Quaternary
Extinctions

Quaternary Extinctions

A PREHISTORIC REVOLUTION

PAUL S. MARTIN
RICHARD G. KLEIN

EDITORS

The University of Arizona Press
Tucson & London

About the Editors

PAUL S. MARTIN has contemplated Pleistocene events of arid regions from his desert laboratory on Tumamoc Hill west of Tucson, Arizona, for more than twenty-five years. He has studied the herpetofauna of eastern Mexico, the Pleistocene fossil pollen record of Arizona, and the potential for fossil packrat middens in studies of climatic changes as well as the causes of late Pleistocene extinctions. His interest in the extinction of late Pleistocene large animals has taken him to fossil sites in the four corners of the world. A geosciences professor, he has been on the faculty of the University of Arizona in Tucson since 1957.

RICHARD G. KLEIN has done research on Early Man in the Soviet Union, Spain, and especially South Africa, where he has analyzed fossil assemblages ranging from early Pliocene to historic in age and including archaeological and "natural" occurrences. His primary interest is in animal bones as indicators of environmental change and of collector behavior. He taught at the University of Wisconsin-Milwaukee, Northwestern University, and the University of Washington before joining the University of Chicago as a professor of anthropology.

Third printing 1995

THE UNIVERSITY OF ARIZONA PRESS

Copyright © 1984
The Arizona Board of Regents
All Rights Reserved

This book was set in 10/11 V-I-P Century Oldstyle.
⊗ This book is printed on acid-free, archival-quality paper.
Manufactured in the U.S.A.

Library of Congress Cataloging in Publication Data

Main entry under title:

Quarternary extinctions.

    1. Paleontology, Quaternary. 2. Extinct animals.
I. Martin, Paul S. (Paul Schultz), 1928–
II. Klein, Richard G.
QE741.Q29 1984   560'.1'78   83-18053

ISBN 0-8165-1100-4

Contents

ASIA AND AFRICA: MODEST LOSSES

AUSTRALIA, NEW ZEALAND, AND THE ISLAND PACIFIC: SEVERE LOSSES

AN OVERVIEW

A Word from the Editors

TO SOME SCHOLARS THE QUESTION of why species become extinct is a trivial one. The answer is clear. The world is ever-changing and what constitutes a successful adaptation to environment at one time is certain to represent an unsuccessful adaptation later. Thus extinction is part of evolution and is the fate of every species sooner or later.

There is no quarreling with this basic perspective, but to many scholars and to the public, especially, it is still interesting and important to ask why a species became extinct when it did. The question is particularly compelling when large numbers of species appear to have become extinct simultaneously, as for example, in the case of dinosaurs at the end of the Cretaceous, or in the case of many large mammals at or near the end of the Pleistocene. The terminal Pleistocene extinctions are especially fascinating because of the possibility that they were entirely promoted by prehistoric people. The principal alternative is that they were caused by a major climatic change. Third is the possibility that they were part of a massive evolutionary turnover, which coincidentally happened late in prehistoric time. It is the attempt to choose between these alternatives or combinations of them that is the central theme of this book.

As numerous chapters in this book show, there is no shortage of stimulating ideas on just how natural or human impacts could have affected various now-extinct species. Just as important, the chapters reflect the ongoing development of the data and dating necessary to evaluate or test these ideas. On one hand, important advances have been made in our knowledge of Pleistocene climatic change, revealed regionally by geomorphology and palynology and globally by the nearly continuous fossil record on the deep-sea floor. On the other hand, our understanding of later Pleistocene faunal and human history has increased significantly. Perhaps the greatest advances have been made in Australia; in years to come we can expect yet more detailed information on the chronological relationships between Australian large mammal extinctions, the arrival of people, and late Pleistocene climatic change. Already, the comparison of the Australian record with that of other continents, particularly North America, is helping to eliminate some extinction scenarios and to promote others. Future research will further narrow the possibilities. While the causes of Pleistocene extinctions remain debatable, this book shows that the debate need be neither endless nor fruitless.

Late Pleistocene extinction is a matter of potential interest to many specialists, but vertebrate paleontologists in particular must deal with it. In this book the field is represented by E. Anderson, Gilbert, Gingerich, Guilday, Guthrie, Saunders, Steadman,

Lundelius, Graham, Webb, Vereshchagin, L. Martin, Horton, Merrilees, Olson, and James. Since the extinctions may impinge on prehistory in one way or another, it is not surprising to find considerable interest among archaeologists. An archaeological outlook is advanced by Agenbroad, Grayson, Whittington, Dyke, Gruhn, Bryan, Haynes, Dewar, Klein, A. Anderson, Cassels, Mead, Meltzer, Trotter, and McCulloch. Geologists are drawn to the problem, such as Agenbroad, Haynes, Marcus, Berger, Liu Tung-sheng, and Li Xing-gou. Botanists, such as Kershaw, may contribute by general study of Pleistocene plant extinctions, or through paleoecological and paleoclimatic studies as in the case of King, Phillips, and Hope. Zoologists may be involved through anatomy, biogeography or paleontology: they include Murray, Tchernov, McDonald, and P. Martin. It will be seen that many authors wear several hats, which is as it should be.

Geographically, North America receives the greatest attention, with sixteen chapters devoted to it. Nevertheless, an equal number are largely or entirely devoted to other parts of the world. E. Anderson's *Who's Who* covers the more important extinct animals of the late Quaternary on a world basis. We are gratified with the contributions from Australia, New Zealand, Hawaii, and Madagascar where massive losses of animals must be dealt with. The less momentous extinctions of Afro-Asia extend over a much longer time. South America is largely neglected in this book, except for Gruhn and Bryan's illuminating treatment of the Taima-taima site in Venezuela. Fortunately South American extinct faunas have been recently reviewed by Marshall, Webb, and Simpson, all referred to in Martin's chapter.

The parent volume to this book is *Pleistocene Extinctions: The Search for a Cause,* edited by P. S. Martin and H. E. Wright, Jr., published by Yale University Press in 1967. Long out of print and increasingly obsolete in content, it seemed desirable to us to attempt a fresh approach rather than a revision of *Pleistocene Extinctions.* While he maintains a keen judicial interest in the subject, H. E. Wright, Jr. wished to be replaced.

Without the benefit of any preceding symposium or conference, informal requests for manuscripts for this book were sent to various colleagues, beginning with authors of *Pleistocene Extinctions.* J. E. Guilday, C. V. Haynes, E. L. Lundelius, Jr., and N. K. Vereshchagin again responded, joined by forty new contributors who willingly shared the burden of chapter authorship and, in some cases, content review. From the level of response, we suspect that twice as many qualified authors could have been recruited had the number of invitations been doubled. Beyond the concept, little material in *Quaternary Extinctions* can be traced to *Pleistocene Extinctions.*

Thanks is extended to various people for assistance: in Australia Jeanette Hope and R.S.V. Wright led us to potential contributors; in New Zealand Roger Green served a similar role. John Guilday helped to recruit Elaine Anderson for the important task of writing the bestiary; Dave Webb recommended Richard Kiltie. Professor Liu of the Academia Sinica kindly responded to an invitation for help with a chapter, initially extended while he was visiting the Australian National University. Larry Marshall moved to the University of Arizona at a critical time, bringing his special knowledge of paleontology when it was most needed.

Finally, we express our gratitude to all chapter authors for their enthusiastic and skilled efforts.

The University of Arizona Press encouraged us to attempt this book, helped guide us in its organization, and devoted itself to details of editing and production. In particular we thank Marshall Townsend, Elizabeth Shaw, Rose Houk, Kim Vivier, and Harrison Shaffer for their efforts.

PAUL S. MARTIN
RICHARD G. KLEIN

Historical Background and the Beasts Themselves

Historical Background
and the Beasts Themselves

It is less than 200 years since fossils have been seen for what they are—evidence of extinction and a transformed world. The discovery was shattering for those who invested philosophical faith in a creator of a perfect, immutable world. Thomas Jefferson was among those who found it difficult to accept the notion that animals might indeed be extinct. He mistook the claw of an extinct ground sloth to be that of a giant, and perhaps living, American lion. By failing to find such a beast, or any other living monsters in the wilderness west of the Mississippi that might resolve the matter of the great claw and other mysterious large bones that had reached the White House, Lewis and Clark helped provide the additional negative evidence. The fact of extinction had to be reckoned with. The animals were gone. The reverberations of that fact linger still.

It was bones of mammoth, mastodont, and ground sloth, not dinosaurs or titanotheres, that first forced recognition of the problem. Along with the fossils came the discovery of glaciation and evidence of an ice age climate. Other revolutionary discoveries followed. By the mid-nineteenth century it became apparent to Darwin and Lyell that Boucher de Perthes, a Frenchman and a catastrophist at that, had made a remarkable discovery: human artifacts were associated with bones of the extinct ice age fauna. Confronted with the extinctions, nineteenth-century geologists and natural historians struggled mightily to explain what might have caused them. A new development overshadowed the field; by the latter half of the century the remains of dinosaurs, titanotheres, and other strange ancient beasts long antedating the Pleistocene, eclipsed much of the excitement generated by the ice age megafauna.

Many of the historical events have been forgotten, and in his introductory chapter Don Grayson shows how discoveries of the time caused dynamic shifts in nineteenth-century thinking.

By the twentieth century, late Pleistocene extinctions, along with all others, were commonly regarded as the other side of the coin of evolution. Extinction no longer needed to be marveled at or even to be explained. Given evolutionary turnover, extinction is inevitable. Like death for the individual, nothing is more certain in the future of a species than its ultimate removal.

This book is an effort to refuel interest in what some may regard as a properly dead issue. Current models have a long history, as Grayson's chapter indicates. Is the problem of late Pleistocene extinction again worthy of a new inquiry? The foremost new technique to transform traditional approaches to the problem is radiocarbon dating, and the full impact of this development is yet to be felt.

Understanding the nature of the late Pleistocene requires a "who's who" of the extinct beasts, an ice age field guide. Elaine Anderson's mission for this book is to bring to life the missing animals of the last 100,000 years. More than 100 genera of mammals had to be resurrected. While her "who's who" is much larger than the bestiary of Pleistocene animals that Martin and Guilday offered in Pleistocene Extinctions in 1967, it still does not cover all the extinct genera of terrestrial vertebrates known from the late Pleistocene. New ones, freshly minted by their authors, continue to appear, especially from oceanic islands.

Finally, should anyone have doubts regarding the abundance of Pleistocene bones, Larry Agenbroad's maps of fossil mammoth sites should be illuminating. The maps suggest how common mammoths were in parts of North America. To find their bones it would appear prudent to proceed to central Alaska, the Yukon, or to central Texas, and to spurn most of eastern Canada.

Generating isopleth maps such as Larry Agenbroad's requires detailed spot maps from state museums or regional geological studies. O. P. Hay (1924) was the last paleontologist to attempt nationwide maps for the extinct American megafauna. David Horton, in a chapter to follow, appears to be the first Australian to attempt to map the fossil distribution of extinct Pleistocene giant marsupials. Perhaps Agenbroad's and Horton's efforts will inspire others to map more of the neglected extinct species of the last Ice Age.

Armed with an historical perspective of the theoretical issues, a guide to the more important animals lost, and a notion of how common they can be as fossils, one can more closely examine Pleistocene extinctions.

Nineteenth-Century Explanations of Pleistocene Extinctions: A Review and Analysis

DONALD K. GRAYSON

THE PAST FEW DECADES HAVE SEEN INTENSE INTEREST in the vertebrate extinctions that occurred toward the end of the Pleistocene in many parts of the world. In this chapter, I review an earlier period of great interest in the same set of extinctions, the decades from the first widespread acceptance of the reality of extinction to just beyond the acceptance of a tremendous human antiquity, an antiquity inferred from the association of human artifacts with the remains of the extinct animals themselves.

Establishing the Reality of Extinction

Although we are now surrounded by the reality of extinction, two centuries ago it was not at all clear that organisms could actually disappear from the earth. Before the late eighteenth century, the possibility that organisms could become extinct seemed remote, and numerous well-reasoned and elegant denials of the possibility are found in the literature of that time. Extinction, as Rudwick (1972) has noted, seemed to imply some imperfection in the Creation, some original flaw in the design of things.

For most of those who dealt with the history of life before the late 1700s, the idea of extinction was an affront to the guardianship that the Creator maintained over his creations and an attack on the principle of plenitude, of the fullness of the Creation and all that that implied (Lovejoy 1964). The continuity suggested by the Great Chain of Being—"the sacred phrase of the eighteenth century" (Lovejoy 1964:184)—implied that removing a link of that chain threatened the whole with destruction:

> Vast chain of being, which from God began,
> Natures aethereal, human, angel, and man,
> Beast, bird, fish, insect what no eye can see,
> No glass can reach! from Infinite to thee,
> from thee to Nothing!—On superior pow'rs
> Were we to press, inferior might on ours:
> Or in the full creation leave a void,
> Where, one step broken, the great scale's destroy'd:
> From Nature's chain whatever link you strike,
> Ten or ten thousandth, breaks the chain alike.
> (Pope, *Essay on Man,* 1733–1734)

5

Up to the 1790s this intellectual framework was still strongly in place. Thomas Jefferson's discussion of the remains of *Megalonyx* from West Virginia, an animal he thought was a carnivore but that was immediately recognized to be more closely allied to the sloths (Wistar 1799), illustrates the framework well. Jefferson was no great paleontologist, as Simpson (1942; see also Osborn 1935) has noted, and he undoubtedly pursued the study of large mammal bones for reasons as nationalistic as they were scientific (Hindle 1974). Nonetheless, his approach to the question of whether this large creature still existed displays the pervasiveness of the feeling that extinction could not occur. Although such an animal had never been seen alive in North America, Jefferson (1799:255–266) argued that

> In fine, the bones exist: therefore the animal has existed. The movements of nature are in a never-ending circle. The animal species which has once been put into a train of motion, is probably still moving in that train. For if one link in nature's chain might be lost, another and another might be lost, till this whole system of things should evanish by piece-meal. . . . If this animal has once existed, it is probable on this general view of the movements of nature that he still exists.

The similarity between Jefferson's argument and Pope's elegant lines are obvious. Jefferson's denial of the possibility of extinction is, in turn, similar to many other denials that appeared in the contemporary literature (Burkhardt 1977, Rudwick 1972).

Many arguments for the reality of extinction made during the 1700s were based on fossil invertebrates that seemed to have no living counterpart. Given the prevailing theoretical framework, which denied the possibility of extinction, such arguments were easily countered by noting discoveries of existing populations of organisms that had first been described on the basis of fossil shells (Burkhardt 1977). It was Georges Cuvier who turned this position around by basing arguments for extinction on animals so large that their possible future discovery on the hoof could not be used to counter those arguments. Regarding such large quadrupeds, he argued, "there is little hope of one day finding those which we have seen only as fossils" (1812a:39). Beginning in 1796 with ground sloths (*Megatherium*) and elephants (*Mammuthus*) (Cuvier 1796a, 1796b), Cuvier's detailed demonstration that no living representatives of these animals existed soon ended opposition to the reality of extinction (see the excellent discussion in Rudwick 1972).

It proved easy to reconcile the power of the Creator with the fact of extinction. Indeed, soon after Jefferson's denial of the possibility of extinction, Blumenbach (1865 [1806]), Parkinson (1811), and others argued that it was a measure of God's omnipotence to have created a world in which pieces could be removed without the whole tumbling into a shambles:

> some good and very learned men have regarded the loss of a single link, in the chain of creation, as inadmissable: it implying, they say, such a deviation from the first plan of creation, as might be attributed to a failure in the original design. But such an inference does by no means follow; since that plan, which prevents the failure of a genus, or species, from disturbing the general arrangement, and œconomy of the system, must manifest as great a display of power and wisdom, as could any fancied chain of beings, in which the loss of a single link would prove the destruction of the whole (Parkinson 1804:469–470).

Many of the animals Cuvier analyzed during the last decades of the eighteenth and first few decades of the nineteenth centuries we now know to have become extinct near the end of the Pleistocene. Indeed, as Cuvier discussed, many of the European animals he had described had long been known to have occurred in Europe, but were previously felt to belong to species that still existed somewhere. It was one of Cuvier's many contributions to demonstrate that most of these animals no longer existed anywhere. By the time he published the first edition of *Ossemens Fossiles* in 1812, the magnitude of ancient European extinctions was well known.

Recognition of the magnitude of ancient mammalian extinction was also rapid in North America. In 1818, the first American edition of the English translation of Cuvier's *Discours Préliminaire* (Cuvier 1812a) was published along with a general overview of North American paleontology by Samuel Latham Mitchill (Cuvier and Mitchill 1818). Mitchill's review was descriptively weak, but a thorough review of North American mammals, including extinct species, appeared seven years later. Able to draw not only on the work of Cuvier but also on that of Cooper (1824), Mitchill (1824), Wistar (1818) and others, Harlan (1825) described ten species of extinct North American terrestrial vertebrates (see Simpson 1942 for a discussion of Harlan's work). The list included *Megalonyx, Megatherium* (now *Eremotherium*), *Tapirus, Mastodon, Elephas,* (now *Mammuthus*), *Cervus americanus* (now *Cervalces scotti*), *Bos bombifrons* (now *Bootherium bombifrons*), and *Bos latifrons* (now *Bison latifrons*). Only thirteen years after the initial publication of *Ossemens Fossiles,* the list of extinct North American mammals had become lengthy.

The responses of those who pondered such lists show clearly that the magnitude of both European and American Pleistocene extinctions was becoming well understood. "It is impossible," Darwin noted in 1836, "to reflect on the state of the American continent without astonishment. Formerly it must have swarmed with great monsters; now we find mere pigmies compared with the antecedent, allied races" (1871a [1845]: 22). The astonishment shown by Cuvier and Darwin at the diverse array of extinct Pleistocene beasts was shared by many other early nineteenth-century naturalists. While Bewick (1807) suggested that elephants might still exist in North America, and Prichard (1902) explored Patagonia for living ground sloths, the scope of the problem to be solved was delineated during the first few decades of the nineteenth century.

Early Extinction Chronologies

Cuvier and other late eighteenth- and early nineteenth-century earth historians grappling with the issue of extinction, and with the broader story of the history of vertebrates, did not confine their interests to animals that we now know to have been Pleistocene in age. Cuvier, for instance, dealt with Tertiary fossils as well. Given that vertebrate remains cutting across millions of years were under investigation, it must be asked if early nineteenth-century earth historians treated the extinction of animals now known to have been late Pleistocene in age as a discrete set of chronologically linked creatures.

My use of the term "Pleistocene" in an early nineteenth-century setting is, of course, anachronistic. Lyell (1833) defined four subdivisions of the Tertiary on the basis of their fossil mollusk content, the two most recent of which he called the Older and Younger Pliocene. In 1839, he substituted the term Pleistocene for Newer Pliocene, substituting Pliocene for what he had previously called the Older Pliocene. Subsequently, Forbes (1846:403) redefined the term Pleistocene to refer to "that section of geological time which was typically distinguished by the presence of severe climatal conditions through a great part of the northern hemisphere, and during which those marine accumulations... which have been called 'northern drift' were formed," a redefinition that Lyell (1873) eventually accepted (see reviews in Wilmarth 1925 and Berry 1968).

After Lyell's initial definition of the Pleistocene, more than a century passed before the stratigraphic and chronologic relationships between the mammals that became extinct at the end of the Pleistocene and Pleistocene deposits themselves became known with any precision (Martin 1967b). In addition, the mechanisms that produced Pleistocene deposits in glaciated areas were not generally agreed upon until the second half of the nineteenth century (see Flint 1971 for a short historical review). Nonetheless, natural historians of this time tended to treat the extinction of Pleistocene mammals as a coherent problem. Although the origin of Pleistocene deposits was not accurately known, deposits resulting from Pleistocene glaciation were well recognized and widely discussed. Such deposits were often attributed to the results of flooding, and debate centered primarily on how the flooding might have occurred, not on whether a distinct set of deposits was involved. Indeed, Buckland's term for these deposits, diluvium, referring to "those extensive and general deposits of superficial loam and gravel" (1823:2), was widely accepted and continued in use long after the advent of glacial theory. But no matter how they were referred to or how they were explained, they were recognized as a set of deposits differing in age from those above and below them, and the general relationships of the deposits to such mammals as woolly mammoth (*Mammuthus primigenius*) and woolly rhinoceros (*Coelodonta antiquitatis*) was recognized as well.

Old and New World diluvium was temporally correlated on the basis of general morphological similarity and because of its stratigraphically superficial position. Such a correlation also implied the temporal correlation of fossils associated with diluvium. Extinct Pleistocene mammals found outside glaciated areas, and not associated with diluvium, were correlated through association with extinct mammals that were found in the diluvium, and by the fact they they, too, were found in strata close to the earth's surface. More generally, the animals were seen to form a rough chronological group because they either belonged to extinct species of extant genera, or belonged to extinct genera with closely related survivors, *and* were not found stratigraphically associated with more ancient animals belonging to genera now completely unknown. Thus, Cuvier (1812a:17) noted that "the most celebrated of the unknown species, which belong to known genera or to genera very near those that are known, such as the elephants, rhinoceroses, hippopotamuses, and fossil mastodons are not found with most ancient genera. It is only in the alluvium ["terrains de transport"] that they are found." In this fashion, the principles of stratigraphic correlation allowed the reconstruction of an entire fauna, although the chronological placement of some animals routinely found in bog deposits, including the Irish "elk" (*Megalocerus giganteus*) and the American mastodont (*Mammut americanum*), posed difficulties throughout the century. By the time the concept of extinction became accepted, it began to be accepted as well that there was a set of mammals *generally* associated with diluvium that must have become extinct at *roughly* the same period of time the diluvium was deposited.

To be sure, there were arguments that some or all of the animals lived on well after the time of deposition of the diluvium. Nonetheless, it was well recognized that the problem involved explaining the relatively recent extinction of a set of large quadrupeds. I will, therefore, use the convenient phrase "Pleistocene extinctions" to refer to the target of nineteenth-century explanations: both modern and nineteenth-century approaches defined the explanatory target in similar ways.

Nineteenth-Century Explanations of Pleistocene Extinctions

It did not take long for speculations on the causes of Pleistocene extinctions to begin once it became accepted that such extinctions had occurred. By the early 1800s, three major external causes of those extinctions had found adherents: rapid geological change, slow and natural changes across the surface of the earth, and human activities. While internal causes (for instance, racial senility) were occasionally called upon, they did not play a major role in the interpretation of Pleistocene extinctions during the nineteenth century.

Rapid Geological Change

The hypothesis that Pleistocene extinctions had resulted from one of a series of recurring episodes of rapid geological change had a large following prior to the middle of the nineteenth century (e.g. Cuvier 1812a, Buckland 1823, d'Orbigny and Gente 1851). Such approaches were catastrophic in that they called upon causes differing in both kind and intensity from those observed in modern times (see discussion and references in Grayson 1979).

Only after careful consideration of the kinds of processes that he saw operating in the modern world did Cuvier (1812a) conclude that such slowly operating forces could not account for some of the major changes evidenced by the geological record. Inclined strata containing salt water mollusks, alternating sequences of terrestrial and marine deposits, frozen mammoth carcasses in Siberia, and a wealth of other well-described phenomena could, he felt, be explained only by calling upon forces working more powerfully and more rapidly than any that had ever been seen in operation. As regards extinction, Cuvier (1812a) concluded that the species of elephant, rhinoceros, hippopotamus, and other quadrupeds he had so carefully described had become extinct as a result of some major, rapid change, some revolution, that had affected the surface of the earth.

Unlike later scientists who called upon universal catastrophes to account for Pleistocene extinctions, Cuvier called upon more localized events. Although he suggested that Siberian mammoths may have been terminated by rapid refrigeration, he emphasized localized floods, caused by changing relationships between land and sea, to explain the extinction of mammoths and of the extinct quadrupeds with which they were so often associated. "If there is anything established in geology," Cuvier (1812a:110) wrote, "it is that the surface of our globe was the victim of a great and sudden revolution that can date to no more than five or six thousand years ago," a revolution that saw low-lying areas inundated and the former seabed exposed, followed by the return of these waters to the modern sea. Recolonization of flooded areas then occurred from regions in which extinction had not occurred. However, such modern animals as elephants and rhinoceroses could not have descended from the extinct European forms he had described because those species had been totally exterminated. Had that not been the case, why do we not find elephants in places like Mexico or South America, he asked, places that seem well suited for them (Cuvier 1812b, Vol. 2:140)?

Cuvier thus called on rapid, but local and transient, geological change as the cause of Pleistocene extinctions, change that not only accounted for the extinctions but also provided an argument against Lamarck's transformism (see Rudwick 1972, Burkhardt 1977, and Grayson 1983 for discussions of this dispute). The causes of such revolutions were unknown to him but were, in theory, knowable.

In his writing, Cuvier showed little concern for the relationship between his arguments and theological issues. One searches in vain in his work for any attempt to reconcile science and religion or to explain geological phenomena through divine deeds. This process began, however, as soon as Cuvier's work crossed the English Channel.

The first major summary of vertebrate paleontology written in England after Cuvier had completed his initial studies was published by James Parkinson (1811). The last of a three-volume discussion of organic fossils, the first volume of which was published in 1804, this work depended heavily on Cuvier's published accounts of ancient vertebrates. Parkinson, however, was intimately concerned with theological issues and was adding another contribution to British natural theology. Indeed, to Parkinson, the fact of extinction itself afforded "a direct proof of the Creator of the universe continuing a superintending providence over the works of his hands" (1811:xiv).

Although the bulk of his discussion was descriptive, Parkinson's conclusions directly addressed the issue of the relationship between the fossil record and the Mosaic account of earth history. He found that this entire record, from the earliest known rocks to the late appearance of human beings, was fully in accord with Genesis. "So close indeed is this agreement, that the Mosaic account is thereby confirmed in every respect, except as to the age of the world, and the distance of time between the completion of different parts of the creation" (1811:451). Not even this apparent chronological problem was crucial, however, since it disappeared as long as the Mosaic day was taken figuratively. Cuvier's localized revolution that had caused the last episode of extinction became the Noachian deluge; the sporadic and repeated appearance of new kinds of organisms during earth history, including the human race, provided evidence for the direct and repeated intervention of the Creator during that history.

It remained for William Buckland, "the teacher of a whole generation of geologists" (Burchfield 1975:8), to solidify and popularize this alteration of Cuvier's approach. Buckland was explicit about his view of "the absurdity, if not impiety of dissolving that union, by which Philosophy becomes associated in its natural and just office, as the faithful auxiliary and handmaid of Religion" (Buckland 1820:28). For Buckland, one of the wonders of geology was that it not only provided evidence for the design of the earth by the Omnipotent Architect, but that it also provided clear evidence that the Creator had continually exerted his will through secondary laws that guided the history of the earth toward beneficial ends. As it had for Parkinson, the study of geology showed that the Mosaic account of the Creation and of subsequent events was correct in all its major points, including "the two great points... of the low antiquity of the human race, and the universality of a recent deluge" (1820:24).

Buckland's *Reliquiae Diluvianae* (1823) was written explicitly to deal with the second of these two great points. In it, he attempted to

> throw new light on a period of much obscurity in the physical history of our globe; and, by affording the strongest evidence of an universal deluge, leads us to hope, that it will no longer be asserted, as it has by high authorities, that geology supplies no proofs of an event in the reality of which the truth of the Mosaic records is so materially involved (1823:iii).

Thus, Buckland wished to show that the paleontology and geology of western Europe specifically, and of other parts of the world in general, supported the biblical accounts of the Noachian flood. Although his aims and concerns were much the same as Parkinson's, his ammunition and skills in this area were more powerful. He proceeded by examining the paleontology and geology of a number of cave and fissure sites from western Europe, in particular from England. In addition, he reviewed a wide range of evidence for diluvial action from many parts of the world, with most attention paid to Europe. From this review, he concluded that "we may for the present rest satisfied with the argument that numberless phenomena have already been ascertained, which without the admission of an universal deluge, it seems not easy, nay, utterly impossible to explain" (1823:228). Interestingly, he had reached an identical conclusion several years before, prior to gaining much of the evidence presented in *Reliquiae Diluvianae* to support it (Buckland 1820:38). Buckland's arguments were carefully and tightly made; the volume, with its detailed consideration of taphonomy and geology, was and remains an important contribution and a pleasure to read. Indeed, much of the evidence that Buckland amassed in support of his diluvial arguments was later to be accounted for by glacial theory. Not only did Buckland live to see this transition, but he himself later abandoned his flood in favor of Agassiz's equally catastrophic glaciers (e.g. Buckland 1842, Woodward 1883, E. C. Agassiz 1885).

The relationship of Pleistocene mammals to diluvial deposits left little doubt in Buckland's mind as to the cause of their extinction:

> How is it possible to explain the general dispersion of all these remains, but by admitting that the elephants as well as all the other creatures whose bones are buried with them, were the antediluvian inhabitants of the extensive tracts of

country over which we have been tracing them? and that they were all destroyed together, by the waters of the same inundation which produced the deposits of loam and gravel in which they are embedded (1823:183–184).

Like Parkinson, Buckland relied heavily on Cuvier as an authority and greatly transformed Cuvier's last revolution, producing a universal deluge where Cuvier had posited only localized incursions of the sea. All three men saw the human species as a recent phenomenon. For Cuvier, the recency of people was a fact to be noted, although it was also a fact of importance in denying Lamarck's suggestion that human activities might have caused some extinctions. For Parkinson, the absence of human remains in association with the extinct mammals was a fact that had to be explained, since it was in apparent contradition with the pre-Deluge appearance of people presented in the Mosaic account. For Buckland, this recency was one of the "two great points" in which earth history and revelation agreed, since the book of God's words implied that it was in paleontologically poorly known Asia that people existed before the flood. For all three, however, Pleistocene extinctions were to be explained by a sudden and major change in the nature of the earth's surface, a change that had involved massive flooding, and a change that had occurred no more than five or six thousand years previously.

When Buckland abandoned Noah's flood for Agassiz's glaciers as a cause of the deposition of diluvium, he was by no means abandoning the catastrophists' position. For Agassiz, those glaciers and the phenomena associated with them were every bit the extinction-causing catastrophe that the Deluge had been argued to have been.

Agassiz early became convinced that glaciers provided the explanatory mechanism for Pleistocene extinctions. In the late 1830s, he wrote to Buckland that "since I saw the glaciers I am quite of a snowy humour, and will have the whole surface of the earth covered with ice, and the whole prior creation dead by cold" (E. C. Agassiz 1885:289). In his famous announcement of glacial theory, the *Discours de Neuchâtel,* delivered in 1837, Agassiz made his position explicit. Agassiz (1837) drew an analogy between the temperature histories of individual organisms and of the earth itself. Just as an organism begins life with the generation of heat, lives that life in a kind of energy equilibrium, and ends that life by the generation of "a glacial cold," so too has been the history of the temperature of the earth. At the beginning, the earth's temperature was high, but as time passed this temperature decreased in stepwise fashion, each plunge separated by a period of stasis.

> The temperature of the earth maintained itself without considerable oscillation during a given geological epoch, as has happened during our epoch, then decreased suddenly at the end of each epoch, with the disappearance of the organized beings characteristic of that epoch, to rise again with the appearance of a new creation at the beginning of the following epoch, but to a value less than that of the preceeding epoch (1837:30).

Thus, not only Pleistocene extinction but each preceding episode of mass extinction had been caused by a sudden drop in temperature. As for Pleistocene mammals themselves, "a Siberian winter established itself... on ground previously covered by rich vegetation and occupied by great mammals, similar to forms that today occupy the warm regions of India and Africa. Death enveloped all nature in a shroud" (1837:xxiv). Extinction was so thorough that the present creation is completely separate from the preceding one, and even if living species seem identical to those revealed by the fossil record, it cannot be assumed that the latter are ancestral to the former.

Agassiz developed these arguments in more detail in *Études sur les glaciers* (1840), adding the observation that people did not appear on the earth until after the last catastrophe had occurred. Few scientists have written so eloquently on the causes of extinction:

The appearance of this great cover of ice must have brought with it the extinction of all organic life on the surface of the globe. The ground of Europe, recently covered by tropical vegetation and occupied by herds of great elephants, enormous hippopotamuses, and gigantic carnivores, found itself entombed under a vast mantle of ice that covered fields, lakes, seas, and plateaus alike. To the movement of a powerful creation succeeded the silence of death. Springs dried up, streams ceased to flow, and the sun's rays, in rising over those frozen expanses (if they still reached there), were met only by the whistling of the northern winds and by the thunder of crevasses as they spilt the surface of this vast ocean of ice (1840:314).

Agassiz's position on this issue did not waver. As time passed, his observations in the New World led him to extend the scope of his glaciers to the southern hemisphere as well as the northern (Agassiz 1887). Nor, really, was there any reason to expect that position to have changed, since his mechanism of extinction was an important part of his view of the development of life. Always strongly against transformationist theories, be they Lamarck's, Chambers's, or Darwin's, Agassiz saw the unfolding of life as progressing toward humankind in discontinuous steps (see, for instance, the excellent discussion in Bowler 1976). New creations following mass extinctions were an important part of this view, and the extinction of Pleistocene mammals by the "Siberian winter" must be seen in this context. Thus, Darwinian evolutionists working toward the end of Agassiz's career were right in thinking that one of Agassiz's reasons for searching for glaciation on a global scale was to strike a blow against Darwinian theory by demonstrating that glaciers had eradicated life in the not-too-distant past. In writing Darwin in 1866, for instance, Asa Gray noted that Agassiz had given a talk in which he argued that the entire American continent had at one time been covered by ice, and concluded, "so here is the end of the Darwin theory" (Darwin and Seward 1903:160). Gray noted that "I said last winter that Agassiz was bent on covering the whole continent with ice, and that the motive of the discovery he was sure to make was that there should be no coming down of life from the Tertiary or post-Tertiary period to ours" (Darwin and Seward 1903:160).

Nearly thirty years after the *Discourse,* Agassiz once again summarized his explanation of Pleistocene extinctions, an explanation that had to be countered not only by Darwinians and other transformationists, but also by nonprogressive uniformitarians:

The long summer was over. For ages a tropical climate had prevailed over a great part of the earth, and animals whose home is now beneath the Equator roamed over the world from the far South to the very border of the Arctics. The gigantic quadrupeds, the Mastodons, Elephants, Tigers, Lions, Hyenas, Bears, whose remains are found in Europe from its southern promontories to the northernmost limits of Siberia and Scandinavia, and in America from the Southern States to Greenland and the Melville Islands, may indeed be said to have possessed the earth in those days. But their reign was over. A sudden intense winter, that was also to last for ages, fell upon our globe; it spread over the very countries where these tropical animals had their homes, and so suddenly did it come upon them that they were embalmed beneath masses of snow and ice, without time even for the decay which follows death (1866:208).

By the time of his death in 1873, Agassiz was alone among major naturalists in forwarding an explanation of Pleistocene extinctions based solely on sudden geological change. Although the history of catastrophist approaches to the geological record as a whole is beyond the scope of this chapter (see Gillispie 1951; Glass et al. 1959; Greene 1959; Rudwick 1972), reaction to the methods and arguments of catastrophists played a major role in structuring other explanations of Pleistocene extinctions during the nineteenth century. It is to those explanations that I now turn.

Slow, Natural Changes Across the Surface of the Earth

The early champion of slow, natural changes across the surface of the earth as a cause of Pleistocene extinctions was Charles Lyell. As the author of the uniformitarian synthesis, Lyell rejected catastrophist explanations of all sorts. All that was required to counter catastrophist explanations of Pleistocene extinctions was to show that extinct Pleistocene mammals had been found both above and below diluvium, and to then argue that these extinctions must have occurred over long periods of time. This was precisely what Lyell did (e.g. Lyell 1830, 1844, 1845, 1853). Thus, while catastrophists stressed the general stratigraphic relationship of extinct Pleistocene mammals to diluvial deposits in order to establish that they had been terminated suddenly by the event that formed those deposits, Lyell used specific examples to show that certain animals had, in fact, survived after that event (e.g. Lyell 1853).

Prior to about 1860, Lyell maintained that Pleistocene extinctions were due to slow, natural changes across the surface of the earth, changes due to the same kind of processes that operate today. He suggested that the extinction of large Pleistocene mammals resulted from "physiological laws which render warm-blooded quadrupeds less capable, in general, of accommodating themselves to a great variety of circumstances, and consequently, of surviving the vicissitudes to which the earth's surface is exposed in a great lapse of ages" (1833:140). Before 1860 he repeatedly argued that

> causes more general than the intervention of man have occasioned the disappearance of the ancient fauna from so many extensive regions... all the species, great and small, have been annihilated one after another in circumstances in the organic world which are always in progress, and are capable in the course of time of greatly modifying the physical geography, climate, and all other conditions on which the continuance upon the earth of any living being depends (1859a:164).

We must not, Lyell argued, be surprised at the fact of extinction. Not only can extinction be shown to be occurring in modern times (Lyell 1832), as any uniformitarian approach requires, but in addition

> the possibility of the existence of a certain species in a given locality, or of its thriving more or less therein, is determined not merely by temperature, humidity, soil, elevation, and other circumstances of like kind, but also by the existence or non-existence, the abundance or scarcity, of a particular assemblage of other plants and animals in the region.
>
> If we show that both these classes of circumstances, whether relating to the animate or inanimate creation, are perpetually changing, it will follow that species are subject to incessant vicissitudes and if the result of these mutations, in the course of ages, be so great as to materially affect the general condition of *stations,* it will follow that the successive destruction of species must now be part of the regular and constant order of nature (1832:141; emphasis in original).

Thus, the extinction of species is a predictable, natural, and ongoing phenomenon, one that can be expected to occur slowly during the course of ages. Since Pleistocene extinctions seem to have occurred over a lengthy period of time as a result of vicissitudes that affected the earth's surface, these extinctions are readily incorporated into the uniformitarian world view.

Lyell was not often precise about the nature of the mechanisms that he felt caused the extinction of Pleistocene quadrupeds. However, in the first edition of the *Principles of Geology* he presented an argument for the extinction of the woolly mammoth that is quite precise and that influenced Darwin's early views on the causes of Pleistocene extinctions. After arguing that the vast numbers of mammoth remains in Siberia indicated that the preserved animals had died over a long period of time (and thus did not require catastrophist explanations) and that their mode of deposition implied that the animals "had continued to exist in Siberia after the winters had become cold," Lyell (1830:96) concluded that the extinction of these huge mammals could be explained by supposing climatic change at the time of extinction to have been

> extremely slow, and to have consisted not so much in a diminution of the mean annual temperature, as in an alteration from what has been termed an "insular" to an "excessive" climate, from one in which the temperature of winter and summer were more nearly equalized to one wherein the seasons were violently contrasted.

As I shall discuss, Lyell's explanation of the extinction of the Siberian mammoth through changing seasonal extremes of temperature represented an attempt to provide a gradualist account of the disappearance of an animal that until then seemed best explained in a catastrophic fashion.

As the years passed, Lyell continued to adhere to the general notion that the cause of Pleistocene extinctions was to be found in the operation of slow, natural changes, but it became increasingly difficult for him to conceive of the precise nature of the mechanisms involved. The apparently global nature of the extinctions required a global cause, and, prior to about 1860, the only reasonable cause available seemed to be climatic change. Yet, Lyell increasingly came to feel that the geological record could not support the hypothesis that climatic change supplied the required mechanism. "The disappearance of the ancient fauna," was, he noted, "the more remarkable, as many of the species had a very wide range, and must therefore have been capable of accommodating themselves to considerable variations of temperature" (1849:259–260).

Lyell, of course, could not accept Agassiz's "sudden intense winter" as a cause of the extinctions, and Agassiz's arguments certainly must have tarnished climatic explanations for him. Against the hypothesis that "the cold of the glacial period was so intense and universal as to annihilate all living creatures throughout the globe" (Lyell 1853:139), Lyell consistently made two kinds of arguments. First, he examined the faunal associates of the extinct organisms, and pointed out that those associates were often still extant in the same area. For instance, he noted that "from Canada to South Carolina" (1846:406), the bones of extinct quadrupeds are found alongside the same molluscan species found in those areas today; such facts, he concluded, imply that neither atmospheric nor oceanic temperatures have changed significantly since the time of extinction (Lyell 1844, 1846). Second, he noted that in Europe, North America, and South America, the mammals that became extinct had been discovered both beneath and above glacial deposits, demonstrating that they lived on after the deposition of glacial debris (e.g. Lyell 1844, 1845, 1846, 1853, 1863, 1873). Here, Lyell was using the same argument he had used earlier to refute those who asserted that the extinctions had occurred as a result of rapid flooding; now, with "diluvium" replaced by "drift," the argument is used to refute the notion that Pleistocene extinctions were caused by the cold temperatures of the glacial period.

These coupled arguments were meant to remove two major supports from the refrigeration hypothesis: it could not have been *that* cold, since extant taxa are found alongside extinct ones, and, even, if it were that cold, the chronology of extinction does not support refrigeration as a cause, since many of the animals can be shown to have lived on after the deposition of drift. Thus, "the coldness of climate which probably coincided in date with the transportation of the drift, was not as some pretend the cause of their extinction" (Lyell 1844:323).

In short, Lyell's uniformitarian approach to Pleistocene extinctions saw those extinctions as having occurred slowly and one-by-one. He suggested that slow, natural changes across the surface of the earth had caused the extinctions, but tended to shy away from more precise statements as to cause. He argued strongly against the hypotheses offered by others to account for those extinctions—against the sudden and catastrophic changes required by Cuvier's and Buckland's floods and Agassiz's glaciers. For similar reasons, as I shall discuss, he also argued against a human role in those extinctions during much of his life. As time passed, the exact nature of the changes that could have caused Pleistocene extinctions seemed to become less, not more, apparent to him. In fact, prior to about 1860 he came to agree with Darwin (1871a [1845]) that the causes of extinction were probably so complex that "it is the height of presumption for any geologist to be astonished that he cannot render an account of them" (Lyell 1849:260).

Explaining Surprising Sympatric Relationships

For nineteenth-century naturalists, one of the more surprising attributes of European Pleistocene faunas related to the climatic affinities of the modern relatives of the mammals making up those faunas. During the earliest decades of the century, the known extinct mammals seemed to have primarily southern relatives. As the decades wore on, however, numerous clearly northern mammals were added to the European faunal list, ultimately creating an extremely curious montage of northern and southern forms. Those calling upon rapid change to account for Pleistocene extinctions dealt less with this matter than did those calling upon slow change across the earth's surface. In part, this was because truly substantial evidence for the sympatric occurrence of northern and southern forms, and especially for the presence of extant arctic and alpine species in lowland western Europe, did not begin to accumulate until after catastrophic approaches had begun to wane. Those who stressed the uniformity of geological processes dealt with these apparent sympatric relationships in some detail, and, for some, the nature of their explanations of these relationships helped determine how they accounted for the extinction of Pleistocene mammals. I step aside from the essentially chronological development of my review in order to examine these issues in coherent fashion.

Both Cuvier and Buckland spent much more time demonstrating that they were dealing with animals that had lived and died locally than they spent pondering the possible climatic meaning of those animals. The detailed arguments they presented to establish that these animals represented a native fauna were essential, given the many previous arguments that these animals were recent importations by people, and in particular by the Romans, or, more importantly, that they had been rafted from the south on the waters of the Deluge (e.g. Kirwan 1799). As Buckland noted,

> To the question which here so naturally presents itself as to what might have been the climate of the northern hemisphere when peopled with genera of animals which are now confined to the warmer regions of the earth, it is not essential to the point before us to find a solution; my object is to establish the fact, that the animals lived and died in the regions where their remains are now found, and were not drifted thither by the diluvian waters from other latitudes (1823:44).

In addition, the fact that so many of the animals were extinct clearly implied that their adaptations may have differed from those of their extant relatives, as the frozen carcasses of woolly rhinoceros and woolly mammoth discovered in Siberia in the late eighteenth century so plainly demonstrated. Even so, both Cuvier and Buckland leaned toward the hypothesis that some of the animals implied a previously warmer climate. For both, however, the fact that these animals represented a native fauna was of paramount importance.

At first, Lyell also leaned toward the view that some of the extinct mammals implied a previously warmer climate, although he underscored the fact that extinct mammals need not have had the same temperature requirements as their extant southern relatives (Lyell 1832). In Siberia, certainly, the presence of the woolly mammoth implied a warmer climate since it seemed impossible for this area to have supported sufficient vegetation for the sustenance of this giant herbivore had that not been the case. To this point, Lyell's early position did not clash with those of Cuvier and Buckland. He could not, however, agree with the catastrophist position that the degree of preservation of frozen carcasses required instantaneous refrigeration, as both Cuvier and Buckland had suggested, and he substituted a gradualist argument.

It was clear, Lyell (1830) asserted, that mammoths continued to exist in Siberia long after the winters had become extremely cold, as indicated by the fact that "their bones are found in icebergs, and in the frozen gravel, in such abundance as could only have been supplied by many successive generations" (1830:96). Having established gradual extinction, he then suggested that the demise of these animals was due to a slow shift from an insular to an excessive climate. This shift ultimately thinned their numbers, after which their final loss was ensured by the arrival of other animals better adapted to the changed conditions. Thus, while agreeing with Cuvier and Buckland that the Siberian mammoth must have thrived when the area was warmer, Lyell was able to accommodate this position, and frozen mammoth carcasses, with gradual extinction and faunal replacement.

The publications of Cuvier and Buckland had a tremendous impact on the pace of fossil gathering in western Europe. Throughout this area, cave after cave was excavated and superficial gravels examined for the remains of ancient creatures. The results were remarkable. By the 1840s and 1850s, numerous clearly arctic or alpine mammals, including musk oxen *(Ovibos moschatus)*, pikas *(Ochotona* sp.), marmots *(Marmota marmota)*, and lemmings *(Lemmus lemmus)*, began to be reported in stratigraphic association with species whose modern representatives were either temperate or southern in distribution (for contemporary historical reviews, see Owen 1846; Lartet 1864; Dawkins 1869, 1872). In the valley of the Thames, for instance, woolly mammoth, woolly rhinoceros, musk ox, reindeer *(Rangifer tarandus)*, hippopotamus *(Hippopotamus amphibius)*, and cave lion *(Felis leo spelaea)* had all been found by 1855 in stratigraphic contexts that seemed to indicate contemporaneity (Morris 1850, Lyell 1853, Owen 1856).

These accumulating discoveries no longer allowed the climatic assessment of a group of animals that might have had the same climatic meaning. Although one could still argue around the southern affinities of the hippopotamus by suggesting that it might have been cold-adapted (e.g. Prestwich 1860), these suggestions were no longer convincing to most. There were simply too many animals of northern and southern affiliation involved to allow the hypothesis of adaptational change to stand, as Geikie (1872), among many others, noted. If adaptational shifts could not explain this situation, what could?

For Agassiz (1866), the presence of these northern and southern animals was welcome information. He interpreted the presence of these two apparently conflicting sets of animals as representing faunal replacement that was in line with his model of climatic change. The southern set, including elephants, mastodonts, lions, and hyenas, represented the tropical fauna of the long summer; the northern taxa, including marmot, reindeer, and musk ox, provided clear evidence of the geological winter. Two successive faunas, one tropical and one arctic, accounted for the presence of these animals in European Pleistocene sediments.

Most other earth scientists, however, preferred very different approaches. While Agassiz was arguing for rapid climatic change and faunal replacement as the cause of the mixed nature of European Pleistocene faunas, three gradualist explanations for this phenomenon emerged: seasonal migration, secular migration, and climatic equability.

Arguments depending on the seasonal migratory habits of animals usually did not stand alone, but were instead combined with hypotheses of secular movement. In his earlier writings, however, W. Boyd Dawkins (e.g. Dawkins 1869, 1871; Dawkins and Sanford 1866) did feel that seasonal movements alone provided a fully adequate explanation for the apparent sympatry of northern and southern mammals during the Pleistocene. "It is incredible," he noted (1871:392), "that the climate suited for the well-being of the hippopotamus could at the same time have been adapted for the reindeer, the lemming, or the musk-sheep; for we have no reason to believe that the powers of resisting heat or cold possessed by those animals differed from those which they now possess." Pointing out the well-known lengthy migrations of such Siberian and northern North American mammals as the reindeer, elk, and wolf, he argued that "oscillation to and fro of the animals according to the seasons" (1871:393) accounted for the mixed nature of the European Pleistocene mammalian fauna. In the winter, he suggested, northern mammals moved south, while in the summer southern mammals moved north, traversing territory that was then exposed as a result of uplift but is now covered with water. Not only did the elevation of land eliminate barriers to population movement, but the increased land mass itself accounted for the greater seasonal temperature extremes that characterized the Pleistocene and drove the migrations during those times when northern and southern mammals seasonally occupied the same territory (Dawkins 1871).

The hypothesis of seasonal migrations was not attractive to James Geikie. He argued that even if summer temperatures at the time had been warm enough to "woo the hippopotamus northwards" (1872:167), this animal simply could not have migrated very far. Not only was it too bulky a brute, as he put it, to accomplish lengthy seasonal moves, but such movements would also imply too great a difference from the habits of its modern relatives: here, he was hoisting Dawkins on his own actualist petard. Geikie (1872) also argued that there were too many bulky southern mammals involved, including several species of elephants and rhinoceroses, to support the seasonal hypothesis. In its place, Geikie (1872) substituted a very different argument.

The Pleistocene, he noted, was apparently characterized by periods of both glacial advance and glacial retreat. The periods of glacial retreat, or interglacials, were times of remarkably equable climate, with greatly dampened seasonal swings of temperature. During these mild periods, Europe was occupied by such warmth-loving creatures as the elephant, rhinoceros, and hippopotamus. When colder conditions returned, the southern herbivores retreated and were replaced by animals that could endure the changed conditions. Extinction of the southern forms occurred because of those changed conditions; reoccupation of Europe by surviving southern mammals has not occurred because the land bridges that had connected Africa with Europe were not reformed after the last episode of glaciation. Why, then, were the bones of northern and southern mammals found intermingled in Pleistocene sediments? Stratigraphic mixture of faunas of very different ages caused such intermingling: "we cannot therefore infer from the occurrence of the horns of a reindeer and the remains of a hippopotamus, in juxtaposition in a Pleistocene deposit, that these animals have lived under similar conditions" (1881:138; see also Geikie 1877).

For Geikie, then, secular migrations followed by stratigraphic mixture accounted for the co-occurrence of northern and southern animals in European Pleistocene deposits. The animals may have been stratigraphically contemporaneous, but they certainly had not been sympatric. Indeed, they could not have been, since the climatic conditions that had allowed the existence of one set of these animals had caused the extinction or extirpation of the other.

By 1874 Dawkins had lost faith in a solely seasonal explanation of the surprising sympatric relationships. While objecting to Geikie's argument that stratigraphic mixture accounted for the co-occurrence of northern and southern mammals (Dawkins 1874:397), he also argued that "it must not... be supposed that the southern animals migrated from the Mediterranean area as far north as Yorkshire in the same year, or the northern as far south as the Mediterranean" (1880:113). Only secular movements could account for the sympatric occurrence of such animals across so wide an area. Dawkins (1874, 1880) now suggested that both seasonal and secular movements, keyed to seasonal and long-term climatic fluctuations respectively, accounted for the mixed nature of the European Pleistocene mammalian faunas. Northern and southern animals would not only meet in some areas during seasonal movements, but they would also meet as they migrated into new territory during periods of major climatic change. And, as for Geikie, these major climatic changes explained the extinction or extirpation of many of the Pleistocene mammals.

Dawkins was not the first or the only earth historian to offer these explanations for the stratigraphic co-occurrence of northern and southern mammals during the Pleistocene. Lyell, in fact, had used the seasonal/secular hypothesis many years before:

> Whenever there is a continuity of land from polar to temperate and equatorial regions, there will always be points where the southern limits of an arctic species meets the northern range of a southern species; and if one or both have migratory habits... they may each penetrate mutually far into the respective provinces of the other. There may also have been several oscillations of temperature during the periods which immediately preceeded and followed the more intense cold of the glacial epoch (1853:147; see also Lyell 1859a, 1863, 1873).

There was, however, another way to reconcile the mixture of northern and southern animals in the Pleistocene deposits of Europe while retaining the assumption that modern relatives of these animals provided an accurate guide to their adaptations. If the musk ox required cold, and the hippopotamus required warmth, and the stratigraphic evidence implied that they had coexisted, than a straightforward reading of all this information could imply that glacial climates had not, as most felt, been marked by severe winters, but had instead been equable.

That reading was, in fact, made by the French paleontologist Edouard Lartet. "In a word," Lartet (1867:191) argued, "there must have been cooler summers for the reindeer and musk-ox; and, on the other hand, warmer winters for the hippopotamus and other species whose analogs are today found withdrawn toward the tropical regions." Not only did this hypothesis account for the co-occurrence of northern and southern mammals, but it also accounted for their subsequent extinction. After the return of a continental climate to Europe, brought about by the retreat of the glacial seas (the same seas used by Lyell and others to account for a variety of Pleistocene geological phenomena, including low elevation drift), arctic and alpine animals would have retreated northwards or upwards, while southern animals would have retreated toward the tropics. Those that did not or could not retreat would have become extinct because of the increased continentality of the climate.

Nearly forty years before, Lyell had suggested that the Siberian mammoth may have become extinct as a result of the increased continentality of Siberian climate. Now, Lartet, working within the framework of glacial theory, introduced equability to account for the mixed nature of an entire fauna, and the loss of equability to account for the extinction of members of that fauna. To adopt this hypothesis, Lartet (1867) realized, it was necessary to abandon the notion of a glacial epoch with severe winters, and to abandon the notion that some Pleistocene extinctions had been caused by the onset of colder conditions. This, Lartet noted, would be difficult for many to do.

By assuming that northern and southern mammals had lived together year-round in Europe, Lartet's hypothesis eliminated the need for Dawkins's migrations. Dawkins (1869) took little time in responding. During the time those mammals had occupied Britain, he argued, the evidence suggested Britain was part of the continent. Given that this was the case, the expansive seas needed to produce an insular climate could not have existed: the climate must have been continental. The immediate issue revolved around the timing of both geological and paleontological phenomena. For Lartet, the animals had coexisted at a time when the glacial seas were sufficiently advanced to create an insular climate; for Dawkins, this coexistence had occurred when those seas were low and the climate, therefore, continental. Only migration, Dawkins concluded, could account for the facts.

In a subsequent paper, written in 1869 or 1870 but published four years after his death, Lartet (1875) continued to argue that an equable glacial climate accounted for both the mixed European Pleistocene mammalian fauna and for the subsequent extinction or extirpation of many of those mammals. Now, however, he cited the work of Gaston de Saporta in his support. Saporta had described and discussed a series of French Pleistocene floras that contained mixed assemblages of northern and southern plants that no longer co-occurred, a situation analogous to that provided by the mammalian data (e.g. Saporta 1870, 1876: for a discussion of Saporta's work and a review of his publications, see Zeiller 1895). The implications of the botanical co-occurrences seemed clear to Saporta: only a humid, equable climate would have allowed such an association. In addition, he felt that these floras were contemporaneous with such Pleistocene mammals as the mammoth, hippopotamus, and rhinoceros, and with times of glacial advance. "Glacial expansion is a phenomenon without direct relation by itself with the rigor of cold" (1876:107), but was instead induced by a combination of the uplift of mountains and increased humidity. It was, therefore, no contradiction to find that the climate at the time of glacial advance was equable. It was only toward the end of the Pleistocene that European climate became colder, drier, and continental, a shift that saw the retreat of glaciers toward their current modest proportions, and the extinction or extirpation of southern mammals from Europe.

While Saporta (1870, 1876) was quite clear on what he saw as the meaning of the floral data, that information could be interpreted in many ways. How others interpreted Saporta's floral assemblages depended in large part on how they explained the climatically mixed faunal assemblages, and, more generally, on how they viewed the march of Pleistocene geological and climatic change. For Lartet (1875), Saporta's plants were fully in line with the equable climate suggested by sympatric northern and southern mammals during the Pleistocene. For Edouard Dupont, whose work in Belgian caves had provided many examples of boreal mammals in Pleistocene sediments and who agreed with Lartet that equability explained the mammalian data (see Dupont 1871: 16–17), Saporta's plants in part conformed with the equability hypothesis. Dupont felt, however, that "an important element seemed to be missing from the beds explored by Saporta" (in Saporta 1876:110), since Saporta had found no subarctic plants in those beds. The problem lay not in the equability hypothesis, but instead with the data, which seemed incomplete. For Dawkins (1880), these floras were early Pleistocene, and simply showed one stage in the retreat of tropical plants from Europe, a retreat that began well into the Tertiary. For Geikie (1881), Saporta's plants were interglacial, fully in accord with the secular faunal movements he had hypothesized. The problem was a stratigraphic one. Lacking clear stratigraphic and chronologic resolution, the temporal position of the plants could be aligned with many conflicting interpretations of the Pleistocene record.

It was also possible to attack the equability argument on the grounds that Lartet's data resulted from stratigraphic mixture, not from sympatry. Alexander Anderson

(1875) employed this approach, just as Geikie (1874) had against Dawkins (1871). Anderson's contribution was written late in 1870 in Canada; Lartet probably never saw it, since he died in January 1871 (Fischer 1873).

Unlike hypotheses invoking seasonal or secular population movements, the equability of glacial climates as an explanation for the co-occurrence of northern and southern mammals in European Pleistocene sediments never gained much popularity during the nineteenth century. Since that was the case, the loss of equability as a cause of Pleistocene extinctions also did not gain much popularity. In his review of the causes that had been advanced to explain mammalian extinctions, Osborn (1906) did not mention loss of climatic equability, nor is Lartet's hypothesis to be found in Reynolds's (1922) otherwise thorough classification of the late nineteenth-century explanations of the distribution of the hippopotamus in Europe during the Pleistocene.

In sum, by the middle of the nineteenth century, accumulating records for the co-occurrence of northern and southern mammals in western European Pleistocene sediments presented a biogeographic picture that demanded explanation. For Agassiz, the picture was an attractive one, since it could be interpreted as implying the rapid replacement of southern by northern faunas, and thus be used to support his geological and biological theoretical positions. For others, however, rapid change of this sort would not do. Three major hypotheses accounting for these stratigraphic co-occurrences emerged: seasonal migration, secular migration, and climatic equability. Although, as the approach of Lyell (1853, 1859a, 1863, 1873) demonstrates, it was not necessary to account for Pleistocene extinctions with the same mechanism used to account for the stratigraphic co-occurrences, Dawkins, Geikie, and Lartet all argued that the same climatic shifts that explained the co-occurrences also explained the extinction or extirpation of many of these mammals in Europe.

Human Causation: Pleistocene Overkill

During the first half of the nineteenth century, most of those who championed the operation of slow, natural changes across the surface of the earth as the cause of Pleistocene extinctions had difficulties suggesting precisely what those changes might have been. A different set of naturalists had no such problem. They suggested that some or all of these extinctions had been caused by human predation. Martin (1967a) has aptly labeled this postulated phenomenon "Pleistocene overkill." I will use this term to refer to the notion that people were the sole or chief cause of Pleistocene extinctions, and I will also examine multivariate hypotheses that saw people as playing a lesser, but still important, role in these extinctions, also using the term "overkill" to indicate the human component.

Early versions of the overkill hypothesis predate the nineteenth century. These early arguments tended to be applied to extinct or extirpated mammals whose remains had been discovered in peat bogs or other superficial deposits thought to be relatively recent in age. The American mastodont provides an excellent example.

In combining misinterpretations of extinct mammals with the simple statement that people caused the demise of those mammals, George Turner's arguments characterize eighteenth-century approaches to overkill explanations. Turner (1799) agreed with earlier conclusions (Hunter 1768) that the American mastodont had been carnivorous. He argued that the numerous fossil remains at Bigbone Lick, Kentucky, must represent the remains of meals of this giant carnivore. He inferred from the size of the animal and the nature of its prey that it must have been a saltatorial predator: "as the immense volume of the creature would unfit him for coursing after his prey through thicket and woods, Nature had furnished him with the power of taking it with a mighty leap" (1799:517–518). Clearly, such an animal would have been "at once the terror of the forest and of man," and from this it follows that the extinction of the mastodont may have been due to people, who may "have made the extirpation of this terrific disturber a common cause" (1799:518).

Early nineteenth-century actualist attempts to attribute Pleistocene extinctions to human activities depended on two major assumptions, one or both of which was contested by the opponents of the overkill hypothesis.

First, it had to be assumed that people had the ability to cause the extinction of large quadrupeds. While it was recognized that modern people could cause great changes in the distribution of larger vertebrates, that they could cause their total extinction was a point of contention. Cuvier, for instance, doubted that ancient people could have caused Pleistocene extinctions, since "modern peoples... have continually driven back the noxious animals but have succeeded in exterminating none" (1812a:47; 1825:78). Cuvier's position was part of a larger argument directed against Lamarck (1801, 1809). In attempting to demonstrate that members of modern species could have been modified in descent from more ancient ones, Lamarck argued that if the paleontological record really documented any extinctions, they were probably only those few that had been caused by human activities (Rudwick 1972, Burkhardt 1977, Grayson 1983).

As the nineteenth century advanced, however, arguments against the ability of people to cause vertebrate extinctions quickly faltered in the face of indisputable evidence that they could and did cause such changes. Even though Lyell, for instance, noted that "we often... form an exaggerated estimate of the extent of power displayed by man in extirpating some of the inferior animals" (1830:161), he was nonetheless deeply impressed with the impact of humans on the distribution of modern plants and animals. He noted that "we wield the sword of extermination as we advance" (1832:155), and that the effects of people are such that "we must at once be convinced, that the annihilation of a multitude of species has already been effected, and will continue to go on hereafter, in a still more rapid ratio, as the colonies of highly civilized nations spread themselves over unoccupied lands" (1832:155). Indeed, as the century advanced, it became so clear that people could have marked effects on the distribution and abundance of animals that for some the argument became not whether people could cause extinctions, but whether any historically documented cases of vertebrate extinctions could be shown to be due to any other causes (Owen 1860).

Agreement that people could cause the extinction of large vertebrates did not automatically result in an increase in the popularity of overkill hypotheses, since the second assumption required by these approaches was much more substantial than the first. Even if human activities could cause extinction, in order to maintain that people had caused the extinction of Pleistocene vertebrates one had to assume that they had coexisted.

Frequent criticisms of this second assumption by influential scientific figures are to be found in the first half of the nineteenth century. Cuvier (1812a) noted the absence of human remains in deposits that had yielded extinct Pleistocene vertebrates, providing yet another rebuttal to Lamarck. For Buckland, the absence of human traces in such deposits was one of the "two great points" in which geology and Moses agreed. Both left open the possibility that the remains of early peoples might be found: Cuvier (1812a) suggested that such people may have occupied only narrowly circumscribed regions, while Buckland (1823) felt that the vestiges of our earliest ancestors were to be found in Asia (see Grayson 1983).

Many subsequently accepted demonstrations of the association of human remains with extinct Pleistocene mammals made prior to the 1850s were ignored or rejected by major earth scientists (Daniel 1950, Oakley 1964, Grayson 1983). Prior to 1860, Lyell routinely rejected such arguments. In the second volume of the *Principles of Geology,* he reviewed suggested associations between human remains and the remains of extinct Pleistocene mammals in French caves, and rejected them all:

Must we infer that man and these extinct quadrupeds were contemporaneous inhabitants of the south of France at some former epoch? We should unquestionably have arrived at this conclusion if the bones had been found in an undisturbed *stratified* deposit... but we must hesitate before we draw... inferences from evidences so equivocal as that afforded by the mud, stalagmites, and breccias of caves, where the signs of *successive* deposition are wanting (1832:225–226; emphasis in original).

For nearly three decades, Lyell remained convinced that the appearance of people on earth postdated the extinction of Pleistocene mammals. His uniformitarian model of earth history saw the human species as an exception to the general rule that the paleontological record did not illustrate the progressive development of life on earth. To explain away the apparent progression shown by that record, Lyell relied on the incompleteness of our knowledge of the past (Bowler 1976). Regarding human history, however, he was guided by the belief that our species played a special role on earth, a role indicated in part by the very recency of our appearance. To support such recency, he depended heavily on the same sort of negative evidence that he rejected for other organisms (see, for instance, his scientific journals on the species question [Wilson 1970:262]). In addition, he dismissed positive evidence for ancient human beings, evidence he would have quickly seized for any other life form (Bartholomew 1973, Bowler 1976, Grayson 1983). Thus, even though he stressed the tremendous impact people could have on animal abundance, and even though he lacked a convincing uniformitarian explanation for Pleistocene extinctions, he had no choice but to reject overkill approaches: acceptance would have required people on earth at a much earlier time than he was willing to admit. After 1859, as I shall discuss, his position on these issues changed sharply.

Although Lyell, Cuvier, and others were unwilling to make one or both of these assumptions during the early 1800s, a number of naturalists were not, and the idea that people had caused the demise of numerous Pleistocene mammals gained some popularity during the early decades of the nineteenth century. This was especially true in North America, where glacial deposits were much less widespread than in Europe, and the relationship between the time of extinction and these deposits therefore less evident. In fact, by 1849, Lyell was able to note, without approval, that the hypothesis that American Pleistocene extinctions were caused "by the arrows of the Indian hunter, is the first idea presented to the mind of almost every naturalist" (1849:259).

The Reverend John Fleming, an important Scottish zoologist, was one of those naturalists. Fleming's early arguments (1824, 1826), directed against Buckland's catastrophism (for a discussion of the debate, see Page 1969, 1972; see also Buckland 1825), are instructive since they deal explicitly with both assumptions required to defend the overkill hypothesis within an actualist framework early in the nineteenth century.

Fleming began his argument by noting that "the progress of society is exerting, and has exerted, a powerful influence on the geographical distribution of British animals," to the point that some species have "perished from off the land" (1824:295). To establish this point, Fleming surveyed historic changes in the distribution and abundance of various mammals and birds in the British Isles. He noted that "Eagles, Ravens, and Bustards have entirely disappeared from the more cultivated districts" (1824:291), and that numerous other vertebrates had suffered similar or more severe fates. Fleming concluded that people had done great damage to the native British fauna, and the data indicated that the same was true for continental Europe. Fleming thus established a major point in his uniformitarian argument: people could cause vertebrate extinction. Indeed, Lyell (1832) made heavy use of Fleming's arguments in his own attempt to establish that extinctions have occurred during modern times.

Fleming developed the rest of his argument quickly. European Pleistocene mammals had a distribution similar to those of mammals extirpated by people in recent times. Situations in which the remains of these animals are found indicate that no great change in the physical nature of the earth has taken place since their extinction. These animals also occur in paleontological sites side-by-side with those that survived and "seem well suited to the climate" (1824:302). Thus, Fleming rejected both catastrophist arguments that called for marked, rapid changes in the nature of the earth to account for extinction, as well as arguments that called upon slow climatic change.

In arguing for the contemporaneity of people and extinct Pleistocene mammals, Fleming could have taken two approaches. First, he could have argued that the animals were old, dating to and predating the time of diluvial deposition. In this case, any associated human remains would have had a much greater antiquity than was generally supposed. Second, he could have argued that the animals had survived for a long time after the deposition of the diluvium.

Fleming took the second approach, arguing that "the relics of these ancient animals occur in postdiluvian strata" (1826:211). The discovery of such extinct mammals as the Irish "elk" and woolly rhinoceros in peat and marl deposits in western Europe, "the recent formation of which is not disputed by any class of geologists" (1824:296), demonstrates the recent extinction of such animals. At least some of these extinctions, then, occurred prior to recorded history but well after the deposition of diluvium.

By approaching the age of extinct mammals in this fashion, Fleming was able to argue that Buckland's universal deluge could not have caused the extinctions because at least some of the animals survived the flood. This point was, in fact, a major thrust of Fleming's thesis. In addition, the association of extinct mammals with human remains became less contentious, because the associations no longer implied an unacceptable human antiquity.

Perhaps this is why Fleming did not go to great lengths to establish the contemporaneity of people and extinct animals. In Fleming's view, such an association meant little in terms of human antiquity since the late survival of extinct Pleistocene quadrupeds could easily be encompassed within the "long term of nearly 6000 years" (1824:290) that he, in accord with general scientific opinion, allowed for that antiquity. Fleming's arguments concerning these associations are markedly weak. Rather than examining them in detail, he merely cited two published examples. In the first, he noted that the remains of an elephant (presumably *Mammuthus primigenius)* had been found in the same kind of deposits at roughly the same depth and in roughly the same area as a "copper battle-axe" (1824:298). In the second, he noted that human bones had been found beneath the remains of extinct Pleistocene mammals in a fissure at Köstritz, East Germany (an association that the analyst of the Köstritz remains, Ernst von Schlotheim, denied, as Buckland [1823] had carefully pointed out). In neither case did Fleming provide stratigraphic evidence for the validity of the associations, nor did he explore alternative explanations for them. In fact, his arguments were so unconvincing that the editor of the *Edinburgh Philosophical Journal,* in which Fleming's papers appeared, appended demurrers to each, noting that "we do not yet possess any authentic instances of human remains occurring in the beds that contain bones of elephants, rhinoceros, etc." (in Fleming 1824:302).

Nonetheless, Fleming felt the point had been well made. He concluded that "man was an inhabitant of this country at the time these animals, now extinct, flourished, his bones and instruments having been found in similar situations with their remains" (1824:303). Given this contemporaneity, and given that there is no reason to believe that ancient peoples did not carry out "extirpation operations" (1826:236) against numerous species of vertebrates, just as modern peoples do, Fleming's answer to Buckland followed logically: "we must refer the extinctions of these early quadrupeds to the destructive influence of the chace" (1826:235); "the weapons of the huntsman completed the extinction of these animals, from the first ages the object of his persecution" (1824:304). Not only did the cause of these extinctions differ from the causes asserted by Buckland and other catastrophists, but the rate of extinction differed as well. The extinctions had not occurred all at once, but had taken place slowly and one-by-one, since the "process of extirpation is gradual" (1826:237).

Although Fleming and Lyell disagreed about the causes of extinction, their methods and conclusions were quite similar. Both established that people could cause extinction, although the reasons for establishing this fact differed; both argued that at least some of the extinctions occurred after the deposition of diluvium; and both maintained that the extinctions occurred slowly. Their arguments were uniformitarian in method and conclusion. The real difference revolved around one point: Lyell had much stricter standards for accepting the contemporaneity of people and extinct mammals than did Fleming, and, at this time, rejected such contemporaneity.

Fleming exemplifies early overkill theorists not only in the ease with which he accepted the contemporaneity of people and extinct mammals, but also because he was not a major theoretician in a wide range of matters relating to earth history. His dispute with Buckland, in fact, shows both his zoological strengths and his geological weaknesses. Lamarck (1801, 1809) was willing to suggest that some few large Pleistocene quadrupeds might have become extinct as a result of human activities. Playfair (1802), in the course of an argument directed against catastrophists de Luc (e.g. 1793, 1794a, 1794b, 1795) and Kirwan (1799), suggested that people may have played a role in causing the extinction of some Pleistocene quadrupeds. But prior to the late 1850s, the overkill hypothesis was rarely adopted by earth scientists well versed in the geological and paleontological records. Evidence was simply insufficient to support both the assumptions that an actualist approach to the overkill hypothesis required, and, as the decades passed, to support the argument that people and extinct Pleistocene mammals had coexisted.

The importance of the demonstration of the contemporaneity of people and extinct Pleistocene mammals to the overkill hypothesis cannot be overestimated. Prior to the 1850s, overkill approaches were adopted primarily by those who were not influential theoreticians in a wide variety of matters relating to earth history. After the 1850s, many of the most substantial synthesizers and earth historians included an overkill component in their explanations of Pleistocene extinctions. And, it was during this decade that the reality of associations between human remains and Pleistocene mammals was finally accepted by virtually all Quaternary scientists.

The development of Lyell's explanations of Pleistocene extinctions is instructive in this regard. As part of his pre-1860 denial of a human role in Pleistocene extinctions, Lyell strongly questioned the argument that humans and the animals involved had been contemporaneous. Given the general similarity of Lyell's approach to Fleming's, given that Lyell was extremely impressed with the ability of modern peoples to cause extirpation and extinction, and given that Lyell could not specify precise causes for the extinctions, it is reasonable to speculate that the lack of convincing evidence for this contemporaneity was the major reason for Lyell's total rejection of the overkill hypothesis during the first half of the century.

But by 1860 Lyell had changed his mind. In a journal entry made in March he noted that the antiquity of man "throws great light on extermination of animals, and in Denmark, of trees" (Wilson 1970:356). And, in 1863, while still convinced that the causes of Pleistocene extinctions lay largely in causes "more general and powerful than the agency of man," he now argued that "the growing power of man may have lent its aid as the destroying cause of many Pleistocene species" (1863:374).

This shift was a response to archaeological research in France and England during the 1850s that demonstrated to the satisfaction of most western scientists that people and extinct Pleistocene mammals had, indeed, coexisted. The establishment of a proper order of human antiquity, in which the association of human remains with extinct Pleistocene mammals played a major role, was a complex affair that has yet to be analyzed in its entirety. Empirical arguments came from considerations of the linguistic and physical divergence of human groups as well as from the archaeological record itself. It was the archaeological evidence that was decisive, since only archaeology provided a direct access to chronological questions. The argument that the human race had great time depth on earth became successful during the middle 1800s. Darwin immediately provided a theoretical framework in which a tremendous human antiquity could be understood, and questions of human antiquity quickly became caught up in discussions of the larger issue of human evolution (see Grayson 1983).

For those who were struggling with the matter of Pleistocene extinctions, the crucial demonstration resulted in largest part from the work of a French civil servant, Jacques Boucher de Perthes, who, beginning in the late 1830s, accumulated evidence for the contemporaneity of people and extinct mammals in the Somme River Valley (Boucher de Perthes 1847, 1857, 1864).

In conducting this work, Boucher de Perthes realized that if he were to establish contemporaneity, he would have to gather evidence with care and from multiple sites: "it is not lightly that one can put in doubt a generally admitted order of things and a system established on the basis of long experience. An isolated fact proves nothing. . . " (Boucher de Perthes 1847:227). As a result, he paid close attention to stratigraphy and to the association of artifacts with deposits that were demonstrably diluvial or prediluvial. In addition, he depended heavily on the opinions of geologists who had worked in the area for assessments of the ages of the strata in which he was working. This care was made even more necessary by the fact that, unlike Lyell and Fleming, Boucher de Perthes felt that extinct Pleistocene mammals were no younger than diluvial in age, thus implying a great antiquity for any associated human materials.

Boucher de Perthes was, in fact, a committed catastrophist. He argued that there had not only been "a last deluge, that of the scriptures and of tradition" (1847:244), but that this had been preceded by other, yet more terrible catastrophes, that resulted in the immediate dissolution of all forms of life. This was standard catastrophism of a form still acceptable in France, though rejected by most contemporary English scientists. However, it was not standard to argue that these earlier catastrophes had destroyed "the human species, as well as all of the races of which fossil debris is found" (1847:244–245), and that these species were renewed after each catastrophe had been terminated. Bold enough to hypothesize such a series of catastrophes and associated effects on people, Boucher de Perthes suggested that the earliest members of the human species may have walked the earth thousands of centuries ago. Indeed, this view of human antiquity was mild compared to the millions of centuries he assigned to that antiquity in the draft of the first volume of *Antiquités Celtiques et Antédiluviennes* (Aufrère 1940).

Boucher de Perthes's stratigraphic excavations were conducted in and near Abbeville, in the valley of the Somme. Here, he or his workers found in and below diluvial deposits a series of objects, including "haches diluviennes," that he attributed to human workmanship. In or above the strata that contained these objects were the bones of extinct Pleistocene mammals, including those of woolly mammoth, woolly rhinoceros, and hippopotamus, "mammals whose races, actually destroyed or foreign to our climate, pertain to the diluvian epoch" (Boucher de Perthes 1847:244). The geological relationship of artifacts and bones was so clear as to allow no conclusion other than that man and extinct beast had existed contemporaneously. Although Boucher de Perthes had not found the bones of the makers of his artifacts (indeed, many of his artifacts were not Paleolithic in age [see Aufrère 1940], and many were not even artifacts [see Boucher de Perthes 1847, chapter 23]), he was certain that they would be found. In addition, he was equally certain that they would differ in form from those of modern peoples: "post-diluvian men are no more descendants of ante-diluvian men than are today's elephants descended from those found in the clysmien strata, and if some day the bones of these ante-diluvian men are discovered, nuances of form will be found which will prove what I assert" (Boucher de Perthes 1847:245).

In understanding the context of Boucher de Perthes's speculations, it is important to recognize the similarity between his approach to earth history and that of some of the late eighteenth-century French Enlightenment philosophes. For instance, Paul Henri

Thiry, Baron d'Holbach presented a model of earth history in his *Système de la Nature* (1770) which, though developed in less detail and set in a militantly atheistic framework, is similar in several ways to that presented by Boucher de Perthes. Holbach (1770, Vol. 2:29) suggested that "there may have been, perhaps, men on the earth from all eternity, but in different periods they may have been destroyed, together with their monuments and their sciences; those who survived these periodical revolutions, each time formed a new race of men...." As did Boucher de Perthes, Holbach suggested that the world would have been a different place after such a revolution, and that, as a result, the new races of people would have been different in form from those before them. In fact, he suggested that "primitive man differed, perhaps, more from modern man than the quadruped differs from the insect" (1770, Vol. 1:85). Catastrophes, tremendous human antiquity, and periodic destruction and subsequent appearance of morphologically distinct races of people all formed part of Holbach's system, and formed as well the basis of Boucher de Perthes's earth, and within it, human history. It might be fruitful to explore the possibility that the origins of at least some of Boucher de Perthes's ideas on human history lie within the writings of Holbach and the members of his circle (for a discussion of Holbach's work and his circle, see Wickwar 1935, Naville 1943, Gay 1966, and Kors 1976). In a very real sense, Boucher de Perthes's work represents Holbach's system as it applies to human history put to the empirical test. Seen in this light, Boucher de Perthes's great desire to find the bony remains of the makers of his ancient stone tools, a desire that led to the Moulin-Quignon debate (Boucher de Perthes 1864; see Oakley 1964 and Grayson 1983 for access to the voluminous contemporary literature on this debate), becomes even more understandable. Only such remains would have allowed him to demonstrate the differences in "nuances of form" that his system led him to expect.

It is also true that Boucher de Perthes's early approach was an anti-transformationist one. His repeated catastrophic extinctions of life forms had the effect of denying the possibility of descent with modification for those forms that had been totally destroyed. These totally destroyed forms explicitly included the antediluvian peoples whose remains he had found (e.g. Boucher de Perthes 1847:578–580). Working at a time, and in a place, that had produced both Lamarck and Étienne Geoffroy Saint-Hilaire, Boucher de Perthes had built into his initial theoretical structure a denial of any transformationist possibilities relating to the discovery of his ancient peoples (for a discussion of the development of Boucher de Perthes's views, see Grayson 1983).

Boucher de Perthes was confident that his arguments would ultimately be accepted. "How many things which were improbable a half-century ago," he asked, "are today proven, and how many others that are rejected as absurd today will, before another half-century passes, be recognized as logical and incontestable" (1847:267). Although his confidence in himself was not misplaced, his concerns were also appropriate, and his work was at first ignored or rejected. At the time, for instance, Darwin looked at Boucher de Perthes's work and "concluded that the whole was rubbish" (F. Darwin 1911:200); Lyell, in commenting on his pre-1850s rejection of apparent associations between human remains and extinct Pleistocene mammals, frankly explained that "I can only plead that a discovery which seems to contradict the general tenor of previous investigation is naturally received with much hesitation" (1863:68).

But Boucher de Perthes persisted (Boucher de Perthes 1857, 1864; see Aufrère 1940, Daniel 1950, Oakley 1964, Grayson 1983). In the late 1850s, a succession of geologists and archaeologists visited his excavations and left favorably impressed (Evans 1860; Prestwich 1860, 1873). Shortly thereafter, Darwin noted that "the high antiquity of man has recently been demonstrated by a host of eminent men, beginning with M. Boucher de Perthes, and this is the indispensable basis for understanding his origin. I shall, therefore, take this conclusion for granted..." (187lb:3). Darwin even felt that Boucher de Perthes "has done for man something like what Agassiz did for glaciers" (F. Darwin 1911:200), while Huxley (1869) compared him to Columbus.

To understand the initial rejection of Boucher de Perthes's work, it is essential to recall that he embedded his evidence for great human antiquity in an extreme catastrophist model of earth history, a model that had more in common with the outmoded speculations of Holbach and his associates than to any approach that enjoyed the general favor of contemporary scholars. No wonder, then, that British and continental scientists alike did not receive his views favorably. From their viewpoint, Boucher de Perthes was a provincial amateur publishing an outlandish system of earth history. If Robert Chambers, whose *Vestiges of the Natural History of Creation* was published three years before the first volume of *Antiquités Celtiques,* was on the outskirts of the scientific community (Ruse 1979), Boucher de Perthes was in some respects in a different century altogether. How could Darwin feel it was anything but rubbish, especially in Chambers's decade? Geikie's explanation for the rejection was certainly in large part correct:

> Perhaps one of the reasons why the French discoveries were so long passed over by English scientific men was the general conclusion arrived at by Boucher de Perthes, that the flint implements and mammalian remains were entombed together by the Noachian deluge. By geologists in this country the idea of a general deluge had long been discredited... so deeply had uniformitarian doctrines been imbibed, that *debâcles* and deluges of any kind had come to be looked upon with considerable disfavour (Geikie 1881:123, emphasis in original; see also Evans 1860).

It was more than the debâcles that dictated the response to Boucher de Perthes's work, it was his entire theoretical framework, coupled with the generally accepted notion that people were of relatively recent origin.

It is, then, no surprise that verification of Boucher de Perthes's claims for great human antiquity was largely an offshoot of the careful excavation of Brixham Cave in England, an excavation conducted by geologists for geological reasons, but that provided strong evidence for the contemporaneity of people and extinct mammals (see the excellent discussion by Gruber 1965). Before Boucher de Perthes's ancient humans could be accepted, they had to be extricated from his model of earth history, and it was this extrication that Brixham Cave allowed.

Lyell was one of the geologists who visited Boucher de Perthes's excavations in 1859 and came away convinced by the evidence (Lyell 1859b). In 1860, he wrote to a friend that

> I have been very busy lately with the proofs afforded by the flint implements found in the drift of the valley of the Somme at Amiens and Abbeville... of the high antiquity of man. That the human race goes back to the time of the mammoth and rhinoceros (Siberian) and not a few other extinct mammals is perfectly clear... (K. Lyell 1881:341; see also Wilson 1970:lvii).

Shortly thereafter, Lyell transformed this new conviction into the *Geological Evidences for the Antiquity of Man* (Lyell 1863), which provided a lengthy summary of the evidence for the coexistence of people and extinct Pleistocene mammals. Such evidence made it possible to incorporate an overkill component in a uniformitarian approach to earth history. As I have discussed, Lyell had long stressed the importance of biotic interactions in causing extinction (e.g. Lyell 1832), and had long stressed the impact of humans on flora and fauna. A human role in causing Pleistocene extinctions provided him with the most powerful cause of those extinctions he could have had. And, in *Antiquity of Man,* he accepted the view that people may have helped cause the extinction of some of Europe's Pleistocene mammals.

While I have examined Lyell's approach to Pleistocene extinctions in detail, the ideas of other naturalists evolved similarly. Famed anatomist and paleontologist Richard Owen provides an excellent example. In his survey of British fossil mammals and birds,

Owen (1846) noted that since only a part of the Pleistocene fauna had become extinct, catastrophist explanations for those extinctions had to be rejected. He also felt a strong analogy could be made between the extinction of Pleistocene mammals and the reduction in numbers of animals that accompanied human occupation of both islands and continents. While maintaining that it would be wise to be cautious in accepting the negative evidence that people and Pleistocene mammals had not coexisted in Europe, and therefore could not have played a role in the extinction of those mammals, he also argued that

> the saber-toothed Machairodus, the great Spelaean Tiger, Hyaena, and Bear, together with the gigantic pliocene Pachyderms, became extinct here and elsewhere, as it would seem, before the creation of Man, —which would indicate that the extirpating cause, if it were exterior to their own constitution, had been due to changes of the configuration and climate of the great continents over which they ranged (1846:xxxiii–xxxiv).

After noting that neither rapid nor gradual changes in climate seemed sufficient to account for the extinction of the mammoth, he concluded that

> with regard to many of the large Mammalia, especially whose which have passed away from the American and Australian continents, the absence of sufficient signs of extrinsic extirpating cause or convulsion, makes it almost as reasonable to speculate... on the possibility that species like individuals may have had the cause of their death inherent in their original constitution, independently of changes in the external world, and that the term of their existence, or the period of exhaustion of the prolific force, may have been ordained from the commencement of each species (1846:270).

By 1860 Owen's ideas had changed. After again denying the possibility that the extinctions had occurred as a result of "exceptional cataclysmal changes," he noted that "all hitherto observed causes of extirpation point either to continuous slowly operating geological causes, or to no greater cause than the, so to speak, spectral appearance of mankind on a limited tract of land not before inhabited" (1860:399). Regarding the woolly mammoth and woolly rhinoceros, "recent discoveries indicate that... a rude primitive race of mankind may have finished the work of extermination begun by antecedent and more general causes" (1860:401). Owen (1860) leaves no doubt as to what those recent discoveries were. After favorably mentioning the overkill hypothesis, he enumerated a number of instances in which associations between people and extinct Pleistocene mammals had been established, an enumeration that began with and focused on the work of Boucher de Perthes.

Owen later abandoned the caution he showed in 1860 in attributing aspects of Pleistocene extinctions to people. For instance, in discussing the extinction of Australian marsupials, he went further than did most influential nineteenth-century naturalists in adopting an overkill position for both Europe and Australia: "the extirpating cause of the *Felis spelaea* together with the huger herbivores (*Elephas, Rhinoceros, Megaceros, Bos primigenius*), represented by remains in British caverns, may be inferred to have operated in relation to the analogous evidence in Australia. That cause I conceive to have been prehistoric man" (Owen 1883:643; see also Owen 1870, 1872, 1884). He did not, however, apply this argument to South America. Although fully aware of the magnitude of extinction there (he had initially described many of the extinct mammals involved), he doubted that people would have persecuted so useful an animal as the horse, and generalized this doubt to other extinct genera as well (Owen 1869). Indeed, Owen's concept of directed evolution saw the horse as having evolved expressly for human use (Owen 1868), and it would have been inconsistent for him to suggest that the very reason for the existence of horses could also have been the reason for their extinction.

Owen's broad acceptance of the overkill hypothesis may have been facilitated by his study of extinct moas, for he had long maintained, with good reason, that these giant birds had not become extinct until very recent times, and that this extinction was due to human predation (Owen 1849, 1862a, 1862b). Indeed, he even argued that cannibalism among New Zealand natives may have resulted from the loss of moas as a source of food (Owen 1849). Gideon Mantell, whose son had collected the bones described by Owen (1849), immediately denied the cannibalism hypothesis by pointing out that his son had also excavated mounds that contained the bones of "Man, Moa, and Dog... promiscuously intermixed" (1850:175). Clearly, moa hunting and cannibalism had been contemporaneous, not consecutive, practices. Mantell nonetheless agreed with Owen on the cause of moa extinction.

Many other nineteenth-century naturalists followed suit (e.g. Le Conte 1879; Lubbock 1872, 1890; Lydekker 1896), although some took longer than others to incorporate overkill into an explanation of Pleistocene extinctions. Alfred Russel Wallace was one of the recalcitrants. Like Darwin, Wallace was well aware of the great differences between Pleistocene and Holocene faunas in many parts of the world, and of the explanatory problems posed by the disappearance of numerous Pleistocene mammals.

> We live in a zoologically impoverished world, from which all the hugest, and fiercest, and strangest forms have recently disappeared; and it is, no doubt, a much better world for us now they have gone. Yet it is surely a marvellous fact, and one that has hardly been sufficiently dwelt upon, this sudden dying out of so many large mammalia, not in one place only but over half the land surface of the globe (1876:150).

In 1876 Wallace argued that the cause of Pleistocene extinctions "lies in the great and recent physical change known as the 'Glacial Epoch'" (1876:151), a view he repeated nearly two decades later (Wallace 1892). Not until after the turn of the century did he become attracted to the overkill hypothesis and incorporate that hypothesis into his explanation of Pleistocene extinctions. "What we are seeking for," Wallace (1911:264) noted, "is a cause which has been in action over the whole earth during the period in question, and which was adequate to produce the observed result. When the problem is stated in this way, the answer is very obvious. It is, moreover, a solution which has often been suggested, though generally to be rejected as inadequate. It has been so with myself, but why I can hardly say." Wallace then quoted the use of the overkill hypothesis by Lyell (1863), and concluded

> Looking at the whole subject again, with the much larger body of facts at our command, I am convinced that the rapidity of... the extinction of so many large Mammalia is actually due to man's agency, *acting in co-operation with those general causes* which at the culmination of each geological era has led to the extinction of the larger, the most specialised, or the most strangely modified forms (1911:264; emphasis in original).

While Wallace noted that he did not understand why he rejected the overkill hypothesis for so long, the reason seems evident. Wallace's major biogeographic works (Wallace 1876, 1880, 1892) relied heavily on "recent changes of climate as dependent on changes of the earth's surface, including the causes and effects of the glacial epoch" as factors that could "explain the dispersal of all kinds of organisms, and thus bringing about the actual distribution that now prevails" (Wallace 1905:100). In *The Geographical Distribution of Animals* (1876) and *Island Life* (1880, 1892), Wallace suggested that glacial phenomena were found in most parts of the earth. The events of the glacial epoch, then, provided a global mechanism that could explain many aspects of the modern distribution of organisms. If this were true, then certainly Pleistocene distributions, including Pleistocene extinctions, could be explained in the same way. Given this framework, why consider yet another cause for those extinctions? To do so might suggest that the events of the Pleistocene were not as efficacious as Wallace had argued.

But as the true extent of Pleistocene glaciation and associated phenomena became better and better known, it became evident that "the ice sheet had very definite limits" (Wallace 1911:262) and could not, in fact, provide the global cause he realized was necessary. It was only after he concluded that glaciation and its accompanying effects could not account for Pleistocene extinctions that he reevaluated the overkill hypothesis. The "much larger body of facts at our command" clearly referred not to any new data on the antiquity of people, an antiquity that Wallace had long asserted was very great (e.g. Wallace 1864, 1891), or to any new data on the impacts of people on Pleistocene faunas, but instead to accumulating information that implied the effects of glaciation were not as widespread as he once believed.

Not everyone was willing to accept a human role in the extinction of European Pleistocene mammals. With almost ironic similarity to Lyell's much earlier criticism, Searles Wood noted that as far as the extinction of European and Asian large mammals was concerned "the favorite hypothesis seems to have been that they owed their extinction to the attacks of Post-glacial man, whose implements are not infrequently found with their remains..." (Wood 1872:155). For Wood, this hypothesis made little sense. In Africa and Asia, he noted, it was only recently, and with the aid of firearms, that any decrease in the number of large mammals had occurred.

> Are we to suppose that thousands of years before... the scattered tribes of men who managed to exist along the shores and rivers of Europe, and of Northern and Central Asia, exterminated with their feeble weapons of bone and flint the gigantic pachydermata and felines of the Post-glacial period? Fancy attacking a rhinoceros, whose hide will turn a rifle bullet, with a flint hatchet or a bone skewer! (1872:155–156).

Wood then approached the problem in much the same way Dawkins had. He noted the apparently incongruous presence of northern and southern mammals in Europe during the Pleistocene, pointed out Dawkins's discussion of this incongruity, and argued that a combination of seasonal and secular migrations accounted for the stratigraphic co-occurrence of such animals. Having explained the co-occurrence, he then argued that the extinctions were due to climatic change.

Overkill also played no role in the approaches of Dawkins (1871, 1874, 1880), Geikie (1872, 1877, 1881), and Lartet (1867, 1875). As with Wood, all three were impressed by the apparent ecologic incompatibility of such animals as the musk ox and hippopotamus, and all three focused on climatic explanations of these seemingly incongruous relationships. Having derived what they felt were fully adequate climatic accounts for those relationships, and at the same time for the subsequent extinction or extirpation of the animals involved, there was no need to call upon overkill. As was the case with Wallace (1876, 1880, 1892), to have called upon overkill would have weakened the explanatory mechanism they had employed to account for other aspects of European Pleistocene faunas. This, I suspect, is why Saporta included overkill as a likely cause of Pleistocene extinctions in 1868, but, after the development of the equability hypothesis to account for the mixed nature of European Pleistocene floras and faunas, mentioned only equability as a cause of those extinctions in 1874 (Saporta 1868, 1876).

A Note on Darwin

Darwin's early thoughts on mammalian extinction were heavily influenced by Lyell, as Herbert (1974) and Kohn (1980) have noted. In 1835, while on board the *Beagle,* Darwin rejected catastrophic explanations for the extinction of South American mammals and followed Lyell in suggesting that those extinctions must have occurred over a long period of time: "with respect to the *death* of species of Terrestrial mammalia in the S. Part of S. America I am strongly inclined to reject the action of any sudden debacle. –Indeed the very numbers of the remains renders it to me more probable that they are owing to a succession of deaths after the ordinary course of nature" (Darwin MSS, Vol. 42, University Library, Cambridge; cited in Herbert 1974:236; emphasis in original).

But what was the "ordinary course of nature" above and beyond the succession of deaths? What was the cause of these extinctions? During the late 1830s, and after having become a transmutationist, Darwin dealt with two major possible causes of extinction. First, and following Lyell, he drew an analogy between the length of life of an individual and the length of existence of a species. "And that at present can be said with certainty," he wrote in 1837, "is that, as with the individual, so with the species, the house of life has run its course, and so is spent" (1839:212). Similarly, and at about the same time, he noted to himself that he was "tempted to believe animals created for a definite time—not extinguished by change of circumstances" (Red Notebook, p. 129; Herbert 1980:66). This approach, which explicitly likened the generation of species to the generation of individuals (Notebook B, p. 63; de Beer 1960a: 49), provided Darwin with a nonenvironmental cause for extinctions (Herbert 1980) and with a means of keeping the number of species on earth in equilibrium through time (Notebook B, p. 37; de Beer 1960a: 46).

In his analysis of Darwin's early approach to transmutation, Gruber (1974) suggested that one of Darwin's reasons for adopting a nonenvironmental cause for extinctions related to his reaction to catastrophist accounts of ancient extinctions. More convincing, however, is Kohn's suggestion that Darwin was impressed by the lack of evidence for adaptational causes for the extinction of *Macrauchenia* in South America (Kohn 1980). Certainly, the rejection of catastrophist accounts did not exclude adaptational explanations of extinction, as Lyell had already shown. Indeed, Darwin soon rejected racial senility as implausible (Kohn 1980) and began to search for extrinsic, adaptational causes. Thus, early in Notebook B (written between July 1837 and February 1838), Darwin suggested that "death of species is a consequence (contrary to what would appear from America) of non-adaptation of circumstances" (Notebook B, pp. 38–39; de Beer 1960a:46). In subsequent notebooks dealing with transmutation, written between 1838 and 1839, Darwin elaborated the adaptational theme. Here again the influence of Lyell is noteworthy. In Notebook C he generalized Lyell's equability argument to both Old World and New World Pleistocene extinctions: "whatever destroyed great Pachyderms in S. America destroyed great Edentata or American form.... The climates having grown more extreme both in N. and S. America, is only common cause I can conceive of destruction of great animals in Europe and America" (Notebook C, p. 37; de Beer 1960b:86; see also de Beer 1960c:162).

Darwin's notebooks went unpublished until this century. Of the senility and equability arguments, only the former was published by Darwin (1839). In 1845, we find Darwin still close to Lyell in rejecting both catastrophist and overkill explanations:

> What, then, has exterminated so many species and whole genera? The mind is at first irresistably hurried into belief of some great catastrophe; but thus to destroy animals, both large and small, in southern Patagonia, in Brazil, on the cordillera of Peru, in North America up to Behring's Straits, we must shake up the entire framework of the globe.... Did man, after his first inroad into South America, destroy, as has been suggested, the unwieldy Megatherium and other Edentata? We must at least look to some other cause for the extinction of the fossil mice and other small quadrupeds (1871a [1845]: 223–224).

Climatic change did not seem to provide a mechanism, and certainly climatic change or any other phenomena associated with the glacial epoch did not do so. His South American data, Darwin noted, fully agreed with Lyell's North American observations: the extinct mammals were found stratigraphically above ice age deposits (Darwin 1846).

Nonetheless, Darwin (1846) also felt that even though the precise causes of Pleistocene extinctions were unknown, we should feel no more surprise when we are confronted with the fact of extinction than when confronted with the fact that a species is rare. Darwin addressed the general concept of extinction at length in the *Origin of Species* (1859, 1958 [1872]). In this discussion, he stressed the complexity of extinction and the importance of understanding competitive relationships within faunas if any instance of extinction is to be understood. He concluded, as he had earlier, that

> Whenever we can precisely say why this species is more abundant in individuals than that; why this species and not another can be naturalized in a given country; then, and not till then, we may justly feel surprise why we cannot account for the extinction of any particular species or group of species (1859:322).

Given this frequently expressed position (e.g. Darwin 1839, 1871a [1845], 1859, 1871b, 1958 [1872]), it is understandable that Darwin did not forward precise explanations for Pleistocene extinctions in his later works. The statements he did make were noncommittal. For instance, in an 1877 letter to Wallace, Darwin responded to Wallace's suggestion that the extinction of Pleistocene mammals was caused by the effects of the Pleistocene itself by saying that "I cannot feel quite easy about the glacial period and the extinction of large mammals, but I must hope you are right" (Darwin and Seward 1903:13). Why he felt uneasy about the glacial period probably related both to his earlier geological observations in South America and to Agassiz's use of the sudden winter as an anti-Darwinian weapon. Why he hoped Wallace was right probably related to Darwin's diplomacy. How he felt about the alternative that human activities were to blame he did not say.

The Nineteenth Century: An Overview

The period of time I have examined here is not an arbitrary one. The nineteenth century began as the reality of extinction was being firmly established, in large part on the basis of detailed analyses of mammals we now know to have been Pleistocene in age. Some sixty years later, the contemporaneity of people and extinct Pleistocene mammals became generally accepted, and the development of the impacts of this acceptance on explanations of Pleistocene extinctions is to be seen during the ensuing decades of the century.

As with numerous other aspects of earth history, the extinction of Pleistocene mammals became a subject of debate between and among catastrophist and uniformitarian scientists during the early decades of the nineteenth century. The greatest empirical debates centered on the rate at which the extinctions had occurred and on the relative antiquity of human beings and the extinct mammals.

Early in the century, catastrophists maintained that the extinctions took place rapidly and that people appeared on earth after the extinctions were complete. The uniformitarians at this time agreed with the catastrophists on human antiquity: people *were* recent. But the uniformitarian world view required that the extinctions occurred slowly, a point which both Lyell and Fleming made by pointing out that members of extinct Pleistocene taxa had been found both above and below diluvial deposits. The point for them then became *how* slowly.

For Lyell, the extinctions were sufficiently ancient and human history sufficiently recent that the latter could not have caused the former. At the least, such a view could not be entertained until strong stratigraphic evidence for the association between the two had been collected. Lyell's views on the relative antiquity of people and extinct

Pleistocene mammals led him to employ much more rigid standards for these associations than were employed by such early overkill theorizers as Fleming. Rejecting evidence of the sort he would have readily employed for any other organism, Lyell initially concluded that while Pleistocene extinctions had occurred slowly, they nonetheless had been completed prior to the arrival of people on earth. Prior to about 1860, Lyell concluded that Pleistocene extinctions had been caused by slow and natural changes across the earth's surface.

But even in the early 1800s, all it took to incorporate the overkill hypothesis into a strictly uniformitarian world view were two related steps: a sufficient extension of the slowness of extinctions to bring them into the time when people existed, and a relaxing of the criteria of geological association to the point where associations otherwise judged unacceptable became acceptable. It was this approach that Fleming took, and which enabled him to become one of the earliest natural historians to develop the overkill hypothesis in detail while remaining within a strict uniformitarian framework.

Both Lyell and Fleming were reacting to, and arguing against, the catastrophist approaches of Cuvier and Buckland, who had maintained in convincing fashion that Pleistocene extinctions were due to either localized or universal flooding. Later, Lyell had to contend with Agassiz's equally catastrophic glaciers. The arguments he used against the one, he could and did use against the other. Evidence for slow extinction could counter a catastrophist explanation no matter what the mechanism involved.

By the mid-1800s, catastrophism was no longer a preferred method for the interpretation of earth history. While individual catastrophists still existed, and while uniformitarian geologists still took time to deprecate such approaches, by the 1850s Pleistocene extinctions were primarily explained through the application of uniformitarian methods. The major change in the results of the application of these methods must be attributed to the work of Boucher de Perthes, whose demonstration of the association between human artifacts and extinct Pleistocene mammals gained wide acceptance during the late 1850s. This demonstration provided the crucial piece of evidence needed to incorporate the overkill hypothesis into uniformitarian explanations of Pleistocene extinctions at a time when these extinctions were proving more difficult to explain in any other way. Boucher de Perthes corrected the mistakes made by both Lyell and Fleming in their approaches to the overkill hypothesis. By working in well-stratified deposits in an open setting, as opposed to excavating the stratigraphically more complex and often mixed deposits of caves, he showed that Lyell was wrong in assuming a time depth for people less than that for the extinctions. In addition, he provided the strong association that Fleming needed to make his arguments more convincing, and he supplied the demonstration that allowed a more convincing incorporation of overkill into uniformitarian explanations of Pleistocene extinctions. Yet, the fact that he embedded his evidence for great human antiquity within a thoroughly catastrophist and outdated model of earth history delayed the acceptance of his evidence until other work led to a more detailed examination of his claims.

After the acceptance of Boucher de Perthes's demonstration, the popularity of the overkill hypothesis increased rapidly. Lyell, Owen, Wallace, and many other nineteenth-century naturalists adopted the hypothesis, either as a sole explanation of Pleistocene extinctions, or, more frequently, as part of a multivariate explanation of those extinctions. The shift came rapidly, with explicit recognition of Boucher de Perthes's role. The shift did not come, however, because anyone had provided a detailed and convincing argument in favor of a human role in Pleistocene extinctions. In fact, the most detailed argument in favor of overkill during the nineteenth century was that presented by Fleming, an argument best understood in the context of Fleming's debate with Buckland over the proper interpretation of diluvium, and an argument that was not cited after Boucher de Perthes's demonstration of a great human antiquity. Instead, the overkill hypothesis gained adherents because other hypotheses seemed inadequate. At the same time, a human role in the extinctions helped lessen the impact of the new realization that people had coexisted with Pleistocene mammals: extinction due to human activities was very clearly a part of the modern world. Thus, Boucher de Perthes, an extreme catastrophist, had provided the actualists, and among them the uniformitarians, with the evidence they needed to attribute Pleistocene extinctions, at least in part, to human activities, while remaining actualists.

Those who remained convinced that vicissitudes of the earth's surface had been of sufficient magnitude to account for the extinctions continued to reject the overkill hypothesis. Agassiz's "sudden intense winter," part of his creationist model of earth history, provided such a mechanism for him. Dawkins, Geikie, Lartet, and other earth scientists also had explanations that accounted for Pleistocene extinctions while explaining numerous other phenomena as well. As did Agassiz, these scientists also rejected the overkill hypothesis. The individual histories of the explanations of some of these men display the process well. Saporta, for instance, accepted a possible human role in the extinctions until the development of the equability hypothesis allowed him to explain both Pleistocene extinctions and many other phenomena with a single approach. And, Wallace rejected a human component in the extinctions until the power of glacial theory no longer seemed adequate to him. Had they lived, neither Darwin nor Lyell would have been surprised that, at the close of the nineteenth century, there was still no agreement as to what had caused Pleistocene extinctions.

Acknowledgments

I thank Peter J. Bowler and Robert C. Dunnell for very helpful reviews of an early version of this paper, and Joseph T. Gregory and Jacob W. Gruber for assistance provided a number of years ago.

References

Agassiz, E. C. 1885. *Louis Agassiz. His life and correspondence.* Volume 1. New York: Houghton Mifflin.

Agassiz, L. 1837. Discours prononcé à l'ouverture des séances de la Societé Helvétique des Sciences Naturelles, à Neuchâtel le 24 Juillet 1837. *Societé Helvétique des Sciences Naturelles, Actes,* 22me session: V–XXXII.

———. 1840. *Études sur les glaciers.* Neuchâtel: Jent and Gassman.

———. 1866. *Geological sketches.* Boston: Ticknor and Fields.

———. 1887. *Geological sketches.* Second series. New York: Houghton Mifflin.

Anderson, A. C. 1875. Further remarks on the reindeer; and on its assumed coexistence with the hippopotamus. In *Reliquiae acquitanicae; being contributions to the archaeology and paleontology of Périgord and the adjoining provinces of southern France,* by E. Lartet and H. Christy, edited by T. R. Jones, pp. 153–160. London: Williams and Norgate.

Aufrère, L. 1940. Figures des préhistoriens. 1. Boucher de Perthes. *Préhistoire* 7:1–134.

Bartholomew, M. 1973. Lyell and evolution: an account of Lyell's response to the prospect of an evolutionary ancestry for man. *British Journal for the History of Science* 6:261–303.

Berry, W. B. N. 1968. *Growth of a prehistoric time scale based on organic evolution:* San Francisco: W. H. Freeman.

Bewick, T. 1807. *A general history of quadrupeds.* Newcastle upon Tyne: T. Bewick and S. Hodgson.

Blumenbach, J. F. 1865. Contributions to natural history. Part 1 (1806). In *The anthropological treatises of Johann Friedrich Blumenbach,* translated by T. Bendyshe, pp. 277–324. London: Anthropological Society of London.

Boucher de Perthes, J. 1847. *Antiquités celtiques et antédiluviennes. Memoire sur l'industrie primitive et les arts à leur origine* (Vol. 1). Paris: Treuttel and Wertz.

————. 1857. *Antiquités celtiques et antédiluviennes. Memoire sur l'industrie primitive et les arts à leur origine* (Vol. 2). Paris: Treuttel and Wertz.

————. 1864. *Antiquités celtiques et antédiluviennes. Memoire sur l'industrie primitive et les arts à leur origine* (Vol. 3). Paris: Treuttel and Wertz.

Bowler, P. J. 1976. *Fossils and progress.* New York: Science History Publications.

Buckland, W. 1820. *Vindicae geologicae; or the connexion of geology with religion explained.* Oxford: Oxford University Press.

————. 1823. *Reliquiae diluvianae; or, observations on the organic remains contained in caves, fissures, and diluvian gravel, and on other geological phenomena, attesting the action of an universal deluge.* London: John Murray.

————. 1825. Professor Buckland's reply to some observations in Dr. Fleming's remarks on the distribution of British animals. *Edinburgh Philosophical Journal* 12:304–319.

————. 1842. On the former existence of glaciers in Scotland and in the north of England. *Proceedings of the Geological Society of London* 3:332–337; 345–348.

Burchfield, J. D. 1975. *Lord Kelvin and the age of the earth.* New York: Science History Publications.

Burkhardt, R. W., Jr. 1977. *The spirit of system. Lamarck and evolutionary biology.* Cambridge: Harvard University Press.

Cooper, W. 1824. On the remains of the *Megatherium* recently discovered in Georgia. *Annals of the Lyceum of Natural History of New York* 1:114–124.

Cuvier, G. 1796a. Memoire sur les espèces d'elephans tant vivantes que fossiles. *Magasin Encyclopédique,* 2me année, 3:440–445.

————. 1796b. Notice sur le squelette d'une très-grande espèce de quadrupède inconnue jusqu'à présent, trouvé au Paraquay, et déposé au Cabinet d'Histoire naturelle de Madrid. *Magasin Encyclopédique,* 2me année, 1:303–310.

————. 1812a. Discours préliminaire. In *Recherches sur les ossemens fossiles des quadrupèdes, ou l'on rétablit les caractères de plusieurs espèces d'animaux que les révolutions du globe paroissent avoir détruites.* Paris: Deterville.

————. 1812b. *Recherches sur les ossemens fossiles des quadrupèdes, ou l'on rétablit les caractères de plusieurs espèces d'animaux que les révolutions du globe paroissent avoir détruites.* Paris: Deterville.

————. 1825. *Discours sur les révolutions de lu surface du globe.* Paris: Dufour and d'Ocagne.

Cuvier, G. and S. L. Mitchill. 1818. *Essay on the theory of the earth, with mineralogical notes and an account of Cuvier's geological discoveries by Professor Jameson, to which are now added, observations on the geology of North America; illustrated by the description of various organic remains, found in that part of the world.* New York: Kirk and Mercein.

Daniel, G. 1950. *A hundred years of archaeology.* London: G. Duckworth.

Darwin, C. 1839. *Journal of research into the geology and natural history of the various countries visited by* H. M. S. Beagle, *under the command of Capt. Fitzroy, R. N., during the years 1832 to 1836.* London: Colburn.

————. 1846. *Geological observations on South America, being the third part of the geology of the* Beagle, *under the command of Capt. Fitzroy, R. N., during the years 1832 to 1836.* New York: Smith, Elder.

————. 1859. *On the origin of species by means of natural selection, or the preservation of favoured races in the struggle for life.* London: John Murray.

————. 1871a. *Journal of researches into the natural history and geology of the countries visited during the voyage of* H. M. S. Beagle *round the world under the command of Capt. Fitz Roy, R. N.* (1845 edition). New York: Hafner.

————. 1871b. *The descent of man, and selection in relation to sex,* Volume 1. London: John Murray.

————. 1958. *The origin of species by means of natural selection, or the preservation of favoured races in the struggle for life* (Sixth edition, 1872). New York: New American Library.

Darwin, F. 1911. *The life and letters of Charles Darwin,* Volume 2. New York: Appleton.

Darwin, F. and A. C. Seward. 1903. *More letters of Charles Darwin,* Volume 2. London: John Murray.

Dawkins, W. B. 1869. On the distribution of British Postglacial mammals. *Quarterly Journal of the Geological Society of London* 25:192–217.

————. 1871. On Pleistocene climate and the relation of the Pleistocene mammalia to the glacial period. *Popular Science Review* 10:388–397.

————. 1872. *The British Pleistocene mammalia. Part V: British Pleistocene Ovidae.* Ovibos moschatus *Blainville.* London: Palaeontographical Society.

————. 1874. *Cave hunting, researches on the evidence of caves respecting the early inhabitants of Europe.* London: Macmillan.

————. 1880. *Early man in Britain and his place in the Tertiary period.* London: Macmillan.

Dawkins, W. B. and W. A. Sanford. 1866. Introduction. In British Pleistocene Felidae, Volume 1, in *A Monograph of the British Pleistocene mammalia.* London: Palaeontographical Society.

de Beer, G. 1960a. Darwin's notebooks on transmutation of species. Part I. First notebook (July 1837—February 1838). *British Museum (Natural History) Historical Series* 2 (2).

————. 1960b. Darwin's notebooks on transmutation of species. Part II. Second notebook (July 1837–February 1838). *Museum (Natural History) Historical Series* 2 (3).

————. 1960c. Darwin's notebooks on transmutation of species. Part IV. Fourth notebook (October 1838–10 July 1839). *British Museum (Natural History) Historical Series* 2 (5).

de Luc, J.-A. 1793. Geological letters, addressed to Professor Blumenbach. Letter 1. *British Critic* 2:231–238; 351–358.

————. 1794a. Geological letters. To Professor Blumenbach. Letter IV. *British Critic* 4:212–218; 328–336.

————. 1794b. Geological letters. To Professor Blumenbach. Letter V. *British Critic* 4:447–459; 569–578.

————. 1795. Geological letters. To Professor Blumenbach. Letter VI. *Britih Critic* 5:197–207; 316–326.

d'Orbigny, C. and A. Gente. 1851. *Géologie appliquée aux arts et à l'agriculture, comprenant l'ensemble des révolutions du globe.* Paris: M. A. Gente.

Dupont, E. 1871. *Les temps antéhistoriques en Belgique. L'homme pendant les ages de la pierre dans les environs de Dinant-sur-Meuse.* Brussels: Mucquardt.

Evans, J. 1860. On the occurrence of flint implements in undisturbed beds of gravel, sand, and clay. *Archaeologia* 38:280–307.

Fischer, P. 1873. The scientific labors of Edward Lartet. *Annual Report of the Smithsonian Institution for 1872:* 172–184.

Fleming, J. 1824. Remarks illustrative of the influence of society on the distribution of British animals. *Edinburgh Philosophical Journal* 11: 287–305.

————. 1826. The geological deluge, as interpreted by Baron Cuvier and Professor Buckland, inconsistent with the testimony of Moses and the phenomena of nature. *Edinburgh Philosophical Journal* 14:205–239.

Flint, R. F. 1971. *Glacial and Quaternary geology.* New York: John Wiley.

Forbes, E. 1846. On the connexion between the distribution of the existing flora and fauna of the British Isles, and the geological changes which have affected their area, especially during the era of the Northern Drift. *Great Britain Geological Survey Memoir* 1:336–432.

Gay P. 1966. *The Enlightenment: an interpretation,* Volume 1. New York: Norton Library.

Geikie, J. 1872. On changes of climate during the Glacial epoch. Fifth Paper. *Geological Magazine* 9:164–170.

————. 1877. *The great ice age and its relation to the antiquity of man.* Second edition. London: Daldy, Isbister.

————. 1881. *Prehistoric Europe, a geological sketch.* London: Stanford.

Gillispie, C. C. 1951. *Genesis and geology.* Cambridge: Harvard University Press.

Glass, B. P., O. Temkin, and W. L. Strauss, Jr. (editors). 1959. *Forerunners of Darwin.* Baltimore: Johns Hopkins.

Grayson, D. K. 1979. Vicissitudes and overkill: the development of explanations of Pleistocene extinctions. In *Advances in archaeological method and theory,* edited by M. B. Schiffer, Volume 3, pp. 357–403. New York: Academic Press.

————. 1983. *Establishment of human antiquity.* New York: Academic Press.

Greene, J. C. 1959. *The death of Adam.* New York: New American Library.

Gruber, H. E. 1974. A psychological study of scientific creativity. Book one, in *Darwin on man,* by H. W. Gruber and P. H. Barrett. New York: E. P. Dutton.

Gruber, J. W. 1965. Brixham Cave and the antiquity of man. In *Context and meaning in cultural anthropology,* edited by M. E. Spiro, pp. 373–402. New York: Free Press.

Harlan, R. 1825. *Fauna Americana: being a description of the mammiferous animals inhabiting North America.* Philadelphia: Finley.

Herbert, S. 1974. The place of man in the development of Darwin's theory of transmutation: Part 1. To July 1837. *Journal of the History of Biology* 7:217–258.

————. 1980. *The Red Notebook of Charles Darwin.* Ithaca and London: British Museum (Natural History) and Cornell University Press.

Hindle, B. 1974. *The pursuit of science in revo-*

lutionary America. New York: Norton Library.

[Holbach, P. H. T., Baron d']. 1770. *Système de la nature. Ou les loix de monde physique et du monde moral.* Two volumes. London.

Hunter, W. 1768. Observations of the bones, commonly supposed to be elephant's bones, which have been found near the river Ohio in America. *Philosophical transactions of the Royal Society* 58:34–45.

Huxley, T. H. 1869. The anniversary address of the President. *Quarterly Journal of the Geological Society of London* 25:xxviii–liii.

Jefferson, T. 1799. A memoir on the discovery of certain bones of a quadruped of the clawed kind in the western part of Virginia. *Transactions of the American Philosophical Society* 4:246–260.

Kirwan, R. 1799. *Geological essays.* London: Bremner.

Kohn, D. 1980. Theories to work by: rejected theories, reproduction, and Darwin's path to natural selection. *Studies in History of Biology* 4:67–170.

Kors, A. C. 1976. *D'Holbach's coterie: an enlightenment in Paris.* Princeton: Princeton University Press.

Lamarck, J.-B.-P.-A. 1801. *Système des animaux sans vertèbres, ou tableau général des classes, des ordres, et des genres de ces animaux; présentant leurs caractères essentials et leur distribution, d'après la considération de leurs rapports naturels et de leur organisation, et suivant l'arrangement établi dans les galeries du Muséum d'Histoire naturelle, parmi leurs dépouilles conservées; précédé du discours d'ouverture du cours de zoologie, donné dans le Muséum National d'Histoire Naturelle, l'an VIII de la République.* Paris.

———. 1809. *Philosophie zoologique ou exposition des considérations relatives à l'histoire naturelle des animaux; a la diversité de leur organization et des facultés qu'ils en obtiennent; aux causes physiques qui maintiennent en eux la vie et donnent lieu aux mouvements qu'ils exécutent; enfin, à celles qui produisent, les unes le sentiment, et les autres l'intelligence de ceux qui en sont doués.* Paris: Dentu [and] l'Auteur.

Lartet, E. 1864. Sur un portion de crâne fossile d'Ovibos musqué *(O. moschatus, Blainville),* trouvée par M. le Dr. Eug. Robert dans le diluvium de Précy (Oise). *Comptes Rendus Hebdomadaires des Séances de l'Academie des Sciences* 58:1198–1201.

———. 1867. Note sur deux têtes de carnassiers fossiles (Ursus et Felis), et sur quelques débris de rhinocéros, provenant des découvertes faites par M. Bourguignat dans les cavernes du Midi de la France. *Annales des Sciences Naturelles: Zoologie et Paleontogie,* Series 5, 8: 157–194.

———. 1875. Notes on the reindeer and hippopotamus. In *Reliquiae acquitanicae; being contributions to the archaeology and paleontology of Périgord and the adjoining provinces of southern France,* by E. Lartet and H. Christy, edited by T. R. Jones, pp. 147–152. London: Williams and Norgate.

Le Conte, J. 1879. *Elements of geology.* New York: Appleton.

Ledieu, A. 1885. *Boucher de Perthes. Sa vie, ses oeuvres, sa correspondence.* Abbeville: E. Caudron.

Lovejoy, A. O. 1964. *The great chain of being.* Cambridge: Harvard University Press.

Lubbock, J. 1872. *Pre-historic times, as illustrated by ancient remains and the manners and customs of modern savages.* Second edition. New York: Appleton.

———. 1890. *Pre-historic times, as illustrated by ancient remains and the manners and customs of modern savages.* Fifth edition. New York: Appleton.

Lydekker, R. 1896. *A geographical history of mammals.* Cambridge: Cambridge University Press.

Lyell, C. 1830. *Principles of geology, being an attempt to explain the former changes of the earth's surface by reference to causes now in operation,* Volume 1. London: John Murray.

———. 1832. *Principles of geology, being an attempt to explain the former changes of the earth's surface by reference to causes now in operation,* Volume 2. London: John Murray.

———. 1833. *Principles of geology, being an attempt to explain the former changes of the earth's surface by reference to causes now in operation,* Volume 3. London: John Murray.

———. 1844. On the geological position of the *Mastodon giganteum* and associated fossil remains at Bigbone Lick, Kentucky, and other localities in the United States. *American Journal of Science* 46:320–323.

———. 1845. *Travels in North America; with geological observations on the United States, Canada, and Nova Scotia,* Volume 1. London: John Murray.

———. 1846. On the newer deposits of the southern states of North America. *Quarterly Journal of the Geological Society of London* 2:405–410.

———. 1849. *A second visit to the United States of North America,* Volume 1. New York: Harper and Brothers.

———. 1853. *Manual of elementary geology.*

Fourth edition. New York: Appleton.
———. 1859a. *Manual of elementary geology.* Sixth edition. New York: Appleton.
———. 1859b. On the occurrence of works of human art in Post-Pliocene deposits. *British Association for the Advancement of Science Report* 29:93–95.
———. 1863. *The geological evidences of the antiquity of man, with remarks on theories of the origin of species by variation.* Philadelphia: G. W. Childs.
———. 1873. *Geological evidences of the antiquity of man, with an outline of glacial and post-Tertiary geology and remarks on the origin of species, with special reference to man's first appearance on earth.* Fourth edition. London: John Murray.
Lyell, K. 1881. *Life, letters and journals of Sir Charles Lyell, Bart.,* Volume 2. London: John Murray.
Mantell, G. 1850. *A pictorial atlas of fossil remains, consisting of coloured illustrations selected from Parkinson's "Organic remains of a former world," and Artis's "Antediluvian phytology."* London: Bohn.
Martin, P. S. 1967a. Pleistocene overkill. *Natural History* 76: 32–38.
———. 1967b. Prehistoric overkill. In *Pleistocene extinctions: the search for a cause,* edited by P. S. Martin and H. E. Wright, Jr., pp. 75–120. New Haven: Yale University Press.
Mitchill, S. L. 1824. Observations on the teeth of the *Megatherium* recently discovered in the United States. *Annals of the Lyceum of Natural History of New York* 1:58–60.
Morris, J. 1850. On the occurrence of mammalian remains at Brentford. *Quarterly Journal of the Geological Society of London* 6:201–204.
Naville, P. *Paul Thiry d'Holbach et la philosophie scientifique au XVIIIe siècle.* Paris: Gallimard.
Oakley, K. A. 1964. The problem of man's antiquity. *British Museum (Natural History), Geology* 9(5).
Osborn, H. F. 1906. The causes of extinction of mammalia. *American Naturalist* 40:769–795; 829–859.
———. 1935. Thomas Jefferson as a paleontologist. *Science* 82:533–538.
Owen, R. 1846. *A history of British fossil mammals, and birds.* London: Van Voorst.
———. 1849. On *Dinornis,* an extinct species of tridactyle struthious birds, with descriptions of portions of the skeleton of five species which formerly existed in New Zealand. *Transactions of the Zoological Society of London* 3:235–275.
———. 1856. Description of a fossil cranium of the musk-buffalo [*Bubalus moschatus,* Owen; *Bos moschatus* (Zimm. & Gmel.), Pallus; *Bos pallasii,* De Kay; *Ovibos Pallasii,* H. Smith & Bl.] from the "lower-level drift" at Maidenhead, Berkshire. *Quarterly Journal of the Geological Society of London* 12:124–137.
———. 1860. *Paleontology, or a systematic summary of extinct mammals and their geological relations.* Edinburgh: A. and C. Black.
———. 1862a. On *Dinornis* (Part VI): containing a description of the leg of *Dinornis* (Palapteryx) *struthoides* and of *Dinornis gracilis,* Owen. *Transactions of the Zoological Society of London* 4:141–147.
———. 1862b. On *Dinornis* (Part VII): containing a description of the bones of the leg and foot of *Dinornis elephantopus,* Owen. *Transactions of the Zoological Society of London* 4:149–158.
———. 1868. *On the anatomy of vertebrates,* Volume 3. London: Longmans and Green.
———. 1869. On fossil remains of equines from Central and South America referred to *Equus tau,* Ow., and *Equus arcidens,* Ow. *Philosophical Transactions of the Royal Society* 159:559–573.
———. 1870. On the fossil mammals of Australia.—Part III. *Diprotodon australis,* Owen. *Philosophical Transactions of the Royal Society* 160:519–578.
———. 1872. On the fossil mammals of Australia.—Part VII. Genus *Phascolomys:* species exceeding the existing ones in size. *Philosophical Transactions of the Royal Society* 162:241–258.
———. 1883. Pelvic characters of *Phylacoleo carnifex. Philosophical Transactions of the Royal Society* 174:639–643.
———. 1884. Evidence of a large extinct monotreme (*Echidna Ramsayi* Ow.) from the Wellington Breccia Cave, New South Wales. *Philosophical Transactions of the Royal Society* 175:273–375.
Page, L. E. 1969. Diluvialism and its critics in Great Britain in the early nineteenth century. In *Toward a history of geology,* edited by C. J. Schneer, pp. 256–271. Cambridge: M. I. T. Press.
———. 1972. Fleming, John. In *Dictionary of scientific biography,* Volume 5, edited by C. C. Gillispie, pp. 31–32. New York: Scribner's.
Parkinson, J. 1804. *Organic remains of a former world. An examination of the mineralized remains of the vegetables and animals of the antediluvian world; generally*

termed *extraneous fossils. The first volume; containing the vegetable kingdom.* London: Robson and others.

————. 1811. *Organic remains of a former world. An examination of the mineralized remains of the vegetables and animals of the antediluvian world; generally termed extraneous fossils. The third volume; containing the fossil starfish, echini, shells, insects, amphibia, mammalia, &c.* London: Sherwood, Neely and Jones, and others.

Playfair, J. 1802. *Illustrations of the Huttonian theory of the earth.* Edinburgh: Creech.

Prestwich, J. 1860. On the occurrence of flint-implements, associated with remains of animals of extinct species in beds of a late geological period, in France at Amiens, and in England at Hoxne. *Philosophical Transactions of the Royal Society* 150:277–317.

————. 1873. Report on the exploration of Brixham Cave. With descriptions of the animal remains by George Bush, Esq., and of the flint implements by John Evans, Esq. F. R. S. *Philosophical Transactions of the Royal Society* 163:471–572.

Prichard, H. 1902. *Through the heart of Patagonia.* New York: Appleton.

Reynolds, S. H. 1922. The hippopotamus. In *A monograph of the British Pleistocene mammalia.,* 3(1):1–38. London: Palaeontographical Society.

Rudwick, M. J. S. 1972. *The meaning of fossils.* New York: American Elsevier.

Ruse, M. 1979. *The Darwinian revolution.* Chicago: University of Chicago Press.

Saporta, G. de. 1866. La végétation du globe dans les temps antérieurs à l'homme. *Revue des Deux Mondes* 74:315–360.

————. 1870. Les anciens climats. *Revue des deux mondes* 80:208–238.

————. 1876. Sur le climat présumé de l'époque Quaternaire dans l'Europe centrale, d'après des indices tirés de l'observation des plantes. *Compte Rendu de la Congrès International d'Anthropologie et d'Archéologie Préhistoriques,* 7ᵉ Session, 1874, pp. 83–111.

Simpson, G. G. 1942. The beginnings of vertebrate paleontology in North America. *Transactions of the American Philosophical Society* 86(1):130–188.

Turner, G. 1799. Memoir on the extraneous fossils, denominated mammoth bones: principally designed to shew, that they are the remains of more than one species of nondescript animal. *Transactions of the American Philosophical Society* 4:510–518.

Wallace, A. R. 1864. The origin of human races and the antiquity of man deduced from the theory of "natural selection." *Journal of the Anthropological Society of London* 2:clviii–clxxxvii.

————. 1876. *The geographical distribution of animals, with a study of the relations of living and extinct faunas as elucidating past changes of the earth's surface,* Volume 1. New York: Harper and Brothers.

————. 1880. *Island life, or the phenomena and causes of insular faunas and floras, including a revision and attempted solution of the problem of geological climates.* London: Macmillan.

————. 1891. *Natural selection and tropical nature, essays on descriptive and theoretical biology.* London: Macmillan.

————. 1892. *Island life, or the phenomena and causes of insular faunas and floras, including a revision and attempted solution of the problem of geological climates.* Second edition. New York: Macmillan.

————. 1905. *My life, a record of events and opinions.* London: Chapman and Hall.

————. 1911. *The world of life, a manifestation of creative power, directive mind, and ultimate purpose.* New York: Moffat, Yard.

Wickwar, W. H. 1935. *Baron d'Holbach. A prelude to the French revolution.* London: Allen and Unwin.

Wilmarth, M. G. 1925. The geologic time scale classification of the United States Geological Survey compared with other classifications. *United States Geological Survey Bulletin.* 769.

Wilson, L. G. 1970. *Sir Charles Lyell's scientific journals on the species question.* New Haven and London: Yale University Press.

Wistar, C. 1799. An account of the bones deposited, by the President, in the museum of the Society, and represented in the annexed plates. *Transactions of the American Philosophical Society* 4: 526–31.

————. 1818. An account of two heads found in the morass, called the Big Bone Lick, and presented to the Society, by Mr. Jefferson. *Transactions of the American Philosophical Society,* new series 1:375–380.

Wood, S. V., Jr. 1872. On the climate of the Post-glacial period. *Geological Magazine* 10:153–161.

Woodward, H. B. 1883. Dr. Buckland ᵃrd the glacial theory. *Midland Naturalist* 6:225–229.

Zeiller, R. 1895. Le Marquis G. de Saporta, sa vie et ses travaux. *Revue générale de Botanique* 7:353–388.

2

Who's Who in the Pleistocene: A Mammalian Bestiary

ELAINE ANDERSON

Illustration by Peter Murray, based on a painting by Barbara J. Hoopes in *Zoobooks,* copyright 1980, Wildlife Education Ltd. Other drawings in chapter by Peter Murray.

SINCE MARTIN AND GUILDAY WROTE "A Bestiary for Pleistocene Biologists" in 1967, our knowledge of Pleistocene animals has been augmented by the publication of hundreds of papers and several books discussing faunas, species, regions, and problems. Now the Pleistocene mammals of Europe (Kurtén 1968), North America (Kurtén and Anderson 1980), and Africa (Maglio and Cooke 1978) are fairly well known, but those from other areas, especially South America and southern Asia, are poorly known and the literature is widely scattered. In this chapter, I have attempted to characterize some extinct genera, extinct species of surviving genera, and extant genera that are well represented in Pleistocene faunas. For well-known genera the stratigraphic and geographic ranges, number of recognized species, major faunal sites, habitat, distinguishing characters, food, habits, associations with man, cause(s) of extinction, and radiocarbon dates, if available, are given. For other genera only the stratigraphic and geographic ranges are given. Most of the mammals discussed are large ones, though not all can be called megafauna. In many papers megafauna has included everything larger than a rabbit; yet the ecological requirements and the animal's effect on the environment of, for example, *Capromeryx minor,* a small pronghorn weighing about 10 kg, were certainly much different from those of *Glossotherium harlani,* an ox-sized ground sloth, or *Mammut americanum,* the mastodont which stood about 2.7 m high at the shoulder.

Table 2.1. Correlation of Plio-Pleistocene Land Mammal Ages

MY B.P.		North America	South America	Europe	Africa
					Late Pleistocene
0.125		Rancholabrean	Lujanian	Steinheimian	
					Middle Pleistocene
0.4					
			Ensenadan	Cromerian	0.7
0.7	Pleistocene	Irvingtonian		1.0	
1.0					Early Pleistocene
			1.5		
				Late Villafranchian	1.8
1.8		1.9		1.9	
1.9					
2.0			Uquian	Middle Villafranchian	
					Villafranchian
2.5	Pliocene		2.5	2.5	
			Chapadmalalan		
3.0		Blancan		Early Villafranchian	
			3.0		
3.5					

The Pleistocene epoch began about 1.9 million years ago (mya) and ended about 10,000 years ago. Table 2.1 attempts to correlate the land mammal ages (a relative chronology showing faunal succession, each age characterized by the presence or absence of certain mammalian genera) in North America, South America, and Europe; land mammal ages have not been designated for Africa, so the Pleistocene is divided into early, middle and late faunas.

Extinction in the Pleistocene was a gradual process occurring throughout the epoch. The disappearance of many large forms culminated in the early Pleistocene in Africa and in the late Pleistocene–early Holocene in Eurasia and North and South America. Extinction of island faunas occurred mainly in the early Holocene or later. Although many causes of Pleistocene extinctions have been advanced, the reasons for most extinctions remain unknown. An informal survey of colleagues for *Pleistocene Mammals of North America* (Kurtén and Anderson 1980) revealed that the majority did not believe that a single factor, such as overkill, was responsible for the disappearance of many genera; instead a mosaic of factors, including man, combined to cause extinction. They stressed that too little is known about the animals themselves, how they were affected by climatic and vegetational changes, and the hunting behavior of Early Man.

Table 2.2 lists all mammalian genera (except pinnipeds, cetaceans, and groups for which the Pleistocene record is unknown) identified in Pleistocene faunas. As noted, the record is incomplete for many mammals and many regions; much work remains to be done. For accounts of extant genera of mammals, the reader is referred to Walker (1975).

Acknowledgments

I would like to thank Björn Kurtén, H. Gregory McDonald, Larry G. Marshall, and Paul S. Martin for providing data on unfamiliar genera, species and faunas.

Table 2.2. Mammalian Genera Recorded From Pleistocene Faunas
†Extinct genus

MONOTREMATA
  Tachyglossidae—Echidnas
    Zaglossus
  Ornithorhynchidae—Platypus
    Ornithorhynchus
MARSUPIALIA
  Didelphidae—American Opossums
    Caluromys
    Chironectes
    Didelphis
    Lestodelphys
    Lutreolina
    Marmosa
    Metachirus
    Micoureus
    Monodelphis
    Philander
    †Sparassocynus
    Thylamys
    †Thylophorops
  Dasyuridae—Marsupial "Carnivores"
    Antechinomys
    Antechinus
    Dasycercus
    Dasyuroides
    Dasyurus
    Phascogale
    Phascolosorex
    Planigale
    Sarcophilus
    Sminthopsis
  Myrmecobiidae—Numbat
    Myrmecobius
  †Thylacinidae—Marsupial "Wolf"
    †Thylacinus
  Peramelidae—Bandicoots
    Chaeropus
    Isodon
    Perameles
  Thylacomyidae—Rabbit-eared Bandicoots
    Macrotis
  Phalangeridae—Phalangers
    Phalanger
    Trichosurus
  Petauridae—Gliders
    Gymnobelideus
    Petaurus
    Pseudocheirus
    Schoinobates
  Burramyidae—Pygmy Possums
    Acrobates
    Burramys
    Cercartetus
  Macropodidae—Kangaroos, Wallabies
    Aepyprymus
    Bettongia
    Caloprymnus
    †Fissuridon
    Hypsiprymnodon
    Lagorchestes
    Macropus

    Onychogalea
    Petrogale
    Potorous
    †Prionotemnus
    †Procoptodon
    †Propleopus
    †Protemnodon
    Setonix
    †Sthenurus
    †Synaptodon
    Thylogale
    †Troposodon
    Wallabia
  Tarsipedidae—Honey Possum
    Tarsipes
  Vombatidae—Wombats
    Lasiorhinus
    †Phascolonus
    †Ramsayia
    Vombatus
  †Diprotodontidae—Diprotodonts
    †Diprotodon
    †Nototherium
    †Sthenomerus
    †Zygomaturus
  †Palorchestidae—Palorchestids
    †Palorchestes
  †Thylacoleonidae—Marsupial "Lion"
    †Thylacoleo
    Phascolarctidae-Koala
    Phascolarctos
INSECTIVORA
  Soricidae—Shrews
    †Beremendia
    Blarina
    Crocidura
    Cryptotis
    Neomys
    †Nesiotites
    Notiosorex
    †Petenyia
    Sorex
    Soriculus
    Suncus
  Talpidae—Moles
    Condylura
    Desmana
    Scalopus
    Scapanus
    Talpa
  Erinaceidae—Hedgehogs
    Erinaceus
  †Nesophontidae—Extinct W.I. Shrews
    †Nesophontes
  Solenodontidae—Solenodon
    Solendon
  Macroscelidae—Elephant Shrews
    Elephantulus
    Macroscelides
PRIMATES
  Lemuridae—Lemurs

Table 2.2. Mammalian Genera Recorded From Pleistocene Faunas
†Extinct genus
(continued)

†Archaeolemur	Chaetophractus
†Hadropithecus	Chlamyphorous
Lemur	Dasypus
Lepilemur	†Eutatus
†Megaladapis	Euphractus
Indriidae—Indris	†Holmesina
†Archaeoindris	†Kraglievichia
Avahi	†Pampatherium
Indri	†Propraopus
†Mesopropithecus	Tolypeutes
†Neopropithecus	Zaedyus
†Palacopropithecus	†Glyptodontidae—Glyptodonts
Propithecus	†Daedicuroides
Daubentoniidae—Aye-Aye	†Doedicurus
Daubentonia	†Glyptodon
Cercopithecidae—Old World Monkeys	†Glyptotherium
Cercocebus	†Hoplophorous
†Cercopithecoides	†Lomaphorus
Cercopithecus	†Neothoracophorus
Colobus	†Panochthus
†Dinopithecus	†Paxhaplous
†Gorgopithecus	†Megalonychidae—Megalonychid Ground Sloths
Macaca	†Acratocnus
Papio	†Habanocnus
†Parapapio	†Megalocnus
Presbytis	†Megalonyx
Theropithecus	†Mesocnus
Cebidae—New World Monkeys	†Micronus
Aloutta	†Miocnus
Cebus	†Neomesocnus
Pongidae—Apes	†Parocnus
Pongo	†Paulocnus
CHIROPTERA	†Synocnus
Mormoopidae—Leaf-chinned Bats	†Megatheriidae—Megatheres
Mormoops	†Eremotherium
Phyllostomatidae—Amer. Leaf-nosed Bats	†Megatherium
Desmodus	†Nothrotheriops
Leptonycteris	†Nothrotherium
Tonatia	†Mylodontidae—Mylodonts
Vespertilionidae—Vespertilionid Bats	†Glossotherium
Antrozous	†Lestodon
Barbastella	†Mylodon
Eptesicus	†Scelidotherium
Histiotus	CARNIVORA
Lasionycteris	Mustelidae—Weasels, Otters, Badgers
Lasiurus	Aonyx
Miniopterus	†Baranogale
Myotis	†Brachyprotoma
Nyctalus	Conepatus
Nycticeius	Eira
Pipistrellus	Enhydra
Plecotus	Galictis
Vespertilio	Gulo
Molossidae—Free-tailed Bats	Lutra
Eumops	Lyncodon
Tadarida	Martes
Rhinolophidae—Horseshoe Bats	Meles
Rhinolophus	Mellivora
EDENTATA	Mephitis
Dasypodidae—Armadillos	Mustela
Cabassous	Spilogale

Table 2.2. Mammalian Genera Recorded From Pleistocene Faunas
†Extinct genus
(continued)

Taxidea
Vormela
Canidae—Dogs, Wolves, Foxes
  Alopex
  Canis
  Cerdocyon
  Chrysocyon
  Cuon
  Dusicyon
  Fennecus
  Lycaon
  Nyctereutes
  Otocyon
  †Protocyon
  Urocyon
  Vulpes
Procyonidae—Raccoons
  Ailurus
  Bassariscus
  †Brachynasua
  Nasua
  Procyon
Ursidae—Bears
  Ailuropoda
  †Arctodus
  Tremarctos
  Ursus
Felidae—Cats
  Acinonyx
  †Dinofelis
  Felis
  †Homotherium
  Lynx
  †Machairodus
  †Meganteron
  Panthera
  †Smilodon
Viverridae—Civets
  Atilax
  Crossarchus
  Cynictis
  Genetta
  Herpestes
  Ichneumia
  Mungos
  Suricata
  Viverra
Hyaenidae—Hyenas
  †Chasmaporthetes
  Crocuta
  †Euryboas
  Hyaena
  Proteles
RODENTIA
  Aplodontidae—Sewellel
  Aplodontia
  Sciuridae—Squirrels
    Cynomys
    Eutamias

Glaucomys
Marmota
Sciurus
Spermophilus
Tamias
Tamiasciurus
Geomyidae—Pocket Gophers
  Geomys
  Heterogeomys
  †Nerterogeomys
  Pappogeomys
  Thomomys
Heteromyidae—Kangaroo Rats, Pocket Mice
  Dipodomys
  Liomys
  Microdipodops
  Perognathus
  †Prodipodomys
Castoridae—Beavers
  Castor
  †Castoroides
  †Trogontherium
Cricetidae—Voles, Hamsters
  Akodon
  Andinomys
  Arvicola
  †Atopomys
  Baiomys
  Calomys
  Clethrionomys
  Cricetulus
  Cricetus
  Dicrostonyx
  Dolomys
  Eligmodontia
  Ellobius
  Euneomys
  Gerbillus
  Holochilus
  Kunsia
  Lagurus
  Lemmus
  †Majoria
  Meriones
  Microtus
  †Mimomys
  Myopus
  Nectomys
  Neofiber
  Neotoma
  Ochrotomys
  Ondatra
  Onychomys
  Oryzomys
  Oxymycterus
  Peromyscus
  Phenacomys
  Phyllotis
  Pitymys
  †Pliomys

Table 2.2. Mammalian Genera Recorded From Pleistocene Faunas

†Extinct genus

(continued)

†Pliophenacomys
†Pliopotamys
†Predicrostonyx
†Proneofiber
Reithrodon
Reithrodontomys
†Rhinocricetus
Scapteromys
Sigmodon
Synaptomys
Tatera
†Tyrrhenicola
Zapodidae—Jumping Mice
  Napaeozapus
  Zapus
Dipodidae—Jerboas
  Allactaga
  Jaculus
Spalacidae—Mole-rats
  †Prospalax
  Spalax
Myoxidae—Dormice
  Dryomys
  Eliomys
  Myoxus
  †Hypnomys
  †Leithia
  Muscardinus
Muridae—Old World Rats and Mice
  Apodemus
  Arvicanthus
  †Coryphomys
  Dendromus
  Micromys
  Mus
  †Parapodemus
  †Rhagamys
  Steatomys
  †Stephanomys
Pedetidae—Springhares
  Pedetes
Thryonomyidae—Cane Rats
  Thryonomys
Bathyergidae—African mole-rats
  Heterocephalus
Hystricidae—Old World Porcupines
  Hystrix
Erethizontidae—New World Porcupines
  Coendou
  Erethizon
Ctenomyidae—Tuco-tucos
  Ctenomys
Abrocomyidae—Chinchilla Rats
  Abracomys
Echimyidae—Spiny Rats
  Carterodon
  Cercomys (= Tricomys)
  Euryzygomatomys
  †Heteropsomys (= Boromys, Brotomys)
  Proechimys

Capromyidae—Hutias, Coypu
  †Aphaetreus
  †Hexolobodon
  †Isolobodon
  †Macrocepnomys
  Myocastor
Chinchillidae—Chinchillas
  Lagostomus
Caviidae—Guinea Pigs, Cavies
  Cavia
  Dolichotis
  Galea
  Microcavia
Octodontidae—Viscacha Rats
  †Alterodon
  †Pithanotomys
†Heptaxodontidae—Extinct W.I. Rats
  †Amblyrhiza
  †Clidomys
  †Elasmodontomys (= Heptaxodon)
  †Quemisia
  †Speoxemus
  †Spirodontomys
Hydrochoeridae—Capybaras
  Hydrochoerus
  †Neochoerus
PHOLIDOTA
  Manidae—Pangolins
    Manis
    Phataginus
LAGOMORPHA
  Ochotonidae—Pikas
    Ochotona
    †Ochotonoides
    †Prolagus
  Leporidae—Rabbits, Hares
    Brachylagus
    †Hypolagus
    Lepus
    Oryctolagus
    Pronolagus
    Sylvilagus
†LITOPTERNA
  †Macrauchenidae—Macrauchenids
    †Macrauchenia
    †Windhausenia
†NOTOUNGULATA
  †Toxodontidae—Toxodonts
    †Mixotoxodon
    †Toxodon
  †Mesotheriidae—Typotheres
    †Mesotherium
PERISSODACTYLA
  Equidae—Horses, Asses, Zebras
    Equus
    †Hipparion
    †Hippidion
    †Onohippidium
  Tapiridae—Tapirs
    Tapirus

Table 2.2. Mammalian Genera Recorded From Pleistocene Faunas
†Extinct genus
(continued)

Rhinocerotidae	Rangifer
Ceratotherium	†Sangamona
†Coelodonta	Antilocapridae—Pronghorns
Diceros	Antilocapra
Dicerorhinus	†Capromeryx
†Elasmotherium	†Stockoceros
Rhinoceros	†Tetrameryx
†Chalicotheriidae—Chalicotheres	Bovidae—Cattle, Sheep, Buffalo, Antelope
†Ancyclotherium	Aepyceros
ARTIODACTYLA	Alcelaphus
Suidae—Hogs, Pigs	Ammotragus
†Hippohyus	Antidorcas
Hylochoerus	Antilope
†Kolpochoerus	Beatragus
†Metridiochoerus	Bison
†Notochoerus	†Bootherium
Phacochoerus	Bos
Potamochoerus	Boselaphus
†Stylochoerus	Bubalus
Sus	†Bucapra
Tayassuidae—Peccaries	Capra
Catagonus	Cephalophus
†Mylohyus	Connochaetes
†Platygonus	Damaliscus
Tayassu	†Damalops
Hippopotamidae—Hippopotamus	†Duboisia
Hexaprotodon	†Euceratherium
Hippopotamus	Gazella
Camelidae—Camels, Llamas	†Hemibos
†Camelops	Hemitragus
Camelus	Hippotragus
†Hemiauchenia	Kobus
Lama	Madoqua
†Palaeolama	†Makapania
†Titanotylopus	†Megalotragus
Vicugna	†Menelikia
Giraffidae—Giraffe, Okapi	†Myotragus
Giraffa	†Numidocapra
Okapi	Oreamnos
†Sivatherium	Oreotragus
Cervidae—Deer	Oryx
Alces	Ourebia
Blastoceros	Ovibos
Blastocerus	Ovis
Capreolus	†Parmularius
†Cervalces	†Pelorovis
Cervus	†Platybos
†Choritoceros	†Praeovibos
Dama	†Proboselaphus
†Eucladoceros	Raphicerus
Hippocamelus	Redunca
†Libralces	Rupicapra
Mazama	Saiga
†Megaloceros	†Sivacobus
†Morenelaphus	†Sivadenota
†Navahoceros	†Soergelia
Odocoileus	†Spirocerus

Table 2.2. Mammalian Genera Recorded From Pleistocene Faunas
†Extinct genus
(continued)

Sylvicapra	Trichechus
†Symbos	PROBOSCIDEA
Syncerus	†Mammutidae—Mastodonts
Taurotragus	†Mammut
†Thalerceros	†Gomphotheriidae—Gomphotheres
Tragelaphus	†Anancus
TUBULIDENTATA	†Cuvieronius
Orycteropodidae—Aardvarks	†Haplomastodon
Orycteropus	†Stegomastodon
†Plesiorycteropus	Elephantidae—Elephants, Mammoths
HYRACOIDEA	Elephas
Procaviidae—Hyraxes	Loxodonta
†Gigantohyrax	†Mammuthus
Procavia	†Stegodontidae—Stegodonts
SIRENIA	†Stegodon
Dugongidae—Dugongs, Sea Cows	†DEINOTHERIOIDEA
†Hydrodamalis	†Deinotheriidae—Deinotheres
Trichechidae—Manatees	†Deinotherium

Abbreviations

Ast.–Astian	**Amer.**–America (in text)	**U.S.**–United States
Bl.–Blancan	**Ariz.**–Arizona	**W.Af.**–West Africa
Chapad.–Chapadamalan	**Aust.**–Australia	**W.I.**–West Indies
Crom.–Cromerian	**Calif.**–California	**Wyo.**–Wyoming
Ens.–Ensenadan	**Cent. Am.**–Central America	
Hemph.–Hemphillian	**Eng.**–England	**C**–central
Holoc.–Holocene	**Eur.**–Europe	**E**–Early
Irv.–Irvingtonian	**Fla.**–Florida	**E**–east, eastern
Luj.–Lujanian	**Ger.**–Germany	**EC**–east-central
Mioc.–Miocene	**Id.**–Idaho	**L**–Late
Olig.–Oligocene	**Isl.**–Island	**M**–Middle
Paleoc.–Paleocene	**Kans.**–Kansas	**N**–north, northern
Pleist.–Pleistocene	**Mex.**–Mexico	**S**–south, southern
Plioc.–Pliocene	**Mo.**–Missouri	**SE**–southeast
R–Recent	**Neb.**–Nebraska	**SW**–southwest
RLB–Rancholabrean	**New Mex.**–New Mexico	**W**–west, western
Sang.–Sangamonian	**N.Af.**–North Africa	**gm**–grams
Steinh.–Steinheimian	**N.Am. (Amer.)**–North America	**ht.**–height
Uq.–Uquian	**Penn.**–Pennsylvania	**kg**–kilogram
Villafr.–Villafranchian	**S.Af.**–South Africa	**lt.**–length
Weich.–Weichselian	**S.Am. (Amer.)**–South America	**m**–meter
	S. Car.–South Carolina	**mm**–millimeter
Af.–Africa	**Tenn.**–Tennessee	**sp.**–species (1)
Alb.–Alberta	**Tex.**–Texas	**spp.**–species (2+)
Am.–America (geog. range)	**Va.**–Virginia	**wt.**–weight

The Bestiary
(† = extinct genus)

MARSUPIALIA

†*Sparassocynus*   Mercerat, 1898   Didelphidae
Extinct opossum. L Mioc.–Uq. S.Am.

†*Thylophorops*   Reig, 1952   Didelphidae
Extinct opossum. L Plioc.–Uq. S. Am

†*Thylacinus*   Temminck, 1838   Thylacinidae
"Tasmanian wolf." L Mioc.–R (now extinct) Aust.; Pleist.–R (now extinct) Tasmania;
Plioc.–Pleist. New Guinea. Largest carnivorous marsupial; about the size of *Canis latrans*. Extinction probably due to competition with the dingo and hunting by man (Marshall 1981).

†*Fissiuridon*   Bartholomai, 1973   Macropodidae
Extinct kangaroo. Pleist. Aust.

†*Priotemnus*   Stirton, 1955   Macropodidae
Extinct kangaroo. Plioc.–Pleist. Aust.

†*Procoptodon*   Owen, 1874   Macropodidae
Extinct short-faced kangaroo. Pleist. Aust.

†*Propleopus*   Longman, 1924   Macropodidae
Extinct rat kangaroo. Plioc.–Pleist. Aust.

†*Protemnodon*   Owen, 1874   Macropodidae
Extinct kangaroo. Mioc., Plioc.–Pleist. Aust.; Pleist. Tasmania; Plioc. New Guinea.

†*Sthenurus*   Owen, 1874   Macropodidae
Giant short-faced kangaroo. Plioc.–Pleist. Aust.; Pleist. Tasmania. It had a massive head, short incisors, reduced diastema, complex molars, and a short tail. Height 3 m. Browsing habits.

†*Synaptodon*   De Vis, 1889   Macropodidae
Extinct kangaroo. Pleist. Aust.

†*Troposodon*   Bartholomai, 1967   Macropodidae
Extinct kangaroo. Plioc.–Pleist. Aust.

†*Phascolonus*   Owen, 1872   Vombatidae
Giant wombat. Plioc.–Pleist. Aust.; Pleist. Tasmania; its remains are common at Lake Calabonna. Size of a hog. Had a single pair of rodentlike ever-growing incisors in each jaw, apparently used to remove bark or to cut broadleafed plants. Cheek teeth rootless, crowns lacking enamel; dentine wore to two ridges on each tooth. Terrestrial and fossorial. (Churcher 1980, Kurtén 1972a)

†*Ramsayia*   Tate, 1951   Vombatidae
Extinct wombat. Pleist. Aust.

†*Diprotodon*   Owen, 1838   Diprotodontidae
Diprotodont. Plioc.–Pleist. Aust.; Pleist. Tasmania (King Isl.); remains are common at Lake Calabonna. Largest known marsupial (shoulder ht. 2 m, lt. 3.2 m); graviportal build, resembled an oversized wombat. Three upper incisors present in each jaw half, the first pair chisel-like; in the lower jaw only a single pair of procumbent incisors

present. Long diastema separates incisors from molars. Molars quadrate, transversely bilophodont with lophs forming shallow crescentic ridges. Analysis of the stomach contents preserved in some skeletons reveals that diprotodont fed on *Salsola* and other Chenopodiaceae, typical plants of the saline plains. (Churcher 1980, Kurtén 1972a)

†*Nototherium*   Owen, 1845   Diprotodontidae
Diprotodont. Plioc.–Pleist. Aust.; Plioc. New Guinea.

†*Sthenomerus*   De Vis, 1883   Diprotodontidae
Diprotodont. Pleist. Aust.

†*Zygomaturus*   Owen, 1858   Diprotodontidae
Diprotodont. L Mioc., Pleist. Aust.; Pleist. Tasmania.

†*Palorchestes*   Owen, 1874   Palorchestidae
Giant kangaroo. L Mioc.–Pleist. Aust.; Pleist. Tasmania. Largest of the giant kangaroos (ht. ca. 3.5 m, skull lt. 384 mm).

†*Thylacoleo*   Owen, 1858   Thylacoleonidae
"Marsupial lion." Plioc.–Pleist. Aust.; Pleist. Tasmania. About the size of a small extant lion (head and body lt. 1.2 m). Massive skull with strong post-orbital bar and large crests for attachment of muscles. It had a peculiar dentition with enlarged middle incisors that may have functioned as canines, small canines, nubbinlike anterior premolars, a pair of large shearing carnassials in each jaw that form scissorlike blades (nothing similar in any other mammal), and reduced posterior molars. Whether *Thylacoleo* was a harmless fruit eater or a voracious carnivore is uncertain. It had opposable digits on the manus and pes and large, sharp compressed claws. (Churcher 1980, Kurtén 1972a)

INSECTIVORA

†*Beremendia*   Kormos, 1934   Soricidae
Beremend shrew. Ast.–L Crom. Eur., China.

†*Nesiotites*   Bate, 1945   Soricidae
Extinct Mediterranean shrew. 3 spp. Pleist. Balerics, Corsica, Sardinia.

†*Petenyia*   Kormos, 1934   Soricidae
Petenyi's shrew. Villafr.–Crom. Eur.

†*Nesophontes*   Anthony, 1916   Nesophontidae
Extinct West Indian shrews. 3–6 spp. Pleist.–Holoc. (now extinct). Cuba, Puerto Rico. Became extinct after the arrival of the Spaniards and introduction of *Rattus* and *Mus*. Ranged in size from a mouse to a chipmunk.

PRIMATES

†*Archaeolemur*   Filhol, 1895   Lemuridae
Extinct lemurs. 3 spp. Pleist.–postglacial bog deposits, Madagascar. Skull about 50 percent larger than in living indriids; braincase relatively larger, facial region shortened; orbits face forward.

†*Hadropithecus*   Lorenz, 1899   Lemuridae
Extinct lemur. 1 sp. Very rare, known only from the caves of Andrahomana and adjacent deposits in S Madagascar. Pleist. Similar to *Archaeolemur* but with front limbs longer than hind and different cheek teeth. Profile more simian than other lemurs.

Magaladapis, the largest of the extinct Madagascar lemurs

†*Megaladapis*   Major, 1893   Lemuridae
Giant lemur. Pleist.–E Holoc. Remains of the largest known lemur have been found in
bog deposits in Madagascar. Size of a small adult human; maximum skull lt. 300–310
mm; skull shows disproportionate elongation of facial region, a small cranium with a
brain only one-half the size of *Archaeolemur,* and small orbits. Extinction apparently
occurred after the arrival of man.

†*Archaeoindris*   Standing, 1908   Indriidae
Extinct indriid. Known from 1 specimen found in E Holoc. deposit of Ampasambazimba.
Skull (lt. 260 mm) nearly as large as *Megaladapis.* Large powerful jaws, large orbits;
probably diurnal.

†*Mesopropithecus*   Standing, 1908   Indriidae
Extinct indriid. Known only from the type locality, Ampasambazimba, Madagascar.
Pleist. Skull lt. ca. 100 mm, orbits proportionately smaller than in extant indriids.

†*Neopropithecus*   Lambert, 1936   Indriidae
Extinct indriid. 2 spp. W, C Madagascar. Pleist. Closely related to the extant *Prop-
ithecus.* Small size, gracile build.

Daubentonia   E. Geoffroy, 1795   Daubentoniidae
Aye-aye. 1 living sp., 1 extinct sp.; both species have been found in E Holoc. deposits
in dry SW Madagascar. Found today only in heavily forested areas. Extinct *D. robusta*
was about 50 percent larger than living aye-ayes. Has long attenuated middle finger,
large ever-growing rodentlike incisors. Teeth are used for ornaments.

†*Cercopithecoides*   Mollett, 1947   Cercopithecidae
Extinct monkey., 1 sp. E Pleist Af.; common in S.Af. fossil sites. Larger than modern
colobines; pronounced sexual dimorphism.

†*Dinopithecus*   Broom, 1936   Cercopithecidae
Extinct monkey. E Pleist. S.Af.

†*Gorgopithecus*   Broom and Robinson, 1949   Cercopithecidae
Extinct baboon. E Pleist. S.Af. Braincase longer than *Papio,* short muzzle, narrow
rostrum. No sexual dimorphism.

†*Parapapio*   Jones, 1937   Cercopithecidae
Extinct baboon. E Pleist., mainly S.Af. and Angola. Most primitive papionin genus.
Difficult to distinguish from *Macaca;* probably more arboreal than *Papio.*

Theropithecus   I. Geoffroy, 1843   Cercopithecidae
Gelada baboon. 1 sp. Rocky and mountainous areas of Ethiopia. E Pleist.–R, E and S
Af. Associated with Acheulean cultures, hunted by man. Replaced by *Papio*. Nostrils
open at side of nose (terminal in other baboons).

CHIROPTERA

Desmodus   Wied-Neuwied, 1826   Phyllostomatidae
Vampire bats. 1 sp. N Mexico S to C Argentina. Pleistocene range included Florida,
California, Texas, Mexico, and Cuba. Feed solely on fresh blood. Wt. 15–50 gm.

Tonatia   Gray, 1827   Phyllostomatidae
Round-eared bats. 5 extant spp. Cent. and S. Am. An extinct sp. is known from
Jamaica. Wt. 9–30 gm.

EDENTATA

Dasypus   Linnaeus, 1758   Dasypodidae
Nine-banded armadillo. ca. 5 spp. Kans.-Mo. S to Mexico and into S.Am., M Bl.–R,
N.Am., Ens.–R, S.Am. The extinct N. Amer. species *D. bellus* (Simpson) has been
identified in about forty faunas. It was about twice as large (lt. 1.2 m) but otherwise
identical to the living *D. novemcinctus* Linn. Availability of insects throughout the year
and lack of moist soil for digging were probably limiting factors to the distribution of *D.
bellus*. Slaughter (1961) noted that its presence in a fauna indicates winters no more
severe than found in north-central Texas today. No associations with man are known.
Deteriorating climate was probably a factor in its extinction, and it was replaced by *D.
novemcinctus*. (Kurtén and Anderson 1980, Slaughter 1961)

†*Eutatus*   Gervais, 1867   Dasypodidae
Extinct armadillo. Uq.–Luj. S.Am.

†*Holmesina*   Simpson, 1930   Dasypodidae
North American pampathere. Irv.–L RLB in SE U.S. to as far N as Kansas (Kanopolis)
at about seventy-five localities. *Holmesina* was similar in appearance to the living
Dasypus novemcinctus but was much larger (lt. 2 m, ht. 1 m), and had three free
moveable bands, armor on top of the head and feet, and rings of scutes on the tail. It
probably lived in a lowland environment along the coast and in river valleys. Edmund
(pers. comm.) suggests that they were vegetarians and ground their food with their
flat-crowned teeth. One species, *H. septentrionalis* Simpson, is recognized. No associations with man are known; a terminal date is 9,880 yr B.P. at Hornsby Springs, Fla.
(Kurtén and Anderson 1980)

†*Kraglievichia*   Castellanos, 1927   Dasypodidae
Pampathere. L Bl.–E Irv. (Florida); Chapad. Argentina. Ancestral to *Pampatherium*
and *Holmesina*. About the size of *D. bellus*.

†*Propraopus*   Ameghino, 1881   Dasypodidae
Extinct armadillo. Ens.–Luj. S.Am.

†*Pampatherium*   Gervais and Ameghino, 1880 (=*Chlamytherium*,   Dasypodidae
                                      Chlamydotherium)
South American pampathere. Uq.–Luj. S.Am. *Pampatherium* is closely related to
Holmesina. The conservative pampathere lineage is characterized by an increase in
size and progressive "molarization" of the anterior teeth (Edmund in Hibbard 1978).
At Arroio Touro Passo, Brazil, charcoal associated with an extinct fauna that included *Pampatherium* has been dated 11,040±190 yr B.P. (I-9628). (Hibbard et al.
1978, Marshall et al. MS)

†*Glyptodon*   Owen, 1838   Glyptodontidae
South American glyptodont. Uq.–Luj. S. Am. Glyptodonts were a primarily S. Amer.
group of heavily armored edentates whose body and upper limb segments were encased
in an immense turtlelike carapace made up of thick polygonal plates covered with horny
scales. The top of the head was covered with a bony casque, and the relatively short tail
was enclosed in a sheath of overlapping rings with prominent conical projections on the
top and sides. The skeleton was greatly modified to support the massive carapace, and
axial mobility was limited. These clumsy, overspecialized mammals survived until the
end of the Lujanian (13,000+ yr B.P. at Taima-taima, Venezuela; 11,040±190 yr B.P.
[I-9628] at Arroio Touro Passo, Brazil).

   Several other genera of Pleistocene glyptodonts have been described from S. Amer.,
among them *Daedicuroides, Doedicurus, Hoplophorus* (=*Sclerocalyptus*), *Lomaphorus,
Neothoracophorus, Panochthus,* and *Paxhaplous,* but for the most part they are inade-
quately distinguished from each other and are badly in need of revision. (Gillette 1973
and pers. comm., Marshall et al. MS)

†*Glyptotherium*   Osborn, 1903   (=*Boreostracon, Brachyostracon*)   Glyptodontidae
North American glyptodont. L Plioc.–L RLB N.Am. Three species of *Glyptotherium*
are now recognized in the U.S. The terminal species, *G. floridanus* (Simpson) has
been found in Florida, South Carolina, and Texas. Carapace lt. was ca. 1.5 m, total
lt. 2 m, ht. ca. 1.2 m, and it weighed ca. 1 ton. Glyptodonts inhabited tropical and
subtropical regions where they browsed on lush vegetation growing along water-
ways. Their remains are often associated with those of capybaras (*Hydrochoerus,
Neochoerus*). Gillette (1973) believes that since *Glyptotherium* was so specialized
and lived in such a restricted environment that local populations were easily wiped
out by climatic change. There are no associations with man. (Gillette 1973, Kurtén and
Anderson 1980)

†*Megalonyx*   Harlan, 1825   Megalonychidae
Megalonychid ground sloths. L Hemph.–L RLB N.Am. The Rancholabrean megalonychid, *M. jeffersonii* (Demarest) has been found at some 75 localities in the
eastern two-thirds of the U.S., the West Coast N to Alaska, and inland to Idaho. It was
about the size of an ox (lt. 2.5–3.0 m), larger and more robust than *Nothrotheriops*
and had plantigrade hind feet. A browser, *Megalonyx* lived in forests and woodlands.
There are no definite associations with man; recently McDonald (pers. comm.) reported a tentative association at Warm Mineral Springs, Sarasota Co., Fla., dated
10,980±160 yr B.P. (Kurtén and Anderson 1980, McDonald 1977 and pers. comm.)

Megalocnus, the largest of the extinct West Indian ground sloths

†West Indian ground sloths   Megalonychidae
Ten genera of megalonychid ground sloths (*Megalocnus* Leidy, 1868, Cuba; *Acratocnus*
Anthony, 1918, Puerto Rico, Hispaniola; *Mesocnus* Matthew, 1918, Cuba; *Microcnus*

Matthew, 1919, Cuba; *Miocnus* Matthew, 1919, Cuba; *Parocnus* Miller, 1929 Haiti; *Paulocnus* Hooijer, 1962, Curacao; *Synocnus* Paula Couto, 1967, Haiti; *Neomesocnus* Arredondo, 1961, Cuba; *Habanocnus* Mayo, 1978, Cuba) have been described from the West Indies. These poorly known sloths are greatly oversplit and badly in need of revision. The age of the material is late Rancholabrean–early Holocene. Little is known about the appearance or habits of these small edentates although *Acratocnus* may have been semiarboreal. *Megalocnus,* the largest of these insular sloths, reached the size of a black bear. Bones of *Acratocnus* and *Parocnus* have been found in association with fragments of pottery and pig bones. (Hooijer 1962, 1963; Paula Couto 1967)

†*Eremotherium*   Spillman, 1948   Megatheriidae
Hermit megathere. Luj. S.Am.; L RLB SE U.S., Cent. Am. One of the largest known ground sloths was *Eremotherium;* the largest species, *E. mirabile* (Leidy) (= *E. rusconii, E. carolinensis*) reached a length of about 6 m and weighed more than 3 tons. Sexual dimorphism was pronounced, with males about 50 percent larger than females. A browser, *Eremotherium* lived in a savanna habitat. There are no known associations with man, and a terminal date for its extinction is 13,900 yr B.P. at the Talara tar seep, Peru. (Churcher 1966, Gazin 1956, Hoffstetter 1952)

†*Megatherium*   Cuvier, 1795   Megatheriidae
Megathere. Uq.–Luj. S.Am. *Megatherium,* a giant ground sloth (lt. 6 m), differs from *Eremotherium* in having four digits on the forefoot with the II, III, and IV bearing well-developed claws. *Eremotherium* has three digits on the forefoot. The premaxillae of *Megatherium* form a narrow thick bar solidly fused to the maxillae. The skull is small compared to the huge body, and there is a spoutlike symphysis that probably housed a long, flexible tongue. The teeth, 5/4, are quadrate, ever-growing prisms with transverse crests. Megatheres were probably browsers pulling branches to their mouth while standing semierect, supported by the massive tail. *Megatherium* survived to the end of the Lujanian. There are no known associations with man. (Edmund pers. comm., Scott 1962)

Nothrotheriops, the Shasta ground sloth, best known for its dung deposits from caves of arid North America

†*Nothrotheriops*   Hoffstetter, 1954   Megatheriidae
North American nothrothere. L Irv.–L RLB. The North American genus *Nothrotheriops* was found in Florida in the late Irvingtonian and ranged from N Mexico to S Alberta in the Rancholabrean. *N. shastensis* (Sinclair) was described (as *Nothrotherium shastense*)

from Potter Creek Cave, Calif.; a large sample was recovered from San Josecito Cave, Mex., and a mummified specimen was found in Aden Crater, New Mex. *Nothrotheriops* was the smallest N. Amer. ground sloth, but was twice as large as *Nothrotherium* and weighed between 135–180 kg. Remains of *Nothrotheriops* and large amounts of its dung have been found at a number of caves in the SW. A browser, it fed on roots, stems, seeds, flowers, and fruits of desert plants. No archaeological material has been found in direct association with the bones or dung. The youngest dated sample at Rampart Cave, Ariz. is 10,780 yr B.P. (Hansen 1978; Long and Martin 1974; Martin 1975; Martin, Sabels, and Shutler 1961; Paula Couto 1971)

†*Nothrotherium*   Lydekker, 1889   Megatheriidae
South American nothrothere. Ens.–Luj. S.Am. Formerly the genus *Nothrotherium* included both North and South American nothrotheres, but Paula Couto (1971) has shown them to be generically distinct. The North American specimens are *Nothrotheriops*. *Nothrotherium* differs from *Nothrotheriops* in its smaller size and characters of the skull and hind limb bones. It may have been semiarboreal. No associations with man are known. (Paula Couto 1971)

†*Glossotherium*   Owen, 1840   (=*Paramylodon*)   Mylodontidae
Big-tongued sloth. L Bl.–L RLB N.Am. (from Florida to Washington, from C Mexico to Idaho); Luj. S.Am. (Brazil, Chile, Colombia, Ecuador, Paraguay, Peru, Uruguay, Venezuela). At Rancho La Brea, Calif. more than 75 individuals have been recovered. The N. Amer. species, *G. harlani* (Owen), was powerfully built with enormous forelimbs and huge claws. It was covered with coarse, shaggy hair and had pebblelike dermal ossicles embedded in the skin. The head was elongated, but the rostrum was not inflated. *Glossotherium* probably fed on grass and small shrubs and used its claws to dig roots. Clumsy and slow-moving, *Glossotherium* inhabited open country. There is no association with man at any N. Amer. site; it survived until 9,880 yr B.P. at Hornsby Springs, Fla. In S. Amer., two subgenera (*Glossotherium* and *Oreomylodon*) and four species are recognized. A terminal date is 11,040±190 yr B.P. (I-9628) at Arroio Touro Passo, Brazil. (Hoffstetter 1952, Marshall et al. MS, Owen 1842, Stock 1925)

†*Lestodon*   Gervais, 1855   Mylodontidae
Broad-faced sloth. Ens.–Luj. S. Am. (Pampian of Argentina and Uruguay, as far N as Rio Grand de Sul, Brazil, and Tarija Valley, Bolivia). Several species have been described, but only *L. armatus* is now considered valid. *Lestodon* differs from *Mylodon* in having a greatly expanded anterior portion of the skull and having the first tooth in both jaws enlarged, triangular in shape, almost tusklike. *Lestodon* was a gigantic sloth, intermediate in size between *Mylodon* and *Megatherium*. (McDonald pers. comm.)

†*Mylodon*   Owen, 1840   Mylodontidae
Mylodon. Ens.–Luj. S.Am. (Argentina, Chile, Ecuador, Paraguay, Uruguay). *Mylodon* is distinguished from *Glossotherium* by a fusion of the premaxillae and nasals to form a bony ring. The skull is short and broad with a sharply truncated muzzle. The teeth, numbering 5/4, are lobate-shaped and rootless; the anterior one is caniniform and separated from the rest of the series by a diastema. Massively built, *Mylodon* stood about 130 cm over the back. A mummified specimen found in Ultima Esperanza Cave, Chile, had coarse, straight yellowish hair; embedded in the skin was a pavement of dermal ossicles. *Mylodon* fed on grass and shrubs. Remains of *Mylodon* from several sites have been radiocarbon dated: Ultima Esperanza Cave, 13,500±470 yr B.P. (NZ-1680); Palli Aike Cave, Chile, 8,639 yr B.P. (C-485); Fell's Cave, Chile, 10,080±160 yr B.P. (I-5146); Gruta del Indio, Argentina, 10,740± yr B.P. (A-1351). At Fell's Cave the remains were associated with *Homo*. (McDonald pers. comm., Marshall et al. MS, Scott 1962)

†*Scelidotherium*   Owen, 1840   (=*Scelidodon*)   Mylodontidae
Scelidothere. Ens.–Luj. S.Am. (Chile, Argentina, Uruguay, Bolivia, Peru, E Brazil).

Four subgenera and several species are recognized. *Scelidotherium* had an elongated skull with a tubular muzzle and weak dentition. It was as large as *Glossotherium* and *Mylodon*. Remains of *Scelidotherium* have been found in the Talara tar seeps, Peru, dated about 13,900 yr B.P. and at Arroio Touro Passo, Brazil dated 11,040±190 yr B.P. (I-9628); these remains are associated with *Homo*. (McDonald pers. comm, Marshall et al. MS)

CARNIVORA

†*Baranogale*   Kormos, 1934   Mustelidae
Extinct banded polecat. E–M Pleist. Eur. Very common. Resembled *Vormela*.

†*Brachyprotoma*   Brown, 1908   Mustelidae
Short-faced skunk. Irv.–E Holoc. EC U.S. 1 sp. *Brachyprotoma* was the size of a small *Spilogale* and probably fed on hard-shelled insects. Short skull and jaws, premolars crowded, toothrow curved. At many sites it is asscicated with a northern or boreal fauna. There is no association with man at any site, and its extinction is unexplained. (Kurtén and Anderson 1980)

Canis   Linnaeus, 1758   Canidae
Wolves, coyotes, dogs, jackals. 8 spp. N.Am., Cent. Am., Eur., Asia, Af., Aust. Bl.–R N.Am.; Villafr.–R Eur.; Pleist.–R Asia; E Pleist.–R Af.; R Aust. The extant genus *Canis* is well represented in Pleistocene faunas, and a number of species have been described. The best known extinct species, *C. dirus* Leidy, the dire wolf, ranged from S Alberta to Perú during the late Pleistocene. At Rancho La Brea a minimum of 1,646 individuals have been found, and large samples are also known from San Josecito Cave and the Talara tar seeps. The species probably arose in S. Amer. and spread northward in the Sangamonian. It exceeded the living gray wolf, *C. lupus*, in size and had a heavier build, sturdier limbs with relatively shorter lower limb segments, a larger broader head, and a smaller braincase than that species. The teeth of *C. dirus* are more powerful than those of any other *Canis*. Dire wolves were hunters and scavengers. There is no direct association with man at any site, though the bones have been found at the same level as artifacts. Extinction was probably due to competition with *C. lupus*; terminal dates include 9,400±760 yr B.P. at Brynjulfson Cave, Mo.; 9,860 ±550 yr B.P. at Rancho La Brea, and 10,690±360 yr B.P. at La Mirada, Calif.

Remains of the domestic dog, *C. familiaris* Linn., are rare in Pleistocene faunas. Ancestry from *C. lupus* is indicated. Domestication had occurred by Magdalenian times in Europe, and the dog probably accompanied man across Beringia. Its remains have been recovered at Old Crow River Loc. 11A (probable age 20,000 yr B.P., Beebe 1978). At Jaguar Cave, Id., dated 10,370±350 yr B.P., two distinct size classes of dogs were found. Dog remains are highly variable and can be confused with those of wolves and coyotes. (Beebe 1978, Kurtén and Anderson 1980, Nowak 1979)

Cuon   Hodgson, 1838   Canidae
Dhole. 1 sp. Parts of Russia, Siberia, most of Asia. M–L Pleist. Eur.; Pleist.–R Asia; RLB Beringia and Mexico (San Josecito Cave). Wt. 14–21 kg.

†*Protocyon*   Giebel, 1855   Canidae
Extinct bush dog. Uq.–Luj. S.Am., E Irv. N.Am. (Rock Creek, Tex.). Probably related to the living *Speothos*. Short-legged, short broad face, slender high-cusped cheek teeth. Highly predacious. Size of *Canis lupus*.

†*Arctodus*   Leidy, 1854   Ursidae
Short-faced bears. Irv.–L RLB N.Am. Uq.–Luj. S.Am. 2 spp. In the late Ranchola-brean *A. simus* (Cope) ranged from Alaska to Mexico except in Florida and SE U.S. It was a long-legged, short-bodied animal with a short face and broad muzzle. Highly predacious, *Arctodus* was the most powerful predator of the American Pleistocene. At

Lubbock Lake, Tex., a few foot bones were found with possible cut marks on them (E. Johnson pers. comm. This material has been dated at 12,650±350 yr B.P.). The cause of extinction of *Arctodus* is believed to have been competition with invading brown and grizzly bears (*Ursus arctos*). Remains of the two species have been found together at Little Box Elder Cave, Wyo. (Kurtén 1967, Kurtén and Anderson 1980)

Ursus spelaeus, the extinct cave bear of Europe (left) and *Tremarctos*, the spectacled bear, extinct in North America but surviving in South America

Tremarctos   Gervais, 1758   Ursidae

Spectacled bear. 1 sp. Mountainous regions of S.Am. Bl.–RLB N.Am.; Uq.–R S.Am. The extant S. Amer. genus *Tremarctos* was represented in the N. Amer. Pleistocene (and perhaps in the Blancan) by *T. floridanus* (Gidley), a widespread species that may have survived until the early Holocene (Devil's Den, Fla.). Its remains are common in sinkhole and river deposits in Florida; it was also found in E U.S. and W Mexico. The extinct species is distinguished from *T. ornatus* Cuvier, the extant spectacled bear, by much larger size and heavier proportions. There was a tendency towards reduction of the premolars and elongation of the back molars. These herbivorous bears filled the same niche as *Ursus spelaeus,* the European cave bear. There is no known association with man (Kurtén 1966).

Ursus   Linnaeus, 1758   (=*Euarctos, Thalarctos*)   Ursidae
Brown and grizzly bears, black bear, polar bear, cave bear. N.Am., Eurasia. Bl.–R
N.Am.; Ast.–R Eur.; Pleist.–R Asia. The best known Pleistocene bear is *U. spelaeus*
Rosenmüller and Heinroth, the cave bear, an endemic European species that lived
during the Steinheimian. Remains of this animal are common in cave deposits where
enormous numbers of complete skeletons have been found. *U. spelaeus* differs from the
living *U. arctos* by its unusually large size (rivaling the Alaskan brown bear), short legs,
large head with a strongly vaulted forehead, and tubercular molars. Cave bears were
exclusively vegetarian in habits. They denned in caves, apparently with sexual separation, for some caves contain the remains of males, others, females and young. The great
accumulation of bones in caves represents animals that died during hibernation (most
are young, very old, or diseased animals). There is no real evidence for a Neandertal
cave bear cult (see Kurtén 1976). Although cave bears were occasionally hunted, it does
not seem possible that man hastened their extinction. Cave bears were still relatively
common at the end of the Mousterian about 30,000 years ago, but by the end of the
Pleistocene only a few caves in central Europe supported relict populations. Paleolithic
drawings of bears mostly represent *U. arctos,* the brown bear, and there are no stone
implements or cave bear bones with cut marks on them.

The earliest known appearance of *U. arctos* Linn. is from the middle Pleistocene of
China (Choukoutien); the species reached Europe in the early Steinheimian and N.
Amer. in the late Rancholabrean. *U. americanus* Pallas, the extinct American black bear,
is present from Irvingtonian times onward. Polar bear (*U. maritimus* Phipps) fossils are
very rare; the species dates back to the late Steinheimian in Europe and the early
Holocene in N.Amer. (Kurtén 1972b, 1976)

Acinonyx   Brookes, 1828   Felidae
Cheetah. 1 sp. Villafr.–Crom. Eur.; Villafr.–R (now extinct) Asia; ?E Pleist.,
L Pleist.–R Af.; Bl.–L RLB N.Am. Until recently the genus *Acinonyx* was thought to
be endemic to the Old World where it was represented by *A. pardinensis* Croiset and
Jobert, a gigantic form reaching the size of a lion; it was known from the Villafranchian of
Europe, India, and China, and in the late Pleistocene by *A. jubatus* Schreber, the extant
cheetah. An extinct cheetah, *A. trumani* (Orr), has been recognized in the late Wiscon-
sinan of Wyoming (Natural Trap Cave) and Nevada (Crypt Cave); and Adams (1977) has
referred the Blancan and Irvingtonian species *Felis studeri* and *F. inexpectata* to
Acinonyx and has placed these three species in a distinct subgenus *Miracinonyx*. The
New World cheetahs are closely related to pumas (*F. concolor*) and differ from the Old
World animals in having less inflated bullae and frontal sinuses and fully retractile claws.
Cheetahs are distinguished from other felids in having elongated slim limbs, a small
head, and lithe body. Like the living animal, Pleistocene cheetahs were superbly adapted
to a cursorial mode of life and inhabited grasslands and steppes. Remains of cheetah are
rare, and there are no known associations with man in either the Old World or the New.
(Adams 1979, Kurtén 1968)

†*Dinofelis*   Zdansky, 1924   Felidae
Sabertooth. Villafr. Eur., Asia; Plioc.–E Pleist. Af.; Bl. N.Am. Large sabertooth with
unserrated upper canines.

Felis   Linnaeus, 1758   Felidae
Small cats (puma, ocelot, jaguarundi, margay, wild cats, serval, jungle cats, etc.); ca. 25
spp. Nearly worldwide except Australian region, Madagascar, and W.I. Bl.–R N.Am.;
Uq.–R S.Am.; L Villafr.–R Eur.; L Plioc.–R Asia; E Pleist.–R Af. Characterized by a
short rostrum, skull highly arched in frontal region, usually a long tail. Size ranges
from smaller than a house cat to ca. 100 kg (*F. concolor*). Remains of *F. silvestris* and
F. concolor are common in late Pleistocene cave deposits.

Panthera leo spelaea, extinct cave lion of Europe (above); *Smilodon,* New World sabertooth (left); *Homotherium,* scimitar cat (right); all extinct

†*Homotherium*   Fabrini, 1890   (=*Dinobastis*)   Felidae

Scimitar cat. Villafr.–L Pleist. Eur. (L Weich. Britain); L Plioc.–M Pleist. China, Java; E Pleist. Af.; Bl.–L RLB N.Am. Several spp. are recognized. The Old World species *H. latidens* (Owen) was about the size of a lion; a rare animal, it was probably solitary in habits. *H. serum* (Cope), the closely related N. Amer. species, was less common than the sabertooth or the American Pleistocene lion. At Friesenhahn Cave, Tex., skeletons of both adults and juveniles have been found along with hundreds of milk molars of juvenile mammoths, probably the scimitar cat's favorite prey. *Homotherium* had elongated forelimbs and short hindlimbs which resulted in a sloping back and a long head carried high on a long neck. The relatively short scimitarlike sabers are serrated fore and aft, and the carnassials are reduced to thin, slicing blades. The animal probably hunted by ambush. There are no associations with man, and extinction was due to the disappearance of its prey. (Kurtén and Anderson 1980)

†*Machairodus*   Kaup, 1833   Felidae
Sabertooth. L Mioc.–E Pleist. Eur., Asia, Af.; Hemph. N.Am. Typical Pliocene sabertooths. Upper canine large, bladelike, coarsely serrated; it protruded beneath the chin
when the mouth was closed because the flange was lost; lower canine incisiform greatly
reduced in size. Relatively long face, elongated limbs. Size of a lion or tiger.

†*Meganteron*   Croiset and Jobert, 1828   Felidae
Sabertooth. Villafr. Eurasia; Bl. N.Am.; E Pleist. Af. Short face, long narrow unserrated upper canines, shortened limbs. Size of a puma. Ancestral to *Smilodon*.

Panthera   Oken, 1816   Felidae
Lion, leopard, tiger, jaguar. 4 spp. Villafr.–L Steinh. Eur.; L Plioc.–R Asia; Irv.–R
N.Am.; Ens.–R S.Am.; E Pleist.–R Af. The four large cats were present in the
Pleistocene, and though larger than their extant counterparts, most are now considered
only subspecifically distinct. *P. leo spelaea* Goldfuss, the European cave lion, and *P. l.
atrox* Leidy, the American Pleistocene lion, were considerably larger, perhaps weighing
half again as much as a living lion (wt. 181–227 kg). Hemmer (1974) estimates head and
body length for *P. l. spelaea* as 1.4–2.7 m and 1.6–2.5 m for *P. l. atrox*; for the extant lion
it is 1.8–2.4 m. *P. l. atrox* is absent from pre-Sangamonian faunas south of Alaska, but
from the Sangamonian to the end of the Wisconsinan it was found primarily in open
country from Alaska to Peru, being absent only in the East and peninsular Florida. Like
modern lions, Pleistocene ones probably hunted in prides. Lions have been pictured at
Les Combarelles, and in collections at Bordeaux a piece of reindeer scapula is engraved
with the tail of a lion showing the tuft. At Jaguar Cave, Id., bones of *P. l. atrox* have been
found at the same level as hearths, but they have no cut marks on them; the hearths
have been dated at 10,370±350 yr B.P.
    Though not as abundant as the lion, Pleistocene leopards were widespread by the
middle Pleistocene and have been found in Europe where they survived until the end of
the epoch, in the Middle East, China, Java, and Africa. Leopards have a long body, short
legs, a small head, and a very long tail; weight is ca. 90 kg. They feed on medium-sized
prey which are hunted by stalking or ambush.
    Pleistocene tigers are known from China. Tigers differ from lions in having much
larger canines in relation to premolars, a different shaped nasal opening, and other skull
characters. Largest of the living cats, tigers have a head and body length of 1.8–2.8 m
and weigh 227–272 kg. Lions and tigers may have had a common ancestry in the late
Pliocene of Africa.
    The jaguar, *P. onca* (Linnaeus), has been found in Irvingtonian and in over thirty
Rancholabrean faunas in N. Amer. and Ensenadan and Lujanian faunas in S. Amer. The
early Pleistocene form may be conspecific with the contemporaneous Palearctic jaguar,
P. gombaszoegensis (Kretzoi). Jaguars are smaller than lions (wt. 68–136 kg) and have
shorter, stockier limb bones. The Wisconsinan jaguar, *P. o. augusta* Leidy, was 15 to 20
percent larger than its living counterpart. Lion and jaguar remains have seldom been
found together—the jaguar being found in forests and along streams, the lion in open
country. No direct associations with man are known. (Hemmer 1974, Kurtén and
Anderson 1980)

†*Smilodon*   Lund, 1842   Felidae
New World sabertooth. E Irv.–L RLB N.Am.; Ens.–Luj. S.Am. At Rancho La Brea,
Calif. thousands of specimens have been recovered. An extensive collection also is
known from the Talara tar seeps, Peru. Three species are currently recognized:
S. gracilis Cope, E Pleist. on both continents; *S. fatalis* (Leidy) (=*S. californicus,
S. floridanus*) L Irv.–L RLB from Alberta to western S.Am.; and *S. populator* Lund
(=*S. neogaeus*) Luj. from Argentina, Brazil, Ecuador, Peru, and Uruguay, a larger
species with shortened distal limb segments. *Smilodon* had long slender canines with
large coarse serrations, a huge gape (ca. 100°), and reduced coronoid processes.
S. fatalis was about the size of an African lion but had heavier forequarters, lighter

hindquarters, and a bobbed tail. It was not as fleet as a lion. *Smilodon* preyed on large, slow-moving animals which it killed by stabbing in vital areas. A few bones from Rancho La Brea have marks on them which may have been made by man. Extinction was probably due to the disappearance of its large prey. Sabertooths were not adapted to catch relatively small animals. Bone collagen of *Smilodon* from the First National Bank Site, Nashville, Tenn., has been dated 9,410±155 yr B.P. (I-6125). (Guilday 1977, Kurtén and Anderson 1980, Martin, L. D. 1980, Merriam and Stock 1932)

†*Chasmaporthetes*   Hay, 1921   Hyaenidae
Hunting hyena. Bl.–E Irv. N.Am.; E Pleist. Af.; Villafr. Eur.; L Plioc.–?E Pleist. Asia. Only hyenid to reach N.Am. Sectorial dentition, toothrow curved. Cheetahlike predator.

Crocuta   Kaup, 1828   Hyaenidae
Spotted hyena. 1 sp. Crom.–Steinh. Eur.; E Pleist.–R Af.; L Mioc.–L Pleist. Asia. The spotted hyena ranged from South Africa to Europe and E to China in the Pleistocene but now is found only in Africa S of the Sahara. Enormous numbers of its bones have been found in European caves (Kirkdale, Tornewton, Kent's Cavern, Eng.) that were used as dens for thousands of years. Bone remains from hyena dens have sometimes been mistaken for human artifacts (see Sutcliffe 1970). Pleistocene *Crocuta* was larger than the extant animal (wt. 59–82 kg), and a subspecies, *C. c.spelaea* Goldfuss, characterized by shortened, thickened foot bones and very long humeri and femora is recognized. *Crocuta* is a highly specialized scavenger with the most powerful jaws in proportion to size of any mammal. The premolars are strong bone-crushing teeth, the carnassials are elongated to form slicing blades, but the canines are weak. Studies have shown that *Crocuta* is an efficient predator capable of bringing down animals as large as zebra and wildebeest. *Crocuta* became extinct in Eurasia at the end of the ice age. There are few representations of the spotted hyena in Paleolithic art; an ivory sculpture from La Madeleine, Dordogne, is notable. (Kurtén 1968)

†*Euryboas*   Schaub, 1941   Hyaenidae
Hunting hyena. L Mioc.–E Pleist. Af.; Villafr. S Eur. Inhabited open areas. Teeth slender and sharp, tooth row straight. Diurnal, cursorial hunter. Essentially an African form; evolved independently from *Chasmaporthetes*.

Hyaena   Brumich, 1772   Hyaenidae
Striped and brown hyenas. 2 spp. L Mioc.–R Af.; Villafr.–L Steinh. Eur.; Pleist.–R Asia. The genus *Hyaena* was widespread during the Pleistocene, and several extinct species have been described. The short-faced hyena, *H. brevirostris* Aymard, lived from late Villafranchian to middle Cromerian times in Europe, China (Choukoutien), and Java. It was the size of a lion (shoulder ht. 1 m, head and body lt. 1.5 m). Its extinction was due to competition with *Crocuta,* the spotted hyena. The extinct African species, *H. abronia,* was ancestral to the living striped hyena, *H. hyaena* Linnaeus (L Plioc.–R). By the late Pleistocene it had spread to Europe and Palestine; now it ranges throughout S and SW Asia and the northern half of Africa. The Pleistocene form was somewhat larger but otherwise similar to the living animal (wt. 27–54 kg). Carrion is its main food. The extant brown hyena, *H. brunnea* Thunberg, is known from middle Pleistocene faunas in S. Af., its present range. It is an advanced species that some workers believe is generically distinct and should be placed in *Hyaenicititherium,* the ancestral genus. It has relatively large canines, and its cheek teeth are less specialized for carrion-feeding than other hyenas. (Hendey 1978, Kurtén 1968)

Proteles   Geoffroy, 1824   Hyaenidae
Aardwolf. 1 sp. Sandy plains and bush country of S and E Af. L Plioc.–E-M Pleist. Transvaal. An extinct species, *P. transvaalensis* Hendey, from Swartkrans and Kromdraai still had functional carnassials. *Proteles* is a hyenalike animal that has weak jaws, vestigial teeth, and sharp canines. It feeds on termites, insect larvae, eggs, and occasionally mice. Wt. 10–15 kg. (Hendey 1974)

Castoroides, the extinct giant beaver of the North American Pleistocene

RODENTIA

†*Nerterogeomys* Hay, 1927   Geomyidae
Early pocket gopher. 4 spp. Kans., Tex., Ariz. E Bl.–E Irv. Size of *Geomys*.

†*Prodipodomys* Hibbard, 1939   Heteromyidae
Extinct kangaroo rats. 2 spp. Bl.–E Irv. W U.S. Collateral lineage with *Dipodomys*.

†*Castoroides* Foster, 1838   Castoridae
Giant beaver. 1 sp. Bl.–L RLB N.Am. The largest rodent in N. Amer. during the
Pleistocene was *Castoroides*, an animal the size of a black bear. It ranged from Alaska
south to Florida and from Nebraska eastward but was most common just south of the
Great Lakes. The giant beaver inhabited lakes and ponds bordered by swamps, and its
habits were more like those of a muskrat *(Ondatra)* than a beaver *(Castor)*. There is no
evidence that *Castoroides* built dams or felled trees. It fed on coarse swamp vegetation.
Characteristic are the enormous blunt-tipped, longitudinally fluted, convex incisors;
short legs, small front feet, large probably webbed hind feet, and a relatively long
narrow tail. There is no association with man at any site. Extinction was probably due
to competition with *Castor* and reduction and disappearance of its preferred habitat.
(Kurtén and Anderson 1980)

†*Trogontherium* Fischer, 1809   Castoridae
Eurasian giant beaver. Villafr.–E L Pleist. Eurasia; its remains are common at Choukoutien. The Eurasian giant beaver was smaller than *Castoroides*, being only slightly larger
than a modern beaver. It may have had a prehensile upper lip and fed while swimming.
(Kurtén 1968)

†*Atopomys* Patton, 1965   Cricetidae
Extinct vole. L Irv. Tex. and Appalachian Mtns. Rootless, cementless teeth.

†*Majoria* Thomas, 1915   Cricetidae
Extinct Madagascar mouse. Pleist. Madagascar.

†*Mimomys* Major, 1902   Cricetidae
Extinct voles. ca. 8 spp. Ast.–Crom. Eur.; Bl. N.Am. (includes *Ophiomys, Cosomys,
Ogmodontomys*; (C. A. Repenning MS). Ancestral to *Arvicola*.

†*Pliomys* Meheley, 1914   Cricetidae
Extinct snow vole. Plioc.–Crom. Eur.; E Irv. Alaska. Related to *Dolomys*.

†*Pliophenacomys* Hibbard, 1938   Cricetidae
Extinct voles. 3 spp. E Bl.–E Irv. Kans., Neb. Teeth have closed roots, lack cement.

†*Pliopotamys*   Hibbard, 1938   Cricetidae
Extinct muskrat. Bl.–E Irv. N. Am. Ancestral to *Ondatra,* probably to *Neofiber.*

†*Predicrostonyx*   Guthrie and Matthews, 1971   Cricetidae
Extinct lemming. 1 sp. Irv. Alaska, treeless tundra. Ancestral to *Dicrostonyx.*

†*Proneofiber*   Hibbard and Dalquest, 1973   Cricetidae
Extinct water rat. 1 sp. E Irv. Tex. (Gilliland). Size of the primitive muskrat, *Ondatra annectens.*

†*Rhinocricetus*   Kretzoi, 1941   Cricetidae
Extinct dwarf hamster. L Villafr.–Crom. C Eur. Intermediate in size between *Cricetus* and *Cricetulus.*

†*Tyrrhenicola*   Major, 1905   Cricetidae
Extinct Tyrrhenian vole. Crom.–E Holoc. cave breccias of Corsica and Sardinia. May be a subgenus of *Pitymys.*

†*Prospalax*   Meheley, 1908   Spalacidae
Primitive mole-rat. Ast.–Crom. C Eur., Mediterranean region.

†*Hypnomys*   Bate, 1919   Myoxidae
Extinct Baleric dormice. 2 spp. Mallorca and Minorca. Steinh.–R (now extinct). Relict form. Intermediate in size betwen *Myoxus* and *Leithia.*

†*Leithia*   Lydekker, 1896   Myoxidae
Maltese dormice. Crom.–Steinh. Malta and Sicily. Giant form, size of a hamster.

†*Coryphomys*   Schaub, 1937   Muridae
Extinct East Indian mouse. Pleist. East Indies.

†*Parapodemus*   Schaub, 1938   Muridae
Extinct wood mouse. E Plioc.–Crom. Eur. Ancestral to *Apodemus.*

†*Rhagamys*   Major, 1905   Muridae
Extinct Mediterranean field mouse. 1 sp. Crom.–E Holoc. Corsica and Sardinia. Related to *Apodemus.*

†*Stephanomys*   Schaub, 1938   Muridae
Extinct field mouse. M Plioc. Eur.; L Plioc.–Pleist. Asia.

†*Heteropsomys*   Anthony, 1916   (=*Boromys, Brotomys*)   Echimyidae
Extinct spiny rats. Pleist.–Holoc. Puerto Rico, Haiti, Cuba, Dominican Republic. Size of *Proechimys.* Common in pre-Columbian times; extinction probably due to introduced predators, diseases, and hunting by man.

†*Aphaetreus*   Miller, 1922   Capromyidae
Montane hutia. Pleist.–Holoc. Mountains of Haiti, Dominican Republic.

†*Hexlobodon*   Miller, 1929   Capromyidae
Extinct Hispaniolan hutia. 1 sp. Pleist.–Holoc. Haiti. Size of *Capromys.* Exterminated by man.

†*Isolobodon*   J. A. Allen, 1916   Capromyidae
Extinct hutia. Pleist.–E Holoc. West Indies. Remains numerous in caves (preyed on by giant owl, *Tyto ostolaga*) and in middens. Size of *Plagiodontia.*

†*Alterodon*   Anthony, 1920   Octodontidae
Extinct West Indian degu. Pleist. W.I.

†*Pithanotomys*   Ameghino, 1887   Octodontidae
Extinct degu. Plioc.–Ens. S.Am.

Extinct West Indian heptaxodont rodents   Heptaxodontidae
Heptaxodont rodents (Family Heptaxodontidae, considered a subfamily of the Di-

nomyidae, false pacas, by some workers) inhabited the West Indies during the late Pleistocene; three genera *(Amblyrhiza, Elasmodontomys, Quesmia)* survived to about the sixteenth century. These terrestrial rodents are characterized by massive skulls and stout bodies; each cheek tooth has from four to seven laminae with nearly parallel crests arranged obliquely to the long axis of the skull. Six genera and seven species have been described: †*Amblyrhiza* Cope, 1868, Anguilla, skull lt. 400 mm; †*Clidomys* Anthony, 1920, Jamaica; †*Elasmodontomys* Anthony, 1916 (incl. †*Heptaxodon)*, Puerto Rico, skull lt. 125 mm; †*Quesmia* Miller, 1929, Haiti, Dominican Republic, size of *Agouti*; †*Spirodontomys* Anthony, 1920, Jamaica; †*Speoxenus* Anthony, 1920, Jamaica. Whether all these genera are valid is uncertain. (Verona 1974, Walker 1975)

Hydrochoerus Brisson, 1762 (=*Xenohydrochoerus*)   Hydrochoeridae
Capybara. 1 sp. Uq.–R S.Am.; Pleist. Cent. Am., W.I.; Irv.–RLB Florida. The extinct N. Amer. species, *H. holmesi* Simpson, had larger cheek teeth than the living species. Today capybaras are found in Panama and N S. Amer. east of the Andes. The capybara is the largest living rodent (wt. 27–50 kg, ht. at shoulder 50 cm). They have a robust build with a large broad head, a truncated muzzle, and small eyes set relatively far back on the head. The ever-growing cheek teeth are made up of transverse lamellae joined together by cement; the incisors are white and slightly grooved. Capybaras are semiaquatic and feed on grass and aquatic plants. (Kurtén and Anderson 1980)

Neochoerus, an extinct giant capybara of North and South America

†*Neochoerus* Hay, 1926 (=*Pliohydrochoerus*)   Hydrochoeridae
Giant capybara. Uq.–Luj. S.Am.; Irv.–RLB Fla., Tex., S. Car., perhaps Ariz. The N. Amer. species, *N. pinckneyi* (Hay), was about 40 percent larger in size with teeth 30 percent larger than those of the extant *Hydrochoerus*. The relationship between *Neochoerus* and *Hydrochoerus* is unknown at this time. No associations with man are known. (Kurtén and Anderson 1980, Marshall et al. MS)

LAGOMORPHA

†*Ochotonoides* Teilhard de Chardin and Young, 1931   Ochotonidae
Extinct Chinese pika. Pleist. Asia (Choukoutien)

†*Prolagus* Pomel, 1853   Ochotonidae
Sardinian pika. M Mioc.–E Holoc. Eur. *P. sardus* (Crom.–E Holoc.) Sardinia, Corsica. Larger than *Ochotona*; lineage shows increase in size through time. Adept climbers.

†*Hypolagus* Dice, 1917   Leporidae
Early rabbit. Hemph.–Bl. N.Am.; L Plioc.–Pleist. Eurasia. ca. 6 spp.

LITOPTERNA

†*Macrauchenia*   Owen, 1840   Macrauchenidae
Macrauchenids. Uq.–Luj. S.Am. One of the last members of the endemic S.Amer.
order Litopterna (L Paleoc.–L Pleist.) was *Macrauchenia*. It had elongated limbs, three
functional toes, a long neck and snout, and high-crowned cheek teeth. It was about the
size of a camel. Position of the nasal opening, shifted backward onto the top of the skull,
characteristic. *Macrauchenia* was probably amphibious in habits. No associations with
man are known. Extinction of the litopterns is generally attributed to the invasion of
advanced placental carnivores from N. Amer. who ate them or their food. (Kurtén
1972a, Romer 1966)

†*Windhausenia*   Kraglievich, 1930   Macrauchenidae
Macrauchenid. L Plioc.–Luj. S.Am.

The last members of the last South American orders to become extinct, *Toxodon*
(Notoungulata, left) and *Macrauchenia* (Litopterna, right and skull, center)

NOTOUNGULATA

†*Mixotoxodon*   Toxodontidae
Toxodont. Luj. S.Am.

†*Toxodon*   Owen, 1837   Toxodontidae
Toxodonts. Chapad.–Luj. S.Am. Probably the most common mammal in the S. Amer.
Pleistocene was *Toxodon,* a widespread notoungulate that has been described as a giant
guinea pig built like a short-legged rhinoceros with hippolike habits. The dorsal position
of the nasal opening suggests a large snout. Unlike other notoungulates, *Toxodon* has
a diastema between the incisors and the cheek teeth. There are three hoofed toes
on each foot. Toxodonts were amphibious. At Arroio Touro Passo, Brazil, dated
11,040±190 yr B.P. (I-9628) *Toxodon* was a member of the fauna that also included

man. Predation by invading carnivores and competition with invading ungulates spelled doom to the toxodonts; perhaps relict populations were hunted by man. (Kurtén 1972a, Marshall et al. MS, Romer 1966)

†*Mesotherium* Serres, 1867   Mesotheriidae
Typothere. Uq.–Luj. S.Am. especially the Argentine pampas. Highly specialized with a broad depressed stout skull; rodentlike incisors, no canines, ever-growing cheek teeth.

PERISSODACTYLA

Equus Linnaeus, 1758   Equidae
Horses, asses, zebras. ca. 8 spp. Bl.–L RLB N.Am.; M Villafr.–L Steinh. Eur.; Plioc.–R Af.; L Plioc.–R Asia; now worldwide by introduction. The taxonomy of Pleistocene equids is in a state of confusion. Scores of species, many based on isolated teeth, have been described. Studies by Dalquest (1978) and Churcher and Richardson (1978) on N. Amer. and African equids respectively have somewhat clarified the picture. Several subgenera are recognized: *Dolichohippus*—extinct N. Amer. Blancan species, *E. capensis* Plioc.–L Pleist. S.Af.; and the extant Grevy zebra, *E. grevyi; Hippotragus*—zebras, *E. burchelli, E. zebra,* and *E. quagga* (became extinct in 1880); *Equus*—true horses, *E. przewalskii,* the only living wild species, and *E. caballus,* the modern horse; *Asinus* and *Hemionus*—asses and donkeys, includes many extinct species and the living *E. hemionus* and *E. asinus.* The domestic horse is probably descended from the tarpan and Przewalski's horses.

Equus is monodactyl with lateral digits absent and lateral metapodials reduced to splints; the hooves lack the medial split seen in *Hipparion.* Cheek teeth are strongly hypsodont, and canines are usually absent in females. *Equus* has a relatively long neck, head, and tail, with a mane on the neck; some species have a forelock. All are swift runners that inhabit plains, savannas, and in some cases mountainous regions.

Horses and asses were dominant members of many faunas of the steppes, where their remains are often the most numerous animals found. Horses were frequently pictured by Stone Age artists in Europe. Both Przewalski's horse and the tarpan have been recognized at Lascaux. Direct association between man and horse is uncommon in N. Amer., however, in Europe horses were hunted extensively by Paleolithic man. Terminal dates for the extinction of *Equus* in N. Amer. are 10,370±350 yr B.P. at Jaguar Cave, Id., 8,150 yr B.P. at Beverly Pit near Edmonton, Alberta, and ca. 8,000 yr B.P. at Pashley, Alb. Why some Pleistocene equids became extinct while others survived is unknown. (Churcher and Richardson 1978, Dalquest 1978, Kurtén and Anderson 1978)

†*Hipparion* de Cristol, 1832   Equidae
Three-toed horses. Many spp. Mioc.–M Plioc. N. Am.; M Mioc.–L Pleist. Af.; M Mioc.–E Villafr. Eurasia. The genus *Hipparion* arose in N. Amer. in the Miocene and spread rapidly to populate large areas of Eurasia and Africa as well as their original homeland. *Hipparion* reached its heyday in the Pliocene, then succumbed to competition with the better-adapted *Equus. Hipparion* survived in Africa where populations persisted well into the Pleistocene. The *Hipparion* lineage is characterized in having functional II, III (largest), and IV digits, a deep median slit on the terminal phalanx (hoof), increasingly hypsodont cheek teeth, large incisors, canines present in both sexes, and prominent ectostylids on the lower permanent cheek teeth. *Hipparion* was smaller than extant asses or zebras. A grazer, it lived in herds on the savannas. *H. libycum* Pomel is known from Pleistocene faunas in North, East and South Africa; it is the most advanced species, with extremely hypsodont teeth. The extinction of this widespread species was probably due to the explosive radiation of bovids that resulted in a proliferation of antelope on the savannas. (Churcher and Richardson 1978)

†*Hippidion*   Owen, 1869   Equidae
Extinct South American horse. Widely distributed in S. Amer., especially in Bolivia and
Argentina. Uq.–Luj. Small, short-legged, heavily built; long slim nasal bones. Mountain
dweller. It and *Onohippidium* probably derived from *Pliohippus* rather than *Equus*.

†*Onohippidium*   Moreno, 1891   Equidae
Extinct South American horse. S. Amer. Uq.–Luj. Closely related to *Hippidion*. Both
differ from *Equus* in having extremely long, slender nasal bones attached to the skull
only at the posterior end, leaving a long slit in the side of the skull, small orbits situated
lower down, and relatively short, wide metapodials. Also characteristic is a large head,
relatively short neck and heavy limbs.

Tapirus   Brisson, 1762   Tapiridae
Tapirs. 4 spp. Cent. and S Amer., SE Asia. Bl.–L RLB N.Am.; Uq.–R S.Am.; Plioc.–
Pleist. Asia; Pleist. E. Indies; Ast.–E M Pleist. Eur. Tapirs were fairly common in
Rancholabrean times in N. Amer., and a few specimens have been found in earlier
deposits. Most of the specimens come from Florida, but remains of tapirs have been
found as far N as Pennsylvania and W to California. Several species have been described, but relationships are unclear and revision is needed. *T. veroensis* Sellards (=*T.
tennesseae, T. excelsus*) was slightly larger than the living Neotropical species. These
primitive perissodactyls are heavily built, with four toes on the front feet and three toes
on the hind feet. The teeth are brachydont; tapirs feed on soft vegetation. Generally
solitary, tapirs are semiaquatic and live in moist forests. No associations with man are
known; its extirpation in N. Amer. was probably due to climatic change. (Kurtén and
Anderson 1980)

Ceratotherium   Gray, 1867   Rhinocerotidae
White rhinoceros. 1 sp. Plioc.–R Af. Originating in Africa, *Ceratotherium* split off from
the *Diceros* stock in the Pliocene. The earliest known species, *C. praecox* Hooijer and
Patterson, was ancestral to the living *C. simum* (Burchell) identified in East African
deposits dated 3 my. The white rhino is now found in savannas and brushy habitats in
S. Af., S Sudan, Uganda, and Zaire. It has a square upper lip, high-crowned cheek teeth,
a shoulder height of 1.6–2.0 m and weighs 2.3–2.6 tons. A grazer, it feeds on grasses
and low shrubs. Populations have been greatly reduced, but the species is protected in
S. Af. (Hooijer 1978, Walker 1975)

†*Coelodonta*   Bronn, 1831   Rhinocerotidae
Woolly rhinoceros. 1 sp. M–L Steinh. Eurasia. At Starunia in the Carpathian foothills
three individuals were found preserved in salt and petroleum deposits, and frozen
mummies have been found in Siberia. *C. antiquitatas* Blumenbach was adapted to live in
a cold climate and had a thick woolly pelage, thick skin, and a layer of subcutaneous fat.
It has two nasal horns supported by robust nasal bones and a bony nasal septum; the
laterally flattened anterior horn was longer and more slender than the posterior one.
The horns were probably used as weapons and to scrape away snow. The head was
carried low, accentuating the shoulder hump. A specimen from Starunia measures
358 cm in length plus a 60-cm tail. An inhabitant of the steppe-tundra, the woolly rhino
fed on grasses and low plants. *Coelodonta* is represented in Paleolithic cave paintings
and engravings (Lascaux, Rouffignac). Why *Coelodonta* did not enter Alaska during the
ice age as did so many cold-adapted Eurasian animals is unknown. Two specimens from
the Indigirka and Yana rivers, NE Siberia, have been radiocarbon dated at 38,000 and
33,000 yr B.P., respectively. (Kurtén 1972b)

Diceros   Gray, 1821   Rhinocerotidae
Black rhinoceros. 1 sp. Plioc.–R Af.; Plioc. Eur., SW Asia. The extant species, *D.
bicornis* Gray, is recognized at the 4-my level at Kanam West and Afar, E. Af., making it
one of the oldest surviving mammal species. Today the black rhino is found in E and S
Africa. Shoulder height is 1.4–1.5 m, weight is 1.0–1.8 tons; it has a prehensile upper

The last rhinos from northern Eurasia, *Elasmotherium*
(foreground) and *Coelodonta* (background)

lip and two median horns composed of keratin are located on the nasal and frontal
bones. A browser, it feeds on twigs and leaves. *Diceros* is more aggressive and less
sociable than *Ceratotherium*. Since the black rhinoceros is valued for its skin and horns,
its populations have been reduced in many areas. (Hooijer 1978, Walker 1975)

Dicerorhinus   Gloger, 1841   Rhinocerotidae
Sumatran rhinoceros. 1 sp. SE Asia. Rare, may be extinct. L Olig.–Pleist. Eur.; Mioc.
Af.; L Olig.–R Asia. Several extinct spp. recognized, including *D. hemitoechus* Falconer
which survived in Europe until the end of the Steinheimian. The smallest living rhino,
D. sumatrensis (shoulder ht. 1.1–1.5 m, wt. ca. 1 metric ton), has two horns and the
folded skin is covered with bristlelike hair.

†*Elasmotherium*   Fischer, 1808   Rhinocerotidae
Giant rhinoceros. 1 sp. Pleist. steppes of Eurasia. Larger than any living rhinoceros;
enormous (lt. 2 m) horn on forehead; complicated high-crowned cheek teeth. Grazer.

Rhinoceros   Linnaeus, 1758   Rhinocerotidae
Indian rhinoceros. 2 spp. Tropical Asia. L Plioc.–R Asia. One nasal horn; loose folds of
skin give it an armored appearance. Greatly reduced in numbers due to hunting and loss
of habitat. Wt. 2,000–4,000 kg.

†*Ancyclotherium*   Gaudry, 1863   Chalicotheriidae
Chalicothere. Plioc. SE Eur., Iran; Plioc.–E Pleist. Af. (Omo, Olduvai, Makapansgat).
Chalicotheres are an aberrant group of perissodactyls that had claws instead of hooves.

Their main area of evolution was Asia N of the Tethys. These large animals inhabited wooded savanna and fed on leaves and soft vegetation; they may have been tree-top browsers. The cause of their extinction is unknown but competition with giraffes and proboscideans for the same food source may have been a factor. (Butler 1978)

ARTIODACTYLA

†*Hippohyus*   Falconer and Cautley, 1840–45   Suidae
Extinct Asian hog. E Plioc.–Pleist. Asia. Retains features found in ancestral pigs. The zygomatic arches are widely separated from the cranium, the orbits are elevated, and the masseter muscles are well developed. Grazer.

Hylochoerus   Thomas, 1904   Suidae
Forest hog. 1 sp. L. Pleist.–R Af. These medium to large suids (shoulder ht. 76–96 cm, wt. 160–275 kg) have a poor fossil record because they occupied forest habitat. The extant species, *H. meinertzhageni* (Thomas), has been identified in late Pleistocene faunas in Kenya and is found in dense forests in Cent. Af. today. It has greatly expanded zygomatic arches, a broad occiput, large outflaring canines, and is covered with long, coarse black pelage. *Hylochoerus* has a facial gland and lacks facial protuberances. (Cooke and Wilkinson 1978)

†*Kolpochoerus*   E. C. N. van Hoepen, 1932   (=*Mesochoerus*)   Suidae
Extinct bush pig. 5 spp. Plioc.–M Pleist. from N to S Af. The zygomatic arches are expanded laterally and droop, the molars are brachydont, the canines are shorter and stouter than those of *Hylochoerus,* and sexual dimorhism is pronounced. Medium to large size. Both *Kolpochoerus* and *Potamochoerus* are derived from a common ancestor that possessed basic *Sus*-like characters. (Cooke and Wilkinson 1978)

†*Metridiochoerus*   Hopwood, 1926   Suidae
Hopwood's extinct hog. 3 spp. Plioc.–M Pleist. E and S Af. (Olduvai I–IV, Omo, Swartkrans). *Metridiochoerus* has broad parietal and occipital regions, elevated orbits, reduced premolars, and hypsodont molars, especially M3. Size medium to large. (Cooke and Wilkinson 1978)

†*Notochoerus*   Broom, 1925   Suidae
Broom's extinct hog. 3 spp. Plioc.–E Pleist. Af. (Omo, Hadar, Makapansgat). *Notochoerus* has reduced premolars with only P3 and P4 present in adults, hypsodont third molars, and upper canines that are dorsoventrally flattened and flare outward. Size large, twice as large as extant warthog. (Cooke and Wilkinson 1978)

Phacochoerus   Cuvier, 1817   Suidae
Warthog. 1 sp. Plioc.–R Af. 3 extinct spp. The widespread extant warthog inhabits the savannas of Africa; during the Pleistocene its range was larger and included areas that are now desert. Medium size (shoulder ht. 64–73 cm, wt. 75–100 kg); has elevated orbits, broad thickened zygomatic arches, large outflaring upper canines, and hypsodont molars. Males have large facial protuberances and lack a facial gland. These diurnal, omnivorous suids are hunted for their tasty flesh. (Walker 1975)

Potamochoerus   Gray, 1854   Suidae
African bush pig. 1 sp. Fossil remains of the extant genus *Potamochoerus* are rare in sub-Saharan Africa. *P. porcus* Linn. now inhabits bush and forest areas near water in Africa and Madagascar. The skull is similar to that of *Sus,* but the zygomatic arches are more expanded laterally and may be inflated; the canines in males are protected by prominent flanges. These nocturnal, secretive suids are omnivorous and use their snouts to root. (Walker 1975)

†*Stylochoerus*   E. C. N. and H. E. van Hoepen, 1932   Suidae
Van Hoepen's extinct hog. 1 sp. M Pleist. (Koobi Fora, Olduvai, Kanjera, Vaal River gravels, S. Af.). The distinctive canines of *Stylochoerus* are oval in cross section and have a core of cancellous osteodentine; the upper canines arise at a 40° angle from the palatal plane. It was twice the size of the living warthog. (Cooke and Wilkinson (1978)

Sus   Linnaeus, 1758   Suidae
Wild boar, domestic pig. ca. 5 spp. Eurasia, N. Af., Japan, Malaysia, introd. U.S.; domestic pig has a worldwide distribution. E Plioc.–R. Remains of *Sus* have been found in Villafranchian faunas in Europe and N. Africa, and it is common in the late Pleistocene, especially in interglacial and interstadial faunas because frozen ground and deep snow were and are limiting factors to its distribution. *Sus* is a medium- to large-size suid, the zygomatic arches are not inflated, the upper and lower canines are in full contact with each other thus wearing to sharp edges, the molars are brachydont and bunodont. Wild boars are hunted for food and sport. (Cooke and Wilkinson 1978, Kurtén 1972b)

Catagonus   Ameghino, 1904   Tayassuidae
Chaco peccary. 1 sp. Ens.–R S. Am. The S. Amer. genus *Catagonus* was thought to have become extinct in the Lujanian. In the 1970s, however, live Chaco peccaries were discovered in the Chaco region of Paraguay and Bolivia! The living animals are referred to as *C. wagneri* Rusconi. Closely related to *Platygonus, Catagonus* differs from that genus in having larger teeth, longer tooth rows, a shorter diastema, a four-cusped P4 that is nearly as large as M1, and a more quadrangular M3. A large head and much larger rostrum separate *Catagonus* from the extant *Tayassu*. An inhabitant of the scrub-thorn grass refugium, the Chaco peccary is a diurnal browser. The *Catagonus* lineage is more conservative than that of *Platygonus*, probably a result of having fewer enemies. (Wetzel 1977, Wetzel et al. 1975)

†*Mylohyus*   Cope, 1889   Tayassuidae
Long-nosed peccary. 2 spp. Bl.–L RLB E and C U.S.; whether the genus also inhabited S. Amer. is presently uncertain since taxonomy of S. Amer. peccaries needs revision. The Blancan species *M. floridanus* Kinsey, was probably ancestral to *M. nasutus* (Leidy), a widespread Irvingtonian to late Wisconsinan/early Holocene species. It was a long-legged cursorial animal about the size of a small white-tailed deer. An inhabitant of open areas and forest edges, the long-nosed peccary was less common than *Platygonus,* and its remains are usually found singly. It has not been found in association with man at any site, although cultural materials and the bones of *Mylohyus* were removed from Hartmann's Cave, Penn. in the 1800s. Loss of habitat and competition with *Ursus americanus,* the black bear, were major factors in the extinction of *Mylohyus*. (Kurtén and Anderson 1980)

†*Platygonus*   Le Conte, 1848   Tayassuidae
Flat-headed peccary. 4 spp. Bl.–L RLB N. Am. *Platygonus* was probably the most common medium-sized animal in North America during the Pleistocene. Remains of *P. compressus* Le Conte (M-L RLB) has been found from coast to coast, as far north as the Yukon (Old Crow) and south to Mexico. It was about the size of the European wild boar, *Sus scrofa.* It had longer legs than the living peccaries, no dew claws, and a more specialized dentition that included razor-sharp canines, their major offensive and defensive weapons. *Platygonus* has not been found in direct association with Early Man at any site, and there is no evidence that man played any part in its extinction. Loss of habitat due to climatic and vegetational changes, a low reproductive rate, and direct competition with and predation by *Ursus americanus* were probably major factors in its extinction. A terminal date for its extinction is 11,900±750 yr B.P. at Mosherville, Penn. (Guilday et al. 1971, Kurtén and Anderson 1980)

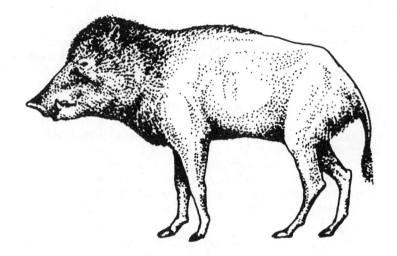

Platygonus, the extinct flat-headed peccary of North America

Tayassu   Fisher, 1814   Tayassuidae
Peccaries. 2 spp. SW U.S., Cent. and S.Am. Luj.–R S.Am.; R N.Am. Wt. 16–30 kg.

Hippopotamus   Linnaeus, 1758   Hippopotamidae
Hippopotamus. 1 sp. E Pleist. India; Pleist. interglacials Eur.; Pleist. Egypt to S. Af.
The extant genus *Hippopotamus* is now restricted to river areas in east and central
Africa. The European form, sometimes regarded as a distinct species, *H. antiquus*
Desmarest, differs from the living *H. amphibius* Linn. only in larger size. Dwarf hippos
have been found in Crete, Malta, Cyprus, Sicily (several spp. have been described), and
Madagascar *(H. lemerlie)*; their small size and sturdy build were adaptations for foraging
in rough terrain. The Madagascar hippo became extinct about 1,000 years ago, probably
through hunting and habitat destruction. Neolithic rock etchings of *Hippopotamus* have
been found in the Tibesti Mountains of N.Af., and in ancient Egypt the hippopotamus
was venerated as the goddess Thoeris.
    The only living amphibious artiodactyl, the hippopotamus has a barrel-like body, short
stocky legs, broad snout, enormous mouth, protruding eyes, and nostrils located on the
top of the snout. The incisors (never more than two pair, tetraprotodont) and canines
are tusklike and ever-growing. Hippos weigh between 3.0–4.5 tons and are 3.7–4.5
m long. They feed on water plants and grass. Valued for their hides, fat and ivory,
hippo populations have been reduced in many areas. (Coryndon 1978, Kurtén 1968,
Walker 1975)

Hexaprotodon   Falconer and Cautley, 1836   (=*Choeropsis*)   Hippopotamidae
Pygmy hippopotamus. 1 sp. Plioc.–Pleist. Pakistan, Burma, Java; Plioc.–R Af. (several
extinct spp.). The extant *H. liberiensis* is found in wet forests and coastal plains in W.Af.
from Guinea to the Ivory Coast. It differs from *Hippopotamus* in having nonprotruding
orbits on the side of the head, a rounder head, smaller size (shoulder ht. 0.75–1.0 m, wt
160–240 kg), and incisors 2/1 (2/2 in *Hippopotamus*). The primitive extinct species had
incisors 3/3 (hexaprotodont). When disturbed, *H. liberiensis* seeks shelter in forests
rather than in the water. Hexaprotodont hippos were dominant in the late Tertiary; their
decline and the rise of *Hippopotamus* are correlated with the proliferation of bovids
which took over the habitat and food of *Hexaprotodon* but did not compete with the
amphibious *Hippopotamus*. (Coryndon 1978)

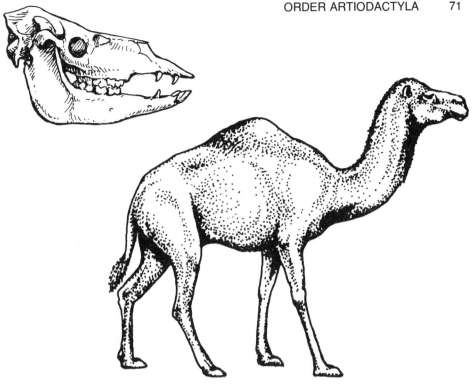

Camelops, the extinct western camel of North America, with skull (not to scale)

†*Camelops* Leidy, 1854   Camelidae
Western camel. Bl.–L RLB W half U.S. 6 spp. (probably not all valid). The Rancholabrean species, *C. hesternus* (Leidy), has been found in some thirty local faunas, and there are many isolated occurrences; about forty individuals have been found at Rancho La Brea. *C. hesternus* probably resembled the living dromedary, *Camelus dromedarius,* but its limbs were about 20 percent longer. It has a longer, narrower muzzle and thicker lips than *Camelus,* a high shoulder region, a middorsal hump, and steeply sloping hindquarters. Primarily a grazer, this camelopine travelled in large herds. *C. hesternus* has been found in association with man at Burnet and Jaguar caves, Tule Springs, Casper, Lubbock Lake, Sandia, and Clovis, but its remains are rare in Paleo-Indian sites because *Camelops* was less susceptible to Indian hunting techniques. A terminal date is 10,080±170 yr B.P. (RL-208) at Casper, Wyo. (Frison et al. 1978, Kurtén and Anderson 1980)

Camelus Linnaeus, 1758   Camelidae
Camels. 2 spp. Central Asia, Arabia. Pleist. Asia, N.Af. *C. dromedarius,* the one-humped camel, is known only in the domesticated form which has been widely introduced. Wild population of *C. bactrianus,* the two-humped camel, may still exist in Cent. Asia. Wt. 450–690 kg.

†*Hemiauchenia* M. Gervais and Ameghino, 1880 (=*Tanupolama*)   Camelidae
Long-legged llama. 5 spp. Hemh.–L RLB N.Am.; Uq.–Luj. SE S.Am. In North America remains of *H. macrocephala* (Cope) have been found from coast to coast; it survived until the late Wisconsinan. Larger than the extant *Lama glama,* it was a highly cursorial grazer. Its remains have not been found in association with man at any site. (Kurtén and Anderson 1980, Webb 1974)

Lama   F. Cuvier, 1800   Camelidae
Guancoes (wild), llamas and alpacas (domesticated). 3 spp. S and W S.Am., semidesert habitat to 5,000 m. Uq.–R, 2 extinct spp. Wt. 48–96 kg.

†*Palaeolama*   P. Gervais, 1867   Camelidae
Stout-legged llamas. 3 spp. Uq.–Luj. N S.Am.; Irv.–L RLB S U.S. *P. weddelli* (P. Gervais) is known from Bolivia, Ecuador and Peru; *P. aequatorialis* Hoffstetter is known from the Pacific Coast of S. Amer., and its remains are common in the Talara tar seeps, Peru. *P. mirifica* (Simpson) is known from Florida (15 sites), Texas, and California (Emery Borrow Pit). *Palaeolama* is distinguished from *Hemiauchenia* in having a complex P4 with multiple fossetids, low-crowned cheek teeth, and stocky limbs. It was adapted to live in rough terrain and fed on shoots and leaves of bushes and trees as well as grass. It has not been found in association with man at any site. (Kurtén and Anderson 1980, Webb 1974)

†*Titanotylopus*   Barbour and Schultz, 1934   Camelidae
Giant camel. 3 spp. Hemph.–Irv. W N.Am. Ht. ca. 3.5 m, long massive limbs, large hump, relatively small braincase, well developed P3 in both jaws.

Vicugna   Lesson, 1842   Camelidae
Vicunas. 1 sp. Andes of S.Am. Ens.–R. Distinguished from *Lama* by its ever-growing incisors and short, high-domed skull. More nearly resembles *Palaeolama* than *Lama*. Wt. 35–65 kg.

Giraffa   Brisson, 1756   Giraffidae
Giraffe. 1 sp. Mioc.–R Af.; E Plioc. Greece, Turkey, Pakistan; Plioc.–L Pleist. Asia. *Giraffa* first appears in the fossil record in the late Miocene of Kenya and by the middle Pleistocene was found over most of the African savanna; they were extirpated from N. Africa in Neolithic times and today inhabit the bush veld S of the Sahara. Several extinct African species have been described (*G. jumae* Leakey, *G. gracilis* Arambourg, *G. pygmae* Harris), but whether they are all valid is uncertain since the range of variation seems to fall within the extremes of the living species, *G. camelopardalis* Brisson. It is the tallest extant mammal (shoulder ht. 3.5 m, total ht. 5.0–5.5 m). The greatly elongated limbs and neck enable giraffes to browse on foliage above the reach of other large herbivores. Paired ossicones (with a median third one often present), located on the frontoparietal suture, are straight, blunt-ended, skin-covered, fully vascularized and present in both sexes. Giraffes are hunted for their skins, the natives using snares and pitfalls to capture them. (Churcher 1978)

Okapia   Lankester, 1901   Giraffidae
Okapi. 1 sp. Equatorial rainforests of Zaire. E-M Pleist., R.

†*Sivatherium*   Falconer and Cautley, 1832   (=*Libytherium*)   Giraffidae
Sivatheres. Pleist. S Eurasia and Af. Sivatheres were gigantic giraffids that had a massive body, nonelongated neck and limbs, shortened skull with retracted nasals, and large low-crowned teeth. The ossicones are highly variable, often antlerlike; usually two pair were present in males. Sivatheres lived in open woodlands or bush veld and probably had grazing or low browsing habits. In Africa their remains have been found at several Acheulean (Lower Paleolithic) sites. (Churcher 1978)

Alces   Gray, 1821   Cervidae
Moose, elk (in Europe), Crom.–R Eur.; Pleist.–R Asia; L Irv.–R N.Am. An extinct species, *A. latifrons* (Johnston), (Crom. N Eurasia, E RLB Beringia), was even larger than the living moose and had long-beamed, small-palmed antlers. The only extant species, *A. alces* (Linnaeus) first appears in late Steinheimian faunas in Europe and the Wisconsinan in N. Amer. It is the largest living cervid (shoulder ht. 1.4–1.9 m, wt. up to

Sivatherium, an antlered giraffe of the Lower Paleolithic of Africa

825 kg) and has massive short-beamed, large-palmed antlers. Moose inhabit timbered and muskeg boreal regions where they browse on trees, shrubs, and aquatic plants. (Kurtén and Anderson 1980)

Blastoceros Fitzinger, 1860   (=*Ozotoceros*)   Cervidae
Pampas deer. 1 sp. Brazil to N Patagonia. Uq.–R.

Blastocerus Wagner, 1844   Cervidae
Swamp deer. 1 sp. Luj.–R S.Am.; Sang. Fla. (Sabertooth Cave). The extant species, *B. dichotomus* Illiger, is about the size of a mule deer and has stout antlers with a bifurcating beam; the hooves are widely spreading, an adaptation for walking on soft ground. They inhabit wet savannas with tall grass and scattered trees. Their skins and antlers are used by natives, but the flesh is not favored.

Capreolus Gray, 1821   Cervidae
Roe deer. 1 sp. Crom.–R Eur.; Pleist.–R Asia. Pleistocene roe deer were larger but otherwise identical to extant *C. capreolus* (Linnaeus), an inhabitant of open forests throughout Eurasia south to N China. Roe deer stand 65–77 cm at the shoulder and weigh 15–30 kg; the antlers are erect, roughened at the base, and seldom more than three tined. (Kurtén 1968)

†*Cervalces* Scott, 1885   Cervidae
Stag-moose. 1 sp. RLB–E Holoc. Alberta, Saskatchewan, and EC U.S. *C. scotti* (Lydekker) has been identified at more than twenty Wisconsinan faunas, and two complete skeletons have been found in New Jersey bogs. It was about the size of a moose with long limbs and complexly palmated antlers. It probably inhabited muskegs. No associations with man are known, and the cause of its extinction was probably competition with *Alces alces* and a restricted geographic range. (Kurtén and Anderson 1980)

Cervus Linnaeus, 1758   Cervidae
Red deer, Wapiti. ca. 13 spp. Villafr.–R Eur., Asia; Irv.–R N.Am.; several extinct
species are recognized. *C. elaphus* Linnaeus, wapiti and red deer, was common
throughout its range in the late Pleistocene and is represented in Paleolithic art. The
species presently is found in open forests and meadows in Eurasia and N. Amer. It is
characterized by large size (shoulder ht. 1.2–1.5 m, wt. 225–450 kg); males have
magnificent spreading antlers. Wapiti are prized game animals, and their remains,
though not as numerous as *Odocoileus,* are often found in archaeological sites. (Kurtén
and Anderson 1980)

†*Choritoceros* Hoffstetter, 1963   Cervidae
Bolivian deer. 1 sp. Luj. Bolivia. This small extinct deer is characterized by peculiar
antler form.

Dama Frisch, 1775   Cervidae
Fallow deer. 1 sp. Villafr.–R Eur., Asia. The living species, *D. dama* (Linnaeus), first
appears in the Eemian interglacial throughout much of Europe, and its remains are
common in caves. By the Weichselian its range was restricted to the Mediterranean
region, and its Recent range is Asia Minor; the species has been widely introduced in
Europe and the U.S. Fallow deer stand about 1 m high at the shoulder and weigh 40–80
kg; their distinctive antlers are flattened and palmate with numerous points. These deer
are highly gregarious. (Kurtén 1968)

Eucladocerus, brush-antlered deer of the Eurasian middle Pleistocene

†*Eucladoceros* Falconer, 1868   Cervidae
Bush-antlered deer. Villafr.–M Crom. Eur.; M Pleist. China. Greatly branched antlers
with 12 tines on each side. Size of a moose.

Hippocamelus Leuckart, 1816   Cervidae
Andean deer. 2 spp. Andes of S.Am. 3,300–5,000 m. Luj.–R. Wt. 45–65 kg.

†*Libralces* Azzaroli, 1952   Cervidae
Extinct moose. L Villafr. Eur. *(L. gallicus)*. Had long-beamed, small-palmed antlers;
nasals moderately reduced. Lived in savannas or open forests.

Mazama   Rafinesque, 1817   Cervidae
Brocket deer. ca. 10 spp. S Mexico to Paraguay. Luj.–R. Wt. 16–21 kg.

†*Megaloceros*   Brooks, 1828   (=*Megaceros*)   Cervidae
Giant deer. M Crom.–E Holoc. Eur., N. Af., N Asia, China. *M. giganteus* Blumenbach,
the Irish elk, is commonly found in bog deposits in Ireland. It rivaled a large moose,
Alces alces, in size and was one of the largest known deer. Characteristic are the
enormous antlers which in some individuals had a 4-m spread; the main part was formed
by great palmation with several tines on the rim. Antler size increased 2½ times faster
than body size. The antlers were used in ritualized behavior and selection pressure for
large antlers was intense. The Irish elk was adapted to live in grassy, sparsely wooded
country, and extinction resulted when *Megaloceros* could not adapt to the subarctic
tundra or heavy forest. (Gould 1979, Kurtén 1968)

†*Morenelephas*   Carette, 1922   Cervidae
Moreno's deer. Uq.–Luj. S. Am.

†*Navahoceros*   Kurtén, 1975   Cervidae
Mountain deer. 1 sp. RLB Mexico, N. Mex., Wyo. The genus *Navahoceros* was derived
for stout-limbed deer that had been included in the genus *Sangamona*. *N. fricki* (Schultz
and Howard) weighed about 225 kg, making it slightly larger than *Sangamona,* or
between a mule deer and a wapiti in size. It had three-tined antlers, stocky limbs, and
very short metapodials. *Navahoceros* was adapted to a climbing mode of life. At Burnet
Cave, N. Mex., its bones were found at the same level as artifacts, thus it survived to
about 11,500 yr B.P. (Kurtén 1979)

Odocoileus   Rafinesque, 1832   Cervidae
White-tailed and mule deer. 2 spp. Bl.–R N. Am.; Luj.–R S. Am. The New World genus
Odocoileus was the most common cervid in late Blancan and Pleistocene faunas in N.
Amer. These medium-sized deer have large dichotomously forked antlers on a steeply
rising beam, a narrow face, relatively small teeth, and slender limbs. Both *O. virginianus* (Zimmerman), the white-tailed deer, and *O. hemionus* (Rafinesque), the mule
deer, are found from S Canada to S. Amer. in a variety of habitats where cover is
sufficient for concealment. These deer were important game animals for Paleo-Indians
and are hunted extensively today. (Kurtén and Anderson 1980)

Rangifer   H. Smith, 1827   Cervidae
Caribou, reindeer. 1 sp. Crom.–R Eurasia; L Irv.–R N. Am. The circumboreal genus
Rangifer probably originated in Beringia or NE Asia; the earliest N. Amer. record is
Cape Deceit, Alaska, the earliest European record is Süssenborn, Ger., both middle
Pleistocene in age. During the late Pleistocene, caribou were found far south of their
present range. They were hunted extensively by Early Man on both continents, and
their remains are relatively common in archaeological sites. A flesher made from a
caribou tibia from Old Crow River Loc. 14N was dated 27,000+3,000/−2,000 yr B.P.
making it one of the oldest known New World artifacts. *Rangifer* is the only cervid in
which both sexes have antlers. Superbly adapted to arctic conditions, caribou continue
to thrive. (Kurtén and Anderson 1980)

†*Sangamona*   Hay, 1920   Cervidae
Stilt-legged deer. 1 sp. L RLB. Remains of *Sangamona* have been found in Wisconsinan-
age faunas in EC U.S. (Frankstown, Whitesburg, Cavetown, Alton, Peccary, Brynjulfson). *S. fugitiva* Hay weighed between 135–180 kg, had unusually long limbs with
stiltlike metapodials, and weak development of ribs on the molars. The antlers are
unknown. It was a cursorial animal with grazing habits. No associations with man are
known; it survived until 9,440±760 yr B.P. at Brynjulfson Cave, Mo. (Kurtén 1979)

Antilocapra   Ord, 1918   Antilocapridae
Pronghorn. 1 sp. Deserts and grasslands SW Canada, W U.S., N Mexico. L RLB–
R. Wt. 36–60 kg.

†*Capromeryx* Matthew, 1902   Antilocapridae
Small four-horned pronghorns. 4 spp. Bl.–RLB N.Am. An endemic N.Amer. genus,
Capromeryx ranged from California to Florida, S to Mexico. *C. minor* Taylor
(=*Breameryx*), the diminutive pronghorn, has been found at Rancho La Brea, McKittrick, Black Water Draw, and several other sites. It stood about 56 cm at the shoulder
and weighed about 10 kg. The paired horncores rose from a common pedestal, the
anterior prong is short and spurlike, the posterior prong long and slender. A grazer, it
lived on the plains. It has not been found in association with man at any site.(Kurtén and
Anderson 1980)

†*Stockoceros*   Frick, 1937   Antilocapridae
Stock's pronghorns. 2 spp. RLB Mexico, Ariz., New Mex., Neb. Frick (1937) described
Stockoceros as a subgenus of *Tetrameryx*, but later workers have given it full generic
rank. It was intermediate in size between *Antilocapra americana* and the extinct *Capromeryx minor* and had symmetrically forked horncores arising from a common base
with separate sheaths covering each prong. Remains of *Stockoceros* have been found at
the same level as artifacts at Burnet Cave, dated 11,500 yr B.P. (Frick 1937, Kurtén and
Anderson 1980)

†*Tetrameryx*   Lull, 1921   Antilocapridae
Large four-horned pronghorns. 5 spp. Irv.–L RLB Tex., Calif., Mexico. Another group
of Pleistocene pronghorns have been placed in the genus *Tetrameryx* which is in need of
revision. These pronghorns were about the size of *Antilocapra americana* and had
paired horncores with the posterior one longer than the anterior one. Like the other
pronghorns, *Tetrameryx* was a plains grazer. There are no known associations with man.
(Kurtén and Anderson 1980)

Aepyceros   Sundevall, 1847   Bovidae
Impala. 1 sp. C and S Af. E-M Pleist., R. Wt. 65–75 kg.

Alcelaphus   Blainville, 1816   Bovidae
Hartebeest. 2 spp. Senegal and Sudan S to S.Af. L Pleist.–R. Wt. 160–180 kg.

Ammotragus   Blyth, 1840   Bovidae
Aoudad. 1 sp. L Pleist.–R N Af. in rough, rocky, waterless areas. Only wild sheep
indigenous to Af. Wt. 50–115 kg.

Antidorcas   Sundevall, 1847   Bovidae
Springbok. 1 sp. Treeless veldt of S and SW Af. E Pleist.–R. Several extinct
spp. recognized. *A. bondi* (Chelmer, Rhodesia, Swartkrans, Vlakkraal, Florisbad,
Border Cave, S.Af. dated ca. 36,000 yr B.P.) has small, extremely hypsodont teeth.
Wt. 32–36 kg.

Antilope   Pallas, 1766   Bovidae
Blackbuck. 2 sp. West Pakistan and India. L Plioc.–R Asia. Wt. 37 kg.

Beatragus   Heller, 1912   (sometimes incl. in *Damaliscus*)   Bovidae
Hunter's antelope. 1 sp. Kenya, Somaliland in grassy plains and shrubby areas. E
Pleist., R.

Bison   H. Smith, 1827   Bovidae
Bison, wisent. 2 extant, several extinct spp. The earliest known *Bison* come from
Pliocene deposits in China and were established in Europe by the middle Pleistocene.
Bison reached N. Amer. in the early Rancholabrean, perhaps somewhat earlier, and it is

Bison latifrons, the extinct long-horned bison of the North American late Pleistocene

regarded as an indicator fossil for Rancholabrean faunas. Early bison were forest animals, later becoming plains dwellers with corresponding changes in their social profile. (Guthrie 1980).

Bison differs from *Bos* in having a shorter, broader forehead, more tubular orbits, a less reduced posthorn region, and parietals visible in the dorsal aspect. However, some workers consider *Bison* and *Bos* to be congeneric (Van Gelder 1978).

Bison priscus (Bojanus), the steppe bison, was a widespread and common animal that congregated in large herds on the steppes of Eurasia from the Cromerian to the end of the Steinheimian. Paleolithic artists accurately portrayed it on cave walls (Altamira) as having long horns, massive humped shoulders, and slender hindquarters. *B. priscus* reached Alaska in Illinoian times and survived in Beringia until the late Wisconsinan (11,910±180 and 12,460±220 yr B.P. at Old Crow River Loc. 11). The Beringian population was the source of two distinct southward migrations. The first in the Illinoian gave rise to the giant-horned (tip-to-tip horncore span 213 cm) *B. latifrons* (Harlan); the second in the late Sangamonian–early Wisconsinan gave rise to *B. bison* (Linnaeus). *B. latifrons* is common in Sangamonian faunas and survived until the late Wisconsinan (21,500±700–31,300±2,300 yr B.P. at Rainbow Ranch, Id.). Its extinction resulted from competition with *B. bison* and genetic swamping.

Later N. Amer. bison have been known under a variety of specific names, but recent studies (Wilson 1974, 1975) have shown the presence of only one species, *B. bison,* with *B. b. antiquus* and *B. b. occidentalis* recognized as extinct subspecies. Characteristic of this lineage is a north-south cline and a reduction in horncore size through time. *B. b. antiquus* (horncore span 88 cm) was widespread south of the ice; *B. b. occidentalis* (horncore span 75 cm) had a more northern distribution. These bison were most numerous in the Great Bison Belt, the shortgrass prairies from Alberta to Texas. Both subspecies were extensively hunted by Paleo-Indians and provided the major protein source for the Clovis-Folsom people. In N. Amer. the plains bison, *B. b. bison* (horncore span 58 cm), and the wood bison, *B. b. athabascae* (horncore span 66 cm), survive mainly in national and state parks. In central Europe the forest-dwelling wisent, *B. bonasus* (Linnaeus), a postglacial immigrant from America, is barely holding its own. (Guthrie 1980; Kurtén and Anderson 1980; Van Gelder 1978; Wilson 1974, 1975)

†*Bootherium*   Leidy, 1852   Bovidae
Leidy's ox. 1 sp. L RLB N.Am. Status of genus uncertain (see *Symbos*). Smaller than *Ovibos;* posterior part of skull slopes steeply downward as it does in sheep and goats.

Bos   Linnaeus, 1758   Bovidae
Aurochs, cattle, yak. ca. 6 spp. in Asia and *B. taurus,* domestic cattle, which have a worldwide distribution. One ancestor of modern cattle was the auroch, *B. primigenius* Bojanus. This extinct species first appears in the Holsteinian interglacial in Europe; aurochs are rare in Pleistocene deposits though they have been found in N. Africa as well as Europe and were pictured by Paleolithic man at Lascaux and Teyjat. Characteristic of the auroch lineage was a decrease in size through time; the Holsteinian form was gigantic with enormous horns. By the early Holocene aurochs were very common, and they survived until Viking times in Scandinavia; the last known specimen died in Poland in 1627. Originally aurochs inhabited overgrown grasslands and open forests, but in historic times they sought refuge in deep forests.

In the Pleistocene, the yak, *B. grunniens* Linnaeus, was widespread in northern Asia and reached Beringia (Fairbanks), but did not spread south in the New World. An inhabitant of the high plateaus and mountains of central Asia, the sturdy, sure-footed yak weighs about 500 kg. It has been domesticated in Tibet for centuries. (Kurtén 1968)

Boselaphus   Blainville, 1816   Bovidae
Nilgai. 1 sp. Peninsular India. Pleist.–R Asia. Largest of the Indian antelope; wt. 200 kg.

Bubalus   H. Smith, 1827   Bovidae
Water buffalo. 1 sp. Egypt to the Philippines; widely domesticated. Pleist.–R Asia. Spread of horns, 1.2 m along the outer edge, exceeds that of all other living bovids. Wt. 725–815 kg.

†*Bucapra*   Rütimeyer, 1877   Bovidae
Extinct goat. L Pleist. S. Asia.

Capra   Linneaus, 1758   Bovidae
Goats, ibex. 5 spp. Rugged mountain country from Spain to India, N to Siberia. Pleist.–R. Domestication took place in SW Asia 8,000 to 9,000 years ago. Wt. 50–120 kg.

Cephalophus   H. Smith, 1827   Bovidae
Duikers. ca. 10 spp. E Pleist., R, Af. Both sexes horned. Wt. 5–65 kg.

Connochaetes   Lichtenstein, 1814   Bovidae
Gnus. 2 spp. E Pleist.–R. Open grassy plains E and S Af. Wt. 230–275 kg.

Damaliscus   Sclater and Thomas, 1894   Bovidae
Topi, blesbok. 6 spp. E Pleist.–R. Af. Wt. 114–136 kg.

†*Damalops*   Pilgrim, 1939   Bovidae
Extinct hartebeest. Plioc.–L Pleist. E and S Asia. Related to *Alcelaphus, Rabaticeras,* and *Damaliscus.*

†*Duboisia*   Stremme, 1911   Bovidae
Dubois' antelope. Pleist. S Asia, E. Indies.

†*Euceratherium*   Furlong and Sinclair, 1904 (=*Preptoceras, Aftonius*)   Bovidae
Shrub oxen. Irv.–L RLB N.Am. *Euceratherium* was one of the first bovids to reach N. Amer. (in the Irvingtonian), and it survived until the end of the Wisconsinan. Remains of *E. collinum* (Furlong and Sinclair) have been found from northern Calif. to the state of Mexico in Mexico and east to Illinois. These heavily built oxen were about four-fifths the size of modern bison or slightly larger than a living musk ox. *Euceratherium* has

strongly hypsodont teeth with well-developed styles on the upper molars. The massive horncores are characteristic. These grazing animals probably inhabited the foothills. At Burnet Cave, N. Mex., bones of *Euceratherium* were found at the same level as artifacts. (Kurtén and Anderson 1980)

Gazella Blainville, 1816 Bovidae
Gazelles. ca. 12 spp. Plains of N and E Af., Arabia, Israel, C Asia, India. E Plioc. Eur.; E Plioc.–R Asia; Pleist.–R Af. Wt. 14–75 kg.

†*Hemibos* Falconer, 1865 Bovidae
Extinct water buffalo. L Pleist. Asia, Palestine. Ancestral to *Bubalus.*

Hemitragus Hodgson, 1841 Bovidae
Tahrs. 3 spp. Himalayas, Nilgai Hills of S India, Arabia; Pleist.–R Asia; Pleist. Af. Wt. to 100 kg.

Hippotragus Sundevall, 1846 Bovidae
Sable and roan antelopes. 2 spp. Af. S of the Sahara. Plioc. Eur., Asia; Pleist.–R Af. Several extinct species are recognized: *H. gigas* Leakey has been found at Olduvai I–III, Omo, East Turkana, Elandsfontein and Florisbad; one of the early populations may have been ancestral to the roan and the sable.

Kobus A. Smith, 1840 Bovidae
Waterbuck, kob, lechwe. ca. 6 spp. Af. S of the Sahara and Nile Valley in swampy areas, never far from water. E Pleist.–R. Several extinct species are recognized. Wt. up to 270 kg.

Madoqua Ogilby, 1837 Bovidae
Dik-dik. ca. 6 spp. Arid areas from Ethiopia to SW Af. E Pleist.–R. Small, wt. 3–5 kg.

†*Makapania* Wells and Cooke, 1956 Bovidae
Extinct ovibovine. E Pleist. S.Af. (Makapansgat). Moderate to large size, with horncores emerging almost transversely from the skull. Related to *Megalovis,* a European Villafranchian ovibovine.

†*Megalotragus* Broom, 1909 Bovidae
Extinct alcelaphine. E-L Pleist. E and S Af. Larger than any living alcelaphine (gnu, hartebeest, topi); long, curved horns.

†*Menelikia* Arambourg, 1941 Bovidae
Extinct kob. E-M Pleist. E Af. Medium-sized reduncine with extensive internal hollowing of the frontals; horncores with transverse ridges.

†*Myotragus* Bate, 1909 Bovidae
Cave goat. Villafr.–E Holoc. Baleric Islands. Remains of *M. balearicus* Bate, the late Pleistocene species, are common in the island caves. At Muleta Cave, Mallorca, more than 500 individuals have been recovered, and it is believed that over a long period of time the animals tumbled off a ledge to their deaths. This aberrant species is perhaps related to the chamoislike bovids. The cannon bones are short, and there is some fusion between the carpals and tarsals and the phalanges, an adaptation for climbing in rough terrain. Shoulder height was about 50 cm. The middle pair of lower incisors grew into huge chisel-like teeth with open roots, the cheek teeth are high-crowned, an adaptation for dealing with coarse, grit-covered vegetation. *Myotragus* survived into historic times and was hunted by man, who was a factor in the extinction of this highly specialized insular animal. (Kurtén 1968)

†*Numidocapra* Arambourg, 1949 Bovidae
Extinct caprid. M Pleist. N.Af. (Aïn Hanech, Algeria). Long thick horncores; short braincase which is strongly inclined.

Oreamnos   Rafinesque, 1817   Bovidae
Mountain goat. 1 sp. Rocky areas above timberline in the Rockies and coastal ranges of
N.Am. M RLB–R, rare as fossils, 1 extinct sp. Closely related to *Rupicapra*. Wt.
75–140 kg.

Oreotragus   A. Smith, 1834   Bovidae
Klipspringer. 1 sp. Rocky hills of E and S Af., N Nigeria. E, L Pleist.–R. Habits similar
to *Oreamnos* and *Rupicapra*. Wt. 11—16 kg.

Oryx   Blainville, 1816   Bovidae
Oryx, gemsbok. 4 spp. Arid plains and deserts of Arabia, N, E and S Af. E Pleist.–R.
Wt. up to 210 kg.

Ourebia   Laurillard, 1841   Bovidae
Oribi. 1 sp. Af. S of the Sahara. M Pleist., R. Wt. 14–21 kg.

Ovibos   Blainville, 1816   Bovidae
Musk ox. 1 sp. E Pleist.–Holoc. Eur.; Pleist.–Holoc. N Asia; E RLB–R N.Am. During
the Pleistocene *Ovibos* had a circumboreal distribution and was quite common over
much of its range. Musk oxen were extirpated in Eurasia about 3,000 years ago and
today are found in N Canada and Greenland and have been reintroduced into Alaska.
Musk oxen are compactly built with short legs and a slight hump over the shoulders; the
downward sweeping horns are closely set, nearly meeting at the midline to form an
almost solid boss. The dentition is less hypsodont than that of *Bison*. Gregarious, musk
oxen are found in herds numbering up to 100. Man is their chief enemy. (Kurtén and
Anderson 1980)

Ovis   Linnaeus, 1758   Bovidae
Sheep. 6 spp. Dry uplands and mountains of N.Am., USSR, China, Kashmir, Pakistan
and the Middle East. Villafr. Eur., Asia; by the late Pleistocene sheep had spread into
mountainous regions of N.Af., Eurasia, and N.Am. (earliest record, Illinoian, near
Fairbanks, Alaska). Wt. 75–200 kg.

†*Parmularius*   Hopwood, 1934   Bovidae
Extinct alcelaphine. E-L Pleist. E Af. Medium-sized; horncores with basal swelling.

†*Pelorovis*   Reck, 1928   Bovidae
Long-horned buffalo. 2 spp. E-L Pleist. Af. This African genus of gigantic-horned buffalo
ranged stratigraphically from the early (Olduvai, Koobi Fora) to the middle and late
Pleistocene in North, East and South Af. *P. oldowayensis* Reck was probably ancestral
to *P. antiquus* Duvernoy. Compared to *Syncerus caffer*, the extant African buffalo,
Pelorovis, had much longer horncores (lt. of span 1.8 m) that are less dorsoventrally
flattened and lack basal bosses, and much shorter metapodials. At Olduvai, remains of
Pelorovis were associated with stone tools. (Gentry 1978)

†*Platybos*   Pilgrim, 1939   Bovidae
Extinct Asian ox. L Pleist. S Asia.

†*Platycerabos*   Barbour and Schultz, 1941   Bovidae
Flat-horned ox. L Irv. Neb. Large bovid resembling *Bos* more than *Bison*. Known from
one specimen.

†*Praeovibos*   Staudinger, 1908   Bovidae
Extinct musk ox. M Pleist. C Eur.; E RLB (Illinoian) Alaska, Yukon Terr. Larger,
longer-legged, and more slender than *Ovibos*.

†*Proboselaphus*   Matsumoto, 1915   Bovidae
Extinct nilgai. E-L Pleist. E Asia.

Raphicerus   H. Smith, 1827   Bovidae
Steenbok. 3 spp. E and S Af. E Pleist., R. Wt. 7–14 kg.

Redunca   H. Smith, 1827   Bovidae
Reedbuck. 3 spp. Af. S of the Sahara. E Pleist.–R; several extinct spp. Wt. 23–91 kg.

Rupicapra   Blainville, 1816   Bovidae
Chamois. 1 sp. Mountains of Eur. and Asia Minor. L Pleist.–R Eur. Wt. 24–50 kg.

Saiga   Gray, 1843   Bovidae
Saiga antelope. 1 sp. Pleist. Eurasia, N N.Am. During the Pleistocene the range of *Saiga* extended from England to Alaska and the Northwest Territory. Rough terrain, glaciers, and deep snow were barriers to its southward advance. As late as 1600 A.D. *Saiga* was still found in the steppe zones of eastern Europe, but today it inhabits only the cold, dry plains of central Asia. Standing 75–80 cm at the shoulder and weighing 36–69 kg, *Saiga* has a large head, downward pointing nostrils, and an inflated mobile muzzle containing convoluted choanae and mucous glands for warming and moistening the dry air. Only the males have horns. Formerly, saiga antelope were hunted extensively for their horns valued in the Chinese pharmaceutical trade; they are now strictly protected and are the most numerous hoofed animal in their range. (Kurtén and Anderson 1980, Sokolov 1974)

†*Sivacobus*   Pilgrim, 1939   Bovidae
Extinct waterbuck. L Pleist. S Asia.

†*Sivadenota*   Pilgrim, 1939   Bovidae
Extinct antelope. L Pleist. S Asia.

†*Soergelia*   Schaub, 1951   Bovidae
Primitive musk ox. M Pleist. Eurasia, N.Am. Size of a steer, heavily built, massive limbs. Occupies an intermediate position between *Euceratherium* and *Ovibos*.

†*Spirocerus*   Boule and Teilhard de Chardin, 1929   Bovidae
Extinct antelope. E-L Pleist. Asia.

Sylvicapra   Ogilby, 1837   Bovidae
Gray duiker. 1 sp. Af. Often at high elevations. M Pleist., R. Wt. 14–17 kg.

†*Symbos*   Osgood, 1905   Bovidae
Woodland musk ox. Irv.–L RLB N.Am. Remains of *Symbos* have been found from Alaska south to Mississippi and from the Pacific to the Atlantic coasts; most of the finds are, however, from eastern Beringia and east-central U.S. Most workers now believe that *S. cavifrons* (Leidy) is the only valid species, and *Bootherium sargenti* Gidley is a female *S. cavifrons*. *Symbos* was taller and more slender than the extant musk ox, *Ovibos moschatus* (Zimmerman). Diagnostic of *Symbos* is the pitted, rough-basined surface between the horncores which are situated higher on the skull than those of *Ovibos*. Woodland musk oxen were grazers inhabiting the plains and woodlands and were adapted to warmer conditions than *Ovibos*. No direct associations with man are known. A terminal date for its extinction is 11,100±400 yr B.P. at Kalamazoo, Michigan. (Kurtén and Anderson 1980)

Syncerus   Hodgson, 1847   Bovidae
African buffalo. 1 sp. Af. S of the Sahara. L Pleist.–R. Massively built, wt. 600–900 kg.

Taurotragus   Wagner, 1855   Bovidae
Eland. 2 spp. Plains and moderately hilly areas of C and S Af. M Pleist.–R. Largest of the African antelope, wt. 900 kg.

†*Thaleroceros*   Reck, 1935   Bovidae
Extinct kob. M Pleist. Af. Known only from Olduvai Bed IV. Moderate to large size with massive horncores.

Tragelaphus   Blainville, 1816   Bovidae
Bushbuck, nyala, kudu, sitatunga. 6 spp. Af. S of the Sahara. E-M Pleist., R. Wt. 120–270 kg.

TUBULIDENTATA

Orycteropus   E. Geoffroy, 1795   Orycteropodidae
Aardvark. 1 sp. Af. S of the Sahara where sufficient quantities of ants and termites are available; also feeds on the fruit of a species of Cucurbitaceae. E Pleist.–R. Fossorial. Wt. 50–70 kg.

†*Plesiorycteropus*   Filhol, 1895   Orycteropodidae
Madagascar aardvark. E Holoc. forests of Madagascar. ^{14}C date 1,035±50 yr B.P. Most isolated tubulidentate both geographically and morphologically. Differs from other aardvarks in having a short skull, low glenoid articulation, and teeth greatly reduced or absent. Limb bones indicate digging, jumping, and climbing abilities. Fed exclusively on ants. (Patterson 1978)

HYRACOIDEA

†*Gigantohyrax*   Kitching, 1965   Procaviidae
Giant hyrax. 1 sp. Plio-Pleist. S.Af. (Makapansgat). Differs from *Procavia* by larger size (1.5x) and dental characters.

Procavia   Storr, 1780   Procaviidae
Hyrax. 1 sp. (*P. capensis* Storr). Af., extreme SW Asia in rocky, scrub-covered areas. Plio-Pleist. S.Af.; 2 extinct spp.—the large *P. transvaalensis* Shaw and the small *P. antiqua* Broom. The hypsodont dentition resembles that of a rhinoceros. Wt. 1.4–2.0 kg.

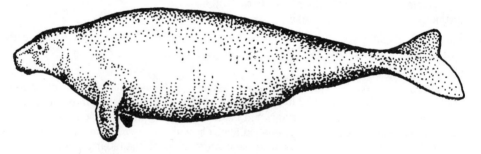

Hydrodamalis, Steller's sea cow, a genus of large mammal that became extinct in historic time

SIRENIA

†*Hydrodamalis*   Retzius, 1794   Dugongidae
Sea cow. The genus *Hydrodamalis* inhabited shallow waters of the North Pacific Ocean from the late Pliocene to 1768 when the last specimen was killed by meat-hungry Russian hunters in the Bering Sea. *H. gigas* (Zimmerman) was about 7.5 m long and weighed about 10 tons; Bering Sea populations were slightly smaller. The species is characterized by a small head, small hooklike forelimbs lacking phalanges, no hind limbs, dense ribs, and a horizontally flattened tail. In place of teeth, *Hydro-damalis* had a horny plate covering the anterior palate and mandible; they fed on kelp and marine algae. The skin was rough and barklike, beneath it was a thick layer of blubber. There are no known Pleistocene associations with man, though Domning (1970) postulates that Paleo-Indians crossing Beringia may have hunted these inoffen-

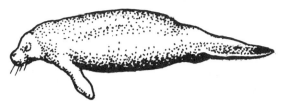

Trichechus, the manatee, a living marine mammal suffering severe population reduction

sive, easy-to-kill mammals in coastal waters. Later Russian crews decimated the relict population between 1743 and 1763; the last known specimen was killed in 1768. (Domning 1970, Kurtén and Anderson 1980)

Trichechus Linnaeus, 1758   Trichechidae
Manatees. 3 spp. Found along the coasts and in sluggish coastal rivers of SE U.S. to N S.Am., the Amazon and Orinoco rivers, and W.Af. Bl.–R SE U.S.; remains, especially the dense ribs, are common in Pleist. deposits in Florida. Fusiform, nearly hairless body, small head, six cervical vertebrae, rounded paddle-shaped tail, forelimbs modified into paddles, hindlimbs vestigial or absent. Maximum lt. 4.5 m, wt. up to 600 kg. (Kurtén and Anderson 1980)

PROBOSCIDEA

†*Mammut* Blumenbach, 1799   (=*Mastodon*)   Mammutidae
Mastodont. E Mioc.–E Pleist. Af., Eurasia; M Mioc.–E Holoc. N.Am.; absent from S.Am. In the Rancholabrean *Mammut americanum* (Kerr), the highly variable American species, was found from Alaska to Florida, but finds are most common in the eastern forests. Their preferred habitat was open spruce woodlands, spruce forests, and pine parklands (see Saunders 1977 for a discussion of pollen analysis, ecology and thanatocoenoses of *Mammut*). Mastodonts fed on twigs and cones of conifers, leaves, coarse grasses, and swamp plants. Standing 2.7–3.0 m at the shoulder with a total body length of 4.5 m, mastodonts were more heavily built than mammoths. Tusks were present in the upper jaws of both sexes; males had vestigial lower tusks that were usually lost by maturity. The cheek teeth have large low cusps arranged one behind the other to form low ridges separated by open valleys; two to three teeth were present in each jaw half at a given time. Increasing dryness resulting in vegetational changes was a factor in the extinction of mastodonts. Relict populations would then have been subject to other environmental stresses, including hunting pressure. Accepted terminal dates range between 9,000 and 12,000 yr B.P. (Kurtén and Anderson 1980, Saunders 1977)

†*Anancus* Aymard, 1855   Gomphotheriidae
Straight-tusked gomphotheres. L Mioc.–M Pleist. Af.; L Plioc.–E Pleist. Eur.; L Plioc.–Pleist. Asia. Tendency towards reduction and finally loss of lower tusks; upper tusks are extremely long (3.5 m) and nearly straight. Browsing habits.

†*Cuvieronius* Osborn, 1923   Gomphotheriidae
Andine gomphothere. Bl.–E RLB S N.Am.; Ens.–L Luj. S.Am., especially common in the southern Andes. One species is recognized in S.Am. Characteristic of the genus is a long skull, low brow, simple molars with secondary trefoils and no cement, and long straight upper tusks with a spiral band of enamel. At Tagua-Tagua, Chile, charcoal associated with the remains of *Cuvieronius, Equus, Blastocerus,* and *Homo* has been dated at 11,380±320 yr B.P. (GX-1205). (Marshall et al. MS, Simpson 1980)

†*Haplomastodon* Hoffstetter, 1950   Gomphotheriidae
Pampean gomphothere. 1 sp. Luj. of tropical S.Am., especially Brazil and Ecuador. The skull and brow are intermediate in size between those of *Stegomastodon* and

Extinct Pleistocene Proboscideans and Deinotheres. Left, top to bottom: *Cuvieronius, Stegomastodon, Haplomastodon, Anancus,* and *Stegodon;* Right, top to bottom: *Mammut, Deinotherium, Mammuthus (Archidiskodon) meridionalis* and *Mammuthus primigenius*

Cuvieronius. The molars are similar to those of *Cuvieronius,* and as Simpson (1980) notes they are more often confused with than distinguished from each other. The tusks are stout and upturned, and juvenile specimens have a straight enamel band. At Taima-Taima, Venezuela, butchered bones of a juvenile were found in direct association with artifacts of the El Jobo complex. Four radiocarbon dates obtained on wood samples associated with the bones average 13,500 yr B.P., and the authors (Bryan et al. 1978) point out that a big game hunting complex of an entirely different technological tradition was flourishing in northern Venezuela at least 1,500 years before the start of the Clovis complex in southwestern U.S. (Bryan et al. 1978, Marshall et al. MS, Simpson 1980)

†*Stegomastodon*   Pohlig, 1912   Gomphotheriidae
Stegomastodont. Bl.–E Irv. N.Am.; Ens.–L Luj. N S.Am. *Stegomastodon* has a short skull and mandible, a high brow, and large, uniformly curved upper tusks lacking enamel; lower tusks are absent. The molars are very complex with numerous conical cusps and cement between the cones; they wear to at least a double trefoil pattern. In N. Amer. *Stegomastodon* succumbed to competition with the invading *Mammuthus* and became extinct in the early Irvingtonian, but in S. Amer. where neither *Mammuthus* nor *Mammut* penetrated, *Stegomastodon* survived throughout the late Pleistocene. Remains of *Stegomastodon* at Arroio Touro Passo, Brazil, dated 11,040±190 yr B.P. (I-9628) and at Muaco, Venezuela, dated 14,300±500 yr B.P. (M-1068) and 16,375±400 yr B.P. (0-999), may be associated with Early Man. (Kurtén and Anderson 1980, Marshall et al. MS)

Elephas   Linnaeus, 1758   Elephantidae
Elephant. 1 extant sp., 9 extinct spp. E Plioc.–L Pleist. Af.; L Plioc.–L Pleist. Eur.; L Plioc.–R Asia. The extant genus *Elephas* shows more resemblances to *Mammuthus* than to *Loxodonta* and is the most diverse genus in the family. Two species of dwarf elephants, *E. falconeri* Busk (L Pleist. Mediterranean Islands, shoulder ht. 90–140 cm) and *E. celebensis* Hooijer (M-L Pleist. Celebes, Java) are recognized; they are derived from *E. namadicus* Falconer and Cautley and *E. planifrons* Falconer and Cautley respectively, and their small stature is attributed to adverse environmental conditions. The extant species, *E. maximus* Linn., is known from early Holocene times onward in India and SE Asia where it lives in a variety of habitats from dense jungles to open grassy plains. It is a small- to medium-sized species (lt. 5.5–6.4 m, ht. at shoulder 2.5–3.0 m, wt. 2.5 tons). It has narrow molars with 22–27 plates on M3 and coarsely folded enamel; the upper tusks are gently curved, almost parallel-sided. The skull is high with an expanded parietal region, and the forehead is foreshortened; the top of the head is the highest point. These gregarious animals feed on grass, vines, leaves, shoots, and fruit. Populations have been greatly reduced in many parts of its range. (Coppens et al. 1978, Maglio 1973, Walker 1975)

Loxodonta   F. Cuvier and G. Saint-Hillaire, 1825   Elephantidae
African elephant. 1 extant sp., 2 extinct spp.: *L adaurora* Maglio and *L. atlantica* (Pomel). M Plioc.–R Af., an endemic African genus. Fossils of the living African elephant, *L. africana* (Blumenbach), are known from the end of the early Pleistocene in Kenya and middle and late Pleistocene faunas in South Africa and Chad; it was probably derived from *L. adaurora*. Its Recent range was formerly most of Africa south of the Sahara in savannas, river valleys, dense forests, and desert scrub, but populations have been extirpated or greatly reduced in many areas. The African elephant is the largest living terrestrial mammal (lt. 6.0–7.5 m, ht. at shoulders 3.0–4.0 m, wt. 5–7 tons). The molars of *Loxodonta,* though highly specialized, are relatively conservative compared to other genera of proboscideans. There are 8–15 plates on M3 with enamel 2–5 mm thick; characteristic are the broad, lozenge-shaped wear surfaces. The upper tusks

are gently curved upward, not twisted; lower tusks are absent. *Loxodonta* differs from *Elephas* by its larger size, much larger ears, a more convex forehead, and having the highest point over the shoulders. These social animals congregate in herds and feed on grasses, branches, and fruit. (Maglio 1973, Coppens et al. 1978, Walker 1975)

†*Mammuthus* Burnett, 1830   Elephantidae
Mammoths. L Plioc.–E Pleist. Af.; E Pleist.–E Holoc. Eur., Asia; E Irv.–E Holoc. N. Am. The genus *Mammuthus* (previously called *Elephas, Primelephas, Loxodonta, Paleoloxodon, Archidiskodon,* etc., see Maglio 1973) was relatively common over much of its range and was the dominant member of many Pleistocene faunas. Species taxonomy of mammoths is in need of revision.

The best known species is *M. primigenius* Blumenbach, the woolly mammoth, a late Pleistocene inhabitant of the tundra and northern taiga regions of Eurasia and N. Amer. Its appearance is well known from numerous cave paintings and engravings as well as frozen carcasses found in Siberia and Alaska. Standing about 2.8 m high at the shoulder, the woolly mammoth had a high domed head, a humped sloping back, a relatively short tail, small ears, and a rather short trunk ending in two fingers at the tip. It was covered with long, probably black hair and a thick undercoat beneath which was an insulating layer of fat. It had large, greatly curved tusks, and the molars were the most complex of any mammoth, with extremely thin, closely appressed plates (20–27 on M3, 11–13/100 mm). A macrograzer, the woolly mammoth fed on grasses and tundra plants.

In N. Amer. south of the ice, the Sangamonian-Wisconsinan species was *M. jeffersonii* (Osborn) (often called *M. columbi* or *M. imperator,* see Kurtén and Anderson 1980). It was larger than *M. primigenius* and stood between 3.2–3.4 m at the shoulder, had large lyrate or incurved tusks, a long broad cranium, a shallow jaw with a slightly downturned chin, and very complex molars (isolated teeth have been mistakenly identified as being from *M. primigenius,* the northern species). *M. jeffersonii* inhabited open prairies especially in the West and fed primarily on grasses.

Remains of both species have been found at kill sites and hunting camps, and mammoths were hunted by Early Man. At Murray Springs, Dent, Blackwater Draw, and Miami (Texas), evidence indicates that Clovis Paleo-Indians killed entire family groups (Saunders 1980). Mammoths survived until about 11,000 years ago. Their extinction was due to overspecialization, climatic changes, and increasing hunting pressure by man. (Kurtén and Anderson 1980, Maglio 1973, Saunders 1980)

†*Stegodon* Falconer, 1857   Stegodontidae
Stegodonts. M Plioc.–L Pleist. Asia, Plioc. Af.; primarily an Asiatic group of mammutid proboscideans. Several spp. Pygmy stegodonts have been found on Sulawesi, Timor, and Flores. Masticatory apparatus parallels that of elephants. Low crowned teeth have thick enamel which becomes highly folded and scalloped with wear. Males had gigantic tusks situated close together so that the trunk had to be held at the side. (Coppens et al. 1978)

DEINOTHERIOIDEA

†*Deinotherium* Kaup, 1829   Deinotheriidae
Deinotheres. L Mioc.–M Plioc. Eur., Asia; L Mioc.–E Pleist. Af. (Omo, Olduvai, Chemoigut Beds, Hadar). Although often placed with the Proboscidea, deinotheres are related to moeritheres and barytheres and probably constitute a separate order. Deinotheres are large, long-legged, short-necked terrestrial mammals. Diagnostic are mandibular tusks that curve downward. The skull is low and elongated with high occipital condyles and well-developed paroccipital processes. The teeth are brachydont; deinotheres fed on soft vegetation in their woodland habitat. (Harris 1978)

References

Adams, D. B. 1979. The cheetah: native American. *Science* 205:1155–58.

Beebe, B. 1978. Two new Pleistocene mammal species from Beringia. *Abst. 5th Biennial Meeting AMQUA, 2–4 Sept. 1978. Edmonton, Alberta.* p. 159.

Bryan, A. L., R. M. Casamiquela, J. M. Cruxent, R. Gruhn, and C. Ochsenius. 1978. An El Jobo mastodon kill at Taima-Taima, Venezuela. *Science* 200:1275–77.

Butler, P.M. 1978. Chalicotheriidae. In V. J. Maglio and H. B. S. Cook (eds.). *Evolution of African Mammals.* Cambridge, Mass., Harvard Univ. Press. pp. 368–70.

Churcher, C. S. 1966. The insect fauna from the Talara tar seeps, Peru. *Canad. J. Zool.* 44:985–93.

———. 1978. Giraffidae. In V. J. Maglio and H. B. S. Cooke (eds.). *Evolution of African Mammals.* Cambridge, Mass., Harvard Univ. Press. pp. 509–35.

———. 1980. Marsupials. In R. W. Fairbridge and D. Jablonski (eds.). Encyclopedia of Earth Sciences, Vol. 7, The Encyclopedia of Paleontology. Stroudsburg, Penn., Dowden, Hutchinson and Ross. pp. 445–61.

Churcher, C. S. and M. L. Richardson. 1978. Equidae. In V. J. Maglio and H. B. S. Cooke (eds.). *Evolution of African Mammals.* Cambridge, Mass., Harvard Univ. Press. pp. 379–422.

Cooke, H. B. S. and A. F. Wilkinson. 1978. Suidae and Tayassuidae. In V. J. Maglio and H. B. S. Cook (eds.). *Evolution of African Mammals.* Cambridge, Mass., Harvard Univ. Press. pp. 435–82.

Coppens, Y, V. J. Maglio, C. T. Madden, and M. Beden. 1978. Proboscidea. In V. J. Maglio and H. B. S. Cooke (eds.). *Evolution of African Mammals.* Cambridge, Mass., Harvard Univ. Press. pp. 336–67.

Coryndon, S. C., 1978. Hippopotamidae. In V. J. Maglio and H. B. S. Cooke (eds.). *Evolution of African Mammals.* Cambridge, Mass., Harvard Univ. Press. pp. 483–95.

Dalquest, W. W. 1978. Phylogeny of American horses of Blancan and Pleistocene age. *Acta Zool. Fennici* 15:191–99.

Domning, D. P. 1970. Sirenian evolution in the North Pacific and the origin of Steller's sea cow. *Proc. 7th Ann. Conf. Biol., Sonar and Diving Mammals, Stanford Res. Inst.* 970:217–20.

Frick, C. 1937. Horned ruminants of North America. *Bull. Amer. Mus. Nat. Hist.* 69:1–669.

Frison, G. C., D. N. Walker, S. D. Webb, and G. M. Zeimans. 1978. Paleo-Indian procurement of *Camelops* on the northwestern plains. *Quatern. Res.* 10(3):385–400.

Gazin, C. L. 1956. Exploration for remains of giant ground sloths in Panama. *Ann. Rept. Smithsn. Inst.* Publ. 4272:341–54.

Gentry, A. W. 1978. Bovidae. In V. J. Maglio and H. B. S. Cooke (eds.). *Evolution of African Mammals.* Cambridge, Mass., Harvard Univ. Press. pp. 540–72.

Gillette, D. D. 1973. A review of North American glyptodonts (Edentata, Mammalia): osteology, systematics, and paleobiology. Ph.D. dissertation, So. Meth. Univ., Dallas.

Gould, S. J. 1979. The misnamed, mistreated and misunderstood Irish elk. *Ever Since Darwin.* New York and London: W. W. Norton and Co. pp. 79–90.

Guilday, J. E. 1977. Sabertooth cat, *Smilodon floridanus* (Leidy), and associated fauna from a Tennessee cave (40Dv40), the First American Bank Site. *J. Tenn. Acad. Sci.* 52(3):84–94.

Guilday, J. E., H. W. Hamilton, and A. D. McCrady. 1971. The Welsh Cave peccaries *(Platygonus)* and associated fauna, Kentucky Pleistocene. *Ann. Carnegie Mus.* 43(9):249–320.

Guthrie, R. D. 1980. Bison and man in North America. *Canad. J. Anthropol.* 1(1):55–73.

Hansen, R. M. 1978. Shasta ground sloth food habits, Rampart Cave, Arizona. *Paleobiology* 4(3):302–19.

Harris, J. M. 1978. Deinotherioidea and Barytherioidea. In V. J. Maglio and H. B. S. Cooke (eds.). *Evolution of African Mammals.* Cambridge, Mass., Harvard Univ. Press. pp. 315–32.

Hemmer, H. 1974. Zur Artgeschichte des Löwen *Panthera (Panthera) leo* Linnaeus. *Veroff. Zool. Staatssamml.* Munich. 17:167–280.

Hendey, Q. B. 1974. New fossil carnivores from the Swartkrans Australopithecine site (Mammalia, Carnivora). *Ann. Transvaal Mus.* 29(3):27–51.

———. 1978. Late Tertiary Hyaenidae from Langebaanweg, South Africa, and their relevance to the phylogeny of the family. *Ann. S. Afr. Mus.* 76(7):265–97.

Hibbard, C. W., R. J. Zakrzewski, R. E. Eshelman, G. Edmund, C. D. Griggs, and C. Griggs. 1978. Mammals from the Kanopolis local fauna, Pleistocene (Yarmouth) of Ellsworth County, Kansas.

Contrib. Mus. Paleontol. Univ. Mich.
25(2):11–44.

Hoffstetter, R. 1952. Les mammiféres Pléistocéne de la Republique de L' Equateur.. *Mem. Soc. Geol. France.* No. 66. 391 pp.

Hooijer, D. A. 1962. A fossil ground sloth from Curacao, Netherlands Antilles. *Proc. Koninkl. Nederl. Akad. Wetensch. Amsterdam Ser. B* 65(1):46–60.

———. 1963. Mammalian remains from an Indian site on Curacao. *Stud. fauna Curacao and other Caribbean islands.* 14(64):119–22.

———. 1978. Rhinocerotidae. In V. J. Maglio and H. B. S. Cooke (eds.). *Evolution of African Mammals.* Cambridge, Mass., Harvard Univ. Press. pp. 371–78.

Kurtén, B. 1966. Pleistocene bears of North America, 1. Genus *Tremarctos,* spectacled bears. *Acta Zool. Fennica* 115:1–120.

———. 1967. Pleistocene bears of North America, 2. Genus *Arctodus,* shortfaced bears. *Acta Zool. Fennica* 117:1–60.

———. 1968. *Pleistocene Mammals of Europe.* London, Weidenfeld and Nicolson. 317 pp.

———. 1972a. *The Age of Mammals.* New York, Columbia Univ. Press. 250 pp.

———. 1972b. *The Ice Age.* New York, G. P. Putnam's Sons. 179 pp.

———. 1976. *The Cave Bear Story.* New York, Columbia Univ. Press. 163 pp.

———. 1979. The stilt-legged deer, *Sangamona* of the North American Pleistocene. *Boreas* 8:313–21.

Kurtén, B. and E. Anderson. 1980. *Pleistocene Mammals of North America.* New York, Columbia Univ. Press. 443 pp.

Long, A. and P. S. Martin. 1974. Death of American ground sloths. *Science* 186:638–40.

McDonald, H. G. 1977. Description of the osteology of the extinct gravigrade edentate, *Megalonyx,* with observations on its ontogeny, phylogeny and functional anatomy. M.A. thesis, Univ. Fla. Gainesville.

Madden, C. in press. The Proboscidea of South America. *Geol. Soc. Amer. Abst.*

Maglio, V. J. 1973. Origin and evolution of the Elephantidae. *Trans. Amer. Phil. Soc. n.s.* 63(3):1–149.

Maglio, V. J. and H. B. S. Cooke (eds.). 1978. *Evolution of African Mammals.* Cambridge, Mass., Harvard Univ. Press. 641 pp.

Marshall, L. G. 1981. The families and genera of Marsupialia. *Fieldiana, Geology n.s.* 8, Publ. 1320, pp. 1–65.

Marshall, L. G., A. Berta, R. Hoffstetter, R. Pascual, O. A. Reig, M. Bombin, and A. Mones. MS. Geochronology of the continental mammal-bearing Quaternary of South America.

Martin, L. D. 1980. Functional morphology and the evolution of cats. *Trans. Neb. Acad. Sci.* 8:141–54.

Martin, P. S. 1975. Sloth droppings. *Nat. Hist.* 84(7):74–81.

Martin, P. S. and J. E. Guilday. 1967. A bestiary for Pleistocene biologists. In P. S. Martin and H. E. Wright (eds.). *Pleistocene Extinctions: The Search for a Cause.* New Haven and London, Yale Univ. Press. pp. 1–62.

Martin, P. S., B. E. Sabels, and D. Shutler. 1961. Rampart Cave coprolite and ecology of the Shasta ground sloth. *Amer. J. Sci.* 259(2):102–27.

Merriam, J. C. and C. Stock. 1932. The Felidae of Rancho La Brea. *Carnegie Inst. Washington Publ.* 422:1–232.

Nowak, R. 1979. North American Quaternary *Canis. Mon. Mus. Nat. Hist. Univ. Kans.* 6:1–154.

Owen, R. 1842. Description of the skeleton of an extinct giant sloth, *Mylodon robustus* Owen. London, Taylor. 176 pp.

Patterson, B. 1978. Pholidota and Tubulidentata. In V. J. Maglio and H. B. S. Cooke (eds.). *Evolution of African Mammals.* Cambridge, Mass., Harvard Univ. Press. pp. 268–78.

Paula Couto, C. de. 1967. Pleistocene edentates of the West Indies. *Amer. Mus. Novitates* 2304:1–55.

———. 1971. On two small Pleistocene ground sloths. *An. Acad. Brasil Cienc.* (1971) (supp.) 43-499–513.

Repenning, C. A., MS. Biochronology of the microtine rodentia of the United States. In Woodburne, M. O. (ed.). Cenozoic Mammals: Their Temporal Record, Biostratigraphy and Biochronology. Berkeley: Univ. California Press.

Romer, A. S. 1966. *Vertebrate Paleontology.* 3rd ed. Chicago, Univ. Chicago Press. 468 pp.

Saunders, J. J. 1977. Late Pleistocene vertebrates of the western Ozark Highland, Missouri. *Rept. Inv. Ill. State Mus.* 33:1–118.

———. 1980. A model for man-mammoth relationships in Late Pleistocene North America. *Canad. J. Anthropol.* 1(1):87–98.

Scott, W. B. 1962. *A History of Land Mammals in the Western Hemisphere.* New York, Hafner Publ. Co. 786 pp.

Simpson, G. G. 1945. The principles of classification and a classification of mammals. *Bull. Amer. Mus. Nat. Hist.* 85:1–350.

———. 1980. *Splendid Isolation. The Curious History of South American Mammals.* New Haven and London, Yale Univ. Press. 266 pp.

Slaughter, B. H. 1961. The significance of *Dasypus bellus* Simpson in Pleistocene local faunas. *Texas J. Sci.* 13(3):311–15.

Sokolov, V. E. 1974. *Saiga tatarica. Mammal. Species.* 38:1–4.

Stock, C. 1925. Cenozoic gravigrade edentates of western North America with special reference to the Pleistocene Megalonychidae and Mylodontidae of Rancho La Brea. *Carnegie Inst. Washington Publ.* 331:1–206.

Sutcliffe, A. J. 1970. Spotted hyena: crusher, gnawer, digester, and collector of bones. *Nature* 227:1110–13.

Van Gelder, R. G. 1977. Mammalian hybrids and generic limits. *Amer. Mus. Novitates* 2635:1–25.

Varona, L. S. 1974. Catálogo de los Mamíferos vivientes y extinguidos de las Antilles. Acad. Cienc. Cuba, Havana. 139 pp.

Walker, E. P. 1975. *Mammals of the World.* 3rd ed. Baltimore, The Johns Hopkins Press. 2 vol. 1,500 pp.

Webb, S. D. 1974. Pleistocene llamas of Florida with a brief review of the Lamini. In S. D. Webb (ed.). *Pleistocene Mammals of Florida.* Gainesville, Univ. Fla. Presses. pp. 170–213.

Wetzel, R. M. 1977. The Chacoan peccary, *Catagonus wagneri* (Rusconi). *Bull. Carnegie Mus. Nat. Hist.* 3:1–36.

Wetzel, R. M., R. E. Dubois, R. L. Martin, and P. Myers. 1975. *Catagonus,* an "extinct" peccary, alive in Paraguay. *Science* 189:379–81.

Wilson, M. V. 1974. The Casper Site local fauna and its fossil bison. In G. C. Frison (ed.). *The Casper Site: a Hell Gap Bison Kill on the High Plains.* New York, Academic Press. pp. 125–71.

———. 1975. Holocene fossil bison from Wyoming and adjacent areas. M.A. thesis, Univ. Wyo., Laramie.

3

New World Mammoth Distribution

LARRY D. AGENBROAD

FIVE FIELD SEASONS of excavations at the Hot Springs Mammoth Site, South Dakota, provided geological and paleontological evidence (Agenbroad, chap. 4; Laury 1980) that the site is unique as a mammoth-selective trap and in a geological-hydrologic setting that provided unusually good preservation of the faunal remains. A search of the literature for reports on mammoth deposits yielded nearly 1,500 localities for more than 3,100 individual mammoth in the New World.

Between 1923 and 1927 Oliver P. Hay, of the Carnegie Institute, published a series of works on the Pleistocene vertebrate fauna of North America. These publications summarized what was known of selected Pleistocene megafauna up to approximately 1930. Since that time, the industrial, agricultural, municipal, and transportation expansion that has taken place on this continent has led to multiple new discoveries of ice age fauna.

In updating the record of only one genus of Pleistocene megafauna, the mammoth, I have come to appreciate the magnitude of Hay's effort. With very few exceptions, the data used here are only from published works. No attempt was made to go through the collections of various state, county, educational, and private institutions for catalogued and as yet unpublished material. Nor was an effort made to incorporate data from private collections. With these limitations in mind, the reader will understand that while the results presented here are representative of the abundance and distribution of mammoth sites and occurrences, they are far from comprehensive.

Several states or regions have excellent summary publications for mammoth and other Pleistocene animals (e.g. Skeels 1962, West and Dallman 1980, Mehl 1962, Harington and Shackleton 1978). Many have no recorded discoveries of mammoth since Hay's works. I hope that this chapter will stimulate renewed interest in mammoth occurrences and distribution within regions and will generate improved compilations, so that a more representative update and compilation of mammoth occurrence and distribution in the New World will be possible.

Accepting the limitations, I feel that the general conclusions as to distribution and abundance of mammoth based on a minimum count are valid. Determining the distribution and abundance of individual species of mammoth is less satisfactory partly for the following reasons: 1) lack of published reports; 2) general lack of species information in published reports; 3) uncertainty regarding taxonomic criteria for identifying species; 4) overproliferation of species names (e.g. splitting) in certain reported data.

Table 3.1. Family Elephantidae, Subfamily Elephantinae, in the Pleistocene

(Age in Years Before Present)

Continent	Early Pleistocene (1.8 my–700,000)	Middle Pleistocene (700,000–130,000)	Late Pleistocene (130,000–10,000)	Recent (10,000–)
Africa	*Loxodonta atlantica* *Elephas recki*	*Loxodonta atlantica* *Elephas recki* *Loxodonta africana*	*Elephas iolensis* *Loxodonta africana*	*Loxodonta africana*
Europe	*Mammuthus meridionalis*	*Mammuthus armeniacus* *Elephas namadicus* *Elephas falconeri*	*Elephas namadicus* *Mammuthus primigenius*	
Asia	*Elephas planifrons* *Elephas hysudricus* *Elephas celebensis*	*Elephas platycephalus* *Elephas namadicus*	*Elephas maximus* *Elephas namadicus (?)* *Mammuthus primigenius*	*Elephas maximus*
North America	*Mammuthus meridionalis*	*Mammuthus imperator*	*Mammuthus columbi* *Mammuthus primigenius* *Elephas exilis**	

SOURCE: Data from Maglio (1973), as modified by Agenbroad (1980).

*Included in *M. imperator* by Maglio

For taxonomic purposes, I have modified (Agenbroad 1980) Maglio's (1973) system for the Elephantidae. Table 3.1 compares Pleistocene elephants in North America with those on other continents.

One work not previously cited should be noted. Madden (1981) has undertaken a systematic revision of mammoths in which he concludes there are six valid species in North America. These six (oldest to youngest) are given as *M. hayi, M. imperator, M. (Parelephas) exilis, M. (Parelephas) columbi, M. (Parelephas) jacksoni, M. mammonteus.* Although this work is detailed and exhaustive, it proposes terminology not in use in the published literature, and Maglio's system has been utilized in this chapter.

Taxonomy and Species Identification

Historically, classification of mammoth in the New World has been, and unfortunately remains, confused. Maglio (1973:61–63) points out that because of morphologic gradation of dentition from primitive to derived forms and because of overlapping stages, one species cannot be separated from another on the basis of dental criteria alone (See also Osborn 1942, Skeels 1962, Mawby 1967, Saunders 1970, Miller 1971, Davis et al. 1973, Harington and Shackleton 1978). The problem can be traced to the time of Hay. The "woolly mammoth" in particular has been identified uncritically.

Concerning species identification, Kurtén and Anderson (1980:350) "provisionally recognize four North American species or stages—*Mammuthus meridionalis, Mammuthus columbi, Mammuthus jeffersonii,* and *Mammuthus primigenius.*" They note the fact that Osborn (1922, 1942) referred to *M. jeffersonii* as more progressive than *M. columbi* and more primitive than *M. primigenius.* Kurtén and Anderson have chosen to adopt Osborn's definition of *M. jeffersonii* for what most investigators have reported as *M. columbi,* and to use *M. columbi* for what has been referred to as *M. imperator* in the literature. Addressing the confusion of generic and specific names in North American proboscideans, Maglio (1973:62) states:

The great confusion associated with the name *M. columbi* resulted in part from Falconer's inadequate holotype specimen and from Osborn's (1922) selection of two neotype specimens (AMNH 13707) both of which are very close to *M.*

imperator, if not actually identical to it. Osborn concluded that the holotypes of *imperator* and *columbi* were probably conspecific, although in later publications he retained both names. For the more progressive elephant material that had previously been referred to *M. columbi,* Osborn proposed the specific name *jeffersonii.* Although Osborn was correct in considering Falconer's original holotype specimen as inadequate for species diagnosis, there is little evidence that his neotype accurately reflects the true characters of the original. Thus, it is probably best at present to retain Leidy's name *imperator* for the more primitive of these mammoths and Falconer's name *columbi* for the more progressive stages. This also conforms with the most common usage of these names.

In my opinion, there appears to be a temporal gradation from *M. imperator* to *M. columbi* to *M. jeffersonii* to *M. primigenius* without distinct characteristics separating these proposed species. Since the publication of volume II of *Proboscidea* (Osborn 1942), most researchers have used *M. columbi* as the intermediate stage in North American mammoth development, between *M. imperator* and *M. primigenius.* Though Kurtén and Anderson (1980) are technically correct in their use of *M. jeffersonii,* I shall adopt the nomenclature used by Maglio and the numerous publications since 1942 which refer to *M. columbi.*

In several summary papers for a given geographic area there are no data on the species, nor is there information on the geologic context in which the remains occurred. Such information is vital and should be included when available. The majority of reported remains are simply assigned to *Mammuthus* (sp.), with no attempt at refinement of the classification. For the purpose of this chapter, I have used the species designation provided by the author of the publication cited unless the terminology is archaic, in which case it has been placed within the system presented by Maglio (1973).

The number of individual animals represented at a given site or locality is lacking in the majority of published reports. For this reason, the number of individuals presented in this chapter must be considered to be a minimal count.

New World Mammoth Distribution

Figures 3.1 and 3.2 illustrate the distribution of mammoth sites and the number of individual mammoths, respectively, as reported in the literature. Multiple animals from a given site generate a more detailed configuration for the map of the number of individuals (fig. 3.2) than for the map of the number of localities (fig. 3.1).

As can be seen from the maps, the heartlands for mammoth and mammoth localities are in Alaska, in the prairie provinces of Canada, and in a band of southwestern and central states in the United States. In general ecologic terms the mammoth population of the New World reflects the grassland and arctic-steppe (Matthews 1976) environments. It must be kept in mind that this pattern also reflects all reported remains, regardless of the age of the deposit. Mammoth records from the domain of the Laurentide Ice Sheet over Hudson Bay presumably all predate the last glaciation; most at its periphery postdate the last glaciation.

Some areas not designated on the maps probably have mammoth remains, but the information has not been gathered or has not been published. For example, Arkansas was not represented in Hay's compilation of mammoth remains, nor have I found any subsequent published data on mammoth occurrence there. Neighboring states provide multiple occurrences. Certain geographic areas have greater paleontological visibility due to the activity and interests of persons working there.

In the general pattern of distribution, Alaska has the greatest frequency of individual occurrences, followed by Texas and the central plains. Mammoths occur frequently in the Pleistocene sediments of the west coast of the United States, then swinging northeast from Arizona to Minnesota. Florida is an isolated, high-frequency area.

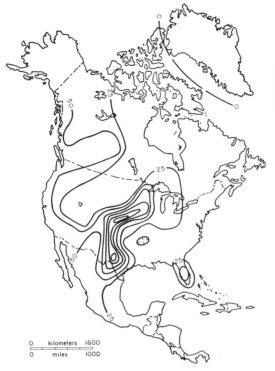

Figure 3.1. Minimum reported localities of *Mammuthus* in the New World. Contour interval = 25 localities.

Figure 3.2. Minimum number of undifferentiated individuals of *Mammuthus* in the New World. Contour interval = 25 individuals south of Alaska and Yukon; 200 individuals in Alaska and Yukon.

Mexico has numerous reported finds; vague reports for Mexico indicate a greater mammoth frequency than the published data reveal. The most southerly accepted occurrence of mammoth in the New World is in El Salvador (Stirton and Gealey 1949).

It appears that the Appalachian and the northern Rocky Mountain regions of the United States are relatively devoid of mammoth remains. This phenomenon may be due to scarcity of suitable sediments for preservation, rather than to a natural distribution of animals. A similar pattern is evident in the case of mastodont (Gross 1951:113). It has often been stated that mammoth were more or less excluded from the forested regions of the eastern portion of the continent, which was the preferred range of American mastodont. The "exclusion" was of an adaptive nature, rather than a competitive nature, however. Both mammoth and mastodont are not uncommon off the east coast, on the present continental shelf (Whitmore et al. 1967), which was exposed during the Pleistocene.

An attempt to plot the geographic range for four species of mammoth (*meridionalis, imperator, columbi,* and *primigenius*) was made from the published occurrences. Figures 3.3 through 3.6 give the general ranges and frequencies for these species. It is apparent that Columbian mammoth most closely reflect the general pattern of site distribution. They are also the most numerous of identified species. Imperial mammoth reflect the southern reaches of the Columbian distribution; however, imperial mammoth have been reported with high frequency as far north as British Columbia. *M. primigenius* is relatively restricted to the northeastern United States, Canada, and Alaska, although occurrences have been reported in much of the western United States and in Mexico.

Mammuthus meridionalis

Mammoth migration into the New World apparently took place more than 1.7 million years ago. The earliest form, *Mammuthus meridionalis,* has been reported from a number of localities in Canada, the United States, and Mexico (fig. 3.3). Three of these localities provide dates ranging from 1.36 million years ago at Bruneau, Idaho (Malde and Powers 1962), 1.5 million years at San Francisco (Hall 1965), to 1.7 million years in deposits of the Wellsch Valley, Saskatchewan (Harington and Shackleton 1978). Other, nondated occurrences are Aguascalientes, Mexico (Mooser and Dalquest 1975); Arkalon, Kansas; Crete and Angus, Nebraska; and Holloman, Oklahoma (Maglio 1973). Schultz et al. (1978) have questioned the Idaho date as well as the validity of the examples of the holotypes and suggest that these examples be assigned to *M. imperator.* They also suggest that mammoth remains from Gilliland, Texas, and Butler County, Nebraska, represent valid *M. meridionalis* specimens.

Figure 3.3 presents an envelope containing reported, early Pleistocene *Mammuthus meridionalis* localities. The embayment in the envelope in the Montana-Utah region follows the similar embayment in Figures 3.1 and 3.2, and 3.4 through 3.6. In general, the locations of the earliest (most primitive) mammoth known in the New World display the same pattern as that produced for the general data (figs. 3.1 and 3.2) and that for *M. columbi* (fig. 3.5), and reflect a western and plains adaptation.

Mammuthus imperator

The imperial mammoth (*M. imperator*) distribution (fig. 3.4) is similar to that for all mammoth (fig. 3.1) but follows a more southerly and southeasterly pattern. The heartland of the *M. imperator* range extends from northern Mexico through Texas and Nebraska and into Minnesota. Relatively high frequencies of this species have been reported in California, British Columbia, and Florida. Most of the eastern portion of the continent is apparently devoid of imperial mammoth, and they are sparsely represented in the mountain west.

Figure 3.3. Reported distribution of *Mammuthus meridionalis,* with available radiometric ages.

Figure 3.4. Reported distribution of *Mammuthus imperator.* Contour interval = 10 individuals.

Mammuthus columbi

Columbian mammoth (*M. columbi*), including *M. jeffersonii,* appear to be wide-spread, with the greatest density of specimens in a bifurcate configuration from Arizona northeast to the Great Lakes and from the central plains (Nebraska) to Texas (fig. 3.5); Florida and the west coast also have relatively high populations. As noted earlier, the *M. columbi* distribution closely duplicates the map of locality distribution for all reported mammoth localities (fig. 3.1).

Mammuthus primigenius

Distribution of woolly mammoth (*M. primigenius*) (fig. 3.6) is concentrated in an area from Hudson Bay through the Great Lakes region and the northern plains of the United States across northern Canada, and the highest concentration is in Alaska. It should be noted that *M. primigenius* is reported from many southwestern states and Mexico. It is possible that some of these southerly identifications are in error. *M. primigenius* distribution is often correlated with tundra environment; tundra-boreal forest; or even cold, loess-steppe environments (Harington and Shackleton 1978). This distribution rather closely reflects the maximum extent of glaciation and glacial and near-glacial environments for the continent.

Environment of Deposition or Entrapment

In a sample of 856 localities drawn from several geographic regions for which information had been given as to the geologic environment in which the mammoth remains were found, 73 percent represented aqueous environments such as stream channels (or former stream channels), flood plains, marsh or bog environments, springs, or lacustrine deposits. Table 3.2 gives a relative breakdown on modes of occurrence of the mammoth remains. These data probably represent the environment of deposition, accumulation, and preservation more closely than they represent habitat of the living animal.

Multiple mammoth occurrences have been reported at several localities. Note-worthy among these is Friesenhahn Cave, Texas, where "very large number(s) of immature elephant bones" were found (Evans 1961), as well as "441 isolated teeth of young individuals...as compared to only 14 complete and fragmentary teeth of adult elephants" (Graham 1976). The geologic nature of this cave, a vertical solution/collapse feature in limestone, coupled with the associated carnivores, indicates "that young elephants were the preferred and principal diet of the great cat, *Dinobastis*" (Evans 1961), which used the cave as a den. Evans estimates that several hundred young elephants are present in the cave fill. This mammoth locality represents a prey accumulation, selective for but not exclusively of immature animals.

Hay (1924) wrote that "at a place in Kansas, over 200 teeth of Columbian mammoth" were collected and "tons of broken bones and tusks" were found. The locality was a small, 2–3-rod (10–15 m, 33–50 ft.) diameter basin in a small ravine in the Niobrara chalk. Some of these remains were assigned to *M. primigenius*. Judging from the description given, the accumulation appears to represent a unique erosional feature in which faunal remains accumulated.

The Hot Springs Mammoth Site in South Dakota represents a spring-fed karst feature in which the remains of more than thirty Columbian mammoth have been preserved (Agenbroad 1976, 1978; Dutrow 1980; Agenbroad et al. 1978; Laury and Agenbroad 1980). The accumulation is natural, mammoth-selective, and attritional, and it represents as much as a 300-to-500-year period of the infilling of the depression (Laury 1980). Considering the major multiple-mammoth localities in the literature in

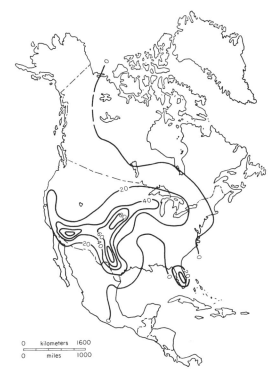

Figure 3.5. Reported distribution of *Mammuthus columbi.* Contour interval = 20 individuals.

Figure 3.6. Reported distribution of *Mammuthus primigenius.* Contour interval = 5 individuals south of Alaska; 200 individuals in Alaska.

Table 3.2. General Environment of Deposition/Preservation of Mammoth

Type of Environment	Number of Localities	Relative Percent of Total
Fluvatile		57.3
streams, rivers	243	
gravel/sand pits near streams	245	
Lacustrine/Oceanic		7.2
lake beds	12	
shoreline/beachlines	35	
offshore (marine)	14	
Marsh/Bog/Swamp		6.5
clay	30	
bog	25	
Spring		3.2
spring deposits	27	
Glacial		8.7
drift/outwash	74	
Other		17.1
construction	62	
wells	24	
farming (fields)	17	
loess	13	
phosphate/salt mines	15	
sinkholes	5	
caves	3	
tar pits	2	
lava tubes	1	
lava flows	1	
salt licks	1	
ash deposits	1	
alkali pockets	1	
TOTAL	851	100.0

Table 3.3 New World Multiple-Mammoth Localities

Locality	Minimum Number of Individuals	Type of Deposit
New Siberian Is., Alaska	+1,250	—
Friesenhahn, Texas	"Several hundred"	Carnivore Den
Fairbanks, Alaska	+113	Alluvium
Penndennis, Kansas	+50	Erosional/Depositional
Hot Springs, South Dakota	+30	Sinkhole w/Spring
Trinity River, Dallas, Texas	+28	Alluvial Gravels
Lamb Springs, Colorado	+24	Spring Deposit
Charleston, South Carolina	20	Phosphate Beds
Blackwater Draw, New Mexico	15	Human Kill Site
Lehner Ranch, Arizona	13	Human Kill Site
Dent, Colorado	12	Human Kill Site
Frankstown, Pennsylvania	7	Cave Fill
Colby, Wyoming	7	Human Kill Site
Bradenton, Florida	7	River Sand
Dutton, Colorado	+5	Pond
Selby, Colorado	+5	Pond
Miami, Texas	5	Human Kill Site
San Francisco, California	5	Alluvial Fan
Murray Springs, Arizona	4	Human Kill Site
Slanton, Texas	4	Lake Beds
Silverspring, Florida	3	Spring/Pond

which numbers of individuals are represented (Table 3.3), the Hot Springs Site is unique in North America, representing a natural trap for a large, living, adult population of mammoth.

A large number of mammoth have been found in the alluvium of the second terrace of the Trinity River, near Dallas, Texas (Lull 1921; Slaughter 1960), where the remains have been recovered from sand and gravel pits. Slaughter reports "fifteen semi-articulated skeletons of *Mammuthus columbi* in a single pit." Lull reports thirteen skulls from the pits in 1921.

A large number of reported, multiple-mammoth occurrences from single localities reflect the campsite and butchering activity of human predation. Such sites (see Table 3.4) cluster at about 11,200 years ago, are characterized by Clovis projectile points, and are distributed in basically the same pattern as the high-frequency trend in Figure 3.5. Clovis association with mammoth remains are also coincident with, or very close to, the terminal dates for mammoth occurrences.

Chronology

Seventy-eight localities provide absolute dates for mammoth remains or for associated materials. The majority of mammoth localities have not yielded absolute dates. Most dating has been done for sites which have potential cultural affiliation. Table 3.4 gives the chronologic information for dated mammoth localities.

The age range and distribution of *M. meridionalis* are provided in Figure 3.3, based on K-Ar dates of basalt, tuff (Evernden et al. 1964, Hall 1965), or paleomagnetic data (Harington and Shackleton 1978). *M. imperator* has absolute dates from five localities. The Channel Islands off Santa Barbara, California, provide dates of 15,820; 29,700; and greater than 37,000 years ago. The Lindsay mammoth site in Montana provides dates of 10,700; 10,980; and 11,926 yr B.P. Wiseton, Saskatchewan, and Santa Isabel Iztapan (Mexico), yield dates of 10,600 and 11,003 yr B.P., respectively. Domebo, Oklahoma, provides dates of 11,045 and 11,220 years ago. While sparse, the dates suggest that the imperial mammoth was extant until approximately 10,600 years ago in Canada, possibly finding its way into Paleo-Indian menus around 11,000 years ago.

Localities providing temporal information for *M. columbi* give a range from 26,075 to 8,815 years ago. [It is likely that the minimal date is more representative of the material dated at that site (bone collagen) than it is of a terminal date for Columbian mammoth.] The youngest dates on mammoth may be on "acid insoluble residues" rather than a high yield of bone collagen. While an early Holocene age of mammoth survival is entirely possible, the age needs to be carefully replicated. A series of nine of the most recent dates for *M. columbi* (excluding the Sandy Mammoth date, Table 3.4) gives an average of 11,424 yr B.P. as a terminal date for the species. In man-mammoth associations the dates cluster near 11,200 ± 100 yr B.P. *M. primigenius* occurrences provide sixteen reported radiocarbon dates ranging from greater than 35,000 to 7,670 yr B.P.

Overlapping age ranges for the species prevent drawing meaningful conclusions as to either geographic or temporal relationships. The paucity of radiometric dates from paleontological sites and the clustering of dates from sites with cultural components also tend to mask any temporal or geographic patterns that might exist.

Dates such as those from Sandy, Utah; Kyle, Saskatchewan; and Stein Ranch and Manhattan, Montana, seem to reflect the dating of bone or bone fractions with low collagen yields and increased opportunity of contamination and are probably younger than the demise of the species. Lake Mills and Stiles, Wisconsin; La Paloma Ranch, Texas; Hudson Hope, British Columbia; Seward, Alaska; and Kassler, Colorado, apparently represent some of the most recent occurrences of mammoth in the New World.

Table 3.4. Reported Radiometric Dates for New World Mammoth (*Mammuthus*) Localities

(*Denotes Locality with Reported Cultural Affiliation at <15,000 yr B.P.)

Country	State or Province	Locality	Date, yr B.P.	Lab No.	Species	Reference
Canada	Alberta	Empress	20,400 ± 320	CGS-1387	M. columbi / M. primigenius	Harington 1978
	British Columbia	Babine Lake	43,800 ± 1830	GSA 1687	M. columbi	Harington 1978
			42,900 ± 1860	GSC 1657	M. columbi	Harington 1978
			34,000 ± 690	GSC 1754	M. columbi	Harington 1978
		Hudson Hope	7670 ± 150	I-2244	Mammuthus (sp.)	Harington and Shackleton 1978
		Taylor	27,400 ± 580	GSC-2034	M. primigenius	Harington and Shackleton 1978
	N.W. Territories	Tununuk	19,440 ± 290	I 8578	M. primigenius	Harington 1978
		Melville Is.	21,900 ± 320	GSC 1760	Mammuthus (sp.)	Harington 1978
	Ontario	Woodbridge	45,000 ± 900	GSC 1181	M. primigenius	Harington 1978, Lowden et al. 1971
	Saskatchewan	Kyle	12,000 ± 200	S-246	M. primigenius	McCallum and Wittenberg 1968
			8680 ± 400	A-619	M. columbi	Haynes et al. 1971
		Wiseton	10,600 ± 140	S-232	M. imperator	Rutherford et al. 1973
	Yukon	Old Crow	25,750 + 1800 − 1500	GX-1568	M. primigenius	Irving and Harington 1973
			+ 39,900	I-4428	Mammuthus (sp.)	Bonnichsen 1978
			29,100 + 3000 − 2000	GX-1567	M. primigenius	Irving and Harington 1973
	Yukon	Whitestone River	30,300 ± 2000	I 3576	M. primigenius	Harington 1978
		Gold Run Creek	32,250 ± 1750	I 4226	M. primigenius	Harington 1978
United States	Alaska	Fairbanks	21,300 ± 1300	L-601	M. primigenius	Péwé 1975
			15,380 ± 300	SI-453	M. primigenius	Stuckenrath and Mielke 1970
			12,622 ± 750	W-401	Mammuthus (sp.)	Hester 1960
		Kotzebue	26,900 + 2400 − 3400	AU-90	Mammuthus (sp.)	Hopkins et al. 1976
		Ester Creek	>35,500		M. primigenius	Tucek 1943
		Dry Creek	10,690 ± 250	SI-1561	Mammuthus (sp.)	Thorson and Hamilton 1977
		Dome Creek	32,700 ± 980	ST-1632	M. primigenius	Gillespie 1970
		Seward Peninsula	+ 10,200 ±	L-137	Mammuthus (sp.)	Hopkins 1963
		Sullivan Creek	>38,800	W-1132	Mammuthus (sp.)	Ives et al. 1964
		Canyon Creek	39,360 (avg)	DIC-1819	M. primigenius	Weber et al. 1981
		Tyone River	29,450 ± 610	SI 355	Mammuthus (sp.)	Thorson et al. 1981
		Lost Chicken Creek	26,760 ± 300	I 10649	M. primigenius	Péwé 1975
			20,500 ± 390	I 8582	M. primigenius	Harington 1978
			10,370 ± 160	I 9998	M. primigenius	Harington 1978
		Baldwin Peninsula	10,050 ± 150	AU-90	M. primigenius	Harington 1978
			26,900 ± 3400		M. primigenius	Reeburgh and Young 1976
	Arizona	Hurley	21,210 ± 770	A-988	M. columbi	Haynes et al. 1971
		Naco	9250 ± 300	A 9-10	M. columbi	Hester 1960

State	Location	Date	Sample No.	Species	Reference
California	*Lehner Ranch	11,260 ± 360	AVG.	M. columbi	Damon et al. 1964
		11,240 ± 190	A-42	M. columbi	Damon and Long 1962
	Winslow	22,360 ± 500		M. columbi	Saunders 1970
	Cerros Negros	12,000 ± 300	A-854	M. columbi	Haynes 1968
	*Murray Springs	11,230 ± 340	A-805	M. columbi	Haynes and Haas 1974
	China Lake	18,600 ± 450	UCLA-1800	M. columbi	Davis 1978
		11,800 ± 800	UCLA-106	M. columbi	Libby and Fergusson 1962
	Santa Rosa	29,700 ± 3000	L-290R	M. imperator	Broecker and Kulp 1957, Hester 1960
		15,820 ± 280	L-244	M. imperator	Orr 1956, Hester 1960
		>37,000	UCLA-749	M. imperator	Lorenzo 1978
		11,800 ± 800	UCLA-106	M. exilis	Damon and Long 1962
	San Francisco	1,500,000		M. meridionalis	Hall 1965
	Rancho La Brea	13,890 ± 280	Y-354b	M. imperator, M. columbi	Hester 1960
Colorado	*Dent	15,390 ± 230	Y-355b	M. imperator, M. columbi	Hester 1960
		11,200 ± 500	I-622	Mammuthus (sp.)	Haynes 1964, Frison 1978, Butler 1981
	*Lamb Springs	13,140 ± 1000	M-1464	Mammuthus (sp.)	Crane and Griffin 1968, Stanford et al. 1981
	*Selby and Dutton	11,710 ± 150	SI-2877	M. columbi	Stanford 1979, Graham 1981
	Kassler	10,200 ± 350	W-401	M. columbi	Scott 1963
Idaho	*Owl Cave	10,920 ± 150	WSU-1786	Mammuthus (sp.)	Miller and Dort 1978
		12,850 ± 150	WSU-1281	Mammuthus (sp.)	Butler 1972
	American Falls	12,250 ± 200	WSU-1259	Mammuthus (sp.)	Sheppard and Chatters 1976
		21,500 ± 700	WSU-1423	Mammuthus (sp.)	Sheppard and Chatters 1976
		31,300 ± 2300	WSU-1424	Mammuthus (sp.)	Sheppard and Chatters 1976
	Bruneau	1,360,000	KA-1188	M. meridionalis	Evernden et al. 1964
Kentucky	Big Bone Lick	10,600 ± 250	W-1358	Mammuthus (sp.)	Levin et al. 1965
Michigan	Midland Co.	24,000 ± 4000	M-2145	M. columbi	Kapp 1970
	Jackson Co.	12,200 ± 700	M-507	M. columbi	Crane and Griffin 1958
Montana	*Lindsay	11,925 ± 350	S-918	M. imperator	Frison 1978, Rutherford et al. 1973
		10,980 ± 225	I-9220	M. imperator	Frison 1978
		10,700 ± 290	WSU-652	M. imperator	Frison 1978
		9490 ± 135	I-7028	M. imperator	Rutherford et al. 1973
	Stein Ranch	8890 ± 300	A-584	Mammuthus (sp.)	Haynes et al. 1971
	Manhattan	6050 ± 750	A-587	Mammuthus (sp.)	Haynes et al. 1971
	Sun River	11,500 ± 300	W-1753	M. columbi	Marsters et al. 1969
Nevada	Tule Springs	11,500 ± 500	UCLA-636	M. columbi	Haynes 1967
		>40,000	UCLA-517	M. columbi	Haynes 1967
New Mexico	Hermit Cave	12,270 ± 450	W-499	M. columbi	Schultz 1968
		11,850 ± 350	W-498	Mammuthus (sp.)	Schultz 1968, Hester 1960
		12,270 ± 450	W-499	Mammuthus (sp.)	Schultz 1968
	*Clovis	11,170 ± 360	A-481	M. columbi	Hester 1972
		+30,000		Mammuthus (sp.)	Crane 1955
	Sandia	25,000		Mammuthus (sp.)	Hibben 1955, Crane 1956
		+20,000	M-247	Mammuthus (sp.)	Hester 1960, Crane 1956

Table 3.4. Reported Radiometric Dates for New World Mammoth (*Mammuthus*) Localities
(*Denotes Locality with Reported Cultural Affiliation at < 15,000 yr B.P.)
(continued)

Country	State or Province	Locality	Date, yr B.P.	Lab No.	Species	Reference
	New York	*Lucy	14,300 ± 650	M-1434	*Mammuthus* (sp.)	Crane and Griffin 1958
		McCullum Ranch	15,750 ± 760	A-375	*Mammuthus* (sp.)	Damon et al. 1964
	Oklahoma	Malloy Farm	12,100 ± 400	I-838	*Mammuthus* (sp.)	Buckley et al. 1968
		*Domebo	11,220 ± 500	SI-172	*M. imperator*	Leonhardy 1966
			11,045 ± 647	SM-695	*M. imperator*	Leonhardy 1966
		Bartow	11,990 ± 170	A-582	*Mammuthus* (sp.)	Haynes et al. 1971
	South Dakota	Hot Springs	21,000 ± 700	GX-5356-A	*M. columbi*	Agenbroad 1978
			26,075 ± 975	GX-5895-A	*M. columbi*	Agenbroad and Laury 1979
	Texas	Lewisville	>37,000	O-235	*M. columbi*	Brannon et al. 1957
			>38,000	UCLA-110	*M. columbi*	Crook and Harris 1961
		Plainview	9800 ± 500	L-303	*M. columbi*	Hester 1960
	Texas	Toyah	12,140 ± 140	I-7088	*Mammuthus* (sp.)	Buckley 1976
		Grosbeck Creek	16,775 ± 565	Socony-Mobil Oil	*M. columbi*	Dalquest 1965
		*Bonfire	>10,230 ± 160	TX-153	*Mammuthus* (sp.)	Dibble and Lorrain 1968
		Laubach	13,970 ± 310	TX-1138	*Mammuthus* (sp.)	Lundelius and Davidson 1975
			28,340 ± 1710	TX-1419	*Mammuthus* (sp.)	Lundelius and Davidson 1975
		*Lubbock Lake	12,650 ± 250	LL-308	*Mammuthus* (sp.)	Johnson, C. 1974, Johnson, E. 1974
		Hueco Mountains	33,000	A-1721	*Mammuthus* (sp.)	Van Devender 1979 (pers. comm.)
		La Paloma Ranch	10,700 ± 4300	TX-2194	*Mammuthus* (sp.)	Valastro et al. 1979
			9400 ± 4700	TX-2197B	*Mammuthus* (sp.)	Valastro et al. 1979
			8080 ± 480	TX-2197A	*Mammuthus* (sp.)	Valastro et al. 1979
	Utah	City Creek	14,150 ± 800	RL486	*M. columbi*	Madsen et al. 1976
		Silver Creek	18,150 ± 950	UCR-331	*M. columbi*	Madsen et al. 1976
		Sandy	8815 ± 100	SI-2341a	*M. columbi*	Madsen et al. 1976
	Wisconsin	Lake Mills	9065 ± 90	WISC-704	*M. primigenius*	Bender et al. 1977
		Stiles	9335 ± 90	WISC-786	*M. primigenius*	Bender et al. 1977
	Wyoming	*Colby	11,200 ± 200	RL-392	*M. columbi*	Frison 1978
		*U.P.	11,280 ± 200	I-499	*M. columbi*	Irwin et al. 1962
		Natural Trap Cave	10,920 ± 300	A-366	*Mammuthus* (sp.)	Martin and Gilbert 1978
		Rawhide Butte	10,550 ± 350	A-366	*Mammuthus* (sp.)	Damon et al. 1964
Mexico	Mexico	*Santa Isabel Iztapan	11,003 ± 500	C-205	*M. imperator*	Hester 1960, Libby 1955
			+16,000	C-204	*M. imperator*	Hester 1960
	Puebla	Tlapacoya	33,500 + 3200 − 2300	GX-1103	*Mammuthus* (sp.)	Mirambell 1978
			24,200 ± 4000	A-794-B	*Mammuthus* (sp.)	Haynes et al. 1971
			23,150 ± 950	GX-0950	*Mammuthus* (sp.)	Mirambell 1978
			21,700 ± 500	I-4449	*Mammuthus* (sp.)	Lorenzo 1978
		Valsequillo	20,780 ± 800	W-1897	*Mammuthus* (sp.)	Kelley et al. 1978

Figure 3.7. Carbon-14 isochrons of reported post-15,000 B.P. *Mammuthus* localities. The dotted line represents the southern limit of the Wisconsin glacial (after Flint 1971) and the proposed ice-free Canadian corridor (Haynes 1964, Dumond 1980). Contour interval = 1,000 years.

    To visualize the extinction model for mammoth, all radiocarbon dates of less than 15,000 yr B.P. were plotted on a map of North America. Exceptions were Sandy, Utah; Kyle, Saskatchewan (minimum date); and Stein Ranch and Manhattan, Montana, for reasons discussed above. There were forty-eight dated localities with dates of less than 15,000 years. The resulting plot (fig. 3.7) indicates a "ripple effect" of extinction with terminal dates of greatest antiquity (+12,000 yr B.P.) centered in an area from Texas northwest to California and Utah. This pattern is also reflected in the northeastern United States and Canada. Surrounding the center of +12,000-year dates is an envelope of +11,000-year dates extending from Mexico through the United States and into central Canada. Dated sites of less than 11,000 years antiquity are scattered outside the core area described.

    The isochrons proposed in Figure 3.7 are discordant with the chronology modeled by Martin (1973, 1974), in which a sequential extinction wave is proposed, initiated at some point in Canada and sweeping toward the south. Using reported C-14 dates, from which Figure 3.7 is derived, it appears that the extinction of mammoth began in a central area south of Canada and spread peripherally, much like ripples from an object falling into the center of a pond. Greatest antiquity (post-15,000 yr B.P.) lies in California, Utah, New Mexico, and Texas.

    The fact that fourteen of the forty-eight dates have, or are thought to have, cultural affiliations reflects human predation in 29 percent of the radiocarbon-dated mammoth sites of less than 15,000 years antiquity. Such a high visibility of human predation suggests the growing role of *Homo sapiens* in the extinction of mammoth. Also, if one considers only the data from which Figure 3.7 was generated, it appears that the earliest dates for successful mammoth kills (at least as reported in the literature) occur in the same geographic region as the highest concentration of mammoth (fig. 3.2), and that younger kill sites diffuse in all directions from that center.

The center of mammoth range and concentration is also coincident with the "threshold" and "entryway" of man into what is now the United States, as shown by the proposed ice-free corridor delineated in Figure 3.7. Heaviest hunter impact would have been felt initially in this threshold and entry area, and, as game was depleted or driven out of the area by hunting pressure, the spread of extinction would have been in all directions, as proposed in the model outlined above. Alternative models, such as those proposing maximum thermal effects and concomitant range deterioration due to de-creased precipitation, air mass distribution, and so forth, might also be postulated to account for the isochron pattern.

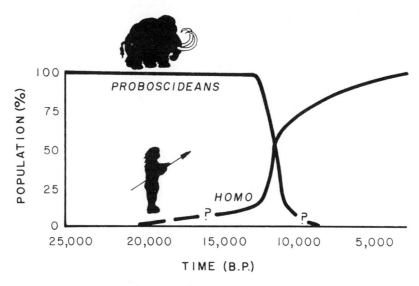

Figure 3.8. Population and temporal relationships of mammoth and men in the late Pleistocene of the New World.

Reported radiometric ages show that there have been mammoth in the New World at least since 1.7 million years ago. Temporal ranges for imperial, Columbian, and woolly mammoth overlap, but all species became extinct approximately 10,000 years ago. Sites which display man-mammoth association tend to cluster at or near 11,200 yr B.P., approximately 1,200 years prior to the most recent credible mammoth occurrences. Figure 3.8 shows the coincidence of the archaeological visibility of *Homo sapiens* in the New World and the extinction of *Mammuthus*. It would indeed be an extraordinary coincidence if cultural impacts had little or nothing to do with the timing of New World mammoth extinction.

References

Agenbroad, L. D. 1976. Mammoth Site of Hot Springs, South Dakota. Caxton Printers Ltd., 20 pp.

———. 1978. Excavations at the Hot Springs Mammoth Site: A Pleistocene Animal Trap. *Transactions of the Nebraska Academy of Sciences* 6:127–30.

———. 1980. Quaternary Mastodon, Mammoth and Men in the New World. *Canadian Journal of Anthropology* 1(1):99–101.

Agenbroad, L. D., and R. L. Laury. 1979. Excavation and Sedimentology of the Mammoth Site of Hot Springs, South Dakota. *National Geographic Society Research Reports* 1976–78. Washington, D.C.

Agenbroad, L. D., R. L. Laury, B. Dutrow, J. Mead, and P. Bjork. 1978. Hot Springs Mammoth Site, South Dakota, and

Hudson-Meng Paleoindian Bison Kill, Nebraska. *Guidebook and Roadlogs for Rocky Mountain-Plains Field Conference: Friends of the Pleistocene.* 40 pp.

Bender, M. M., R. A. Bryson, and D. A. Baerreis. 1977. University of Wisconsin Radiocarbon Dates XIV. *Radiocarbon* 19:127–137.

Bonnichsen, R. 1978. Critical Arguments for Pleistocene Artifacts from the Old Crow Basin, Yukon: A Preliminary Statement *in* A. Bryan (ed.), Early Man in America from a Circum-Pacific Perspective. *Occasional Papers of the Department of Anthropology, University of Alberta.* 1:102–118.

Brannon, H. R., Jr., A. C. Daughtry, D. Perry, L. H. Simmons, W. W. Whitaker, and M. Williams. 1957. Humble Oil Company Radiocarbon Dates I. *Science* 125:147–150.

Broecker, W. S., and J. L. Kulp. 1957. Lamont Natural Radiocarbon Measurements IV. *Science* 126:1324–1334.

Buckley, J. D. 1976. Isotopes Radiocarbon Measurements XI. *Radiocarbon* 18:172–189.

Buckley, J. D., M. A. Trautman, and E. H. Willis. 1968. Isotopes Radiocarbon Measurements VI. *Radiocarbon* 10:246–294.

Butler, B. R. 1972. The Holocene or Postglacial Ecological Crisis on the Eastern Snake River Plain. *Tebiwa* 15:49–63.

Butler, W. B. 1981. Eastern Colorado Radiocarbon Dates. *Southwestern Lore* 47:12–31.

Crane, H. R. 1955. Antiquity of the Sandia Culture: Carbon-14 measurements. *Science* 122:689–690.

———. 1956. University of Michigan Radiocarbon Dates I. *Science* 124:664–672.

Crane, H. R., and J. B. Griffin. 1958. University of Michigan Radiocarbon Dates II. *Science* 1127:1098–1105.

———. 1968. University of Michigan Radiocarbon Dates XII. *Radiocarbon* 10:61–114.

Crook, W. W., Jr., and R. K. Harris. 1961. Significance of a New Radiocarbon Date from the Lewisville Site. *Bulletin of the Texas Archeological Society* 32:327–330.

Dalquest, W. W. 1965. New Pleistocene Formation and Local Fauna from Hardeman County, Texas. *Journal of Paleontology* 39:63–72.

Damon, P. E., and A. Long. 1962. Arizona Radiocarbon Dates III. *Radiocarbon* 4:239–249.

Damon, P. E., C. V. Haynes, and A. Long. 1964. Arizona Radiocarbon Dates V. *Radiocarbon* 6:91–107.

Davis, E. L. 1978. Association of People and Rancholabrean Fauna at China Lake, California *in* Bryan (ed.), Early Man in America from a Circum-Pacific Perspective. *Occasional Papers of the Department of Anthropology, University of Alberta* 1:183–217.

Davis, L. C., R. E. Eshelman, and J. C. Prior. 1973. A Primary Mammoth Site with Associated Fauna in Pottawattamie County, Iowa. *Proceedings of the Iowa Academy of Science* 79:62–65.

Dibble, D. S., and D. Lorrain. 1968. Bonfire Shelter: A Stratified Bison Kill Site, Val Verde County, Texas. *Texas Memorial Museum Miscellaneous Papers* 1:9–138.

Dumond, D. E. 1980. The Archaeology of Alaska and the Peopling of America. *Science* 209:984–991.

Dutrow, B. L. 1980. Population Structure of a Late Pleistocene Mammoth Assemblage, Hot Springs, South Dakota. *American Quaternary Association Abstracts and Program* 6:67.

Evans, G. L. 1961. The Friesenhahn Cave. *Bulletin of the Texas Memorial Museum* 2:7–22.

Evernden, J. F., D. E. Savage, G. H. Curtis, and G. T. James. 1964. Potassium-Argon Dates and the Cenozoic Mammalian Chronology of North America. *American Journal of Science* 262:145–198.

Flint, R. F. 1971. *Glacial and Quaternary Geology.* John Wiley and Sons, Inc., New York. 892 pp.

Frison, G. C. (ed.) 1978. *Prehistoric Hunters of the High Plains.* Academic Press, New York. 457 pp.

Gillespie, J. M. 1970. Mammoth Hair: Stability of Alpha-Keratin Structure and Constituent Proteins. *Science* 170:1100–1101.

Graham, R. W. 1976. Pleistocene and Holocene Mammals: Taphonomy and Paleoecology of the Friesenhahn Cave Local Fauna, Bexar County, Texas. Ph.D. dissertation, University of Texas, Austin.

———. 1981. Preliminary Report on Late Pleistocene Vertebrates from the Selby and Dutton Archeological/Paleontological Sites, Yuma County, Colorado. *Contributions to Geology, University of Wyoming* 20:33–56.

Gross, H. 1951. Mastodons, Mammoths and Man in America. *Texas Archaeological and Paleontological Society Bulletin* 22:101–123.

Hall, N. T. 1965. Late Cenozoic Stratigraphy between Mussel Rock and Fleishhacker Zoo, San Francisco Peninsula *in* Schultz and Smith (eds.), *Northern Great Basin and California: Guide Books for Field Conference I,* 7th INQUA Conference, 151–158.

Harington, C. R. 1978. Quaternary Vertebrate Faunas of Canada and Alaska and Their Suggested Chronologic Sequence. *Syllogeus* No. 15, National Museum of Canada, Ottawa, pp. 1–105.

Harington, C. R., and D. M. Shackleton. 1978. A Tooth of *Mammuthus primigenius* from Chestermere Lake near Calgary, Alberta, and the Distribution of Mammoths in Southwestern Canada. *Canadian Journal of Earth Sciences* 15:1272–1283.

Hay, O. P. 1923. *The Pleistocene of North America and its Vertebrated Animals from the States East of the Mississippi and from the Canadian Provinces East of Longitude 95 degrees*. Carnegie Institute of Washington, D.C. 322:499 pp.

————. 1924. *The Pleistocene of the Middle Region of North America and its Vertebrated Animals*. Carnegie Institute of Washington, D.C. 322A:385 pp.

————. 1925. Extinct Proboscideans of Mexico. *Pan American Geologist* 44: 21–37.

————. 1927. *The Pleistocene of the Western Region of North America and its Vertebrated Animals*. Carnegie Institute of Washington, D.C. 322B:346 pp.

Haynes, C. V. 1964. Fluted Projectile Points: Their Age and Dispersion. *Science* 145: 1408–1413.

————. 1967. Quaternary Geology of the Tule Springs Area: Clark County, Nevada *in* Wormington and Ellis (eds.), Pleistocene Studies in Southern Nevada. *Nevada State Museum Anthropological Papers* 13:15–104.

————. 1968. Preliminary Report on the Late Quaternary Geology of the San Pedro Valley, Arizona. *Southern Arizona Guidebook III: Arizona Geological Society* 76–96.

Haynes, C. V., Jr., D. C. Grey, and A. Long. 1971. Arizona Radiocarbon Dates VIII. *Radiocarbon* 13:1–18.

Haynes, C. V., Jr., and H. Haas. 1974. Southern Methodist University Radiocarbon Date List I. *Radiocarbon* 16:368–380.

Hester, J. J. 1960. The Late Pleistocene Extinctions. *American Antiquity* 26:58–87.

————. 1972. Blackwater Locality No. 1: A Stratified Early Man Site in Eastern New Mexico. Fort Burgwin Research Center, Taos. 238 pp.

Hibben, F. C. 1955. Specimens from Sandia Cave and Their Possible Significance. *Science* 122:688–689.

Hopkins, D. M. 1963. Geology of Imuruk Lake Area, Seward Peninsula, Alaska. *USGS Bulletin* 1141:C1–C101.

Hopkins, D. M., R. E. Giterman, and J. V. Matthews, Jr. 1976. Interstadial Mammoth Remains and Associated Pollen and Insect Fossils, Kotzebue Sound Area, Northwestern Alaska. *Geology* 4:169–172.

Irving, W. N., and C. R. Harington. 1973. Upper Pleistocene Radio-carbon Dated Artifacts from the Northern Yukon. *Science* 179:335–340.

Irwin, C. H., Irwin, and G. Agogino. 1962. Wyoming Muck Tells of Battle: Ice Age Man vs. Mammoth. *National Geographic* 121:828–838.

Ives, P. C., B. Levin, R. D. Robinson, and M. Rubin. 1964. U.S. Geological Survey Radiocarbon Dates VII. *Radiocarbon* 6: 37–76.

Johnson, C. 1974. Geologic Investigations at the Lubbock Lake Site *in* C. Black (ed.), History and Prehistory of the Lubbock Lake Site. *The Museum Journal* 15:70–105.

Johnson, E. 1974. Zooarchaeology and the Lubbock Lake Site *in* C. Black (ed.), History and Prehistory of the Lubbock Lake Site. *The Museum Journal* 15:107–122.

Kapp, R. O. 1970. A 24,000 Year Old Jefferson Mammoth from Midland County, Michigan. *Michigan Academician* 3:95–99.

Kelley, L., E. Spiker, and M. Rubin. 1978. U.S. Geological Survey, Virginia, Radiocarbon Dates XIV. *Radiocarbon* 20:283–312.

Kurtén, B., and E. Anderson. 1980. *Pleistocene Mammals of North America*. Columbia University Press. 442 pp.

Laury, R. L. 1980. Paleoenvironment of a Late Quaternary Mammoth-bearing Sinkhole Deposit, Hot Springs, South Dakota. *Geological Society of America Bulletin* 91:465–475.

Laury, R. L., and L. D. Agenbroad. 1980. The Hot Springs Site: A Unique Late Quaternary Mammoth Trap in South Dakota. *American Quaternary Association Abstracts and Program* 6:125.

Leonhardy, F. C. 1966. Domebo: A Paleo-Indian Mammoth Kill in the Prairie-Plains. *Contribution of the Museum of the Great Plains* 1:1–53.

Levin, B., P. C. Ives, C. L. Oman, and M. Rubin 1965. U.S. Geological Survey Radiocarbon Dates VIII. *Radiocarbon.* 7: 372–398.

Libby, W. F. 1955. *Radiocarbon Dating*. 2nd ed. University of Chicago Press. 175 pp.

Libby, W. F., and G. J. Fergusson. 1962. UCLA Radiocarbon Dates I. *Radiocarbon* 4:109–114.

Lorenzo, J. L. 1978. Early Man Research in the American Hemisphere: Appraisal and

Perspectives *in* Bryan (ed.), Early Man in America from a Circum-Pacific Perspective. *Occasional Papers of the Department of Anthropology, University of Alberta* 1:1–9.

Lull, R. S. 1921. Fauna of the Dallas Sand Pits. *American Journal of Science* 2:159–176.

Lundelius, E. L., and B. Davidson. 1975. Late Pleistocene Vertebrates from Laubach Cave, Texas. *Abstracts with Programs, South Central Section, Geological Society of America* 7:211–212.

McCallum, K. J., and J. Wittenberg. 1968. University of Saskatchewan Radiocarbon Dates V. *Radiocarbon* 10:369.

Madden, Cary T. 1981. Mammoths of North America, Ph.D. dissertation, University of Colorado, Boulder, 271 pp.

Madsen, D. B., D. R. Currey, and J. H. Madsen. 1976. Man, Mammoth, and Lake Fluctuations in Utah. *Utah State Historical Society Antiquities Section Selected Papers* 5:1–58.

Maglio, V. J. 1973. Origin and Evolution of the Elephantidae. *Transactions of the American Philosophical Society: New Series* 63:1–149.

Malde, H. E., and H. A. Powers. 1962. Upper Cenozoic Stratigraphy of Western Snake River Plain, Idaho. *Geological Society of America Bulletin* 73:1197–1120.

Marsters, B., E. Spiker, and M. Rubin. 1969. U.S. Geological Survey Radiocarbon Dates X. *Radiocarbon* 11:210–227.

Martin, L. D., and B. M. Gilbert. 1978. Excavations at Natural Trap Cave. *Transactions of the Nebraska Academy of Sciences* 6:107–116.

Martin, P. S. 1973. The Discovery of America. *Science* 179:969–974.

———. 1974. Palaeolithic Players on the American Stage: Man's Impact on the Late Pleistocene Magafauna *in* Ives and Barry (eds.), *Arctic and Alpine Environments*. Metheun & Co., Ltd., London.

Matthews, J. V., Jr. 1976. Arctic-steppe—An Extinct Biome. *Abstracts of the Fourth Biennial Meeting American Quaternary Association*, Tempe, Arizona, pp. 73–77.

Mawby, J. E. 1967. Fossil Vertebrates of the Tule Spring Site, Nevada *in* Wormington and Ellis (eds.), Pleistocene Studies in Southern Nevada. *Nevada State Museum Anthropology Papers* 13:105–129.

Mehl, M. G. 1962. Missouri's Ice Age animals. *Educ. Ser. Missouri Geol. Surv. Water Resources* 1:1–104.

Miller, S. J., and W. Dort, Jr. 1978. Early Man at Owl Cave: Current Investigations at the Wasden Site, Eastern Snake River Plain, Idaho *in* A. Bryan (ed.), Early Man in America from a Circum-Pacific Perspective. *Occasional Papers of the Department of Anthropology, University of Alberta* 1: 129–139.

Miller, W. E. 1971. Pleistocene Vertebrates of the Los Angeles Basin and Vicinity (Exclusive of Rancho La Brea). *Bulletin of the Los Angeles County Museum of Natural History* 10.

Mirambell, L. 1978. Tlapacoya: A Late Pleistocene Site in Central Mexico *in* A. Bryan (ed.), Early Man in America from a Circum-Pacific Perspective. *Occasional Papers of the Department of Anthropology, University of Alberta* 1:221–230.

Mooser, O., and W. W. Dalquest. 1975. Pleistocene Mammals from Aguascalientes, Central Mexico. *Journal of Mammology* 56:781–820.

Orr, P. C. 1956. Radiocarbon Dates from Santa Rosa Island, I. *Santa Barbara Museum of Natural History Bulletin* 2:1–10.

Osborn, H. F. 1922. Species of American Pleistocene Mammoths, *Elephas jeffersonii*, New Species. American Museum Novitiates 41:1–16.

———. 1942. *Proboscidea: A Monograph of the Diversity, Evolution, Migration and Extinction of the Mastodons and Elephants of the World.* American Museum of Natural History, New York. 1630 pp.

Péwé, T. L. 1975. Quaternary Stratigraphic Nomenclature in Unglaciated Central Alaska. *U.S. Geological Survey Professional Paper* 862:1–32.

Reeburgh, W. S., and M. S. Young. 1976. University of Alaska Radiocarbon Dates I. *Radiocarbon* 18:1–15.

Rutherford, A. A., J. Wittenberg, and K. J. McCallum. 1973. University of Saskatchewan Radiocarbon Dates VI. *Radiocarbon* 15:193–211.

Saunders, J. J. 1970. The Distribution and Taxonomy of *Mammuthus* in Arizona. M. S. thesis, University of Arizona. 115 pp.

Schultz, C. B. 1968. The Stratigraphic Distribution of Vertebrate Fossils in Quaternary Eolian Deposits in the Mid-Continent Region of North America. *in* Schultz and Frye (eds.), Loess and Related Eolian Deposits of the World. *Proceedings of the 7th INQUA Congress* 115–138.

Schultz, C. B., L. D. Martin, L. G. Tanner, and R. G. Corner. 1978. Provincial Land Mammal Ages for the North American Quaternary. *Transactions of the Nebraska Academy of Sciences* 5:59–64.

Scott, G. R. 1963. Quaternary Geology and Geomorphic History of the Kassler Quadrangle,

Colorado. *U.S. Geological Survey Professional Paper.* 421-A:1–67.

Sheppard J. C., and R. M. Chatters. 1976. Washington State University Natural Radiocarbon Measurements II. *Radiocarbon* 18:140–149.

Skeels, M. A. 1962. The Mastodons and Mammoths of Michigan. *Papers of the Michigan Academy of Science, Arts, and Letters* 47:101–133.

Slaughter, B. H. 1960. A New Species of Smilodon from a Late Pleistocene Alluvial Terrace Deposit on the Trinity River. *Journal of Paleontology* 34:486–493.

Stanford, D. 1979. The Selby and Dutton Sites: Evidence for a Possible Pre-Clovis Occupation of the High Plains *in* Humphrey and Stanford (eds.), *Pre-Llano Cultures of the Americas: Paradoxes and Possibilities.* Anthropological Society of Washington, Washington, D.C., pp. 101–123.

Stanford, D., W. R. Wedel, and G. R. Scott. 1981. Archaeological Investigations of the Lamb Spring Site. *Southwestern Lore* 47:14–27.

Stirton, R. A., and W. K. Gealey. 1949. Reconnaissance Geology and Vertebrate Paleontology of El Salvador, Central America. *Geological Society of America Bulletin* 60:1731–1754.

Stuckenrath, R., and J. E. Mielke. 1970.

Smithsonian Institution Radiocarbon Dates VI. *Radiocarbon* 12:193–204.

Thorson, R. M., and T. D. Hamilton. 1977. Geology of the Dry Creek Site: A Stratified Early Man Site in Interior Alaska. *Quaternary Research* 7:149–176.

Thorson, R. M., E. J. Dixon, Jr., G. S. Smith, and A. R. Batten. 1981. Interstadial Proboscidean from South-Central Alaska: Implications for Biogeography, Geology and Archeology. *Quaternary Research* 16:404–417.

Tucek, S. 1943. Perennially Frozen Ground in Alaska. Its Origin and History. *Geological Society of America Bulletin* 54:1433–1548.

Valastro, S., Jr., E. M. Davis, and A. G. Varela. 1979. University of Texas Radiocarbon Dates XIII. *Radiocarbon* 21:257–273.

Weber, F. R., T. D. Hamilton, D. M. Hopkins, C. A. Repenning, and H. Hass. 1981. Canyon Creek: A Late Pleistocene Vertebrate Locality in Interior Alaska. *Quaternary Research* 16:167–180.

West, R. M., and J. E. Dallman. 1980. Late Pleistocene and Holocene Vertebrate Fossil Record of Wisconsin. *Geoscience Wisconsin* 4:25–45.

Whitmore, F. C., Jr., K. O. Emery, H. B. S. Cooke, and D. J. P. Swift, 1967. Elephant Teeth from the Atlantic Continental Shelf. *Science* 156:1477–1481.

A Close Look
at Significant Sites

A Close Look
at Significant Sites

Ice Age fossils form intriguing deposits that may be found in remarkable places. The five localities reported here are all quite extraordinary—even for the Pleistocene, for which the extraordinary may seem commonplace. Agenbroad's record of thirty mammoth in an ancient warm spring brings to mind an uninvited party trespassing in a private hot tub. The elephantine frolickers turned serious and then trumpeted frantically as they gradually realized that the walls of the natural "tub" were too high and too slippery to climb.

The fossil mammoth site Agenbroad portrays allows us to picture how death may have come to some of America's largest native herbivores. Experienced in archaeology of early man sites, Agenbroad sought evidence of stone artifacts. None were found in situ with the splendidly preserved mammoth bones of Wisconsin glacial age. Agenbroad attributes spiral fractures of bones present in the deposit to breakage before mineralization, perhaps the result of a freshly entrapped mammoth trampling the carcass of an earlier victim. He discusses the problem of identifying an archaeological site by bone breakage alone. If prehistoric people were in America when this site was active, why didn't they use it for their own purposes? Appreciating its value, residents of Hot Springs, South Dakota, are endeavoring to protect it within a permanent building where researchers and the public can share their interest in the site.

Gruhn and Bryan write of another spring deposit also attractive to proboscideans, in this case mastodonts, once inhabiting what is presently the cactus-clad coast of northern Venezuela. Again it seems that water, and perhaps mineralized earth, was the attraction. The cause of mortality may not be natural. While the presence of bones of large mammals in springs is not unusual, the association with artifacts is of great interest. If sound—and Gruhn and Bryan are quite convincing on that point—the dating of their site at slightly but significantly over 11,000 years suggests the oldest evidence yet for hunters of "elephants" in America. Paleontologists will rightly insist that a mastodont is not an elephant any more than a camel is a giraffe. Both are in different families. However, the mastodonts had tusks, trunks, ponderous feet, and an elephantine build. The fact that they were hunted at all will seem remarkable enough to some. The thought that the fossil record of South America is much richer in evidence of early archaeological associations than many have believed is indeed provocative, as Gruhn and Bryan well realize. Have the earliest hunters been overlooked in North America? Or did the hunters somehow reach South America first? The continent invites much more work on its extinct fauna and possible archaeological associations.

A different fate was in store for cursorial animals in northern Wyoming. Twenty-four meters below the mouth of a bell-shaped limestone cavern a variety of bones are all that is left of large Pleistocene herbivores and carnivores that fell to their doom. Did big cats in hot pursuit across snowy terrain tumble in as both they and their equine prey failed to swerve away in time to avoid the aptly named "Natural Trap"? Did the trap have some special lure of its own that led curious animals moving slowly and suspiciously ever closer to the edge to be swallowed up? Also entombed in this natural catacomb were mammoth, antelope, and mountain sheep of unusual size. From the bones, Gilbert and L. Martin infer the nature of southern Wyoming's fauna at the height of the last Ice Age and find support for a climatic model of megafaunal extinction.

Five hundred feet above the Colorado River in the western end of the Grand Canyon just inside the Grand Wash Cliffs is another remarkable catacomb. Rampart Cave is a horizontal cave, and the bones it contained, mainly of ground sloth, were embedded in an even more unusual deposit, the dry dung of the animal. According to radiocarbon dates on dung, Shasta ground sloth extinction took place close to 11,000 years ago. From middens of plant material left by packrats, Phillips describes the displaced and ecologically anomalous woodland found in this part of the canyon when the ground sloths were last alive. He discovered that the first known collector of sloth dung was not a human but a rodent, a packrat around 9,000 years ago who added a bolus of sloth dung to its nest, inadvertently providing an unbiased sample for Phillips to date. The age was in accord with the model of extinction by 11,000 B.P., based on other samples found in situ. According to Phillips, the sloths disappeared at a time when the natural vegetation would seem to have favored their continued existence.

Finally, at the famous Rancho La Brea tar pits, whose fauna is superbly displayed at the Page Museum in Hancock Park, Los Angeles, Marcus and Berger summarize two decades of radiocarbon dating. The tar-impregnated bones were originally held to be hopelessly contaminated and thus unsuitable for radiocarbon analysis. Subsequent methods of extracting the tar revealed that this natural impregnation had preserved bone amino acids in sufficient quantities for chromatographic separation and dating. Thus what at first appeared a handicap turned out to be an asset. Under ordinary circumstances buried bone yields very little amino acid residue, too little for routine ^{14}C dating. Under extraordinary circumstances at Rancho La Brea, amino acids were better preserved, and more than 100 ^{14}C dates, many on amino acids, trace the tar-pit fauna from 40,000 to around 11,000 years ago. There were no signs of reduction in diversity or numbers predating extinction, and no evidence of human bones in association with the extinct giant fauna.

While few Pleistocene bone beds are as spectacular as these, all late Pleistocene deposits have the potential for yielding new clues regarding the cause or causes of the extinctions which the bone beds predate. Long neglected because the study of the late Pleistocene megafauna was regarded as an unpromising source of novelty for taxonomically or stratigraphically minded paleontologists, the old bone quarries as well as new ones are being shown to yield far more new and useful information than had been imagined a generation ago. The following five chapters illustrate what paleoecologists may discover in the late Pleistocene, a lost world that inevitably challenges their investigative skills and provokes their efforts at explaining what happened to the big mammals.

4

Hot Springs, South Dakota
Entrapment and Taphonomy of Columbian Mammoth

LARRY D. AGENBROAD

DISCOVERY OF THE HOT SPRINGS MAMMOTH SITE was accidental. During earth-moving operations of the initial phase of construction for a housing development on the south side of Hot Springs, South Dakota, bulldozers and graders exposed teeth, tusks, skulls, and postcranial skeletal elements of mammoth.

Two weeks of salvage excavations and stabilization were carried out in 1974. Testing in the summer of 1975 proved the site to be paleontologically significant, and formal excavations were undertaken during the summers of 1976, 1977, 1978, and 1979. In those field seasons the general extent, content, and make-up of the site and its geologic history were outlined and refined. It soon became evident that the deposit represents a unique trap for a large population of late Pleistocene mammoth.

Geology of the Site

Location and Physiographic Setting

The Hot Springs Mammoth Site is located within the city limits of Hot Springs, South Dakota. An extension of existing streets would place the locality near the intersection of Nineteenth Street and Evanston Avenue, in the southern portion of the city.

Located in Red Valley, a topographic low eroded in the Spearfish Shale, which encircles the Black Hills interior, the site is a local topographic high. The small hill on which it is located is due, in fact, to the relative resistance to erosion of the sediments containing the fauna, as opposed to the less resistant Spearfish Shale. The site remains as an example of reversed topographic expression, the sedimentary fill of a former sinkhole having become the top of a small hill by differential erosion. To the east and southeast, prominent ridges of resistant sandstones are exposed. To the north, northwest, and southwest, hills formed of limestones and older sedimentary units rise in response to the anticlinal structure of the Black Hills (fig. 4.1).

Origin

Solution and removal of gypsum and anhydrite from underlying formations, particularly the Minnelusa Formation, resulted in post-solution collapse. The collapse caused breccia pipes to be formed above the collapse areas, and such pipes extend

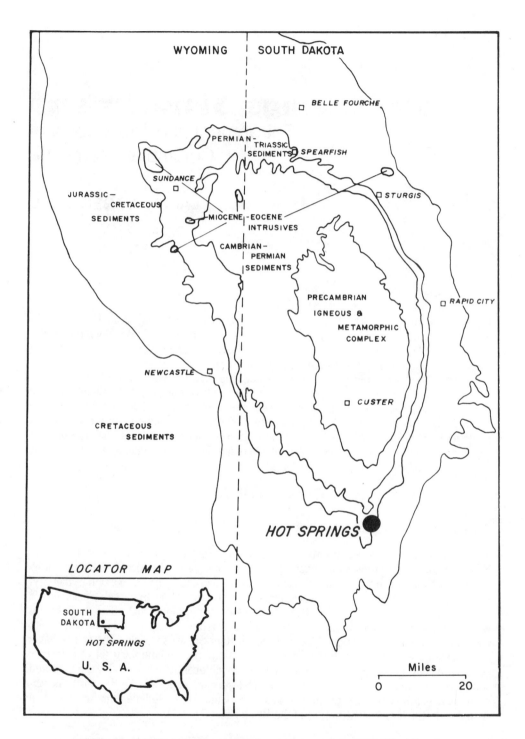

Figure 4.1. Simplified location map and geologic map of the Black Hills, Wyoming and South Dakota. The mammoth site is located within the city limits of Hot Springs, South Dakota.

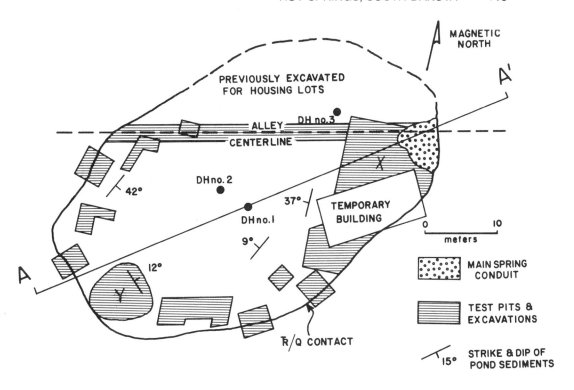

Figure 4.2. Map of the sinkhole at Hot Springs, South Dakota. Shaded areas are test pits and excavations made during the 1974–79 field seasons. Major excavations were concentrated in the conduit area and alley right-of-way. The temporary building houses in situ exhibits. DH = drill hole, Ŕ = Triassic, Q = Quaternary, X = area of disarticulated bone in high-energy environment, Y = area of articulated and semiarticulated bone in low-energy environment.

upward to the modern surface, penetrating as much as 300 meters (Laury 1980) of overlying sediments. The Hot Springs Mammoth Site was formed in one of these breccia pipes (fig. 4.2).

In addition to the physical formation of the breccia pipe and resulting karst feature, a critical factor in producing an animal trap from what would otherwise be just another sinkhole was the presence of an artesian spring. Groundwater in the Minnelusa Formation used the conduit made by the breccia pipe and formed the spring. Spring effluent created a standing body of water within the steep-walled sink. The water was warm; using biological and sedimentary evidence, Laury (1980) estimated it to be 35°C (95°F). The pool of water served as the attraction for mammoth, trapping a large number of individuals whose remains are incorporated in the sediments that ultimately filled the sinkhole.

Artesian and thermal springs are not uncommon in the Black Hills Uplift (Rahn and Gries 1973). The snowpack and runoff are able to infiltrate into limestones, dolomites, and sandstones which are dipping from the uplifted crystalline core. The hydraulic gradient of these aquifers and confining beds produces sufficient hydrostatic pressure to generate artesian springs wherever structural features such as faults or breccia pipes allow water to penetrate overlying formations.

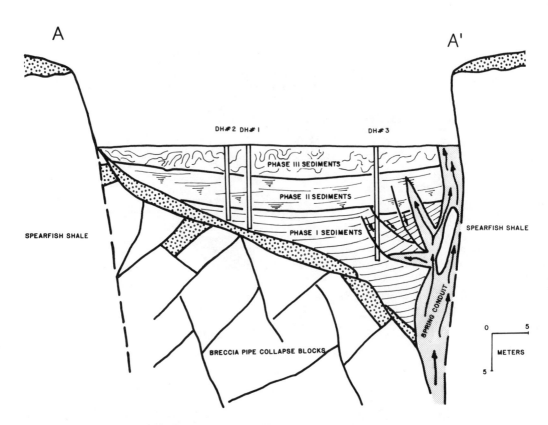

Figure 4.3. Diagrammatic cross-section A–A′ (see fig. 4.2) of the Hot Springs Mammoth Site, based on drill holes and site excavations. DH = drill hole.

Sedimentary History

Following sinkhole collapse, filling began, and three subaqueous, sedimentary episodes (fig. 4.3) can be recognized (Laury 1980). Phase I represents the initial sedimentary sequence of relatively coarse-grained deposits. Phase II is an intermediate unit of laminated, fine-grained sediments. Phase III is the terminal infilling by subaqueous and colluvial deposits.

Phase I sediments reflect gravels, probably derived from surficial gravels of the fourth terrace above Fall River (Kempton 1980) shortly after the collapse of the breccia pipe. Fall River is a spring-generated stream which originates just west of Hot Springs, flows through the city, and joins the Cheyenne River approximately ten miles southeast of Hot Springs. Remnants of this gravel surface are still preserved on partially modified land immediately adjacent to the Hot Springs site. Contemporaneous deposits of rhythmic laminations of clayey silt and silty sand were being deposited in the central portions of the sink. Subaqueous, sediment gravity flows and microfaulting are common phenomena in Phase I sediments (Laury 1980).

Sediments in Phase II are thin laminae of sand and silts, with very few large clasts other than mammoth bone. Faunal remains and tracks occur in all phases of the sedimentary fill. Based on the sequence of mammoth tracks (fig. 4.4) within the deposits, Laury (1980) suggests that the water level was approximately 15 feet (5 m) deep during much of the pond history, at least in the first two phases.

Figure 4.4. A vertical section of a mammoth footprint preserved in the laminated Phase II pond-fill sediments of the sinkhole. Contorted Phase III sediments can be observed above the datum line. The track has been outlined in ink for clarity; the arrow indicates bone exposed at a lower elevation in the wall of the excavation trench. (Photograph by R. Laury)

Phase III sediments suggest weak spring effluent and final desiccation of the deposit. These sediments are finer in grain size and highly bioturbated. It is probable that the spring effluent had been diverted to a lower elevation due to downcutting of Fall River, allowing desiccation of the sinkhole and ending its effectiveness as a mammoth trap.

Excavation of the Site

Methodology

Excavation of the site followed standard archaeological and paleontological techniques as applied at Murray Springs (Hemmings 1970) and Lehner Ranch (Haury et al. 1959), Arizona; Boney Springs, Missouri (Saunders 1977); and Hudson-Meng, Nebraska (Agenbroad 1978). Bones were mapped in situ, both vertically and horizontally, as encountered in the fill. Horizontal provenance was obtained with a string grid and transferred to metric graph paper. A vertical datum gave levels on individual bones. A level was taken on the central portion of the bone except in instances in which one end was vertically much higher or lower.

Remains were mapped, field-numbered, and stabilized in place, or removed and taken to the laboratory for further stabilization, reconstruction, and identification. A portion of the remains was left in situ after excavation, for subsequent development of the site. By 1982 community leaders in the city of Hot Springs were engaged in a major effort at permanent preservation and public interpretation of the mammoth site.

In addition to hand tools, a backhoe/front-end loader was used in trenching, removal of overburden, and backfill operations. A contracted power shovel gave us greater depth penetration for trenches cut for stratigraphic information. The South Dakota Geological Survey provided a truck-mounted drill rig to give us three test holes in the fall of 1978. The information from those holes provided data on the total depth and configuration of the sedimentary fill within the karst depression.

Fauna

Two weeks of salvage paleontological excavations were conducted at the site in the initial season. These efforts were undertaken to assess the extent of the deposit and its paleontological significance. The results of the initial efforts indicated that at least four to six individual mammoth were represented in the area of construction disturbance. The mammoth remains were considered to be *Mammuthus columbi*. This identification was verified by a graph of the Index of Hypsodonty (Cooke 1947, 1960) to the length-lamella ratio of the first molars (Dutrow 1977). Associated remains indicated the presence of at least one specimen each of bear, coyote, camel, and peccary, and one large, unidentified raptor.

Initially, bones were exposed in two concentrations, labeled X and Y in Figure 4.2. Subsequent testing proved mammoth bone to be present in every area of the site, especially at the periphery of the deposit, within 6 meters (laterally) of the contact with Spearfish Shale.

Area X contained a dense concentration of disarticulated bone in a sandy matrix, suggesting a high-energy environment such as spring conduit. Area Y contained articulated and semiarticulated remains in laminated silts and silty clay, suggesting a low-energy, pond environment.

Each season saw an increase in the inventory of individual mammoth virtually unaccompanied by other species of megafauna. Thirty mammoth had been found by the end of the 1979 field season. Invertebrates were recovered, such as ostracods, gastropods, and pelecypods (Mead 1978). Microvertebrate remains were recovered in screen wash of the spring conduit excavations (Bjork 1978). Pollen and macroflora were sought without success.

The detailed mapping of vertical walls of the trenches and excavations allowed analysis and interpretations of the stratigraphy and sedimentation of the sinkhole fill deposits (Laury 1980). This study verified the model proposed in 1976, that is, an artesian spring in the northeast corner of the deposit, which fed a pond environment on the west and southwest of the sinkhole. One unexpected result was the discovery of mammoth tracks preserved (in vertical section) in the sediments. More than twenty such features (fig. 4.4) were recognized in the excavation walls and trenches. These tracks were present in all levels of the sedimentary fill.

Taphonomy of the Site

The Hot Springs Mammoth Site represents a unique combination of spring and lacustrine environments which formed as a result of the special conditions provided by a karst feature in a particular geologic environment. The Hot Springs assemblage represents a thanatocoenosis (death assemblage) of mammoth in the unusual circumstance of an untransported or minimally transported accumulation, in a springfed pond within a karst depression.

Death, decomposition, disarticulation, minor downslope, subaqueous transport, and deposition took place within the confines of a small bowl-like depression with no transport mechanism available except those which occurred within that depression. Those processes were artesian spring discharge, overbank flow, and subaqueous gravity movements. One other factor that could account for emplacement of some elements of the assemblage would be bloat and floating of decomposing carcasses.

Data for the interpretation of the assemblage at this site are derived primarily from the detailed sedimentation and stratigraphic studies of Laury (1980) and the biological/paleontological evidence in the form of distribution, abundance, and degree of disarticulation represented in the partial recovery of the bone bed by excavation.

Origin of the Hot Springs Thanatocoenosis

As Saunders (1977) has stated, "Taphonomy begins at death, and the first event to be considered in formational analysis is the cause of death." In the case of the Hot Springs mammoth, death came about in one of two ways: by drowning or by starvation.

An explanation is called for in both cases, as elephants in general are excellent swimmers. The special circumstances present at Hot Springs were in the form and nature of the trap. The wall rock of the sinkhole is the Spearfish Formation, a red, sandy shale of Permo-Triassic age, from 250 to 700 feet thick (Rahn and Gries 1973). The contact between the Spearfish and the sinkhole fill materials was observed to be no less than 60°; in some cases, such as near the conduit, the walls were overhanging. It has been estimated by various evidence that the depth of water in the sinkhole was 4 to 5 meters (Laury 1980), at least in early stages of the history of the pond. A mammoth, attracted into the sinkhole by water or bankside vegetation, was trapped on entering the pond, whether the entry was made purposely or accidentally, such as by sliding the last few feet on submoistened Spearfish Shale. Once immersed, the mammoth had little or no hope of extricating itself because of the steepness of the walls and the slipperiness of the submoistened shale. The doomed animal could only swim until exhausted or position itself near the wall on whatever slump block, shelf, or sediment allowed foothold. If drowning did not claim the victim, starvation ultimately would.

This model of entrapment is supported by the distribution and condition of the faunal remains. It became apparent in the initial seasons of excavation that most of the bones were to be found within 6 meters of the Spearfish wall rock. Conversely, the central portions of the pond fill seemed to be relatively devoid of bone, as indicated by backhoe trenching and test pits dug for stratigraphic and sedimentation information and mapping. In addition, the fact that most of the bone was partially disarticulated argues for pond-side decomposition of the carcass and disarticulation of the skeleton. The bones became units within the sedimentary sequence by sliding and rolling down subaqueous sedimentary slopes, until they reached their final resting place and were covered by the sediments.

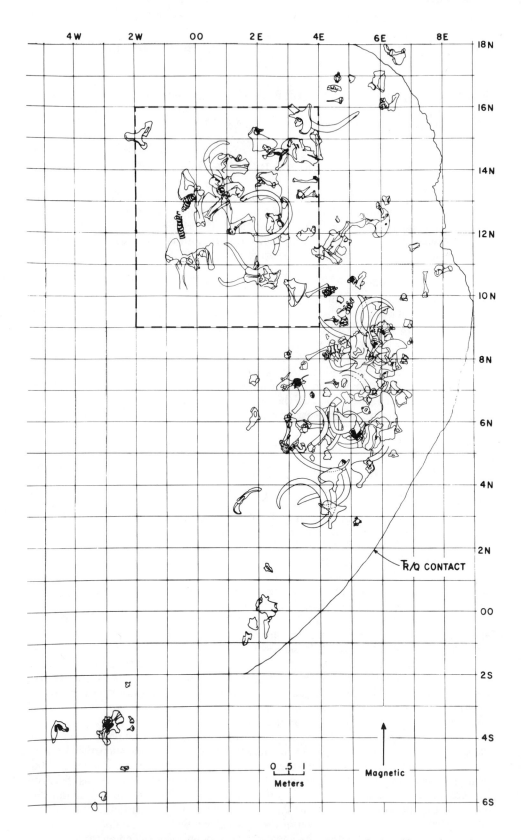

Figure 4.5. Bone map of the east edge of the sinkhole at the Hot Springs Mammoth Site showing areas of articulated bone (surrounded by dashed line) and disarticulated bone near spring conduit.

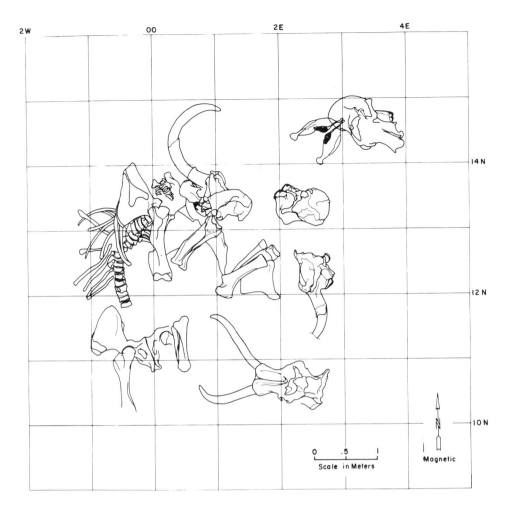

Figure 4.6. Detail of Figure 4.5 showing outlined area of articulated bone (small bones, isolated tusks, etc., have been deleted). Articulated specimen not completely excavated. (Map by B. Dutrow)

Alternatively, a case can be made for partial and completely articulated skeletal units being incorporated as units. The discovery of several articulated feet suggests deposition and burial (some in deep-water portions of the pond) as complete skeletal components. Presumably ligaments and other tissue held these small bones in articulation until after burial. In the final week of the 1979 field season a completely articulated skeleton of a mature mammoth was discovered. The animal was in the deepest excavation area of the site (excluding drill holes and the deep trench of 1979). The skeleton was just west of the major spring conduit (fig. 4.5) and had apparently sunk in that location after having been afloat in the pond. The position of the skeleton (fig. 4.6) suggests that it sank on its back, with the skull rotating 160° to 180° due to the weight of the tusks. Apparently the carcass sank in this location prior to final decomposition since small bones, such as hyoids, were still in anatomical position, and a gland stone (bile stone) was found in the approximate location of the organs in which it was produced.

Test pits along the periphery of the pond fill exposed other partially articulated specimens. These consisted mainly of vertebral columns and, in some cases, portions of the appendicular skeleton.

Thirty-one mastodonts *(Mammut americanum)* at Boney Springs, Missouri, died because of drought and nutritional deficiency, according to Saunders (1977). Based on Kurtén's population work (1953) and the age structure of the population, Saunders concludes that there was a mass mortality.

At Hot Springs the trapped animals represent young adults and mature animals (Dutrow 1980). Scarcity of juveniles and aged animals argues against a catastrophic or mass accumulation of a single herd of gregarious mammoth (Voorhies 1969). The sedimentary evidence of superimposed mammoth tracks, preserved in vertical section of excavation walls, indicates continuing entrapment of occasional animals. Bones are also incorporated throughout the sedimentary sequence, which may represent several hundred years in duration.

Predators apparently played a very minor part in bone disarticulation. Although coyote, bear, and raptor remains have been recovered from the mammoth site, the minimum number of individuals represents only one animal in each case. It is quite possible that these remains reflect the chance accumulation of carnivore and scavenger elements as a result of overbank flow, rather than an accumulation of animals actually present in the sinkhole. It is equally possible that the decomposing carcasses did attract scavengers. The mammoth bones have not, as yet, produced evidence of carnivore action such as gnawing.

Bone from the site does give other evidence of postmortem or death processes which have been interpreted in various ways by other investigators. A small but conspicuous percentage of bones displays spiral fractures—usually considered to be "green bone" breaks. It should also be noted that there is an absence of evidence of subaerial bone decomposition and weathering; exceptions are the cases in which bones were exposed during the erosion that postdated the destruction of the trap, after it ceased to function due to infilling by sedimentary processes and the decline of hydrologic activity.

At other late Pleistocene sites Irving and Harington (1973), Stanford (1979), and others have interpreted the presence of green bone breaks to be representative of human activity. Human butchering of Hot Springs mammoth is very unlikely. Bonnichsen (1979) has undertaken an extensive experimental study on the fracturing of bone and has examined detailed photographs of spiral-fractured, long-bone fragments from the Hot Springs site (figs. 4.7–4.11). His conclusion concurs with my own: the bones are spirally fractured, indicative of the breaking of fresh bone, or of bone broken shortly postmortem. The processes that would provide such breakage are limited to only two, exclusive of human activity: 1) torsional stress, as caused by trying to extricate a limb mired in mud, muck, or quicksand—stress possibly even enhanced by an accompanying accidental fall; or 2) the possibility of trampling of recently deceased animals by newly

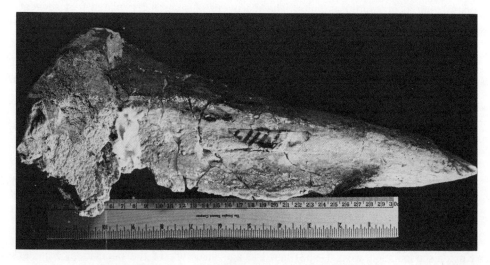

Figure 4.7. Spiral fracture of a mammoth scapula fragment.

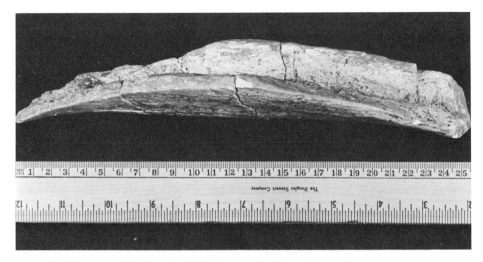

Figure 4.8. Spiral fracture of a mammoth long-bone fragment.

Figure 4.9. Spiral fracture of an unidentified mammoth bone.

entrapped individuals. The intricate intermixing of bones, as evidenced in their excavation, suggests abundant opportunity for limbs to provide "anvil" associations which could have been stepped on by living animals—with the mass of an elephant—to provide sufficient force to cause spiral fractures. Sediment-loading is considered to be inconsequential as a factor in production of spiral fractures in this deposit.

Additional fracturing of bone was noted in several locations within the excavated portions of the bone bed. These fractures are a consequence of displacement of units of the fill materials due to compaction and desiccation upon the cessation of the hydrologic activity of the site or other post-depositional movements. Small fracture zones with up

Figure 4.10. Spiral fracture of a mammoth humerus.

Figure 4.11. Detail of the termination of the spiral fracture in Figure 4.9.

to several centimeters of relative displacement were observed to cut across bone accumulations adjacent to the spring conduit. In one case, a skull was fractured and the opposing elements displaced slightly. Several pelves and long bones, as well as tusks, displayed similar breakage in an adjacent area.

Additional postmortem agents in the distribution of faunal elements are at least twofold. The absence of long bones in sufficient quantity to match crania, scapulae, and pelvic bones suggests that these elements were concentrating in other portions of the deposit. As a field hypothesis, we coined an "index of rollability," suggesting that long bones such as femora, radii, ulane, and humeri, being rather cylindrical in shape, might roll on the steep, subaqueous slopes to come to rest in deeper, unexcavated portions of the deposit. This field hypothesis was partly substantiated in the 1979 field season by the discovery of increasing numbers of long bones in the bone concentrations of the deep portions of the excavation. Conversely, ribs, scapulae, pelves, and so forth, which have a low pivotability were found throughout the excavations. Skull emplacement follows a different process. Though weighted by molars and ivory, the skull is buoyed by the cancellous material of the maxilla, nasal regions, and even the cranial vault. Visual evidence of a floating proboscidean cranium was provided on the 1979 Friends of the Pleistocene field trip to Agate Basin, Wyoming. George Frison and his crew were mascerating the skull of a recently deceased zoo elephant in a stock pond near the Agate Basin archaeological site. The skull floated in the pond and moved about in response to wind direction. Even with the attachment of weights and forced immersion, the skull continued to float in the pond. Such a mechansim could account for the abundance of skulls located in the deep excavation of 1979 (fig. 4.5).

Chronology

Faunal remains include the bear and coyote, which are extant, and extinct Pleistocene camel and peccary. Both are typical of late Pleistocene deposits—and both became extinct at the end of the Wisconsin glacial along with the mammoth.

Several unsuccessful attempts were made to obtain sufficient bone collagen from scrap mammoth bone for radiocarbon dating. Ultimately, bone apatite was resorted to, with a date of $21,000 \pm 700$ yr B.P. (Gx–5356–A) from scrap bone throughout the excavation to that time. During the following field season, expendible bone from a stratigraphic horizon within the excavation was accumulated and submitted for a more controlled age determination. The date of $26,075 +975/-790$ yr B.P. (Gx–5895–A) also on bone apatite from a stratigraphic horizon approximately one half of the depth of the fill of the deposit gives a chronologic base. Bone apatite dates should be considered minimal.

Site Significance

The significance of the site is due to 1) its unique geological, hydrologic origin; 2) the fact that it was a mammoth-selective trap; 3) the large number of individuals represented by the excavation of approximately 15 percent of the deposit; 4) the accumulation of a large sample of local mammoth populations over a relatively short time span; 5) the contribution to Quaternary geology and paleontology of the Black Hills, South Dakota, the local geographic region, and to the continent as a whole; and, 6) a large, late Pleistocene mammoth death assemblage predating human predation.

In a sample of 856 mammoth localities (of at least 1,427) in the literature (Agenbroad, chap. 3, this volume), only five localities (0.5 percent) record mammoth remains in sinkholes. Friesenhahn Cave, Texas (Evans 1961, Graham 1976), is a vertical solution/collapse feature in limestone which contains evidence of several hundred immature elephants. The cave is considered to be a secondary accumulation representing the remains of young mammoth which were taken back to the den of the great cat, *Dinobastis*. Sellards (1938) reports at least five elephants associated with artifacts in shallow sinkholes from Roberts County, Texas. Other reported sinkhole occurrences are single-animal occurrences, with the exception of the Hot Springs site. In summary, Hot Springs is the only large-population, sinkhole-occurrence of mammoth independent of predator activity in North America.

Paleontologically, the Hot Springs site represents the largest single-locality, primary occurrence of mammoth remains in South Dakota and in the entire north-central United States. Penndennis, Kansas (Hay 1924), has a large population from a fluvial deposit. Similarly, Dallas, Texas, has numerous recorded mammoth from the alluvial terraces along the Trinity River (Slaughter 1960). Alluvial deposits imply secondary as well as primary accumulation of faunal remains. The Hot Springs locality ranks fifth in reported multiple-mammoth localities in the New World, and third outside Alaska.

Excavations to date exposed only 10 to 15 percent of the total deposit. The recovery of teeth and tusks indicates that at least thirty individuals are represented by the sample and suggests that two or three times that number remain to be discovered in the undisturbed fill. Age-structure analysis indicates a death population of subadult and mature animals. This information, plus the presence of mammoth remains and tracks in all levels of the stratified fill, indicates an autochthonous, attritional accumulation of mammoth throughout the effective history of this natural trap. The data suggest occasional rather than catastrophic entrapment, death, and deposition of individuals from the mammoth population in the southern Black Hills during the terminal Wisconsin.

The presence of spiral fractured bone in an environment allowing little or no transport of faunal elements indicates spiral fracturing of mammoth limb bones by living animals, as in torsional stress, or in the very early postmortem history of deceased animals. The site provides evidence of spiral fracturing in a strictly paleontological setting, with no evidence of human activity, at a time interval in which some investigators of other deposits are citing spiral fractures alone as evidence of human activity.

The absence of predator or scavenger tooth marks makes the deposit unusual. Accumulation of a large sample of mature to subadult mammoths in a karst feature is unique. The deposit represents a live-trap mechanism rather than a depositional accumulation of deceased individuals, such as an alluvial deposit in a fluvial system. No attributes of the deposit portend the coming total extinction of American mammoths at or near 11,000 years ago.

Acknowledgments

Many individuals have contributed to the excavation development and interpretation of the Hot Springs Mammoth Site. The sedimentology was done by R. Laury of Southern Methodist University (S.M.U.). Barbara Dutrow served as field foreperson and cartographer for the 1975–79 field work and compiled her M.S. thesis at S.M.U. on the postcranial attributes of the mammoth population. Pamela Kempton completed an M.S. thesis at S.M.U. on the geomorphic history of Fall River. Phil Bjork and others from South Dakota School of Mines and Technology were active in microfaunal analysis and support. James Mead, of the University of Arizona, has served as my assistant director and studied the microfauna and mollusks of the site since the initial efforts in 1974. Graphics were supplied, in part, by B. Dutrow and D. Meier. Funding was provided by the National Geographic Society, the Center for Field Research (Earthwatch), High Plains Center, and Geological Society of America. Indispensable support of every kind was provided by the Hot Springs Mammoth Site, Inc., as well as by other interested persons.

References

Agenbroad, L. D. 1978. *The Hudson-Meng Site: An Alberta Bison Kill in the Nebraska High Plains.* University Press of America. 230 pp.

Bjork, P. R. 1978. A Preliminary Report on the Microvertebrates of the Mammoth Site. *Guidebook and Roadlogs for Rocky Mountain-Plains Field Conference: Friends of the Pleistocene,* Hot Springs, South Dakota, pp. 37–38.

Bonnichsen, R. 1979. Pleistocene Bone Technology in the Beringian Refugium. National Museum of Man, Mercury Series, Archaeological Survey of Canada, Paper No. 89. Ottawa. 297 pp.

Cooke, H. B. S. 1947. Variation in the Molars of the Living African Elephant and a Critical Revision of the Fossil Proboscidea of South Africa. *American Journal of Science* 245:434–517.

————. 1960. Further Revision of the Fossil Elephantidae of Southern Africa. *Palaeontologia Africana* 8:46–58.

Dutrow, B. L. 1977. Preliminary Post-Cranial Metric Analysis of Mammoths from the Hot Springs Mammoth Site, South Dakota. *Transactions of the Nebraska Academy of Sciences* 4:27–31.

————. 1980. Metric Analysis of a Late Pleistocene Mammoth Assemblage, Hot Springs, South Dakota. M.S. thesis, Southern Methodist Univ., Dallas, Texas.

Evans, G. L. 1961. The Friesenhahn Cave. *Bulletin of the Texas Memorial Museum* 2:7–22.

Graham, R. W. 1976. Pleistocene and Holocene Mammals, Taphonomy and Paleoecology of the Friesenhahn Cave Local Fauna: Bexar County, Texas. Ph.D. dissertation, University of Texas, Austin.

Haury, E. W., E. B. Sayles, and W. W. Wasley. 1959. The Lehner Mammoth Site, Southeastern Arizona. *American Antiquity* 25 (1):2–34.

Hay, O. P. 1924. *The Pleistocene of the Middle Region of North America and Its Vertebrated Animals.* Carnegie Institute of Washington, D.C., 322A:385 pp.

Hemmings, E. T. 1970. Early Man in the San Pedro Valley, Arizona. Ph.D. dissertation, University of Arizona, Tucson.

Irving, W. N. and C. R. Harington. 1973. Upper Pleistocene Radio-carbon Dated Artefacts from the Northern Yukon. *Science* 179:335–340.

Kempton, P.D. 1980. Quaternary Terrace Development Along the Fall River, Hot Springs Area, South Dakota. M.S. thesis, Southern Methodist University, Dallas Texas.

Kurtén, B. 1953. On the Variation and Population Dynamics of Fossil and Recent Mammal Populations. *Acta Zoologica Fennica* 76: 1–122.

Laury, R. L. 1980. Paleoenvironment of a Late Quaternary Mammoth-Bearing Sinkhole Deposit, Hot Springs, South Dakota. *Geological Society of America Bulletin.* Part 1, Vol. 91, No. 8, pp. 465–475.

Mead, J. I. 1978. Freshwater Molluscs from the Hot Springs Mammoth Site, South Dakota. *Guidebook and Roadlogs for Rocky Mountain-Plains Field Conference: Friends of the Pleistocene,* Hot Springs, South Dakota, pp. 32–36.

Rahn, P. H., and J. P. Gries. 1973. Large Springs in the Black Hills, South Dakota and Wyoming. *South Dakota Geological Survey Report of Investigations.* No. 107, 46 pp.

Saunders, J. J. 1977. Late Pleistocene Vertebrates of the Western Ozarks Highland, Missouri. *Illinois State Museum Reports of Investigations,* No. 33, Springfield.

Sellards, E. H. 1938. Artifacts Associated with Fossil Elephants. *Bulletin of the Geological Society of America.* 49:999–1010.

Slaughter, B. H. 1960. A New Species of *Smilodon* from a Late Pleistocene Alluvial Terrace Deposit on the Trinity River. *Paleontology* 34:486–493.

Stanford, D. 1979. The Selby and Dutton Sites: Evidence for a Possible Pre-Clovis Occupation on the High Plains, *in* Humphrey, R. L., and Stanford, D., eds. *Pre-Llano Cultures of the Americas: Paradoxes and Possibilities.* Washington, D.C., Anthropological Society of Washington, pp. 101–123.

Voorhies, M. R. 1969. Taphonomy and Population Dynamics of an Early Pliocene Vertebrate Fauna, Knox County, Nebraska. *Contributions to Geology: Special Paper No. 1.* University of Wyoming, Laramie.

5

The Record of Pleistocene Megafaunal Extinction at Taima-taima, Northern Venezuela

RUTH GRUHN AND ALAN L. BRYAN

THE COASTAL ZONE of north-central Venezuela today is a hot, dry expanse of plains and lowlands covered with open scrub and cacti, the most notable fauna being herds of varicolored goats introduced with the Spanish conquest. In late Pleistocene times, however, the zone was populated by mastodonts, giant ground sloth, glyptodonts, horses, and early man.

Archaeological and paleontological research on the coastal plain of northern Venezuela has been conducted by J. M. Cruxent since the early 1950s (Cruxent and Rouse 1956). Abundant megafaunal remains of late Pleistocene age have been recovered from clayey sand deposits at two waterhole sites, Muaco (Royo y Gómez 1960, Cruxent 1961) and Taima-taima (Cruxent 1967, Bryan et al. 1978, Ochsenius and Gruhn n.d.); and from the gravels in an arroyo at Cucuruchú (Cruxent 1970). All three of these sites are located close to the present coastline and within several kilometers of each other, in a hilly region about 10 to 15 km east of the city of Coro (fig. 5.1). The most detailed stratigraphic work was carried out at Taima-taima in 1976 (Bryan et al. 1978, Ochsenius and Gruhn n.d.).

The site of Taima-taima is now a waterhole in a small basin at an elevation of approximately 23 meters above sea level. The waterhole drains by means of a small shallow stream which crosses several low bedrock shelves on its way to the sea, only about half a kilometer to the north. The waterhole is formed by an outcropping of a soft, aquiferous Miocene marine sandstone, through which the artesian water, derived from the base of the San Luís Mountains about 10 km to the south, seeps. The archaeological and paleontological evidence indicates that the waterhole was in existence in late Pleistocene times and served as an attraction for megafauna and man.

Stratigraphy and Paleoenvironments

Excavations for four seasons by Cruxent over an area of 150 square meters before 1976 exposed many bones, but unfortunately most were destroyed by vandals before identification and analysis. Although more limited in area (80 square meters), the 1976 excavations provided detailed evidence which made possible a refinement of the stratigraphic distribution of the megafaunal remains. Three separate faunal assemblages were determined and correlated with major stratigraphic horizons (fig. 5.2).

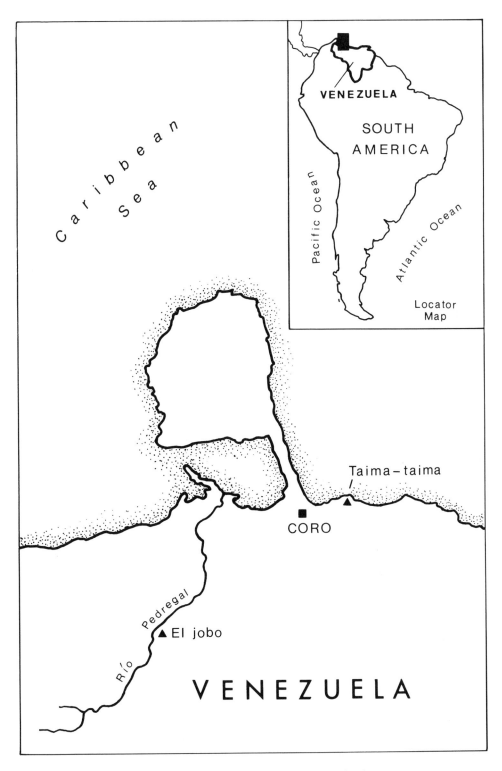

Figure 5.1. Map of north-central Venezuela, showing location of Taima-taima and El Jobo type site.

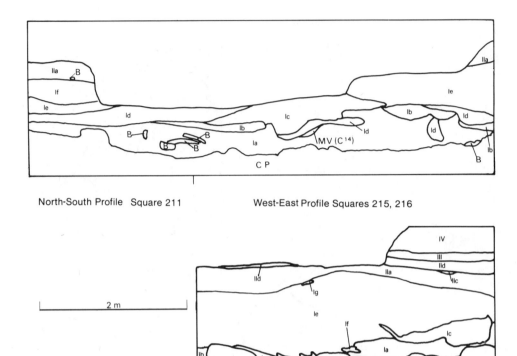

North-South Profile   Square 211   West-East Profile Squares 215, 216

North-South Profile (East Wall) Squares 216, 217

Figure 5.2. Description of stratigraphic features:

Unit I. Fine sand, often convoluted, reduced where saturated oxidized near top.

Ia. Convoluted dark grey clayey sand.
Ib. Light grey sand.
Ic. Brownish grey sand.
Id. White sand.
Ie. Yellowish to reddish brown sand.
If. Red sand.
Ig. Dark red sand.
B   Bone
CP   Cobble pavement and Miocene sand
MV   Masticated vegetation

Unit II. Laminated sand.
IIa. Yellowish red sand with reduction streaks, occasional pebbles.
IIc. Brick red sand (laterite).
IId. Yellowish white sand with oxidation streaks.

Unit III. Black organic clay.

Unit IV. Brown colluvial sand.

    The floor of the waterhole in the excavated area consisted of waterworn, limestone cobbles and pebbles tightly impacted in a pavement over the compact Miocene sand. The cobbles and pebbles were all derived from the same local formation, a fossiliferous Miocene limestone which outcrops on the sides of the basin: apparently blocks and fragments of this limestone were let down by erosion onto the floor of the basin and became impacted and rounded in the water hole. Numerous broken and worn bones of megafauna were exposed embedded in this cobble pavement, forming the first faunal horizon.

Overlying the cobble pavement was a deposit of grey clayey sand approximately 0.75–1.0 m in thickness. This zone, designated Unit 1, displayed local convoluting and particle-sorting due to the upward seepage of water. Its upper surface, above the present water table, was oxidized, but the lower third of the deposit was still saturated at the time of excavation. Organic material as well as bone was preserved by the continuously waterlogged condition of the lower part of Unit I. A stratigraphic series of fifteen radiocarbon dates on Unit I range between 12,580 ± 150 yr B.P. and 13,390 ± 130 yr B.P. (Bryan and Gruhn n.d.). In 1976 the butchered remains of a juvenile mastodont in association with an El Jobo projectile point and a utilized jasper flake were exposed near the base of Unit I, and four radiocarbon dates on a mass of sheared twigs believed to be the stomach contents of the mastodont indicate a date of 13,000 years ago for the kill (Bryan et al. 1978, Ochsenius and Gruhn n.d.). Bones and teeth of other megafauna forming the second faunal horizon were also recovered from the grey clayey sand of Unit I.

The top of Unit I is marked by a prominent disconformity indicating an erosional interval. On the surface of the disconformity appeared remnants of a paleosol, with a thin line of pebbles and many weathered bone fragments. The faunal assemblage on the Unit I/II disconformity, forming the third faunal horizon, is the last evidence of megafauna at Taima-taima. This horizon is not directly dated, although Bryan and Gruhn (n.d.) speculate that an anomalous date of 11,860 ± 130 yr B.P. (IVIC-655) obtained from wood recovered in previous excavations may actually be derived from a water-logged root originating from a tree which grew on the Unit I/II land surface. The red sands of the overlying Unit II are sterile of organic remains and undated, but an overlying organic black clay deposit (Unit III) has yielded six stratigraphically consistent radiocarbon dates ranging from 10,290 ± 90 yr B.P. to 9650 ± 80 yr B.P. The final stratigraphic unit at Taima-taima, a deposit of colluvial brown sand (Unit IV), is sterile of organic remains and is undatable.

Direct evidence for the paleoenvironment at different horizons in the Taima-taima stratigraphic sequence is sparse. Plant remains recovered from the lower part of the grey clayey sand of Unit I include seeds of Portulacaceae (*Portulaca oleracea* and *P. venezuelensis*) and *Coccoloba uvifera* ("uva de playa"), all known from the vicinity of Taima-taima at present. The sheared twigs recovered from the vicinity of the butchered juvenile mastodont unfortunately could not be identified, but thorns were notable on the twigs; and it is likely that genera now in the area, such as *Prosopis, Cercidium,* and *Caesalpinia,* are represented. The sparse evidence from the plant remains in Unit I, then, suggests a vegetation pattern like that of the present (Ochsenius n.d.). Plant remains from the organic black clay deposit of Unit III could not be identified due to an inadequate sample of the deposit still exposed in 1976, but the black clay is believed to represent an interval of ponding after colluvial deposition and weathering of the red sand of Unit II. The old land surface of the Unit I/II disconformity may indicate a drier climatic interval, but it is also possible that the spring flow migrated temporarily to another part of the basin.

The Faunal Sequence

The bones recovered from Taima-taima in 1976 were identified by Rodolfo M. Casamiquela after the remains had been removed from the site at the close of excavations. Due to inadequate control of the water level during excavation, most of the bones were in poor condition by the time he examined the collection. Casamiquela had only a few weeks at his disposal to analyze the collection in Coro, and no comparative collection at hand. In view of these restrictions, he regards the classification (detailed in Casamiquela n.d.) as only tentative, with certain taxonomic assignations requiring reanalysis and refinement.

The Cobble Pavement

Bones recovered from the cobble pavement in the 1976 excavations were water-worn and often broken. The remains which could be definitely assigned to this horizon were exclusively mastodont, representing individuals of all ages. Casamiquela tentatively classified the molars as pertaining to the genera *Stegomastodon* and *Haplomastodon*. Elements present include tusk fragments, molars, mandibular fragments, vertebrae, ribs, fragments of scapulae and pelves, femora, tibiae, humeri, and podials. It is notable that three femora showed clear evidence (in the form of multiple, crisscrossing, linear abrasion scars on the upper surface) of human use as anvils: two of these were embedded in the cobble pavement in the vicinity of the butchered juvenile mastodont situated at the base of Unit I and must have been used by the successful hunters expediently as chopping blocks.

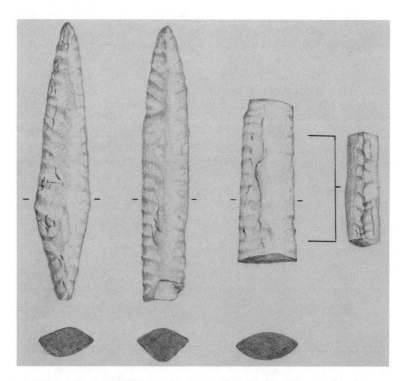

Figure 5.3. Three projectile points from the El Jobo type site in the Río Pedregal valley. The midsection is similar to the specimen found within the body cavity of the juvenile mastodont at Taima-taima. The edge of the point shown at the far right was produced by using the flake arrises on one surface as platforms for the detachment of pressure flakes on the opposing surface (scale is 1:1; illustration by R. Will).

Unit I

Numerous bones were recovered in the 1976 excavations of the grey clayey sand of Unit I. Mastodont remains were in highest frequency. A partially articulated skeleton of a juvenile mastodont, with clear evidence of butchering and an El Jobo projectile point (fig. 5.3) in the pubic cavity, was the major find, situated at the base of Unit I (figs. 5.4, 5.5). The head and right forelimb had been completely removed by the successful hunters. Other mastodont remains from Unit I included adult or juvenile mandibles,

Figure 5.4. Excavations in progress at Taima-taima in 1976. in the foreground is the skeleton of the juvenile mastodont near the base of Unit I; in the background is an inverted glyptodont carapace on the Unit I/II disconformity.

molars, tusk fragments, and fragmentary humeri, ulnae, vertebrae, and scapulae. On the basis of the molar morphology Casamiquela identified *Haplomastodon* and perhaps *Stegomastodon.* In addition to the mastodont remains, the Unit I sand yielded three fragmentary glyptodont scutes (*Glyptodon* and ?Sclerocalyptonae *incertae sedis*) and three molars of *Equus.* A fragmentary tooth of *Glossotherium* and a molar tooth of an ursid, probably *Pararctotherium,* were also recovered from Unit I, in addition to a small scapula and rib which may pertain to a felid.

Unit I/II Disconformity

On the old land surface represented by the Unit I/II disconformity were numerous eroded bone fragments—molars, vertebrae, ribs, fragmentary limbs, scutes—representing a variety of animals. Notably absent, however, is mastodont. The absence can hardly be a sampling error, as mastodonts have bigger bones to leave more identifiable fragments than the taxa which are represented. The most numerous remains identifiable are those of horse (*Equus* and *Hippidion*), glyptodont (*Glyptodon* and ?Sclerocalyptonae *incertae sedis*), and *Macrauchenia.* A major find, requiring very careful excavation and removal *en bloc,* was an inverted glyptodont carapace, with fragmentary remains of the ilium and the vertebral column in the interior (see fig. 5.4). In addition, scant remains of a mylodontid and an artiodactyl were found, plus five fragments of tortoise carapace (*Geochelone*).

Discussion

The 1976 excavations at Taima-taima indicated three faunal assemblages correlated with major stratigraphic horizons: bones impacted into the cobble pavement, bones enclosed within the grey clayey sand of Unit I, and eroded bone fragments exposed on the Unit I/II disconformity. Mastodont remains are the predominant element in the two earlier faunal assemblages; in the assemblage from the Unit I/II disconformity mastodont is conspicuously absent. To be considered as factors involved in an explanation of the faunal sequence are environmental changes and human activity.

Figure 5.5. Bones mapped at Taima-taima in 1976. Dashed line indicates profiles depicted on Figure 5.2.

All bones are of mastodont unless otherwise noted. Bones with heavy outline on map relate to the juvenile mastodont skeleton or occur at the same level in Unit I; other bones were embedded in the underlying cobble pavement; the El Jobo point is the small triangle in the pubic cavity.

Map No.	Identification	Map No.	Identification
29.	unidentified pelvic or scapular fragment	89.	small long bone
30.	patella	90.	left femur—used as anvil
31.	basal fragment of cranium, well-worn	91.	unidentified
32.	terminal fragment of tusk, juvenile	92.	glyptodont scute
33.	atlas	93.	unidentified foot bone
34.	8th right thoracic rib	94.	unidentified foot bones
35.	7th right thoracic rib	95.	phalange, left
36.	unidentified fragment	96.	unidentified
37.	fragment of rib	97.	17th right thoracic rib
38.	phalange	98.	fragment of pelvis
39.	unidentified fragment	99.	scapula of carnivore
40.	unidentified rib	100.	vertebra, juvenile
41.	unidentified rib	101.	8th left thoracic rib
42.	9th left thoracic rib	102.	rough stone
43.	15th left thoracic rib—one cut mark	103.	11th right thoracic rib
44.	14th left thoracic rib	104.	fragment of left ulna—used as anvil
45.	11th left thoracic rib	105.	left radius
46.	left humerus—six cut marks	106.	10th right thoracic rib
47.	left scapula	107.	14th right thoracic rib
48.	left femur	108.	13th right thoracic rib
49.	fragment of ischium of small mammal	109.	right tibia
50.	fragment of rib	110.	right fibula
51.	left ulna	111.	right femur
52.	19th right thoracic rib	112.	left pubis and ischium
53.	16th right thoracic rib	113.	right pubis and ischium
54.	18th left thoracic rib	114.	small rib
55.	midsection of left humerus	115.	epiphysis of left ilium
56.	left fibula	116.	rough stone
57.	left tibia	117.	small fragment of bone
58.	right femur—used as anvil	118.	left ilium
59.	fragment of vertebra	119.	small scapula
60.	left mandible	120.	small long bone
61.	adult molar	121.	unidentified
62.	flaked pebble	122.	right ilium
63.	fragment of vertebra	123.	1st caudal vertebra
64.	10th left thoracic rib	124.	1st sacral vertebra
65.	juvenile molar	125.	4th lumbar vertebra
66.	flaked pebble	126.	3rd lumbar vertebra
67.	left mandible, juvenile (part of no. 84)	127.	2nd and 3rd sacral vertebrae
68.	rough stone	128.	fragment of vertebra
69.	flaked pebble	129.	2nd lumbar vertebra
70.	adult molar	130.	19th thoracic vertebra
71.	rough stone	131.	18th thoracic vertebra
72.	juvenile molar	132.	17th thoracic vertebra
73.	flaked pebble	133.	16th left thoracic rib
74.	flaked pebble	134.	14th and 15th thoracic vertebrae
75.	flaked pebble	135.	16th and 20th thoracic vertebrae
76.	fragment of tusk, juvenile	136.	fragment of vertebra
77.	midsection of left femur—used as anvil	137.	fragment of vertebra
78.	molar of ursid	138.	fragment of vertebra
79.	fragment of vertebra	139.	12th thoracic vertebra
80.	left mandible, juvenile	140.	13th thoracic vertebra
81.	12th right thoracic rib	141.	midsection of right femur
82.	phalange, left	142.	18th right thoracic rib
83.	mandible, juvenile	143.	1st lumbar vertebra
84.	right mandible, juvenile (part of no. 67)	144.	4th sacral vertebra
85.	midsection of infantile right tibia	145.	17th left thoracic rib—one cut mark
86.	fragment of vertebra	146.	19th left thoracic rib
87.	unidentified	147.	fragment of rib
88.	unidentified	148.	13th left thoracic rib
		149.	fragment of femur

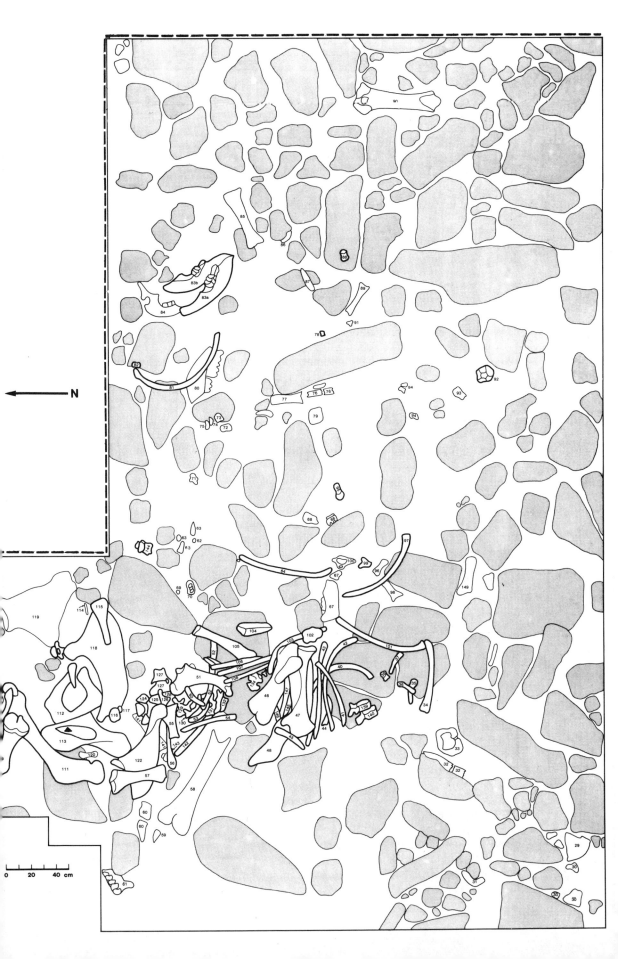

N

0   20   40 cm

Consideration of absolute chronology is also a factor in the discussion. Mastodont bones collected from the cobble pavement in 1976 for radiocarbon analysis have been undatable due to lack of collagen; nevertheless Bryan and Gruhn (n.d.) speculate that an anomalous organic carbon date of $14,400 \pm 435$ yr B.P. (IVIC-191-2) on bone collected in previous excavations, reportedly from the grey sand of Unit I, may actually be a bone imbedded in the grey sand but derived from the cobble pavement. The grey clayey sand of Unit I, with its assemblage of mastodont plus glyptodont, horse, mylontid, ursid, and felid, is securely dated between 13,400 yr B.P. and 12,500 yr B.P. The varied faunal assemblage from the Unit I/II disconformity, distinguished by the absence of mastodont, may be dated at about 11,800 yr B.P.; it is certainly older than 10,200 yr B.P., the date for the black clay of Unit III, which is separated from it by the red sand deposit of Unit II and a weathering horizon at the top of Unit II.

The admittedly sparse paleoenvironmental data from Taima-taima give no evidence of significant differences in the local environment of late Pleistocene times from that of the present. Indeed, Ochsenius (n.d.) argues for a semiarid climate and thorn forest cover in the coastal zone of north-central Venezuela in the late Pleistocene, notwithstanding plausible counter-evidence in the form of the presence of megafauna until 10,000 years ago. All taxa including the proboscideans might have been adaptable to semiarid conditions, much as similar taxa are to the present semiarid environments of east and south Africa.

Further detailed research in aspects of the paleoenvironment reflected in late Pleistocene deposits of north-central Venezuela will be necessary to give a full picture of the late Pleistocene habitat. A most crucial paleoenvironmental horizon to study will be that corresponding to the Unit I/II disconformity at Taima-taima, when mastodont has disappeared from the area although other Pleistocene taxa continue. We suspect that drier climatic conditions may have prevailed at this time, but at present there is no adequate data base for postulating local environmental change as the cause of extinction.

The hand of man is certainly evident at Taima-taima, with a definite mastodont kill at the edge of the water hole about 13,000 years ago documented in the 1976 excavations, as well as evidence of other kills at the site found in previous excavations. Projectile points of the El Jobo complex were also recorded with mastodont remains, including deliberately grooved mastodont bones (cf. Rouse and Cruxent 1963b:pl. 4) at the nearby water hole of Muaco, which has two radiocarbon dates on burned bone of $16,375 \pm 400$ yr B.P. (M-1068) and $14,300 \pm 500$ yr B.P. (0-999) (Cruxent 1961). The animals may have been ambushed at these water holes, or the hunters may have struck elsewhere and tracked the wounded animals to water. A third probable kill site with El Jobo points in association with mastodont remains is Cucuruchú, in a side canyon in a gorge near the sea a few kilometers east of Taima-taima.

Artifacts of the El Jobo complex are also known from surficial sites on the middle terraces of the Río Pedregal, about 80 km south-southwest of Taima-taima. This is an area with abundant fine quartzite, a flakeable material commonly used by El Jobo hunters; many of the archaeological sites from this type area are clearly quarry workshops. It is notable that assemblages of stone artifacts on the higher benches of the Río Pedregal valley include no projectile points—only flakes, cores, choppers, uniface scrapers, and large, thick bifaces. An evolution of these industries (Camare complex, Las Lagunas complex) into the El Jobo complex has been postulated (Rouse and Cruxent 1963a:28–33, Bryan 1973), although it is undemonstrable without a detailed study of the geomorphology of the Río Pedregal terraces and the geochronology of the sites. On typological grounds, however, it is possible to speculate that the local precursors of the El Jobo complex in north-central Venezuela innovated bifacial, lanceolate projectile points from their existing lithic repertoire.

The El Jobo complex, then, would mark the local development of a proboscidean-hunting complex certainly by 13,000 years ago and quite possibly as early as 16,000 years ago. This development in north-central Venezuela must be entirely independent of the emergence of the Clovis or Llano complex of North America, with its markedly different lithic technology and age of only 11,500–11,000 yr B.P. Indeed, a focus on proboscidean hunting may be seen as an independent cultural development which took place in several different areas as responses to available economic resources in the Old World as well as in the New World.

Once the technology was evolved in north-central Venezuela, mastodonts may have yielded to hunting pressures. Even if the mastodonts were gone by 11,800 years ago, however, various other Pleistocene megafauna—glyptodont, horse, *Macrauchenia,* mylodont—continued to range the Taima-taima area beyond that time, apparently able to cope, at least for some time, with human predation. The dating and cause of final extinction of these taxa in north-central Venezuela is yet to be determined.

Acknowledgments

We are grateful to the Centro do Investigaciones del Paleoindio y Cuaternario Sudamericano (CIPICS) for supporting the 1976 excavations at Taima-taima.

References

Bryan, Alan L. 1973. Paleoenvironments and cultural diversity in Late Pleistocene South America. *Quaternary Research* 3(2):237–256.

Bryan, A. L., R. M. Casamiquela, J. M. Cruxent, R. Gruhn, and C. Ochsenius. n.d. An El Jobo mastodon kill at Taima-taima, Venezuela. *Science* 200:1275–1277.

Bryan, Alan L., and Ruth Gruhn. 1980. The radiocarbon dates. *In*: Ochsenius, C., and R. Gruhn (editors), Taima-taima: Final Report on the 1976 Excavations. *Monografías Científicas* 3, Programa CIPICS, Universidad Francisco de Miranda, Coro, Venezuela. In press.

Casamiquela, Rodolfo M. n.d. An interpretation of the fossil vertebrates of the Taima-taima site. *In*: Ochsenius, C., and R. Gruhn (editors), Taima-taima: Final Report on the 1976 Excavations. *Monografías Científicas* 3, Programa CIPICS, Universidad Francisco de Miranda, Coro, Venezuela. In press.

Cruxent, José M. 1961. Huesos quemados en el yacimiento prehistórico de Muaco. IVIC Depto. de Antropología Boletín Informativo 2:20–21.

———. 1967. El Paleo-Indio en Taima-taima, Estado Falcón, Venezuela. *Acta Científica Venezolana* Suppl. 3:3–17.

———. 1970. Projectile points with Pleistocene mammals in Venezuela. *Antiquity* 44:223–225.

Cruxent, José M., and Irving Rouse. 1956. A lithic industry of Paleo-Indian type in Venezuela. *American Antiquity* 22:172–179.

Ochsenius, Claudio. n.d. A brief paleoecological interpretation of the site of Taima-taima and its surroundings. *In*: Ochsenius, C., and R. Gruhn (editors), Taima-taima: Final Report on the 1976 Excavations. *Monografías Científicas* 3, Programa CIPICS, Universidad Francisco de Miranda, Coro, Venezuela. In press.

Ochsenius, Claudio, and Ruth Gruhn (editors). n.d. Taima-taima: Final Report on the 1976 Excavations. *Monografías Científicas* 3, Programa CIPICS, Universidad Francisco de Miranda, Coro, Venezuela. In press.

Rouse, Irving, and José M. Cruxent. 1963a. Some recent radiocarbon dates for western Venezuela. *American Antiquity* 28:537–540.

———. 1963b. *Venezuelan Archaeology.* Yale University Press. New Haven.

Royo y Goméz, José. 1960. El yacimiento de vertebrados Pleistocenos de Muaco, Estado Falcón, Venezuela, con industria lítica humana. *International Geological Congress*, 21st report, Part 4, pp. 154–157. Copenhagen.

6

Late Pleistocene Fossils of Natural Trap Cave, Wyoming, and the Climatic Model of Extinction

B. MILES GILBERT AND LARRY D. MARTIN

THE PRESENCE OF FOSSILS in Natural Trap Cave has been known for some time. In 1970 L. L. Loendorf made a small collection. In 1971 Loendorf, accompanied by W. B. Vincent and G. Middaugh, excavated under the entrance, and additional work was done in 1972. These excavations were reported by Rushin (1974). Further excavations were conducted jointly by the University of Missouri at Columbia and the University of Kansas from 1974 to 1980.

The late Pleistocene fauna of Wyoming is not well known. Descriptions of only a few other sites, including Little Box Elder Cave (Anderson 1968, 1970; Kurtén and Anderson 1974), Bell Cave (Zeimans and Walker 1974), Horned Owl Cave (Guilday et al. 1967) and Chimney Rock Animal Trap (Hager 1972), have been published. Anderson (1974) summarized the fauna from these sites and other late Pleistocene localities in adjacent states. The Hell Gap site (Roberts 1970) and numerous other published archaeological sites also provide information on the late Pleistocene through early Holocene of Wyoming (Frison 1978). Many of these published sites are clustered in the southeastern corner of the state. Natural Trap Cave was the first major Pleistocene locality to be developed in northern Wyoming, and the recovery of more than 30,000 specimens representing a wide variety of vertebrates has established it as one of the major late Pleistocene sites in North America.

Natural Trap Cave is an 85-foot-deep karst sinkhole in the Madison (Mississippian) limestone on the western slope of the Big Horn Mountains in north-central Wyoming (fig. 6.1). It has been trapping animals at least since Sangamon times. Excavations of the remains of those animals and sediments demonstrated that they were serially deposited and are clearly stratified. They document environmental changes from the last interglacial through the last full glacial advance and represent the longest and most extensive continuous record of late Pleistocene biota in the northern Rockies (Martin and Gilbert 1978).

The fauna from Natural Trap Cave is notable for the abundance of many animals, such as the cheetah, *Miracinonyx trumani* (Adams 1979); wolf, *Canis lupus;* musk ox, *Symbos;* and pronghorn, *Antilocapra,* which are rare in other Pleistocene localities. Natural Trap Cave is the only, and by far the largest, sample of biota which is in many respects more similar to the steppe-tundra fauna of the Alaskan Pleistocene (Guthrie 1968) than it is to other fossil faunas south of the continental ice sheet. The fact that it is a stratified deposit composed of dated levels enhances the usefulness of this locality in

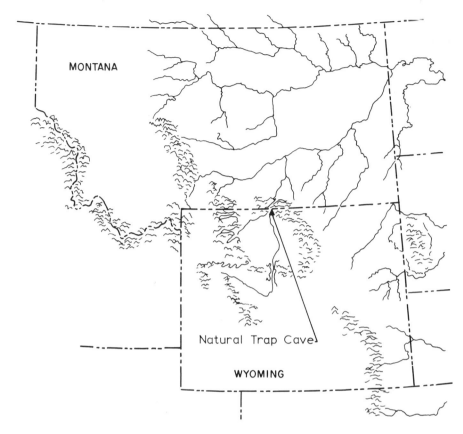

Figure 6.1. Location of Natural Trap Cave in north-central Wyoming. (Base map from Anderson 1974)

tracing the environmental and evolutionary changes in this community during the late Pleistocene and early Holocene and permits us to examine in detail some of the parameters assumed by the climatic model of extinction.

Martin and Neuner (1978) have evaluated the various assumptions inherent in the overkill model for the megafaunal extinction. We may now examine the assumptions connected with the climatic model. These assumptions include: 1) that climatic change should be reflected by vegetational change; 2) that in any given area both large and small mammals should be affected; 3) that the climatic change at the end of the Pleistocene should be unlike (in quality and magnitude) that associated with the numerous ice-age interglacials; and 4) that animals which survived the extinction should reflect climatic change through changes in their morphology and distribution.

The greatest problem in comparing the modern environment with previous inter-glacials in North America is in finding sediments, faunas, and floras which are unequivo-cally interglacial. This may be done in marine cores, but it is difficult to derive the nature of the continental climates from the changes in oxygen isotopes in the shells of marine invertebrates. Very few terrestrial faunas or floras are known which may be confidently assigned to an interglacial period. Part of the problem is a lack of long continuous sequences and the difficulty in dating continental deposits earlier than 40,000 yr B.P.

The duration of the Sangamon has been subject to various interpretations, but one of the longer durations suggested (all of isotope stage 5: Heusser and Shackleton 1979) implies a time from about 125,000 to 75,000 yr B.P. Almost all of this range should be encompassed in the lower levels of the present excavation at Natural Trap Cave. These excavations now extend to a depth of 11 feet with potentially as much as 20 feet of additional deposits. Near the base of the excavation we have found three superimposed layers of volcanic ash separated by distinct layers of cave sediments. The upper of these ashes has been dated by the fission track method on glass at 110,000 yr. B.P. (Gilbert et al. 1980), and faunas have been collected below this date as well as below the lowermost (undated) ash. The sequence continues upward into a thin Holocene layer. Although the Holocene is much better represented in nearby shelter caves (Chomko and Gilbert in press), Natural Trap Cave and the surrounding area has a sequence which includes at least the last (Sangamon) interglacial and the Pleistocene-Holocene boundary. This range provides an opportunity to test all of the previously suggested assumptions for the climatic extinction model.

Vegetational Change

Vegetational change at the end of the Pleistocene seems to have been roughly contemporaneous throughout North America. Wells (1974) has shown that the modern vegetational aspects of the Chihuahuan desert region began to appear in the woodrat midden record about 12,000 yr B.P., and the last remnants of the Pleistocene vegetational pattern may have lingered as late as 9000 yr B.P. (Wells and Hunzicker 1976). Van Devender and Spaulding (1979) would extend some aspects of the later xeric woodlands in what are now deserts until about 8000 yr B.P., and this date would fall into the youngest interval of dates generally given for extinctions. The climatic change in the eastern United States began with the first retreat of the ice sheet in the Great Lakes region, starting about 12,000 yr B.P. (Wright 1968), and the boreal woodland was replaced by a mixed coniferous-deciduous forest about 10,500 yr B.P. (Maxwell and Davis 1972). In some areas remnants of the Pleistocene vegetational pattern persisted until about 9000 yr B.P. (Wright 1968, Martin and Webb 1974). The fossil record at Natural Trap Cave shows that Pleistocene fauna and vegetation persisted until at least 12,000 yr B.P. Little cave infilling occurred afterward. This sequence correlates closely with the time of most marked climatic change found elsewhere. Excavations by Chomko (1980) in nearby rockshelters indicate an absence of extinct or extralocal species of mammals at least as far back as 8500 yr B.P.

The dates for the last appearance of extinct mammalian species published in Kurtén and Anderson (1980, Table 19.6) show the duration of the end of the Pleistocene extinction (fig. 6.2). More than half of the dates fall in the interval of 12,000 to 9,000 years before present. This interval is somewhat longer than that proposed by the overkill model of extinction, which requires a completion of the extinction during the time occupied by Clovis hunters, that is, the interval of 12,000 to 11,000 years before the present (Haynes 1967, 1970). The relatively short interval for Clovis artifacts accounts for less than 20 percent of the dated last appearances of megafauna. If we take into account likely errors (the last *dated* occurrence is probably not the last *real* occurrence, and the radiocarbon date might be in error), the importance of the interval 12,000 to 11,000 years in extinction, even with new dates, is no more likely to improve than that of the interval 11,000 to 10,000 yr B.P. In fact it is much less likely to improve because it would require two assumptions: 1) the *dated* last appearance in the interval 12,000 to 11,000 yr B.P. really are *last* appearances, and 2) all dates in the interval 12,000 to 11,000 yr B.P. are correct (19 percent of the dates), and all dates later than this interval (61 percent of the dates) are systematically too young. The latter conclusion would require a major reevaluation of radiocarbon dating as a reliable tool in

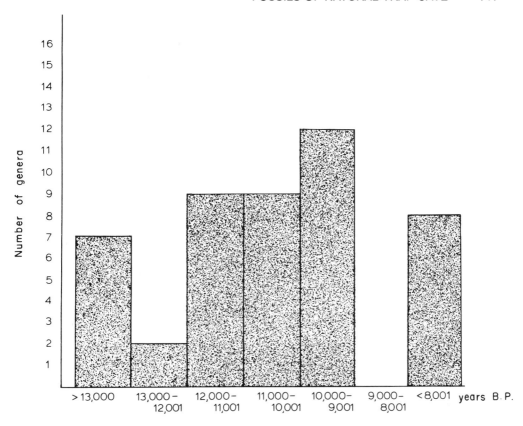

Figure 6.2. Last records of genera becoming extinct at the end of the Pleistocene.
(After Kurtén and Anderson 1980, Table 19.6)

deciding the time of the extinction, and while it might be a correct interpretation, it would mean that no correlation between the dating of Clovis occupation and extinction could be reasonably supported by using radiocarbon dates.

Because a major change of vegetation will affect the associated vertebrate fauna in a variety of ways, it is probably unproductive to seek a single direct cause of extinction in the climatic model. In general, organisms are well-adapted to the environments in which they live, and change of any sort is more likely to be detrimental than helpful. One major change in the composition of the flora at Natural Trap Cave occurred at about 20,000 yr B.P., when the percentage of C_3 grasses dropped from around 90 percent to some 70 percent of the phytoliths in the cave, and C_4 grass percentages went up. C_4 grasses are even more important in the modern flora. Studies by Leslie Marcus on bone from Natural Trap Cave indicate that very little if any C_4 grass was being consumed. This conclusion is consistent with results that Marcus has developed elsewhere including Rancho La Brea (personal communication, 1980). The change from C_3 to C_4 grass dominance may have had a profound effect on grazing mammals at the end of the Pleistocene (Wilson 1973, Martin and Neuner 1978).

W. Johnson has studied the pollen from Natural Trap Cave and has given us some of his preliminary results, showing that pine was an important pollen component at around 12,000 yr B.P. with no *Juniperus*. Pine is not present in the near vicinity now but the adjacent slopes are covered by *Juniperus*. The pollen record is thus also consistent with a change to the modern vegetation subsequent to 12,000 yr B.P.

Types of Mammals Involved

The mammals we find in the Pleistocene levels at Natural Trap Cave include three categories. First, mammals which are extinct: *Martes nobilis* (extinct pine marten), *Arctodus simus* (short-faced bear), *Canis diris* (dire wolf), *Panthera atrox* (American lion), *Miracinonyx trumani* (American cheetah), *Mammuthus* sp. (mammoth), *Equus* sp. (four kinds of extinct horses), *Camelops* sp. (American camel), *Symbos cavifrons* (woodland musk ox), *Bison antiquus* (fossil bison), *Ovis catclawensis* (extinct bighorn sheep). Second, mammals which are not extinct but which did not occur near the cave at the time of appearance of white settlers in the region: *Brachylagus* (pygmy rabbit), *Lepus arcticus* (Arctic hare), *Ochotona* sp. (pika), *Discrostonyx* (collared lemming), *Gulo gulo* (wolverine). Third, mammals which still occur or have occurred nearby during historic times: *Sorex merriami* (Merriam's shrew), *Myotis* sp. (bat), *Sylvilagus* sp. (cottontail rabbit), *Lepus* sp. (jackrabbit), *Marmota flaviventris* (yellow-bellied marmot), *Spermophilus* sp. (ground squirrel), *Eutamias minimus* (chipmunk), *Perognathus* sp. (pocket mouse), *Microtus montanus* (montane vole), *Microtus ochrogaster* (prairie vole), *Lagurus curtatus* (sagebrush vole), *Canis lupus* (gray wolf), *Antilocapra americana* (pronghorn).

The first category contains eleven taxa (all medium or large mammals), and the second contains five taxa (all medium or small mammals). The large mammal fauna has been decreased by some 70 percent, and the small mammal fauna has been changed by local extirpation of about 25 percent of its members. Extinction and extirpation seem to have been simultaneous in the Natural Trap vicinity. Within a local area they have the same community result, and we infer that climatic factors were responsible.

Uniqueness of the Present Climate

Most workers have used the present climate as a model for past interglacials, and this assumption has implications for extinction. Most climatic models of extinction include the inference that the interglacials were significantly different (more equable) than the Holocene. While this interpretation appears discordant with the deep sea cores, where the isotope record indicates that ice retreats at least as great as the most recent one have occurred in the past (Heusser and Shackleton 1979), there is no proof that the North American continental air circulation or the overall climate was precisely the same as at present (Bryson and Murray 1977). Evidence exists that the ice-age climates were becoming progressively colder and drier in North America during the Pleistocene (Schultz et al. 1972), and no evidence exists for the present distribution of deserts and prairies at any previous time during the ice age. The oxygen isotope curves are not in agreement as to amplitude (Shackleton 1980), and insolation data (Johnson 1980) suggest that our present situation is unique for the Pleistocene. Slaughter (1967) has pointed out that Pleistocene small mammal faunas contain combinations of species of small mammals which are not presently sympatric, and Miller (1975) has argued the same thing for terrestrial and freshwater mollusks. Either the present climate is more seasonal than previous interglacials, or we must not have sampled the small mammal or molluscan faunas of any interglacial.

Natural Trap Cave has a sequence which includes the last interglacial and the change to modern climate. This record can be studied in terms of two kinds of paleobotanical data. One data set is composed of opal phytoliths. These are tiny biogenic structures formed by living plants. They are characteristic and may be identified at least to within a family of plants. Although a wide variety of plants produce phytoliths, their study is relatively new. We have used only those produced by grasses in the present analysis, although shrubby plant phytoliths are also present. Pearsall (Gilbert et al. 1980) has divided the grass phytoliths into three categories—festucoid, panicoid and chloridoid—and calculated relative abundances. The categories are shown in Figure 6.3.

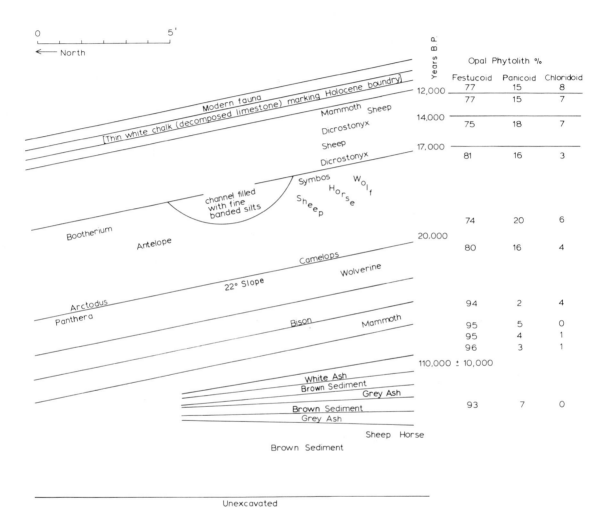

Figure 6.3. Schematic diagram of Natural Trap stratigraphy showing associated fauna, phytolith frequencies, and dates.

The sediments near the 110,000 yr B.P. date contain 90 percent or more festucoid phytoliths and no chloridoid phytoliths. This evidence suggests a cooler growing season and a wetter climate than currently exists near Natural Trap Cave (Pearsall, personal communication, 1981). Near the upper levels of Natural Trap Cave the percentage of festucoid phytoliths goes down and there begin to be significant levels of chloridoid phytoliths, but at no Pleistocene level is there a phytolith composition consistent with the modern vegetation.

W. Johnson (personal communication) informs us that the pollen associated with the Sangamon date at Natural Trap Cave contains much more pine than the pollen from near the 20,000 yr B.P. date (full glacial), which was indicative of open country. However, the record of Sangamon pollen is also unlike that of the modern vegetation. Thus both the phytolith and the pollen records suggest that the present vegetation is not similar to that of the previous interglacial.

Changes in Morphology

Reduction in size of fauna at the end of the Pleistocene was common and is well documented (Davis 1977; Wilson 1974a,b,1979). There seems to be general agreement that this size reduction was the result of climatic change (Bergmann's rule). If this is so, and if the modern climate is typical of an interglacial, we would expect to find similar size reduction in previous interglacials followed by increases in size in the glacials. This phenomenon has not been observed in any lineage of North American mammals of which we are aware. This lack may be due to our failure to recognize interglacial faunas, but even so we should still be able to find examples of Pleistocene large mammals which show as much size reduction when compared with full-glacial examples as do their modern counterparts. Perhaps individuals from the interglacials are entirely missing from our Pleistocene mammalian record. This question can be addressed in part by using fossil *Ovis* (bighorn sheep) from Natural Trap Cave. The *Ovis* from Natural Trap Cave are on an average 15 percent larger than the modern samples from Wyoming and Montana. The oldest members of the *Ovis* samples are from below a 110,000 yr B.P. date and may reasonably be thought to be from the Sangamon interglacial. Other examples are associated with the following radiocarbon dates: DiCarb-1689, 21,370 +830/−920; DiCarb-1687, 20,250±275; DiCarb-1686, 17,870 ± 230; DiCarb-1690, 17,620 +1490/−1820; DiCarb-1689, 14,670 +670/−730; Loendorf, personal communication, 12,770 ± 900 and 10,930 ± 300. No size reduction is evident during this period; all the size reduction in bighorn sheep seems to have occurred after 12,000 yr B.P. This pattern seems to hold for other animals in Natural Trap Cave whose modern counterparts are smaller, including wolves (*Canis lupus*), wolverines (*Gula gula*), pronghorn (*Antilocapra*) and bison (*Bison*).

In general, a higher percentage of the fauna from Natural Trap Cave survived the end-Pleistocene extinction than is normal for an ice-age fauna. This phenomenon may be due to the area's relatively open grassland nature and to the fact that the modern vegetation is also an open grassland, although of a different character than the Pleistocene steppe-tundra. The Pleistocene vegetational structure is reflected in the cursorial nature of the large mammal fauna. This fauna is dominated by the long-legged pursuit predators, *Miracinonyx trumani* (Martin et al. 1977, Adams 1979), *Panthera atrox* (Martin and Gilbert 1978), *Canis lupus,* and *Arctodus simus* (Martin and Gilbert 1978). *Antilocapra* is also common, as are several kinds of horses.

Almost all of the large herbivores are long-legged. The extinct sheep, *Ovis catclawensis,* was not only much larger in body size than the living bighorn, but its legs were longer and thus better adapted for steppes and open country than for the rocky habitat associated with the living species.

Camelops and *Mammuthus* are not common in Natural Trap Cave although several individuals of each have been recovered. Interestingly, browsing forms like the

American mastodont, *Mammut americanum,* or the ambush predator, *Smilodon,* have not yet been found in Natural Trap Cave, thus providing additional evidence for open terrain. The total fauna bears a strong resemblance to the fauna associated with the steppe-tundra in Alaska (Guthrie 1968), which is itself mostly a northern extension of the western montane fauna that occurred south of the continental ice and which is characterized by *Camelops, Arctodus simus,* and *Panthera atrox.* This fauna apparently occupied a suite of montane conifer parklands which varied somewhat in floral content but was a generally uniform floral type over a vast area. It seems likely that many of the grasses in this parkland were C_3 grasses (Wells 1976). Approximately 10,000 years ago those parklands disappeared and were replaced by deserts and treeless steppes of the sort that are so extensive in the present western United States. Within the modern steppes the dominant grasses are C_4 grasses. Apparently a similar floral turnover occurred in the Natural Trap area, and the latest record of extinct animals that we have from there date from 12,000 to 11,000 yr B.P. The work by Leslie Marcus on animal bone from Natural Trap Cave supports the thesis that Pleistocene herbivores in this area were markedly dependent on C_3 grasses. However, not all of the herbivores using C_3 grasses became extinct. Bison, bighorn sheep, and pronghorn all survived. In these animals climatic change was not reflected in extinction but in decrease in size.

Apparently in the study area and elsewhere the end of the Pleistocene was marked by the establishment of climatic conditions essentially similar to those found in the same areas today. Without exception this change has resulted in a decrease in animal diversity both through redistribution of taxa and through extinction. It would appear that floral diversity within a given area may also be reduced, although extinction was not a factor with plants. In general, Pleistocene communities can be characterized as being more heterogeneous and complex than are the communities occupying the same regions today. This greater complexity is likely due to decreased seasonality (Taylor 1965, Holman 1980) at the end of the Pleistocene. It appears that seasonality of the modern type became established in North America by about 10,000 years ago, and at this time the complex Pleistocene floral assemblages broke up to form less heterogeneous biomes, resulting in the expansion of prairies, deserts, and deciduous forests. This same change apparently did not take place at Natural Trap Cave during the Sangamon interglacial.

Although Clovis points have been found in the study area, they have not been found in a stratified context. We presume that man has been in this area at least since 12,000 yr B.P. and perhaps longer (Chomko 1979), but the correlation of extinction is as good with climatic changes at Natural Trap Cave as with human arrival and is probably better. This same observation seems to hold for the rest of North America. Insofar as we can test them, the assumptions of the climatic model of extinction seem to be valid, and it remains an attractive alternative to overkill.

Acknowledgments

This research has been supported by National Science Foundation Grant DEB75–21234, the National Geographic Society, the University of Kansas General Research Fund, the University of Missouri Research Council, and Earthwatch. We thank the National Park Service and the U.S. Bureau of Land Management for their cooperation, which has made the excavation of Natural Trap Cave possible. R. Hoffmann and S. Chomko read the manuscript and made useful suggestions. The figures were made by M. A. Klotz and Judy Vali.

References

Adams, D. B. 1979. The cheetah: native American. *Science* 205:1155–1158.

Anderson, E. 1968. Fauna of the Little Box Elder Cave, Converse Co., Wyo. *The Carnivora: Univ. Colo. Study Earth Sci. Ser.* (6):1–59.

———. 1970. Quaternary evolution of the genus *Martes* (Carnivora, Mustelidae). *Acta Zoologica Fennica* 130:1–132.

———. 1974. A survey of the late Pleistocene and Holocene mammal faunas of Wyoming. *In* Wilson, M., editor, Applied Geology and Archaeology: The Holocene History of Wyoming. *Geol. Surv. Wyo. Report* 10:78–87.

Bryson, R., and J. Murray. 1977. *Climates of Hunger.* Univ. Wisc. Press, Madison.

Chomko. S. A. 1978. Late Pleistocene biostratigraphy and prehistory of Prospects Shelter, Wyoming. Paper read at the 36th Plains Conference, Denver.

———. 1979. Late Pleistocene environments and man in the Big Horn Mountains. Paper read at the 44th Society for American Archaeology meeting, Vancouver.

———. 1980. The late Pleistocene/Holocene mammalian record in the northern Big Horn Mountains. Paper read at the 38th Plains Conference, Iowa City.

Chomko, S. A., and B. M. Gilbert. In press. The late Pleistocene/Holocene mammalian record in the northern Bighorn Mountains, Wyoming. *In* Semken, Holmes, and Graham, Russel, editors, Late Pleistocene/Holocene Environmental Changes in the High Plains: The Vertebrate Record. *Illinois State Museum, Reports of Investigations,* Springfield.

Davis, S. 1977. Size variation in the fox *Vulpes vulpes* in the palearctic region today and in Israel during the Late Quaternary. *J. Zool.* 182:343–351.

Frison, G. 1978. *Prehistoric Hunters on the High Plains.* Academic Press, New York.

Gilbert, B. M., D. Pearsall, and J. Boellstorff. 1980. Post-Sangamon record of vulcanism and climatic change at Natural Trap Cave, Wyoming. *Abstracts and Program,* sixth biennial meeting, American Quaternary Assoc., p. 26.

Guilday, J., H. Hamilton, and E. K. Adam. 1967. Animal remains from Horned Owl Cave, Albany Co., Wyo. *Contrib. Geol. Univ. Wyo.* 6:97–99.

Guthrie, R. D. 1968. Paleoecology of the large-mammal community in interior Alaska

during the late Pleistocene. *Am. Mid. Nat.* 79:346–363.

Hager, M. W., 1972. A late Wisconsinan-Recent vertebrate fauna from the Chimney Rock Animal Trap, Larimer Co., Colo. *Contrib. Geol. Univ. Wyo.* 11:63–71.

Haynes, C. V. 1967. Carbon-14 dates and early man in the New World. *In* Martin, P. S., and Wright, H. E., editors, *Pleistocene Extinctions: The Search for a Cause.* Yale Univ. Press, New Haven.

———. 1970. The geochronology of man-mammoth Sites and their bearing on the origin of the Llano complex. *In* Dort, W., and Jones, J. K., editors, *Pleistocene and Recent Environments of the Central Great Plains.* Univ. Kansas Press, Lawrence.

Heusser, L. E., and Shackleton, N. J. 1979. Direct marine-continental correlation: 150,000 year oxygen isotope-pollen record from the North Pacific. *Science* 204:837–839.

Holman, J. A. 1980. Paleoclimatic implications of Pleistocene herpetofaunas of eastern and central North America. *Trans. Neb. Acad. Sci.* 8:131–140.

Johnson, R. G. 1980. Ending of Quaternary extremes: a new high latitude insolation criterion. *Abstracts and Program,* sixth biennial meeting of the American Quaternary Assoc., p. 115.

Kurtén, B., and Anderson, E. 1974. Association of *Ursus arctos and Arctodus simus* (Mammalia:Ursidae) in the late Pleistocene of Wyoming. *Breviora* 426:1–6.

———. 1980. *Pleistocene Mammals of North America.* Columbia Univ. Press, New York.

Martin, L. D., and Gilbert, B. M. 1978. Excavations at Natural Trap Cave. *Trans. Nebraska Acad. Sci.* 6:107–116.

Martin, L. D., and Neuner, A. M. 1978. The end of the Pleistocene in North America. *Trans. Nebraska Acad. Sci.* 6:117–126.

Martin, R. A., and Webb, S. D. 1974. Late Pleistocene mammals from the Devil's Den fauna. *In* Webb, S. D., editor, *Pleistocene Mammals of Florida,* pp. 114–145. Univ. of Florida Press, Gainesville.

Martin, L. D., Gilbert, B. M., and Adams, D. B. 1977. A cheetah-like cat in the North American Pleistocene. *Science* 195:981–982.

Martin, L. D., Gilbert, B. M., and Chomko, S. A. 1979. *Dicrostonyx* (Rodentia) from the late Pleistocene of northern Wyoming. *J. Mamm.* 60:193–195.

Maxwell, J. A., and Davis, M. B. 1972. Pollen

evidence of Pleistocene and Holocene vegetation on the Allegheny Plateau, Maryland. *Quat. Res.*, 2:506–530.

Miller, B. B. 1975. A sequence of radiocarbon-dated Wisconsinan nonmarine molluscan faunas from southwestern Kansas-northwestern Oklahoma. *In* Smith, G. R., and Friedland, N. E., editors, Studies on Cenozoic paleontology and stratigraphy. *Papers on Paleontology* No. 12, Mus. of Paleontology, Univ. of Michigan, pp. 9–18.

Mosimann, J. E., and Martin, P. S. 1975. Simulating overkill by Paleoindians. *American Scientist* 63:304–313.

Roberts, M. F. 1970. Late glacial and postglacial environments in southeastern Wyoming. *Paleogeog., Paleoclimate, Paleoecol.* 8(1):5–19.

Rushin, C. J. 1974. Test excavation in Natural Trap Cave, Wyoming. Ms. at Dept. Forestry, Univ. Montana, Missoula, 56 pp.

Schultz, C. B., Tanner, L. G., and Martin, L. D. 1972. Phyletic trends in certain lineages of Quaternary mammals. *Bull. Univ. Nebraska State Mus.* 9:183–195.

Shackleton, N. J., 1980. The oxygen-isotope record as an ice-volume monitor. *Abstracts and Programs* sixth biennial meeting of the American Quaternary Assoc., p. 172.

Slaughter, R. H. 1967. Animal ranges as a clue to late Pleistocene extinction. *In* Martin, P. S., and Wright, H. E., ediors, *Pleistocene Extinctions: The Search For a Cause.* Yale Univ. Press, New Haven.

Taylor, D. W. 1965. The study of Pleistocene nonmarine mollusks in North America. *In* Wright, H. E., Jr., and Frey, D. G., editors, *The Quaternary of the United States,* pp. 597–611. Princeton Univ. Press, Princeton.

Van Devender, T., and Spaulding, W. G. 1979. Development of vegetation and climate in the southwestern United States. *Science* 204:701–710.

Wells, P. V. 1974. Post-glacial origin of the present Chihuahuan Desert. *In* Wauer, R. E., and Rigkind, D. H., editors, *Transactions of the Symposium on the Biological Resources of the Chihuahuan Desert Region, United States and Mexico.* No. 2. National Park Service, Washington, D. C.

———. 1976. Macrofossil analysis of wood rat (*Neotoma*) middens as a key to the Quaternary history of arid America. *Quat. Res.* 6:223–248.

Wells, P. V., and Hunzicker, J. H. 1976. Origin of the creosote bush deserts of southwestern North America. *Annals Missouri Bot. Gardens* 63:843–861.

Wilson, J. W., III. 1973. Photosynthetic pathways and spatial heterogeneity on the North American plains: suggestion for the cause of Pleistocene extinction. Paper read at the annual meeting of the Society of Vertebrate Paleontologists.

Wilson, Michael. 1974a. History of the *Bison* in Wyoming with particular reference to early Holocene forms. *In* Wilson, M., editor, Applied geology and archaeology: the Holocene history of Wyoming. *Geol. Surv. Wyo. Report* 10:91–99.

———. 1974b. The Casper local fauna and its fossil bison. *In* Frison, George G., editor, *The Casper Site.* Academic Press, New York.

———. 1979. The Elnora bison bone bed. *Archaeological Survey, Alberta Occasional Papers* 14. Edmonton.

Wright, H. E., Jr. 1968. The roles of pine and spruce in the forest history of Minnesota and adjacent areas. *Ecology* 49:937–955.

Zeimens, G., and Walker, D. N. 1974. Bell Cave, Wyoming: preliminary archeological and paleontological investigations. *In* Wilson, M., editor, Applied geology and archaeology: the Holocene history of Wyoming. *Geol. Surv. Wyo. Report* 10:88–90.

7

Shasta Ground Sloth Extinction
Fossil Packrat Midden Evidence From the Western Grand Canyon

ARTHUR M. PHILLIPS, III

RAMPART CAVE, located at the western end of the Grand Canyon in Mohave County, northwestern Arizona, is a site of extraordinary interest to the Pleistocene paleoecologist. Until a fire in 1976 and 1977, the cave's deposit of dung of the extinct Shasta ground sloth (*Nothrotheriops shastensis*) was perfectly preserved and virtually undisturbed since the last ground sloth departed from the cave more than 11,000 years ago.

A number of studies have been carried out in Rampart Cave, mostly centering on attempts to determine the date of the demise of the sloth (Martin et al. 1961, Long et al. 1974, Hansen 1978). In visits to Rampart Cave beginning in 1972 I became interested in the potential of ancient packrat middens as a source of paleoecological information independent of that contained in sloth dung balls. Since Shasta ground sloths were selective browsers, an analysis of their diet (Hansen 1978) will not reveal all the plants in their environment. More than fifty late Pleistocene and early Holocene packrat middens containing evidence of displaced vegetation were found in Rampart Cave, in nearby Vulture Cave (Mead and Phillips 1981), and in numerous small rockshelters on both sides of the Colorado River within 3 km of Rampart Cave. With radiocarbon dates ranging from 8500 yr B.P. to 30,000 yr B.P., these middens provide a record of vegetation dynamics and climatic change during and immediately following the last full glacial episode (Phillips and Van Devender 1974, Phillips 1977). Of these middens, twelve were radiocarbon dated between 10,000 and 12,000 yr B.P. (Table 7.1) and thus span the time of local extinction of the Shasta ground sloth which occurred both at Rampart Cave and elsewhere around 11,000 years ago (Thompson et al. 1980). These fossil packrat middens provide evidence about the flora and climate of the lower Grand Canyon at the time of the ground sloth's extinction.

Packrat midden analysis was developed in the early 1960s by P. V. Wells (Wells and Jorgensen 1964, Wells 1966, Wells and Berger 1967, Wells 1976). The method has been further refined by Van Devender (1973, 1977) and others at the University of Arizona. Middens of *Neotoma* spp. (packrats or woodrats), hardened by urine and protected from moisture in caves, rockshelters, and rock crevices in the arid and semiarid Southwest, preserve a record of the local flora at the time of their deposition. Seeds, leaves, twigs, spines, and other plant parts are preserved in the indurated, urine-cemented blocks and are readily released when the middens are soaked in water.

Table 7.1. Lower Grand Canyon Packrat Midden Localities and ^{14}C Dates Between 10,000 and 12,000 yr B.P.

DA, Desert Almond Canyon; IC, Iceberg Canyon; VC, Vulture Canyon; WR, Window Rock Canyon

Midden	Elevation Meters	Feet	Slope Exposure	Lab. No.	Radiocarbon Date yr B.P.	Material Dated
DA 1 A+B	490	1600	N	A-1380	10,100 ± 200	*Nolina* leaves
DA 2 A	490	1600	N	A-1426	10,930 ± 460	*Nolina* leaves, *Prunus* seeds, *Fraxinus* twigs
DA 3 B	520	1700	N	A-1422	11,190 ± 150	*Juniperus* twigs
DA 4	585	1920	N	A-1423	11,990 ± 490	*Juniperus* twigs
DA 5 A+B	565	1850	SW	A-1456	10,450 ± 420	*Mortonia* leaves, *Juniperus* twigs, *Fraxinus* twigs and seeds
DA 6 A+B	570	1860	S	A-1427	10,910 ± 450	*Fraxinus* twigs, *Nolina* leaves
VC 2 B	460	1520	SE	A-1567	10,250 ± 290	*Juniperus* twigs
VC 4	495	1620	SE	A-1566	10,610 ± 320	*Juniperus* seeds and twigs
VC 14	645	2120	SE	A-1587	11,870 ± 190	*Juniperus* seeds and twigs
WR 1 H	465	1525	N	A-1314	11,310 ± 380	*Juniperus* twigs
WR 2 C	465	1525	N	A-1352	10,250 ± 200	*Juniperus* seeds and twigs
IC 1	425	1400	NW	A-1322	11,010 ± 400	*Juniperus* twigs

In desert areas old middens can be recognized by the presence of extralocal woodland species such as juniper. The macrofossils obtained from the middens can be identified, often to species level, by comparison with modern reference material.

Each midden generally contains between ten and twenty-five identifiable plant taxa. Individual middens reflect differences in habitat and time; taken together, a number of middens provide a considerable species list reflecting the flora of the area in which they were collected (Phillips and Van Devender 1974, Phillips 1977).

Analysis of Rampart Cave Area Middens

The middens considered here were all found in small rockshelters in the Rampart Cave area. Locations of the collection sites are shown in Figure 7.1. Three were collected in Vulture Canyon; six were found in Desert Almond Canyon; two were at the mouth of Window Rock Canyon, a more exposed site; and one was found 15 km west of the Grand Wash Cliffs on the south wall of Iceberg Canyon. Table 7.2 lists the plant species identified from each midden. The modern vegetation of the area consists of a low-elevation, hot desert community of low shrubs and succulents typical of the eastern portions of the Mohave Desert (Phillips 1975).

Forty-nine plant species were recovered from the twelve middens contemporary in ^{14}C age with the extinction of the ground sloth at Rampart Cave. These are displaced woodland and modern desert species; with minor variations, this is the same pattern that persisted for 3,000 years after the time of extinction and had prevailed for at least 18,000 years prior to extinction (Phillips 1977).

The predominant woodland species represented in the middens are *Juniperus* sp. (juniper) and *Fraxinus anomala* (single-leaf ash). Today these trees are commonly found only at higher elevations in the Grand Canyon, and near Rampart Cave their lowest occurrence is 1,000 m above the cave and several kilometers back from the rims. Other extralocal woodland species found in some of the middens include *Cercocarpus intricatus* (little-leaf mountain mahogany), *Ostrya knowltonii* (Knowlton hophornbeam), *Rhamnus betulaefolia* (birchleaf buckthorn), *Rhus trilobata* (squawbush), and *Symphoricarpos* sp. (snowberry). Extralocal high-desert species not found today at low

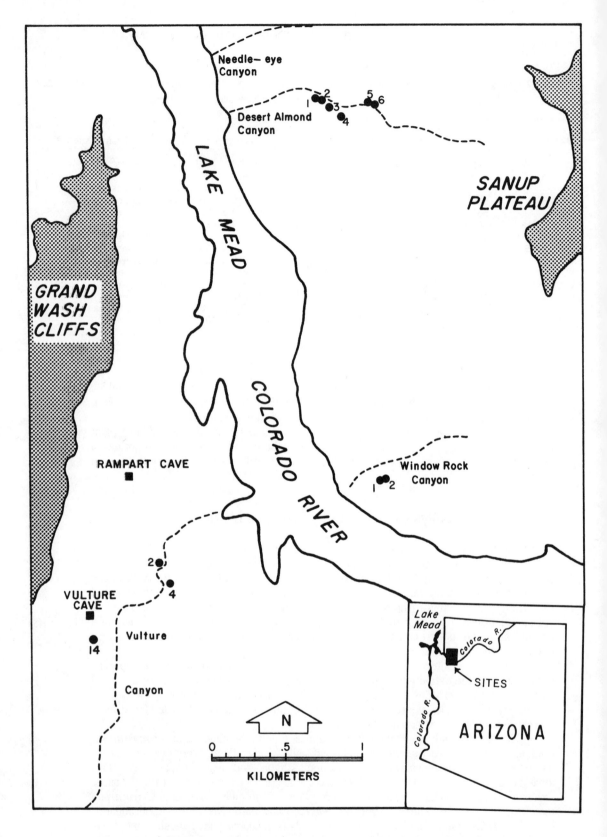

Figure 7.1. Late Pleistocene packrat midden sites (numbered) in the lower Grand Canyon.

Table 7.2. Macrofossils Recovered From Lower Grand Canyon Packrat Middens Dating 10,000 to 12,000 yr B.P.

Relative abundances: 5 = abundant, 4 = very common, 3 = common, 2 = occasional, 1 = rare, x = present (relative abundance not calculated)

	Sites											
	Desert Almond						Vulture Canyon			Window Rock		Iceberg Canyon
Species	1	2	3	4	5	6	2	4	14	1	2	1
EXTRALOCAL												
Atriplex confertifolia	–	–	–	–	–	–	–	–	3	–	–	–
Cercocarpus intricatus	–	–	–	3	–	–	–	–	1	–	–	–
Chrysothamnus sp.	–	–	–	1	3	1	–	–	–	–	–	–
Coleogyne ramosissima	–	–	1	1	1	1	–	1	–	1	x	1
Cryptantha virginensis	–	–	–	–	1	–	–	–	–	–	–	–
Fraxinus anomala	4	4	4	4	3	5	3	3	4	1	–	3
Juniperus sp.	2	2	5	5	5	4	5	5	5	5	x	5
Opuntia erinacea	–	–	–	–	–	–	–	–	–	–	–	1
Ostrya knowltonii	–	–	–	–	–	–	–	–	2	–	–	–
Rhamnus betulaefolia	–	–	–	–	–	1	–	–	–	–	–	–
Rhus trilobata	–	–	–	1	1	2	–	–	–	1	–	–
Salvia dorrii	–	–	–	–	–	–	1	–	–	–	–	1
Symphoricarpos sp.	–	–	1	1	–	1	–	–	4	–	x	1
Yucca baccata	1	–	1	1	–	1	–	–	–	1	x	–
RELICTUAL												
Amsonia tomentosa	–	–	–	1	1	–	–	2	–	–	–	1
Nolina microcarpa	5	4	4	5	–	5	–	–	–	–	–	–
Opuntia whipplei	–	–	–	3	1	4	4	3	1	3	x	–
Prunus fasciculata	4	4	–	1	2	2	–	–	–	4	x	–
Vitis arizonica	–	–	–	–	–	–	–	–	–	–	x	–
UNCHANGED												
Acacia greggii	3	–	–	–	3	1	–	–	–	–	–	2
Agave utahensis	–	–	2	1	–	3	–	–	1	–	x	5
Allionia incarnata	–	–	–	–	–	–	–	2	–	–	–	–
Anemone tuberosa	–	–	–	–	–	1	–	–	–	1	–	1
Argemone sp.	1	–	–	–	2	2	–	1	–	1	–	–
Crossosoma bigelovii	–	–	–	1	–	3	–	–	3	1	x	–
Echinocactus polycephalus	1	–	1	–	1	5	1	1	–	1	x	3
Echinocereus sp.	1	–	3	2	–	3	1	1	–	–	x	–
Encelia farinosa	2	–	–	–	2	–	–	2	–	2	x	–
Ephedra sp.	3	5	3	3	5	3	3	4	1	5	x	3
Ferocactus acanthodes	1	–	–	–	–	–	–	1	–	–	–	–
Franseria confertiflora	–	–	–	–	1	–	–	–	–	–	–	–
Gutierrezia microcephala	1	–	3	4	–	3	2	1	1	–	x	3
Larrea tridentata	–	–	–	–	–	–	–	–	1	–	–	–
Lycium andersonii	–	–	–	–	–	–	–	–	–	–	x	–
Mentzelia sp.	–	–	–	1	–	1	–	–	–	–	x	–
Mortonia scabrella var. utahensis	1	–	1	4	4	4	–	–	–	–	–	–
Oenothera cavernae	–	–	–	–	–	–	–	–	–	1	–	–
Opuntia basilaris	2	1	1	–	1	2	–	–	2	1	–	2
Penstemon eatonii	2	–	–	–	–	1	–	–	–	–	–	–
Phoradendron californicum	–	–	–	–	1	1	–	–	–	–	–	–
Physalis sp.	3	3	1	–	4	3	2	3	1	2	x	1
Sphaeralcea sp.	1	–	–	–	3	1	1	2	–	3	x	–
ANNUAL												
Astragalus nuttallianus	–	–	–	–	–	–	–	–	–	1	–	–
Caucalis microcarpa	–	–	–	–	–	–	–	–	–	–	x	–
Cirsium sp.	–	–	–	–	–	1	–	2	–	1	x	–
Galium sp.	–	–	–	–	1	–	–	–	–	–	x	–
Lepidium sp.	1	–	–	–	–	–	1	3	1	1	x	–
Phacelia crenulata	1	–	–	–	1	–	–	–	–	1	x	–
Thysanocarpus amplectens	–	–	–	–	–	–	–	–	–	1	–	1

elevations near Rampart Cave include *Atriplex confertifolia* (shadscale), *Chrysothamnus* sp. (rabbitbrush), *Coleogyne ramosissima* (blackbrush), and *Yucca baccata* (banana yucca). A number of species now reaching their lower elevational limits in the area were, according to the midden record, more widespread in the late Pleistocene. These include *Nolina microcarpa* (beargrass), *Opuntia whipplei* (Whipple cholla), and *Prunus fasciculata* (desert almond).

The remaining species, more than 50 percent of those identified, are low-elevation Mohave and Sonoran desert species that are found in the modern flora of the Rampart Cave area (Phillips 1975). These include *Acacia greggii* (catclaw acacia), *Agave utahensis* (Utah agave), *Echinocactus polycephalus* (woolly-headed barrel cactus), *Encelia farinosa* (brittlebush), *Ephedra* spp. (Mormon tea), *Opuntia basilaris* (beavertail cactus), *Physalis* sp. (ground cherry), and *Sphaeralcea* sp. (globemallow). Plant nomenclature follows Lehr (1978).

Desert Almond Canyon

Desert Almond Canyon is across the Colorado River (Lake Mead) from Rampart Cave and approximately 2 km downstream. It is a steep side canyon heading in an amphitheater on the North Rim at the Sanup Plateau. The midden sites are concentrated in a narrow section of the canyon 0.5 km east of the river.

Juniper and single-leaf ash are the predominant species recovered from most of the middens, which appear to record an open woodland dominated by these two trees with an understory of woodland and desert shrubs, cacti, and herbs 10,000 to 12,000 years ago. Other middens in the canyon document the presence of woodland trees until at least 8500 yr B.P. The abundance of beargrass, desert almond, and *Mortonia scabrella* var. *utahensis* (sandpaper bush) is notable in most of these middens; these species are absent or less common in the middens from other sites. All three species are represented by a few scattered plants at their regional lower elevational limits in protected microhabitats in Desert Almond Canyon today. The late Pleistocene middens show a more widespread distribution in the past.

Vulture Canyon

Located just upstream from Rampart Cave on the same side of the Colorado River as the cave, Vulture Canyon has proved to be a rich source of ancient middens. Most of them are older than the time period considered here. Vulture Cave, the source of fifteen late Pleistocene packrat middens reported by Mead and Phillips (1981), is located in the upper part of the canyon.

Juniper and single-leaf ash also predominate in these middens, with a mixture of understory woodland and desert plants similar to that of Desert Almond Canyon. Two of the middens in Vulture Canyon, VC 2 and VC 4, are near the canyon floor and show a more xerophytic assemblage than VC 14, a midden which is 0.5 km farther up the canyon and 185 m higher. Knowlton hophornbeam, shadscale, and abundant snowberry indicate the differing ecological position of VC 14.

Window Rock

Window Rock Canyon is on the north side of the Colorado River directly across from Vulture Canyon. The two middens found here were close together in small rockshelters on the south side of the canyon at the point where the cliffs open up to the Grand Canyon. This exposed site lacks the protection afforded by side canyon localities.

Juniper predominates in both middens; both lack single-leaf ash, perhaps because of their exposure. Seeds of desert almond are very common; it probably grew below the shelters along the canyon floor. A few seeds of *Vitis arizonica* (canyon grape) were found in one of the middens; perhaps a small seep issued from the base of the Muav limestone at the mouth of the canyon 10,000 years ago. Canyon grape is found in a few such places near Rampart Cave today.

Iceberg Canyon

Iceberg Canyon is 20 km downstream along the pre–Lake Mead course of the Colorado River from Rampart Cave and 15 km west of the Grand Wash Cliffs, which mark the western end of the Grand Canyon. A single midden was found on the south wall of the narrow, steep-walled canyon, a somewhat protected site whose exposure is probably comparable to the larger side canyons in the Grand Canyon. At 425 m, this is the lowest elevation at which a Pleistocene packrat midden has been found in the vicinity of Rampart Cave.

As with the Grand Canyon sites, juniper twigs are the most abundant extralocal fossils in the midden, establishing the existence of a late Pleistocene woodland along the Colorado River outside the Grand Canyon. Two additional displaced woodland species are present: single-leaf ash, represented by numerous twigs and seeds; and snowberry, with a single seed. According to the midden record, the flora of Iceberg Canyon at the time of ground sloth extinction was comparable to that around the midden sites near Rampart Cave.

Packrat Middens and the Pleistocene Flora

Laudermilk and Munz (1938) and Martin et al. (1961) concluded that Shasta ground sloths were eating essentially modern vegetation at the time of their extinction 11,000 years ago and thus that modern vegetation and climate prevailed in the Rampart Cave area by about 12,000 yr B.P. The presumption was that since ground sloths survived after the displacement of Pleistocene woodlands by the Holocene hot desert, climatic change at the end of the ice age could be ruled out as the cause for their extinction. The prevalence of desert plants such as *Yucca, Ephedra,* and *Sphaeralcea* in their diets was shown as evidence that they were well adapted to the browse provided by the desert habitat.

A study by Hansen (1978) confirms the earlier conclusion that the Shasta ground sloth ate desert plant species, most of which still occur near Rampart Cave. Taken from throughout the dung profile, Hansen's samples show *Sphaeralcea ambigua* (desertmallow) as consistently the most abundant plant fragment, followed by *Ephedra nevadensis* (Nevada Mormon tea), *Atriplex* spp., catclaw acacia, cactus, and *Phragmites australis* (common reed). In the upper sections of the profile these species always constitute 90 percent or more of the dung. Single-leaf ash was the only extralocal species to appear regularly, and in all levels it was less than 2 percent. Juniper was present at 0.5 percent or less. Of the seventy-two species identified by Hansen, 68 percent are present in the modern flora near the cave; 17 percent are extralocal woodland or high-desert species; 10 percent are aquatics such as common reed, most of which would have been expected in the modern pre–Lake Mead flora along the Colorado River or in springs around Columbine Falls; and 5 percent are relict species present but rare in the modern flora. It seems clear that the Shasta ground sloth was selectively browsing desert vegetation and ingested woodland plants only accidentally or incidentally. Were it not for the packrat middens, a comingling of desert and woodland elements (Table 7.2) might not be suspected.

Throughout the Southwest the midden record shows a mixture of woodland and desert species at low elevations through the last full glacial (Van Devender and Spaulding 1979). The terminal date for displaced woodland was consistently found to be about 8000 yr B.P. synchronously in all the Southwestern hot deserts (Van Devender 1977).

Figure 7.2a. Vegetation below Rampart Cave 15,000 years ago. (Drawing by Pamela Scott Lungé)

Figure 7.2b. Modern vegetation below Rampart Cave. (Drawing by Pamela Scott Lungé)

The pattern of vegetation change and plant community structure at Rampart Cave is consistent with these observations (Phillips 1977). Figure 7.2 shows my interpretation of the flora around Rampart Cave at the end of the full glacial compared with the present.

Ground sloths were absent from the Rampart Cave record during the last full glacial, 14,000 to 24,000 years ago (Long et al. 1974). They returned to feed on desert shrubs nearly to the exclusion of woodland species, a fact which led researchers to the erroneous conclusion through dung studies that the woodland plants had retreated from the Inner Gorge (Phillips 1977). In fact, sloths may have gone extinct under their most favorable climatic and floristic conditions, when their preferred food plants were still flourishing. The hot, dry climatic conditions of the Altithermal were still 3,000 years in the future (Mehringer 1967).

The packrat middens found within Rampart Cave deserve special mention. Of the separate middens located within the cave, none was dated within the time period considered here. Six were radiocarbon dated at 1,000 or more years prior to ground sloth extinction; one dated approximately 1,500 years after the terminal date for ground sloth occupation. During the period from 24,000 to 14,000 yr B.P., when sloths were absent from the cave, a layer of unconsolidated packrat den materials about 15 cm thick was deposited on top of the pre-full glacial sloth dung; this layer was subsequently buried in dung after the return of the sloths some 3,000 years prior to their final departure. The four packrat middens in the cave dating between 14,000 and 12,000 years old are all indurated middens found in the far corners and passageways of the cave where the sloths would not disturb them.

One midden postdating sloth extinction was found on top of the deposit, perhaps indicating a reemergence of packrats from their protected corners. Three radiocarbon dates were obtained on plant materials in the midden, ranging from 9770 to 9520 yr B.P. Incorporated in the midden was a sloth dung ball which was dated at 11,140±250 yr B.P. (A-1453). This date is significantly older than the dates on plant materials in the midden and apparently represents a dung ball collected from the top of the deposit and incorporated into the midden by packrats long after the demise of the sloths. This interpretation is in accord with the conclusion of Long et al. (1974) and Thompson et al. (1980) that the Shasta ground sloth did not survive into the early post glacial period. The date agrees very closely with the dates on undisturbed dung balls from the top of the deposit, representing the termination of use of the cave by ground sloths.

The flora of the Rampart Cave area at the time of the last ground sloth was a mixture of woodland species now found above the Grand Canyon rims near Rampart Cave, high-desert species found at or just below the rims and relictual at low elevations, and low-desert species of which the modern flora is exclusively composed.

Climatic inferences are difficult to draw in an area as complex as the Grand Canyon; however, a few general trends are apparent. Woodland species present in the Inner Gorge indicate wetter Pleistocene conditions and probably cooler summers than at present. The high diversity of desert species not or rarely found today in the woodlands on the rims suggests mild winter temperatures in the Pleistocene, perhaps only slightly cooler than today. The abundance of annual species associated only with a winter rainy season and the absence of annuals indicative of summer rains indicate that in the Pleistocene, as today, the balance was strongly in favor of winter rainfall and that summer rains have been scanty and unreliable for at least the past 30,000 years (Phillips 1977). While slight warming and drying of the Grand Canyon climate probably began about 14,000 years ago, woodland species prevailed until about 8,500 years ago, some 2,000 years after the ground sloths disappeared. At about that time, woodland species were extirpated at low elevations, perhaps in response to climatic change toward the modern warm, dry conditions.

"Disharmonious" or "anomalous" associations may be inferred in the case of midden records of *Encelia farinosa* and *Ferocactus acanthodes* with *Juniperus* and *Fraxinus anomala* (Table 7.2). These species rarely associate locally, if at all, in the modern flora. If the dissolution of such communities is held to be a key to the extinction of late Pleistocene mammals (e.g. Graham and Lundelius this volume) it should be noted that such associations persist beyond the terminal dates of the Shasta ground sloths. Few changes are evident in the 2,000 years of midden record (Table 7.2) embracing, according to Thompson et al. (1980), the time of ground sloth extinction. If a significant change in climate or plant associations was responsible for the extinction of the Shasta ground sloths, the contemporaneous plant record of packrat middens does not disclose the nature of the event.

The mild climate in the lower Grand Canyon, suggested by the presence of both woodland and desert plant species, and the availability of a varied diet, indicated by the apparent increase in desert browse species, make it seem probable that extinction of the Shasta ground sloth occurred at a time in the late Pleistocene which should have been nearly ideal for its continued existence.

Acknowledgments

The research reported here was carried out as part of a Ph.D. dissertation in the Department of General Biology at the University of Arizona under the direction of Drs. Paul S. Martin and Willard Van Asdall. Dr. Thomas Van Devender taught me packrat midden analysis, Jim I. Mead discovered Vulture Cave, and many others helped at various times with the field work. Pamela S. Lungé prepared Figures 7.1 and 7.2. My wife, Dr. Barbara G. Phillips, offered untiring assistance and encouragement through all phases.

Middens and plants were collected under permits from Lake Mead National Recreation Area and Grand Canyon National Park. Radiocarbon dates were provided by the University of Arizona Radiocarbon Laboratory under the direction of Dr. Austin Long. Financial assistance was received from National Science Foundation grants GB-27406 and DEB 75-13944 to Paul S. Martin, and a Grand Canyon Natural History Association research grant to the author.

References

Hansen, R. M. 1978. Shasta ground sloth food habits, Rampart Cave, Arizona. *Paleobiology* 4:302–319.

Laudermilk, J. D., and P. A. Munz. 1938. Plants in the dung of *Nothrotherium* from Rampart and Muav Caves, Arizona. *Carnegie Inst. Wash. Publ.* 487:271–281.

Lehr, J. H. 1978. A catalogue of the flora of Arizona. Desert Botanical Garden, Phoenix.

Long, A., R. M. Hansen, and P. S. Martin. 1974. Extinction of the Shasta ground sloth. *Geol. Soc. Am. Bull.* 85:1843–1848.

Martin, P. S., B. E. Sabels, and D. Shutler, Jr. 1961. Rampart Cave coprolite and ecology of the Shasta ground sloth. *Am. J. Sci.* 259:102–127.

Mead, J. I., and A. M. Phillips, III. 1981. The late Pleistocene and Holocene fauna and flora of Vulture Cave, Grand Canyon, Arizona. *Southwest Nat.* 26:257–288.

Mehringer, P. J., Jr. 1967. Pollen analysis of the Tule Springs area, Nevada. *Nev. State Mus. Anthro. Pap.* 13(Part 3):130–200.

Phillips, A. M., III. 1975. Flora of the Rampart Cave area, lower Grand Canyon, Arizona. *J. Ariz. Acad. Sci.* 10:148–159.

———. 1977. Packrats, plants, and the Pleistocene in the lower Grand Canyon. Ph.D. dissertation, University of Arizona, Tucson. 123 pp.

Phillips, A.M., III, and T. R. Van Devender. 1974. Pleistocene packrat middens from the Lower Grand Canyon of Arizona. *J. Ariz. Acad. Sci.* 9:117–119.

Thompson, R. S., T. R. Van Devender, P. S. Martin, T. Foppe, and A. Long. 1980. Shasta ground sloth (*Nothrotheriops shastense* Hoffstetter) at Shelter Cave, New Mexico: environment, diet, and extinction. *Quat. Res.* 14:360–376.

Van Devender, T. R. 1973. Late Pleistocene plants and animals of the Sonoran Desert: a survey of ancient packrat middens in southwestern Arizona. Ph.D. dissertation, University of Arizona, Tucson. 179 pp.

———. 1977. Holocene woodlands in the Southwestern deserts. *Science* 198:189–192.

Van Devender, T. R., and W. G. Spaulding. 1979. Development of vegetation and climate in the southwestern United States. *Science* 204:701–710.

Wells, P. V. 1966. Late Pleistocene vegetation and degree of Pluvial climatic change in the Chihuahuan Desert. *Science* 153:970–975.

———. 1976. Macrofossil analysis of wood rat (*Neotoma*) middens as a key to the Quaternary vegetational history of arid America. *Quat. Res.* 6:223–248.

Wells, P. V., and R. Berger. 1967. Late Pleistocene history of coniferous woodland in the Mohave Desert. *Science* 155:1640–1647.

Wells, P. V., and C. D. Jorgensen. 1964. Pleistocene wood rat middens and climatic change in Mohave Desert: a record of juniper woodlands. *Science* 143:1171–1174.

8

The Significance of Radiocarbon Dates for Rancho La Brea

LESLIE F. MARCUS AND RAINER BERGER

THE REMARKABLE RANCHOLABREAN FOSSIL DEPOSITS in metropolitan Los Angeles continue to yield new information about the biota of southern California during the late Pleistocene. An extensive carbon-14 dating program has dated the major fossil-bearing accumulations in some detail and allows estimates for the lastest extinction times of some taxa in the Los Angeles Basin. Much of the processed floral material from the excavation of Pit 91 has been thoroughly studied (Warter 1976, 1979, 1980). Scott Miller has reviewed the insects at La Brea (Miller and Peck 1979, Doyen and Miller 1980, Miller 1983), additional small and large mammals have been described (W. Miller 1968, Akersten et al. 1979), and the diatoms of Pit 91 were being studied in late 1982. This activity represents a reawakening of research at La Brea.

The majority of the Rancho La Brea fossils come from deposits located within the twenty-three-acre area of Hancock Park (figs. 8.1, 8.2). This locality serves as the type for the Rancholabrean Land Mammal Age (Savage 1951, Savage and Downs 1954). Some smaller deposits are present to the east and southeast, as shown in Figure 8.1, but are covered by man-made structures. The small collections from these sites have not been studied. In 1975, during excavation for the Page Museum, which houses and displays the Natural History Museum of Los Angeles County (LACM) collection on the Rancho La Brea site, the first articulated skeletons of some of the extinct animals were unearthed (Duque and Barnes 1975). As of 1982, these fossils have yet to be studied.

The stratigraphy of the deposits has been summarized by Woodard and Marcus (1973, 1976). The vertebrate-containing deposits are underlain by a bed of sand, containing marine invertebrate fossils; this sand represents the last recorded marine incursion in the area. As far as is known, the main bone accumulations are restricted to the upper 30 feet of sediments. The bones and other fossils are found in sands and clays usually permeated with petroleum. The radiocarbon dates range from greater than 40,000 years (based on plants) to 4500 ± 80 yr B.P. (based on a wooden artifact). The collagen dates of bones for some elements of the extinct fauna range from $>36,000$ yr B.P. to approximately 11,000 yr. B.P., while one human skeleton is dated at 9000 ± 80 yr B.P.

History of Collecting

Rancho La Brea was recognized as a major fossil-bearing deposit in 1905, though the asphalt was commercially mined earlier. The site has been known as a place of asphalt accumulation since early historic times (see references and discussion in Stock

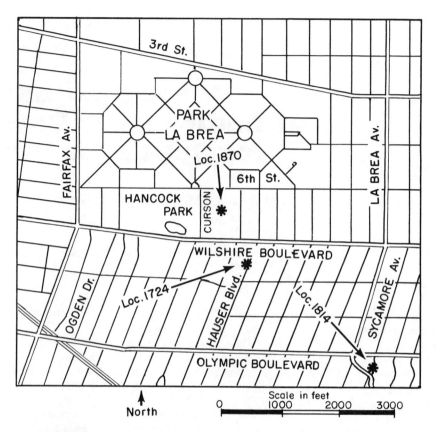

Figure 8.1. Location of Hancock Park Scientific Monument (Rancho La Brea) and nearby fossil-bearing brea localities (Loc.) of the Natural History Museum of Los Angeles County.

1956). Excavations were undertaken by the University of California intermittently until 1912; major excavations were terminated by that institution in 1913. The Natural History Museum of Los Angeles County (LACM) then conducted the largest exploration and collection activity to date between 1913 and 1915 (figs. 8.3, 8.4). The LACM carried on a number of additional explorations and excavations, most notably in 1929, 1945, and from 1969 to 1982 (locality LACM 6909, originally designated Pit 91) (fig. 8.5). Other institutions in the Los Angeles area also made collections before 1913 (Howard 1960, 1962). Beginning with the major University of California excavations, each area of excavation was divided into 3-foot squares which were assigned letter and number grid coordinates. A record was made of the depth to individual bone accumulations or skeletal elements whenever possible in the earlier excavations. A photographic record was made of much of this activity, and the photographs were placed on file at the LACM. The earlier collecting concentrated on well-preserved larger mammals, though good records of birds are available for some of the deposits. Many tiny fossils representing smaller individuals seem to have been missed. Much poorly preserved material was cut through because the bone was too fragile to be saved or was discarded during the frequent cave-ins that plagued the LACM excavators.

The excavations of the 1970s and 1980s, known as the Rancho La Brea Project (G. Miller and Singer 1970, Shaw 1982), have been done on a finer scale than earlier ones, with an emphasis on recording all macroscopic organic material and retaining subsamples for sedimentology and microfossil examination. Diatoms from some of these

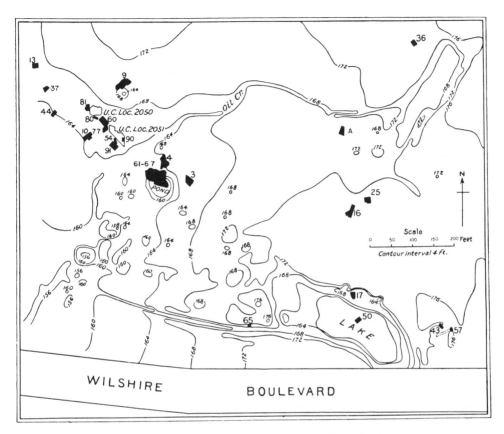

Figure 8.2. Topographic map of Rancho La Brea showing locations of principal excavations. Modified after Marcus 1960.

samples have been examined. Ostracoda and pollen have not yet been studied in detail. Detailed cross-sections, maps and photographs have been made for each grid square. Cave-ins have been prevented by shoring of the pit walls (G. Miller 1971).

Nature of the Collection

The bulk of the earlier collections consists of bones of large mammals. Bird bones are also very abundant. Vertebrate tissues other than bone and teeth are not usually preserved. Gretchen Sibley (pers. comm.) has recognized hair which probably came from some of the large mammals. The bone collagen has an amino acid composition comparable to that of Recent vertebrates (Ho 1966). The impregnation of the bone by petroleum seems to exclude water, which slows both the rate of hydrolysis of the protein and racemization of the amino acids. Plant fossils are excellently preserved, and leaves, seeds, cones, wood, and pollen are present. Insect chitin is well preserved, and there is an extensive collection of mostly disarticulated insects. Freshwater mollusk shells are also well preserved.

The largest collections of Rancho La Brea fossils are housed at the George C. Page Museum (a branch of the LACM) and the Museum of Paleontology of the University of California at Berkeley (UCMP). Small collections have been exchanged with

Figure 8.3. Pit 61–67. Looking southeast from the northwest wall of Pit 61, April 16, 1915. A typical La Brea Pleistocene fauna came from this large excavation. Several human artifacts were also found in this deposit. A three-foot-by-three-foot grid system was in use, but shoring was limited; collecting was oriented toward large specimens. Major cave-ins occurred in February of 1915; "timbering went down on the night of the 21st, and the pit this morning has an average depth of 8′ of muck" (Wyman field notes for Feb. 23, 1915).

other institutions, and many museums of the world have some Rancho La Brea material and frequently display composite mounted skeletons. Most exchange was done by the University of California; the LACM collections have been maintained essentially intact.

A complete catalog of the bird collection for the earlier LACM excavations is on magnetic tape (available on request from the senior author) and records more than 80,000 specimens. The mammal collection is incompletely cataloged, and additional taxa are being recognized in the process of recurating the early collections and in the examination of the newer collections (W. Miller 1968, Akersten et al. 1979, George Jefferson, pers. comm.). A partial catalog of the mammal collection is also available on tape.

Bones were largely disarticulated in the original deposits. Reconstruction of the distribution and association of remains is possible through use of the grid records kept with the specimens and recorded in the catalogs. Figure 8.6 is an example of the distribution of bones of the major extinct mammals (excluding uncataloged *Bison*) at the

Figure 8.4. Pit 91. Looking from the west end of a bone deposit left intact for future excavation, August 10, 1915. This is the deposit that was reopened in 1969.

9-foot level in Pit 3. In the excavation of Pit 91 the coordinates and orientation of each specimen have been recorded so that its position and relation to other specimens can be recovered. These data await detailed taphonomic study.

The Fauna

The deposit is best known for the large number of extinct genera and species of megafauna. Based on the catalogs of the LACM, it is estimated that parts of more than 3,400 individuals of coyote or larger-sized extinct mammals are preserved (Marcus 1960). The 5,845 individuals of birds (Howard 1962) represent several living as well as extinct species. The collection of mammals is dominated by large carnivores. The dire wolf (*Canis dirus*—48 percent) and sabertooth cat (*Smilodon floridanus*—30 percent) together account for approximately 78 percent of the individuals. In decreasing order of abundance, lesser numbers of coyotes (*Canis latrans*—7 percent), bison (*Bison antiquus*—5 percent), horse (*Equus occidentalis*—4 percent), large cats (*Panthera atrox*—2 percent), mylodont sloths (*Glossotherium harlani*—2 percent), camels

163

Figure 8.5. Pit 91. Looking west, in 1970, six months after the pit was reopened. A large bone mass was re-exposed approximately 10 feet below the surface datum. Shoring and sumps were installed to prevent flooding and caving. All material was screened for small fossils, and subsamples were saved for micro-fossils, sedimentology and other future studies. (Photo by Armando Solis)

(*Camelops hesternus*—1 percent), brea antelopes (*Capromeryx minor*—1 percent), and nothrothere sloths (*Nothrotheriops shastensis*—1 percent) are present. There are a few (fewer than twenty) individuals of short-faced bears (*Arctodus* sp.), grizzly or brown bears (*Ursus arctos*), black bears (*Ursus americanus*), mammoths (*Mammuthus columbi*), mastodonts (*Mammut americanus*), the Jefferson ground sloths (*Megalonyx jeffersonii*), mountain lions (*Felis concolor*), deer (*Odocoileus hemionus*) (Miller, G., in press), and the American antelope (*Antilocapra americanum*), which are not included in the percentages above. A tapir (*Tapirus* sp.) and peccary (*Platygonus* sp.) are represented by one individual each. The giant *Bison latifrons* and slender-limbed camel *Hemiauchenia* sp. were more recently recognized in the collections from Rancho La Brea (W. Miller 1968). W. Miller and Brotherson (1979) give further evidence for the presence of *B. latifrons* in response to the criticism (Schultz and Hillerud 1977) of Miller's earlier identification of this taxon. The jaguar (*Panthera onca*) has also been identified (Jefferson 1983). The census figures for smaller mammals have not been recorded, and counts may not be reliable for the older collections.

The dominant birds are Falconiformes (approximately 60 percent), owls (14 percent), fowl (13 percent), and perching birds (12 percent). Lesser numbers of water and shore birds and other groups make up the remainder of the census (Howard 1962).

The smaller collections of amphibians and reptiles have been summarized by Brattstrom (1953a, 1953b, 1958). Swift (1979) has recognized three species of freshwater fish in the collection.

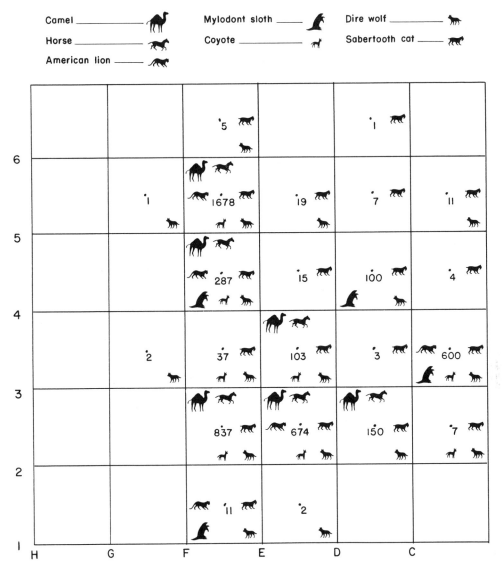

Figure 8.6. Plot of number of bones in Pit 3 at a depth of 9 feet in each grid square with indication of presence of large-sized species (*Bison* not included).

Additional uncataloged and unstudied mammal material is available. The figures presented in Marcus (1960) and Howard (1962) do not include later excavations or the UCMP collections, which are significant but smaller. Some of the early collections such as the Los Angeles High School collection are also excluded. The census by Marcus (1960) based on the available catalogs may give an underestimate of the numbers of larger mammals collected by LACM from 1913 to 1915. For example, G. Miller's (1968) careful analysis of all collections of *Smilodon* at the LACM located 2,100 individuals based on skulls or fragmentary skulls. This is more than twice as many individuals as reported by Marcus (1960) from the LACM catalogs. The discrepancy is accounted for in part by the incomplete cataloging of broken and juvenile specimens, and collections made before 1913.

A part of the insect fauna from the earlier collections was studied by Pierce (see Stock 1956 for references). Pierce named a number of supposedly extinct taxa, almost all of which are now recognized as synonyms of living species (S. Miller and Peck 1979, Doyen and S. Miller 1980, S. Miller et al. 1981, and Gagne and S. Miller 1982), although identifications of some taxa are difficult due to the fragmentary nature of the material. Seven orders of insects have been recorded. In terms of ecological groups, aquatic and semiaquatic species, carrion and dung scavengers, ground beetles, and assorted herbivorous species can be recognized. Bone damage by dermestid beetle larvae has also been found (S. Miller and Reynolds, in prep.). In addition, millipedes, arachnids, ostracods, isopods, and mollusks (both freshwater and terrestrial) are known but await study.

Evidence for Man at Rancho La Brea

As of 1982, there has been no published comprehensive study of the evidence for man at Rancho La Brea. Separate reports include the description of the Rancho La Brea skull by Kroeber (1962, published posthumously by Heizer 1962) and descriptions of atlatl foreshafts (Woodward 1937), a cogstone (Salls 1980), and possible artifacts (G. Miller 1969). A description of the human fossils has been prepared (Bromage and Shermis 1981). The partial human skeleton, radiocarbon-dated at 9000±80 yr B.P. (Berger et al. 1971), from Pit 10 was associated with a fauna that is believed not to include the extinct mammals typical of the late Pleistocene association, although extinct birds were in the same deposit with the human skeleton. Reynolds (in prep). discusses the details of the deposit and the relationships of the dated human skeleton and dated extinct *Equus* in the pit. The bones of the extinct large mammals show little to no pit wear and are in a different matrix than the human skeleton and associated bones that do show slight to extreme pit wear.

Woodward (1937) gave an inventory of the artifacts from Pits 61 and 67 (thought to be a single deposit and labeled 61–67 hereafter; see fig. 8.3). These consisted of worked antler, bone, shell and wood, and a fragment of a cog stone. A "quartz knife object which may be a crude knife blade" was found in one of the University of California (UC) excavations. Manufacture and wear pattern are evident on many of the tools, and "the assemblage compares favorably to others attributed to Shoshonean-speaking peoples who inhabited this area from ca. 10,000 B.P. to present" (Oswalt 1979). A domestic dog is also recorded from Pit 61–67 but was originally named as a separate species, *Canis petrolei* (Reynolds 1979).

Although the artifacts occurred over a considerable depth range in Pit 61–67, this deposit had been mined for asphalt before scientific study began and was used as a dump for the muck removed in the excavation of Pit 4. A number of marine shells were also found. One or more shells (number indicated in parentheses) occurred in Pit 3 (1 at 15 ft), Pit 4 (1 each at 5 ft and 18 ft), Pit 10 (6 between 9 ft and 13.5 ft), Pit 12 (1 at 5.5 ft), Pit 60 (1 at 12.5 ft), Pit 61–67 (44 between 8.5 ft and 19 ft or in caved walls or muck), Pit 81 (1 with no data), Pit UC2051 (1 with no data), and 4 others with "no data" or in the "Pond Muck" (census provided by R. Reynolds). It has been suggested that the mollusks were introduced by Indians of the area (Reynolds, pers. comm.), and the occurrence of artifacts in Pit 61–67 and the human skeleton in Pit 10 supports that suggestion for the majority of them. The shells from Pit 60, and the well-dated Pit 3 are more difficult to explain and might have been introduced from below through a vent or by unknown causes. Until all of the artifacts are thoroughly described and illustrated, it is difficult to evaluate them as evidence for the antiquity of man at La Brea.

One tibia of *Smilodon* with cuts attributed to human activity by G. Miller (1969) has been dated at 15,200±800 yr B.P. (UCLA 1292L from Pit 4 at 11.5 feet deep). Three other *Smilodon* bones (two from Pit 4 and one from Pit 77), one *Panthera atrox* bone (from Pit 3), and one *Bison* cf. *antiquus* bone from Pit 4 show similar markings. Preliminary experiments by G. Miller suggest that the cuts may have been made by stone tools. The bone from Pit 3 (the only well-stratified pit) would have the same age (within statistical error) as UCLA 1292L, based on its depth (13 feet) in that deposit.

An artifact made from elk antler (LACM HC6315) has also been found in the collection. It comes from Pit 77 at a depth of 10–11.5 feet.

In summary, the human remains and objects most accepted as artifacts are from deposits with mixed stratigraphy and give no clearcut evidence for man at La Brea prior to 9000±80 yr B.P. Questionable artifacts are dated at 15,200±800 yr B.P. or earlier.

Stratigraphy and Age of the Deposits

Using original field notes of the LACM excavators, test cores, and nearby well logs, Woodard and Marcus (1973, 1976) inferred that fluviatile processes played an important part in the accumulation of the bone deposits. They assigned the late Pleistocene sediments at Rancho La Brea to the Palos Verdes Sand, which they subdivided into members *A, B,* and *C* from older to younger. The bone-bearing deposits are contained largely within member *C,* comprising the upper 18–26 feet of the Palos Verdes Sand, and extend into the uppermost part of Member *B.* Unit *B* is a massive black asphaltic sand with gravel and clay lenses. Member *C* was further subdivided into submembers *a, b,* and *c.*

Valentine and Lipps (1970) interpreted marine fossils occurring in the lower part of Unit *B* (labeled *unit D* in their study) as belonging to a biozone correlated with dated marine deposits and marine terraces representing the last interglacial. However, the correlation of Valentine and Lipps may be in doubt. The assignment to this biozone was based on the presence of *Crassinella branneri* in the fauna, which was believed to be a warm-water indicator. *C. branneri* has now been synonomized with *C. pacifica,* which ranges from Laguna Beach, California, to Peru (Emerson 1980). The marine incursion recorded in the lower part of Unit *B* may represent the last interglacial or could be earlier.

Additional evidence is required before a date can be established for the last marine incursion in the northwestern Los Angeles Basin. If there should be an original hiatus representing emergence and erosion of the marine deposits (member *B*), before alluviation and accumulation of the bone-bearing deposits (member *C*), the marine biozone may not aid in setting a useful lower age limit for the bone deposits.

Submember *b* of Unit *C* consists of asphaltic sand with clay and gravel lenses. It contained the majority of the fossil bone concentrations. Bone is also found in the underlying, largely clay submember *a.* Submember *b* was interpreted as a continuous stratigraphic unit based on both the available radiocarbon dates and its occurrence at a similar depth level over much of the area (Woodard and Marcus 1973).

Dates for wood and asphalt from Rancho La Brea became available soon after radiocarbon-dating technology became widespread (the earlier dates are discussed in Howard 1960). A more extensive dating program was initiated when it became possible to date bone itself (Berger et al. 1964). The hypothesis of stratification of Woodard and Marcus (1973) and the speculations that different deposits were accumulated at different times (Marcus 1960, Howard 1962) could now be tested. More than 108 radiocarbon date determinations have been made on eighty-two samples from Rancho La Brea or proximate brea localities (Table 8.1). Seven laboratories have dated samples. The majority of dates are for bone collagen in the form of amino acids, which have been determined by the University of California at Los Angeles (UCLA) and Queens College (QC) radiocarbon laboratories. A simpler technique based on collagen in the form of gelatin has also been used (Longin 1971), and a few bone carbonate and bone apatite dates have been produced. A number of wood dates are available, but in most cases there was less care taken by the collectors in recording and retaining grid-coordinate and depth data for plant materials. One leaf date is available as well. In some cases the wood and bone dates are not concordant for similar stratigraphic levels, and there may be considerable difference among bone dates at the same depth.

Table 8.1. Rancho La Brea Date List

Depth (feet)	Coordinates	Age* (yr B.P.)	Average Error (yr)	dC-13	Material Dated	Species	Lab	Reference and Remarks
					Pit 3			
1–4.5	E-4	13,820	840	−23.04	BAmi-femur	*Smilodon*	QC401	
6	E-3	13,035	275		BAmi-femur	*Smilodon*	QC279	
6	E-3	13,745	275	−23.89	BAmi-humerus	*Smilodon*	QC414	
7	C-4	12,650	160		BAmi-femur	*Smilodon*	UCLA1292B	Berger and Libby 1968
11.5	C-4	14,400	2100		BAmi-femur	*Smilodon*	UCLA1292E	Berger and Libby 1968
12	E-2	14,500	190		BAmi-femur	*Smilodon*	UCLA1292C	Berger and Libby 1968
12	D-2	14,440	300		Wood	Cypress (trunk)	LJ55	Hubbs et al. 1960
		15,390	230		(same)	Cypress	Y354B	Deevey et al. 1959
		14,500	210		(same)	Cypress	Y354A	Deevey et al. 1959
		14,110	420		(same)	Cypress	Y355A	Deevey et al. 1959
		13,890	280		(same)	Cypress	Y355B	Deevey et al. 1959
		15,200	150		(same)	Cypress (limb)	QC422A	
		14,500	140		(same)	Cypress (root)	QC422B	
		> 28,000			Asphalt—from cavity in trunk		LJ89	Hubbs et al. 1960
14	E-5	14,350	175		BAmi-ulna	*Mammut*	UCLA1292T	Berger et al. in press
15	C-3	14,430	200		BAmi-tibia	*Mammut*	UCLA1292AA	Berger et al. in press
22	E-5	21,400	560		BAmi-femur	*Smilodon*	UCLA1292A	Berger and Libby 1968
		20,500	900		(same)	*Smilodon*	UCLA1292J	Berger and Libby 1968
22–25	E-3	9860	550		BCarb-humerus	*Canis dirus*	GEO	
26	E-4	19,300	395	−21.84	BAmi-femur	*Smilodon*	UCLA1292K	Berger and Libby 1968
		19,555	820		(same)	*Smilodon*	QC283	
					Pit 4			
4.5–8.5		19,800	300		BAmi-femur	*Smilodon*	UCLA1292R	Berger et al. in press
5	G-3	33,700	1600		Wood		UCLA773A	Berger and Libby 1966
5		35,300	2500		Wood		QC426	
8	D-2+4	26,700	900		BAmi-femur	*Smilodon*	UCLA1292G	Berger and Libby 1968
9	A-5	22,000	1200	−23.60	BAmi-femur	*Equus*	QC412	
10.5	D-3+4	13,500	170		BAmi-scapula	*Bison latifrons*	UCLA1292Q	Berger et al. in press
		27,000	1600		BAmi-axis	*Bison antiquus*	UCLA1292Z	Berger et al. in press
11.5	C-2	15,200	800		BAmi-tibia	*Smilodon*	UCLA1292L	Berger and Libby 1968
15	F-4+5	26,995	4000	−25.69	BAmi-femur	*Smilodon*	QC386	
15.5	D-2	28,000	1400		BAmi-femur	*Smilodon*	UCLA1292D	Berger and Libby 1968
18	F-4+5	35,500	2200		BAmi-femur	*Smilodon*	UCLA1292S	Berger et al. in press
18.5	B-5	12,760	150		BAmi-femur	*Smilodon*		Berger et al. in press
20–22	F-4+5	> 36,000			BAmi-femur	*Smilodon*	UCLA1292M	Berger et al. in press
23.5	B-5	29,600	1100		BAmi-femur	*Smilodon*	UCLA1292O	Berger et al. in press

Depth	Grid	Date	±	δ	Element	Species	Sample No.	Reference
Pit 9								
8.5		13,300	160		Wood		UCLA773D	Berger and Libby 1966
		13,430	210		(same)		QU724	
		13,120	230		(same)		QC429	
10.5		>38,600			Wood		QC423	
		>40,000			(same)		UCLA773B	Berger and Libby 1966
		>40,000			Wood		UCLA773F	Berger and Libby 1966
16		34,285	1675		(same)		QC424	
Pit 10								
4–5.5	F-11	15,700	530		BAmi-femur	*Equus*	UCLA1292CC	Berger et al. in press
5.5		5270	155		BCol-femur	*Ursus arctos*	QC916R	
6–9		9000	80		BAmi-femur	*Homo*	UCLA1292BB	Berger et al. 1971
Pit 13								
11	E-11	14,950	430		BAmi-femur	*Smilodon*	UCLA1292F	Berger and Libby 1968
13	G-10	15,360	480		BAmi-femur	*Smilodon*	QC339	
14.5	F-10	15,300	200		BAmi-femur	*Smilodon*	UCLA1292I	Berger and Libby 1968
20–23	F-11	14,310	920	−23.42	BAmi-femur	*Bison a.*	QC420	
Pit 16								
3–6		>32,850			BAmi-metatarsal	*Bison a.*	QC277II	
		12,275	775		BAmi-metatarsal	*Bison a.*	QC371	
4.5		>40,000		−24.66	Wood		UCLA773G	Berger and Libby 1966
6.5		19,485	275		(same)		QC427	
		18,430	500		(same)		QC427R	
8–12		10,710	320		BCarb-humerus	*Canis dirus*	GEO	
12		33,870	1350		Wood		QC428	
		>40,000			(same)		UCLA773E	Berger and Libby 1966
		>37,310			(same)		QU725	
		>38,780			(same)		QU767	
12–14		24,400	535		BAmi-metacarpal	*Bison a.*	QC278	
Pit 60								
8–9	C-10	27,900	2700		BAmi-femur	*Smilodon*	QC280	
9–9.5	F-11	>28,850		−23.68	BAmi-metapodial	*Equus*	QC365	
		24,460		1.14	BApa-(same)	*Equus*		
9–12		23,700	600		BAmi-femur	*Smilodon*	UCLA1292H	Berger and Libby 1968
12		7600	195		BApa-rib	*Glossotherium*	QC361	
		23,420	350		BAmi-(same)	*Glossotherium*	QC361	
14	C-12	24,900	3360		BAmi-tibia	*Equus*		
		22,330	1060		BApa-(same)	*Equus*	QC410	

Table 8.1. Rancho La Brea Date List

(continued)

Depth (feet)	Coordinates	Age* (yr B.P.)	Average Error (yr)	dC-13	Material Dated	Species	Lab	Reference and Remarks
					Pit 61–67			
(SE caved wall)								
10	D-16	4450	200		Wood	(atlatl shaft)	LJ121	Hubbs et al. 1960
15–20	H-10	12,000	125		BAmi-femur	Smilodon	UCLA1292X	Berger et al. in press
16–18.5	F-10	12,200	200		BAmi-femur	Smilodon	UCLA1292Y	Berger et al. in press
		11,640	135		BAmi-femur	Smilodon	QC302A	
		11,980	260	−22.00	BAmi-(same)	Smilodon	QC302B	
18–20	B-9	11,130	275	−2.35	BAmi-femur	Smilodon	QC413	
					BApa-(same)	Smilodon		
					Pit 77			
9–11	G-11	28,200	980		BAmi-femur	Smilodon	UCLA1292W	Berger et al. in press
13–15		33,100	600		BAmi-femur	Smilodon	UCLA1292U	Berger et al. in press
18.5–21	F-10	31,300	1350		BAmi-femur	Smilodon	UCLA1292V	Berger et al. in press
caved		29,470	1150		Wood		QC425	
		37,000	2660		(same)		UCLA773C	Berger and Libby 1966
					Pit 81			
caved		14,415	3250	−0.44	BApa-tibia	Equus	QC405	
		10,940	510	−23.76	BAmi-(same)	Equus		
					Pit 91			
6–7	L-5	30,800	600		BAmi-sacrum	Smilodon	UCLA1718	Berger et al. in press
6–8		8850	455		BApa-radius	Equus	QC384	
7.2–7.5	M-3+4	32,600	2800		BAmi-femur	Smilodon	UCLA1738D	Berger et al. in press
8–8.3	L-10	25,100	1100		BAmi-tibia	Smilodon	UCLA1738F	Berger et al. in press
8.1–8.5	F-11	29,100	1200		BAmi-femur	Smilodon	UCLA1738C	Berger et al. in press
8.5–8.8	I-6	25,100	850		BAmi-humerus	Smilodon	UCLA1738A	Berger et al. in press
8.8–9	N-11	33,000	1750		BAmi-humerus	Smilodon	UCLA1738B	Berger et al. in press
8.8–9	F-7	35,735	4050		Wood		QC658	
					Pit 2050			
10.8	D-10	30,470	1090		Wood		QC349A	
		30,870	1650		Wood		QC349B	
					Pit 2051			
5	N-21	22,890	500	−23.00	BAmi-rib	Glossotherium	QC443	

Level	Grid	Material	Species	Date (BP)	±	δ13C	Lab No.	Reference
6	I-3	BCol-femur	*Equus*	26,140	2200		QC430	
		BApa-(same)	*Equus*	6160	530			
6.8	G-5	BApa-humerus	*Smilodon*	20,410	2450		QC436	
		BAmi-(same)	*Smilodon*	13,950	1570			
8	T-22	BAmi-rib	*Glossotherium*	20,450	460	−24.15	QC390	
11	J-13	BAmi-rib	*Smilodon*	22,355	3400	−23.42	QC431	
15	H-19	BAmi-femur	*Smilodon*	23,850	1200		QC440	
15	D-13	BAmi-rib	*Glossotherium*	19,480	550		QC381	
16	D-7	BApa-humerus	*Smilodon*	20,900	2700	−24.17	QC435	
		BAmi-(same)	*Smilodon*	18,475	320			
16	Q-19	BApa-ulna	*Smilodon*	>29,760			QC438	
		BAmi-(same)	*Smilodon*	28,250	1030			
17	D-7	BApa-metacarpal	*Camelops*	17,630	1400	3.87	QC442	
		BAmi-(same)	*Camelops*	20,300	1750			

LACMNH LOC 1814

Material	Date (BP)	Lab No.
Wood	>46,500	QC504 (near Sycamore and La Brea)

LACMNH LOC 7247

Material	Species	Date (BP)	±	δ13C	Lab No.
BApa-metacarpal	*Bison*	23,630	4560	2.88	QC570

Page Museum Salvage—LGB 1383

Material	Date (BP)	±	Lab No.
BApa-bone fragments	6400	140	QC684

Samples from UC Berkeley

Material	Species	Date (BP)	±	Lab No.	Reference
Wood	Cypress	23,300	510	UCLA737A	Berger and Libby 1966
Leaves	*Quercus agrifolia*	32,350	1400	UCLA737B	Berger and Libby 1966

Miscellaneous La Brea Samples of Unknown Provenience

Material	Date (BP)	±	Lab No.	Reference
Wood	16,250	2000		Douglas 1952
Wood	16,400	2000		Douglas 1952
Asphalt	>34,000		LJ344, LJ345, LJ346	

BAmi = Bone Collagen dated as amino acids prepared following Ho et al. 1969

BCol = Bone Collagen changed to gelatin following Longin 1971

BCarb = Bone Carbonate

BApa = Bone Apatite

*All dates were computed with a half-life of 5568 years C-14. All averages or statistical tests were computed using the methods outlined in Long and Rippeteau (1974). A 5 percent significance level was used throughout.

Additional details on the nature of the separate deposits have been published by Woodard and Marcus (1973) and Howard (1960, 1962). All of the UCLA, QC and QU (University of Quebec) dates will be published in greater detail in the journal *Radiocarbon*.

A major concern for any series of radiocarbon dates is contamination. The asphalt environment of the Rancho La Brea deposits required special pretreatment to remove the petroleum contaminants. For plant materials, a variety of organic solvents were used in a Soxhlet distilling apparatus until petroleum was no longer extracted (Berger and Libby 1968). The wood was then burned to produce carbon dioxide. Bone has been dated on both organic and inorganic components. Two dates from Geochron Laboratories were on carbonate fractions and appear too young from the context of other nearby dates. Bone apatite dates obtained by the Queens College Radiocarbon Laboratory are also suspect. Apatite CO_2 carbon delta 13 values in parts per thousand relative to the PDB standard (Craig 1953) range from -2.35 to 3.87. Apatite from fresh bone is usually more negative, and values near zero suggest exchange with groundwater. Radiocarbon dates based on carbonate and apatite are not considered further in this chapter.

The potential for contamination was greatest for the organic bone components. It is known that collagen, the principal protein of bone, does not exchange carbon with the environment (Hassan and Hare 1978). However, sources of contamination might be organic preservatives; other organic materials, such as petroleum, introduced since deposition; or bacterial activity. The Rancho La Brea bones were treated following a specially developed method of Ho et al. (1969) to produce free amino acids. These were then burned to form carbon dioxide. Analyses of amino acids (Ho 1965, 1966, 1967; Ho et al. 1969) in Rancho La Brea bone give results comparable to Recent bone collagen, and there is no evidence for contamination by other proteins or organic substances. Depending on the preservation of the bone, between 75 and 300 grams of cleaned bone powder were used for preparation of amino acids. Some samples had to be supplemented with infinite-age carbon dioxide if they proved too small. This supplementation led to a decrease in the upper age limit of the method. The carbon 13 values for the amino acids ranged from -21.84 to -25.69 (Table 8.1). This is the range of values one would expect for animals whose diets consisted essentially of C3 plants (De Niro and Epstein 1978).

The oldest radiocarbon dates are for wood from Pits 9 and 16 dated at $>40,000$ yr B.P. A date of $>46,000$ yr B.P. was obtained for a wood specimen from a small brea deposit approximately one mile from Hancock Park (at the corner of Sycamore Avenue and La Brea Boulevard: LACMVP 1814, fig. 8.1), where a small fossil collection was made. The oldest finite bone collagen date is $35,500\pm2200$ yr B.P. for Pit 4, though there were several dates obtained (e.g., sample QC 365 in Pit 60) with limits younger than 40,000 yr B.P. when insufficient material was available for a finite age determination. It is worthwhile here to review the separate deposits and then summarize the overall picture.

Pit 3 had bones preserved throughout the 26-foot-deep excavation. It appeared to be divided into an upper and lower unit. This interpretation is supported by stratigraphic and field-note information (Wyman 1915, Woodard and Marcus 1973). A tree was rooted in a blue-clay deposit at 12 feet below datum, and the bones near the tree have a comparable age. The tree has now been dated by three laboratories. The dates are significantly heterogeneous, though the overall range of dates is from $13,890\pm280$ yr B.P. to $15,390\pm230$ yr B.P. Dates in the upper 15 feet range from $12,650\pm160$ to $15,390\pm230$ yr B.P. This part of the deposit may have accumulated over a relatively short period. There is a hiatus between the upper and lower parts of this deposit reflected in the sedimentary interpretations (Woodard and Marcus 1973). The date from 22 feet is significantly older than the date from 26 feet (see Table 8.1).

Pit 4 is the most complicated to interpret and is also the most difficult to explain by the hypothesis of stratification of the deposits. The deposit clearly consisted of at least three bone masses apparently related to three separate accumulations, designated

Vents I (grids A–C/4–6), II (grids D–E/4–5) and III (grids B–C/2–3) (Woodard and Marcus 1973). In Vent I a date of 12,760±150 yr B.P. was obtained at 18.5 feet between two much older dates in stratagraphic order. Two of three specimens marginal to Vent II (grids F/4–5) give the oldest bone dates for Rancho La Brea. There is only one date of 15,200±800 yr B.P. available for the more shallow Vent III. It is from a *Smilodon* tibia with peculiar oblique parallel grooves; G. Miller (1969) has suggested that this tibia is an artifact. Three out of four dates near the center of Pit 4 (grids D/2–4) from 8 feet to 15.5 feet are not significantly different from each other, averaging 27,060±680 yr B.P. The fourth, at 10.5 feet, gives a quite discordant date of 13,500±1700 yr B.P. The only wood date with grid coordinates from Pit 4 is at a shallow depth of 5 feet and nearest to Vent II. It is not significantly different in age from the oldest finite bone date in that vent. The wood may have been washed in and the deposit may have been reworked (either during the Pleistocene or by excavators), while fissures, petroleum vents or other causes may explain the heterogeneity. One marine shell (a human artifact?) occurred in Pit 4 at 18 feet.

Pit 9 was the deepest bone-bearing deposit with bone occurring below 30 feet. Only wood dates are available for Pit 9. There was a considerable amount of water in the deposit, and the bone was poorly preserved. Attempts to obtain sufficient collagen for a bone date have failed. This deposit contained more bones and individuals of mammoth and mastodon than any other deposit. The upper part yielded a wood date of 13,300±110 yr B.P. at 8.5 feet (average for three determinations by three laboratories), and non-finite or relatively old dates were obtained at 10.5 and 16 feet.

Pit 10 contains the only known human remains from Rancho La Brea. A human femur gave a date of 9000±80 yr B.P. (Berger et al. 1971, Berger 1975). The part of the deposit bearing the human skeleton contained artifacts and a mammal fauna of more recent aspect. This deposit was apparently intermixed with an older deposit containing more typical Rancho La Brea faunal elements, including *Equus* dated at 15,700±530 yr B.P. A date of 5270±155 yr B.P. was obtained for a grizzly or brown bear bone. This is the only pit that contains *Ursus arctos*. Extinct birds may be associated with the human skeleton.

Pit 13 appeared to contain a bone deposit completely capped by clay. All four dated bones from depths of 11 to 23 feet yielded ages which indicate that this deposit may have been active over a short period of time. The dates are not significantly different, and the average age is 15,220±170 yr B.P. The age of Pit 13 is comparable to part of the 11-to-15-foot interval of Pit 3. Pit 13 bones show an unusual amount of "pit wear" or abrasion, apparently caused by movement of the bones against one another in a plastic gritty matrix.

Pit 16 was a small circular deposit 27 feet deep. It yielded the largest collection of birds and the fewest large mammals in comparison to the other LACM major deposits. Coyotes are especially abundant in this pit. One wood specimen gave very discordant results among labs, and the other dates for Pit 16, like some of those for Pit 4, are especially difficult to explain in terms of stratification. There may have been some mixing in this deposit.

Pit 60 was excavated to 15.5 feet and was not large. Dates for this deposit range from 23,700±600 to 27,900±700 yr B.P. at 8–9 feet and >28,850 yr B.P. at 9–9.5 feet. Thus Pit 60 seems to represent a restricted period of time prior to the deposition of the lower part of Pit 3. The dates are significantly heterogeneous.

Pit 61–67 represented a series of connected pockets of asphalt reaching to a maximum depth of 20 feet. The age range for the fossils (11,130±275 yr B.P. to 12,200±200 yr B.P.) represents the shortest time span for any dated deposit at Rancho La Brea. The dates are, however, significantly different from each other, the youngest being the deepest. There were a number of cave-ins during excavation of this deposit.

Many human artifacts consisting of worked shell, bone, antler and wood were found in 61–67. A dated wooden artifact (exact depth and coordinates unknown) gives a date of 4450±200 yr B.P. for some of the human activity.

Pit 77 was excavated to 20.5 feet. Many large mammals but relatively few birds were recovered from Pit 77. It yielded the most consistent older bone dates (28,200±980 yr B.P. to 33,100±600 yr B.P.) for Rancho La Brea, though an older wood date is available for a specimen without stratigraphic information. There is a significant discordance for the two dates from different laboratories for this specimen.

Pit 81 has been dated only once. The majority of this bone deposit was removed intact as a block and is on display at the Page Museum. An *Equus* bone (whose exact position within the deposit is unknown) yielded the youngest bone date, 10,940±510 yr B.P., for the extinct fauna.

Pit 91 (LACM 6909) is a continuation of a deposit set aside for future study by the LACM in its original 1913–1915 excavations (see figs. 8.4, 8.5). Bone deposits were relocated at about 6 feet below the datum of 169–170 feet above sea level. The excavation has reached about 12 feet. The six bone dates available range from 25,100±1100 yr B.P. to 33,000±1750 yr B.P. over a small stratigraphic interval and probably represent at least two bone concentrations. This deposit is most comparable in age to Pits 60 and 77 on the basis of bone dates.

Pit UC 2050 represented the smaller of the two main UC excavations. A piece of wood available in the UC collections yielded two dates averaging 30,600±910 yr B.P. No bone specimens with stratigraphic information had been dated as of 1982 from this deposit.

Pit UC 2051 was the largest UC excavation and actually consisted of several separate bone accumulations (Stoner 1913). The deposit was earlier called Locality 1059 by UC. Pits 1 and 2 of LACM were in the sides of the large UC excavation, and the dates from these pits (Pit 1 represents D-13 and Pit 2 is T-22 on the UC grid) are included with those for Pit UC 2051. While there is no apparent stratification in the dates, the range of dates—13,950±1570 to 26,410±2200 yr B.P.—puts this deposit in the time period overlapping Pit 60 and Pit 3.

The miscellaneous dates for the UC collection and for material without detailed locality information are all consistent with the other Rancho La Brea dates, but further details are not available. The first dates for Rancho La Brea by Douglas (1952) are not thought to be reliable (Howard 1960).

The radiocarbon dates for Rancho La Brea are listed in Table 8.1 and summarized in Figure 8.7. Pits 61–67, 13 and the upper part of 3 represent the latter part of the Wisconsinan Age. The lower part of Pit 3 and parts of Pit UC 2051 are a little older. The middle time range of bone accumulation activity is represented by Pits 60, 91 and UC 2051 while Pits 77 and 91 represent the older bone-bearing deposits (ignoring the more confusing Pit 4 with the oldest dates, and Pits 16 and 9, which may be old). Only Pit 3 is clearly stratified into an older and younger part, but internally these parts are not stratified in detail. The older date for Pit 10 represents a date comparable to the upper parts of 3 or 13. There are no bone dates for La Brea between 19,300 yr B.P. and 15,700 yr B.P., though this range is slightly narrowed when standard errors are taken into consideration. This period was near the late Wisconsinan glacial maximum (Clague et al. 1980) and may have been a time of erosion. A gap of 3,600 years or longer for seventy dates distributed at random over a 30,000-year period would occur by chance with a probability between .01 and .05. This result is based on a Monte Carlo simulation.

There does not seem to be a simple relation between the age of the separate deposits and their geographic locations (see fig. 8.2). Most of the deposits fall along a northwest-southeast trend. If the controlling factor in deposition was stream activity and alluviation from the Santa Monica Mountains in an area of active petroleum seepage, then the separate times of activity of the individual deposits may represent a coincidence

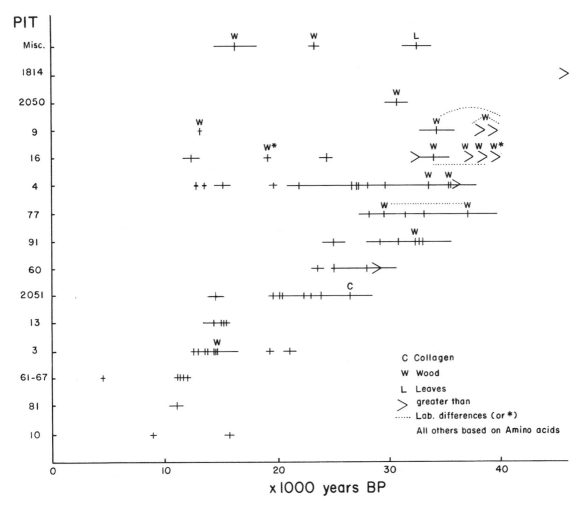

Figure 8.7. Distribution of all available radiocarbon dates by pit locality. Horizontal lines give range of standard errors for dates from Table 8.1.

of sand accumulation and formation of shallow tar pools in stream beds or locally ponded areas. The capping of some of the deposits by clay would represent periods of relative quiescence or when the stream channel had moved laterally (Akersten 1979).

The loss of a coherent stratigraphy in some of the deposits may represent natural cut and fill, that is, superimposed accumulation of fossils in stream-channel sands where petroleum had collected. Erosion of earlier formed deposits followed by new seepage, new entrapment and aggregation of younger fossils would produce new accumulations adjacent to earlier fossils. These activities in one area might be separated by many thousands of years. Petroleum seepage in vents through deposits would further complicate the picture. It is clear that no single simple explanation will account for all of the fossil distribution and dates in the different deposits, and each has to be approached individually. Only Pit 91 (figs. 8.4, 8.5) has been documented with sufficient attention to detail to allow a future accurate or complete reconstruction of accumulation at a part of the Rancho La Brea site.

The rate of amino acid racemization has been determined for Rancho La Brea bone by using a small piece of radiocarbon-dated UCLA 1738F from Pit 91 (Kvenvolden and Peterson 1972, Bada et al. 1973). Bada and Helfman (1975) determined that the racemization rate was slow for aspartic acid and suggested that this phenomenon was due to the lack of water in the petroleum-impregnated bone. Additional samples have been analyzed, and it appears that the rate of racemization is low for Rancho La Brea bone relative to other localities (McMenamin et al. 1982). If the rate is consistent, then this analysis may be a useful technique for dating Rancho La Brea bones. Since a much smaller sample (less than a gram rather than hundreds of grams) is required for this technique than for radiocarbon dating, some deposits whose ages are unknown could be dated. It may be possible to apply the method to bird bones as well.

Some of the earlier pit census data (Marcus 1960, Howard 1962) and morphometric analyses may be reinterpreted in terms of the age analysis of the deposits. However, it now appears that the Pit 3 collection should be divided into two parts for census and statistical comparisons. The census differences noted by Marcus (1960) are not clarified by the available dates. The only obvious difference is a smaller percentage of coyotes in Pit 77. Howard (1962) constructed several indices based on the bird census, which appeared to be useful for determining the relative ages of the deposits. Several of the deposits with significant bird accumulations have not yet been dated, however. For the dated pits, Pit 77 has the most individuals of extinct species, followed by Pit 61–67 (the youngest pit). On this criterion, Pits 13, 60 and 77 fall into the correct age order. Pit 10 of the dated pits has the smallest percentage of individuals of extinct species. Pits 3, 4, 13, 16, 61–67 and 77 were considered "older" in Howard's analysis. It is unfortunate that dates are not available for the "younger" pits (37,28,36, *A* and *Acad*) included in Howard's study. Based on thirteen of what Howard called "critical" extinct species forming part of the nucleus of the Rancho La Brea Pleistocene avifauna, Pits 3, 61–67 and 16 appear to be younger than Pits 4, 13, 77 and 60.

The statistical analyses of Menard (1947) on *Smilodon* metapodials and Nigra and Lance (1947) for *Canis dirus* metapodials demonstrated differences among the pits, some of which can be shown to be statistically significant. *Smilodon* metapodials from Pits 13, 3 and 67 are significantly larger on the average than those in Pit 77. Perhaps *Smilodon* was becoming larger before it became extinct? For *Canis dirus,* specimens from Pits 13 and 3 average smallest, and those from Pit 4 are the largest. Those from Pits 61–67 and 77 are nearly the same intermediate size. Thus, a different trend in size is observed in *C. dirus* from that in *Smilodon.*

Evidence for Extinction

The first suite of bones selected for dating was chosen to test the hypothesis of stratification within deposits and determine the range of activity for the separate deposits. The dense shaft bone of an individual femur of *Smilodon floridanus* provided sufficient material for one date (or in some cases two) following the method of Ho et al. (1969). An attempt was made to find specimens with the youngest and oldest possible dates from pits with larger collections. The latest date for *Smilodon* is 11,130±275 yr B.P. from an 18–20-foot depth in Pit 61–67. Two others are nearly as young but significantly different (see Table 8.1). A second suite of specimens from a variety of species is being dated to corroborate the earlier series, for which *Smilodon* could not be obtained from some deposits. This series has produced the youngest date for *Equus,* 10,940±540 yr B.P., from Pit 81 at an unknown depth. These two youngest dates are not significantly different and together average 10,900±240 yr B.P. *Bison antiquus* dated from Pit 13 (12,275±775 yr B.P.) is also not significantly different in age from these two, and the three together average 11,200±230 yr B.P. If this is taken as the youngest date for the Rancho La Brea extinct fauna, then these three individuals died more than 10,810 years ago (95 percent confidence level). Separately the youngest *Equus, Smilodon* and *Bison* give upper limits of 10,050 yr B.P., 10,680 yr B.P. and 11,000 yr B.P. The best estimate of the date for extinction of the Pleistocene fauna in the vicinity of Rancho La Brea is thus somewhat later than 11,000 yr B.P. Note that the young date for *Canis dirus* (9860±550 yr B.P. from Pit 3) reported in Kurtén and Anderson (1980) and Table 8.1 is a carbonate date and is thought to be too young.

The next youngest date is for a human skeleton in Pit 10 at 9000±80 yr B.P. associated with a largely Recent mammalian fauna and a small percentage of individuals of extinct birds (Howard and Miller 1939, Howard 1962). The human skeleton and the Recent bones show slight pit wear, while the extinct mammal fossils (including the dated horse bone) are preserved in a distinctive matrix, have a different color and texture, and do not show pit wear (R. Reynolds, pers. comm.). The youngest date on a large mammal, 5270±155 yr B.P., is on bone collagen of *Ursus arctos,* the living grizzly bear.

Extinction for the large mammals does not appear to be spread over a period of time, as all of the more common extinct species are preserved in pits spanning the entire time of recorded activity at Rancho La Brea. Thus, it appears that extinction of the mammalian megafauna occurred over a relatively short time span.

The evidence for extinction of birds is less clear-cut because Pit 10 preserves eleven of the nineteen extinct species and seven of the eleven extinct genera at La Brea (Howard 1962) in the same matrix and state of preservation as the fossils associated with the human skeleton dated at 9000±80 yr B.P. The other "younger" pits, that is, 37, 36 and 28 (the less well-documented Pits *A* and *Acad* are excluded) preserve four, sixteen and seven extinct species and one, ten and four extinct genera respectively, although some of the species assignments are questionable. These numbers were obtained from an examination of the complete bird catalog and are slightly larger than those published in Howard (1962). Pits 28, 36 and 37 have not been dated. Pit 36 does have remains of several extinct mammalian species, and Pits 28 and 37 each have a few bones of *Equus* (recorded in the mammal catalog—those from Pit 37 appear to be from modern *Equus asinus*). Thus some of the bird extinctions may have occurred later than 11,000 yr B.P. and, based on their associations in Pit 10, perhaps later than 9000 yr B.P. Using the census data in Howard (1962), but a less complete date list than included here, Grayson (1977) concluded that the eleven genera included in the "younger" deposits (including "pits" *A* and *Acad*) "became extinct during very late Pleistocene times." The evidence cited above for Pit 10 suggests that many of the extinct genera of birds discussed by Grayson became extinct later. Additional dates are needed to test the hypothesis that the so-called "younger" pits are indeed younger. The lack of large mammal bones makes these deposits difficult to date. The extinction of the birds at Rancho La Brea may have occurred more gradually than the extinction of the mammals but probably still over a rather short period of time.

Accumulation of the Fossil Deposit

The high ratio of carnivores to herbivores is explained by the "death trap of the ages" model: trapped herbivores attracted carnivores which were in turn trapped by the asphalt. This is the only model that has been proposed to explain the unusual excess of carnivores. Dire wolves, sabertooth cats and other large cats combined outnumber large prey mammals by more than six to one, while carnivorous and carrion-feeding birds also dominate the avian census. In Africa, Recent carnivorous mammals typically make up less than 4 percent of the biomass of communities of large-sized mammals (Beland and Russel 1979).

Small outpourings of oil are seen at the site today, and insects, reptiles, amphibians and birds may be trapped in them. In times of more active seepage, larger pools of sticky liquid asphalt would have easily accumulated. Unwary mammals could be caught and would attract carnivores or carrion-feeders who in turn would be caught. Streams from the nearby Santa Monica Mountains built up the alluvial floodplain deposits. Ponds or marshes apparently occurred in the area as they did in historic times. Faults provided fissures through which petroleum from underlying late Tertiary sediments migrated upwards and laterally into the porous stream gravel and sands. In turn, pools of liquid asphalt formed in shallow depressions in the surface alluvium. A model of shallow

petroleum accumulation, with trapped animals subsequently buried by alluviation processes, repeated cycles of petroleum seepage, new entrapments and accumulations is
suggested from the detailed analysis of the 1970s excavations (Maloney and Daigh 1971,
Maloney and Warter 1974, Maloney and Akersten 1976, Maloney 1970, Akersten 1979).

The Mythology of Rancho La Brea

A number of myths have been perpetuated about La Brea. The appeal to fantasy
was strong even for the most serious researcher: "These tar traps during their active
periods appear verily to have been ulcerous spots on the face of Nature, where death
mocked the comparative serenity of life in the open.... The cries and struggles of the
wounded, the not infrequent stench of offal, and the fierce combats of those not yet
mired may well have made of the individual trap a quagmire whose horrors and iniquities
are now veiled with the inevitable passing of geological time" (Stock 1929, quoted in
Savage and Downs 1954).

For the more than 450 large herbivores (*Capromeryx* or larger) censused in the
LACM collections (Marcus 1960), an entrapment event on the average no more frequent than once every fifty years would easily explain the numbers of individuals preserved over the last 25,000 to 30,000 years of the Pleistocene. Even if the total
numbers of individuals were greater by a factor of ten than the collections censused,
entrapments would have been relatively rare events, less often than once every five
years. An oversimplified Poisson model assuming constant probability for entrapment
leads to the conclusion that if the census represented only one percent of the animals
trapped, two or more entrapments would occur only about once every other year. It is
difficult to provide a more realistic model at this time, but this scenario was not the one
envisioned by some of the earlier, more dramatic authors.

Another common belief about the deposits is that there must have been a great
amount of mixing or "churning" (i.e., vertical and horizontal redistribution of specimens
in the deposit). After excavation and exploration were completed, the open pits were
left to fill naturally by rain and groundwater (the water table is high in the area) and to be
covered by petroleum seeping from below. The large gas bubbles escaping through the
petroleum and water that are to be seen today in the man-made lake (shown on the
cover of *Science* for 30 May 1969) were unlike the appearance of many of the deposits at
the time of original excavation. Many of the original deposits were covered with an
oxidized and hardened asphalt cap, commonly outcropping above ground level. This was
occasionally removed during excavation by blasting to expose the more easily worked
bone-bearing matrix below. Narrow deposits of asphalt filled chimneys, pipes and fissures occurred, and some specimens no doubt moved vertically and horizontally through
these channels. Some of the deposits show clear stratification, and some show mixing,
while others represent a relatively short period of time. The two published descriptions
of the earlier major excavations (Stoner 1913, Woodard and Marcus 1973—based in part
on the field notes of Wyman, which lacked detailed geological observations) and the
more recent observations for Pit 91 depict perhaps the different local features encountered in the separate deposits. The earlier reconstructions may have been prejudiced by
the excavation techniques used and possibly by the beliefs of the excavators.

It has been claimed that the collections have an excess of juvenile and aged
individuals, as well as a large percentage of diseased or maimed individuals for some
species (Merriam 1911, based on an early unpublished census). An age-distribution
study of *Smilodon* by G. Miller (1968), using all available material from the earlier LACM
collections, showed that all age classes were present and that the age distribution was
consistent with that found for living populations of the African lion today. Howard (1945)
found that 23 percent of 750 individuals of the turkey *Parapavo* represented by tar-
sometatarsi were from individuals less than one year old. Howard suggested that over
the years *Parapavo* bred in the area. Additional studies are required for the other

species. Pathological specimens are abundant at Rancho La Brea, but the collection is large and represents an unusual situation for sampling greater absolute numbers of normal and pathological individuals from once-living populations than does any other known deposit. This is especially true for carnivores. A better test of the speculations about proportion of young and old or diseased and maimed would be to compare the collections at Rancho La Brea to the accumulated data from other Rancholabrean sites or comparable Recent collections.

Late Pleistocene Environment at Rancho La Brea

Johnson (1977a, b) has summarized the evidence for climatic conditions in southern California during Rancholabrean time. The more than 25,000-year period recorded at Rancho La Brea covers periods of Wisconsinan glacial maxima and interstadials. Johnson suggests that "full-glacial winters in coastal California may have resembled modern winters, but summers may have been cooler and perhaps more moist than now (but still relatively dry)." The present Mediterranean climate of this area has wet and cool winters with nearly rainless summers.

Evidence for the climate at La Brea comes mostly from an analysis of the plant remains preserved there. The detailed study by Warter (1976, 1979, 1980) of the "stream-drift" preserved in Pit 91 (radiocarbon dates for Pit 91 range from 25,100±850 yr B.P. to 33,000±1750 yr B.P.) records a variety of communities preserved in what was apparently a stream-fed marshy channel. Some of the plant fossils undoubtedly rafted in from some distance during floods. Wood and cones from nearby cypress and pine trees formed part of the abundantly represented closed-cone pine association that was believed "to have been growing in the immediate vicinity of Rancho La Brea" (Warter 1976, p. 35). A smaller percentage of a chaparral and foothill woodland association was presumably washed down from the nearby Santa Monica Mountains. *Sequoia* wood (Coast redwood) preserved at Rancho La Brea is assumed to have drifted in from the more sheltered valleys of the Santa Monica Mountains. Aquatic and emergent plants associated with pond snails, tiny clams and ostracods indicate the presence of standing water in the form of freshwater ponds or marshes. Perhaps small ponds existed adjacent to the main stream channel during the time these sediments were laid down. The diatom assemblage also supports this interpretation (Jon Sperling, pers. comm.).

A riparian or streambank assemblage was also identified and was also presumably drifted in from riparian situations further upstream, closer to the foothills. Finally, a miscellaneous group of drier-adapted herbs was identified on the basis of seeds and pollen and perhaps represents elements of several of the associations mentioned above.

The dates for Pit 91 may represent an interstadial when Rancho La Brea would have been nearer the ocean than during a glacial maximum. This might account for the coastal plant association, typical today of the Monterey Peninsula. "This southward extension (200+ miles) of coastal pine and cypress forest, and the presence of upland vegetation at lower elevations suggests 25–30+ inches of glacial winter rainfall, and cool but dry summers" (Warter 1980). During glacial maxima, Rancho La Brea most likely would have been a few miles farther from the coast. A flora for the last glacial maximum for Rancho La Brea has not been studied in detail. Note that the period from 15,700–19,300 yr B.P. is almost dateless (see fig. 8.7).

Carbon 13 analyses of bone collagen give values in the range of C3 plants or C3 plant consumers (De Niro and Epstein 1978). The carbon delta 13 values are for *Equus* −23.6 and −23.8, for *Bison antiquus* −23.4 and −24.7, and for *Glossotherium* −23.0 and −24.2, while the values for their potential predator *Smilodon* range from −21.8 to −25.7. The herbivores represent both grazers and browsers. There is a strong correlation between the percentage of C4 grass species and normal July daily minimum temperatures in North America (Teeri and Stowe 1976). The nearest sampled location to Rancho La Brea was San Luis Obispo, California, which has 18 percent C4 species.

Arizona and the Sonoran Desert localities included in Teeri and Stowe have 57 percent and 82 percent C4 grass species respectively.

The three species of fish preserved in Pit 91 "definitely indicate permanent stream conditions" (Swift 1979). A juvenile rainbow trout, *Salmo gairdnerii*; an adult arroyo chub, *Gila orcutti*; and spines of the three-spined stickleback, *Gasterosteus aculeatus*, represent species common and native to the Los Angeles Basin today. The "exclusively small salmonids and adult stickleback suggest a winter or early spring deposition" (Swift 1979).

Warm, relatively dry summers and wetter winters suggest that the streams were more active in the winter. Akersten (1979) suggested that sand was accumulated in the stream, while clay was deposited lateral to the channels: "As summer approached, the streams dried and asphalt (now warm and less viscous than during the winter) seeped through the channel sands to form shallow puddles. When concealed by dust and leaves, these puddles would trap an occasional herbivore which would in turn lure predators and scavengers to their fates."

Acknowledgments

Many individuals have contributed to producing the data on which this chapter is based. Richard Pardi, assisted by Carlos Perdomo and Marion Newman, ran all of the later bone dates at the Queens College Radiocarbon Laboratory. Richard Reynolds helped with the selection of datable bones, did much of the early pretreatment and unselfishly shared his vast knowledge of Rancho La Brea with us. Claude Hilar-Marcell's laboratory at the University of Quebec, Montreal, provided the carbon 13 analyses. The Natural History Museum of Los Angeles County and the University of California Museum of Paleontology kindly loaned specimens for dating. Computer graphics for Figure 8.6 were done at the Health Sciences Computing Facility at the University of California Los Angeles, and all figures were redrawn by Raymond Gooris.

Conversations with G. Davidson Woodard, Richard Reynolds, William Akersten, William K. Emerson, Walter Newman, Richard Pardi, Scott Miller and George Jefferson contributed to our understanding of the complex problems at Rancho La Brea. We are responsible for any distortion of the ideas or facts that they contributed. All of the above-named individuals, as well as Hildegarde Howard, Theodore Downs, George Miller, Nancy Neff and Janet Warter, provided useful comments on a draft of this chapter.

This study was partially supported by National Science Foundation and City University of New York PSC BHE grants. Excavations of Pit 91 were partially supported by grants from the LACM Foundation and the National Science Foundation.

References

Akersten, W. A. 1979. Genesis of Rancho La Brea fossil deposits. *Abstracts,* So. Calif. Acad. Sci. Annual Meeting, May 5–6, p. 28.

Akersten, W. A., R. L. Reynolds and A. E. Tejada-Flores. 1979. New mammalian records from the late Pleistocene of Rancho La Brea. *Bull. So. Calif. Acad. Sci.* 78(2):141–143.

Axelrod, D. I. 1967. Quaternary extinctions of large mammals. *University of California Publications in Geological Science* 74:1–42.

Bada, J. L., and P. M. Helfman. 1975. Amino acid racemization dating of fossil bones. *World Archaeology* 7:160–173.

Bada, J. L., K. A. Kvenvolden and E. Peterson. 1973. Racemization in amino acids in bones. *Nature* 245(5424):308–310.

Bada, J. L., R. A. Schroeder and G. F. Carter. 1974. New evidence for the antiquity of man in North America deduced from aspartic acid racemization. *Science* 184:791–793.

Bada, J. L., R. A. Schroeder, R. Protsch and R. Berger. 1974. Concordance of collagen-based radiocarbon and aspartic-acid racemization ages. *Proc. Nat. Acad. Sci.* 71(3):914–917.

Beland, P., and D. A. Russel. 1979. Ectothermy in dinosaurs: paleoecological evidence from Dinosaur Provincial Park, Alberta. *Can. J. Earth Sci.* 16:250–255.

Berger, R. 1975. Advances and results in radiocarbon dating: early man in America. *World Archaeology* 7(2):174–184.

Berger, R., and W. F. Libby. 1966. UCLA radiocarbon dates V. *Radiocarbon* 8:467–497.

———. 1968. UCLA radiocarbon dates VIII. *Radiocarbon* 10(2):402–416.

Berger, R., A. B. Horney and W. F. Libby. 1964. Radiocarbon dating bone and shell from their organic components. *Science* 144:999–1001.

Berger, R., R. Protsch and R. Reynolds. In press. UCLA Radiocarbon Dates X. *Radiocarbon*.

Berger, R., R. Protsch, R. Reynolds, C. Rozaire and J. R. Sackett. 1971. New radiocarbon dates based on bone collagen of California paleoindians. In Stross, F. H. (ed.), The Application of the Physical Sciences to Archaeology, *Contributions of the University of California Archaeological Research Facility* 12:43–49.

Brattstrom, B. H. 1953a. The amphibians and reptiles from Rancho La Brea. *Transactions of the San Diego Society of Natural History* 11(14):365–392.

———. 1953b. Records of Pleistocene reptiles from California. *Copeia* 1953(3):174–179.

———. 1958. New records of cenozoic amphibians and reptiles from California. *Bull. So. Calif. Acad. Sci.* 57(1):5–13.

Bromage, T. G., and S. Shermis. 1981, The La Brea woman (HC 1323): descriptive analysis. In *Contributions to Western Archeology, Occasional Papers of the Society for California Archeology* 3:59–75.

Campbell, K. E., Jr. 1979. The non-Passerine Pleistocene avifauna of the Talara tar seeps, northwestern Peru. *Royal Ontario Museum Life Sci., Cont.* 118:1–203.

Churcher, C. S. 1959. Fossil *Canis* from the tar pits of La Brea, Peru. *Science* 130:564–565.

Clague, J. J., J. Armstrong and W. H. Mathews. 1980. Advance of the late Wisconsin Cordilleran ice sheet in southern British Columbia since 22,000 BP. *Quaternary Research* 13:322–326.

Craig, H. 1953. The geochemistry of the stable carbon isotopes. *Geochimica and Cosmochimica Acta* 3:53–92.

De Niro, M. J., and S. Epstein, 1978. Carbon isotope evidence for different feeding patterns in the *Hyrax* species occupying the same habitat. *Science* 201:906–908.

Deevey, E. S., L. J. Gralenski and V. Hoffren. 1959. Yale natural radiocarbon measurements IV. *American Journal of Science, Radiocarbon Supplement* 1:144–172.

Douglas, D. L. 1952. Measuring low-level radioactivity. *General Electric Review* 55:16–20.

Doyen, J. T., and S. E. Miller. 1980. Review of Pleistocene darkling ground beetles of the California asphalt deposits (Coleoptera: Tenebrionidae, Zopheridae). *Pan-Pacific Entomologist* 56(1):1–10.

Duque, J., and L. G. Barnes. 1975. *Smilodon*, is this how you looked? *Terra* 14(1):18–24.

Emerson, W. K. 1980. Invertebrate faunules of late Pleistocene age, with zoogeographic implications, from Turtle Bay, Baja California Sur, Mexico. *The Nautilus* 94(2):67–89.

Gagne, R. J., and S. E. Miller. 1982. *Protochrysomyia howardae* from Rancho La Brea, California, Pleistocene, new junior synonym of *Cochliomyia macellaria* (Diptera: Calliphoridae). *Bull. So. Calif. Acad. Sci.* 80:95–96.

Grayson, D. K. 1977. Pleistocene avifaunas and the overkill hypothesis. *Science* 195:691–693.

Hassan, A. A., and P. E. Hare. 1978. Amino acid analysis in radiocarbon. *Archaeological Chemistry* II, pp. 109–116. American Chemical Society.

Hassan, A. A., J. D. Termine and C. V. Haynes. 1977. Mineralogical studies on bone apatite and their implications for radiocarbon dating. *Radiocarbon* 19(3):364–374.

Heizer, R. F. 1962. "Prefatory remarks" to "The Rancho La Brea skull," posthumous paper by A. L. Kroeber. *American Antiquity* 27(3):416–417.

Ho, T. Y. 1965. The amino acid composition of bone and tooth proteins in late Pleistocene mammals. *Proc. Nat. Acad. Sci.* 54(1):26–31.

———. 1966. The isolation and amino acid composition of the bone collagen in Pleistocene mammals. *Comp. Biochem. Physiol.* 18:353–358.

———. 1967. Relationship between amino acid contents of mammalian bone collagen and body temperature as a basis for estimation of body temperature of prehistoric mammals. *Comp. Biochem. Physiol.* 22:113–119.

Ho, T. Y., L. F. Marcus and R. Berger. 1969. Radiocarbon dating of petroleum-impregnated bone from tar pits at Rancho La Brea, California. *Science* 164:1051–1052.

Howard, H. 1945. Observations on young tar-sometatarsi of the fossil turkey *Parapavo californicus* (Miller). *The Auk* 62:596–603.

———. 1960. Significance of carbon-14 dates for Rancho La Brea. *Science* 131:712–714.

———. 1962. A comparison of avian assemblages from individual pits at Rancho La Brea, California. *Contrib. Sci. Natur. Hist. Mus. Los Angeles County* 58:1–24.

Howard, H., and A. H. Miller. 1939. The avifauna associated with human remains at Rancho La Brea. Carnegie Inst. Wash., *Pub. No.* 514:39–48.

Hubbs, C. L., G. S. Bien and H. E. Suess. 1960. La Jolla natural radiocarbon measurements. *American Journal of Science, Radiocarbon Supplement* 2:197–223.

———. 1962. La Jolla natural radiocarbon measurements II. *Radiocarbon* 4:204–238.

Jefferson, G. T. 1983. First record of jaguar from the late Pleistocene of California. *Bull. So. Calif. Acad. Sci.*

Johnson, D. L. 1977a. The California ice-age refugium and the Rancholabrea extinction problem. *Quaternary Research* 8:149–153.

———. 1977b. The Late Quaternary climate of coastal California: evidence for an ice age refugium. *Quaternary Research* 8:154–179.

Kroeber, A. L. 1962. The Rancho La Brea skull. *American Antiquity* 27(3):416–417.

Kurtén, B., and E. Anderson. 1980. *Pleistocene mammals of North America*. Columbia University Press, New York, 442 pp.

Kvenvolden, K. A. and E. Peterson, 1972. Amino acids in late Pleistocene bone from Rancho La Brea, California (Abstract). *Geol. Soc. Amer. Abstracts with Programs* 5(7):704–705.

Long, A., and B. Rippeteau, 1974. Testing contemporaneity and averaging radiocarbon dates. *American Antiquity* 39:205–215.

Longin, R. 1971. New method of collagen extraction for radiocarbon dating. *Nature* 230:241–242.

McMenamin, M. A. S., D. J. Blunt, K. A. Kvenvolden, S. E. Miller, L. F. Marcus, and R. R. Pardi. 1982. Amino acid geochemistry of fossil bones from the Rancho La Brea asphalt deposit, California. *Quaternary Research* 18:174–183.

Maloney, N. J. 1970. Late Pleistocene sedimentation and asphalt entrapment, Rancho La Brea (Abstract). *Abstracts, So. Calif. Acad. Sci.* Annual Meeting, May 5–6, p. 28.

Maloney, N. J., and Akersten, W. A. 1976. Formation of calcareous sandstone at asphalt-groundwater contacts in fluvial sediments, Rancho La Brea, California. *Geological Society of America Cordilleran Section, Abstracts with Programs* 8(3):393.

Maloney, N. J., and J. Daigh. 1971. Sediment facies in Pit 91, Rancho La Brea Tar Pits, California. *Geological Society of America Cordilleran Section, Abstracts with Programs* 3(2):156–157.

Maloney, N. J., and J. K. Warter. 1974. Probable origin of the deposits in Pit 91, Rancho La Brea Tar Pits, California. *Geological Society of America Cordilleran Section, Abstracts with Programs* 6(3):212.

Maloney, N. J., J. J. Criscione and L. L. Bramlett. 1973. Fluvial sedimentation at Rancho La Brea *Geological Society of America Cordilleran Section, Abstracts with Programs* 5(1):77.

Marcus, L. F. 1960. A census of the abundant large Pleistocene mammals from Rancho La Brea. *Contrib. Sci. Natur. Hist. Mus. Los Angeles County* 38:1–11.

Masters, P. M., and J. L. Bada. 1978. Amino acid racemization dating of bone and shell. *Archaeological Chemistry* II, pp. 117–138. American Chemical Society.

Menard, H. W. 1947. Analysis of measurements in length of the metapodials of *Smilodon*. *Bull. So. Calif. Acad. Sci.* 46(3):127–135.

Merriam, J. C. 1911. The fauna of Rancho La Brea. Part 1. Occurrence. *Memoirs of the University of California* 1(2):199–213.

———. 1914. Preliminary report on the discovery of human remains in an asphalt deposit at Rancho La Brea. *Science* 90(1023):198–203.

Miller, G. J. 1968. On the age distribution of *Smilodon californicus* Bovard from Rancho La Brea. *Contrib. Sci. Natur. Hist. Mus. Los Angeles County* 131:1–17.

———. 1969. Man and *Smilodon*: a preliminary report on their possible coexistence at Rancho La Brea. *Contrib. Sci. Natur. Hist. Mus. Los Angeles County* 163:1–8.

———. 1971. Some new improved methods for recovering and preparing fossils as developed on the Rancho La Brea Project. *Curator* 14(4):293–307.

———. In press. A new subgenus and two new species of *Odocoileus* (Mammalia, Cervidae) with a study of variation in antlers. University of Oregon Press.

Miller, G. J., and H. Singer. 1970. The Rancho La Brea Project: 1969–1970. *Los Angeles County Museum of Natural History Quarterly* 9(1):26–30.

Miller, S. E. 1979. Pleistocene insects of Rancho La Brea and other California asphalt deposits. *Abstracts, So. Calif. Acad. Sci.* Annual Meeting, May 5–6, p. 46.

———. 1983. Late Quaternary insects at Rancho La Brea and McKittrick, California. *Quaternary Research.*

Miller, S. E., and S. B. Peck. 1979. Fossil carrion beetles of Pleistocene California asphalt deposits, with a synopsis of Holocene California Silphidae (Insecta; Coleoptera: Silphidae). *Trans. San Diego Soc. Natural History* 19(8):85–106.

Miller, S. E., R. D. Gordon and H. F. Howden. 1981. Reevaluation of Pleistocene scarab beetles from Rancho La Brea, California (Coleoptera: Scarabaeidae). *Proc. Entomol. Soc. Wash.* 83(4):625–630.

Miller, W. E. 1968. Occurrence of a giant bison, *Bison latifrons,* and a slender-limbed camel, *Tanupolama,* at Rancho La Brea. *Contrib. Sci. Natur. Hist. Mus. Los Angeles County* 147:1–9.

———. 1971. Pleistocene vertebrates of the Los Angeles basin and vicinity (exclusive of Rancho La Brea). *Natural History Museum of Los Angeles County Science Bulletin* 10:1–124.

Miller, W. E., and J. D. Brotherson. 1979. Size variation in foot elements of *Bison* from Rancho La Brea. *Contrib. Sci. Natur. Hist. Mus. Los Angeles County* 323:1–19.

Moore, I., and S. E. Miller. 1978. Fossil rove beetles from Pleistocene California asphalt deposits (Coleoptera: Staphylinidae). *The Coleopterists Bulletin* 32(1):37–39.

Nigra, J. O., and J. F. Lance. 1947. A statistical study of the metapodials of the dire wolf group. *Bull. So. Calif. Acad. Sci.* 46(1):26–34.

Oswalt, S. S. 1979. Artifacts at Rancho La Brea. *Abstracts, So. Calif. Acad. Sci.* Annual Meeting, May 5–6, p. 42.

Purdue, J. R. 1980. Clinal variation of some mammals during the Holocene in Missouri. *Quaternary Research* 13:242–258.

Reynolds, R. L. 1979. Occurrence of domestic dogs at Rancho La Brea. *Abstracts, So. Calif. Acad. Sci.* Annual Meeting, May 5–6, p. 43.

Salls, R. 1980. The La Brea cogged-stone. *The Masterkey* 54(2):53–59.

Savage, D. E. 1951. Late Cenozoic vertebrates of the San Francisco Bay region. *Univ. Calif. Publ. Bull. Dept. Geol. Sci.* 28(10):215–314.

Savage, D. E., and T. Downs. 1954. Cenozoic land life of southern California. *Bulletin of the California Division of Mines* 170:43–58.

Schultz, C. B., and J. M. Hillerud. 1977. The antiquity of *Bison latifrons* (Harlan) in the Great Plains of North America. *Trans. Nebr. Acad. Sci.* 4:103–116.

Shaw, C. A. 1982. Techniques used in excavation, preparation and curation of fossils from Rancho La Brea. *Curator* 25:63–77.

Stock, C. 1956. Rancho La Brea: a record of Pleistocene life in California. *Natural History Museum of Los Angeles County Science Series* 20:1–81.

Stoner, R. C. 1913. Recent observations on the mode of accumulation of the Pleistocene bone deposits of Rancho La Brea. *University of California Publications, Bulletin of the Department of Geology* 7(20):387–396.

Swift, C. C. 1979. Freshwater fish of the Rancho La Brea deposit. *Abstracts, So. Calif. Acad. Sci.* Annual Meeting, May 5–6, p. 44.

Teeri, J. A., and L. G. Stowe. 1976. Climatic patterns and the distribution of C-4 grasses in North America. *Oecologia* 23:1–12.

Valentine, J. W., and J. H. Lipps. 1970. Marine fossils at Rancho La Brea. *Science* 169:277–278.

Warter, J. K. 1976. Late Pleistocene plant communities—evidence from the Rancho La Brea tar pits. In Latting, J. (ed.), *Plant Communities of Southern California,* California Native Plant Society Spec. Publ. 2, pp. 32–39, Berkeley.

———. 1979. The environment of Pit 91, Rancho La Brea, as interpreted by plant remains. *Abstracts, So. Calif. Acad. Sci.* Annual Meeting, May 5–6, p. 44.

———. 1980. Late Pleistocene environment of the Los Angeles Basin based on plant remains from the Rancho La Brea Tar Pits (Abstract). *Bulletin of the Ecology Society of America* 61(2):107.

Woodard, G. D., and L. F. Marcus. 1973. Rancho La Brea fossil deposits: a re-evaluation from stratigraphic and geological evidence. *Journal of Paleontology* 47(1):54–69.

———. 1976. Reliability of late Pleistocene correlation using C-14 dating: Baldwin Hills-Rancho La Brea, Los Angeles, California. *Journal of Paleontology* 50(1):128–132.

Woodward, A. 1937. Atlatl dart foreshafts from the La Brea pits. *Bulletin of the Southern California Academy of Sciences* 36(29):41–60.

Wyman, L. E. 1915. Field notes on the excavation of Rancho La Brea, July 1913 to September 1915. Unpublished, on file at the George C. Page Museum, Los Angeles, California.

The Theoretical Marketplace: Geologic-Climatic Models

The Theoretical Marketplace:
Geologic-Climatic Models

Two leading hypotheses or models that have been developed to account for late Quaternary extinction emphasize either climatic or cultural impact. This part and the following one reflect this division. In both sections seventeen authors display their models and techniques, looking at Pleistocene extinctions from various angles.

In the marketplace of competing ideas what is excluded from a model may be as troublesome to the thoughtful "shopper" as what is featured. Prudent shoppers will also ask "What can I do with this model that I can't with another?" and of course the inevitable: "Can this model be disproved?"

If climatic models ultimately prevail in accounting for Pleistocene extinction, they are likely to be derived from examples offered in this section. Webb starts with a megascopic view of the last ten million years. His valuable detailed stratigraphic inventory of extinct North American mammals from the Miocene on will be viewed with envy by the Australians, among others, where such a record is barely taking shape. In it Webb finds five major extinction episodes, the greatest being the late Hemphillian nearly five million years ago, which exceeded the generic losses of the Rancholabrean by twenty genera. There can be no argument that the great majority of extinctions in the last ten million years were not man-caused; Webb notes a general relationship of extinctions to termination of glacial cycles.

Gingerich views faunal turnover from an even longer time span, the entire Cenozoic. He finds especially high origination rates in the Pleistocene. Thus the very high rate of late Pleistocene (Rancholabrean) extinctions may be driven by the evolutionary exuberance and intercontinental intrusions of the previous two million years. At least in North America late Pleistocene extinctions may have been bound to happen, like a great earthquake in a seismically active region following a long quiescent episode.

While authors of the next four papers are thoroughly familiar with the long-term view, they have specifically examined the late Pleistocene in search of climatic or ecological explanations for the extinctions of the Rancholabrean. In general they adopt the concept of loss of equability advanced in the past two decades by Hibbard, Axelrod, Slaughter, and Guilday. Graham and Lundelius develop evidence from plants as well as animals for an ice age ecosystem in which boreal shrews coexisted with temperate cricetids and other creatures now allopatric in their ranges. Guilday illustrates this specifically for the Appalachian Mountains where the range of fifty-seven of seventy-five nonvolant mammals shifted. According to Guilday, large mammal taxa did not simply vanish from a scene of ecological composure but instead from one of ferment. Graham and Lundelius emphasize that, at least at the species level, extinctions were not exclusive to large mammals. Some small mammal species became extinct, implying that even if man had not been on the scene large mammal extinctions would have occurred as a result of biotic reorganization 10,000 to 12,000 years ago, with a coevolutionary disequilibrium playing a major role.

Guthrie expands on the same idea, identifying a large variety of unusual or even anomalous associations in the Wisconsinan biotic communities. The world of the last Ice Age was compositionally different from any we know now, and not simply because many large mammals became extinct at its end. Guthrie theorizes that the vegetation mosaic of the ice ages was more fine grained. The changes in the Holocene impacted caecalid digesters and noncervid ruminants in particular. Even the expansion of grasslands in the Holocene need not mean an expansion of niches for grazers. In another aspect of seasonality, the matter of breeding season, Kiltie looks at ways in which reproductive output of large mammals could have been depressed at the start of the Holocene. He finds that nearly all recent species with gestation times of more than one year occur in tropical or subtropical areas. Many late Pleistocene species that became extinct undoubtedly had long gestation periods.

Finally King and Saunders review the fossil record of the American mastodont in eastern North America, highlighting the Wisconsinan-age plant assemblages that they have recovered from bone-rich spring deposits in Missouri. They relate mastodont extinction to a trend toward dwarfism accompanied by shrinkage and breakup of its preferred habitat.

Other authors besides the ones included here support geologic-climatic models of extinction, for example, Gilbert and L. Martin and Horton. One requirement or "test" of many climatic models would seem to be that the beginning of the Holocene must have been climatically different, and not merely a replay of the beginning of the Sangamon. Here our ignorance of interglacial periods in general and the Illinoian/Sangamon boundary in particular looms large.

9

Ten Million Years of Mammal Extinctions in North America

S. DAVID WEBB

THE UNAIDED EYE, THE LIGHT MICROSCOPE, and the electron microscope "see" remarkably different images of the same object. One's view is governed greatly by the scale of one's perceptions. The human mind, however, knowing that these images must be reconciled, is up to the task. Similarly, different time scales will feature different aspects of the same chapter of earth history. In this essay I present land mammal extinctions in the perspective of the last ten million years, two orders of magnitude greater than is usual in Pleistocene extinction studies. It is not surprising that this longer scale renders different features. Understandably, this longer view must be reconciled with views based on a shorter scale.

In the past 300 million years, a succession of large tetrapod groups have reigned over the earth and then vanished with apparent abruptness. Bakker (1977), for example, recognized eight dynasties of this sort and dealt in some detail with the demise of the dinosaurs in the late Cretaceous. Each decline spanned several million years. I view the decline of the large-mammal dynasty during the late Cenozoic on about the same scale as these older extinction events.

In the early Miocene, mammals had reached a peak of ordinal and familial diversity in the world (Lillegraven 1972). The proportion of large mammals (those with body weights greater than 5 kg) was on the increase. Bourliére (1975) explains "the bimodal distribution of body weights among terrestrial mammals" and recognizes as large mammals those with average adult body weight of more than 5 kg. In North America during the middle Miocene (12 to 15 million years ago), major radiations were in progress among native ungulates like the horses, camels, and pronghorn. Presumably the burgeoning of such ungulate groups was an evolutionary response to the expansion of savanna and steppe in North America at the expense of forest habitats (Gregory 1971; Webb 1977 and 1983a).

In the late Miocene, however, a series of declines began that continued through the late Pleistocene extinctions. The first sharp decrease in the number of large-mammal genera in North America came at the end of the Clarendonian about nine million years ago (mya). On a long time scale these late Cenozoic extinctions appear to form a string of related events culminating in the best-documented one about ten thousand years ago. See Tedford (1970) and Tedford et al. (in press) for principles and practices concerning the mammalian ages utilized here.

This postulated relationship between the late Pleistocene extinctions and the broader late Cenozoic decline can be illustrated by the extinction of the horses (Equidae) in North America. It is sometimes stated that this family became extinct on this continent at the end of the Pleistocene. In fact only one genus with a few species was extirpated in the late Pleistocene of North America (Kurtén and Anderson 1980). Another genus, *Nannipus*, and about four other equid species became extinct about two million years earlier in the latest Pliocene (late Blancan). Before that, however, three even greater apparent extinction events occurred in the long history of this native North American family. The first decline occurred in the late Clarendonian (nine mya) when two browsing genera and one grazing genus were lost. Two other equal declines occurred in the medial Hemphillian (about seven mya) and again in the late Hemphillian (about five mya). The stair-stepped late Cenozoic decline of the Equidae is summarized in Figure 9.1. It is true that the final demise of the family came with the late Rancholabrean extinctions, but this was merely the last, and by no means the worst, episode in the late Cenozoic decline of the Equidae.

My primary purpose here is to summarize the patterns of late Cenozoic extinctions in North American land mammals. A further purpose is to consider the possible mechanisms of these extinctions. In the framework of this long time scale, secular climatic deteriorations appear to be the only persistent cause sufficient to explain the late Cenozoic decline of land mammals.

Figure 9.2 represents the fossil record of late Cenozoic land mammal extinctions in North America, using nine biochronologic divisions. This figure summarizes the known stratigraphic ranges of all North American land mammal genera from Table 9.1. The greatest extinction episode, the late Hemphillian (about five mya), involved sixty-two genera, thirty-five of which were large mammals. The Rancholabrean episode did not devastate as many *land mammal genera*, but it did affect thirty-nine *large land mammal genera*, four more than in the late Hemphillian episode. Next, in descending order of their effects on number of mammal genera, were the Clarendonian, the late Blancan, and the early Hemphillian extinctions. These three intervals also had a major impact on large land mammals, with twenty to twenty-four genera making their last appearances in each interval. The least important was the late Irvingtonian. The six late Cenozoic extinctions appear separated by roughly two-million-year intervals of relatively stable existence for land mammal genera.

A major problem in comparing extinction sets of different ages concerns the varying scale of the stratigraphic record. As Figure 9.2 indicates, the older intervals are longer (more crudely defined) than the younger intervals. With increasing antiquity our capability at radiometric dating of short-term events, those occurring within a few millenia of each other, vanishes. In addition, the sedimentary record becomes exponentially less complete (Sadler, 1981). There are two opposite interpretations of the relationship between apparent extinctions and the stratigraphic record from which they are deduced. One interpretation assumes that all the extinctions in one stratigraphic interval occurred at one "instant" (e.g. "the dinosaurs died at the end of the Cretaceous"). The opposite assumes a uniform rate of extinction and prorates them over the entire interval. If extinctions occurred episodically, this second approach would dilute their impact in longer stratigraphic intervals. This dilemma can be resolved only by continuing efforts to refine the entire stratigraphic enterprise.

The consequences of this scaling dilemma are great. In the present study one cannot be sure whether or not the late Hemphillian extinctions were as great a cataclysm as the Rancholabrean extinctions. In both more than forty genera disappear. For the Rancholabrean, many carbon dates at many well-stratified sites indicate that these extinctions were closely synchronized within a few thousand years. If prorated, they are still impressive, for that age is only 0.5 my long. The late Hemphillian, on the other hand, cannot be dated with the precision of the Rancholabrean, and many fewer fossiliferous sections are available. Possibly the late Hemphillian extinctions all occurred

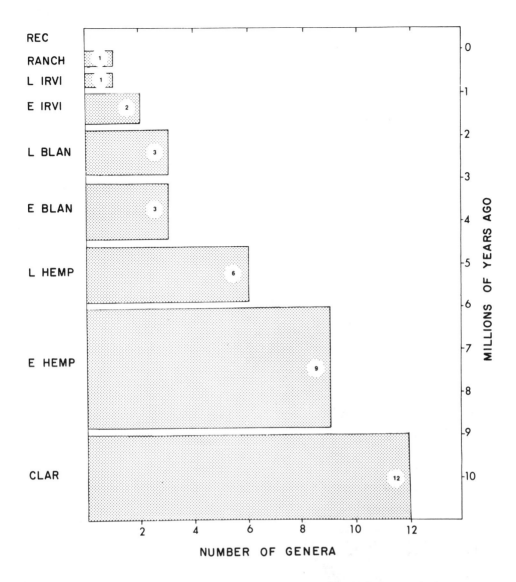

Figure 9.1. The decline of horses in North America. From twelve genera in the Clarendonian, three extinction episodes of the late Miocene reduced the number to three. The late Pleistocene saw the final extermination of *Equus* from North America. The North American land mammal ages on the left are in ascending order: Clarendonian (CLAR); early and late Hemphillian (E HEMP and L HEMP); early and late Blancan (E BLAN and L BLAN); early and late Irvingtonian (E IRVI and L IRVI); Rancholabrean (RANCH); and Recent (REC). The height of each block represents approximate duration of each age. See Tedford (1970) for a thorough analysis of these ages and their usage.

near the end of that interval within a few thousand years, in which case it was at least as great an extinction episode as the Rancholabrean. But, if the extinctions are prorated through their stratigraphic interval, late Hemphillian extinctions span 1.5 my and occur at only one-third the rate of the Rancholabrean. One wonders whether the more ancient extinction episodes were as strongly pulsed as the Rancholabrean episode.

Extinction and Climate

Most discussions of mammal extinctions emphasize either human hunting or climatic deterioration. The human hunting hypothesis is particularly attractive because

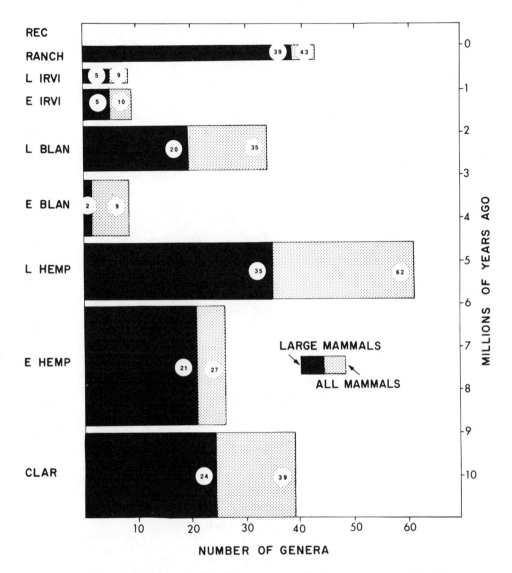

Figure 9.2. Late Cenozoic extinction episodes in North American land mammals. The effect on small mammals (less than 5 kg body weight) was much greater in the late Hemphillian episode than in the late Rancholabrean. (For abbreviations see Figure 9.1 legend.)

of its widespread efficacy (across diverse habitats and various climatic regimes), and because of its resemblance to the dire effects that Recent human cultures are inflicting on many surviving species of large mammals.

In the broader scale of the late Cenozoic, however, man appears too late to be the major cause. The genus *Homo* evolved midway in the late Cenozoic glacial ages but arrived in the New World only for the last in a long string of extinctions. Hunting cultures surely played a role in this last great extinction pulse. Just as surely, however, man could not have figured in the preceding extinction pulses. Thus, by elimination, *climatic deterioration* becomes the primary causal hypothesis for late Cenozoic large mammal extinctions.

Table 9.1. Stratigraphic Ranges of Late Cenozoic Mammals in North America

(*over 5 kg adult body weight; s.l. = sensu lato)

Taxon	Clar	E Hemp	L Hemp	E Blan	L Blan	E Irvi	L Irvi	Ranch	Rec
Didelphidae									
Didelphis									
Erinaceidae									
Metechinus									
Lanthanotherium									
Soricidae									
Meterix									
Limnoecus									
Parydrosorex									
Alluvisorex									
Hesperosorex									
Sorex									
Planisorex									
Microsorex									
Blarina									
Notiosorex									
Cryptotis									
Paracryptotis									
Anouroneomys									
Talpidae									
Achyloscaptor									
Mystipterus									
Domninoides									
Gaillardia									
Lemoynea									
Condylura									
Scalopus									
Scapanus									
Parascalops									
Neurotrichus									
Dasypodidae									
*Kraglievichia									
*Holmesina									
*Dasypus									
Glyptodontidae									
*Glyptotherium									
Megalonychidae									
*Pliometanastes									
*Megalonyx									
Megatheriidae									
*Nothrotheriops									
*Eremotherium									
Mylodontidae									
*Thinobadistes									
*Glossotherium									
Ochotonidae									
Hesperolagomys									
Ochotona									

Table 9.1. Stratigraphic Ranges of Late Cenozoic Mammals in North America
(*over 5 kg adult body weight; s.l. = sensu lato)

(continued)

Taxon	Clar	E Hemp	L Hemp	E Blan	L Blan	E Irvi	L Irvi	Ranch	Rec
Leporidae									
Hypolagus	■	■	■	■	■				
Notolagus		■	■	■	■				
Alilepus		■	■						
Pratilepus				■	■				
Nekrolagus				■					
Brachylagus								■	■
Alurolagus				■					
Sylvilagus					■	■	■	■	■
Lepus					■	■	■	■	■
Romerolagus									■
Aplodontidae									
Liodontia	■								
Aplodontia								■	■
Mylagaulidae									
*Mylagaulus	■	■							
*Epigaulus		■							
Sciuridae									
Sciurus	■	■	■	■	■	■	■	■	■
Tamias	■	■	■	■	■	■	■	■	■
Eutamias								■	■
Tamiasciurus							■	■	■
Spermophilus	■	■	■	■	■	■	■	■	■
Ammospermophilus	■	■	■	■	■	■	■	■	■
Cynomys					■	■	■	■	■
Marmota							■	■	■
*Paenemarmota			■	■					
Glaucomys							■	■	■
Sciuropterus	■								
Cryptopterus					■				
Microsciurus									■
Syntheosciurus									■
Castoridae									
Hystricops	■	■							
Eucastor	■	■							
Dipoides	■								
Procastoroides			■	■	■				
*Castoroides					■	■	■	■	■
*Paradipoides							■		
*Castor			■	■	■	■	■	■	■
Eomyidae									
Ronquillomys	■	■							
Kansasimys		■							
Leptodontomys	■	■							
Heteromyidae									
Perognathoides	■	■							
Perognathus	■	■	■	■	■	■	■	■	■
Diprionomys	■	■							
Cupidinimus	■	■							
Liomys								■	■
Eodipodomys	■								

Table 9.1. Stratigraphic Ranges of Late Cenozoic Mammals in North America

(*over 5 kg adult body weight; s.l. = sensu lato)

(continued)

Taxon	Clar	E Hemp	L Hemp	E Blan	L Blan	E Irvi	L Irvi	Ranch	Rec
Heteromyidae *(cont.)*									
Prodipodomys									
Dipodomys									
Microdipodops									
Heteromys									
Geomydidae									
Parapliosaccomys									
Pliosaccomys									
Pliogeomys									
Geomys									
Thomomys									
Nerterogeomys									
Cratogeomys									
Heterogeomys									
Zygogeomys									
Pappogeomys									
Orthogeomys									
Macrogeomys									
Zapodidae									
Macrognathomys									
Pliozapus									
Zapus									
Napaeozapus									
Cricetidae									
Copemys									
Peromyscus									
Ochrotomys									
Calomys									
Neotoma									
Paraneotoma									
Oryzomys									
Baiomys									
Prosigmodon									
Sigmodon									
(new genus)									
Paronychomys									
Onychomys									
Galushamys									
Gnomomys									
Tregomys									
Reithrodontomys									
Symmetrodontomys									
Pliotomodon									
Repomys									
Nelsonia									
Neotomodon									
(13 recent genera)									
Arvicolidae									
Microtoscoptes									
Goniodontomys									
Promimomys									
Mimomys									
Pliopotamys									
Ondatra									

Table 9.1. Stratigraphic Ranges of Late Cenozoic Mammals in North America
(*over 5 kg adult body weight; s.l. = sensu lato)
(continued)

Taxon	Clar	E Hemp	L Hemp	E Blan	L Blan	E Irvi	L Irvi	Ranch	Rec
Arvicolidae *(cont.)*									
Pliolemmus					—				
Nebraskomys				—					
Atopomys							—		
Synaptomys					——————————————————				
Propliophenacomys			—						
Pliophenacomys				—					
Phenacomys							————————————————		
Proneofiber						—			
Neofiber							————————————————		
Allophaiomys						———————			
Pitymys (s.l.)							————————————————		
Microtus (s.l.)							————————————————		
Lemmus								—————————	
Lagurus								—————————	
Clethrionomys							————————————————		
Dicrostonyx								—————————	
Arborimus									—
Hydrochoeridae									
*Neochoerus					————————————————————————				
*Hydrochoerus					————————————————————————				
Erethizontidae									
*Erethizon					——————————————————————————————————				
*Coendou					——————————————————————————————————				
Canidae									
*Tomarctus	—								
*Cynarctus	—								
*Carpocyon	———								
*Epicyon	———————								
*Aelurodon	——								
*Osteoborus		—————————							
*Borophagus		———————————————————						
*(new genus)	———————								
*Leptocyon	—								
*Vulpes							————————————————		
*Urocyon		——							
*Alopex								—————————	
*Canis									
*Cerdocyon		———————							
*Cuon							————————————————		
*Speothos									—
Amphicyonidae									
*Ischyrocyon	——								
*Pseudocyon	—								
Ursidae									
*Hemicyon	——								
*Indarctos		——							
*Agriotherium			———						
*Plionarctus			—						
*Arctodus					————————————————————————				
*Thalarctos								—————————	
*Tremarctos					———				
*Ursus				—					

Table 9.1. Stratigraphic Ranges of Late Cenozoic Mammals in North America

(*over 5 kg adult body weight; s.l. = sensu lato)

(continued)

Taxon	Clar	E Hemp	L Hemp	E Blan	L Blan	E Irvi	L Irvi	Ranch	Rec
Procyonidae									
*Bassariscus									
*Parailurus									
*Procyon									
*Nasua									
*Arctonasua									
*Paranasua									
Mustelidae									
Mustela (s.l.)									
Martes (s.l.)									
Leptarctus									
Brachypsalis									
Sthenictis									
Plionictis									
Buisnictis									
Trigonictis									
Simocyon									
Sminthosinis									
Ferinestrix									
*Pliotaxidea									
*Taxidea									
*Eomellivora									
*Plesiogulo									
*Gulo									
Tisisthenes									
Osmotherium									
Pliogale									
Spilogale									
Canimartes									
Brachyopsigale									
Brachyprotoma									
Mephitis									
*Conepatus									
*Lutra									
*Lutravus									
*Satherium									
*Enhydriodon									
*Enhydra									
Eira									
Galictis									
Nimravidae									
*Barbourofelis									
Felidae									
*Pseudaelurus									
*Nimravides									
*Machairodus									
*Ischyrosmilus									
*Dinobastis									
*Megantereon									
*Smilodon									
*Homotherium									
*Dinofelis									
*Metailurus									

Table 9.1. Stratigraphic Ranges of Late Cenozoic Mammals in North America
(*over 5 kg adult body weight; s.l. = sensu lato)
(continued)

Taxon	Clar	E Hemp	L Hemp	E Blan	L Blan	E Irvi	L Irvi	Ranch	Rec
Felidae *(cont.)*									
*Lynx									
*Felis									
*Panthera									
*Acinonyx									
Hyaenidae									
*Chasmaporthetes									
Mammutidae									
*Pliomastodon									
*Mammut									
Gomphotheriidae									
*Gomphotherium									
*Amebelodon									
*Platybelodon									
*Rhynchotherium									
*Stegomastodon									
*Cuvieronius									
Elephantidae									
*Mammuthus									
Tapiridae									
*Tapiravus									
*Tapirus									
Rhinocerotidae									
*Peraceras									
*Aphelops									
*Teleoceras									
Equidae									
*Archaeohippus									
*Hypohippus									
*Megahippus									
*Merychippus									
*Protohippus									
*Pseudhipparion									
*Neohipparion									
*Cormohipparion									
*Hipparion									
*Nannippus									
*Calippus									
*Pliohippus									
*Astrohippus									
*Dinohippus									
*Dolichihippus									
*Equus (s.l.)									
Toxodontidae									
*Mixotoxodon									
Tayassuidae									
*Prosthennops									
*Macrogenis									
*Platygonus									
*Mylohyus									
*Tayassu									

Table 9.1. Stratigraphic Ranges of Late Cenozoic Mammals in North America

(*over 5 kg adult body weight; s.l. = sensu lato)

(continued)

Taxon	Clar	E Hemp	L Hemp	E Blan	L Blan	E Irvi	L Irvi	Ranch	Rec
Merycoidodontidae									
*Ustatochoerus									
Protoceratidae									
*Synthetoceras									
*Kyptoceras									
Camelidae									
*Miolabis									
*Aepycamelus									
*Blancocamelus									
*Protolabis									
*Nothotylopus									
*Procamelus									
*Titanotylopus									
*Megatylopus									
*Camelops									
*Hemiauchenia									
*Palaeolama									
*Alforjas									
Gelocidae									
*Pseudoceras									
*new genus									
Dromomerycidae									
*Cranioceras									
Yumaceras (s.l.)									
Moschidae									
*Blastomeryx									
*Longirostromeryx									
Cervidae									
*Bretzia									
*Odocoileus									
*Navahoceros									
*Cervus									
*Sangamona									
*Cervalces									
*Alces									
*Rangifer									
*Mazama									
Antilocapridae									
*Plioceros									
*Texoceros									
*Sphenophalos									
*Proantilocapra									
*Merycodus									
*Ilingoceros									
*Hexameryx									
*Hexobelomeryx									
*Ceratomeryx									
*Capromeryx									
*Osbornoceros									

Table 9.1. Stratigraphic Ranges of Late Cenozoic Mammals in North America
(*over 5 kg adult body weight; s.l. = sensu lato)
(continued)

Taxon	Clar	E Hemp	L Hemp	E Blan	L Blan	E Irvi	L Irvi	Ranch	Rec
Antilocapridae *(cont.)*									
*Ottoceros									
*Hayoceros									
*Stockoceros									
*Tetrameryx									
*Antilocapra									
*new genus									
Bovidae									
*Neotragocerus									
*Euceratherium									
*Bootherium									
*Bison									
*Platycerabos									
*Praeovibos									
*Ovibos									
*Symbos									
*Soergelia									
*Saiga									
*Ovis									
*Neorhaeodus									
*Oreamnos									

The late Cenozoic mammal extinctions coincide broadly with the period of glacial climates. The oldest glaciations in the northern hemisphere appeared nearly ten million years ago as alpine glaciers in southern Alaska (Denton and Armstrong 1969). More importantly, in the southern hemisphere two major ice caps had been emplaced in Antarctica, and this had worldwide effects, such as lowered sea levels and cooling of bottom waters, especially at middle latitudes (Shackleton and Kennett 1975; Schnitker 1980). Donn and Shaw (1977) regard the late Miocene as the time when mean annual temperatures dropped to freezing at high northern latitudes, "accounting for the late Cenozoic glacial age."

Figure 9.3 provides a set of correlations between late Tertiary oxygen-isotope curves from benthic forams in deep-sea cores and North American land mammal ages by way of paleomagnetic chronology and radiometric dates. There is a suggestive relationship between the three sharp reversals in the bottom-water oxygen-isotope curve between five and ten million years ago and the three major extinction episodes in North America during that same interval. Obviously more refined correlations are needed and are rapidly being developed. Nonetheless, it is fair to note a general late Miocene coincidence between sharp cooling trends in the deep sea and the onset of severe extinction episodes on the land.

It is perhaps to be expected that the great majority of the late Cenozoic extinctions on temperate land masses were concentrated in the late Tertiary; for that is when the biota, previously adjusted to more equable conditions, first felt the severity of glacial conditions in polar regions. I have discussed elsewhere the dramatic faunal disruption in temperate North America, especially the loss of browsing mammals, that followed the replacement of predominantly savanna by predominantly steppe vegetation (Webb 1977, Webb 1983b). Leopold (1967) found the same late Miocene replacement with respect to

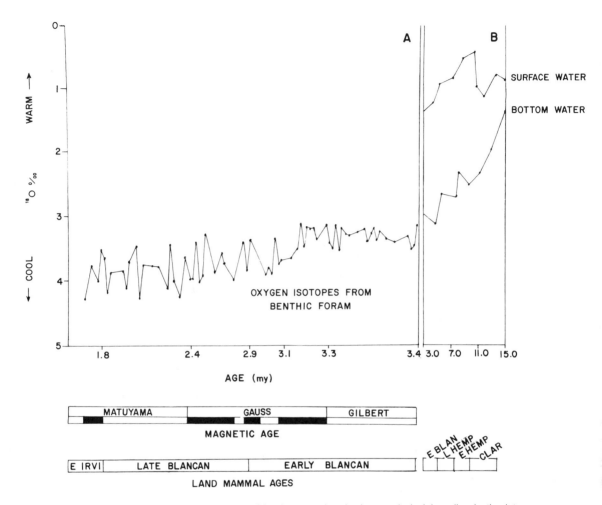

Figure 9.3. Correlations between times of land mammal extinctions and glacial cooling in the late Cenozoic. Compare with Figure 9.2 for times of greatest land mammal extinctions. Notable features are the reversals of cooling trends at about nine, six, and four million years ago. Oxygen isotopes from benthic foram, *Globocassidulina subglobosa*. Oxygen-isotope curves (A) above 3.0 my from Shackleton and Kenneth 1975; (B) below 3.4 my from Shackleton and Opdyke 1977. Note time scale change at this break. (For abbreviations see Figure 9.1 legend.)

vascular plants in North America and Europe and concluded that "the bulk of recorded late Cenozoic extinctions are of late Tertiary, not Quaternary, age." The data for land mammal extinctions corroborate this view, and suggest that the whole biota was most severely reorganized at the beginning of the late Cenozoic ice ages.

Extinctions and Glacial Terminations

A corollary of the climatic hypothesis of land mammal extinctions is that greater extremes of temperature, moisture availability and seasonality suddenly occurred to produce misadaptations of plants and animals in many different localities and habitats.

Much evidence indicates that extremes of instability and maximum seasonality came during glacial terminations (Dreimanis 1977, Porter 1980). If land mammal extinctions can be shown to coincide with glacial terminations, then the climatic explanation becomes more plausible.

One of the interesting features of the oxygen-isotope curves from Pleistocene deep-sea cores is their saw-toothed (asymmetrical) pattern, showing more rapid heating than cooling. The widest excursions with the steepest slopes occur during glacial terminations (e.g. Shackleton and Opdyke, 1977). One plausible explanation for this pattern was recently advanced by Kukla et al. (1981) to the effect that the cyclical nature of glacial history is determined by earth-sun orbital cycles and that the most important effect is the onset of an interglacial interval, driven primarily by the perihelion. On the continents at temperate latitudes, conditions during the onset of an interglacial were particularly extreme and unstable. The ameliorating effects of glacial masses were lost as they retreated, and the rates of retreats accelerated (Birchfield et al. 1981, Dreimanis 1977, Wright 1970). Thus, the most inequable climates on land might be expected at high latitudes in continental interiors during glacial terminations. Such conditions of climatic instability (the inequability of Axelrod 1967) and drought would have the most devastating effects on land biota.

It is, therefore, intriguing to note that late Cenozoic land mammal extinctions in North America may have been massed at times of glacial terminations. The most familiar case is the late Rancholabrean extinction episode which coincides with the end of the last glacial cycle. Several students of late Pleistocene vertebrates have noted the retreat of many Recent species from much wider Rancholabrean distributions. These often include the breakup of enriched ("disharmonious") faunas, often with the northward retreat of some taxa and the southward retreat of others. Indications include greater drought, colder winters, and hotter summers (Hibbard 1970, Holman 1976, Graham 1976, Guthrie this volume, Lundelius 1967, Graham and Lundelius this volume). Such range restrictions have a suggestive correlation with the late Rancholabrean extinctions and with the end of the Wisconsin glaciation.

Three other mammal extinction episodes from the late Cenozoic of North America evidently correlate with glacial terminations. The late Hemphillian extinction episode apparently coincides with the end of the Messinian glacio-eustatic cycle. As discussed more fully elsewhere (Webb 1983b, Schultz 1977), this extinction episode is dated in the Texas panhandle at about 5.5 mya, and therefore probably correlates with the terminal Messinian transgression cycle, involving a change of at least 40 meters in sea level. The episode falls stratigraphically near the end of the mid-Hemphillian (Webb et al. 1978; Peck et al. 1979). If the Messinian sea level cycle is correctly interpreted as a major glacio-eustatic cycle (Hsu et al. 1977, Adams et al. 1978, Schnitker 1980), then the major mid-Hemphillian extinction episode correlates with the termination of the late Miocene glacial cycle about five mya.

The late Blancan extinction event can be placed rather concisely near the end of a glacial cycle as Repenning (1980) notes. Several magnetochronologic studies (Lindsay et al. 1975, Opdyke et al. 1977, Johnson et al. 1975, Neville et al. 1979) consistently place it in the basal part of the Matuyama magnetozone just below the Olduvai Event. Meanwhile deep-sea oxygen-isotope and magnetochronology records have shown that this same time (about 1.8 mya) was the termination of an early series of intense glacial cycles in the northern hemisphere. These glacials "of a magnitude of at least two-thirds that of the late Pleistocene glacial maxima" extended from 2.4 to 1.8 mya (Shackleton and Opdyke 1977) (see fig. 9.3). A deep core from Searles Lake in California shows that the end of the first long pre-Pleistocene wet period ended just below the Olduvai Event (Liddicoat et al. 1980). Thus, the late Blancan extinctions may be correlated tentatively with the dry interval at the termination of the late Tertiary glacial interval of the northern hemisphere.

The late Irvingtonian is the third early extinction event that may be correlated with a glacial termination. The Cudahy local fauna (a late Irvingtonian sample) occurs in the lower Brunhes normal magnetochron, well above an ash dated 0.6 mya and below other parts of the section that are Rancholabrean (Kurtén and Anderson 1980). A glacial correlation of the Cudahy local fauna is supported by the small mammal species and molluscs of northern aspect (Hibbard 1970), and thus it ought to be assigned to oxygen-isotope stage 12 (Repenning 1980). From these relationships one may conclude that the Irvingtonian extinctions probably coincided with the end of glacial stage 12.

It remains to be shown why certain glacial terminations apparently produced major extinction events and others did not. In his minutely detailed review of the European history of continental glaciations, Kukla (1977) noted that there were probably 11 glacial terminations (22 oxygen-isotyope stages) during the past one million years, and that the duration of each stage was generally comparable to that of the latest glacial cycle. Kukla (1977) further suggested that, as measured by deep-sea oxygen-isotope data, the magnitude of glacial stages 2, 6, 12, 16, and 22 appeared greater than the others. It is interesting to note that stages 2 and 12 correspond respectively to the Rancholab-rean and the Irvingtonian extinction events. Also, according to Kukla's (1977) review, the major Villafranchian extinctions in Europe fell at about stage 22, and Bonifay (1980) finds the extinctions of the first cold-adapted European fauna to coincide with stage 16 or about the end of the first Biharian faunas. Thus it is possible that *the terminations of more intense glacial stages effectively triggered large mammal extinction episodes in the late Cenozoic.*

Mammal Extinctions as a Latitudinal Function

It has long been evident that glacial cycles and their climatic effects were harsher at higher latitudes. If, as is argued above, the climatic changes associated with glacial terminations initiated land mammal extinctions, then would not these extinctions be diminished at lower latitudes? A search for such a relationship between lower latitudes and increased survival of land mammal genera may be regarded as a test of the climatic hypothesis.

First, it is important to support the premise that land areas at lower latitudes were less severely affected by climatic changes during glacial cycles. Few continental Pleis-tocenes sites yielding climatic data are available for North America below 20° N latitude (Peterson et al. 1979). In the absence of detailed local evidence, especially in the Central American lowlands, it is best to cite the broad results from the adjacent sea surface, as produced by the CLIMAP project (1976). About 17,000 years ago, during the maximum extent of the Wisconsinan glaciation, surface water temperatures in the Gulf of Mexico and at the straits of Florida were only about 2° C cooler than during present winters. At even lower latitudes, the glacial decrease measured by the proxy climatic data of foraminiferal samples was also only about 2° Celsius. At latitudes between 30°N and 40°N, however, sea surface temperatures along the Atlantic coast were about 10° Celsius lower than present winter temperatures. The lowland American tropics thus apparently suffered little cooling during the last glacial interval, and the same may be presumed for earlier cycles.

A broad method of testing latitudinal effects on extinctions is to ask how many of the land vertebrates that are counted among the Rancholabrean extinctions in temper-ate North America actually survive elsewhere at lower latitudes. The largest contingent survives today in tropical America, including *Tapirus, Tremarctos* (the spectacled bear), various species of *Felis* and *Panthera* (including jaguars, jaguarundis, and ocelots), *Geochelone* (giant tortoises, now confined in the New World to the Galapagos Islands), and above the generic level, sloths, capybaras, peccaries, and llamas. A newly discovered example is the peccary *Catagonus*, now living in the Gran Chaco and closely related to the extinct *Platygonus* (Wetzel 1977). Other Rancholabrean "refugees" at low latitudes in the Old World include cheetahs, horses, and elephants (Reed 1970). Thus, nearly half the groups that are counted as Rancholabrean extinctions at temperate latitudes have tropical survivors.

The same pattern of tropical survival applies to earlier extinction episodes. A newly discovered example is the progressive cricetid rodent, *Repomys*, known from California and Nevada in the late Hemphillian and Blancan, but now represented by *Nelsonia* at high latitudes in central Mexico (May 1981). Most such histories are difficult to trace, however, because fossil distributions at low latitudes are poorly known. Larger mammal taxa that evidently survived the Blancan extinction event at low latitudes are gomphotheres, *Stegomastodon* (into the early Irvingtonian in southern Arizona and southern California), hippidion horses, tropical deer such as *Mazama*, peccaries near *Tayassu*, tropical procyonids, and the tropical mustelids *Galictis* (related to Blancan and early Irvingtonian *Trigonictis*), and *Pteronura* (near the Blancan *Satherium*). Several taxa, including *Geochelone, Nasua, Hydrochoerus, Cuvieronius,* and *Glyptotherium*, show a stepwise tropical retreat. Whereas they ranged through high temperate latitudes during Blancan time, they retracted southward to the Gulf Coastal Plain and Mesoamerica during Irvingtonian and Rancholabrean time. The Blancan, Irvingtonian, and Rancholabrean climatic events not only produced extinctions, but they also swept many North American mammals southward. The related effects were total extinction in some cases and tropical restriction in others.

The number of species in the Recent North American mammal fauna shows an impressive inverse correlation with latitude (McCoy and Conner 1980). Increasing from latitude 38°N, the number of species decreases markedly; below that latitude the number of species increases more gradually toward the equator to 25°N latitude. Farther south the numbers decrease, presumably because the rapidly decreasing area of Central America has an overriding effect. This latitudinal species gradient is equally apparent in large and small animals. The 38th parallel coincides closely with the southern limits of the last system of Pleistocene glaciers. Presumably the recurrent effects of late Cenozoic glacial climates produced this strong latitudinal decline in mammal species above 38°N in temperate North America. The tropical North American fauna includes a much greater percentage of old native groups than the temperate fauna, which is more heavily loaded with Asiatic immigrant groups. This pattern indicates the effectiveness of large mammal extinctions correlated with latitude.

Possible Effects of Immigrants

Elsewhere I have elaborated the view that major extinction episodes can be produced by either of two fundamentally different mechanisms: one resulting from biological interactions following immigration, the other from environmental stress accompanying climatic change (Webb 1969). I have shown how the two basic types of extinction can be distinguished in the fossil record (Webb 1983b). The role of immigrants and climatic change with respect to the late Cenozoic land mammal extinctions provides an interesting comparison.

Figure 9.4 records the number of large land mammal genera in North America, divided into new immigrants (allochthonous genera) and natives (autochthonous genera). In the Clarendonian, few immigrant genera entered North America. The number increased slightly in the early Hemphillian and still more in the late Hemphillian. While the absolute numbers do not change greatly between these two substages, the immigration rate for the late Hemphillian is approximately twice that of the early Hemphillian.

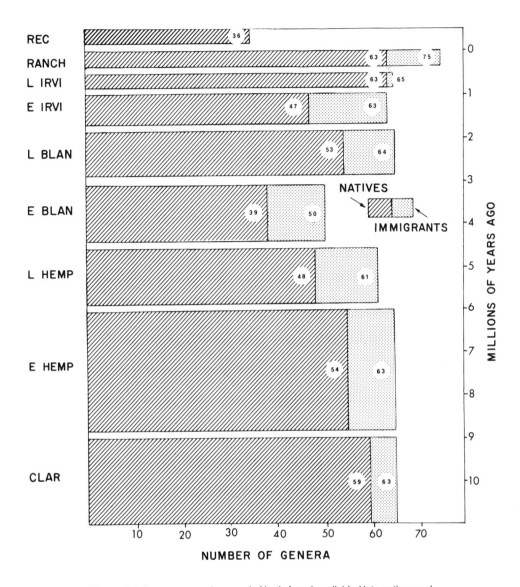

Figure 9.4. Large mammal genera in North America, divided into natives and immigrants. Times of high immigrant numbers do not correlate with times of low native numbers. Instead, immigrant rates seem independently to undergo a general increase through the late Cenozoic. (For abbreviations see Figure 9.1 legend.)

The early Blancan rates also remain high. And thereafter immigration rates increase still more to the very high rates of the Irvingtonian and the Rancholabrean. The record suggests that immigrations generally tracked extinctions. In effect, immigrant taxa filled the vacuums produced by prior extinctions.

The Blancan rush of immigrants exemplifies the largely noninteractive relationship between allochthons and autochthons. At the beginning of that stage the North American land mammal fauna was about as depauperate as at present. Then immigrants swelled the fauna with diversely adapted types. From South America came large hystricognath rodents and at least six large edentate genera unlike anything in this continent (Webb 1978). From Asia by way of Beringia came such diverse mammals as aquatic cricetid rodents, omnivorous carnivores such as *Ursus* and *Parailurus*, browsing Cervidae, and the large mesic flying squirrel *Cryptopterus* (Repenning 1967, 1980; Robertson 1976; Tedford and Gustafson 1977). Most of these immigrants had become well established before the next great extinction episode in the late Blancan.

During Irvingtonian and especially during Rancholabrean time, the number of trans-Beringian immigrants rose markedly. This greatly unbalanced the rate of faunal interchange between Asia and North America which previously had been balanced or, in the Hemphillian, unequal in the opposite direction (Repenning 1967, Webb 1969). Furthermore, these Pleistocene immigrants from Asia were virtually all scrub, steppe, or steppe-tundra inhabitants, reflecting the evolutionary importance of that vast Asiatic terrain during the Pleistocene. The new immigrants were predominantly small grazers (microtine rodents), large grazers (bovids and mammoths), and large browsers (moose and elk). Possibly the last major waves of Asian immigrants bore a causal relationship to the final great extinction episode. Gingerich (this volume) favors such a view. On the other hand, it is difficult to show a closely synchronous relationship between immigrations and extinctions in the late Pleistocene of North America. Immigrations continue generally to increase throughout the late Cenozoic record, often in the wake of extinctions.

Another biological interaction of some interest is the possible complementary relationship between large mammal and small mammal diversity. Webb (1969) showed an apparent correlation between late Cenozoic decreases in large mammalian herbivore genera and increases in small mammalian herbivore genera. Figure 9.5 shows clearly the broad shift from a preponderance of large mammal genera in the late Miocene to a preponderance of small mammal genera at present. There was also a decline at the end of the Hemphillian that was restored by late Blancan time. From over 60 percent of genera in the Clarendonian, large mammals (those >5 kg) have declined to less than 40 percent of genera today. Much of this diversity shift involves loss of large grazing herbivores in the late Miocene, as exemplified by the Equidae. In their place, during the Pleistocene came a dramatic increase in small grazers, derived from various stocks of Cricetidae (Repenning 1980). These data suggest that the replacement of ungulate diversity by rodent diversity occurred not because of direct competition, but as a consequence of most major extinctions more adversely affecting large ungulates. In the late Cenozoic the net result was that small grazing herbivores filled the vacancies left by large grazing herbivores.

Conclusions

During the last ten million years, six major extinction episodes have devastated the land mammal fauna of North America. The greatest of these, according to available records, was the late Hemphillian (nearly five million years ago) when more than sixty genera of land mammals (of which thirty-five were large, weighing more than 5 kg) disappeared from this continent. The late Rancholabrean extinction pulse (about 10,000 years ago) was the next greatest; over forty genera became extinct, of which nearly all were large mammals. If these events are prorated by the durations of their respective ages, the Rancholabrean is the most severe. Three lesser, but still major, extinction episodes, each involving a score of genera, occurred in the late Clarendonian (about nine million years ago), the early Hemphillian (about six million years ago), and the late Blancan (1.9 million years ago). The sixth was the late Irvingtonian.

Some evidence shows that these extinction episodes were correlated with terminations of glacial cycles, when climatic extremes and instability are thought to have reached their maxima. The best documented correlation is that between the late Rancholabrean and the Wisconsinan termination. Correlation between the late

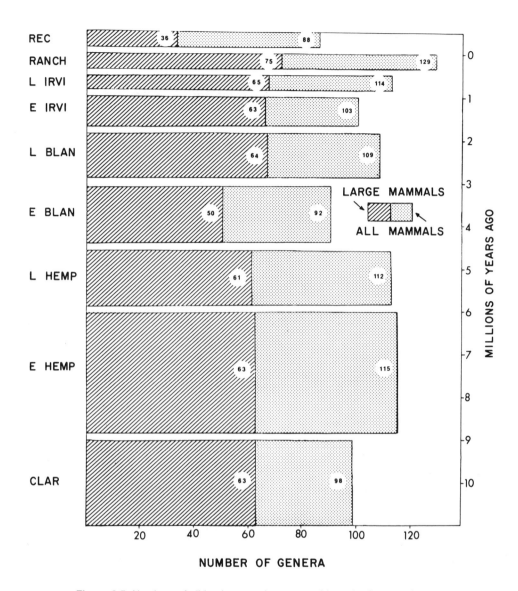

Figure 9.5. Numbers of all land mammal genera and large land mammal genera during the late Cenozoic in North America. From about 65 percent during the Clarendonian, the large mammal contribution to the total land mammal genera has declined to about 40 percent at present. The relative decline continued more or less regularly through the late Miocene and then again during the late Pleistocene. (For abbreviations see Figure 9.1 legend.)

Hemphillian extinctions and the end of the Messinian (latest Miocene) and also between the late Blancan and termination of the latest Tertiary glacial phase may also indicate such a relationship.

In addition to the temporal correlation between the glacial terminations and large mammal extinctions, there is an inverse latitudinal correlation. In temperate North America during the late Cenozoic, large mammal genera are more severely extirpated at high latitudes than at low latitudes. About half the families of large vertebrates that became extinct in temperate North America during the late Pleistocene survive at low latitudes, mainly in the American tropics.

The net result of the late Cenozoic extinctions has been a major transformation in the character of North American land mammal fauna. In the late Miocene (late Clarendonian and early Hemphillian), climatic deteriorations destroyed the large browsers of North America as the extensive midcontinental savanna gave way to steppe (Gregory 1971). And in the latest Miocene (late Hemphillian) mixed feeders and even grazers, including peccaries, horses, rhinos, antilocaprids, and gomphotheres, were decimated. By late Pliocene (late Blancan) time, the diversity of all groups was again severely reduced. A net trend toward replacement of large herbivores by small herbivores has been documented (Webb 1969). With the climatically induced extinctions of old native ungulate groups, rodents have risen to become generically the most diverse herbivores in North America.

Acknowledgments

I am deeply grateful to the following scholars who have freely shared their special knowledge about late Cenozoic mammals and their stratigraphic distribution: Michael Voorhies, Eric Gustafson, Rick Zakrzewsky, Bruce MacFadden, Charles Repenning, and Richard Tedford. Repenning, Tedford, and MacFadden, also provided valuable criticism of this chapter. Support from the National Science Foundation (DEB 78-10672) to study late Cenozoic vertebrates in Florida has aided this work (Florida State Museum Contribution to Vertebrate Paleonotology Number 199).

References

Adams, G. C., Benson, R. H., Kidd, R. B., Ryan, W. B. F., and Wright, R. C. 1978. The Messinian salinity crisis and evidence of late Miocene eustatic changes in the world ocean. Nature 269: 383–386.

Axelrod, D. I. 1967. Quaternary extinctions of large mammals. Univ. Calif. Publ. Biol. Sci. 74:1–42.

Bakker, R. T. 1977. Tetrapod mass extinctions—a model of the regulation of speciation rates and immigration by cycles of topographic diversity. pp. 439–468 In A. Hallam (ed.), Patterns of Evolution as Illustrated by the Fossil Record. Elsevier Publ. Co., Amsterdam and Oxford.

Birchfield, G. E., Weertman, J., and Lunde, A. T. 1981. A paleoclimate model of northern hemisphere ice sheets. Quat. Res. 15: 126–142.

Bonifay, M. F. 1980. Relations entre les donnees isotopiques, oceaniques et l'histoire des grandes faunes Europeenes Plio-Pleistocenes. Quat. Res. 14:251–262.

Bourliére, F. 1975. Mammals, small and large: the ecological implications of size. pp. 1–8 In E. B. Golley et al. (eds.), Small Mammals: Their Productivity and Population Dynamics. Cambridge Univ. Press, Cambridge, London, New York and Melbourne.

CLIMAP Project Members. 1976. The surface of the Ice Age Earth. Science 191:1131–1144.

Denton, G. H. and Armstrong, R. L. 1969. Miocene-Pliocene glaciations in southern Alaska. Am. J. Sci. 267:1121–1142.

Donn, W. L. and Shaw, D. M. 1977. Model of climate evolution based on continental drift and polar wandering. Geol. Soc. Am. Bull. 88:390–396.

Dreimanis, A. 1977. Late Wisconsin glacial retreat in the Great Lakes region, North America. Ann. New York Acad. Sci. 288: 70–79.

Graham, R. W. 1976. Late Wisconsin mammalian faunas and environmental gradients of the eastern United States. Paleobiology 2: 343–350.

Gregory, J. T. 1971. Speculations on the significance of fossil vertebrates for the antiquity of the Great Plains of North America. Abh. Hessisches Landesamtes Bodenforsch. 60:64–72.

Hibbard, C. W. 1970. Pleistocene mammalian local faunas from the Great Plains and central lowland provinces of the United States. pp. 395–433 In W. Dort and J. K. Jones (eds.), Pleistocene and Recent Environments of the Central Great Plains. Univ. Press Kansas. Spec. Publ. 3.

Holman, J. A. 1976. Paleoclimatic implications of "ecologically incompatible" herpetological species (late Pleistocene: southeastern United States). Herpetologica 32:290–295.

Hsü, K. J., Montadert, L., Bernoulli, D., Cita, M. B., Erickson, A., Garrison, R. E., Kidd, R. B., Mèlierés, F., Müller, C., and Wright, R. 1977. History of the Mediterranean salinity crisis. Nature 267:399–403.

Johnson, N. M., Opdyke, N. D., and Lindsay, E. H. 1975. Magnetic polarity stratigraphy of Pliocene-Pleistocene terrestrial deposits and vertebrate faunas, San Pedro Valley, Arizona. Geol. Soc. Am. Bull. 86:5–12.

Kukla, G. J. 1977. Pleistocene land-sea correlations. I. Europe. Earth Sci. Rev. 13:307–374.

Kukla, G. J., Berger, A., Lotti, R., and Brown, J. 1981. Orbital signature of interglacials. Nature 290:295–300.

Kurtén, B. and Anderson, E. 1980. Pleistocene Mammals of North America. Columbia Univ. Press, New York. 442 pp.

Leopold, E. B. 1967. Late-Cenozoic patterns of plant extinction. pp. 203–246 *In* P. S. Martin and H. E. Wright, Jr. (eds.), Pleistocene Extinctions. Yale Univ. Press, New Haven.

Liddicoat, J. S., Opdyke, N. D., and Smith, G. I. 1980. Paleomagnetic polarity in a 930-m core from Searles Valley, California. Nature 286:22–25.

Lillegraven, J. A. 1972. Ordinal and familial diversity of Cenozoic mammals. Taxon. 21:261–274.

Lindsay, E. H., Johnson, N. M., and Opdyke, N. D. 1975. Preliminary correlation of North American land mammal ages and geomagnetic chronology. *In* Studies on Cenozoic Paleontology and Stratigraphy. Claude W. Hibbard Memorial 3, Univ. Michigan Pap. Paleontol. 12:111–119.

Lundelius, E. L. 1967. Late Pleistocene and Holocene faunal history of central Texas. pp. 287–319 *In* P. S. Martin and H. E. Wright (eds.), Pleistocene Extinctions. Yale Univ. Press, New Haven.

McCoy, E. D. and Connor, E. F. 1980. Latitudinal gradients in the species diversity of North American Mammals. Evolution 34:193–203.

May, S. R. 1981. *Repomys* (Mammalia: Rodentia, gen. nov.) from the late Neogene of California and Nevada. J. Vert. Paleont. 1(2):219–230.

Neville, C., Opdyke, N. D., Lindsay, E. H., and Johnson, N. M. 1979. Magnetic stratigraphy of Pliocene deposits of the Glenns Ferry Formation, Idaho, and its implications for North American mammalian biostratigraphy. Am. J. Sci. 279:503–526.

Opdyke, N. D., Lindsay. E. H., Johnson, N. M., and Downs, T. 1977. The paleomagnetism and magnetic polarity stratigraphy of the mammal-bearing section of Anza-Borrego State Park, California. Quat. Res. 7:316–329.

Peck, D. M., Missimer, T. M., Slater, D. H., Wise, S. W., Jr., O'Donnell, T. H. 1979. Late Miocene glacial-eustatic lowering of sea level: evidence from the Tamiami Formation of South Florida. Geology 7:285–288.

Peterson, G. M., Webb, T., III, Kutzbach, J. E., Van Der Hammen, T., Wijmstra, T. A., and Street, F. A. 1979. The continental record of environmental conditions at 18,000 yr B.P.: an initial evaluation Quat. Res. 12:47–82.

Porter, S. C. 1980. Rapid deglaciation of alpine regions at the end of the last glaciation. Am. Quat. Assoc. 6th Bien. Mtg. Abst. 157.

Reed, C. A. 1970. Extinction of mammalian megafauna in the Old World late Quaternary. BioScience 20:284–288.

Repenning, C. A. 1967. Palearctic-Nearctic mammalian dispersal in the late Cenozoic. pp. 288–311 *In* D. M. Hopkins (ed.), The Bering Land Bridge. Stanford Univ. Press, Palo Alto.

————. 1980. Faunal exchanges between Siberia and North America. Canadian J. Anthro. 1:37–44.

Robertson, J. S. 1976. Latest Pliocene mammals from Haile XVA, Alachua County, Florida. Bull. Florida State Mus., Biol. Sci. 20(3):111–186.

Sadler, P. M. 1981. Sediment accumulation rates and the completeness of stratigraphic sections. J. Geol. 89:569–584.

Schnitker, D. 1980. Global paleoceanography and its deep water linkage to the Antarctic glaciation. Earth Sci. Rev. 16:1–20.

Schultz, G. E. 1977. Guidebook for field conference on late Cenozoic biostratigraphy of the Texas Panhandle and adjacent Oklahoma. Dept. Geol. Anthrop. Spec. Publ., Canyon, Texas 1:1–160.

Shackleton, N. J. and Kennett, J. P. 1975. Paleo-temperature history of the Cenozoic and the initiation of Antarctic glaciation; oxygen and carbon isotyope analyses in DSDP sites 277, 279, and 281. *In* J. P. Kennett et al. (eds.), Initial Reports of the Deep Sea Drilling Project. U.S. Govt. Print. Off. 29:743–755.

Shackleton, N. J. and Opdyke, N. D. 1973. Oxygen isotope and paleomagnetic stratigraphy of equatorial Pacific core V 28–239; oxygen isotope temperatures and ice volumes on a 10^5–10^6 year scale. Quat. Res. 3:39–55.

———. 1977. Oxygen isotope and paleomagnetic evidence for early Northern Hemisphere glaciation. Nature 270:216–219.

Tedford, R. H. 1970. Principles and practices of mammalian geochronology in North America. *In* Proceedings of the North American Paleontology Convention, September 1969. Part F: 666–703.

Tedford, R. H. and Gustafson, E. P. 1977. First North American record of the extinct panda, *Parailurus*. Nature 265:621–623.

Tedford, R. H., Galusha, T., Skinner, M. F., Taylor, B. E., Fields, R. W., MacDonald J. R., Patton, T. H., Rensberger, J. M., and Whistler, D. P. In press. Faunal succession and biochronology of the Arikareean through Hemphillian interval (late Oligocene through late Miocene epochs) North America. *In* M. O. Woodburne (ed.), Vertebrate Paleontology as a Discipline in Geochronology. Univ. California Press, Berkeley.

Webb, S. D. 1969. Extinction-origination equilibria in late Cenozoic land mammals of North America. Evolution 23:688–702.

———. 1976. Mammalian faunal dynamics of the Great American Interchange. Paleobiology 2:220–234.

———. 1977. A history of savanna vertebrates in the New World. Part I: North America. Ann. Rev. Ecol. Syst. 8:355–380.

———. 1978. A history of savanna vertebrates in the New World. Part II: South America and the Great Interchange. Ann. Rev. Ecol. Syst. 9:393–426.

———. 1983a. The rise and fall of the late Miocene ungulate fauna in North America. *In* M. Nitecki (ed.), Coevolution Symposium. Univ. Chicago Press, Chicago.

———. 1983b. Large-scale community succession by two kinds of rapid faunal turnover. *In* J. Van Couvering and W. Berggren (eds.), Uniformitarianism, Catastrophes, and Earth History. Princeton Univ. Press, Princeton.

Webb, S. D., Wise, S. W., and Wright, R. C. 1978. Late Miocene glacio-eustatic cycles in Florida: marine and fluvio-estuarine sequences. Geol. Soc. Am. Abst. 10:513.

Wetzel, R. M. 1977. The extinction of peccaries and a new case of survival. Ann. New York Acad. Sci. 288:538–544.

Wright, H. E., Jr. 1970. Vegetational History of the Central Great Plains. pp. 157–173 *In* W. Dort, Jr. and J. K. Jones, Jr. (eds.), Pleistocene and Recent Environments of the Central Great Plains. Univ. Press Kansas. Spec. Publ. 3.

10

Pleistocene Extinctions in the Context of Origination-Extinction Equilibria in Cenozoic Mammals

PHILIP D. GINGERICH

SOME OF THE MOST INTERESTING MAMMALS of the Pleistocene are now extinct. The rhino-sized marsupial *Diprotodon* of Australia, the large koala-like primate *Megaladapis* of Madagascar, the antlered giraffe *Sivatherium* and robust hominid *Australopithecus* of Africa, the woolly rhinoceros *Coelodonta* of Eurasia, the camel-like litoptern *Macrauchenia* of South America, and the giant beaver *Castoroides,* Jefferson's ground sloth *Megalonyx,* and mastodon *Mammut* of North America are representative of animals no longer with us. The Pleistocene sabertooth cats, *Smilodon, Dinobastis,* and *Homotherium,* are among the most conspicuous of extinct carnivores.

Viewed in the context of the entire Cenozoic era, are Pleistocene extinctions unusual? Thinking more broadly, how are the disappearances of old taxa (extinction) related to the evolutionary first appearances of new taxa (origination)? How is our understanding of both of these processes colored by problems of perception in comparing relatively recent events with older ones receding and fading into the geological past?

Patterns of Origination and Extinction in Cenozoic Mammals

Simpson (1953) summarized the history of diversification of carnivore genera in a matrix reproduced here as Table 10.1, based on his 1945 classification of mammals. Simpson's matrix shows the number of genera that first appeared in each subdivision of the Cenozoic (rows), and when each of these genera disappeared (columns). Simpson's data are now somewhat out of date (many of the Paleocene and Eocene genera he included in Carnivora have since been transferred to other orders) and the geological time scale has been revised; but his matrix shows several surprising things relating to the origination and extinction of Cenozoic mammals.

Simpson assigned durations to each subdivision of the geological time scale, and used the distribution of data in Table 10.1 to derive the survivorship curve shown in Figure 10.1. Survivorship was then used to predict the number of genera originating in the Pleistocene that should survive to the Recent (Table 10.2). Of thirty-four genera originating in the Pleistocene, Simpson predicted that thirty-three should survive until the Recent. Only nineteen did so, leading Simpson (1953) to offer an explanation involving unusually high mortality in the Pleistocene:

Table 10.1. Distribution of Genera of Carnivora (Except Pinnipedia)

(Figures are numbers of known genera)

Last Known Appearance

First Known Appearance		Paleocene			Eocene			Oligocene			Miocene			Pliocene			Pleisto-cene	Recent	Total First Known Appearance
		L	M	U	L	M	U	L	M	U	L	M	U	L	M	U			
Paleocene	L	5	0	0	1	..	..	..	..	..	..	..	..	..	..	..	..	..	6
	M	..	12	0	1	..	..	..	..	..	..	..	..	..	..	..	..	..	13
	U	..	..	5	3	..	..	..	..	..	..	..	..	..	..	..	..	..	8
Eocene	L	..	..	..	8	5	3	..	..	..	..	..	..	..	..	..	..	..	16
	M	..	..	..	..	7	4	..	..	..	..	..	..	..	..	..	..	..	11
	U	..	..	..	..	..	13	6	1	0	0	2	..	..	..	..	..	..	22
Oligocene	L	..	..	..	..	..	..	9	1	4	5	2	0	1	..	..	..	..	22
	M	..	..	..	..	..	..	..	3	..	..	..	..	..	..	..	..	..	3
	U	..	..	..	..	..	..	..	0	3	1	..	..	..	..	..	..	..	4
Miocene	L	..	..	..	..	..	..	..	..	20	0	3	2	..	..	..	..	..	25
	M	..	..	..	..	..	..	..	..	..	9	3	7	1	1	1	..	..	22
	U	..	..	..	..	..	..	..	..	..	..	8	8	1	0	0	2	19
Pliocene	L	..	..	..	..	..	..	..	..	..	..	..	23	5	4	1	7	40
	M	..	..	..	..	..	..	..	..	..	..	..	4	2	0	2	8
	U	..	..	..	..	..	..	..	..	..	..	..	4	2	3	9
Pleistocene		..	..	..	..	..	..	..	..	..	..	..	..	15	19	34
Total last known appearances		5	12	5	13	12	20	15	5	4	28	14	14	41	11	11	19	33	262

SOURCE: Simpson (1953: Table 11)

Among the carnivores, survival to Recent agrees sufficiently with expectation for genera that appeared before late Pliocene, but it is much lower than expectation for late Pliocene and Pleistocene genera. The discrepancy was largely, perhaps wholly, caused by the unusually high mortality of the Pleistocene.

Another interpretation is possible. The average number of genera appearing in each of six subdivisions of the Miocene and Pliocene was 20.5 (Table 10.2). If only 20.5 genera had made first appearances in the Pleistocene, 98 percent survivorship would yield twenty expected survivals, almost equal to the nineteen actual survivals observed. Viewed from this perspective, the striking feature of Simpson's tables is not the number of last appearances in the Pleistocene but rather the great number of first known appearances in the Pleistocene. In the Miocene and Pliocene, Simpson's subdivisions average 19.8 last known appearances of carnivore genera (compared with nineteen observed in the Pleistocene) and 20.5 first known appearances (compared with a surprising thirty-four in the Pleistocene).

Pleistocene Originations and Extinctions

The three orders of mammals that include the most genera with a good fossil record are Rodentia (about 560 genera), Artiodactyla (about 500 genera), and Carnivora (about 260 fissiped genera). Romer (1966) most completely summarizes available data on the stratigraphic ranges of genera. His data can be used to calculate rates of origination and extinction for genera of each of the three orders through the course of the Cenozoic. Rodents, artiodactyls, and carnivores span a full range of ecological types, they are essentially worldwide in distribution, and they have a good fossil record. Hence, patterns of origination and extinction in these three orders should accurately

Table 10.2. Expected and Actual Generic Survivorship to the Recent

Time	Genera Appearing	Percentage of Approximate Expectation of Survival to Recent	Expected Survivals	Actual Survivals
		CARNIVORA		
Early Miocene	25	0	0	0
Middle Miocene	22	2	0	0
Late Miocene	19	15	3	2
Early Pliocene	40	23	9	7
Middle Plicene	8	37	3	2
Late Pliocene	9	90	8	3
Pleistocene	34	98	33	19

SOURCE: Simpson (1953: Table 12)

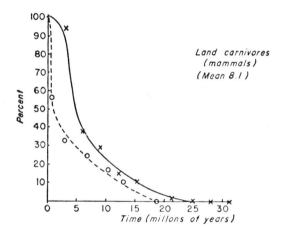

Figure 10.1. Survivorship curve for genera of carnivorous mammals showing the percentage of all genera (ordinate) surviving to a given age (abscissa). Solid line is based on genera now extinct. Dashed line is based on surviving genera. (From Simpson 1953, Figure 5)

represent the general Cenozoic mammalian pattern. Changes and additions to Romer's data published since 1966 alter some details, but the large sample size available for each of the three orders makes it unlikely that general patterns would be altered significantly. Pseudo-origination and pseudo-extinction, by evolution of one genus directly into another, cannot be estimated accurately, but these should inflate both origination and extinction curves by a constant proportion.

Analysis of Romer's data for fissiped Carnivora using Simpson's method (1953, see Table 10.1) shows that 55 genera originated and 12 genera became extinct in the early Pleistocene. According to Romer, no genera made evolutionary first appearances in the late Pleistocene, but 24 genera became extinct. In other words, as Simpson's data and discussion imply, the number of originations of carnivore genera in the early Pleistocene was approximately twice the number of extinctions in the late Pleistocene. Similar analysis of Artiodactyla shows that 145 genera originated in the early Pleistocene and 72 genera became extinct in the late Pleistocene. Analysis of Rodentia yields 176 early Pleistocene originations and 67 late Pleistocene extinctions. In both large mammals and small, carnivores and herbivores, it appears that Pleistocene originations exceeded Pleistocene extinctions by a wide margin (Gingerich 1977).

Cenozoic Rates of Origination and Extinction

The standard method of calculating rates requires that the number of originations or extinctions in any geological interval be divided by the estimated length of the interval in millions of years (cf. Simpson 1953). Rates of origination and extinction for rodents, artiodactyls, and fissiped carnivores combined are listed in Table 10.3. All three orders were represented in the fossil record by early Eocene time and, as shown by the great increase in total genera (Table 10.3), all three underwent a great diversification subsequent to their origin. Results of calculating rates of origination and extinction in this way are illustrated graphically in Figure 10.2A. Assuming the duration of both early and late Pleistocene (in Romer's sense) to be about 1 my, the rate of extinction seen in the late Pleistocene is 163 genera per million years. More striking, however, is the much higher rate of origination in the early Pleistocene, when new genera of rodents, artiodactyls, and carnivores appeared at a rate of 376 genera/million years. The high rate of generic origination in the early Pleistocene greatly exceeds the rate of late Pleistocene extinction. Comparisons with rates of origination and extinction earlier in the Cenozoic show that late Pleistocene rates of extinction for these three groups exceeded

**Table 10.3. Rates of Origination and Extinction in Cenozoic Genera of
Rodentia, Artiodactyla, and Fissiped Carnivora**

Rates are calculated as total number of originations or extinctions in an interval divided by the duration of the interval or by the total number of genera known from the interval. Data on generic ranges are from Romer (1966). Temporal durations are estimated from Berggren (1972) consistent with Romer's subdivision of epochs. The time scale, especially placement of the Miocene-Pliocene boundary, has been revised subsequent to Romer's work, but this should not affect general patterns evident in his data.

Epochs		Approximate Duration (my)	Total Genera	Originations			Extinctions		
				Total Number	Rate Per Million Years	Rate Per Total Genera	Total Number	Rate Per Million Years	Rate Per Total Genera
Paleocene	Early	3	0	0	0.0	0.00	0	0.0	0.00
	Middle	3	3	3	1.0	1.00	1	0.3	0.33
	Late	6	4	2	0.3	0.50	1	0.2	0.25
Eocene	Early	5	27	24	4.8	0.89	12	2.4	0.44
	Middle	3	47	33	11.0	0.70	19	6.3	0.40
	Late	8	118	89	11.1	0.75	76	9.5	0.64
Oligocene	Early	6	132	90	15.0	0.68	63	10.5	0.48
	Middle	2	111	42	21.0	0.38	42	21.0	0.38
	Late	6	115	46	7.7	0.40	46	7.7	0.40
Miocene	Early	6	238	169	28.2	0.71	155	25.8	0.65
	Middle	2	146	63	31.5	0.43	45	22.5	0.31
	Late	3	168	67	22.3	0.40	57	19.0	0.34
Piocene	Early	6	334	223	37.2	0.67	178	29.7	0.53
	Late	4	256	100	25.0	0.39	110	27.5	0.43
Pleistocene	Early	1	522	376	376.0	0.72	85	85.0	0.16
	Late	1	439	2	2.0	0.00	163	163.0	0.37

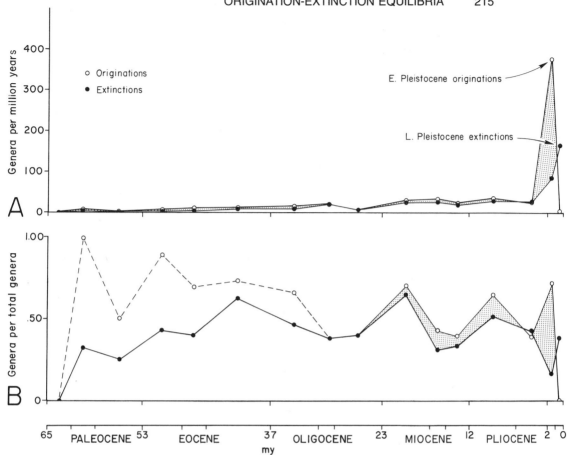

Figure 10.2. Patterns of change in the origination and extinction of genera of Rodentia, Artiodactyla, and fissiped Carnivora, expressed in (A) terms of genera per million years or (B) genera per total genera for each subdivision of the Cenozoic time scale. Note the close correlation of origination and extinction throughout most of the Cenozoic. Note also the unusually high rate of early Pleistocene originations preceding late Pleistocene extinctions, regardless of whether this is expressed in terms of rate per million years or percentage of the total fauna. Based on data from Romer (1966) (see Table 10.3, this chapter). Early Pliocene of Romer is now regarded as late Miocene by most authors, i.e., the Miocene-Pliocene boundary is now placed at 5 mya.

early and late Pliocene rates by a factor of four or five. Early Pleistocene originations of these groups exceeded Pliocene rates by a factor of more than ten, and earlier Cenozoic rates by an even wider margin.

Given the great overall increase in total number of rodent, artiodactyl, and carnivore genera during the Cenozoic, it is possible that high rates of Pleistocene originations and extinctions are an artifact of increasing diversity, or of an increasingly well-sampled fossil record, or both. This bias can be corrected by calculating rates of generic origination and extinction as proportions of the total number of genera under study in any given time period. Standardized rates, in genera per total genera, are also listed in Table 10.3. These standardized rates are plotted in Figure 10.2B. As before, origination rates appear to exceed extinction rates in the early Cenozoic during the explosive initial radiation of the three orders studied, but during most of the later Cenozoic origination and extinction proceeded at about the same rate. When corrected for total known diversity in each subdivision of the time scale, the late Pleistocene extinction rate is slightly below average for the whole of the Cenozoic. The standardized early Pleistocene rate of originations, on the other hand, is exceeded only by abnormally high middle Paleocene and early Eocene origination rates. Both Pleistocene rates would increase by a factor ranging from two to eight relative to other subdivisions of the time scale if the durations of subdivisions were also taken into account.

Origination-Extinction Equilibria and Pleistocene Extinctions

One notable feature of the origination and extinction curves shown in Figure 10.2 is the way origination and extinction follow each other. When origination rate is high, extinction rate tends to be high as well. Paleocene and Eocene local faunas have diversity values characteristic of extant mammalian faunas (Rose 1982), suggesting that there is little reason to expect steep secular trends in overall Cenozoic mammalian diversity. High rates of origination are thus necessarily accompanied by high rates of extinction, and vice versa, because these two processes are in dynamic equilibrium (MacArthur and Wilson 1963, 1967; Webb 1969). The actual level of the equilibrium, in a sense the diversity capacity (MacArthur and Wilson's *saturation*), at any time is determined by the environment. As long as saturation levels remain constant, faunas can experience unlimited substitution of one genus or species for another, i.e. turnover may be high or low, but net diversity can only fluctuate within narrow limits.

The relative height of any origination/extinction pair of points in Figure 10.2 is a measure of the amount of faunal turnover in a given time interval. Assuming that origination and extinction are in dynamic equilibrium constrained by environmental diversity capacity, the difference in height between origination and extinction in any origination/extinction pair is a measure of change in the level of environmental saturation. Where origination rate greatly exceeds extinction rate we can infer an increase in saturation, and where extinction exceeds origination we can infer a decrease in this quantity. Thus the early Miocene and the early Pliocene (*fide* Romer 1966, "early Pliocene" of Romer is now regarded as late Miocene by Berggren 1972 and most authors) appear to have been intervals of high mammalian faunal turnover compared to the late Oligocene or middle and late Miocene (middle Miocene).

The fact that origination and extinction track each other so closely suggests little change in saturation through the Miocene and Pliocene. However, as we have seen, during the early Pleistocene mammalian originations greatly exceeded extinctions, suggesting a marked increase in saturation level or diversity capacity. Given that the extinction rate remains in close equilibrium with origination rate, the high rate of late Pleistocene extinctions can be viewed as a natural equilibration ending a one- to two-million-year interval of high saturation that followed an unusually high rate of early Pleistocene originations. In this context, what requires explanation is not late Pleistocene extinctions but the very high rate of early Pleistocene originations.

Coincidence of an unusually high origination rate with a low early Pleistocene extinction rate (fig. 10.2) produced a Pleistocene fauna of exceptionally high diversity. Diversity usually has a positive correlation with temperature, productivity, and stability (Klopfer 1959, Fischer 1960, Pianka 1966), but the Pleistocene was a time of climatic cooling and instability (Flint 1971). Hence these factors cannot explain the increased diversity. An explanation is possibly to be found in the fragmentation and diversification of habitats accompanying successive Pleistocene glaciations. Spatial heterogeneity has a positive effect on faunal diversity within communities (MacArthur and MacArthur 1961) and between communities (Kurtén 1969, Valentine and Moores 1970, Flessa 1975), and this appears to be a plausible explanation for abnormally high origination rates and low extinction rates in the early Pleistocene.

In North America, Pleistocene glaciations apparently compressed latitude-parallel life zones (Dillon 1956, Delcourt and Delcourt 1981), which were fragmented longitudinally by large rivers fed by glacial meltwater. One would expect greater endemism and at the same time predict higher rates of species differentiation in such a complex environmental mosaic. Fluctuating climates facilitate dispersal, and immigration undoubtedly complemented speciation in augmenting origination rates and diversity. Both voles and shrews exhibited higher species densities during the Wisconsinan glaciation than is evident at the same localities today (Graham 1976). This trend toward greater diversification of shrews and voles in local faunas can presumably be extrapolated to other mammalian groups as well.

A large proportion of late Pleistocene extinctions could represent the natural result of faunal equilibration associated with return to nonglacial geographic and climatic conditions toward the end of the Pleistocene. Retreat of continental ice sheets would reduce environmental heterogeneity, and former high levels of Pleistocene mammalian diversity might no longer be supportable.

Origination-Extinction Equilibria Within the Pleistocene— A Test of the Environmental Heterogeneity Hypothesis

High rates of origination at the beginning of the Pleistocene and high rates of extinction at the end of the Pleistocene may reflect increased diversity capacity during this epoch caused by increased environmental heterogeneity associated with continental glaciation. This hypothesis predicts that *within* the Pleistocene, glacial intervals should have more diverse faunas than interglacials. Or, stated in terms of origination (or appearance in a restricted geographic area) and extinction (disappearance), rates of origination should exceed extinction by a significant amount during glacial intervals and extinctions should approach or exceed originations during interglacials.

Origination and Extinction Within the Pleistocene

Kurtén and Anderson (1980) provide the most extensive compilation available of temporal ranges for North American Plio-Pleistocene mammals. The Blancan Land Mammal Age is divided into four successive intervals, and the Irvingtonian and Rancholabrean are divided into three each. Irvingtonian-1 coincides with the Nebraskan glacial interval, Irvingtonian-2 corresponds to the Aftonian interglacial, and Irvingtonian-3 includes both Kansan glacial and Yarmouthian interglacial intervals. Rancholabrean-1 corresponds to the complex Illinoian glacial interval, Rancholabrean-2 represents the Sangamonian interglacial, and Rancholabrean-3 corresponds to Wisconsinan glaciation (Kurtén and Anderson 1980).

Rates of appearance and disappearance for all North American species of Plio-Pleistocene mammals are summarized in Table 10.4, and these data are shown graphically in Figure 10.3. As noted for Cenozoic mammals in general, rates of appearance and disappearance track each other closely and appear to be in dynamic equilibrium. Appearances almost always exceed disappearances by a small margin. Whether calculated in terms of species per million years or species per total species, there are three peaks indicating a high rate of appearance of new species. These peaks correspond to glacial intervals in Irvingtonian-1, Irvingtonian-3, and Rancholabrean-3, respectively. The rate of disappearance of species in glacial intervals, which should be low, appears to have remained near the average for Plio-Pleistocene mammals when considered in terms of species per total species.

Table 10.4. Rates of Appearance and Disappearance in North American Species of Plio-Pleistocene Mammals

Rates are calculated as total number of appearances or disappearances in an interval divided by the duration of the interval or by the total number of species known from the interval.

Land-Mammal Age	Glacial (G) or Interglacial (I)	Approximate Duration (my)	Total Species	Appearances			Disappearances		
				Total Number	Rate Per Million Years	Rate Per Total Species	Total Number	Rate Per Million Years	Rate Per Total Species
Blancan-1	Preglaciation	0.3	35	28	93.3	0.80	12	40.0	0.34
Blancan-2	Preglaciation	0.4	100	77	192.5	0.77	50	125.0	0.50
Blancan-3	G (Montane)	0.5	91	41	82.0	0.45	25	50.0	0.27
Blancan-4	I (Plio-Pleistocene)	0.8	103	37	46.3	0.36	49	61.3	0.48
Irvingtonian-1	G (Nebraskan)	0.4	100	46	115.0	0.46	36	90.0	0.36
Irvingtonian-2	I (Aftonian)	0.3	87	23	76.7	0.26	18	60.0	0.21
Irvingtonian-3	G/I (Kansan/Yarmouthian)	0.3	163	94	313.3	0.58	32	106.7	0.20
Rancholabrean-1	G (Illinoian)	0.2	158	27	135.0	0.17	8	40.0	0.05
Rancholabrean-2	I (Sangamonian)	0.2	208	58	290.0	0.28	22	110.0	0.11
Rancholabrean-3	G (Wisconsinan)	0.1	289	103	1030.0	0.36	77	770.0	0.27

SOURCE: Kurtén and Anderson (1980) for species ranges, temporal framework, and glacial/interglacial sequence

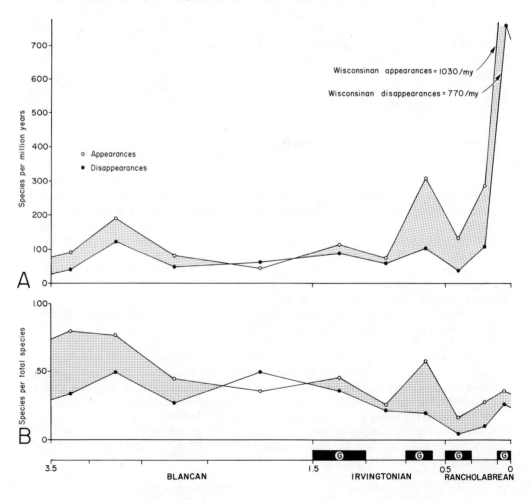

One glacial interval, Rancholabrean-1, is characterized by unusually low appearance and disappearance rates, and in this sense it does not conform to expectation. There are several possible reasons for this. Rancholabrean-1 represents the interval of Illinoian glaciation, which is itself complex, including at least one major interstadial (Frye et al. 1965, Schultz and Martin 1970). The first appearance of the taxon defining the beginning of the Rancholabrean, *Bison*, is not well dated, and it may have immigrated into North America sometime during the Illinoian rather than at the beginning of this glacial interval (Kurtén and Anderson 1980, p. 5). Finally, the distinction between Irvingtonian and Rancholabrean faunas now appears less sharp than it was originally assumed to be (Kurtén and Anderson 1980, p. 37).

Correspondence of three out of four Pleistocene peaks of high appearance rate with glaciation provides some corroboration of the environmental heterogeneity hypothesis outlined above to explain high rates of Pleistocene origination of new taxa, but the evidence available from mammal distributions within the Pleistocene is still inadequate for a definitive test. Finer subdivisions of the Pleistocene mammalian record and better correlation with complex glacial cycles will be required before a definitive test of the environmental heterogeneity hypothesis can be attempted.

Extinction of Large Mammals

It is worth noting that the equilibrium between origination and extinction in large mammals follows a pattern similar to that of mammals in general through most of the Pleistocene (Table 10.5). However, as Martin (1967), Van Valen (1970), Webb (this volume), and others have noted, extinction affected large mammals very differently at the end of the Pleistocene. The rate of disappearance of large mammals in Rancholabrean-3 greatly exceeded their rate of appearance (fig. 10.4). Large mammals (Artiodactyla, Perissodactyla, and Proboscidea) disappeared in Rancholabrean-3 at a rate of 290 species/million years more rapidly than new species appeared in this interval. Stated in terms of species per total species, 56 percent of large mammalian species disappeared in Rancholabrean-3 and were not replaced by new large mammals. Viewed in the context of origination-extinction equilibria in Cenozoic mammals (fig. 10.2), or viewed in the context of appearance-disappearance equilibria for Pleistocene mammals in general (fig. 10.3), the disappearance of 56 percent of the large mammal fauna without replacement is unusual. This aspect of Pleistocene extinctions cannot reflect simple equilibration to previously existing levels of environmental saturation. It could represent a temporary lag in replacement (we are, after all, only 10,000 years removed from Wisconsinan glaciation), or it could possibly be a consequence of increasing human predation decimating large mammal populations that will never be replaced.

Conclusions

Pleistocene extinctions are real and significant, but they should not be viewed in isolation from the other half of the dynamic equilibrium that controlled Pleistocene faunal diversity: originations. Mammalian generic origination rates throughout most of the

Figure 10.3. Patterns of change in the appearance and disappearance of all species of North American Plio-Pleistocene mammals, expressed in (A) terms of species per million years or (B) species per total species. Note the general correlation of rates of first appearance with rates of disappearance. Wisconsinan first appearances ("originations") exceeded Wisconsinan disappearances ("extinctions") by 360 species per million years or, expressed as genera per total genera, by 9 percent. (Based on data in Kurtén and Anderson 1980) (see Table 10.4 this chapter.)

Table 10.5. Rates of Appearance and Disappearance in North American Species of Large Plio-Pleistocene Mammals (Artiodactyla, Perissodactyla, Proboscidea)

Rates calculated as in Table 10.4.

Land-Mammal Age	Glacial (*G*) or Interglacial (*I*)	Approximate Duration (my)	Total Species	Appearances Total Number	Appearances Rate Per Million Years	Appearances Rate Per Total Species	Disappearances Total Number	Disappearances Rate Per Million Years	Disappearances Rate Per Total Species
Blancan-1	Preglaciation	0.3	2	2	6.7	1.00	0	0.0	0.00
Blancan-2	Preglaciation	0.4	16	14	35.0	0.88	5	12.5	0.31
Blancan-3	*G* (Montane)	0.5	15	4	8.0	0.27	0	0.0	0.00
Blancan-4	*I* (Plio-Pleistocene)	0.8	19	4	5.0	0.26	7	8.8	0.37
Irvingtonian-1	*G* (Nebraskan)	0.4	23	11	27.5	0.48	5	12.5	0.22
Irvingtonian-2	*I* (Aftonian)	0.3	27	9	30.0	0.33	6	20.0	0.22
Irvingtonian-3	*G/I* (Kansan/Yarmouthian)	0.3	33	12	40.0	0.36	3	10.0	0.09
Rancholabrean-1	*G* (Illinoian)	0.2	40	10	50.0	0.25	3	15.0	0.08
Rancholabrean-2	*I* (Sangamonian)	0.2	52	15	75.0	0.29	11	55.0	0.21
Rancholabrean-3	*G* (Wisconsinan)	0.1	52	11	110.0	0.21	40	400.0	0.77

SOURCE: Kurtén and Anderson (1980)

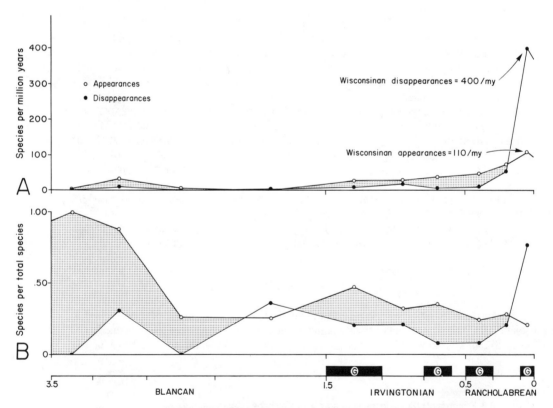

Figure 10.4. Patterns of change in the appearance and disappearance of large herbivorous species of North American Plio-Pleistocene mammals, expressed (A) in terms of species per million years or (B) species per total species. Note that Wisconsinan disappearances ("extinctions") exceeded Wisconsinan first appearances ("originations") by 290 species per million years, or 56 percent of the total number of species. (Based on data in Kurtén and Anderson 1980) (see Table 10.5 this chapter).

Cenozoic were equal to or only slightly greater than extinction rates until the Pleistocene. In the early Pleistocene, origination rates greatly exceeded extinction rates, whether these are calculated per million years or as a percentage of total genera. The rate of late Pleistocene extinctions was the highest of any subdivision of the Cenozoic, but this may be in part an artifact of a more completely known Pleistocene fossil record. When viewed in terms of genera per total genera, the rate of late Pleistocene extinctions was about average for earlier parts of the Cenozoic.

Given the equilibrium observed between origination and extinction throughout the Cenozoic, the high rate of late Pleistocene generic extinctions may be viewed as a natural sequel to an unusually high rate of originations in the early Pleistocene.

The high rate of early Pleistocene generic originations may be a response to increased environmental heterogeneity associated with the onset of continental glaciation in the early Pleistocene. If this explanation has merit, it should be possible to correlate levels of generic or species diversity (saturation), and the originations (appearances) and extinctions (disappearances) that control saturation, with glacial and interglacial stages within the Pleistocene: glacial intervals should have high standing diversity with increased rates of origination at the beginning balanced by increased rates of extinction at the end. If there is any lag between environmental change and its effect on diversity, glacial intervals might be expected to show high rates of origination and interglacial intervals high rates of extinction. This hypothesis is tested using the temporal ranges of North American species of Plio-Pleistocene mammals published by Kurtén and Anderson (1980).

Throughout the Pleistocene, as throughout the Cenozoic, rates of origination and extinction appear to be closely related. When all North American species are considered, the rate of appearance of new species in the Pleistocene exceeds the rate of disappearance of species, even during the late Rancholabrean (Wisconsinan) glacial interval so well known for its fauna of unusual mammals now extinct.

When large herbivorous North American species are considered in isolation, the rate of disappearance (extinction or emigration) of species during the Wisconsinan greatly exceeds their rate of appearance (origination or immigration). Some 56 percent of Wisconsinan artiodactyls, perissodactyls, and proboscidians disappeared from the North American fauna without replacement. Late Wisconsinan extinction of large mammals is certainly real, but much of the "unusually high mortality of the Pleistocene" cited by Simpson (1953) is best regarded as a natural consequence of high faunal turnover caused by major oscillations in climate and environmental heterogeneity.

Patterns within the Pleistocene are inadequate to test the general environmental heterogeneity hypothesis advanced here to explain high rates of early Pleistocene originations. Three out of four glacial intervals of the Pleistocene exhibit high rates of species appearances, as predicted, but the remaining interval has an unusually low rate. The rates of disappearance of Pleistocene species in different intervals show no obvious correlation with glacials or interglacials. The tendency for high rates of species appearances to coincide with glaciations is suggestive, but better sampling of appearances and disappearances at the beginning and end of glacial and interglacial stages is required. Diversification patterns within the Pleistocene do not yet provide an adequate test of the influence that increased environmental heterogeneity during continental glaciation may have had on the origination and extinction of genera and species of Pleistocene mammals.

Acknowledgments

I thank William R. Farrand, Karl W. Flessa, Paul S. Martin, Leigh Van Valen, S. David Webb, Neil A. Wells, and Dale A. Winkler for comments on various drafts of this manuscript. Karen Klitz drew Figures 10.3 and 10.4.

References

Berggren, W. A. 1972. A Cenozoic time-scale—some implications for regional geology and paleobiogeography. Lethaia 5:195–215.

Delcourt, P. A. and H. R. Delcourt. 1981. Vegetation maps for eastern North America: 40,000 yr B.P. to the present. *In* R. C. Romans (ed.), *Geobotany II* (Plenum Publ., New York), pp. 123–165.

Dillon, L. S. 1956. Wisconsin climate and life zones in North America. Science 123:167–176.

Fischer, A. G. 1960. Latitudinal variations in organic diversity. Evolution 14:64–81.

Flessa, K. W. 1975. Area, continental drift, and mammalian diversity. Paleobiology 1:189–194.

Flint, R. F. 1971. *Glacial and Quarternary Geology* (John Wiley, New York, 892 pp.).

Frye, J. C., H. B. Willman, and R. F. Black. 1965. Outline of glacial geology of Illinois and Wisconsin. *In* H. E. Wright and D. G. Frey (eds.), *The Quarternary of the United States* (Princeton University Press, Princeton, 922 pp.), pp. 43–61.

Gingerich, P. D. 1977. Patterns of evolution in the mammalian fossil record. *In* A. Hallam (ed.), *Patterns of Evolution as Illustrated by the Fossil Record* (Elsevier Publ. Co., Amsterdam, 591 pp.), pp. 469–500.

Graham, R. W. 1976. Late Wisconsin mammalian faunas and environmental gradients of the eastern United States. Paleobiology 2:343–350.

Klopfer, P. H. 1959. Environmental determinants of faunal diversity. American Naturalist 43:337–342.

Kurtén, B. 1969. Continental drift and evolution. Scientific American 220(3):54–64.

Kurtén, B. and E. Anderson. 1980. *Pleistocene Mammals of North America* (Columbia University Press, New York, 442 pp.).

MacArthur, R. H. and J. W. MacArthur. 1961. On bird species diversity. Ecology 42:594–598.

MacArthur, R. H. and E. O. Wilson. 1963. An equilibrium theory of insular zoogeography. Evolution 17:373–387.

———. 1967. *The Theory of Island Biogeography* (Princeton University Press, Princeton, 203 pp.).

Martin, P. S. 1967. Prehistoric overkill. *In* P. S. Martin and H. E. Wright (eds.), *Pleistocene Extinctions, The Search for a Cause.* (Yale University Press, New Haven, 453 pp.), pp. 75–120.

Pianka, E. R. 1966. Latitudinal gradients in species diversity: a review of concepts. American Naturalist 100:33–46.

Romer, A. S. 1966. *Vertebrate Paleontology* (University of Chicago Press, Chicago, 468 pp.).

Rose, K. D. 1981. The Clarkfordian land-mammal age and mammalian faunal composition across the Paleocene-Eocene boundary. University of Michigan Papers on Paleontology 26:1–196.

Schultz, C. B. and L. D. Martin. 1970. Quaternary mammalian sequence in the central Great Plains. *In* W. Dort and J. K. Jones (eds.), *Pleistocene and Recent Environments of the Central Great Plains* (University of Kansas Press, Lawrence, 433 pp.), pp. 341–353.

Simpson, G. G. 1953. *The Major Features of Evolution* (Columbia University Press, New York, 434 pp.).

Valentine, J. W. and E. M. Moores. 1970. Plate-tectonic regulation of faunal diversity and sea level: a model. Nature 228:657–659.

Van Valen, L. 1970. Evolution of communities and late Pleistocene extinctions. Proceedings North American Paleontological Convention 1(E):469–485.

Webb, S. D. 1969. Extinction-origination equilibria in late Cenozoic land mammals of North America. Evolution 23:688–702.

Coevolutionary Disequilibrium and Pleistocene Extinctions

RUSSELL W. GRAHAM AND ERNEST L. LUNDELIUS, JR.

THE END OF THE PLEISTOCENE WAS MARKED by global climatic changes and by a major faunal extinction that was most severe for terrestrial mammals. The demise of many members of a fauna is not unique to the late Pleistocene; extinction appears to have been the rule rather than the exception in geologic history. However, because the late Pleistocene is not so far back in geologic time, we can study the extinctions in greater detail, providing a model for the possible causes of preceding extinctions.

The time of extinction of individual species can be more accurately established by radiometric dating techniques, especially radiocarbon for the last 50,000 years, than for earlier periods, and the fossil accumulations have not been as badly degraded by erosion and later diagenesis. Precise timing of earlier extinction events is hampered by the lack of datable materials, higher standard errors for absolute dates, and the incompleteness of the fossil record. The late Pleistocene extinction is also associated with one of the most extensive and rapid environmental changes in the earth's history and, unlike earlier extinctions, is associated with the emergence of man as a dominant and efficient predator.

The apparently "instantaneous" nature of the late Pleistocene extinction has led to the development of two dominant theories concerning its cause. One theory, prehistoric overkill, advocates that the abrupt and massive elimination of late Pleistocene species was a consequence of the predatory habits of Upper Paleolithic hunters (Martin 1967, 1973, this volume; Mosimann and Martin 1975). The actual impact of human predation on the late Pleistocene biota is debatable, and many facets of the extinction event are not adequately explained by the "overkill" hypothesis (Graham 1979; Grayson 1977; L. Martin and Neuner 1978; Webb 1977, 1978).

Environmental change has also been proposed as a general theory to account for late Pleistocene extinctions. Within this framework many independent models have been offered (Guilday 1967; Lundelius 1967, 1972, 1976; L. Martin and Neuner 1978; Slaughter 1967; Webb 1978; Wilson in Kurtén and Anderson 1980). Most environmental models focus on the general nature of environmental change. None considers the consequences of a disequilibrium created by the disruption of coevolutionary interactions between plants and animals. These coevolved relationships are important to the stability and perpetuation of biological communities. The purpose of this chapter is to evaluate the role of coevolutionary disequilibrium in the late Pleistocene extinction. Obviously, it is naïve to isolate a single factor as the sole reason for extinction, but it is essential to thoroughly understand all the factors that might have contributed to the extinction before we can arrive at a comprehensive understanding of the phenomenon.

Late Pleistocene Biotic Communities

Late Pleistocene communities were characterized by the coexistence of species that today are allopatric and presumably ecologically incompatible. These associations have been referred to as "disharmonious" in terms of the modern distributional patterns (Semken 1974). However, this does not imply that the biota was not adapted to prevailing environments at the time. Disharmonious associations have been documented for late Pleistocene floras (Cushing 1965, 1967; Davis 1967; Van Devender and Spaulding 1979; Webb and Bryson 1972; West 1967; Wright 1970, 1981), terrestrial invertebrates (Campbell 1980; Coope 1979; Kavanaugh 1980; Lehmkuhl 1980; Matthews 1974; Miller 1976; Morgan and Morgan 1979, 1980), lower vertebrates (Holman 1976), birds (Guilday et al. 1977), and mammals (Dalquest 1965; Dalquest et al. 1969; Graham 1976a, 1979; Graham and Semken 1976; Guilday et al. 1978; Hibbard 1960; Lundelius 1967, 1974, 1976; Semken 1974; Slaughter 1975).

In fossil assemblages, the assumption that species with apparently disparate environmental requirements were contemporaneous must take into account the possibility of mixing of remains of different faunas. Slow accumulation of a deposit through a period of rapid environmental change could produce such associations. Most of the presently available evidence suggests that individual stratigraphic units are deposited in too short a time in relation to the rate of environmental change for this to be a likely cause. This is readily apparent in palynological studies, which show disharmonious associations in environments of rapid deposition such as lakes (Cushing 1965, 1967).

Failure to distinguish between stratigraphic units during excavation or analysis could lead to a false association of taxa. Spurious associations can also result from redeposition of fossils from older to younger strata. These possibilities must be considered separately for each local fauna; they frequently can be detected by careful methods of excavation. The widespread occurrence of disharmonious faunas in Pleistocene deposits also indicates that these associations were much too common to be spurious in all cases. In addition, if these associations are caused by sedimentary mixing, their frequency should be about the same for all time periods; but disharmonious associations are rare in Holocene faunas, and in stratified faunas they usually disappear at the Pleistocene/Holocene contact.

There are no modern analogues to late Pleistocene biotas or environments (Cushing 1965; Graham 1979; Hibbard 1960; Lundelius 1974; Matthews 1974, 1976; Wright 1981). Late Pleistocene disharmonious floras and faunas generally had higher numbers of species than their modern counterparts. In modern floras and faunas there is a positive correlation between increased species diversity and decreased climatic variability as measured by winter-summer differences in mean temperature (MacArthur 1975). Thus the degree of diversity in late Pleistocene disharmonious biotas suggests that they existed during times when the climate was equable and seasonal extremes in temperature and effective moisture were reduced (Graham 1976a, 1979). Although the late Pleistocene climates were more equable than those of today, they still supported environmental gradients which, like modern ones, created biotic associations that can be grouped into biogeographical provinces (Graham 1979, Guilday et al. 1978, L. Martin and Neuner 1978). Because of the disharmonious nature of the late Pleistocene biota, the species composition of these provinces was quite different from modern ones (Cushing 1965, Graham 1979, Wright 1981).

Changes in the late Pleistocene biota probably were initiated by global changes in climate. The biological response to these complex and time transgressive environmental changes was not a synchronous, unidirectional shift. Instead the biosphere underwent a mosaic pattern of evolution. Organisms did not migrate as intact and immutable community units; each species responded individually to environmental change based upon its own limits of tolerance for temperature, moisture, and other environmental variables (Cushing 1965; Davis 1976; Graham 1976a, 1979; F. King and Graham 1981). The evolution of modern community patterns from those of the late Pleistocene is a consequence of intricate biological and biophysical interactions of individual species. To some extent, the composition of modern communities is also a matter of chance due to historical accidents of species distributions and migration routes in the past.

These individualistic readjustments to environmental change have significant paleoecological and biological implications. If each species migrated independently, community composition would not have been stable over long periods of time. Therefore, modern communities cannot be used as direct analogues in the reconstruction of Pleistocene environments. These species-specific changes also reduced the predictability of the structure and composition of the evolving communities. The degree of predictability would have been a function of the magnitude of environmental change. Fisher's (1958) model of adaptation to environmental deterioration predicts that a random environmental change of a given magnitude is more likely to improve the level of adaptation of the generalist rather than that of the specialist (Pianka 1974). However, the probability of adaptive improvement for both specialists and generalists decreases with increasing magnitudes of environmental change (fig. 11.1).

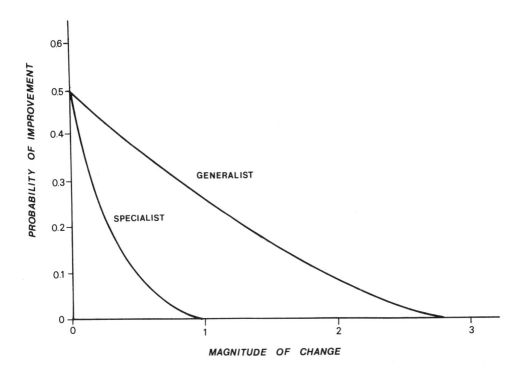

Figure 11.1. Fisher's (1958) theoretical model of the probability of improvement of fitness for generalists and specialists in response to random environmental changes of given magnitudes. (Modified after Pianka *Evolutionary Ecology,* Harper & Row, 1974)

The evolution of the North American grassland communities at the end of the Pleistocene clearly illustrates the magnitude of these changes. The Quaternary fossil record for grasses is meager, but genetic studies of modern grasses can be used to establish phylogenetic relationships, infer past distributions, and suggest migration routes for the flora (Stebbins 1975). From these studies it has been concluded that the patterns of grassland communities of the central and western United States were established during the early Holocene (Stebbins 1981).

Bluestems (*Andropogon* or *Schizachyrium*), Indian grass (*Sorghastrum*), grama grasses (*Bouteloua*), and buffalo grasses (*Buchloe*), species which now dominate the plains, are polyploids that appear to have close diploid relatives either in the southern United States or in northern Mexico (Stebbins 1981). The polyploid complexes to which these species belong have distribution patterns which suggest youthfulness (Stebbins 1971); thus these species may have originated during the Holocene (Stebbins 1981). *Stipa comata* and *Stipa spartea* probably immigrated from Asia during the early Pleistocene; *Stipa viridula, S. agropyron,* and *Elymus* spp. were derived from the western mountains of the United States. *Stipa viridula* probably is a late Pleistocene or post-Pleistocene immigrant.

Many of the herbaceous species (*Artemisia*, Chenopodiaceae, Gramineae) of the modern Great Plains grassland communities were elements of the late Pleistocene "forest" and "tundra" communities of the eastern United States. Pollen diagrams for the Midwest as well as the eastern United States show increases in *Artemisia*, Chenopodiaceae, Gramineae, and Cyperaceae, which are frequently associated with the late Pleistocene spruce zone. Although the total pollen influx and pollen percentages for shrubs and herbs are low in these zones (Birks 1981, King 1981, Wright 1981), it is apparent that these nonarboreal plants were integral components of the late Pleistocene communities. These species are either insignificant or absent in modern spruce forests. Many elements of modern grassland communities may have been derived from these eastern environments. Distributional studies of modern insects (Ross 1970), birds (Mengel 1970), and mammals (Hoffmann and Jones 1970) also support an eastern forest origin for many species in the grassland communities.

Late Pleistocene vertebrate faunas from the Great Plains also reflect disharmonious associations of species (Graham 1981, in press; Harris 1977; Hibbard 1970; Lundelius 1967; Lundelius et al. in press). Northern species such as *Blarina brevicauda* (short-tailed shrew), *Sorex cinereus* (masked shrew), *Sorex palustris* (water shrew), *Zapus hudsonius* (meadow jumping mouse), *Spermophilus richardsoni* (Richardson's ground squirrel), and *Microtus pennsylvanicus* (meadow vole) had distributions extending into the southern plains of Texas and New Mexico. These species coexisted with southern taxa such as *Sigmodon hispidus* (cotton rat), *Geochelone wilsoni* (Wilson's tortoise), and *Dasypus bellus* (giant armadillo). Eastern species such as *Blarina carolinensis* (Carolina shrew), *Synaptomys cooperi* (bog lemming), *Sciurus* spp. (tree squirrels), and *Tamias striatus* (eastern chipmunk) were also present in the late Pleistocene communities of the southern Great Plains.

The evolution of modern floral and faunal community patterns for the North American grasslands was a complex interaction of many biological processes. Numerous species of the modern grassland communities were residents of the late Pleistocene grasslands. Some floral and faunal species immigrated into the central grasslands of the United States from the eastern forest, western mountains, northern subarctic, and southern savannas. Other species were added to the Holocene grassland communities by autochthonous evolution (Stebbins 1975). Individual species were eliminated from the ancestral late Pleistocene communities by extirpation and extinction. The culmination of all of these events was reached in the late Pleistocene/early Holocene when the modern grassland communities emerged.

Coevolutionary
Disequilibrium and Extinction

Broadly defined, coevolution refers to the common evolution of multiple taxa that do not exchange genes but do share close ecological relationships, and in which reciprocal selective forces act to make the evolution of either taxon partially dependent upon the evolution of the other (Ehrlich and Raven 1964, Pianka 1974). In a more restricted sense, coevolution refers primarily to the interdependent evolutionary interactions between plants and animals. These interactions may be beneficial to one organism and detrimental to the other or they may be mutually beneficial. For example, a plant may invest a given amount of energy in the production of secondary chemical substances that deter the vast majority of herbivore predators. Evolution of this strategy by the plant may insure protection from predation. However, if an herbivore can evolve a physiological means of processing the chemical deterrent, then the herbivore can gain an uncontested food supply. Coevolution of this nature may favor herbivore specificity for a single species or a few closely related species of plants.

In other cases of coevolutionary relationships, plants and animals form cooperative or mutually beneficial relationships. In many of these cases a balance is established between the evolutionary gains for the plant and animal involved in the interaction. Environmental change or evolution of new adaptive strategies, as well as temporal and spatial migrations of species, may disrupt this equilibrium. The consequences of the coevolutionary disequilibrium may vary from establishment of a new equilibrium or new interaction sphere involving other species to extirpation or extinction of one of the organisms.

The grazing succession exemplified by the modern diverse herbivore megafauna of the African savannas is a highly coevolved system. Herbivore species migrate throughout the savannas, each grazing on a specific plant species, plant part, or select group of plants. This grazing activity stimulates the growth and development of other plant species or plant parts that will be consumed by subsequent migratory waves of herbivores. Development of the food resource for one species depends on the grazing activities of the preceding species (Bell 1971, Gwynne and Bell 1968, McNaughton 1976, Vesey-Fitzgerald 1960). The productivity patterns of the vegetation, and consequently the migrational behavior of the herbivores, is directly related to the annual climatic cycle (McNaughton 1976, Vesey-Fitzgerald 1960).

To preserve this grazing succession it is essential to retain the tightly coevolved, sequential structure of the plant-animal interactions. Consequently, the herbivores must follow a specific migrational sequence and each plant species must be of nutritional value at the time it is to be used by the herbivore. Changes in the migratory behavior of herbivore species, adjustments in the productivity schedules or distributions of plant species, or alterations in the climatic patterns will disrupt this succession.

Coevolved grazing sequences partition the environment by well-defined niche differentiation, allowing the coexistence of many large herbivores. Such ecological separation has also been documented for browsing ungulate communities in Tsavo East National Park, Kenya (Bell 1971, Leuthold 1978, McNaughton 1976, Vesey-Fitzgerald 1960, Western and Ssemakula 1981). A coevolved grazing succession must have existed in the late Pleistocene savanna and grassland environments of North America (Graham 1981, Lundelius 1972). Other late Pleistocene communities, including arctic-steppe, tundra, and open forests or parklands may have had similar, though possibly abbreviated, grazing sequences, as indicated by the presence of such grazers as *Mammuthus* (mammoth), *Equus* (horse), *Camelops* (camel), and *Bison*.

Increased patchiness of the environment also may allow an area to support a highly diverse vertebrate fauna (see Guthrie this volume). The diversity of the late Pleistocene mammalian fauna of the eastern United States may have been maintained by a vegetational mosaic (Brown and Cleland 1968). Fossil pollen diagrams for that area show high percentages of prairie vegetation (*Artemisia*, Gramineae, Chenopodiaceae), along with a high spruce peak (Cushing 1965, 1967; Davis 1967; Kapp et al. 1969; King 1973, 1981; McAndrews 1966; Van Zant 1979; Whitehead 1979, 1981; Wright 1981). These data suggest that some of the late Pleistocene forests were more open, and perhaps more patchy, than those today.

This type of vegetational mosaic would have supported a primarily browsing mammalian fauna, but would have accommodated a grazing element. The late Pleistocene mammalian faunas of the eastern United States are characterized by browsing elements (*Mammut* [mastodon], *Cervalces* [stag-moose], *Symbos* [woodland musk ox], *Mylohyus* [long-nosed peccary], *Castoroides* [giant beaver], *Tapirus* [tapir], and *Megalonyx* [ground sloth]) as well as grazers (*Equus, Bison, Platygonus* [peccaries], *Mammuthus, Spermophilus* [ground squirrels], and possibly *Glossotherium* [ground sloth]). These diverse faunal elements were obviously closely related to, and coevolved with, the vegetational communities. In fact, the open nature of the late Pleistocene forests may have been maintained by highly coevolved seed and sapling predation systems. The high density and low diversity characteristic of many modern plant associations such as the spruce forest may be the result of escape from predation by plants due to the extinction of specific plant predators (Janzen 1970, Janzen and Martin 1982).

If the late Pleistocene biota responded to environmental changes as communities rather than as individual species, then the large herbivore communities should have been able to track the vegetational shifts. Stability of the community patterns would have allowed the integrity of the grazing sequence to be retained. Furthermore, given the increased continentality of the late Pleistocene/early Holocene environment (Graham 1976a, Hibbard 1960), such a zonal model would require the expansion of the grassland communities at the expense of the forest communities. Models of island biogeography (MacArthur and Wilson 1967) would predict different extinction rates for the expanding and contracting habitat islands. Assuming zonal continuity, expanding grasslands should have lower extinction rates than contracting forest habitats. This was not the case in North America, where numerous grazing taxa became extinct.

Alternatively, if each species responded individually to environmental change at the end of the Pleistocene, then new community patterns would emerge. The species composition of these new communities would be totally different from that of their late Pleistocene predecessors. Consequently, even though a physiognomic association such as the grasslands might be expanding, species composition of the communities would be continually changing. Genetic studies of modern grass species, palynological data, and distributional studies of modern fauna, as well as invertebrate and vertebrate paleontological data, demonstrate that each species responded independently to environmental changes at the end of the Pleistocene. The changes in the flora would have drastically disrupted the grazing sequence of the late Pleistocene grasslands and savannas.

With the breakdown of the grazing sequence structure, or any previously coevolved system, the niche differentiation of the large herbivore species would have lost its clear definition. As a consequence, both interspecific and intraspecific competition would have been increased by these environmental and biological changes (Guilday 1967). Based on the Lotka-Volterra competition equations, there are four possible outcomes of competition: three lead to the extinction of one species, and the fourth leads to the stable coexistence of both species (Pianka 1974). Competition in the large herbivore communities of the late Pleistocene would have driven species with reduced fitness to extinction, selected for species best adapted to the new community patterns, and redefined niche differentiation for sympatric species that survived the extinction event.

The impact of competitive exclusion on a large herbivore grazing succession is clear in two theoretical studies (Soulé et al. 1979, Western and Ssemakula 1981) of extinction equilibria on future game preserves in East Africa. Both models consider game parks as habitat islands so their predictions are based on island biogeographic principles. However, the outcome of each model is significantly different. Soulé's paradigm, referred to as the "true island" model, predicts that 50 percent of the large mammals (~ fifteen ungulate species) of the Serengeti will become extinct in the first 250 years of isolation, while 75 percent of the mammalian megafauna (~ sixteen ungulate species) will become extinct during the same time interval at Nairobi National Park, and that within a thousand years only 15 percent (~ three species) of the megafauna will remain. In contrast, the "stable ecosystem" model proposed by Western and Ssemakula predicts that only 3 percent (one species) and 50 percent (twenty-one species), respectively, of the megafauna will become extinct at the same two localities within 250 years.

These differences arise in part from assumptions about the slope of the species-area curves used in each model (Western and Ssemakula 1981:9–12). A more fundamental difference between the two models is the relative importance placed on competition and habitat loss as mechanisms of extinction. The "true island" model considers competition to be the dominant force, whereas the "stable ecosystem" model emphasizes habitat loss.

The "stable ecosystem" model may be relevant for short-term stable environments, but this model does not apply to the magnitude and nature of change experienced in the late Pleistocene and perhaps in the future. The "true island" model may not be directly applicable to late Pleistocene environmental changes because individualistic response along an environmental gradient would have reduced the effectiveness of the boundaries of the habitat islands. However, this model does underscore the significance of competition in the late Pleistocene extinction. The rates of extinction suggested by the "true island" model are compatible with the extinction rates in North America during the late Pleistocene. The differences in these models serve to highlight the importance of considering the effects of long-term (geologic time) as well as short-term (ecologic time) changes in planning policies for megafaunal preserves.

Diverse megaherbivore communities also support a number of generalized feeders. Presumably the impact of the late Pleistocene environmental changes would have been less severe for these species because they would have been able to shift to new resources more easily than the specialists. Even for generalist herbivores, however, the interaction between the secondary compounds of plants and the detoxification systems of animals is critical.

The unpredictability of change due to the individualistic response of the late Pleistocene biota would have required adjustments in feeding strategies even for generalists. Evolution of new floral communities would have forced generalist herbivores to develop new strategies for searching for and sampling food items (Freeland and Janzen 1974) and to modify mixed foliage intake to minimize damage from secondary compounds or to balance nutrient intake (Janzen 1977). There would have to be an adjustment of enzyme systems, which sometimes transform substances that initially are relatively innocuous into ones that are actually toxic (Brattsten et al. 1977). Given the complexity of these systems it is possible that some herbivore species literally poisoned themselves into extinction.

Predictive Hypotheses

If coevolutionary disequilibrium was a factor in the late Pleistocene extinction event, then it should be possible to make certain predictions based on this hypothesis about the pattern of extinctions during the Pleistocene. Four of these predictions are contrary to those based on the "overkill" hypothesis, and thus may be used to compare the predictive success of the two hypotheses. In all four cases patterns observed in the fossil record appear to better agree with the coevolutionary disequilibrium predictions than with the corresponding "overkill" predictions.

Prediction 1

If disruption of coevolutionary interactions contributed significantly to the late Pleistocene extinction event, then the extinction should not have been restricted to a particular adaptive zone, trophic class, size category, or taxonomic group; herbivores, however, should form the largest percentage of the extinct fauna. In contrast, the "overkill" hypothesis, as conceived by Martin (1967), predicts that large mammals should be the largest portion of the fauna to become extinct on the continents, and that small mammals and other taxonomic groups, including birds, should form an insignificant portion, except in the case of commensals and scavengers.

Kurtén and Anderson (1980, Table 19.2:358) tabulate the number of species that became extinct within the three land mammal ages of the Pleistocene. In order to make consistent comparisons between land mammal ages, the data were used to analyze chronological trends in mammalian body size for Quaternary extinctions. Table 11.1 lists only species that became extinct at the end of the Rancholabrean, and it does not include taxa that became extinct during earlier intervals of this land mammal age. The species included in Table 11.1 were derived from the biostratigraphic ranges given by Kurtén and Anderson (1980, Appendix 2:407–417), but the categorization of them by body size, trophic class, and adaptive zone was determined by us. These data were then used in analyzing the composition of the mammalian fauna that became extinct at the end of the Pleistocene.

Analysis of Pleistocene extinctions of mammal species by body size illustrates several chronological trends (fig. 11.2). Throughout the Pleistocene, extinction rates tended to increase for large (182 kg to 1.9 tons) and very large (> 2 tons) mammal species and to decrease for small (1 to 907 g) mammal species. The extinction rate for medium-sized (908 g to 181 kg) mammal species appears to have increased only slightly. Although Pleistocene extinction rates showed directional trends with regard to body size, large and very large mammals did not dominate the late Pleistocene extinction event as small mammals did in the early and middle Pleistocene extinctions (fig. 11.2).

The large and very large mammalian species together comprise about 40 percent of the extinct fauna; large mammalian species comprised about 33 percent and very large species 6 percent of the total. Small mammals make up about 28 percent of the Rancholabrean extinct mammalian fauna and medium-sized species account for about 32 percent. Therefore, the late Pleistocene extinction encompasses all size categories of mammals, and three of the four size groups (small, medium, and large) contributed about equally to the extinction.

Some authors (Klein this volume, Martin this volume) believe that analyses such as these which are based on the number of species may be misleading because of the taxonomic problems involved in the definition of a fossil species. This may be true even if higher taxonomic units such as genera are used in the quantitative analysis. For instance, several systematic studies published in the last five years have drastically altered the numbers of recognized genera of extinct Pleistocene mammals. Gillette and Ray (1981), in their systematic revision of glyptodonts, reduced the number of recognized genera from five (*Glyptodon, Xenoglyptodon, Boreostracon, Brachyostracon,* and *Glyptotherium*) to one (*Glyptotherium*). The same study only reduced the number of recognized species from eight to five. Thus the percentage change in extinct genera (80 percent) was significantly greater than the percentage change in species (37.5 percent). Because there are generally fewer genera than species in the Linnaean hierarchy, any alterations in the number of genera will almost always be of greater relative magnitude than changes in the number of species.

Table 11.1. Size Category, Trophic Class, and Adaptive Zone of Mammal Species That Became Extinct in North America at the End of the Pleistocene

KEY: *Size:* S = small, 1–907 g, M = medium, 908 g–181 kg, L = large, 182 kg–1.9 tons, VL = very large, more than 2 tons;
Trophic Class: SA = sanguivore, IN = insectivore, BH = browsing herbivore, GH = grazing herbivore, CA = carnivore, OM = omnivore, GR = granivore;
Adaptive Zone: FL = flying, TR = terrestrial, FO = fossorial, AQ = aquatic

Extinct Species	Size	Trophic Class	Adaptive Zone
Desmodus stocki	S	SA	FL
Myotis rectidentis	S	IN	FL
Myotis magnamolaris	S	IN	FL
Plecotus tetralophodon	S	IN	FL
Tadarida constantinei	S	IN	FL
Holmesina septentrionalis	L	IN	TR
Dasypus bellus	L	IN	TR
Glyptotherium floridanum	L	BH	TR
Glyptotherium mexicanum	L	BH	TR
Megalonyx jeffersoni	L	BH	TR
Eremotherium rusconii	L	GH	TR
Nothrotheriops shastensis	L	BH	TR
Glossotherium harlani	L	BH	TR
Martes nobilis	M	CA	TR
Brachyprotoma obtusata	M	OM	TR
Canis dirus	M	CA	TR
Bassariscus sonoitensis	M	OM	TR
Tremarctos floridanus	L	OM	TR
Arctodus pristinus	VL	CA	TR
Arctodus simus	VL	CA	TR
Smilodon fatalis	L	CA	TR
Homotherium serum	L	CA	TR
Panthera atrox	L	CA	TR
Acinonyx trumani	L	CA	TR
Felis amnicola	S	CA	TR
Tamias aristus	S	GR	TR
Thomomys orientalis	S	BH	FO
Thomomys microdon	S	BH	FO
Castoroides ohioensis	M	BH	AQ
Peromyscus nesodytes	S	GR	TR
Peromyscus anyapahensis	S	GR	TR
Peromyscus imperfectus	S	GR	TR
Peromyscus cochrani	S	GR	TR
Hydrochoerus holmesi	M	BH	AQ
Neochoerus pinckneyi	M	BH	AQ
Sylvilagus leonensis	M	BH	TR
Equus hemionus	L	GH	TR
Equus conversidens	L	GH	TR
Equus giganteus	L	GH	TR
Equus occidentalis	L	GH	TR
Equus complicatus	L	GH	TR
Equus fraternus	L	GH	TR
Equus scotti	L	GH	TR

SOURCE: Data modified from Kurtén and Anderson (1980) with the following alterations: *Synaptomys australis* is treated as a subspecies of *S. cooperi;* while extirpated from North America, *Cuon alpinus* is not extinct; *Equus tau* is considered a *nomen nudum; Cuvieronius* is not known with certainty after the early Wisconsin (Lundelius 1972); and *Mammuthus columbi* is used instead of *M. jeffersoni* for the late Pleistocene mammoth species contemporaneus in North America with *M. primigenius.*

Table 11.1. Size Category, Trophic Class, and Adaptive Zone of Mammal Species That Became Extinct in North America at the End of the Pleistocene

(continued)

Extinct Species	Size	Trophic Class	Adaptive Zone
Equus niobrarensis	L	GH	TR
Equus lambei	L	GH	TR
Tapirus copei	L	BH	TR
Tapirus veroensis	L	BH	TR
Mylohyus nasutus	M	BH	TR
Platygonus compressus	M	GH	TR
Camelops huerfanensis	L	GH	TR
Camelops hesternus	L	GH	TR
Hemiauchenia macrocephala	M	GH	TR
Paleolama mirifica	M	GH	TR
Navahoceros fricki	L	BH	TR
Sangamona fugitiva	L	BH	TR
Alces latifrons	L	BH	TR
Cervalces scotti	L	BH	TR
Capromeryx minor	M	GH	TR
Capromeryx mexicana	M	GH	TR
Tetrameryx shuleri	M	GH	TR
Stockoceros conklingi	M	GH	TR
Stockoceros onusrosagris	M	GH	TR
Oreamnos harringtoni	M	GH	TR
Euceratherium collinum	L	GH	TR
Symbos cavifrons	L	BH	TR
Bootherium bombifrons	L	BH	TR
Bison priscus	L	GH	TR
Bison latifrons	L	GH	TR
Mammut americanum	VL	BH	TR
Mammuthus columbi	VL	GH	TR
Mammuthus primigenius	VL	GH	TR

Even excluding such statistical considerations, the species, not the genus, is the fundamental taxonomic unit in the interaction of biological systems (Mayr 1970:39). Thus extinction must be considered to occur at the level of the species, even if doubt remains about the validity of some species.

The taxonomic validity of fossil species such as the bats, *Myotis magnamolaris* and *M. rectidentis*, and the chipmunk, *Tamias aristus*, has been questioned because each is known from only one locality. However, some living rodent species, including *Peromyscus simulatus* (Jico deer mouse), *P. altilineatus* (Todos Santos deer mouse), *P. ochraventer* (El Carrizo deer mouse), *Neotoma palatina* (Bolanoz wood rat), *N. varia* (Turner Island wood rat), *Rheomys mexicanus* (Mexican water mouse), *R. raptor* (Goldman's water mouse), *Microtus oaxacensis* (Oaxacan vole), and *M. umbrosus* (Zempoaltepec vole), are known from only a few isolated specimens in restricted geographic areas (Hall 1981). Other species, including *Sorex merriami* (Merriam's shrew), *S. saussurei* (Saussure's shrew), *Myotis volans, Perognathus flavus* (silky pocket mouse), and *Peromyscus difficilis* (Zacatecan deer mouse), are known from only a single late Pleistocene locality but have a wide geographic distribution today. Therefore, the limited occurrence of a species in the fossil record reflects either the limited distribution of living populations of this species at a particular time or inadequate sampling of past biotas by the fossil record.

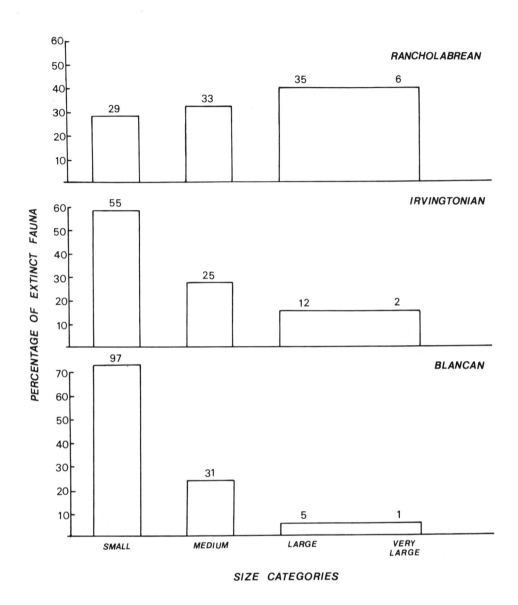

Figure 11.2. Extinctions of mammal species by body size for the three Land Mammal Ages (LMA) of the Pleistocene. Kurtén and Anderson (1980) consider the Blancan LMA as late Pliocene and early Pleistocene; other authors (Marshall et al. 1982) consider it to be strictly Pliocene. Integers are the exact number of species for each size class. Data were taken directly from Kurtén and Anderson (1980:358; Table 19.2) and were *not* calculated from Table 11.1. (Courtesy of Illinois State Museum)

If the limited Pleistocene record of a given species is the result of low abundance or restricted distribution of the living population, then the probability of extinction for that species would have been greater than average. The probability of extinction also would have been high for those extinct small mammal species that were restricted to

Table 11.2. Composition of Extinct Fauna by Adaptive Zone and Size
(Percentages calculated from data in Table 11.1)

Size*	Number of Species (%) by Adaptive Zone			
	Flying	Terrestrial	Fossorial	Aquatic
Small	5(100)	6(10)	2(100)	0(0)
Medium	0(0)	14(23)	0(0)	3(100)
Large	0(0)	41(67)	0(0)	0(0)

*Small = 0–1 kg; Medium = 1.1–200 kg; Large = >200 kg

islands during the late Pleistocene. Examples are *Peromyscus nesodytes* (Santa Rosa mouse), known only from Rancholabrean deposits on Santa Rosa Island, and *P. anyapahensis* (Anacapa mouse), which occurs only on Anacapa Island off the coast of California. These extinctions are in accord with predictive models of island biogeography (MacArthur and Wilson 1967).

Other extinct small mammal species had wide stratigraphic and geographic distributions in the fossil record. *Desmodus stocki* (extinct vampire bat) ranged from the Sangamon to the Wisconsin, and its remains have been found from Florida through Texas and Mexico to northern California. *Sylvilagus leonensis* (extinct pygmy marsh rabbit) occurs in late Pleistocene faunas in Mexico and Florida. From these data it is clear that the extinction of small mammal species is more than an artifact of taxonomy.

Analysis of the composition of the extinct mammalian fauna by habitat adaptation and size categories demonstrates that the late Pleistocene extinction event encompassed a variety of adaptive zones (Table 11.2). Small mammals made up 100 percent of the extinct flying and fossorial fauna. Among aquatic mammals, only three medium-sized species became extinct. No scansorial or arboreal species became extinct. Except for very large mammals, the relative number of extinct species of terrestrial mammals decreases with decreasing body size.

Seven trophic classes are represented by extinct mammal species (Table 11.3). In all trophic classes except granivores and sanguivores, mammals from at least two size classes became extinct. Among browsing and grazing herbivores and carnivores, the majority of the extinct species were large mammals. The largest percentage of extinct omnivores were medium-sized mammals.

Table 11.3. Composition of Extinct Fauna by Trophic Level and Size
(Percentages calculated from data in Table 11.1)

Size[†]	Number of Species (%) by Trophic Level*						
	IN	BH	GH	GR	CA	SA	OM
Small	4(67)	2(10)	0(0)	5(100)	1(11)	1(100)	0(0)
Medium	0(0)	5(24)	9(35)	0(0)	2(22)	0(0)	2(67)
Large	2(33)	14(66)	17(65)	0(0)	6(67)	0(0)	1(33)
			9[††](35)				

*IN = insectivore, BH = browsing herbivore, GH = grazing herbivore, GR = granivore, CA = carnivore, SA = sanguivore, OM = omnivore
[†]small = 0–1 kg; medium = 1.1–200 kg; large = 200 kg
[††]number of species of *Equus*

In terms of both absolute and relative numbers of extinct species, the large browsing and grazing herbivores were in the majority in the late Pleistocene extinct mammalian fauna (fig. 11.3). The large mammal browsing herbivore trophic class is not dominated by a single taxonomic group or ecological adaptation. Instead, this trophic class includes a diverse array of taxa, including sloths, proboscideans, musk oxen, cervids, and others, which appear to represent a broad spectrum of habitat types. Although the diversity of taxa, as well as ecotypes, for the large grazing herbivores is lower since several of the seventeen extinct species were horses (Table 11.3), a wide variety of adaptive forms remains, including proboscideans, bison, and antelope, among others, and even within the horses different adaptive forms exist (such as kiangs, asses, zebrines, and caballines). The late Pleistocene mammalian extinction transcended not only size classes, but also a host of ecological adaptations. The late Pleistocene extinction event was not restricted to mammalian taxa. As with the mammals, extinctions of birds, reptiles, and amphibians encompassed a variety of size classes and ecological adaptations. A majority of the extinct nonmammalian species were primary consumers.

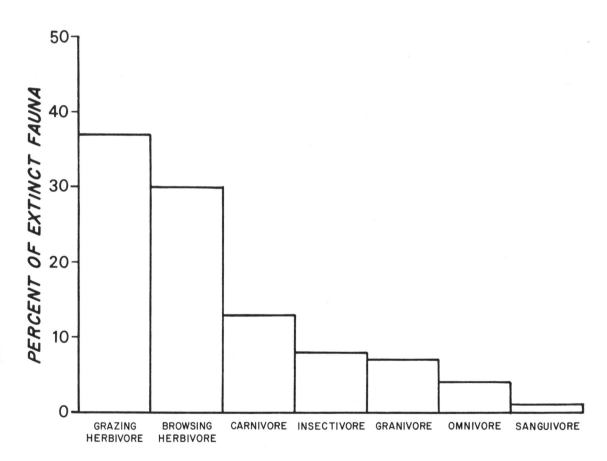

Figure 11.3. Percentage species composition of extinct mammalian fauna grouped by trophic class. Calculations based on data in Table 11.1. (Courtesy of Illinois State Museum)

Prediction 2

If the individualistic nature of the reorganization of the late Pleistocene biota was an important factor in the extinction, as it must be for the coevolutionary disequilibrium model, then the time of the extinction event for a specific geographic area should have been correlated with the time of the reorganization of communities, especially the vegetation, in that area. This reorganization was dependent on environmental changes and independent of the appearance of human cultures. The "overkill" hypothesis predicts correlation of the extinction event with the first appearance of efficient human predators, which may have been independent of environmental changes.

Specific examples of biological reorganization and the evolution of community patterns have been given for the Great Plains of the United States. However, the general process of individualistic response of species to changing environments can be extended to late Pleistocene community evolution throughout the world (Birks 1973; Cushing 1965; Graham 1979; Hopper 1979; Livingstone 1967, 1975; Pregill and Olson 1981; West 1964; Wright 1981). These changes were not synchronous on a global scale, but for those areas for which we have adequate data the evolution of new communities is correlated with the regional late Pleistocene extinction event.

In North America disharmonious associations occur in most late Pleistocene faunas and floras. Although different species are involved in various geographic areas, the phenomenon is the same everywhere. Disharmonious mammal faunas from northern Mexico and central Texas are distinctly different from those of other areas of North America. In San Josecito Cave (Nuevo León, Mexico) *Sorex cinereus* and *Synaptomys cooperi* are associated with *Liomys irroratus* (spiny mouse), *Microtus mexicanus, Cryptotis mexicana* (Mexican shrew), and *Sorex saussurei*; the present-day ranges of these two groups do not overlap. In central Texas southern species such as *Liomys irroratus, Microtus mexicanus, Cryptotis mexicana,* and *Sorex saussurei* are absent but northern species such as *Microtus pennsylvanicus, Sorex cinereus, Synaptomys cooperi,* and *Mustela erminea* (ermine) are disharmonious with southern taxa such as *Baiomys taylori* (pygmy mouse), *Perognathus merriami* (Merriam's pocket mouse), and *Sigmodon hispidus.*

The late Pleistocene fauna from Cave Without a Name (Kendall County, Texas) has disharmonious pairs such as *Mustela erminea* and *Perognathus hispidus* (hispid pocket mouse), *Sorex cinereus,* and *Sigmodon hispidus* (Lundelius 1967). The Schulze Cave fauna (Edwards County, Texas) contains such disharmonious pairs as *Sorex vagrans* (vagrant shrew) and *Tamias striatus, Sorex cinereus* and *Baiomys taylori, Mustela erminea* and *Perognathus merriami* (Dalquest et al. 1969). In Friesenhahn Cave (Bexar County, Texas) *Tamias striatus, Synaptomys cooperi,* and *Blarina carolinensis* are associated with *Perognathus hispidus, Dipodomys* sp. (kangaroo rats), and *Cynomys ludovicianus* (black-tailed prairie dog) (Graham 1976b).

Dicrostonyx torquatus (collared lemming), *Ovibos moschatus* (musk ox), *Rangifer tarandus* (caribou), *Sorex arcticus* (arctic shrew), and *Microtus xanthognathus* (yellow-cheeked vole) are disharmonious with many species in late Pleistocene faunas from the northern Plains and upper Midwest. Many of these northern species appear to have been distributed in a relatively narrow band just south of the ice sheet during the late Pleistocene. Extinct species such as *Mammuthus primigenius* (woolly mammoth) also may have been restricted to this late Pleistocene faunal province (Graham 1979).

Numerous Pleistocene faunas in the Appalachian Mountains have associations of species which do not occur together today. Baker Bluff Cave (Tennessee) has *Microtus xanthognathus* in association with species such as *Blarina brevicauda, Sorex fumeus* (smoky shrew), *Tamias striatus, Neotoma floridana* (eastern woodrat), and *Spilogale putorius* (spotted skunk). *Neotoma floridana* is associated with *Phenacomys intermedius* (spruce vole), *Synaptomys borealis* (northern bog lemming), and *Rangifer tarandus,* as well as with *Microtus xanthognathus*. The Pleistocene fauna from Eagle Cave (Pendleton County, West Virginia) (Guilday and Hamilton 1973) also has *Microtus xanthognathus*, in this case associated with *Microtus pinetorum* (pine vole) and *Synaptomys cooperi*; other disharmonious pairs from this cave are *Microtus chrotorrhinus* (rock vole) and *Neotoma floridana* and *Sorex arcticus* and *Neotoma floridana*.

Late Pleistocene faunas from Florida, considered a late Pleistocene refugium for southern species, also exhibit disharmonious pairs of northern and southern taxa. The Devil's Den local fauna (Martin and Webb 1974) contains *Synaptomys cooperi* and *Microtus pennsylvanicus*, neither of which has a present-day range overlapping that of *Neofiber alleni* (Florida water rat). Forest-, river-, and plains-dwelling species appear to have been sympatric in the late Pleistocene faunas from Melbourne, Vero, Ichtucknee River, and Seminole Field (Webb 1974). The Ladds local fauna (Bartow County, Georgia) is composed of a mixture of northern (*Sorex cinereus, S. fumeus, Synaptomys cooperi, Zapus hudsonius,* and *Martes pennanti* [marten]), southern (*Neofiber alleni, Conepatus leuconotus* [hog-nosed skunk], and *Panthera onca* [jaguar]) mammalian and herpetile species, none of which overlap today (Holman 1976, Ray 1967). Northern species such as *Clethrionomys* (red-backed vole) and *Synaptomys borealis* have been found in other late Pleistocene faunas of Georgia (Voorhies 1974). Many extinct species which appear to have been endemic to the coastal plain environments of the late Pleistocene also are abundant in the faunas from Florida.

The late Pleistocene disharmonious mammalian faunas of Alaska and Siberia are similar in composition (Harington 1978, Péwé 1975, Sher 1968, Vangengeim 1967), but have some significant differences. Some genera, such as *Sus* (wild pigs), *Hyaena* and *Crocuta* (hyenas), and *Coelodonta* (woolly rhinoceros) were indigenous to the Old World; other species such as *Mammut* (mastodonts), *Symbos* and *Bootherium* (musk oxen), and *Camelops* (camels) were restricted to the New World. Another group of large mammals (*Saiga* [saiga antelope], *Rangifer, Ovibos, Bison, Equus, Cervus* [wapiti], *Alces* [moose], *Ovis* [sheep], *Alopex* [arctic fox], and *Canis lupus* [wolf]) and small mammals (*Spermophilus, Marmota* [marmots], *Microtus, Lepus* [hares], *Gulo* [wolverine], and *Mustela* [weasels]) ranged widely throughout Siberia, Alaska, and the Bering land bridge.

In the late Pleistocene faunas of Alaska and the Yukon Territory, tundra species such as *Dicrostonyx, Lemmus,* and *Ovibos* are associated with woodland and plains species such as *Cervus, Taxidea* [badger], and *Bison* (Guthrie 1966, 1968a, 1968b; Harington and Clulow 1973; Irving and Harington 1973; Kurtén and Anderson 1980). These tundra species also occur in other late Pleistocene faunas in the same area with Old World species such as *Saiga tatarica* and *Cuon alpinus* (dhole) which no longer inhabit the tundra (Kurtén and Anderson 1980). The late Pleistocene fauna of northeastern Siberia has been characterized as a mixture of species exclusively from the far north, such as *Alopex lagopus* (arctic fox), *Dicrostonyx, Lemmus, Ovibos,* and *Ovis canadensis* (bighorn sheep), along with representatives of the southern Siberian fauna (*Felis tigris* [tiger], *Saiga, Cervus,* and *Equus* [horse]) which today do not occur as far north as 60° N latitude (Okladnikov 1970). Okladnikov (1970) believes that the late Pleistocene fauna of northern Siberia may have contained as many as eight arctic species and twenty-six southern species.

In general, modern community patterns began to evolve from these disharmonious associations between 12,000 and 10,000 years ago throughout North America. This interval of time also spans the majority of the North American late Pleistocene extinctions. Lags in the evolution of modern community patterns in the northern Great Plains (Graham 1981, Lundelius et al. in press) can be correlated with the late survival of extinct taxa such as *Camelops* (Frison et al. 1978). Early Holocene alterations in the vegetation of the southwestern United States may not be associated with an extinction; however, the initiation of these community reorganizations 11,000 years ago (Van Devender and Spaulding 1979) does overlap with the late Pleistocene extinction event.

The disharmonious late Pleistocene biotas were not confined to North America but occurred throughout most of the world (Lundelius 1976). Disharmonious mammalian faunas are found in all parts of Australia from which fossil assemblages are known (Lundelius 1983). As in North America, zonation can be seen in the Pleistocene faunas of Australia. Those from southeastern Australia contain such disharmonious pairs as *Burramys parvus* (mountain pgymy possum) and *Gymnobelideus leadbeateri* (Leadbeater's possum), *Burramys parvus* and *Melomys cervinipes* (fawn-footed mosaic-tailed rat), *Acrobates pygmaeus* (pygmy glider) and *Cercartetus lepidus* (little pygmy possum). Although all of these species pairs are now allopatric, they are all characteristic of humid environments. In contrast, faunas from the rest of Australia have disharmonious pairs such as *Potorous platyops* (broad-faced potoroo) *and Petaurus breviceps* (sugar glider), *Antechinus swainsoni* (dusky marsupial mouse) and *Perameles bougainville* (barred bandicoot), *Phascolarctos cinereus* (koala) and *Caloprymnus campestris* (desert rat kangaroo). These species pairs, which are now allopatric, are characteristic of very different moisture regimes.

In Australia, as elsewhere, individualistic biotic reorganization was associated with a major megafaunal extinction. However, in Australia the extinction event and the emergence of new community patterns began at least 18,000 years ago (Hope et al. 1977, Lundelius in press). A similar pattern of environmental change and extinction appears to have occurred throughout the South Pacific islands, though exact timing of events for various geographic areas may not have been the same everywhere. The appearance of man on the Australian continent predated the beginning of the late Pleistocene extinction event by 12,000 to 32,000 years (Hope and Hope 1976). The extinction of *Sarcophilus harrisii* (Tasmanian devil) and *Thylacinus cynocephalus* (Tasmanian wolf) late in the Holocene may have been the result of the introduction of the dingo rather than of the general change in environment or of human predation (Archer 1974, Calaby and White 1967).

There are numerous records of disharmonious floras (West 1964, 1968) and faunas (Sutcliffe 1962, Sutcliffe and Kowalski 1976) in the British Isles during the late Pleistocene. At Tornewton Cave (Devon) tundra and boreal species (*Rangifer tarandus, Dicrostonyx torquatus, Clethrionomys glareolus, Microtus oeconomus* [tundra vole], *Microtus gregalis* [gregarious vole], and *Mustela erminea*) are associated with *Crocuta crocuta* (spotted hyena) in the Upper Paleolithic Reindeer Stratum (Sutcliffe and Zeuner 1962, Sutcliffe and Kowalski 1976). Other strata ("Diluvium," Bear, and Glutton) also contain disharmonious mixtures of tundra-steppe, boreal, and forest species (Sutcliffe and Kowalski 1976:66, Table 8). These disharmonious associations are reminiscent of the boreal, steppe, and deciduous components found in late Pleistocene cave faunas from the eastern United States (Graham 1976a).

The present-day distribution of *Apodemus sylvaticus*, the wood mouse, does not overlap with that of either *Dicrostonyx* or *Lemmus* (brown lemmings) in the British Isles (Van Den Brink 1968). However, the wood mouse appears to have been contemporaneous with one or both of these lemming genera in the late Pleistocene deposits at Langwith, Dowel, and Dog Holes caves (Sutcliffe and Kowalski 1976). The reorganization of the flora and fauna of the British Isles is associated with the late Pleistocene extinction.

In southern France, Würm I faunas have *Crocuta crocuta* associated with *Cervus elaphus* and *Felis lynx* (lynx), *Hippopotamus amphibius* (hippopotamus) associated with *Ursus arctos* (brown bear) and *Cervus elaphus*. In the same area Würm II faunas contain such disharmonious pairs as *Crocuta crocuta* and *Rangifer tarandus* and *Crocuta crocuta* and *Gulo gulo* (Pillard 1972). Similar associations of now allopatric taxa are known from late Paleolithic sites in Cantabrian Spain (Freeman 1973).

Paleoenvironmental data for the Quaternary of Africa are sparse, but certain patterns relating extinction to environmental change are beginning to emerge. Fifty-six mammalian genera became extinct during the Pleistocene, one of the highest extinction rates for any of the Cenozoic epochs in Africa (Maglio 1978). Thirty-three (59 percent) of these fifty-six genera became extinct during the early Pleistocene, twelve more (21 percent) disappeared during the middle Pleistocene, and another eleven (20 percent) became extinct during the late Pleistocene. With the exception of *Dicerorhinus kirchbergensis* (Kirchberg's rhinoceros), which may be a special case, all of the late Pleistocene extinctions apparently occurred at the terminal Pleistocene/early Holocene boundary, 12,000 to 10,000 years B.P. (Klein this volume).

The structure of African large mammal herbivore communities of the late Pleistocene was quite different from the structure of historic communities. For instance, *Raphicerus melanotis* (cape grysbok) and *Pelea capreolus* (Vaal rhebuck) are not sympatric today with *Hippotragus equinus* (roan antelope), *Alcelaphus buselaphus* (Bubal hartebeest), *Tragelaphus strepsiceros* (greater kudu), or *Taurotragus oryx* (Livingstone's eland) (Dorst and Dandelot 1969), even though all these species occur in the Southern Savanna Woodland biotic province of South Africa (Rautenbach 1978). However, many of these species were sympatric in late Pleistocene faunas from South African caves. *Pelea capreolus* and *Raphicerus melanotis* were sympatric with *Tragelaphus strepsiceros* and *Taurotragus oryx* at Die Kelders I between 80,000 and 25,000 years B.P. (Klein 1975) and at Boomplaas A between 12,000 and 9000 years B.P. (Klein 1978). At the Klasies River Mouth sites (Klein 1976), *Pelea capreolus* and *Raphicerus melanotis* are associated with *Hippotragus equinus* and *Acephalus buselaphus*. *Taurotragus oryx* is found with *Pelea capreolus* and *Raphicerus melanotis* at Byneskranskop I (Klein in Avery 1982:345, Table 49). At Nelson Bay Cave, *Pelea capreolus* and *Raphicerus* sp. are associated with *Taurotragus oryx* during the late Pleistocene (Klein 1972).

Because species were extirpated from these disharmonious faunas in an individualistic manner, many of the disharmonious associations persisted into the Holocene. However, as in North America, "modernization" of these biotas was essentially contemporaneous with the profound climatic change at the end of the Pleistocene, 10,000 to 12,000 years ago. Small mammal communities also reflect major environmental changes at this critical time (Avery 1982).

These environmental changes were not restricted to South Africa, but occurred throughout the African continent (Klein 1980, this volume). Pollen evidence also reflects a major individualistic reorganization of the African flora in response to climatic change 10,000 to 12,000 years ago (Livingstone 1967, 1975).

Environmental change has been considered an incomplete explanation for the late Pleistocene extinction, at least for Africa, since the extinct species survived earlier periods of change (Klein this volume). However, the character of environmental changes at the end of the Pleistocene, when the climate became more continental, was not the same as that of the changes during earlier parts of the Pleistocene (Graham 1979, Klein this volume). Furthermore, dramatic extinctions were associated with environmental changes in Africa during the early and middle Pleistocene as well as during earlier epochs of the Cenozoic (Maglio 1978). Therefore, it is not necessary to invoke man as the dominant agent of the late Pleistocene extinction, although man may have played a more significant role in Africa than in North America. Man's long evolutionary history in Africa also suggests that he had only a minor impact on the late Pleistocene extinction event.

Late Pleistocene paleoenvironments are not as well documented for Central and South America as for other areas. The late Pleistocene extinction of South American mammals occurred at the end of the Lujanian Land Mammal Age, 8000 to 10,000 years B.P. (Marshall et al. 1982). Palynological and paleoclimatological studies (Bradbury et al. 1981; Colinvaux 1972; Simpson 1975; Simpson and Haffer 1978; van der Hammen 1963, 1974; Vuilleumier 1971), as well as geomorphological analyses (Bigerella and Andrade 1965, Bigerella and Becker 1975, Journaux 1975), of portions of South America and related areas such as the Galapagos Islands document major environmental changes between 8,000 and 10,000 years ago. The changes in climates and floras in South

America appear to have been exactly the opposite of those in North America. In North America, grasslands expanded at the expense of forest habitats; in South America the forests expanded and the grasslands diminished.

Prediction 3

If late Pleistocene alterations of the vegetational patterns disrupted the coevolutionary equilibrium of plant-animal interactions, then surviving remnants or ecomorphs of the late Pleistocene extinct fauna should be associated with floral groups analogous to those of the late Pleistocene. The "overkill" hypothesis predicts survival of the late Pleistocene extinct fauna in areas not invaded by early man.

Some survivors of the late Pleistocene extinction are phylogenetically related to or ecologically equivalent to extinct taxa. According to the coevolutionary disequilibrium model, these species should inhabit environments which are similar, at least floristically, to those of the late Pleistocene. Although these environments probably are not exact replicates, they reflect the coevolved link between vegetational changes and megaherbivore community structure.

Elements of the modern South American mammalian fauna (llamas, tapirs, peccaries, spectacled bears, capybaras, deer, jaguars, armadillos, and several kinds of tortoises) are similar to the late Pleistocene Panamerican Intermingled Savanna fauna (Marshall et al. 1982; Webb 1976, 1977, 1978). This fauna occurred throughout South America, Central America, Mexico, and the southern United States. Spectacled bear, tapir, capybaras, and some others prefer wooded environments. Other species, including llamas, armadillos, and Pampas deer, are adapted to open environments and now live in the South American Pampas. The mammals of the Pampas are associated with plants such as diploid species of *Hordeum,* a genotype which is ancestral to many North American grassland species (Stebbins 1975, 1981), but which is no longer found in North America, except for *H. californicum.* Diploid ancestors of the North American polyploid complexes of *Andropogon, Sorghastrum,* and *Panicum* also survive sympatrically with *Hordeum* on the Pampas. These diploid ancestors probably were extirpated from the Great Plains sometime during the Pleistocene (Stebbins 1975). Thus, portions of the Pampas flora may be a vestige of grassland environments that extended throughout the Americas during the late Pleistocene. The relative stability of these coevolved systems supports remnants or ecomorphs of an extinct fauna.

The mammalian fauna of Africa has frequently been referred to as a living Pleistocene fauna because the diversity of the megafauna is similar to that of many extinct late Pleistocene faunas. However, the late Pleistocene fauna of Africa was considerably more diverse than the African fauna of historic times (Bigalke 1978). The survival of the diverse present-day African fauna, especially the grazing sequence of the Serengeti, may seem enigmatic in light of the coevolutionary disequilibrium model of extinction. However, several explanations for their survival are compatible with this model. The magnitude of the environmental change, which would have determined the predictability of the biotic reorganization, may not have been as great in Africa, and possibly in South America, as it was in other parts of the world. This may account for the lower, late Pleistocene extinction rates for Africa compared to other continents. The survival into the Holocene of savanna environments in Africa would have allowed the continued support of a diverse, though altered, grazing sequence. These savanna environments were widespread in the New World during the late Pleistocene, but today are restricted to areas like the South American Pampas. Therefore, the coevolutionary disequilibrium model is consistent with observed patterns.

The times of extinction and the survival of specific taxa do not agree with the predictions of the "overkill" hypothesis. *Homo sapiens* and pre-*sapiens* species were associated with megafauna for a long period in Africa, but the late Pleistocene extinction in Africa culminated between 10,000 and 12,000 years ago (Klein this volume). The late Pleistocene extinction event also is not synchronous with the first appearance of man in Europe, Asia, Australia, and possibly North and South America. The "overkill" hypothesis does not explain the survival of taxa closely related to extinct species that would have been a good subsistence resource (Graham 1979). However, these survivals are readily explained by the coevolutionary disequilibrium model of extinction.

Prediction 4

If the individualistic reorganization of communities and the subsequent coevolutionary disequilibrium increased competition between herbivores, then selection would favor survival of species preadapted to the evolution of new vegetational communities or species that could rapidly adapt to new vegetational associations. In either case, the modern distributions of the surviving large herbivores should be correlated with specific vegetational associations, and as a consequence of competition these species should also exhibit divergence and habitat partitioning.

The "overkill" hypothesis predicts that surviving large herbivores were not hunted by early man, resided in areas inaccessible to early man, or possessed adaptive strategies to protect them from predation by early man. In the first instance, many of the surviving large herbivores (including *Bison, Odocoileus,* and *Ovis*) were hunted extensively by members of post-Clovis cultural complexes, but these predation pressures did not result in their extinction. In the second instance, little, if any, paleontological or archaeological evidence suggests that large segments of the surviving "Big Game" resided in refugia not accessible to early man. In the third instance, some species, such as *Ovibos moschatus,* have defensive behavioral patterns that would increase their vulnerability to predation and "overkill" by man, but nonetheless survived the late Pleistocene extinction. Furthermore, extinction caused by factors such as "overkill" which are independent of the biological and physical environment should result in many empty niches in the modern environment for large herbivores (Martin 1975) and, as a consequence, in low competition between surviving large herbivores.

Species freed from some interspecific competition frequently exhibit "ecological release" and are able to exploit a wider range of habitats. However, reduction of competition does not necessarily cause an increase in the variety of foods eaten (MacArthur 1972, Pianka 1974). In Canada and the United States only twelve species of "Big Game" survived the extinction event, and one of these, *Dicotyles tajacu,* may be a recent immigrant into the United States (Table 11.4). All of these species have relatively wide geographic distributions today; most of these species, however, are adapted to particular habitats and their ecologies are significantly different.

Table 11.4. Native "Big Game" of the United States and Canada

Cervus canadensis (Elk)	*Antilocapra americana* (Antelope)
Odocoileus hemionus (Mule Deer)	*Bison bison* (Bison)
Odocoileus virginianus (White-Tailed Deer)	*Oreamnos americanus* (Mountain Goat)
Alces alces (Moose)	*Ovibos moschatus* (Musk Ox)
Rangifer tarandus (Caribou)	*Ovis canadensis* (Mountain Sheep)
Tayassu tajacu (Peccary)*	*Ovis dalli* (Dall's Sheep)

*Late Holocene immigrant

Habitat specificity is evident in the biology of all four species: *Ovibos moschatus* (musk ox), *Rangifer tarandus, Alces alces* (moose), and *Cervus canadensis* (elk) that inhabit the boreal environments of North America. *Ovibos moschatus* and *Cervus canadensis* are allopatric herbivores that feed on similar food types such as grasses, herbs, forbs, sedges, twigs, and bark. *Alces alces* is sympatric with *Cervus canadensis* and allopatric with *Ovibos moschatus*. Moose are specifically adapted to freshwater aquatic environments and browse on twigs, bark, saplings, and aquatic vegetation in forests and brushy habitats. During the winter, when *Alces alces* and *Cervus canadensis* are sympatric in Riding Mountain National Park in Manitoba, they show strong preferences for different vegetational associations (Rounds 1981).

Rangifer tarandus is sympatric with all three species, but specific anatomical adaptations reflect its divergent ecology. Caribou feed primarily on lichens and mosses, although other food items such as grasses, herbs, and willows are included in their diet (Edwards and Ritcey 1959, Edwards et al. 1960, Kelsall 1968). Caribou may exhibit selective and preferential ingestion of particular plant species and plant parts (Kelsall 1968) which is reminiscent of the selectivity characteristic of grazing sequences. In fact, *Rangifer tarandus* in Wells Gray Park, British Columbia, is dependent on a single lichen genus, *Alectoria* (Edwards et al. 1960). North American caribou have a strong seasonal migratory pattern from their summer home in the tundra into the spruce forest during the winter. During this annual cycle caribou utilize different food types depending on the habitats and amount of snowfall that their migratory routes transect (Fuller and Keith 1981).

The four large herbivore species that inhabit the modern boreal environments appear to have partitioned the environment by specialization and distribution. These patterns must have begun to emerge at the end of the Pleistocene, since most of the area occupied by modern boreal environments was under glacial ice during the late Pleistocene. In addition, *Rangifer tarandus* and *Alces alces* may have immigrated to positions south of the continental ice sheets in North America during the Wisconsin.

Although preferential plant ingestion, habitat specialization, and allopatric distributions may result from competition, observation of these phenomena does not necessarily prove competition. One theoretical perspective in modern ecology suggests that herbivores are not limited by food (Slobodkin et al. 1967). This theory suggests that foliage in natural communities is seldom visibly overconsumed, and that the "greenness" of the world is evidence. These theories ignore the significance of plant-animal coevolutionary interactions and their effectiveness in deterring herbivorous predators. The chemical warfare of plants clearly reduces the effective supply of plant foods and, in conjunction with seasonally restricted vegetational productivity, serves to narrow the herbivore food niche. Sufficient differentiation of niches has evolved in megaherbivore communities to reduce interspecific competition (Harris 1972; Lamprey 1963, 1964).

Rangifer tarandus, Ovibos moschatus, and *Antilocapra americana* (pronghorn) are gregarious herd herbivores which survived the late Pleistocene extinction. These species are known from late Pleistocene faunas, but at low frequencies. This may reflect their actual paucity in the biocoenosis or may result from taphonomic or sampling factors independent of their actual relative abundance in the fauna. However, the high relative frequencies in late Pleistocene faunas of other presumably gregarious taxa such as *Camelops, Equus, Hemiauchenia* (llamalike camel), *Mammuthus,* and *Platygonus* suggest that taphonomy cannot be the complete explanation. Instead, it is possible that *Rangifer, Ovibos,* and *Antilocapra* existed in low numbers in the highly diverse megafaunas of the late Pleistocene. With the extinction of numerous herd herbivores, other preadapted species such as *Rangifer, Ovibos,* and *Antilocapra* were able to invade the newly formed communities and expand their population sizes in response to lower diversity. Thus the late Pleistocene environmental changes not only influenced the evolution of new communities and the extinction of a diverse fauna, but may also have directed the sociobiological evolution of some surviving species.

Conclusions

The late Pleistocene extinction was a global event correlated with the formation of new biotic communities. Extinction and community evolution were closely linked in each geographic area, but occurred at different times in different areas. However, the critical period of environmental change for most continents seems to have been 10,000 to 12,000 years B.P.

Environmental changes at the end of the Pleistocene caused a major biotic reorganization. Instead of simple shifts of biotic zones, individual species responded to these environmental changes in accord with their own tolerance limits. This individualistic response of each species reduced the predictability of the composition and structure of the new communities. In coevolved systems these changes would disrupt coevolutionary relationships between plants and animals, thus creating a disequilibrium in the system.

A coevolutionary disequilibrium creates detoxification problems for some plant predators which have to adapt to new vegetational associations; it is possible that some herbivore species may have poisoned themselves into extinction. For most species, especially herbivorous megamammals, a coevolutionary disequilibrium would reduce niche differentiation and consequently heighten competition among the herbivore species. Competition in the herbivore communities of the late Pleistocene would have driven species with reduced fitness to extinction, selected for species best adapted to the new community patterns, and redefined niche differentiation for species that survived the extinction event. These processes explain the extinction of certain herbivore species and the survival of others. The coevolutionary disequilibrium model of extinction is not restricted to specific taxonomic groups, size categories, or ecological types.

The coevolutionary disequilibrium model of extinction is independent of direction of environmental change or expansion and contraction of particular habitat types or physiognomies. This model depends on the magnitude of environmental change and the disruption of coevolved systems caused by individualistic reorganization. Greater magnitudes of environmental change reduce the predictability of biotic organization and increase the probability of extinction. Environmental change can therefore influence the extinction of taxa in both expanding and contracting habitats if the coevolved systems are significantly disrupted. Extinctions can be caused by different environmental changes in various geographic areas. Thus, changes from one environmental regime to another and vice versa can cause extinction by the same mechanism, coevolutionary disequilibrium.

The high density and low diversity characteristic of many modern plant associations such as the spruce forest may be explained by the disruption of coevolved systems. Extinction of specific plant predators would release certain species of plants from predation pressure and allow them to expand their populations. Thus population controls would shift from predation pressure to interspecific competition with other plant species. Conversely, preadapted herbivores were able to expand from small populations into gregarious herds. Changes in population structure may have directed sociobiological evolution of certain herbivore species.

If individualistic reorganization and coevolutionary disequilibrium are integral components of the late Pleistocene extinction process, then it can be asked why insects were not adversely affected. Insects have probably coevolved with plants to a greater degree than any other group. However, the Quaternary record for Coleoptera (other insects are rarely preserved in Quaternary deposits) shows extremely low rates of extinction or speciation. This probably is due to the fact that plants other than grasses evolved very slowly during the Quaternary. Most coevolved insects are adapted to a specific plant species, and therefore the insects were able to track the shifting distributions of individual plant species; other organisms, which were adapted not to single species but to specific plant associations, were not able to do this.

The coevolutionary disequilibrium model is directly applicable to plant-animal systems, but similar arguments could be applied to other coevolved relationships such as predator-prey systems involving animal-animal interactions. A coevolutionary disequilibrium model best depicts the observed patterns of the late Pleistocene extinction and probably can be extended to earlier extinction events. Although it is naive to propose a single cause for a major extinction event, coevolutionary disequilibrium must have played a major role in the late Pleistocene extinction.

Acknowledgments

We owe many people a debt of thanks for helping us formulate our ideas and for assistance in putting them into print. James E. King's comments and suggestions about interpretations of pollen spectra were invaluable in our reconstructions of paleovegetation. We also gained valuable insights about the late Pleistocene extinction event at an informal discussion held at the University of Minnesota's Biological Research Station. We thank Paul Martin, Dave Meltzer, and Jim Mead for organizing this session and Herb Wright for hosting the round table and for acting as the discussion leader. Wright's penetrating questions helped clear away a lot of cobwebs and pointed to new directions for inquiry. We thank all others who participated in these discussions. Judy Lundelius, Melissa Winans, and Ann Maginnis helped render our thoughts into intelligible English and turned out the final typescript: Mary Ann Graham drafted the illustrations. Financial support was provided by the Geology Foundation of the University of Texas at Austin. This paper is contribution No. 71 of the Archeological and Quaternary Studies Program of the Illinois State Museum.

References

Archer, M. 1974. New information about Quaternary distribution of the Thylacine (Marsupialia, Thylacinidae) in Australia. *J. R. Soc. Western Australia,* v. 57, pp. 43–50.

Avery, D. M. 1982. Micromammals as paleoenvironmental indicators and an interpretation of the late Quaternary in the southern Cape Province, South Africa. *Ann. South Afr. Mus.,* v. 85.

Bell, H. V. 1971. A grazing ecosystem in the Serengeti. *Sci. Amer.,* v. 225, pp. 86–93.

Bigalke, R. C. 1978. Present-day mammals of Africa, pp. 1–16, *in* Maglio, V. J., and Cooke, H. B. S., Editors, *Evolution of African Mammals.* Cambridge, Harvard Univ. Press.

Bigarella, J. J. and de Andrade, G. O. 1965. Contribution to the study of the Brazilian Quaternary. *Geol. Soc. Amer. Spec. Paper,* v. 184, pp. 433–451.

Bigarella, J. J. and Becker, R. D. 1975. Topics for discussion. International Symposium of the Quaternary (Southern Brazil, July 15–31, 1975). *Bol. Paranaense Geocienc.,* v. 33, pp. 171–275.

Birks, H. J. B. 1973. *Past and Present Vegetation of the Isle of Skye—A Palaeoecological Study.* Cambridge, Cambridge Univ. Press.
———. 1981. Late Wisconsin vegetational and climatic history at Kylen Lake, northeastern Minnesota. *Quat. Res.,* v. 16, pp. 322–355.

Bradbury, J. P., Leyden, B., Salgado-Nobourian, M., Lewis, W. M., Jr., Schubert, C., Binford, M. W., Frey, D. G., Whitehead, D. K., and Neibejabrn, F. H. 1981. Late Quaternary environmental history of Lake Valencia, Venezuela. *Science,* v. 214, pp. 1299–1305.

Brattsten, L. B., Wilkinson, C. F., and Eisner, T. 1977. Herbivore-plant interactions: Mixed function oxidases and secondary plant substances. *Science,* v. 196, pp. 1349–1352.

Brown, J. and Cleland, C. 1968. The late-glacial and early post-glacial faunal resources in midwestern biomes newly opened to human adaptation, pp. 114–122, *in* Bergstrom, R. E., Editor, The Quaternary of Illinois, *Univ. Ill. College Agr. Spec. Publ.,* no. 14.

Calaby, J. H. and White, C. 1967. The Tasmanian devil (*Sarcophilus harrisii*) in northern Australia in recent times. *Austral. J. Sci.,* v. 29, pp 473–475.

Campbell, J. M. 1980. Distribution patterns of Coleoptera in eastern Canada. *Canad. Entomol.,* v. 112, pp. 1161–1175.

Colinvaux, P. A. 1972. Climate and the Galapagos Islands. *Nature,* v. 240, pp. 17–20.

Coope, G. R. 1979. Late Cenozoic fossil Coleoptera: evolution, biogeography, and ecology. *Ann. Rev. Ecol. Syst.,* v. 10, pp. 247–267.

Cushing, E. J. 1965. Problems in the Quaternary phytogeography of the Great Lakes Region, pp. 403–416, *in* Wright, H. E., Jr., and Frey, D. G., Editors, *The Quaternary of the United States.* Princeton, Princeton Univ. Press.

———. 1967. Late-Wisconsin pollen stratigraphy and the glacial sequence in Minnesota, pp. 59–88, *in* Cushing, E. J., and Wright, H. E., Jr., Editors, *Quaternary Paleoecology.* New Haven, Yale Univ. Press.

Dalquest, W. W. 1965. New Pleistocene formation and local fauna from Hardemann County, Texas. *J. Paleontol.,* v. 39, pp. 63–79.

Dalquest, W. W., Roth, E., and Judd, F. 1969. The mammal fauna of Schulze Cave, Edwards County, Texas. *Bull. Fla. St. Mus.,* v. 13, pp. 206–276.

Davis, M. B. 1967. Glacial climate in the northern United States: a comparison of New England and the Great Lakes Region, pp. 11–44, *in* Cushing, E. J., and Wright, H. E., Jr., Editors, *Quaternary Paleoecology.* New Haven, Yale Univ. Press.

———. 1976. Pleistocene biogeography of temperate deciduous forests. *Geosci. and Man,* v. 13, pp. 13–26.

Dorst, J. and Dandelot, P. 1969. *A Field Guide to the Larger Mammals of Africa.* Boston, Houghton Mifflin Co.

Edwards, R. and Ritcey, R. 1959. Migrations of caribou in mountainous areas in Wells Gray Park, British Columbia. *Canad. Field Natur.,* v. 73, pp. 21–25.

Edwards, R. Y., Soos, J., and Ritcey, R. W. 1960. Quantitative observations on epidendric lichens used as food by caribou. *Ecology,* v. 41, pp. 425–431.

Ehrlich, P. R. and Raven, P. H. 1964. Butterflies and plants: a study in coevolution. *Evolution,* v. 18, pp. 586–608.

Fisher, R. A. 1958. *The Genetical Theory of Natural Selection.* Oxford, Clarendon Press.

Freeland. W. J. and Janzen, D. H. 1974. Strategies in herbivory by mammals: the role of plant secondary compounds. *Amer. Nat.,* v. 108, pp. 269–289.

Freeman, L. G. 1973. The significance of mammalian faunas from Paleolithic occupations in Cantabrian Spain. *Amer. Antiqu.,* v. 38, pp. 3–44.

Frison, G. C., Walker, D. N., Webb, S. D., and Zeimens, G. M. 1978. Paleo-Indian procurement of *Camelops* on the Northwestern Plains. *Quat. Res.,* v. 10, pp. 385–400.

Fuller, T. K. and Keith, L. B. 1981. Woodland caribou population dynamics in northeastern Alberta. *J. Wildlife Man.,* v. 45, pp. 197–213.

Gillette, D. D. and Ray, C. E. 1981. Glyptodonts of North America. *Smiths. Contrib. Paleobiol.,* v. 40.

Graham, R. W. 1976a. Late Wisconsin mammal faunas and environmental gradients of the eastern United States. *Paleobiology,* v. 2, pp. 343–350.

———. 1976b. Pleistocene and Holocene mammals, taphonomy, and paleoecology of the Friesenhahn Cave local fauna, Bexar County, Texas. Univ. of Texas, Austin, unpubl. doctoral dissertation.

———. 1979. Paleoclimates and late Pleistocene faunal provinces in North America, pp. 49–69, *in* Humphrey, R. L., and Stanford, D., Editors, *Pre-Llano Cultures of the Americas: Possibilities and Paradoxes.* Washington, D.C., Wash. Anthro. Soc.

———. 1981. Preliminary report on late Pleistocene vertebrates from the Selby and Dutton archeological/paleontological sites, Yuma County, Colorado. *Contrib. Geol.,* v. 20, pp. 33–56.

———. In press. Late Pleistocene/Holocene environmental changes in the southwestern Plains of the Unites States: the mammalian record, *in* Graham, R. W., and Semken, H. A., Jr., Editors, Late Pleistocene/Holocene environmental changes in the Great Plains of the United States: the mammalian record. *Ill. St. Mus. Sci. Pap.*

Graham, R. W. and Semken, H. A. 1976. Paleoecological significance of the short-tailed shrew (*Blarina*), with a systematic discussion of *Blarina ozarkensis. J. Mamm.,* v. 57, pp. 433–449.

Grayson, D. K. 1977. Pleistocene avifaunas and the overkill hypothesis. *Science,* v. 195, pp. 691–693.

Guilday, J. E. 1967. Differential extinction during late Pleistocene and recent times, pp. 121–140, *in* Martin, P. S., and Wright, H. E., Jr., Editors, *Pleistocene Extinctions: The Search for a Cause.* New Haven, Yale Univ. Press.

Guilday, J. E. and Hamilton, H. W. 1973. The late Pleistocene small mammals of Eagle Cave, Pendleton County, West Virginia. *Ann. Carn. Mus.,* v. 44, pp. 45–58.

Guilday, J. E., Hamilton, H. W., Anderson, E., and Parmalee, P. W. 1978. The Baker Bluff Cave deposit, Tennessee, and the late Pleistocene faunal gradient. *Bull. Carn. Mus. Nat. Hist.,* v. 11, pp. 1–67.

Guilday, J. E., Parmalee, P. W., and Hamilton, H. W. 1977. The Clark's Cave bone deposit and the late Pleistocene paleoecology of the central Appalachian Mountains of Virginia. *Bull. Carn. Mus. Nat. Hist.,* v. 2, pp. 1–86.

Guthrie, R. D. 1966. The extinct wapiti of Alaska and Yukon Territory. *Canad. J. Zool.,* v. 44, pp. 45–47.

———. 1968a. Paleoecology of the large mammal community in interior Alaska during the late Pleistocene. *Amer. Midl. Nat.,* v. 79, pp. 346–363.

———. 1968b. Paleoecology of a late Pleistocene small mammal community from interior Alaska. *Arctic,* v. 21, pp. 223–244.

Gwynne, M. D. and Bell, R. H. V. 1968. Selection of vegetation components by grazing ungulates in the Serengeti National Park. *Nature,* v. 220, pp. 390–393.

Hall, E. R. 1981. *The Mammals of North America.* Vol 2. New York, John Wiley and Sons.

Harington, C. R. 1978. Quaternary vertebrate faunas of Canada and Alaska and their suggested chronological sequence. *Syllogeus,* v. 15, pp. 1–105.

Harington, C. R. and Clulow, F. V. 1973. Pleistocene mammals from Gold Run Creek, Yukon Territory. *Canad. J. Earth Sci.,* v. 9, pp. 1039–1051.

Harris, A. H. 1977. Biotic environments of the Paleoindian, pp. 1–12, *in* Johnson, E., Editor, Paleoindian Lifeways. *The Museum Jour.,* v. 17, Texas Tech. Univ.

Harris, L. D. 1972. An ecological description of a semi-arid East African ecosystem. *Range Science Dept. Series,* No. 11. Colorado St. University.

Hibbard, C. W. 1960. An interpretation of Pliocene and Pleistocene climates in North America. *Ann. Rept. Mich. Acad. Sci., Arts, and Lett.,* v. 62, pp. 5–30.

———. 1970. Pleistocene mammalian local faunas from the Great Plains and Central Lowland Provinces of the United States,

pp. 394–433, *in* Dort, W., Jr., and Jones, J. K., Jr., Editors, *Pleistocene and Recent Environments of the Central Great Plains.* Lawrence, Univ. Press of Kans.

Hoffmann, R. S. and Jones, J. K., Jr. 1970. Influence of late glacial and post-glacial events on the distribution of recent mammals of the northern Great Plains, pp. 355–395, *in* Dort, W., Jr., and Jones, J. K., Jr., Editors, *Pleistocene and Recent Environments of the Central Great Plains.* Lawrence, Univ. Press of Kans.

Holman, J. A. 1976. Paleoclimatic implications of "ecologically incompatible" herpetological species (late Pleistocene: southeastern United States). *Herpetologica,* v. 32 pp. 290–294.

Hope, J. H. and Hope, G. S. 1976. Palaeoenvironments for man in New Guinea, pp. 29–54, *in* Kirk, R. L., and Thorne, A. G., Editors, *The Origins of the Australians.* Canberra, Austral. Inst. of Aboriginal Stud.

Hope, J. H., Lampert, R. J., Edmondson, E., Smith, M. J., and Van Tets, G. F. 1977. Late Pleistocene faunal remains from Seton Rock Shelter, Kangaroo Island, South Australia. *J. Biogeogr.,* v. 4, pp. 363–385.

Hopper, S. D. 1979. Biogeographical aspects of speciation in the southwest Australian flora. *Ann. Rev. Ecol. Syst.* v. 10, pp. 399–422.

Irving, W. N. and Harington, C. R. 1973. Upper Pleistocene radiocarbon-dated artefacts from the northern Yukon. *Science,* v. 179, pp. 335–340.

Janzen, D. H. 1970. Herbivores and the number of tree species in tropical forests. *Amer. Nat.,* v. 104, pp. 501–528.

———. 1977. Promising directions of study in tropical plant-animal interactions. *Ann. Missouri Bot. Gard.,* v. 64, pp. 706–736.

Janzen, D. H. and Martin, P. S. 1982. Neotropical anachronisms: the fruits the Gomphotheres ate. *Science,* v. 215, pp. 19–27.

Journaux, A. 1975. Recherches géomorphologiques en Amazonie Brésilienne. *Cent. Nat. Rech. Sci. Cent. Géomorphol. de Caen Bol.,* v. 20, pp. 3–68.

Kapp, R. O., Bushouse, S., and Foster, B. 1969. A contribution to the geology and forest history of Beaver Island, Michigan. *Proc. 12th Conf. Great Lakes Res.,* (1969), pp. 225–236.

Kavanaugh, D. H. 1980. Insects of western Canada, with special reference to certain Carabidae (Coleoptera): present distribution patterns and their origins. *Canad. Entomol.,* v. 112, pp. 1129–1144.

Kelsall, J. P. 1968. The migratory barren-ground caribou of Canada. *Canad. Wildlife Serv. Mon.* No. 3.

King, F. B. and Graham, R. W. 1981. Effects of ecological and paleoecological patterns on subsistence and paleoenvironmental reconstructions. *Amer. Antiq.*, v. 46, pp. 128–142.

King, J. E. 1973. Late Pleistocene palynology and biogeography of the western Ozarks. *Ecol. Mon.*, v. 43, pp. 539–565.

———. 1981. Late Quaternary vegetational history of Illinois. *Ecol. Mon.*, v. 51, pp. 43–62.

Klein, R. G. 1972. The Quaternary mammalian fauna of Nelson Bay Cave (Cape Province, South Africa): its implications for megafaunal extinctions and for environmental and cultural change. *Quat. Res.*, v. 2, pp. 135–142.

———. 1975. Middle Stone Age man-animal relationships in South Africa: Evidence from Die Kelders and Klasies River Mouth. *Science*, v. 190, pp. 265–267.

———. 1976. The mammalian fauna of the Klasies River Mouth sites, southern Cape Province, South Africa. *S. Afr. Archaeol. Bull.*, v. 31, pp. 75–98.

———. 1978. A preliminary report on the larger mammals from the Boomplaas Stone Age site, Congo Valley, Oudtshoorn District, South Africa. *S. Afr. Archaeol. Bull.*, v. 33, pp. 66–75.

———. 1980. Environmental and ecological implications of large mammals from upper Pleistocene and Holocene sites in southern Africa. *Ann. S. Afr. Mus.*, v. 81, pp. 223–283.

Kurtén, B. and Anderson, E. 1980. *Pleistocene Mammals of North America*. New York, Columbia Univ. Press.

Lamprey, H. F. 1963. Ecological separation of the large mammal species in the Tarangire Reserve, Tanganyika. *East Africa Wildlife J.*, v. 1, pp. 63–92.

———. 1964. Estimations of large mammal densities, biomass and energy exchange in the Tarangire Game Reserve and Masai steppe in Tanganyika. *East Africa Wildlife J.*, v. 2, pp. 1–46.

Land, L. S., Lundelius, E. L., Jr., and Valastro, S. 1980. Isotopic ecology of deer bones. *Palaeogeogr. Palaeoclimatol. Palaeoecol.*, v. 32, pp. 143–159.

Lehmkuhl, D. M. 1980. Temporal and spatial changes in the Canadian insect fauna: patterns and explanation. The prairies. *Canad. Entomol.*, v. 112, pp. 1145–1160.

Leuthold, W. 1978. Ecological separation among browsing ungulates in Tsavo East

National Park, Kenya. *Oecologia*, v. 35, pp. 241–252.

Livingstone, D. A. 1967. Postglacial vegetation of the Rowenzori Mountains in Equatorial Africa. *Ecol. Mon.*, v. 37, pp. 25–52.

———. 1975. Late Quaternary climatic change in Africa. *Ann. Rev. Ecol. Syst.*, v. 6, pp. 249–280.

Lundelius, E. L., Jr. 1967. Late-Pleistocene and Holocene faunal history of central Texas, pp. 288–319, *in* Martin, P. S., and Wright, H. E., Jr., Editors, *Pleistocene Extinctions: The Search for a Cause*. New Haven, Yale Univ. Press.

———. 1972. Fossil vertebrates from the late Pleistocene Ingleside fauna, San Patricio County, Texas. *Univ. Texas Bur. Econ. Geol. Rept. of Invest.* No. 77.

———. 1974. The last fifteen thousand years of faunal change in North America, pp. 141–160, *in* Black, C. C., Editor, History and Prehistory of the Lubbock Lake Site, *The Museum Jour.*, v. 15, Texas Tech. Univ.

———. 1976. Vertebrate paleontology of the Pleistocene: an overview. *Geosci. and Man*, v. 13, pp. 45–59.

———. 1983. Climatic implications of late Pleistocene and Holocene faunal associations in Australia. *Alcheringa*.

Lundelius, E. L., Jr., Graham, R. W., Anderson, E., Guilday, J., Holman, J. A., Steadman, D., and Webb, S. D. In press. Terrestrial vertebrate faunas, *in* Porter, S. C., Editor, *Late-Quaternary Environments of the United States: The Late Pleistocene*. Minneapolis, Univ. of Minn. Press.

McAndrews, J. H. 1966. Postglacial history of prairie, savanna, and forest in northwestern Minnesota. *Torrey Bot. Club*, Mem. 22.

MacArthur, J. W. 1975. Environmental fluctuations and species diversity, pp. 74–80, *in* Cody, M. L., and Diamond, J. M., Editors, *Ecology and Evolution of Communities*. Cambridge, Harvard Univ. Press.

MacArthur, R. H. and Wilson, E. O. 1967. *The Theory of Island Biogeography*. Princeton, Princeton Univ. Press.

McNaughton, S. J. 1976. Serengeti migratory wildebeest: facilitation of energy flow by grazing. *Science*, v. 191, pp. 92–94.

Maglio, V. J. 1978. Patterns of faunal evolution, pp. 603–619, *in* Maglio, V. J., and Cooke, H. B. S., Editors, *Evolution of African Mammals*. Cambridge, Harvard Univ. Press.

Marshall, L. G., Webb, S. D., Sepkoski, J. J., Jr., and Raup, D. M. 1982. Mammalian evolution and the great American interchange. *Science,* v. 215, pp. 1351–1357.

Martin, L. D. and Neuner, A. M. 1978. The end of the Pleistocene in North America. *Trans. Nebr. Acad. Sci.,* v. 6, pp. 117–126.

Martin, P. S. 1967. Prehistoric overkill, pp. 75–120, *in* Martin, P. S., and Wright, H. E., Jr., Editors, *Pleistocene Extinctions: The Search for a Cause.* New Haven, Yale Univ. Press.

———. 1973. The discovery of America. *Science,* v. 179, pp. 969–974.

———. 1975. Vanishings, and future, of the prairie. *Geosci. and Man,* v. 10, pp. 39–49.

Martin, R. A. and Webb, S. D. 1974. Late Pleistocene mammals from the Devil's Den fauna, Levy County, pp. 114–145, *in* Webb, S. D., Editor, *Pleistocene Mammals of Florida.* Gainesville, Univ. of Fla. Presses.

Matthews, J. V. 1974. Quaternary environments of Cape Deceit (Seward Peninsula, Alaska): evolution of a tundra ecosystem. *Bull. Geol. Soc. Amer.,* v. 85, pp. 1353–1384.

———. 1976. Arctic-steppe: an extinct biome. *Amer. Quat. Assoc. Abstracts, Fourth Biennial Meeting,* Arizona State Univ., Tempe, pp. 73–77.

Mayr, E., 1970. *Populations, Species, and Evolution.* Cambridge, Harvard Univ. Press.

Mengel, R. M. 1970. The North American central Great Plains as an isolating agent in bird speciation, pp. 279–340, *in* Dort, W., Jr., and Jones, J. K., Jr., Editors, *Pleistocene and Recent Environments of the Central Great Plains.* Lawrence, Univ. Press of Kans.

Miller, B. B. 1976. The late Cenozoic molluscan succession in the Meade County, Kansas area, pp. 73–85, *in* Guidebook 24th Ann. Mtg. Midwest Friends of the Pleist.: stratigraphy and faunal sequence —Meade County, Kansas. *Guidebook Series I.* Lawrence, *Kans. Geol. Surv.*

Morgan, A. V. and Morgan, A. 1979. The fossil Coleoptera of the Two Creeks Forest Bed, Wisconsin. *Quat. Res.,* v. 12, pp. 226–240.

———. 1980. Faunal assemblages and distributional shifts of Coleoptera during the late Pleistocene in Canada and the northern United States. *Can. Entomol.,* v. 112, pp. 1105–1128.

Mosimann, J. E. and Martin, P. S. 1975. Simulating overkill by Paleoindians. *Amer. Sci.,* v. 63, pp. 304–313.

Okladnikov, A. P. 1970. *Yakutia Before its Incorporation into the Russian State.* Michael,

H. N., Editor, Montreal, McGill-Queens Univ. Press.

Péwé, T. L. 1975. Quaternary geology of Alaska. *U.S. Geol. Surv. Prof. Pap.* No. 835, pp. 1–45.

Pianka, E. R. 1974. *Evolutionary Ecology.* New York, Harper and Row.

Pillard, B. 1972. La fauna des grands mammiferes du Würmien II de la grotte de l'Hortus (Valfaunes, Herault), pp. 163–205, *in* de Lumley, H., Editor, La grotte de l'Hortus, *Laboratoire de paleontologie humaine et de prehistoire,* Memoire No. 1, Marseille, Université de Provence.

Pregill, G. K. and Olson, S. L. 1981. Zoogeography of West Indian vertebrates in relation to Pleistocene climatic cycles. *Ann. Rev. Ecol. Syst.,* v. 12, pp. 75–98.

Rautenbach, I. L. 1978. A numerical reappraisal of the southern African biotic zones, pp. 175–187, *in* Schlitter, D. A., Editor, Ecology and taxonomy of African small mammals, *Bull. Carn. Mus. Nat. Hist.,* v. 6.

Ray, C. E. 1967. Pleistocene mammals from Ladds, Bartow County, Georgia. *Bull. Georgia Acad. Sci.,* v. 25, pp. 120–150.

Ross, H. H. 1970. The ecological history of the Great Plains: evidence from grassland insects, pp. 225–240, *in* Dort, W., Jr., and Jones, J. K., Jr., Editors, *Pleistocene and Recent Environments of the Central Great Plains.* Lawrence, Univ. Press of Kans.

Rounds, R. C. 1981. First approximation of habitat selectivity of ungulates on extensive winter ranges. *J. Wildlife Man.,* v. 45, pp. 187–196.

Semken, H. A. 1974. Micromammal distribution and migration during the Holocene. *Amer. Quat. Assoc. 3rd Biennial Meeting,* Univ. of Wisc., Madison, p. 25.

Sher, A. V. 1968. Fossil saiga in northeastern Siberia and Alaska. *Int. Geol. Rev.,* v. 10, pp. 1247–1260.

Simpson, B. B. 1975. Pleistocene changes in the flora of the high tropical Andes. *Paleobiology,* v. 1, pp. 173–294.

Simpson, B. B. and Haffer, J. 1978. Speciation patterns in the Amazonian forest biota. *Ann. Rev. Ecol. Syst.,* v. 9, pp. 497–518.

Slaughter, B. H. 1967. Animal ranges as a clue to late-Pleistocene extinction, pp. 155–168, *in* Martin, P. S., and Wright, H. E., Jr., Editors, *Pleistocene Extinctions: The Search for a Cause.* New Haven, Yale Univ. Press.

———. 1975. Ecological interpretations of Brown Sand Wedge local fauna, pp. 179–192, *in* Wendorf, F., and Hester, J. J., Editors, Late Pleistocene Environments of the Southern High Plains. *Fort Burgwin Res. Cent.*, Pub., no. 9.

Slobodkin, L. B., Smith, F. E., and Hairston, H. G. 1967. Regulation in terrestrial ecosystems and the implied balance of nature. *Amer. Nat.*, v. 101, pp. 109–124.

Soulé, M. E., Wilcox, B. A., and Holtby, C. 1979. Benign neglect: a model of faunal collapse in game reserves of East Africa. *Biol. Conserv.*, v. 15, pp. 259–272.

Stebbins, G. L. 1971. *Chromosomal Evolution in Higher Plants.* London, E. Arnold.

———. 1975. The role of polyploid complexes in the evolution of North American grasslands. *Taxon,* v. 24, pp. 91–106.

———. 1981. Coevolution of grasses and herbivores. *Ann. Missouri Bot. Gard.*, v. 68, pp. 75–86.

Sutcliffe, A. J. 1962. A note on some late Pleistocene mammalian remains from Lummoton Quarry. *Trans. Proc. Torquay Nat. Hist. Soc.*, v. 13, pp. 4–7.

Sutcliffe, A. J. and Kowalski, K. 1976. Pleistocene rodents of the British Isles. *Bull. Brit. Mus. Nat. Hist.: Geol.*, v. 27, pp. 31–147.

Sutcliffe, A. J. and Zeuner, F. E. 1962. Excavations in the Torbryan Caves, Devonshire. I Tornewton Cave. *Proc. Devon. Archaeol. Exploration Soc.*, v. 5, pp. 127–145.

Tiezen, L. L. and Imbamba, S. K. 1980. Photosynthetic systems, carbon isotope discrimination and herbivore selectivity in Kenya. *African J. Ecol.*, v. 18, pp. 237–242.

Van Den Brink, F. H. 1968. *A Field Guide to the Mammals of Britain and Europe.* Boston, Houghton Mifflin Co.

Van der Hammen, T. 1963. A palynological study of the Quaternary of British Guiana. *Leidse Geol. Meded.*, v. 29, pp. 125–180.

———. 1974. The Pleistocene changes of vegetation and climate in tropical South America. *J. Biogeogr.*, v. 1, pp. 3–26.

Van Devender, T. R. and Spaulding, W. G. 1979. Development of vegetation and climate in the southwestern United States. *Science,* v. 204, pp. 701–710.

Vangengeim, E. A. 1967. The effect of the Bering Land Bridge on the Quaternary mammalian faunas of Siberia and North America, pp. 281–287, *in* Hopkins, D. M., Editor, *The Bering Land Bridge.* Palo Alto, Stanford Univ. Press.

Van Zant, K. 1979. Late glacial and postglacial pollen and plant macrofossils from West Okoboji, northwestern Iowa. *Quat. Res.*, v. 12, pp. 358–390.

Vesey-Fitzgerald, D. F. 1960. Grazing succession among East African game animals. *J. Mamm.*, v. 41, pp. 161–172.

Voorhies, M. R. 1974. Pleistocene vertebrates with boreal affinities in the Georgia Piedmont. *Quat. Res.*, v. 4, pp. 85–93.

Vuilleumier, B. S. 1971. Pleistocene changes in the fauna and flora of South America. *Science,* v. 173, pp. 771–780.

Webb, S. D. 1974. Pleistocene llamas of Florida, with a brief review of the Lamini, pp. 170–213, *in* Webb, S. D., Editor, *Pleistocene Mammals of Florida.* Gainesville, Univ. of Fla. Presses.

———. 1976. Mammalian faunal dynamics of the great American interchange. *Paleobiology,* v. 2, pp. 220–234.

———. 1977. A history of savanna vertebrates in the New World: Part I: North America. *Ann. Rev. Ecol. Syst.*, v. 8, pp. 355–380.

———. 1978. A history of the savanna vertebrates in the New World. Part II: South America and the great interchange. *Ann. Rev. Ecol. Syst.*, v. 9, pp. 393–426.

Webb, T. and Bryson, R. A. 1972. Late and post-glacial climatic change in the northern Midwest: quantitative estimates derived from fossil pollen spectra by multivariate statistical analysis. *Quat. Res.*, v. 2, pp. 70–115.

West, R. G. 1964. Inter-relations of ecology and Quaternary paleobotany. *British Ecol. Soc. Jubilee Symp., J. Ecol.*, v. 52 (supplement), pp. 47–57.

———. 1967. The Quaternary of the British Isles, pp. 1–87, *in* Rankama, K., Editor, *The Quaternary,* Vol. 2, New York, Interscience.

———. 1968. *Pleistocene Geology and Biology.* New York, John Wiley and Sons.

Western, D. and Ssemakula, J. 1981. The future of the savanna ecosystems: ecological islands or faunal enclaves. *African J. Ecol.*, v. 19, pp. 7–19.

Whitehead, D. R. 1979. Late-glacial and postglacial vegetational history of the Berkshires, western Massachusetts. *Quat. Res.*, v. 12, pp. 333–357.

———. 1981. Late Pleistocene vegetational changes in northeastern North Carolina. *Ecol. Mon.*, v. 51, pp. 451–471.

Wright, H. E., Jr. 1970. Vegetational history of the central Great Plains, pp. 157–172, *in* Dort, W., Jr., and Jones, J. K., Jr., Editors, *Pleistocene and Recent Environments of the Central Great Plains.* Lawrence, Univ. Press of Kans.

———. 1981. Vegetation east of the Rocky Mountains 18,000 years ago. *Quat. Res.*, v. 15, pp. 113–125.

12

Pleistocene Extinction and Environmental Change
Case Study of the Appalachians

JOHN E. GUILDAY

TWO BASIC ASSUMPTIONS ABOUT THE TERRESTRIAL EXTINCTION pattern associated with the late Pleistocene/early Holocene climatic transition ($\pm$ 12,000 yr B.P.) are: 1) only large mammals were involved, or were affected to a greater degree than were smaller vertebrates and 2) many died out without a niche replacement (P. S. Martin 1967). The implications are that large mammals (big game) met their demise at the hands of human hunters who harvested beyond the capacity of their prey to recoup, that little if any ecological pressure was being exerted upon these now extinct creatures, and that their adaptive niches remained while the show moved on without them.

Both assumptions need qualification. While each is partially true, they are weighted so that they appear, at first glance, to offer strong presumptive evidence for the human overkill hypothesis. If only "big game" disappeared from an otherwise Elysian scene, it would indeed appear that man was the primary factor behind the major wave of early postglacial extinctions.

The taxonomic details of the sweeping faunal change that occurred in the wake of the last glaciation are well known (see E. Anderson this volume), and the differential extinction of *large* mammals is striking indeed when emphasis is placed solely upon biological extinction—the abrupt termination of phyletic lines. But to imply that small vertebrates (nongame) were relatively unaffected, presumably because they simply were not tasty, overstates the case and denigrates the effect of the drastic environmental changes that were coincident with these extinctions and with the interjection, in the Americas, of a new and sophisticated hunter—the Paleo-Indian—into an already stressed ecosystem.

Exact timing of the inception of the process or processes leading to the extinctions is not of major importance if the extinction process was cumulative rather than catastrophic, but the apparent concordance of terminal dates for the late Pleistocene megafauna does demand an explanation and suggests some prime mover (or remover). It is possible to visualize a primitive hunting overload that could result in the extinction of the prey species but only under conditions not directly related to hunting pressure per se, but rather to the overall ecological health of that species. Under such circumstances, hunting pressure becomes just one of many interrelated "causes" acting in consort to drive the species to extinction, none of which could be singled out as the ultimate, or even the primary, cause.

The ecological richness of late Pleistocene, midlatitude North America and west-ern Eurasia (the two areas most extensively studied) compared to their relative ecologi-cal homogeneity during the succeeding Holocene is well illustrated by vertebrate faunas containing both large and small mammals of these two time periods.

I will use the mid-Appalachian Plateau and Ridge and Valley portions of the Ap-palachian Mountains to illustrate. The area, bounded by latitudes 36°N and 45°N, in-cludes most of Pennsylvania, West Virginia, western Virginia, western Maryland, and extreme eastern parts of Ohio, Kentucky, and Tennessee, approximately 475,000 square kilometers of relatively low (usually less than 1,000 m) northeast/southwest trending ridges, intermontane valleys, and dissected rolling plateaus. During Holocene times, prior to the ecological destruction of the past 300 years, the area was covered by a mixed, oak/chestnut (*Quercus/Castanea*) dominant, closed-canopy, largely deciduous, mast forest. Sufficient topographical relief was present to provide isolated, low order, relict stations for northern species of plants and animals that had lingered since glacial times, as well as relatively isolated river-bottom meadows and mountain glades not extensive enough to support endemic faunas. From a mammalian and avian standpoint (leaving aside aquatic habitats), the area was one vast deciduous forest. All of it, except the extreme northern portion and a few isolated high altitude areas which tend toward Canadian conditions, is included within the Carolinian biotic province (Dice 1943).

The mammalian fauna consisted primarily of forest product consumers (mast feeders, browsers), insectivores, and carnivores. Grazing animals were restricted to small rodents and lagomorphs (*Microtus, Synaptomys, Sylvilagus*) capable of exploiting the restricted, often temporary, areas of grassland. The presence of buffalo, *Bison bison,* other than as a transient, has not been satisfactorily established. Primary mast feeders were largely arboreal—five species of sciurids, turkey (*Meleagris gallopavo*), ruffed grouse (*Bonasa umbellus*), and passenger pigeon (*Ectopistes migratorius*), which occurred in phenomenal numbers reminiscent of the Great Plains buffalo herds in terms of biomass. Under the closed-canopy, the relatively unproductive forest floor was gleaned by various small rodents in sylvan microhabitats. The Holocene mammalian fauna of the area totaled an estimated fifty-one species. Of these only six were large (>20 kg in weight): white-tailed deer (*Odocoileus virginianus*) and elk (*Cervus elaphus*), both mast/browsers; the specialized predators, mountain lion (*Felis concolor*) and timber wolf (*Canis lupus*); the omnivorous black bear (*Ursus americanus*); and the transient grazer *Bison bison.* Despite the size of the area, major ecological niches were limited in a generally ecologically homogeneous deciduous mast forest with a long winter resting period under intermittent snow cover.

The Fauna

The late Pleistocene fauna of this same area was more diverse. It is relatively well known from fossil sites in caves and rarer bog and saltlick deposits, and is summa-rized below.

Talpidae: Three moles. Hairy-tailed mole (*Parascalops breweri*), star-nosed mole (*Condylura cristata*), and, marginally, eastern mole (*Scalopus aquaticus*). They are rare as fossils and occur in the area today.

Soricidae: Eight shrews. Arctic shrew (*Sorex arcticus*), a boreal forest species now locally extinct. The ranges of the rock shrew (*Sorex dispar*) and the water shrew (*Sorex palustris*) have since shrunk to relict status. The pygmy shrew (*Microsorex hoyi*), common (about 10 percent of all soricids) in late Pleistocene cave deposits, is now rare and local. The masked shrew (*Sorex cinereus*) and short-tailed shrew (*Blarina brevicauda*) were larger; they have since undergone size reduction within the area, probably climatically induced (Bergmann's rule). Least shrew (*Cryptotis parva*) is known only from a few cave deposits where it may have been intrusive. Smoky shrew

(*Sorex fumeus*) is still common throughout the area today. Thus at least six of the seven species of soricids definitely known from the late Pleistocene of the area were affected in some way by post-Wisconsinan environmental changes.

Vespertilionidae: Nine species of bats show no appreciable morphological changes. The ranges of gray myotis (*Myotis grisescens*) and big-eared bat (*Plecotus* sp.) were more extensive, judging from fossil deposits. Data are insufficient to comment on any other species.

Leporidae: Two species. The snowshoe hare, (*Lepus americanus*) has increased in size in the area since late Pleistocene times, in keeping with the present north-south size cline of the species. Formerly widespread, it is now of relict distribution in the area. New England cottontail (*Sylvilagus transitionalis*) was identified from Clark's Cave, Virginia, while eastern cottontail (*S. floridanus*) and *S. transitionalis* occur in the area today. Their Pleistocene distributions are poorly known.

Sciuridae: Eight squirrels. The ground squirrels, least chipmunk (*Eutamias minimus*) and thirteen-lined ground squirrel (*Spermophilus tridecemlineatus*), are locally extinct and found farther west today. Eastern chipmunk (*Tamias striatus*), red squirrel (*Tamiasciurus hudsonicus*), and flying squirrels (*Glaucomys volans* and *Glaucomys sabrinus*) were represented by larger forms than now occur in the area. The ranges of *Tamiasciurus hudsonicus* and the northern flying squirrel (*G. sabrinus*) have contracted since late Pleistocene times, while that of the gray squirrel (*Sciurus carolinensis*) has expanded. The fox squirrel (*S. niger*) is a Holocene immigrant from the south and west. Only the common woodchuck (*Marmota monax*) seems to have been little affected by postglacial changes.

Cricetidae: Three cricetid rodents. Eastern woodrat (*Neotoma floridana*), deer mouse (*Peromyscus maniculatus*), and white-footed mouse (*P. leucopus*) have been relatively unaffected by postglacial changes. Four new cricetids—hispid cotton rat (*Sigmodon hispidis*), eastern harvest mouse (*Reithrodontomys humulis*), marsh rice rat (*Oryzomys palustris*), and golden mouse (*Ochrotomys nuttalli*) have since appeared along the southern edge of the area.

Arvicolidae: Ten voles. Labrador collared lemming (*Dicrostonyx hudsonius*), heather vole (*Phenacomys intermedius*), northern bog lemming (*Synaptomys borealis*), and yellow-cheeked vole (*Microtus xanthognathus*) are locally extinct and found in boreal areas to the north today. Ranges of rock vole (*Microtus chrotorrhinus*) and southern red-backed vole (*Clethrionomys gapperi*) were more extensive than they are today. The relative abundance of woodland vole (*M. pinetorum*) (more now) and southern bog lemming (*Synaptomys cooperi*) (fewer now) appears to have changed. Meadow vole (*M. pennsylvanicus*) and *S. cooperi* were represented by smaller individuals than now occur in the area. Only the aquatic muskrat (*Ondatra zibethicus*) appears to have been unaffected by post-Pleistocene changes.

Zapodidae: Two jumping mice. Woodland jumping mouse (*Napaeozapus insignis*) has undergone a size reduction, and its range has become restricted. Meadow jumping mouse (*Zapus hudsonius*) is relatively common in fossil deposits.

Erethizontidae: One species, *Erethizon dorsatum*, the porcupine, has undergone a range reduction.

Castoridae: Two beaver species. *Castoroides ohioensis* is now extinct; *Castor canadensis* is still widespread and common.

Canidae: Three species. *Canis dirus,* the dire wolf, is extinct; *Canis latrans,* coyote, is locally extinct; the red fox (*Vulpes vulpes*) has had a complicated Holocene history. Initially it receded to the north. It was absent from the area (based on Holocene archaeological faunas) until colonial land clearing, at which time it reinvaded from the north; it now shares the area with the gray fox (*Urocyon cinereoargenteus*), an early Holocene invader from the south.

Ursidae: Three bears. Giant short-faced bear (*Arctodus simus*) is extinct; the grizzly bear (*Ursus arctos*) has retreated to the west; black bear (*Ursus americanus*) survives. It is possible that the extinct Florida cave bear (*Tremarctos floridanus*) occurred in the southern part of the discussion area.

Procyonidae: One species. The raccoon, *Procyon lotor*, rare in late Pleistocene deposits from the area, is now common and widespread.

Felidae: At least two cats. The felid record is unsatisfactory. Only jaguar (*Felis onca augusta*) and bobcat (*Lynx rufus*) are definitely recorded from late Pleistocene sites. It is possible that mountain lion (*Felis concolor*), lion (*Felis leo*), American cheetah (*Acinonyx trumani*), sabertooth (*Smilodon floridanus*), and Canada lynx (*Lynx canadensis*) were present. Only *F. concolor, L. rufus,* and, marginally, *L. canadensis* occurred during the Holocene.

Mustelidae: Nine species. The marten (*Martes americana*), ermine (*Mustela erminea*), and least weasel (*Mustela nivalis*) have restricted ranges today compared to their widespread late Pleistocene distributions. The short-faced skunk (*Brachyprotoma obtusata*) is extinct. Fisher (*Martes pennanti*), mink (*Mustela vison*), long-tailed weasel (*Mustela frenata*), and striped skunk (*Mephitis mephitis*) are known from both late Pleistocene and Holocene sites throughout the area. The wolverine (*Gulo gulo*) probably inhabited the area during the late Pleistocene but now occurs only farther north. Badger (*Taxidea taxus*), a steppe form, has retreated to the west. The eastern spotted skunk (*Spilogale putorius*) has apparently since invaded from the south.

Tayassuidae: Two peccaries. Long-nosed peccary (*Mylohyus nasutus*) and flat-headed peccary (*Platygonus compressus*), both common and widespread in late Pleistocene deposits, are now extinct.

Cervidae: Five deer. Fugitive deer (*Sangamona fugitiva*) and stag-moose (*Cervalces scotti*) are extinct. Caribou (*Rangifer tarandus*) has retreated to the north. White-tailed deer (*Odocoileus virginianus*) and wapiti (*Cervus elaphus*) (extirpated, now reintroduced) still inhabit the area. The Pleistocene and Holocene record for moose (*Alces*) in the area is unsatisfactory.

Bovidae: Four species. Woodland musk ox (*Symbos cavifrons*), Harlan's musk ox (*Bootherium bombifrons*), and bison (*Bison antiquus*) are extinct. *Ovibos moschatus* has retreated to the north. Only *Bison bison* occurred marginally in the Holocene.

Tapiridae: One tapir (*Tapirus* cf. *veroensis*), formerly widespread and common, now extinct.

Equidae: At least one horse, *Equus complicatus*, has been recorded, now extinct.

Dasypodidae: One armadillo, the extinct *Dasypus bellus*, has been sparingly reported from as far north as central West Virginia (Organ-Hedricks Cave).

Mylodontidae: Harlan's ground sloth (*Glossotherium harlani*) recorded from Big Bone Lick, Kentucky, may have occurred in the western sector of the discussion area; there is no additional fossil evidence.

Megalonychidae: The extinct Jefferson's ground sloth (*Megalonyx jeffersonii*) was abundant and widespread.

Mammutidae: The mastodont (*Mammut americanum*) was common and widespread; now extinct.

Elephantidae: One, possibly two mammoths (*Mammuthus* sp.), now extinct.

In additon, at least four birds no longer inhabit the area; the ptarmigan (*Lagopus* cf. *mutus*) and the spruce grouse (*Dendragapus canadensis*) are now found farther north, the sharp-tailed grouse (*Pedioecetes phasianellus*) farther northwest, and the magpie (*Pica pica*) farther west. The whooping crane (*Grus americanus*) (Natural Chimneys, Virginia) is no longer found in the East, but its disappearance may be due to ecological disruption during the European colonial period. The passenger pigeon (*Ectopistes migratorius*) has become extinct during historic times. One amphibian from a late

Pleistocene deposit (New Paris No. 4, Pennsylvania) has been referred, on the basis of its small size, to a present boreal subspecies, *Bufo americanus copei*.

Eighteen large and one small species of mammal have become biologically extinct in the mid-Appalachians since late Pleistocene times; three large and ten small species of mammals are locally extinct (eight have receded to the north, five to the west). Of those mammalian species that still occur in the area, four are rare and local boreal relicts; nine others have become relatively less common or have undergone some measure of ecological adjustment expressed in range reductions. Nine species of mammals have undergone a Holocene size reduction, while four have increased in size within the area during the last 11,000 years, paralleling, for the most part, modern latitudinal size clines (Guilday et al. 1964, p. 180).

In summary, at least fifty-seven (twenty large, thirty-seven medium to small) of the seventy-five species known from the late Pleistocene of the mid-Appalachians (exclusive of bats), for which there is an adequate paleontological record, appear to have been affected by post-Pleistocene changes that can be monitored by paleontological deposits: biological and local extinctions, altitude and habitat adjustments, morphological changes. The fossil record of most of the eighteen remaining species is too poor to offer any pertinent data.

A Perspective

The above faunal summary was presented, not to beg the issue of the very real disproportionate percentage of large mammal compared to small mammal extinctions that occurred coincident with postglacial change, but rather to place these extinctions in proper perspective. Large mammal taxa did not simply vanish from a scene of ecological composure, but instead disappeared during a time of great ecological ferment when biotas were adjusting, dissolving, and reforming under new climatic parameters to emerge into the Holocene in greatly altered aspect.

Plant and animal taxa were forced to adjust on an individual basis. Not only did selected species become extinct, but so did "ecological niches." This must be inferred indirectly by the dated fossil remains of plants and animals that formed these biological communities.

The ecological adaptations of the extinct forms can only be inferred from analogies drawn from their closest living relatives. One of the advantages of studying a time period as late as the Pleistocene-Holocene transition is that the plants, and most of the smaller vertebrates—particularly the ecologically sensitive small mammals—are still living, and their ecological requirements can be directly observed.

The biological response to postglacial changes in eastern North America was a patterned one. Pollen studies (summarized in Watts 1979, Delcourt and Delcourt 1981) demonstrate that at the height of the Wisconsinan glaciation, and immediately following from at least 18,000 yr B.P. to around 10,500 yr B.P., the environment was boreal in character. Initially there was a periglacial tundra belt, probably less than 100 km wide, along the northern portion of the discussion area. The remainder of the area was an extended spruce/jackpine/fir, birch (*Picea/Pinus banksiana/Abies, Betula*) parkland with an understory of woody shrubs, grasses, sedges, and herbs. There is no modern counterpart. Northern hardwood deciduous trees were initially poorly represented. This broad belt of coniferous forest and parkland stretched from the Dakotas to the Atlantic Coast, and from near the glacial margin south to central Tennessee (Delcourt and Delcourt, 1981). Postglacial warming brought a complete change in plant covertype, from open coniferous to the present closed-canopy deciduous forest.

During late Pleistocene times the area supported many species of mammals that are now allopatric and found in four distinct modern biotic zones—tundra (*Dicrostonyx hudsonius*), boreal forest (*Phenacomys intermedius*), midwestern prairies (*Spermophilus tridecemlineatus*), and temperate mast forest (*Glaucomys volans*). None of these species is sympatric today, yet each, except for the rare tundra element, was common and apparently broadly distributed throughout the area during late Pleistocene times.

Grazers, absent from the Holocene fauna except for a few micrograzers, were well represented in this boreal parkland by large elephants, horses, and bovids (*Symbos, Bootherium, Ovibos, Bison*) and possibly the cervid, *Sangamona* (Kurtén and Anderson 1980). A variety of large browsers and specialized herbivores (*Cervalces, Rangifer, Tapirus, Mammut, Mylohyus, Platygonus, Megalonyx*, possibly *Glossotherium*, as well as the surviving *Odocoileus* and *Cervus*) occurred in the area, suggesting a greater variety of trophic niches; boreal forest and grassland forms were both represented, as well as a variety of now ostensibly temperate forest taxa in this unique, extinct, coniferous parkland.

The terrestrial mid-Appalachian mammalian fauna crashed about 10,000 to 12,000 years ago. As the ecologically diverse late Pleistocene parklands with their broad spectrum of housekeeping possibilities changed to closed-canopy deciduous forest, large grazers could no longer sustain themselves. A few micrograzers persisted in small, often ephemeral, meadow situations within the mast forest and survive to this day but large mammals were reduced from at least 31 percent of the fauna in the late Pleistocene to 12 percent in the Holocene. Most view the fate of the specialized larger carnivores as tied directly to the fate of their modal prey species.

Thus, the broad belt of ecologically diverse, predominantly coniferous parkland that extended from at least Wyoming east to the Atlantic Coastal Plain, and probably viable in some form throughout the Pleistocene, disintegrated as a biological unit within a relatively short period as meteorological patterns changed and the continental glaciers receded, its component species either becoming extinct or regrouping themselves into assemblages that continued to polarize up to the present day. Pleistocene local faunas assigned to both pre-Wisconsinan glacial and interglacial periods are equally diverse in composition and their taxonomic makeup, especially of the reptiles and amphibians, suggests climatic equability (Holman 1980) without a hint of Holocene polarity.

The deterioration and disintegration of the eastern and western segments of this parkland were, in some respects, mirror images of one another. In the western segment, the Great Plains tree cover disappeared almost completely except for scattered fire-break ridges (Wells 1965) and along river courses where corridor woodlands persisted. In the eastern segment, as closed-canopy deciduous forest evolved, grasslands became restricted primarily to river valley corridors. Neither of these corridors, wooded in the Plains, grassed in the East, was extensive enough to support more than a few large mammals on a sustained basis.

Evidence for the extinction of habitats or ecological niches is Martin and Neuner's observation (1978) that on the Great Plains those mammalian taxa that became extinct were those that were relatively most common in deposits older than ca. 12,000 yr B.P. (equids, camelids, long-horned bison, tayassuids, antilocaprids—three genera, one of which was no larger than a jackrabbit—large edentates and proboscideans) while those taxa that were then rare not only survived but in some cases [(*Bison bison*), pronghorn (*Antilocapra americana*)—the largest of the antilocaprids] increased to phenomenal numbers. They were able to expand and flourish in spite of human predation. Their earlier rarity suggests that ecological conditions did not favor them and that they were existing marginally in a parkland environment, while the "big game" that flourished in a former ecosystem reminiscent of East Africa today dwindled to extinction as that ecosystem, with its varied ecological niches, was transformed into the relatively featureless modern prairie of the North American Midlands.

Thus, ecological changes, both in the East and in the West, acted, at the very least, to reduce the geographical distribution of adversely affected species, resulting in a reduction in total population size. The smaller the population, the greater the potential for eventual extinction through environmental attrition, as total reproductive capacity diminished.

This same picture of climatic change and a lessening of ecological diversity was also characteristic of other parts of the continent and of the world during the Pleistocene/Holocene interchange. It is this that makes the riddle so complex. All of this ecological fermentation served as a backdrop behind the spread and increasing technological sophistication of man-the-hunter. Only in a few cases, such as the extinction of island megafaunas by man, as in New Zealand and Madagascar, is the case clearcut.

A source of ambiguity, at least in North America, is the low archaeological visibility of human and extinct megafaunal interaction. If man slaughtered all this game, where is the evidence of such activity, commensurate with that found in European (Lumley 1976, Vereshchagin 1959, Kurtén 1968) or African sites (Klein this volume)? This might be explained in the East by poor preservation; rapid erosion, acidic soil, and abundant phreatic water may well have destroyed evidence of any such activity. Evidence of predation by man on the caribou (*Rangifer tarandus*) at ca. 12,000 yr B.P. is documented from Dutchess Quarry Cave, New York (Funk et al. 1970). The one eastern archaeological site with a dated faunal sequence extending ostensibly back into late glacial times, the Meadowcroft Rockshelter in Pennsylvania (Adovasio et al. 1980), has no extinct species. Western North American sites from that time period, where preservation is much better, are primarily opportunistic kill sites of elephants (Clovis) and long-horned bison (Folsom). No large concentrations of Paleo-Indian housekeeping debris are known (Haynes 1980), and it is obvious that human populations at that time were small and nomadic. P. S. Martin (1967) has treated this positively by postulating that such a nonshow is compatible with a catastrophic extinction of large mammals by an expanding wave of paleo-hunters as they progressed south from the Alaska area into lands teeming with big game that had no time to respond defensively to the new predator, and that man had little difficulty in harrying the unsophisticated beasts to extinction. The swifter the job was done, the narrower the archaeological window, therefore the fewer the sites that might be expected to record the event. This is, of course, unanswerable and untestable.

Saunders (1980), from a study of the age composition of *Mammuthus columbi* kill groups taken by Clovis hunters, concludes that these groups were demographically normal, suggesting that these elephants were not under ecological stress. He cites this as corroborating evidence for man-the-exterminator. But it could also be argued that the same data suggest that they were also under little stress from hunting (except on an individual basis) as well. Samples are so small, however, that the entire argument seems moot.

Large mammals are inherently more vulnerable to environmental changes simply because they are large and require a greater expanse of primary habitat to sustain themselves because of greater individual demands for food or space to play out their reproductive and defensive strategies, cover, flight, herding, etc. (Guilday 1967). Greater demands are placed upon the habitat by an elephant than by a small rodent, although each may be a grazing or browsing specialist. Vulnerability is directly correlated with size. Given this, it follows that any deteriorating ecosystem will lose its ability to support larger organisms first because of their need for a larger share of primary habitat. Small mammals, with lesser individual demands, smaller home ranges, and with greater reproductive capacity, can survive in smaller areas of primary habitat. They are able to sustain higher numbers in relict situations that may serve as extinction buffers, and can recover more rapidly from local catastrophic events whose effects loom larger as a population dwindles and fragments. Taken to its logical conclusion this leads to a pattern of differential extinction—the largest mammals being the most affected.

In a deteriorating ecosystem, species must either 1) adjust their ranges, following, if geographically possible, migrating resources. This apparently took place in the Sahara whose late Pleistocene fauna was essentially Recent East African (Klein this volume); 2) adapt in place to new parameters, not a viable alternative in such a rapid and devastating climatic change as that seen in at least North America from ca. 14,000 yr B.P. to 10,000 yr B.P.; or 3) become extinct. As areas of primary adaptation shrink geographically, numbers lessen and other factors such as heightened interspecific competition and environmental attrition (disease, predation, local catastrophes such as harsh winters, floods, drought, etc.) whittle them down. Under this scenario, large mammals, precisely those forms designated as "big game," are differentially eliminated, even in the absence of man.

The actual circumstances surrounding the demise of individual species on a global basis certainly, as pointed out by most authors, varied with individual circumstances. It is possible, indeed probable, that human predation was the event that forced the apparently geologically concordant extinction of so many large mammals at the end of the Pleistocene. This is true, however, only if those same species had previously been reduced numerically and geographically to relict status from which they may, or may not, have expanded again under more favorable conditions, given the absence of human predation.

Some species may well have persisted to this day if they had not been harried to extinction. P. S. Martin et al. (1961) and Thompson et al. (1980) made a plausible case for the extinction of the sloth, *Nothrotheriops shastensis*, by man, even in the absence of direct archaeological evidence. Slow, relatively inoffensive, and conspicuous, it was apparently adapted (judging from dung analysis) for a desert browsing niche, now open but unoccupied. But the once varied herds of big game on the now ecologically monotonous Great Plains, and the equally diverse eastern megafauna in a region covered in the Holocene by an equally monotonous closed-canopy temperate forest, suggest a major culling and recombination of those once diverse faunas as ecological possibilities (adaptive niches) ceased to exist.

In summary, late Pleistocene climatic changes and the dispersal of man into new areas took place simultaneously. Both obviously had an effect upon the ecosystems they encountered. What effect one would have had in the absence of the other is untestable, therefore their relative importance must remain unknown. Did one set up the punch which the other delivered, or was it vice versa? In any event their combined effect was devastating, and the world is much the poorer.

Acknowledgments

Mid-Appalachian research was supported by National Science Foundation, National Geographic Society, and Carnegie Museum of Natural History. I wish to thank the reviewers of this paper, Drs. Mary R. Dawson, Hugh H. Genoways, and Joseph F. Merritt; Allen D. McCrady and Harold W. Hamilton for support both in the field and in the laboratory; Elizabeth Hill for typing the manuscript; and Alice M. Guilday, who is always there.

References

Adovasio, J. M., J. D. Gunn, J. Donahue, R. Stuckenrath, J. Guilday, K. Lord, and K. Volman. 1980. Meadowcroft rockshelter—retrospect 1977: part 2. North American Archaeologist 1(2):99–136.

Delcourt, Paul A. and Hazel R. Delcourt. 1981. Vegetation maps for eastern North America: 40,000 yrs. B.P. to the present, pp. 123–165. *In* Robert C. Romans (ed),

Geobotany II. Plenum Press, New York, 263 pp.

Dice, L. R. 1943. The Biotic Provinces of North America. Univ. Michigan Press, Ann Arbor, 78 pp.

Funk, R. E., D. W. Fisher, and E. M. Reilly, Jr. 1970. Caribou and Paleo-Indian in New York State: A presumed association. Amer. J. of Sci. 268:181–186.

Guilday, J. E. 1967. Differential extinction during late-Pleistocene and Recent times. *In* P. S. Martin and H. E. Wright (eds.), Pleistocene Extinctions: The Search for a Cause. New Haven and London: Yale Univ. Press, pp. 121–140.

Guilday, J. E., P. S. Martin, and A. D. McCrady. 1964. New Paris No. 4: A Pleistocene cave deposit in Bedford County, Pennsylvania. Bull. Nat. Speleol. Soc. 26(4):121–194.

Haynes, C. Vance. 1980. The Clovis culture. Canadian Journal of Anthropology 1(1):115–121.

Holman, J. A. 1980. Paleoclimatic implications of Pleistocene herpetofaunas of eastern and central North America. Trans. Nebraska Acad. Sciences 8:131–140.

Kurtén, Björn. 1968. Pleistocene Mammals of Europe. Weidenfeld and Nicolson, London. 317 pp.

Kurtén, Björn and Elaine Anderson. 1980. Pleistocene Mammals of North America. Columbia Univ. Press, New York. 442 pp.

Lumley, Henry de. 1976. La Prehistoire Francaise, Vol. 1. Editions du Centre National de la Récherche Scientifique, Paris. 759 pp.

Martin, Larry D. and A. M. Neuner. 1978. The end of the Pleistocene in North America. Trans. Nebraska Acad. Sciences 6:117–126.

Martin, P. S. 1967. Prehistoric overkill. *In* P. S. Martin and H. E. Wright (eds.), Pleistocene Extinctions: The Search for a Cause. New Haven and London: Yale Univ. Press, pp. 75–120.

Martin, P. S., B. E. Sabels, and D. Shutler. 1961. Rampart Cave coprolite and ecology of the Shasta ground sloth. Amer. J. of Sci. 259(2):102–127.

Saunders, J. J. 1980. A model for man-mammoth relationships in late Pleistocene North America. Canadian Journal of Anthropology 1(1):87–98.

Thompson, R. S., T. R. Van Devender, P. S. Martin, T. Foppe, and A. Long. 1980. Shasta ground sloth (*Nothrotheriops shastense* Hoffstetter) at Shelter Cave, New Mexico: environment, diet, and extinction. Quaternary Research 14:360–376.

Vereshchagin, N. K. 1959. The Mammals of the Caucasus. A History of the Evolution of the Fauna. Israel Program for Scientific Translations, Jerusalem, 1967. 816 pp.

Watts, W. A. 1979. Late Quaternary vegetation of Central Appalachia and the New Jersey Coastal Plain. Ecological Monographs 49(4):427–469.

Wells, Philip V. 1965. Scarp woodlands, transported grassland soils, and concept of grassland climate in the Great Plains region. Science 148(3667):246–249.

13

Mosaics, Allelochemics and Nutrients

An Ecological Theory
of Late Pleistocene
Megafaunal Extinctions

R. DALE GUTHRIE

WHEN I CAME TO ALASKA IN THE EARLY 1960s as a young paleobiology professor, freshly bathed in such concepts as chronofaunas, geofloras and biotic provinces, the accepted interpretation of Arctic and Subarctic vegetation changes during the Pleistocene was that the tundra-taiga treeline had been raised and lowered altitudinally by fluctuations in temperature during the glacials and interglacials. At this time geofloras were thought to have maintained associations for millions of years, and the Clementsian ecological theory of climatically determined climax communities and fixed plant associations held sway. Coming from the Midwest I was familiar with the then current ideas about telescoping vegetational provinces—shifting the classic vegetation zones north and south during the Pleistocene glacials and interglacials.

Likewise, being from the Plains I was most familiar with the osteology of bison, horse and mammoth, the dominant plains grazers.

When I began to study the thousands of fossils collected from Alaskan Pleistocene deposits, I found that 95 percent were from bison, horse, and mammoth. How could these grazing animals thrive in a tundra landscape? Clearly the ideas about Alaskan Pleistocene vegetation needed some adjustment (Guthrie 1968), as did my paleoecological models.

Subsequent work has shown that during the Pleistocene a large, complex grassland existed in Alaska and northeast Canada that stretched across most of northern and central Asia and eastward to cover much of Europe. It had many facies and was marked by considerable local diversity in plants, insects, and mammals. Yet despite the heterogeneity, there was such a meaningful integrity throughout this steppe area, especially among the mammals, that a new name was justified. I have referred to it as the mammoth steppe (Guthrie 1980, 1982).

The most curious aspect of this mammoth steppe is its peculiar mixture of biota, without counterpart among modern biomes. This is why the species associations we find there appear so disharmonious. Many forms appeared in Alaska which are now associated with southern grasslands: badgers (*Taxidea*), ferrets (*Mustela*), bison (*Bison*), and grama grass (*Bouteloua*). Eurasiatic forms like saiga antelope (*Saiga*), horses (*Equus*), and lion (*Panthera*) were mixed with the North American short-faced bears (*Arctodus*) and camels (*Camelops*). Other species stretch the imagination: sabertooth cats (*Homotherium*), yaks (*Bos*), and bonnet-horned musk oxen (*Symbos*). All these appeared in a heterogeneous mix with large mammals present in Alaska today: moose (*Alces*), caribou (*Rangifer*), musk oxen (*Ovibos*), sheep (*Ovis*) and others (Guthrie 1968, Péwé 1975, and Harington 1977).

At the end of the last glaciation many of these animals became extinct in Alaska, and the complex grassland that had existed in the north throughout most of the Pleistocene disappeared. The landscape then took its present form, dominated by a more mesic zonation of boreal forest and tundra. No ranges of significant size remained for saiga, horses, asses, mammoth, and the other species which became extinct. The ranges of bison, wapiti, badgers, and others contracted further to the south.

Thus, in the north there is a direct causal relationship between ecological changes in vegetation patterns and the mammalian extinction peak at the end of the last glaciation. I had assumed that this ecological phenomenon was unique to the north, but now I feel that the same sorts of events (with local variations) occurred throughout the Holarctic and in other parts of the world at about the same time.

The peculiar mixtures of fauna and flora in Pleistocene deposits were once explained simply as a periglacial phenomenon: local ice conditions creating a uniquely cool environment far south where biota from several biomes could mix. It has become apparent, however, that peculiar mixtures of floral and faunal elements occurred not only in central European and central North American periglacial regions, but also in far northern glacial refugia and thousands of kilometers south of the glacial margins (e.g. Davis 1976). Most of today's familiar biotas are without Pleistocene analogue. In a relatively short time many species became extinct and there was a resorting of biota on a global scale.

The difficulty most paleoecologists have had with extinction theories based on climatically induced events is their unsatisfactory ambiguity and diffuseness in the proximate coup. Several competing theories involve young born out of season, cold weather, drought, decline in food resources; some even invoke all of these (Axelrod 1967). But if climatically induced ecological changes are responsible for such a major phenomenon, a central biotic explanation is likely.

In addition to a new theory accounting for the Pleistocene extinctions, I present a model for the network of ecological events at the late Pleistocene—early Holocene boundary, events of which the megafaunal extinctions were but a part.

The Model—In a Nutshell

The ecological model of late Pleistocene events proposed here is divided into three headings:

1. Climatic changes in seasonal regimes decreased diversity, increased zonation of plant communities, and caused a shift in net antiherbivory defense strategies.
2. This change in the plant community resulted in a shorter, less diverse growing season for ungulates, decreasing net annual quality and quantity of resources available to many large mammal species.
3. These restrictions in available resources decreased local faunal diversity, body size, distributional ranges, and frequently resulted in extinctions.

The changes affecting the late Pleistocene extinctions have roots deep in time. Paleoecologists generally agree that the almost worldwide closed-canopy forests of the early Tertiary were slowly (albeit irregularly) dismantled into more open, herb-dominated communities (savannas and steppes) in a trend that has continued to the present (see Wolfe 1975, Kennett and Shackleton 1976, Webb 1977). The exact climatic causes for this trend are uncertain, primarily because the forces responsible for maintaining steppes and savannas are themselves debatable. Though paleoecologists might argue about proximate mechanisms behind this trend, I think most agree that it has to do with increasing seasonality, particularly the seasonal distribution of moisture throughout the year (Sinclair 1975, Andrews and Van Couvering 1975). There has also been general cooling throughout the Tertiary as well.

We limit ourselves here to the nature of the biotic repercussions of the climatic events. Concerning large mammals, it is most important to focus on the "open ground" communities of savannas and steppes because the great diversity of ungulates is primarily a matter of adaptations to these environments. Harris (1978) considered the gradient of vegetation zones from evergreen broadleaf forests to desert shrub and concluded that "throughout the tropics the climatic variable that most comprehensively affects plant growth is the intra-annual variability expressed in terms of the length of the dry season." Actually, in more poleward regions, cool seasons could be substituted for (or added to) dry seasons, and one could then generalize that vegetation provinces characteristically reflect the *length and variability of the annual growing season*. The growth season for plants is defined by a combination of temperature, soil fertility (nutrient availability), and moisture. For other biota, the length of growth season is more complex.

The climatic changes occurring throughout the Cenozoic can be viewed biotically as a trend toward decreasing length of the growth seasons (and the intra-annual variability within those growth seasons). At first this trend to shorter growth seasons promoted vegetational mosaics and, as it continued, increased vegetation zonation. Accompanying this zonation (Leopold 1967) was a decrease in local floristic diversity —not simply a reduction in numbers of species, but declines in diversity of growth habit and diversity of antiherbivory defenses and other features.

Vegetational changes during the Cenozoic which affected ungulates include the opening worldwide of a diverse forest cover due to the lengthening of the season unfavorable to plant growth. Shrubs increased, taking the place of trees. Herbs and other plants without investments in woody tissue evolved to accommodate the shorter growing season. As the trend continued, large land areas became covered with a vegetation mosaic of trees, shrubs, and herbs. Because large ungulates cannot digest much lignin or woody tissue, nor can they climb trees to reach more nutritious foliage, they were (and are) infrequent in dense forests. In the Tertiary large ungulates increased in diversity and biomass with the spread of shrubs, herbs, and trees. The increasing volume and diversity of plants suitable for ungulates reached a zenith sometime during the late Miocene and Pliocene when ungulates reach unbelievable diversity. The quantitatively expensive antiherbivory defenses of the long-growing-season plants yielded to an increasing dominance of the qualitatively cheaper compounds found in plants with a brief growing season but rich substrate (plants with a less conservative life history). This also favored ungulates, particularly ruminants.

As this climatic and vegetational trend continued, it moved past an "ungulate optimum." As the growth season shortened, reductions occurred in the diversity from the ungulates' viewpoint; as the open-ground vegetation patterns began to simplify, so did ungulate diversity.

A more homogeneous vegetation array increased ungulate competition and resulted in the elimination of some species (through geographic displacement and extinctions). As ungulate diversity declined, the vegetation biomass, quality, or both, increased for a few species and these particular species experienced an increase in body *quality*: large body size and large social organ size. The timing of the peak in quality varied; for many species (mammoths, bison, horses, etc.) it occurred during the mid-Pleistocene.* But the increasing trend toward shorter growth seasons and the

---

*This is a familiar biotic phenomenon of the value peak being somewhere in the middle of a scale rather than at the ends. Grime (1979) calls this a "humpbacked relationship" where plant species diversity increases in relation to increasing environmental stress to a point, then abruptly begins to decline.

resulting plant zonation and reduced resources eventually caught up with these giants and they experienced a decline in body size. This brings us to the end of the Pleistocene.

I argue that the Pleistocene-Holocene change was an unparalleled jolt across a threshold toward more constricted growth seasons which sharply reduced resources available for ungulate growth. This shift resulted in major repercussions:

1. Many complex plant associations, vegetational communities, and even biomes were fractionated into homogeneous zones at the beginning of the postglacial.
2. As a result of decreased vegetational mosaics (local diversity) and increased vegetational zonation, large mammal biomass declined.
3. Most large mammals experienced postglacial dwarfing.
4. Related to this dwarfing was a decrease in size of antlers, tusks, horns, and other social weaponry.
5. Some mammalian species experienced an increase in numbers and major range expansions, presumably invading portions of newly vacated niches (e.g. bison).
6. Many animal species that occurred together in the Pleistocene no longer overlap in their distribution; allopatric distributions increased.
7. As a result, local species diversity generally decreased.
8. Some animal species not related to the megafauna and its predators became extinct (e.g. birds).
9. Many large and small mammal species underwent range reduction, in effect regional extinction.
10. Due to earlier adaptations to plant diversity and antiherbivory defenses, some ungulate groups were affected by vegetational changes more than others.

The direct result of Pleistocene climatic changes toward shorter, less variable (intra-annual) growth seasons was an increase in local plant competition and a geographic decline in the range of climatic tolerance for individual plant species. Each plant species became more restricted to local habitats nearer its climatic optimum where its competitive edge was keenest, resulting in a marked vegetational zonation and locally reduced species diversity. Since local herbivore diversity depends in part on vegetational diversity, of both species and growth form, herbivore diversity was also reduced.

It is axiomatic that two sympatric species never use all the same resources in exactly the same way. Such competition would result in the demise of one or both. As ecological evidence accumulates, we are better able to understand how the living large mammals partition their food resources (e.g. Jarman 1974; Bell 1969; Janis 1975; McNaughton 1976, 1978; Tieszen et al. 1979; Jewell 1974; Child and Richter 1969; Ferrar and Walker 1974). When living sympatrically, each ungulate often has its special plant form or part on which it can do better than its competitors. Complex altitudinal, successional, habitat, and seasonal dimensions must be added to these interactions. Generally, however, more monotonous herbaceous array and growth habitat reduces the potential for herbivore species diversity, and a high vegetational diversity allows for higher herbivore diversity. The decline of local vegetational heterogeneity and the new dominance of extensive areas by a few plant species will cause contraction of herbivore ranges and many local extinctions. The sum of many local extinctions added up to taxonomic loss of large herbivores and their predators. A few Holarctic megafaunal species expanded in numbers to dominate some of the new plant zones: caribou-reindeer (*Rangifer*) in the northern tundras, moose (*Alces*) in the boreal forest, deer (*Odocoileus* and *Capreolus*) in the deciduous forest, bison (*Bison*) and cattle (*Bos*) in the parklands and shortgrass prairies and plains, red deer or wapiti (*Cervus*) as a "chink-filler" between these species, and brown bear (*Ursus arctos*) as a predator–scavenger–berry eater.

I don't mean to imply that these changes affected only ungulate groups. Grayson (1977) and Graham and Lundelius (this volume) have reviewed some of the evidence that the effects were widespread; however, the impact of the changes is spectacular among the ungulates. This group had radiated to take advantage of the mosaic vegetation of the various Pleistocene "savannas." Some groups and some areas were particularly hard hit; those which relied on a great diversity of plant forms in their diet, such as the large monogastric caecalids—the mastodont, mammoth, rhinos, large edentates, asses, and horses. These groups were adapted to eating high fiber/low protein plants and plant parts that do not rely on the more toxic antiherbivory defenses. Low percentages of milk protein, slow growth rate, later reproductive maturity, longer gestation periods, and other aspects of a slow growth pattern allowed these conservative herbivores to thrive on the low-protein end of the range resource spectrum. These species did not possess the large ruminant's ability to degrade a wide array of toxins within the gut or to synthesize most of the essential amino acids, fatty acids, and vitamins. Rather, they depended on a relatively nutrient-deficient, but abundant, energy staple supplemented by vegetational odds and ends to provide a balanced nutritional mix. Without the appropriate local vegetational diversity to provide this mix the more conservative species were outcompeted. These caecalid-monogastric ungulates were hardest hit in regions which already had a moderately short growth season (e.g. temperate and arctic environments).

Plaids vs. Stripes: Postglacial Zonation and the Decline in Local Diversity

It has been assumed that by studying a local Pleistocene fauna and flora one could easily reconstruct the climate from modern analogues. But, increasingly, many Quaternary paleontologists are frustrated in their attempts to relocate past distributions of contemporary biotic provinces. The literature is replete with references to "disharmonious species mixtures" (Graham 1976, Graham and Lundelius, this volume), "defunct species associations" (Slaughter 1967), Pleistocene communities "without modern or extant counterparts" (Matthews 1979), no "modern analogues," or "no equivalent." Neither glacial nor interglacial age Pleistocene biotic provinces closely match those which occurred during the Holocene. Obviously we have been asking the wrong question. It is not "Where were today's biotic provinces during the late Pleistocene?", but rather "Why were the late Pleistocene biotic provinces so different from those today?" The postglacial and modern communities are highly peculiar derivatives of the late Pleistocene.

A review of some evidence for the diversity of Holarctic Pleistocene communities and their characteristic quality of nonequivalency with modern biomes provides key elements in understanding Pleistocene seasonality not available from other data. It is the biota, rather than physical evidence, which is particularly sensitive to seasonal shifts.

Most paleoecologists argue that biotic evidence from the far north indicates a complex mixture of tundra and steppe elements with no modern analogue (Kowalski 1967, Guthrie 1968, Matthews 1976). They envision a vast "mammoth steppe" (Guthrie 1968, 1980) reaching from England to northwestern Canada, with an amazingly diverse ungulate assemblage dominated by grazers. This habitat predominated as the norm throughout most of the Pleistocene but was interrupted by the brief interglacials which acounted for about 10 percent of the Pleistocene (Davis 1976).

Matthews (1975) has shown that insect species now found in the High Arctic of North America existed sympatrically during the last glacial with species now found only far to the south. Vereshchagin (1977) reviews a similar phenomenon in the Berelekh (high Siberian arctic) mammoth cemetery which dates between 11,000 and 13,000 yr B.P. There were mixtures of today's beetle species from tundra, forest, mountain tundra, aquatic areas, cold steppe and arid Mongolian habitats far to the south. The same is true for plant species in Siberia, judging from macrofossils (Tikhomorov 1958).

From pollen studies Giterman (1975) concluded that a woodless vegetation of a pronounced complex structure existed (i.e. "that mesophytic and xerophytic are distributed in common"). Many of the mammalian species in Alaska common in glacial sediments are now found only far to the south, [e.g. badgers (*Taxidea*), wapiti (*Cervus*), and ferrets (*Mustela*)]. However, species of animals now restricted to the extreme north coast of Alaska and Canada—like musk oxen (*Ovibos*) and lemmings (*Dicrostonyx*—were present much farther south throughout unglaciated Alaska and the periglacial border into Idaho and Pennsylvania.

The taiga, which now occupies the glaciated area, existed during the glacials in midcontinent North America, but in a peculiar community which Davis (1976) refers to as "different than any known today."

In Pleistocene fossil localities in Kansas, Hibbard (1960) observed mixtures of typically boreal and southern mammals, and concluded that "present day climates with their seasonal extremes of temperature and aridity are geologically atypical, even in the Pleistocene." Others have reported strange mixtures of tapir (*Tapirus*) and jaguar (*Panthera*) with caribou (*Rangifer*) in Tennessee (Guilday et al. 1978). Hoffman and Jones (1970) discuss the relict populations of eastern mammals in the west, indicating east-west mixtures during the last glacial of faunas now separate.

Taylor (1965), working with mollusks from the midcontinent, deduced that all previous interglacials had mild, frost-free winters and cooler summers. He concluded that the late Wisconsin and early postglacial were characterized by aridity and strong seasonal contrasts. Ashworth (1977) describes the late Wisconsin beetle assemblages from the Great Plains as "a mixture of faunal elements which has no modern analogue." For example, species now western were then 2,500 kilometers east of their present limits. Holman (1977), investigating the Pleistocene distributions of herpetofaunas, found that most fossil assemblages contain forms climatically incompatible today.

Graham and Semken (1976) discuss the characteristic sympatric association of three "phena" of *Blarina brevicauda* during the Wisconsin, which are now allopatric. They commented on the increased diversity of midcontinental glacial faunas due to decreased seasonal extremes. Graham (1976), in a study of Wisconsin-age small mammal communities, concludes that the "relative frequencies of boreal deciduous and steppe species are more equal in late Wisconsin faunas than in modern faunas which are predominantly composed of only one group." The same conclusion can be drawn about large mammal communities.

In the midcontinent, botanical evidence indicates that vast open plains are a postglacial event (Wells 1970). The separation of short, mid, and tallgrasses into vertical belts seems to be unique to the postglacial. Earlier interglacials and glacials reveal a fauna and flora more like that of a heterogeneous savanna. Anomalous mixtures of pollen have been interpreted as indicative of a more mosaic floral environment (Mehringer et al. 1970, Webb and Bryson 1972, and Wright 1970). Chomko (1982) describes a fauna from the northern Bighorn Mountains in Wyoming that existed before 13,500 B.P., consisting of seventeen genera of small mammals, "many of which are today allopatric."

Spaulding (1978), studying the late Pleistocene history of the vegetation in southern Nevada, argues that the Pleistocene record contains no analogues of modern vegetation. He further concludes that the modern communities in his study area "appear to be unique to the last 10,000 years." Likewise, in the Southwest, Van Devender (1976) and Cole (1982a,b) found unusual mixtures of plants "with no modern counterparts" in packrat middens. Slaughter (1967) talks about the common "disjunct" plant associations in the Southwest without modern counterpart.

Although we normally think of the American southwestern deserts as having a long history in something close to their present form, these desert assemblages seem to be Holocene products. Wells (1977) comments, "As recently as 8000 to 12,000 B.P., the deserts as we know them today either did not exist or were so drastically restricted as to elude detection thus far." Many of the "desertic" faunal and floral species have been identified in the fossil record, but they appear as parts of more mosaic Pleistocene communities. This observation is primarily based on a series of studies by Van Devender (1976, 1978) and colleagues (Van Devender et al. 1977) who found anomalous mammalian, herpetofaunal, and plant assemblages in many fossil packrat (*Neotoma*) middens. They conclude that late Pleistocene communities were complex mixtures of species not separated into the communities seen today. Cole (1982b) deduces from *Neotoma* middens in the western Sierra Nevada that the present-day Mediterranean climatic regime at the altitudes for which he has data is a Holocene event.

Guilday et al. (1978) have discussed the faunal disharmony in the Northeast as being a strange mixture of arctic and southern elements. This disharmony led them to question our current concepts of biome integrity. They found deposits at Welsh Cave, Kentucky, dated 12,950 yr B.P., that contain badger (*Taxidea taxus*), thirteen-lined ground squirrel (*Spermophilus tridecemlineatus*), mammoth (*Mammuthus* sp.), horse (*Equus*), and grizzly (*Ursus arctos*), combined with boreal forest rodents and insectivores. These diverse species suggest a prairie savanna or parkland in what is natively now a closed deciduous forest (Guilday et al. 1978). Likewise, Schwert and Morgan (1980) concluded from late glacial insect assemblages from northwestern New York that these environmental conditions may have no modern analogue in North America today.

For the Southeast Aufenberg and Milstead (1965) state "The available record of Pleistocene reptiles in Florida indicates that this area possessed a sufficiently equable climate so that many cold-sensitive types existed there throughout the Pleistocene until the Wisconsin, perhaps even until near its closing." Webb (1977), in his review of the paleoecology of southeastern North America, doubts that we have ever before experienced the current degree of continentality. He proposes that this continentality is responsible for the extinction of the broad Pleistocene savannas and the heterogeneous fauna of that area.

Late glacial deposits from Ladds Quarry, Bartow County, Georgia, also exhibit disharmonious mixtures (Holman 1976, Ray 1967, Wetmore 1967) of what are now northern species [spruce grouse (*Canachites canadensis*), masked shrew (*Sorex cinereus*), fisher (*Martes pennanti*), and bog lemming (*Synaptomys cooperi*)], mixed with what are now southern species [cotton rat (*Sigmodon hispidus*), round-tailed muskrat (*Neofiber alleni*), and hog-nosed skunk (*Conepatus leuconotus*)]. Holman (1976) refers to these as "climatically incompatible."

The British Isles are informative because their paleoecology is so well studied. Sparks and West (1972), discussing plants, state that "a redistribution of plant species occurred to produce combinations which we no longer know today." And about the climate, they write, "There are, in fact, no exact homologues on the surface of the earth today of the climates Britain experienced in the cold glacials. . . . The character of the stadial floras is an intriguing one. We do not see a flora of the type that is now characteristic of the northern tundra. Instead we see a flora of much more diverse phytographic affinities." They identified Pleistocene assemblages of species found today only in diverse environments (i.e. in mountains, maritime, far northern regions, continental, and even Mediterranean situations). Some plant species present in Europe throughout the Pleistocene became extinct at the beginning of the Holocene, and are now found only in east Asia and North America. Mayhew (1977) has shown that the more complex faunal assemblages of the Pleistocene were not simply a north-south mixing of extant forms, but also had east-west components. For example, the extant ground squirrels (*Spermophilus major*) now confined to central Asia, had a range extending as far west as England during the last glaciation. Likewise Coope (1974) has identified from the late Pleistocene of England a beetle which now lives in Tibet.

From interstadial sediments Coope and Angus (1975) showed strange mixtures of insects associated with a temperate climate regime and "cold adapted" mammoth fauna.

In Europe, as in North America, the strange assemblages of arctic, heliophytic, and thermophilous plants occur together in late glacial deposits (Iversen 1954). Reid (1915) was the first to document the broad Pleistocene overlap of various plant species which do not now overlap in their distributions. This led Leopold (1967) to conclude that the glacial habitat and climate has "no modern counterpart." This disharmony extends also to the vertebrate fossils. Kowalski (1967) discussed the peculiar mixture of the arctic collared lemming (*Dicrostonyx*) with the hamster of the southern steppes (*Cricetulus*) in central Europe. The African hyena (*Crocuta crocuta*) roamed the British Isles together with arctic reindeer (*Rangifer tarandus*), and leopards (*Panthera pardus*) were associated with musk oxen (*Ovibos*) in southwestern Europe. The Mediterranean Basin was occupied by a megafauna now confined to the southern African savannas. And these were mixed with native fauna now considered desert species (Axelrod 1967). Likewise, during the Würm, there were species in North Africa that are now considered northern in their adaptations [e.g. bears (*Ursus*), marten (*Martes*), and others]. Uerpmann (1981), describing the unusual faunal association of the Levant in the eastern Mediterranean, says "it is difficult to imagine the environmental conditions which made these faunal conditions possible. They certainly do not exist in any recent type of environment." Also Suc (1976) talks of a strange Pleistocene fauna in France, living in two concurrent extremes, *"un mileu plus sec et un mileu plus humide."*

In eastern Europe Bader (1978) discusses the environment of the Sungir Site as a curious Upper Paleolithic mixture of anomalous "temperate" floral assemblages and "cold-loving" faunal assemblages with no modern counterpart. Further east Tsukada and Sugita (1982), working in Japan, find peculiar Pleistocene plant combinations and propose that the vegetation patterns cannot be interpreted with zonal integrity and that the present condition is a Holocene event.

Although these major environmental changes have been identified throughout the Holarctic, recent evidence suggests that they were global. Salgadov-Labourau (1980) provides palynological evidence that the Amazon Basin was without rainforest during the last glacial. These data and others suggest "some considerable variety or mosaic in the ancient landscape allowing genera characteristic of rain forests to persist alongside savanna plants (Colinvaux 1980). Campbell (1973) documents a well-watered savanna in the late Pleistocene of eastern Peru that is now desert.

What does this outstanding diversity of Pleistocene biotas mean? One problem is that our concepts of woodlands, grasslands, savanna, tundra, etc., based on current biomes, simply do not fit the Pleistocene. Webb (1977, 1978) proposed that the Pleistocene was characterized by "savanna" vegetation patterns. But the complex mixtures of biota present in Pleistocene deposits indicate more than savanna—they indicate a diverse and complex association of steppe species *and* a complex association of woody plants *and* even more complex mosaics of the two. The latter mosaic association we recognize under the broad category of savannas, but there is evidence of Holocene community simplification or zonation across the entire spectrum of plant and animal associations. The Holocene zonation was not only a decrease in savannas, but also a reduction in the diversity of all growth forms among both herbaceous and woody species.

I argue that this Pleistocene species diversity is primarily the character of the growing season combined with the kinds, degrees, and variations of seasonal stress. Given a long and internally varied growth season, plant competition can result in temporal or phenological species displacement rather than spatial elimination. Thus a long and varied growth season on a productive substrate increases "sympatric tolerance." This diversity normally occurs in a moderate to fine-grain mosaic. The complex vegetative mosaic is then reflected in the animals: insect species emerge and reproduce in a sequential pattern, tracking flowering and seed production. The diversity is seen among birds and mammals which also experience more heterogeneous associations.

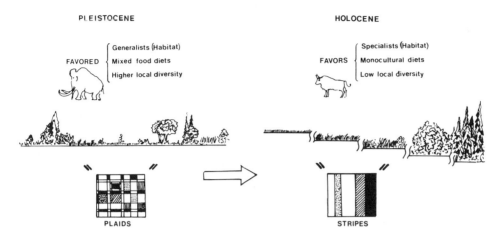

Figure 13.1. The longer, more intra-annually variable growth season of the Pleistocene favored a higher local diversity of plant species which favored the large herbivore species that used mixed food diets and made use of a broad range of habitats. The switch to the shorter, less varied growing seasons of the Holocene resulted in a greater biotic zonation, favoring simplified faunal assemblages that are more precisely adapted to a specific habitat and that can thrive on a more monocultural diet.

It is quite probable that the Pleistocene Holarctic soils were more fertile than those of today (Guthrie 1982). The longer growth seasons during the last half of the Tertiary and the Pleistocene seem to have been internally varied as well (Guthrie 1982). It is probably no coincidence that the remnants of the earlier Quaternary large mammal fauna live in special, more equatorial conditions with long, internally varied, growth seasons. The high Serengeti savanna is, of course, a remnant of the Tertiary pattern of long, varied, growth seasons. The high altitude East African savannas have two sharply defined wet seasons and two dry seasons each year. The Holocene relicts of the once widely distributed Pleistocene northern equids, American camelids, peccaries, tapirs, proboscidians, lions, hyenas, and a number of other species, are now found far to the south or on high alpine plateaus in temperate areas where the growth seasons are comparatively long.

The Holocene's less fertile soil and a less varied and shortened growth season increased plant competition. The resulting competitive displacement sorted species more geographically—a shift from plaids to stripes (fig. 13.1). The shortened growth season heightened the competitive edge of some species. In a local area some species could exclude others that could potentially grow in the same circumstances but could not survive the new competitive pressure.

Late Glacial–Holocene Megafaunal Dwarfing:
Its Ecological Significance

At certain times of the year, known as growth dormancy periods, wild animals generally do not grow (Wood et al. 1962). Somatic growth* in any one year is limited to a "growth season." Generally reduced growth rate is the result of diminished resources

---

*By somatic growth I refer to the body frame which is mainly protein—connective tissue, muscle, nerves, skin, bones, etc. Energy (i.e. carbohydrates and fat resources in the diet) may greatly affect body dimensions and weight, but our main concern here is with the body frame size.

(Lodge and Lamming 1967) which can be due either to a suboptimal volume of food, suboptimal nutrient levels, or both. A seasonal peak occurs when both volume and nutrient level of available forage exceeds the physiological capacity of the herbivores who feed upon it. In a seasonal environment this peak occurs sometime during the height of new plant growth. The duration of this peak, marked by the high quality of growth resources, can vary widely. I argue that it is the duration of this high quality peak in growth resources which is critical in dwarfing, not some abstract yearly energy average or maximum or minimum quality.

The volume of the forage is limited in the early spring, but after the peak of plant growth, volume is usually not critical. At this point it is the deteriorating nutrient quality, particularly the protein content of the forage, along with the increasing levels of plant defenses (fiber, tannin, resins, etc.) which limit the somatic growth of herbivores. Young plant shoots generally contain the lowest levels of such antiherbivory compounds. At this same time the main protein components, phosphorus and nitrogen, and the easily digestible energy sources, the soluble carbohydrates or easily fermentable fibers, are at the highest levels. As plant growth continues, the quantity of available nutrients and easily digestible carbohydrates decline while fiber, minerals, and antiherbivory compounds increase. Thus a mature plant may look healthy and green, but as forage it is usually poor in quality for a large mammal's somatic growth. Plants well past maturity usually contain levels of available protein far below optimum for continued ungulate growth and reproduction (fig. 13.2).

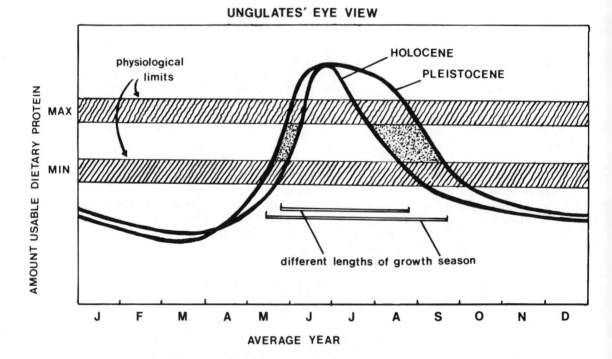

Figure 13.2. A model of the average "protein year" as seen by average Pleistocene and Holocene ungulates. The amount of usable plant protein available to an herbivore has a striking seasonal cycle, normally falling well below the animal's requirements for growth, reproduction, and even maintenance levels for part of the year, then rising above its physiological capacity to make full use of the protein. Evidence from the biota points to a longer seasonal peak in this protein availability during the Pleistocene.

Thus the peak of quality forage occurs early in the spring-summer growth season when new plant growth is most abundant. I stress the importance of usable protein levels because it is surely the most limiting factor to somatic growth (White 1978, Mattson 1980). Later in summer the mature forage is rich in carbohydrates (energy convertible to fat) but low in the protein necessary for continued somatic growth. Thus, most Holarctic large mammals postpone fat accumulation until late summer at which time somatic growth has slowed or ceased.

The somatic growth season for ungulates differs considerably from that of plants. I refer to the ungulate growth season as the average period of the year in which the animals are near their maximum potential for daily somatic growth. Conceptually this is a workable unit, although, like the "plant growth season," the quantification has somewhat fuzzy edges.

Although the matter of ungulate growth season is critical to body size, it has not been widely recognized and cannot be found in an atlas of physiographic zones. For example, it is often possible for northern ungulates to have a greater average net utilizable growth resource than their kin to the south, and this usually occurs. I think the "growth season" better accounts for Bergmann's law than any other factor. Klein (1970a, b) points out that plants in the north have higher protein levels as a result of adaptations for more rapid growth. Northern plants are also more nutritious due to the absence of nighttime catabolic periods. Additionally phenological variations are greater in varied topography due to the low sun angle in the far north. Northern ungulate species can move about in this phenologically varied vegetative landscape and piece together a rich diet for a relatively long growth season.

Virtually all large mammal herbivores have adapted to maximizing the seasonal length of nutrient availability by several strategies: (1) migrating topographically to catch the same species of plants in early stages of growth (as the phenological stages vary with topography); (2) moving within the same topographic area from one plant community to another in a patchy vegetational environment; (3) foraging on a phenological succession of young plants as they appear in the same location over the growth season; (4) selecting among species, morphs within populations, and plant parts least defended by antiherbivory devices (or selecting those plants and parts with defenses one is phylogenetically adapted to overcome) and thus have higher levels of available protein and carbohydrates; (5) selectively mixing species high in utilizable protein but with specific toxic chemical defenses, thereby keeping the general level of any one toxin low, which ultimately results in (6) grazing varied assortments of plants that will have complementary essential nutrients.

Only rarely does the volume of plant resources limit herbivore growth during the comparatively short seasonal resource peak. More commonly there is a surplus of suitable forage, and somatic growth is curtailed and competition comes into play at the beginning and end of the seasonal resource peak. Competition and quantity of forage resources is usually extremely important on winter or dry-season ranges (Mattson 1980). The growth season can be squeezed by intraspecific competition for the high quality plant resources, and it can be affected by changes in the timing and assortment of high quality plant growth.

When diversity of local plant species decreases or plant communities of a mosaic nature separate into broad geographic units characterized by more homogenous growth forms, these changes affect ungulates in several ways. Homogenization of forage reduces the ungulate's ability to piece together a diet with high levels of nutrients for a long period. With increased plant zonation, the nutrient "pulse" in any one plant community is simplified and shortened. Species that can tolerate a shorter growing season thrive, but even they may experience distributional range restrictions because of interspecific competition in plant communities to which they are less well adapted. Species adapted to a longer growth season begin to decline in body size and become more juvenile in appearance (e.g. reduction of adult social organs).

Also, with increasing zonation plants are likely to have similar classes of antiherbivory defenses. This new situation will be detrimental for an herbivore which requires dietary diversity to balance essential nutrients and to dilute the toxic secondary plant substances. For this mixed feeder much of the new array of vegetation may be unusable. Instead of taking bites from a local diverse flora, it must move farther away to find the proper selection of plants.

More spectacular than the general postglacial decline in body size was a reduction in size of some body parts. Social organs (antlers, horns, and other parts of animal dress) are particularly susceptible to reduced nutrient regimens (both developmentally and phylogenetically). Geist (1977) refers to these as "organs of low growth priority." Wildlife managers use social organs as ready indices of the quality of game ranges. The large and well-developed social organs of Pleistocene animals are circumstantial evidence of the large quantities of nutrient resources that were available for allocation to reproduction (acquisition of high stature for mating privileges). If the shortened growth seasons at the glacial-postglacial boundary reduced the net amount of available protein, resources for survival would take evolutionary precedent over the more costly devices for early or expanded reproduction in any one season.

The reason small herbivorous mammals did not necessarily* experience a Holocene decline in body size is that it takes small herbivores less than one "growth season" to reach mature size, so a shortened growth season affects only the last litter(s) of the season; also, other selection vectors are probably more directly important in determining small-mammal body size than nutrients (i.e. competition, predation, etc.); a scale phenomenon of very small home ranges means that these herbivores were already adapted to a comparatively simple environment (at least simplicity on a different scale); and by being smaller their scale of dietary selectivity is different from that of a large mammal.

The point at which the length of time to maturity is less than a growth season seems to change with mammals about the size of a hare or fox, and this seems to be the point in the size continuum where Holocene dwarfing did not occur or occurred less dramatically. Holocene dwarfing among carnivores is more complicated than that of herbivores, but it did occur (Kurtén 1968).

The occurrence of this Holocene dwarfing has been known for some time, but only in the last few years has it been recognized as a general evolutionary trend (Kurtén 1968, Marshall and Corruccini 1978). Specific examples are common in the literature (Kurtén 1959, 1965a, 1965b, 1965c, 1968; Hooijer 1950, 1977; Schultz et al. 1972; Harris and Mundel 1974; Wetzel 1977; Wen-Chung 1963; Davis 1981).

Dwarfing occurred not only in those species which survived but also in those that became extinct. Virtually all mammoths associated with Clovis points in the New World are diminutive and have reduced tusks. Saunders (1978) has plausibly interpreted this as an indication that these animals were all females, however, I contend that the general evidence suggests dwarfing instead. In Europe, Asia, and North America new late-glacial sites are revealing dwarfed mammoth. The spectacular Gonnersdorf open-air site in western Germany (Poplin 1976), which dates between 11,000 and 12,000 yr B.P., contains both bones and drawings on slate of dwarfed woolly rhino (*Coelodonta*), dwarfed woolly mammoth (*Mammuthus*), and dwarfed horses (*Equus*). Likewise the Berelekh mammoth graveyard from northern Siberia (Vereshchagin 1977) has the bones of small mammoth bulls with such vestigial tusks that they were first described as females. At the same site dwarfed bison also appear. In Alaska we have dated wood found immediately below a small mammoth jaw at just under 13,000 yr B.P. Russian workers (e.g. Sher 1967, Garutt 1964, Vereshchagin 1977) have documented a general size reduction in the extinct megafauna in Siberia at the end of the last glacial.

---

*Tchernov (1981) does show minor size declines in the Holocene small mammals of the Middle East.

I have compared (Guthrie, unpublished Dry Creek report) the Rancholabrean-aged fauna from Central Alaska with their living descendants. All have undergone size reductions: moose (*Alces*), musk oxen (*Ovibos*), caribou (*Rangifer*), and Dall sheep (*Ovis*). From a study in progress of the Dry Creek archaeological site we know that this size reduction was taking place sometime around 11,000 to 10,000 yr B.P.

My studies of horn growth on Dall sheep taken from the wild show that, like Dall sheep in the wild, they exhibit obligate growth dormancy throughout the winter, even when they are fed food of the highest quality. During the peak season of late spring/early summer growth, their horn growth rate does not increase over that of wild sheep. The diet of the Dall sheep in the wild is extremely rich and abundant during this peak growth season (Whitten 1975, Winters 1980). Thus I have not been able to produce a change in the maximum growth rate or in the growth dormancy period. However, by supplementing the diet of the captive sheep so that they are continually maintained at the high quality level (ca. 20 percent protein) of the peak growth season, these captive sheep do grow for a *longer period of each each summer* than their counterparts on the wild range. They are much larger than the sheep in the wild population from which they were taken, and the captive sheep are almost as large as Pleistocene Dall sheep.

Dwarfing recorded at the end of the Pleistocene does not seem to be due to decreasing the number of growth years (the survivorship curves do not show this, and young age cohorts of Pleistocene species are larger bodied than their living counterparts). Growth rates are difficult to reconstruct, but evidence inferred from living large mammals suggests that growth rates of the large Pleistocene mammals may have been only slightly greater than those of modern animals. The rate difference would account for some dwarfing, but information from living species suggests that the key to the dwarfing is the length of the ungulate growth season—the duration of the seasonal peak in usable high quality forage.

Late Glacial Regional Extinctions

The major *regional* extinctions of extant species occurred at the same time as the *blanket* or geologic extinctions of mammoths and others, that is, at the beginning of the Holocene. This simultaneity suggests a common cause.

Extant species that experienced broad regional extinctions, such as wapiti (*Cervus*), bison, (*Bison*), and horse (*Equus*), are common in late glacial sediments throughout unglaciated Siberia and Alaska, thousands of kilometers north of their historic relatives. Restocking has been either inappropriate, unsuccessful, or limited to confined habitats of a few square miles. Alaska has virtually no natural grazing lands with available winter forage for these species. The exposed alpine winter ranges are already occupied by mountain sheep (*Ovis*), and even they were more widespread during the glacials; however, the habitat does not exist for restocking mountain sheep in most of their Pleistocene ranges. The same could be said for proposals to reintroduce tapirs (*Tapirus*) and caribou (*Rangifer*) in the Ozarks of the central United States.

These regional large mammal extinctions seem to have occurred because of habitat reductions—the change in character of plant communities. Diverse, geographically extensive habitats disintegrated into regionally uniform communities, which changed the competitive balance. Although insects and small mammals underwent few blanket extinctions, they certainly experienced major regional extinctions. Schwert and Ashworth (1982), for example, argue that the Arctic-Subarctic beetles south of the continental ice sheet became extinct at the end of the last glaciation. The areas in the northeastern and north-central part of the continent were later recolonized from the Alaskan-Yukon refugium.

The best organisms for studying regionalized extinctions and their causes are probably small mammals. Few documented blanket extinctions of small mammals exist, but considerable distributional changes occurred during the glacial-postglacial transition.

These postglacial changes usually involve decreases in distributional extent, decreases in local diversity, and large scale geographic separation of species (many species sympatric in the Pleistocene local faunas are now allopatric).

Traveling from south to north in eastern North America one crosses the distributional margins of several small mammals. Guilday et al. (1978) showed that this was not the case in the late glacial; most local faunas in a wide belt from Pennsylvania to Tennessee included the same small mammal species. This led them to agree with Hibbard's (1960) interpretation of heightened seasonality during the postglacial. It also points to increasing north-south zonation of habitat.

Semken (1974) cites the Howard Ranch local fauna in Texas (Dalquest 1965) as containing the boreal watershrew (*Sorex palustrus*), the southern cotton rat (*Sigmodon hispidus*), and the western prairie dog (*Cynomys ludovicianus*), all allopatric today. Graham's (1976) study of the vole and shrew diversity of Pleistocene localities in the eastern United States reveals the same phenomenon as in the West—that the Pleistocene small mammal diversity was universally greater than it is today. I found the case to be the same in central Alaska (Guthrie 1968), and others have also remarked on this in Europe (e.g. Kowalski 1967). Graham (1976) concludes that Pleistocene faunas were "mixtures of elements from several biomes and were not characteristic of expanded or contracted modern biomes."

One of the dramatic postglacial distributional changes is that of the yellow- or chestnut-cheeked vole (*Microtus xanthognathus*), which is now confined to the far north. Hallbert et al. (1974) argued that the community composition in which the chestnut-cheeked vole is found in the Wisconsin-aged deposits of the eastern United States is indicative of a type of parkland with no modern analogue. Harington (1977) agrees and proposes that it is in this peculiar parkland that the extinct musk oxen *Symbos cavifrons* was most heavily concentrated.

Lundelius (1967, 1974) and Graham and Lundelius (this volume) have championed the postglacial climatic equability decrease. Arguing from the perspective of the general decrease in both small and large mammal species diversity and sympatry Lundelius says "ecotones, which are presumably more patchy than the major biotic provinces they divide, were of much wider extent during the Pleistocene than they are now. The fact that fossil assemblages of this kind are very widespread in North America suggests that the major biotic communities of the Pleistocene were more like modern ecotones in their habitat diversity."

Ecology of the Blanket Extinctions

It is a well-known axiom among range ecologists (e.g. Coustan 1972) that because a community of ungulate species must share dietary and habitat resources, they experience different life histories and their susceptibility to environmental change is quite different. This is most apparent in the late Pleistocene extinctions. Although most large mammal species experienced reductions in body size and distributional range, the blanket extinctions occured within some ungulate groups more than others. To see how this might have happened, we should review the dietary adaptations of various ungulate groups.

Most ruminant artiodactyls can tolerate the chemical antiherbivory defenses of dicotyledonous plants because they degrade them through "composting" in the rumen (Janis 1975). Small ruminants are highly selective and choose mostly young dicot leaves. These, more so than monocots, have a high percentage of soluble carbohydrates and easily fermentable fibers (hemicellulose). These small ruminants do not have as well-developed rumen retention structures as do the large grazing ruminants (Hoppe 1978). Because of the limited amount of food in the small rumen at any one time, these smaller ruminants generally select a diverse high quality diet so that the concentration of any one antiherbivory chemical is relatively low. Thus they normally thrive on a diverse, low-growth form, dicot community of forbs and woody plants. This degree of selectivity precludes much volume of forage and hence the smaller body size.

The shift to a more simplified vegetation pattern in the late Pleistocene would have affected this artiodactyl group somewhat—with the small to medium-sized antilocaprids on the American Great Plains being most influenced. The antilocaprids were reduced to a single species, *Antilocapra americana*, the extant pronghorn. Two deer genera (*Sangamona* and *Navahoceros*) were probably outcompeted by *Odocoileus*.

Among the larger cervid dicot eaters, *Cervalces*, *Megaceros*, and *Alces*, only *Alces* survived. The strategy of these large deer has been to switch from the small deer's selective diet of bark and terminal buds during the off-season* to a high volume diet of the large woody twigs themselves. Bare twigs, like standing dead grass, were an abundant resource, but like grass, twigs may present nutritional difficulties (Bryant and Kuropat 1980).

In the course of their evolution some large ruminants have shifted away from dicots and have become grazers. To capitalize on the grazing diet of more slowly fermentable fiber (cellulose), they had to tolerate longer fermentation times (Hoppe 1978). They have adapted to this with a rumen of larger capacity, with rumen retention structures and other features, because grasses are defended from herbivores with phytoliths (Field 1976a, b) and with fiber that is slow to degrade. No severe handicap is posed by eating more monotonous diets, as long as available protein and other nutrients are sufficient to nurture the rumen flora. To keep protein levels sufficiently high, ruminants normally avoid the fibrous growth of mature tall and midgrass stems. Instead, they concentrate on grass leaves, shortgrasses, or young and regrowth stages of midgrasses and tallgrasses (McNaughton 1976). Except in the far north, this group of large ruminant grazers was essentially unaffected by the late Pleistocene extinctions in the Holarctic. *Bos* and *Bison* thrived on the simplified floras of the Holocene. One advantage a ruminant has over a monogastric is that it recycles considerable body nitrogen in the form of urea via the saliva into the rumen to feed the rumen flora. So, on a no-growth resource it can get by on a limited amount of food with moderate to high fiber at low protein levels (Janis 1975). It does this by retaining the fibrous food in the rumen.

Although large monogastrics like horses and elephants have advantages to their digestive strategies, they cannot do well on limited quantities of high fiber food. Their "strategy" is to pass as much food volume through as possible, breaking down the soluble carbohydrates in the simple stomach and fermenting fiber in the caecum. In general they have a much faster gut transit time than large grazing ruminants because the monogastrics quickly pass through the high fiber parts. They do not possess the sievelike retention device (reticulo-omasal orifice) of large ruminants (Janis 1975). As an example of this difference, Borman et al. (1982) have shown that in the summer horses spend 19.28 hours per day grazing while cows spend 10.42 hours. Likewise, horses produce 5.6 kg of feces per day, while cows produce 3.9 kg.

One of the important features of the rumen is its function as a detoxifier for secondary plant compounds through the action of the diverse bacterial and protozoan flora and associated alkaline conditions (Freeland and Janzen 1974). Because the caecalid's fermentation chamber is positioned after the small intestine, many plant toxins consumed by monogastrics are assimilated into the bloodstream before they are degraded. The main adaptation of large monogastrics to this lack of a broad-spectrum detoxifying rumen is to select as forage staples those plants and plant parts which

---

*It is difficult to refer to the period outside the season of major growth as the "non-growth season" because often some growth occurs during this time. Also, it is not always "winter" that describes this season, because in some areas it is the dry season. So I will use the more general term "off-season" to refer to this period.

depend on structural defenses (phytoliths and fiber) rather than chemical toxins. However, another inherent feature of the monogastric frustrates this adaptive avoidance of chemical toxins: the gut flora is unable to synthesize all the organic nutrients necessary for growth and maintenance. Ruminants, with the aid of the complex rumen flora, require few essential amino acids, fatty acids and vitamins; they are manufactured in the rumen.

Large monogastrics being adapted to an abundant, year-round diet of grasses that is energy-rich but (seasonally) nutritionally poor (and in the case of mastodonts probably an off-season diet of twigs of deciduous trees), must supplement their diet with a variety of more nutritionally rich dicotyledonous plants (forbs, woody plants, or both). Janis (1975) believes the generally greater variety of plants in the diet of perissodactyls than in that of artiodactyls is evidence of this. Olivier (1979) proposes the same argument to account for the greater variety of plant species in the diet of proboscidians. He argues that large monogastrics normally have an abundant staple which is low in toxins and is supplemented with more toxic dicots when nutritional unbalances occur. Malpas (1977) has shown that when elephants are unable to supplement their staple, body conditions are adversely affected.

These more toxic, supplemental dicots can be degraded by monogastrics. Monogastrics use "microsomal" degradation (Freeland and Janzen 1974), that is, they use conjugation with waste products or their own tissue once the toxins have reached the bloodstream (normally detoxification of this type occurs mainly in the liver and kidneys). However, this system depends on a detoxifier whose action is rate limited (i.e. the more toxin, the more time it takes to detoxify). Thus a great amount of a single toxin quickly exceeds the degradation capacity. Monogastrics seem to circumvent this (Freeland and Janzen 1974) by selecting a variety of plants with different toxins. As long as the volume of each toxic plant species is kept low, each can be degraded by processes which are rate limited.

The optimal diet for monogastric ungulates thus seems to be one which includes one or several nutritionally poor but allelochemically undefended plant staple(s) and a diverse variety of more allelochemically well-defended species that are nutritionally rich or at least nutritionally complementary to the staple. I think this concept is critical to an explanation of why the Holarctic monogastric ungulates virtually all became extinct about 11,000 years ago. From a plant's view, protections against mammalian herbivores can be seen as a cost-benefit proposition. A plant must produce some defenses if it is to live a long time on marginal resources, where it is subject to attack by herbivores (Janzen 1975).

Herbivore ecology studies have undergone a major change over the last few years (see Harborne 1978, Rosenthal and Janzen 1979). Less than a decade ago, herbivory studies were mainly concerned with available energy and nutrients. We are now realizing that plants are not passively vulnerable to herbivores, but rather, have a complex array of defense compounds which vary with the taxa, polymorph, habitat, soil fertility, age, past browsing, part of the plant, season, and other factors. Thus the plant species or plant parts an herbivore may utilize depends on how well that herbivore is adapted to each plant's defenses. The chances of any one herbivore being adapted to eating any one random plant species as a dietary staple are low. Energy and nutrients are usually available to an herbivore; the main problem is how to avoid the plant defenses, dilute their effectiveness, or detoxify them.

Plant Defenses

Without reviewing the entire spectrum of plant antiherbivory defenses or nutrient availability, some general patterns in herbivory ecology can be related to the extinction problem. These patterns arise from attempts to examine plant foods from an ungulate's view. One of the most striking insights recently developed in studies of herbivory ecology is that herbivores generally do not select plants on the basic gradient of nutrient quality, but rather, the primary emphasis in selection is to avoid secondary plant constituents—the antiherbivory defenses (see Bryant and Kuropat, 1980, for a review). In other words, herbivores like most of the plants that are defended least. A complication to this rule is that some groups of herbivores are adapted to specific kinds of plant defenses, and other groups are adapted to other plant defenses.

According to Rhoades and Cates (1976), there are two general categories of plant allelochemic defenses. Put simply, the first group is comprised of chemically toxic compounds which interfere with the herbivore's internal metabolism. These compounds, mainly nitrogen-based, include alkaloids, glucosinolates, cynogenic substances, nonprotein amino acids, various toxic proteins, and peptides. These are termed "qualitative" defenses. In a relative sense, these qualitative defenses may be less costly for the plant as they are to some extent recyclable, produced in small quantities, and depend upon getting into the herbivore's bloodstream where they then act on a very specific tissue or system. They normally occur in plants on nitrogen-rich fertile soil or in plants with nitrogen-fixing symbiotics. The second group of plant defenses includes mainly carbon-based compounds such as lignin, phenolic resins, and various terpenoid compounds. These "quantitative" and comparatively more "expensive" defenses are incorporated into the plant's structure where they reduce the digestibility of the plant tissue. They seem to be more common in plants on infertile soil.

Rhoades and Cates contend that the two types of defenses are not distributed randomly. Plants adapted to less seasonal environments, particularly those which enjoy a long and predictable growing season, are well defended by the "expensive" defenses. Plants which thrive in a short and nutritionally rich growing season have more inexpensive defensive compounds (Rhoades 1979). Janzen (1975), Rhoades (1979), and Klein (1977) have outlined a gradient along which plants increasingly produce the more expensive, broad spectrum, quantitative defenses. Before we discuss antiherbivory defenses and extinctions further, let me present four general plant categories which take into account seasonal variations in nutrient quality as well as seasonal differences in antiherbivory defenses.

Graminoids

Most monocots depend mainly on silica for defense though they also use fiber, particularly in coarse stems. During the flush of early growth these plants are poorly defended against herbivores and are also quite nutritious (Whitten 1975, White 1978). As they mature, their main defense is structural coarseness and the translocation of nutrients to the root system, leaving the above-ground plant parts very low in nutrients. Na, Ca, N, P, K, and other elements are all generally quite low in late growth season, dry season, or in winter grasses (Chapin 1980). In addition to absolute deficiencies, many of the remaining nutrients occur in imbalanced ratios relative to a mammal's needs. However, a considerable volume of vegetation remains in the late growth season grasses and throughout the off-season. Thus the energy present in the form of cellulose and hemicellulose is comparatively undefended and available to many large grazers adapted to such a diet and able to supplement this nutrient-deficient dietary staple with complementary nutrients from other sources.

Grasses have a short nutrient pulse during the early part of the growth season and sometimes in the end; however, different species often have different phenological peaks (Brown and Trlica 1977). A grazer can combine the nutrient peaks of different grass species to prolong the length of its own growth season. Grazing specialists tend to be mammals of large size, mainly monogastric caecalids (equids, rhinocerids, and proboscidians) but also some large ruminants (e.g. bison and wildebeest).

Forbs

Forbs, particularly the more ruderal ones, are often defended by nitrogen-based allelochemic compounds. (By forbs, I mean most nongraminoid herbs.) When young, these plants are usually high in nutrients, but as they mature so do their defenses, while their nutrient content declines (nutrient content remains relatively high in many perennials, particularly legumes). Forbs seldom leave nutrients or energy above ground in the form of vegetative growth during the off-season. More than any other plant group, forbs are adapted to a local succession of phenological growth peaks, flowering, setting seed, etc. For a selective feeder able to detoxify nitrogen-based allelochemics, these plants offer a rich resource during the growing season, but not much during the off-season. Large mammal forb specialists tend to be the medium to small ruminant classes in the range of Thompson's gazelles (McNaughton 1976). In North America the pronghorn fits this category (Koerth et al. 1982).

Deciduous Woody Plants

Like evergreens, these plants are primarily defended by carbon-based allelochemics, but they differ from evergreens in many ways. Like graminoids and forbs, deciduous woody plants are pulsers—experiencing a nutrient flush and burst of growth at the beginning of the growth season. New leaves are nutrient rich and relatively undefended; older leaves accumulate antiherbivory compounds and are relatively low in nutrient content. The off-season stems and twigs of new growth are high in allelochemics (Bryant and Kuropat 1980), while the woody growth on the upper parts of more mature plants is less well defended. Generally these plants are found on fertile soils and translocate nutrients into storage reserves during the off-season. A gradient of defense toxicity within this group relates to several ecological and life history parameters. Felt leaf willow, *Salix alaxensis,* and alder, *Alnus crispa,* are examples of the extremes (Bryant and Kuropat 1980). Many willows are more ruderal and can be destroyed by fire or broken by ungulates; they leave few nutrients in winter woody stems, and mature plants are poorly protected with allelochemics against browsing in winter. Alder, however, a "nitrogen-fixer," is high in nutrients and allelochemics, particularly the more toxic resins (see Bryant and Kuropat, 1980, for a more complete review of woody plants' winter defense strategies).

In winter the more undefended species of this category are similar to standing dead grasses as the mature woody browse of willow, aspen, and others provides abundant energy, but relatively poor nutrition. These plants serve as an off-season staple for most living large mammal browsers (i.e., most cervids); undoubtedly they were a winter staple for some now extinct browsing species (e.g. the mastodon). Like grasses, the nutrient content of each of these deciduous species during the off-season is not only low, but imbalanced, particularly for monogastrics, and is usually supplemented with complementary dietary items for the browser to maintain body functions (Olivier 1979).

Most large mammal herbivores that use the new leaves from this group as a growth staple are ruminants with rapid growth rates (cervids, medium-sized antelope). Those which use the large diameter woody stems in the off-season tend to be large in body size, much like the large mammal grazers which use grass stems during the same off-season period [moose (*Alces*) and probably some extinct deer, (e.g. *Megaceros*)]. Ungulates which selectively use the small stems, terminal buds, and bark are small in body size (e.g. roe deer, *Capreolus*).

Evergreen Woody Plants

As a general rule large mammalian herbivores avoid these plants. They usually grow on infertile soils and are severely nutrient limited (alpine, arctic, subarctic) or moisture limited (desert). These are reviewed by Chapin (1980). The evergreens' strategy is a conservative one with long-lived photosynthetic tissue and roots which maximize nutrient use. Consequently, they have no major seasonal flush in nutrients, and nutrients are exposed to herbivores the year-round. Being nutrient limited, rather than carbon limited, they allocate considerable carbon energy toward the more toxic resinous allelochemic defenses. Unlike deciduous species, these plants are tolerant of nutrient stress and seldom have sufficient reserves of nutrients to withstand the damage that would result from substantial browsing by large mammals. Their strategy is not to discourage browsing, but to prevent it.

At this point it is important to emphasize the existence of a gradient of effectiveness in the carbon-based antiherbivory compounds from rather benign to highly toxic. Insoluble carbohydrates like cellulose (which both ruminants and caecalids can digest if given time) mean only special effort and adaptations on the herbivore's part. Lignin cannot be eaten in large quantities by ruminants because it slows rumen function and hinders the digestion of cellulose and hemicellulose. Within reasonable limits caecalids can bypass both lignin and tannin-lignin complexes around the caecum. Ruminants can destroy tannins in the rumen, again within reasonable limits. So although the above compounds are deterrents to ungulate digestion, they can be effectively ingested, as long as quantities are limited. At the more toxic end of the scale are the antimicrobial resins (terpenes, phenolic resins) which greatly hamper digestive functions, reducing the herbivore's production of proteins, fatty acids, and vitamins, and interfering with the assimilation of these compounds and sodium absorption. It is these resins which make most evergreens unusable to large herbivores.

In light of these four plant categories, we can now return to our herbivores and examine the effects of declining plant diversity on the spectrum of ungulate adaptations.

Adaptations of Large Herbivores to Antiherbivory Defenses

Large Ruminant Grazers

Because of the ability of the rumen to synthesize a balanced diet of amino acids, fatty acids, vitamins, etc., large ruminants such as bison and cattle can flourish on a monotonous summer range of just a few plant species. They take advantage of the growth season nutrients, but as they are not highly selective feeders and have correspondingly large mouths (Field 1976a), they do not experience extremely sharp seasonal peaks and have relatively conservative life histories. They can also subsist on winter range comprised of only a few plant species as long as nutrient levels do not fall too low (Peden et al. 1974). These large ruminants seem to have flourished during the late glacial–Holocene transition (*Bison* in North America, *Bos* in Eurasia) except in deserts and the subarctic and arctic where the evergreens and the nutrient-impoverished tussock grasses were becoming predominant. This is probably why bison was driven from its traditional homeland in the northern parts of Eurasia and in Beringia (Guthrie 1980). Like the African buffalo (*Syncerus*), bison and cattle are grass leaf eaters and generally avoid grasses with large stem fractions (Field 1976a).

Medium to Small Ruminant Grazers

With their smaller-sized rumen, members of this group require relatively high quality food year-round. They have to be selective feeders, especially during the seasonal flush of early plant growth. At this time they must feed on a variety of forbs and/or the leaves of woody plants to obtain the high quality diet they need for growth (Whitten 1975, Winters 1980). Small selective grazers are characterized by small mouth parts (Field 1975) allowing them to obtain small plants and precise plant parts. These ungulate species were greatly restricted in range at the beginning of the Holocene. At the peak of the last glacial phase saiga's (*Saiga*) range extended from Canada to western Europe; at the beginning of the Holocene it was rapidly confined to central Asia. Sheep (*Ovis*) underwent considerable local range restrictions. As is the case with bison, sheep were driven from their ranges by vegetational changes. It can be assumed that the encroachment of a simpler plant community and an increasing preponderance of plant species with defenses for which these ruminants are not well adapted (particularly winter browse of resinous, stress-tolerant dicots, and nutrient-poor northern Holocene graminoids) were responsible for their distributional restriction.

Some of the camelids may belong in this group. Their living counterparts seem to be mainly grazers, and their teeth are hypsodont, suggesting a rapid wear rate from siliceous grasses. This group also experienced a major range contraction as well as considerable extinctions.

Large Caecalid Grazers

These monogastric-caecalid grazers, like the large ruminant grazers, capitalize on flushes of nutrients available during the prime season of plant growth, but also like the large ruminant grazers, they are unable to selectively feed on the plant parts that are highest in nutrients and the least defended. This is not to say that they are unselective feeders; they are highly selective (Field 1976b) but at a quite different scale when compared with smaller ungulates. Members of this group often eat the entire grass plant (seeds, leaves, stems). The stems dilute the leaf nutrients somewhat, but a large caecalid's specialty lies in utilizing large volumes of plant matter of low to modest nutrient quality. Thus these caecalids have extremely conservative life histories, even more conservative than their ruminant counterparts. The caecalid's ability to process grass stems does give them definite advantages in the off-season. The undigestible stem fibers of many winter grasses provide more bulk than a ruminant can handle, but the large monogastric caecalid can process more low-quality grasses than the ruminant by shunting the undigestible plant parts past the caecum. Thus, given ample quantities of grasses, the caecalid can live on grasslands where ruminants cannot (Janis 1975).

This grazing strategy presents two potential problems. First, the seasonal peak in plant quality must be long enough to maintain the slow but lengthy annual growth of the animal. Secondly, since the nutrients in both the growth season and off-season grass staples are insufficient for a monogastric caecalid, these staple grasses must be supplemented with other plant species containing complementary nutrients. Monogastric ungulates thus require greater dietary diversity (Field 1976b, Olivier 1979) which they obtain by ranging across a diverse vegetational habitat to select supplemental food from a variety of dicots as far along the "least defended" gradient as possible (Jewell 1974). Their optimum habitats seem to be very mixed vegetational mosaics with considerable local heterogeneity.

The large caecalid monogastric mammals are primarily co-evolved with "conservative" plants which depend on a less seasonal environment and on the less toxic end of the expensive defenses. Monogastrics with their complex hypsodont teeth, a caecal-bypass which allows rapid gut transit time of indigestible components, conservative growth strategy, low reproductive rates, etc., have allowed them to adapt to use plant staples high in silica, fiber, etc. Virtually all the large Holarctic caecalid ungulates became extinct at the beginning of the Holocene.

Forb-feeding Ungulates

Few ungulates in the Holarctic are exclusively forb feeders. Forbs do provide complementary nutrient supplements for some caecalids and important seasonal boosts in nutrients for many ruminants. This is particularly true of the smaller ruminants which select the least defended and highest quality dicots (Hoppe 1978). Most of these ruminant dicot specialists are rather specific in their feeding habitats (Jewell 1974, Ferrar and Walker 1974, Child and Richter 1969); likewise, these ruminants are rather specialized in selecting different plant parts (Bell 1969), and in choosing plants according to the degree of previous grazing (Bell 1969, McNaughton 1976).

The small gazelles in Africa seem to be forb specialists, at least during some seasons. The only Holarctic equivalent may have been the antilocaprids which seem analogous in body build, dental character, and habitat. Several species of antilocaprids became extinct during the end of the last glacial.

Large and Small Ruminant Browsers

These browsers take advantage of the pulse of nutrients during the plant growth season and consequently tend to be unconservative in their life histories (they grow and mature rapidly, and females generally give birth to more than one young every year). These species characteristically use woody twigs as their winter staple, particularly the deciduous species which are abundant, poorly defended, and comparatively digestible, but nutrient poor. They therefore require nutrient supplements from a relatively high diversity of plant browse species. Field (1975) found that among the African bovids the browsing eland (*Taurotragus*) required a protein diet, both rich and diverse. This meant it had to be nomadic.

This group experienced some blanket extinctions throughout the Holarctic at the beginning of the Holocene. The giant deer, *Megaceros,* was distributed across Eurasia at midlatitudes, from Ireland to the eastern Soviet Union and Japan, even on the Mediterranean islands. Judging from its dentition, *Megaceros* had dietary adaptations somewhat similar to *Alces.* In the New World the cervids (*Cervalces, Navahoceros,* and *Sangamona* became extinct. So did *Symbos* and perhaps other ovibovines (the systematics are controversial). In the Holocene, *Alces* and *Ovibos* retreated northward from their more extensive glacial ranges. We know from studies of *Alces* (review by Bryant and Kuropat 1980) and *Ovibos* (Jingfors 1980) that in high quality populations their preferred browse is mature willow of several species low in allelochemics. Even so, they must supplement these with other plant species to form a complete diet (Bryant pers. comm.).

Large Caecalid Browsers

No large caecalid browsers now live in the Holarctic; like the large caecalid grazers, they all became extinct. The tapir was driven southward into Central America. The two groups which most precisely fit this category are the ground sloths and mastodont. For the mastodont, growth season nutrient demands were modest but spread over a long time. We know this because the volume of food it required would have precluded precise selection of high quality plant parts. Rather, the best strategy would have been to avoid quantity doses of allelochemics by feeding on the poorly defended end of the gradient of deciduous plants. During the growth season mastodonts would have eaten leaves and branches from the youngest available plant growth. During the off-season, the major diet would have been twigs and stems from the upper branches of mature stands (the least defended and hence comparatively digestible, but yet lowest in nutrient percentages and utilizable balance of nutrients). These may have been the staples of the mastodont diet, but from what we know of living proboscidians, mastodonts would have required multiple supplements from different plant species to achieve a complete diet (Olivier 1979). In other words, mastodont needed the right kind of plant diversity.

It is a telling point that most of the extinct species and their living relatives do not fit exactly into these categories. Each must range across vegetational habitats and use different plant forms and parts to assemble a diet sufficiently high in energy and nutrients, low in allelochemics, and of proper nutritional balance to live and reproduce—which is to say that these species required a quite heterogeneous diet seasonally or all year.

There seems to be a special interaction in both the rumen and caecum of a wide variety of toxins from a number of plant forms or taxa. Instead of having an additive effect, some toxins have a buffering effect, or synergistic relationship in the mammalian gut (Klein 1977, Bryant and Kuropat 1980). The toxicity of an allelochemic compound seems then to be related to what other plant species or plant parts are in the gut (Klein 1977). Most wild ungulate herbivores must not only eat a mixed diet, it must be the right kind of mixture. We know little about the details of these interactions, but the concept is critical to my argument for heterogeneity being a requisite in the diet of these extinct species.

The stress from an increasingly seasonal environment throughout the late Cenozoic seems to have created both an evolutionary and ecological trend in the vegetation toward a sharper peak in the soil and plant nutrient pulses and toward more toxic defenses (nitrogen-based compounds and the more toxic carbon-based compounds)—those defenses to which monogastrics are least adapted. These trends would have favored ruminants that can detoxify many of these compounds in the rumen and capitalize on the brief but high nutrient pulse with their more radical life history pace. This may well be the major reason for the irregular replacement of monogastrics by ruminants during the latter part of the Cenozoic. This shorter "flush" growth season has undoubtedly affected other animals as well. A shift toward more toxic plant defenses and increased peak nutrient and energy availability of stressed plants is often accompanied by small mammal outbreaks and plagues (Rhoades 1979). Webb (1969) has documented a major increase in Holarctic rodent taxa since the Pliocene at the expense of all ungulate taxa. Small ruminants and many rodents (particularly ground squirrels and microtines) do well in the face of a shift toward toxic plant defenses. Much the same holds true for lagomorphs. Like microtines, they thrive on a "stressed" vegetation, and population outbreaks often occur during environmental stresses such as drought. Although all these small mammals are monogastric caecalids, their situation is quite different from the large caecalids, like horses and mammoths. Primarily because of their size, small mammals can select the parts of plants relatively low in defenses and extremely high in nutritional quality. As a result, they have unconservative life histories: rapid growth rates, prolific reproduction, short time to maturity, etc. This has allowed them to gain an expanding role in the mammalian herbivore community with the increasing seasonality of the late Cenozoic. Characteristically, they experienced no major late Pleistocene extinctions.

So not only has the shortened growth season affected herbivores by changing local vegetation diversity and zonation, it seems also to have been felt in the changing balance of antiherbivory defenses (fig. 13.3).

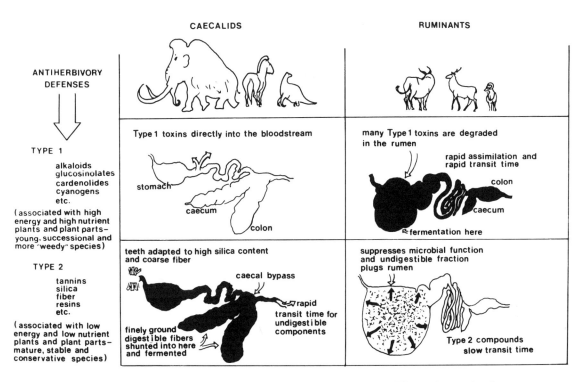

Figure 13.3. Monogastric ungulates and ruminants are adapted to two different classes of antiherbivory compounds. Monogastrics are adapted to "quantitative" defenses (type 2) of more conservative plant species. These are mainly defended by compounds which are indigestible or react with other material proteins and carbohydrates in turn making them undigestible. Ruminants are adapted to "qualitative" defenses (type 1). These are characteristic of less conservative species. Their toxins attack the organism's metabolic function. It is probable that the irregular decline in monogastric ungulates and their replacement in the late Tertiary and Pleistocene by ruminants is due in part to the shift in vegetative changes from more quantitative to qualitative defenses, as part of an increase in less conservative species. This continued shift probably characterized the Pleistocene-Holocene transition, when monogastrics became extinct.

The Role of Vegetational Change in Large Mammal Extinctions

At this point we return to the matter of extinctions, and take a closer look at the climatic-vegetational events of the late glacial in relation to the various herbivory adaptations of the different ungulates. As vegetational zonation intensified, the selective feeders which depended on a complex mixture of secondary habitats and a diversity of plant species (the mastodont, mammoth, horses and asses, camels, peccaries, antilocaprids, ovibovines, and some deer species) could be expected to decrease in *quality*. It is probable that sometime during the first stages of these late Pleistocene vegetational changes a few of these ungulate species actually increased in numbers as the local volume of their staple food increased (e.g. mammoth and ground sloth in the western United States). But as potential dietary diversity continued to decrease, the quality of forage, in terms of the variety of supplemental species, became more critical, and the numbers of animals as well as their quality declined.

These ecological repercussions probably have several facies, all of which we many never know exactly. But, in light of the above discussion, it is worthwhile summarizing three which must have played a part. These are, it must be remembered, parts of the same syndrome of repercussions.

1. *Duration of the plant protein peak, the growth season, probably changed as a direct effect of the late Pleistocene-Holocene climatic changes.* For a large herbivore the shortened season allowed less time to piece together a high quality diet. Judging from the decreases in body size this seems to have hit almost every large species, but it would have affected the more conservative species to a greater extent.

2. *Declines in local plant diversity and the balance of nutrients.* As the plant communities became more zoned, there were fewer optimal "plaid" mixtures of plants for the species requiring nutritional diversity in their diet. We know that today the large monogastric ungulates range across many plant habitats in the African savanna to obtain a forage mix. Few Pleistocene plants became extinct (Kershaw this volume); rather, as the vegetation started to shift into simpler plant zones, this geographical sorting of plants into broad "stripes" meant that monogastrics had to trek further for an optimal mixture. Within these "stripes," there was a simplified diversity of plant forms (from the shorter growth season and less intra-annual seasonal diversity). This "intra-stripe" shift can be seen in the change of the competitive balance. The large bovids, *Bison* and *Bos,* came to dominate the now "grazophilic" prairies, parklands, and plains. Many of the small selective feeding ruminants of the grasslands died out. Pronghorn (*Antilocapra americana*) came to dominate the plains with bison. They appear complementary to bison, thriving on the browse and forbs exposed when bison intensively graze the shortgrass plains. All the other antilocaprids and even camels were squeezed out in this new environment. *Alces* (moose) and *Odocoileus* (American deer) do comparatively well on a short nutrient pulse diet of browse, particularly summer regrowth which is low in antiherbivory toxins but nutrient rich.

3. *Shift in plant defenses.* Not only were nutrients no longer sufficient for the groups which became extinct, they were increasingly locked-up in plant taxa for whose defenses these extinct lines were not adapted (judging from their living counterparts elsewhere in the world). Most extinct animals had been adapted to a major energy staple which was not well defended allelochemically. This is true in both the growth season and off-season. This adaptation was mainly a tolerance for *stems*—coarse straw-like stems of tall and midgrasses among the large caecalid grazers and woody stems used by large caecalid and ruminant browsers. These had to be supplemented from plant parts and species which were better defended. So long as a diversity of the latter species existed, nutrient complements could be obtained by diluting a variety of different toxins (most of which could be detoxified in modest quantities). Without that special kind of diversity, these herbivores would be forced to take more of some toxins than they could detoxify, reducing general digestibility, hampering normal metabolic function, reproductive success, and, in sum, decreasing interspecific competitive abilities.

The stresses of a shortened Holocene growth season (probably in combination with a nutritionally poorer substrate and less intra-annual variability) seem to have altered plant antiherbivory defenses further in the scale of stress tolerance. The stressed dicots would have responded in species composition, through genetic change, and developmentally by increasing the more toxic defenses (particularly late growth season and off-season defenses). Some herbivores profited by this switch, but others were decidedly maladapted to this class of plant defenses. In some areas the severe stress-tolerant plant species became dominant with their "long-lived organs, evergreen habit, slow growth rates, slow turnover of carbon and mineral nutrients, and infrequent flowering and presence of mechanisms which allow the intake of resources during temporarily favorable conditions" (Grime 1979), as well as their extensively developed antiherbivory defenses which virtually no ungulate can penetrate. These include some of the dominant vegetation of the tundra and coniferous forest rim of the northern Holarctic and much of the dominant vegetation along the Holarctic's southern arid margin.

The climatic stresses had moved beyond the Tertiary crest which was optimal for ungulates and in a jerky march were now promoting plant forms uncongenial to most ungulate adaptations.

I proposed earlier that the major brunt of the extinctions was borne by those ungulates with conservative life histories that do not do well in short growing seasons. Using milk quality and gestation length as a broad indicator of conservative life strategies and an inability to capitalize on a short nutrient flush, one can almost contend that every ungulate with a gestation period longer than nine months and milk protein percentages below 4.0 became extinct everywhere throughout the Holarctic. The gestation periods of living proboscidians and rhinocerids are well over a year; horses, asses, hemionids, Asian and South American camelids also about a year; and living Brazilian tapirs about thirteen months (Table 13.1). Representatives of these groups make up the majority of the postglacial megafaunal extinctions.

Also, these large caecal digesters (nonartiodactyl monogastrics) and camelids all have extremely low percentages of protein in the milk (the nutrient most indicative of growth potential during the first growth season). Table 13.2 shows that the cervids, which dominate the Holarctic ungulate megafauna in the Holocene, have extremely high percentages of milk protein.

It is unlikely that the extinct proboscidians, equids, rhinocerids, and camelids differed little from their modern analogues in percentage of milk protein or gestation length since these are conservative evolutionary features. Likewise there is only a small environmental component to the variation of milk protein and gestation length. Changes in range quality or the female's physiological condition will produce changes in the amount of milk or the timing of ovulation.

The significance of the low percentage of milk protein and the long gestation period in these species is part of an adaptive syndrome geared to exploit nontoxic, low quality, bulk staples and a wide diversity of toxic supplements. Also, equids, rhinos, and proboscidians rely on rapid gut transit in a monogastric, caecal-diverticulum gastrointestinal tract (Janis 1975). The rumen system of camels seems to be anatomically and physiologically unusual and more monogastric in many respects. Milk protein and gestation length are not the only traits associated with the more conservative ungulate dietary adaptations. Growth rates, age at weaning, age at sexual maturity, potential longevity, and litter size of these species all reflect the same basic conservative patterns. Because of their conservative strategy, they rely on a long growing season and long life expectancy.

The vegetational changes greatly favored the less conservative ruminant system, but even so, they imposed growth limitations on almost every ungulate species, creating the widespread phenomenon of postglacial dwarfing.

The postglacial and particularly postagriculture rise to dominance in the northern and temperate latitudes of medium to small cervids is illustrative of this shortened ungulate growing season. The cervids are the "weeds" among the Holarctic ungulates: they "live fast and die young," being adapted to a short growth season of high nutrient pulses and high mortality by having rapid growth rates, high percentages of milk protein, early weaning, and multiple births. They have specialized in using high quality resources for a short season and avoiding, as much as possible, large quantities of toxic forage. During the shortened postglacial growth seasons they outcompeted the "oak" end of the ungulate spectrum. In addition to the rumen detoxification system, which can also make up for nutrient deficiencies and imbalances, their size allows greater selectivity for the least defended plants and plant parts and those which are compatible in the gut.

An additional angle to the nutritional effects of a changed vegetational pattern involves ovulation, which is partly regulated by body condition (usually fat levels). As an adaptive regulator for the female, ovulation fails to occur once body condition falls below

Table 13.1. Ungulate Gestation Lengths

Species	Gestation Period (Months)								
	6	8	10	12	14	16	18	20	22
Caribou *Rangifer tarandus*									
Mule deer *Odocoileus hemionus*				Cervids and Bovids					
Moose *Alces alces*									
Wapiti *Cervus canadensis*									
Bison *Bison bison*									
Horse *Equus caballus*				Equids					
Ass *Equus asinus*									
Lama *Lama glama*						Camelids			
Arabian camel *Camelus dromedarius*									
Bactrian camel *Camelus bactrianus*								Large Caecalids	
White rhino *Ceratotherium*									
Black rhino *Diceros bicornis*									
Indian elephant *Elephas maximus*									
African elephant *Loxodonta africana*									
	6	8	10	12	14	16	18	20	22

a certain level. Because monogastric ungulates have a comparatively long gestation period, they must ovulate in the spring or early summer to have their young the following spring(s) when growth nutrients are at their peak. This timing means that an animal having experienced a poor growing season or poor off-season would be in a lowered body condition and would be less likely to ovulate. Recruitment, therefore, is reduced. Ruminants having a shorter gestation period ovulate in late summer or early winter when body condition is at its annual zenith. A shortened growth season and/or increased nutritional stress during the off-season that affected winter body condition would theoretically affect ovulation physiology to a greater extent in monogastric ungulates than in ruminants.

Thus mammoths, mastodonts, horses, asses, camels, sloths, peccaries, and others were replaced not by new, ecological equivalents entering the system, but by the formerly uncommon ruminants in the system. Moose (*Alces*) are nowhere common in the Pleistocene record, yet during the Holocene they became the most abundant large mammal over a vast area of Eurasia and North America. The same occurs in the small and medium-sized deer: roe deer (*Capreolus*), and white-tailed and mule deer (*Odocoileus*). Although not abundant in the late Pleistocene fossil record, in the Holocene they dominate local faunas and now number in the millions. Bison are a frequent Pleistocene fossil throughout the midlatitudes in North America, but they too did not reach dominance until the formation of the great American steppes in the Holocene (Guthrie 1980).

Mastodonts, judging from their distribution and anatomy (and from aspects of the ecology of extant elephants), were probably mixed browsers, concentrating on bark, leaves, and limbs of deciduous trees and shrubs less than 2 to 3 cm in diameter for their dietary staple. The Holocene vegetational changes probably decreased the heterogeneous habitat for which mastodont was adapted. The multiple resources and hence balance of nutritional supplements and toxicity avoidance that mastodonts required occurred in the same community mix. Their consequent numerical reduction and subsequent decreased use of browse may have allowed the cervid browsing specialists to expand by shifting the competitive edge in their favor. It is possible that in addition to this direct decline in suitable forage, the sheer increase in numbers of cervid browsers hastened

Table 13.2. Ungulate Milk Protein

Species	Protein in Milk (%)											
	1	2	3	4	5	6	7	8	9	10	11	12
Reindeer *Rangifer tarandus*												
Red deer *Cervus elaphus*												
White-tailed deer *Odocoileus virginianus*												Cervids
Black-tailed deer *Odocoileus hemionus*												
Moose *Alces alces*												
Roe deer *Capreolus capreolus*												
Wild pig *Sus scrofa*												
Water buffalo *Bubalus bubalis*												
Giraffe *Giraffa camelopardalis*						Some Other						
Musk oxen *Ovibos moschatus*						Artiodactyls						
Peccary *Tayassu tajacu*												
Hippopotamus *Hippopotamus amphibius*												
Bison *Bison bison*												
Lama *Lama glama*				Camelids								
Bactrian camel *Camelus bactrianus*												
Arabian camel *Camelus dromedarius*												
Indian elephant *Elephas maximus*				Large								
Horse *Equus caballus*				Caecalids								
Ass *Equus asinus*												
Black rhinoceros *Diceros bicornis*												
	1	2	3	4	5	6	7	8	9	10	11	12

the mastodont's extinction. For a shrub to be of substantial nutritive value it must contain leaves, buds, and twig tips. Moose and white-tailed deer both "cream" browse by taking the nutritionally best and least defended portions. Once "creamed" by these cervids, a shrub begins to fall below the level of nutrients needed for proboscidians' (in this case mastodont) growth and reproduction. As moose or deer populations become denser they remove a higher percentage of nutrients from the browse. African elephant growth resource minimums range around 10 percent crude protein (Laws et al. 1975), and mastodont requirements must have been similar. Once released from their competitive disadvantage, the cervids would have made it difficult for mastodont to survive and reproduce by reducing the available nutrients and nutrient balance below a critical level.

I think that parallel competition occurred at the same time between other ruminants and monogastrics, bison and horses in particular. Horses can tolerate a higher percent of fiber in grass than can bison (Janis 1975). Grass seeds which tend to be destroyed in the rumen composting process are usually isolated high on an undigestible coarse stem of mature plants. This coarse stem is avoided by most ruminants because it clogs the rumen with relatively undigestible fiber. Horses take advantage of the seeds' placement on the stems; they are able to pass the stem quickly through the gastrointestinal tract in essentially undigested form. At the same time, horses can ingest, masticate, and digest the seeds which are high in nutrient quality and easily assimilated. The few seeds that are uncrushed by the teeth pass through unharmed and sprout in a fertilized manure pile. There seems to be a long co-evolutionary relationship between several monogastric ungulates and grass.

Horses also seem to require a wide variety of plant species, probably because their monogastric digestive system is a poor detoxifier and a little of several kinds of secondary plant substance is better than a lot of one kind (Janis 1975). Obtaining a wide variety of plant species in an increasingly zoned grassland would become more difficult for horses. The other large caecalids would have faced the same problem. Bison, however, as ruminants, did not have this problem. They could survive among the more homogeneous stands of midgrasses and could thrive on the shortgrass plains (Guthrie 1980). During the European colonization of the American plains, horses did poorly on the shorter grasses.

Once grazed, grass responds not by another rapid attempt at seed formation but by increased low leafy growth more conducive to the grassleaf grazing specialist, bison. In the Holocene, however, bison seem to have shifted toward the horse and mammoth niche by consuming moderately large quantities of stems (Guthrie 1980). They probably recycle more urea and handle more coarse fiber than, say, domestic cattle (Peden et al. 1974). Bison have also experienced a slight increase in molar complexity (Guthrie 1980), paralleling the complex teeth of horse and mammoth, for the rapid processing of greater quantities of high fiber, low-nutrient food.

Most ungulates have seasonally specific dietary specializations, eating certain plants or plant parts one season and switching to quite a different selection during another season. This would have been particularly true for the omnivorous monogastric peccaries. The extinct *Platygonus* and *Mylohyus* undoubtedly capitalized on the local plant diversity of the more complex Pleistocene communities.

The best evidence we have about the diets of the large Pleistocene fossil monogastrics comes from studies of fossil ground sloth dung (e.g. Laudermilk and Munz 1934, Long and Martin 1974). The most detailed study was done by Hansen (1978) on the plant microhistological leaf structure of dung from Rampart Cave in Grand Canyon, Arizona, where pellets were taken in a stratified series with ^{14}C dates. Hansen found no evidence of a natural disaster that might have reduced the forage available to the Shasta ground sloth, *Nothrotheriops shastensis;* in addition, he saw no sign of mineral, energy, or protein declines before the sloth became extinct. However, Hansen's data do show some striking trends in the sloth diet with regard to antiherbivory compounds and floristic variety just before the sloth's extinction (fig. 13.4).

More and more Mormon tea (*Ephedra nevadensis*) was incorporated along with their major staple, globemallow (*Sphaeralcea ambigua*), finally surpassing it. Mormon tea is either shunned or used as a low priority food item by extant large herbivores, while globemallow is a preferred food item. As these two differ little in protein and energy (Hansen 1978), this disparity in dietary preference is probably due to the more effective antiherbivory defense compounds in Mormon tea. Several studies analyzing sloth dung from different areas support Hansen's assessment of the sloth as an "opportunistic feeder." This probably precludes its having specialized detoxification adaptations as in the koala (*Phascolarctos*). Also, the diversity of genera in each coprolite ($\bar{x} = 6.8 \pm 2$) fits the pattern of a diverse feeder.

The percentage of dietary supplements markedly declined. As discussed earlier, these supplements are important for a monogastric's growth and reproduction. Hansen also notes that at least a dozen plant taxa present in the sloth dung are absent in the region today. Over half of these were already missing from the upper layer of sloth dung dated at $10,7000 \pm 200$ yr B.P. In a *matrix of similarity* comparing the plant genera in the sloth's diet through the eleven levels of the cave, Hansen's figures show a definite unidirectional dietary shift from bottom to top. These changes in diet, declines in dietary supplements, and declines in late glacial local plant diversity are consistent with the model proposed here.

The diversity of a "mosaic" community is greater than the sum of the individual components. The more complex base allows additional animal species to survive, species which competitively exploit the changing array of habitats and diets. Some of the "additional" ungulate species in such a mosaic community are usually larger monogastric

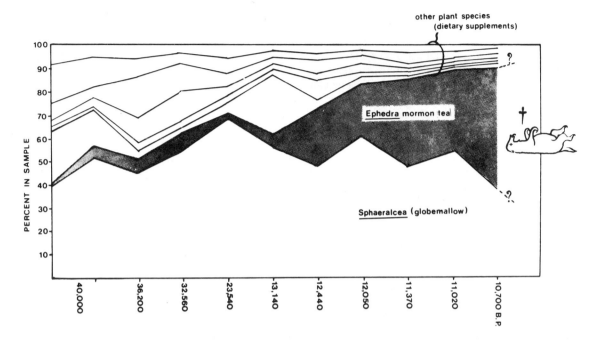

Figure 13.4. A representation of Hansen's (1978) data on the diet of the Shasta ground sloth *(Nothrotheriops shastense).* Two important things are occurring between 13,000 and 10,000 B.P. The first is a replacement of the staple globemallow *(Sphaeralcea)* by Mormon tea *(Ephedra).* The second and most important trend is the declining role of other supplemental plant species in the diet. These supplements are important for a nonruminant because they can seldom synthesize all of the necessary amino acids, fatty acids, vitamins, etc. from one or two staple plant species.

animals which can use as their staple the less sought after, high-fiber, low-digestibility and low-nutrient plants, and use for their nutritional supplement parts of other ungulate's dietary specialty.

Another aspect of the diverse mosaic community of large mammals is the presence of some species which exist at comparatively low densities. These species often use parts of niches of several more common species and can only do so in mosaic vegetation patterns. This may explain the erratic or uncommon distribution in late Pleistocene deposits of several older stocks such as the cervids *(Sangamona, Navahoceros,* and *Cervalces),* ovibovines *(Symbos* and *Eucheratherium),* antilocaprids *(Capromeryx, Tetrameryx, Stockoceros),* and some camelids *(Palaeolama* and *Hemiauchenia).* Some of these were once thought by some paleontologists to be contaminants from older sediments, but subsequent work has shown they lingered, probably at low densities, through late Wisconsin times.

Ecological vs. Overkill Theories

One factor that has frustrated an ecological account of the many biotic changes in the late Pleistocene is the emphasis and attention given to *mean annual* values in temperature, rainfall, etc. No other concept in paleoecology has so distorted our picture of the past. Axelrod (1967) began to articulate some of the limitations and hazards of relying on mean annual values, arguing that one finds the highest species diversity in mild climates, without marked seasonal extremes. Such equable climates allow species which cannot tolerate cold winters to live beside those species which cannot tolerate hot summers. As the climate shifts to a more continental regime marked by hot summers *and* cold winters, species diversity declines.

However, neither Axelrod's "equability" concept nor later versions (e.g. Webb 1977, 1978; Graham 1979) adequately explain the extremely sharp peak of extinctions of so many large mammals over such a short time as occurred in the late Pleistocene. It is unlikely that these extinctions are due directly to increasing seasonal extremes such as the colder winters which accompanied a Holocene trend toward "continentality." Winter temperatures *normally* are not the direct critical factor affecting the physiological conditions for survival of large mammals. Both the quality and availability of winter (or dry season) food is much more immediately limiting (Mattson 1980). Snow depth can affect locomotion and food access, but in a more "continental" Holocene climate such as Axelrod proposes snow cover would be reduced. The diverse Alaskan mammoth fauna of the Pleistocene steppes undermines the theory of more severe winter temperatures in the Holocene causing the extinctions. All the Alaskan cryogenic evidence points to *colder* winter temperatures during glacials. Also, high summer temperatures do not *directly* limit large mammal distributions to any great extent.

There is also the issue of aridity. Many of the dominant large mammals involved in the late Pleistocene extinctions were grazers. This fact is important because herb communities (the grasslands which support grazers) require some aridity to retain their dominance over woody plants. For the most part herb communities at temperate latitudes increased during the climatic changes in the late Pleistocene. Thus, one cannot simply argue that the aridity of a more continental climate during the late Pleistocene itself is the cause of the extinctions. Also, evidence points to a change *toward* a more mesic environmental mean in Alaska at the end of the late Pleistocene (Guthrie 1968).

Additionally there are not really as many "northern" and "southern-adapted" species of large mammals as there are plants. Large mammals are able to thrive in an astonishing range of climatic conditions. Lions (*Panthera leo*), horses (*Equus* sp.), bison (*B. priscus*), camels (*Camelops* sp.), and badgers (*Taxidea* sp.) were in Alaska during the peak of the last glaciation not because the climate was temperate, but because they were animals that could adapt to climatic extremes as long as appropriate food was available throughout the year. Neither winter nor summer extremes are usually the *direct* factor limiting most large mammals at temperate, or even at most other latitudes.

The concept of climatic equability does not recognize the different sorts of growing seasons that can occur under climatic conditions we would conventionally term "equable." Some 14,000 years ago, when the diverse Pleistocene fauna lived in interior Alaska, the climate was different from that of today, but I do not think this difference can be accurately expressed in degrees of equability. Information about *mean annual* climatic conditions can be equally irrelevant and even misleading to paleoecological investigations if the importance of seasonality, of the intra-annual climatic texture, is not understood. And this seasonality cannot be expressed simply in July and January temperatures. It matters when the rain falls, when spring begins, and so on. At midlatitudes particularly, changes in seasonal character greatly affect the competitive balance between species long before the gross temperature thresholds of most species are reached.

We should also recognize the ideas proposed by Lundelius (1967) regarding the decrease in ecotonal gradients which occurred at this same time. The reduction in complex "ecotonal" assemblages seems to have been a real phenomenon, but going a step farther, describing the changes in terms of "ecotonal gradients" demonstrates our misplaced faith in the permanence of modern biotic provinces. To speak of ecotones presupposes ecological zones of marked identity and longevity whose borders comprise the ecotones. This assumption that coniferous forests, grasslands, tundra, etc., as we know them today, are primordial units is what I have called the "fallacy of the primeval." It is the idea that the present biomes each have a long, complex evolutionary history of fine tuning and are much as they have been since, well, since time primeval. The bulk of evidence is quite the contrary. The biotic communities that we recognize so clearly in personal experience and often assume as basic in our biological studies appear as novel assortments when one approaches them from the fossil record of the Pleistocene.

How do the rise in human numbers, colonization of the New World, and large mammal extinctions relate to this ecological change during the late Pleistocene? As I see it, some essential points of the human overkill argument are: (1) when the megafauna became extinct their respective niches remained, unoccupied; (2) at the time of the extinctions of the large grazing mammals, grasslands habitat occupied by these grazers was expanding not decreasing; (3) the human expansion across Eurasia, into North America, and then South America, is paralleled by a chronological succession of faunal extinctions; and (4) although ecological changes occurred during the previous interglacials, these changes were not accompanied by the sort of peak in megafaunal extinctions which characterizes the end of the Pleistocene.

Martin (1973) has argued that the large mammal extinctions were extinctions *without replacement,* implying that the habitat niches which the extinct animals previously used continued to exist after the occupants' demise, but were vacant. Habitat niches, as they appear in textbooks and to some extent in our mental imagery, are static entities—an invisible space, occupied or unoccupied. I think niches are worthwhile models, but they must be portrayed with a seasonal dimension. Given that large mammals have different requirements and adaptations throughout the seasons, a specific vegetational change does not uniformly affect all the seasonal features of a particular niche. For example, spring and summer resources may be plentiful, but if autumn resources are severely reduced, growth may not be maintained and fat accumulations may be insufficient to see the user-animals through the winter. These animals, and eventually the species, may become extinct. In this case, the spring and summer resources may remain unused, or they may be occupied by other species.

I would argue that we are seeing niches occupied by other species in many instances of the late Pleistocene extinctions. Seasonal niche components that survived their prior occupants were subsequently invaded by the previously subordinate members of the large mammal community. Some of these formerly marginal faunal species became more numerous during the Holocene. Other seasonal resources of the extinct species may remain unused. Janzen and Martin (1982) discuss the seasonally available fruits of some trees which seem to have depended on gomphotheres for dispersal and other trees defending themselves against no known herbivore. This does not mean that an entire niche is left unfilled.

In the late stages of the Wisconsin glaciation North American grasslands began to expand across the midcontinent in a roughly triangular area, the base backed up against the Rocky Mountains, the apex heading toward Ohio and Pennsylvania. Although this is when mammoths, horses, camels, etc., became extinct, grasslands seem to have been expanding at their greatest rate (ca. 12,000 to 10,000 B.P.). Why did these grazers become extinct?

The problem is that while numerous grazing species do imply the existence of some form of grasslands, this does not work in reverse. Widespread grasslands *do not* necessitate the existence of many grazing species. We have good indications that the Pleistocene grasslands were much more complex vegetational entities than those we know today. They do not seem to have been zoned into "cards of a suit" but instead were well-shuffled, "mixed hand" mosaic communities that could support a surprising diversity of grazers—the Holocene grasslands could not. Successful introduction of blackbuck, horses, Barbary sheep, and other exotic species into the North American landscape has been accomplished after 10,000 years of its alteration by hunters, farmers, domestic livestock, fires, and predator changes and does not mean that the niches these animals have found today were available at, say, 10,000 B.P.

Martin's implication that the end of the last glaciation was a time of expanding grasslands *and hence* a time of increased range quality and potentially high numbers of ungulates is correct in part. But one might rather ask the question, for whom did the range quality increase? For the more conservative megafauna which relied on a long growing season and a diverse vegetation to provide a balanced diet, the new Holocene buffet was ultimately a disaster.

It is difficult to refute the human overkill model with theoretical discussions of predator-prey relations because human entry into North America was a novel situation, and certainly predators do sometimes "overeat" their prey. Most recent extinctions, however, are due to habitat reductions rather than human overhunting. This is true of the large mammal species which were reduced in midcontinent North America after European colonization. The bears, wolves, bison, bobcats, pumas, wapiti, and others could be easily restocked in a matter of years—but they are not because the "habitat" in which we can tolerate them living, amongst sheep ranches, row crops, and suburbia, has all but been eliminated. This is not to say that humans cannot cause considerable destruction through overhunting. Human hunters with projectile weapons would have been quite a new predator in North America, and I think it is probable that humans could have caused some of the extinctions. Once boats came into use along the Northwest coast, the Steller's sea cow (*Hydrodamalis*) was apparently killed off immediately and survived only near isolated islands uninhabited by humans (Jones 1968). My arguments, and those of others who argue for ecological causes to the extinctions, do not diminish the heuristic appeal of the overkill hypothesis and its eloquent presentation by Martin and others. In fact the correlation is so good between the timing of the extinctions and the early Clovis culture in North America and the boom of the Duyktai in the Arctic that the burden of proof has probably shifted from the overkill theorists to those who would argue for nonhuman causes.

It is possible to argue that both the megafaunal extinctions and the expansion of humans are features of the same climatic event, an event that opened the door in the Arctic to human expansion while at the same time bringing the environmental changes that led to the extinctions. Exactly what held humans south of the unglaciated Asian High Arctic until the late Pleistocene is unclear. The ungulate prey species were there in great diversity. There is evidence, however, for some major seasonal changes in the late Pleistocene which are a good clue to what may have made the far north more inhabitable.

From several lines of evidence (Guthrie 1982) we can reconstruct the winter landscape of the far north during the past glacial as an extremely windy, treeless, cold, and virtually snowless plain. But during the period beginning some time after 14,000 B.P. (around 12,000 B.P. seems to have been an especially critical time), winter snows became deeper and winter winds subsided. Shrublands and woodlands began to colonize and increase at a rapid rate. This more habitable winter combined with ample wood for tools, cooking, and marrow rendering may have been the key to human access.

However, these climatic events in the far north were the same ones that were destroying the Arctic rangelands of the mammoth fauna. From Beringia the early peoples quickly established themselves south into the Americas, again, at the same time that the quality of the rangelands was decreasing there for mammoth and others (at this time the mammoth seem to have been undergoing a major reduction in body size and social organs). With this perspective one can propose that both the human colonization of the Arctic (and hence the Americas) and the extinctions were causally related primarily to climatic change and were only related secondarily to each other.

So if it was not hunting that caused the extinctions, what was so special about the late Pleistocene–Holocene climatic transition? One can rightfully criticize the catastrophists' scenes of the first spring the climate changed: frozen mammoths with daisies in their teeth and eddies of rotting horse carcasses in the arroyos. Martin (1973) challenged paleoecologists to show the sort of vegetational or climatic change which could be responsible for such dramatic postglacial extinctions. He implicitly argued that the Holocene is just another interglacial like previous ones and that no unique ecological events occurred from 13,000 to 10,000 B.P. other than the spread of humans.

But a growing body of evidence attests that the intensifying "seasonality" at the end of the last glacial was unique in both extent and in pace. *Biotically* the Holocene is not like other interglacials. Hibbard (1960) was probably the first to emphasize this point. Looking at his local faunas scattered throughout Kansas and the central part of North America, he could see that the preceding interglacials harbored cold sensitive forms and complex communities. In a later paper he concluded that interglacial summers were milder, that freezing winter temperatures were rarely reached, and that the vegetation pattern was more like savanna than steppe (Hibbard 1970).

Likewise, in the Old World the interglacial climate and biota do not seem to be analogous to that of the Holocene. Béné and Singer (1975) comment on the peculiar woodland and steppe mixtures of European interglacial faunas. These same interglacial faunal mixtures are described by Kahlke (1961).

Even as far north as Baffin Island, Canada, Andrews (1977) found that there is "no precise modern analogue" to the pollen assemblages from the last interglacial and that the last interglacial must have been warmer than the Holocene optimum by 2°–4°C summer temperature. The last interglacial must indeed have been quite different from the Holocene as it allowed hippopotamus to live in central England, sweet gum to grow in Toronto, and turkeys on the Yukon.

The pedology of the American Central Plains suggests a unique postglacial continental climate. Ruhe (1970) concludes that the "present rigorous semi-arid regime of the Central Plains dates from Brody time 11,400 to 9,100 years ago." Sangamon climates were more moist and may have been warmer than those of today. Wells (1970) has argued from botanical data that the steppes of North America are a recently derived biome nonexistent in past interglacials. Although Kapp (1970) cautions that the prairies themselves are not unique to the postglacial period, he finds considerable Sangamon pine pollen on the Great Plains and suggests scattered pine groves intermixed with prairies.

The generally low number of species of insects (Ross 1970), birds (Mengel 1970), and grasses endemic to the American steppes (Wells 1970) argue that the postglacial extent and even existence of the American steppes may be a Holocene phenomenon. The present extent of the American deserts may also be unique to the Holocene (Axelrod 1979).

Haynes (1982) proposes that the ecological-climatic change at the end of the Pleistocene was so major that the different hydrologic regimes caused a continent-wide geostratigraphic marker.

The shift at the beginning of the Holocene occurred with rapidity. Wright (1970), in a synthesis of data relating to the Central Plains, argued for a relatively steep gradient from about 12,000 to 11,000 B.P. which caused dissolution of the continent-wide boreal forest. Though the Holarctic climatic events began earlier and continued later, this critical millenium was near the threshold of a major ecological change.

The great magnitude of these ecological events is not mirrored in the physical data of deep-sea cores (Greenland ice bores, etc.) because the physical data and the marine biota reflect mean annual differences in climate whereas the terrestrial biota is most sensitive to changes in seasonality.

Late Pleistocene extinctions exhibit a global pattern. The phenomenon was mainly Holarctic and more dramatic in the northern portions (Alaska and Siberia), while the rest of the world was affected more lightly. Any climatic change produced by insolarity shifts would be more extreme at higher latitudes. The continents are skewed asymmetrically toward the northern half of the globe. Central and South America (Webb 1978) and Africa (Klein this volume) also underwent minor large mammal extinctions centered about the same time period as in the Holarctic (ca. 11,000 yr B.P.).

Looking at the extinction problem through the eyes of a young paleontologist in the early 1960s, I encountered my first important lesson—that the present can be used to understand the past only with sensitive discretion. In fact, much of the past may have no modern analogue. Most of us find it difficult to imagine the exotic animals of the Pleistocene, at home, here, in our familiar landscapes. It is strange enough to picture woolly mammoth and lion on plains now crossed by Interstate 80. Perhaps it is just too great a conceptual shift to picture our general surroundings a few thousand years ago, the landscape itself, as dramatically different from that of today. But I believe most Quaternary paleoecologists have come to realize that their particular local area was indeed quite different. And many of us are beginning to conclude that these local oddities add up to a radically different late Pleistocene environment throughout the Holarctic—and perhaps the entire globe.

References

Andrews, J. T. 1977. Inferred climates of the last interglacial and early Wisconsin glaciation. Baffin Island N.W.T., Canada: biostratigraphic evidence. (Abstracts). 10th International Quaternary Association Congress, Birmingham.

Andrews, P. J. and J. A. H. VanCouvering. 1975. Paleoenvironments of the East African Miocene. pp. 62–103. *In* Approaches to Primate Paleobiology. F. S. Szalay (ed.). Basel: Karger.

Ashworth, A. C. 1977. Late Wisconsin beetle assemblages from North America. p. 21. (Abstracts). 10th International Quaternary Association Congress, Birmingham.

Aufenberg, W. and W. W. Milstead. 1965. Reptiles in the Quaternary of North America. pp. 557–568. *In* The Quaternary of the United States. H. E. Wright, Jr. and D. G. Frey (eds.). Princeton Univ. Press, Princeton, 922 pp.

Axelrod, D. I. 1967. Quaternary extinctions of large mammals. Univ. of California. Publ. in Geol. Sci. 74:1–42.

———. 1979. Age and origin of the Sonoran Desert vegetation. Occ. Pap. Calif. Acad. Sci. No. 132. pp. 1–74.

Bader, O. N. 1978. Sungir: An Upper Paleolithic Site. Nauka, Moscow. 272 pp.

Bell, R. H. V. 1969. The use of the herb layer by grazing ungulates in the Serengeti. pp. 25–46. *In* Animal Populations in Relation to Their Food Resources. A. Watson (ed.). Blackwell, Oxford.

Béné, E. L. and R. Singer. 1975. Clacton-on-Sea, ein interglazialer Biotope und eine interglaziale Biozonose. Quatär Paläontologie I: 183–186.

Borman, M. D., D. E. Johnson, and L. R. Rittenhouse. 1982. Comparisons of grazing time and fecal output between lactating and nonlactating mares and cows. (Abstract). 35th Annual Meeting Society of Range Management, Calgary, Alberta.

Brown, F. L. and M. J. Trlica. 1977. Simulated dynamics of blue gramma production. Jour. Appl. Ecol. 14:215–224.

Bryant, J. P. and P. J. Kuropat. 1980. Selection of winter forage by subarctic browsing vertebrates: the role of plant chemistry. Ann. Rev. Ecol. and Syst. 11:261–285.

Campbell, K. E. 1973. The Pleistocene avifauna of the Talara tar seeps, northwestern Peru. Ph.D. Dissertation, Univ. of Florida.

Chapin, F. Stuart, III. 1980. The mineral nutrition of wild plants. Ann. Rev. Ecol. and Syst. 11:233–260.

Child, G. and W. von Richter. 1969. Observations on ecology and behavior of Lechwe, puku, and waterbuck along the Chobe River, Botswana. *Zeit. für Saügtier.* 34:275–295.

Chomko, S. A. 1982. Late Pleistocene–Holocene faunal successions in the Northern Bighorn Mountains, Wyoming. p. 80. (Abstracts). 7th Biennial Meeting American Quaternary Association, Seattle.

Cole, K. L. 1982a. Late Quaternary vegetation

in the eastern Grand Canyon. Science 217:1142–1144.

———. 1982b. Pleistocene packrat middens from the western Sierra Nevada, California. p. 82. (Abstracts). 7th Biennial Meeting of American Quaternary Association, Seattle.

Colinvaux, P. 1980. Pollen from the late glacial of tropical South America: vegetation and climate at first settlement. Quart. Rev. of Arch. 2:1–7.

Coope, G. R. 1974. Tibetan species of dung beetle from the late Pleistocene of England. Nature (London) 245:335–336.

Coope, G. R. and P. B. Angus. 1975. An ecological study of a temperate interlude in the last glaciation based on fossil Coleoptera from Islesworth, Middlesex. Jour. Anim. Ecol. 44:365–391.

Coustan, K. J. 1972. Winter foods and range use of three species of ungulates. Jour. Wildl. Mgmt. 36:1068–1076.

Dalquest, W. W. 1965. New Pleistocene formation and local fauna from Hardeman County, Texas. Jour. Paleont. 39:63–79.

Davis, M. B. 1976. Pleistocene biogeography of temperate deciduous forests. Geoscience and Man 13:13–26.

Davis, S. J. M. 1981. The effects of temperature change and domestication on body size of late Pleistocene to Holocene mammals of Israel. Paleobiology 7:101–114.

Ferrar, A. A. and B. H. Walker. 1974. An analysis of herbivore/habitat relationships in Kyle National Park, Rhodesia. Jour. So. Afr. Wildl. Mgmt. Assoc. 4:137–147.

Field, C. R. 1975. Climate and food habits of ungulates on Galana Ranch. E. Afr. Wildl. Jour. 13:203–220.

———. 1976a. Palatability factors and nutritive values of the food of buffalos (*Syncerus caffer*) in Uganda. E. Afr. Wildl. Jour. 14:181–201.

———. 1976b. The savanna ecology of Kidepo Valley National Park. E. Afr. Wildl. Jour. 14:1–15.

Freeland, W. J. and D. H. Janzen. 1974. Strategies in herbivory by mammals: the role of plant secondary compounds. Amer. Nat. 108:269–289.

Garutt, V. E. 1964. *Das mammut (Mammuthus primigenius)* Neue Brehm-Bücherei, Wittenberg-Lutherstadt, Ziemsen.

Geist, V. 1977. A comparison of social adaptations in relation to ecology in gallinaceous birds and ungulate societies. Ann. Rev. Ecol. and Syst. 8:193–208.

Giterman, R. E. 1975. Palynologische Charakteristik der unterpleistozänen ablagerungen von Unterlauf der Kolyma. Quatär Paläontologie 1:7–11.

Graham, R. W. 1976. Late Wisconsin mammalian faunas and environmental gradients of the eastern United States. Paleobiology 2:343–350.

———. 1979. Paleoclimates and late Pleistocene faunal provinces of North America. pp. 49–69. *In* Pre-Llano Cultures of the Americas. R. L. Humphrey and D. Stanford (eds.). The Anthropological Society of Washington. 150 pp.

Graham, R. W. and H. A. Semken. 1976. Paleoecological significance of the short-tailed shrew (genus: *Blarina*) with a systematic discussion of *Blarina ozarkensis*. Jour. Mammal. 57:433–449.

Grayson, D. K. 1977. Pleistocene avifaunas and the overkill hypothesis. Science 195:691–692.

Grime, J. P. 1979. Plant Strategies and Vegetation Processes. John Wiley & Sons, New York. 222 pp.

Guilday, J. E., H. W. Hamilton, E. Anderson, and P. W. Parmalee. 1978. The Baker Bluff Cave Deposit, Tennessee, and the late Pleistocene faunal gradient. Bull. Carnegie Mus. Nat. History. No. 11. 67 pp.

Guthrie, R. D. 1968. Paleoecology of the large mammal community in interior Alaska during the late Pleistocene. Amer. Midl. Nat. 79:346–363.

———. 1980. Bison and man in North America. Canadian Jour. of Anthropology 1:55–75.

———. 1982. Mammals of the mammoth steppe as paleoenvironmental indicators. *In* The Arctic Mammoth Steppe. D. M. Hopkins, J. V. Matthews, C. Schweger, S. Young (eds.). Academic Press, New York.

Hallbert, G. R., H. A. Semken, and L. C. Davis. 1974. Quaternary record of *Microtus xanthognathus* (Leach). The yellow-cheeked vole, from northwestern Arkansas and southwestern Iowa. Jour. Mammal. 55:640–645.

Hansen, R. M. 1976. Foods of free-roaming horses in southern New Mexico. Jour. Range Mgmt. 29:347.

———. 1978. Shasta ground sloth food habits, Rampart Cave, Arizona. Paleobiology 4:302–319.

Harborne, J. B. 1978. Biochemical Aspects of Plant and Animal Evolution. Academic Press, New York. 435 pp.

Harington, C. R. 1977. Pleistocene mammals of the Yukon Territory. Ph.D. Dissertation. Univ. of Alberta, Edmonton. 1060 pp.

Harris, D. R. 1978. The distribution, diversity,

and development of tropical savanna environments. Wenner-Gren Foundation Symposium '79. Burg-Wartenstein, Austria.

Harris, A. H. and P. Mundel. 1974. Size reductions in bighorn sheep (*Ovis canadensis*) at the close of the Pleistocene. Jour. Mammal. 55:678–680.

Haynes, C. V. 1982. Pleistocene–Holocene boundary in the United States: alluvial stratigraphy and geochronology. p. 97: (Abstract). 7th Biennial Meeting American Quaternary Association, Seattle.

Hibbard, C. W. 1960. An interpretation of Pliocene and Pleistocene climates in North America. Mich. Acad. Sci., Arts, Lett., 62nd Annual Report. pp. 5–30.

———. 1970. Pleistocene mammalian local faunas from the Great Plains and the central lowland provinces in the United States. pp. 395–433. *In* Pleistocene and Recent Environments of the Central Great Plains. W. Dort, Jr. and J. K. Jones, Jr. (eds.). Univ. Press of Kansas, Lawrence.

Hibbard, C. W., C. E. Ray, D. E. Savage, D. W. Taylor, and J. E. Guilday. 1965. Quaternary mammals of North America. pp. 509–525. *In* The Quaternary of the United States. H. E. Wright, Jr. and D. G. Freyes (eds.). Princeton Univ. Press, Princeton. 922 pp.

Hoffman, R. S. and J. K. Jones, Jr. 1970. Influence of late glacial and post glacial events on the distribution of Recent mammals on the northern Great Plains. pp. 355–396. *In* Pleistocene and Recent Environments of the Central Great Plains. W. Dort, Jr. and J. K. Jones, Jr. (eds.). Univ. Press of Kansas, Lawrence.

Holman, J. A. 1976. Paleoclimatic implication "ecologically incompatible" herpetological species (late Pleistocene: southeastern United States). Herpetologica 32:290–295.

———. 1977. Herpetofaunal evidence for the Pleistocene climatic equability hypothesis in North America. p. 211. (Abstracts). 10th International Quaternary Association, Birmingham.

Hooijer. D. A. 1950. The study of specific advances in the Quaternary. Evolution 4:360–361.

———. 1977. Pleistocene remains of *Panthera tigris* (Linnaeus) subspecies from Wanhsien, Szechwan China, compared with fossil and Recent tigers from other localities. Amer. Anthr. Novit. 1346:1–17.

Hoppe, P. P. 1978. Rumen fermentation in African ruminants. pp. 1–149. Proc. 13th International Congress of Game Biologists.

Iversen, J. 1954. The late glacial flora of Denmark and its relation to climate and soil. Denmark's Geol. Unders. Sev. II. 80:87–119.

Janis, C. 1975. The evolutionary strategy of the Equidae and the origins of the rumen and caecal digestion. Evolution 30:757–774.

Janzen, D. H. 1975. Ecology of plants in the tropics. Arnold, London. 361 pp.

Janzen, D. H. and P. Martin. 1982. Neotropical anachronisms: fruits the gomphotheres left behind. Science 215:19–27.

Jarman, J. P. 1974. The social organization of antelope in relation to their ecology. Behavior 48:215–266.

Jewell, P. A. 1974. Managing animal populations. pp. 87–119. *In* Conservation in Practice. A. Warren and F. B. Goldsmith (eds.). John Wiley & Sons, New York. 689 pp.

Jingfors, K. T. 1980. Habitat relationships and activity patterns of a reintroduced musk oxen population. M.S. Thesis, Univ. of Alaska, Fairbanks. 113 pp.

Jones, R. E. 1968. A *Hydrodamalis* skull from Monterey Bay, California. Jour. Mamm. 48:143–144.

Kahlke, H. D. 1961. Revision der Saügetierfaumen der Klassischen deutchen Pleistözn-Fundstellen von Susseborn, Mosbach, und Taubach. Geologie 10:493–532.

Kapp, R. O. 1970. Pollen analysis of pre-Wisconsin sediments of the Great Plains. pp. 143–155. *In* Pleistocene and Recent Environments of the Central Great Plains. W. Dort, Jr., and J. K. Jones, Jr. (eds.). Univ. Press of Kansas. 433 pp.

Kennett, J. P. and N. J. Shackleton. 1976. Oxygen-isotopic evidence for the development of the psychronosphere 38 M years ago. Science 260:513–515.

Klein, D. R. 1970a. Food selection by North American deer and their response to overutilization of preferred plant species. pp. 25–44. *In* Animal Populations in Relation to their Food Resources. A. Watson (ed.). Blackwell, Oxford.

———. 1970b. Tundra ranges north of the boreal forest. Jour. Range Mgmt. 23:8–14.

———. 1977. Winter food preferences of snowshoe hares (*Lepus americanus*) in interior Alaska. pp. 266–275. Proc. 13th International Congress of Game Biologists.

Koerth, B. H., L. J. Krysl, B. F. Sowell, and E. C. Bryant. 1982. Estimating seasonal diet quality of pronghorn antelope from fecal analysis. (Abstract). 35th Annual

Meeting Society of Range Management, Calgary, Alberta.

Kowalski, K. 1967. The Pleistocene extinction of mammals in Europe. pp. 343–364. *In* Pleistocene Extinctions: The search for a cause. P. S. Martin and H. E. Wright, Jr. (eds.). Yale Univ. Press, New Haven.

Kurtén, B. 1959. Rates of evolution in fossil mammals. Cold Spring Harbor Symp. on Quant. Biol. 34:205–215.

———. 1965a. The Carnivora of the Palestine caves. Acta Zool. Fenn. 107:1–104.

———. 1965b. On the evolution of the European wild cat *Felis silvestris* Schreber. Acta Zool. Fenn. 111:1–26.

———. 1965c. The Pleistocene Felidae of Florida. Bull. Florida State Mus. 9:215–273.

———. 1968. Pleistocene Mammals of Europe. Aldine Publ. Co., Chicago. 317 pp.

Lamprey. H. F. 1963. Ecological separation of large mammal species in the Tarangire Game Reserve, Tanganyika. E. Afr. Wildl. Jour. 1:63–92.

Laws, R. M., I. S. C. Parker, and R. C. B. Johnstone. 1975. Elephants and their Habitats. Oxford Univ. Press, London. 376 pp.

Laudermilk, J. D. and P. A. Munz. 1934. Plants in the dung of *Nothrotherium* from Gypsum Cave, Nevada. Carnegie Inst. Wash. Publ. 453:31–37.

Leopold, E. B. 1967. Late Cenozoic patterns of plant extinction. pp. 203–246. *In* Pleistocene Extinctions: The search for a cause. P. S. Martin and H. E. Wright, Jr. (eds.). Yale Univ. Press, New Haven.

Lodge, G. A. and G. E. Lamming. 1967. Growth and Development of Mammals. Plenum Press, New York. 527 pp.

Long, A. and P. S. Martin. 1974. Death of the American ground sloth. Science 186:638–640.

Luick, J. R., R. G. White, A. M. Gau, and R. Jenness. 1974. Compositional changes in the milk secreted by grazing reindeer. 1. Gross composition and ash. Jour. of Dairy Science 57:1325–1333.

Lundelius, E. L., Jr. 1967. Late Pleistocene and Holocene history of central Texas. pp. 288–319. *In* Pleistocene Extinctions: The search for a cause. P. S. Martin and H. E. Wright, Jr. (eds.). Yale Univ. Press, New Haven.

———. 1974. The last fifteen thousand years of faunal change in North America. pp. 141–160. *In* History and Prehistory of the Lubbock Lake Site. Black, C. G. (ed.). The Mus. Jour. 15:1–160.

MacGintie, H. E. 1968. Climate since the late Cretaceous. pp. 61–79. *In* Zoogeography. C. L. Hubbs, (ed.). Amer. Assoc. Adv. Sci. Publ. 51. 509 pp.

McNaughton, J. S. 1976. Serengeti migratory wildebeest: facilitation of energy flow by grazing. Science 191:92–95.

———. 1978. Serengeti ungulates: feeding selectivity influences the effectiveness of plant defense guilds. Science 199:238–250.

Malpas, R. C. 1977. Diet and the condition of growth of elephants in Uganda. Jour. Appl. Ecol. 14:489–504.

Marshall, L. G. and R. S. Corruccini. 1978. Variability, evolutionary rates, and allometry in dwarfing lineages. Paleobiology 4:101–119.

Martin, P. S. 1973. The discovery of America. Science 179:969–974.

Martin, P. S. and H. E. Wright, Jr. 1967. (eds.). Pleistocene Extinctions: The Search for a Cause. Yale Univ. Press, New Haven. 453 pp.

Martin, R. A. and S. D. Webb. 1974. Late Pleistocene mammals from Devil's Den Fauna, Levy County. pp. 114–145. *In* Pleistocene Mammals of Florida. S. D. Webb. (ed.). Univ. of Fla. Press, Gainesville. 270 pp.

Matthews, J. V., Jr. 1975. Insects and plant macrofossils from two Quaternary exposures in the Old-Crow-Porcupine region, Yukon Territory, Canada. Arctic and Alpine Research 7:249–259.

———. 1976. Arctic steppe—an extinct biome. pp. 73–77. (Abstracts). American Quaternary Association 4th Biennial Meeting, Tempe.

———. 1979. Tertiary and Quaternary environments: historical background for an analysis of the Canadian insect fauna. *In* Canada and its Insect Fauna. H. V. Danks (ed.). Mem. Entom. Soc. Canada No. 108, 573 pp.

Mattson, W. J. 1980. Herbivory in relation to plant nitrogen content. Ann. Rev. Ecol. and Syst. 11:119–162.

Mayhew, D. F. 1977. New information on British Quaternary lagomorphs and larger rodents. p. 296. (Abstracts). 10th International Quaternary Association Congress, Birmingham.

Mehringer, P. J., Jr., J. E. King, and E. H. Lindsay. 1970. A record of Wisconsin-age vegetation and fauna from the Ozarks of western Missouri. pp. 173–185. *In* Pleistocene and Recent Environments of the Central Great Plains. W. Dort, Jr. and J. K. Jones, Jr. (eds.). Univ. Press of Kansas, Lawrence.

Mengel, R. M. 1970. The North American Central Plains as an isolating agent in bird speciation. pp. 89–119. *In* Pleistocene and Recent Environments of the Central Great Plains. W. Dort, Jr. and J. K. Jones, Jr., (eds.). Univ. Press of Kansas, Lawrence.

Olivier, R. C. D. 1979. Ecology and behavior of living elephants: bases for assumptions concerning the extinct woolly mammoth. Wenner-Gren Foundation Symposium '81. Burg-Wartenstein, Austria.

Peden, D. G., G. M. Van Dyne, R. W. Rice, and R. M. Hansen. 1974. The trophic ecology of *Bison bison* on short grass prairie. Jour. App. Ecol. 11:489–498.

Péwé, T. L. 1975. Quaternary Geology of Alaska. U.S. Geol. Prof. Paper 835. 212 pp.

Poplin, F. 1976. Les grandes vertébrates de Gönnersdorf fouilles. 1968. Franz Steiner Verlag GMBH Wiesbaden, D.D.R.

Ray, E. E. 1967. Pleistocene mammals from Ladds, Barlow County, Georgia. Bull. Georgia Acad. Sci. 25:120–150.

Reed, C. A. 1970. Extinction of mammalian megafauna in the Old World Late Quaternary. Bioscience 20:284–288.

Reid, C. 1915. The plants of the late glacial deposits of the Lea Valley. Geo. Soc. London Quat. Jour. 155–163.

Rhoades, D. F. 1979. Evolution of plant chemical defense against herbivores. pp. 1–48. *In* Herbivores: Their interaction with secondary plant metabolites, G. A. Rosenthal and D. H. Janzen (eds.). Academic Press, New York.

Rhoades, D. F. and R. G. Cates. 1976. Toward a general theory of plant antiherbivory chemistry. Rec. Adv. Phytochem. 10:168–213.

Rosenthal, G. A. and D. H. Janzen. 1979. Herbivores: Their interaction with secondary plant metabolites. Academic Press, New York. 718 pp.

Ross, H. H. 1970. An ecological history of the Great Plains: Evidence from the insects. pp. 225–240. *In* Pleistocene and Recent Environments of the Central Great Plains. W. Dort, Jr. and J. K. Jones, Jr. (eds.). Univ. Press of Kansas, Lawrence. 433 pp.

Ruhe, R. V. 1970. Soils, paleosols, and environments. pp. 37–52. *In* Pleistocene and Recent Environments of the Central Great

Plains. W. Dort, Jr., and J. K. Jones, Jr. (eds.). Univ. Press of Kansas, Lawrence.

Salgadov-Labourau, M. L. 1980. A pollen diagram of the Pleistocene-Holocene boundary of Lake Valencia, Venezuela. Rev. of Paleobot. and Palyn. 30:297–312.

Saunders, J. J. 1978. *Mammuthus* in the Clovis horizon of North America. pp. 99–107. (Abstracts). American Quaternary Association. 5th Biennial Meeting, Edmonton, Alberta.

Schultz, C. B., L. G. Tanner, and L. D. Martin. 1972. Phyletic trends in certain lineages of Quaternary mammals. Bull. Univ. Nebraska State Mus. 9:183–195.

Schwert, D. P. and A. C. Ashworth. 1982. A model for the postglacial development of the Arctic-Subarctic beetle (Coleoptera) fauna of North America. p. 161 (Abstract) 7th Biennial Conference American Quaternary Association, Seattle.

Schwert, D. P. and A. V. Morgan. 1980. Paleoenvironmental implications of late glacial insect assemblages from northwestern New York. Quat. Res. 13:93–110.

Semken, H. A. 1974. Micromammal distribution and migration during the Holocene. American Quaternary Association, 3rd Biennial Meeting, Madison. 1974:25.

Shaul, D. M. B. 1962. The composition of milk in wild animals. The International Zoological Yearbook 4:333–342.

Sher, A. V. 1967. Saiga remains mined in the northeast of Siberia and in Alaska. Commission on Studies of the Quaternary Period, USSR, Bulletin 33. pp. 97–112.

———. Mammals and stratigraphy of the Pleistocene of the extreme northeast of the USSR and North America. Nauka, Moscow. 310 pp. (translation, Int. Geol. Rev. 16:1–284, 1974).

Sinclair, A. R. E. 1975. The resource limitation of trophic levels in tropical grassland ecosystems. Jour. Anim. Ecol. 44:497–520.

Skinner, M. F. and O. C. Kaisen. 1947. The fossil *Bison* of Alaska and preliminary revision of the genus. Bull. Amer. Mus. Nat. History. 89:127–256.

Slaughter, B. H. 1967. Animal ranges as a clue to late Pleistocene extinction. pp. 155–168. *In* Pleistocene Extinctions: The search for a cause. P. S. Martin and H. E. Wright, Jr. (eds.). Yale Univ. Press, New Haven.

———. 1975. An ecological interpretation of the Brown Sand Wedge Local Fauna, Black Water Draw, New Mexico: A hypothesis concerning the late Pleistocene extinction.

pp. 179–192. *In* Late Pleistocene Environments of the Southern High Plains. F. Wendorf and J. J. Hester (eds.). Fort Bergwin Res. Center 9:1–290.

Slaughter, B. H. and B. R. Hoover. 1963. Sulphur River formation and the Pleistocene mammals of Ben Franklin local fauna. Southern Methodist Univ. Grad. Res. Center Jour. 31:132–148.

Sparks, B. W. and R. G. West. 1972. The Ice Age of Britain. Metheuen Ltd., London. 279 pp.

Spaulding, W. G. 1978. The changing vegetation of a southern Nevada mountain range. p. 177. (Abstracts). American Quaternary Association 5th Biennial Meeting, Edmonton, Alberta.

Starkel, L. 1977. Shifting of landscape zones or the creation of new ones? p. 435. (Abstracts). 10th International Quaternary Association Congress, Birmingham.

Suc, J. P. 1976. La vegetation au Pleistocene inférieur en Languedoc Méditerrannéen Roussillon et Catagogne. *In* La Préhistoire Francaise. H. Lumley (ed.). Paris.

Taylor, D. W. 1965. The study of Pleistocene nonmarine mollusks in North America. pp. 597–612. *In* The Quaternary of the United States. H. E. Wright, Jr. and D. G. Frey (eds.). Princeton Univ. Press, Princeton.

Tchernov, E. 1981. The impact of the postglacial in the fauna of Southeast Asia. pp. 197–216. *In* Beiträge zur Umweltgeschichte des Vorderen Orients. W. Frey and H. P. Uerpmann (eds.). Ludwig Reichert, Wiesbaden.

Tieszen, L. L., D. Hein, S. A. Quortrup, J. H. Troughton and S. K. Imbamba. 1979. Use of ^{13}C values to determine vegetation selectivity in East African herbivores. Oecologia 37:351–360.

Tikhomorov, B. A. 1958. Natural conditions and vegetation in the mammoth epoch in Northern Siberia. Problems in the North 1:168–188.

Tsukada, M. and S. Sugita. 1982. Coniferous forests during the last maximum glacial period in Japan. (Abstract). 7th Biennial Meeting American Quaternary Association, Seattle.

Uerpmann, H. P. 1981. The major faunal areas of the Middle East during the late Pleistocene and early Holocene. pp. 99–106. *In* Colloques Internationaux du C.N.R.S. No. 598. Paris.

Van Devender, T. R. 1976. The biota of hot deserts in North America during the last glaciation: The packrat midden record. pp. 62–63. (Abstracts). 4th Biennial Meeting American Quaternary Association, Tempe.

———. 1978. Early Holocene and late Pleistocene amphibians and reptiles in Sonoran Desert packrat middens. Copeia 78:464–475.

Van Devender, T. R. and J. I. Mead. 1976. Late Pleistocene and modern plant communities of Shinumo Creek and Peach Springs Wash, lower Grand Canyon, Arizona. Jour. Ariz. Acad. Sci. 11:16–22.

Van Devender, T. R., A. W. Phillips III, and J. I. Mead. 1977. Late Pleistocene reptiles and small mammals from the lower Grand Canyon of Arizona. The Southwestern Naturalist. 22:49–66.

Vereshchagin, N. K. 1977. Berelekh "cemetery" of mammoths. pp. 5–50. *In* Mammoth Fauna of the Russian Plain and Eastern Siberia. O. A. Skarlato (ed.). Nauka, Leningrad.

Webb, S. D. 1969. Extinction-origination equilibrium in late Cenozoic land mammals of North America. Evolution 23:688–702.

———. 1977. A history of savanna vertebrates in the New World. Part I. Ann. Rev. Syst. and Ecol. 8:355–380.

———. 1978. A history of savanna vertebrates in the New World. Part II. Ann. Rev. Syst. and Ecol. 9:393–426.

Webb, T. III and R. A. Bryson. 1972. Late and postglacial climatic changes in the northern midwest U.S.A.: Quantitative estimates derived from fossil pollen spectra by multivariate statistical analysis. Quat. Res. 2:70–115.

Wells, P. V. 1970. Vegetational history of the Great Plains: A postglacial record of coniferous woodland in southeastern Wyoming. pp. 185–202. *In* Pleistocene and Recent Environments of the Central Great Plains. W. Dort, Jr. and J. K. Jones, Jr. (eds.). Univ. of Kansas Press, Lawrence.

———. 1977. Quaternary vegetation history of arid America. p. 499. (Abstracts). International Quaternary Association, Birmingham.

Wen-Chung, P. 1963. On the problem of the change in body size in Quaternary mammals. Sci. Sinica 12:231–235.

Wetmore, A. 1967. Pleistocene Aves from Ladds, Georgia. Bull. Georgia Acad. Sci. 25:151–153.

Wetzel, R. M. 1977. The Chacoan peccary *Catogonus wagneri* (Rusconi). Bull. Carnegie Mus. Nat. Hist. 3:1–36.

White, T. C. R. 1978. The importance of a relative shortage of food in animal ecology. Oecologia 33:71–86.

Whitten, K. R. 1975. Habitat relationships and population dynamics of Dall sheep (*Ovis dalli dalli*) in Mt. McKinley National Park. M.S. Thesis, Univ. of Alaska. Fairbanks. 177 pp.

Winters, J. 1980. Summer habitat and food utilization by Dall sheep and its relation to body and horn size. M.S. Thesis. Univ. of Alaska, Fairbanks. 211 pp.

Wolfe, J. A. 1975. Some aspects of plant geography of the northern hemispheres during late Cretaceous and Tertiary. Ann. Missouri Bot. Garden 62:264–279.

Wood, A. J., I. Mct. Cowan, and H. C. Nordan. 1962. Periodicity of growth in ungulates as shown by deer of the genus *Odocoileus*. Canadian Jour. Zool. 40:593–603.

Wright, H. E., Jr. 1970. Vegetational history of the Central Great Plains. pp. 157–173. *In* Pleistocene and Recent Environments of the Central Great Plains. W. Dort, Jr. and J. K. Jones, Jr. (eds.). Univ. Press of Kansas. 433 pp.

14

Seasonality, Gestation Time, and Large Mammal Extinctions

RICHARD A. KILTIE

AS THE GLACIERS RETREATED, the climate in regions outside the tropics became less "equable" than ever in the Pleistocene. Winters became colder, summers hotter, and precipitation more seasonally variable, so that the time favorable for production of offspring became more restricted. Larger mammals with longer gestation periods and "inflexible" mating habits then produced young at unfavorable times of the year. The decline in reproductive success of the populations culminated in species extinctions. This is the scenario that Slaughter (1967) and Axelrod (1967) have suggested to explain the widespread extinctions of large mammals at the end of the Pleistocene.

Two potential weaknesses exist with this explanation. One concerns the accuracy of the statement that environmental seasonality increased over most of the world as glaciers retreated. Such an assertion contradicts the widespread tendency to picture unglaciated areas in glacial times as experiencing extreme arctic weather (e.g. Reed 1969), but as Axelrod (1967) especially emphasized, mean temperature and equability may vary independently. Some evidence cited by Slaughter (1967), Axelrod (1967), and others (Hibbard et al. 1965; Taylor 1965; Leopold 1967; Bryson et al. 1970; Martin and Neuner 1978; Holman 1976, 1980; R. D. Guthrie, cited in Stanley 1980; Guthrie this volume; Graham and Lundelius this volume) supports the suggestion of more extreme weather cycles in early postglacial times. This evidence, however, is not overwhelming (Martin 1967, Mehringer 1967, Van Valen 1969, McDonald this volume), and a consensus is not yet apparent with respect to global changes in seasonality, per se, at the end of the Pleistocene (see Wright 1976, Lamb 1977, Pearson 1978, Hecht et al. 1979, Frakes 1979).

The other potential weakness of Slaughter's and Axelrod's hypothesis is the assumption that larger species would be less able to adapt their mating habits to increasingly seasonal environments than smaller species. This assumption might be reasonable if we could be sure that larger species had no refuges to migrate to and if the effect of environmental change was rapid and extreme. However, it has not been made clear why the reproductive habits of larger species would not have been amenable to adaptive change if environmental cycles *gradually* became more extreme or less regular (over, say, a thousand years).

Here I review three potentially negative influences on mammalian reproductive success ("fitness") when seasons favorable for birth become shorter or less regular. These influences generally do not depend on a rapid change in the severity or regularity of the seasons. They also tend to be more marked for species with longer gestation periods than for those with shorter gestations. Because mean gestation length is proportional to about the one-fourth power of adult body weight (Kihlström 1972, Sacher and Staffeldt 1974, Blueweiss et al. 1978, Zeveloff and Boyce 1980), this is equivalent to asserting that the influences should be more adverse for larger species than for smaller ones. Hence they may provide some insight into the question of megafaunal extinctions.

I will first describe these factors in terms of idealized examples and then apply them to the observed pattern of late Pleistocene extinctions. Throughout, my focus is on herbivores, which suffered the most unusual number of extinctions at this time (Hibbard et al. 1965, Martin and Guilday 1967, Kurtén 1968, Webb 1969, Kurtén and Anderson 1980).

1. *Direct vs. correlated breeding cues in changing or unpredictable environments.* Suppose that an environment consists of two seasons affecting survival of females and their newborn young. The favorable season is F days long and the unfavorable season is $365-F$. Assume that gestation period is fixed for each species. If the favorable season (F) is longer than the gestation period, individuals of a species will always be able to produce at least one litter each year if they initiate breeding by "direct" cues—that is, by conditions such as temperature or food availability that actually make a season favorable for birth. On the other hand, if F is less than the gestation period, breeding has to be initiated by cues that do not directly affect reproductive success but which are correlated in time with the conditions that directly influence survival of mother and young (Sadleir [1969] provides a more detailed classification of relationships between "innate" breeding cycles and environmental cycles.) Virtually any regularly recurring environmental conditions could be used as correlated breeding cues; temperature, rainfall, or photoperiod frequently serve this purpose (Sadleir 1969, Spinage 1973). Because gestation period is positively related to body size, the likelihood of needing to use correlated breeding cues is greater for larger species.

Such observations served as the basis for Slaughter's and Axelrod's suggestions that species with long gestation periods should be more severely affected than those with short gestation periods as seasons favorable for birth become more restricted. Confronted with such an environmental change, many large species would have to make one of two kinds of behavioral changes: Species that previously depended on correlated cues to initiate mating would have to alter their use of such cues to produce young only in the new, restricted favorable periods. Species that previously were able to use direct cues, but now have gestation periods greater than the shortened favorable period, would have to switch to indirect cueing. In both cases, and perhaps especially in the second, populations of the species would experience reduced fitness until more appropriate breeding habits were evolved. Populations of large species would experience depressed reproductive rates, hence higher risks of extinction (MacArthur and Wilson 1967) for a longer time than small species while such adjustments are made because of their lower birth rates and longer gestation times (Smith 1954, Bonner 1965, Pianka 1970, Fenchel 1974, Blueweiss et al. 1978, Western 1979).

Species with gestation periods sufficiently short that direct cues could still be employed in the more seasonal environment would bear no such costs because their system for initiating mating would allow immediate and perfect adaptation to the new conditions (see Nichols et al. 1976). Use of direct cues thus allows a rapid, "organismic" response to environmental change, while use of correlated cues requires a slower, "population" response through genetic change (Wilson 1975).

Essentially the same arguments lead to the inference that species using correlated cues should be more severely affected by decreases in the *predictability* of favorable periods, even if the environment's average conditions do not change. If the favorable period in any year begins later or ends earlier than normal, or if the favorable season is completely displaced, some individuals of such species might bear young before or after conditions are favorable and thus lower the population's reproductive output. Species using direct cues would not suffer unless the favorable period became shorter than the gestation period. Because many extratropical regions apparently became drier in post-glacial times and because precipitation is commonly less predictable where it is less frequent (e.g. Sadleir 1969, Leuthold and Leuthold 1975, Low 1978), it is possible that seasonal precipitation cycles became less predictable at the end of the Pleistocene (see also Bryson et al. 1970). The main way for large placental species to adapt to such unpredictability would have been to breed "conservatively," i.e. so that young were produced at the most reliably favorable time.

I should point out here that mating habits based solely on direct cues are probably comparatively rare among extratropical placentals, large or small (Negus et al. 1977). The reason is obvious. If the favorable period is longer than the gestation time plus time to next conception, a second litter can be produced in any season by producing one litter as soon as conditions become favorable. This can only be done if the arrival of favorable conditions is anticipated by the initiation of mating with correlated cues. Many small species appear to combine correlated cues (such as photoperiod) with direct cues; the former are used to initiate breeding and the latter determine the end of the mating season (e.g. see Sadleir 1969). As long as some species at least partially employ direct cues and others respond solely to correlated cues, the generalization that those using only correlated cues (larger species) should be more severely affected by increasingly seasonal or unpredictable environments should still hold.

2. *Uncertainty in the length of gestation period.* Although each species can be characterized by a mean gestation period, it should not be forgotten that there must also be intraspecific variation in gestation period. Such variation is relevant to the question of adaptation to seasonal environments because even if a species uses mating cues that perfectly correlate with future conditions affecting maternal/offspring survival, there can still be an element of uncertainty in the timing of birth with respect to season, which can reduce mean fitness.

Figure 14.1 shows a plot of range and standard deviation of gestation period versus mean gestation period for as many cases as I could find in the literature with seven or more individual measurements of gestation length (Kiltie 1982). Not surprisingly, both measures of variation are positive functions of the mean. Although there is some scatter about the regression lines, no dramatic divergences occur from the overall positive relationship. Furthermore, a multiple regression analysis (Kiltie 1982) failed to find any significant effects of neonatal brain and body size, neonatal brain and body advancement, adult brain and body size, or litter size on variation in gestation time independent of the effects of these factors on *mean* gestation time (see Sacher and Staffeldt 1974). Variation in gestation time, therefore, does not appear amenable to great adaptive reduction without a concomitant reduction in mean gestation time. If uncertainty of gestation time decreases average reproductive success in seasonal environments, the effect should generally be greater for large species with large-brained neonates because they must have long gestation periods (Sacher and Staffeldt 1974).

Figure 14.2 develops this possibility more explicitly. For each mean gestation length, it shows the proportion of births that would be expected to fall within a favorable period of F days if (1) gestation period is normally distributed about the mean with a standard deviation described by the regression equation in Figure 14.1 and (2) all females conceive on the date that would have them bear young at the midpoint of the favorable time, given the mean (expected) gestation period. Clearly, as the length of the favorable period decreases or as the mean gestation period increases, there is eventually some decline in the proportion of a population whose births fall within the favorable period and hence in the reproductive success of the population. However, except in extremely seasonal environments (e.g. when $F = 7$ days), the effect of variation in gestation period is likely to be significant only for large species with long gestation

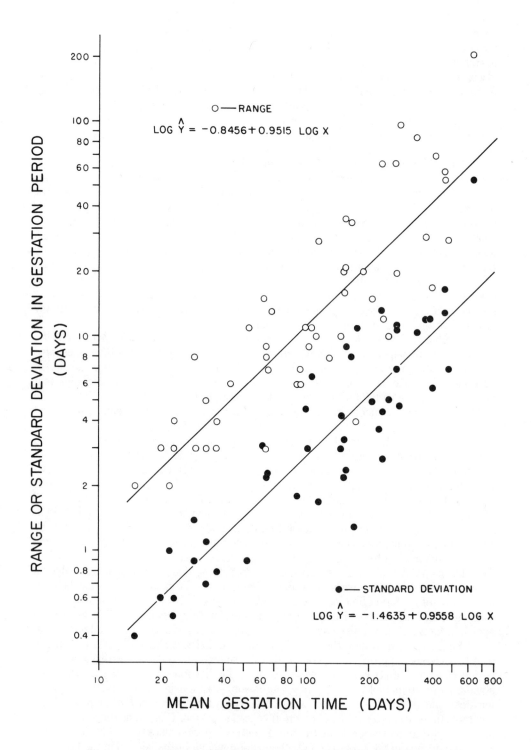

Figure 14.1. Logs of observed range and standard deviation of gestation period plotted against the log of mean gestation period. For the regression of observed range, $N = 51$, $r = 0.85$, and $P < 0.001$. For the regression of standard deviation, $N = 46$, $r = 0.89$, and $P < 0.001$ (from Kiltie, 1982).

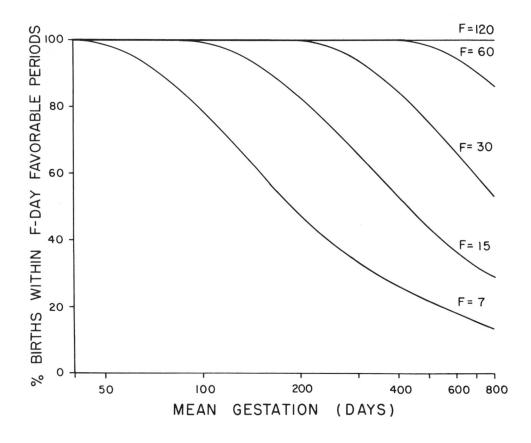

Figure 14.2. Percent of births expected to fall within varying periods of favorability (*F*) as a function of mean gestation period under assumptions described in the text.

periods, especially those with gestation lasting more than about a year. Failure to predict perfectly the midpoint of the favorable period (due to variation in the time of mating or in the occurrence of the favorable period) would worsen the problem.

I should note here that *F* would not necessarily be equal for all species in a particular habitat. Ultimately the favorable period, which I have only vaguely defined, must depend on the biology and ecology of any particular species. The large young of large species undoubtedly can survive equivalent exposure to low temperature or other harsh conditions longer than small young of small species (Geist 1974, Schmidt-Nielsen 1979). However, large species' young are by no means invulnerable to neonatal mortality due to harsh conditions (e.g. Dauphiné and McClure 1974, Bergerud 1975, Bunnel 1982, Thompson and Turner 1982). Furthermore, as Axelrod (1967) and many others have pointed out, small mammals can "escape" harsh conditions by seeking out microhabitats that make conditions favorable over a longer time than might otherwise be possible. Nests also provide a protective environment for young of smaller species, but no herbivores larger than pigs build such structures (Delaney and Happold 1979). The only comparable option for escape by large mammals is migration to more favorable regions (see Churcher 1980), but the evolution and maintenance of this behavior in response to environmental change probably has considerable cost (Cohen 1967), more than for microhabitat selection or nest-building. The need to migrate itself can serve to restrict birth seasons (Dauphiné and McClure 1974, Bunnell 1982).

Large species sometimes also appear to have shorter birth seasons because offspring need considerable time to reach a size sufficient for survival in an approaching unfavorable season (e.g. see Dauphiné and McClure 1974, Bergerud 1975, Guthrie 1980) or because of the need to synchronize births to "saturate" predators (e.g. Estes 1976); small mammals do not seem to use this strategy, and they should not be as severely affected by the problem of maturation time because this trait is inversely related to body size (Blueweiss et al. 1978, Western 1979, see also McNab 1980). Finally, we may note that in habitats in which the length or severity of the favorable season varies from year to year, very small species with sufficiently short gestation periods can take advantage of unpredictable lengthening of the favorable period, while species using correlated cues cannot. To the extent that these factors tend to constrict periods favorable for birth more for large species than for small ones in a given habitat, it becomes even more likely that intraspecific variation in gestation time will more severely reduce the reproductive success of large species.

3. *Time lost in synchronizing reproductive cycles to environmental cycles.* If individuals of a species time their mating activities so that young are produced only in certain seasons, some time may be lost that could otherwise have been spent in production of young. This lost time, which is not necessarily equal for all species in a given environment, can reduce the reproductive rate of the population.

The basic nature of this problem (which was foreshadowed by Kurtén 1953) can be illustrated by two examples. Consider first a species with an 11-month minimum interval between litters. This interval, which I will abbreviate *MIBL*, represents the gestation time plus the minimum time from birth to conception of the next litter. Suppose that this species lives in an environment in which $F = 1$ month and the same month every year is the favorable one. If individuals of this species are able to time their breeding activities perfectly so that offspring are always produced in this month, one month will pass between the birth of one litter and conception of the next (fig. 14.3a). Young can be produced every year.

Now consider a species with a 13-month *MIBL* (fig. 14.3b). Under the same conditions as described for the 11-month *MIBL* case, this species will wait 11 months between the birth of one litter and the conception of the next. By timing its breeding so that young are produced only in favorable seasons, the 13-month *MIBL* species will thus suffer a comparatively greater decrease in maximum reproductive rate than will the 11-month *MIBL* species, as long as there are not other differences between the two species.

Still under the assumption that the favorable period is one month each year, Figure 14.4 shows the percentage of potential "reproductive time" lost by synchronizing reproductive cycles to the environmental cycle (i.e. waiting time ÷ [waiting time + *MIBL*]) for all *MIBL* values from 1 to 48 months. Under less harshly seasonal conditions ($F > 1$ month), the percent potential time lost is less for all species if they breed so as to "predict" favorable times perfectly; however, the overall form of the relationship remains the same. The relative time lost is for species with *MIBL* at or just under an integer multiple of 12 months. It is greatest for species with *MIBL* just greater than an integer multiple of 12 months. In comparing species with *MIBL* values that are a given number of months greater than an integer multiple of 12 (e.g. 4, 16, 28, 40 months) we see that the percentage of time lost is less for progressively greater values. Thus the relationship is more complex between gestation time (which sets a lower limit on *MIBL*) and this "cost" of seasonal reproduction than for the previous two factors discussed.

Of course, many adaptations in behavior and life history could evolve in association with seasonal breeding habits and compensate for or reduce the reproductive delay (see Giesel 1976; Stearns 1976, 1977; Boyce 1979; Horn 1978; among many others). In general, however, small mammals appear to have more possibilities for such adaptation

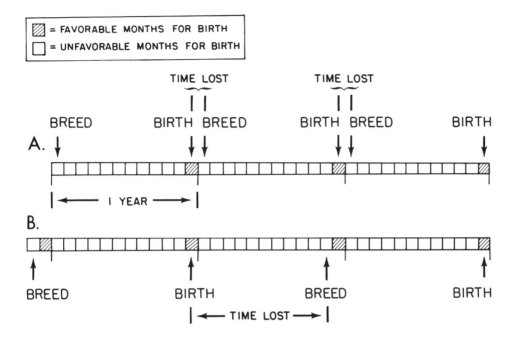

Figure 14.3. Comparison of percent time lost in synchronizing reproductive cycles to environmental cycles for two *MIBL* values when $F = 1$. The lost time is the interval between the completion of minimal parental duties with respect to one offspring (assuming litter size is one) and the conception of the next. No time would be lost in a continuously favorable environment (where $F = 12$). (A) $MIBL = 11$ months. One month is lost for each litter produced. (B) $MIBL = 13$ months. Eleven months are lost per litter.

than large ones. First, as mentioned, small species may have a variety of behavioral methods for escaping unfavorable conditions and lengthening the period in a year that is favorable for giving birth. Second, even when body size is factored out, great variation is found among small species in life history traits such as gestation time, development of young at birth, litter size, growth rate, metabolic rate, and interval from birth to subsequent conception; variation in all these traits may have adaptive bases (e.g. Spencer and Steinhoff 1968, Millar 1977, Case 1978, McNab 1980, Tuomi 1980). Large ungulate species, in contrast, generally have comparatively advanced young at birth, litter sizes of one or at most two, high growth rates, and high metabolic rates (Kurtén 1953, Geist 1974, Sacher and Staffeldt 1974, Case 1978, Robbins and Robbins 1979, McNab 1980, Bunnell 1982). The main way for them to evolve shorter gestation periods or to increase litter size would be to sacrifice advancement of young at birth (Sacher and Staffeldt 1974, Case 1978, Zeveloff and Boyce 1980), but such adjustments would likely have negative effects on neonatal survival in the face of harsh weather or predation.

The period from birth to subsequent conception is a variable trait in extant species of large mammals and is probably relatively more amenable to adaptive influences. Some species like domestic horse (*Equus caballus*) (Stabbenfeldt and Hughes 1977, Evans 1977), Burchell's zebra (*E. burcheli*) (Klingel 1969), alpaca (*Lama pacos*) (San-Martin et al. 1968), vicuña (*Vicugna vicugna*) (Koford 1957) and other tropical ungulates (Dittrich 1974, Gardner 1971) show intervals from birth to subsequent conception that are less than or equal to the period of the normal estrous cycle. Other species (which are not highly seasonal breeders) have long intervals from birth to subsequent conception and

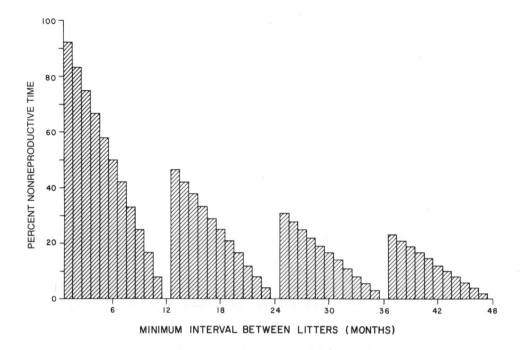

Figure 14.4. Percent reproductive time lost (defined as waiting time ÷ [waiting time + *MIBL*]) for *MIBL* = 1 to 48 months when *F* = 1 month.

great intraspecific variation; representatives of this group include black rhinoceros (*Diceros bicornis*) (Goddard 1967), Bubal hartebeest (*Alcelaphus buselaphus*) (Gosling 1969), African buffalo (*Syncerus caffer*) (Grimsdell 1973, Sinclair 1977), African elephant (*Loxodonta africana*) (Hanks 1972, Laws et al. 1975), giraffe (*Giraffa camelopardalis*) (Leuthold and Leuthold 1978), and defassa waterbuck (*Kobus defassa*) (Spinage 1969). The quality of available nutrition appears to be the main factor influencing intraspecific variation in time to resumption of estrous cycles in this group (Laws et al. 1975, Sinclair 1977, Rattray 1977), but to my knowledge the reasons for differences among species in "vulnerability" to environmental influences on postpartum conception intervals have never been clearly analyzed. In any case, *MIBL* obviously cannot be less than the gestation period.

One factor that almost surely would tend to compensate for time lost in synchronizing breeding cycles to environmental cycles for large species would be the improvement in female survivorship from less frequent exposure to birth-related mortality (for examples of such mortality see Bergerud 1971, Estes and Estes 1979). However, it has been shown in a variety of ungulate species that fertility tends to decrease with age (Hickey 1960, Low 1969, Caughley 1971, Laws et al. 1975, Chaplin 1977, Giesel 1979, Frisch 1980). Thus, individuals that delay breeding to older age classes may on average produce fewer young than those breeding earlier. Furthermore, if all other factors remain equal, seasonally delayed reproduction causes an increase in generation time, which leads to a reduction in reproductive rate (May 1976). Such a decrease will reduce the rate at which a population recovers from any unusual decrease in density and thereby increase its risk of extinction.

In summary, then, although Figure 14.4 makes it appear that very small species would potentially bear the greatest reduction in reproductive rate when they breed seasonally (relative to that possible in a perpetually favorable environment), most species in this group can probably negate the effect through compensatory adaptations. Breeding delay in seasonal environments appears more likely to cause a real decrease (either directly or indirectly) in the fitness of large ungulates with long gestation periods and small litter sizes. The decrease in fitness should be greatest for species with $MIBL$ just greater than an integer multiple of 12 months and least for those with $MIBL$ just under an integer multiple of 12 months. If there are any groups of small mammals that do not have the capacity to compensate for reproductive delay (very small ungulates?), they could also suffer reproductive costs as great or greater than those with $MIBL$ just over one year.

Late Pleistocene Extinctions and Survival With Respect to Gestation Period

Without specific information on maternal/offspring survival as a function of season and other life history data, or exact information on how environmental seasonality changed, if at all, it is impossible to state whether or not a given species "should" have become extinct at the end of the Pleistocene on the basis of any of the factors that I have described. All three factors might have contributed to late Pleistocene extinctions. However, because the first factor should have affected any species with gestation greater than a few months, it cannot have contributed to the *preponderant* demise of very large species (except by virtue of the fact that larger species' longer generation times imply slower adaptation to new breeding cues). Because the second and third factors appear to become important only when gestation period exceeds one year, and hence apply only to very large species, these factors would seem to hold greater potential in helping to explain megafaunal extinctions. Here I want to examine the possibility of differential extinction with respect to this "critical" one-year gestation period.

Table 14.1 lists all extant species of terrestrial mammals that have average gestation periods greater than one year and that do not appear to have any form of delayed fertilization or delayed implantation. These species are in the Elephantidae, Equidae, Tapiridae, Rhinocerotidae, Camelidae, and Giraffidae. The only families that include species with both supra- and infra-annual gestation periods are the Equidae and Camelidae. The best documented cases for bactrian camel (*Camelus bactrianus*) and dromedary camel (*C. dromedarius*) indicate mean gestation periods of about 13 months, while the smaller New World species guanaco (*L. guanacoe*), alpaca (*L. pacos*), llama (*L. glama*), and vicuña (*Vicugna vicugna*) average about 10 to 11 months (Novoa 1970, Schmidt 1978). The African equids, ass (*E. asinus*), Grevy's zebra (*E. grevyi*), and Burchell's zebra (*E. burchelli*) are reported to have gestation periods slightly greater than one year (Asdell 1964, Wackernagel 1965, Anon. 1959), while the Eurasian horse species *E. caballus* and *E. przewalskii* usually require 330 to 340 days (Veselovsky and Volf 1965, Mohr 1971, Evans 1977, Asdell 1964). The few reports available for the remaining extant equids, kiang (*E. kiang*), mountain zebra (*E. zebra*), and onager (*E. hemionus*), indicate 11 to 12 months mean gestation periods (Asdell 1964, Anon. 1959, MacClintock 1976, Groves 1974, David 1966).

Figure 14.5 shows the cumulative range in historical times of the species with average gestation periods greater than one year. Almost all are limited to tropical or subtropical regions, areas not subject to extreme temperature fluctuations or to snowfall during the year. These areas instead tend to be characterized by seasonal cycles in rainfall. These tropical areas are apparently characterized by longer seasons favorable for birth because all tropical species with supra-annual gestation periods are not as highly seasonal in their breeding habits as extratropical species with infra-annual gestation. The bactrian camel, *Camelus bactrianus,* is the only apparent exception to this pattern of tropical distribution and not extremely seasonal breeding among species with

**Table 14.1. Extant Species of Terrestrial Mammals With
Average Gestation Period Greater Than One Year**

Species	Approx. Average Gestation (months)
ELEPHANTIDAE	
Loxodonta africana, African elephant	22
Elephas maximus, Asian elephant	21–22
EQUIDAE	
Equus asinus, African ass	12–13
E. burchelli, Burchell's zebra	12–13
E. grevyi, Grevy's zebra	13
TAPIRIDAE	
Tapirus terrestris, Brazilian tapir	13
T. bairdii, Baird's tapir	13
T. pinchaque, Mountain tapir	13
T. indicus, Malayan tapir	13
RHINOCEROTIDAE*	
Rhinoceros unicornis, Indian rhinoceros	16
R. sondaicus, Javan rhinoceros	?
Dicerorhinus sumatrensis, Sumatran rhinoceros	?
Diceros bicornis, Black rhinoceros	15
Ceratotherium simum, White rhinoceros	?
CAMELIDAE	
Camelus bactrianus, Bactrian camel	13–14
C. dromedarius, Dromedary camel	12–13
GIRAFFIDAE	
Okapia johnstoni, okapi	14–15
Giraffa camelopardalis, giraffe	15–16

*The periods shown for *R. unicornis* and *D. bicornis* have been well documented by Lang et al. (1977) and Anon. (1967). They are about two months shorter than the figures usually listed in compendia like Anon. (1959) and Asdell (1964) for these and other rhinoceros species. Probably the others are also in the vicinity of fifteen to sixteen months.

SOURCE: Primary references are Asdell (1964), Anon. (1959), Altman and Dittmer (1962), and Frazer and Huggett (1974).

supra-annual gestation periods. This species currently inhabits arid regions of central Asia and may breed highly seasonally in the wild, although it has not been well studied (Novoa 1970, Schmidt 1973).

All of the ungulate species besides *C. bactrianus* that survived until Recent times in extratropical regions have gestation periods less than one year, including some (e.g. moose, *Alces alces,* and wapiti, *Cervus canadensis*) large enough to be called "megafauna." All breed so that young are produced over periods of about two months or less. It is noteworthy that all except *C. bactrianus* also appear to have *MIBL* of less than one year. The larger ruminants achieve this by having less mature young (Geist 1974, Robbins and Robbins 1979) after a shorter gestation period (Kurtén 1953, Sacher and Staffeldt 1974) than would be expected on the basis of size-related variation in other groups. The horses (*E. caballus* and *E. przewalskii*) can have short intervals between birth and subsequent estrus—this "foal heat" usually occurs less than one month after birth (Evans 1977), and usually results in conception (Tyler 1972). The Asiatic asses *E. kiang* and *E. onager* seem likely candidates for postpartum conception as well, although they have not been well studied (Groves [1974] states without data that *E. kiang* and *E. hemionus* females foal every other year).

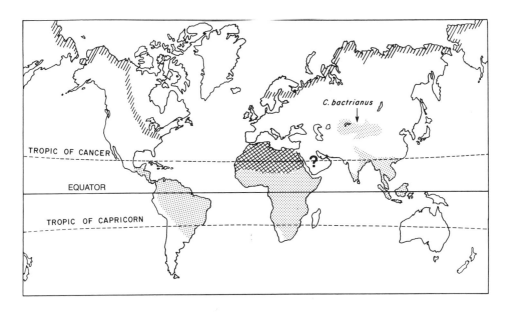

Figure 14.5. The approximate distribution of all species with gestation period greater than one year in Recent preindustrial times is shown by the stipple. Cross-hatch indicates the survival of species with supra-annual gestation period in North Africa after megafaunal extinctions in extratropical regions (e.g. see Axelrod 1967); these North African populations became extinct before modern industrial times. The question mark shows the suspected site of origin for *Camelus dromedarius,* which is now extinct in the wild (Zeuner 1963, Bulliet 1975). The label shows the distribution of *C. bactrianus,* which is unique among Recent species with supra-annual gestation periods in being wholly outside the tropics. Single hatching indicates the northernmost distribution of woolly mammoths *(Mammuthus primigenius)* during the Pleistocene (Kurtén 1968).

What about the species that became extinct at the end of the Pleistocene? Obviously one can never rule out exceptional reproductive traits in such species. However, it seems reasonable to suppose that the extinct Proboscideans, Tapiridae, Rhinocerotidae, and the larger Camelidae and Equidae had gestation periods greater than one year because their closest relatives of equal or smaller size today have supra-annual gestation periods. On the basis of size alone, I would also include the extinct species of Megatheriidae, Mylodontidae, Macraucheniidae, and Toxodontidae among those that likely had supra-annual gestation periods. (This inference for the ground sloths is strengthened by the observation that extant tree sloths have very long gestation periods for their size in association with low metabolic rates [McNab 1980]). Thus the extinction of these species may have been promoted by the factors described here if extratropical environments became more highly seasonal as the glaciers retreated.

Gestation may or may not have exceeded one year in the smaller species of extinct equids, camelids, and ground sloths and the larger bovids and cervids. The smaller extinct species of bovids, cervids, and tayassuids almost certainly did not approach the one-year period. Conceivably, smaller species with gestation periods less than about six months (e.g. the tayassuids, perhaps also the pronghorn *Capromeryx*

[*Breameryx*] *minor*) could have been adversely affected by a combination of the first and third factors (switching from direct to correlated cues; comparatively long waiting time between pregnancies) if the environment became more seasonal. The current absence of very small ungulates (less than about 10 kg) from extratropical regions* may also reflect the influence of heightened seasonality (see also Boyce 1979).

Except for the bactrian camel, possession of an infra-annual gestation period appears to have been a necessary, but not a sufficient, condition for survival into the Holocene in extratropical regions. Thus the factors discussed here clearly cannot account for all extinctions of herbivorous mammals in extratropical regions at the end of the Pleistocene. They also cannot account for extinctions of carnivores, which undoubtedly had infra-annual gestation periods, or for megafaunal extinctions in the New World tropics. It is therefore worth emphasizing that these ideas are not incompatible with other reasons that have been suggested as possibly having contributed to the extinction of species during the time of glacial retreat, such as vegetation change or predatory "overkill." Indeed, long gestation periods are part of a suite of traits contributing to low reproductive rates in large mammals, which make it difficult for them to recover numerically and genetically from any environmental change that causes population reduction (Guilday 1967, Van Valen 1969, Mertz 1971, Terborgh 1974, Southwood et al. 1974, Stanley 1979, many others).

Additional paleoclimatic research is clearly critical to the ultimate usefulness of the arguments presented here. If this research confirms an increase in extratropical environmental seasonality at the end of the Pleistocene and at other times of similar megafaunal extinctions (Martin and Neuner 1978, Webb this volume), these ideas may contribute to our understanding of such extinctions. If it is found that environments of megafaunal populations were more harshly seasonal before the end of the Pleistocene, when these populations were still healthy, we must wonder what adaptations these species evolved to overcome these problems. In either case, it would also be desirable to have more information about the reproductive biology and ecology of bactrian camels. As the only Recent species with supra-annual gestation periods to survive in extratropical regions, bactrian camels could be the exception that either proves or disproves the suggestion that heightened environmental seasonality presents special problems to species with very long gestation periods.

Summary

It has been suggested that extratropical regions became more harshly seasonal, or less equable, at the end of the Pleistocene and that large mammal species with long gestation periods became extinct because they had "inflexible" mating habits causing them to bear young at unfavorable times (Slaughter 1967, Axelrod 1967). One potential problem with this scenario is that it assumes that climatic change was too rapid for natural selection to alter such species' mating habits so that young would be born only in favorable seasons. Three factors could have depressed reproductive output of mammal species with long gestation periods, even if their mating habits had adapted to shorter seasons (on average) favorable for birth. In principle, a reduction in reproductive "fitness" could have stemmed from (1) use of indirect cues to time mating if favorable seasons also became less predictable; (2) inherent variation in the length of gestation

---

*There are two apparent exceptions to this generalization: Musk deer (*Moschus moschiferus*) and Chinese water deer (*Hydropotes inermis*) are Asian cervids that occur in north-temperate habitats and that may weigh less than 10 kg as adults (Walker et al. 1975). *Hydropotes*, however, has rodentlike litter sizes of four to seven according to Walker et al. (1975). *Moschus*, with more typically ungulate litters of one to two offspring, thus stands out as the main exception, for which I can offer no special explanation.

causing more young to be born outside shorter favorable seasons; and (3) time "lost" in synchronizing reproductive cycles to environmental cycles, which causes a reduction in reproductive rate.

Although the first factor could conceivably have contributed to late Pleistocene extinctions of very large species, it cannot primarily account for the differential extinctions of these species because many smaller species also use indirect cues for timing breeding. The second and third factors apply more specifically to large species, especially those with average gestation periods greater than about one year. Nearly all Recent species with gestation periods greater than one year occur in tropical or subtropical areas (the only sure exception is *Camelus bactrianus*) and many (although not all) of the large species that became extinct at the end of the Pleistocene undoubtedly had similarly long gestation periods. Except for *C. bactrianus*, then, infra-annual gestation periods appear to have been a necessary, but not sufficient, condition for survival into the Holocene in extratropical regions.

Acknowledgments

I thank David Webb, Brian McNab, Jerry McDonald, Steve Thompson, and John Robinson for helpful comments on a preliminary version of this paper. I also thank the Department of Zoology, University of Florida, for support through a postdoctoral fellowship.

References

Altman, P. L. and Dittmer, D. S., Editors. 1962. *Growth including reproduction and morphological development.* Washington, D.C., Federation of Experimental Biology, 608 pp.

Anonymous. 1959. Mammalian gestation periods. *International Zoo Yearbook,* v. 1, pp. 157–160.

———. 1967. Tabulated data on the breeding biology of the black rhinoceros *Diceros bicornix* compiled from reports in the yearbook. *International Zoo Yearbook,* v. 7, p. 166.

Asdell, S. A. 1964. *Patterns of mammalian reproduction (second edition).* Ithaca, N.Y., Comstock, 670 pp.

Axelrod, D. I. 1967. Quaternary extinctions of large mammals. *Univ. Calif. Publ. Geol. Sci.,* v. 74, pp. 1–42.

Bergerud, A. T. 1971. The population dynamics of Newfoundland caribou. *Wildl. Monogr.,* v. 25, pp. 1–155.

———. 1975. The reproductive season of Newfoundland caribou. *Can. J. Zool.,* v. 53, pp. 1213–1221.

Blueweiss, L., Fox, H., Kudzma, V., Nakashima, D., Peters, R., and Sams, S. 1978. Relationships between body size and some life history parameters. *Oecologia (Berl.),* v. 37, pp. 257–72.

Bonner, J. T. 1965. *Size and cycle: an essay on the structure of biology.* Princeton, Princeton Univ. Press, 219 pp.

Boyce, M. S. 1979. Seasonality and patterns of natural selection for life histories. *Am. Nat.,* v. 114, pp. 569–583.

Bryson, R. A., Baerreis, D. A., and Wendland, W. M. 1970. The character of late-glacial and post-glacial climatic changes, pp. 53–74 *in* Dort, W., Jr. and Jones, J. K., Jr. (Editors), *Pleistocene and Recent environments of the central great plains.* Lawrence, Univ. Press of Kansas, 433 pp.

Bulliet, R. W. 1975. *The camel and the wheel.* Cambridge, Mass., Harvard Univ. Press, 327 pp.

Bunnell, F. L. 1982. The lambing period of mountain sheep: synthesis, hypotheses, and tests. *Can. J. Zool.* v. 60, pp. 1–14.

Case, T. J. 1978. On the evolution and adaptive significance of postnatal growth rates in the terrestrial vertebrates. *Q. Rev. Biol.,* v. 53, pp. 243–282.

Caughley, G. 1971. Demography, fat reserves and body size of a population of red deer *Cervus elaphus* in New Zealand. *Mammalia,* v. 35, pp. 369–383.

Chaplin, R. E. 1977. *Deer.* Poole, Dorset (U.K.), Blandford, 218 pp.

Churcher, C. S. 1980. Did the North American

mammoth migrate? *Can. J. Anthropol.* v. 1, pp. 103–106.

Cohen, J. 1967. Optimization of seasonal migratory behavior. *Am. Nat.,* v. 101, pp. 5–17.

Dauphiné, T. C. and McClure, R. L. 1974. Synchronous mating in Canadian barren-ground caribou. *J. Wildl. Manage.,* v. 38, pp. 54–66.

David, R. 1966. Breeding the Indian wild ass *Equus hemionus khur* at Ahmedabad Zoo. *International Zoo Yearbook,* v. 6, pp. 197–198.

Delaney, M. J. and Happold, D. C. D. 1979. *Ecology of African mammals.* London, Longman, 434 pp.

Dittrich, L. 1974. Postpartum conception in African antelope. *International Zoo Yearbook,* v. 14, pp. 181–182.

Estes, R. D. 1976. The significance of breeding synchrony in the wildebeest. *E. Afr. Wildl. J.,* v. 14, pp. 135–152.

Estes, R. D. and Estes, R. K. 1979. The birth and survival of wildebeest calves. *Z. Tierpsychol.,* v. 50, pp. 45–95.

Evans, J. W. 1977. Anatomy and physiology of reproduction in the mare, pp. 351–380 *in* Evans, J. W., Borton, A., Hintz, H. F. and Van Vleck, L. D. (Editors), *The horse.* San Francisco, Freeman, 766 pp.

Fenchel, T. 1974. Intrinsic rate of natural increase: the relationship with body size. *Oecologia (Berl.),* v. 14, pp. 317–326.

Frakes, L. A. 1979. *Climates throughout geologic time.* Amsterdam, Elsevier, 310 pp.

Fraser, J. F. D. and Huggett, A. St. G. 1974. Species variations in foetal growth rates of eutherian mammals. *J. Zool., Lond.,* v. 174, pp. 481–509.

Frisch, R. E. 1980. Fatness, puberty, and fertility. *Natural History,* v. 89, no. 10, pp. 16–27.

Futuyma, D. J. 1979. *Evolutionary biology.* Sunderland, Mass., Sinauer, 565 pp.

Gardner, A. L. 1971. Postpartum estrus in a red brocket deer, *Mazama americana,* from Peru. *J. Mammal.,* v. 52, pp. 623–624.

Geist, V. 1974. On the evolution of reproductive potential in moose. *Naturaliste Can.,* v. 101, pp. 527–537.

Giesel, J. T. 1976. Reproductive strategies as adaptations to life in temporally heterogeneous environments. *Ann. Rev. Ecol. Syst.,* v. 7, pp. 57–80.

——. 1979. Associations between age specific mortality and fecundity rates in mammals. *Exp. Geront.,* v. 14, pp. 189–192.

Goddard, J. 1967. Home range, behaviour, and recruitment rates of two black rhinoceros populations. *E. Afr. Wildl. J.,* v. 5, pp. 133–150.

Gosling, L. M. 1969. Parturition and related behaviour in Coke's hartebeest, *Alcelaphus buselaphus cokei* Günther. *J. Reprod. Fert.,* supp. 6, pp. 265–286.

Grimsdell, J. J. R. 1973. Reproduction in the African buffalo, *Syncerus caffer,* in western Uganda. *J. Reprod. Fert.,* supp. 19, pp. 303–318.

Groves, C. P. 1974. *Horses, asses and zebras in the wild.* Hollywood, Fla., Ralph Curtis Books, 192 pp.

Guilday, J. E. 1967. Differential extinction during late-Pleistocene and Recent times, pp. 121–40 *in* Martin, P. S. and Wright, H. E., Jr. (Editors), *Pleistocene extinctions: the search for a cause.* New Haven, Yale Univ. Press, 453 pp.

Guthrie, R. D. 1980. Bison and man in North America. *Can. J. Anthropol.,* v. 1, pp. 55–74.

Hanks, J. 1972. Reproduction of elephant, *Loxodonta africana,* in the Luangwa valley, Zambia. *J. Reprod. Fert.,* v. 30, pp. 13–26.

Hecht, A. D., Barry, R., Fritts, H., Imbrie, J., Kutzbach, J., Mitchell, J., and Saven, S. M. 1979. Paleoclimatic research: status and opportunities. *Quat. Res.,* v. 12, pp. 6–17.

Hibbard, C. W., Ray, C. E., Savage, D. W., Taylor, D. W., and Guilday, J. E. 1965. Quaternary mammals of North America, pp. 509–25 *in* Wright, H. E., Jr. and Frey, D. G. (Editors), *The Quaternary of the United States.* Princeton, Princeton Univ. Press, 922 pp.

Hickey, F. 1960. Death and reproductive rate in relation to flock culling and selection. *N. Z. J. Agric. Res.,* v. 3, pp. 332–44.

Holman, J. A. 1976. Paleoclimatic implications of "ecologically incompatible" herpetological species (Late Pleistocene: southeastern United States). *Herpetologica,* v. 32, pp. 290–95.

——. 1980. Paleoclimatic implications of Pleistocene herpetofaunas of eastern and central North America. *Trans. Nebraska Acad. Sci.,* v. 8, pp. 131–140.

Horn, H. S. 1978. Optimal tactics of reproduction and life-history, pp. 411–29 *in* Krebs, J. R. and Davies, N. B. (Editors), *Behavioural ecology: an evolutionary approach.* Sunderland, Mass., Sinauer, 494 pp.

Kihlström, J. E. 1972. Period of gestation and body weight in some placental mammals. *Comp. Biochem. Physiol.,* v. 43A, pp. 673–80.

Kiltie, R. A. 1982. Intraspecific variation in the

mammalian gestation period. *J. Mammal-ogy,* v. 63, pp. 646–652.

Klingel, H. 1969. Reproduction in the plains zebra, *Equus burchelli boehmi:* behaviour and ecological factors. *J. Reprod. Fert.,* supp. 6, pp. 339–45.

Koford, C. B. 1957. The vicuña and the puna. *Ecol. Monogr.,* v. 27, pp. 153–219.

Kurtén, B. 1953. On the variation and population dynamics of fossil and Recent mammal populations. *Acta Zool. Fenn.,* v. 76, pp. 1–122.

———. 1968. *Pleistocene mammals of Europe.* Chicago, Aldine, 317 pp.

Kurtén, B. and Anderson, E. 1980. *Pleistocene mammals of North America.* New York, Columbia Univ. Press, 392 pp.

Lamb, H. H. 1977. *Climate: present, past and future (v. 2, Climatic history and the future).* London, Methuen, 835 pp.

Lang, E. M., Leutenegger, M., and Tobler, K. 1977. Indian rhinoceros *Rhinoceros unicornis* births in captivity. *International Zoo Yearbook,* v. 17, pp. 237–38.

Laws, R. M., Parker, I. S. C., and Johnstone, R. C. B. 1975. *Elephants and their habitats.* Oxford, Clarendon, 376 pp.

Leopold, E. B. 1967. Late-Cenozoic patterns of plant extinction, pp. 203–46 *in* Martin, P. S. and Wright, H. E., Jr. (Editors), *Pleistocene extinctions: the search for a cause.* New Haven, Yale Univ. Press, 453 pp.

Leuthold, B. M. and Leuthold, W. 1978. Ecology of the giraffe in Tsavo East National Park, Kenya. *E. Afr. Wildl. J.,* v. 16, pp. 1–20.

Leuthold, W. and Leuthold, B. M. 1975. Temporal patterns of reproduction in ungulates of Tsavo East National Park, Kenya. *E. Afr. Wildl. J.,* v. 13, pp. 159–69.

Low, B. S. 1978. Environmental uncertainty and the parental strategies of marsupials and placentals. *Am. Nat.,* v. 112, pp. 197–213.

Lowe, V. P. W. 1969. Population dynamics of the red deer (*Cervus elephus* L.) on Rhum. *J. Anim. Ecol.,* v. 38, pp. 425–57.

MacArthur, R. H. and Wilson, E. O. 1967. *The theory of island biogeography.* Princeton, Princeton Univ. Press, 203 pp.

MacClintock, D. 1976. *A natural history of zebras.* New York, Scribner's, 143 pp.

McNab, B. K. 1980. Food habits, energetics, and the population biology of mammals. *Am. Nat.,* v. 116, pp. 106–24.

Martin, L. D. and Neuner, A. M. 1978. The end of the Pleistocene in North America. *Trans. Nebraska Acad. Sci.,* v. 6, pp. 117–26.

Martin, P. S. 1967. Prehistoric overkill, pp. 75–120 *in* Martin, P. S. and Wright, H. E., Jr. (Editors), *Pleistocene extinctions: the search for a cause.* New Haven, Yale Univ. Press, 453 pp.

Martin, P. S. and Guilday, J. E. 1967. A bestiary for Pleistocene biologists, pp. 1–62 *in* Martin, P. S. and Wright, H. E., Jr. (Editors), *Pleistocene extinctions: the search for a cause.* New Haven, Yale Univ. Press, 453 pp.

May, R. M. 1976. Estimating *r:* a pedagogical note. *Am. Nat.,* v. 110, pp. 496–99.

Mehringer, P. J., Jr. 1967. The environment of extinction of the late-Pleistocene megafauna in the arid southwestern United States, pp. 247–66 *in* Martin, P. S. and Wright, H. E., Jr. (Editors), *Pleistocene extinctions: the search for a cause.* New Haven, Yale Univ. Press, 453 pp.

Mertz, D. B. 1971. The mathematical demography of the California condor population. *Am. Nat.,* v. 105, pp. 437–453.

Millar, J. S. 1977. Adaptive features of mammalian reproduction. *Evolution,* v. 31, pp. 370–86.

Mohr, E. 1971. *The Asiatic wild horse* Equus przewalskii *Poliakoff, 1881.* London, J. A. Allen, 124 pp.

Negus, N. C., Berger, P. J., and Forslund, L. G. 1977. Reproductive strategy of *Microtus montanus. J. Mammal.,* v. 58, pp. 347–53.

Nichols, J. D., Conley, W., Batt, B., and Tipton, A. R. 1976. Temporally dynamic reproductive strategies and the concept of *r-* and *K*-selection. *Am. Nat.,* v. 110, pp. 995–1005.

Novoa, C. 1970. Reproduction in Camelidae. *J. Reprod. Fert.,* v. 22, pp. 3–20.

Pearson, R. 1978. *Climate and evolution.* London, Academic Press, 274 pp.

Pianka, E. R. 1970. On *r* and *K* selection. *Am. Nat.,* v. 104, pp. 592–97.

Rattray, P. V. 1977. Nutrition and reproductive efficiency, pp. 553–75 *in* Cole, H. H. and Cupps, P. T. (Editors), *Reproduction in domestic animals (third edition).* New York, Academic Press, 665 pp.

Reed, C. A. 1969. They missed the ark. *Ecology,* v. 50, pp. 343–46.

Robbins, C. T. and Robbins, B. L. 1979. Fetal and neonatal growth patterns and maternal reproductive effort in ungulates and subungulates. *Am. Nat.,* v. 114, pp. 101–16.

Sacher, G. A. and Staffeldt, E. F. 1974. Relation of gestation time to brain weight for placental mammals: implications for the theory of vertebrate growth. *Am. Nat.,* v. 108, pp. 593–615.

Sadleir, R. M. F. S. 1969. *The ecology of reproduction in wild and domestic mammals.* London, Methuen, 321 pp.

San-Martin, M., Copaira, M., Zuniga, J., Rodriguez, R., Bustinza, G., and Acosta, L. 1968. Aspects of reproduction in the alpaca. *J. Reprod. Fert.*, v. 16, pp. 395–99.

Schmidt, C. R. 1973. Breeding seasons and notes on some other aspects of reproduction in captive camelids. *International Zoo Yearbook*, v. 13, pp. 387–90.

Schmidt-Nielsen, K. 1979. *Animal physiology: adaptation and environment.* Cambridge, Cambridge Univ. Press, 560 pp.

Sinclair, A. R. E. 1977. *The African buffalo: a study of resource limitation of populations.* Chicago, Univ. Chicago Press, 355 pp.

Slaughter, B. H. 1967. Animal ranges as a clue to late-Pleistocene extinction, pp. 155–68 *in* Martin, P. S. and Wright, H. E., Jr. (Editors), *Pleistocene extinctions: the search for a cause.* New Haven, Yale Univ. Press, 453 pp.

Smith, F. E. 1954. Quantitative aspects of population growth, pp. 274–94 *in* Boell, E. (Editor), *Dynamics of growth processes.* Princeton, Princeton Univ. Press, 307 pp.

Southwood, T. R. E., May, R. M., Hassell, M. P., and Conway, G. R. 1974. Ecological strategies and population parameters. *Am. Nat.*, v. 108, pp. 791–804.

Spencer, A. W. and Steinhoff, H. W. 1968. An explanation of geographic variation in litter size. *J. Mammal.*, v. 49, pp. 281–86.

Spinage, C. A., 1969. Reproduction in the defassa waterbuck *Kobus defassa ugandae* Neumann. *J. Reprod. Fert.*, v. 18, pp. 445–57.

———. 1973. The role of photoperiodism in the seasonal breeding of tropical African ungulates. *Mammal Review*, v. 3, pp. 71–84.

Stabbenfeldt, G. H. and Hughes, J. P. 1977. Reproduction in horses, pp. 401–31 *in* Cole, H. H. and Cupps, P. T. (Editors), *Reproduction in domestic animals (third edition).* New York, Academic Press, 665 pp.

Stanley, S. M. 1979. *Macroevolution.* San Francisco, Freeman, 332 pp.

Stanley, V. 1980. Paleoecology of the Arctic-steppe mammoth biome. *Curr. Anthropol.*, v. 21, pp. 663–66.

Stearns, S. C. 1976. Life history tactics: a review of the ideas. *Q. Rev. Biol.*, v. 51, pp. 3–47.

———. 1977. The evolution of life-history traits: a critique of the theory and a review of the data. *Ann. Rev. Ecol. Syst.*, v. 8, pp. 145–71.

Taylor, D. W. 1965. The study of Pleistocene nonmarine mollusks in North America, pp. 597–612 *in* Wright, H. E., Jr. and Frey, D. G. (Editors), *The Quaternary of the United States.* Princeton, Princeton Univ. Press, 922 pp.

Terborgh, J. 1974. Preservation of natural diversity: the problem of extinction prone species. *BioScience*, v. 24, pp. 715–22.

Thompson, R. W. and Turner, J. C. 1982. Temporal geographic variation in the lambing season for bighorn sheep. *Can. J. Zool.*, v. 60, pp. 1781–1793.

Tuomi, J. 1980. Mammalian reproductive strategies: a generalized relation of litter size to body size. *Oecologia (Berl.)*, v. 45, pp. 39–44.

Tyler, S. 1972. The behaviour and social organization of the New Forest ponies. *Anim. Behav. Monogr.*, v. 5, pp. 85–196.

Van Valen, L. 1969. Late Pleistocene extinctions. *Proc. N. Am. Paleont. Conv., 1969*, pp. 469–85.

Veselovsky, Z. and Volf, J. 1965. Breeding and care of rare Asian equids at Prague Zoo. *International Zoo Yearbook*, v. 5, pp. 28–37.

Wackernagel, H. 1965. Grant's zebra, *Equus burchelli boehmi*, at Basel Zoo, a contribution to breeding biology. *International Zoo Yearbook*, v. 5, pp. 38–41.

Walker, E. P., Warnick, F., Hamlet, S. E., Lange, K. I., Davis, M. A., Uible, H. E., and Wright, P. A. 1975. *Mammals of the world*, third edition. Baltimore, Johns Hopkins Univ. Press, 1,500 pp. (2 vols.).

Webb, S. D. 1969. Extinction-origination equilibrium in late Cenozoic land mammals of North America. *Evolution*, v. 23, pp. 688–702.

Western, D. 1979. Size, life history and ecology in mammals. *Afr. J. Ecol.*, v. 17, pp. 185–204.

Wilson, E. O. 1975. *Sociobiology: the new synthesis.* Cambridge, Mass., Belknap/Harvard Univ. Press, 697 pp.

Wright, H. E., Jr. 1976. The dynamic nature of Holocene vegetation: a problem in paleoclimatology, biogeography, and stratigraphic nomenclature. *Quat. Res.*, v. 6, pp. 581–96.

Zeuner, F. E. 1963. *A history of domesticated animals.* London, Hutchinson, 560 pp.

Zeveloff, S. I. and Boyce, M. S. 1980. Parental investment and mating systems in mammals. *Evolution*, v. 34, pp. 973–82.

15

Environmental Insularity and the Extinction of the American Mastodont

JAMES E. KING AND JEFFREY J. SAUNDERS

It should follow that the *incognitum* in former times has been a very general inhabitant of the globe. And if this animal was indeed carnivorous, which I believe cannot be doubted, though we may as philosophers regret it, as men we cannot but thank Heaven that its whole generation is probably extinct.

William Hunter, before the Royal Society of London, February 25, 1768.

FROM NUMEROUS DESCRIPTIVE STUDIES of abundant fossil remains, we have come to know a single but highly variable species of *Mammut americanum,* the American mastodont. As pointed to by Osborn (1936:131), the origin, early history, and progressive development of this species (and its paleobiology) remain relatively obscure. The American mastodont is still a bit of the *incognitum* it was two hundred years ago.

The reasons for this obscurity are several and include our general lack of knowledge about Pleistocene forest-dwelling animals and particularly the habits of the presumably reclusive mastodont. This contrasts with its singular abundance as fossils and thus, by extension, an unwelcome burden to collections and curators: "Ho-hum, another mastodont for the attic, or basement, or for trade." Yet we are particularly intrigued by this animal, for the very reason of the contradiction explicit in the justaposition of the terms "familiarity" and "obscurity."

The American mastodont was a great favorite of Thomas Jefferson (who in essence charged Captains Meriwether Lewis and William Clark to learn of its existence in the "West") and many of his learned contemporaries; in fact it provided a basis for the early development of vertebrate paleontology in America as it emerged from the myths, prejudices, and superstitions of the eighteenth and nineteenth centuries (Green 1959). We intend to argue a better understanding of this animal's "place in nature," as Huxley (1863) phrased it, and to demonstrate reasonable and probable agents, including catastrophic disruption of habitat and resulting floristic insularity, in its extinction.

Phyletic History and Attributes

Mammut americanum was the last surviving species of an old and exceedingly conservative family of Proboscidea—the Mammutidae—that can be traced in North America (arriving here as an immigrant) from the Middle Miocene *(Miomastodon merriami,* Thousand Creek, Nevada, and Pawnee Creek, Colorado, Osborn 1936:154) and from somewhat earlier, the Lower Miocene, in Europe *(Miomastodon depereti,* Sables de l'Orléanis, France, Osborn 1936:154). Farther back, the lineage is only somewhat more shadowed. *Paleomastodon beadnelli,* from the Lower Oligocene of the Fayum, Egypt, is, in terms of dental form and cranial morphology, a likely though uncertainly demonstrated ancestor for the Mammutidae (Osborn 1936, Simpson 1945).

Mammutid hallmarks are several: a broad and low cranium, abbreviated mandibular rami housing persistent though variable lower tusks, relatively simple upper tusks that are both upturned and out-turned and, notably, simple low-crowned cheek teeth consisting of a series (two, three, four, or four plus) of lophs/lophids each composed of two major "nipple like" cusps (mastodont=breast or nipple tooth) arranged side by side. The transverse valleys separating the lophs/lophids are open and unobstructed; in addition there is a prominent and diagnostic longitudinal median sulcus or furrow running the length of each cheek tooth that separates and accentuates the two major cusps. It is the sharply cuspate crown of the mastodont cheek tooth that early suggested carnivorous habits for this enigmatic animal, a view that persisted even after 1767 when Peter Collinson, seconded by Benjamin Franklin by 1768, pointed out that an elephantlike build implied a ponderous, woefully inept, thus preposterous carnivore (Green 1959:103, 105).

Mammut americanum in North America

Remains of the American mastodont have been recovered throughout the United States, including Alaska (Fairbanks area, Frick 1937), as well as in the Yukon (Harington 1977), Manitoba (Osborn 1936), and Ontario (Dreimanis 1967). It ranged southward to at least central Mexico (Valsequillo, Irwin-Williams 1967). In addition, mastodont molars have been dredged by fishermen from numerous localities on the Atlantic continental shelf where specimens have been recovered as far as 300 km from the present shoreline (Whitmore et al. 1967).

The earliest records of *Mammut americanum* are in faunas referred to the Blancan Land Mammal Age (Kurtén and Anderson 1980). During this time mastodont distribution is centered in the Pacific Northwest (White Bluffs, Washington, and Grandview and Hagerman, Idaho), but the species occurs in Florida (Santa Fe River IB) as well. The Hagerman fauna is large and has been the focus for much research, including radiometric dating which indicates that the fauna is 3.5 million years old (Bjork 1970). This date marks the earliest appearance of *Mammut americanum* (*senso stricto*). During the succeeding Irvingtonian Land Mammal Age, commencing ca. 1.5 million years ago, mastodont distribution continues to include the Pacific Northwest (Delight, Washington) but in the east expands to include Pennsylvania (Port Kennedy Cave) and Maryland (Cumberland Cave) in addition to Florida (Inglis IA). There is one record from Nebraska (Mullen) referred to this middle Pleistocene interval. By the Rancholabrean Land Mammal Age, commencing about 400,000 years ago (Kurtén and Anderson 1980), the range of *Mammut americanum* had expanded from Alaska to Florida but was most numerous, increasingly so with time, in the eastern forests centered about the Great Lakes (Illinois, Indiana, Michigan, Ohio, New York, and Ontario) and along the Atlantic Coast.

In the Great Lakes, *Mammut americanum* is generally associated with other animals, predominantly browsers, indicative of forested areas: castorids (*Castor, Castoroides*), cervids (*Cervus, Odocoileus, Cervalces*), and bovids (*Symbos* and *Bootherium*). Outside this region, in western marginal areas (California, Arizona, Texas, western Missouri) and along the Atlantic seaboard, including Florida, *Mammut americanum* is often associated with grazers indicative of open forests or woodlands: elephantids (*Mammuthus*), equids (*Equus*), and bovids (*Bison*).

Dreimanis (1968) focused on the extinction of late Pleistocene *Mammut americanum* from the viewpoint of strict environment dependency. Our analysis is based on twelve years of research in an interdisciplinary endeavor into the geology, paleontology, paleoecology, and cultural history of the western Missouri Ozarks (Wood and McMillan 1976). In the Ozarks, from artesian spring sites occurring in terraces along the Pomme de Terre River, we systematically excavated abundant mastodont remains (seventy-one individuals represented) associated with microstratigraphy, radiometric age determinations, pollen and plant macrofossils, as well as molluscan and insect fossils. These springs, which arise from depth, form boglike environments where they emerge at the surface (Haynes 1976, 1980). They have actively accumulated the remains of large megafauna and their contemporary pollen and plant macrofossils (King 1973, Saunders 1977). The six localities (no. 15 on fig. 15.1) occur along an approximately 11-km stretch of the river (since inundated by a Corps of Engineers impoundment) and together comprise a punctuated record of geological and biological history spanning the last 60,000 years. Five sites contained Pleistocene deposits, four of which had a faunal assemblage dominated by mastodont remains. The importance of this long fossil record is that, while holding space constant, the faunal and floral changes through time can be examined in detail, permitting new insights into the microevolution and paleobiology of this Pleistocene proboscidean.

The Western Missouri Record

The western Missouri pollen record includes the mid-Wisconsinan interstadial (Altonian), the full-glacial (Farmdalian), and late full-glacial (Woodfordian) periods. The localities are discussed chronologically, the oldest first.

Mid-Wisconsinan Interstadial

Jones Spring. Jones Spring fossils were concentrated in two peat lenses associated with artesian spring activity and in sediments comprising the spring conduit and feeder as well as in adjacent host alluvium. The lower peat, which was the main fossiliferous horizon in Jones Spring, was a 15-m diameter lens of brown peat more than 1-m thick, overlying a 4-m by 7-m conduit filled with roughly concentric fossiliferous strata of gravelly sand, and mixed sand and blue chert gravel, in addition to white sand occupying the spring's center or feeder. This lower peat lens was separated into an upper dark brown sandy zone and a lower light brown gravelly zone. Fossils also occurred at the basal contact of the lower peat lens with gray clay of the host alluvium. The upper peat lens was less extensive and less organic. It contained fragmented, often rounded fossils suggesting forceful redeposition from the lower peat lens (Haynes 1980). Apparently after the lower (main) peat unit was deposited, renewed spring discharge forced an intrusive mixture of sediments and fossils from the lower peat into a higher (younger) stratigraphic position within the spring. Because of its nonprimary position, this upper peat unit is of little concern to this discussion.

A radiocarbon age determination on the top of the lower peat indicates it to be >40,000 yr B.P. (Tx–1627, Haynes 1980). A small *Juniperus* log collected from the basal contact of this unit with underlying gray clay has been dated at 48,900±900 yr B.P. (QL–962, Haynes 1980). The Jones Spring dates are consistent with placement in the early mid-Wisconsinan interstadial interval, perhaps correlated with the Altonian substage in Illinois (Haynes 1976). This substage dates to the interval from approximately 28,000 to 75,000 years ago, characterized by alternating episodes of glacial advance and retreat (Willman and Frye 1970).

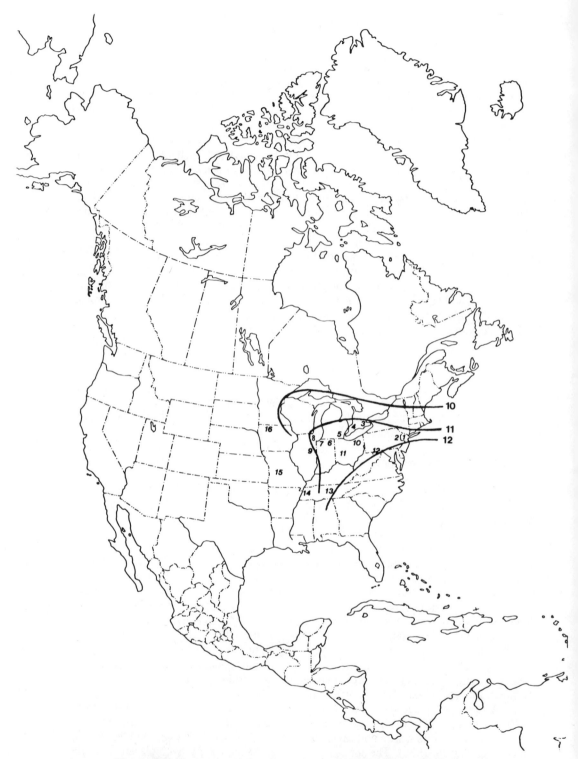

Figure 15.1. The retreat of the spruce/pine boundary in eastern North America between 12,000 and 10,000 radiocarbon years ago plotted from fossil pollen data. These curves, an expanded adaptation from those presented by Bernabo and Webb (1977) and Davis (1976), are based on the following numbered sites and sources: (1 and 2) Watts 1979; (3) Mott and Farley-Gill 1978; (4) Karrow et al. 1978; (5) Kerfoot 1974; (6) Williams 1974; (7) Bailey 1972 (8 and 9) King 1981; (10) Shane 1975; (11) Shane 1976; (12) Maxwell and Davis 1972; (13) Delcourt 1979; (14) Delcourt et al. 1980; (15) King 1973, and (16) Van Zant 1979.

Only the upper part of the lower peat deposit (dark brown sandy peat) contained preserved pollen; the lower portion of the deposit (light brown gravelly peat) did not. However, plant macrofossils did occur throughout the deposit permitting continuous, although limited, paleobotanical interpretations.

Seeds of maple (*Acer*), dogweed (*Cornus* cf. *alternifolia* and *C. florida*), wild plum (*Prunus americana*), wild cherry (*P.* cf. *serotina*), hazel (*Corylus americana*), hawthorn (*Crataegus* sp.), and hornbeam (*Carpinus carolina*) from the basal portion of the peat, associated with QL–962, indicate a definite temperate deciduous tree component around the spring. Other macrofossils from the zone include wood of juniper (*Juniperus* cf. *virginiana*), oak (*Quercus* sp.), maple (*Acer*), honey locust (*Gleditsia* sp.), ash (*Fraxinus* sp.), hickory (*Carya* sp.), and osage orange (*Maclura pomifera*).

Pollen spectra recovered from the upper portion of the lower peat contained 20 to 30 percent pine, 10 percent oak, high percentages of sedge (Cyperaceae), up to 5 percent each of ragweed (*Ambrosia*), grass (Gramineae), and other composites (King 1973). These data suggest cooler climate supporting both deciduous species and pine. A jack pine cone *(Pinus banksiana)* was recovered from this zone. A possible modern analogue would be to areas of the southern lower penninsula of Michigan, where the deciduous species occur on the richer soils while the sandy soils, especially those near Lake Michigan and along the "tension zone" to the north (Potzger 1948), support pine. Open areas of herbs occur throughout the region (Braun 1950).

The Jones Spring faunal assemblage is listed in Table 15.1. Although there is some indication of earlier and later periods of fossil accumulation, distinct, temporally separate faunas cannot be demonstrated with certainty on stratigraphic separation. Fossil wood and root stock fragments associated with QL–962 from the base of the lower peat lens are entirely of deciduous species, whereas pollen from the upper portion of the fossiliferous sequence (dated >40,000 yr B.P.) includes pine. These are interpreted as indicating an open pine-parkland (King 1973). The Jones Spring faunal accumulation probably continued over perhaps an appreciable interval of time. At the beginning of this interval, at least by 48,900 years ago, deciduous species appear to have dominated the immediate site area, which was succeeded before 40,000 years ago by open pine-parkland. This floral readjustment is interpreted as a response to shifting climatic regimes whereby a dryer, warmer climate gave way to a moister, probably appreciably cooler climate after 48,900 years ago but before 40,000 years ago. Associated with this changing climatic phase is the disappearance of alligator from western Missouri, the evolutionary shift from *Bison latifrons* to *Bison antiquus* (Saunders 1983), and a probable increase in the number of mastodonts relative to mammoths in the fauna.

The late phase of the mid-Wisconsinan interstadial as recorded at Trolinger and Koch springs is also associated with abundant mastodont remains. The pollen and plant macrofossils continue to indicate an open pine-dominated parkland with jack pine (King 1973). Radiocarbon dates on these assemblages are 31,880±1340 yr B.P. (Tx–1412) from the middle of the Koch Spring section and >38,000 yr. B.P. (Tx–1457) from its base, and between 34,300±1200 yr B.P. (A–1080) and 29,340±900 yr B.P. (A–1000) at Trolinger Spring (Haynes 1980). These two assemblages appear to be similar in vegetation; they are probably contemporaneous and represent the terminal phase of the cool mid-Wisconsinan interstade.

Trolinger Spring. Two distinct fossil faunas occur in Trolinger Spring (Saunders 1983), designated Trolinger Spring I (earlier) and Trolinger Spring II (later). Each reflects a distinctly different environment.

Trolinger Spring I fossils (Table 15.2) were concentrated in three units comprising the conduit complex in Trolinger Spring. Trolinger Spring I sediments have been dated at >55,000 yr. B.P. (QL–1428, Haynes et al. in press).

Although the Trolinger Spring I fauna lacks mastodont, it does include bear, mammoth, horses, deer, and bison, all representative of savanna-steppe conditions. There is no associated pollen evidence, but fossil wood specimens of deciduous species

**Table 15.1. Late Pleistocene Vertebrate Fauna From
Jones Spring, Hickory County, Missouri**

Taxon	Number of Specimens	Minimum Number of Individuals
Class Reptilia		
Order Chelonia		
Chrysemys scripta	1	1
Terrapene carolina putnami	18	3
cf. *Terrapene carolina putnami*	>150	16
Trionyx sp.	1	1
Order Crocodilia		
Alligator mississipiensis	8	2
Class Aves		
Order Aseriformes		
cf. *Anas carolinensis*	2	1
cf. *Aythya collaris*	1	1
indeterminate duck sp.	1	1
Class Mammalia		
Order Edentata		
Glossotherium harlani	6	1
Order Rodentia		
Geomys sp.	1	1
Castoroides ohioensis	1	1
cf. *Microtus* sp.	1	1
Order Carnivora		
Procyon lotor	1	1
Smilodon cf. *floridanus*	1	1
Order Proboscidea		
Mammut americanum	244	25
Mammuthus jeffersonii	81	12
Order Perissodactyla		
Equus complicatus	154	10
Equus calobatus or *Equus hemionus*	7	1
Tapirus veroensis	2	1
Order Artiodactyla		
Camelops sp.	17	4
Odocoileus virginianus	3	2
Bison latifrons	36	5
Bison latrifrons and/or *Bison antiquus*	50	4
Symbos or *Bootherium*	25	4

SOURCE: Data are from Saunders (1983).

recovered from these sediments suggest that the spring was surrounded by open woodland or savanna. The fauna may be contemporaneus with pollen from adjacent Kirby Spring which indicates deciduous savanna and which is associated with dates of >37,000 yr B.P. (Gx–2718) and >57,000 yr B.P. (QL–1426) (King 1973, Haynes et al. in press).

Trolinger Spring II fossils (Table 15.3) were concentrated in three strata overlying or set within Trolinger Spring I sediments. Trolinger Spring II sediments and fossils are dated between 22,000 to 39,000 years ago (Haynes et al. in press). The actual accumulation of fossils probably took place between 32,000 and 39,000 years ago and may have been restricted to a short interval within this period. Haynes (1976) tentatively correlated the terrace containing Trolinger Spring with the Altonian substage of

Table 15.2. Late Pleistocene Vertebrate Fauna FromTrolinger Spring I

Taxon	Number of Specimens	Minimum Number of Individuals
Class Reptilia		
Order Chelonia		
Chrysemys scripta	1	1
Class Mammalia		
Order Carnivora		
Ursus americanus amplidens	3	1
Order Proboscidea		
Mammuthus jeffersonii	15	5
Order Perissodactyla		
Equus complicatus	65	4
Equus cf. *scotti*	5	2
Order Artiodactyla		
Odocoileus sp.	3	3
Bison sp.	26	5

SOURCE: Data from Saunders (1983).

the Wisconsinan glacial interval, 28,000 to 75,000 years ago. We suggest that Trolinger Spring II fauna is correlated with the upper part of this substage.

Trolinger Spring II sediments were highly organic and set within dominantly inorganic Trolinger Spring I sediments. The fossils represent a browsing fauna including mastodont, stilt-legged deer (*Sangamona fugitiva*), and woodland musk ox (*Symbos cavifrons*) (Saunders 1983). Pollen associated with these fossils (King 1973) indicates that the Trolinger Spring II fauna occupied an open pine-parkland.

Koch Spring. The Koch Site was initially excavated in 1840 by Albert C. Koch, proprietor of the St. Louis Museum, and subsequently reinvestigated by us in 1971 and 1978. During 1971 a north-south trench through the center of the spring exposed a once-fossiliferous brown peat lens under 2.5 to 3 m of gray alluvial clay containing chert gravel lenses and overlying interbedded clay, sand, and gravel (Haynes 1980). The entire central part of the spring had been disturbed by Koch's and possibly later excavations.

Table 15.3. Late Pleistocene Vertebrate Fauna From Trolinger Spring II

Taxon	Number of Specimens	Minimum Number of Individuals
Class Mammalia		
Order Insectivora		
Blarina brevicauda	2	1
Order Rodentia		
Peromyscus spp.	2	2
Synaptomys sp.	1	1
Order Proboscidea		
Mammut americanum	323	15
Order Artiodactyla		
Sangamona fugitiva	2	1
Symbos cavifrons	56	4

SOURCE: Data are from Saunders (1983).

Radiocarbon dates of 31,880±1340 yr B.P. (Tx–1412) from the middle of the peat lens, and >38,000 yr B.P. (Tx–1457) from organic gray clay below the peat (Haynes 1980) confirm its mid-Wisconsinan age. As previously mentioned, these dates suggest a degree of contemporaneity between the fossiliferous peat in Koch Spring and Trolinger Spring II.

The Koch Spring fauna is listed in Table 15.4. According to accounts of his contemporaries (Harlan 1843), Koch recovered a fauna including the remains of numerous mastodonts (Harlan mentions more than 300 teeth), associated with the remains of sloth (*Megatherium* sp. and *Glossotherium harlani*), deer, elk, and ox (?*Bison* sp., ?*Symbos cavifrons*) from this locality. Koch dispersed his collection here and abroad in 1843, with much of the *Mammut americanum* material, including a nearly complete mounted skeleton, purchased by the British Museum (Natural History). Lydekker (1886:16–27) lists ninety-two mastodont specimens, exclusive of the mounted skeleton, as being "from Missouri" and purchased "about 1844." This still leaves much of Koch's collection unaccounted for and its present location(s) unknown. Our excavations corroborated the occurrence of *Mammut* remains (a few fragments of bone and tusk) in the mixed pebbly clay and peat in the disturbed zone overlying the conduit area (Haynes 1980). No fossil remains referrable to other taxa were recovered. From this we infer that Koch's and possibly later excavators' methods resulted in nearly complete recovery of what must have initially been an extensive bone bed.

Table 15.4. Late Pleistocene Vertebrate Fauna From Koch Spring, Hickory County, Missouri

Taxon	Number of Specimens	Minimum Number of Individuals
Class Mammalia		
Order Edentata		
Megatherium sp.	?	?
Glossotherium harlani	49+	3
Order Proboscidea		
Mammut americanum	300+	38 (est.)
Order Artiodactyla		
Odocoileus? (deer)	?	?
Cervus? (elk)	?	?
Bison? and/or *Symbos?* (ox)	?	?

SOURCE: Data from Harlan (1843) with generic inferences by Jeffrey J. Saunders.

Pollen analysis indicates that an open pine-parkland existed during accumulation of most of the peat at Koch Spring, but spruce pollen does appear near the top of the profile (King 1973). Haynes (1980) suggests that the peat at Koch Spring represented a dry to wet cycle within the Altonian substage of Frye and Willman (1973), possibly the Cherrytree stadial of Terasmae and Dreimanis (1976).

Wisconsinan Full-Glacial

There are no spring deposit faunas known from the maximum Wisconsinan Full-Glacial interval in western Missouri. For whatever reason(s), the faunal record during this interval (and presumably earlier glacial episodes as well) is commonly recovered from solution cavity fill (e.g. sinkholes, fissures, and caves), or from other nonspring depositional environments (e.g. colluvial and alluvial valley fill).

Such sites dating to this interval (ca. 25,000 to 16,000 years ago) in the mid-continent that contain mastodont remains include Nonconnah Creek, Tennessee (17,195±505 yr B.P., Lackey 1977) and possibly Barnhart (undated, R. W. Graham pers. comm.) and Crankshaft caves (Parmalee et al. 1969, published dates certainly much too young) in Missouri.

Between 25,000 and 20,000 years ago the pollen record from western Missouri indicates a dramatic change in vegetation and climate. In pollen diagrams spanning this period from both Trolinger and Boney springs, there is an abrupt shift from pine and herbaceous pollen to spruce pollen dominance. Prior to this shift, spruce had been essentially absent; by about 23,000 B.P., however, spruce first appears and within a few centimeters vertically percentages as high as 60 to 90 percent occur in the sediments of the two springs. This shift from pine and herbs to spruce reflects the climatic shift from cool interstadial to cold full-glacial conditions (King 1973) and correlates with the boundary between the Plum Point interstadial and the Nissouri stadial or the main Wisconsinan glacial stage of the eastern Great Lakes (Dreimanis and Karrow 1972, Terasmae and Dreimanis 1976).

The pollen data from both Trolinger and Boney springs indicates that the early part of the full-glacial must have been the coldest period during the last 60,000 years in western Missouri. The pollen during this period is dominated almost exclusively by spruce, 70 to 90 percent, with few other species represented. This spruce-dominated flora is located 400 km south of the maximum extent of Wisconsinan glaciation and is similar in composition to the early late-glacial spruce zones in the Great Lakes region (Wright 1968).

Late Full-Glacial

Boney Spring. The final accumulation of mastodonts and associated pollen records in western Missouri is represented by the Boney Spring fossil assemblage (King 1973; Saunders 1977). This locality contained the first reported spruce and proboscidean assemblage this far south of the maximum glacial limit in the mid-continent (Mehringer et al. 1968, 1970).

Boney Spring deposits are a sequence of primarily alluvial clays and peat. The fossil material was concentrated in two strata, gray alluvial clay overlying a granular tufa filling the spring feeder. An extensive bone bed, unparalleled in western Missouri, was contained in the gray alluvial clay. Immediately below the bones at the base of this unit (4.75 m below the present surface) were wood fragments of spruce and larch and clumps of moss, probably remnants of an earlier extensive moss mat. In addition to dispersed vertebrate remains, the granular tufa filling the spring feeder, representing primarily calcareous overgrowths on decayed moss, contained seeds, wood, ostracod shells, and insect fragments. Radiocarbon dates from the bone bed fall into two groups. Four age determinations on spruce wood recovered from immediately below the bone bed are remarkably consistent: 16,450±200 yr B.P. (I-4236), 16,580±220 yr B.P., (I-3922), 16,490±290 yr B.P. (Tx-1477), and 16,540±170 yr B.P. (Tx-1478). Two radiocarbon dates on organic debris filling the pulp cavities of two mastodont tusks from the uppermost portion of the bone bed are similarly consistent: 13,700±600 yr B.P. (M-2211) and 13,550±400 yr B.P. (A-1079). In addition, a radiocarbon date on moss filling the spring feeder near but below the level of the bone is 16,190±400 yr B.P. (Tx-1629), establishing both feeder activity and moss decay at that time. Taphonomic analysis strongly suggests that faunal accumulation occurred during the latest portion of this interval, i.e. approximately 13,500 years ago during a stressful period of drought (Saunders 1977).

The Boney Spring fauna is listed in Table 15.5. The ostracod fauna has a modern aspect and is comparable to present-day faunas of other Ozark springs. The fossil insects are notable for the possible occurrence of *Olophrum,* a staphylinid beetle with a

Table 15.5. Late Pleistocene Fauna From Boney Spring, Benton County, Missouri

Taxon	Number of Specimens	Minimum Number of Individuals
Class Ostracoda		
Potamocypris smaragdina	200	- -
cf. *Potamocypris illinoisensis*	4	- -
Cypridopsis sp.	50	- -
Candona crogmaniana	35	- -
Candona sigmoides	1	- -
Candona cf. *fluviatillis*	10	- -
Limnocythere reticulata	2	- -
Class Pterygota		
cf. *Pterostichus*	2	- -
indeterminate Dytiscidae	- -	- -
cf. *Helophorous*	- -	- -
cf. *Olophrum*	1	- -
indeterminate Scarabaeidae	- -	- -
indeterminate Curculionidae	2	- -
indeterminate Chrysomelidae	- -	- -
Class Osteichthyes		
indeterminate fish	11	2
Class Amphibia		
Order Anura		
Bufo sp.	10	4
Rana catesbeiana	1	1
Rana sp.	8	2
Order Urodela		
Ambystoma opacum	1	1
Class Reptilia		
Order Squamata		
Eumeces cf. *fasciatus*	1	1
Carphophis amoenus	4	1
Diadophis punctatus	1	1
Lampropeltis triangulum	1	1
Storeria sp.	2	1
Thamnophis proximus or *Thamnophis sauritus*	1	1
Thamnophis sp.	6	1

present-day boreal distribution. Otherwise they are common to Missouri today. Similarly, the nine genera and eleven species of herptofauna are found today in Missouri. The mammalian fauna includes twenty-one genera and twenty-two species. Three genera (*Glossotherium, Castoroides,* and *Mammut*) and five species (*Glossotherium harlani, Castoroides ohioensis, Mammut americanum, Equus* sp., and *Tapirus veroensis*) are extinct. *Napaeozapus insignis* is extant but does not presently occur in Missouri. *Microtus pennsylvanicus* possibly occurs in Missouri today but does not inhabit the Pomme de Terre River valley. In the Pomme de Terre River valley, *Synaptomys cooperi* is near the southern limit of its distribution. In sum, the late Pleistocene fauna of western Missouri 13,500 years ago had a strong modern aspect.

Pollen recovered from dated mastodont tusk pulp cavity fillings is dominated by *Picea* but contains larger amounts of deciduous tree pollen (oak, willow, alder, poplar, elm, and ironwood) and lower spruce values, 26 to 36 percent, than occur in the earlier full-glacial section at Trolinger Spring.

Table 15.5. Late Pleistocene Fauna From Boney Spring, Benton County, Missouri
(continued)

Taxon	Number of Specimens	Minimum Number of Individuals
Class Mammalia		
Order Insectivora		
Blarina brevicauda	28	4
Cryptotis parva	2	2
Scalopus aquaticus	30	2
Order Edentata		
Glossotherium harlani	87	4
Order Lagomorpha		
Sylvilagus floridanus	1	1
Order Rodentia		
Sciurus cf. *niger*	1	1
Marmota monax	4	1
Tamias striatus	7	2
Glaucomys volans	4	1
Geomys bursarius	4	1
Castoroides ohioensis	3	2
Peromyscus cf. *leucopus*	21	6
Neotoma floridana	3	2
Synaptomys cooperi	5	2
Microtus pennsylvanicus	3	1
Microtus ochrogaster and/or		
Microtus pinetorum	16	4
Microtus sp.	13	- -
Napaeozapus insignis	1	1
Order Carnivora		
Procyon lotor	1	1
Order Proboscidea		
Mammut americanum	717	31
Order Perissodactyla		
Equus sp.	1	1
Tapirus veroensis	5	2
Order Artiodactyla		
Odocoileus sp.	8	1

SOURCE: Data from Saunders (1977).

The Boney Spring assemblage represents a fauna associated with the terminal phase and probably the collapse of the full-glacial vegetation in Missouri. The complete disappearance of spruce, including refugia or "islands," probably occurred within the following thousand years. Spruce was definitely gone from central Tennessee by 12,500 years ago (Delcourt 1979) and from northeastern Kansas by 11,300 years ago (Grüger 1973). Its distribution was retreating to the north and east, and thermophilous deciduous forest developed in its place.

Discussion

Late-Glacial Setting: 12,000 to 10,000 B.P.

With the collapse of the Pleistocene spruce forest to the south, distribution of mastodonts appears to center in the northeast. During the brief interval between 12,000 and 10,000 B.P., a number of important changes occurred. The ice retreated north of the Great Lakes. The early pioneering plant communities of spruce woodland and tundra were replaced by a brief episode of pine and then deciduous species. The first undisputable evidence of man in the New World appears, on occasion associated with extinct megafauna. And the American mastodont became extinct.

The Laurentide ice sheet began to retreat northward somewhat before 15,000 B.P. in response to gradual climatic warming. The vegetational evidence of this warming first occurs in the pollen records from western Missouri and Tennessee (Delcourt 1979, Delcourt et al. 1980), areas well south of the maximum ice extent. In these parts of the continental interior, the pollen of deciduous trees began to appear in spruce-dominated pollen zones about 16,000 B.P. Further north, closer to the ice margin, a belt of spruce woodland and tundra, much like the modern forest-tundra ecotone in northern Canada, developed on the newly deglaciated surfaces. Sites in Iowa (Van Zant 1979), central Illinois (King 1981), northeastern Indiana (Williams 1974), Ohio (Shane 1976), Ontario (Mott and Farley-Gill 1978), eastern Pennsylvania and northern New Jersey (Watts 1979), and Maryland (Maxwell and Davis 1972) all contain low influx spruce pollen zones that are comparable with the modern forest-tundra transition (Ritchie and Lichti-Federovich 1967). This spruce woodland moved northward, following the retreating ice, and by approximately 12,000 B.P. it formed an ecosystem centered in the southern Great Lakes. Glaciers existed to the north of this area, while to the southwest (Missouri, Illinois) and south (Tennessee) vegetational succession had already progressed to deciduous forest associations. Between 13,500 and 12,000 B.P. spruce disappears from western Missouri, central Illinois, and Tennessee.

By this time the Great Lakes region contained a vegetation that was more or less uniform in composition containing species, primarily spruce, that formerly had much wider, continental, geographical distributions. As Dreimanis (1968) noted, this area contains the largest concentration of fossil mastodont remains of any area in North America; together with the Atlantic Coastal region, it appears to form the terminal locus of distribution of the American mastodont. Between 12,000 and 10,000 radiocarbon years ago, the spruce woodland in this critical region was replaced by various mixtures of pine and deciduous tree species.

To the north in Canada the Laurentide ice continued to decay and many large areas were inundated by proglacial lakes (Chapman and Putnam 1973, Karrow et al. 1975). By the time these areas drained and began to support spruce forests, the mastodonts were gone.

The late-glacial spruce pollen zones are quite different from modern pollen spectra from Canada's spruce forests. The fossil zones contain higher spruce pollen percentages, and they lack evidence for many of today's commonly associated species. These fossil pollen zones appear to represent forest communities that were composed almost exclusively of spruce, forming plant communities unlike those at present (Wright 1968). These differences have been explained by differential rates of migration between species (Davis 1976, Wright 1968). Spruce, the first to migrate on to the newly exposed land surfaces, assumed dominance primarily due to lack of competitors. These late Pleistocene spruce forests may have formed a critical available environment for the mastodont, especially with the continual restriction of its habitat.

The shift from spruce to pine dominance in pollen diagrams, marking the collapse of the unique late Pleistocene spruce forests, forms a prominent marker horizon throughout eastern North America (Ogden 1967). Studies plotting species migrations based on pollen data from the eastern half of the continent (Bernabo and Webb 1977, Davis 1976) demonstrate that this marker horizon is a time-transgressive front moving northward (fig. 15.1). Between 12,000 and 10,000 B.P. the spruce/pine pollen boundary moved rapidly across the Great Lakes and Atlantic coastal regions.

Plant Remains Associated
With Late-Glacial Mastodonts

Although mastodont remains are widely distributed, with more than 600 occurrences of their fossils reported from the southern Great Lakes alone (Dreimanis 1968), few records have associated palynological or plant macrofossil data. Those that do portray a boreal environment.

The majority of the pollen evidence associated with mastodonts is from the late-glacial terminal period of the animals' existence, 13,000 to 10,000 B.P. The botanical evidence associated with these mastodonts indicates varying amounts of spruce dominance with some fir, larch, and pine pollen. Many of them also exhibit small percentages, less than 15 percent, of deciduous tree pollen.

In central lower Michigan, Oltz and Kapp (1963) analyzed pollen from alveolar cavities in the Smith Mastodont skull and found 42 percent spruce and 35 percent pine; other taxa were less than 5 percent. This specimen was dated at 10,700±400 yr B.P. Pollen associated with the Pitt Mastodont from the same general area contained 27 percent spruce and 59 percent pine, suggesting a slightly later age for this undated specimen (Oltz and Kapp 1963). Stoutamire and Benninghoff (1964) found the pollen and plant macrofossils associated with the Pontiac Mastodont, dated at 11,900±350 yr B.P. to be dominated by spruce with low amounts of larch, pine, and some deciduous species. The authors note that less pine pollen is associated with the Pontiac specimen than with the 1,200-year younger Smith Mastodont. The spruce/pine pollen boundary had apparently passed through southern Michigan in the intervening years (i.e. between 12,000 and 10,700 years ago). Unfortunately the Michigan mastodont specimens for which mid-Holocene radiocarbon dates were reported in the 1950s lack paleoecological data that could now be used to place them chronologically. Those late radiocarbon dates, run on tusk and bone using the solid carbon technique, are considered too young in light of present knowledge; they are excluded from our discussion.

Numerous late Pleistocene mastodonts with associated pollen data have been recovered from Ohio. Pollen with the Orleton Farms Mastodont (Sears and Clisby 1952) included 45 percent spruce, 39 percent pine, 3 percent fir, 3 percent oak, and 6 percent grass (percentages corrected to include non-tree pollen). The date on this specimen of 8420 yr B.P. (Thomas 1952) is a solid carbon date and clearly too young for the associated pollen. The similar percentages of spruce and pine do, however, suggest a late time within the late-glacial period. Other specimens from Ohio, including the Johnston Mastodont, 10,190±160 yr B.P.; the Coles Mastodont, 9460±305 yr B.P. (date not directly associated with the animal, authors consider it a minimum age); the Pontius Farms Mastodont, 13,180±520 yr B.P. (Ogden and Hay 1967); and the Novelty Mastodont, 10,654±188 yr B.P. (Ogden and Hay 1965), all exhibit spruce pollen dominance with lesser amounts of pine, fir, sedge, grass, and deciduous elements.

In central Indiana Whitehead et al. (1982) report pollen evidence for an open white spruce (*Picea glauca*)-dominated boreal forest, with early indications of the immigration of hardwood taxa associated with mastodont, caribou, and giant beaver in Christensen Bog. Four radiocarbon dates bracket the fauna between 12,000 to 14,000 yr B.P. From north-central Indiana, the Wells Mastodont, dated 12,000±450 yr B.P. (Gooding and Ogden 1965), has associated pollen dominated by spruce, 35 to 60 percent, along with fir, birch, ash, and small amounts of other deciduous taxa.

Elsewhere botanical remains associated with mastodonts from New York, New Jersey, and Ontario also suggest coniferous species (Dreimanis 1968). From southwest Virginia a late Pleistocene fauna of *Mammut americanum, Megalonyx, Equus, Rangifer,*

Symbos, Bootherium, Bison, and possibly *Cervalces* is associated with spruce and pine pollen indicating a "boreal parkland" (Ray et al. 1967). A date of 13,450±420 yr B.P. (SI–461) has been reported for this assemblage (Guilday et al. 1975).

One of the few mastodonts with botanical data recovered from outside the northeast is the Manis Mastodont from the Olympic Peninsula of Washington dated at 12,000±310 yr B.P. (Gustafson et al. 1979). They report that the associated pollen is dominated by freshwater plants, cattails and sedge, and characterize the regional vegetation as shrub grassland including willow (*Salix* sp.), buffalo-berry (*Shepherdia canadensis*), and blackberries or raspberries (*Rubus* sp.).

Other late Pleistocene pollen records associated with extinct fauna (not mastodont) occur in Pennsylvania and Michigan. In the New Paris No. 4 cave deposit in Pennsylvania, the pollen, 50 to 60 percent pine and 10 percent spruce, suggests an open pine-parkland associated with a large small mammal fauna and *Mylohyus* about 11,300 yr B.P. (Guilday et al. 1964). Near Kalamazoo, Michigan, Benninghoff and Hibbard (1961) report a musk ox, *Symbos cavifrons,* dated at 13,000±600 yr B.P. associated with 87 percent spruce pollen.

The distribution of late Pleistocene mastodonts along with their associated pollen evidence indicates that they were mostly eastern North American solitary forest dwellers. The pollen data indicate that the last mastodonts died out in a woodland environment of pine and spruce with lesser amounts of hardwood species. Mastodont fossils have not been found in any deposits dominated by deciduous forest pollen which prevailed after 9000 to 10,000 B.P.

Morphological Response to Environment in Wisconsinan Mastodonts

Pine- versus spruce-adapted mastodonts. Based on a study of dentitions, it was suggested by Saunders (1977:112–113) that mastodonts, although probably preferring a spruce-dominated forest or woodland landscape, nevertheless adapted to a pine-dominated vegetation through selection for a rugged cheek tooth morphology. As noted by Leidy (1869:242) during the last century, two distinct varieties occur as morphological extremes among large series of *Mammut americanum* cheek teeth. A "smooth variety" is distinguished by possessing transverse valleys that are uninterrupted by only weak or moderate cristae descending from the anterior and posterior faces of the primary, pretrite cusp. The enamel is relatively smooth and the cingula are only weakly to moderately developed. In the "rugged variety" the transverse valleys are interrupted in their bottoms by the juncture of moderately to strongly developed cristae descending from the pretrite cusp. In the extreme condition these cristae descend from the secondary, posttrite cusp as well. The walls of the valleys, as well as the labial and lingual surfaces of the crown, bear strong plications producing numerous vertical corrugations of the enamel surface. In addition the enamel is choerodont, and the cingula are well developed.

Cheek teeth of *Mammut americanum* from Trolinger Spring are of the "rugged variety," with rugose enamel and generally interrupted valleys (Saunders 1977:32, 34–35). They are in direct contrast with the "smooth variety" of *M. americanum* cheek teeth from Boney Spring that are characterized by smooth enamel and uninterrupted valleys (Saunders 1977:54–61). An example of each type of tooth is illustrated in Figure 15.2. *Mammut americanum* from Trolinger Spring occupied an open pine-parkland while at Boney Spring this taxon occupied a forest of spruce mixed with deciduous species. Mastodont cheek teeth from Trolinger Spring also show greater wear, relative to their progression through the jaws, than do those from the Boney Springs mastodonts. This is apparently due to the more rapid wear of teeth in the mastication of pine or pine-associated plants. This suggests that *Mammut americanum* fared less favorably (although successfully) in an open pine-parkland and that optimum conditions for

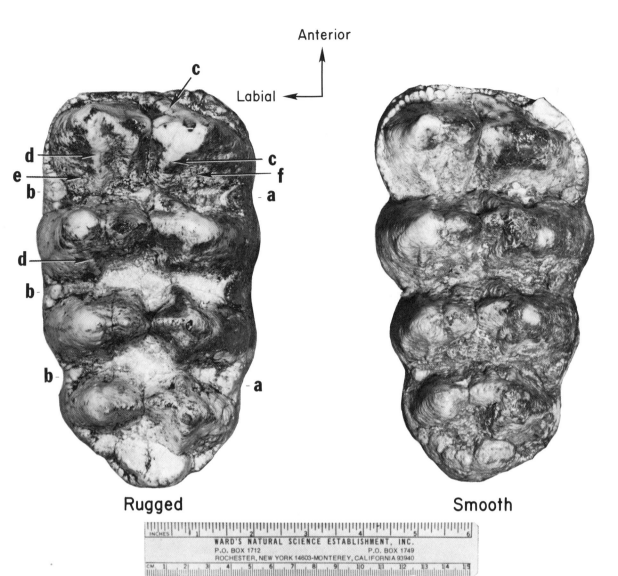

Figure 15.2. Crown views of right M3/'s of *Mammut americanum* from Trolinger (left; EHL179) and Boney (right; 114BS71) springs, illustrating, respectively, examples of the rugged and smooth varieties of cheek teeth in this taxon. Both teeth are in full eruption and in early wear and both are from mature individuals inferred to have been 32 African Equivalent Years of age at their death. See Saunders 1977 (pp. 28–36, 50–61) for full discussion of tooth morphology and the use of teeth for assigning discrete ages to individual mastodonts.

The rugged variety (left) exhibits greater development of the lingual (a) and labial (b) cingula. In addition the cristae (c) descending from the pretrite cusps are stronger, and the incipient cristae (d) on the posttrite cusps are more developed. The enamel is choerodont (e) and plications (f) are developed.

In the symbol at the top, the vertical arrow indicates anterior, the left-facing arrow indicates labial. EHL179 (EHL = Everett H. Lindsay) and 114BS71 (BS = Boney Spring) are field numbers assigned to each specimen.

M. americanum in the late Pleistocene apparently were spruce forests or woodlands. In this we agree with the assessment of Kurtén and Anderson (1980), although our work does not support their view that *Mammut americanum* was morphologically homogeneous in space and time. Rather, the data indicate that *Mammut americanum* adapted to less favorable high pine habitat by selection for a more rugged cheek tooth morphology in partial compensation for accelerated wear of teeth in this environment.

Size trends: mid-Wisconsinan versus late-Wisconsinan mastodonts. In addition to cheek teeth character, absolute size of teeth also appears to correlate with environment. Saunders (1977:113) suggested that *Mammut americanum* from deposits correlated with late Wisconsinan stadial conditions in western Missouri were larger, on the average, than *M. americanum* from deposits correlated with mid-Wisconsinan interstadial conditions. When length and width measurements are compared, Trolinger Spring samples have smaller cheek teeth than Boney Spring samples. Measurements of the Jones Spring cheek teeth (M1–M3) indicate they are either smaller than, as small as, or only slightly larger than those from Trolinger Spring; they are definitely smaller than the large molars from Boney Spring. Table 15.6 presents the size relationships of M2/2 (upper and lower second molars) of *Mammut americanum* from Jones, Trolinger, and Boney springs, Missouri.

Table 15.6. Comparison of Data on Second Molars of Mid-Wisconsinan and Late Wisconsinan *Mammut americanum* From Western Missouri

KEY: Measurements (mm) and Statistics: n = number of specimens; OR = observed range; x̄ = mean; s = standard deviation; CV = coefficient of variation; M2/ = upper 2nd molar; M/2 = lower 2nd molar.

| | Mid-Wisconsinan | | Late Wisconsinan |
| | Early | Late | |
	Jones Spring	Trolinger Spring	Boney Spring
M2/length			
n	28	8	34
OR	102–132	106–124	110–137
x̄	112.536	115.000	123.206
s	8.053	6.347	7.368
CV	7.16	5.52	1.26
M2/width			
n	28	8	34
OR	79–105	84–96	83–104
x̄	88.000	90.750	93.324
s	7.741	3.882	6.079
CV	8.80	4.28	6.51
M/2 length			
n	16	13	37
OR	105–127	101–124	101–135
x̄	113.313	113.923	120.081
s	6.770	8.200	8.015
CV	5.97	7.20	6.67
M/2 width			
n	16	13	37
OR	78–94	79–95	81–99
x̄	85.125	85.538	89.189
s	5.265	5.651	4.415
CV	6.18	6.61	4.95

Length and width measurements of M2/2 are plotted in Figures 15.3 and 15.4 and include measurements of mastodont teeth from an additional locality in Missouri (Kimmswick, Jefferson County, eastern Missouri) as well as from a site in the Mississippi River valley of western Tennessee (Nonconnah Creek, Shelby County). Kimmswick is a Clovis culture-mastodont association in eastern Missouri (Graham et al. 1981). The artifacts associated with the remains of at least two individual mastodonts include several diagnostic fluted projectile points. No organic samples suitable either for radiocarbon dating or paleoenvironmental reconstruction have so far been recovered from the Kimmswick excavations. Elsewhere in North America, Clovis assemblages date to the interval between 11,300 and 11,000 years ago (Saunders 1980), and presumably the mastodonts from the Clovis levels at Kimmswick also date to this interval. Along Nonconnah Creek, a tributary to the Mississippi River, pollen and plant macrofossils associated with a mastodont have been dated at 17,195±505 yr B.P. (Lackey 1977, Delcourt et al. 1980). Saunders has examined this mastodont and recorded measurements of the cheek teeth, including the M2/'s (upper second molar). The associated pollen, 75 percent spruce, 4 percent pine, 1 to 2 percent fir and larch, and 7 to 16 percent oak, are interpreted as representing a spruce forest containing deciduous species. The plant macrofossils include such temperate deciduous taxa as beech (*Fagus grandiflora*), tulip-tree (*Liriodendron tulipifera*), hickory (*Carya* sp.), hazel (*Corylus* sp.), black walnut (*Juglans nigra*), and oak (*Quercus* sp.). This mixed spruce and deciduous forest had occurred there since at least 23,000 years ago and represents a mid-continent, full-glacial vegetation considerably south of Wisconsinan ice.

The teeth represented in Figures 15.3 and 15.4 and in Table 15.6 from Jones, Trolinger, and Boney springs were analyzed using t-tests to determine any differences in mean length and width of M2/2. Comparing the size of M2/ in Jones Spring and Boney Spring, the specimens from Jones were an average of 10.7 mm shorter (t=5.60, p<0.001) and 5.3 mm narrower (t=3.03, p<0.01). The Trolinger sample was also shorter, on the average by 8.2 mm, than the Boney specimens (t=2.90, p<0.01); there was, however, no significant difference in the width of Trolinger and Boney M2/. The mean length and width of M2/ did not differ in the Jones and Trolinger samples.

In general, the M/2 demonstrate a pattern similar to the upper teeth. Jones Spring specimens were shorter by an average of 6.8 mm (t=2.95, p<0.01) and narrower by an average of 4.1 mm (t=2.90, p<0.01) than those from Boney Spring. The Trolinger samples were also smaller than the Boney by 6.2 mm in length and 3.7 mm in width; these differences are, however, less significant (t=2.37 and 2.38, respectively; both p<0.05). As with the upper teeth, the M/2 from Jones and Trolinger springs did not differ in either length or width.

Summarizing the trends in the data, Table 15.6 and Figures 15.3 and 15.4 show that: 1) differences in means of length and width of M2/2 between mid-Wisconsinan *Mammut americanum* samples (Jones Spring and Trolinger Spring II) are not significant and, 2) mastodonts from the mid-Wisconsinan interval were significantly smaller, based on means of M2/2 lengths and M/2 width, than those from the full-glacial (Boney Spring). Furthermore, although sample sizes do not compare and are inadequate, there is some indication that mastodonts of the Wisconsinan maximum full-glacial (Nonconnah Creek, Tennessee) were larger than mid-Wisconsinan and fit well within the size limits of the Boney Spring mastodonts; and mastodonts from the late-glacial (Kimmswick Clovis levels, Missouri) were as small as mid-Wisconsinan mastodonts. These data suggest that mid-continent mastodonts were of a relatively constant size during the mid-Wisconsinan interstadial, that they increased in size during the full-glacial, and that they may have begun decreasing in size in the late-glacial. It was during this last trend that the mastodonts became extinct. A similar trend, based on forelimb stature, has been suggested by Heintz and Garutt (1965), on the basis of small sample size (n=6), for Siberian mammoths (*Mammuthus primigenius*) during this same dated interval: 47,500 to 11,200 years ago. They interpret this trend as a morphological response of mammoths to favorable (large size) and less favorable (small size) environments. The same relationship exists in our data on mid-continent mastodonts. Mid-Wisconsinan mastodonts associated with pine-parkland possess rugged cheek teeth and are small, whereas later mastodonts associated with Wisconsinan full-glacial spruce forest have smooth cheek teeth and are large. In addition, data suggest that mastodonts during the terminal late-glacial to early post-glacial interval were again trending toward smaller size. Maintenance of small size as well as trends toward smaller size appear to be correlated with less favorable habitat. Extreme conditions could result in dwarfism, as

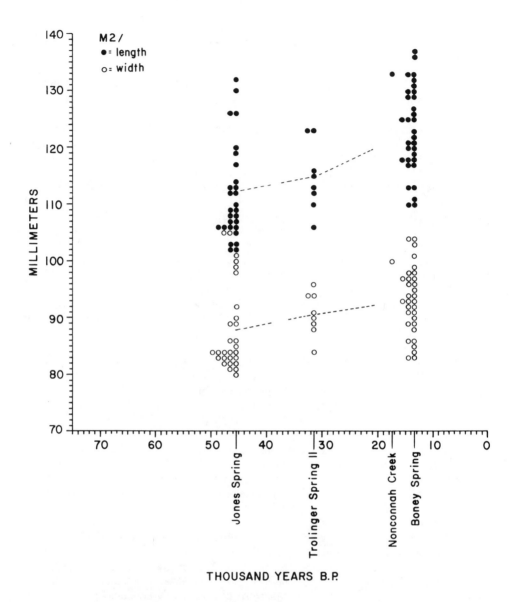

Figure 15.3. Length and width measurements of *Mammut americanum* upper second molars (M 2/'s) from western Missouri and western Tennessee.

seen in insular elephants from the Mediterranean region (Sondaar and Boekschoten 1967, Sondaar 1977), as well as mammoths from the Channel Islands, California (Stock and Furlong 1928, Johnson 1978). The result, given that favorable environments did not reappear, would be extinction of the mastodonts.

Sample Trends: Mid-Wisconsinan versus Late-Wisconsinan Mastodonts

Mid-Wisconsinan. Our data pertaining to mid-Wisconsinan mastodont sample trends are primarily age structures inferred for mastodonts found in Jones and Trolinger springs. Saunders (1977), following the method of Laws (1966) for African elephants,

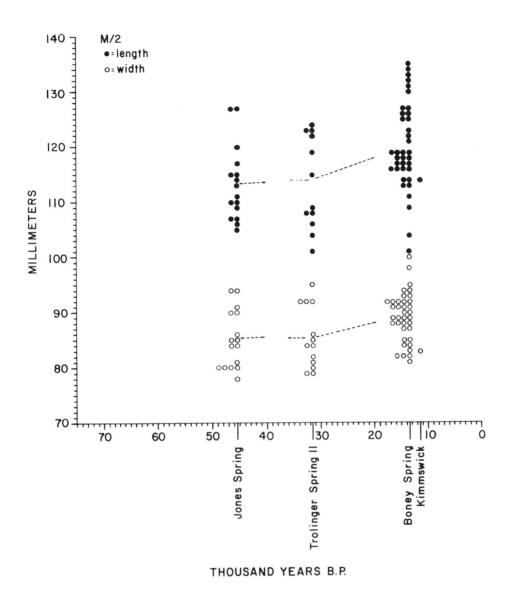

Figure 15.4. Length and width measurements of *Mammut americanum* lower second molars (M/2's) from western and eastern Missouri.

assigned discrete ages to individual mastodonts in the western Missouri record based on the eruption and progression of cheek teeth through the lower jaw. These ages, in African-equivalent years (AEY), are compiled as histograms in Figure 15.5.

The mean inferred age of the Jones and Trolinger springs samples (25.4 and 24.8 AEY, respectively) closely agreed. Recruitment, or the percentage of suckling individuals inferred in the sample (= those through 3 AEY, Sikes 1971:169–170) was 12 percent in the Jones Spring sample and 13 percent in the Trolinger Spring sample. The percentage of immature individuals in the samples (= those less than 15 AEY) was 36 percent for Jones Spring but fell to 23 percent for Trolinger Spring. Finally, prime individuals (= those 25 through 39 AEY, a period during the life span marked by the

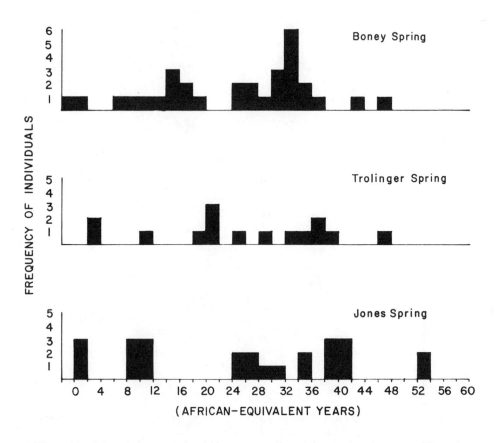

Figure 15.5. Inferred age structure of *Mammut americanum* from springs in western Missouri.

first appearance of wear on the erupting M/3 and continuing until M/2 is eliminated) comprised 44 percent of the Jones Spring sample and 47 percent of the Trolinger Spring sample.

Late-Wisconsinan. The Boney Spring fossil assemblage contains the only large sample of late-Wisconsinan *Mammut americanum* known from western Missouri. The mean inferred age calculated for this sample was 25.7 AEY. Recruitment was 3 percent. Immature individuals made up 27 percent of the sample, while those individuals in their prime accounted for 53 percent.

Table 15.7 summarizes individual inferred age data for mid-Wisconsinan and late-Wisconsinan *Mammut americanum* from western Missouri. Unexpectedly, the mean inferred age indicated for each sample was very similar. Recruitment indicated for mid-Wisconsinan samples was constant but fell appreciably in the late-Wisconsinan sample. The percentage of immature individuals was high in the earlier mid-Wisconsinan sample, fell appreciably in the later mid-Wisconsinan sample, and then rose slightly in the late-Wisconsinan sample. The percentage of prime individuals in the samples from the mid-Wisconsinan showed a slight increase in the later sample, compared to the earlier, and then increased appreciably in the late-Wisconsinan sample. While the mean inferred age for each sample was similar, nevertheless a probably significant increase occurred in the number of prime individuals in the late-Wisconsinan sample relative to the mid-Wisconsinan samples. Indication of decreased recruitment during the late-Wisconsinan is also probably significant.

Table 15.7. Comparison of Individual Age Data on Mid-Wisconsinan and Late-Wisconsinan *Mammut americanum* From Western Missouri

| | Mid-Wisconsinan | | Late Wisconsinan |
| | Early | Late | |
	Jones Spring	Trolinger Spring	Boney Spring
Sample size	25	15	30*
Mean age (AEY)	25.4	24.8	25.7
Recruitment (%)	12	13	3
Immatures (%)	36	23	27
Prime (%)	44	47	53

*Although thirty-one individuals are represented in the Boney Spring sample, one, a fetus, is not included in the calculations.

We interpret these trends in the western Missouri record to be evidence of mastodonts in poor condition in the late-Wisconsinan by 13,500 years ago, relative to their good or better condition indicated for the mid-Wisconsinan interval. The increase in the proportion of prime individuals entering the fossil record during the late-Wisconsinan was a direct consequence of the disruption and disappearance of coniferous habitat. As the coniferous habitat was collapsing, it appears from the Boney Spring data that mastodonts were undergoing self-regulation, evidenced by low recruitment and high modal age class analogous to the African evidence (Laws and Parker 1968), during a time of environmental stress. Because an environment more favorable to mastodonts failed to return to western Missouri after 13,500 years ago, this self-regulation was in vain, and mastodonts became locally extinct.

The Implications of a Mastodont-Coniferous Forest Linkage in the Terminal Late-Wisconsinan

The pollen data from western Missouri (King 1973), northeastern Kansas (Grüger 1973), and central Tennessee (Delcourt 1979) suggest that pine was not present in sufficient numbers to form discrete late-glacial pine communities in these western and southern areas of the Midwest. The dominant conifer was spruce. There was also no late-glacial pine in central Illinois (King 1981). Pine-dominated pollen zones are present in late-glacial pollen diagrams from the southern Great Lakes, but the data indicate that they were of short duration. In northern Illinois, this late-glacial pine zone lasts from 10,900 to 10,300 B.P., only 600 years (King 1981); in northwestern Indiana pine persisted from 11,800 to 10,000 B.P. (Bailey 1972); in northeastern Indiana from 10,500 to 9500 B.P. (Williams 1974); in southern lower Michigan from 11,000 to just after 10,000 B.P. (Kerfoot 1974); and in northeastern Ohio the pine zone is dated between 11,000 and 9000 B.P. (Shane 1975), but the pine decline there began before 9000 B.P. Throughout the southern Great Lakes region pine dominance was brief, and pine was succeeded by deciduous trees shortly after 10,000 years ago. To the north, where the modern boreal forest was in the early stages of development, vast areas were still either ice-covered, under large proglacial lakes, or supporting shrub-tundra vegetation (Ritchie 1976, Prest 1969).

To the south the late glacial spruce-pine woodland had been replaced by deciduous forest, a biome that we believe mastodonts could not adapt to perhaps because of the suddenness with which environmental events occurred. The rapid warming that terminated the Pleistocene occurred much faster than any other temperature change since the warming at the beginning of the last interglacial, stage 5e of the deep sea record

(Shackleton and Opdyke 1973). The fact that mastodonts survived that rapid temperature shift apparently without effect must be related to the sequence and rate of environmental events at that time.

The spread of dates of this pollen data indicates that as the coniferous vegetation was declining it formed a mosaic of temporary island communities interspersed with early Holocene deciduous forests. These islands would have provided short-term habitats for mastodont concentrations. The mid-Wisconsinan data from western Missouri indicate that mastodonts had adapted to pine-dominated vegetation at that time through general small size and rugged cheek teeth. During the terminal late-glacial period, some evidence shows that mastodonts again decreased in size, perhaps reflecting incipient dwarfing in response to insular distributions. It is probably no coincidence that as these coniferous islands vanished so did the mastodont.

We infer that a mastodont-coniferous vegetation linkage was vital for the maintenance of healthy and vigorous mastodont populations during the late Wisconsinan. When pine collapsed into islandlike distributions, this linkage was effectively broken and mastodonts, retreating into these islands, would be expected to decline rapidly as a consequence of natural self-regulation. When these islands disappeared, the mastodonts vanished in a manner that was heralded at Boney Spring, Missouri, 13,500 years ago. Unlike earlier episodes of changing climates and vegetation, i.e. the mid-Wisconsinan fluctuations, which proceeded apace with mastodont adaptive plasticity, events of the late-glacial moved at a rate that mastodont adaptability could not accommodate. Considering mastodont habitats, adaptive morphologies and strategies, and sample trends from the western Missouri record, an environmental insularity hypothesis can explain the extinction of *Mammut americanum*. The fossil records from Jones, Trolinger, and Boney springs provide temporally discrete views of western Missouri mastodont populations. We feel that this record has broad implications, that is, it illuminates the time-transgressive events and conditions involved in the extinction of *Mammut americanum*.

This hypothesis is testable and vulnerable. In our study area in western Missouri it is demonstrated to be null if samples of *Mammut americanum* are found that postdate 13,000 years B.P. Furthermore, because the hypothesis is time-transgressive into the Great Lakes region, it is also demonstrated to be null if local, healthy (non-relict, non-stressed) mastodont samples are found which survived the collapse of the late Pleistocene vegetation, i.e. are found to be associated with early postglacial deciduous forests prevailing after 9000 to 10,000 years B.P.

Acknowledgments

Many people have contributed to our endeavor with data and ideas. We especially wish to acknowledge B. Dawson-Saunders, R. W. Graham, C. R. Harington, F. B. King, and R. B. McMillan. In addition, studies and analyses by C. V. Haynes, collaborator in the western Missouri project, provided the geological framework and the radiocarbon assays reported here. The research reported here was supported in part by the National Park Service, the National Science Foundation, the Kansas City District of the U.S. Army Corps of Engineers, and the Illinois State Museum Society.

References

Bailey, R. E. 1972. Late- and postglacial environmental changes in northwestern Indiana. Doctoral Dissertation. Indiana University, Bloomington, Indiana.

Benninghoff, W. S. and C. W. Hibbard. 1961. Fossil pollen associated with a late-glacial woodland musk ox in Michigan. *Papers of the Michigan Academy of Science, Arts, and*

Letters 46:155–159.

Bernabo, C. J. and T. Webb III. 1977. Changing patterns in the Holocene pollen record of northeastern North America: a mapped summary. *Quaternary Research* 8:64–96.

Bjork, P. R. 1970. The Carnivora of the Hagerman local fauna (late Pliocene) of southwestern Idaho. *Transactions of the American Philosophical Society*. New Series, Volume 60, Part 7, 54 pp.

Braun, E. L. 1950. *Deciduous forests of eastern North America*. Blakiston, Philadelphia, Pennsylvania. 596 pp.

Chapman, L. J. and D. F. Putnam. 1973. *The physiography of southern Ontario*. Second edition with revisions. University of Toronto Press, Toronto, Canada. 366 pp.

Curtis, J. T. 1959. *Vegetation of Wisconsin: an ordination of plant communities*. University of Wisconsin Press, Madison, Wisconsin. 657 pp.

Davis, M. B. 1976. Pleistocene biogeography of temperate deciduous forests. *Geosciences and Man* 13:13–26.

Delcourt, H. R. 1979. Late Quaternary vegetation history of the eastern highland rim and adjacent Cumberland Plateau of Tennessee. *Ecological Monographs* 49:255–280.

Delcourt, P. A., H. R. Delcourt, R. C. Brister, and L. E. Lackey. 1980. Quaternary vegetation history of the Mississippi embayment. *Quaternary Research* 13:111–132.

Dreimanis, A. 1967. Mastodons, their geologic age and extinction in Ontario, Canada. *Canadian Journal of Earth Sciences* 4:663–675.

———. 1968. Extinction of Mastodons in eastern North America: testing a new climatic-environmental hypothesis. *Ohio Journal of Science* 68:337–352.

Dreimanis, A. and P. F. Karrow. 1972. Glacial history of the Great Lakes–St. Lawrence region, the classification of the Wisconsin(an) stage, and its correlatives. 24th International Geological Congress 24 (Section 12):5–15.

Frick, C. 1937. Horned ruminants of North America. *Bulletin of the American Museum of Natural History* 64, 669 pp.

Frye, J. C. and H. B. Willman, 1973. Wisconsinan climatic history interpreted from the Lake Michigan lobe deposits and soils. *In* Black, R. F., R. P. Goldthwait, and H. B. Willman (eds.), *The Wisconsinan Stage*. Geological Society of America Memoir 136, pp. 135–152.

Gooding, A. M. and J. G. Ogden III. 1965. A radiocarbon dated pollen sequence from the Wells Mastodon site near Rochester, Indiana. *Ohio Journal of Science* 65:1–11.

Graham, R. W., C. V. Haynes, D. L. Johnson, and M. Kay. 1981. Kimmswick: A Clovis-mastodon association in eastern Missouri. *Science* 213:1115–1117.

Green, J. C. 1959. *The death of Adam: evolution and its impact on western thought*. The Iowa State University Press, Ames, Iowa. 382 pp.

Grüger, J. Studies on the late Quaternary vegetation history of northeastern Kansas. *Geological Society of America Bulletin* 84:239–250.

Guilday, J. E., H. W. Hamilton, and P. W. Parmalee. 1975. Caribou (*Rangifer tarandus* L.) from the Pleistocene of Tennessee. *Journal of the Tennessee Academy of Science* 50 (3):109–112.

Guilday, J. E., P. S. Martin, and A. D. McCrady. 1964. New Paris No. 4: A late Pleistocene cave deposit in Bedford County, Pennsylvania. *Bulletin of the National Speleological Society* 26:121–194.

Gustafson, C. E., D. Gilbow, and R. D. Daugherty. 1979. The Manis Mastodon site: early man on the Olympic Peninsula. *Canadian Journal of Archaeology* 3:157–164.

Harington, C. R. 1977. Pleistocene mammals of the Yukon territory. Doctoral Dissertation. University of Alberta, Edmonton, Canada. 1066 pp.

Harlan, R. 1843. Description of the bones of a new fossil animal of the order Edentata. *American Journal of Science* (Series 1) 44:69–80.

Haynes, C. V. 1976. Late Quaternary geology of the lower Pomme de Terre valley. *In* W. R. Wood and R. B. McMillan (eds.), *Prehistoric man and his environments: a case study in the Ozark highland*. Academic Press, New York. pp. 47–61.

———. 1980. Late Quaternary geochronology of the lower Pomme de Terre River, Missouri. Report of findings prepared for the Kansas City District, U.S. Army Corps of Engineers.

Haynes, C. V., M. Stuiver, H. Hass, J. E. King, F. B. King and J. J. Saunders. In press. Mid-Wisconsinan radiocarbon dates from mastodon- and mammoth-bearing springs, Ozark Highland, Missouri. *Radiocarbon*.

Heintz, A. E. and V. E. Garutt. 1965. Determination of absolute age of the fossil remains of mammoth and woolly rhinoceros from the permafrost in Siberia by the help of radiocarbon (C[14]). *Norsk Geologisk Tidsskrift* 45:73–79.

Huxley, T. H. 1959. *Man's place in nature.* (Reprint of the 1863 edition). The University of Michigan Press, Ann Arbor, Michigan.

Irwin-Williams, C. 1967. Associations of early man with horse, camel, and mastodon at Hueyatlaco, Valsequillo (Puebla, Mexico). *In* P. S. Martin and H. E. Wright, Jr. (eds.), *Pleistocene extinctions: the search for a cause.* Yale University Press, New Haven, Connecticut. pp. 337–347.

Johnson, D. L. 1978. The origin of island mammoths and the Quaternary land bridge history of the northern Channel Islands, California. *Quaternary Research* 10:204–225.

Karrow, P. F., T. W. Anderson, A. H. Clarke, L. D. Delorme, and M. R. Sreenivasa. 1975. Stratigraphy, paleontology and age of Lake Algonquin sediments in southwestern Ontario, Canada. *Quaternary Research* 5:49–87.

Kerfoot, W. C. 1974. Net accumulation rates and the history of Cladoceran communities. *Ecology* 55:51–61.

King, J. E. 1973. Late Pleistocene palynology and biogeography of the western Missouri Ozarks. *Ecological Monographs* 43:539–565.

———. 1981. Late Quaternary vegetational history of Illinois. *Ecological Monographs* 51:43–62.

Kurtén, B. and E. Anderson. 1980. *Pleistocene mammals of North America.* Columbia University Press, New York. 442 pp.

Lackey, L. E. 1977. Mastodon and associated full glacial fauna and flora from Memphis, Tennessee. *Abstracts,* Geological Society of America, Southeastern Section, Winston-Salem, North Carolina. pp. 156–157.

Laws, R. M. 1966. Age criteria for the African elephant, *Loxodonta a. africana. East African Wildlife Journal* 4:1–37.

Laws, R. M. and I. S. C. Parker. 1968. Recent studies on elephant populations in East Africa. A symposium of the Zoological Society of London 21:319–359.

Leidy, J. 1869. The extinct mammalian fauna of Dakota and Nebraska, including an account of some allied forms from other localities, together with a synopsis of the mammalian remains of North America. *Philadelphia Academy of Natural Sciences Journal,* Vol. 7 (2nd series), pp. 1–472.

Lydekker, R. 1886. *Catalogue of the fossil Mammalia in the British Museum (Natural History),* Pt. 4. British Museum (Natural History), London.

Maxwell, J. A. and M. B. Davis. 1972. Pollen evidence of Pleistocene and Holocene vegetation in the Allegheny Plateau, Maryland. *Quaternary Research* 2:506–530.

Mehringer, P. J., Jr., C. E. Schweger, W. R. Wood, and R. B. McMillan. 1968. Late Pleistocene boreal forest in the western Ozark highland? *Ecology* 49:568–569.

Mehringer, P. J., Jr., J. E. King, and E. H. Lindsay. 1970. A record of Wisconsin-age vegetation and fauna from the Ozarks of western Missouri. *In* W. Dort, Jr., and J. K. Jones, Jr. (eds.), *Pleistocene and recent environments of the central Great Plains.* University of Kansas Press, Lawrence, Kansas. pp. 173–183.

Mott, R. J. and L. D. Farley-Gill. 1978. A late-Quaternary pollen diagram from Woodstock, Ontario. *Canadian Journal of Earth Sciences* 15:1101–1111.

Ogden, J. G. III. 1967. Radiocarbon and pollen evidence for a sudden change in climate in the Great Lakes region approximately 10,000 years ago. *In* E. J. Cushing, and H. E. Wright, Jr., (eds.), *Quaternary paleoecology.* Yale University Press, New Haven, Connecticut. pp. 117–127.

Ogden, J. G. III and R. J. Hay. 1965. Ohio Wesleyan University natural radiocarbon measurements II. *Radiocarbon* 7:166–173.

———. 1967. Ohio Wesleyan University natural radiocarbon measurements III. *Radiocarbon* 9:316–332.

Oltz, D. F., Jr., and R. O. Kapp. 1963. Plant remains associated with mastodon and mammoth remains in cental Michigan. *American Midland Naturalist* 70:339–346.

Osborn, H. F. 1936. *Proboscidea. A monograph of the discovery, evolution, migration and extinction of the mastodonts and elephants of the World.* Vol. 1—Moeritherioidea, Deinotherioidea, Mastodontoidea. The American Museum of Natural History. New York. 802 pp.

Parmalee, P. W., R. D. Oesch, and J. E. Guilday. 1969. Pleistocene and recent vertebrate faunas from Crankshaft Cave, Missouri. *Illinois State Museum Reports of Investigations* No. 14. Springfield, Illinois.

Potzger, J. E. 1948. A pollen study of the Tension Zone of lower Michigan. *Butler University Botanical Studies* 8:161–177.

Prest, V. K. 1969. Retreat of the Wisconsin and recent ice in North America. Geological Survey of Canada, map 1257A, Ottawa, Canada.

Ray, C. E., B. N. Cooper, and W. S. Benninghoff. 1967. Fossil mammals and pollen in a late Pleistocene deposit at Saltville,

Virginia. *Journal of Paleontology* 41:608–622.

Ritchie, J. C. 1976. The late-Quaternary vegetation history of the western interior of Canada. *Canadian Journal of Botany* 54:1793–1818.

Ritchie, J. C. and S. Lichti-Federovich. 1967. Pollen dispersal phenomena in arctic-subarctic Canada. *Review of Palaeobotany and Palynology* 3:255–266.

Saunders, J. J. 1977. Late Pleistocene vertebrates of the western Ozark Highland, Missouri. *Illinois State Museum Reports of Investigations* No. 33, Springfield, Illinois. 118 pp.

———. 1980. A model for man-mammoth relationships in late Pleistocene North America. *Canadian Journal of Anthropology* 1:87–98.

———. 1983. Mitigation of the adverse effects upon the local paleontological resources of the Harry S. Truman Dam and Reservoir, Osage River Basin, Missouri. Report of findings prepared for the Kansas City District, U.S. Army Corps of Engineers.

Sears, P. B. and K. H. Clisby. 1952. Pollen specta associated with the Orleton Farms Mastodon site. *Ohio Journal of Science* 52:9–10.

Shackleton, N. J. and N. D. Opdyke. 1973. Oxygen isotope and paleomagnetic stratigraphy of Equatorial Pacific Core V28–238: oxygen isotope temperatures and ice volumes on a 10^5 year and 10^6 year scale. *Quaternary Research* 3:39–55.

Shane, L. C. K. 1975. Palynology and radiocarbon chronology of Battaglia Bog, Portage County, Ohio. *Ohio Journal of Science* 75:96–102.

———. 1976. Late-glacial and postglacial palynology and chronology of Darke County, west-central Ohio. Doctoral Dissertation. Kent State University, Kent, Ohio.

Sikes, S. K. 1971. *The natural history of the African elephant*. Weidenfeld and Nicolson, London. 397 pp.

Simpson, G. G. 1945. The principles of classification and a classification of the mammals. *Bulletin of the American Museum of Natural History* 85, 350 pp.

Sondaar, P. Y. 1977. Insularity and its effect on mammal evolution. *In* M. K. Hecht, P. C. Goody, and B. M. Hecht (eds.), *Major patterns in vertebrate evolution*. Plenum Press, New York. pp. 671–707.

Sondaar, P. Y. and G. J. Boekschoten. 1967. Quaternary mammals in the South Aegean Island arc; with notes on other fossil mammals from the coastal regions of the Mediterranean. I. *Proc. Kon. Ned. Akad. v. Wetensch.*, Amsterdam, Ser. B, 70:556–576.

Stock, C. and E. L. Furlong. 1928. The Pleistocene elephants of Santa Rosa Island, California. *Science* 68:140–141.

Stoutamire, W. P. and W. S. Benninghoff. 1964. Biotic assemblage associated with a mastodon skull from Oakland County, Michigan. *Papers of the Michigan Academy of Science, Arts, and Letters* 49:47–60.

Terasmae, J. and A. Dreimanis. 1976. Quaternary stratigraphy of southern Ontario. *In* W. C. McHaney (ed.), *Quaternary stratigraphy of North America*. Dowden, Hutchinson and Ross, Inc., Stroudsburg, Pennsylvania. pp. 51–63.

Thomas, E. S. 1952. The Orleton Farms Mastodon. *Ohio Journal of Science* 52:1–5.

Van Zant, K. 1979. Late-glacial and postglacial pollen and plant macrofossils from Lake West Okoboji, northwestern Iowa. *Quaternary Research* 12:358–380.

Watts, W. A. 1979. Late Quaternary vegetation of central Appalachia and the New Jersey coastal plain. *Ecological Monographs* 49:427–469.

Whitehead, D. R., S. T. Jackson, M. C. Sheehan, and B. W. Leyden. 1982. Late-glacial vegetation associated with caribou and mastodon in central Indiana. *Quaternary Research*. 17:241–257.

Whitmore, F. C., K. O. Emery, H. B. S. Cooke, and D. J. P. Swift. 1967. Elephant teeth from the Atlantic continental shelf. *Science* 156:1477–1481.

Williams, A. S. 1974. Late-glacial–postglacial vegetational history of the Pretty Lake region, northeastern Indiana. *Professional Paper* 686-B, U.S. Geological Survey, Washington, D.C.

Willman, H. B. and J. C. Frye, 1970. Pleistocene statigraphy of Illinois. *Illinois State Geological Survey Bulletin* 94, Urbana, Illinois. 204 pp.

Wood, W. R. and R. B. McMillan. 1976. *Prehistoric man and his environments: a case study in the Ozark highland*. Academic Press, New York, New York. 271 pp.

Wright, H. E., Jr. 1968. The roles of pine and spruce in the forest history of Minnesota and adjacent areas. *Ecology* 49:934–955.

The Theoretical Marketplace:
Cultural Models

The Theoretical Marketplace: Cultural Models

Those who appeal to prehistoric human activity as a cause for most or all late Quaternary extinction must acknowledge the role of natural processes in all earlier extinctions. The issue is whether, in the absence of prehistoric people, a great many more large animals would have survived to modern times than did.

If people, not climate, caused the extinctions, then the fossil record of large animals cannot be relied upon as indisputable evidence of climatic change. At least some "tropical" animals, such as the tapir, alligator, and capybara, viewed in some quarters as evidence of mild winters, may have abandoned temperate latitudes because of human, not climatic, impacts.

One test of the overkill model is stratigraphic. While Haynes acknowledges that the evidence is too circumstantial to convict early man of megafaunal extinction, he notes that when sedimentary units lend themselves to meticulous examination the coincidence of the first evidence of hunters and the last bones of extinct megafauna is impressive.

Martin lists the extinct fauna of five continents and a variety of oceanic islands, looking at the timing and magnitude of megafaunal extinctions in the last 100,000 years. Magnitude of extinction seems inversely related to abundance of kill sites, an attribute of "blitzkrieg." Prehistoric human impact is easier to envision on oceanic islands, although many have yet to reveal extinct faunas, and few of the ones that are known are well dated.

McDonald emphasizes the increase in land area, plant productivity, and megafaunal biomass at the time of North American extinctions. He believes habitat patches were becoming more fine grained, the reverse of what Guthrie proposed. In McDonald's view megafaunal extinctions are difficult to explain without an anthropogenic assist. Mead and Meltzer review North American radiocarbon dates on extinct animals, proposing an objective system for screening radiocarbon dates. They ask whether the extinct fauna survived into the Holocene. When quality dates only are considered, there is little good evidence of Holocene survivals, although it

343

is not at all clear that all extinctions occurred as close to 11,000 years ago as Mosimann and Martin's blitzkrieg model requires. Mastodonts in particular may have lingered at least another thousand years.

The Mosimann-Martin model is reviewed by Whittington and Dyke. They favor a smaller annual destruction rate of animals by hunters and a "frontless" distribution of hunters. The result is similar to Budyko's model of mammoth extinction in the U.S.S.R. However, the time to extinction is increased, which should also require a higher visibility of kill sites or processing sites than is now known in the North American archaeological record.

Late Pleistocene bird extinction is examined by Steadman and Martin, who regard most of the lost genera as victims of extinction by association with the large mammals. Avian losses were quite small compared with the large mammals of the continent, and represent a small fraction of the known late Pleistocene fossil birds.

A major test requirement of cultural extinction models is chronology. One can view the scarcity of kill sites in America, Australia, and Madagascar as supporting evidence for sudden extinction. However, such a model requires virtually simultaneous extinction of the entire megafauna, a requirement that may be hard to meet, especially in Australia. So the effort at propounding the overkill theory brings the investigator back to searching for deposits or remains that can be well dated. What is or is not "well dated" is less easy to agree upon.

16

Stratigraphy and Late Pleistocene Extinction in the United States

C. VANCE HAYNES

IN REFLECTING UPON THE RESULTS OF LATE QUATERNARY stratigraphic studies and radiocarbon dating over the past twenty-five years, one aspect of the extinction of the megafauna is more impressive than all others. This is its statigraphic abruptness. The more common large mammals of the late Pleistocene megafauna, especially horses, camels, and mammoths, do not occur in any primary context above an abrupt stratigraphic contact. Wherever the record is sufficiently complete, that is, with little missing time (hiatuses), this is a marked stratigraphic break representing an erosional episode between 12,500 and 11,500 B.P. (Haynes 1968a). During this interval of a millenium or less, the last of the glacial-age floodplains are abandoned as many, if not most, streams change regimen from net aggradation to net degradation and incise their channels.

When net aggradation resumes within a few centuries of 11,500 B.P., streams are generally smaller than before. Channel deposits of sand and gravel, commonly reworked deposits, become buried by fine-grained alluvium via overbank deposition (vertical accretion) or slopewash.

It is in such sequences (fig. 16.1) that the stratigraphic evidence of Pleistocene extinction is best recorded and fixed chronologically by radiocarbon dating (Haynes 1967). With only a few questionable exceptions most of the main elements of the Rancholabrean megafauna are absent from the stratigraphic record above the contact between the channel facies (unit β_1) and the top strata (unit β_2). The youngest in situ remains of mammoths, mastodonts, camels, horses, tapirs, dire wolves, and others, occur at the basal contact of unit β_2 or on adjacent terraces where preserved by colluvial or overbank deposition. It is probably significant that in many cases the remains are associated with Paleo-Indian artifacts that, without exception, represent the Clovis culture (fig. 16.2).

In addition to reduced stream size, upon deglaciation water tables experienced a net lowering, until, in the southwestern United States, many low order streams became ephemeral or intermittent and many springs underwent reduced discharge or dried up completely. Many traditional watering places for Pleistocene megafauna may have disappeared at the onset of the Holocene. Animals that concentrated about the remaining water places were more vulnerable to predation. Adding Clovis hunters to the scene may have been the determining factor in bringing about extinction 11,000 years ago (Jelinek 1967).

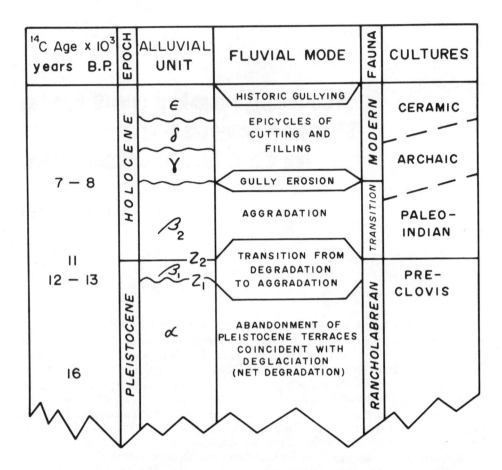

Figure 16.1. Correlation chart of some late Pleistocene-Holocene events in North America. Alluvial unit α represents terminal Pleistocene streams with discharges several times greater than streams of unit β_1 which represents the beginning of Holocene aggradation following a pronounced interval of degradation represented by unconformity Z_1. Along this much reduced channel, elements of Rancholabrean fauna spent their final days as Clovis hunters pursued them 11,000 years ago. The depositional contact Z_2, between β_1 and β_2, marks the abrupt end of Pleistocene extinction, but it is localized for widespread recognition as the Pleistocene-Holocene boundary. This is better represented by Z_1. Unit β_2, the first significant post Pleistocene aggradation, is silty sand or fine sandy silt commonly lacking coarser channel facies and is probably slope-washed loess. Subsequent alluvial units (γ, δ, and ϵ) are derived from slope wash and reworking of earlier alluvial units. Pre-Clovis inhabitation of America south of the Wisconsin ice border has not been conclusively demonstrated. Radiocarbon ages of erosional intervals indicate the range of time during which degradation occurred. The transition from Rancholabrean fauna to the modern fauna is represented by extinct forms of bison in unit β_2 as the last vestige of the Pleistocene megafauna.

The Evidence

The type Clovis site (Blackwater No. 1) in eastern New Mexico is in a spring-fed depression that during the late Pleistocene overflowed into Blackwater Draw, a tributary of the Brazos in west Texas (Haynes and Agogino 1966, Haynes 1975). Rancho-

Figure 16.2. At the Murray Springs Paleo-Indian site, Arizona, the skeleton of an adult female mammoth (*M. columbi*), killed or scavenged by Clovis hunters, rests on an erosional surface that truncates marl (white area) deposited in a lake (unit α) that dried up 12,000 years ago and underlies a black algal mat (black band near base of the wall on the left) that began to form about 11,000 years ago. The black mat (lower β_2) marks the end of a full Rancholabrean fauna in southeastern Arizona. The channel facies (unit β_1) that followed the 11,000- to 12,000-year-old erosional episode (contact Z_1) was exposed in the upper right corner by gullying in the early 1960s.

labrean fauna was relatively abundant, but by 11,000 B.P. spring discharge had decreased, and Clovis hunters attacked and killed mammoths and bison around spring feeders in a shallow pond that subsequently dried up before becoming wet again under more arid conditions (Haynes 1975, Hester 1972, Warnica 1966). Figure 16.3 shows the classic stratigraphy as seen in 1949. The Clovis artifacts and the bones of their prey

Figure 16.3. The classic Paleo-Indian stratigraphy at the Blackwater Draw Clovis site consists of gray sand (unit α), probably more than 15,000 years old, separated from the overlying diatomite (lower β_2) by an erosional contact (unconformity Z_1) upon which Clovis kills of mammoth and extinct bison have been found with associated artifacts. Channel facies (unit β_1) are not known from Blackwater Draw, but there are springlaid (brown sand wedge) of equivalent age containing extinct fauna and Clovis artifacts. The diatomite, containing Folsom artifacts with bison (*B. antiquus*) bones, is overlain by the "carbonaceous silt" (upper β_2) containing Plano projectile points and bones of extinct bison (*B. occidentalis*). This is unconformably overlain by the "jointed sand" (unit γ), containing Archaic artifacts and modern bison (*B. bison*). The mammoth, horse, camel, and dire wolf are gone from the stratigraphic record after unconformity Z_1. (1949 photograph of the west wall of the south gravel pit by geologist Glen L. Evans)

lie either in a spring-laid sand wedge, unit C of the local sequence (unit β_1 of the generalized sequence, fig. 16.1), between 11,000 and 12,000 years old or on the erosional contact between an older (15,000 B.P.), more extensive, spring-laid sand, unit B_1 (unit α), and an organic diatomaceous earth, unit D (lower β_2), deposited during the final pond phase between 11,000 and 10,000 years ago (Haynes 1967). The Pleistocene horse, mammoth, and camel were gone from Blackwater Draw before deposition of the diatomite began approximately 11,000 years ago. Of the large mammals only bison persisted as manifested by their skeletons associated with Folsom and Agate Basin artifacts in the diatomite. A smaller but still extinct form of bison occurred in a carbonaceous silt, unit E (upper β_2), overlying the diatomite, and modern bison occurred in overlying units.

In the Texas panhandle at the Lubbock Lake site, about 180 km downstream from Blackwater No. 1, only bison remain after 11,000 B.P. The extinction is marked by the contact between Stratum 1B, and fluvial sand, and Stratum 2A, organic diatomite (Stafford 1981).

Farther west, in Arizona's San Pedro Valley, there are four (Naco, Lehner, Murray Springs, and Escapule), and possibly two more (Leikem and Navarette), stratified Clovis sites where mammoth bones and artifacts occur in association. They overlie an erosional surface dated 10,900 ± 40 yr B.P. on the basis of twenty-three radiocarbon analyses on charcoal from Clovis camp fires. The occupational surface, in places containing articulated bones of horse, bison, camel, and dire wolf as well as mammoth (fig. 16.2), is overlain by a buried black mat of dark brown to black organic, clayey silt or sand yielding radiocarbon dates as old as 10,800 yr B.P. In addition to these archaeological sites, Rancholabrean megafaunal elements occur below the black mat at at least five other localities in the San Pedro Valley. Sediments above the basal contact have yielded only bones of living species, mostly bison (Haynes 1968b; Agenbroad and Haynes 1975).

Before the black mat was deposited, the tributary streams at the Clovis sites were spring fed and probably perennial. During the brief interval of Clovis activity they may have nearly dried up, causing animals to concentrate at springs and seeps. Earlier, between 26,000 and 12,000 B.P., many tributary valleys were occupied by ponds or lakes, some spring fed, around which the Rancholabrean fauna was relatively abundant. These sites might also be expected to attract early hunters, but despite careful excavation at three vertebrate fossil localities and periodic examination of several others, these older deposits, in the process of eroding, have not yielded artifacts.

At the Tule Springs site in southern Nevada abundant bones of Rancholabrean fauna were associated with ponds and marshes between 30,000 and 12,000 B.P. By 12,000 B.P. these dried up and were subsequently dissected by ephemeral streams that after 11,000 B.P. deposited clayey silts and sands in which no horse, mammoth, or camel bones occur other than by redeposition. The only unquestionable artifacts occurred in the post-11,000 B.P. sediments (Shutler 1967 and others) which represent several cycles of arroyo cutting and filling typical of southwestern channels during the Holocene (Haynes 1968b).

The Domebo site, a Clovis mammoth kill in southwestern Oklahoma, occurs in gray organic sediments (unit C_2 of Albritton 1966) under saturated, anaerobic conditions that preserved the bones as well as rooted tree stumps dated approximately 11,000 yr B.P. by radiocarbon (Leonhardy 1966). No extinct fauna occur in the 1 m of overlying gray, clayey silt (unit C) or 10 m of brown silt (unit B). Apparently the water table has fluctuated above the level of the bone bed throughout most of Holocene time.

Rodgers Shelter in the lower Pomme de Terre Valley of the Ozark Highlands, Missouri, contains a nearly complete, radiocarbon-dated, Holocene succession which began to accumulate approximately 11,000 years ago after a period of degradation that occurred between then and 13,000 B.P. The shelter sediments yielded bones of a variety of large and medium-sized mammals hunted by Paleo-Indians (McMillan 1976); all represent extant species.

The alluvial terraces of the Pomme de Terre River contain a stratigraphic record extending from 13,000 B.P. back to probably Sangamon times (Haynes 1976, 1978; Brakenridge 1981). Spring deposits associated with various terraces contain abundant remains of mastodont and other typical Rancholabrean big game animals without associated artifacts (Saunders 1977). The brief episode of degradation between Boney Spring alluvium and Rodgers alluvium marks Pleistocene extinction in the Ozarks.

Near Kimmswick, 32 km south of St. Louis, Missouri, a 6-m alluvial terrace of probable late Wisconsin age contains late Pleistocene mammal bones. The terrace is overlain by relatively thin clayey silts containing Clovis artifacts associated with mastodont bones, the only known Clovis-mastodont association. Overlying colluvial silts and gravels contain Archaic artifacts but no bones (Graham et al. 1981), probably because of weathering. At Modoc Rock Shelter, 52 km downstream from Kimmswick, the floodplain of the Mississippi River 11,000 years ago was at least 3 m below its present level and only a modern fauna was found in the alluvial deposits filling the shelter (Fowler 1959). In the Mississippi Valley Pleistocene extinction evidently occurred after postglacial degradation and at or before the beginning of Holocene aggradation.

Stratified sites with extinct fauna in the eastern United States are scarce, and those controlled by radiocarbon dates are even more so. If we assume that Clovis points were associated with extinct Pleistocene fauna in the East at the same time as in the West, then Pleistocene extinction at the Thunderbird site in the Shenandoah Valley of Virginia occurred after a period of net degradation and abandonment of a gravel terrace (Q1). This probably represented Wisconsin-age discharge and onset of a change to net aggradation by overbank deposition ("Clovis clay") which buried the Clovis level and eventually the late Wisconsin terrace (Gardner 1974).

This interpretation is consistent with that of Carbone (1974) who points out a late Pleistocene fauna at Saltville, Virginia (385 km southwest of Thunderbird) dated 13,460 ±420 yr B.P. (SI–461) (Guilday 1971), and another late Pleistocene fauna dated 11,300± 1000 yr B.P. (Y-727) at the New Paris No. 4 sinkhole in Pennsylvania 143 km to the north (Guilday and others 1964). By 9340±1000 yr B.P. (M-1291) a completely modern fauna had become established at Hosterman's Cave (Guilday 1967) approximately 105 km northeast of the New Paris No. 4. At Thunderbird extinction was apparently complete by the end of deposition of the "Clovis clay" (Gardner 1974).

The easternmost stratified Clovis site, Shawnee-Minisink on the Delaware River at Stroudsburg, Pennsylvania, is, like Thunderbird, on a late Wisconsin terrace buried by early Holocene alluvium (Crowl and Stuckenrath 1977; McNett et al. 1977). The earliest occupation surface, dated 10,590 ± 300 yr B.P. (W-2994) and 10,750 ± 600 yr B.P. (W-3134), was overlain by Archaic horizons, all of which were truncated during a mid-Holocene episode of degradation (McNett personal communication). As at Thunderbird, no faunal remains were found, but the stratigraphic correlation seems obvious.

Stratified deposits at the Marmes rockshelter on the Palouse River, Washington, in northwestern United States contain artifacts and human bones dated 10,130 ± 300 yr B.P. (W-2218) (Kelley et al. 1978). While elk bones are larger than modern, the associated fauna is Holocene (Fryxell et al. 1968, Fryxell personal communication). Near Squium, 415 km to the northwest, mastodont bones occur in deposits dated approximately 12,000 ± 310 yr B.P. and overlying Vashon till (Gustafson et al. 1979, Mehringer personal communication). At Marmes it appears, therefore, that Pleistocene extinction occurred during the erosional hiatus between late Wisconsin lacustrine sediments and the earliest postglacial aggradation of the Palouse River.

Discussion

In this brief review of selected localities in the United States that have provided an opportunity for rigorous stratigraphic radiocarbon control, the large extinct Pleistocene mammals commonly found in Rancholabrean faunas were gone by 10,500 B.P. While Mead and Meltzer (this volume) do not exclude their survival to later times on the basis of ^{14}C dates throughout North America, Marcus and Berger (this volume) report no younger ^{14}C dates on amino acids extracted from bone of the extinct fauna at Rancho La Brea. Stratigraphically, 11,000 B.P. corresponds with the basal contact of the first postglacial episode of stream aggradation (vertical accretion) that appears to have occurred throughout North America south of the Wisconsin ice border. This contact (fig. 16.1, Z_2), commonly erosional, is the surface occupied by Clovis people wherever their artifacts are found in primary stratigraphic position whether it is in the east, west, or central United States.

From a stratigraphic viewpoint Clovis appears as a fully developed culture during a relatively brief erosional hiatus of no more than one millenium, and the time depth for Clovis is probably 500 radiocarbon years or less. This, plus the fact that pre-Clovis artifacts have not been found below the hiatus, has led to the hypothesis that the Clovis culture may have been the first to enter the New World south of the Wisconsin ice border (Haynes 1964, 1966, 1970, 1980; Martin 1973). If people were here earlier, they were not very successful in hunting megafauna or in populating the continent. They must have been indifferent to hunting or scavenging large mammals because there is so little firm evidence of associated artifacts, and what exists is controversial.

The extinction of the Pleistocene megafauna between 11,500 and 10,500 B.P. coincides so closely with the age of the Clovis culture that a possible cause-and-effect relationship has not escaped notice (Haynes 1966; Martin 1966, 1967). However, as Jelinek (1967) points out, other factors suggest that man the predator was not the sole cause of extinction. In North America one of the most abrupt climatic changes "and certainly the one about which we know the most came at the end of the Pleistocene about 11,000 years ago" (Wright 1976, p. 10). Whereas this may have caused stress to some species, the late Pleistocene fauna survived the Sangamon. Upon deglaciation, more, not fewer, favorable habitats may have become available to the large herbivores, a view amplified by McDonald (this volume) from Mehringer (1967) and counterargued by Guthrie (this volume).

Another factor may have been sources of water. The paleohydrological data from many of the sites provide ample evidence that local water tables declined with deglaciation as streams underwent net degradation even in nonglacial catchments (Haynes 1973, 1975, 1978). In the Southwest many low order streams changed from effluent to influent between 11,500 and 10,500 B.P. Therefore, many traditional watering places dried up, at least seasonally during deglaciation, and water table fluctuations approached an epicyclic steady state during the same period that extinction occurred and Clovis hunters appeared. When the factor of fauna unadjusted to the presence of *Homo sapiens* is added, a situation was created where watering places became places of death rather than sustenance. Jelinek (1967) alludes to the possibility that man might have placed already-stressed animals under further stress by denying access to traditional water sources.

Other possibilities that could have had significant effect are the use of poison, annihilation of entire family units in the case of mammoth (Saunders 1980), and advantageous use of individual behavior (Haynes 1980).

The evidence is strictly circumstantial and indeed may never be adequate to unequivocally convict man of megafaunal extinctions, even if overhunting was responsible. Nevertheless, the stratigraphic coincidence of the first visible evidence of hunters and the last skeletons of the late Pleistocene extinct megafauna is intriguing.

References

Agenbroad, L. D. and C. V. Haynes, Jr. 1975. *Bison bison* remains at Murray Springs, Arizona. *The Kiva,* v. 40, pp. 309–313.

Albritton, C. C., Jr. 1966. Stratigraphy of the Domebo Site. *In* Domebo: a Paleo-Indian mammoth kill in the Great Plains. *Museum of the Great Plains, Contribution* No. 1, pp. 11–13.

Brakenridge, G. R. 1981. Late Quaternary floodplain sedimentation along the Pomme de Terre River, southern Missouri. *Quaternary Research,* v. 15, pp. 62–76.

Carbone, V. A. 1974. The paleoenvironment of the Shenandoah Valley. *In* The Flint Run Paleoindian complex: a preliminary report 1971–1973 seasons, W. M. Gardner, ed.

Occasional Publication No. 1, Catholic University, Washington, D.C., pp. 84–99.

Crowl, G. H. and R. Stuckenrath. 1977. Geological setting of the Shawnee-Minisink Paleoindian archaeological site. *In* Amerinds and their paleoenvironments in northeastern North America, W. S. Newman and Bert Salouen, eds. *New York Academy of Science Annals,* v. 288, pp. 218–222.

Fowler, M. L. 1959. Summary report of Modoc Rock Shelter. *Illinois State Museum Report of Investigations* 8.

Fryxell, R., T. Bielicki, R. D. Daugherty, C. E. Gustafson, H. T. Irwin, and B. C. Keel. 1968. A human skeleton from sediments of mid-Pinedale Age in southeastern Washington. *American Antiquity,* v. 33, pp. 511–515.

Gardner, W. M. 1974. The Flint Run Paleoindian complex: a preliminary report 1971–1973 seasons. *Occasional Publication* No. 1, Catholic University, Washington, D.C.

Graham, R. W., C. V. Haynes, D. L. Johnson, and M. Kay. 1981. Kimmswick: a Clovis-mastodon association in eastern Missouri. *Science,* v. 213, pp. 1115–1117.

Guilday, J. E. 1967. The climatic significance of the Hosterman's Pit local fauna, Centre County, Pennsylvania. *American Antiquity,* v. 32, pp. 231–232.

———. 1971. The Pleistocene history of the Appalachian mammal fauna. Res. Div. Mongr. Virginia Polytechnic Inst. State University, v. 4, pp. 233–262.

Guilday, J. E., P. S. Martin, and A. D. McCrady. 1964. New Paris No. 4: a Pleistocene cave deposit in Bedford County, Pennsylvania. *Bull. Nat. Speleol. Soc.,* v. 26, pp. 121–194.

Gustafson, C. E., D. Gilbow, and R. D. Daugherty. 1979. The Manis mastodon site: Early Man on the Olympic Peninsula. *Canadian Journal of Archaeology* No. 3, pp. 157–164.

Haynes, C. V., Jr. 1964. Fluted projectile points: their age and dispersion. *Science,* v. 145, pp. 1408–1413.

———. 1966. Elephant-hunting in North America. *Scientific American,* v. 214, pp. 104–112.

———. 1967. Carbon-14 dates and Early Man in the New World. *In* Pleistocene extinctions: the search for a cause, P. S. Martin and H. E. Wright, Jr., eds., Yale University Press, New Haven, pp. 267–286.

———. 1968a. Geochronology of late-Quaternary alluvium. *In* Means of correlation of Quaternary successions, R. B. Morrison and H. E. Wright, Jr., eds., Proceedings VII INQUA Congress, University of Utah Press, Salt Lake City, v. 8, pp. 581–631.

———. 1968b. Preliminary report on the late Quaternary geology of the San Pedro Valley, Arizona. *Southern Arizona Guidebook III,* Arizona Geol. Soc., pp. 79–96.

———. 1970. Geochronology of man-mammoth sites and their bearing upon the origin of the Llano Complex. *In* Pleistocene and Recent environments of the central Plains, W. E. Dort and A. E. Johnson, eds., University of Kansas Press, pp. 77–92.

———. 1973. Geochronology and paleohydrology of the Murray Springs Clovis site, Arizona. *Geol. Soc. Am. Abstracts with Programs,* v. 5, p. 659.

———. 1975. Pleistocene and Recent stratigraphy. *In* Late Pleistocene environments of the southern High Plains, F. Wendorf and J. J. Hester, eds., *Fort Burgwin Research Center Publ.* No. 9, pp. 57–96.

———. 1976. Late Quaternary geology of the lower Pomme de Terre Valley. *In* Prehistoric man and his environments: a case study in the Ozark Highland, W. R. Wood and R. B. McMillan, eds., Academic Press, New York, pp. 47–61.

———. 1978. Geochronology of the lower Pomme de Terre River. Paper presented at 43rd Annual Meeting, Society of American Archaeology, Tucson.

———. 1980. The Clovis culture. *Canadian Journal of Anthropology,* v. 1, pp. 115–121.

Haynes, C. V., Jr. and G. A. Agogino. 1966. Prehistoric springs and geochronology of the Clovis site, New Mexico. *American Antiquity,* v. 31, pp. 812–821.

Hester, J. J. 1972. Blackwater No. 1: a stratified, Early Man site in eastern New Mexico. *Fort Burgwin Research Center Publ.* No. 8, 238 pp.

Jelinek, A. H. 1967. Man's role in the extinction of Pleistocene faunas. *In* Pleistocene extinctions: the search for a cause, P. S. Martin and H. E. Wright, Jr., eds., Yale University Press, New Haven, pp. 193–200.

Kelly, L., E. Spiker, and M. Rubin. 1978. U.S. Geological Survey, Reston, Virginia, Radiocarbon Dates XIV. *Radiocarbon,* v. 20, pp. 283–312.

Leonhardy, F. C. 1966. Domebo: a Paleo-Indian mammoth kill in the Great Plains. *Museum of the Great Plains, Contribution* No. 1, 53 pp.

McMillan, R. B. 1976. Rodgers Shelter: a record of cultural and environmental change. *In* Prehistoric man and his environments: a case study in the Ozark Highland, W. R. Wood and R. B. McMillan, eds., Academic Press, New York, pp. 212–232.

McNett, C. W., Jr., B. A. McMillan, and S. B. Marshall. 1977. The Shawnee-Minisink site. *In* Amerinds and their paleoenvironments in northeastern North America, W. S. Newman and B. Salwen, eds., *Annals of the New York Academy of Science,* v. 288, pp. 282–296.

Martin, P. S. 1966. Africa and Pleistocene overkill. *Nature,* v. 212, pp. 339–342.

———. 1967. Prehistoric overkill. *In* Pleistocene extinctions: the search for a cause, P. S. Martin and H. E. Wright, Jr., eds.,

Yale University Press, New Haven, pp. 75–120.

———. 1973. The discovery of America. *Science*, v. 179, pp. 969–974.

Mehringer, P. J., Jr. 1967. The environment of extinction of the late-Pleistocene megafauna in the arid southwestern United States. *In* Pleistocene extinctions: the search for a cause, P. S. Martin and H. E. Wright, Jr., eds., Yale University Press, New Haven, pp. 247–266.

Saunders, J. J. 1977. Late Pleistocene vertebrates of the western Ozark Highland, Missouri. *Illinois State Museum Report of Investigations* No. 33, 118 pp.

———. 1980. A model for man-mammoth relationships in late Pleistocene North America.

Canadian Journal of Anthropology, v. 1, pp. 87–98.

Shutler, Richard, Jr. 1967. Archaeology of Tule Springs. *In* Pleistocene studies in southern Nevada, H. M. Wormington and D. Ellis, eds., *Nevada State Museum Anthropological Papers,* n. 13, 411 pp.

Stafford, T. W., Jr. 1981. Alluvial geology and archaeological potential of the Texas southern High Plains. *American Antiquity,* v. 46, pp. 548–565.

Warnica, J. M. 1966. New discoveries at the Clovis site. *American Antiquity* v. 31, pp. 345–357.

Wright, H. E., Jr. 1976. Pleistocene ecology —some current problems. *Geoscience and Man,* v. 13, pp. 1–12.

17

Prehistoric Overkill: The Global Model

PAUL S. MARTIN

TOWARD THE END OF THE ICE AGE, in the last interglacial before the last great ice advance, the continents were much richer in large animals than they are today. Mammoth, mastodonts, giant ground sloths, and seventy other genera, a remarkable fauna of large mammals, ranged the Americas. Australia was home to more than a dozen genera of giant marsupials. Northern Eurasia featured woolly mammoth, straight-tusked elephant, rhinoceros, giant deer, bison, hippopotamus, and horse. New Zealand harbored giant flightless birds, the moas, rivaled by the elephant birds of Madagascar. Even Africa, whose modern fauna is often viewed as saturated with large mammals, was richer before the end of the last glacial episode by 15 to 20 percent. In America, the large mammals lost in the ice age did not disappear gradually. They endured the last as well as the previous glaciations only to disappear afterward, the last "revolution in the history of life" (Newell 1967). Armed with modern techniques of analysis, one turns to deposits of the late Pleistocene and the prospect of determining the cause of this prehistoric revolution.

Late Pleistocene sediments yielding bones of extinct megafauna are widespread. Outcrops of late Pleistocene age are commonly more accessible than those of earlier times; as a result many more fossil mammoths have been found than brontosaurs. Dating of the late Pleistocene fossil faunas can be attempted by various geochemical techniques, especially radiocarbon analysis, an essential tool in attempting correlations, now improved by the use of accelerators. Under favorable circumstances the ages of late Pleistocene sedimentary units can be determined to within less than one hundred years. Fossil pollen and associated plant and animal remains of living species disclose the natural environment at the time of the extinctions. In a few favorable cases frozen paunch contents or ancient coprolites preserved in dry caves reveal what some extinct beasts actually ate. Mummified carcasses and cave paintings may show what others looked like. In the Old World the bones found in Paleolithic hunters' camps suggest the importance of large mammals, living and extinct, as prey. Taphonomy, the study of post-mortem processes affecting fossils, may disclose the cause of local mortality. While complete life histories of the extinct animals are unattainable, helpful clues may be found in the habits of living relatives. Many opportunities to study the late Pleistocene extinctions remain unexploited.

Given the great interest among paleontologists in earlier geological revolutions, one might expect to find an especially keen interest in the last one. For various reasons, this has not been the case, perhaps because of the peculiar attributes of the late Pleistocene extinctions. Certainly the extinctions at the end of the Pleistocene are quantitatively dwarfed by those at the Cretaceous/Tertiary boundary and the Permian/Triassic boundary. Late Pleistocene sediments, deposited during the time of the last major episode of extinctions, are neglected by many geologists who seek older events. Conversely, and for the opposite reason, ecologists have often ignored the late Pleistocene megafauna, so unlike that of historic time. While archaeologists have pioneered the study of late Pleistocene deposits and chronology, they may abandon the field at the most interesting moment when ancient artifacts give way to culturally "sterile" sediments that yield only the bones of an extinct fauna. Fortunately, the disciplinary fences that "Balkanized" late Pleistocene research earlier in this century are crumbling and some superior modern studies of megafaunal bone beds have appeared in recent years, such as those of Missouri's mastodont deposits, South Dakota's mammoth Hot Springs, Wyoming's Natural Trap, New Zealand's moa deposits, and California's tar pits. These and other examples of key sites are examined elsewhere in this volume.

Late Pleistocene extinctions have proved refractory. As Grayson shows in Chapter 1, the fascinating nineteenth-century efforts of Cuvier, Buckland, Fleming, Lyell, Darwin, Wallace, Owen, Agassiz, and others seeking to account for or to explain Pleistocene extinctions led to no solution and no consensus. Despite radiocarbon dating and other new techniques, will twentieth-century investigators manage to do any better?

I think so, perhaps because I remain optimistic about the possibility of devising a comprehensive late Pleistocene extinction model that will incorporate all significant pieces of evidence from the many parts of the globe. By judicious application of radiocarbon dating, rigorous chronological tests can be leveled at any hopeful model. Furthermore, recent discoveries regarding anomalous amounts of iridium in sediments of the Cretaceous-Tertiary boundary (Alvarez et al. 1980) suggest that at least one earlier "revolution in the history of life" may have been far less gradual than commonly thought. The possibility of cosmic events being involved has helped rejuvenate interest in all geological extinctions. As mammalian extinctions of the last 100,000 years become better known, which is the purpose of this book, the present theoretical difficulty in explaining them should vanish.

The Meaning of Late Pleistocene Extinctions

From the chronology, intensity, and character of late Pleistocene biotic changes in different land masses, I will review late Pleistocene extinctions on a global scale, revising an earlier attempt (Martin 1967). Large mammals of the continents were the group most obviously affected. Outside Africa most genera of large mammals lost in the Pleistocene were lost in the last 100,000 years, with an increasing number toward the end of that interval. The large mammals or "megafauna" can be defined in various ways. Since most genera of continental herbivores which became extinct in the late Pleistocene approximated or exceeded 44 kg (100 lbs.) in adult body weight, animals of this size or larger will be assigned to the "megafauna"; those under 44 kg will be considered "small."

Extinctions of interest within the last 100,000 years were not limited to large mammals. On oceanic islands during the Holocene a large number of genera of small birds and mammals disappeared. "Oceanic" islands are those surrounded by deep water beyond the continental shelf, remaining separate from the continent even during marine regressions of glacial age. Even when such islands are in full view of the mainland, they will not support a balanced mainland terrestrial fauna.

Late Pleistocene extinctions represent losses of native fauna within the last 100,000 years up to historic time. Those who end the Pleistocene around 10,000 years ago, recognizing the "Recent" as a separate time unit, may prefer to speak of "late Quaternary extinctions" which I accept as synonymous. It appears that one's understanding of the cause of the extinctions is not improved by boundary definitions.

"Extinction" has a special meaning in physics and psychology; in geology and biology it has classically been applied to the permanent disappearance of a taxonomic lineage, as in "extinction is forever." A different meaning is found in MacArthur and Wilson's (1967) accidental extinctions, the departure or extirpation of undifferentiated island populations of plants and animals, which may be followed by later reinvasion. While MacArthur and Wilson did not attempt to account for geological extinctions, their concept was soon applied to fossils, for example by Webb (1969).

Stenseth (1979) has referred to the two types as *local* extinctions if they are repeatable phenomena that operate in ecological time, and as *global* extinctions if they are irreversible and operate in geological time. Along with most authors in this book, I will mainly be analyzing global extinctions. Partial exceptions may occur as in the "extinction" of the horse (*Equus*) in America, the musk ox (*Ovibos*), in Eurasia, and the spiny anteater (*Zaglossus*) in Tasmania. These genera are each extant in other parts of the globe; conceivably they could return. However, their regional extirpation was seen only once in the Pleistocene fossil record when it was accompanied by extinctions of many other animals. The sweeping range reductions cited do not appear to be repeatable phenomena operating in ecological time.

While data on species changes may be preferable when available, I employ the fossil genus as a practical measure of Ice Age extinctions. "The genus is the smallest taxonomic unit for which geological range data are readily available, and the genus is probably the unit most consistently defined by mammalian paleontologists" (Gingerich 1977). Detailed species records available for Europe (Kurtén 1968) and North America (Kurtén and Anderson 1980) show that no sizable loss of Pleistocene fauna occurred without registering an accompanying loss of genera. An important characteristic that distinguishes the late Pleistocene from earlier mammalian extinctions, at least since the Hemphillian (late Miocene), is that higher taxonomic categories are affected, including families and even certain mammalian orders.

The Pleistocene embraces the last 1.7 million years (my), equivalent to or slightly younger than the top of the Olduvai Event in the paleomagnetic timescale (Haq et al. 1977). In North America the land mammal ages of the Rancholabrean and the Irvingtonian are included. The major continental glaciations fall within this time period, as does most of the evolution of the genus *Homo*. During the last 100,000 years modern man, *Homo sapiens,* replaced *Homo erectus* and spread over the world.

Most groups of organisms escaped massive extinction in the late Pleistocene. For that matter certain groups that display striking range changes and even local extirpation in response to early Pleistocene climatic changes, such as European vascular plants (Leopold 1967) and beetles (Coope 1979), suffered little or no global extinction of genera at any time in the Pleistocene (see Kershaw this volume). To take another tack, the plants eaten by extinct ground sloths (Hansen 1978) and by woolly mammoth did not become extinct themselves. The episode was far less revolutionary than the large mammal losses would portend.

Important Pleistocene generic extinctions among freshwater ostracods are few or unknown (Benson pers. corres.). Marine extinctions of planktonic organisms involve the last appearance of species or populations of diatoms, calcareous nannoplankton, radiolarians, and foraminifers. The marine stratigraphic record displays more last appearances in the 0.9 my prior to the Quaternary than during the last 1.7 my itself (Berggren et al. 1980). The twenty-three last appearances of planktonic marine organisms in the 2.6 my since the Gauss/Matuyama boundary all predate the late Pleistocene losses of land megafauna. Whether of whales, clams, or phytoplankton, no marine extinctions match or accompany the conspicuous loss of large land animals at the end of the Pleistocene. The Quaternary shows no change that can be compared with the late Permian or late Cretaceous when the marine record most forcefully documents the loss of numerous taxa.

One important event in the paleoclimatic record of the oceans does correlate closely with many late Pleistocene extinctions on land. The last 100,000 years of glacial expansion, as recorded by oxygen-isotope ratios in deep sea cores from the Atlantic and the equatorial Pacific, terminated abruptly around 12,000 years ago (Broecker and van Donk 1970, Ruddiman and McIntyre 1981). A very rapid ice melt caused a rapid rise of sea level. Large mammal extinctions in North and South America are coeval with Termination I as plotted by Ruddiman and McIntyre. Detailed land fossil records show a major movement of plant and animal species at the time, especially into formerly glaciated terrain. American megafaunal extinctions occurred during a time of rapid climatic change as seen in fossil pollen and small animal records. It is natural that the paradigm of climatic change which works so well in explaining so many biogeographic events of the Cenozoic should also be applied to the late Pleistocene extinctions. However, the marine record of the last half million years displays many gradual glacial expansions and sudden melts or terminations. Evidently Termination II occurred about as rapidly 125,000 years ago as Termination I, without triggering extinctions of large animals.

A realistic global extinction model must account for differences in extinction intensity between continents of the Southern Hemisphere—Africa, Australia, and South America. Heavy losses occurred in Australia predating by thousands of years the time of Termination I (see Horton this volume). In contrast South American extinctions, also very heavy, coincided with Termination I. African generic extinctions were unusual in being heavier early rather than late in the Pleistocene; only relatively minor losses occurred from 10,000 to 12,000 years ago.

The point I raise is that the proxy data from ocean sediments indicate that climatic changes accompanying late Pleistocene deglaciation were not unique but occurred repeatedly throughout the Ice Age. Apart from Africa, continental extinctions of large land mammals are unremarkable except during the time of Termination I. Then they coincide with heavy losses in the New World and not in Africa, while postdating massive extinctions in Australia. The extinction pattern does not obviously track changes in climate throughout the Pleistocene. Furthermore, the extinctions vary in time and intensity within the late Pleistocene.

Unlike changes in climate, the global spread of prehistoric *Homo sapiens* was confined to the late Pleistocene. Prehistoric man's role in changing the face of the earth can conceivably include major depletions of fauna, an anthropogenic "overkill." Overkill is taken to mean human destruction of native fauna either by gradual attrition over many thousands of years or suddenly in as little as a few hundred years or less. Sudden extinction following initial colonization of a land mass inhabited by animals especially vulnerable to the new human predator represents, in effect, a prehistoric faunal "blitzkrieg" (Mosimann and Martin 1975). A close look at the late Pleistocene extinction pattern will help establish ways in which the extinctions might be conceived of happening. A look at various land masses will establish whether extinction coincides with human invasion.

The Late Pleistocene Extinction Pattern

The following eight attributes of late Pleistocene extinction seem especially noteworthy:

1. *Large mammals were decimated.* In North America thirty-three genera of large mammals disappeared in the last 100,000 years (or less), while South America lost even more (Table 17.1). Not since the late Hemphillian, several million years before the Pleistocene, was there a comparable loss of large terrestrial mammals in America. Africa lost about eight genera in the last 100,000 years while Australia lost a total of perhaps nineteen genera of large vertebrates, not all of them mammals. Europe lost three genera by extinction and nine more by range shrinkage.

2. *Continental rats survived; island rats did not.* In the Pliocene and early Pleistocene many genera of small mammals were lost in continental North America. In the late Pleistocene there were significant changes in range and in body size of small mammals. However, the episode of late Pleistocene large mammal extinction was unaccompanied by equally heavy loss of small mammals. Oceanic islands are another matter. On the West Indies and oceanic islands in the Mediterranean small endemic genera of mammals experienced very heavy prehistoric loss.

3. *Large mammals survived best in Africa.* From continent to continent the loss of megafauna was highly variable. Many more genera of large mammals were lost from the late Pleistocene in America and Australia than from Africa (Table 17.1). No late Pleistocene families were lost from Asia or Africa to match the New World loss of families and orders.

4. *Extinctions could be sudden.* In parts of New Zealand at the end of the Holocene giant birds, the moas, disappeared within 300 years or less. Radiocarbon dates on Shasta ground sloth dung from caves in the southwestern United States indicate extinction of different populations of ground sloths around 11,000 yr B.P. (Thompson et al. 1980). Evidently North American mammoth disappeared quite suddenly around the same time (Haynes 1970). Those extinct North American genera that can be dated by radiocarbon disappeared roughly between 15,000 and 8,000 years ago. It is not certain that any one genus actually died out before another (Martin 1974).

5. *Regional extinctions were diachronous.* Moas in New Zealand and elephant birds in Madagascar were still alive eight millenia after the mammoths and mastodonts were gone from America. The latter were predeceased by the mammoths of northern Europe and China. The European mammoths in turn survived the diprotodonts of Australia by many thousands of years. Since late Pleistocene extinctions occur on one land mass at a

Table 17.1. Late Pleistocene Extinct and Living Genera of Terrestrial Megafauna (>44 kg adult body weight) of Four Continents*

	Extinct (last 100,000 years)	Living	Total	% Extinct
Africa	7	42	49	14.3
North America	33	12	45	73.3
South America	46	12	58	79.6
Australia†	19	3	22	86.4

*Includes genera living elsewhere, i.e. *Equus* lost in North and South America and *Tapirus* lost in North America.

†Australian data include large extinct reptiles, *Meiolania*, *Megalania*, and *Wonambi* and extinct and living large birds, *Genyornis*, *Dromiceius*, and *Casuarius*.

SOURCE: Data from Tables 17.2, 17.3, 17.6, 17.7, 17.8, and 17.9.

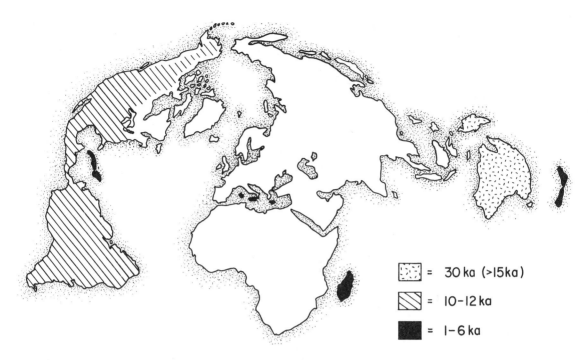

Figure 17.1. Major late Pleistocene extinction events on the continents and larger islands, excluding Afro-Asia; ka = thousand years. Heavy large mammal extinction occurred some time before 15,000 years ago in Australia and New Guinea. North and South America were especially affected 10,000 to 12,000 yr B.P.; the Greater Antilles and Mediterranean islands experienced extinction in the early or mid-Holocene, and Madagascar and New Zealand in the late Holocene. The sequence and intensity of loss tracks the spread of prehistoric cultures.

different time and intensity than on another, they are unlikely to be explained by a sudden extraterrestrial catastrophe *à la* the asteroid impact model of Alvarez et al. (1980). Late Pleistocene extinctions were time transgressive, following the sequence shown in Figure 17.1.

6. *Extinctions occurred without replacement.* The terrestrial extinctions seen throughout the late Pleistocene were not necessarily preceded or accompanied by significant faunal invasion. There was no mixing of long-separated or isolated faunas, sometimes claimed as the cause of geologic extinctions, unless man is considered to be the crucial element in the new community.

7. *Extinctions followed man's footsteps.* In America, Australia, and on oceanic islands, very few megafaunal extinctions of the late Pleistocene can be definitely shown to predate human arrival. Conversely, the survival into historic time of large, easily captured animals such as the dodo of Mauritius, the giant tortoises of Aldabra and the Galapagos, and Steller's sea cow on the Commander Islands of the Bering Sea, occurred only on remote islands undiscovered by or unoccupied by prehistoric people. The sequence of extinction during the last 100,000 years follows human dispersal, spreading out of Afro-Asia into other continents and finally reaching various oceanic islands.

8. *The archaeology of extinction is obscure.* In Eurasia and Africa, where comparatively few late Pleistocene extinctions occurred, the bones of large extinct mammals are commonly found in Paleolithic sites of various ages. In America where there are many late Pleistocene extinctions, very few archaeological sites have yielded bones of extinct

mammals; most of these have been of mammoth, and the mammoth sites are roughly equivalent in age (Haynes 1970, 1980). In Australia no megafaunal "kill sites" have been found. While the lack of kill or processing sites has often been regarded as a serious obstacle to assigning prehistoric man much of a role in the extinction process, few would be expected if man's impact were truly swift and devastating (Mosimann and Martin 1975).

Two features that would be discordant with the model of overkill include: (1) extinction of animals too small or ecologically resistant to be vulnerable to human impact and (2) extinction of late Pleistocene faunas under circumstances that preclude an anthropogenic effect, i.e. before prehistoric human arrival. Some features that I regard as diagnostic of late Pleistocene extinctions may not be exclusive to it. For example, the end of the Hemphillian could have seen a wave of extinctions happening suddenly (see Webb this volume). In addition, earlier Cenozoic extinctions were also probably not synchronous on different continents. Finally, it is possible that large mammals have always been more susceptible to extinction than small ones. Nevertheless, severe late Pleistocene extinctions appear to have been uniquely the fate of large mammals of the continents. Taken together, the eight features of the late Pleistocene are compatible with a model of overkill. Taken together they are not features one would necessarily encounter if phyletic replacement, faunal turnover, or climatic change were primary causes of late Pleistocene losses.

Excluding Antarctica, which is not known to have suffered any late Pleistocene extinctions, and limiting treatment of Asia to its European segment only, I will examine extinction events on all major land masses. Beginning with North America the sequence will proceed from continents that suffered heavily to those that did not, ending with a sketch of events on oceanic islands, especially those known to have experienced significant prehistoric faunal losses.

North American Extinction

It has long been known that the kinds of large mammals native to North America were greatly diminished by the end of the Pleistocene. The fate of small mammals could not be determined on the basis of ample evidence until the technique of screen-washing led to the recovery of abundant bones of small mammals from various sediments. The new records yielded many early and very few late Pleistocene extinctions of small mammals (Table 17.2). Kurtén and Anderson's (1980) review of Pleistocene mammals of North America enables biogeographers to examine the Pleistocene and Pliocene extinction record with regard to taxonomic group, size of animal, temporal sequence, and other useful details. Their Blancan age begins 3.5 million years ago (mya), the Irvingtonian 1.7 mya, and the Rancholabrean 0.7 mya.

From Kurtén and Anderson (1980) I have revised earlier tables of extinct Plio-Pleistocene genera for both large and small mammals (Martin 1967). Since it probably exceeded 44 kg, I have transferred *Procastoroides*, a large beaver two-thirds the size of extinct giant beavers to the "large mammal" category and moved the diminutive pronghorn antelope *Capromeryx*, of which one species weighed about 10 kg, as well as its larger relative, *Stockoceros*, to the "small mammal" category.

None of these changes appreciably alters the pattern presented previously for extinction of large mammals (Martin 1967). Ten genera of large mammals were lost in the Blancan, seven in the Irvingtonian, and three in the Rancholabrean prior to Wisconsin glacial times (Table 17.3). Thus twenty genera of large mammals represent the total of all known megamammal extinctions over three million years prior to the last glaciation. They are followed at the end of the Wisconsin by the loss of thirty-three genera of

Table 17.2. Extinct Small Mammals of the Plio-Pleistocene in North America (exclusive of bats)

Scientific Name (Common Name or Nearest Living Relative)	Blancan			Irvingtonian			Rancholabrean		
	Early	Middle	Late	Early	Middle	Late	Illinoian	Sangamon	Wisconsin
	MY 3.5			MY 1.7			MY 0.7		
Paracryptotis (shrews)									
Ferinestrix (wolverines)									
Parailurus (red pandas)									
Pliogeomys (pocket gophers)									
Symmetrodontomys (mice)									
Cosomys (voles)									
Notolagus (rabbits)									
Pratilepus (rabbits)									
Aluralagus (rabbits)									
Nekrolagus (rabbits)									
Planisorex (shrews)									
Sminthosinus (grisons)									
Businictis (skunks)									
Dipoides (beaver)									
Bensonomys (mice)									
Ogmodontomys (voles)									
Pliophenacomys (voles)									
Nebraskomys (voles)									
Pliolemmus (voles)									
Pliopotamys (muskrats)									
Trigonictis (grisons)									
Canimartes (grisons)									
Satherium (otters)									
Enhydriodon (sea otters)									
Paenemarmota (giant marmot)									
Cryptopterus (flying squirrels)									
Ophiomys (voles)									
Hypolagus (rabbits)									
Osmotherium (skunk)									
Protocyon (dog)									
Nerterogeomys (pocket gopher)									
Prodipodomys (kangaroo rat)									
Etadonomys (kangaroo rat)									
Proneofiber (water rat)									
Pliomys (vole)									
Tisisthenes (weasel)									
Predicrostonyx (lemming)									
Atopomys (vole)									
Mimomys (vole)									
*Coendou (porcupine)									
Paradipoides (beaver)									
*Heterogeomys (hispid pocket gopher)									
*Cuon (dhole)									
Capromeryx (pronghorn)									
Stockoceros (pronghorn)									
Brachyprotoma (short-faced skunk)									
TOTAL EXTINCT, N = 46	10	10	8	6	3	3	1	1	4

*Living genus outside North America
SOURCE: Kurtén and Anderson 1980

Table 17.3. Extinct Large Mammals of the Plio-Pleistocene in North America

Scientific Name (Common Name or Nearest Living Relative)	Blancan			Irvingtonian			Rancholabrean		
	Early	Middle	Late	Early	Middle	Late	Illinoian	Sangamon	Wisconsin
	MY 3.5			MY 1.7			MY 0.7		
Megatylopus (camel)									
Bretzia (false elk)									
Ceratomeryx (pronghorn)									
Meganteron (dirktooth)									
Ischyrosmilus (sabertooth)									
Dinofelis (false sabertooth)									
Procastoroides (beaver)									
Nannippus (gazelle-horse)									
Blancocamelus (camel)									
Rhynchotherium (gomphothere)									
Borophagus (plundering dog)									
Chasmaporthetes (hunting hyena)									
Stegomastodon (gomphothere)									
Titanotylopus (camel)									
Kraglievichia (pampathere)									
Hayoceros (pronghorn)									
Platycerabos (flat-horned ox)									
Soergelia (Soergel's ox)									
Praeovibos (musk oxen)									
*Blastocerus (marsh deer)									
Holmesina (pampathere)									
Glyptotherium (glyptodont)									
Megalonyx (megalonychid) (ground sloth)									
Eremotherium (giant ground sloth)									
Nothrotheriops (Shasta ground sloth)									
Glossotherium (mylodont ground sloth)									
*Tremarctos (spectacled bear)									
Arctodus (short-faced bear)									
Smilodon (sabertooth)									
Homotherium (scimitartooth)									

megamammals including all North American members of seven families and one order, the Proboscidea. The loss of various species of horses and camels terminated the dynasty of the Equidae and Camelidae in the land of their origin. The changes can be visualized by plotting the cumulative loss of genera of large mammals for the three mammalian age-stages. The minor break in slope suggests a minor extinction episode in the late Blancan, with a major break in slope only at the end of the Rancholabrean (fig. 17.2).

The losses of megamammals well dated by radiocarbon occurred around 11,000 radiocarbon years ago (Martin 1974, Mead and Meltzer this volume); uncertainty exists in the case of those genera that are so rare or rarely dated that their extinction chronology cannot be established with confidence (Mead and Meltzer this volume). When extinct mammals can be located within rock-stratigraphic units, the boundary between living and extinct fauna is abrupt (see Haynes this volume). Stafford's work on the Llano Estacado of west Texas provides another example. In a sequence of alluvial and colluvial deposits alternating with erosion surfaces, bones of large mammals are found throughout the depositional units. Extinct fauna of mammoth, horse, and camel

Table 17.3. Extinct Large Land Mammals of the Plio-Pleistocene in North America

(continued)

Scientific Name (Common Name or Nearest Living Relative)	Blancan			Irvingtonian			Rancholabrean		
	Early	Middle	Late	Early	Middle	Late	Illinoian	Sangamon	Wisconsin
	MY 3.5			MY 1.7			MY 0.7		
*Acinonyx (cheetah)									
Castoroides (giant beaver)									
*Hydrochoerus (capybara)									
Neochoerus (capybara)									
*Equus (horses, asses, onager)									
*Tapirus (tapir)									
Mylohyus (peccary)									
Platygonus (peccary)									
Camelops (camel)									
Hemiauchenia (llama)									
Palaeolama (stout-legged llama)									
Navahoceros (mountain deer)									
Sangamona (fugitive deer)									
Cervalces (stag-moose)									
Tetrameryx (pronghorn)									
*Saiga (saiga)									
Euceratherium (shrub-ox)									
Symbos (woodland musk ox)									
Bootherium (woodland musk ox)									
*Bos (yak)									
Mammut (American mastodont)									
Cuvieronius (Cuvier's gomphothere)									
Mammuthus (mammoth)									
TOTAL EXTINCT, N = 53	3	0	7	3	2	2	2	1	33

*Living genus outside North America
SOURCE: Kurtén and Anderson 1980

are found in the lower units, the most recent being about 11,000 years old (fig. 17.3). The large bones deposited in younger units above unit one (fig. 17.3) are exclusively those of bison.

When ideal material for radiocarbon dating of an extinct genus is available, such as charcoal with mammoth remains in archaeological sites (Haynes 1970, 1980, this volume), collagen-rich bone from Rancho La Brea (Marcus and Berger this volume), or Shasta ground sloth dung in desert caves (Thompson et al. 1980), the youngest remains of the extinct animal approximate 11,000 radiocarbon years. Whether or not prehistoric people were in the Americas earlier, 11,000 yr B.P. is the time of unmistakable appearance of Paleo-Indian hunters using distinctive projectile points. It also appears to be the time when the better dated extinct genera such as *Mammuthus, Nothrotheriops,* and *Equus* disappeared quite suddenly. The circumstances are in accord with the "blitzkrieg" model.

The small mammal extinction chronology is entirely different. The plot of cumulative extinction of small genera displays no break in slope (fig. 17.2). Unlike the large mammals, more genera of small mammals were lost during the Blancan than in the

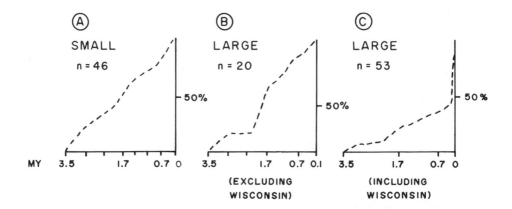

Figure 17.2. Cumulative extinctions in percentage of genera of North American mammals during the past three million years. A relatively constant rate of extinction through time will plot as a straight line; the extinction record of small mammals is an example (A). A punctuated or concentrated episode of extinction will plot as a break in slope. The extinction record of large mammals shows two breaks, one at the end of the Blancan (B) and one at the end of the Rancholabrean (C). The latter event involves many more losses (data from Tables 17.2 and 17.3).

Irvingtonian and Rancholabrean combined, twenty-eight compared with eighteen. Only four genera of small mammals disappeared during the Wisconsin. At least two of these, the extinct antelopes, may be regarded as possible human prey and the third, *Cuon* of Alaska, as a carnivore dependent on the megafauna. The extinction of birds in the Rancholabrean involves mainly scavengers and commensals plus a few large species that may have been human prey such as the flightless duck *Chendytes* (Steadman and Martin this volume; for opposing view see Grayson 1977). The only genus of small mammals to become extinct around the time of the megafaunal crisis that cannot be readily linked to human impact, or to some side effect of the extinction of the megafauna, is the spotted skunk relative, *Brachyprotoma*. It is noteworthy that in the Lancian, at the time of late Cretaceous dinosaur extinction, nine genera of small mammals (mouse to fox terrier in size) disappeared, including four genera of multituberculates, four of marsupials, and one placental (Clemens et al. 1981). No genera of continental rats or mice became extinct in the late Pleistocene.

Significant size differences in the intensity of extinction can be examined in more detail by comparing species losses within mammalian orders. Percent extinction of species was derived from Kurtén and Anderson (1980, their Table 19.3). The number of species extinctions are plotted for each land mammal age, without regard to absolute time (fig. 17.4). The edentates, odd- and even-toed ungulates, and elephants represent orders of mammals of large size. The insectivores, rodents, rabbits, and the majority of the carnivores represent small- to medium-size animals that were under 44 kg adult body weight. While small animals may have been hunted avidly, they would be much less vulnerable to sustained human predation. It massive small mammal extinctions accompanied the loss of large ones, as in the late Hemphillian, the assumption of any anthropogenic effect in the late Pleistocene might be questioned (see Graham and Lundelius this volume).

For each of the four orders of predominantly large mammals, extinction of species reached a maximum value within the last mammalian age, the Rancholabrean. The carnivores, which include some large as well as many small- and medium-size species, show an intermediate response. Finally, there were relatively few losses in the Rancholabrean among the orders of exclusively small mammals. The relative differences in

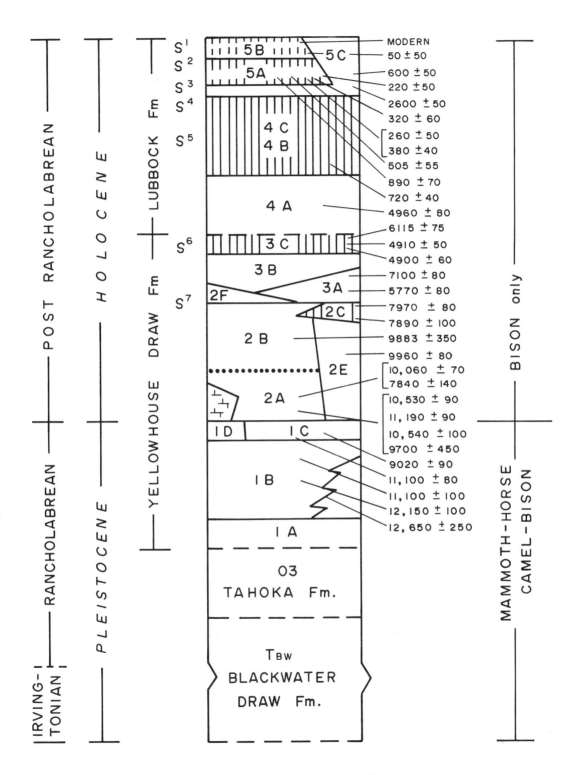

Figure 17.3. Part of the lithostratigraphic sequence in west Texas. Bones of large mammals, living and extinct, occur in units 1 and those below. Only bones of the living genus, *Bison,* occur in units 2 and above. Reproduced by permission of the Society for American Archaeology, from *American Antiquity* 46(3), Stafford, 1981.

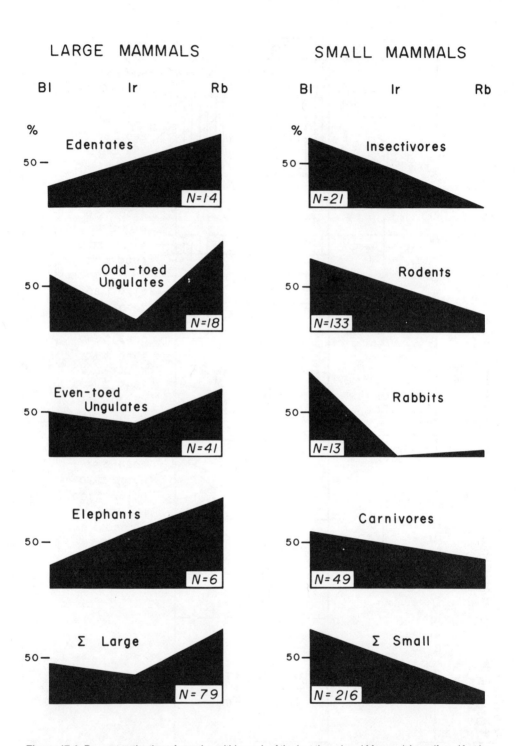

Figure 17.4. Percent extinction of species within each of the last three Land Mammal Ages (from Kurtén and Anderson 1980). Four orders of large mammals and four of small mammals are shown. N = total number of species that became extinct. Since the Blancan (Bl) lasted longer than the Irvingtonian (Ir) and the latter longer than the Rancholabrean (Rb), a decline in percent extinctions from Blancan to Rancholabrean would be expected if losses were relatively constant. The small mammals show such a result. The extremely high percentage of large mammal species extinctions (well over 50 percent) in the Rancholabrean is anomalous (see text).

extinction intensity are in accord with the model of human impact as a size selective force, operating effectively on the larger (and more slowly reproducing) fauna and only in the Rancholabrean.

If Ice Age climatic changes were important in determining the extinction of American large mammals, as many believe, it is not obvious why earlier glaciations and interglacial warmups were unaccompanied by faunal losses. A recent pollen study of a long core from Clear Lake, California, supports the view that while the Sangamon was even warmer than the Holocene (Adam et al. 1981), no extinctions occurred then. The lack of extinctions accompanying pre-Wisconsin glaciations was contrary to expectations of O. P. Hay, who mistakenly assigned an early Pleistocene extinction to camels (Nelson and Madsen 1980) and other extinct genera (Martin 1967). If man's role in the process was negligible or quite incidental, it would be an amazing coincidence that extinctions of most large mammals occurred when they did. The only significant difference between the transition to the Sangamon and the transition to the Holocene is the presence of early hunters in North America.

It can be argued that by limiting analysis to the late Pliocene and Pleistocene I have taken too narrow a view. Webb (1969, this volume) has shown that the end of the Miocene was marked by a time of even more extensive loss than in the late Pleistocene, but it is not known how rapidly late Miocene (Hemphillian) extinctions occurred.

According to Webb (this volume), the late Hemphillian saw the loss of two genera of bears; four of mastodonts; eight perissodactyls, including the family Rhinocerotidae; and sixteen artiodactyls, including the family Protoceratidae. The loss of twenty-six genera of large mammals is less than but certainly comparable to late Pleistocene losses. On the other hand, and unlike the late Pleistocene, numerous small mammalian genera became extinct. Late Hemphillian extinctions included nine genera of insectivores; fifteen of rodents, including two rodent families; and eight genera of small- to medium-size carnivores (Table 9.1 of Webb this volume). The extinction of thirty-two genera of Hemphillian small mammals may be compared with the loss of only four genera of small mammals in the late Pleistocene (the Wisconsin in Table 17.4). Thus there is no difficulty in distinguishing late Hemphillian extinctions from the late Pleistocene in one crucial attribute—body size. If climatic change was involved in bringing about both late Pleistocene, Blancan, and Hemphillian extinctions in North America, the differences in the effect on the small mammals remain to be explained.

Zoologists interested in living mammals rather than the dead commonly ask: "If ancient hunters were such potent destroyers, how did the bison and the other large genera native to North America manage to survive?"

**Table 17.4. Generic Extinction of North American Mammals
Since the Mid-Hemphillian**

The Wisconsin, a segment of the Rancholabrean equivalent to the late Pleistocene, is shown separately to emphasize the late occurrence of large mammal extinctions within the Rancholabrean.

	Small (<44 kg)	Large (>44 kg)	Total	% Large	Time 10^6 yr
Late Hemphillian	32	26	58	44.8	>3.5
Blancan	28	10	38	26.3	3.5–1.7
Irvingtonian	12	7	19	36.8	1.7–0.7
Rancholabrean	6	36	42	85.7	0.7–0.0
Wisconsin	4	33	37	89.2	0.1–0.0
Total	78	79	157		

SOURCE: Data from Tables 17.2 and 17.3 and Webb (this volume).

Living Megafauna of America

The sudden loss of large mammals must mean that in some way large size was a decided handicap. Nevertheless, certain survivors such as moose, elk, and musk ox were much larger than many of the smaller extinct mammals such as the long-nosed peccary, extinct pronghorn (*Capromeryx*), and extinct llamas. To restate the question, how did some large mammals manage to survive when many others disappeared?

Virtually all of the twelve surviving genera of large mammals were of Eurasian origin, arriving late in the Pleistocene. The one notable exception is the pronghorn, *Antilocapra.* "It is noteworthy that most of the Eurasian invaders of North America, the moose, wapiti, caribou, musk ox, grizzly bears, and so on—were able to maintain themselves, perhaps because of their long previous conditioning to man" (Kurtén 1971, p. 221). The survivors are known for behaviors that should have reduced their vulnerability to hunters, such as the constant and seasonally erratic movements of *Rangifer* and *Bison,* the solitary or forested haunts of *Cervus* and *Alces,* and the rigorous arctic or subalpine habitats embraced in the ranges of *Ovis, Oreamnos,* and *Ovibos.* The archaeological record indicates that all were hunted in the North American Holocene long after megafaunal extinctions. Perhaps bison, caribou, and the other surviving large mammals served as a reliable resource for the hunters during critical years when those animals destined for extinction were themselves too scarce to provide a stable food supply. If so, the extant large mammals unwittingly helped exterminate those on the threshold of extinction, in effect maintaining hunter populations during the last few years before the extinctions of other megafauna.

A second group of New World survivors remains to be considered. North of Mexico, North America lost its camelids, tapir, capybara, spectacled bear, and mountain deer (see Table 17.5). In each case close relatives survived in South or Central America.

Table 17.5. Extinct North American Large Mammals Surviving in South America

North American Extinct Species (North of Mexico)		Living South American Relative	
*Palaeolama	(llamine)	Lama	(guanaco)
*Hemiauchenia	(llamine)	Lama	(guanaco)
Tapirus	(tapir)	Tapirus	(tapir)
Hydrochoerus	(capybara)	Hydrochoerus	(capybara)
*Platygonus	(plains peccary)	Catagonus	(Paraguay peccary)
*Navahoceros	(mountain deer)	Hippocamelus	(Andean deer)
Blastocerus	(Florida marsh deer)	Blastocerus	(marsh deer)
Tremarctos	(bear)	Tremarctos	(spectacled bear)

*Extinct genus

The converse is not known; no late Pleistocene genus that became extinct in South America managed to survive further north. The disappearance of "tropical" species from North America north of Mexico, such as tapir and capybara, has been attributed to climatic change (Marshall et al. 1982). One difficulty with a climatic explanation is the well-known evidence of warming conditions in the interval from 10,000 to 12,000 years ago. The range of semitropical species should have been expanding, not contracting, at the time of the extinction of the "tropical" element.

It is possible to suggest a cultural explanation. The hunters first entering America would of necessity have mastered the severe, treeless subarctic tundra of the Bering platform (see Colinvaux 1981). They should have accommodated readily to the more open woodland or savanna habitats within warm temperate and subtropical parts of

North America, ultimately encountering and exterminating peripheral populations of tapir and capybara and other "tropical" species in those areas. However, the hunters would have been less effective in the dense tropical forest habitats also occupied by tapir and capybara. Heavy forest may have required more ecological adjustment than any of the more open habitats previously encountered.

After the human invaders penetrated the tropical forests of northern South America, their progeny "rediscovered" open habitats, South American savanna, steppe, and subalpine tundra (paramo). The populations that first reached the steppes and inter-Andean basins of South America were no longer part of a lineage as experienced in subarctic and montane steppe as their ancestors. Cold climate adaptations would have had to be relearned. Thus the South American faunas of large mammals, the mountain deer and llamoids, may have experienced less hunting pressure than their North American relatives.

I have left for last the intriguing matter of Holocene (postglacial) survival. To the best of my knowledge all securely dated archaeological deposits of the last 10,000 years on the American continent have yielded bones exclusively of living mammalian genera. New World prehistoric art of the Holocene portrays no extinct animals. Large numbers of bison remains are found in Holocene deposits of the High Plains, the North American range shared in the late Pleistocene by mammoth, horse, camel, and ground sloth. Numbers of guanaco bones are found in Andean and Patagonian archaeological sites in South America, the late Pleistocene domain of mastodonts, glyptodonts, ground sloths, and toxodonts. Even the recently discovered peccary, *Catagonus,* is a "living fossil" of the Holocene, not of the Pleistocene. Bones of a near relative were known from Holocene archaeological sites before the living Paraguay population was described (Wetzel et al. 1975). Extinction of genera of large American mammals had ended by at least 8,000 years ago, despite continuing human population growth in the late Holocene triggered by agricultural economies based on production of maize, beans, and squash.

The persistence of surviving large mammals that endured the disaster of the late Pleistocene may be explained by the increasingly sedentary pattern of subsequent human occupation. Apart from local incursions, such as that of Athabaskan-speaking people late in prehistoric time, the main cultural developments were regional. Locally distinct linguistic groups developed. Intertribal conflicts led to the development of inter-tribal buffer zones, which may have provided prey species with a natural refuge. For example, the tribal boundary between the Chippewa and the Sioux supported deer (*Odocoileus virginianus*) which vanished only when a U.S. government–enforced peace treaty was established in the nineteenth century (Hickerson 1965). In the Amazon basin intertribal warfare and periodic taboos against huntable large animals ensured continuity of the larger prey species (Ross 1978). A biological analogue is suggested in the behavior of certain wolf packs which maintain a territorial spacing sufficient to provide refuge for deer, their natural prey (Mech 1977). For whatever reason, prehistoric people witnessed little or no extinctions of genera in the last 10,000 years.

In historic time the introduction of European diseases and other cultural disruptions shattered the traditional society of native Americans. According to C. Martin (1978) the spiritual game "bosses" were blamed for the crisis and, abandoning traditional practice, a new overkill began with Indians hunting and trapping wantonly. While C. Martin found the concept of prehistoric overkill to be flawed and theatrical, his historical analysis may help in modeling the late Paleolithic. What would happen to a stable society of Eurasian mammoth and reindeer hunters when they first set foot in the vast hunters' paradise sweeping south from Alaska? Any preexisting hunting taboos might not have been extended to protect mastodonts, ground sloths, and giant peccaries of the new environment. The main difficulty in proceeding further is not the theatrical quality of the stage, which must involve the retreat of glaciers, the penetration of a continent by Paleolithic hunters, the drama of the extinctions, and at least the proposition that these events are linked. The problem is that the critical changes may have run their course too rapidly to be embraced by the fossil record. The first centuries following historic contact of native Americans and Europeans offer an analogue. Presumably overkill of the ninth millenium was also a time of rapid and catastrophic change in human behavior, with the relaxation of any preexisting restraints on excessive hunting.

To summarize, the large mammals surviving in the New World appear more gracile than the extinct ones (e.g. cervids compared with ground sloths), to be unpredictable in their movements (caribou, bison, antelope), or cryptic denizens of heavy cover (moose, spectacled bear, tapir). Many North American megafaunal survivors were themselves immigrants from the Old World where, despite a much longer history of being hunted and despite severe reductions in range, they also managed to survive. If extinctions were sudden, as seems to be the case, it is not surprising that the fossil record is insufficient to reveal many details.

South America

Separated from other continental faunas through much of the Tertiary, the evolution of South American mammals and their subsequent fate following land bridge contact with North America has fascinated biogeographers (Patterson and Pascual 1972; Webb 1978a, 1978b; Simpson 1980; Marshall 1981a; Marshall et al. 1982). The Pleistocene faunas of South America were much richer than those of historic time. Extinctions earlier in the Cenozoic were also numerous, and the possibility of explaining them has invited considerable interest, despite Simpson's pessimistic view (1953, p. 303; 1980) that there are too many possible causes to determine exactly what led to the demise of any one species.

Following Marshall (1981a) and Marshall et al. (MS.), I have listed all small (Table 17.6) and large (Table 17.7) extinct genera of South American Pleistocene mammals according to their land mammal ages. The Lujanian and upper Ensenadan combined can be roughly correlated with the Rancholabrean of North America, the lower Ensenaden and upper Uquian with the Irvingtonian, and the lower Uquian, Chapadmalalan and upper Montehermosan with the Blancan.

While the small mammal record of the South American Pleistocene is not as well known as that of North America and Europe, there were thirty-two extinctions of genera of small- and medium-sized mammals, including certain marsupials, armadillos, rodents, and carnivores. As in North America, these are largely confined to the early Pleistocene (Table 17.6). Contrary to Darwin's curious comment of over 100 years ago, there was no appreciable loss of late Pleistocene small mammals.

Darwin (1855, p. 224) wrote:

> Did man, after his first inroad into South America, destroy, as has been suggested [by Owen], the unwieldly *Megatherium* and the other Edentata? *We must at least look to some other cause for the destruction of the little Tucutuco at Bahia Blanca, and of the many fossil mice and other small quadrupeds in Brazil* [ital. added].

The many fossil extinctions that accompanied those of the giant ground sloths and other large Edentata in the Lujanian were of large or moderately large mammals (Table 17.7) plus some birds (Campbell 1979). No fossil mice are involved.

Late Pleistocene faunas are much more abundant and thus better samples than those of earlier times, so that the earlier land mammal ages will undoubtedly gain proportionally to the Lujanian as they become better known. Nevertheless, the total number of mammalian genera known from the Lujanian, 120, is not vastly more than that known from the Ensenadan, 106, or even the Uquian with 84 (see Marshall 1981a, pp. 171–172). The Lujanian extinctions stripped South America of 21 genera and 4 families of large edentates; 1 genus of giant rodent; 3 genera of large carnivores; the last 2

Table 17.6. South American Small Mammal (< 44 kg) Extinctions

C = Chapadmalalan; U = Uquian; E = Ensenadan; L = Lujanian (after Marshall 1981a).

	C	U	E	L
Duration (million years)	1.0	1.0	0.7	0.3
MARSUPIALIA				
Argyrolagidae				
Argyrolagus	—			
Microtragulus	———————			
Didelphidae				
Paradidelphys	—			
Thylatheridium	———————			
Thylophorops	———————			
Sparassocynidae				
Sparassocynus	———————			
EDENDATA				
Dasypodidae				
Doellotatus	—			
Chorobates	—			
Plaina	—			
Ringueletia	—			
Macroeuphractus	—			
NOTOUNGULATA				
Hegetotheriidae				
Paedotherium	———————			
RODENTIA				
Abrocomidae				
Protabrocoma	—			
Caviidae				
Cardiomys	—			
Caviodon	—			
Caviops	—			
Dolicava	—			
Orthomyctera	—————————————			
Palaeocava	—			
Chinchillidae				
Lagostomopsis	—			
Cricetidae				
Cholomys		———		
Dankomys		———————		
Echimyidae				
Eumysops	———————			
Myocastoridae				
Isomyopotamus	—			
Tramyocastor		———		
Octodontidae				
Actenomys	—			
Eucoelophorus	———————			
Megactenomys	?	———		
Pithanotomys	—————————————			
Plateaomys			———	
Pseudoplateomys	—————————————			
CARNIVORA				
Canidae				
Protocyon		———————		
Mustelidae				
Stipanicicia		———		
Procyonidae				
Brachynasua			———	
Cyonasua	——	?		
ARTIODACTYLA				
Tayassuidae				
Argyrohyus	—			
Total extinctions, N = 34	17	14	5	0

Table 17.7. South American Large Mammal (>44 kg) Extinctions

C = Chapadmalalan; U = Uquian; E = Ensenadan; L = Lujanian (after Marshall 1981).

Duration (million years)	C 1.0	U 1.0	E 0.7	L 0.3
EDENTATA				
Dasypodidae				
Kraglievichia	—			
Pampatherium		—	—	
Propraopus	?	?	—	
Glyptodontidae				
Chlamydotherium			—	—
Daedicuroides			—	
Doedicurus			—	—
Glyptodon			—	—
Hoplophorus			?	—
Lomaphorus			—	—
Neothoracophorus			—	—
Neuryurus			—	
Panochthus			—	—
Paraglyptodon	—	?		
Plaxhaplous			—	—
Plohophoroides	—			
Sclerocalyptus			—	—
Trachycalyptus	—			
Urotherium	—			
Megalonychidae				
Diheterocnus	—			
Nothropus			—	—
Nothrotherium			—	—
Ocnopus				—
Megatheriidae				
Eremotherium				—
Megatherium			—	—
Plesiomegatherium	?	?		
Mylodontidae				
Glossotherium	—			
Glossotherium			—	—
Lestodon			—	—
Mylodon		?	—	—
Scelidodon			—	—
Scelidotherium	—	?		
Scelidotherium			—	—
LITOPTERNA				
Macraucheniidae				
Macrauchenia			—	—
Promacrauchenia	—			
Windhausenia			—	—
Proterotheriidae				
Brachytherium	—			
NOTOUNGULATA				
Hegetotheriidae				
Tremacyllus	—			
Mesotheriidae				
Mesotherium			—	
Pseudotypotherium	—			
Toxodontidae				
Mixotoxodon				—

Table 17.7. South American Large Mammal (>44 kg) Extinctions

C = Chapadmalalan; U = Uquian; E = Ensenadan; L = Lujanian (after Marshall 1981a).

(continued)

	Duration (million years)	C 1.0	U 1.0	E 0.7	L 0.3
Toxodon		———	———	———	———
Xotodon		———			
RODENTIA					
Dinomyidae					
Telicomys		———			
Hydrochoeridae					
Chapalmatherium		———			
Hydrochoeropsis			———		
Neochoerus			———	———	———
Nothydrochoerus			———		
Protohydrochoerus		———	?		
CARNIVORA					
Felidae					
Smilodon				———	———
Smilodontidion			———		
Procyonidae					
Chapalmalania		———			
Ursidae					
Arctodus				———	———
ARTIODACTYLA					
Camelidae					
Astylolama					———
Eulamaops					———
Hemiauchenia				———	———
Palaeolama				———	———
Protauchenia					———
Cervidae					
Agalmaceros					———
Charitoceros					———
Morenelaphus				———	———
Paraceros					———
Tayassuidae					
Brasiliochoerus (Catagonus)				———	———
Platygonus				———	———
Selenogonus			?		
PERISSODACTYLA					
Equidae					
Equus				———	———
Hippidion				———	———
Onohippidium				———	———
PROBOSCIDEA					
Gomphotheriidae					
Cuvieronius				———	———
Haplomastodon					———
Notiomastodon				?	———
Stegomastodon				———	———
Total extinctions N = 72		14	7	3	46

genera of the ungulate order Litopterna, and the last 2 of the ungulate order Notoungulata; 4 genera of mastodonts and the order Proboscidea; 3 genera of horses and the family Equidae; and 11 genera of artiodactyls including peccaries, camelids, and deer (Table 17.7). While the number of large mammalian genera lost in the Lujanian may be inflated by taxonomic splitting, especially among the glyptodonts, the number as it stands greatly exceeds the combined large mammal losses of the previous 2.7 million years: 14 genera in the Chapadmalalan, 7 genera in the Uquian and 3 genera in the Ensenadan. The 46 Lujanian extinctions considerably exceed the large mammal extinctions of the late Pleistocene of North America. While South America became a refuge for some large mammals that had ranged widely in North America, such as tapir, camelids, and spectacled bear, it became the graveyard of far more. No other continent lost as many mammals in the late Pleistocene (Table 17.1).

Through radiocarbon dating it can be shown that at least the more common members of the South American extinct faunas survived into the late glacial in deposits 15,000 to 8,000 years old (Marshall et al. MS.). The age of extinction coincides closely with the arrival of prehistoric hunters in South America. Human impact has been held accountable for the event, at least in part (Patterson and Pascual 1972, Webb 1976, Marshall 1981a).

The impact of North American mammalian invaders in South America has been widely discussed. While hystricomorph rodents, procyonid carnivores, and platyrrhine primates had managed to reach South America much earlier, around three million years ago a land bridge was established and South America's "splendid isolation" ended (Simpson 1980). Felids, canids, cricetine rodents, proboscideans, and ungulates of North American stock spread into South America to encounter a long isolated fauna. Apart from endemic rodents and bats, most South American mammals were entirely different from the new invaders even at the ordinal level. Some South American land mammals moved north such as the opossum, porcupines, and especially the edentates, of which a few had managed to enter North America several million years before the land bridge was established. It is tempting to make the "great interchange" the cause of most South American extinctions of Pleistocene age (e.g. Preston 1962, Newell 1967). If competition from North American invaders can satisfactorily account for the Lujanian extinctions, there would be less need to pursue other explanations for them.

Marshall and Hecht (1978) and Marshall (1981a) noted that two of the native South American ungulate orders were in decline since the Miocene, long before land bridge time. One South American ungulate of camel-like build, *Macrauchenia*, is associated in Lujanian faunas with *Palaeolama*, from North America, which would not be expected if they were competitors. Some of the invaders may have displaced giant native South American rodents (Simpson 1980), and rodent losses of early Pleistocene time are appreciable (Table 17.6). On the other hand, another South American group, the ground sloths and glyptodonts, increased in generic diversity during the first two million years after land bridge time. They spread north and became common in North America in apparent disregard of the established Holarctic large herbivores. Few clear-cut ecological replacements can be seen among the invaders from North America which Patterson and Pascual (1972) viewed as insinuators into South America. Introduced mammals can be highly effective at displacing natives, and the extinction of certain endemic mammal faunas may occur rapidly following invasion by aliens (Simpson 1953). The possibility of competition and early Pleistocene faunal loss following North American invasion of a saturated South American megafauna has been defended by Webb (1976, 1978b) and Marshall et al. (1982). However, this hypothesis has no bearing on the late Pleistocene megafaunal extinctions which did not happen until several million years later. As most authors note, it is of particular interest that extinctions affected animals of northern origin as well as the South American autochthons. Of the forty-six extinct Lujanian genera in South America, twenty represented lineages of North American origin. Had competition or turnover caused the extinctions, one would expect to see them occur much earlier along with the loss of native hystricomorphs and marsupials in the Chapadmalalan or Uquian.

While there may be disagreement about the early Pleistocene, most authors no longer attribute Lujanian extinctions to the impact of the great American interchange. The groups that survived such as peccaries, llama, deer, and tapir were smaller and presumably more fecund than the large and robust gomphotheres, ground sloths, and toxodonts. These attributes should have endowed the survivors with a greater potential for escaping human predation than those of their extinct contemporaries. It is difficult to imagine an assemblage of large mammals that would have been easier to track, hunt, and exterminate than the numerous kinds of ground sloths and glyptodonts that disappeared down to the last species, leaving as living relatives only the much smaller tree sloths and anteaters.

The time of South American megafaunal extinction has been reviewed by Marshall et al. (MS.). Direct dates on dung, hide, or hair of extinct animals are the most desirable; dates on other organic material in association with bones are invariably not as satisfactory; dates on bone may be suspect unless the analyst can defend the result by demonstrating that uncontaminated collagen or amino acids, not "acid insoluble residues," were the source of the carbon (Taylor 1980). Ground sloth extinction in South America may be slightly younger than in North America (Long and Martin 1974). No ground sloth dung younger than 10,000 years old has been found. A debatable mid-Holocene survival of extinct ground sloths in Patagonia is advanced on stratigraphic, not radiocarbon evidence (Moore 1978).

The date of human arrival in South America is unclear. MacNeish (1979) has reported a ground sloth kill and butchering site at Pikimachay Cave, Peru at 14,100 B.C. ± 1200 with older occupation to 20,000 B.C. Other excavators of cave sites in the region have not recovered occupations this old or found evidence of man associated with large extinct mammals (Rick 1980, Lynch 1980, Hurt et al. 1976). As in North America, deposits rich in remains of late Pleistocene extinct fauna such as the tar seeps of Talara, Peru, the mineral springs at Araxa, Brazil, or the dung deposit of Mylodon Cave, Chile, have not yielded unimpeachable evidence of human artifacts in abundant association with extinct animals.

For this reason I proposed a rapid overkill or blitzkrieg of the South American extinct fauna (Martin 1973). As the man-mammoth sites in North America are all about 11,000 years old (Haynes 1980), and as most archaeologists infer migration by land from the north, the South American sites should be younger. My age interpretation conflicts with South American claims of a greater human antiquity (MacNeish 1976, 1979; Bryan et al. 1978; Gruhn and Bryan this volume). A compromise might be reached by theorizing that scattered bands pursuing mainly small game and plants spread slowly south from North America many thousands of years before the arrival of skilled big game hunters, to coexist peacefully with the ground sloths and toxodonts prior to their extinction at the hands of later invaders. However, Gruhn and Bryan (this volume) claim to have spear points with a 13,000-year-old mastodont site; thus the issue is far from resolved. If prehistoric hunters are found associated with South American megafauna significantly before 11,000 B.P., the result would be entirely discordant with the blitzkrieg model. Claims of early South American archaeological sites are proliferating. Establishing their validity will require successful replication by disinterested parties.

Climatic change has been proposed as the cause for South American and other extinctions (Axelrod 1967). Using Africa as a model, Guilday (1967) attributed the losses to habitat reduction. Recent pollen and paleolimnological work in Venezuela (i.e. Bradbury et al. 1981) supports earlier biogeographic evidence for an expansion of tropical forest at the expense of savanna 8,000 years ago. While this is close to the time of the Lujanian extinctions and may be viewed as supporting evidence by those who favor a climatic model, it would appear to entail increased rather than reduced equability as a mechanism initiating extinction, the reverse of climatic models applied to North America (Graham and Lundelius this volume).

Finally, the climatic model which attributes large mammal extinctions in temperate latitudes to the cessation of warm winters and the onset of severe freezing temperatures in the Holocene will not operate in frost-free tropical latitudes. As noted by Grayson (this volume), the model was first proposed by Lyell and independently restated by Hibbard (1960) and Slaughter (1967). Much has been made of the assumed vulnerability of giant tortoises (*Geochelone, Testudo*) to freezing temperatures, at least in Kansas. However, tortoises of large size disappeared throughout the American tropics, including the West Indies, where temperatures must have remained mild even during a glaciation. The only environment favorable for the survival of large tortoises appears to have been those remote and arid oceanic islands such as Aldabra and parts of the Galapagos that escaped much human settlement or heavy tortoise harvesting by either prehistoric or historic voyagers.

Australia

Australia resembles the Americas in magnitude of its megafaunal extinctions. While it lost fewer genera, it had fewer to lose; the proportions are similar. Australia may have lost almost as many *species* of large animals as North America (see Table 29.1 in Horton this volume; Tables 27.1 and 27.2 in Murray this volume). Only one genus and four species of terrestrial mammals exceeding 44 kg survived. Australia differs from North America in the lack of allochthonous genera. There was no Pleistocene enrichment comparable to the Eurasian invasion of North America by bison, moose, musk ox, caribou, mammoth, etc., in the ice ages. Australia (incluing New Guinea) is the only known Pleistocene refuge for large marsupials. Apart from bats, the only placentals to invade were murid rodents in the Tertiary, the dingo in the Holocene and, of course, prehistoric people in the late Pleistocene. Thus the concept of phyletic or ecological turnover which can be invoked to account for American megafaunal extinctions in the late Pleistocene and earlier (Webb 1969, this volume) is less promising in the case of Australia.

The late Pleistocene megafauna of Australia includes roughly thirteen extinct genera of large marsupials (Table 17.8). Of the six families that were represented in the Pleistocene (Marshall 1981b), three became extinct. To these may be added the Tasmanian wolf, *Thylacinus,* and the pig-footed bandicoot, *Chaeropus,* two Australian genera lost in historic times. The Australian megafauna included a rhino- or hippo-sized form, *Diprotodon,* as well as *Zygomaturus,* a smaller relative in the same endemic family. Australia lacked elephant-sized mammals represented in America by mammoths, mastodonts (including gomphotheres), and giant ground sloths (*Eremotherium, Megatherium*). Overall the extinct genera of Australian megafauna appear to be smaller on the average than their American counterparts (Table 17.8). Two "browsing antelopines," *Propleopus* and *Protemnodon,* were somewhat under 44 kg adult body weight as well as the smaller species of *Sthenurus* (see Murray this volume). Other members of the extinct assemblage could be roughly compared in build with extinct American megafauna, the extinct giant wombats with capybaras, *Palorchestes* with the tapir, and the smaller extinct macropods, *Sthenurus* and *Procoptodon,* with small ground sloths.

Late Pleistocene extinction of Australian large animals was not restricted to the marsupials. There was a giant varanid lizard, *Megalania,* larger than the Komodo dragon; a giant horned tortoise, *Meiolania*; an extinct ostrichlike bird, *Genyornis*; and an extinct snake the size of a python, *Wonambi.* Four species of kangaroos, *Macropus giganteus, M. fuliginosus, M. rufus,* and *M. robustus,* represent the full complement of

Table 17.8. Australian Pleistocene Megafauna

Excludes *Glaucodon* of the early Pleistocene, after Marshall (1981b). All except *Macropus* are extinct. *Nototherium* and *Prionotemnus* may predate the Pleistocene; *Sthenomerus* and *Synaptodon* are very poorly known and *Phascolomis* may be congeneric with living *Vombatus*.

Family	Genus	# of Extinct Species	Estimated Weight (kg)	Ecological Equivalent
Thylacoleonidae	*Thylacoleo*	2	200	big cat
Vombatidae	*Phascolonus*	1	500	giant capybara
	Ramsayia	1		giant capybara
	Phascolomis	2		giant capybara
Palorchestidae	*Palorchestes*	1		tapir
Diprotodontidae	*Diprotodon*	2	2000	rhino
	Zygomaturus	1	1000	small rhino
	Nototherium	1		
	Sthenomerus	1		
Macropodidae	*Propleopus*	1	30	browsing antelopine
	Protemnodon	3	40	browsing antelopine
	Troposodon	1		
	Macropus	4–8	150	(giant kangaroos)
	Sthenurus	10	30–60	very small ground sloths
	Procoptodon	3–4	300	small ground sloths
	Fissuridon	1		
	Prionotemnus	1		
	Synaptodon	1		

living "large" mammals to which one may add the giant birds, the emu, and cassowary. Apart from Antarctica no continent is more impoverished in its native terrestrial megafauna (Table 17.1).

As in North and South America, Australia apparently experienced no appreciable extinction of small mammals in the late Pleistocene. For many years the small marsupial mouse, *Burramys,* described in 1895 by Broom from its Pleistocene bones, was an apparent exception. Recently it was added to the modern fauna when it was found to be alive near the tree line in the Australian Alps (Walker 1975, p. 67). Since forty of the forty-eight genera of living Australian marsupials are known to have Pleistocene or, in some cases, Pliocene fossil records, it seems unlikely that further work in the late Pleistocene will uncover many undescribed extinct genera of small size. As in America late Pleistocene extinction is restricted to animals of large size.

One remarkable feature of the Australian Pleistocene megafauna is the relative scarcity of carnivores (Hecht 1975). They are limited to three species: the fiercely clawed large "lion," *Thylacoleo*; the large "dog," *Thylacinus*; and the giant varanid lizard, *Megalania*. Only the thylacine could have functioned as a cursorial predator. Presumably because the Australian large herbivores did not coevolve with an array of fleet predators like the canids, hyaenis, and felids of Africa, most of them, at least the extinct ones, appear to have been at best sprinters rather than distance runners.

The first human inhabitants should have found the large mammals relatively easy to pursue, especially the lumbering diprotodontids with their small plantigrade feet, the stumpy giant wombats, and the ponderous, extinct kangaroos that were less gracile than their living relatives. Morphologies of the extinct genera can be judged from reconstructions by Archer (1981) and Murray (1978, this volume). Gill (1955) has suggested psychological factors that may have made them vulnerable to the hunters. Both their metabolism and their brain size were reduced compared with that of large placental mammals on other continents (Murray pers. corres.).

Within their lineages, the surviving mammals appear selected for either bounding speed, gracile body form, smaller body size, or nocturnal habits. Relatively slow moving and easily captured animals such as the living wombats and the koala may have managed to survive by virtue of cryptic behavior, including feeding at night. The koala, *Phascolarctos,* lost its Pleistocene range in the Murray-Darling basin and in western Australia even though suitable habitat persists there. I suggest a local extirpation by hunters exploiting habitat islands that were not easily reinvaded by surviving populations located in the more extensive forests of eastern Australia. A larger or less cryptic mammal than the koala would not have survived. Aboriginal effectiveness at trimming local populations can be gauged by flourishing faunas of the larger marsupials found only on those islands adjacent to the continent which were unoccupied by aborigines in historic time (Abbott 1980).

"Only twice in human history were entire continents, Australia and America, colonized suddenly" (Jones 1979). The arrival of a potent and deadly species, *Homo sapiens,* landing on a continent that had previously known few large cursorial carnivores, and none of these in the Order Carnivora, seems uniquely favorable for overkill. The loss of the less fleet and more conspicuous large mammals seems inevitable and unremarkable.

While the overkill hypothesis was advanced more than a hundred years ago in the time of Owen (1877), and was recast several times in this century, especially by Merrilees (1968) and Jones (1968), it has not found widespread endorsement (see Horton 1980 and this volume). For example, White and O'Connell (1979) concluded: "Whatever the value of overkill models proposed for North America and elsewhere, they are inapplicable in the Australian situation."

The divergence of opinion stems in part from a failure to find extinct animal remains in the known archaeological sites and the question of the age of the last of the extinct fauna. With the advent of radiocarbon dating it became apparent that Australian megafaunal extinctions had occurred before the late Pleistocene extinction in North America (Martin 1967). It is still not clear how closely the extinctions track prehistoric cultural events.

In the 1960s Mulvaney's excavation of Kenniff Cave established human occupation of Australia by at least 14,000 yr B.P. Australian archaeology soon began to take great strides back into the late Pleistocene. Ground stone axes in caves in Arnhem Land were dated to over 20,000 yr B.P. (C. White 1971); cave wall designs and other evidence of human occupation at a chert source found in total darkness within Koonalda Cave in the Nullarbor Plain were dated at over 20,000 yr B.P. (Wright 1971); a dozen bone points up to an estimated 29,500 yr B.P. were reported from Devils Lair, western Australia (Dortch 1979, p. 270); and clay ovens or hearths at Lake Mungo exceeded 30,000 years in radiocarbon age (Barbetti and Allen 1972). An articulated human skeleton at Lake Mungo was dated at 28,000 years (Bowler and Thorne 1976). These and other impressive prehistoric finds discovered in Australia are reviewed in Mulvaney (1975). A few years later Jones (1979) reported a total of twenty-nine sites predating 15,000 radiocarbon years. Australian early man sites in excess of 15,000 years even include Tasmania during the glacial age (Kiernan, et al. 1983) and stand on much firmer ground than those claimed in the Americas at the same time or earlier.

While Australian archaeology had been extended into the late Pleistocene to at least 30,000 years, the new discoveries were lacking in one item that apparently had been and still is thought by some to be inexplicable in its absence. If human impact was involved in the extinctions, where are the sites with bones of the large extinct mammals themselves? Apart from occasional scraps such as three tooth fragments of *Sthenurus* at Seton's Cave, the bone record associated with human occupation is commonly limited to members of the living fauna. The fact that observations "on the association of man and the Pleistocene giant fauna are few and unsatisfactory" (Calaby 1971) was and is held against the overkill thesis. Undocumented claims of bones of extinct megafauna associated with artifacts were noted (Calaby 1976). Cuts on extinct macropodid bone at Lancefield were attributed to a scavenger, presumably *Thylacoleo,* rather than to human activity (Horton and Wright 1981). Bowdler (1977) noted: "In Australia we have an overkill hypothesis, but we are still waiting for a kill site."

Deposits of extinct Australian mammal remains can be extremely rich, such as at Lancefield, Mammoth Cave, Wellington Cave, and Victoria Cave. Unfortunately, none of these deposits has proved easy to date by radiocarbon, and when dates have been obtained the results may be ambiguous (e.g. Wells 1978). Occasional fragments of *Sthenurus* (Hope et al. 1977), including a mandible and humerus at Cloggs Cave (Flood 1980), may indicate survival of a few local populations into the late glacial. However, ambiguities and evidence for redeposition of bone of extinct genera have been encountered in caves studied elsewhere (Milham and Thompson 1976, Lundelius and Turnbull 1978, Balme 1978). It is not yet clear that the age of any of the reported late glacial survivors is securely established. Even the bone bed at Lancefield, which many authors have accepted as 26,000 years old, rests on the association of two charcoal dates. Sample association problems are suggested by plant material at the bottom of the flooded swamp deposit which yielded a modern ^{14}C date (Gillespie et al. 1978). It appears that the bone-rich deposits of late Pleistocene mammals known in Australia are not readily dated. No ideal source of organic carbon comparable to American sloth dung or bone amino acid residues have been found to help eliminate the ambiguity in age determination. Most, perhaps all, extinct Australian genera appear to be older than the youngest Rancholabrean faunas in North America. In America a large number of extinct faunas are known to be less than 30,000 years in ^{14}C age (Martin 1974, Mead and Meltzer this volume). Carbon-14 dating of extinct American megafauna into the late glacial has proved far easier than it appears to be in Australia.

Unless survival of the extinct Australian megafauna can be firmly dated to within the last 30,000 years, the possibility of man's sudden impact, an Australian blitzkrieg, cannot be discarded. I illustrate a chronological model of events for both continents (fig. 17.5). Ample evidence of human presence in Australia exists back to 30,000 years ago. In contrast, such evidence beyond 12,000 years ago in the New World is sparse at best. Evidence for the occurrence of extinct mammals in the New World from 30,000 to 11,000 years ago is also abundant. In Australia, however, such evidence seems relatively sparse and scrappy. The timing of extinction on both continents invites a careful, thoughtful comparison. If the model shown in Figure 17.5 survives chronological tests, a strong case could be made for sudden overkill or "blitzkrieg" in both continents.

The lack of kill sites in Australia is not necessarily remarkable. As mentioned, clear-cut associations of human artifacts with extinct fauna are few even in America. The model of sudden human impact does not predict the discovery of kill sites. One might expect more to be preserved in America since New World extinctions are evidently younger and thus stratigraphically more accessible. Given the rapid rate at which highly vulnerable and previously unhunted prey might be destroyed by the first invading human hunters, the chance of finding appreciable evidence of man's deadly passage is small (Mosimann and Martin 1975). This theoretical view will apply to Australia no less than to America, especially if the Australian fauna was poorly adapted to cursorial and diurnal human predation.

The blitzkrieg concept was designed to explore maximum human impact on large mammals engendered by the sudden arrival of hunters. The model can be rejected in Australia if a "late" extinction chronology proves valid, that is, if the large extinct mammals are found to have maintained sizable populations long after initial human invasion of the continent, exactly as Horton has concluded (this volume). If any extinct faunal kill sites actually are found in Australia, they should be over 30,000 years in age to conform with the blitzkrieg model. With active work underway the chronological uncertainty may eventually be clarified; the major difficulty is the matter of validity in radiocarbon dating of bone samples that have lost most of their bone collagen.

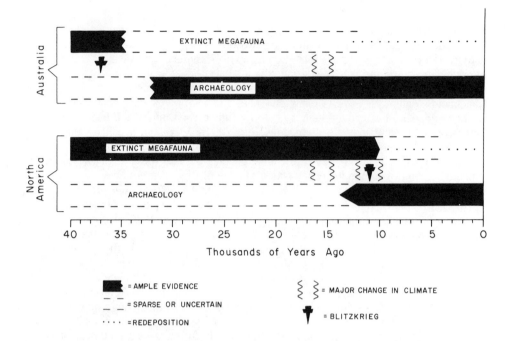

Figure 17.5. A comparison of evidence of extinct faunas and prehistoric artifacts in Australia and North America. Radiocarbon-dated deposits of extinct megafauna in Australia are sparse or in an uncertain context over the last 30,000 years. In America extinct megafauna is abundantly dated up to 11,000 years ago and in an uncertain context afterward.

Radiocarbon-dated archaeological sites in Australia are numerous and well dated back to 30,000 years ago in a variety of sites and uncertain beyond that. In America archaeological sites are numerous and well dated only to 12,000 years ago and uncertain or questionable before. The record provides a test of sudden overkill (blitzkrieg). The discovery of abundant, well-dated extinct faunas in Australia 15,000 to 30,000 years ago, or abundant, well-dated hunting cultures in America at the same time would be discordant with the blitzkrieg model.

Africa

"As the land of big game par excellence, Africa more than any other continent has retained something of the richness and strangeness of the mammal fauna that dominated the world before the Pleistocene extinctions took their toll" (Bigalke 1978). The African record of Pleistocene extinction is less familiar to biogeographers. Its most unusual feature is the sizable number of extinctions of large mammals in the Lower Pleistocene, a time when less change is seen in the Americas. Unlike the Americas and Australia, the African fauna escaped being greatly altered by late Pleistocene extinctions. If Africa provided our only Pleistocene record, its fossils would appear to be in accord with classical ideas about ice-age extinctions tracking climatic changes, i.e. more loss early on.

Following new developments in magnetostratigraphy, Butzer and Isaac (1975) place the Plio-Pleistocene boundary at the time of the Olduvai Event, around 1.8 million years ago. The Lower-Middle Pleistocene boundary was set to coincide with the Matuyama-Brunhes reversal 0.7 million years ago. Finally, the Middle-Upper Pleistocene boundary falls at the beginning of the last interglacial/marine transgression about 125,000 years ago. Table 17.9, adapted from Maglio (1978) and Klein (MS.), displays extinctions of the Lower, Middle, and Upper Pleistocene. The African Lower Pleistocene correlates with the Irvingtonian of North America; the Middle and Upper Pleistocene in Africa correlate with the Rancholabrean of North America.

Thirty seven genera of African mammals disappeared in the Pleistocene, including two families (Chalicotheriidae, Gomphotheriidae) and an order (Deinotherioidea) represented by a single genus in each (Table 17.9). Ten genera of small mammals, mainly rodents and insectivores, were lost in the early Pleistocene. While I suspect that the small mammal faunas of later times escaped any appreciable extinction, the African Pleistocene is too poorly known to be certain of this.

There was a considerable faunal enrichment during the African Pleistocene. "This epoch gained more than it lost..." (Maglio 1978, p. 613). Competition has been cited in accounting for losses. Black (1978, p. 423) implies competition from the Hippopotamidae to account for the Pleistocene extinction of Anthracotheres. Churcher and Richardson (1978, p. 402) relate the disappearance of the last three-toed equids to the explosive radiation of the Bovidae. Coppens et al. (1978, p. 364) note that the disappearance of *Mammuthus* in Africa around two million years ago "may relate to competition with *Elephas.*" In view of the proliferation of the Bovidae in the lower Pleistocene, bovid extinctions need occasion no surprise. Gentry's characterization (1978, p. 568) of *Beatragus hunteri* as a "species naturally on the verge of extinction at the present time" implies competitive replacement. Competition of *Homo erectus* is mentioned as a possibility for explaining the disappearance of robust species of *Australopithecus,* although Howell (1978, p. 273) notes "their disappearance is also part of a larger pattern of mammalian faunal turnover and replacement within the continent around a million years ago."

Replacements are not difficult to find for losses within the Artiodactyla, the order which underwent the most evolutionary radiation throughout the Pleistocene. While the Bovidae lost three genera in the Lower and three in the Middle Pleistocene, it had earlier gained fifteen genera in the Pliocene and eleven in the Early Pleistocene. The loss of five suids in the Early and Middle Pleistocene was preceded by the Plio-Pleistocene arrival of seven. The artiodactyls were not the only group to experience appreciable extinctions. In the early Pleistocene the primates and the carnivores lost six genera each. The machairodonts and the hyaenids may have suffered competition from *Panthera* and new types of canids (Klein this volume). The most likely competitor to replace the various giant baboons would be *Homo.* Lampbrecht (1980) speculates that climatic change and a reduction in favorable forest habitats one million years ago allowed rapid expansion of the genus *Homo.*

The alleged depredations of *Australopithecus* on his fellow primates proposed by Dart to explain the nature of bone accumulations in South African caves has been discounted (Brain 1981); it is more likely that *Australopithecus* was a savenger. If the extinct sabertooths (*Dinofelis, Meganteron,* and *Homotherium*) and hyaenids (*Percrocuta, Leecyaena,* and *Euryboas*) were also scavengers or predators of "easy" prey, the early Pleistocene hominids were among their potential competitors. Based on dental morphology as well as phylogeny, the early Pleistocene hominids also were the potential competitors of the six genera of cercopithecoid primates lost in the Lower Pleistocene. A competitive crisis might even be extended to the suids (Hutely and Kappelman 1980). As the Early Pleistocene hominids improved their root-gathering techniques, they may have reduced ecological opportunities for those pigs whose dentition suggests a comparable food niche.

Extinctions around one million years ago reduced the hominid lineages to one, *Homo erectus.* A long interval of typological stability followed. The extinction of *Elephas recki,* baboons of the subgenus *Simopithecus,* and several large pigs occurred much later, toward the end of the Acheulean. According to Isaac (1977), "The time span during which the numerous extinct elements of the Acheulean died out is not yet documented in any rock-stratigraphic sequence in eastern or southern Africa." Until the African Pleistocene extinctions are better dated, any attempt at explanation must remain highly conjectural (Klein this volume). However, the hominid shadow falls across the African lower Pleistocene at a time when the Americas experienced fewer mega-

Table 17.9. Extinct African Genera of Large Mammals

| | Absolute Age | | |
Lower Pleistocene 1,800,000–700,000	Middle Pleistocene 700,000–130,000	Upper Pleistocene 130,000 to present	Living
PRIMATES			
Cercopithecidae			
Dinopithecus			_Pan_
Parapapio			_Mandrillus_
Gorgopithecus			_Gorilla_
Cercopithecoides	_(Simopithecus, subgenus)_		_Theropithecus_
Paracolobus			
Hominidae			
Australopithecus			_Homo_
CARNIVORA			
Hyaenidae			
Percrocuta			_Hyaena_
Leecyaena			_Crocuta_
Euryboas			
Felidae			
Dinofelis	_Machairodus_		_Acinonyx_
Megantereon			_Panthera_
Homotherium			
PROBOSCIDEA			
Gomphotheriidae			
Anancus			
Elephantidae			
*Mammuthus		_*Elephas_	_Loxodonta_
DEINOTHERIOIDEA			
Deinotheriidae			
Deinotherium			
TUBULIDENTATA			_Orycteropus_
PERISSODACTYLA			
Chalicotheriidae			
Ancyclotherium			
Equidae		_Hipparion_	_Equus_
Rhinocerotidae			_Diceros_
			Ceratotherium

faunal extinctions and when, for example, American saber cats flourished. While the possibility of a sudden overkill of large herbivores of the type I have modeled for Australia and America seems very remote, the effect of ancient human predation may nevertheless be traced in the character of the evolving fauna. The highly diverse and cursorial nature of the surviving African animals, the Pleistocene radiation of the Bovidae, and the lack of any species of very slow-moving, ponderous herbivores comparable to the American ground sloths and glyptodonts, or to the Australian diprotodonts, may reflect the coevolutionary history of Pleistocene man and African mammals.

The extinct genera of large mammals known from the Acheulean are found mainly in archaeological sites, leading Clark (1959) and Martin (1966) to suggest that they ultimately became victims of human predation at the end of the Acheulean. However, the outstanding feature of the African Pleistocene is the astonishing number of large

Table 17.9. Extinct African Genera of Large Mammals

(continued)

Absolute Age			Living
Lower Pleistocene 1,800,000–700,000	Middle Pleistocene 700,000–130,000	Upper Pleistocene 130,000 to present	Living
ARTIODACTYLA			
Suidae			
Notochoerus	Metridiochoerus		Phacochoerus
Potamochoeroides	Kolpochoerus		Sus
	Stylochoerus		Potamochoerus
			Hylochoerus
Hippopotamidae			Hippopotamus
			Hexaprotodon
Camelidae		*Camelus	(introduced)
Cervidae		Megaloceros	Cervus
Giraffidae	Sivatherium		Giraffa
			Okapia
Bovidae			
Simatherium	Menelikia	Megalotragus	Tragelaphus
"Hemibos"	Rabaticeras	Pelorovis	Boocerus
Makapania	Numidocapra	Parmularius	Taurotragus
			Syncerus
			Cephalophus
			Kobus
			Redunca
			Hippotragus
			Oryx
			Addax
			Damaliscus
			Alcelaphus
			Beatragus
			Connochaetes
			Aepyceros
			Litocranius
			Gazella
			Capra
			Ammotragus
TOTAL 21	9	7	42

SOURCE: Adapted from Maglio (1978) and Klein (MS.)

*genus survived elsewhere

mammals which survived, forty-two genera in thirteen families, in contrast to the extinction of nine genera of large mammals in the middle and seven in the upper Pleistocene (Table 17.9)

While the survival of many kinds of large animals in Africa has been used as an argument against overkill (Eiseley 1943), I would use it as an argument against climate, or more strictly against a combination of climatic and cultural causes as independent variables, both contributing to faunal extinction. The record of extinction in Africa, seven genera within the last 125,000 years (Table 17.9), of which *Megaloceros* and *Hipparion* were known only from North Africa and *Parmularius* and *Megalotragus* were known only from South Africa (see Klein this volume), is slight compared with Australia and America. This is not to imply that climatic changes in Africa are less significant than on other continents. Fossil pollen records show that Africa experienced important late Pleistocene climatic change (Scott 1979, Livingstone 1975).

Table 17.10. Pleistocene Extinct Small Mammals (< 44 kg) of Europe

Genus / Common Name	Astian	Villafranchian Phases					Middle Pleistocene					Late Pleistocene				
		a	b	c	d	e	A	1 (i)	B	1 (ii)	C	2	D	3	F	4
Parailurus, panda																
Glirulus, dormouse																
Pliosciuropterus, squirrel																
Nyctereutes, raccoon-dog																
Procamptoceras, chamois																
Gazella, gazelle																
Dolichopithecus, monkey																
Soriculus, shrew																
Baranogale, polecat																
Prospalax, mole rat																
Parapodemus, field mouse																
Vormela, polecat																
Enhydrictis, tayra																
Rhinocricetus, hamster																
Hypolagus, rabbit																
Petenyia, shrew																
Pannonictis, tayra																
Allophaiomys, lemming																
Beremendia, shrew																
Mimomys, vole																
Trogontherium, beaver																
Pliomys, vole																
Ochotona, pika																
Aonyx, otter																
Dicrostonyx, vole																
Allactaga, jerboa																
Extinctions (N = 26)	0	1	1	1	1	2	1	2	0	5	3	1	4	0	0	4

SOURCE: After Kurtén (1968); European Pleistocene subdivisions in this table and Table 17.11 follow Kurtén (1968).
*genera surviving elsewhere

I suggest that human hunters existed in a very different relationship to the African fauna than did those of Australia and America. Successful faunal survival in late prehistoric time can be attributed to the role of tsetse flies in limiting the encroachment of pastoralists (Lambrecht 1967) and perhaps even earlier resident human populations. Human disease may well have aided faunal survival in Africa, preventing the excessive buildup of human populations in the game-rich savannas, as was the case in historic time (Lambrecht 1980).

To summarize, the majority of large mammal extinctions in Africa were concentrated early rather than late in the Pleistocene, the reverse of the record in the Americas. Despite an amazing proliferation of bovids and some immigration, there was comparatively little late Pleistocene loss of genera of mammals. The extinction of *Megalotragus, Parmularius,* and a few others can be attributed to human hunters. The rich late Pleistocene archaeological record provides an enviable sample of the hunted populations and suggests improved techniques by hunters after the Middle Stone Age (Klein this volume). The distinction between African and American Pleistocene extinctions is seen in the difference between gradually developing hominids evolving for millions of years with large animals on one continent, compared with the onslaught of a highly advanced hunting society at the peak of its power suddenly arriving on the other. Had America rather than the Old World been the center of human origins, the late Pleistocene record of extinction might well have been reversed.

Table 17.11. **Pleistocene Extinct Large Mammals (>44 kg) of Europe**

Genus	Common Name	Astian	Villafranchian Phases					Middle Pleistocene	1 (i)	B	1 (ii)	Late Pleistocene	2	D	3	F	4
			a	b	c	d	e	A	(i)	B	(ii)	C	2	D	3	F	4
Agriotherium,	hyena bear																
Zygolophodon,	mastodont																
Hipparion,	horse																
Hesperoceras,	goat																
Deperetia,	antelope																
Megalovis,	giant sheep																
Euryboas,	hunting hyena																
Gazellospira,	antelope																
*Syncerus,	buffalo																
Meganteron,	dirk-tooth																
Gallogoral,	goral																
*Tapirus,	tapir																
Anoglochis,	deer																
Anancus,	mastodont																
*Acinonyx,	cheetah																
*Lycaon,	hunting dog																
Leptobos,	ox																
Eucladoceros,	deer																
Euctenoceros,	deer																
Archidiskodon,	elephant																
Praeovibos,	musk ox																
Soergelia,	ox																
*Bubalus,	water buffalo																
*Hyaena,	hyena																
*Hemitragus,	tahr																
Homotherium,	sabertooth																
*Dicerorhinus,	rhinoceros																
*Equus,	horse																
*Cuon,	dhole																
*Hippopotamus,	hippo																
*Ovibos,	musk ox																
*Crocuta,	hyena																
Megaloceros,	giant deer																
Mammuthus,	mammoth																
Palaeoloxodon,	elephant																
Coelodonta,	woolly rhino																
*Saiga,	saiga antelope																
Extinctions (N = 37)		0	2	2	0	2	5	2	0	0	3	4	2	1	0	1	13

SOURCE: After Kurtén (1968)
*genera that survived elsewhere

Europe

In the absence of a suitable checklist of its Pleistocene animals, I will not attempt to summarize late Pleistocene changes throughout Eurasia; instead, I will concentrate on Europe (including the islands of the Mediterranean, to be discussed shortly). I have tabulated European late Pliocene and Pleistocene generic extinctions following Kurtén (1968). In the last three million years continental Europe experienced twenty-six extinctions of genera of small mammals (Table 17.10) and thirty-seven of large mammals (Table 17.11). As on other continents, there was no appreciable loss of small mammals at the end of the Pleistocene. The disappearance from Europe of pika, otter, *Dicrostonyx* and jerboa at the end of the last glaciation (Kurtén 1968) reflects range changes within living genera.

Unlike Africa, there was no marked extinction of large mammals in the lower Pleistocene. The loss of thirteen large genera at the end of the Pleistocene (Table 17.11) is impressive, although less so than the late glacial decline of megafauna in America. The European event is inflated by the nine genera which were not actually lost but locally extirpated in Europe itself to survive elsewhere in Asia, in Africa or, in the case of the musk ox, in America. Only three European genera completely disappeared throughout their range. These were the giant deer, *Megaloceros*; a woolly rhino, *Coelodonta*; and the woolly mammoth, *Mammuthus*. The extinct genus of straight-tusked elephant, *Palaeoloxodon*, has been synonymized with the living genus *Elephas* (Maglio 1973). The scimitar tooth, *Homotherium*, may not be indigenous to the late Pleistocene of Britain or the continent (Stuart 1982). The genera of large mammals surviving in Europe at least into historic time include elk (*Alces*), wisent (*Bison*), auroch (*Bos*), ibex (*Capra*), red deer (*Cervus*), fallow deer (*Dama*), lion (*Panthera*), reindeer (*Rangifer*), wild boar (*Sus*), brown bear (*Ursus*), and wolf (*Canis*).

In the Pliocene and early Pleistocene of Europe warm temperate forest elements were lost in the European flora, such as Taxodiaceae, *Nyssa, Carya,* and *Pterocarya* (Leopold 1967, West 1980). Increased continentality of the climate may also account for the early Pleistocene extinction of tapir, lesser panda, raccoon-dog, and mastodonts. In the glacial ages the climate of western Europe became extremely cold and dry as the ameliorating effect of the Gulf Steam was deflected south. North of the Alps Europe was treeless, its biota a mixture of tundra and steppe elements. Based on the argument that fossil beetles track climatic changes more rapidly than plant communities as disclosed by their fossil pollen, Coope (1975) has modeled several sudden and severe warmups in the European late glacial climate between 16,000 and 10,000 B.P. Despite the severe climatic changes, no loss of plant genera is evident in the late Pleistocene.

Unlike America, the decline of large mammals in Europe appears to be gradual and sequential. For example, cave bear (*Ursus speleus*) extinction began in the mid-Weichselian and continued into the Magdalenian and even the Mesolithic (Kurtén 1976, p. 142). In Europe proboscidean extinction advanced gradually from south to north. In the Levant the elephants, *Stegodon* and *Elephas,* were last recorded in the Acheulean, the latter in abundance (Farrand 1977). Rhinoceros persisted into the Levalloiso-Mousterian, considerably after the last proboscideans (Farrand 1977, p. 9; Tchernov this volume). One of the youngest fossil records of elephant in the region is a piece of tusk from Tabun Ec, not identified to species according to Bate (1937). This would be somewhere in Isotope Stage 5, 90,000 yr B.P. according to Farrand (pers. corres.).

In Spain at the other end of the Mediterranean numerous elephant bones and early Paleolithic remains are found together in the Acheulean (Freeman 1978). The survival of *Elephas (Palaeoloxodon)* can be traced into the Mousterian. *Elephas* was apparently not replaced by the more cold-adapted mammoth: "*Mammuthus primigenius* is rare [in Spain], even in Last Glacial sites, and it would be difficult to estimate its latest appearance with any assurance" (Richard Klein pers. corres.). The abundance of proboscideans in late glacial age sites (14,000–10,000 yr B.P.) in the New World and at higher latitudes within parts of the Old World stands in marked contrast with their late glacial scarcity or absence along the southern margin of Europe. There is an additional difference. In the eastern Mediterranean late Pleistocene fossils of large animals (including occasional proboscideans) are typically found in archaeological sites (Tchernov this volume). In the New World late Pleistocene large mammals are found in natural deposits until the ninth millenium B.C., when mammoth especially is seen in cultural deposits.

Radiocarbon-dated records of woolly mammoth (*Mammuthus primigenius*) include evidence of bone in the construction of a hut at the site of Krakow, Poland, 20,000 years ago (Kozlowski et al. 1974). Around this time the remains of at least 516 individual mammoth were used in construction of Paleolithic huts in the Dnepr-Desna region of the Ukraine (Klein 1973, p. 53). The youngest mammoth bones at Molodova in the western Ukraine were around 13,400 years old (Coles and Higgs 1969). According to Stuart

(1982, p. 164): *"Mammuthus primigenius* and *C. antiquitatis* are conspicuously absent from the numerous assemblages which can be attributed with varying precision to the end of the Late Devensian ('late glacial'), post-dating the main glaciation of about 18,000 to 15,000 B.P."* While artistic representations of mammoth are known from Gonnersdorf and El Castillo with a sporadic representation in other latest Pleistocene cultural-stratigraphic units (G. Clark, pers. corres.), mammoth are scarce or absent during late glacial time in the classic Magdalenian and other sites of the Dordogne (Alimen 1967). The megafauna of most late Paleolithic sites is either of reindeer (*Rangifer*) or red deer (*Cervus*).

The late glacial fossil fauna of Denmark includes horse, bison, and elk (*Alces*); there are no records of mammoth or straight-tusked elephants. The absence of proboscidians is curious for several reasons. The fossil record of Denmark of interstadial age (24,000 years or older) was very rich in fossil elephants, mostly of mammoth with a few straight-tusked elephants. Eighty-five localities have been mapped by Mohl in Berglund et al. (1976) and forty are shown in Figure 17.6. Lime-rich following glaciation, the flat Danish landscape should have supported numbers of mammoths after the last ice retreat, as it did during the earlier stadials. Evidently no proboscideans inhabited Denmark after the last ice retreat. Freshly deglaciated portions of North America in New York, Michigan, Minnesota, and Alberta were accumulating fossils. One hundred and sixty late glacial mastodont localities are known from the lower peninsula of Michigan alone (Martin 1967, Skeels 1962) and sixty-three localities are known in southeastern Ontario (Dreimanis 1967).

The radiocarbon dates on mammoth in Europe reviewed by Berglund et al. (1976), in Russia by Orlova (1979), and in China by Liu and Li (this volume) show many more dates older than 20,000 yr B.P. than younger; this is the reverse of the case in America where numerous late glacial dates have been obtained (Martin 1974, Agenbroad this volume). For Norway and Sweden eight radiocarbon dates on mammoth are all 19,000 years or older. There is one exception, the Lockarp tusk which was dated 13,300 yr B.P. The youngest radiocarbon date on mammoth in western Europe accepted by Berglund et al. (1976) is LY – 877, 12,170 ± 210 yr B.P. from Praz Rodent, Switzerland. Three additional dates from adjacent parts of France are slightly older. "Only a small number of mammoths survived in Europe as far as the late Weichselian time" (Berglund et al. 1976). The few that can be dated came from localities close to melting glaciers of the Alps or Scandinavia. While there is an unverified date of 11,000 years on mammoth from Kostenki, Ukrainian mammoth extinction occurred by 13,000 yr B.P. according to Klein (1973) who notes (p. 59) that woolly rhino lasted until 11,000 yr B.P. and bison until 10,000 yr B.P. In north central Eurasia Shimkin (1978) reports mammoth to be present in Upper Paleolithic horizons to about 14,000 yr B.P. It appears that woolly mammoth were rare or absent in Europe (including European Russia) after 13,000 yr B.P. at a time when late glacial age mammoth and mastodont were still present in apparently undiminished numbers in North America. The decline or extinction of mammoth first in Europe and later in America is in accord with a later and sudden appearance of big game hunters in America.

The disappearance of mammoth, woolly rhino, and musk ox from the late Pleistocene of Europe unavoidably coincided with late glacial climatic changes. The return of forested conditions during the early Holocene in Europe would have been unfavorable for tundra- or steppe-adapted species large or small. The extinction of giant elk in Ireland is typically held to be the effect of such a change in climate (e.g. Gould 1974). For these and other reasons the extinction of large mammals in the late glacial of Europe has often been attributed to climatic change.

On the other hand, the failure of *Hippopotamus* and certain other warm temperate mammals common in the last interglacial of England (Stuart 1976) to return there during the Holocene is not in accord with the climatic model. Furthermore, the vicissitudes of ice age climatic change swept over Europe many times. The relatively gradual attrition

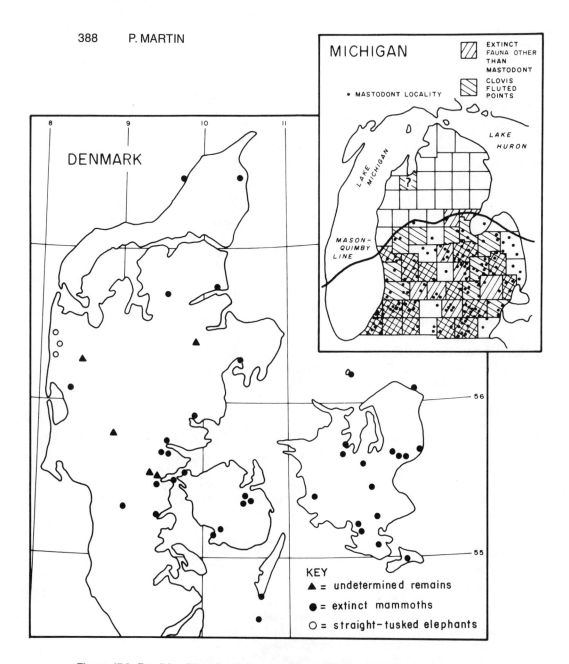

Figure 17.6. Fossil localities of extinct mammoth, straight-tusked elephants, and undetermined remains, from an unpublished map by U. Mohl, Copenhagen Museum. All records predate the late glacial, 14,000 years or older, compared with fossil localities of extinct mastodont in Michigan, that are mainly of late glacial age, 14,000 years or less. European proboscideans largely disappeared before American mastodonts and mammoth.

of large mammals from Europe after the last interglacial compared with the rapid late glacial extinctions in America matches the more gradual, climatically restricted deployment of Paleolithic hunters in the Old World compared with their sudden appearance in the New World. "It seems fairly certain that modern man has played a dominant role in the wiping out of many species, although perhaps by indirect influences as much as by actual hunting" (Kurtén 1968, p. 273).

Rather than choosing between climatic or cultural factors, many authors prefer a synthesis of both. At first this judicious approach promises an ideal solution. At the time of the extinctions climatic changes were having an unmistakable effect on the late glacial environment of Europe. The late glacial environment also saw the spread of Paleolithic hunters whose material culture was abundant and highly developed (Jelinek 1975, Klein 1973). The problem is that despite the changing climates and considerable cultural activity, the extinction record in Europe is not that impressive. While nine genera of large mammals were locally extirpated, only three genera *Mammuthus, Coelodonta,* and *Megaloceros* were completely lost throughout their range, compared with the much larger loss in the same interval in North and South America. If both climatic and cultural impacts acted together as a common cause, one might expect proportionally as much late glacial extinction of megafauna in the Old World as in the New, with fewer survivors among the genera of large mammals in Europe than in America.

Although continental Europe as well as Africa escaped massive late Pleistocene extinctions, the islands just beyond the continental shelf did not. In the Mediterranean thirteen endemic extinct genera of large and small mammals (Table 17.12) are listed by Kurtén (1968) from the Balearics, Sicily, Malta, Sardinia-Corsica, Crete, and Cyprus (see fig. 17.7). In addition giant land tortoises (Bate 1914) were present. Although some

Figure 17.7. Oceanic islands of the Mediterranean Sea showing numbers of genera of extinct endemic mammals (after Sondaar, see text); at least some extinctions were Holocene in age. Glacial sea-level depression shown inside modern coast.

Mediterranean islands were in contact with the mainland in the Tertiary, in the Pleistocene the ones listed in Table 17.12 were surrounded by water deep enough to prevent entry of the mainland fauna even during times of lowered sea levels. They received an unbalanced (not representational) endemic fauna (Sondaar 1977). Their mammalian faunas originated by arrival of mainland species swimming or rafting to the deep water islands in Simpson's "sweepstakes" method (Simpson 1940). In isolation the elephants, hippo, and deer underwent dwarfing and evolved hypsodont teeth, presumably for

Table 17.12. Extinct Endemic Mammals of the Mediterranean

Genus Common Name	Mallorca	Menores	Sardinia	Corsica	Sicily	Malta	Crete	Cyprus
*Nesiotites, shrews	x	x	x	x				
Aonyx (Nesolutra), otters				x		x		
Elephas (*Paleoloxodon), dwarf elephants			x		x	x	x	x
*Praemegaceros, dwarf deer			x	x	x		x	
Cervus, dwarf red deer					x	x		
Hippopotamus, dwarf hippo					x	x	x	x
?Neomorhaedus, goral			x					
*Myotragus, cave goat	x	x						
*Leithia, dormouse					x	x		
*Hypnomys, dormouse	x	x						
*Tyrrhenicola, vole			x	x				
*Rhagamys, field mouse			x	x				
*Prolagus, pika			x	x				

SOURCE: Kurtén (1968)

*extinct genus

ingestion of coarse grass. "Low gear locomotion" (Sondaar, 1977) is evident in the case of the goat-antelope of the Balearics and the endemic hippo of Cyprus which became less aquatic than its ancestors; Boekschoten and Sondaar (1972) assigned it to a new genus, *Phanuoris*. Eleven islands of the Aegean Archipelago alone have endemic faunas of dwarf elephant, hippo, and deer (Dermitzakis and Sondaar 1978). Unable to migrate, the island endemics were exposed to, and at least some patently survived, the full effect of late Pleistocene climatic changes. In the eastern Mediterranean during the full glacial at 18,000 yr B.P. winter sea surface temperatures dropped more than 6°C below their present value (Thunnell 1979). Under a climatic model insular extinctions might be expected then or subsequently, with the dwarf "giant deer" and dwarf elephants vanishing along with their continental cousins during the late glacial.

Kurtén (1968), however, assigns the extinct fauna of the Mediterranean islands to the Holocene. The age of the youngest goat-antelopes (*Myotragus*) of Mallorca, dated by collagen of a limb bone, was 4900±390 yr B.P. (Burleigh and Clutton-Brock 1980). Previous Holocene as well as Pleistocene dates of ca. 32,000 to 14,000 yr B.P. were obtained by W. H. Waldren, discoverer of Son Muleta Cave. In the eastern Mediterranean Bachmayer et al. (1976) list dates on dwarf elephants of Tilos of 4390±600 and 7090±680 yr B.P. Pending verification, these dates indicate an intriguingly late survival for the last proboscideans in Europe. Ironically the last European elephants appear to have been dwarfs occupying oceanic islands, an environment inevitably viewed by biogeographers as especially prone to the hazards of natural extinction. Natural extinction of Tertiary-age island endemics is known in the Mediterranean following tectonic movements (Sondaar 1977). In the Holocene, however, prehistoric man and his commensals (e.g. *Rattus*) are regarded as the main agent of extinction of endemics on Sardinia (Azzaroli 1981) and on other Mediterranean islands (Sondaar 1977, p. 698).

Unlike late Pleistocene deposits on the continent, where Paleolithic artifacts are found in abundant association with mammoth and other large mammal bones (e.g. Klein 1973, Shimkin 1978), most of the extinct island mammals of the Mediterranean have not been found in a clear-cut archaeological context. An exception may occur in the coexistence of Bronze-Age cultures and extinct goat-antelope bones on Mallorca (Waldren and Kopper 1967). On other Mediterranean islands as in America, absence of the extinct fauna in an archaeological context should not be held against the possibility of sudden prehistoric overkill. Apparently most, perhaps all, island endemics of the Mediterranean survived the rigors of the last Ice Age, only to succumb to those of the Holocene. Holocene extinctions on islands in the Mediterranean would be quite mysterious without the likelihood of overkill. The Mediterranean islands appear to be especially strategic places to study extinction as well as evolution of insular forms.

Oceanic Islands

Turning from the faunally rich continents to impoverished oceanic islands, where the megafauna was limited to giant flightless birds, land tortoises, or dwarfed forms of those continental mammals that managed to colonize, may seem anticlimatic. However, the youngest deposits yielding evidence of late Pleistocene extinctions are to be found here, along with the greatest opportunity for examining details of the extinction process. The fact that most oceanic islands within tropical latitudes were discovered and occupied by prehistoric voyagers long before the advent of European navigators is often overlooked. Both the islands of the Mediterranean and the West Indies were inhabited by the mid-Holocene. While New Guinea and Australia were occupied much earlier, at least by 30,000 years ago, voyagers did not reach Polynesia and Micronesia until between 6,000 and 1,000 years ago in a gradient of decreasing age from west to east and from warmer to colder waters (Bellwood 1978, Kirch 1980). In the Indian Ocean Madagascar was unoccupied by prehistoric man until about 1,000 years ago (see Dewar this volume), a remarkably late date coeval with the Polynesian arrival in New Zealand.

While extinct land birds and mammals have long been known in Madagascar, New Zealand, and the West Indies, the discovery of an extinct fauna of Holocene age on other oceanic islands has just begun, as in the case of the birds of the Hawaiian Islands (see Olson and James this volume). Unlike the continents and the islands of the continental shelf, the subfossil faunas of oceanic islands are relatively rich in extinct endemic genera of small mammals and land birds. The subfossil record shows native genera of land mammals to have been more common on oceanic islands relatively close to the continent, as in the case of the West Indies, than those more distant, as the Galapagos.

In the West Indies a very rich extinct fauna of rodents, edentates, and insectivores (Table 17.13) has long been known. Nevertheless, new discoveries of fossil endemic "giants" are still being reported (Morgan et al. 1980). The extinct avifauna includes numerous species of raptorial birds, some of great size (Olson 1978). Presumably through failure to find organic material suitable for radiocarbon dating, the chronology of West Indian extinction is based largely on stratigraphic interpretations.

Prehistoric extinct island endemics that may well have fallen victim to overkill have not necessarily all been found within early archaeological sites. In Madagascar, for example, remains of the extinct giant lemurs have not been found associated with artifacts (see Dewar this volume). In New Zealand, Scarlett (1968) has claimed that the Owlet Nightjar (*Megaegotheles*) was already extinct before human arrival. At Palliser Bay on the North Island of New Zealand Leach and Leach (1979) encountered a rich archaeological site extending from 1200 to 1600 A.D. While midden refuse was rich in fish, shellfish, and marine mammals, they failed to find many moas or other extinct birds. One of the few moa records proved to be of a species evidently traded in from the South Island. Older natural deposits near Palliser Bay yielded many extinct fossil birds. For these reasons Leach and Leach abandoned human impact as a cause of moa extinction, at least locally, in favor of a climatic model. Nevertheless, absence of evidence is not evidence of absence. At Palliser Bay the possibility of "blitzkrieg" cannot be eliminated. We know that the extinct endemic New Zealand avifauna survived into the late Holocene. In a few years or less, before an equilibrium based on other food resources

Table 17.13. Extinct Late Pleistocene—Holocene Genera of West Indian Land Mammals

Number of fossil species in each genus is shown for various islands.

Genera	Weight in Kg >10	1–10	<1	Cuba	Hispañola	Jamaica	Puerto Rico	Lesser Antilles
INSECTIVORA, shrews								
Nesophontes		x		4	3		1	
PRIMATES, monkeys								
Xenothrix		x				1		
EDENTATA, sloths								
Megalocnus	x			1				
Mesocnus	x			2				
Parocnus	x				1			
Neomesocnus	x			1				
Neocnus	x			1				
Acratocnus	x						1	
Synocnus	x				1			
Miocnus	x			1				
RODENTIA, rodents								
Alterodon		x				1		
Macrocapromys		x		1				
Hexolobodon		x			1			
Isolobodon		x			1		1	
Heteropsomys			x	2	2		2	
Elasmodontomys		x					1	
Clidomys		x				2		
Spirodontomys		x				1		
Speoxenus		x				1		
Amblyrhiza	x							1
Quemisa		x			1			
TOTAL genera/species				8/13	7/10	5/6	5/6	1/1

SOURCE: After Varona (1973)

could be established, local moa and other bird populations may have been rapidly destroyed. There is no guarantee that the time of extinction will inevitably be discovered by archaeologists. Evidently a much more gradual attrition occurred on the South Island where moas are commonly found in an archaeological context (see A. Anderson this volume).

Occasionally the possibility of human destruction can be eliminated. A case in point involves the island of Aldabra in the Indian Ocean. Taylor et al. (1979) report Pleistocene faunas in solution pits in limestone. Besides giant tortoises (*Geochelone gigantea*), the fossil record includes crocodiles, lizards, and land snails and an endemic duck (Harrison and Walker 1978) not found on Aldabra historically. The island was entirely submerged by at least two marine transgressions in the Pleistocene, and all terrestrial animals were presumably swept away. The tortoises managed to reestablish themselves by drifting back to Aldabra from Madagascar. Unlike the Mascarenes, where land tortoises were destroyed along with the dodo and the solitaire within the last 500 years, and unlike the Comoro Islands and Madagascar, where giant tortoises presumably were obliterated earlier in the Holocene, the giant tortoises of Aldabra escaped complete destruction at the hand of man.

Human impact can also be rejected as a possible cause of extinction of a species of horned tortoise, *Meiolania platyceps*. Horned tortoise bones found on Lord Howe Island between Australia and New Zealand occur in Pleistocene calcarenite (Sutherland and Ritchie 1977) dated at over 20,000 years (Squires 1963). When Lord Howe Island was discovered by a British vessel in 1788 it lacked an indigenous human population. Subsequent archaeological work has not uncovered evidence of prehistoric occupation. While the early historic accounts indicated an abundance of sea turtles and birds so tame they could be knocked over, no mention was made of the tortoises which must have then been extinct. Early Holocene extinction may have accompanied sea level transgression of most of their habitat on the Lord Howe–Balls Pyramid platform (A. Ritchie pers. comm.). Lord Howe is unique among the wet tropical Pacific islands in that it was not inhabited prehistorically. Its biota is proving to be remarkably rich in endemic terrestrial plants and invertebrates as well as land birds (Recher and Ponder 1981). A major conservation effort is underway (Recher and Clark 1974). Lord Howe Island may serve as an example of the degree of endemism that was attained elsewhere on Pacific islands prior to prehistoric human incursions of the Holocene.

I have dwelt on the case of Aldabra and Lord Howe to illustrate geological extinctions on islands under circumstances which eliminate prehistoric man as a cause. A record of natural extinctions proposed for land birds in the West Indies is not as clear cut. Pregill and Olson (1981) have concluded that the rise in sea level and extensive glacial-age aridity were the major cause of extinctions.

Work early in the twentieth century led most authors to regard the extinct West Indian fauna of giant hystricomorph rodents and dwarf ground sloths as archaeologically associated (Allen 1972, Harrington 1921). At least one extinct capromyid rodent, *Isolobodon portoricensis*, was probably transported between Hispañola, Puerto Rico, and the Virgin Islands by prehistoric people (Miller 1918, Reynolds et al. 1953). The Bahama Bank suffered the loss of 90 percent of its surface area at the end of the Ice Age, and certainly may have lost terrestrial fauna as a result. On the other hand, twenty extinct genera of endemic mammals of the Greater Antilles (Table 17.13) would not have been subject to much loss of habitat from the rise in sea level. It is not clear that any of the Greater Antillean extinct endemic mammals disappeared as early as 11,000 years ago, the time of faunal catastrophe on the continent. At least some extinct genera survived to a later date to be incorporated in archaeological middens. Unless it can be shown that the others definitely disappeared before human arrival, most West Indian extinctions may be attributed to "blitzkrieg" or to more gradual impacts accompanying the growth of sizable prehistoric populations. The human population of Hispañola exceeded one million by the time the Spanish arrived (Sauer 1966).

Conceivably some species thought to represent prehistoric losses actually survived into the historic period to succumb during the first two centuries of European contact. Historic extinctions may have occurred without attracting the attention of the naturalists of the day. For example, on Ascension Island in the South Atlantic, bones of a flightless rail the size of a Virginia rail were recently found in abundance in two extinct volcanic fumeroles and described as a new species, *Atlantisia elpenor* (Olson 1973). Presumably this was the species captured and eaten by Peter Mundy and his shipmates in 1656. Its extermination soon after is attributed to introduced mammals (Olson 1977). Without Mundy's account the late survival of *Atlantisia* might be disputed. Historic extinctions are less certain on St. Helena where two genera of endemic rails "probably survived until shortly after the island's discovery in 1502" (Olson 1977).

The only extinct genus of terrestrial vertebrate found in the extensively sampled lava tube caves of the Galapagos is the muskrat-size rodent, *Megaoryzomys*. Its remains are common among the abundant subfossils of the late Holocene which are rich in Darwin's finches and other land birds (Steadman 1981). Archaeological remains are negligible; evidently the Galapagos were not permanently occupied by prehistoric people. Thus prehistoric man's role in the extinction process can be discounted. As in the case of the rails of St. Helena. *Megaoryzomys* escaped historic notice. Whether its extinction occurred before or after historic contact in the sixteenth century is uncertain.

In the context of overkill those oceanic islands that were not occupied by prehistoric people are of particular interest. If Holocene extinctions of endemics on oceanic islands are mostly anthropogenic, as I propose, extinctions of land birds and mammals on Lord Howe Island, the Galapagos, and Ascension should have been much less than on those islands heavily populated prehistorically, such as Hawaii and the West Indies. The subfossil deposits of Hawaii and the Galapagos in particular suggest the difference in avifaunal survivals to be expected in archipelagos with and without heavy pre-historic human occupation. The former are known to be rich in extinct subfossils (Olson and James this volume), the latter relatively poor, with only one extinct genus (Steadman 1981).

Anthropogenic extinction of small terrestrial animals on oceanic islands need not be the direct result of human hunting. In the Pacific the prehistoric introduction of rats, pigs, and domestic fowl, and perhaps new diseases such as avian malaria need to be considered. Finally, most forested islands should have been highly vulnerable to habitat destruction. The Polynesians cultivated field crops. Traces of their impact may be inferred from the introduction of land snails. Six species on Tikopea in the Solomons indicate that the environment "... was significantly modified by humans during the prehistoric period, probably as a result of agriculture practices" (Christensen and Kirch 1981).

In the case of Fiji the potential impact of human arrival can be inferred in another way. In the swamp of Waitabu on Lakeba six meters of sediment was deposited in 2,000 years. According to Hughes and Hope (in Brookfield et al. 1979) the lower half of the deposit yielded a mean erosion index within the watershed of 2,600 tons per square kilometer per year. The presence of charcoal throughout the core led Hughes and Hope to believe that the massive erosion was accompanied by repeated fires. Only fern spores were recovered, so the degree of change in the natural vegetation is unknown. It appears that prehistoric human impact on Fiji, and perhaps on most Pacific island ecosys-tems, may have been much greater than has been suspected. Even islands not likely to have been cultivated suffered an anthropogenic change in natural vegetation. In the Chathams east of the South Island of New Zealand prehistoric Polynesians not engaged in farming were nevertheless at least party responsible for destruction of forest and its replacement by bracken grassland (Dodson and Kirk 1978).

To date, Aldabra provides one of the few examples of population extinctions on an oceanic island quite independent of human presence. With rare exceptions the extinction of island endemic fauna is not known to predate human arrival. In the case of New Zealand a wave of bird extinctions accompanies the expansion of prehistoric Polynesians (Trotter and McCullough this volume). In Hawaii over half of the native land birds of the prehistoric period survived into historic time (Olson and James this volume). In the Mediterranean virtually 90 percent of the endemic mammalian genera were lost in the Holocene, apparently at the hands of Neolithic invaders (Sondaar 1977). The West Indies lost 21 endemic genera of mammals. On all oceanic islands the subfossil record invites much more careful attention. The Holocene, a time of little change on the continents, is much more dynamic on oceanic islands. The opportunity to test various extinction models is especially promising.

Conclusions

Late Pleistocene extinctions can be studied in detail by judicious application of radiocarbon dating. In Africa large mammal extinctions were comparatively few, seven in the last 100,000 years; in Australia nineteen genera and more than fifty species disappeared within the last 40,000 years; in America more than seventy genera of large mammals disappeared, most, perhaps all, within the last 15,000 yr. B.P. Many terres-trial mammals, some birds, a few reptiles, and virtually no plants or invertebrates became extinct. All continents except Antarctica were affected, America and Australia to a much greater degree than Africa and Asia (fig. 17.8).

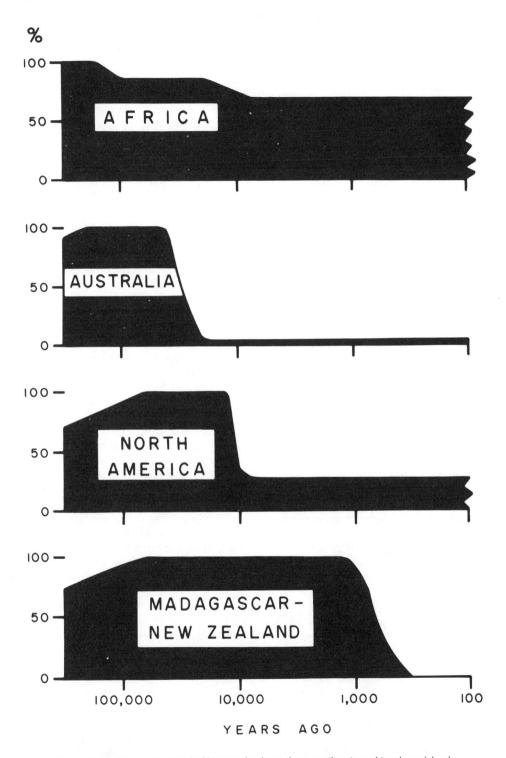

Figure 17.8. Percent survival of large animals on three continents and two large islands; time in common logarithms. The intensity and sequence of extinctions follow cultural development: extinction was less severe in Africa compared with America, Australia, New Zealand, and Madagascar which suffered heavy losses.

On the continents small- or medium-size animals survived the late Pleistocene with minor losses at worst. Genera of small mammals had suffered appreciable extinction in the early Pleistocene or Pliocene, at least in America and Europe. Small mammals and birds suffered severe losses in the last few thousand years only on oceanic islands. The subfossil faunas of certain oceanic islands disclose a much larger endemic fauna than has been anticipated by biogeographers.

Timing of late Pleistocene extinctions is diachronous. Australia was stricken before North and South America. The last mammoths of China and England died out before those of America. The animals which became extinct on the continents disappeared before the Holocene. Extinctions continued throughout the Holocene on oceanic islands, first on archipelagos close to the continents such as the West Indies and islands of the Mediterranean, and later on more remote islands such as Hawaii, New Zealand, and Madagascar. The lack of synchroneity between the extinctions on different continents and their variable intensity, for example, heavier in America than Africa, appears to eliminate as a cause any sudden extraterrestrial or cosmic catastrophe.

Regional models that correlate climatic change with the extinctions are often encountered; regional extinctions often coincide with changes in climate. Attempts at a synthetic treatment that incorporate the intensity, timing, and character of late Pleistocene extinction into a worldwide model related to climatic changes are more difficult. On a global scale the late Pleistocene extinction patterns appear to track the prehistoric movements or activities of *Homo sapiens* much more closely than any widely agreed-upon pattern of especially severe global climatic change in the late Pleistocene. The moderate loss of animals in Afro-Asia can be related to the gradual history of human spread which allowed an ecological equilibrium to develop. The loss of the great majority of large animals from North America, South America, and Australia can be related to sudden and severe human impact on these continents, which were previously unexposed to evolving hominids.

Oceanic island extinctions can be correlated closely with arrival of prehistoric seafarers. While they did not necessarily hunt all the small mammals or island birds and other animals to extinction, the side effects of the arrival of prehistoric colonizers were severe and involved fire, habitat destruction, and the introduction of an alien fauna. Most native endemic fauna succumbed rapidly at the hands of prehistoric people.

The model of human impact of "overkill" as a unique cause of late Pleistocene extinctions has attracted considerable interest and criticism. The model will not serve to explain earlier extinction events, such as those of the late Miocene (Hemphillian). Details within the chronology of large animal extinctions and human arrivals have been challenged. Disputes have arisen about quality of radiocarbon dates, and even about the size of a "large" mammal. A controversy exists in America regarding when hunters first arrived.

A conceptual difficulty has centered on the failure of the fossil record of many regions to disclose ample evidence of extinct faunas in kill sites in any other cultural context. Continents in which extinction of higher categories is severe, such as America and Australia, provide much less evidence of human contact with the lost fauna than continents with far fewer losses, such as Africa and Asia. The lack of kill sites can be accommodated by blitzkrieg, a special case of faunal overkill that maximizes speed and intensity of human impact and minimizes time of overlap between the first human invader and the disappearance of native fauna (Mosimann and Martin 1975). Paleontologists have noted that an effective new invader need not overlap in the fossil record with the species it replaces (Simpson 1953, p. 301). In the case of the Paleolithic invaders of America, Janzen (1983) proposes that the native carnivores acted to enhance the hunter's impact, a "fifth column" that increased the rate of extinction.

A rapid rate of change can be anticipated whenever a land mass was first exposed to *Homo sapiens*. The vulnerable species in a fauna first encountering prehistoric people will suddenly vanish from the fossil record. By virtue of the youth of its archaeological sites, New Zealand provides an exception. It yields abundant evidence of temporal overlap between prehistoric people and an extinct megafauna, the moas, whose extinction occurred in this millenium (see Cassels, Trotter and McCullough, and A. Anderson this volume). Furthermore the North Island, environmentally more suitable for rapid prehistoric Polynesian settlement, has yielded much less evidence of "moa hunting" than the South Island.

How is the overkill model to be used? Outside Europe and Africa it means that prehistorians in search of kill sites should devote themselves to the first centuries following the arrival of early hunters and not be discouraged if the search proves difficult. In America this means increased attention to deposits of the ninth millenium B.C. The Australians should search for much older sites, as indeed they are. Success could be fatal to the model of sudden overkill if archaeological sites yielding extinct fauna prove to be variable in age, that is, scattered over thousands of years. Success at finding much evidence of killing or processing of the extinct fauna is not predicted by the blitzkrieg version of overkill.

Australia may offer the best opportunity for refutation of prehistoric blitzkrieg. Horton (1980, this volume) finds that some faunal extinctions occurred more than 10,000 years after the absolute minimum date for human arrival. This is especially intriguing since, although the first Australians may not have been specialized big game hunters to the same degree as the first Americans, the extinct Australian large marsupials appear to have been especially ill-equipped to escape human predators (Murray this volume). Oceanic islands, such as New Zealand and Hawaii, should provide an opportunity for the archaeological investigation of prehistoric extinctions under circumstances attending human colonization. Many oceanic islands, such as the Solomons, New Caledonia, and the Canaries, appear to be worthy of prospecting by those who wish to discover extinct prehistoric faunas yet unknown.

One striking advantage of the overkill model over others is its testability. The organism lost must be of a size or in a location potentially vulnerable to human impacts. The pattern and radiocarbon dating of late Pleistocene extinction can be studied on a global scale independent of the pattern and timing of prehistoric cultural changes. Extinctions and cultural changes can then be compared for degree of fit. Serious anomalies, if they emerge, will refute the model. To date those that have been advanced appear to be conceptual and not evidential discrepancies, such as the claims of a New World population long before 11,000 yr B.P. (no unequivocal evidence of an earlier invasion of big game hunters has been established) or the lack of kill sites in Australia and on some islands (archaeological associations will be rare or absent if extinction occurred rapidly, i.e. blitzkrieg). The late Pleistocene extinction pattern bears an intriguing relationship to prehistoric events. The model of overkill and its special case, blitzkrieg, offer a challenging alternative to other explanations of the last geological revolution in the history of life.

Acknowledgments

I thank University of Arizona colleagues and staff, especially Karl W. Flessa, Austin Long, James H. Brown, Susan Archias, Christopher Duffield, and the Tumamoc Hill group, including Julio L. Betancourt, Faith L. Duncan, Suzanne K. Fish, Robert D. McCord, Vera Markgraf, Jim I. Mead, David W. Steadman, Robert S. Thompson, Tinco E. A. Van Hylckama, and Thomas R. Van Devender. Various chapter authors responded with editorial or technical aid, especially Larry D. Agenbroad, Elaine Anderson, Alan L. Bryan, Richard Cassels, Robert E. Dewar, C. Vance Haynes, Helen James, Don K. Grayson, the late John Guilday, Dale Guthrie, Richard A. Kiltie, Richard Klein, E. L. Lundelius, Jr., Leslie F. Marcus, Storrs L. Olson, and S. David Webb. My effort at reviewing Australia was guided by Jeanette H. Hope, David R. Horton, Rhys Jones, Peter Murray, Alex Ritchie, and Richard V. S. Wright; New Zealand by Roger Green and Richard Cassels; South America by Larry G. Marshall; Africa by Richard G. Klein; Europe by Bjorn Kurtén, Paul Y. Sondaar, and Anthony J. Stuart; Pacific Islands by Patrick V. Kirch; and the West Indies by Karl P. Koopman. Finally, extra thanks goes to Deborah A. Gaines for typing and editing.

References

Abbott, I. 1980. Aboriginal man as an exterminator of wallaby and kangaroo populations on islands round Australia. *Oecologia* 44:347–354.

Adam, D. P., J. D. Sims, and C. K. Throckmorton. 1981. 130,000-yr continuous pollen record from Clear Lake, Lake County, California. *Geology* 9:373–377.

Alimen, F. 1967. The Quaternary of France. *In* The Quaternary (K. Rakama, ed.), pp. 89–238. Interscience Publishers.

Allen, G. M. 1972. Extinct and vanishing mammals of the western Hemisphere. Cooper Square Publishers.

Alvarez, L. W., W. Alvarez, F. Asaro, and H. V. Michel. 1980. Extraterrestrial cause for the Cretaceous-Tertiary extinction. *Science* 208:1095–1108.

Archer, M. 1981. A review of the origins and radiations of Australian mammals. *In* Ecological biogeography of Australia (A. Keast, ed.), pp. 1437–1488. W. Junk, The Hague.

Axelrod, D. L. 1967. Quaternary extinctions of large mammals. University of California, *Publication in Geological Science* 74:1–42.

Azzaroli, A. 1981. Cainozoic mammals and the biogeography of the island of Sardinia, western Mediterranean. *Palaeogeography, Paleoclimatology, Palaeoecology* 36:107–111.

Bachmayer, F., N. Symeonidis, R. Seeman, and H. Zapfe. 1976. Die Ausgrabungen in der Zwergelefantenhole "Charkadio" auf den insel Tilos. *Ann. Naturhistor. Mus. Wien* 80:113–144.

Balme, J. 1978. An apparent association of artifacts and extinct fauna of Devil's Lair, Western Australia. *The Artefact* 3:111–116.

Barbetti, M. and H. Allen. 1972. Prehistoric man at Lake Mungo, Australia, by 32,000 years B.P. *Nature* 240:46–48.

Bate, D. M. A. 1914. A gigantic land tortoise from the Pleistocene of Menorca. *Geol. Mag.* 1:100–107.

———. 1937. The Stone Age of Mount Carmel (Vol. I, Part II, Palaeontology). Oxford, Clarendon.

Bellwood, P. 1978. Man's conquest of the Pacific. Oxford University Press.

Berggren, W. A., L. H. Burkle, M. B. Cita, H. B. S. Cooke, B. M. Funnell, S. Gartner, J. D. Hays, J. P. Kennett, N. D. Opdyke, L. Pastouret, N. J. Shackleton, and Y. Takayanagi. 1980. Towards a Quaternary time scale. *Quaternary Research* 13: 277–302.

Berglund, Bijorn E., Soren Hakansson, and Erik Lagerlund. 1976. Radiocarbon-dated mammoth (*Mammuthus primigenius* Blumenbach), finds in South Sweden. *Boreas* 5:177–191.

Bigalke, R. C. 1978. Present-day mammals of Africa. *In* Evolution of African mammals (Vincent J. Maglio and H. B. S. Cooke, eds.), pp. 1–16. Harvard University Press.

Black, C. C. 1978. Anthracotheriidae. *In* Evolution of African mammals (Vincent J. Maglio and H. B. S. Cooke, eds.), pp. 423–434. Harvard University Press.

Boekschoten, G. J. and P. Y. Sondaar. 1972. On the fossil mammalia of Cyprus. Proceedings, series B. Koninklijke Nederlandse Akademie van Wetenschappen: 75(4): 306–338.

Bowdler, S. 1977. The coastal colonization of Australia. *In* Sunda and Sahul, Prehistoric studies in southeast Asia, Melanesia and Australia (J. Allen, J. Golson and R. Jones, eds.), pp. 205–246. Academic Press, London.

Bowler, J. M. and A. G. Thorne. 1976. Human remains from Lake Mungo: discovery and excavation of Lake Mungo III. *In* The origin of the Australians (D. L. Kirk and A. G. Thorne, eds.). Australian Institute of Aboriginal Studies, Canberra.

Bradbury, J. P., B. Leyden, M. Salgado-Labouriau, W. M. Lewis, Jr., C. Schubert, M. W. Binford, D. G. Frey, D. R. Whitehead, and F. H. Weibezahn. 1981. Late Quaternary environmental history of Lake Valencia, Venezuela. *Science* 214:1299–1305.

Brain, C. K. 1981. The hunters or the hunted?

An introduction to African cave taphonomy. University of Chicago Press.

Broecker, W. S. and J. van Donk. 1970. Insolation changes, ice volumes, and the O^{18} record in deep-sea cores. *Reviews of Geophysics and Space Physics* 8:169–197.

Brookfield, H. C., M. Latham, M. Brookfield, B. Salvat, R. F. McLean, R. D. Bedford, P. J. Hughes, and G. S. Hope. 1979. Lakeba: environmental change, population dynamics and resource use, pp. 93–110 *in* The UNESCO/UNFPA Population and Environment Project in the eastern islands of Fiji. Man and the Biosphere (MAP) Programme, Project 7: Ecology and rational use of island ecosystems. Canberra.

Bryan, A. L., R. M. Casamiquela, J. M. Cruxent, R. Gruhn, and C. Ochsenius. 1978. An El Jobo mastodon kill at Taima-Taima, Venezuela. *Science* 200:1275–1277.

Burleigh, R. and J. Clutton-Brock. 1980. The survival of *Myotragus balearicus* Bate, 1909, into the Neolithic on Mallorca. *Journal of Archaeological Science* 7:385–388.

Butzer, H. W. and G. L. Isaac (eds.). 1975. After the Australopithecines. Mouton, The Hague.

Calaby, J. H. 1971. Man, fauna, and climate in aboriginal Australia. *In* Aboriginal man and environment in Australia (D. J. Mulvaney and J. Golson, eds.), pp. 81–93. Australian National University Press, Canberra.

———. 1976. Some biogeographical factors relevant to the Pleistocene movement of man in Australia. *In* The origin of the Australians (R. L. Kirk and A. G. Thorne, eds.), pp. 23–28. Australian Institute of Aboriginal Studies, Canberra.

Campbell, K. E., Jr. 1979. The non-passerine Pleistocene avifauna of the Talara tar seeps, Peru. *Life Science Contribution*, Royal Ontario Museum 118:203 pp.

Christensen, C. C. and P. V. Kirch. 1981. Non-marine molluscs from archaeological sites in Tikopia, southeastern Solomon Islands. *Pacific Science* 35:75–88.

Churcher, C. S. and M. L. Richardson. 1978. Equidae. *In* Evolution of African mammals (Vincent J. Maglio and H. B. S. Cooke, eds.), pp. 379–422. Harvard University Press.

Clark, J. D. 1959. The prehistory of southern Africa. Penguin Books, Ltd., 341 pp.

Clemens, W. A., J. D. Archibald, and L. J. Hickey. 1981. Out with a whimper, not a bang. *Paleobiology* 7:293–298.

Coles, J. M. and E. S. Higgs. 1969. The archaeology of Early Man. Faber and Faber, London.

Colinvaux, Paul. 1981. Historical ecology of Beringia: the southland bridge coast at St. Paul Island. *Quaternary Research* 16:18–36.

Coope, G. R. 1975. Climatic fluctuation in northwest Europe since the last interglacial, indicated by fossil assemblages of Coleoptera. *Geol. Journal Spec. Issue* 6:153–168.

———. 1979. Late Cenozoic fossil Coleoptera: evolution, biogeography and ecology. *Ann. Rev. Ecol. Syst.* 10:247–267.

Coppens, Y., V. J. Maglio, C. T. Madden, and M. Beden. 1978. Proboscidea. *In* Evolution of African mammals (Vincent J. Maglio and H. B. S. Cooke, eds.), pp. 336–367. Harvard University Press.

Darwin, C. 1855. Journal of Researches ... Voyage of the *Beagle*. 2 vols. Harper & Bros.

Dermitzakis, M. D. and P. Y. Sondaar. 1978. The importance of fossil mammals in reconstructing paleogeography with special reference to the Pleistocene Aegean archipelago. *Annales Geologiques des pays Helleniques* 29:808–840.

Dodson, J. R. and R. M. Kirk. 1978. The influence of man at a quarry site, Nairn River Valley, Chatham Island, New Zealand. *Journal Roy. Soc. of New Zealand* 8:377–384.

Dortch, C. E. 1979. Devil's Lair, an example of prolonged cave use in southwestern Australia. *World Archaeology* 10:258–281.

Dreimanis, A. 1967. Mastodons, their geologic age and extinction in Ontario, Canada. *Canadian Journal of Earth Sciences* 4:663–675.

Eiseley, L. C. 1943. Archaeological observations on the problem of postglacial extinction. *American Antiquity* 8:209–217.

Farrand, W. R. 1977. Palaeoenvironment of Pleistocene man in the Levant. Eretz-Israel, *Archaeol. Historical and Geographical Studies* 13:1–13. Israel Exploration Society, Jerusalem.

Flood, Josephine. 1980. The moth hunters. Australian Institute of Aboriginal Studies, Canberra, 388 pp.

Freeman, L. G. 1978. The analysis of some occupation floor distributions from earlier and middle Paleolithic sites in Spain. *In* Views of the past (L. G. Freeman, ed.), pp., 57–116. Mouton, The Hague.

Gentry, A. W. 1978. Bovidae. *In* Evolution of African mammals (Vincent J. Maglio and H. B. S. Cooke, eds.), pp. 540–572. Harvard University Press.

Gill, E. C. 1955. The problem of extinction with special reference to Australian marsupials. *Evolution* 9:87–92.

Gillespie, R., D. R. Horton, P. Ladd, P. E. Macumber, T. H. Rich, R. Thorne, and R. V. S. Wright. 1978. Lancefield swamp and the extinction of the Australian megafauna. *Science* 200:1044–1048.

Gingerich, P. D. 1977. Patterns on evolution in the mammalian fossil record. *In* Patterns of evolution (A. Hallam, ed.), pp. 475–478. Elsevier.

Gould, S. J. 1974. The origin and fluctuation of "bizarre" structures: antler size and skull size in the "Irish elk," *Megaloceros giganteus. Evolution* 28:191–220.

Grayson, D. K. 1977. Pleistocene avifaunas and the overkill hypothesis. *Science* 195:691–692.

Guilday, J. 1967. Differential extinction during late-Pleistocene and recent times. *In* Pleistocene extinctions: the search for a cause (P. S. Martin and H. E. Wright, Jr., eds.), pp. 121–140. Yale University Press.

Haq, B. W., W. A. Berggren, and J. A. Van Couvering. 1977. Corrected age of the Pliocene/Pleistocene boundary. *Nature* 269:483–488.

Harrington, M. R. 1921. Cuba before Columbus. Indian notes and monographs, Museum of the American Indian, Heye Foundation, New York.

Harrison, C. J. O. and C. A. Walker. 1978. Pleistocene bird remains from Aldabra Atoll, Indian Ocean. *Journal of Natural History* 12:7–14.

Haynes, C. V. 1970. Geochronology of man-mammoth sites and their bearing on the origin of the Llano Complex. *In* Pleistocene and Recent environments of the central Great Plains (W. Dort, Jr., and J. Knox Jones, eds.), pp. 77–192. University Press of Kansas.

———. 1980. The Clovis culture. *Canadian Journal of Anthropology* 1:115–121.

Hecht, M. K. 1975. The morphology and relationships of the largest known terrestrial lizard, *Megalania prisca* Owen, from the Pleistocene of Australia. *Proc. Roy. Soc. Victoria* 87:239–250.

Hibbard, C. W. 1960. An interpretation of Pliocene and Pleistocene climates in North America. *Mich. Acad. Sci., Arts, and Letters Papers* 62:5–30.

Hickerson, Harold. 1965. The Virginia deer and intertribal buffer zones in the upper Mississippi valley. *In* Man's culture and animals (Anthony Leeds and A. P. Vayda, eds.), pp. 43–66. AAAS Publ. No. 78.

Hope, J. H., R. J. Lampert, E. Edmondson, M. J. Smith, and G. F. Van Tets. 1977. Late Pleistocene faunal remains from Seton Rock Shelter, Kangaroo Island, South Australia. *Journal of Biogeography* 4:363–385.

Horton, D. R. 1980. A review of the extinction question: man, climate and megafauna. *Archaeol. Phys. Anthropol. Oceania* 13:86–97.

Horton, D. R. and R. V. S. Wright. 1981. Cuts on Lancefield bones: carnivorous *Thylacoleo,* not humans, the cause. *Archaeol. in Oceania* 16:73–80.

Howell, F. C. 1978. Hominidae. *In* Evolution of African mammals (Vincent J. Maglio and H. B. S. Cooke, eds.), pp. 154–248. Harvard University Press.

Hurt, W. R., V. Van Der Hammen, and G. Correal U. 1976. The El Abra Rockshelters, Sabana de Bogotá, Colombia. *Indiana University Museum, Occasional Papers and Monographs,* Bloomington.

Hutley, T. and J. Kappelman, 1980. Bears, pigs, and Plio-Pleistocene hominids: a case for the exploitation of below ground food resources. *Human Ecology* 8:371–387.

Isaac, G. 1977. Olorgesailie: Archaeological studies of a middle Pleistocene lake basin in Kenya. University of Chicago Press.

Janzen, D. 1983. The Pleistocene hunters had help. *American Naturalist* 121:598–599.

Jelinek, J. 1975. the evolution of man. Hamlyn, London, 552 pp.

Jones, R. 1968. Geographical background to the arrival of man in Australia. *Archaeol. Phys. Anthropol. Oceania* 3:186–215.

———. 1979. The fifth continent: problem concerning the human colonization of Australia. *Ann. Rev. Anthropology* 8:445–466.

Kiernan, K., R. Jones and D. Ransom. 1983. New evidence from Fraser Cave for glacial age man in south-west Tasmania. *Nature* 301:28–32.

Kirch, P. V. 1980. Polynesian prehistory: cultural adaptation in island ecosystems. *American Scientist* 68:39–48.

Klein, R. G. 1973. Ice age mammals of the Ukraine. University of Chicago Press, 140 pp.

———. (MS.). The large mammals of southern Africa: late Pliocene to Recent. *In* Later Cenozoic environments, ecology, and prehistory in southern Africa (R. G. Klein, ed.). A. A. Balkema. In press.

Kozlowski, J. K., E. Sachse-Kozlowski, H. Kobiak, B. Van Vlioet, and G. Zakrzewska. 1974. Upper Palaeolithic site with dwellings of mammoth bones—Cracow, Spadzista Street B. *Folia Quaternaria* 44, 158 pp.

Kurtén, B. 1968. Pleistocene mammals of Europe. Aldine.

———. 1971. The age of mammals. Columbia University Press.

———. 1976. The cave bear story. Columbia University Press.

Kurtén, B. and E. Anderson. 1980. Pleistocene mammals of North America. Columbia University Press.

Lambrecht, F. 1967. Trypanosomiasis in prehistoric and later human populations, a tentative reconstruction. *In* Diseases in antiquity (D. Brothwell and A. T. Sandison, eds.). Ch. C. Thomas, Springfield, Ill.

———. 1980. Paleoecology of tsetse flies and sleeping sickness in Africa. *Proc. Amer. Philos. Soc.* 124:367–385.

Leach, H. M. and B. F. Leach. 1979. Environmental change at Palliser Bay. *In* Prehistoric man in Palliser Bay. *National Museum Bulletin* 21:229–240, Otago, New Zealand.

Leopold, E. B. 1967. Late-Cenozoic patterns of plant extinction. *In* Pleistocene extinctions: the search for a cause (P. S. Martin and H. E. Wright, Jr., eds.), pp. 203–246. Yale University Press.

Livingstone, O. 1975. Late Quaternary climatic change in Africa. *Adv. in Ecological Research* 6:249–280.

Long, A. and P. S. Martin. 1974. Death of American ground sloths. *Science* 186:638–640.

Lundelius, Jr., E. L. and W. D. Turnbull. 1978. The mammalian fauna of Madura Cave, Western Australia, part III. *Fieldiana, Geology* 38:1–120.

Lynch, T. F. 1980. Guitarrero Cave: Early man in the Andes. Academic Press.

MacArthur, R. H. and E. O. Wilson. 1967. The theory of island biogeography. Princeton University Press.

MacNeish, R. S. 1976. Early man in the New World. *American Scientist* 64:316–327.

———. 1979. The Early Man remains from Pikimachai Cave, Ayacucho basin, highland Peru. *In* Pre-llano cultures of the Americas: Paradoxes and possibilities (R. L. Humphrey and D. Stanford, eds.), pp. 1–47. The Anthropological Society of Washington.

Maglio, V. J. 1973. Origin and evolution of the Elephantidae. *Trans. Amer. Phil. Soc.* 63:1–149.

———. 1978. Patterns of faunal evolution. *In* Evolution of African mammals (V. J. Maglio and H. B. S. Cooke, eds.), pp. 603–619. Harvard University Press.

Marshall, L. G. 1981a. The great American interchange—an invasion induced crisis for South American mammals. *In* Biotic crisis in ecological and evolutionary time (M. H. Nitecki, ed.), pp. 133–229. Academic Press.

———. 1981b. The families and genera of Marsupialia. *Fieldiana, Geology* n.s. No. 8, 65 pp.

Marshall, L. G. and M. K. Hecht. 1978. Mammalian faunal dynamics of the great American interchange: an alternative interpretation. *Paleobiology* 4:203–209.

Marshall, L. G., A. Berta, R. Hoffstetter, R. Pascual, O. A. Reig, M. Bombin, and A. Mones. (MS.). Geochronology of the continental mammal bearing Quaternary of South America. *Paleovertebrata.* In press.

Marshall, L. G., S. D. Webb, J. J. Sepkoshi, and D. M. Raup. 1982. Mammalian evolution and the great American interchange. *Science* 215:1351–1357.

Martin, C. 1978. Keepers of the game. University of California Press, Berkeley.

Martin, P. S. 1966. Africa and Pleistocene overkill. *Nature* 212:339–342.

———. 1967. Prehistoric overkill. *In* Pleistocene extinctions: the search for a cause (P. S. Martin and H. E. Wright, Jr., eds.), pp. 75–120. Yale University Press.

———. 1973. The discovery of America. *Science* 179:969–974.

———. 1974. Paleolithic players on the American stage. *In* Arctic and alpine regions (J. D. Ives and R. Barry, eds.), pp. 669–700. Methuen, London.

Mech, L. D. 1977. Wolf-pack buffer zones as prey reservoirs. *Science* 198:320–321.

Merrilees, D. 1968. Man the destroyer: Late Quaternary changes in the Australian marsupial fauna. *Journal Roy. Soc. Western Australia* 51:1–24.

Milham, P. and P. Thompson. 1976. Relative antiquity of human occupation and extinct fauna at Madura Cave, southeastern Western Australia. *Mankind* 10:175–180.

Miller, G. S. 1918. Mammals and reptiles collected by Theodoor de Booy in the Virgin Islands. *Proc. U.S. Nat. Mus.* 54:507–508.

Moore, D. M. 1978. Post-glacial vegetation in the South Patagonian territory of the giant ground sloth, *Mylodon. Botanical Journal, Linnean Society* 77:177–202.

Morgan, G. S., C. E. Ray, and O. Arredondo. 1980. A giant extinct insectivore from Cuba (Mammalia: Insectivora: Soledontidae). *Proc. Biol. Soc. Wash.* 93:597–608.

Mosimann, J. E. and P. S. Martin. 1975. Simulating overkill by Paleoindians. *American Scientist* 63:304–313.

Mulvaney, D. J. 1975. The prehistory of Australia. Pelican Books.

Murray, P. F. 1978. Australian megamammals: restorations of some late Pleistocene fossil marsupials and a monotreme. *The Artefact* 3:77–79.

Nelson, M. E. and J. H. Madsen, Jr. 1980. The Hay-Romer camel debate: fifty years later. *Contributions to Geology, University of Wyoming* 18:47–50.

Newell, N. D. 1967. Revolutions in the history of life. *Geol. Soc. Amer. Special Paper* 89:63–91.

Olson, S. L. 1973. Evolution of the rails of the South Atlantic islands (Aves: Rallidae). *Smithsonian Contributions to Zoology* 152:1–53.

———. 1977. A synopsis of the fossil Rallidae. *In* Rails of the world, by S. Dillon Ripley, pp. 339–373.

———. 1978. A paleontological perspective of West Indian birds and mammals. *In* Zoogeography in the Caribbean, Academy of Natural Sciences of Philadelphia, *Special Publ.* 13, pp. 99–117.

Orlova, L. A. 1979. the radiocarbon age of mammoth remains excavated on the territory of the USSR. Novosibirsk Izvestiya Sibirskogo Otdeleniya Akademi Nauk SSR No. 6, pp. 89–97 (in Russian).

Owen, R. 1877. Researches on the fossil remains of the extinct mammals of Australia. Erxleben, London.

Patterson, B. and R. Pascual. 1972. The fossil mammal fauna of South America. *In* Evolution, mammals and southern continents (A. Keast, E. C. Erk, and B. Glass, eds.), pp. 247–309. State University of New York Press.

Pregill, G. K. and S. L. Olson. 1981. Zoogeography of West Indian vertebrates in relation to Pleistocene climatic cycles. *In* Ann. Rev. of Ecol. and Syst. 12:75–98.

Preston, F. W. 1962. The canonical distribution of commonness and rarity: Part II. *Ecology* 43:410–432.

Recher, H. F. and S. S. Clark. 1974. A biological survey of Lord Howe Island with recommendations for the conservation of the Island's wildlife. *Biol. Conservation* 6:263–273.

Recher, H. F. and W. F. Ponder, (eds.). 1981. Lord Howe Island: a summary of current and projected scientific activities. *Occasional Reports of The Australian Museum,* Sydney. 72 pp.

Reynolds, T. E., K. F. Koopman, and E. E. Williams. 1953. A cave faunule from western Puerto Rico with a discussion of the genus *Isobbodon. Breviora* 12:1–8.

Rick, J. W. 1980. Prehistoric hunters of the high Andes. Academic Press.

Ross, E. B. 1978. Food taboos, diet and hunting strategy: the adaptation to animals in Amazon cultural ecology. *Curr. Anthropol.* 19:1–36.

Ruddiman, W. F. and A. McIntyre. 1981. Oceanic mechanisms for amplification of the 23,000 year ice-volume cycle. *Science* 212:617–627.

Sauer, C. O. 1966. The early Spanish Main. University of California Press, Berkeley.

Scarlett, R. J. 1968. An owlet-nightjar from New Zealand. *Notornis* 15:254–266.

Scott, L. 1979. Late Quaternary pollen analytical studies in the Transvaal (South Africa). Ph.D. dissertation, University of the Orange Free State, Bloemfontein.

Shimkim, E. M. 1978. The upper Paleolithic in north-central Eurasia: evidence and problem. *In* Views of the past (L. G. Freeman, ed.), pp. 193–315. Mouton, The Hague.

Simpson, G. G. 1940. Mammals and land bridges. *Journal Washington Acad. Sci.* 30:137–163.

———. 1953. The major features of evolution. Columbia University Press.

———. 1980. Splendid isolation. Yale University Press.

Skeels, M. A. 1962. The mastodons and mammoths of Michigan. *Michigan Acad. Sci., Arts, and Letters Papers* 47:101–133.

Slaughter, B. H. 1967. Animal ranges as a clue to late Pleistocene extinctions. *In* Pleistocene extinctions: the search for a cause (P. S. Martin and H. E. Wright, Jr., eds.), pp. 155–167. Yale University Press.

Sondaar, P. Y. 1977. Insularity and its effect on mammal evolution. *In* Major patterns in vertebrate evolution (M. K. Hecht, R. C. Goody, and B. M. Hecht, eds.), pp. 671–707. Plenum Pub. Co.

Squires, D. F. 1963. Carbon-14 dating of the fossil dune sequence, Lord Howe Island. *Australian Journal Sci.* 25:412–413.

Stafford, T. 1981. Alluvial geology and archaeological potential of the Texas southern high plains. *American Antiquity* 46:548–565.

Steadman, D. W. 1981. Vertebrate fossils in lava tubes in the Galapagos Islands. Proceedings 8th International Congress of Speleology (B. F. Beck, ed.), pp. 549–550.

Stenseth, N. C. 1979. Where have all the species gone? On the nature of extinction and the red queen hypothesis. *Oikos* 33:196–227.

Stuart, A. J. 1976. The history of the mammal fauna during the Ipswichian last interglacial in England. *Phil. Trans. Roy. Soc. London* 276:221–250.

———. 1982. Pleistocene vertebrates in the British Isles. Longman, London and New York. 212 pp.

Sutherland, L. and A. Ritchie. 1977. Defunct volcanoes and extinct horned turtles. *In* Lord Howe Island (N. Smith, ed.), pp. 7–12. The Australian National Museum, Sydney.

Taylor, J. D., C. J. R. Braithwaite, J. F. Peake, and E. N. Arnold. 1979. Terrestrial faunas and habitats of Aldabra during the late Pleistocene. *Phil. Trans. Roy. Soc. London* B, 286:47–66.

Taylor, R. E. 1980. Radiocarbon dating of Pleistocene bone: toward criteria for the selection of samples. *Radiocarbon* 22:969–979.

Thompson, R. S., T. R. VanDevender, P. S. Martin, T. Foppe, and A. Long. 1980. Shasta ground sloth (*Northrotheriops shastense* Hoffstetter) at Shelter Cave, New Mexico: environment, diet, and extinction. *Quaternary Research* 14:360–376.

Thunnell, R. C. 1979. Eastern Mediterranean Sea during the last glacial maximum: an 18,000 years B.P. reconstruction. *Quaternary Research* 11:353–372.

Varona, L. A. 1974. Catalogo de los mamiferos vivientes y extinguidos de las Antillas. Academia de Ciencias de Cuba, Havana.

Waldren, W. H. and J. S. Kopper. 1967. A nucleus for a Mallorca chronology of prehistory based on radiocarbon analysis. Deya Archaeological Museum, Deya, Mallorca (Spain), 21 pp.

Walker, E. P. 1975. Mammals of the world (3rd edition). The Johns Hopkins University Press.

Webb, S. D. 1969. Extinction-origination equilibria in late Cenozoic land mammals of North America. *Evolution* 23:688–702.

———. 1976. Mammalian faunal dynamics of the great American interchange. *Paleobiology* 2:220–234.

———. 1978a. A history of savanna vertebrates in the New World. Part II: South America and the great interchange. *Ann. Rev. Ecol. and Syst.* 9:393–426.

———. 1978b. Mammalian faunal dynamics of the great American interchange: Reply to an alternative interpretation. *Paleobiology* 4:206–209.

Wells, R. T. 1978. Fossil mammals in the reconstruction of Quaternary environments with examples from the Australian fauna. *In* Biology and Quaternary environments (D. Walker and J. C. Guppy, eds.), pp. 103–124. Australian Academy of Science, Canberra.

West, R. G. 1980. Pleistocene forest history in East Anglia. *New Phytologist* 85:571–622.

Wetzel, R. M., R. E. Dubis, R. L. Martin, and P. Meyers. 1975. *Catagonus,* an extinct peccary, alive in Paraguay. *Science* 189:379–381.

White, C. 1971. Man and environment in northwest Arnhem Land. *In* Aboriginal man and environment in Australia (D. J. Mulvaney and J. Golson, eds.), pp. 141–157. Australian National University Press, Canberra.

White, J. P. and J. F. O'Connell. 1979. Australian prehistory: New aspects of antiquity. *Science* 203:21–28.

Wright, R. V. S. 1971. The archeology of Koonalda Cava. *In* Aboriginal man and environment in Australia (D. J. Mulvaney and J. Golson, eds.), pp. 104–113. Australian National University Press, Canberra.

18

The Reordered North American Selection Regime and Late Quaternary Megafaunal Extinctions

JERRY N. McDONALD

TWO-THIRDS OF NORTH AMERICA'S LARGE MAMMAL genera, along with larger-bodied avian and reptilian genera, became extinct during the late Wisconsin and early Holocene. This major biological catastrophe has been attributed to various environmental causes, including decreasing equability and increasing continentality of climate (Axelrod 1967; Graham 1979; Lundelius 1967, 1976; Martin and Neuner 1978; Slaughter 1967); desiccating postglacial ecosystems (Guilday 1967; Lundelius 1967, 1976; Webb 1969, 1977); inability of some taxa to adjust reproductive habits to changing climates (Slaughter 1967); ecological disequilibrium initiated by rapid glacial retreat (Dreimanis 1968, Graham 1979, Hester 1967); competitive exclusion (Guilday 1967, Krantz 1970); and overexploitation by human hunters (Edwards 1967; Jelinek 1967; Martin 1967a, 1967b, 1973; Smith 1975. [Uetz and Johnson (1974) and Van Valen (1969) have summarized and evaluated these hypotheses]. Modern ecological and evolutionary theory, however, predicts that most changes in the North American environment during the period of extinction should have resulted in an increase in megafaunal biomass and perhaps even in diversity.

Whatever the proximal cause or causes of the extinctions in North America, the phenomenon was continent-wide, not regional; it was selective principally against large mammals; and it resulted probably from some ultimate, ubiquitous causal factor. The magnitude and rate of faunal simplification further indicates that a significant change occurred in the continent's natural selection regime sufficient to render, quickly and decisively, much of the continent's megafauna "unfit."

Viewing faunal simplification as a consequence of decreased fitness, the extinction cluster can be examined by identifying ecologically significant changes in the natural selection regime that transpired between the maximum late Wisconsin glaciation (ca. 20,000 to 18,000 yr B.P.) and the relative stabilization of vegetation changes about 4000 yr B.P. (Wendland 1978), and analyzing their probable selective influences on the megafauna. When this is done, the overkill hypothesis (Martin 1967a) appears to be more harmoniously reconcilable with the direction and degree of continent-wide environmental changes as the probable ultimate cause of extinctions. Most selective forces in the North American selection regime were relaxing during the period of extinction. Except for the limiting effects of new or intensifying human predation, North America's megafaunal biomass should have increased and its taxonomic diversity should have remained constant or also increased (Mehringer 1967), but this did not happen.

The extinctions which occurred instead are explained most easily as consequences of selection against large body size, or some life history character or characters scaled to large body size, brought about by a fundamental change in the nature of the continent's selection regime—that is, by a change from a late Wisconsin K-selection regime (favoring competitive efficiency in relatively "saturated" environments) to a Holocene r-selection regime (favoring relatively rapid population rebound or expansion into relatively "unsaturated" environments).

In this chapter "late Wisconsin fauna" refers to the North American megafaunal complex of about 20,000 to 18,000 yr B.P. "Holocene fauna" refers to the simplified continental terrestrial megafauna (also exclusive of *Homo*) postdating the extinctions, a process completed by about 8000 yr B.P. The principal taxonomic rank used for analysis is the genus, the lowest rank that can be used with relative confidence owing to the existing imbalance in late Quaternary megafaunal taxonomy.

The Changing Environment of North America

The late Wisconsin stadial reached its maximum intensity about 20,000 to 18,000 yr B.P. (CLIMAP 1976, Davis 1976). After that time, the glacial climate gradually waned (but was interrupted periodically by surges of renewed intensity) until about 14,000 yr B.P., when a more rapid change leading to the interglacial climate began. The early Holocene environment was characterized by continued withdrawal of glaciers, climate warming, and an increasing continentality of climate. The maximum Holocene warming was reached about 7000 yr B.P., and had passed by about 4000 yr B.P. Different regions of the continent experienced differing degrees and patterns of lower-order climate change (Bryson et al. 1970, Mercer 1972, Wendland 1978, Wright 1971).

The North American environment at the late Wisconsin glacial maximum is summarized in Figures 18.1 and 18.2. Some 48 percent of the current continental land area was covered with ice. For analytical purposes, three major terrestrial units are recognized for North America at about 18,000 yr B.P. (fig. 18.1):

1. Mid-latitude North America, situated between the glacial systems and what is today Mexico, was the largest and most compact unit. The climate of this region was probably more equable and temperatures were generally lower than the modern climate of the region. Forest vegetation dominated, being largely or entirely coniferous forest to the north and at higher elevations, with more deciduous elements in a generally mixed forest to the southeast. Woodland/shrubland dominated in the southwest. A narrow band of tundra probably bordered the glacial ice and a block of steppe tundra has been recognized in the northern Great Basin. The large mammal inventory included at least forty-five genera, and possibly man as well, representing (with *Homo*) seven orders.

2. Peninsular North America included what is now Mexico and Central America. The 18,000 yr B.P. climate of this region is not well understood, but cooler climates and vegetation changes correlative with those of mid-latitude North America have been suggested (e.g. Emiliani 1971, Canby 1979). Vegetation was dominated by conifer forests at higher elevations, with broadleaf and mixed forests at lower elevations. A variety of woodlands or shrublands occurred on the Mexican Plateau. True desert was restricted to lower elevations along the Gulf of California. The character of this region's megafauna is less well established than that of mid-latitude North America, but it appears that at least thirty-one genera (thirty-two with *Homo*) were present in the late Wisconsin, representing (with *Homo*) seven orders.

3. Eastern Beringia was the eastern end of a land bridge connecting Asia and North America. The climate was generally cold and dry, influenced by Siberia's intensely continental climate. Vegetation was mainly steppe tundra or tundra, with shrubs or woodland near the southern coast. At least twenty-three genera (twenty-four with *Homo*), representing six orders, were present in the late Wisconsin megafauna of

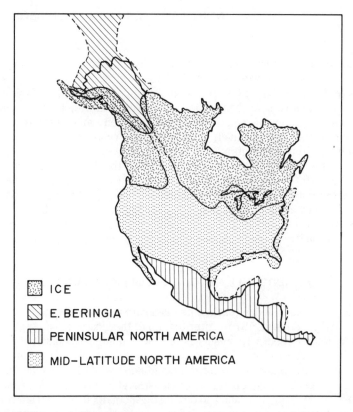

Figure 18.1. Subregions of North America about 20,000 to 18,000 yr B.P. Glacial ice north of modern continental mainland coast not shown. Sea level depression shown for Beringia and North America south of main ice mass only.

eastern Beringia, but exact chronological control is not well established for Alaskan fossils. Even though Beringia was essentially an eastward extension of the Siberian habitat, it was located at an extreme corner of Eurasia and contained Eurasian taxa *that had survived the filtering effect of Siberia's intensely continental climate.* North American taxa dispersing into eastern Beringia were also subject to filtering by a harsh climate and low carrying capacity.

The Holocene climate brought about substantial changes in these patterns. Continental glaciers disappeared and cordilleran glaciers wasted to relict patches. Rising sea levels inundated previously exposed portions of the continental shelf. The most extensive inundation (in Beringia) resulted in the separation of Alaska and Siberia. Continentality increased over mid-latitude North America, whereas eastern Beringia acquired a *more* equable maritime climate. The climate of peninsular North America apparently became warmer and, in the north, drier and more continental. Major vegetation changes (fig. 18.3) included (1) shifts of the previously mid-latitude conifer forests northward, and to higher elevations on deglaciated mountain ranges; (2) a northward shift of the eastern mixed forest; (3) redevelopment and/or expansion of primarily or totally broadleaf forest in southeastern mid-latitude North America; (4) reduction of the conifer forest and expansion of the broadleaf and mixed forests in the North American peninsula; (5) replacement of the intermontane/Mexican Plateau woodland/shrubland with more desertic vegetation; (6) northward shift and expansion of arctic tundra; (7) disappearance of steppe tundra; and (8) development of a mid-latitude grassland. The surviving Holocene megafauna contained no genera that were not present at the late Wisconsin maximum, unless *Homo* was, in fact, not present in North America until after 18,000 yr B.P., and/or *Tayassu* (the peccary) had evolved and was present on the North American peninsula. (Major sources for the interpretation of vegetation changes

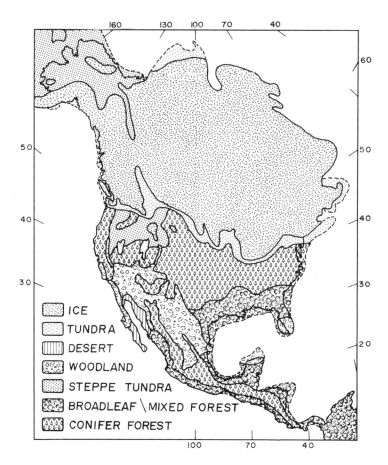

Figure 18.2. Generalized vegetation formations of North America about 20,000–18,000 yr B.P. (After Canby 1979 and McDonald 1981)

presented here include Canby 1979; Davis 1976; Delcourt and Delcourt 1981; Guthrie 1968; Hoffman and Taber 1967; Johnson 1977; Martin and Mehringer 1965; Mehringer 1967, 1977; Péwé 1975; VanDevender and Spaulding 1979; Watts 1970; Wells 1965, 1966; D. Whitehead 1973; Wright 1971; and Yurtsev 1972).

Natural Selection Regimes and the Changing North American Environment

Individual variation within a species' population affords a range of potential phenotypic responses to the collective forces of natural selection. Additionally, spatial variations in the expression of selection throughout a species' range subject populations to different expressions and intensities of selection. *Individual variation and selection gradients combine to provide substantial coevolutionary plasticity in environmental stimulation and adaptive response.* In all but rare or localized species, only a substantive negative shift in some selective force or forces, relative to the tolerance range and adaptive potential of a species' population, should be sufficient to bring about extinction.

Major environmental parameters considered here for their ecological and selective significance include climate; quality, quantity, diversity and accessibility of food and water resources; space; habitat; and predation. Changes in these parameters would have feedback consequences either maintaining, improving, or deteriorating the quality

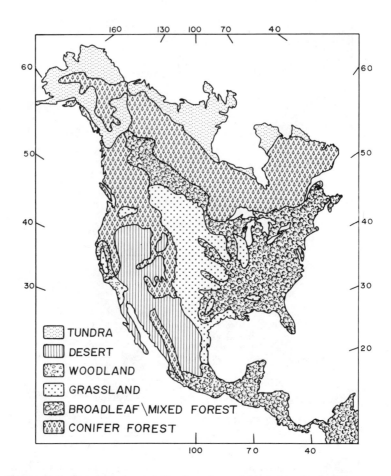

Figure 18.3. Generalized vegetation formations of North America in Holocene. (After Espenshade 1970)

of an environment for a given organism or population. Mammals are relatively euryecious, capable of adapting to or living in a wide range of environments. By virtue of their size, dominance, stamina, greater heat retention per unit volume, general feeding behavior, and mobility, large mammals are among the most ecologically euryecious of all mammals (Gould 1966, Pianka 1978, Stanley 1973). What amounts to a significant environmental change for one taxon need not necessarily be equally significant for all taxa. Generally speaking, larger-bodied mammals should be among the last categories of organisms to be brought to extinction by broad-spectrum environmental changes *unless* selection operated against some character or characters ubiquitous in, and peculiar to, the group.

Biologically, probably the two most important climatic changes that occurred between 18,000 and 7000 yr B.P. were the general increase in ambient temperatures and the net increase of usable moisture over the enlarged ice-free land area. Continentality increased between Alaska and central Mexico. The rain shadow effect of the western mountains, amplified by other drought-producing modifications in the general atmospheric circulation system, probably reached its peak during the 7000 to 4000 B.P. period when the southern and central Great Plains experienced their driest climate of the Holocene (Borchert 1971, Bryson et al. 1970, Dillehay 1974, Wendland 1978, Wright 1970). Equability probably decreased in the central and northern continent, *but remained relatively unchanged* in coastal areas, the southeastern quadrant, and most of the North American peninsula. Alaska's climate became more maritime after the Bering-Chuckchi Platform was inundated (Axelrod 1967, Hopkins 1967).

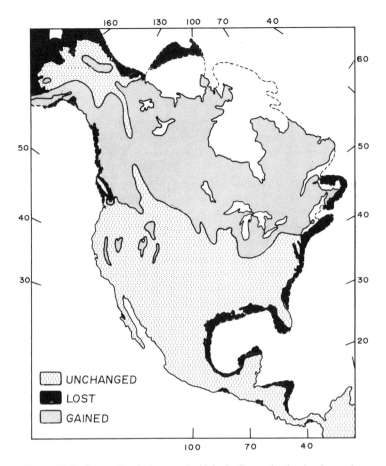

Figue 18.4. Generalized changes in biologically productive land area between the late Wisconsin maximum and the middle Holocene, shown on Lambert's azimuthal equal area projection. Biologically productive land area increased by about 70 percent during this period.

Deglaciation resulted in an increase in continental land area. Allowing for inundation by rising sea levels, the ice-free land area of North America *increased approximately 78 percent* (fig. 18.4). Land area has been positively correlated with species diversity— as land area increases, so, generally, do species diversity (Darlington 1957, Flessa 1975, Preston 1962, MacArthur and Wilson 1967) and biomass, including faunal biomass.

The combination of extensive deglaciation, warmer mean air temperatures, and ample available moisture over much of the continent created a physical environment which substantially increased North America's gross primary productivity (fig. 18.5). The spread of vegetation onto formerly glaciated terrain, development of an extensive mid-latitude grassland, and increase in the absolute abundance of broadleaf flora represent major vegetation developments that considerably increased the continent's potential secondary productivity as well. Alaska, the intermontane region, the Mexican Plateau, and parts of coastal Mexico were the only large regions that might have experienced a decrease in primary productivity. About 18 percent of the Holocene land area experienced decreased productivity. Approximately 10 percent experienced little or no change in productivity. The remaining 72 percent, however, experienced an *increased primary productivity* (cf. figs. 18.2–18.5 and Table 18.1).

Increased primary productivity does not, in itself, guarantee a proportionate increase in the food resources available to terrestrial herbivores and, thereafter, their predators, but the nature of physical and vegetation changes which occurred between

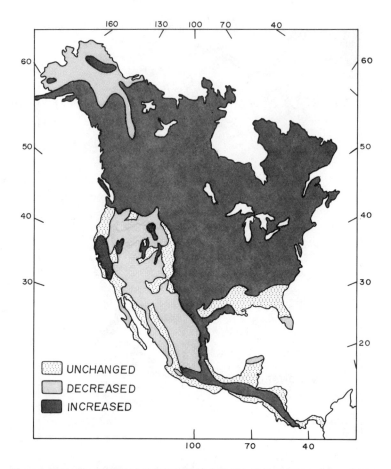

Figure 18.5. Generalized changes in biological productivity between the late Wisconsin maximum and the middle Holocene. I have assumed the following productivity relationships among the vegetation formations used in Figures 18.3 and 18.4; broadleaf/mixed forests > conifer forests > woodlands > grasslands > steppe tundra > tundra > desert (see Table 18.1)

18,000 and 7000 yr B.P. probably significantly increased the forage biomass available to large terrestrial herbivores. The combined increase in grassland, tundra, desert, and woodland vegetation indicates that the area over which low-level forage was distributed also increased. The increased dominance of broadleaf plants over some parts of the continent resulted in a greater supply of generally higher-quality forest browse (Smith 1952, 1957, 1959, Ullrey et al. 1964, 1967, Urness et al. 1971). The absolute frequency of disclimax situations (such as hydroseres and lithoseres, especially on recently de-glaciated terrain) and disruptive events such as floods, wind, and ice storms, certainly must have increased with the greater forest area; this resulted in an absolute increase in the number of seral islands existing at any given time within the forests. Lastly, the increasingly important role of natural and cultural fires in creating and maintaining disclimax communities in forests and ecotones must be acknowledged (Crow 1978, Lewis and Schweger 1973, Rowe and Scotter 1973, Sauer 1956, Stewart 1951, Taylor 1973, Viereck 1973). Large mammal biomass has been shown to increase with precipitation and primary productivity, so a maximum large mammal biomass would be expected where maximum primary productivity and forage availability converged (Coe 1980, Coe et al. 1976, Eisenberg and Seidensticker 1976, Western 1980), that is, in prairies, savannas, or more heavily disturbed forests, all of which must certainly have been more widespread in the Holocene than the late Wisconsin.

Table 18.1. **Estimated Primary Productivity in Different Vegetation Formations**

Odum (1975) (kcal/m²/yr)		Kormondy (1976)* (kcal/m²/yr)		Jones (1979) (g/m²/yr)	
Desert and tundra	200	Desert scrub	315	Desert	70
Grasslands and pastures	2,500	Tundra and alpine	630	Tundra	140
Dry forests	2,500	Temperate grassland	2,250	Boreal forest	500–800
Boreal conifer forest	3,000	Woodland/shrubland	2,700	Mid-latitude	
Moist temperate forest	8,000	Savanna	3,150	deciduous forest	600–1,300
Wet tropical broadleaf		Coniferous forest	3,600	Savanna	700
and subtropical forest	20,000	Temperate forest	5,850	Mediterranean woodland	ca. 925
		Tropical forest	9,000	Seasonal rainforest	1,000
				Tropical rainforest	2,000

*Data from Whittaker 1970.

Corollaries of land area are habitat diversity and patchiness (Darlington 1957, Flessa 1975, Preston 1962, MacArthur and Wilson 1967), both of which should have been increasing in North America during the time of most extinctions (Mehringer 1967). Increased floristic diversity and greater habitat patchiness could have resulted from the absolute increase in the vegetated area.

Eleven carnivore genera (Table 18.2) (and possibly man) represent the large mammal predator complex of the late Wisconsin fauna. Six of the carnivore genera were extinct by the early Holocene, and no new genera from this order were added to the continent's large mammal inventory. The question of human presence in North America before the late Wisconsin is unsettled and being vigorously investigated. One point of view holds that humans were in North America before the onset of the late Wisconsin stadial (Bada et al. 1974, Irving 1978, Morlan 1978, Stalker 1977). Another maintains that evidence does not verify human presence on the continent before the late Wisconsin (Adovasio et al. 1978, Haynes 1967, Martin 1967b, 1973, Miller and Dort 1978, Powers and Hamilton 1978, Reagan et al. 1978). This difference of opinion is critical to the megafaunal extinction problem *only* if there were no cultural differences between the earlier and later human populations. On this matter, proponents of the "earlier" arrival model present evidence that the early Wisconsin humans had a less advanced or less complex material culture than did the later Paleo-Indians. The fact that most large-bodied herbivores in North America about 25,000 to 20,000 yr B.P. and their carnivore predators survived the late Wisconsin maximum further minimizes the importance of human-large mammal interactions insofar as the extinction question is concerned. If humans were present before 18,000 yr B.P., they did not, whether for inability or insignificance, constitute a sufficiently limiting selective force to effect megafaunal extinctions (Martin 1967b). Between 18,000 and 8000 yr B.P., therefore, over half of North America's large-bodied carnivore genera became extinct, whereas the only new, or significantly more successful, predator appears to have been man.

Prediction and Reality

Each change in the North American environment between 18,000 and 7000 yr B.P. represented a relaxation of one or more limiting factors which, theoretically, should have been functioning at maximum limiting intensity during the late Wisconsin maximum glaciation. The megafaunal species' populations should have been reduced to their minimum numbers and honed to their maximum competitiveness by the severe selection regime then existing. They should have increased in numbers thereafter, and either maintained or increased their taxonomic diversity as the Holocene environment emerged.

Table 18.2. Temporal and Spatial Distribution of Large Mammal (>44 kg) Genera Present in the Late Wisconsin of North America

Genus	Appearance in North America*	Late Wisconsin Distribution — Eastern Beringia	Late Wisconsin Distribution — Mid-latitude North America	Late Wisconsin Distribution — Peninsular North America	Holocene Distribution — Eastern Beringia	Holocene Distribution — Mid-latitude North America	Holocene Distribution — Peninsular North America	Holocene Status — Genus Extinct	Holocene Status — Genus Extant, Species Extinct	Holocene Status — Genus and Species Extant
EDENTATA										
Holmesina	Irvingtonian	—	X	X	—	—	—	X	—	—
Glyptotherium	Blancan	—	X	X	—	—	—	X	—	—
Megalonyx	Blancan	X	X	X	—	—	—	X	—	—
Eremotherium	Irvingtonian	—	X	X	—	—	—	X	—	—
Nothrotheriops	Irvingtonian	—	X	X	—	—	—	X	—	—
Glossotherium	Blancan	—	X	X	—	—	—	X	—	—
RODENTIA										
Castoroides	Blancan	—	X	—	—	—	—	X	—	—
Hydrochoerus	Irvingtonian	—	X	X	—	—	X	—	X	—
Neochoerus	Wisconsin	—	X	—	—	—	—	X	—	—
CARNIVORA										
Canis	Blancan	X	X	X	X	X	X	—	—	X
Cuon	Wisconsin	X	—	X	—	—	—	X†	—	—
Tremarctos	Blancan	—	X	X	—	—	—	X†	—	—
Arctodus	Irvingtonian	X	X	X	—	—	—	X	—	—
Ursus	Blancan	X	X	X	X	X	X	—	—	X
Smilodon	Irvingtonian	—	X	X	—	—	—	X	—	—
Homotherium	Irvingtonian	X	X	—	—	—	—	X	—	—
Panthera	Irvingtonian	X	X	X	—	X	X	—	X	—
Acinonyx	Blancan	—	X	—	—	—	—	X†	—	—
Felis	Blancan	—	X	X	—	X	X	—	—	X
Lynx	Blancan	X	X	—	X	X	X	—	—	X
PROBOSCIDEA										
Mammut	Pre-Blancan	X	X	X	—	—	—	X	—	—
Cuvieronius	Blancan	—	X	X	—	—	—	X	—	—
Mammuthus	Irvingtonian	X	X	X	—	—	—	X	—	—

Taxon	Earliest appearance	23	45	31	10	14	11	33	7	13
PERISSODACTYLA										
Equus	Blancan	×	×	×	—	—	—	X[†]	—	—
Tapirus	Irvingtonian	—	×	×	—	—	×	—	×	×
ARTIODACTYLA										
Mylohyus	Blancan	—	×	—	—	—	—	×	—	—
Platygonus	Blancan	×	×	×	—	—	—	×	—	—
Camelops	Blancan(?)	×	×	×	—	—	—	×	—	—
Hemiauchenia	Blancan	—	×	×	—	—	—	×	—	—
Palaeolama	Irvingtonian	—	×	—	—	—	—	—	—	—
Odocoileus	Blancan	—	×	×	—	×	×	×	—	×
Navahoceros	Wisconsin	—	×	×	—	—	—	×	—	—
Sangamona	Wisconsin	—	×	—	—	—	—	—	—	—
Rangifer	Irvingtonian	×	×	—	×	×	—	—	—	×
Alces	Illinoian	×	—	—	×	×	×	×	—	×
Cervalces	Sangamonian (?)	×	×	×	—	—	—	—	—	—
Cervus	Irvingtonian	×	×	×	×	×	×	×	—	×
Capromeryx	Blancan	—	×	—	—	—	—	×	—	—
Tetrameryx	Irvingtonian	—	×	×	—	—	—	×	—	—
Stockoceros	Irvingtonian	—	×	×	—	—	—	×	—	—
Antilocapra	Wisconsin	×	×	×	—	×	×	×	—	×
Saiga	Illinoian	×	—	—	—	—	—	X[†]	—	—
Oreamnos	Sangamonian	—	×	—	×	×	—	—	×	×
Ovis	Illinoian	×	×	—	×	×	×	—	×	×
Euceratherium	Irvingtonian	—	×	×	—	—	—	×	×	—
Symbos/Bootherium	Irvingtonian	×	×	—	—	—	—	×	—	—
Ovibos	Illinoian	×	×	×	×	×	×	—	—	×
Bison	Illinoian	×	×	×	×	×	×	X[†]	×	—
Bos	Illinoian	×	—	—	—	—	—	X[†]	—	—
TOTAL		23	45	31	10	14	11	33	7	13

*Earliest appearances of genera in North America are based upon Adams (1979), Guthrie and Matthews (1971), Hibbard et al. (1965), Kurtén (1975), Kurtén and Anderson (1980), Péwé and Hopkins (1967), Repenning (1967), and Webb (1965, 1974).

[†]These genera survive in South America or Eurasia.

413

Some regional divergence from this predicted response should be expected. Alaska probably should have experienced a reduction of megafaunal biomass and diversity because the large mammal carrying capacity was reduced with the transition from steppe tundra to tundra and forest (Guthrie 1968, Reed 1970); however, increased carrying capacity in the deglaciated areas might have compensated for this. In addition, the connection with Eurasia was severed and a tundra/boreal forest/mountain barrier complex precluded facile colonization from mid-latitude North America. The intermontane–northern Mexico area should also have lost some of its diversity due to the limiting effect of increasing aridity.

The eastern Beringian genera *Bos, Saiga,* and *Alces* could reasonably be expected to have become extinct in North America; they were either rare or localized, and the area they occupied experienced a reduced carrying capacity. On the other hand, these genera could (as *Alces* did) have dispersed into mid-latitude North America and become established there. There is no reason, however, to expect that the genera present in the late Wisconsin intermontane–northern Mexico area collectively should have become extinct. Most of these genera also occurred outside the desiccating area and could have survived in regions where productivity and carrying capacity actually increased (including the southwestern mountains and plateaus); or they could have migrated from the desiccating region as the environment changed, as *Bison* did (Dillehay 1974, McDonald 1981). Considering direction of changes in limiting factors, widespread distribution of most genera, and availability of migration corridors, no ecological reason is apparent (save human hunting) to explain why remaining parts of unglaciated mid-latitude or peninsular North America should have experienced megafaunal depauperization.

The late Wisconsin megafauna of North America includes some forty-nine genera (Table 18.2). All large mammal genera currently occurring naturally in North America, with the exception of the peccary (*Tayassu*) (Kurtén and Anderson 1980, Lundelius 1967), were present in the late Wisconsin fauna. Of these genera, thirty-three (67 percent) apparently became extinct during the late Wisconsin or early-to-middle Holocene. All species in each extinct genus, of course, became extinct and so did some species in surviving genera, some (e.g. *Bison antiquus* and *Ovis catclawensis*) by phyletic replacement (Harris and Mundel 1974, McDonald 1981) as subsequent generations adapted to the newly ordered selection regime. Six genera that became extinct in North America survived in South America or Eurasia. Late/post-Wisconsin extinctions were considerably more extensive than those at the end of previous Quaternary glacial stages (Table 18.3).

As should be expected, mid-latitude North America had, relative to its area and habitat diversity, the largest number of genera (at least forty-five) in its late Wisconsin fauna. The only late Wisconsin North America genera not present were the Palearctic autochthons *Cuon* (unreported but probably present), *Alces* (possibly present), *Bos,* and *Saiga,* the last two rare even in Alaska's fauna. Thirty-two (71 percent) of the mid-latitude genera were lost, and one (*Alces*)was gained, giving a Holocene faunal generic diversity of fourteen, or a 69 percent net loss of this region's late Wisconsin megafaunal diversity. Peninsular North America had the next greatest diversity (thirty-one genera, Table 18.2), also as expected because of its area, tropical-subtropical location, and connection with two large continental areas. Twenty-one genera (68 percent) were lost, one (*Ovis*) was added, leaving eleven large mammal genera (35 percent of previous diversity) on the peninsula. Alaska, the smallest subregion, had only twenty-three genera (Table 18.2). Fourteen (61 percent) of these were lost, one (*Oreamnos*) was added, for a net loss of 56 percent of the region's previous diversity. Theoretically, mid-latitude North America's environment should have permitted the greatest survival of generic diversity, followed by peninsular North America and, lastly, Alaska. However, the opposite of this predicted pattern occurred. Alaska had the least simplification of its fauna (56 percent), followed by peninsular North America (65 percent) and mid-latitude North America (69 percent).

Table 18.3. Periodicity of Quaternary Megafaunal Extinctions in North America

Number of Genera Becoming Extinct	At or near end of
According to Hibbard et al. (1965)	
0	Nebraskan
7	Aftonian
4	Kansan
3	Yarmouth
0	Illinoian
1	Sangamon
33	Wisconsin
According to Kurtén and Anderson (1980)	
2	Middle Irvingtonian (ca. Nebraskan and Aftonian)
2	Late Irvingtonian (ca. Kansan and Yarmouth)
2	Illinoian
1	Sangamon
35	Wisconsin

Most large mammal genera (thirty-eight) in North America were herbivores (78 percent); nineteen genera were Pecoran ruminants (= 50 percent of all herbivores). Three bear genera were probably omnivores to differing degrees. Remaining carnivores were represented by two dog and six cat genera. Extinction came to twenty-seven (71 percent of the herbivores (including 90 percent of the nonruminants and 53 percent of the ruminants), 67 percent of the bears, 50 percent of the dogs, and 50 percent of the cats. The late Wisconsin herbivore:carnivore ratio of about 4.75:1 decreased to about 2.75:1 in the Holocene fauna, suggesting selection for more generalized predatory habits and a potentially increased variety of predatory pressures from carnivores on surviving herbivores.

Among the thirty-three genera that became extinct, twenty-six (79 percent) had been present in North America since at least the Irvingtonian (Table 18.2). Two more genera (6 percent) dated from the Illinoian, and one (3 percent) from the Sangamon (Table 18.2). The remaining four (12 percent) arrived possibly as late as the Wisconsin. Among the surviving sixteen genera, ten (63 percent) arrived before the Rancholabrean, and five (31 percent) before the Wisconsin. At least forty-four (90 percent) of the genera present had, therefore, experienced at least the earlier Wisconsin stadials, and forty-two (86 percent) had experienced a complete glacial-interglacial (Illinoian-Sangamon-Wisconsin) cycle. The fauna was not, therefore, unfamiliar with glacial or interglacial environmental conditions. Extinction was not biased toward either "older" or "younger" member genera.

Incompatibility of Environmental Deterioration Theories and Megafaunal Extinctions

Environmental changes (excluding those associated with humans) were probably not responsible ultimately for North American megafaunal extinctions. The most important and general aspect of this position is that the North American environment improved as large mammal habitat between 18,000 and 7000 yr B.P. Desiccation reduced the large mammal carrying capacity (relative to the late Wisconsin) in only about 15 percent of Holocene North America, with another 10 percent or so added temporarily during the mid-Holocene period of maximum aridity on the southern and central Great

Plains (ca. 7000 to 4000 yr B.P.)—after most extinctions had already occurred. Desiccation producing the central grasslands, however, resulted in increased available forage for large mammals which historically supported a large *Bison–Antilocapra–Canis* megafaunal biomass. Even the hot deserts of North America can support viable populations of large mammals, as evidenced by large numbers of domestic and feral livestock and smaller numbers of exotic game animals from Texas to California (Laycock 1966; McKnight 1958; Norment and Douglas 1977; Seegmiller 1977; Woodward 1976, 1979).

The argument that decreasing equability, increasing continentality, or both caused the megafaunal extinctions is weakened by examining the late Wisconsin-Holocene transition in central Asia and Siberia. This region today has the world's most continental and least equable climate. The Sartan (= late Wisconsin) glacial climate of this region was even more intensely continental than today (Frenzel 1968, Gates 1976), yet it supported a greater variety of megafaunal genera then. If decreasing equability/ increasing continentality caused extinctions in North America, the opposite ecological effect would be expected for central Asia and Siberia. That is, rather than collapsing, megafaunal diversity should have remained stable or increased with the arrival of the Holocene. Many genera which became extinct in North America before 8000 yr B.P. were also present in the glacial continental climate of Siberia (Aigner 1972, Flerow 1967, Reed 1970, Repenning 1967, Vangengeim 1967), which was much more intense than the Holocene continentality in North America. Lastly, even if megafaunal genera in North America were displaced from some regions because of changing climate, they could have migrated to, and survived in, regions with less adverse conditions (as did *Bison*) (Johnson 1977, McDonald 1981).

Evidence concerning the flexibility of mammalian reproduction patterns indicates that adaptive adjustments in breeding and birth seasons, and perhaps even gestation periods, could have been made to adjust to at least moderate environmental changes involving precipitation, temperature, and plant growth regimes (Fraser 1968, Fretwell 1972, Pianka 1978, Snyder 1976). There occurs, in fact, some seasonality in the reproductive patterns of most large mammals, even those living in the tropics and those producing throughout the year in seasonal environments (e.g. *Felis concolor*) in response to variations in critical environmental factors (Asdell 1964, Crandall 1964, Ewer 1973, Fairall 1968, Laws et al. 1975, Leuthold and Leuthold 1975, Sinclair 1977, G. Whitehead 1972). Some species, many possessing clearly seasonal reproductive patterns, are polyestrous, and others (Table 18.4) have a significant range of gestation periods (Asdell 1964, Crandall 1964, Ewer 1973, Kiltie 1982, Walker 1968). The temporal flexibility of reproductive patterns is most clearly shown with large mammals transferred from the northern hemisphere to the southern hemisphere, or vice versa. In numerous instances, viable populations have been established in antipodal environments, an adjustment which necessitates, in part, a six-month (more or less) shift in mating and birth cycles (Groves 1974, Schaller 1977, G. Whitehead 1972). Clinal and regional variations in the onset or duration of reproductive activity, adjusted to seasonal environmental conditions, are also known for species in several genera, including *Equus, Cervus, Odocoileus, Rangifer, Antilocapra,* and *Bison* (Crandall 1964, Hall and Kelson 1959, McDonald 1982, Rue 1968, G. Whitehead 1972).

Since most large herbivores are, in fact, more or less feeding generalists, but with a somewhat narrower band of preferred food items, it is difficult to envision the disintegration of late Wisconsin plant assemblages as being contrary to the foraging interests of the late Wisconsin megafauna, particularly since the overall nutritional level and abundance of forage increased. Where careful observations of large mammal diets have been made or estimated, the variety of food items has frequently been found to be relatively high; specialist preference for a few food categories relatively narrow; and variation in composition and preference from one area to another, and one season or year to another, apparent (i.e. they eat a wide variety of foods, preferring some but in fact eating what is available when available) (Borowski and Kossak 1972, Gebczynska and Krasinska 1972, Hansen and Martin 1973, Meagher 1973, Naumov 1972, Peden 1976, Sinclair 1977, Smith and Malechek 1974, Urness et al. 1971, Woodward and Ohmart 1976). Fox and Morrow (1981) found that feeding specialization is frequently a *flexible* attribute of local populations. Additionally, the "sorting out" of floral assemblages between 18,000 and 7000 yr B.P. made important diet items (e.g. forbs, grasses, and broadleaf woody plants) available in *larger quantities*, thereby improving the diet situation for most, and probably all, large mammals. The current distribution of feral horse and ass populations in western North America clearly illustrates the wide range of habitats that large mammal species, and exotics at that, can viably exploit (McKnight 1958, 1959). Numerous other species confirm the great ecological flexibility, including diet flexibility, of large mammals introduced into exotic environments (Elton 1958, Laycock 1966, McKnight 1964, 1971, 1976).

Lastly, if competitive exclusion at any trophic level was going to produce the extinction of any taxon during the period being considered here, it should have done so during the 20,000 to 18,000 B.P. yr period when selection was most intense and limiting. The fact that most genera that became extinct survived this period of selection by several millenia weakens the argument that competitive exclusion caused the extinctions (Mehringer 1967). Food, water, and space resources are usually sufficiently abundant in self-regulating predator-prey systems to support an adapted herbivore population in all but relatively rare catastrophic situations like drought or ice storms. They do not ordinarily serve as actual, ultimate limiting factors, especially at the species' population level (Pianka 1978, Sinclair 1977, Wiens 1977). Krantz's (1970) argument that human predation increased the survivorship of juveniles and lowered the mean age of prey populations, resulting in a rapid population increase sufficient to eliminate nongame trophic competitors, overlooks density-dependent population regulations which operate in many large-bodied mammals (Ewer 1973, Haney 1969, Laws et al. 1975, McCullough 1979, Snyder 1976, E. Wilson 1975).

K and r Selection and Late Quaternary Extinctions

Late Quaternary mammalian extinctions were biased toward large-bodied terrestrial species. Largeness usually conveys greater individual competitive ability and security, but it also requires more resources. Large body size is correlated positively with several life history characteristics (fig. 18.6), including greater potential and expected life spans, territorial requirements, absolute growth rates, age at sexual maturity and first parturition, gestation time, and weight of young at birth (Table 18.4). Reproductive rates (potential birth, absolute birth, and recruitment) and relative growth rate are negatively correlated with body size (fig. 18.7). These characteristics require that large-bodied organisms possess efficient ecological strategies, such as sedentariness (to reduce energy expenditures), territoriality (to assure control of adequate resources in a resource-limited environment over time), generalized feeding habits (to maximize use of available forage), and individuality or, at least, lower thresholds for density-dependent social responses (to reduce intraspecific competition). Large-bodied organisms have been selected for overall competitive success over long life spans, circumstances which select against high reproductive rates and rapid population growth and turnover. Actual population size and density vary according to available resources and specific adaptive strategies and tolerance levels, but, generally, larger-bodied mammals are less numerous, their population levels are more nearly at the region's carrying capacity, and population size fluctuates less frequently and extremely than that of smaller-bodied mammals.

Table 18.4. Body Weight and Reproduction Data for Selected Mammal Species

	(1) Body Weight[a] (kg)	(2) Age at First Parturition[b] (days)	(3) Gestation[c] (days)	(4) Mean Litter Size[d]	(5) Litter Frequency[e] (per year)	(6) Sources
EDENTATA						
Myrmecophagidae						
Myrmecophaga tridactyla	21/17	—	190	1	1	1, 2, 6, 8
Bradypodidae						
Bradypus tridactylus	4¼/—	—	150 (120–180)*	1	—	1, 8
Choloepus didactylus	—	—	190 (120–263)*	1	—	2, 6, 8
Dasypodidae						
Euphractus sexcinctus	5/—	—	—	2	—	1, 8
Priodontes giganteus	53¼/—	—	—	1 (1–2)	—	1, 5, 8
Dasypus novemcinctus	6½/—	600/728	120*	4 (4–8)	1	4, 6, 8
CARNIVORA						
Canidae						
Canis lupus	55/36	791/1092	63 (60–63)	5 (1–13)	1	1, 2, 4, 5, 6, 7, 8
Canis latrans	11/9	427/728	63 (60–65)	6 (1–19)	1	1, 2, 4, 5, 6, 7, 8
Canis mesomelas	9/—	—	64 (57–70)	4 (1–8)	1	1, 2, 3, 9
Canis adustus	9/—	—	64 (57–70)	2 (1–7)	2	1, 2, 3, 8
Alopex lagopus	5/4	355/365	55 (51–60)	6 (1–14)	1	1, 2, 4, 6, 7, 8, 9
Vulpes velox	3/—	351/365	51	5 (4–7)	1	1, 3, 4, 6, 8, 9
Vulpes vulpes	7/—	352/365	52 (49–56)	5 (1–10)	1	1, 2, 5, 6, 7, 8, 9
Fennecus zerda	1½/—	—	51 (50–51)	2 (1–5)	1	2, 3, 8, 9
Urocyon cinereoargenteus	5/—	363/365	63 (53–63)	4 (1–7)	1	1, 2, 4, 6, 7, 8
Nyctereutes procyonoides	7/—	—	62 (52–79)	6 (1–12)	1	1, 2, 5, 8
Chrysocyon brachyurus	23/—	—	—	2 (2–3)	—	1, 2, 8
Speothos venaticus	6/—	—	65	4 (2–5)	2	2, 3, 5, 8
Cuon alpinus	18/—	—	63 (60–70)	4 (2–7)	—	1, 2, 8
Lycaon pictus	24/—	806/1092	72 (60–80)	7 (2–13)	1	1, 2, 3, 8, 9
Otocyon megalotis	4/—	—	65 (60–75)	3 (2–5)	—	2, 3, 5, 8
Ursidae						
Tremarctos ornatus	125/—	—	255 (240–255)	2 (1–3)	(1)	1, 10
Selenarctos thibetanus	118/—	—	—	2 (1–2)	(1)	2, 5, 8
Ursus arctos						
Grizzly	400/—	1312/1458	220 (180–258)*	2 (1–4)	½–⅓	1, 2, 5, 6, 7, 8, 10
Brown, Eurasia	225/—	1621/1820	165 (151–177)*	2 (1–4)	½–⅓	
Brown, North America	545/320	1671/1820	215 (180–250)*	2 (1–4)	½–⅓	

Ursus americanus	135/120	1302/1458	210 (100–240)*	2 (1–6)	½	1, 2, 6, 7, 8, 10
Thalarctos martimus	450/320	1347/1458	255 (240–270)*	2 (1–4)	½	2, 3, 5, 6, 7, 8
Helarctos malayanus	45/—	—	95 (95–96)	2 (1–2)	—	2, 3, 5, 8
Melursus ursinus	100/—	—	210	2 (1–3)	—	2, 5, 8
Procyonidae						
Bassariscus astutus	1/—	—	—	3 (1–5)	—	2, 4, 6, 8, 9
Procyon lotor	16/—	348/364	63 (60–73)	4 (1–7)	1	2, 4, 6, 7
Nasua nasua	11/—	—	77 (71–77)	2 (2–6)	—	1, 2, 4, 5, 6, 8
Potos flavus	2½/—	—	—	1 (1–4)	—	1, 2, 4, 6, 8
Ailurus fulgens	4½/—	—	100 (90–112)	2 (1–2)	1	1, 2, 5, 8
Ailuropoda melanoleuca	130/105	—	—	1 (1–2)	½	1, 2, 6, 8
Viverridae						
Genetta tigrina	1½/—	—	74 (70–77)	2 (2–3)	2	3, 5, 8
Paradoxurus hermaphroditus	2½/—	—	—	3 (2–6)	2	1, 2, 8
Paguma larvata	4½/—	—	—	2 (1–4)	—	2, 5, 8
Arctictis binturong	11½/—	—	91 (90–92)	2 (2–3)	—	3, 8
Herpestes auropunctatus	2/—	—	60 (32–60)	3 (2–4)	2–3	1, 6, 8
Helogale vetula	.7/—	300/364	52 (35–54)	4 (2–6)	—	1, 2, 3, 8
Hyaenidae						
Crocuta crocuta	70/—	1205/1456	90 (90–110)	1 (1–2)	1	1, 2, 8, 9
Hyaena brunnea	40/—	—	94 (90–98)	2 (2–4)	—	1, 2, 8
Hyaena hyaena	40/—	—	90	3 (1–6)	—	1, 2, 5, 8
Felidae						
Felis silvestris	4/—	—/364	65 (63–68)	4 (3–6)	—	1, 2, 5, 9
Felis pardalis	14/—	—	90	2 (1–3)	—	2, 4, 5, 6
Felis yagouaroundi	8/—	—	63	2 (2–3)	—	4, 6
Felis concolor	70/50	819/1092	91 (82–96)	2 (1–6)	½–¾	1, 2, 4, 5, 6, 7, 8, 9
Lynx caracal	15/—	704/728	70 (60–74)	2 (1–4)	—	1, 2, 8, 9
Lynx lynx	12/—	—	62 (60–90)	2 (1–5)	⅔–1	2, 4, 6, 7, 8
Lynx rufus	9/8	—	55 (49–63)	2 (1–6)	1–2	1, 2, 4, 5, 6, 7
Neofelis nebulosa	20/—	—	—	2 (1–4)	—	2, 8
Panthera leo	200/115	1201/1456	108 (105–113)	3 (1–6)	½–⅓	2, 5, 6, 8, 9, 30
Panthera tigris	230/140	1196/1456	104 (99–109)	3 (1–6)	½	2, 5, 6, 8, 9
Panthera pardus	80/74	—	100 (98–105)	2 (1–4)	—	1, 2, 8
Panthera onca	100/—	829/1092	100 (93–110)	2 (1–4)	½	2, 4, 6, 7, 8
Panthera unica	35/30	—	93	2 (1–4)	(1)	1, 2, 8
Acinonyx jubatus	58/50	750/750	93 (84–95)	2 (2–4)	½	1, 2, 3, 6, 8, 9
PROBOSCIDEA						
Elephantidae						
Loxodonta africana	4500/2800	7139/7139	660 (570–690)	1 (1–2)	⅙–¼	2, 5, 6, 8, 9, 10, 11
Elephas maximus	3650/2500	5372/5372	640 (510–720)	1 (1–2)	⅙–¼	1, 2, 5, 6, 8
PERISSODACTYLA						
Equidae						
Equus przewalskii	300/—	1458/1458	330 (330–335)	1	-	2, 8, 12

Table 18.4. Body Weight and Reproduction Data for Selected Mammal Species

(continued)

	(1) Body Weight[a] (kg)	(2) Age at First Parturition[b] (days)	(3) Gestation[c] (days)	(4) Mean Litter Size[d]	(5) Litter Frequency[e] (per year)	(6) Sources
Equus hemionus	350/275	1073/1093	345 (330–365)	1	½–1	1, 2, 8, 12
Equus asinus	240/—	728/728[f]	365 (348–377)	1	—	1, 2, 6, 8
Equus zebra	250/—	—	360 (300–375)	1	—	2, 8
Equus burchelli	283/—	1088/1093	360 (300–375)	1	—	2, 6, 8, 9, 12
Equus grevyi	350/—	—	390	1	—	2, 5, 8
Tapiridae						
Tapirus indicus	260/—	—	390 (360–405)	1 (1–2)	½	1, 2, 5, 6, 8
Tapirus terrestris	250/—	—	397 (390–400)	1 (1–2)	—	1, 2, 5, 8
Tapirus bairdii	260/—	—	395 (394–400)	1 (1–2)	—	1, 4, 6, 8
Rhinocerotidae						
Rhinoceros unicornis	2000/1600	2126/2126	488 (474–570)	1	⅓–½	1, 2, 5, 6, 8
Rhinoceros sondaicus	2000/—	—	510	1	⅓–½	8
Dicerorhinus sumatrensis	1000/900	—	225 (210–240)	1	—	1, 8
Ceratotherium simum	3000/—	2366/2366	510 (480–540)	1	⅓–½	1, 2, 5, 8, 14
Diceros bicornis	1500/—	2275/2275	495 (450–550)	1	¼–½	2, 5, 6, 8, 9, 13
ARTIODACTYLA						
Suidae						
Potamochoerus porcus	100/—	—	120	4 (1–10)	—	1, 2, 5, 8
Sus scrofa	130/100	662/729	116 (101–130)	4 (1–12)	1	1, 2, 6, 7, 8
Phacochoerus aethiopicus	90/68	—	173 (120–175)	4 (2–4)	—	1, 2, 5, 8, 9
Hylochoerus meinertzhageni	240/200	—	125	6 (2–8)	—	1, 2, 5, 8
Babyrousa babyrussa	80/—	—	140 (125–150)	2 (1–2)	—	1, 2, 5, 8
Tayassuidae						
Tayassu tajacu	23/—	—/729	115 (112–148)	2 (1–2)	—	1, 2, 4, 6, 7, 8
Tayassu pecari	25/—	—	145 (142–148)	2 (1–2)	—	1, 2, 8
Hippopotamidae						
Hippopotamus amphibius	4000/3200	2060/2084	237 (225–257)	1 (1–2)	⅓–½	1, 2, 5, 6, 8, 9
Choeropsis liberiensis	200/162	2060/2084	210 (200–240)	1	⅓–½	1, 2, 5, 6, 8
Camelidae						
Lama guanicoe	80/—	—	315 (300–330)	1	½	2, 5, 6, 8
Vicugna vicugna	50/	—/1093	315 (300–330)	1	½–1	2, 5, 8, 16
Camelus bactrianus	620/—	2184/2184	406 (315–440)	1 (1–2)	½	1, 2, 6, 8
Camelus dromedarius	750/—	2184/2184[f]	390 (315–390)	1	½	1, 2, 5, 15

Tragulidae						
Tragulus meminna	3½/—	—	135 (120–155)	1 (1–2)	—	1, 2, 8
Tragulus napu	3½/—	—	150 (150–155)	1 (1–2)	—	8
Tragulus javanicus	3½/—	—	150 (120–155)	1 (1–2)	—	2, 5, 8
Cervidae						
Moschus moschiferus	13/11	—/364	155 (150–160)	1 (1–2)	1	1, 2, 5, 17, 18
Muntiacus muntjak	17/—	—	180	1 (1–2)	—	2, 5, 8
Elaphodus cephalophus	20/—	—	180	1 (1–2)	—	2, 8, 17
Dama dama	60/—	958/1092	230 (230–246)	1 (1–2)	1	1, 2, 8, 17
Axis axis	75/—	—	225 (210–238)	1 (1–3)	—	2, 17
Axis porcinus	40/—	—	240	1 (1–2)	—	2, 17
Cervus unicolor	270/—	—	240	1 (1–2)	(1)	1, 2, 17
Cervus duvauceli	200/140	—	245 (240–250)	1 (1–2)	—	2, 17
Cervus eldi	100/63	728/728	183[9]	1 (1–2)	(1)	1, 2, 17
Cervus nippon	65/50	719/728	235 (222–250)	1 (1–2)	(1)	1, 2, 17
Cervus elaphus (European)	190/—	—/1092	234 (225–250)	1 (1–2)	(1)	1, 2, 5, 17
Cervus elaphus hanglu	200/—	—	180	1	(1)	1, 2, 17
Cervus canadensis	320/225	1103/1103	255 (210–262)	1 (1–2)	1	2, 4, 6, 7, 17
Odocoileus hemionus	110/—	364/364	203 (199–210)	2 (1–3)	1	1, 2, 4, 6, 7, 17
Odocoileus virginianus	90/—	364/364	200 (195–270)	2 (1–4)	1	2, 4, 6, 7
Mazama americana	18/—	—	—	1 (1–2)	—	1, 4, 8
Hippocamelus antisensis	55/—	—	270	1 (1–2)	(1)	8, 17
Blastocerus dichotomus	90/—	—	270	1	(1)	2, 8
Alces alces	600/410	729/729	245 (226–250)	1 (1–3)	1	1, 2, 6, 7, 17
Rangifer tarandus	140/120	703/728	217 (210–240)	1 (1–2)	1	1, 2, 6, 7, 17
Hydropotes inermis	12/10	—	—	2 (1–7)	—	2, 5, 17
Capreolus capreolus	23/—	—	150*	2 (1–3)	(1)	2, 5, 17
Giraffidae						
Okapia johnstoni	220/220	1464/1464	436 (250–446)	1 (1–2)	½–1/1.25	2, 5, 8
Giraffa camelopardalis	1500/800	1617/1617	435 (420–468)	1 (1–2)	½	1, 2, 6, 8, 9
Antilocapridae						
Antilocapra americana	65/43	709/728	240 (230–240)	2 (1–3)	1	1, 2, 4, 6, 7
Bovidae						
Tragelaphus angasi	120/—	—	—	1	—	1, 2, 8
Tragelaphus buxtoni	218/—	—	—	1	—	8
Tragelaphus spekei	125/—	—	252 (245–258)	1	—	1, 8
Tragelaphus scriptus	40/—	660/728	180 (178–225)	—	—	1, 2, 5, 9
Tragelaphus imberbis	104/65	—	—	1	—	1, 2, 19
Tragelaphus strepsiceros	240/—	728/728	214 (210–240)	1	—	1, 2, 5, 8
Taurotragus oryx	815/—	1150/1456	263 (255–270)	1 (1–2)	—	1, 2, 5, 8, 9

Table 18.4. Body Weight and Reproduction Data for Selected Mammal Species

(continued)

	(1) Body Weight[a] (kg)	(2) Age at First Parturition[b] (days)	(3) Gestation[c] (days)	(4) Mean Litter Size[d]	(5) Litter Frequency[e] (per year)	(6) Sources
Taurotragus derbianus	900/—	1352/1456	260 (250–270)	1	—	1, 8
Boocercus euryceros	220/160	—	—	1	—	2, 8
Boselaphus tragocamelus	165/135	—	245 (245–247)	1 (1–2)	1–1.25	1, 2, 5, 8
Tetracerus quadricornis	19/—	—	183	2 (1–3)	—	2, 8
Bubalus bubalis	770/—	1053/1092	315 (287–340)	1 (1–2)	1	1, 2, 5, 6, 8, 20
Bos gaurus	800/680	—	270	1 (1–2)	1	1, 2
Bos grunniens	525/—	986/1093	258 (255–300)	1	1	1, 2, 5, 6, 8
Syncerus caffer	600/500	1672/1820	340 (330–345)	1	1	2, 6, 8, 9, 20
Bison bison	800/450	1003/1092	275 (270–285)	1 (1–2)	1	1, 2, 3, 5, 6, 7, 21, 22, 23
Bison bonasus	800/—	1003/1092	264 (254–276)	1	1	1, 2, 5, 20, 24
Cephalophus sylvicultor	55/—	—	120	1	—	1, 8
Sylvicapra grimmia	16/13	—	—	1 (1–2)	—	2, 5, 8, 9
Kobus ellipsiprymnus	250/—	—	240	1	—	1, 2
Kobus defasa	160/—	980/1092	250 (243–257)	1	—	1, 2, 9
Kobus leche	70/—	775/775	219 (210–240)	1	—	1, 2, 9
Hippotragus equinus	230/—	—	270 (210–300)	1	—	1, 2, 8
Hippotragus niger	240/—	—	270 (270–281)	1	—	2, 5, 8
Oryx gazella	200/—	—	—	1	—	2, 5
Addax nasomaculatus	120/—	—	315 (300–330)	1	—	1, 2, 8
Damaliscus dorcas	125/—	—	260 (225–300)	1 (1–2)	—	1, 2, 5, 8
Alcelaphus buselaphus	170/125	690/728	220 (214–242)	1	—	1, 2, 5, 8, 9, 25
Connochaetes gnou	250/—	709/728	255 (240–270)	1	—	2, 8
Connochaetes taurinus	250/—	709/728	255 (234–270)	1 (1–2)	—	2, 5, 6, 8, 9, 29
Oreotragus oreotragus	14/—	—	214	1	—	1, 2, 8
Ourebia ourebi	18/—	—	210	1	—	1, 5, 8, 9
Raphicerus campestris	11/—	—	210	1 (1–2)	—	1, 8
Neotragus moschatus	8½/—	—	—	1	—	1, 8
Madoqua kirki	4½/—	355/364	174 (170–180)	1 (1–2)	2	1, 2, 8, 9
Antilope cervicapra	37/—	—	165 (150–180)	1 (1–2)	2	1, 2, 5, 8
Aepyceros melampus	55/—	728/728	191 (171–210)	1 (1–2)	—	1, 2, 5, 8, 9
Litocranius walleri	47/—	—	191	1	—	2, 8
Gazella thomsoni	15/—	550/728	191	1 (1–2)	2	1, 5, 9

Species	Body weight (M/F)[a]	Age at first parturition[b]	Gestation (days)[c]	Litter size[d]	Litters/yr[e]	Sources
Gazella granti	40/—	—	—	1 (1–2)	—	1, 9
Antidorcas marsupialis	34/—	—	170	1 (1–2)	—	2, 5, 8
Pantholops hodgsoni	43/—	—	180	1	—	1, 8, 26
Saiga tatarica	52/40	370/370	145 (145–150)	2 (1–3)	1	1, 2, 5, 8, 26
Nemorhaedus goral	31/31	—	180	1	1	5, 26
Capricornis sumatrensis	110/91	—	225 (210–240)	1 (1–2)	1	1, 2, 5, 8, 26
Oreamnos americanus	110/85	906/1092	178 (176–180)	1 (1–2)	1	2, 4, 6, 7, 26
Rupicapra rupicapra	35/22	711/729	165 (150–210)	1 (1–3)	1	1, 2, 6, 26
Budorcas taxicolor	250/—	—	210 (200–220)	1	1	2, 5, 8, 26
Ovibos moschatus	350/270	1338/1456	246 (240–255)	1	½	1, 2, 4, 5, 6, 8, 26
Hemitragus jemlahicus	90/60	665/728	180 (180–242)	1 (1–2)	1	2, 5, 8, 26
Capra hircus	75/—	—	165 (150–180)	1 (1–2)	1	2, 26
Capra ibex ibex	85/32	1075/1092	165 (150–180)	1 (1–2)	½	1, 2, 5, 6, 26
Capra ibex nubiana	52/—	—	150	1 (1–2)	—	26
Capra ibex sibirica	74/50	660/728	175 (170–180)	1 (1–2)	(1)	1, 2, 26
Capra caucasica	72/—	—	155 (150–160)	1 (1–2)	(1)	1, 26
Capra falconeri	90/37	660/728	154 (153–155)	1 (1–2)	—	1, 2, 5, 26
Pseudois nayaur	64/39	—	160	1 (1–2)	—	1, 2, 8, 26
Ammotragus lervia	90/52	706/728	160 (154–161)	2 (1–2)	—	2, 5, 8, 26
Ovis musimon	40/—	706/728	153 (148–159)	1 (1–3)	—	2, 26, 27
Ovis orientalis	41/—	—	165 (150–180)	2 (1–2)	—	1, 2, 26, 27, 28
Ovis ammon	140/64	—	150	1 (1–3)	—	1, 2, 5, 26, 27
Ovis canadensis	113/80	1075/1092	165 (150–180)	1 (1–2)	1	1, 2, 4, 6, 7, 26, 27
Ovis dalli	86/—	1075/1092	180	1 (1–2)	1	2, 4, 7, 26, 27

SOURCES:: (1) Asdell 1964; (2) Crandall 1964; (3) Ewer 1973; (4) Hall and Kelson 1959; (5) Morris 1965; (6) Palmer 1957; (7) Rue 1968; (8) Walker 1968; (9) Western 1979; (10) Martin and Guilday 1967; (11) Laws et al. 1975; (12) Groves 1974; (13) Hall-Martin and Penzhorn 1977; (14) Owen-Smith 1974; (15) Bullet 1975; (16) Franklin 1974; (17) Whitehead 1972; (18) Green 1978; (19) Leuthold 1974; (20) Sinclair 1977; (21) Lott 1974; (22) McHugh 1972; (23) Meagher 1973; (24) Krasinski and Raczynski 1967; (25) Goslin 1974; (26) Schaller 1977; (27) Geist 1971a; (28) Schaller and Mirza 1974; (29) Estes 1966; (30) Rudnai 1979.

[a]Male weights are left of the slash, female weights to the right. The figures given here represent approximate mean body weights, as determined from the information provided in the sources. If sources provided mean body weights, these values were normally accepted. If ranges only were provided, the median value was normally selected. If maximum body weights were provided, this figure was reduced by 10 percent.

[b]Values to the left of the slash are minimum ages of first potential parturition, determined by adding the age at sexual maturity and the gestation period. Values to the right of the slash are minimum ages of potential first parturition in reproductive patterns adjusted to seasonal peaks.

[c]The primary value approximates the mean gestation period for the species. Values in parentheses indicate the range of days reported.

[d]The primary value is the reported actual mean litter size. Values in parentheses indicate the range of litter sizes reported.

[e]This is the reported actual frequency of litters, not the biologic potential. Assumed or implied frequencies are in parentheses. Most blanks are probably frequencies of one per year.

[f]These values are known only for domestic or feral populations and may not be accurate for wild populations.

[g]"Very short" and needs verification (Crandall 1964).

*Delayed implantation is reported for these species/subspecies.

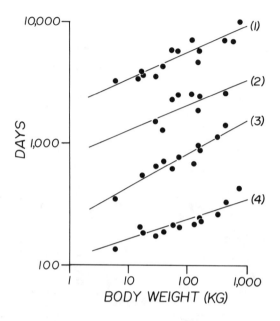

Figure 18.6. Relations between body weight and four other life history parameters in seventeen species of artiodactyls illustrate the scaled nature of life history character complexes. The life history characters shown here are (1) lifespan (y = 2.66X$^{0.22}$), (2) life expectancy at birth (y = 2.31X$^{0.20}$), (3) age at first parturition (y = 1.55X$^{0.27}$), and (4) gestation period (y = 1.60X$^{0.16}$). (After Western 1979)

Smaller-bodied mammals normally exhibit biological and ecological characteristics relatively different from those of larger-bodied mammals. These include shorter potential and actual life spans, smaller territorial requirements, lower absolute and greater relative growth rates, earlier age at sexual maturity and first parturition, shorter gestation time, and lower weight of young at birth. Reproduction rates are normally greater than for larger mammals. Smaller body size often correlates with reduced competitive ability, relatively greater mobility, greater susceptibility to predation, less absolute resource demand, greater sociality, and more specific feeding habits. Selection favors higher reproductive potential instead of overall competitive efficiency of individuals as the prime strategy to maintain population levels, which usually fluctuate and normally increase following periodic catastrophic reductions or colonization of vacant, relatively short-lived habitats.

The characteristics generally exhibited by large and small mammals allow these two groups to be related to the r- and K-selection continuum. Large mammals, exhibiting well-developed competitive qualities that maintain a relatively stable population level at or near a stable environment's carrying capacity, are considered good K-strategists; smaller-bodied mammals, r-strategists, exhibit reproductive qualities that produce a relatively large number of young in a short time to quickly establish or rebuild a depleted population. I suggest that the late Wisconsin megafauna was well adapted to a selection regime in which resources were minimal and intraspecific and interspecific competition were relatively strong. The most fit individuals were those that could successfully seize

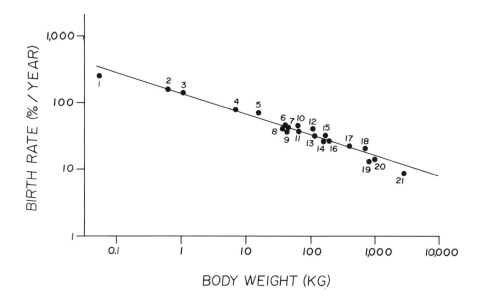

Figure 18.7. Relations between body weight and birth rate in twenty-one species of African mammals, including: (1) Elephant shrew *(Elephantulus rufescens)* (2) Elephant shrew *(Rhynchocyon chrysopygus)* (3) Mongoose *(Mungos mungo)* (4) Dik-dik *(Madoqua kirki)* (5) Thomson's gazelle *(Gazella thomsoni)* (6) Grant's gazelle *(Gazella granti)* (7) Warthog *(Phacochoerus aethiopicus)* (8) Impala *Aepyceros melampus)* (9) Uganda kob *(Adenota kob)* (10) Hyena *(Crocuta crocuta)* (11) Lechwe *(Kobus leche)* (12) Lion *(Panthera leo)* (13) Kongoni *(Alcelaphus buselaphus)* (14) Waterbuck *(Kobus ellipsiprymnus)* (15) Wildebeest *(Connochaetes taurinus)* (16) Zebra *(Equus burchelli)* (17) Buffalo *(Syncerus caffer)* (18) Giraffe *(Giraffa camelopardalis)* (19) Black rhinoceros *(Diceros bicornis)* (20) Hippopotamus *(Hippopotamus amphibius)* (21) Elephant *(Loxodonta africana)*. The regression equation for this line is $y = 2.096X^{-0.327}$. (After Western 1979, 1980)

and control sufficient resources to support themselves and assure the success of their few progeny. (Gould 1966, 1977; Leigh 1975; McCullough 1979; Pianka 1978; Stanley 1973; Van Valen 1973; Western 1979, 1980; and E. Wilson 1975 are useful references for r and K selection and scaled relationships between body size and other life history characteristics.) As the late Wisconsin climate gave way to the Holocene climate—a dramatic event rapid in geologic time *but slow in animal generation time*—the intensity of selection should have relaxed. A relatively K-type regime would have been maintained as large mammal species' populations expanded to keep up with the increased carrying capacity, had not a destabilizing process appeared in the selection regime.

Human hunting, whether a new or substantially improved process, became an element of substantial importance in the selection regime. Humans hunted several species of large game during the very late Wisconsin, using tools to magnify their otherwise limited physical potential. In most predator-prey systems, a positive correlation appears to exist between predator and largest-prey body size, a relationship based ultimately upon the relative size and strength of predator and prey (cf. Hespenheide 1975, Rosenzweig 1966). Considering body size and physical strength only, human hunters should have been limited to considerably smaller prey than they are known to have killed. Tools, however, allowed them to overcome this limitation and to prey not only upon juveniles, aged, or infirmed individuals (groups usually culled by predators) but also healthy individual adults and entire herds of bison and possibly other herding prey. Hunting efficiency suggests that individuals or small populations of large-bodied, relatively sedentary, fearless game animals, adapted to the widespread forested or woodland environments of the late Wisconsin, would have been optimal prey for dispersed small bands of early human predators. Later, as faunal diversity decreased but biomass of surviving taxa increased in the newly opened habitats and niches, cooperative hunting of herds by larger bands—at a time when the human populations and their resource needs had grown—would have been equally optimal.

Not only were humans regularly imposing catastrophic losses upon megafaunal populations (Davis and Wilson 1978, Frison 1978, Hillerud 1970, Johnson 1974, McDonald 1981, Martin 1967a, Wormington 1957), thereby imposing heightened reproductive demands on species with low reproductive potentials, but they were also themselves functioning as r-strategists (cf. Martin 1973). Paleo-Indians of this period probably experienced low intraspecific competition relative to the resources available, and they had access to ample food resources, evidenced by their often excessive and wasteful harvest of some large mammals. They probably had high recruitment rates, permitted by a high quality diet. Initially, they probably lacked strong territorial attachment and floated from resource island to resource island, more easily taking the larger-bodied game populations. The large mammals were more sedentary, territorial, dispersed, and individualistic, qualities rendering them vulnerable to destructive hunting pressures that spread faster than evolutionary adaptive responses. As resources dwindled and competition intensified for what remained, Paleo-Indians probably became more territorial, occupying more productive territories and defending them against encroachment of other populations. Increasing territorialism intensified hunting pressure against remaining large-bodied mammal populations. As these were systematically eliminated by hunters well acquainted with their territory, more r-strategist game biomass appeared within the effectively occupied territories or, more likely, in the buffer zones separating territories of two or more groups of hunting-based human populations (cf. Hickerson 1965).

The practical value of a greater energy return:energy investment ratio in predator-prey systems (Rozenweig 1966) is sufficient reason to suspect that early human hunters preferentially hunted larger game animals, thereby imposing a size bias against the larger herbivore biomass. These hunters would not have passed up smaller game, but, if available, they probably preferred larger game. As the human population spread across the continent, the consequences of hunting practices also spread. Hunting could have brought about the demise of larger-bodied taxa by its selection against the low reproductive rates characteristic of these animals (figs. 18.6–18.9; Table 18.4). Radiocarbon-dated occurrences of several large-bodied herbivore genera indicate that several taxa hunted by Paleo-Indians survived the late Wisconsin maximum and several centuries of human hunting before becoming extinct (fig. 18.10). They were subject, then, to prolonged and apparently continuous hunting pressure. It is useful to relate to figures 18.8 and 18.9 the fact that North America lost all of its Proboscidea, all of its Perissodactyla except the Central American tapir, and many of its larger Artiodactyla and Carnivora. Normally, the extinct taxa were larger than their nearest surviving relative, so the age at first parturition and length of gestation period for extinct forms were ostensibly equal to or greater than those characteristic of surviving genera (figs. 18.8 and 18.9).

Continued human hunting of large-bodied mammals probably had the general effect of lowering the average age of the species' populations (Bombin 1980, Edwards 1967, Krantz 1970). Age-indiscriminant culling had the immediate effect of removing individuals from all age classes while stimulating increased reproduction within the fertile age class, thus accelerating the actual birth rates. In larger-bodied animals, however, with more time required for young females to reach sexual maturity, continued general removal of individuals from all age classes had the net effects of rapidly reducing the population size and lowering its age structure. The threshold of population viability was crossed when the population became so young or so fragmented as to not be able to maintain itself; total collapse was then imminent. The rate of demise would have been even faster if hunters had favored juveniles or females, which was probably the case.

The impact of human hunting on the viability of herbivore populations was compounded by other environmental factors, such as variable and seasonal environments and intensified predation by carnivores. Where critical environmental variables periodically fluctuated beyond the tolerance limits of various individuals (e.g. low precipitation in the desiccating Southwest), reproductive success and survivorship of susceptible young individuals would, as expected, be reduced. This in turn, would reduce the potential viability of the population. Strongly seasonal environments, which *earlier had been occupied successfully* by genera which became extinct, regulated the frequency of births and/or survival of young to a greater extent than did less seasonal environments. Fertile females in species with a gestation period of less than twelve months could theoretically have given birth once every year. The likelihood of annual births probably decreased as the twelve-month gestation period was approached, and biennial births, such as Groves (1974) reports for some equids, were probably more typical. Species with gestation periods greater than twelve months were even more susceptible to the regulating influence of seasonality, which would keep actual birth and recruitment rates below the species' biological potential (Table 18.4, column 2). The downward shift in age structure of surviving populations should have benefitted some of the carnivorous predators, since young individuals constitute one important class of prey, but only if the age reduction was a function of absolutely increased births. If the age structure was reduced because of decreased life expectancy independent of increased births, however, the carnivores would not have benefitted. Krantz (1970) points out that human hunting could have reduced the number of ill and aged individuals in prey populations, and that this, in turn, could have reduced predator survival through the winter season when these adults would normally be culled. With fewer total individuals and relatively fewer old and infirmed adults available as prey, however, intensified predation on juveniles would likely occur, thereby hastening the demise of the prey species.

Reduction of herbivore populations was probably the major immediate cause of most predator extinctions, although herbivore behavioral changes probably also contributed. As the number of herbivores was reduced, competition among predators for remaining resources presumably increased. Smaller or more cursorial predators were more likely able to obtain ample food resources to support and propagate themselves than were larger or more sedentary predators, and thus the larger, heavier built ones disappeared. Increased cursoriality among surviving herbivores also selected against the bulkier and stealth-hunting predators.

Inbreeding probably intensified as breeding populations of herbivores and carnivores became smaller. Numerous physiological and anatomical anomalies that could have greatly reduced the fitness of affected populations are known or suspected to result from inbreeding, including reduced fertility, increased juvenile mortality, greater incidence of stillborn young, irregular estrous cycles, and skeletal and musculature abnormalities (Crandall 1964, Hillman and Carpenter 1980, Miller 1979, Ralls et al. 1979, E. Wilson 1975). An unusually high frequency of skeletal abnormalities has been found in *Bison* from 11,000 to 9000 yr B.P., a period when they were being actively hunted over a large area and were conceivably reduced to dispersed local populations, at least in those areas of most intense hunting (McDonald 1981, Wilson, 1974).

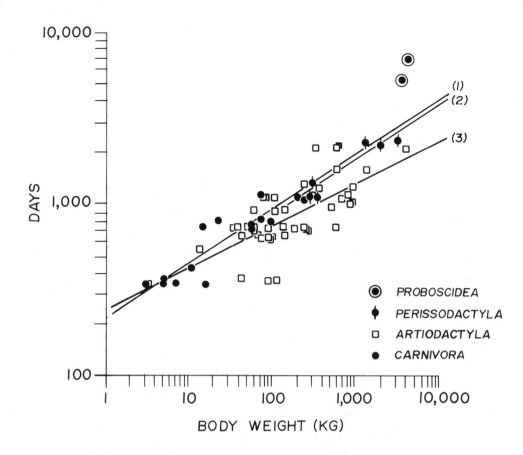

Figure 18.8. Relationship between body weight and age at first parturition among four orders containing large-bodied mammals. Data are from Table 18.4. Data for species with delayed implantation and those known only for domestic or feral populations have been omitted. Regression equations for the orders are: (1) Perissodactyla ($n = 2.29X^{0.319}$, $r = .930$); (2) Artiodactyla ($n = 44$, $y = 2.41X^{0.238}$, $r = .692$); (3) Carnivora ($n = 16$, $y = 2.40X^{0.285}$, $r = .910$).

Table 18.5. Reproductive Potentials and Potential Productivity Among Selected Large Mammal Genera

(1) Genus	(2) ♀ Age @ Sexual Maturity	(3) Gestation Period (days)	(4) Age @ First Parturition	(5) Birth Interval (yr)	(6) Potential # Young/Year	(7) Age @ Decline of Fertility
1. *Elephas*	18 yr.	640	19 yr. 9 mos.	4	0.25	>25*
2. *Elephas*	8 yr.	640	9 yr. 9 mos.	4	0.25	>25*
3. *Equus*	2 yr.	330–360	3 yr.	1	1.0	15
4. *Equus*	2 yr.	330–360	3 yr.	1.5	0.67	15
5. *Equus*	2 yr.	330–360	3 yr.	2	0.5	15
6. *Bison*	2 yr.	275	3 yr.	1	1.0	15
7. *Odocoileus*	5 mos.	200	1 yr.	1	1.8	10

Notes: *Elephas* case 1 is probably more realistic than case 2, which approximates the minimum potential reproductive age reported (Asdell 1964, Crandall 1964, Walker 1968). *Equus* cases 4 and 5 are probably more realistic than case 3 Groves 1974, S. L. Woodward pers. comm. In the reproductive model used to illustrate potential lifetime productivity and descendency, the first birth was considered female, with subsequent births alternate sexes, in all cases except case 7, in which the opposite was true. This permitted maximum productivity over the 25-year model period. The first birth in

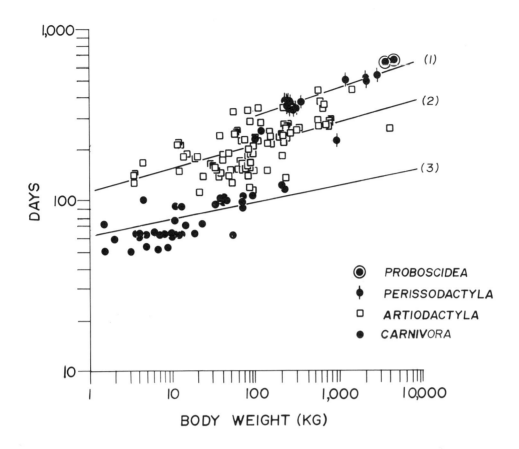

Figure 18.9. Relationship between body weight and gestation period among four orders containing large-bodied mammals. Data are from Table 18.4. Data for species with delayed implantation and those known only for domestic or feral populations have been omitted. Regression equations for the orders are: (1) Perissodactyla ($n = 13$, $y = 2.28X^{0.112}$, $r = .480$); (2) Artiodactyla ($n = 90$, $y = 2.07X^{0.123}$, $r = .590$); (3) Carnivora ($n = 40$, $y = 1.68X^{0.173}$, $r = .720$).

(8) # Young/ 25 Years/♀	(9) Potential # Descendants/25 yrs/♀		(10) Body Weight (kg) ♂ / ♀	(11) Potential Biomass of Descendants (kg)	(12) Biomass Relative to Case 1 Elephas
2	1 ♂	1 ♀	3650/2500	6,150	1.0
4	2 ♂	4 ♀	3650/2500	17,300	2.8
13	575 ♂	786 ♀	300/230	353,280	57.4
9	119 ♂	243 ♀	300/230	91,590	14.9
7	77 ♂	121 ♀	300/230	50,930	8.3
13	575 ♂	786 ♀	800/450	813,700	132.3
16	30,097 ♂	44,405 ♀	90/72	5,905,890	960.3

Odocoileus is usually a single fawn and subsequent births are frequently or usually twins; *Elephas, Equus,* and *Bison* normally give birth only to a single young at a time (Asdell 1964, Crandall 1964, Groves 1974, Morris 1965, Rue 1968).

*I do not have fecundity data for *Elephas,* but Laws et al. (1975) show a sharp decline in *Loxodonta* female fertility at ages >50–55 years.

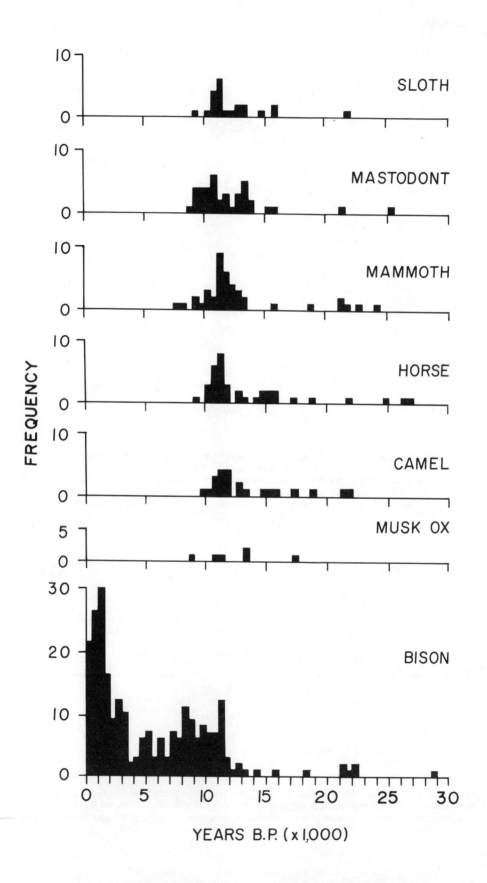

Some aspects of the relative fitness of different reproductive patterns and some implications of these fitness differentials to human hunting populations are modeled in Table 18.5, using data from modern genera. An arbitrarily selected twenty-five year period is used to model population growth and biomass productivity. The differences in growth and productivity among the four genera used would only increase with a longer model period. *Elephas,* a large, long-lived, relative K-strategist, can usefully be compared to the proboscidean (and perhaps edentate) genera of the late Wisconsin fauna of North America. An average and a minimum birth-to-first-parturition period is given for *Elephas* to illustrate the magnitude of potential differences in secondary productivity associated with this critical biological character. *Equus* and *Bison* represent intermediate-size herbivores with substantially shorter birth-to-first-parturition periods than *Elephas,* and modest but significantly different gestation periods of eleven to twelve months and nine months, respectively. *Odocoileus* represents an even smaller-bodied, more r-strategist herbivore with a shorter birth-to-first-parturition period and a significantly higher potential reproductive rate afforded by its twinning tendency.

Theoretically, *Odocoileus* could annually produce more than seven times as many young as *Elephas* (Table 18.5, column 6). A female *Elephas,* over the twenty-five years following her birth, could potentially produce some two to four young while yielding only two to six descendants. *Bison* and *Equus* females could both produce offspring at the same rate if they gave birth annually, but if an *Equus* female produced only at eighteen- or twenty-four-month intervals, which is typical in seasonal environments (Groves 1974, Woodward 1979), then over the twenty-five-year model period its number of offspring could be reduced to between one-half and one-sixth that of a female *Bison,* which typically produces annually (Table 18.5, column 9). The reproductive fitness boundary was lowered under the new selection regime to a level where it passed between *Equus* and *Bison.* This accounts for the heretofore unexplained anomaly of *Equus* (and camelid) extinction and *Bison* survival.

The generic simplification of the North American fauna modeled here resulted in an increase in the potential game biomass and an increase in the rate of biomass formation. A single female deer (*Odocoileus*) potentially could generate in twenty-five years, through her descendants, almost 1,000 times the biomass that a single female *Elephas* could generate (Table 18.5, columns 11 and 12), and do it more energy-efficiently (Petrides et al. 1968). *Bison,* too, would potentially be able to produce anywhere from about 2.3 to 16 times the biomass of *Equus,* depending on the *Equus* birth interval and assuming that *Bison* produced annually. The increased human energy expenditures needed to kill a larger number of smaller game animals to equal previous energy acquisitions could have been offset by the greater absolute abundance of individual game animals. Some game, for example deer, was spread more evenly in space, which increased the probability of repeated hunting successes; other game, such as *Bison,* was unevenly distributed in space but formed large herds containing abundant biomass.

Figure 18.10. Frequencies of radiocarbon-dated occurrences of selected mega-faunal genera present in the late Wisconsin fauna of North America. This histogram contains all acceptable radiocarbon measurements dating the identified taxa. Several dates associated with the taxa were rejected as unacceptable, usually because their validity was questioned in the source, they did not clearly or directly date megafauna, or they were rendered suspect by other information. Occasionally, a date questioned in its source was accepted if other information supported the validity of the date. All measurements presented here are based upon the Libby half-life of 5,568 ± 30 years and have not been calibrated to true calendar years. Multiple dates for a single horizon/stratum have been arithmetically averaged. (Data are from McDonald 1981)

Some Exceptions to the General Pattern

Some irregularity should be anticipated and tolerated along the extinction-survival boundary—that is, some taxa which were below the fit-unfit boundary (i.e. apparently fit) for their orders conceivably could have become extinct; other taxa in their orders that were nearer, on, or above the boundary, (i.e. apparently unfit) could have survived. This could have resulted from rareness, ecological or biological specializations, nonadaptability, chance, etc. Five taxa which were situated below the fit-unfit boundary but nonetheless became extinct were *Breameryx, Capromeryx, Stockoceros, Oreamnos harringtoni,* and *Acinonyx.* The first four names are exceptional in that they are smaller than their nearest surviving relatives. *Breameryx, Capromeryx,* and *Stockoceros* were antilocaprids distributed widely through the Southwest and, based upon modern ecological analogues, could have been foraging specialists living either solitarily or in small groups (Dubost 1979, Eisenberg and McKay 1974, Estes 1974). The combination of a desiccating environment, including the reduction of cover as well as preferred food and water resources, increased predation by carnivores (as their larger prey disappeared, Janzen 1983), and increased competition from other herbivores could have been more instrumental in their extinction than human hunting directly. *Oreamnos harringtoni,* a small mountain goat of localized distribution in the Southwest (Martin and Guilday 1967, Martin and Mead 1980, Mead 1980), could represent a late Wisconsin clinal or insular dwarfing adaptation to resource "islands" that were developing or disappearing in the desiccating broken terrain this species inhabited. If *O. harringtoni* populations did become isolated in patches of suitable habitat that were being fragmented, reduced in area, or extinguished as they shifted to higher elevations on plateaus or mountain peaks (cf. Hoffmann and Taber 1967, Wells and Jorgensen 1964), these populations could have become increasingly vulnerable to extinction by any number of processes (loss of habitat, competitive exclusion, human hunting, carnivore predation, inbreeding, etc.). *Acinonyx,* the cheetah, was uncommon in the late Wisconsin fauna and this rareness, perhaps a consequence of the scarcity of open habitat during the late Wisconsin, rendered it especially vulnerable to extinction, even though open habitat more suited to its continued survival was developing and smaller ungulate prey was becoming more abundant.

Tapirus (with a gestation period > one year) and *Ovibos* (with r < one/fertile female/year, at least at high latitudes) are survivors situated above the fitness boundary. *Tapirus* survived only in a tropical environment that provided abundant cover. *Ovibos* survived in high latitude North America where relative inaccessibility and a harsh arctic environment minimized the impact of human hunting.

Summary

The maximum limiting intensity of reduced continental land area, cooler air temperatures, simplicity of habitats, limited patchiness, lowest total primary productivity, and least availability of food resources were attained in late Quaternary North America about 20,000 to 18,000 yr B.P. Megafaunal biomass was probably lowest at that time while intraspecific and interspecific competition was probably greatest. The limiting intensity of most selective forces was relaxed after 18,000 B.P., and the associated environmental changes should not have produced the cluster of extinctions which in fact occurred.

The appearance of human hunting as a new (or substantially expanded or improved) ecological process represented the only new, important, and ubiquitous factor introduced into the North American selection regime during the very late Wisconsin. As a selective factor hunting was singularly sufficient to change the entire continental megafaunal selection regime from a competitively efficient K-type to a population recovery r-type. Human hunters, behaving initially as relative r-strategists, then becoming more K-strategists as competition for dwindling resources increased, were, directly or indirectly, more responsible than any other selective force in rendering some 67 percent of North America's large mammal genera extinct.

Selection against the large-bodied mammals operated most strongly on their low reproductive rates. Continuous hunting over several centuries reduced the age structure and biomass of species' populations until the threshold of continued population viability was crossed and collapse ensued.

Measures of fitness in the new selection regime included having relatively small body size, a birth-to-first-parturition interval of three years or less if a herbivore or five years or less if a carnivore, a gestation period of less than twelve months, an annual biotic potential equal to or greater than one per fertile female, a ruminant digestive system if an herbivore, and relatively more cursorial and less territorial habits if herbivore or carnivore. The continent's large mammal carrying capacity actually increased, but continued harvesting of large herbivores by human hunters limited their standing crop to less than their biological potential.

Acknowledgments

The ideas and information presented in this paper have benefitted from comments by Elaine Anderson, Richard Kiltie, Peter J. Mehringer, Jr., Charles H. Smith, Jr., and Susan Woodward. Portions of this research were supported by a 1979 Association of American Geographers research grant.

References

Adams, D. B. 1979. The cheetah: native American. *Science* 205:1155–1158.

Adovasio, J. M., J. D. Gunn, J. Donahue, R. Stuckenrath, J. Guilday, and K. Lord. 1978. Meadowcroft rockshelter. In *Early man in America from a circum-Pacific perspective,* ed. A. L. Bryan, pp. 140–180. Univ. Alberta, Dept. Anthropol., Occ. Papers, No. 1.

Aigner, J. S. 1972. Relative dating of north Chinese faunal and cultural complexes. *Arctic Anthropol.* 9:36–79.

Asdell, S. A. 1964. *Patterns of mammalian reproduction.* Ithaca: Cornell Univ. Press. 2nd ed.

Axelrod, D. I. 1967. Quaternary extinctions of large mammals. *Univ. Calif. Publ. Geol. Sci.,* 74.

Bada, J. L., R. A. Schroeder, and G. F. Carter. 1974. New evidence for the antiquity of man in North America deduced from aspartic acid racemization. *Science* 184:791–793.

Bombin, M. 1980. Evolution, man, and Pleistocene megafauna extinctions in North America. *Abstracts, sixth biennial meeting, American Quaternary Association,* pp. 35–36.

Borchert, J. R. 1971. The dust bowl in the 1930s. *Annals. Assoc. Amer. Geog.* 61:1–22.

Borowski, S. and S. Kossak, 1972. The natural food preferences of the European bison in seasons free of snow cover. *Acta Theriol.* 17:151–169.

Bryson, R. A., D. A. Baerreis, and W. M. Wendland. 1970. The character of late glacial and post-glacial climatic change. In *Pleistocene and recent environments of the central Great Plains,* ed. W. Dort, Jr. and J. K. Jones, Jr., pp. 53–74. Univ. Kansas, Dept. Geol., Spec. Publ. 3.

Bulliet, R. W. 1975. *The camel and the wheel.* Cambridge: Harvard Univ. Press.

Canby, T. Y. 1979. The search for the first Americans. *Natl. Geog. Mag.* 156:330–363.

CLIMAP Project Members. 1976. The surface of the ice age earth. *Science* 191:1131–1137.

Coe, M. J. 1980. The role of modern ecological studies in the reconstruction of paleoenvironments in sub-Saharan Africa. In *Fossils in the making: vertebrate taphonomy and paleoecology,* ed. A. K. Behrensmeyer and A. P. Hill, pp. 55–67. Chicago: Univ. Chicago Press.

Coe, M. J., D. H. Cumming, and J. Phillipson. 1976. Biomass and production of large African herbivores in relation to rainfall and primary production. *Oecologia* 22:341–354.

Crandall, L. S. 1964. *Management of wild mammals in captivity.* Chicago: Univ. Chicago Press.

Crow, A. B. 1978. Fire ecology and fire management in the forests of the lower Mississippi River valley. *Geoscience and Man* 19:75–80.

Darlington, P. J. 1957. *Zoogeography: the geographic distribution of animals.* New York: Wiley and Sons.

Davis, L. B. and M. Wilson (eds.). 1978. *Bison procurement and utilization: a symposium.* Memoir 14, *Plains Anthropol.*

Davis, M. B. 1976. Pleistocene biogeography of temperate deciduous forests. *Geoscience and Man* 13:13–26.

Delcourt, P. A. and H. R. Delcourt. 1981. Vegetation maps for eastern North America: 40,000 yr B.P. to the present. In *Geobotany II,* ed. R. C. Romans, pp. 123–165. New York: Plenum Publ. Corp.

Dillehay, T. D. 1974. Late Quaternary bison population changes on the southern plains. *Plains Anthropol.* 19:180–196.

Dreimanis, A. 1968. Extinction of mastodons in eastern North America: Testing a new climatic-environmental hypothesis. *Ohio J. Sci.* 68:257–272.

Dubost, G. 1979. The size of African forest artiodactyls as determined by the vegetation structure. *Afr. J. Ecol.* 17:1–18.

Edwards, W. E. 1967. The late-Pleistocene extinction and diminution in size of many mammalian species. In *Pleistocene extinctions: the search for a cause,* ed. P. S. Martin and H. E. Wright, Jr., pp. 141–154. New Haven: Yale Univ. Press.

Eisenberg, J. F. and G. M. McKay. 1974. Comparison of ungulate adaptations in the New World and Old World tropical forests with special reference to Ceylon and the rainforests of Central America. In *The behaviour of ungulates and its relation to management,* ed. V. Geist and F. Walther, pp. 585–602. Intl. Union Cons. Nature and Nat. Res., No. 24.

Eisenberg, J. F. and J. Seidensticker. 1976. Ungulates in southern Asia: a consideration of biomass estimates for selected habitats. *Biol. Cons.* 10:293–308.

Elton, C. S. 1958. *The ecology of invasions by animals and plants.* London: Chapman and Hall.

Emiliani, C. 1971. The amplitude of Pleistocene climatic cycles at low latitudes and the isotopic composition of glacial ice. In *The late Cenozoic glacial ages,* ed. K. K. Turekian, pp. 183–197. New Haven: Yale Univ. Press.

Epenshade, E. B., Jr. 1970. *Goode's world atlas.* Chicago: Rand McNally & Co. 13th ed.

Estes, R. D. 1966. Behaviour and life history of the wildebeest (*Connochaetes taurinus* Burchell). *Nature* 212:999–1000.

———. 1974. Social organization of the African Bovidae. In *The behavior of ungulates and its relation to management,* ed. V. Geist and F. Walther, pp. 166–205. Intl. Union Cons. Nature and Nat. Res., No. 24.

Ewer, R. F. 1973. *The carnivores.* Ithaca: Cornell Univ. Press.

Fairall, N. 1968. The reproductive seasons of some mammals in Kruger National Park. *Zoologica Africana* 3:189–210.

Flerow, C. C. 1967. On the origin of the mammalian fauna of Canada. In *The Bering land bridge,* ed. D. M. Hopkins, pp. 271–280. Stanford: Stanford Univ. Press.

Flessa, K. W. 1975. Area, continental drift and mammalian diversity. *Paleobiology* 1:189–194.

Fox, L. R. and P. A. Morrow. 1981. Specialization: species property or local phenomenon? *Science* 211:887–893.

Franklin, W. L. 1974. The social behaviour of the vicuna. In *The behaviour of ungulates and its relation to management,* ed. V. Geist and F. Walther, pp. 477–487. Intl. Union Cons. Nature and Nat. Res., No. 24.

Fraser, A. F. 1968. *Reproductive behaviour in ungulates.* New York: Academic Press.

Frenzel, B. 1968. The Pleistocene vegetation of northern Eurasia. *Science* 161:637–649.

Fretwell, S. D. 1972. *Populations in a seasonal environment.* Princeton Univ. Monog. Pop. Biol, No. 5.

Frison, G. C. 1978. *Prehistoric hunters of the High Plains.* New York: Academic Press.

Gates, W. L. 1976. Modeling the ice-age climate. *Science* 191:1138–1144.

Gebczynska, Z. and M. Krasinska. 1972. Food preferences and requirements of the European bison. *Acta Theriol.* 17:105–117.

Geist, V. 1971. *Mountain sheep: a study in behavior and evolution.* Chicago: Univ. Chicago Press.

Gosling, L. M. 1974. The social behaviour of Coke's hartebeest (*Alcelaphus buselaphus cokei*). In *The behaviour of ungulates and its relation to management,* ed. V. Geist and F.

Walther, pp. 488–511. Intl. Union Cons. Nature and Nat. Res., No. 24.

Gould, S. J. 1966. Allometry and size in ontogeny and phylogeny. *Biol. Rev.* 41:587–640.

———. 1977. *Ontogeny and phylogeny.* Cambridge: Belknap Press.

Graham, R. W. 1979. Paleoclimates and late Pleistocene faunal provinces in North America. In *Pre-Llano cultures of the Americas: paradoxes and possibilities,* eds. R. L. Humphrey and D. Stanford, pp. 49–69. Washington, D.C.: The Anthropological Society of Washington.

Green, M. J. B. 1978. Himalayan musk deer (*Moschus moschiferus moschiferus*). In *Threatened deer,* (IUCN Deer Specialist Group symposium), pp. 56–64. Intl. Union Cons. Nature and Nat. Res.

Groves, C. P. 1974. *Horses, asses and zebras in the wild.* Hollywood, Fla.: Ralph Curtis Books.

Guilday, J. E. 1967. Differential extinction during late-Pleistocene and recent times. In *Pleistocene extinctions: the search for a cause,* ed. P. S. Martin and H. E. Wright, Jr., pp. 121–140. New Haven: Yale Univ. Press.

Guthrie, R. D. 1968. Paleoecology of the large-mammal community in interior Alaska during the late Pleistocene. *Amer. Midl. Nat.* 79:346–363.

Guthrie, R. D. and J. V. Matthews, Jr. 1971. The Cape Deceit fauna—early Pleistocene mammalian assemblages from the Alaskan Arctic. *J. Quat. Res.* 1:474–510.

Hall, E. R. and K. R. Kelson. 1959. *The mammals of North America.* New York: Ronald Press Co.

Hall-Martin, A. J. and B. L. Penzhorn. 1977. Behaviour and recruitment of translocated black rhinoceros *Diceros bicornis. Koedoe* 20:147–162.

Haney, J. E. 1969. Studies of white-tailed deer. *Research News* 20: No. 2 (August). Ann Arbor: Univ. Mich.

Hansen, R. M. and P. S. Martin. 1973. Ungulate diets in the lower Grand Canyon. *J. Range Mgmt.* 26:380–381.

Harris, A. H. and P. Mundel. 1974. Size reduction in bighorn sheep (*Ovis canadensis*) at the close of the Pleistocene. *J. Mammal.* 55:678–680.

Haynes, C. V. 1967. Carbon-14 dates and early man in the New World. In *Pleistocene extinctions: the search for a cause,* ed. P. S. Martin and H. E. Wright, Jr., pp. 267–286. New Haven: Yale Univ. Press.

Hespenheide, H. A. 1975. Prey characteristics and predator niche width. In *Ecology and evolution of communities,* ed. M. L. Cody and J. M. Diamond, pp. 158–180. Cambridge: Belknap Press.

Hester, J. J. 1967. The agency of man in animal extinctions. In *Pleistocene extinctions: the search for a cause,* ed. P. S. Martin and H. E. Wright, Jr., pp. 169–192. New Haven: Yale Univ. Press.

Hibbard, C. W., C. E. Ray, D. E. Savage, D. W. Taylor, and J. E. Guilday. 1965. Quaternary mammals of North America. In *The Quaternary of the United States,* ed. H. E. Wright, Jr. and D. G. Frey, pp. 509–525. Princeton: Princeton Univ. Press.

Hickerson, H. 1965. The Virginia deer and intertribal buffer zones in the upper Mississippi valley. In *Man, culture, and animals,* ed. A. Leeds and A. P. Vayda, pp. 43–66. Amer. Assoc. Adv. Sci., Publ. no. 78, Washington, D.C.

Hillerud, J. M. 1970. Subfossil high plains bison. Ph.D. Thesis, Univ. Nebraska (Geology).

Hillman, C. N. and J. W. Carpenter. 1980. Masked mustelid. *Nat. Conservancy News* 30; No. 2 (March-April) 20–23.

Hoffmann, R. S. and R. D. Taber. 1967. Origin and history of Holarctic tundra ecosystems, with special reference to their vertebrate faunas. In *Arctic and alpine environments,* ed. H. E. Wright, Jr. and W. H. Osburn, pp. 143–170. Bloomington: Indiana Univ. Press.

Hopkins, D. M. 1967. The Cenozoic history of Beringia—a synthesis. In *The Bering land bridge,* ed. D. M. Hopkins, pp. 451–484. Palo Alto: Stanford Univ. Press.

Irving, W. N. 1978. Pleistocene archaeology in eastern Beringia. In *Early man in America from a circum-Pacific perspective,* ed. A. L. Bryan, pp. 96–101. Univ. Alberta, Dept. Anthropol., Occ. Papers, No. 1.

Janzen, D. 1983. The Pleistocene hunters had help. *Am. Nat.* 121:598–599.

Jelinek, A. J. 1967. Man's role in the extinction of Pleistocene faunas. In *Pleistocene extinctions: the search for a cause,* ed. P. S. Martin and H. E. Wright, Jr., pp. 193–200. New Haven: Yale Univ. Press.

Johnson, D. L. 1977. The late Quaternary climate of coastal California: evidence for an ice age refugium. *J. Quat. Res.* 8:154–179.

Johnson, E. 1974. Zooarchaeology and the Lubbock Lake site. *The Museum J.* 15:107–122.

Jones, G. 1979. *Vegetation productivity.* London: Longman.

Kiltie, R. A. 1982. Intraspecific variation in the mammalian gestation period. *J. Mammal.* 63:646–652.

Kormondy, E. J. 1976. *Concepts of ecology.* Englewood Cliffs: Prentice-Hall Inc. 2nd ed.

Krantz, G. S. 1970. Human activities and megafaunal extinctions. *Amer. Sci.* 58:164–170.

Krasinski, Z. and J. Raczynski. 1967. The reproductive biology of European bison living in reserves and in freedom. *Acta Theriol.* 12:407–444.

Kurtén, B. 1975. A new Pleistocene genus of American mountain deer. *J. Mammal.* 56:507–508.

Kurtén, B. and E. Anderson. 1980. *Pleistocene mammals of North America.* New York: Columbia Univ. Press.

Laws, R. M., I. S. C. Parker, and R. C. B. Johnstone. 1975. *Elephants and their habitats.* Oxford: Clarendon Press.

Laycock, G. 1966. *The alien animals.* Garden City, N.Y.: The Nat. Hist. Press.

Leigh, E. G., Jr. 1975. Population fluctuations, community stability, and environmental variability. In *Ecology and evolution of communities,* ed. M. L. Cody and J. M. Diamond, pp. 51–73. Cambridge: Belknap Press.

Leuthold, W. 1974. Observations on home range and social organization of lesser kudu, *Tragelaphus imberbis* (Blyth, 1869). In *The behaviour of ungulates and its relation to management,* ed. V. Geist and F. Walther, pp. 206–234. Intl. Union Cons. Nature and Nat. Res., No. 24.

Leuthold, W. and B. M. Leuthold. 1975. Temporal patterns of reproduction in ungulates of Tsavo East National Park, Kenya. *E. Afr. Wildl. J.* 13:159–169.

Lewis, H. T. and C. Schweger. 1973. Paleo Indian use of fire during the late Pleistocene: the human factor in environmental change. *Abstracts, ninth congress, INQUA,* p. 210.

Lott, D. F. 1974. Sexual and aggressive behavior of adult male American bison (*Bison bison*). In *The behaviour of ungulates and its relation to management,* ed. V. Geist and F. Walther, pp. 382–394. Intl. Union Cons. Nature and Nat. Res., No. 24.

Lundelius, E. L., Jr. 1967. Late-Pleistocene and Holocene faunal history of central Texas. In *Pleistocene extinctions: the search for a cause,* ed. P. S. Martin and H. E. Wright, Jr., pp. 287–319. New Haven: Yale Univ. Press.

———. 1976. Vertebrate paleontology of the Pleistocene: an overview. *Geoscience and Man* 13:45–59.

MacArthur, R. H. and E. O. Wilson. 1967. *The theory of island biogeography.* Princeton Univ. Monog. Pop. Biol., No. 1.

McCullough, D. R. 1979. *The George Reserve deer herd: population ecology of a K-selected species.* Ann Arbor: Univ. Michigan Press.

McDonald, J. N. 1981. *North American bison: their classification and evolution.* Berkeley and Los Angeles: Univ. California Press.

———. 1982. Adapatively differentiated reproductive activity patterns in feral ass (*Equus asinus*) populations. Association of American Geographers Program Abstracts, p. 75.

McHugh, T. 1972. *The time of the buffalo.* New York: Knopf.

McKnight, T. L. 1958. The feral burro in the United States: distribution and problems. *J. Wildl. Mgmt.* 22:163–178.

———. 1959. The feral horse in Anglo-America. *Geog. Review* 49:506–525.

———. 1964. Feral livestock in Anglo-America. *Univ. Calif. Publ. Geog.* 16.

———. 1971. Australia's buffalo dilemma. *Annals, Assoc. Amer. Geog.* 61:759–773.

———. 1976. Friendly vermin: a survey of feral livestock in Australia. *Univ. Calif. Publ. Geog.* 21.

Martin, L. D. and A. M. Neuner. 1978. The end of the Pleistocene in North America. *Trans. Neb. Acad. Sci.* 6:117–126.

Martin, P. S. 1967a. Pleistocene overkill. *Nat. Hist.* 76: No. 10 (December) 32–38.

———. 1967b. Prehistoric overkill. In *Pleistocene extinctions: the search for a cause,* ed. P. S. Martin and H. E. Wright, Jr., pp. 75–120. New Haven: Yale Univ. Press.

———. 1967c. Preface. In *Pleistocene extinctions: the search for a cause,* ed. P. S. Martin and H. E. Wright, Jr., pp. v–viii. New Haven: Yale Univ. Press.

———. 1973. The discovery of America. *Science* 179:969–974.

Martin, P. S. and J. E. Guilday, 1967. A bestiary for Pleistocene biologists. In *Pleistocene extinctions: the search for a cause,* ed. P. S. Martin and H. E. Wright, Jr., pp. 1–62. New Haven: Yale Univ. Press.

Martin, P. S. and J. I. Mead. 1980. Extinction of Harrington's mountain goat. *Abstracts, Bulletin Ecol. Soc. Amer.* 61:105.

Martin, P. S. and P. J. Mehringer, Jr. 1965. Pleistocene pollen analysis and biogeography of the southwest. In *The Quaternary of the United States,* ed. H. E. Wright, Jr., and D. G. Frey, pp. 433 451. Princeton: Princeton Univ. Press.

Mead, J. I. 1980. The late Quaternary fauna of the Grand Canyon, Arizona. *Abstracts, sixth biennial meeting, American Quaternary Association,* p. 137.

Meagher, M. M. 1973. *The bison of Yellowstone National Park.* Natl. Park Svc. Scientific Monog. Ser., No. 1.

Mehringer, P. J., Jr. 1967. The environment of extinction of the late-Pleistocene megafauna in the arid southwestern United States. In *Pleistocene extinctions: the search for a cause,* ed. P. S. Martin and H. E. Wright, Jr., pp. 247–266. New Haven: Yale Univ. Press.

———. 1977. Great Basin late Quaternary environments and chronology. In *Models and Great Basin prehistory: a symposium.* ed. D. D. Fowler, pp. 113–167. Desert Res. Inst. Publ. in Social Sci., No. 12, Reno.

Mercer, J. H. 1972. The lower boundary of the Holocene. *J. Quat. Res.* 2:15–24.

Miller, R. I. 1979. Conserving the genetic integrity of faunal populations and communities. *Env. Cons.* 6:297–304.

Miller, S. J. and W. Dort, Jr. 1978. Early man at Owl Cave: current investigations at the Wasden site, eastern Snake River Plain, Idaho. In *Early man in America from a circum-Pacific perspective,* ed. A. L. Bryan, pp. 129–139. Univ. Alberta, Dept. Anthropol., Occ. Papers, No. 1.

Morlan, R. E. 1978. Early man in northern Yukon Territory: perspectives as of 1977. In *Early man in America from a circum-Pacific perspective,* ed. A. L. Bryan, pp. 78–95. Univ. Alberta, Dept. Anthropol., Occ. Papers, No. 1.

Morris, D. 1965. *The mammals: a guide to the living species.* New York: Harper and Row.

Naumov, N. P. 1972. *The ecology of animals.* Urbana: Univ. Illinois Press. (Trans. F. K. Plous, Jr; ed. N. D. Levine.)

Norment, C. and C. L. Douglas. 1977. Ecological studies of feral burros in Death Valley. Univ. Nevada, Las Vegas, Coop. Natl. Park Res. Studies Unit, Contrib. No. 17.

Odum, E. P. 1975. *Ecology.* New York: Holt, Rinehart and Winston. 2nd ed.

Owen-Smith, R. N. 1974. The social system of the white rhinoceros. In *The behaviour of ungulates and its relation to management,* ed. V. Geist and F. Walther, pp. 341–351. Intl. Union Cons. Nature and Nat. Res., No. 24.

Palmer, E. L. 1957. *Palmer's fieldbook of mammals.* New York: E. P. Dutton & Co.

Peden, D. G. 1976. Botanical composition of bison diets on shortgrass plains. *Amer. Midl. Nat.* 96:225–229.

Petrides, G. A., F. B. Golley, and I. L. Brisbin. 1968. Energy flow and secondary productivity. In *A practical guide to the study of the productivity of large herbivores,* ed. F. B. Golley and H. K. Buechner, pp. 9–17. Oxford: Blackwell Scientific Publ.

Péwé, T. L. 1975. Quaternary geology of Alaska. *U.S. Geol. Survey Prof. Paper,* No. 835.

Péwé, T. L. and D. M. Hopkins. 1967. Mammal remains of pre-Wisconsin age in Alaska. In *The Bering land bridge,* ed. D. M. Hopkins, pp. 266–270. Palo Alto: Stanford Univ. Press.

Pianka, E. R. 1978. *Evolutionary ecology.* New York: Harper and Row, 2nd ed.

Powers, W. R. and T. D. Hamilton. 1978. Dry Creek: a late Pleistocene human occupation in central Alaska. In *Early man in American from a circum-Pacific perspective,* ed. A. L. Bryan, pp. 72–77. Univ. Alberta, Dept. Anthropol., Occ. Papers, No. 1.

Preston, F. W. 1962. The canonical distribution of commonness and rarity. Part II. *Ecology* 43:410–432.

Ralls, K., K. Brugger, and J. Ballou. 1979. Inbreeding and juvenile mortality in small populations of ungulates. *Science* 206:1101–1103.

Reagan, M. J., R. M. Rowlett, E. G. Garrison, W. Dort, Jr., V. M. Bryant, Jr., and C. J. Johannsen. 1978. Flake tools stratified below Paleo-Indian artifacts. *Science* 200: 1272–1275.

Reed, C. A. 1970. Extinction of mammalian megafauna in the Old World late Quaternary. *BioScience* 20:284–288.

Repenning, C. A. 1967. Palearctic-Nearctic mammalian dispersal in the late Cenozoic. In *The Bering land bridge,* ed. D. M. Hopkins, pp. 288–311. Palo Alto: Stanford Univ. Press.

Rosenzweig, M. L. 1966. Community structure in sympatric carnivora. *J. Mammal.* 47:602–612.

Rowe, J. S., and G. W. Scotter. 1973. Fire in the boreal forest. *J. Quat. Res.* 3:444–464.

Rudnai, J. 1979. Ecology of lions in Nairobi National Park and the adjoining Kitengela conservation unit in Kenya. *Afr. J. Ecol.* 17:85–95.

Rue, L. L., III. 1968. *Sportsman's guide to game animals: a field book of North American species.* New York: Harper and Row.

Sauer, C. O. 1956. The agency of man on the earth. In *Man's role in changing the face of the earth*, ed. W. L. Thomas, Jr., pp. 49–69. Chicago: Univ. Chicago Press.

Schaller, G. B. 1977. *Mountain monarchs: wild sheep and goats of the Himalaya*. Chicago: Univ. Chicago Press.

Schaller, G. B. and Z. B. Mirza. 1974. On the behavior of the Punjab Urial (*Ovis orientalis punjabiensis*). In *The behaviour of ungulates and its relation to management*, ed. V. Geist and F. Walther, pp. 306–323. Intl. Union Cons. Nature and Nat. Res., No. 24.

Seegmiller, R. F. 1977. Ecological relationships of feral burros and desert bighorn sheep, western Arizona. M. S. Thesis, Arizona State Univ. (Zoology).

Sinclair, A. R. E. 1977. *The African buffalo: a study of resource limitation of populations*. Chicago: Univ. Chicago Press.

Slaughter, B. H. 1967. Animal ranges as a clue to late-Pleistocene extinction. In *Pleistocene extinctions: the search for a cause*, ed. P. S. Martin and H. E. Wright, Jr., pp. 155–167. New Haven: Yale Univ. Press.

Smith, A. D. 1952. Digestibility of some native forages for mule deer. *J. Wildl. Mgmt.* 16:309–312.

———. 1957. Nutritive value of some browse plants in winter. *J. Range Mgmt.* 10:162–164.

———. 1959. Adequacy of some important browse species in overwintering of mule deer. *J. Range Mgmt.* 12:8–13.

Smith, A. D. and J. C. Malechek. 1974. Nutritional quality of summer diets of pronghorn antelopes in Utah. *J. Wildl. Mgmt.* 38:792–798.

Smith, V. L. 1975. The primitive hunter culture, Pleistocene extinctions, and the rise of agriculture. *J. Polit. Econ.* 83:727–755.

Snyder, R. L. 1976. *The biology of population growth*. New York: St. Martin's Press.

Stalker, A. MacS. 1977. Indications of Wisconsin and earlier man from the southwest Canadian prairies. *Annals, N.Y. Acad. Sci.* 288:119–136.

Stanley, S. M. 1973. An explanation for Cope's Rule. *Evolution* 27:1–26.

Stewart, O. C. 1951. Burning and natural vegetation in the United States. *Geog. Review* 41:317–320.

Taylor, D. L. 1973. Some ecological implications of forest fire control in Yellowstone National Park, Wyoming. *Ecology* 54:1394–1396.

Uetz, G. and D. L. Johnson. 1974. Breaking the web. *Environment* 16: No. 10 (December) 31–39.

Ullrey, D. E., W. G. Youatt, H. E. Johnson, P. K. Ku, and L. D. Fay. 1964. Digestibility of cedar and aspen browse for the white-tailed deer. *J. Wildl. Mgmt.* 28:791–797.

Ullrey, D. E., W. G. Youatt, H. E. Johnson, L. D. Fay, and B. E. Brent. 1967. Digestibility of cedar and jack pine browse for the white-tailed deer. *J. Wildl. Mgmt.* 31:448–454.

Urness, P. J., W. Green, and R. K. Watkins. 1971. Nutrient intake of deer in Arizona chaparral and desert habitats. *J. Wildl. Mgmt.* 35:469–475.

VanDevender, T. R. and W. G. Spaulding. 1979. Development of vegetation and climate in the southwestern United States. *Science* 204:701–710.

Vangengeim, E. A. 1967. The effect of the Bering land bridge on the Quaternary mammalian faunas of Siberia and North America. In *The Bering land bridge*, ed. D. M. Hopkins, pp. 281–287. Palo Alto: Stanford Univ. Press.

Van Valen, L. 1969. Late Pleistocene extinctions. *Proc. N. Amer. Paleont. Conv., Part E*, pp. 469–485.

———. 1973. Body size and numbers of plants and animals. *Evolution* 27:27–35.

Viereck, L. A. 1973. Wildfire in the taiga of Alaska. *J. Quat. Res.* 3:465–495.

Walker, E. P. 1968. *Mammals of the world*. Baltimore: The Johns Hopkins Press. 2nd ed.

Watts, W. A. 1970. The full-glacial vegetation of northwestern Georgia. *Ecology* 51:17–33.

Webb, S. D. 1965. The osteology of Camelops. Bull. Los Angeles Co. Mus. Nat. Hist.: Science ser., No. 1.

———. 1969. Extinction-origination equilibria in late Cenozoic land mammals of North America. *Evolution* 23:688–702.

———. 1974. *Pleistocene mammals of Florida*. Gainesville: Univ. Florida Press.

———. 1977. A history of savanna vertebrates in the New World. Part I: North America. *Annual Rev. Ecol. Syst.* 8:355–380.

Wells, P. V. 1965. Scarp woodlands, transported grassland soils, and concept of grassland climate in the Great Plains region. *Science* 148:246–249.

———. 1966. Late Pleistocene vegetation and degree of pluvial climatic change in the Chihuahuan Desert. *Science* 153:970–975.

Wells, P. V. and C. D. Jorgensen. 1964. Pleistocene wood rat middens and climatic change in Mohave Desert: a record of juniper woodlands. *Science* 143:1171–1174.

Wendland, W. M. 1978. Holocene man in North America: the ecological setting and climatic background. *Plains Anthropol.* 23:273–287.

Western, D. 1979. Size, life history and ecology in mammals. *Afr. J. Ecol.* 17:185–204.

———. 1980. Linking the ecology of past and present mammal communities. In *Fossils in the making: vertebrate taphonomy and paleoecology,* ed. A. K. Behrensmeyer and A. P. Hill, pp. 41–54. Chicago: Univ. Chicago Press.

Whitehead, D. R. 1973. Late-Wisconsin vegetational changes in unglaciated eastern North America. *J. Quat. Res.* 3:621–631.

Whitehead, G. K. 1972. *Deer of the world.* London: Constable and Co., Ltd.

Whittaker, R. H. 1970. *Communities and ecosystems.* New York: Macmillan.

Wiens, J. A. 1977. On competition and variable environments. *Amer. Sci.* 65:590–597.

Wilson, E. O. 1975. *Sociobiology: the new synthesis.* Cambridge: Belknap Press.

Wilson, M. 1974. The Casper local fauna and its fossil bison. In *The Casper site: a Hell Gap bison kill on the High Plains.* ed. G. C. Frison, pp. 125–171. New York: Academic Press.

Woodward, S. L. 1976. Feral burros of the Chemehuevi Mountains, California: the biogeography of a feral exotic. Ph.D. Thesis, Univ. California Los Angeles (Geography).

———. 1979. The social system of feral asses (*Equus asinus*). *Z. Tierpsycol.* 49:304–316.

Woodward, S. L. and R. D. Ohmart. 1976. Habitat use and fecal analysis of feral burros (*Equus asinus*), Chemehuevi Mountains, California, 1974. *J. Range Mgmt.* 29:482–485.

Wormington, H. M. 1957. *Ancient man in North America.* Denver Mus. Nat. Hist., Popular ser., No. 4. 4th ed.

Wright, H. E., Jr. 1970. Vegetational history of the central plains. In *Pleistocene and recent environments of the central Great Plains,* ed. W. Dort, Jr. and J. K. Jones, Jr., pp. 157–172. Univ. Kansas, Dept. Geol., Spec. Publ. 3.

———. 1971. Late Quaternary vegetational history of North America. In *The late Cenozoic glacial ages,* ed. K. K. Turekian, pp. 425–464. New Haven: Yale Univ. Press.

Yurtsev, B. A. 1972. Phytogeography of northeastern Asia and the problem of Transberingian floristic interrelations. In *Floristics and paleofloristics of Asia and eastern North America,* ed. A. Graham, pp. 19–54. New York: Elsevier.

19

North American Late Quaternary Extinctions and the Radiocarbon Record

JIM I. MEAD AND
DAVID J. MELTZER

ONE OF THE CRUCIAL ISSUES YET UNRESOLVED in the debate over extinction of the late Pleistocene North American vertebrate fauna is the timing and mode of the extinction process itself. Did a wide variety of taxa become extinct simultaneously? Was the extinction phenomenon a massive, short-term catastrophe or a long, drawn-out process? Was the timing or duration of extinctions the same among the same diverse taxa inhabiting ecologically incompatible environments?

Sir Charles Lyell, for one, felt that the extinction of the "Post-pliocene" fauna was gradual. In sound uniformitarian fashion (in the sense of Grayson 1980), he observed.

> We may presume that the time demanded for the gradual dying out or extirpation of a large number of wild beasts which figure in the Post-pliocene stratigraphy, and are missing in the Recent fauna, was of protracted duration, for we know how tedious a task it is in our times, even with the aid of fire-arms, to exterminate a noxious quadruped.... (1863:374).

Lyell's perception of the problem was colored by his theoretical view of science, as are current statements on the matter. Those inclined to attribute the phenomenon to invasion by Pleistocene hunters argue that extinctions were rapid and synchronous across taxa (e.g. Martin 1974); those inclined to see a multiplicity of causes perceive the process as more gradual (Hester 1967, Hibbard et al. 1965). Neither theoretical position is firmly rooted in an empirical foundation, and thus resolution of the timing of extinctions is unclear.

The problem is compounded by the absence of tight temporally and spatially controlled data on taxa-specific range and population changes through the critical terminal millenia of the Wisconsinan. In the absence of this information the best instrument, and it is only a blunt instrument, for measuring the timing of extinctions is the radiocarbon record. That record can be used in tracking the spatial and temporal distribution of the extinctions, in determining differences or similarities (i.e. the overall synchroneity) in the termination of the extinction of the various taxa, and in providing a baseline for establishing the relation between extinctions and the theories called on to account for them.

For these reasons the ^{14}C record has been previously examined by Hester (1960), Martin (1967, 1974) and most recently in a preliminary statement by Meltzer and Mead (1983). In this chapter we expand on our critical evaluation of the radiocarbon record for the following genera of large extinct North American vertebrates: *Acinonyx, Arctodus, Bootherium, Camelops, Capromeryx, Castoroides, Cervalces, Equus, Euceratherium, Glossotherium, Hemiauchenia, Mammut, Mammuthus, Megalonyx, Mylohyus, Nothrotheriops, Panthera, Platygonus, Sangamona, Smilodon, Stockoceros, Symbos,* and *Tapirus* (terminology follows Kurtén and Anderson 1980).

Methods

As we have noted, (Meltzer and Mead 1983), our data are derived from twenty-three years of the journal *Radiocarbon,* published site reports, and unpublished sources where available. Information was collected on site locality, material dated, potential or actual biases or contaminants in the dates, complimentary geological and pollen data, possible association with archaeological materials, and the validity of the association between the material dated and the extinct genus. We have included dates on extinct fauna that purportedly range in age from 3000 yr B.P. to the earliest limits of the radiocarbon technique, ca. 45,000 yr B.P. The total sample amounted to 150 sites, from north of Mexico and south of about 55° N latitude; this included more than 375 separate ^{14}C determinations.

The sample, composed of nearly all published, radiocarbon-dated extinct fauna in North America, is remarkably small compared to the thousands of extinct vertebrates that have been unearthed on this continent but not radiocarbon dated (see, for example, the samples being dealt with by Dreimanis [1967, 1968] for only one taxa [*Mammut*] in a limited section of the continent; Kurtén and Anderson 1980).

Originally we designed a rigid evaluation system that incorporated a series of criteria for determining the reliability of each radiocarbon date. Since most publications rarely provided the requisite information for each date, we were faced with either working with an extremely small sample or simplifying our rating system We selected the second option and evaluated each radiocarbon date on the material being dated and the strength of the association between the material dated and the target fauna. These two classes of information were the only kinds consistently found in the literature. The rating system (Table 19.1) provides at least a partial control on the idiosyncracies of the dating technique and its application (Stuckenrath 1977) and yields a set of uniformly sorted and reliable dates.

To derive an individual score, the rating in the first category is added to the rating in the second category. The scores can range from a high of 9 (charcoal in strong association) to a low of 2 (carbonate or whole bone in weak association). These numbers are only on an ordinal scale, and the assigned values do not imply any arithmetic relationships. For our study, ratings of 7 or less are considered unreliable, while ratings of 8 or better are considered reliable, thus acceptable indicators of the age of a fauna.

The ranking of the substances in the first category is based on studies of the utility of these various materials for ^{14}C dating (e.g. Haas and Banewicz 1980, Hassan et al. 1977, Haynes 1968, Taylor 1980). It is important to note that after we had examined the data it was apparent that collagen-derived dates were behaving unpredictably, often giving internally inconsistent results. This reinforces the cautions of Taylor (1980) and others on the use of this material in dating; thus we viewed collagen-derived dates, regardless of their score, with some suspicion.

Table 19.1. Rating System Used in Analyzing Radiocarbon Dates Associated With Extinct Fauna

Scores can range from 9 (charcoal, strong association) to 2 (whole bone/terrestrial carbonate, weak association). To derive an individual score choose and add one from Section I and one from Section II. Our test includes rating scores of 8 and 9.

	Score
Section I. MATERIAL DATED	
Derived from Extinct Fauna	
Collagen	5
Body perishables (dung, hide, hair)	5
Apatite	3
Whole bone ("bone date," "bone OM")	1
Derived from Other Organic Material	
Charcoal (elemental)	6
Wood (logs, twigs, leaves)	5
Peat	3
Organic mud (includes gyttja)	3
Soil	3
Shell (freshwater/terrestrial)	2
Terrestrial carbonate (marl)	1
Section II. STRENGTH OF ASSOCIATION	
Strong	3
Unknown or medium	2
Weak	1

Regarding the second category of the rating system, a high score for a particular taxa at a site may not carry over to all taxa represented at the site. For instance, where a ^{14}C date is derived directly from an extinct genus (e.g. a collagen date on the bone), then only that genus is given a high value in category II. Other fauna "associated" with that date but not dated directly are given a lesser value. Our position is predicated on the belief that it is better to assume an association is weak until proven otherwise.

Sites with two or more radiocarbon determinations were dealt with in one of two ways. If the two dates had an identical rating and were statistically indistinguishable, then the dates were averaged and plotted as a single date (tests of statistical identity and averaging of the dates utilized the formulae from Long and Rippeteau [1974]). Where multiple dates on the site had different ratings, were statistically different, or both, the dates were each plotted independently. All dates were thus either plotted individually or plotted as a group mean.

The Christensen Mastodon site (Whitehead et al. 1982) provides a good illustration of the rating system and some of the problems commonly encountered in determining scores. This Indiana fossil locality contains numerous plant and animal remains (including *Mammut* and *Castoroides*) that are the subject of an excellent paleoecological analysis and vegetation reconstruction. Nonetheless, the age of the fauna is ambiguous. A cross section (schematic representation, Whitehead et al. 1982:244, fig. 3) shows the *Mammut* remains in relation to six radiocarbon dates. The bones are distributed through the deposit from about 2 m to approximately 3.5 m below the surface.

Stratigraphically the age of the *Mammut* is between 12,060±100 (ISGS-601) and 14,080±150 (ISGS-502) yr B.P. The younger date, on peat, overlies the bones while the older date, mainly on lake sediments, is below the bones. Both ^{14}C dates are of weak to medium association (are found either in a different unit or not recovered in direct association with the *Mammut* remains). In our rating system they are given scores of 4 and 3 respectively. No dates are directly on or in a strong association with the *Mammut* remains. Only one ^{14}C age of 13,220±100 yr B.P. (ISGS-492) on wood was recovered from the bone-bearing horizon. Again, because of an unclear association, a score of 7 was the highest that could be assigned.

We do not wish to appear unduly critical of this or any other site report, nor have we lost sight of the fact that the reports we have used are likely written for purposes different from our own. Our primary reason in using this rating system is to evaluate the data in terms of how reliable the record may be for an absolute age estimate, given the information available in the published literature. The status of the radiocarbon record is such that without a critical evaluation, one can be led to significant errors of interpretation (Martin 1967).

Results and Discussion

The results are illustrated in Figure 19.1, a histogram of all ^{14}C dates regardless of their score on the critical test. This sample includes 233 dates from the 150 sites (reduced from the initial sample of 375 dates). No sites or clearly erroneous ^{14}C dates are omitted from this figure so that the magnitude and range of the radiocarbon record for extinct late Pleistocene genera can be illustrated.

The structure of this distribution, with the frequency peak occurring between 11,000 and 11,500 yr B.P., reflects the overall distribution of Wisconsinan radiocarbon dates and is not indicative of any phenomena related to the extinction process. While nearly 70 percent of all radiocarbon dates fall between 11,000 and 14,000 yr B.P., this is largely due to sampling biases. Archaeologists and geologists have tended to radiocarbon date late Pleistocene and younger remains when a likelihood exists of determining the last appearance of a taxa or when there are associated cultural materials. In glaciated regions virtually all extinct fauna that can be dated will be younger than 14,000 yr B.P.

As Figure 19.1 illustrates, many radiocarbon determinations are younger than the age often claimed to mark the end of the megafaunal extinction episode (i.e. 11,000 yr B.P.). A fair number of dates are scattered through the early and mid-Holocene. Before any conclusions are drawn about Holocene vertebrate survivals, however, we would caution that this evidence is more apparent than real. When one plots only those dates most likely to be reliable (those with a score of 8 or 9 on the rating system), a different pattern emerges (fig. 19.2). Nearly all the early mid-Holocene dates from the initial sample are absent from this group.

Importantly, all dates younger than 10,000 yr B.P. that remain in this sample are determinations run on bone collagen (fig. 19.2). Taylor has noted the danger of using the term "collagen" simply as a synonym for the acid soluble, insoluble, or undissolved fractions. These fractions will not automatically contain only collagen-derived organics (Taylor 1980:969). This is verified in the pattern here, with the collagen dates differing markedly from dates run on other organic materials, and with the variation apparent in multiple collagen dates. In the case of Sandy, Utah, three "collagen" dates (5985±210, SI-2341b; 7200±190, RL-464; 8815±100, SI-2341a) were obtained from the same mammoth bone.

To better assess the structure and biases of the distribution of these radiocarbon dates, some basic analytical tests were conducted. A chi square statistic was calculated to determine whether the dates rated 2 to 7 were proportional (in frequency) to dates scoring 8 or better. For ease of analysis, the 500-year time brackets on the histograms (figs. 19.1 and 19.2) were combined into 1000-year periods ranging from 6000 yr B.P. to 18,000 yr B.P. (a period that covers the bulk of the radiocarbon dates and the millenia within which extinction likely occurred). The contingency table is shown in Table 19.2. The calculated chi square value for this table is 20.22, a value significant at the 0.05 level (with 11 degrees of freedom), but not at the 0.025 level. The null hypothesis is therefore rejected: the 1000-year time period within which a particular set of dates falls is not independent of the rating of those dates. In essence, while the time brackets do not literally influence the score of a radiocarbon date, within some or all of the time periods there is a disproportionate relationship between scores of 7 or less and 8 or better.

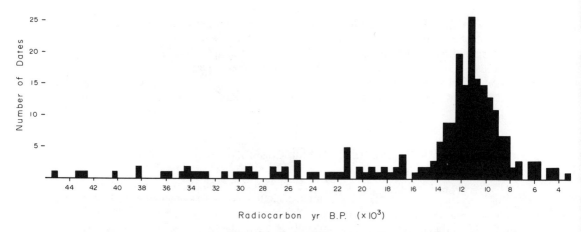

Figure 19.1. Distribution of radiocarbon dates (rating 2–9) for extinct late Pleistocene megafauna.

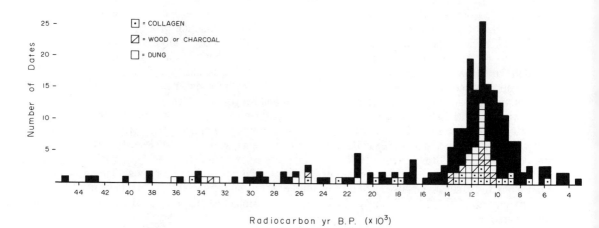

Figure 19.2. Distribution of radiocarbon dates with a score of 8 or 9 (light areas) against the background of all radiocarbon dates.

The chi square statistic used here is 1-tailed, and thus one cannot specify which cells (time periods) contributed most significantly to the chi square value, or the direction in which those cell frequencies are disproportionate. This information, however, can be derived using the procedure for the analysis of residuals outlined in Everitt (1977:46–48). This series of statistics enables one to calculate the adjusted residuals (d_{ij}) for each of the cells, deriving values that have an approximately normal distribution with a mean of 0 and a standard deviation of 1. In a fashion analogous to the analysis of variance, the product identifies which cells differ significantly from the "mean" (in this example contribute most to the chi square value) and, because it is a 2-tailed test, the direction of the difference (either over- or underrepresentation of the frequencies within the cell).

These results are shown in Table 19.3. Four cells within that table are significant at the 5 percent normal deviate ($> \pm 1.96$). For the 9000- to 10,000-year bracket, poorly dated sites are overrepresented ($+2.16$), and well-dated sites are underrepresented (-2.15). In the 11,000- to 12,000-year period, there is an overrepresentation of well-dated sites ($+2.69$) and, commensurately, an underrepresentation of poorly dated

Table 19.2. Frequency Counts Used for Chi Square Analysis of Radiocarbon Dates

Degrees of freedom are 11. X^2 = 20.22. Significant at 0.05.

Rating Score	Radiocarbon Age—Thousands yr B.P.												
	6–7	7–8	8–9	9–10	10–11	11–12	12–13	13–14	14–15	15–16	16–17	17–18	
8–9	0	1	3	3	13	20	10	5	0	0	0	1	= 56
2–7	3	4	11	21	18	21	19	10	5	3	4	2	= 121
Total	3	5	14	24	31	41	29	15	5	3	4	3	= 177

Table 19.3. Calculated Adjusted Residual Scores by Cell

Formula from Everitt 1977. Cells with significant variance are italicized.

Rating Score	Radiocarbon Age—Thousands yr B.P.											
	6–7	7–8	8–9	9–10	10–11	11–12	12–13	13–14	14–15	15–16	16–17	17–18
8–9	−1.18	−0.56	−0.84	−2.15	1.35	2.69	0.35	0.13	1.53	−1.18	−1.37	0.06
2–7	1.18	0.55	0.85	2.16	−1.35	−2.68	−0.35	−0.13	1.53	1.18	1.36	−0.05

sites (−2.68). Based on this table, it is apparent that the ^{14}C determinations for all but two time periods have proportional frequencies of dates with scores of 7 or less and 8 or better.

The significant and disproportionate drop in reliable dates younger than 10,000 yr B.P. is of some importance. It indicates either analytical and sampling difficulties in dating materials younger than 10,000 yr B.P. (a slight possibility at best) or, more probably, that late Pleistocene genera did not survive after 10,000 yr B.P. In other words, despite an overrepresentation of dated sites within this period, very few are reliable dates and, as already mentioned and worth recalling here, those reliable dates (8 or better) are collagen-derived determinations.

These tests indicate that what is needed at this point are additional reliable dates from terminal Pleistocene faunal sites to see if, in fact, any reliable dates fall unequivocally later than 10,000 yr B.P. Certainly sufficient numbers of good dates in earlier millenia (e.g. 11,000 to 12,000 yr B.P.) are available to indicate the genera that survived up to that point. But before the temporal limits of the extinctions are drawn, more reliable dates are needed for the time periods immediately preceding and following 10,000 yr B.P. It will then be safe to infer that the lack of reliably dated faunas indicates their disappearance from the North American landscape and not a sampling or technical problem.

For more specific information on each taxa involved, the latest ^{14}C dates are listed by genus in Table 19.4 and illustrated in Figure 19.3. The table illustrates an apparent lack of synchrony in the terminal dates of the various genera, although this evidence is complicated by the widely varying sample sizes for each taxa and the reliability of the particular ^{14}C determinations. It would be premature to infer an early, full glacial extinction for either *Acinonyx* or *Bootherium* because each is only dated once.

Similarly, while Figure 19.3 appears to illustrate rapid cumulative extinction in the millenia around 10,000 B.P., the shape and details of the curve are subject to revision. The figure includes the latest and most reliable dates for all twenty-three genera regardless of rating. Save for the terminal dates, the cumulative graph bears a close resemblance to the extinction model proposed by Martin (1974, Figure 11.1). At the same time we emphasize that the curve cannot and does not reflect *how* the extinctions actually took place; it is simply the best available age of the latest survivors of the extinction process.

Table 19.4. Youngest ^{14}C dates for Extinct Genera From 150 Sites

If the rating for a particular date is under 7, then, where available, the next youngest date with a score of 8 or better is listed. If that date is based on collagen (*), then the next youngest date with a score of 8 or better and not derived from collagen is listed. See for example *Mammut*.

† = possibly contaminated; ^{14}C date not replicated.

Genus	Number of Sites	Youngest ^{14}C Date	Score	Locality
Acinonyx (American cheetah)	1	$*17,620^{+1490}_{-1820}$	7	a
Arctodus (Short-faced bear)	2	$12,650 \pm 250$	3	b
Bootherium (Musk ox)	1	$17,200 \pm 600$	6	c
Camelops (Camel)	25	8240 ± 960	7	d
		$10,370 \pm 350$	8	e
Capromeryx (Pronghorn)	3	$11,170 \pm 360$	7	f
Castoroides (Giant Beaver)	2	$10,230 \pm 150$	7	g
Cervalces (Stag-moose)	2	$10,230 \pm 150$	7	g
Equus (Horse)	38	8240 ± 960	7	d
		$10,370 \pm 350$	8	e
Euceratherium (Shrub ox)	1	$*8250 \pm 330$	8	h
Glossotherium (Mylodont Ground Sloth)	9	9880 ± 270	2	i
Hemiauchenia (Llama)	3	8527 ± 256	7	j
Mammut (American Mastodont)	52	5950 ± 300	4	k
		$*8910 \pm 150$	8	l
		$10,395 \pm 100$	8	m
Mammuthus (Mammoth)	63	4885 ± 160	6	n
		$*5985 \pm 210$	8	o
		$10,550 \pm 350$	8	p
Megalonyx (Megalonychid Ground Sloth)	10	9380 ± 85	4	q
		$11,500 \pm 500$	7	bb
Mylohyus (Peccary)	3	$*9410 \pm 155$	7	r
Nothrotheriops (Shasta Ground Sloth)	9	9840 ± 160	8	s†
		$10,035 \pm 250$	8	t†
		$10,400 \pm 275$	8	t
Panthera (American Lion)	7	$10,370 \pm 350$	8	e
Platygonus (Peccary)	10	4290 ± 150	4	u
		$*12,950 \pm 550$	8	v
Sangamona (Fugitive Deer)	1	$*9940 \pm 760$	6	w
Smilodon (Sabertooth)	4	$*9410 \pm 155$	8	r
Stockoceros (Pronghorn)	2	8980 ± 300	4	x
		$11,330 \pm 370$	7	y
Symbos (Woodland Musk Ox)	1	$11,100 \pm 400$	4	z
Tapirus (Tapir)	6	9400 ± 250	6	aa

Fossil localities and radiocarbon laboratory numbers:

a Natural Trap Cave, WY, DC–690
b Lubbock Lake, TX, I–246
c Big Bone Lick, KY, W–1617
d Whitewater Draw, AZ, A–184c
e Jaguar Cave, ID, No Lab. Number
f Blackwater Draw, NM, A–481
g Ansonia, OH, I–5843
h Potter Creek, CA, UCR–381
i Hornsby Spring, FL, No Lab. Number

j Gypsum Cave, NV, C–222 (solid carbon)
k Russell Farm, MI, M–347
l Ferguson Farm, Ontario, GSC–614
m Pleasant Lake, MI, BETA–1388
n Kassler, CO, W–288
o Sandy, UT, SI–2341b
p Rawhide Butte, WY, A–366
q Glynn, GA, UGa–79
r First American Bank, TN, I–6125

s Aden Crater, NM, Y–1163a
t Rampart Cave, AZ, L–473a, I–442
u Warren Beach, OH, M–1516
v Welsh Cave, KY, I–2982
w Brynjulfson, MO, ISGS–70
x Ventana Cave, AZ, A–1081
y Shelter Cave, NM, A–1878
z Scotts, MI, M–1402
aa Evansville, IN, W–418
bb Tule Springs, NV, UCLA–636

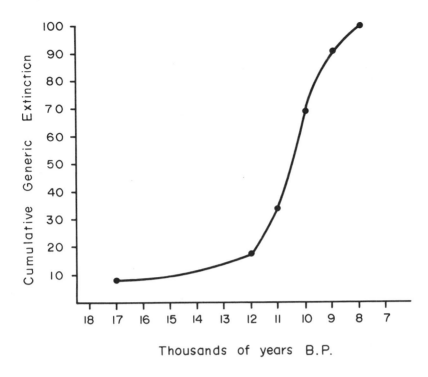

Figure 19.3. Cumulative percentage curve of the terminal dates for all twenty-three genera. Data from the highest rated terminal date listed in Table 19.4.

Certain survivors may have lasted into the early Holocene (*Camelops, Equus, Hemiauchenia*), and perhaps as late as the middle Holocene (*Mammut, Mammuthus, Platygonus*). But of the genera just listed, and for that matter for all genera postdating 10,000 B.P., it can be seen that only three (*Euceratherium, Nothrotheriops, Smilodon*) have scores on our rating system of 8 or better. Of these three exceptions, the *Nothrotheriops* date from Aden Crater (Y-1163a) was based on body tissue thought to have been contaminated (Simons and Alexander 1964). The *Euceratherium* and *Smilodon* dates are each derived from bone collagen whose reliability is, again, suspect.

When one considers only those genera for which we have demonstrably reliable dates (those with a score of 8 or better) that are not derived from bone collagen (*Camelops, Equus, Mammut, Mammuthus, Nothrotheriops, Panthera*), a familiar pattern appears. These reliable dates, predictably on genera for which we have relatively large samples, indicate that late Pleistocene extinctions lasted no later than 10,000 yr B.P. and possibly were complete by 10,800 yr B.P. (assuming one standard deviation error). At least among these genera, the extinction process may contain an element of synchroneity.

For the genera most often thought to have been associated with the Pleistocene hunters, the radiocarbon record contains no surprises. The frequency of radiocarbon-dated sites for *Mammut* and *Mammuthus* reflects the geographical distribution of the fossils of these animals (Table 19.5a, fig. 19.4). *Mammut*, the predominantly eastern taxa, may have become extinct by approximately 14,000 yr B.P. in the western states, although this western sample is extremely small. In the East, however, the mastodon persisted to about 10,300 yr B.P. It appears that the mammoth became extinct at approximately the same time in both the eastern (10,600 yr B.P. and western states (10,500 yr B.P.) (also see Agenbroad, Chapter 3, this volume). There does not appear to be a significant difference between the reliability of mammoth dates and mastodon dates (Table 19.5b).

**Table 19.5. Frequency of *Mammut* and *Mammuthus*
Dated Sites and Radiocarbon Determinations**

	West	East
Mammut	4	48
Mammuthus	51	12

(a) Distribution of *sites* from eastern and western localities by genus

	Score 8–9	Score 2–7
Mammut	7	57
Mammuthus	18	70

(b) Reliability of radiocarbon *dates* by genus

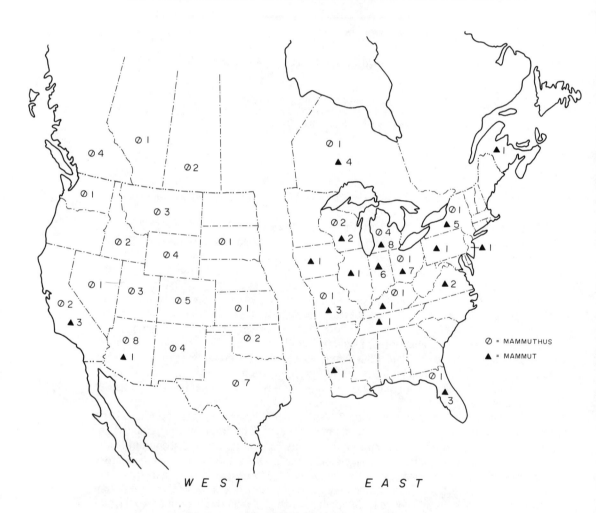

Figure 19.4. States with radiocarbon-dated *Mammut* and
Mammuthus localities. Numbers indicate the quantity of sites of
each genus in that state.

Conclusions

Theories that have been advanced to account for the extinction of the late Pleistocene fauna can be divided into two basic types: those that attribute extinction to the agency of man (Martin 1967, 1973, 1974; Mosimann and Martin 1975), and those that invoke the effects of climatic change (Guilday 1967, Slaughter 1967, Dreimanis 1968). Importantly, the foundation of each is presumed temporal correlation. For example, Long and Martin observe, "the apparently synchronous loss of the Shasta ground sloth with the arrival of big game hunters in Arizona. . . [is] in accord with the [human overkill] model" (Long and Martin 1974:640).

The correlation between the appearance of Pleistocene hunters, the development of Holocene climates, and the disappearance of the Pleistocene fauna remains equivocal. Some archaeologists have questioned the apparent late Pleistocene arrival of man (MacNeish 1976, 1979; Bryan 1978; Bryan et al. 1978). The timing of the major late and postglacial environmental changes has proven to be more complex than previously known (e.g., Wright 1981). And, as has been demonstrated here, the radiocarbon record for the relevant fauna leaves many questions unanswered.

The demonstration from the radiocarbon record that the extinctions were synchronous or gradual and successive will not itself resolve the nearly 200-year-old problem of the extinction of the late Pleistocene megafauna. The evidence from the radiocarbon record will have to be coupled with evidence relating to population dynamics, change, and collapse, and the whole embedded in a viable and testable theory. That will ultimately determine whether there was a wave of synchronous extinctions or whether one genus predeceased another by hundreds or even thousands of years.

Acknowledgments

This chapter is an outgrowth of a paper delivered at a symposium organized by the authors for the 1982 Society for American Archaeology meetings. The symposium included a roundtable discussion of late Pleistocene extinctions organized by Dr. Paul S. Martin and hosted and moderated by Dr. Herbert E. Wright, Jr. We thank the participants in both instances for helping us sharpen our own thoughts on the issue.

In addition to those acknowledged in our preliminary paper, we would like to thank Drs. Donald K. Grayson, C. Vance Haynes, Paul S. Martin, Stephen T. Jackson, and Donald Whitehead for their comments and criticisms. We gratefully acknowledge the support of National Science Foundation Grant DEB79–23804 to Paul S. Martin and the Department of Anthropology, Smithsonian Institution. Finally, we would like to thank Emilee Mead and Suzanne Siegel, who aided in all aspects of the manuscript preparation.

References

Bryan, A. (Editor). 1978. *Early man in America.* Occasional Papers No. 1, Department of Anthropology, University of Alberta, Canada.

Bryan, A., Casamiquela, R. M., Cruxent, J., Gruhn, R. and Ochsenius, C. 1978. An El Jobo mastodon kill at Taima-Taima, Venezuela. *Science* 200:1275–1277.

Dreimanis, A. 1967. Mastodons, geologic age and extinction in Ontario, Canada. *Canadian Journal of Earth Sciences* 4:663–675.

———. 1968. Extinction of mastodons in eastern North America: testing a new climatic-environmental hypothesis. *The Ohio Journal of Science* 68(6):257–272.

Everitt, B. C. 1977. *The analysis of contingency tables.* Chapman and Hall, London.

Grayson, D. K. 1980. Vicissitudes and overkill:

the development of explanations of Pleistocene extinctions. *Advances in archaeological method and theory* 3:357–493. Academic Press, New York.

Guilday, J. 1967. Differential extinctions during late Pleistocene and Recent times. In *Pleistocene extinctions: the search for a cause.* P. Martin and H. E. Wright, Eds., pp. 121–140. Yale University Press, New Haven.

Haas, H. and J. Banewicz. 1980. Radiocarbon dating of bone apatite using thermal release of CO_2. *Radiocarbon* 22:537–544.

Hassan, A., Termine, J. D., and C. V. Haynes. 1977. Mineralogical studies on bone apatite and their implications for radiocarbon dating. *Radiocarbon* 19:364–374.

Haynes, C. V. 1968. Radiocarbon:analysis of inorganic carbon of fossil bone and enamel. *Science* 161:687–688.

Hester, J. 1960. Late Pleistocene extinction and radiocarbon dating. *American Antiquity* 26:58–87.

———. 1967. The agency of man in animal extinctions. In *Pleistocene extinctions: the search for a cause.* P. Martin and H. E. Wright, Eds., pp. 169–192. Yale University Press, New Haven.

Hibbard, C., Ray, C., Savage, D., Taylor, D., and Guilday, J. 1965. Quaternary mammals of North America. In *The Quaternary of the United States.* H. E. Wright and D. Frey, Eds., pp. 509–525. Princeton University Press, Princeton.

Kurtén, B. and E. Anderson. 1980. *Pleistocene mammals of North America.* Columbia University Press, New York.

Long, A. and P. S. Martin. 1974. Death of North American ground sloths. *Science* 186:638–640.

Long, A. and B. Rippeteau. 1974. Testing contemporaneity and averaging radiocarbon dates. *American Antiquity* 39(2):205–215.

Lyell, C. 1863. *The geological evidences of the antiquity of man.* John Murray, London.

MacNeish, R. 1976. Early man in the New World. *American Scientist* 63:316–327.

———. 1979. Earliest man in the New World and its implications for Soviet-American archaeology. *Arctic Anthropology* 16:2–15.

Madsen, D. B., Currey, D. and Madsen, J. 1976. Man, mammoth and lake fluctuations in Utah. *Utah State Historical Society Section, Selected Papers* 5:45–58.

Martin, P. S. 1967. Prehistoric overkill. In *Pleistocene extinctions: the search for a cause.* P. Martin and H. E. Wright, Eds., pp. 75–120. Yale University Press, New Haven.

———. 1973. The discovery of America. *Science* 179:969–974.

———. 1974. Paleolithic players on the American stage: man's impact on the Late Pleistocene megafauna. In *Arctic and alpine environments,* J. Ives and R. G. Barry, Eds., pp. 669–700. Methuen, London.

Meltzer, D. and J. Mead. 1983. The timing of Late Pleistocene mammalian extinctions in North America. *Quaternary Research* 19:130–135.

Mosimann, J. and P. S. Martin. 1975. Simulating overkill by Paleoindians. *American Scientist* 63:304–313.

Simons, E. and H. Alexander. 1964. Age of the Shasta ground sloth from Aden Crater, New Mexico. *American Antiquity* 29:390–391.

Slaughter, R. 1967. Animal ranges as a clue to Late Pleistocene extinctions. In *Pleistocene extinctions: the search for a cause,* P. Martin and H. E. Wright, Eds., pp. 155–168. Yale University Press, New Haven.

Stuckenrath, R. 1977. Radiocarbon: some notes from Merlin's diary. In *Amerinds and their paleoenvironments in northeastern North America,* W. Newman and B. Salwen, Eds. pp. 181–188. Annals of the New York Academy of Sciences, Volume 288, New York.

Taylor, R. 1980. Radiocarbon dating of Pleistocene bone: toward criteria for the selection of samples. *Radiocarbon* 22(3):969–979.

Whitehead, D., Jackson, S. T., Sheehan, M. C., and Leyden, B. 1982. Late-glacial vegetation associated with caribou and mastodon in central Indiana. *Quaternary Research* 17:241–257.

Wright, H. E. 1981. Vegetation east of the Rocky Mountains 18,000 years ago. *Quaternary Research* 15:113–125.

20

Simulating Overkill

Experiments With the
Mosimann and Martin Model

STEPHEN L. WHITTINGTON
AND BENNETT DYKE

THE END OF THE PLEISTOCENE SAW MANY extinctions, especially of the megafauna: a group of large, slow animals with low reproductive rates. The theory that excessive hunting by humans was the major contributing cause to the extinction of these species is known as the overkill hypothesis (Budyko 1967, 1974; Martin 1967, 1973; Johnson 1977). Climatic changes and other natural processes believed to have occurred near the end of the Pleistocene are postulated by proponents of the hypothesis to have played only minor roles in the extinctions (Kroeber 1940, Edwards 1967).

In a 1975 paper, Mosimann and Martin constructed a model based on mathematical equations from Budyko's (1967, 1974) work on the extinction of mammoths in Europe. They wanted to estimate how many human hunters might be sufficient to cause megafaunal extinction, how much time might be required for extinction to occur, and how much evidence might be left in the fossil record. According to their computer simulation, megafaunal extinction could be caused rapidly and with little resultant archaeological evidence by a small number of humans advancing across North America in a "front," a colonizing mechanism first appended to the overkill hypothesis by Martin (1973). A front is a dense, narrow band of population that surrounds a region of much lower density and moves radially away from its point of origin as the population grows. Mosimann and Martin tested various scenarios by changing the values of the prey carrying capacity, the human population growth rate, and the prey destruction rate. Using ranges of values which seemed reasonable to them, they concluded that the simulation demonstrated the feasibility of overkill under a number of scenarios.

The experiments reported here used Mosimann and Martin's computer simulation in a slightly modified form, with a somewhat different set of assumptions about overkill, and with different values more likely to be acceptable to human ecologists and archaeologists for variables. A sensitivity analysis of the modified model was performed, in which the value of one input parameter at a time was varied and its effects on other variables were observed, to determine the basic properties and interactions of the simulation and their implications for the overkill hypothesis.

Our results show that the overkill hypothesis does not depend on the somewhat controversial concept of the front when certain values are used as input, and that the modified model is not invalidated by criticisms of Mosimann and Martin's simulation. In agreement with Wesler (1981), overkill cannot be ruled out as a possible explanation for New World megafaunal extinctions.

Simulation of Overkill in the New World

Mosimann and Martin portrayed events which occurred about 11,500 years ago in the New World. One hundred people (an arbitrary number not based on archaeological evidence) first entered North America in the period immediately preceding the extinction event. These people, bearing a big-game hunting technology, soon found themselves at the southern terminus of an ice-free corridor at about the location of what is now Edmonton, Alberta. They faced a wide-open, extremely rich continent with an abundant megafauna, so they hunted and reproduced at very high rates. As the human population grew and put pressure on resources, it formed an expanding quarter circle of uniform density centered at Edmonton.

The Front

A front was created only when the human population hunted the megafauna to extinction in the inhabited area. Then the humans vacated all or part of that region by advancing away from the Edmonton center into unoccupied territory as far as necessary to find sufficient prey to support the population. Observations of the spread of snails, fish, and large herbivores into previously uninhabited, but habitable, regions have shown that a front consists of a high density population that moves swiftly and radially in a narrow expanding arc from its point of origin, leaving in its wake a much less densely inhabited region (Mead 1961, Riney 1964, Caughley 1970, Zaret and Paine 1973). The region behind the front remained devoid of megafaunal prey because the dense population of hunters made the front impermeable to the megafauna. The front was kept relatively narrow in width because humans everywhere within it still had to be able to hunt to survive.

Since the part of the continent which lay before the front was rich and uninhabited, there was no pressure to control population growth. For the front to absorb the increasing human population, its leading edge, or periphery, was forced to move smoothly away from the Edmonton center in an unbroken band. The front's trailing edge remained stationary because people living there had no reason to advance farther across the continent as long as prey remained within the front. The inhabitants could not move back toward Edmonton because their prey was extinct in that direction.

The pressure exerted by the population growing at a geometric rate to expand into new, unoccupied territory was sufficient to cause the periphery of the region of habitation to move radially at an ever-increasing rate. The distance the periphery swept across the continent in a year increased until limitations, common to all social animals, placed a ceiling on the distance the humans would move. The restraints, which held the radial movement of the front's periphery to twenty miles a year, applied for as long as any megafaunal prey still existed in the area occupied by humans. They were overridden only when the prey was exhausted in the area, at which time the human population in the simulation would advance as far as necessary to find sufficient prey (as when the front was originally formed), causing both the leading and trailing edges of the front to move outward simultaneously, making the front appear to "jump."

The front's sweep across North America reached the eastern and southern coasts of what is now the United States in about 300 years and resulted in an extremely uneven distribution of the human and animal populations, with the vast majority of both living in a coastal band that enclosed a nearly empty continent. Megafaunal extinction occurred within three years after the front reached the coasts. The high density of hunters in the front was more of a factor in causing extinction than was the number of humans on the continent.

Mosimann and Martin's yearly limit of twenty miles for the distance the periphery could move seems conservative, in light of evidence that humans can move much greater distances (Clark 1959). However, the limitation emphasizes that it is unnecessary to postulate long-distance population movements during a year to explain humans reaching the coasts about 300 years after the introduction of the initial population.

Mosimann and Martin felt that the combination of the front's radial movement across the continent and its brief occupation of any locality during its sweep explained both why evidence for megafaunal extinction is so rare and why extinction occurred so quickly. One scenario, which assumed a high human population growth rate, had a total of only 300,000 humans advancing across North America in less than 300 years, hunting to extinction a megafauna whose biomass equaled 93 billion pounds. At such a rate, the passage of the front through any locality would have taken ten years or less, during which time few artifacts would have been deposited. The radial movement of the front across the continent would have been so swift that the New World megafaunal species would not have had time before they became extinct to evolve behavioral adaptations such as those possessed by their Old World counterparts, whose long period of coexistence with humans allowed them to survive human predation (Edwards 1967, Jelinek 1967).

Objections to the Simulation

Major objections to the simulation have come from three directions. The first is that the input values are not reasonable. This criticism especially concerns the human population density (Albini 1975, Hallum 1977), the prey destruction rate (Stauffer 1975, Webster 1981), and the human population growth rate (Hassan 1978, Webster 1981), all of which are considered too high. The second objection concerns the stipulations about the timing of the entry of humans into the New World. Increasing claims of early archaeological sites overshadow the assumption that humans arrived in the area of what is now Edmonton only 11,500 years ago (Rouse 1976, MacNeish 1976, Bryan et al. 1978). The third objection is that the front greatly simplifies what must have happened in reality, since geographical barriers, prey switching, and optimal foraging would probably break up a front or keep one from forming (Webster 1981).

Any of these objections alone appear to be damaging to the overkill hypothesis as it was simulated. Some (Albini 1975, Bryan et al. 1978, Webster 1981) feel that their objections to Mosimann and Martin's simulation make invalid all other overkill models. However, what is fatal to the simulation is not necessarily fatal to all models. Mosimann and Martin's simulation was more a test of a particular colonization model (the front) than of the overkill hypothesis itself, which simply states that humans caused many Pleistocene megafaunal extinctions through excessive hunting. None of the critics has undertaken a program of experimentation with Mosimann and Martin's simulation to study its robustness and to explore whether it is useful and valid in any modified form.

Experimentation With the Simulation

The experiments that follow were made using a simulation program supplied by Mosimann and Martin. The program, written in BASIC, was modified to run on an Exidy Sorcerer microcomputer.

Features of the Simulation

Most of the original features of the model were retained. The human carrying capacity (L) is defined as the maximum density of humans allowed in an inhabited area, and $L=AK/G^*$, where A is the prey replacement rate, K is the prey carrying capacity, G is the prey destruction rate, and $G^*=(1+A)G$. Part of the human population has to migrate during the course of each year into a previously unoccupied area containing megafaunal prey to keep the population density below carrying capacity. As long as any prey remains in an area occupied by humans, however, the humans advance away from the Edmonton center no farther than twenty miles in a year. The human population growth rate begins to decrease (increasing the doubling time) once the population density begins to force the periphery of the inhabited area to move radially twenty miles a year, with the result that distances greater than twenty miles do not have to be traversed. This process results in a logistic growth curve, with the human population growth rate approaching zero. Limiting the periphery's movement may sometimes cause local megafaunal extinctions, even with a decreasing human population growth rate. In such cases, the entire human population will advance as far as necessary to find enough prey to support it, forming a front which appears to jump forward. It is possible for megafaunal extinction not to occur because of the fact that the human growth rate may eventually reach zero, maintaining the density of the human population in the occupied area below its carrying capacity.

The assumption of Mosimann and Martin's simulation that extinction must occur within three years of the date that humans reached the coasts seemed unduly restrictive. A minor change was made to the simulation so that it continues to operate even after humans have reached the coasts, until either the megafauna becomes extinct, or a specified time limit is reached.

Baseline Values

The primary and most important step in setting up the experimentation was to choose "baseline" values: input parameters based upon empirical evidence generally acceptable to archaeologists and human ecologists. The seven input variables are listed in Table 20.1, with their baseline values. The values for six of the seven variables were chosen from the literature on megafaunal extinctions and on human colonization of unoccupied regions. In all simulation runs, initial prey biomass (N) and prey carrying capacity (K) were given identical values, implying that the megafauna of the continent was at carrying capacity and in equilibrium when humans first entered the area. Initial prey biomass and the prey carrying capacity were set at values of 25 a.u./mi² (one animal unit [a.u.] is equivalent to 1,000 pounds of living prey). These correspond to the

Table 20.1. Variables and Baseline Values

Variable	Description of Variable	Baseline Value	Source
N	Initial prey biomass (animal units * per square mile)	25	Mosimann and Martin 1975
K	Prey carrying capacity (animal units per square mile)	25	Mosimann and Martin 1975
A	Prey biomass replacement rate (annual herd growth rate)	0.25	Mosimann and Martin 1975
M(0)	Initial human population size (within 100-mile radius quarter circle centered at Edmonton)	200	Budyko 1967, 1974
C	Human population growth rate	0.0443	Birdsell 1957
L	Human carrying capacity (people per square mile)	1.295	Budyko 1967, 1974
G	Prey destruction rate (annual destruction of animal units per person)	3.862	Derived

*One animal unit (a.u.) is equivalent to 1,000 lb. of living prey

values Mosimann and Martin used in their basic simulation and to the biomass found in many African game parks. The value of the prey biomass replacement rate (A) also was taken from their basic simulation and was set at 0.25. An initial human population size [M(0)] of 200 people gives an initial human density of 0.0255 people/mi^2 which, according to Budyko (1967, 1974), is below the density of populations that do not possess the capability of mass hunting. As will be shown, the value chosen for the initial human population size has relatively little effect on the outcome of megafaunal extinction in the simulation. The human population growth rate (C) was set at 0.0443, which gives a doubling time of sixteen years for the population. While probably too high, this is the maximum doubling time that has been suggested for a human population entering a previously unoccupied but favorable environment (Birdsell 1957). This baseline rate was not retained in all experiments. Human carrying capacity (L) was set at 1.295/mi^2, an upper limit of estimates of human density in Europe at the end of the Upper Paleolithic (Budyko 1967, 1974).

Minimum Kill and the Prey Destruction Rate

The seventh variable is annual destruction of animal units/person, or the prey destruction rate (G). Little agreement is found in the literature concerning reasonable estimates of daily human caloric needs, meat consumption, and wastage (see Lee 1968, 1972; Kemp 1971; Silberbauer 1972; Wheat 1972; Tanaka 1976; Tanno 1976; Webster 1981). The evidence used to support one view over another seems contradictory. Because Mosimann and Martin's choice of very high values for the prey destruction rate has been one of the major areas of contention surrounding their paper (Stauffer 1975, Webster 1981), it was decided to *derive* a baseline value for the rate by running the simulation with the other six variables set at baseline values and determining the minimum value of the prey destruction rate that would cause simulated megafaunal extinction. This threshold value is called the "minimum kill." It equals 3.862 a.u. and represents 3,862 pounds of animal killed per person annually, or 10.6 pounds daily, including waste and nonedible parts. Even if this baseline value exceeds prey destruction rates expected in the real world, it was derived from the interactions of the variables in the simulation and was acceptable for the purpose of examining the relationships between the variables.

After the set of baseline values had been chosen, the sensitivity analysis was performed. The first step in this procedure was to change the value of any one variable away from its baseline value. It was then determined by trial and error what changes, if any, had to be made to the remaining variables to achieve megafaunal extinction.

Effects of Variables on Megafaunal Extinction

The minimum kill was chosen as a convenient value for studying the effects of the other variables on simulated megafaunal extinction. Experimentation revealed that the variables which directly affect the minimum kill are human carrying capacity and those concerned with prey population growth: initial prey biomass, prey carrying capacity, and the prey biomass replacement rate. Figure 20.1 shows how the value of the minimum kill is changed when any of these variables is changed from its baseline value. The point where the solid and dashed lines intersect represents the value of the minimum kill when all other variables are set at baseline values.

As initial prey biomass and prey carrying capacity are increased, the minimum kill also increases by the same proportion. Increasing the prey biomass replacement rate also causes the minimum kill to increase, but not by the same proportion. The effect of increasing human carrying capacity is to decrease the minimum kill by the same proportion. Since changes in the prey biomass replacement rate (A) have relatively smaller effects on the value of the minimum kill than changes of equal proportion in the other variables, A is relatively less important in determining the values of the minimum kill and has relatively less effect on megafaunal extinction than do the other variables.

The broken horizontal and vertical lines in Figure 20.1 indicate that a value for the minimum kill of 1 a.u., for example, can be obtained in more than one way. Setting the prey biomass replacement rate equal to 0.05125 while leaving all other variables at baseline values has the same effect on the value of the minimum kill as setting the value of the human carrying capacity equal to 4.921 people/mi², or setting the values of initial prey biomass and prey carrying capacity equal to 6.5 a.u/mi². Thus, no single "correct" set of interactions between variables can cause the outcome of a simulation run to be megafaunal extinction.

Threshold Values

When all other variables are set at baseline values, a maximum value can be assigned to the prey biomass replacement rate above which megafaunal extinction will not occur. This maximum value has an identical relationship with the prey biomass replacement rate as the minimum kill has with the prey destruction rate: it is a threshold value. The value of the maximum is, in fact, the baseline value of the prey biomass replacement rate, and results from the interactions between all other variables. As with the prey destruction rate and the prey biomass replacement rate, each variable affecting extinction has its own baseline threshold value which will cause extinction.

A possible answer to Eiseley's (1943) question of why some New World species became extinct at the end of the Pleistocene, while closely related forms did not, can be found in the interaction between the prey biomass replacement rate and the minimum kill. Very small variations in the prey biomass replacement rate can mean the difference between simulated megafaunal extinction and survival. With the variables set at baseline values, extinction will occur as long as a prey species having a biomass replacement rate at or below 0.25 is consumed at a rate of 3.862 a.u., while a species with a prey biomass replacement rate above the threshold of 0.25 will survive. The largest animals, which today have the lowest biomass replacement rates, would become extinct first under intensive hunting (Budyko 1967, 1974).

Effects of Variables on Event Timing and Human Population Sizes

The important "events" in Mosimann and Martin's simulation are the arrival of the human population at the coasts and megafaunal extinction. Variables in the simulation which affect the length of time that it takes for these events to occur ("event timing") are initial human population size, human carrying capacity, the prey destruction rate, and the human population growth rate. Figure 20.2 shows that changes in the initial human population size and the prey destruction rate have identical effects on event timing, while changes in human carrying capacity have effects equal in proportion to but the inverse of the effects of those two variables. The human population growth rate has an inverse relationship with event timing and by far the greatest effect on it, since changes in the value of the growth rate cause much larger changes in event timing than do changes of equal proportion in any other variables.

The variables which affect human population sizes are the same as those which affect event timing. Changes in the prey destruction rate and human carrying capacity have large effects of equal proportion on human population sizes (see figs. 20.3 and 20.4). The prey destruction rate has an inverse relationship with the human population size; the prey destruction rate and human carrying capacity also have an inverse relationship. The other variables have only small effects on the population sizes, except that effects on the cumulative population size caused by changes in the human population

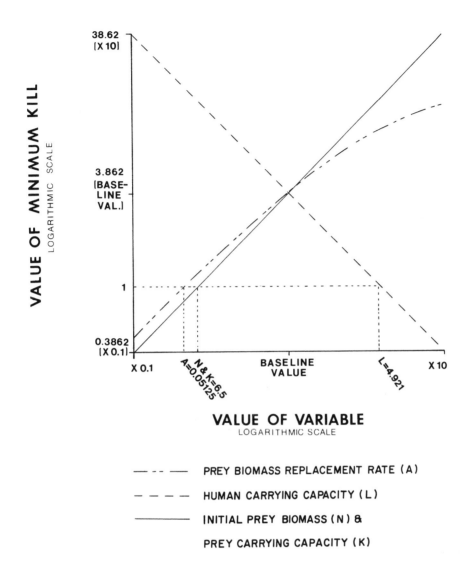

Figure 20.1. Changes in the value of the minimum kill caused by changes in the values of variables which affect it.

growth rate are similar to, but somewhat smaller than, those caused by changes in the prey destruction rate. Like the prey destruction rate, the human population growth rate has an inverse relationship with the cumulative population size.

Of the four variables critical to event timing and human population sizes, the human population growth rate is the most important in determining how quickly events occur, while human carrying capacity and the prey destruction rate play the greatest part in determining the sizes of the population at the times when events occur. Initial human population size plays a minor role in determining the speed with which events occur, sizes of the human population at the times when events occur, and cumulative human population size. A particular size for the human population at the time when an event occurs can result from more than one set of interactions between variables.

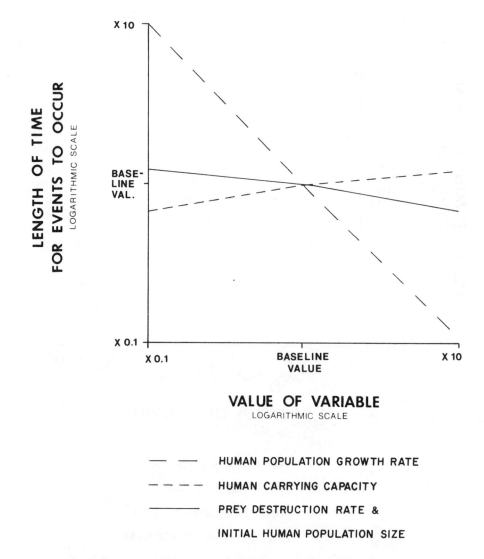

Figure 20.2. Changes in the length of time it takes for events to occur caused by changes in the values of variables which affect it.

Alternate Values for Variables

Because the human population growth rate is the most important factor in the speed with which humans inhabit the entire continent of North America, virtually any length of time for events to occur can be simulated by modifying only the human population growth rate (Table 20.2). The time necessary for humans to reach the coasts based on a 16-year doubling time is somewhat under 250 years, while a population which doubles in size every 160 years takes less than 2,300 years. A human population growth rate of 0.000443, which gives a doubling time of 1,600 years, causes the continent to fill with people in 22,000 years and is consistent with the growing number of archaeological sites dated from before 11,500 years ago (Rouse 1976, MacNeish 1976, Bryan et al. 1978).

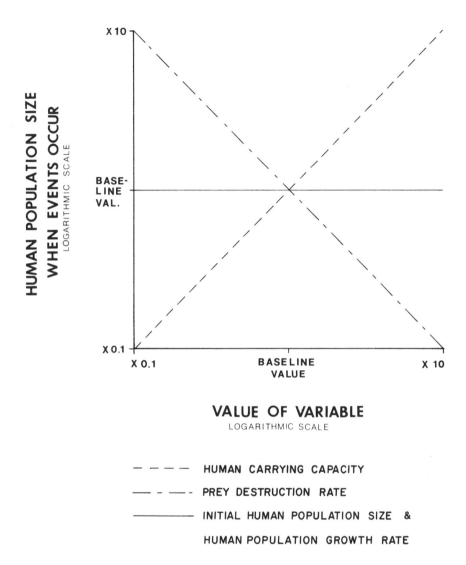

Figure 20.3. Changes in the human population size at the times when events occur caused by changes in the values of variables which affect it.

A simulation can be run using values more likely than those used by Mosimann and Martin to be considered acceptable by human ecologists and archaeologists. With a human carrying capacity of 1.295 people/mi², a prey destruction rate of 3.862 a.u., and all other variables set at baseline values, the resulting human population size when megafaunal extinction occurs is 3,884,860 (Table 20.3). In the real world, human population size may have diminished during the period immediately following megafaunal extinction (Budyko 1967, 1974; Caughley 1970).

Even very low values for the human carrying capacity or for the prey destruction rate can lead to simulated megafaunal extinction, so long as all the other variables which affect extinction (initial prey biomass, prey carrying capacity, and prey biomass replacement rate) are set properly to compensate for the low values. Extinction can be simulated with a human carrying capacity of 0.1295 people/mi², a lower limit of estimates

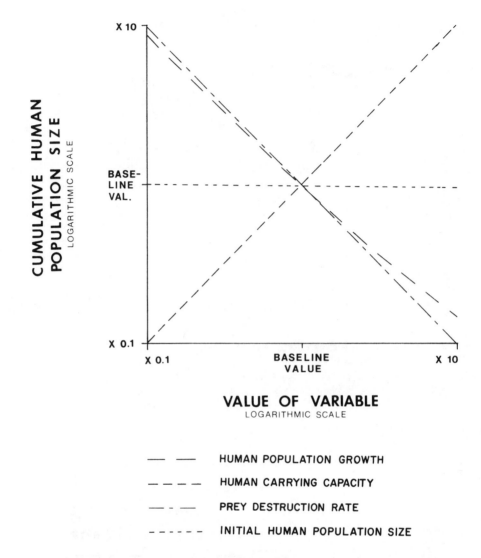

Figure 20.4. Changes in the cumulative human population size caused by changes in the values of variables which affect it.

Table 20.2. Effects of Changes in Human Population Growth Rate (C) on Event Timing and Human Population Sizes

Value of C	Years to Coasts	Population at Coasts	Years to Extinction	Population at Extinction	Cumulative Population
0.0443*	245	3,880,270	327	3,884,860	18,198,400
0.00443	2,243	3,875,010	2,798	3,884,120	122,814,000
0.000443	22,088	3,874,090	26,684	3,882,950	1,060,150,000

*Baseline value

**Table 20.3. Effects of Changes in Human Carrying Capacity (L)
on Event Timing and Human Population Sizes**

Value of L (people /mi²)	Years to Coasts	Population at Coasts	Years to Extinction	Population at Extinction	Cumulative Population
0.1295	192	388,115	269	388,486	1,743,630
1.295*	245	3,880,270	327	3,884,860	18,198,400
12.95	298	38,801,600	381	38,848,600	183,356,000

*Baseline value

of human density in Europe at the end of the Upper Paleolithic (Budyko 1967, 1974). However, either a prey destruction rate of 38.62 a.u. must be associated with that human carrying capacity (Table 20.4) or some other variable must be changed from its baseline value to counter the low value of the carrying capacity. Extinction can occur with a prey destruction rate of only 0.3862 a.u., equivalent to 1.6 pounds of prey per person daily, but the value of some other variable must be changed in compensation. If that variable is human carrying capacity, its value must become 12.95 people/mi², which results in a human population size of 38,848,600 people when extinction occurs. This is more people living in North America than are postulated to have existed in the entire world 10,000 years ago (Deevey 1960, Hassan 1978, Westing 1981).

It is possible to compensate for a low prey destruction rate without increasing the value of human carrying capacity (thereby ending up with an excessively high human population size) by lowering either prey carrying capacity or the prey biomass replacement rate. This could be done if Mosimann and Martin's estimates of values for prey carrying capacity or the prey biomass replacement rate were too high.

With certain values for human carrying capacity, the prey destruction rate, and the human population growth rate, it is possible to simulate extinction that is caused by a human population that kills the megafauna at a low rate and slowly grows to only a small population size by the time extinction occurs.

Model Minus Front

Major objections to Mosimann and Martin's simulation of overkill concern the front, a central concept of their model. Through a critical examination of the concept, the validity of these objections can be determined.

**Table 20.4. Effects of Changes in Human Carrying Capacity (L)
on Minimum Kill [G(min)]**

Value of L (people/mi²)	Value of G(min) (a.u.)
0.1295	38.62
1.295*	3.862
12.95	0.3862

*Baseline value

Population Distributions

The nonuniform distribution of human and animal populations over the North American continent which would result from the existence of a front in the model is certainly an oversimplification. Criticism has been leveled that the unbroken, densely populated front enclosing a region devoid of hunters of megafauna and their prey is too artificial a construct to be acceptable (Albini 1975, Webster 1981). The fact that North America is not a uniform, flat plane with uniformly distributed resources argues against the existence of a front. A human population density high enough to make the front impermeable might not exist everywhere, and barriers such as mountains or deserts could split up any front that might form, allowing prey animals to seep back through into uninhabited regions. In a reply to Albini, Mosimann and Martin admit that the front is a great simplification of reality.

Perhaps the most important result of setting the variables so that the prey destruction rate equals the minimum kill is that no front is ever formed. This is true even for values of the prey destruction rate somewhat greater than the minimum kill. As the human population grows, it forms an expanding inhabited region shaped like a quarter circle, centered at Edmonton and uniform in population density, as in Mosimann and Martin's simulation. The radially moving periphery of the region is arc-shaped, and thus resembles a front, except that the density of hunters at the periphery is no greater than at the point of origin. In this case, the periphery offers no barrier whatsoever to the movement of the megafauna into the region inhabited by humans. Because of the freedom of movement enjoyed by the prey animals, no region is ever vacated by the megafauna or by its hunters. Eventually the periphery reaches the coasts, and a uniformly dense human population coexists everywhere in the continent with a uniformly dense megafaunal population of the same biomass as before humans entered the New World.

This uniform distribution over the entire continent allows certain critical objections to the concept of a front to be avoided. Mountains and deserts, which would prevent the formation of an unbroken front, are not obstacles to megafaunal extinction if there is no densely populated, impenetrable band to be kept intact. As the human population grows, it can simply flow around the barriers into all regions containing the megafauna.

Certainly a population cannot exist everywhere on the continent at one density given a nonuniform distribution of resources, and this is a simplification of what is found in the real world. Even more troubling, it is not necessarily true that the same regions of the continent that could support high densities of the megafauna could also support high densities of humans. The basic requirements for survival for humans and megafauna probably were not the same, and the distribution of humans probably was not determined only by the distribution of the megafauna. However, despite the fact that the distributions of human and megafaunal populations may not always have overlapped, humans could have caused megafaunal extinction without a front wherever the populations did overlap.

A better approximation of reality than uniform population densities would be a model that allows for interactions between megafaunal and human populations whose densities were based on the distribution of various resources. Since this would be a radical departure from Mosimann and Martin's simulation, a reformulation of the model was not undertaken. The major point of this chapter is to present the results of experimentation with their simulation to expose the basic properties and interactions of the variables and what they suggest about the overkill hypothesis.

Population Budding and Prey Switching

Birdsell (1957) has shown evidence that human populations split or bud off into unoccupied areas at some point before the carrying capacity of a region is ever reached. One objection to the overkill simulation which depends upon a front is that the megafauna is hunted to extinction in a region inhabited by humans before the human population moves far enough to relieve the pressure on resources (Webster 1981). In a more realistic simulation, the human population might migrate into an unoccupied region or switch to more easily obtainable prey items as soon as it became energetically inefficient to continue hunting the megafauna. This would occur at some point before the megafauna became extinct and might allow the animal population to replenish itself (Krantz 1970).

In a simulation run in which the prey destruction rate is set so that it is the minimum kill and no front is formed, the budding process is constant, keeping the population from causing local megafaunal extinctions. The entire human population never has reason to vacate an area, which would cause the leading and trailing edges of the front to advance and make the front appear to jump, because the prey biomass never drops to zero as long as the budding proceeds. In fact, as long as the process continues, the prey biomass never reaches a point where its replacement rate fails to keep it at the same level as it was before humans entered the New World. Such a state of affairs in the real world would make prey switching unnecessary.

The budding process could theoretically continue to keep the animal biomass at prehunting levels for an infinite length of time on an unbounded plane, despite a restricted maximum annual radial movement of twenty miles for the front's periphery, since the human population growth rate begins to decrease as soon as the population density forces the maximum movement. However, new and unavoidable constraints on the movement of the periphery appear at the coasts and with the narrowing of the continent in Mexico, which cause the density of the human population to increase. Even after the constraints begin to take effect in the simulation, the megafaunal biomass remains at the same level as it was before humans entered the New World for what may be a greatly extended period of time.

Critical Density

Eventually the human population reaches the value described by Mosimann and Martin as "critical density," equal to AK/G^*, or L (human carrying capacity). At this threshold value, the prey destruction rate outstrips the ability of the megafauna to replace itself. This turns the prey destruction rate (which caused no decrease in prey biomass) into the minimum kill, the minimum prey destruction rate which will cause megafaunal extinction. All over the continent, prey biomass typically drops from prehunting period levels to zero in less than 100 years, no matter how long it has been since the human population first began to expand from the Edmonton center or to reach the coasts. Extinction can occur from 80 to 4,600 years after humans reach the coasts, depending on what value is used for the human population growth rate.

The only certain way to avoid megafaunal extinction is to reduce the human population growth rate to zero before the density reaches the threshold. No matter how slowly the human population continues to grow while still hunting at its old rate, once the threshold is reached the megafaunal population will fall quickly in density and size. This will occur even though the value of the prey destruction rate is only 0.0001 a.u. above what the prey biomass replacement rate will support, an amount presumably imperceptible to a hunter. Even if humans switch to more readily available prey when the animal biomass begins its crash, as long as they continue to occupy the entire continent and kill megafaunal prey when they come upon it, the extinction process will continue.

Budyko's Model

Budyko (1967, 1974) wrote the equations upon which Mosimann and Martin based their simulation. His model contains nothing resembling a front, however, and Wesler (1981) believes that it makes a better argument for the overkill hypothesis than

Mosimann and Martin's. Budyko attempts to demonstrate that extinctions which appear to have been caused by a single event (sudden climatological changes) may actually have been caused by a long-term process of human predation. While human populations with very low growth rates could hunt animals for tens of thousands of years with no perceptible effect, as soon as a threshold human density was reached, sudden massive extinctions would occur. His well-explained model demonstrates the possibility that humans alone could have been the agents of megafaunal extinctions.

Conclusions

By incorporating the concept of the front, Mosimann and Martin's simulation is able to explain the great suddenness with which Pleistocene extinctions occurred. However, the simulation has critics who question the realism of the parameter values used, and the front itself. A more robust model is one in which the general structure of the original simulation is preserved, but which allows for uniform predator population densities so that a front need not be formed. Eliminating the front answers major objections to the simulation and allows the model to remain useful.

A principal motive for including the front as a feature in the original model was Mosimann and Martin's attempt to account for the apparent scarcity of megafaunal hunting sites. Their idea was that kills occurred only within the area defined by the front as it moved swiftly over the landscape, so that archaeological deposition characteristic of long-term habitation did not have time to occur. However, there is growing evidence that humans lived in the New World for thousands of years before megafaunal extinction occurred. Depending on the point at which a human population density threshold is reached, our modified model permits extinction to occur in as few as 250, or as many as 22,000 years (or more). Determination of a reasonable figure will ultimately depend on better archaeological and ecological data. Meanwhile, the overkill hypothesis and the Mosimann and Martin model remain attractive and useful concepts.

Acknowledgments

This chapter would not have been possible if not for the great kindness of Paul S. Martin and James E. Mosimann in making available copies of their computer simulation and output. The comments and criticisms of David L. Webster and James W. Hatch have been invaluable aids to our understanding of the topic of overkill.

References

Albini, Frank A. 1975. Simulating overkill. *American Scientist* 63:500.

Birdsell, Joseph B. 1957. Some population problems involving Pleistocene man. Population studies: animal ecology and demography, *Cold Spring Harbor Symposium on Quantitative Biology.* 22:47–69.

Bryan, A., R. Casamiquela, J. Cruxent, R. Gruhn, and C. Ochsenius. 1978. An El Jobo mastodon kill at Taima-taima, Venezuela. *Science* 200:1275–1277.

Budyko, M. I. 1967. On the causes of the ex- tinction of some animals at the end of the Pleistocene. *Soviet Geography Review and Translation* 8(10):783–793.

———. 1974. *Climate and life.* Academic Press, New York.

Caughley, Graeme. 1970. Eruption of ungulate populations, with emphasis on Himalayan thar in New Zealand. *Ecology* 51:51–72.

Clark, J. D. 1959. *The prehistory of southern Africa.* Penguin Books, Harmondsworth.

Deevey, Edward S., Jr. 1960. The human popu- lation. *Scientific American* 203(3):194–204.

Edwards, William Ellis. 1967. The late-Pleistocene extinction and diminution in size of many mammalian species. In *Pleistocene extinctions: the search for a cause,* edited by Paul S. Martin and Herbert E. Wright, Jr., pp. 141–154. Yale University Press, New Haven.

Eiseley, Loren C. 1943. Archaeological observations on the problem of post-glacial extinction. *American Antiquity* 8:209–217.

Hallum, Sylvia J. 1977. The relevance of Old World archaeology to the first entry of man into new worlds: colonization seen from the antipodes. *Quaternary Research* 8:128–148.

Hassan, Fekri A. 1978. Demographic archaeology. In *Advances in archaeological method and theory.* volume 1, edited by Michael B. Schiffer, pp. 49–103. Academic Press, New York.

Jelinek, Arthur J. 1967. Man's role in the extinction of Pleistocene faunas. In *Pleistocene extinctions: the search for a cause,* edited by Paul S. Martin and Herbert E. Wright, Jr., pp. 193–200. Yale University Press, New Haven.

Johnson, Donald Lee. 1977. The California ice-age refugium and the Rancholabrean extinction problem. *Quaternary Research* 8:149–153.

Kemp, William B. 1971. The flow of energy in a hunting society. *Scientific American* 224(3):104–115.

Krantz, Grover S. 1970. Human activities and megafaunal extinctions. *American Scientist* 58:164–170.

Kroeber, Alfred L. 1940. Conclusions: the present status of Americanistic problems. In *The Maya and their neighbors,* edited by C. L. Hay, R. L. Linton, S. K. Lothrop, H. L. Shapiro, and G. C. Vaillant, pp. 460–487. D. Appleton-Century, New York.

Lee, Richard B. 1968. What hunters do for a living, or how to make out on scarce resources. In *Man the hunter,* edited by Richard B. Lee and Irven DeVore, pp. 30–48. Aldine, Chicago.

———. 1972. The !Kung Bushmen of Botswana. In *Hunters and gatherers today,* edited by M. G. Bicchieri, pp. 327–368. Holt, Rinehart and Winston, New York.

MacNeish, Richard S. 1976. Early man in the New World. *American Scientist* 64:316–327.

Martin, Paul S. 1967. Prehistoric overkill. In *Pleistocene extinctions: the search for a cause,* edited by Paul S. Martin and Herbert E. Wright, Jr., pp. 75–120. Yale University Press, New Haven.

———. 1973. The discovery of America. *Science* 179:969–974.

Mead, A. R. 1961. *The giant African snail.* University of Chicago Press, Chicago.

Mosimann, James E. and Paul S. Martin. 1975. Simulating overkill by Paleoindians. *American Scientist* 63:304–313.

Riney, T. 1964. The impact of introduction of large herbivores on the tropical environment. *International Union for the Conservation of Nature and Natural Resources Publications* (n.s.) 4:261–273.

Rouse, Irving. 1976. Peopling of the Americas. *Quaternary Research* 6:597–612.

Silberbauer, George B. 1972. The G/wi Bushmen. In *Hunters and gatherers today,* edited by M. G. Bicchieri, pp. 271–326. Holt, Rinehart and Winston, New York.

Stauffer, C. E. 1975. Overkill. *American Scientist* 63:380–381.

Tanaka, J. 1976. Subsistence ecology of Central Kalahari San. In *Kalahari hunter-gatherers,* edited by Richard B. Lee and Irven DeVore. Harvard University Press, Cambridge.

Tanno, T. 1976. The Mbuti net-hunters in the Ituri Forest, eastern Zaire. *Kyoto University African Studies* 10:101–136.

Wallace, Alfred Russel. 1911. *The world of life.* Moffat, Yard, New York.

Webster, David L. 1981. Late Pleistocene extinction and human predation: a critical overview. In *Omnivorous primates: gathering and hunting in human evolution,* edited by Robert S. O. Harding and Geza Teleki, pp. 556–594. Columbia University Press, New York.

Wesler, Kit W. 1981. Models for Pleistocene extinction. *North American Archaeologist* 2(2):85–100.

Westing, Arthur H. 1981. A note on how many humans have ever lived. *BioScience* 31(7):523–524.

Wheat, Joe Ben. 1972. The Olsen-Chubbuck site: a Paleo-Indian bison kill. *Memoirs of the Society for American Archaeology* 26(1).

Zaret, Thomas M., and R. T. Paine. 1973. Species introduction in a tropical lake. *Science* 182:449–455.

21

Extinction of Birds in the Late Pleistocene of North America

DAVID W. STEADMAN AND PAUL S. MARTIN

AT THE END OF THE PLEISTOCENE North American birds experienced more extinctions than any other group of organisms except the large mammals. These extinct forms included spectacular species, such as giant teratorns with wingspans of up to sixteen feet, and a variety of cathartid vultures, accipitrid vultures, hawks, and eagles. Birds are less often preserved in continental deposits than mammals, and the Pleistocene is no exception. While the fossil record of large mammalian genera has been relatively stable taxonomically over the past two decades, that of the Pleistocene avifauna has not. Descriptions of new faunas and revisions of old ones are changing its character.

To emphasize rarity of late Pleistocene extinction among small mammals, Martin (this volume, Tables 15.1 and 15.2) groups mammals into "large" (>44 kg) and "small" (<44 kg) categories. By this criterion, all extinct and living late Pleistocene birds are "small," and thus the 44-kg limit is not useful for North American birds. By avian standards the majority of extinct Pleistocene birds are rather large to very large. Late Pleistocene extinction affected flesh-eating birds, particularly carrion feeders, to a greater extent than the carnivorous mammals. Extinction in late Pleistocene mammals involved mainly large forms ("megafauna"), especially large herbivores such as ground sloths, mammoths, mastodonts, horses, and camels. Unlike large mammals, the avian losses typically are only a small fraction of the total bird fauna known from any given late Pleistocene locality. The differences between mammalian and avian extinction suggest different causal mechanisms for each group. How are avian extinctions to be explained?

Synonymy and Chronology

In his *Catalogue of Fossil Birds,* Brodkorb (1963, 1964, 1967, 1971, 1978) listed a total of twenty-eight extinct genera (Table 21.1) from the Pleistocene in North America north of Mexico and excluding the West Indies. Certain of these genera are now considered invalid. Others have not been subjected to rigorous osteological comparisons, and may well prove to be synonymous with living genera. Because of many unresolved problems in the systematics of Pleistocene birds, a discussion of each extinct species is premature. We consider only extinct genera of birds, facilitating comparisons with generic extinctions in mammals.

We generally omit citations for references readily obtainable in the species account in Brodkorb's catalogue. Describers' names will be provided only for those genera described from fossils. Taxonomic judgments are Steadman's. Some genera in Table 21.1 may be eliminated from any discussion of late Pleistocene extinction. The supposed extinct heron, *Palaeophoyx* McCoy, was based on two specimens, one of which has been assigned to the living bittern *Botaurus lentiginosus,* and the other to the living barn owl, *Tyto alba* (Olson 1974a). *Brantadorna* Howard and *Titanis* Brodkorb apparently represent mid- or early Pleistocene extinctions, *Brantadorna* being known only from the Irvingtonian Vallecito Creek fauna, California, and *Titanis* from two sites in Florida, Santa Fe IB (Blancan) and Inglis IA (Irvingtonian). The flightless duck of the Pacific coast, *Chendytes* L. Miller, survived the Pleistocene to become extinct in the Holocene, the only case of its kind. Extinctions of historic time include the formerly widespread passenger pigeon, *Ectopistes,* and the Carolina parakeet, *Conuropsis.*

The supposed occurrence in North America of the eagle, *Hypomorphnus,* was based originally on the proximal end of the tarsometatarsus and a furcular symphysis from Fossil Lake, Oregon, described as *Aquila sodalis* by Shufeldt (1891). These specimens were restudied by Howard (1946), who found the furcular fragment to be from a different eagle than the tarsometatarsal bit, which she tentatively referred to *Hypomorphnus,* a genus now generally synonymized with *Buteogallus* (Blake 1977). In light of the poor quality of the fossils and the unsettled nature of generic level systematics of large accipitrids, we discuss them no further.

Neortyx Holman is known from three elements (nine total specimens) from Reddick and Haile XIB, two Rancholabrean sites in Florida. Another quail, *Colinus,* occurs in each of these faunas, and in great abundance at Reddick. *Neortyx* is within the size range of *Colinus,* a genus that is very similar osteologically to the closely related genera *Callipepla, Lophortyx,* and *Oreortyx,* for which Holman (1961) had few comparative specimens. At present the validity of *Neortyx* is uncertain.

Agriocharis and *Parapavo* L. Miller are now regarded as congeneric with the living turkey *Meleagris* (Steadman 1980). Whatever the status of *Creccoides* Shufeldt (see Olson 1977), the Blanco fauna of Texas is now considered to be Pliocene in age. *Aramides,* the genus of living neotropical wood-rails, has been reported from the late

Table 21.1. Extinct Genera of North American Pleistocene Birds,
As Listed in Brodkorb's *Catalogue of Fossil Birds*

S = a genus believed to be synonymous with a living genus; EP = extinction in the early Pleistocene or Pliocene; H = extinction in the Holocene; see text for details.

Family	Genus	Family	Genus
Phoenicopteridae	*Phoenicopterus*	Phasianidae	*Neortyx*—S
Ardeidae	*Palaeophoyx*—S		*Agriocharis*—S
Ciconiidae	*Ciconia*		*Parapavo*—S
Anatidae	*Anabernicula*	Rallidae	*Creccoides*—EP
	Brantadorna—EP		*Aramides*—S
	Chendytes—H	Phorusracidae	*Titanis*—EP
Vulturidae	*Breagyps*	Charadriidae	*Dorypaltus*
Teratornithidae	*Teratornis*	Columbidae	*Ectopistes*—H
	Cathartornis	Corvidae	*Protocitta*
Accipitridae	*Hypomorphnus*—S		*Henocitta*
	Spizaetus	Icteridae	*Cremaster*
	Morphnus		*Pandanaris*
	Wetmoregyps		*Pyelorhamphus*
	Neophrontops		
	Neogyps		

Pleistocene Seminole Field fauna of Florida; Olson (1974b) has shown that the two specimens involved are of the living *Rallus elegans*. The remaining seventeen genera of extinct birds in Table 21.1 appear to be more defensible taxonomically. All are late Pleistocene in age and are worthy of further consideration in interpreting late Pleistocene extinction.

Avian Extinctions

Of the nineteen late Pleistocene extinct genera, ten are known from the Rancho La Brea tar pits (Table 21.2), in deposits that postdate the last interglacial and lie within range of ^{14}C dating (see Marcus and Berger, this volume). P. Martin (1958, p. 403) proposed that the extinct birds of the late Pleistocene are mostly scavengers or commensals of the mammalian megafauna. We pursue this approach in more detail and find that the loss of many if not all extinct genera of late Pleistocene birds can be attributed to ecological dependencies on large mammals. Other possible explanations include climatic change, climatically induced range contractions, direct human predation, or stochastic turnover.

Table 21.2. Extinct Late Pleistocene Genera of North American Birds

*Indicates a genus that survives today outside North America; Brackets [ ] indicate a genus whose status has been questioned; R = occurrence at Rancho La Brea; see text for details.

Family	Genus	Common Name
Phoenicopteridae	*Phoenicopterus*	flamingo
Ciconiidae,	*Ciconia*	stork—R
Anatidae	Anabernicula	shelduck
Vulturidae	Breagyps	condor—R
Teratornithidae	Teratornis	teratorn—R
	Cathartornis	teratorn—R
Accipitridae	*[Spizaetus]	hawk-eagle—R
	Amplibuteo (formerly *Morphnus*)	eagle—R
	Wetmoregyps	"walking-eagle"—R
	Neophrontops	Old World Vulture—R
	Neogyps	Old World Vulture—R
Falconidae	*Milvago*	caracara
Charadriidae	[Dorypaltus]	lapwing
Burhinidae	*Burhinus*	thick-knee
Corvidae	[Protocitta]	jay
	[Henocitta]	jay
Icteridae	[Cremaster]	hangnest
	[Pandanaris]	cowbird—R
	[Pyelorhamphus]	cowbird

The flamingo, *Phoenicopterus*, is known from deposits of two playa lakes (Fossil Lake, Oregon, and Manix Lake, California), the latter dating to the late Pleistocene. The nearest living populations of flamingos are those of *P. ruber* in the Caribbean region, where they breed locally in widely scattered places, suggesting refugia from human interference. Flamingos are not restricted to tropical climates, as shown by the modern range of *P. ruber* in southern Europe and Asia, and *P. chilensis* in southernmost South America. They are not scavengers. If the Manix Lake flamingo bred in North America, it probably inhabited saline playas of the Great Basin and Mohave deserts at times when they held shallow plankton-rich lakes. The intrusion of human hunters could have fatally disrupted these birds. On their breeding grounds flamingos are extremely vulnerable to nesting site disturbance (Brown 1959). With the arrival of prehistoric people nest robbing and disturbance of nesting birds could have been sufficient to cause a major contraction in the breeding range of American flamingos.

The stork, *Ciconia,* disappeared in the late Pleistocene of North America where it is recorded from California, Idaho, and Florida. Living storks often pursue small prey, as Grayson (1977) notes; others are well known for carrion feeding, e.g. the marabou, *Leptoptilos crumeniferus,* (Kahl 1966) and the wooly-necked stork, *Ciconia (Dissoura) episcopus* (see Mackworth-Praed and Grant 1962, 1970). Some commensal or scavenging role may be suggested in the case of North American storks (see fig. 21.1), including Cuba, where Pleistocene scavengers would have found large carrion in the form of endemic ground sloths and giant tortoises.

The shelduck, *Anabernicula* Ross, which occurs in the late Pleistocene of Oregon, California, Nevada, New Mexico, and Texas, is the only known case of extinction of an anatid genus in the late Pleistocene, evidently close to the time of megafaunal extinction. Migratory and wide ranging birds such as ducks and geese seem especially implausible candidates for extinction by climatic change. A scavenging or commensal niche can also be ruled out in the absence of any modern analogue, and we are left with the bare suggestion of prehistoric human impact. However, in reviewing the habits of living species of the Tribe Tadornini we find nothing that clearly suggests any special vulnerability to human predation.

The only other anatid genus to become extinct in prehistoric time was the flightless duck, *Chendytes,* which occurs abundantly in archaeological sites. According to Morejohn (1976, p. 210), "overharvest by aboriginal man was probably the principal factor contributing to the extinction of this species." *Chendytes* is unique among North American genera of extinct Pleistocene mammals and birds in being highly visible in certain archaeological sites.

The group suffering the greatest late Pleistocene extinction was the raptorial birds—the condors, teratorns, eagles, accipitrid vultures, and caracaras. They make up nine of the nineteen genera in Table 21.2. While feathers may have been sought, extermination by hunting is unlikely. The extinct condor, *Breagyps* L. Miller and Howard, is recorded from three late Pleistocene sites: Rancho La Brea, California; Smith Creek Cave, Nevada; and Shelter Cave, New Mexico (Howard 1971). The huge teratorns include *Teratornis* L. Miller and *Cathartornis* L. Miller. *Cathartornis* occurs only at Rancho La Brea, while *Teratornis* is more widespread. *Teratornis merriami* is known from three late Pleistocene faunas in California (Rancho La Brea, McKittrick, and Carpinteria), as well as the Seminole Field and Bradenton faunas in Florida. Webb (1974) assigns a latest Pleistocene age to Seminole Field, while putting Bradenton very early in the Rancholabrean land mammal age. A related species, *Teratornis incredibilis,* from the late Pleistocene of Smith Creek Cave, Nevada, was the largest known North American flying bird. The hawk-eagles, *Spizaetus,* are recorded from a variety of western late Pleistocene records for the accipitrid "Old World" vultures, *Neophrontops* L. Miller and Smith Creek Cave, Nevada; and Howell's Ridge Cave, New Mexico. *Morphnus woodwardi* is an eagle from Rancho La Brea that Campbell (1979) has assigned to an otherwise South American extinct genus, *Amplibuteo* Campbell. The long-legged "walking eagle," *Wetmoregyps* L. Miller, is known from Rancho La Brea and Carpinteria. Late Pleistocene records for the accipitrid "Old World" vultures, *Neophrontops* L. Miller and *Neogyps* L. Miller, include Rancho La Brea, McKittrick, and Carpinteria, with *Neophrontops* also at Dark Canyon Cave, New Mexico (Howard 1971), and *Neogyps* at Smith Creek Cave, Nevada. A caracara, *Milvago,* is known only from two late Pleistocene faunas in Florida—Itchtucknee River and Arredondo IIA (Campbell 1980).

The extinction of scavengers, such as *Breagyps, Teratornis, Cathartornis, Neophrontops,* and *Neogyps,* can be attributed to their dependency on the megafauna for a diverse and abundant supply of carrion. Presumably some preyed as well on very young, weak, or sick animals. For analogies one turns to those parts of Africa that will sustain a diverse megafauna. According to Houston (1975, p. 55), "Africa now supports more species of accipitrid, or Old World vultures than any other continent, up to seven species being found in any one area." It is not unusual to see five species feeding from the same carcass. One group with long, bare necks reaches far inside a carcass to slice with its sharp bill and withdraw slippery meat on a barbed tongue. These include the griffins, *Gyps africanus* and *G. ruepellii*. A second group that does not gather in large numbers has deep, powerful bills (like those of teratorns) to feed on tougher parts of the carcass such as skin and tendons; these include the lappet-faced vulture, *Torgos tracheliotus,* and the white-headed vulture, *Trigonoceps occipitalis.* Houston (1975) notes a third group of small, thin-billed vultures that takes lizards, insects, and dung as well as carrion and is also partial to village rubbish heaps. The hooded and Egyptian vultures, *Necrosyrtes monachus* and *Neophron percopterus,* belong here. On a strategy of its own, the lammergeyer (*Gypaetus barbatus*) ingests the shattered remains of large limb bones, carcasses, or even tortoises it has dropped from a height in mountain canyons.

In Natal the lappet-faced and white-headed vultures occur in game country; with the elimination of the large native mammals they disappear (Clancey 1964). Boshoff and Vernon (1980) have postulated that the reduced numbers of Cape vultures (*Gyps coprotheres*) in the Cape Province of South Africa is due to the drastic decline of game herds, as well as recent changes in cattle ranching. It seems logical to view the rapid extinction of most of North America's megafauna as having had a comparable effect on New World scavenging birds.

Many hawks and eagles that are generally considered predatory may also feed on carrion. For Africa, Brown (1970, p. 296) lists four eagles (*Aquila rapax, A. heliaca, Haliaeetus vocifer, Terathopius ecaudatus*), four hawks (*Buteo* spp.), and a kite (*Milvus migrans*) as eaters of the carrion of mammals weighing more than 10 kg. In India, Ali and Ripley (1968, pp. 227, 231, 253, 274–279, 286, 291, 325) note occasional to regular carrion feeding in two kites (*Milvus migrans, Haliastur indus*), a hawk (*Buteo rufinus*), six eagles (*Aquila chrysaetos, A. heliaca, A. rapax, A. nipalensis, Haliaeetus albicilla, H. leucoryphus*), and a harrier (*Circus aeruginosus*).

The Rancho La Brea tar deposits have yielded an abundance of scavenging birds. They match the large numbers of bones of scavenging mammals found there, such as the dire wolf (*Canis dirus*). Apparently the niche for all sorts of scavengers was expanded during the late Pleistocene. If the African analogy holds, it seems obvious that *Breagyps, Teratornis, Cathartornis, Neophrontops,* and *Neogyps* shared carrion-feeding duties at Rancho La Brea with other scavengers that survived, such as *Gymnogyps, Cathartes, Coragyps, Aquila, Polyborus,* and *Corvus* (fig. 21.1). Most surviving raptors, eagles as well as vultures, declined with the extinction of other guild members. The only scavenger whose bones increase at Rancho La Brea after the loss of the large mammals and birds is the turkey vulture, *Cathartes aura* (Howard 1962), a consumer of small carrion.

In addition to vultures we propose other raptorial genera as possible obligate scavengers of the Pleistocene megafauna. Grayson (1977, p. 692) has noted that *Wetmoregyps, Morphnus* (=*Amplibuteo*), and *Spizaetus* "most likely relied primarily upon small vertebrates for food, as do the extant congeneric relatives of two of them." However, *Wetmoregyps* is a long-legged "eagle" of unknown ecology; it may well have been a scavenger. The ecology of the extinct genus *Amplibuteo* is also highly speculative at present. *Spizaetus* is tropical in its modern range and is not known to scavenge. We are reluctant to consider any environmental change as sufficient to explain the retreat of *Spizaetus* into the truly tropical parts of Mexico, where it occupies regions dominated by *Ficus, Enterolobium, Inga, Bursera,* and other tropical trees. No tropical plants have been found in the rich paleoflora of Rancho La Brea (Johnson 1977) or at Smith Creek Cave, Nevada, where Pleistocene *Spizaetus,* as well as *Breagyps, Teratornis,* and *Neogyps* have been reported. The associated late Pleistocene fossil plant record at these localities is richer in species of northern rather than southern distribution. Thompson (1978) has shown that the late Pleistocene vegetation in the now treeless area of Smith Creek Cave prior to faunal extinction was a coniferous forest or woodland dominated by bristlecone pine (*Pinus longaeva*).

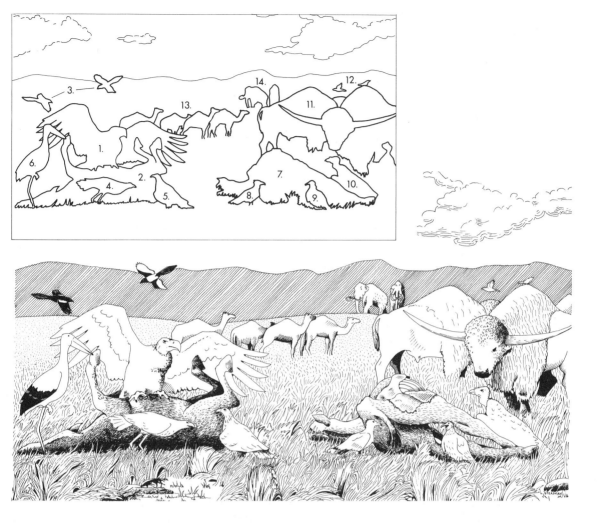

Figure 21.1. A scene from the late Pleistocene of western North America; * = extinct genus or species. On the left, (1) a teratorn, *Teratornis merriami,* stands upon the carcass of a (2) freshly dead horse, *Equus* sp. The teratorn has just scared away (3) two magpies *(Pica pica)* that are flying overhead. Below the teratorn, a (4) long-legged eagle, *Wetmoregyps daggetti,* rushes up to rip a piece of flesh from the carcass. To the right of *Wetmoregyps,* (5) a golden eagle *(Aquila chrysaetos)* pauses to keep an eye on the teratorn before resuming its feeding activities. To the left of the teratorn, (6) a stork, *Ciconia maltha,* awaits its chance to run up to the dead horse and grab a piece of meat or entrails that has been exposed by the teratorn or the eagles. On the right side of the scene is another dead horse (7). This animal has been dead for about a week. The teratorns and eagles have already removed most of the entrails, as well as much of the meat. On the left is an (8) accipitrid vulture, *Neophrontops americanus,* that is keeping its distance from the other scavengers as it looks for an opportunity to feed upon some soft, easy-to-tear flesh that has been exposed. To the right of *Neophrontops* is a (9) black vulture, *Coragyps atratus,* and (10) a large condor *Breagyps clarki.* The large size of *Breagyps* allows it to feed deeply within the body cavity of the horse. Behind *Breagyps* are several (11) long-horned bison, *Bison latifrons,* accompanied by two commensal (12) cowbirds, *Pandanaris convexa.* In the center background, a small herd of (13) camels, *Camelops hesternus,* and two (14) mammoths, *Mammuthus jeffersoni,* are moving through the grasslands. Sooner or later, they too will provide carrion for the avian scavengers.

It seems that Pleistocene populations of *Spizaetus* transcended their modern geographic and ecological range. Whether the extinct populations would have also changed their behavior cannot be determined. The remarkable decline of golden eagles (*Aquila chrysaetos*) and bald eagles (*Haliaeetus leucocephalus*) that accompanied extinction of the teratorns at Rancho La Brea (Howard 1962) suggests that at least locally a number of falconiform birds may have found more opportunity as scavengers in the time of the tar pit megafauna than afterwards. It is possible that *Spizaetus* shared in the generalized scavenging opportunity, one that allowed an extension of their range. All of this depends, of course, upon correct systematic assignment of the fossils currently placed in the genus *Spizaetus*.

Concerning extinction of *Milvago* in North America, Campbell stated (1980, p. 127) that "Many other factors may have been responsible for the extinction of *M. readei*, but the climatic and vegetational changes at the end of the Pleistocene that resulted in the loss of the dry savanna habitat in Florida were probably the most important." We suggest that extinction of large mammals may have been just as important in the North American loss of *Milvago*, a scavenger today confined to Costa Rica, Panama, and much of South America and known to associate with large herbivores. For example, Vuilleumier noted (1970, p. 3), "in Patagonia, *Milvago chimango* flocks are freqent near cattle and horses; I even saw one bird sitting on the flank of a lying horse, pecking from time to time at the skin, perhaps to eat ticks."

A cooling climate could certainly explain range shrinkage of larger raptorial birds that seasonally migrate south from higher latitudes. However, a climatic model that would propose the *complete* extinction of raptors, well known for their broad ecological amplitude, appears unpromising. At least within lower latitudes they appear to be independent of any particular climatic or vegetation type. If the loss of most large mammals drastically reduced the food supply for scavengers, a sequence of scavenger extinctions may be explained. Scavenging and commensal roles are mainly but not entirely the specialty of large raptorial birds. A few extinct genera of smaller birds, including some passerines, are ecologically dependent on large mammals.

Dorypaltus Brodkorb is known from four partial elements (eight total specimens) from the late Pleistocene sites of Arredondo IIA and Haile XIB, Florida. If *Dorypaltus* is indeed a lapwing, its generic distinction (as reported in Brodkorb 1959 and Ligon 1965) is not certain because comparisons were made with only two modern lapwings, the South American *Belonopterus chilensis* and the Eurasian *Vanellus vanellus*. Eight genera and eleven species of lapwings are resident in Africa (Peters 1934) where Mackworth-Praed (1962, 1970) reports that two African lapwings, the black-wing plover (*Stephanibyx melanopterus*) and blacksmith plover (*Hoplopterus armatus*), prefer grassy areas much used by cattle or large native mammals. In India the lapwing, *Vanellus vanellus*, often forages in "wet meadows or grazing land in proximity of cattle" (Ali and Ripley 1969, p. 211). Further speculations on the cause of extinction of *Dorypaltus* may be reserved until its status is clarified.

The thick-knee, *Burhinus*, is recorded from Dark Canyon Cave, New Mexico (Howard 1971), a locality tentatively dated at between 25,000 and 12,500 yr B.P. (Harris 1977), and the Sanborn Formation, Kansas, referred to the Sangamonian interglacial (Feduccia 1980). These fossils appear not to be very different from the living *B. bistriatus*, which reaches its northern limit on the Gulf slope of eastern Mexico with one record from southern Texas (MacInnes and Chamberlain 1963). Feduccia emphasized that modern species of *Burhinus* are characteristic of arid or semiarid tropical areas, and concluded that the same must be true for the fossil species. However, the stone-curlew, *B. oedicnemus*, occurs in temperate Eurasia as far north as the British Isles, Germany, and Poland (Bannerman 1962, p. 28). The Dark Canyon Cave record may be associated with xeric pinyon-juniper woodlands (Van Devender and Spaulding 1979, Thompson et al. 1980) and a much cooler climate than any modern tropical savanna. Possibly the ecological tolerances of American *Burhinus* were broader in the Pleistocene. We found no traits of living *Burhinus* that would suggest an intimate ecological relationship with large mammals.

Late Pleistocene extinction in the largest avian order, the passerines, is limited to members of two families, the corvids and icterids. Loss within these groups is in accord with our dependency model; no other American perching birds are as intimately associated with large mammals.

Protocitta Brodkorb is a supposedly extinct genus of jay. It is known from the late Pleistocene faunas of Reddick and Haile XIB, Florida, and Miller's Cave, Texas, as well as several pre-Rancholabrean faunas. *Protocitta* is very similar, particularly in size (Brodkorb 1972), to living neotropical jays of the genera *Psilorhinus* and especially *Calocitta*. Another extinct jay, *Henocitta* Holman, is known only from a single distal end of a humerus that is insufficient to establish its status firmly. Holman (1959) stated that *Henocitta* is closest to *Psilorhinus*. In his extensive osteology of Pleistocene passerine birds, Hamon (1964) did not compare *Protocitta* and *Henocitta* to *Calocitta* or *Psilorhinus*. We feel that these comparisons should precede detailed ecological speculation. *Calocitta* lives today as far north as central Sonora, whereas *Psilorhinus* reaches its present northern limit in the lower Rio Grande Valley of Texas. The late Pleistocene occurrence of tropical jays in Florida and central Texas would not be surprising and would complement the record of jaguars, ocelots, and other presently "tropical" animals that once occupied this region (see Webb 1974).

If either *Protocitta* or *Henocitta* prove not to be synonymous with living neotropical jays, it is possible that they represent an autochthonous American development of the piapiac or magpie niche, one that collapsed in America with late Pleistocene extinction of large mammals. The piapiac, *Ptilostomus afer,* is a black, rather magpielike African corvid that feeds on insects in close association with large mammals, both wild and domestic (Rice 1963). In the case of American magpies (*Pica pica*), Linsdale (1937) noted that carrion served as their main animal food in winter on the High Plains of North America. Magpies became scarce after buffalo declined in historic time. Winter road kills as well as the spread of domestic livestock apparently restored their food supply and led to their present recovery.

The icterid *Cremaster* Brodkorb is known from only three elements (four total specimens) from Arredondo IIA and Haile XIB, Florida. Until compared to a large series of Mexican forms such as *Amblycercus holosericeus, Cassiculus melanicterus, Dives dives,* and *Icterus gularis,* one cannot safely regard *Cremaster* as an extinct genus. Two other extinct genera of icterids have been described. *Pandanaris* A. Miller is known from Reddick and Haile XIB, Florida, and Rancho La Brea, California. *Pyelorhamphus* A. Miller is from Shelter Cave, New Mexico, the source of a late Wisconsinan biota indirectly associated with the radiocarbon-dated dung of an extinct ground sloth (Thompson et al. 1980). Both *Pandanaris* and *Pyelorhamphus* are cowbirdlike forms that could prove to be congeneric with *Molothrus* (including *Tangavius*). Based on mandibles and rostra instead of only postcranial elements, these genera seem to have a stronger basis than *Cremaster*. As pointed out by P. Martin (1958, p. 403), the extinction of both *Pandanaris* and *Pyelorhamphus* can be explained by proposing a close commensal relationship between them and large Pleistocene herbivores; if not "cowbirds," perhaps they were "mastodont birds."

Aside from the piapiacs already mentioned, other passerine associates of large African mammals include the oxpeckers, *Buphagus* (Family Sturnidae), which are scab feeders, and drongos, *Dicrurus* (Family Dicruridae), which are insect eaters (Rice 1963). Thus it is possible to postulate that a variety of commensal relationships existed independently in the New World, as in Africa, only to be lost with the late Pleistocene disappearance of most American large mammals.

Discussion

Within the late Pleistocene avifauna, human impact may have been the direct cause of extinction only in the case of the flamingo and one anatid, *Chendytes*. The latter, a flightless scoter, survived the extinction of large herbivores presumably to succumb in the Holocene when the coastal Indians of California improved their watercraft to reach the offshore islands where *Chendytes* bred. While details of even historic extinctions are poorly known, the passenger pigeon, *Ectopistes,* and Carolina parakeet, *Conuropsis,* are considered to have been victims of human impacts in historic time.

Grayson (1977) proposed to test the thesis of overkill by comparing the number of extinct birds with extinct mammals, regardless of size. He did this in an unusual way, dividing the number of late glacial extinct genera by the total number of known bird extinctions for the entire Pleistocene. Both birds and mammals yielded extinction ratios of just under 50 percent, which he took as discordant with an overkill model. Grayson's method obscures the fact that virtually all late Pleistocene mammalian extinctions are of large mammals, which is not the case earlier in the Pleistocene (see Martin this volume). Furthermore, unlike the large mammals, most late Pleistocene birds belong to living genera. For example, in the case of song birds, only five genera are listed as extinct, three icterids and two corvids (see Table 21.2) while fifty-one living genera have been reported (see Table 21.3). The late Pleistocene ratio is 10 percent, comparable to late Pleistocene avian extinction estimates of 15 percent previously obtained for Rancho La Brea (Howard 1962) and within the range noted in other late Pleistocene faunas in North America (Selander 1965). Had birds suffered as much extinction as did the *large mammals,* two thirds of the Wisconsin-age avifauna would be extinct.

Birds suffered more generic extinction within the Rancholabrean age than did the small mammals. We agree with Grayson (1977) that in most cases the avian extinctions in continental America are not to be attributed directly to human impact. At the same time we are impressed with the scavenging or commensal behavior that can be attributed to most of the genera listed in Table 21.2. Commensals or scavengers of large mammals are not known among the loons, grebes, gulls, grouse, shorebirds, and woodpeckers, all well represented in the Pleistocene fossil record (see Brodkorb 1963, 1964, 1967, 1971). None of these groups includes extinct genera. No generic extinction occurred among continental owls in the Pleistocene, to match that of the falconiform birds (Miller and de May 1942). We suggest that this is simply because owls, unlike vultures, do not, and did not, scavenge. There is a marked increase in owls and rodent-feeding hawks following megafaunal extinction at Rancho La Brea (Howard 1962; pits 37, 28, 10). This may mark a shift in mammalian biomass with an increase in mice and mouse-eaters after the ground sloths and mastodonts departed.

If climate had played an important role in avian extinctions of the late Pleistocene, one might expect the relatively rich Pleistocene avifaunas of Europe to yield examples, in view of the severe climatic changes that occurred there. However, no generic extinctions of birds are known in Eurasia in the Pleistocene. *Pliogallus* Gaillard, a phasianid listed from the Pleistocene of Hungary in Brodkorb's catalogue, is now considered Pliocene in age. All other birds in the Pleistocene avifaunas of Eurasia are assigned to living genera.

In a few cases avian losses may reflect range contractions of subtropical genera still living in Mexico or southern Texas. In most it appears that late Pleistocene extinct birds fall into groups well known for their scavenging or commensal behaviors. The reason more birds than small mammals became extinct is that very few small mammals (only vampire bats) were ecologically dependent on the megafauna. Late Pleistocene extinct birds of North America constituted 10 to 20 percent of the known fossil genera, far too few to be comparable to the accompanying late Pleistocene megafaunal extinctions, which exceeded 60 percent. No other groups of organisms were appreciably involved in the extinctions.

The avifaunal losses lend no support to any particular causal theory advanced to explain the extinction of the large mammals. Whatever caused megafaunal extinction will indirectly explain known avian extinctions. As in the case of the extinction of mammalian

Table 21.3. Living Genera of Passerines Reported From North American Pleistocene Avifaunas, Derived From Brodkorb (1978)

Question mark indicates uncertainty about identification or age assignment
(i.e. might be Holocene portion of a Pleistocene locality).

Family	Genus	Common Name
Tyrannidae	*Sayornis*	phoebe
	Tyrannus	kingbird
	Contopus	pewee
	Empidonax	empidonax flycatcher
Alaudidae	*Eremophila*	lark
Corvidae	*Gymnorhinus*	pinyon jay
	Aphelocoma	jay
	Cyanocitta	jay
	Perisoreus	gray jay
	Pica	magpie
	Nucifraga	nutcracker
	Corvus	crow, raven
Paridae	*Parus?*	tit
Sittidae	*Sitta*	nuthatch
Troglodytidae	*Troglodytes*	wren
	Salpinctes	wren
	Campylorhynchus?	cactus wren
	Cistothorus	marsh wren
Muscicapidae	*Toxostoma*	thrasher
	Oreoscoptes	sage thrasher
	Mimus	mockingbird
	Sialia	bluebird
	Catharus	thrush
	Turdus	robin
	Chamaea	wren-tit
Bombycillidae	*Bombycilla*	waxwing
Laniidae	*Lanius*	shrike
Vireonidae	*Vireo*	vireo
Coerebidae	*Mniotilta?*	warbler
	Geothlypis	yellowthroat
Tanagridae	*Pyrrhuloxia*	cardinal
	Pheucticus	grosbeak
Icteridae	*Agelaius*	blackbird
	Sturnella	meadowlark
	Quiscalus	grackle
	Euphagus	blackbird
	Molothrus	cowbird
	Dolichonyx	bobolink
Emberizidae	*Pipilo*	towhee
	Calamospiza	lark bunting
	Passerculus	savannah sparrow
	Ammodramus	sparrow
	Pooecetes	vesper sparrow
	Chondestes	lark sparrow
	Amphispiza	sparrow
	Spizella	sparrow
	Junco	junco
	Zonotrichia	sparrow
	Calcarius?	longspur
Passeridae	*Carduelis*	goldfinch
	Carpodacus	finch
	Coccothraustes	evening grosbeak
	Loxia	crossbill

predators, scavengers, or parasites, such as the sabertooth, extinct lion, dire wolf, and an extinct dung beetle, the avian losses appear to depend on the loss of large herbivores.

In terms of total fossil fauna, the extinction of Pleistocene birds is a minor exception to the view (Martin this volume) that small animals on the continent were unaffected by the late Pleistocene extinction phenomenon. Our conclusion can be falsified if future paleontological workers on this continent uncover many more extinct birds of late Pleistocene age, especially birds in groups that cannot be viewed as scavengers or commensals of the large mammals, or be regarded as easy prey for the prehistoric hunters. Bird extinction on oceanic islands is another matter (Cassels and Olson and James this volume).

Acknowledgments

Conversations on Pleistocene birds with Storrs L. Olson and Don K. Grayson and a critical review by Kenneth C. Parkes have been beneficial. Deborah Gaines and Debbie Rollman typed the manuscript. Figure 21.1 is by Lee M. Steadman.

References

Ali, S. and S. D. Ripley. 1968. *Handbook of the Birds of India and Pakistan,* v. 1. Oxford Univ. Press, Bombay, 380 pp.

——. 1969. *Handbook of the Birds of India and Pakistan,* v. 2. Oxford Univ. Press, Bombay, 345 pp.

Bannerman, D. A. 1962. *The Birds of the British Isles,* v. 11. Oliver and Boyd, London, 368 pp.

Blake, E. R. 1977. *Manual of Neotropical Birds,* v. 1. Univ. Chicago Press, 674 pp.

Boshoff, A. F. and C. J. Vernon. 1980. The past and present distribution and status of the Cape vulture in the Cape Province. *Ostrich,* v. 51, pp. 230–250.

Brodkorb, P. 1959. The Pleistocene avifauna of Arredondo, Florida. *Bull. Florida State Mus., Biol. Sci.,* v. 4, pp. 269–291.

——. 1963, 1964, 1967, 1971, 1978. Catalogue of Fossil Birds, parts 1–5. *Bull. Florida State Mus., Biol. Sci.,* v. 7, pp. 179–293; v. 8, pp. 195–335; v. 11, pp. 99–220; v. 15, pp. 163–266; v. 23, pp. 139–228.

——. 1972. Neogene fossil jays from the Great Plains. *Condor,* v. 74, pp. 347–349.

Brown, L. 1959. *The Mystery of the Flamingos.* Country Life Ltd., London, 116 pp.

——. 1970. *African Birds of Prey.* Houghton Mifflin Co., Boston, 320 pp.

Campbell, K. E., Jr. 1979. The non-passerine Pleistocene avifauna of the Talara tar seeps, northwestern Peru. *Life Sci. Contrib. Royal Ontario Mus.,* no. 118, 203 pp.

——. 1980. A review of the Rancholabrean avifauna of the Itchtucknee River, Florida. *In* Campbell, K. E., Jr., ed., Papers in avian paleontology honoring Hildegarde Howard. *Contrib. Sci., Nat. Hist. Mus. Los Angeles Co.,* no. 330, pp. 119–129.

Clancey, P. A. 1964. *The Birds of Natal and Zululand.* Oliver and Boyd, Edinburgh, 511 pp.

Feduccia, A. 1980. A thick-knee (Aves: Burhinidae) from the Pleistocene of North America, and its bearing on ice age climates. *In* Campbell, K. E., Jr., ed., Papers in avian paleontology honoring Hildegarde Howard. *Contrib. Sci., Nat. Hist. Mus. Los Angeles Co.,* no. 330, pp. 115–118.

Grayson, D. K. 1977. Pleistocene avifaunas and the overkill hypothesis. *Science,* v. 195, pp. 691–693.

Hamon, J. H. 1964. Osteology and paleontology of the passerine birds of the Reddick, Florida, Pleistocene. *Florida Geol. Surv., Geol. Bull.,* v. 44, pp. 1–210.

Harris, A. H. 1977. Wisconsin age environments in the northern Chihuahuan Desert: Evidence from the higher vertebrates. pp. 23–52 *In* Wauer, R. H. and D. H. Riskind, eds., Trans. Symp. Biol. Res. Chihuahuan Desert Region, United States and Mexico: *Nat. Park Serv. Trans. Proc. Ser.,* no. 3, 658 pp.

Holman, J. A. 1959. Birds and mammals from the Pleistocene of Williston, Florida. *Bull. Florida State Mus., Biol. Sci.,* v. 5, pp. 1–24.

———. 1961. Osteology of living and fossil New World quails (Aves, Galliformes). *Bull. Florida State Mus., Biol. Sci.,* v. 6, pp. 131–233.

Houston, D. C. 1975. Ecological isolation of African scavenging birds. *Ardea,* v. 63, pp. 55–64.

Howard, H. 1946. A review of the Pleistocene birds of Fossil Lake, Oregon. *Carnegie Inst. Washington,* Pub. 551, pp. 141–195.

———. 1962. A comparison of prehistoric avian assemblages from individual pits at Rancho La Brea, California. *Contrib. Sci., Los Angeles Co. Mus.,* no. 58, pp. 1–24.

———. 1971. Quaternary avian remains from Dark Canyon Cave, New Mexico. *Condor,* v. 73, pp. 237–240.

Johnson, D. L. 1977. The late Quaternary climate and paleoecology of coastal California. *Quat. Res.,* 8:154–179.

Kahl, M. P. 1966. A contribution to the ecology and reproductive biology of the Marabou Stork (*Leptoptilos crumeniferus*) in east Africa. *J. Zool.,* v. 148, pp. 289–311.

Ligon, J. D. 1965. A Pleistocene avifauna from Haile, Florida. *Bull. Florida State Mus., Biol. Sci.,* v. 10, pp. 127–158.

Linsdale, J. M. 1937. The natural history of magpies. *Pacific Coast Avifauna,* no. 25, 234 pp.

MacInnes, C. D. and E. B. Chamberlain. 1963. The first record of the Double-striped Thick-knee in the United States. *Auk,* v. 80, p. 79.

Mackworth-Praed, C. W. and C. H. B. Grant. 1962. *Birds of the Southern Third of Africa,* v. 1. Longmans, Green and Co., Ltd., London, 688 pp.

———. 1970. *Birds of West Central and Western Africa,* v. 1. Longman Group Ltd., London, 671 pp.

Martin, P. S. 1958. Pleistocene ecology and biogeography of North America. pp. 375–420 *in* Hubbs, C. L., ed., *Zoogeography.* American Assoc. Adv. Sci., Washington, D.C., 509 pp.

Miller, L. H. and I. de May. 1942. The fossil birds of California. *Univ. California Pub. Zool.,* v. 47, pp. 47–142.

Morejohn, G. V. 1976. Evidence of the survival to Recent times of the extinct flightless duck *Chendytes lawi* Miller. pp. 207–211 *in* Olson, S. L., ed., Collected papers in avian paleontology honoring the 90th birthday of

Alexander Wetmore. *Smithsonian Contrib. Paleobiol.,* no. 27.

Olson, S. L. 1974a. A reappraisal of the fossil heron *Palaeophoyx columbiana* McCoy. *Auk,* v. 91, pp. 179–180.

———. 1974b. The Pleistocene rails of North America. *Condor,* v. 76, pp. 169–175.

———. 1977. A synopsis of the fossil Rallidae. pp. 339–373 *in* Ripley, S. D., *Rails of the World: A Monograph of the Family Rallidae.* David R. Godine, Boston, 406 pp.

Peters, J. L. 1934. *Check-list of Birds of the World,* v. 2. Harvard Univ. Press, Cambridge, 401 pp.

Rice, D. W. 1963. Birds associated with elephants and hippopotamuses. *Auk,* v. 80, pp. 196–197.

Selander, R. K. 1965. Avian speciation in the Quaternary. pp. 527–542 *in* Wright, H. E. and D. G. Frey, eds., *The Quaternary of the United States.* Princeton Univ. Press, Princeton, 922 pp.

Shufeldt, R. W. 1891. Fossil birds from the *Equus* beds of Oregon. *American Nat.,* v. 25, pp. 818–821.

Steadman, D. W. 1980. A review of the osteology and paleontology of turkeys (Aves: Meleagridinae). pp. 131–207 *in* Campbell, K. E., Jr., ed, Papers in avian paleontology honoring Hildegarde Howard. *Contrib. Sci., Nat. Hist. Mus. Los Angeles Co.,* no. 330.

Thompson, R. S. 1978. Late Pleistocene and Holocene packrat middens from Smith Creek Canyon, White Pine County, Nevada: *Nevada State Mus. Anthro. Pap.,* no. 17, pp. 362–380.

Thompson, R. S., T. R. Van Devender, P. S. Martin, T. Foppe, and A. Long. 1980. Shasta ground sloth (*Nothrotheriops shastense* Hoffstetter) at Shelter Cave, New Mexico: Environment, diet, and extinction. *Quat. Res.,* v. 14, pp. 360–376.

Van Devender, T. R. and W. G. Spaulding. 1979. Development of vegetation and climate in the southwestern United States. *Science,* v. 204, pp. 701–710.

Vuilleumier, F. 1970. Generic relations and speciation patterns in the caracaras (Aves: Falconidae). *Breviora,* no. 355, pp. 1–29.

Webb, S. D. 1974. Chronology of Florida Pleistocene mammals. pp. 5–31 *in* Webb, S. D., ed., *Pleistocene Mammals of Florida.* Univ. Presses of Florida, Gainesville, 270 pp.

Asia and Africa: Modest Losses

Asia and Africa:
Modest Losses

Deposits rich in extinct animal remains in America and Australia rarely yield any archaeological evidence. According to Dewar, this is also true of Madagascar. It is not true, however, in Africa and Eurasia where bone-rich deposits typically are archaeological sites. For over a million years various species of Homo scavenged or hunted the Afro-Asian megafauna. While it is generally agreed that the early scavenger-hunters also harvested plants, the nature of the plant resources is as yet undisclosed.

Large animals were widely hunted in northern Eurasia, although evidence for this is much more abundant in Eastern Europe and the Ukraine than in Siberia, according to Vereshchagin. The Berelich site near the Siberian Arctic Coast is especially bone rich; however, like Agenbroad's Mammoth Hot Springs in South Dakota, no artifacts are directly associated with most of the deposit.

In China mammoth were present toward the end of the last glaciation but not apparently after 20,000 years ago, according to Liu and Li. The disappearance first in China (and the Soviet Maritime Province) and later in the Soviet Arctic seems to have come too soon to be concordant with the retreat of a subarctic-adapted animal as the climate warmed. Archaeological associations are numerous and Liu and Li suspect that Paleolithic man, not climate, drove the mammoth out of China in the last ice age.

Tchernov reviews the changing mammalian faunas of the Quaternary in Israel, where bones of large mammals come mostly from archaeological sites. While some ranges of large mammals shrank considerably in the last 100,000 years, extinctions were few.

Klein surveys the African extinction record and attributes the loss of several species and one genus of South African ungulate to improved hunting methods by stone age people 10,000 to 12,000 years ago. The African record is greatly exceeded by the heavy extinction of large genera in Madagascar immediately east of Africa around one thousand years ago. The association of prehistoric people and extinct fauna (elephant birds, giant lemurs, and a dwarf hippo) in Madagascar is unclear. No "elephant bird hunter" sites are known to match the moa hunter sites of New Zealand. As Dewar notes, the use of the giant eggs as containers may represent quarrying or gathering of subfossil material rather than plundering of fresh nests of the giant birds by the ancient Malagasy. Unless pre-Holocene extinct faunas are found in Madagascar, the comparison with Africa cannot be synchronous. However, the prehistoric fauna was rich, especially in large primates. It is hard to envision an earlier wave of Malagasy extinctions.

While many large mammals disappeared from parts of Africa and Asia, and the range of most large species shrank in the last 100,000 years, comparatively few genera were lost. With the exception of Madagascar, the record of Afro-Asian megafaunal survival, compared with late Pleistocene disappearances in America and Australia, remains the outstanding aspect of the fossil record.

22

Quaternary Mammalian Extinctions in Northern Eurasia

N. K. VERESHCHAGIN AND G. F. BARYSHNIKOV

THE EXTINCTION OF PHYLOGENETIC LINES and of animal species is a complex process that depends on many biological and physical factors. Methodologically, the paleontologist concerned with extinctions works above all with the presence or absence of species at various stratigraphic levels. Additionally, he must know whether species that are linked in phylogenies are really genetically related. Finally, in reconstructing the true history of Quaternary mammalian extinctions, the investigator must be aware of taphonomic problems and must use data from collateral sciences—geology, paleogeography, and archaeology. It must be realized that the latest known record of an extinct species does not record its final extinction, but rather the continued presence of a relatively large population (Efremov 1950). Only correct interpretation of morphological, paleozoogeographical, and taphonomic data can explain extinctions.

There are four major approaches to explaining the causes and tempo of Quaternary mammalian extinctions.

1. Investigation of the influences of natural (ecological) changes—in climate, in weather conditions, in the character of the soil, in landscapes and biotopes.
2. Investigation of lost resistance, morphological plasticity, and physiological adaptation in declining or extinct species, occurring as a result of nonadaptive evolution, of evolutionary inertia, or of orthogenesis and evolutionary dead-ends.
3. Investigation of the influence of predators, epizootic diseases, and biocenotic reorganizations.
4. Study of the destructive activity of man, by direct action or by altering species habitats.

Theoretically, it seems likely that the first, second, and third groups of factors predominated in the early stages of the Quaternary, while the fourth group dominated in the later stages. It is also obvious that the most intensive extinctions occurred as a result of complex combinations of the cited causes. By necessity, we limit ourselves in this chapter to the first and fourth groups.

The colossal size of northern Eurasia and its wide variety of climates and landscapes permitted Pleistocene mammals to survive in some places after they had become extinct in others. This fact complicates the overall picture of extinction but allows a realistic evaluation of the influence of natural factors and of man on separate species and on species complexes. In other words, data on changes in the geographic ranges of Quaternary mammals are useful for establishing the causes of extinction.

Translated by R. G. Klein

The development of species ranges and their dynamics are complex phenomena, but in essence there is the original expansion (the rapid spread of a species from its place of origin), followed by temporary reduction of the range or extinction, and, finally, secondary expansions from relict areas. Flourishing relicts as a rule do not disappear if they are not affected by human activity. In the first stages, extinction is usually linked with rapid or slow shrinkage in species ranges and in overall population size. In this process, regions that were continuously occupied may be broken up into separate areas of occupation. There is a series of cases in which these facts allow a ready understanding and explanation of extinction, especially when the matter is concluded by radical or partial transformation of the environment.

Progressive reduction of a species' range under natural influences often retraces the route of its original spread. Some Central Asiatic steppe species—the yellow lemming, the corsac fox, the kulan, and the saiga—provide an example. Over the last few centuries their ranges have shrunk rapidly from west to east, first as a result of ecological factors (as yet poorly studied) and secondly as a result of human activity in the nineteenth and twentieth centuries.

In this chapter we are concerned only with mammals in the concluding stages of the Quaternary period—the late (Upper) Pleistocene and Holocene. There are too few facts to explain extinctions in the Lower and Middle Pleistocene.

The Evolution of Mammalian Communities in Northern Eurasia During the Quaternary Period

At present about 530 species of terrestrial mammals are known from the Quaternary (including the Villafranchian) of Europe and the USSR. The number of species per order is as follows:

	Number of species known
Insectivora	35
Chiroptera	30
Primates	8
Lagomorpha	30
Rodentia	220
Carnivora	75
Proboscidea	14
Perissodactyla	28
Artiodactyla	100

As many as 182 species of terrestrial mammals are known from the late Pleistocene (Würm/Wisconsin) and Holocene of the USSR. Thirty of them are extinct. The number of species per order is as follows:

	Number of species known	Number extinct
Insectivora	5	—
Chiroptera	10	—
Primates	1	1
Lagomorpha	8	2
Rodentia	82	8
Carnivora	38	5
Proboscidea	1	1
Perissodactyla	6	4
Artiodactyla	31	9
Total	182	30

We must point out that it is not clear that all the "extinct" species truly are extinct. Much of the problem is that some living species may in fact be derived from ones said to be extinct. Domestication introduces complications in this respect. Thus, for example, among the Carnivora, the Volga wolf (*Canis volgensis* M. Pavl.) is only provisionally extinct, insofar as its genes are probably present in numerous breeds of domestic dogs.

Also problematic are the cave hyena of the genus *Crocuta* and the cave lion of the genus *Panthera*. It is possible that these are only cold-adapted northern subspecies of still extant African-Asiatic tropical species.

Some Pleistocene lagomorphs and rodents described as distinct species may in fact be the ancestors of present-day forms. In particular, this conjecture pertains to the Don hare (*Lepus tanaiticus* Gureev)—a probable ancestor of the present day arctic hare (or its collateral branch?)—and to some ground squirrels. In the Perissodactyla the fossil horse (*Equus ferus* Bodd. s. lato) is only provisionally extinct, since one of its subspecies undoubtedly gave rise to the flourishing breeds of domestic horses. Similarly, in the Artiodactyla the aurochs (*Bos primigenius*) is only provisionally extinct, since its genes are certainly present in modern, large-horned cattle.

Given the huge size of the USSR, we must use subregions to list the species of late Quaternary mammals which are known in fossil form (Table 22.1)

Table 22.1. The Species of Late Quaternary Mammals Known in Fossil Form in the USSR

Order and Species	Russian Plain and Crimea	Caucasus	Central Asia	Siberia	Far East
Insectivora					
Erinaceus europaeus L.	H	PH	—	H	—
Erinaceus amurensis Schrenk	—	—	—	—	PH
Desmana moschata L.	PH	—	—	—	—
Talpa caucasica Satunin	—	PH	—	—	—
Sorex sibirica Dukelsky	—	—	—	PH	—
Crocidura russula Güldenstaedt	—	PH	—	—	—
Chiroptera					
Rhinolophus mehelyi Matschie	—	PH	—	—	—
Rhinolophus ferrum equinum Schreber	H	PH	H	—	—
Myotis blythi Tomes	H	PH	H	H	—
Nyctalus noctula Schreber	PH	H	H	H	—
Nyctalus lasiopterus Schreber	PH	H	H	—	—
Eptesicus serotinus Schreber	PH	H	H	—	—
Eptesicus nilssoni Keyserling et Blasius	PH	H	H	PH	—
Vespertilio murinus L.	PH	PH	PH	H	H
Miniopterus schreibersi Kuhl	H	PH	PH	—	H
Primates					
Macaca sp.	—	P	—	—	—
Lagomorpha					
Lepus tanaiticus Gureev	P	—	—	P	—
Lepus timidus L.	?H	—	—	?H	—
Lepus tolai Pallas	—	—	PH	?H	—
Lepus europaeus Pallas	PH	PH	H	H	—
Ochotona daurica Pallas	—	—	—	PH	—
Ochotona alpina Pallas	—	—	—	PH	—
Ochotona azerica Gadziev et Aliev	—	P	—	—	—
Ochotona pusilla Pallas	PH	—	H	—	—

NOTE: P = late Pleistocene; H = Holocene; * = extinct species

Table 22.1. The Species of Late Quaternary Mammals Known in Fossil Form in the USSR
(continued)

Order and Species	Russian Plain and Crimea	Caucasus	Central Asia	Siberia	Far East
Rodentia					
Pteromys volans L.	P	—	—	?H	—
Sciurus vulgaris L.	PH	—	—	H	H
Sciurus anomalus Gmelin	—	PH	—	—	—
Tamias sibiricus Laxmann	H	—	—	H	H
Spermophilus undulatus Pallas	—	—	H	PH	—
Spermophilus glacialis Vinogradov	—	—	—	P	—
Spermophilus relictus Kashkarov	—	—	PH	—	—
Spermophilus severskenis I. Gromov	P	—	—	—	—
Spermophilus suslicus Güldenstaedt	PH	—	—	—	—
Spermophilus muscoides I. Gromov	P	—	—	—	—
Spermophilus musicus Menetrie	—	PH	—	—	—
Spermophilus pygmaeus Pallas	PH	H	H	—	—
Spermophilus superciliosus Kaup	P	—	?	?	—
Spermophilus fulvus Lichtenstein	PH	—	H	—	—
Spermophilus erythrogenys Brandt	—	—	H	PH	—
Marmota bobac Müller	PH	—	H	H	—
Marmota paleocaucasica Baryshnikov	—	PH	—	—	—
Marmota marmota L.	P	—	—	—	—
Marmota baibacina Kastschenko	—	—	H	PH	—
Marmota sibirica Radde	—	—	—	PH	—
Marmota camtschatica Pallas	—	—	—	PH	—
Marmota caudata Geoffroy	—	—	PH	—	—
Castor fiber L.	PH	PH	—	PH	H
Castor canadensis Kuhe	—	—	—	?	—
Hystrix vinogradovi Agryropulo	P	P?	—	—	—
Hystrix leucura Sykes	—	PH	PH	—	—
Dryomys nitedula Pallas	H	PH	H	—	—
Glis glis J.	P	PH	—	—	—
Sicista subtilis Pallas	PH	—	H	H	—
Sicista caucasica Vinogradov	—	?H	—	—	—
Allactaga jaculus Pallas	PH	—	H	H	—
Allactaga elater Lichtenstein	PH	H	H	—	—
Allactaga williamsi Thomas	—	PH	—	—	—
Pygerethmus platyurus Lichtenstein	P	—	H	—	—
Alactagulus pygmaeus Pallas	PH	H	H	—	—
Dipus sagitta Pallas	PH	H	H	H	—
Scirtopoda telum Lichtenstein	PH	H	H	H	—
Paradipus ctenodactylus Vinogradov	—	—	—	PH	—
Nannospalax leucodon Nordmann	?H	—	—	—	—
Spalax microphtalmus Güldenstaedt	?H	PH	—	—	—
Apodemus sylvaticus L.	PH	PH	H	H	—
Apodemus flavicollis Melchior	?H	H	—	—	—
Mus musculus L.	PH	H	H	H	H
Rattus rattoides Hodgson	—	—	PH	—	—
Nesokia indica Gray	—	—	PH	—	—
Ellobius lutescens Thomas	—	PH	—	—	—
Ellobius talpinus Pallas	PH	—	H	—	—
Ellobius tancrei Blasius	—	—	PH	—	—
Allocricetulus eversmanni Brandt	PH	—	H	H	—
Tscherskia albipes Ugnev	—	—	—	—	PH
Cricetulus argyropuloi I. Gromov	—	P	—	—	—
Cricetulus migratorius Pallas	PH	PH	H	H	—
Mesocricetus raddei Nehring	—	PH	—	—	—
Cricetus cricetus L.	PH	PH	H	PH	—
Merionus erythrourus Gray	—	H	PH	—	—
Merionus meridianus Pallas	PH	H	H	—	—

Table 22.1. The Species of Late Quaternary Mammals Known in Fossil Form in the USSR

(continued)

Order and Species	Russian Plain and Crimea	Caucasus	Central Asia	Siberia	Far East
Rombomys opimus Lichtenstein	—	H	PH	—	—
Myospalax myospalax Laxmann	—	—	—	PH	—
Myospalax aspalax Pallas	—	—	—	PH	—
Prometheomys schaposchnikovi Satunin	—	PH	—	—	—
Clethrionomys rufocanus Sunderval	H	—	—	PH	H
Clethrionomys glareolus Schreber	PH	H	—	PH	—
Clethrionomys rutilus Pallas	H	—	—	PH	H
Lagurus lagurus Pallas	PH	—	H	PH	—
Eolagurus luteus Eversmann	P	—	H	PH	—
Dicrostonyx quillielmi Sanford	P	—	—	P	—
Dicrostonyx torquatus Pallas	?H	—	—	?H	—
Lemmus sibiricus Kerr	PH	—	—	PH	—
Myopus schisticolor Lilljeborg	H	—	—	H	—
Arvicola terrestris L.	PH	PH	H	—	—
Pitymys subterraneus Selus-Longchamps	PH	—	—	—	—
Pitymys majori Thomas	—	PH	—	—	—
Pitymys daghestanicus Schidlovskii	—	PH	—	—	—
Microtus gregalis Pallas	PH	—	H	PH	—
Microtus socialis Pallas	?H	PH	H	—	—
Microtus fortis Buchner	—	—	—	PH	H
Microtus maximowiczii Schrenk	—	—	—	H	P
Microtus oeconomus Pallas	PH	—	H	PH	—
Microtus agrestis L.	PH	—	—	PH	—
Microtus arvalis Pallas	PH	PH	—	PH	—
Microtus transcaspicus Satunin	—	—	PH	—	—
Lasiopodomys brandti Radde	—	—	—	PH	—
Chionomys gud Satunin	—	PH	—	—	—
Carnivora					
Nyctereutes procyonides Gray	—	—	—	—	PH
Canis lupus L.	PH	PH	PH	PH	PH
Canis aureus L.	H	PH	PH	—	—
Canis volgensis M. Pavlova	PH	—	—	—	—
Alopex lagopus L.	PH	—	—	PH	—
Vulpes vulpes L.	PH	PH	PH	PH	PH
Vulpes corsac L.	PH	H	H	H	—
Cuon alpinus Pallas	H	PH	H	PH	PH
Selenarctos thibetanus G. Cuvier	—	?	—	—	?H
Ursus arctos L.	PH	PH	H	PH	PH
Ursus spelaeus Rosenmüller et Heinroth	P	PH	—	—	—
Ursus rossicus Borissiak	P	P	—	P	—
Thalarctos maritimus Phipps	—	—	—	PH	—
Martes zibellina L.	?H	—	—	PH	PH
Martes martes L.	PH	PH	—	—	—
Martes foina Erxleben	H	PH	H	—	—
Martes flavigula Boddaert	—	—	—	—	PH
Gulo gulo L.	PH	P	—	PH	PH
Mustela erminea L.	H	H	H	PH	?H
Mustela nivalis L.	PH	PH	H	PH	H
Mustela sibirica Pallas	H	—	—	PH	PH
Mustela altaica Pallas	—	—	H	?H	H
Putorius eversmanni Lesson	PH	H	H	PH	—
Vormela peregusna Güldenstaedt	H	PH	H	—	—
Meles meles L.	PH	PH	H	PH	PH
Lutra lutra L.	PH	H	H	PH	PH

Table 22.1. The Species of Late Quaternary Mammals Known in Fossil Form in the USSR

(continued)

Order and Species	Russian Plain and Crimea	Caucasus	Central Asia	Siberia	Far East
Crocuta spelaea Goldfuss	P	P	?	P	P
Hyaena hyaena L.	—	H	PH	—	—
Panthera leo L.	H	H	—	—	—
Panthera spelaea Goldfuss	P	P	P	P	?
Panthera tigris L.	—	H	H	—	PH
Panthera pardus L.	?	PH	PH	—	PH
Uncia uncia Schreber	—	—	PH	H	—
Felis silvestris Schreber	PH	PH	—	—	—
Felis libyca Forster	H	H	PH	—	—
Felis euptilura Elliot	—	—	—	—	?H
Felis chaus Güldenstaedt	PH	PH	H	—	—
Lynx lynx L.	PH	PH	H	H	PH
Proboscidea					
Mammuthus primigenius Blumenbach	P	P	?	P?	P
Perissodactyla					
Equus ferus Boddaert	PH	PH	PH	PH	PH
Equus przewalskii Poliakov	—	—	?H	?	—
Equus hydruntinus Regalia	P	P	?	—	—
Equus hemionus Pallas	H	H	PH	PH	—
Dicerorhinus kirchbergensis Jaeger	—	P	—	—	—
Coelodonta antiquitatis Blumenbach	PH	?	P	P	P
Artiodactyla					
Sus scrofa	PH	PH	PH	H	PH
Camelus knoblochi Nehring	PH	?	PH	—	—
Moschus moschiferus L.	—	—	—	H	PH

A Review of Some Extinct Pleistocene Mammals

The Mammoth—*Mammuthus primigenius* Blumenbach, 1799

We can now trace the evolution of north Eurasian mammoths from the Lower Pleistocene, when thin-enameled teeth of large elephants appear for the first time. It is possible, however, that they derive from archidiskodons (genus *Archidiskodon*), which convergently developed "mammoth features" as an adaptation to the severe conditions of early glaciations in northeast Siberia.

For stratigraphic purposes, many Soviet paleontologists and geologists accept the hypothesis that mammoths (genus *Mammuthus*) originated from archidiskodon elephants: Gromov's archidiskodon (*Archidiskodon gromovi* Garutt et Alexeeva), southern archidiskodon (*A. meridionalis* Nesti), and Taman archidiskodon (*A. tamanensis* Dubrovo), from the Upper Villafranchian (Khapry and Taman faunal complexes), through the intermediate form of the trogontherii elephant (*A. trogontherii* Pohlig), characteristic of the early Pleistocene (Tiraspol complex) (Gromov 1948, Dubrovo 1964, Gromova 1965, Alekseeva 1977, Garutt 1981).

Vereshchagin believes, however, that it is more likely that the archidiskodons and mammoths split in the latest Pliocene, while the mammoths and the closely related Asiatic elephant *Elephas maximum* L.) separated in the Lower Pleistocene. Thenius (1980) holds a similar opinion.

Table 22.1. The Species of Late Quaternary Mammals Known in Fossil Form in the USSR

(continued)

Order and Species	Russian Plain and Crimea	Caucasus	Central Asia	Siberia	Far East
Dama mesopotamica Brooke	—	P?	—	—	—
Cervus nippon Temminck	—	—	—	—	PH
Cervus elaphus L.	PH	PH	PH	PH	PH
Capreolus capreolus L.	PH	PH	—	—	—
Capreolus pygargus L.	P	H	PH	PH	PH
Megaloceros giganteus Blumenbach	P	P	P	P	—
Alces alces L.	PH	PH	—	PH	PH
Rangifer tarandus L.	PH	—	—	PH	?
Bos primigenius Bojanus	PH	P	PH	P	?
Poephagus baikalensis N. Vereshchagin	—	—	—	P	?
Bison priscus Bojanus	P	P	P	PH	P
Bison bonasus L.	H	?H	—	—	—
Spirocerus kiakhtensis M. Pavlova	—	—	—	P	—
Gazella subgutturosa Güldenstaedt	H	H	PH	—	—
Procapra gutturosa Pallas	—	—	—	PH	—
Parabubalis capricornis V. Gromova	—	—	—	?	—
Saiga tatarica L.	PH	PH	PH	P	—
Ovibos moschatus Zimmermann	P	—	—	PH	—
Naemorhedus caudatus Milne-Edwards	—	—	—	—	PH
Rupicapra rupicapra L.	—	PH	—	—	—
Capra aegagrus Erxleben	—	PH	H	—	—
Capra sibirica Pallas	—	—	PH	PH	—
Capra caucasica Güldenstaedt et Pallas	—	PH	—	—	—
Capra prisca Woldrich	—	P	—	—	—
Ovis orientalis Gmelin	—	PH	—	—	—
Ovis vignei Blyth	—	—	PH	—	—
Ovis ammon L.	?	P	H	PH	—
Ovis nivicola Eschscholz	—	—	—	PH	—

The cold-adapted species of mammoth apparently evolved in northeast Asia. It was already widely distributed throughout the Eurasian periglacial zone in the Middle Pleistocene (Mindel/Riss and Riss). However, the large mammoths of the Khazar fauna of eastern Europe are sometimes regarded as a distinct species, *Mammuthus chosaricus* Dubrovo.

M. primigenius was most widely distributed in the Riss/Würm and especially in the Würm (Valdaj), when mammoths occurred across Eurasia from the Atlantic to the Pacific, and through Alaska into North America. They lived north of the Arctic Circle and the margins of the Arctic basin and spread south to the edges of the Central Asiatic and Mongolian deserts. In Europe they reached Spain, the southern tip of Italy, and the Transcaucasus.

The remains of mammoths are common in Mousterian sites in the Crimea (Chokurcha) and northern Caucasus (Il'skaya, Dakhovskaya), but they do not occur in Upper Paleolithic sites there. By this time the southern limit of mammoth distribution had apparently retreated northward.

Research by Soviet biologists, paleogeographers, and geologists has now produced firm ideas on the biology and ecology of the mammoths. In the north these elephants were well adapted to a dry, cold, and sharply continental climate. The thick cover of hair, the wool on the trunk, and the abundant fat deposits under the skin bear witness to this adaptation. The investigation of stomach contents has provided data on diet.

In summer mammoths fed primarily on herbage—prairie grasses, sedges, cotton grass, and the terminal shoots of shrubs (willow, birch, and alder). They tore the bark off willow and larch with their tusks. In winter, when water bodies froze and snow was absent, mammoths could apparently obtain water by using their tusks to scrape ground ice from the vertical walls of cliffs or from subsurface cracks. This hypothesis seems to be confirmed by lateral wear commonly found on the ends of tusks even in young individuals, and by the frequency of fractured tusks, probably broken in such activity. As yet, there are no data on the winter diet of mammoths, but it probably consisted of dried grass and shoots of leaf-bearing shrubs and conifers (larch, pine, and fir).

Mammoths apparently undertook long southerly treks along the river valleys, especially in the event of heavy snowfalls and droughts. They very often died during flash floods in the valleys and floodplains of the rivers or when they tried to cross the ice of lakes and rivers that was insufficiently solid and masked by early snowfalls.

The bodies of mammoths were buried in earth slides at the base of slopes, in soil flows thawed by the sun, in fluvial silts and sands, and in deltas. Freezing in the sediments, the bodies were preserved for millenia. Rare but excellent burials of carcasses occurred when animals fell into insidious cracks eroded in ground ice by small streams. Soft tissues and even whole bodies were preserved to the present because the mammoths lived in a permafrost environment, with winter temperatures of minus 60–80 degrees centigrade.

In northeast Siberia (Yakutia) frozen mammoth carcasses were buried in windborne loessic silts that form the covering (Sartan) loam deposits. This phenomenon is shown by the presence of thick ice seams that pass to a depth of 30–40 m and press up columns of powdery soil with bones and, in places, disarticulated carcasses of mammoths. These facts, confirming the cold adaptation of the mammoths, are vital for explaining their life, death, and extinction.

One very puzzling aspect of mammoth history is the fact that so few remains are known from the period beyond 50–60,000 years ago, that is, from before the last glacial epoch (Würm, Wisconsin), and also from the period beyond the range of radiocarbon dating. At the same time, taphonomic theory suggests that conditions appropriate for the burial of mammoth skeletons must have been abundant in the preceding Riss/Würm (Dnepr/Valdai) interglacial, when there was extensive ground thawing, erosion, and reconstitution of the draining system (fig. 22.1).

Whenever Pleistocene extinctions are considered, mammoths always receive the most attention. For almost two centuries now, when solid facts and ecological and taphonomic observations were lacking, scholastic discussions or fantastic musings have dominated logical deductions. From the time of Peter the Great and the first printed articles about the mammoth by the state councilor and scholar Tatishchev (Tatishchev and Gmelin 1730), it became popular to think that the mammoth lived in a warmer climate, similar to the African and Asiatic elephants, and that it perished when the climate got colder. This idea has survived right to the present among both scholars and laymen.

In order to explain the massive death and preservation of skeletons and of frozen mammoth carcasses, it has been common to resort to the Deluge hypothesis in its various forms. Usually, the idea is that Siberian rivers transported the carcasses of mammoths from south to north (Middendorf 1860). More rarely, transport is ascribed to a giant wave that was driven through the Himalayas following volcanic eruptions (Pallas 1773) or by the force of a meteor striking the Pacific Ocean. People who are not acquainted with the nature of the buried carcasses and bones believe it is quite possible that the animals perished when an asteroid captured a portion of the atmosphere, leading to rapid deterioration in the climate of Siberia. Howorth (1887) presented a detailed version of the Deluge hypothesis in a broad review of how animals get buried. For its abundant examples of various types of burial in Quaternary deposits, Howorth's century-old account has no equal.

Figure 22.1. Head of the Katanga mammoth (Tajmyr Peninsula), dated to 53,000 years ago. (Photo by N. K. Vereshchagin, 1978)

The first rational ideas on the life of mammoths and on the relationship between the frozen carcasses and the ground ice of the Arctic appeared in the works of the Russian Polar Expedition under the leadership of Toll'. These ideas created the basis for understanding the extinction of our northern elephants. Toll' (1897) himself linked the extinction to the breaking up of a former Arctic continent between Asia and America, which led to a less continental, but colder, climate. As a result, rich pastures disappeared. The geologist and paleontologist Cherskij (1891) thought that greater cold and the development of the taiga zone explained the disappearance of the mammoth fauna in northeast Siberia.

Later reviews of the causes of mammoth extinction were published at the beginning of the twentieth century by the paleontologist Pavlova (1924); the journalist Digby (1926), who visited Yakutia at the beginning of the century; the preparator Pfizenmayer (1926), who excavated the Berezovka mammoth; the geographer Tolmachoff (1929); and others. Basically, they considered the possibility that extinction was a result of human activity, of physical changes in the mammoth's habitat, or of excessive development of the tusks.

Captain Gernet (1930, 2nd ed. 1981) used the glacial theory itself to support the vague and muddle-headed explanation of mammoth extinction from the cold. The geologist Gromov (1948) ascribed mammoth extinctions to climatic deterioration during the Würm. The geophysicist Budyko (1967) saw the same cause and supported it mathematically by hypothesizing small numbers of northern elephants and significant destructive power to primitive hunters.

The paleontologist Pidoplichko (1969) believed that Paleolithic hunters were unquestionably responsible for destruction of the mammoths that lived in the Ukraine and the southern part of the Russian Plain. He drew parallels to the hunts of African aborigines, and pointed out that if there were half a million mammoths in eastern Europe, Paleolithic hunters could have destroyed them in a single millenium.

The ichthyologist Lindberg (1972) naively attempted to explain the extinction of the mammoth by a rise in the level of the Pacific Ocean. The resulting overflow of the Baltic into the Black Sea wiped out the animals on the Russian Plain, while individuals living on the New Siberian Islands were cut off from the mainland and died from the cold.

The paleogeographer Velichko (1973) drew a convincing and detailed picture of a hyperzone of productive tundra-steppe, inhabited by the mammoth fauna, and ascribed the destruction of the mammoth complex and the extinction of the mammoths themselves to landscape changes—thawing of the ground, development of bogs over great expanses, and development of forests at the boundary between the Pleistocene and the Holocene.

The zoologist Vereshchagin (1971b, 1979) believed that the mammoth became extinct primarily as a result of the radical reconstitution of climates and landscapes in northern Eurasia at the end of the last glacial epoch. However, in the southern regions of eastern Europe and Siberia the destructive influence of primitive man could have predominated.

In the opinion of the paleogeographer and frozen-ground specialist Tomirdiaro (1977), the extinction of the mammoth and of the mammoth fauna in northeast Siberia came about at the end of the Würm (Valdaj), when an "ecological catastrophe," linked to climatic amelioration, led to the thawing of the Arctic basin and the melting of ground ice. The fodder-rich tundra-steppe was transformed into a moist, boggy, mossy tundra. There would be no place for mammoths in the present arctic tundra of Eurasia with its dense snow driven by the winds.

The geographer Kvasov (1977) also thinks that one of the reasons for the extinction of the mammoths and the mammoth fauna was the postglacial rise in temperature and humidity, connected with the development of extensive water bodies, the periglacial lakes.

Finally, there is the curious notion of the graphic artist Krause (1977), who compared some of the morphological features of the mammoths (woolly coat, ear structure, fat deposits, etc.) to the same features in other northern animals. As a result, with the direct logic of a person not acquainted with the paleogeographic facts, he declares that the cold adaptation of the mammoths is only a "scientific fiction" and that these beasts, having been adapted to a warm climate, perished from the cold. Such ideas, however, are not new and only confirm the tenacity of naive notions formulated in the eighteenth and nineteenth centuries.

Paleobiologists have repeatedly called attention to human influences on the mammoth population of Eurasia, especially in connection with the investigation of Paleolithic sites. In eastern Europe and in Siberia, massive "cemeteries" of mammoth bones and puzzling constructions made of their skulls and bones have been found (fig. 22.2). Usually, these constructions have been regarded as the foundations of ruined dwellings (Pidoplichko 1969), but some of them could have had ritual meaning or could have been used for musical purposes (Bibikov 1981). With regard to the extinction question, the principal significance of these bone heaps is that they could indicate the existence of intensive hunting. However, there are still no direct data to support the existence of such hunts. For example, traces of blows from tools have not been found on the bones.

At the same time there are numerous actualistic observations showing that primary animal "cemeteries" can form without human help when the carcasses of trampled creatures are concentrated in river meanders and oxbows (Vereshchagin 1972).

Figure 22.2. Ruins of a mammoth bone dwelling. Village of Mezhirich in the Ukraine. (Photo by N. K. Vereshchagin, 1971)

Paleolithic hunters could have transported fully mascerated mammoth bones from such now-dry natural accumulations to the high terraces of rivers for use as construction material. While we do not deny the probable existence of active mammoth hunts conducted with spears, poisoned javelin points, and arrows, or involving drives into natural traps and artificial pits, onto thin ice or even into a marshy bog, we must point out that there is still no undisputed evidence that this happened.

Massive, primary, natural burials of mammoth bones—in situ "mammoth cemeteries"—occur in subaqueous deposits in river valleys and lake basins. Secondary burials, which owe their existence to primitive people, also occur in river valleys and along the edges of lake basins, but they are found most often in the loesses and loesslike loams of the high terraces, both in eastern Europe and in Siberia (fig. 22.3).

Considering the immensity of the mammoth's range, encompassing a wide variety of natural conditions, it is impossible to assert that only one natural factor caused its extinction, for example, climatic amelioration leading to an ecological catastrophe in which boggy tundra and taiga replaced cold tundra. The proponents of this position must then use Upper Paleolithic people and their more abundant Neolithic successors to explain the disappearance of the mammoth in southern Siberia, where steppe dominated.

The same insufficiencies are hidden in explanations that speak of an "evolutionary blind alley," for we still do not know what negative morphological and physiological features could serve to explain mammoth extinction.

Figure 22.3. Ruins of a mammoth bone dwelling. Village of Kostenki on the Don. Paleolithic site of Kostenki XI. (Photo by N. K. Vereshchagin, 1960)

Meanwhile, any reasonable explanation must take into account the latest chronological dates. Table 22.2 shows them for Europe and the U.S.S.R. Together with other similar dates for the extinction of Pleistocene species in Eurasia and North America (Kurten and Anderson 1980), these dates can help in the search for the cause of extinction of the woolly elephant.

When the mammoth became extinct, its distinctive stomach botfly *Cobboldia russanovi* Grunin 1973 also disappeared. This was one of three species of botflies known for these proboscideans.

The Great Cave Bear *Ursus spelaeus* Rosenmüller et Heinroth, 1794

Cave bear history may be traced from the early Pleistocene, when speleoid features began to appear in European *Ursus etruscus* Cuvier. From the intermediate form *U. deningeri* Reich., *U. spelaeus* evolved in the Middle Pleistocene (the Riss) (Erdbrink 1953, Gromova 1965). The karst regions of central Europe and the Mediterranean littoral were the homeland of the great cave bear. In the north the species' range extended to southern England, Belgium, and southern Poland, and it included the Urals, the Crimea, and the Caucasus. Each karst region contained its own cave bear population, with little or no movement between regions. For example, isolated by flat

Table 22.2. Latest Chronological Dates for Mammoth in Europe and the USSR

Country	Site	Date (yr B.P.)
Europe		
Sweden	Lockarp	13,360 ± 95 (Lu–796)
		13,090 ± 120 (Lu–796:2)
		13,260 ± 110 (Lu–865)
France	La Colombière	13,390 ± 300 (Ly–433)
Switzerland	Praz Rodet	12,270 ± 210 (Ly–877)
USSR		
European part	Kunda, Estonian SSR	9,780 ± 260 (TA–12)
	Kostenki 2, Voronezh Oblast'	11,000 ± 200 (GIN–93)
	Timonovka I, Bryansk Oblast'	12,200 ± 300 (IGAN–82)
	Yudinovka, Bryansk Oblast'	13,650 ± 200 (LU–153)
		13,830 ± 850 (LU–103)
	Avdeevo, Kursk Oblast'	13,900 ± 200 (IGAN–78)
Siberia	Yuribej, Gydansk Peninsula	10,000 ± 70 (LU–1153)
	Mamontovaya, Tajmir	11,450 ± 250 (T–297)
	Berelekh, Yakutsk ASSR	10,370 ± 90 (SOAN–327)
		12,240 ± 160 (LU–149)
		13,700 ± 80 (MAG–114)
	Yar Berezovskij, Irtysh R.	12,860 ± 90 (SOAN–1283)

SOURCE: Berglund et al. 1976, Orlova 1979, Vereshchagin 1982a

steppes, a small colony of these animals lived on the Zhigulevsk Highland (Samarsk Dome) in the central Povolzh'e. The cave bears of the Caucasus were also isolated and evolved at a slow rate. Until the very end of the Pleistocene they maintained archaic features of *U. deningeri* (Baryshnikov and Dedkova 1978).

The morphological features of *U. spelaeus* were enormous size and weight (up to 1,000 kg); flat, bunodont molars; a bulging frontal; a powerful sagittal crest; a narrow nasal foramen; and a shortened tibia. These features suggest that the cave bear was less mobile and more vegetarian than the contemporary *Ursus arctos* L. Ecologically the cave bear was closely tied to caves, in which it lived and bore its young. This behavior turned out to be fatal for the species at the end of the Würm (Wisconsin). Practically all of its remains have been found in caves; remains are virtually unknown from alluvium, loess, or covering loams.

Paleolithic tribes regularly hunted the cave bear, and in western Europe there was a strong link between this animal and the life and evolution of man. Similiarly, remains of at least 200 individual bears were found in the Acheulean levels of Kudaro I in the Caucasus.

The cave bear probably became extinct at the same time as the mammoth and the rhinoceros in the late Würm (Magdalenian), since cave bear bones have not been found in Mesolithic sites on the Russian Plain or in the Urals.

Kurtén (1968, 1976) and Gabuniya (1969) review the many hypotheses that have been advanced to explain the extinction of the cave bear. The most popular hypotheses are degeneration as a consequence of inbreeding in isolated populations, imperfect morphological adaptations to a vegetarian diet via nonadaptive evolution, and Paleolithic destruction of the bears in their caves. Like Davitashvili (1969), Gabuniya thinks that cave bears were outcompeted and displaced by brown bears, but this is confusing cause and effect. The omnivorous brown bear may have competed with the herbivorous bear but could not have caused its destruction. In any case, brown bears were rare in the Pleistocene.

We note the fatal attraction of caves to the cave bears. At the end of the glacial epoch, some of the caves that the bears regularly occupied became insidious traps when water levels rose in rivers and streams during the thaw. For example, a succession of caves in the upper reaches of the Kama and Pechora rivers in the Urals became death traps over a period of several centuries (Vereshchagin 1982b).

The bears survived somewhat longer in the western Caucasus and in the upper reaches of the tributaries of the Rion (Kudaro II) than in other regions. Cave bear bones with well-preserved collagen have been found in local caves.

A basic factor in the extinction of this species was the change from a dry, continental climate to a moist one, which made for damp and uncomfortable microclimates in cave refuges.

A closely related species was the small cave bear *Ursus rossicus* Borissiak, which was morphologically even more specialized. It lived in eastern Europe and western Siberia outside karst regions, that is, it was less tied to caves. Nonetheless, like its giant colleague, it became extinct, a fact which seems to point to some evolutionary dead end in the morphology of the cave bear.

The Cave Hyena *Crocuta spelaea* Goldfuss, 1823

Pliocene and Quaternary deposits in the Old World have produced remains of various species of hyenas belonging to two basic phylogenetic branches, represented today by the genera *Crocuta* and *Hyaena*. Representatives of both genera occurred in the Pleistocene of Europe. The striped hyena (*Hyaena hyaena* L.) disappeared in Europe sometime during the Pleistocene, while the cave hyena survived to the end of it (Thenius 1980).

In cranial and skeletal features the cave hyena is very similar to the spotted hyena (*Crocuta crocuta* Erxl.) of Africa. Therefore European paleontologists frequently consider the cave hyena to be a northern subspecies of typical African *Crocuta*. However, this conclusion is debatable.

In Europe, *C. spelaea* appeared in the Cromerian Interglacial, and in the USSR in the Tiraspol' fauna. At the same time, in a series of features, the early forms—for example, the small hyena from Süssenborn—resembled the modern spotted hyena more than the cave hyena (Soergel 1936).

The cave hyena was highly specialized with extremely massive bone-crushing molars and weak canines. Its progressive and regressive features were more highly developed than in modern *Crocuta*. It possessed the awkward construction and elevation of the forepart of the body found also in the cave bear.

The range of the cave hyena covered all of Europe and the subtropical and middle latitudes of northern Asia, reaching to about 56 degrees N in Siberia. It apparently did not penetrate the arctic zone, nor did it occupy high mountains, since its remains have not been found in alpine sites in the Caucasus (Kudaro, Tsona). If the concept of a single African-Eurasian species is correct, then it did not become extinct, but merely suffered a catastrophic shrinkage of the northern, Eurasiatic portion of its range. The cave lion presents a potentially analogous case (see below).

Massive accumulations of cave hyena bones, together with coprolites and bones of prey animals, have been found in many European caves. For example, Kirkdale Cave in England provided abundant remains of hyenas of all ages, from newborn to old, while Tornewton Cave supplied more than 20,000 cave hyena teeth (Kurtén 1968). In the USSR hyena bones are common in caves in the Altaj (Hyena Den Cave) and the Far East (Geographic Society Cave) (Ovodov 1980). The species denned in caves and sometimes died there, vanishing into karst wells (Ageeva et al. 1978).

Like the African spotted hyena, the cave hyena fed primarily on carcasses of large ungulates and thus completed the stage of "secondary biomass" utilization, prior to final destruction by larval insects and bacteria. It is also possible that it hunted rodents, young ungulates, and mammoths.

Reduction in the numbers and range of the cave hyena coincided with the disappearance of herds of steppe and forest-steppe ungulates at the boundary between the Pleistocene and the Holocene, when the plains became boggy and the taiga developed. Like the modern aborigines of Africa, Paleolithic hunters only rarely captured hyenas; fragments of hyena bones, skulls, and teeth are rare in Paleolithic sites, especially in ones dated to the Solutrean-Magdalenian Culture. In the USSR no cave hyena bones have been found in any Mesolithic or Neolithic site.

The cave hyena's extinction is seen by some paleozoologists as a result of excessive specialization and of the anthropological factor—a sharp decline in the number of wild ungulates combined with the growth of agriculture and stock-breeding (Pidoplichko 1951a).

The Cave Lion *Panthera spelaea* Goldfuss, 1810

The history of this gigantic cat is both uncertain and puzzling. Its earliest record in Europe is from the Cromerian Interglacial in the Mosbach fauna. Some European paleontologists (Ryabinin 1919, Hemmer 1967, Kurtén 1968) think that the cave lion was only a subspecies of the modern African *Panthera leo* L. Although these animals are far apart in time and space, they are very similar in cranial and skeletal features. However, there are some differences. The cave lion exhibited even greater specialization of the feline type than the modern lion (Vereshchagin 1971a)

The concept of a single species implies the existence of a single range from the Cape of Good Hope in southern Africa through Scandinavia, Tajmyr (Taymyr), and Beringia in Eurasia, and into North America, to California and Mexico. No other species of mammal has a comparable range (Kurtén and Anderson 1980).

Other scholars (Goldfuss 1821, Leidy 1853, Gromova 1932) believe that the cave lion was a separate species. Besides its adaptation to cold conditions and its typically large size, it possessed cranial and skeletal features reminiscent of both the lion and the tiger (*P. tigris* L.). Its distribution was limited to the northern half of Eurasia, including Europe, from the British Isles through the Crimea, the Caucasus, Siberia, the Far East, Beringia, and even Alaska and the whole southern half of North America, where there was a closely related subspecies, *P. spelaea atrox* Leidy.

In habits and morphology the cave lion was closer to the lion than to the tiger. In the mountains it used caves as refuges, but on the plains it survived without them. It fed on horses, deer, bison, musk ox, and saiga, and in America on horses, bison, camels, and giant sloths. Its existence depended on the abundance of these herbivores.

The cave lion apparently became extinct when populations of ungulates declined sharply at the end of the Würm (Valdaj, Wisconsin). It is interesting that this extinction apparently occurred at the same time in Eurasia and North America, around 10,000 years ago.

Naturally, the proponents of a single African-Holarctic species of lion recognize the extinction of only the north Eurasiatic and North American Pleistocene populations of this remarkable animal. In Europe paleontological finds indicate that a lion of African type survived longest on the northern Black Sea littoral (until 3000 B.C; Bibikova 1973). On the Balkan Peninsula, lion was still present at the beginning of the Christian era, if we believe Herodotus that lions preyed on the camel caravans of Xerxes.

The Woolly Rhinoceros *Coelodonta antiquitatis* Blumenbach, 1799

A characteristic "fellow traveler" of the mammoth, the woolly rhinoceros apparently evolved in the early Pleistocene in the steppes of Mongolia, Transbaikal, and north China. Here the ancestral, early Quaternary form, *C. tologoijensis* Beljaeva, has been found at Tologoi on the River Selenga. Here is also the greatest concentration of woolly rhinoceros fossil remains, dating mainly from the Middle and Upper Pleistocene.

In western Europe the woolly rhinoceros is known from the Mindel (Franken-hausen, Bornhausen), where the molars are less specialized than in later forms (Sickenberg 1962). The woolly rhino became very numerous in the Riss. On the Russian Plain woolly rhinoceros skulls and rarer ones from Merck's rhinoceros (*Dicerorhinus kirchbergensis* Jaeger) occur in the alluvium of the Don and Volga, dated to the second half of the Middle Pleistocene.

The woolly rhinoceros achieved its maximum distribution in the Würm (Valdaj), when it was widespread throughout northern Eurasia, reaching the shores of the North Sea on the west and Italy, the northern Caucasus, and Kirgizia on the south. In north-east Siberia it reached the Kolyma and Anadyr basins and the shores of the Sea of Okhotsk. It is unclear why it was unable to cross the Bering Land Bridge into the New World during the Würm (Wisconsin) continental phase.

A series of morphological and biological features—the low position and inclination of the head, the thick woolly coat, the grassy diet (judging by the stomach contents of the Churapcha specimen from Yakutia)—indicates that this rhinoceros inhabited cold tundra-steppe. There is a paradoxical ecological similarity between the Eurasian tundra-steppe rhinoceros and the white rhinoceros of the African savannas.

It is usually believed that this species became extinct at the same time or earlier than the mammoth, and earlier in Siberia than in western Europe. However, from the collagen* still found in woolly rhinoceros bones in the covering loams of the Ukraine, Pidoplichko (1951b) sometimes established very late dates, on the order of 800–1000 years B.C. "Fresh" examples of rhinoceros bones with the distinct odor of collagen also occur in the Kazan University collections. Other materials from various places, dated by radiocarbon, provide dates on the order of 12–14,000 years B.C. (figs. 22.4, 22.5).

Upper Paleolithic people energetically hunted the woolly rhinoceros. In Upper Paleolithic sites on the Russian Plain rhinoceros bones comprise 1.5–3 percent of all the bones from animals that people exploited (Vereshchagin 1979), while the figure reaches 3–4 percent in sites in Cisbaikal (Ermolova 1978). In the Transbaikal steppes rhinoceros remains are remarkably abundant both in natural burials and in Paleolithic sites. We know little about Paleolithic methods for hunting the rhinoceros or about the impact these hunts may have had on the species. Active hunting with javelins and arrows is shown in the wall art of La Colombière Cave (France). It is probable that the animals were also caught in pits dug across their habitual trails.

Neither depictions of the woolly rhinoceros nor its bones have been found in Mesolithic or Neolithic sites. Thus, the best explanation for its extinction at the end of the Pleistocene and beginning of the Holocene is its morphological inability to adapt to the changing environment. There is a notable analogy to the mammoth in this respect.

The Giant Deer *Megaloceros giganteus* Blumenbach, 1803

Large-antlered deer of the tribe Megalocerini are known in Europe from the Villafranchian and in the USSR from the time of the Khapry fossil mammal complex (Baryshnikov 1981). Through time they increased in body size, while their antlers became more complex and more palmate.

Giant deer apparently first appeared in Europe. An early representative—*Megaloceros giganteus antecedens* Berck. from the Mindel-Riss of Germany (Steinheim) —had broadly palmate antlers. However, these were directed posteriorly, permitting the animals to occupy forest. In the Middle Pleistocene (Riss) *M. giganteus*, with broadly splayed antlers, became abundant (Gromova 1965). Its remains are common

---

*No positive ^{14}C dates on collagen. —EDS.

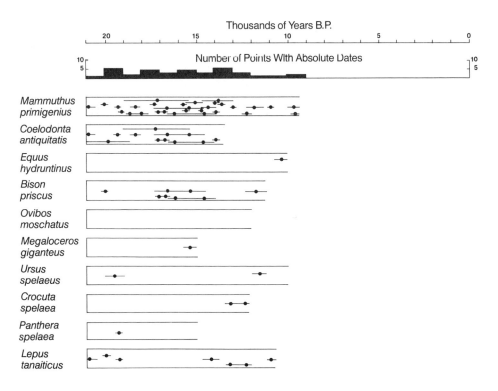

Figure 22.4. The latest radiocarbon dates for remains of Pleistocene mammals in eastern Europe and the Caucasus.

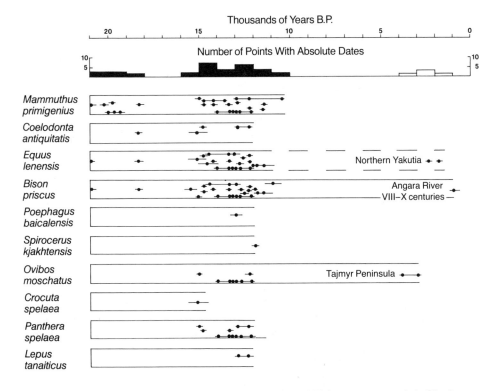

Figure 22.5. The latest radiocarbon dates for remains of Pleistocene mammals in Siberia.

in the Khazar levels of the Volga valley, where they comprise 2–8 percent of all the bones found on sandbars (Vereshchagin 1959). Dwarf forms of large-antlered deer are known from southern Italy and from Mediterranean islands (Bonfiglio 1978).

The giant deer ranged throughout Europe, to Scandinavia and the Timanskij Ridge on the north, to North Africa, the Caucasus, Kazakhstan, and southern Siberia, as far north as Tyumen, and Transbaikal (Kyakhta). Giant deer remains are abundant in Mousterian sites on the Russian Plain: at Starye Duruitory (Moldavia) they represent 2 percent of all bones from economically important species (David 1980), and at Korman IV (Ukraine) they make up 2.2 percent (Tatarinov 1977). In the northern Caucasus they make up 1.3 percent at Il'skaya (Vereshchagin 1959) and 12.0 percent at Barakaevskaya (Baryshnikov 1979). Giant deer remains are very numerous in the Crimea: at Shajtan-Koba they represent 39.4 percent (Gromov 1948). In the Transcaucasus they are rare—only 0.07 percent in Akhshtyr Cave (Vereshchagin 1959).

The antlers of the male giant deer, reaching 3.7 m across, indicate that the species inhabited open landscapes—glades where these alternated with copses on river floodplains. This conclusion is confirmed by comparison with the moose, a forest animal. In the giant deer, the orbits were larger, the teeth more hypsodont, and the lower jaw more massive.

In the late Valdaj (Würm) the range of the giant deer shrank dramatically (fig. 22.6). It disappeared in Siberia, Kazakhstan, and the Caucasus. The latest finds in Siberia date from the beginning of the Upper Paleolithic (Verkholenskaya gora; Ermolova 1978). On the Russian Plain they are rare. Thus, in the Upper Paleolithic levels of Korman IV on the Dnestr, giant deer bones comprise only 0.1 percent of the remains of economically important species (Tatarinov 1977), and at Kostenki VIII on the Don only 0.2 percent (Vereshchagin and Kuz'mina 1977). It was more abundant in the Crimea—up to 7.8 percent of bones from herbivorous mammals (Bibikova and Belan 1979).

In western Europe giant deer bones are rare in sites dating from the Aurignacian and Magdalenian. In France it disappeared in the Allerød (Bouchud 1965), while in the northern part of West Germany it was still present in the Preboreal (Guenther 1960). It is possible that the giant deer survived into the Christian era in Ireland (Mitchell and Parkes 1949) and was destroyed by medieval Anglo-Saxons.*

In eastern Europe the species probably became extinct at the end of the Pleistocene, since it has not been recorded from any Mesolithic or Neolithic site (Paaver 1965, Vereshchagin 1979). Suggestions that some giant deer bones from the Ukraine (Tarasovka in Dneprepetrovsk oblast'; Pidoplichko 1951a) and Siberia (Kamyshlov; Cherskij 1891) date from the Holocene require confirmation.

It is not clear why this species became extinct. The most commonly cited reasons are the colossal energy demands of male antlers, low fertility, the need for relatively warm conditions, the shrinkage of mesophytic meadows during the Würm (Valdaj), and prehistoric hunting.

The Primitive Bison *Bison priscus* Bojanus, 1827

The bison probably originated in southern Asia, since the oldest known representative (subgenus *Eobison*) has been found in Pliocene deposits in India (the Siwaliks) and China. In the USSR remains of the small *B. (Eobison) tamanensis* N. Ver. occur in Upper Pliocene/Lower Pleistocene beds on the shore of the Sea of Azov in the Taman faunal complex.

---

*Most authors place extinction of *Megaloceros* in Ireland at the Allerød. Mitchell and Parkes (1949) are very guarded about two poorly documented claims of Holocene survival. —EDS.

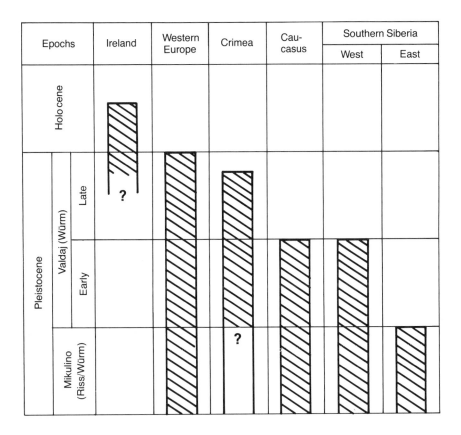

Figure 22.6. Time of extinction of the giant deer.

Early Pleistocene deposits in Europe and northern Asia have provided remains of the larger *B. voigtstedtensis* Fischer (Germany, England) and *B. schoetensacki* Freud. *B. schoetensacki*, considered a forest species by Flerov (1979), ranged as far east as the Vilyuj River in Yakutia. By this time, bison had already reached America.

Bison probably divided into steppe and forest types in the early Pleistocene (Hilzheimer 1918, Flerov 1979). Short-horned forest forms rarely occur in Upper Quaternary deposits. A forest fauna from the Kudaro sites in the Caucasus contains fragmentary bison bones from the Acheulean to the Holocene (Vereshchagin and Baryshnikov 1980).

Considerably more may be said about the evolution of the steppe *Bison priscus*. It is represented by the huge *B. p. gigas* Flerow in the Middle Pleistocene Khazar fauna of the Povolzh'e and Kazakhstan. The hornspan of this subspecies reached 2 m. *B. p. crassicornis* Rich. was the contemporaneous subspecies in Siberia and Alaska. The great breadth and volume of its nasal cavity indicate that it was adapted to the cold climate of the northern forest-steppe (Flerov 1979).

B. p. mediator Hilzh. occupied the extensive Upper Pleistocene periglacial steppes of Europe and western Siberia, while *B. p. occidentalis* Lucas inhabited the tundra-steppe of eastern Siberia, Alaska, and Canada. It was the antecedent steppe forms that made it possible for these bison to penetrate arctic latitudes so extensively—reaching the Gydansk and Tajmyr peninsulas and the Bering Land Bridge. This is confirmed by paleogeographical data and by a comparison of the ecology of the bison with the ecology and distribution of the aurochs, a more mesophylic bovine.

Until the historic period in the Holarctic there existed four geographically isolated forms of bison, distinguished from one another by ecology and morphology. These were the forest bison, *Bison bonasus* L., an inhabitant of the forests of Europe and the Caucasus; a diminished *B. priscus* subsp., which survived in parts of the south Russian Plain (Don basin) until the fifth to the tenth centuries A.D. (Vereshchagin 1971) and in the steppic valley of the Angara and Cisbaikal until the sixth to the seventh centuries (Ermolova 1978); the forest bison, *B. priscus athabascae* Rhoads, which inhabited the woodland and the light-needle coniferous taiga of northeastern Siberia and the taiga of Canada and which survives today only in Canada; and the bison of the North American prairies, *B. bison* L., which flourished until the arrival of Europeans.

As a whole the evolution of forest and steppe adaptations in bison during the Quaternary was a complex process. Fluctuations in bison ranges and populations as a result of natural and anthropogenic factors were equally complex. It is known that bison do not tolerate a prolonged snow cover of more than 40 cm, which limits the restocking of reserves and parks in the USSR. Observations on a thousand head of hybrid bison that have lived in the Caucasus Nature Reserve for the past sixty to seventy years show that bison adaptation to forest and steppe conditions is not absolute. The hybrid individuals have acclimated well to the forest conditions of the western Caucasus. In summer they ascend to the mountain meadows of the alpine zone, but they do not enter the foothill steppe.

The huge range of prehistoric bison in the Eurasian Pleistocene suggests clearly that a natural catastrophe caused their decline in the Holocene. In the northern parts of the bison range the catastrophe was probably increasing warmth and precipitation, a thawing out of the Pleistocene tundra-steppe, and an increase in the depth of the snow cover. The bison populations that survived in the forests and steppes of Europe and the Caucasus and in the steppes of Siberia were almost wiped out by man at the beginning of the twentieth century. Geptner et al. (1961) and Kirikov (1959) review this destruction in detail.

The Musk Ox *Ovibos moschatus* Zimmerman, 1780

The musk ox, which flourished in large numbers in the Pleistocene, is especially interesting in the context of the extinctions problem, since it has survived to the present in parts of the Arctic. As a consequence, we have an actualistic basis for reconstructing its ecology.

The musk ox probably made its first appearance in northeast Siberia. This area has provided a series of ancestral forms belonging to the genus *Praeovibos* (Sher 1971), previously described from European early Pleistocene deposits (Frankenhausen). The oldest known fossils of modern *Ovibos* come from the Mindel gravel of Süssenborn in Germany (Soergel 1942). Pleistocene Eurasiatic musk oxen differ from modern American ones in several cranial features, leading some zoologists to place them in a distinct species, *O. pallantis* H. Smith, 1827 (Ryziewicz 1955). This position is debatable, however. During the Pleistocene, musk ox metapodials changed their proportions: they became shorter and more massive (Kahlke 1963), approaching metapodials of modern musk oxen in massiveness.

During the Middle and Upper Pleistocene the musk ox ranged through most of Europe and northern Asia. In western Europe it spread south to the Dordogne (Les Eyzies) in France, to southeastern Hungary, and to Dobruja in Rumania. In the coldest periods it apparently even reached Spain (Abreda; Estevez 1979). Its southernmost limit on the Russian Plain was at the latitude of Kiev and Volgograd. In Siberia its southern limit was further north, for musk ox bones are known from Paleolithic sites in Cisbaikal (Ermolova 1978). Its northern boundary passed onto the Tajmyr Peninsula and onto the continental shelf of the Arctic basin, where it occurred on the New Siberian Islands.

Modern musk oxen live in herds averaging from twenty to thirty individuals, of which two or three are adult bulls. In summer the males leave the herds. In European Upper Pleistocene sites males outnumber females seven to one. This fact led Soergel (1942) to suggest that carnivores and prehistoric hunters preyed selectively on solitary bulls that had been driven from herds during sexual conflicts.

It is now known that the musk ox does not tolerate a dense snow cover deeper than 30 cm and that it is unable to undertake long migrations. On the other hand, it tolerates temperatures of minus 50 degrees C and harsh winds on Canadian islands and the northeast coast of Greenland. In addition, it can survive on the sparse grass and shrub vegetation of the stony tundra.

Unsuccessful attempts were made in the 1970s to introduce Canadian and Alaskan musk oxen to the southern slopes of the Byrranga Mountains on the Tajmyr Peninsula and to Wrangel Island. This experiment showed that musk oxen could not inhabit large portions of the modern Eurasian arctic tundra.

Beginning in the late Valdaj (Würm III) the range of the musk ox probably retreated progressively northward. A small population survived longest on the Tajmyr Peninsula. Very late radiocarbon dates have been obtained on musk ox skulls and horn sheaths found on the surface in northern Tajmyr: 3800 ± 200 and 2900 ± 95 yr B.P. (Vereshchagin 1971b). In Europe the latest known occurrences are in late glacial deposits in Scandinavia, as exemplified by a skull from Estersund, Sweden, with an age of about 9000 yr B.P. (Borgen 1979).

The principal factor limiting musk ox distribution in the modern Siberian tundra is the dense snow cover, which prevents feeding. Dense snow cover probably accounts for the extinction of musk ox throughout Eurasia in the postglacial. Human activity is responsible for the sharp reduction in musk ox range in North America in recent centuries.

In summary, we can say first that not all the extinct Pleistocene species disappeared at the same time. Some of them apparently retreated into appropriate refugia. Second, the paleontological data provide no basis for the idea that any of the extinct species disappeared as a result of morphological defects.

Local Causes of Extinction and Local Range Displacements in the Pleistocene and Holocene

Modern partial extinctions show clearly how environmental factors affect mammal existence and distribution. Range fluctuations in mesophilic and relatively xerophilic species provide excellent examples.

In the western Ciscaucasus the Caucasian mole (*Talpa caucasica* Satun.) lives in river valleys and on the plain adjacent to the Caucasus foothills. In periodic drought years it becomes extinct over large areas. Extinction is caused by a dramatic reduction in soil invertebrates and probably also by microclimatic deterioration in the burrows—a sharp rise in temperature and a decrease in moisture. The species can disappear completely from sections of the plain when agricultural exploitation and the removal of riverine forests increase aridity and promote the expansion of steppe.

When large expanses of floodplain became more arid in the postglacial epoch, the moles became restricted to relatively moist relict areas surrounded by steppe and even semidesert biotopes. The Karayarsk Oak Forest in the Kura Valley east of Tbilisi is a remarkable example of a mole refugium (Vereshchagin 1959). Relict areas of mole occupation also occur on the Stavropol' Highland, surrounded by dry steppes, and in the alpine meadows of the Armenian Plateau.

The social vole, *Microtus socialis* Pall., a relatively xerophilic species, presents a similar example of range pulsation. This species flourishes in the eastern Transcaucasus (Azerbajdzhan) in semidesert and steppe with 350–800 mm of precipitation per year. A series of dry winters causes the vole to become extinct over large areas. At the same time, its range retreats progressively into the foothills of the Caucasus east of Baku 100–120 km in the course of one or two seasons.

Such local extinction, brought about by seasonal climatic events, can continue over long periods. Complicated by other, nonclimatic factors, it can lead to total extinction. The gradual disappearance of a group of steppe rodents and small carnivores in the Caspian Sea region over the past few centuries is probably another example of such climatically induced local extinctions.

Formozov (1938) described the extinction of the steppe pika (*Ochotona pusilla* Pall.), the yellow lemming (*Eolagurus luteus* Eversm.), the great gerbil (*Rhombomys opimus* Licht.), the northern mole vole (*Ellobius talpinus* Pall.), and the giant mole rat (*Spalax giganteus* Nehr.) in the steppes and semideserts on the northern margin of the Caspian Sea in the middle of the nineteenth and beginning of the twentieth centuries. In the middle of the present century new data appeared on the disappearance of the corsac fox (*Vulpes corsac* L.) and steppe polecat (*Putorius eversmanni* Lesson.) in the Volga-Don steppes and in the steppes north of the Caucasus (Vereshchagin 1959). In this instance, it is possible that a decisive factor was the use of poisons and gasses to destroy small ground squirrels that the carnivores ate. Also perhaps important was the intense exploitation of the carnivores for their skins.

The saiga antelope (*Saiga tatarica* L.) presents another obvious example of fluctuation in range size. This species is adapted to a sharply continental climate and to steppe and semidesert landscapes. It feeds on grasses and wormwood and depends on hard ground for rapid flight. Its heartland is the region of steppe plateaus of central Asia. where its relative, the chiru (*Pantholops hodgsoni* Abel), also occurs. During the Upper Pleistocene the saiga ranged from England on the west to Alaska on the east and from the New Siberian Islands on the north to the central Asiatic deserts on the south. It is possible, however, that a separate species (*Saiga borealis* Tscherskyi) inhabited the arctic steppes of Siberia.

In the historic period the saiga's range has shrunk progressively eastward. In the Middle Ages large numbers of saiga lived in the steppes of the Dnestr and Dnepr basins, and they probably also occurred in Hungary west of the Carpathians. By the 1920s no more than a few hundred animals survived in the Kalmyk steppes (on the Don/Volga interfluve), in the Karakumy adjacent to the Urals, and in the western part of the Bet-Pak-Dala. As a result of this decline in saiga numbers, the subcutaneous botfly, *Pallasiomyia antilopum,* disappeared. It is found today only in an isolated population of Mongolian saiga (Grunin 1957).

In Siberia the major factor in saiga extinction was the disappearance of Pleistocene tundra-steppe and the spread of taiga. On the Russian Plain and in Kazakhstan the extinction of saiga in the nineteenth and early twentieth centuries was directly linked to persecution by nomadic herdsmen and farmers, especially near water sources. The rapid growth in saiga numbers and the restoration of its range in the succeeding thirty to forty years reflect depopulation of the steppe and semidesert and state efforts at conservation—a full ban on hunting. In Kazakhstan today the total saiga population varies around 1.8 to 2 million head, thanks to regulated exploitation and to the shooting of wolves from airplanes.

Overall, the impression we gained is that the steppe group of central Asiatic species, including carnivores (corsac fox and steppe polecat), rodents (yellow lemming and steppe marmot), lagomorphs (steppe pika), and ungulates (kulan, tarpan, and saiga), retreated progressively eastward beginning at the end of the Pleistocene. They disappeared in western Europe very early on and in recent centuries they have also disappeared from eastern Europe. Today they are largely restricted to their central Asiatic homeland—the steppes of Kazakhstan and Mongolia. Their retreat appears to have retraced the route of their Pleistocene expansion.

In Europe and Asia there are numerous well-known examples of mesophilic rodent species that retreated into mountains from plains that became arid and steppic in the postglacial period. In Adam Cave in the semidesert of Dobruja on the right bank of the Danube, Romanian paleontologists have found remains of the snow vole (*Chionomys nivalis* Martins), which lives today in the southern Carpathians. In the Ciscaucasus the closely related Gudarsk vole (*Chionomys gud* Satun.), which presently lives in the high Caucasus, once lived at Pyatigor'e on Mt. Razvalka only 500 to 600 m above sea level in an area of permafrost in teshenite detritus.

The chromosomal species of common vole (*Microtus arvalis* Pall. and *M. subarvalis* Meyer, Orlov et Skholl) are broadly distributed in the mesophilic floodplains of the Russian Plain and in the high Caucasus, but on the southern limit of their range they have a patchy distribution in alpine meadows. The patchy distribution probably came about when populations living in degraded lowlands and intermontane valleys became extinct in the postglacial.

In general, there was considerable mobility in the ranges of mesophilic and xerophilic small mammals during the Pleistocene and Holocene. The discovery of the Binagadinsk Middle Pleistocene fauna in tars of the Aspheron Peninsula in the eastern Transcaucasus shows this mobility clearly.

During the Holocene (?), the corsac fox and the saiga retreated from the Transcaucasus into the Ciscaucasus, while the alpine mole vole and the porcupine retreated from the Transcaucasus onto the Iranian Plateau. Displacements in the ranges of other local species were less remarkable. Thus the boundary of the common vole's range moved 120 to 130 km to the northwest, while that of the golden hamster (*Mesocricetus raddei* Nehr.), which still survives in Dagestan, moved 200 to 250 km.

Other factors led to great variation in the range of the river beaver (*Castor fiber* L.). In the northeastern USSR there are places where the former presence of beavers is obvious from fossil ponds and fossil gnaw marks on branches and tree trunks. On the Enisej, Aldan, and Penzhina rivers, these places are many thousands of kilometers from relict beaver colonies found, for example, in the Tuva Autonomous Region and in the eastern foothills of the Urals along the Konda and Sos'va rivers (Grave 1931, Skalon 1951).

Historically in Siberia beavers were rare in the permafrost zone, because permafrost inhibits the construction of stable burrows in river banks and causes beaver ponds and lairs to freeze over. Beavers probably penetrated far to the north in Siberia during a period when permafrost disappeared. Thus, in our opinion, the relict beaver colonies that survive in the Irtysh Basin and in other parts of Siberia formed in the postglacial epoch no earlier than 8,000 to 9,000 years ago. Their "aboriginality" is therefore relative. Greater "aboriginality," dating from the Mio-Pliocene, may be ascribed to those beaver colonies which occurred in western Asia and Kazakhstan, and which still occur on the Bulugan River.

Until the twentieth century, beaver colonies survived only under the protection of monasteries in European Russia, and elsewhere under the protection of native shamans, that is, also on religious grounds. In the 1950s and 1960s a major state effort to promote the restocking of beavers in the European USSR and in southern Siberia were totally successful and showed clearly that human overexploitation constituted the principal reason for the historic disappearance of beavers from huge areas of eastern Europe, the Caucasus, western and central Asia, and Siberia. In the southern regions of the country beavers were also adversely affected by stockbreeders' destruction of the riverine, gallery forests that provided fodder for these remarkable rodents.

Climate and Landscape Changes
on the Boundary Between the Pleistocene
and Holocene as Causes of the Extinction of the Mammoth Group

As indicated earlier, in the eighteenth, nineteenth, and early twentieth centuries most European scholars believed that the mammoth fauna existed in Siberia under conditions that were much warmer and very different from modern ones. The zoologist Brandt (1865:3) was correct when he wrote that the scanty vegetation of the modern tundras would not sustain gigantic mammoths, rhinoceroses, and bison. He believed that a warmer climate promoted more luxuriant vegetation in the past.

Opinions changed in the first half of the twentieth century. It was supposed that temperatures approximated modern ones. The ornithologist Tugarinov (1928, 1934) believed that the tundra was "not the same as now" when the cave lion and saiga lived there. He postulated that it was "an open landscape with a remarkably xerothermic climate, rather cold, with little winter precipitation" (1928:669). In his opinion, an increase in moisture, a subsequent dry period, and a renewed increase in moisture caused horses, camels, and saigas to retreat from tongues of steppe in eastern Siberia into the region of unbroken steppes and deserts (1934:63–64). Mammoth, rhinoceros, bison, and cave bear did not become extinct as a result of a sharp climatic change or one "in the direction of deterioration." "They were living in marginal conditions to begin with, and a relatively small environmental change was enough to precipitate a crisis."

The excellent ornithologist and paleozoologist Serebrovskij (1935) apparently saw approximately the same landscape zones in the Würm (Valdaj, Wisconsin) as exist in northern Eurasia now; however, he also recognized the existence of a great European ice sheet.

The paleontologist Pidoplichko (1951a, 1969) consistently asserted that mammoths could live in the Ukraine today. He emphasized the lack of sharp climatic and landscape differences between the Pleistocene and the present and discounted such differences as a factor in Pleistocene extinctions. The glaciologist Dajson (1966:138) stated that, from the perspective of mammoth habitation, Pleistocene and modern environments were very similar.

Vague facts and speculative considerations led to such categorical conclusions, since detailed paleoclimatic investigations had not taken place. Different facts were discovered by numerous native paleogeographers, frozen-ground specialists, and geomorphologists working in the arctic and subarctic zones and in the region where permafrost developed during the Pleistocene.

Velichko (1973, 1982) and Tomirdiaro (1977, 1980) obtained very firm information on natural processes during the Pleistocene. From geomorphic observations on the Russian Plain and in northeast Siberia, and from paleogeographic comparisons, both investigators reconstructed the mammoth's environment during the period of peak cold in the Würm (Valdaj). Climate was sharply continental, leading to the development of permafrost up to 1.5 km below the surface and to the formation of subterranean ice vein-walls up to 40 m deep, pressing up columns of earth. Permafrost extended as far south as 46–48 degrees N in Europe. Low winter temperatures were characteristic: around minus 30 degrees C at the latitude of southern England, White Russia, and the Central Russian Highland, judging from paleobotanical data (Velichko 1982). Summer temperatures were not depressed as much. In the periglacial zone, precipitation was no more than 250 to 300 mm per annum.

In Europe the zonal mixed and broadleaf forests were replaced by periglacial vegetation which occupied a wide belt between the Scandinavian ice sheet on the north and the Alpine glacier on the south (Grichuk 1982). Periglacial steppes covered the south Russian Plain and the central Danube lowlands.

Winters with little snowfall and the development of a luxuriant grass cover on hard, dry ground with abundant summer insolation allowed horses, bison, and saiga to

Figure 22.7. Counting bones at the Berelekh mammoth "cemetery" (Yakutia). (Photo by A. V. Lozhkina 1970)

occupy huge expanses of northern Eurasia. The boundary between the Pleistocene and Holocene was characterized by sharp, short climatic oscillations: the Bølling interstadial (12,400–12,000 yr B.P.), the Middle Dryas stadial (12,000–11,800 yr B.P.), and the Allerød interstadial (11,800–11,000 yr B.P.), the Upper Dryas stadial (11,000–10,300 yr B.P.), and so forth, which affected the Pleistocene species decisively. It was precisely in this time range that the massive extinction of the mammoths and their "fellow travellers" occurred in the arctic zone. Testimony to this extinction are the hundreds of thousands of bones from disarticulated skeletons and the occasional frozen carcasses buried in Sartan deposits (late Wisconsin) in northern Yakutia and on the Tajmyr Peninsula (fig. 22.7). Judging from modern examples of mass death among wild and domestic ungulates in the Kazakhstan steppes (Sludskij 1963), the best explanation for such death at the end of the Pleistocene is the frequent occurrence of snowstorms (blizzards) in winter and the transformation of the nutritious Pleistocene tundra-steppe into a boggy, lake-dotted tundra. In subarctic latitudes at this time, taiga and mixed forests advanced rapidly onto open expanses, and a forest fauna developed.

From the paleozoologist's point of view, the most convincing proof that the landscape changed radically on the boundary between the Pleistocene and Holocene is the change from a steppe, mammoth fauna into a forest fauna on the Russian Plain, in the northern Urals (Kuz'mina 1971), in Siberia, and even in the Far East.

Animal Extinction Under Human Influence

Man's ability to destroy economically significant species, especially "harmful" ones, is widely appreciated in our technological age. It has found clear expression in numerous nature-preservation laws, legislative measures, books, pamphlets, and instruction on exploitation and destruction, and we need present no proof. However, the situation was different in the Paleolithic, when there were few people and when technology and economy were at a low level.

The primitive people of the Lower and Upper Paleolithic left abundant evidence of their exploitation of mammals, birds, and fish in western and eastern Europe, the Caucasus, central Asia, Siberia, and the Far East. This evidence comprises hundreds of thousands of artificially broken bones in open-air and cave sites dating from the Paleolithic, Mesolithic, and Neolithic. The mammals taken by primitive hunters included essentially all the large species that were available: carnivores, large rodents, lagomorphs, perissodactyls, artiodactyls, proboscideans, and occasionally even primates. Soergel (1922) and Lindner (1937) have reviewed the hunting methods used in the European Paleolithic and Neolithic. Further reviews may be found in the archaeological and ethnographic literature.

Gromov (1948) was the first to summarize the species composition of Paleolithic faunas in the Soviet Union. Special reviews relevant to ancient hunting are presented by Semenov (1968) and Vereshchagin (1971b). These publications show that on the Russian Plain and in the Caucasus, central Asia, Siberia, and the Far East, man obtained and butchered for food and technical purposes one species of monkey, twenty-eight carnivores, two proboscideans, five lagomorphs, five rodents, four perissodactyls, twenty-three artiodactyls—a total of sixty-seven to seventy species of terrestrial mammals.

Regional differences, linked to zoogeography and terrain, are very obvious. For example, the greatest variety of carnivores and ungulates was obtained on the Russian Plain and in the Caucasus, and the smallest in central Asia. These data are expanding and becoming more precise. One basic pattern is the use of local, immediately available, abundant resources. Taphonomic factors complicate attempts to evaluate objectively the significance of different mammal species in ancient human diets. The bones of large animals—mammoths and ungulates—are better and more readily preserved and thus appear more important than the bones of small and medium-size carnivores and rodents. We have no convincing data on how much meat ancient people ate. Different estimates vary widely. Ethnographic observations indicate that some modern natives of Siberia—Nentsy, Dolganes, Yukagirs, Yakuts, Evenki, and Evenni—can eat 2 to 5 kg or more of reindeer meat in twenty-four hours.

Mowat (1963) published a daily norm of 2 kg of caribou (reindeer) meat for Canadian Eskimos. Soviet investigators accept clearly lower norms for Upper Paleolithic people in eastern Europe: Bibikov (1966), 600 g in twenty-four hours, and Pidoplichko (1969), 800 to 1,000 g. These norms apparently differed sharply between summer and winter. In summer, people ate a great deal of vegetal food: roots of cattails and reeds, parts of umbelliferous plants, fruits and berries, nuts, edible herbage, and even tree bark—for example, willow bast. Pidoplichko (1969:153) thought that vegetal products, fish, and meat of small mammals and birds comprised one third of the entire diet in summer and fall. Over the entire year, he believed, vegetal foods and the meat of small mammals constituted one fourth of the diet, and mammoth meat three fourths. Data from the Kostenki (Don) sites suggest that in the majority of cases Paleolithic people got more meat from horses and hares than from mammoths (Vereshchagin and Kuz'mina 1977).

Given the severe climate of northern Eurasia in the Paleolithic epoch and the brief growing season, Vereshchagin (1971b) assumed that the average person consumed 2 kg of meat in twenty-four hours. At this level of consumption, an Upper Paleolithic population of 15,000 people (a conservative estimate) on the southern half of the Russian Plain would need up to 10,500 tons of meat per year—up to 60,000 horses (100 kg of meat after butchering) and up to 10,000 bison (300 kg of meat after butchering). Considering the difficulty of hunting large mammals with primitive weapons, the prehistoric hunters would have had to work hard all year. Nonetheless, these numbers suggest that Paleolithic people applied considerable pressure to the mammal populations of the plains. However, there is also reason to suspect that the ancient inhabitants of northern Europe often scavenged the bodies of mammoths and other animals that had drowned or frozen. We know this, for example, from the 1980 excavation of bajdzherakhs on the

Figure 22.8. A skull of one of the last mammoths of Berelekh. The tusks were removed by Stone Age hunters. (Photo by N. K. Vereshchagin, 1970)

Berelekh River in northern Yakutia, where there is a huge "mammoth cemetery," formed approximately 13,000 years ago (fig 22.8). (*Bajdzherakh* is a Yakut word for a column and mound of ground pushed out by veins of ground ice during polygonal cracking of the earth.)

Like the recent aborigines of Africa, the prehistoric hunters of northern Eurasia probably also used game killed by carnivores (cave lions and wolves). Such double pressure—from carnivores and settled human tribes—could have been fatal to Upper Pleistocene populations of mammoths and ungulates.

Irregular hunting success and the migratory habits of the mammoths and ungulates must have led Paleolithic people to develop methods for storing meat. They probably dried and cured it in the sun and wind, froze it in winter, and buried it in the permafrost in summer.

In the post-Paleolithic epoch the growth of stockherding and agriculture increased the pressure on exploitable animals, thanks to advances in the technology for capturing wild animals and to domestication of the dog and horse.

The use of the dog and the horse permitted improved methods for reconnoitering and catching large and small game. Numerous petroglyphs observed and described by ethnographers over a huge area, from Scandinavia to Kamtchatka and from the Mediterranean littoral to the Pacific Ocean, tell us about the animal species that were acquired and also about the invention and perfection of bows, harpoons, and boar spears in the Neolithic and the Bronze Age (Kuhn 1956, Savvateev 1970), Vereshchagin and Burchak-Abramovich 1948, Marikovskij 1953, Okladnikov 1959, Dikov 1971, Formozov 1969, Devlet 1980, and others).

Numerous bones of large wild animals (bear, boar, and moose) have been found in Neolithic pile-dwellings on the eastern Baltic littoral. Marine animals (dolphins and pinnipeds) played an important role in the economy of the shoreline settlements (Paaver 1965; Vereshchagin and Rusakov 1979). Neolithic and Bronze Age sites on the shores of the Far Eastern oceans present a similar picture.

Undoubtedly the Neolithic tribes of eastern Europe and Siberia obtained the last, sporadic mammoths, musk oxen, spiral-horned antelopes, Transbaikal buffalo, and Baikal yaks. On the steppes of eastern Europe and southern Siberia the Sarmatian and Scythian tribes were still able to exploit almost untouched herds of tarpan, kulan, saiga, aurochs, and bison.

Medieval manuscripts tell us about the grand battues the Mongol and Tatar cavalry of the Bronze Age conducted against steppe ungulates (bison, red deer, and saiga) (Rashid-Ad-Din 1946, Kirikov 1959). The existence of such hunts is confirmed by the numerous bones found in the eighth- to thirteenth-century Khazar town and fortress of Sarkel on the Don (Vereshchagin 1971a) and by the superb studies of Tsalkin (1962), Bibikova (1953) and Timchenko (1972) on bones excavated from several town sites on the Russian Plain in the middle of the present century. Even with the wide variety of available data it is difficult to evaluate the significance of hunting as a factor in the destruction of several mammalian species. Over the centuries there was a general tendency for the percentage of wild animals present to decline, while the percentage of domestic ones increased. Among the thousands of bones in the collections from medieval sites near the Dnepr River (Ukraine), bones of wild animals (boar, kulan, roe deer, red deer, moose, saiga, bison, and aurochs) comprise from 0.4 to 10.6 percent; on a minimum individual basis they comprise from 5.0 to 29 percent (Timchenko 1972).

The cases of the aurochs and the large cats illustrate what can happen to large mammals as a result of intense hunting, the cutting down of forests, and the plowing of virgin lands. The oldest fossil remains of the aurochs (*Bos primigenius* Bojanus, 1827) come from Lower Quaternary deposits in Europe, western Asia, and the Caucasus. Even then it was a very large animal. The Pleistocene range of the aurochs was extensive but considerably smaller than that of the bison. Besides Europe, remains of aurochs are known from western Asia, the Transcaucasus and Ciscaucasus, central Asia, and southern Siberia as far as Cisbaikal. During the Middle and Upper Pleistocene the aurochs became rather large, but toward the end of the Pleistocene and in the Holocene it split into large and small forms. Both forms of the aurochs were domesticated in the Neolithic.

Ecologically, the aurochs was more mesophilic and warmth-loving than the steppe bison. It inhabited the valleys and floodplains of rivers, especially during droughts or desiccation of interfluves. It penetrated high into mountain ranges. In the medieval forests of Lithuania and Poland, aurochs kept to dense mixed forests, even moist and boggy ones, feeding in part on sprigs and shoots (Geptner et al. 1961).

In the Ukraine, aurochs bones occur in Neolithic, Eneolithic, and Bronze Age sites as late as the tenth or eleventh centuries (Bibikova 1953). In the Caucasus there are Mesolithic and Neolithic depictions of aurochs on rocks. In western and central Asia, the species survived until the Bronze Age in riverine thickets and reed beds along the Euphrates, Amu-Darya, and Syr-Darya. The aurochs disappeared in western and central Europe in the fifteenth century, following the removal of forests. It survived only under feudal protection in some forests of modern Poland and White Russia.

Bones of Pleistocene tigers (*Panthera tigris* L.) have been found in India, China, and the southern part of the Soviet Far East. At the beginning of the twentieth century in the USSR the tiger occurred in Transbaikal, along the courses of central Asiatic rivers, in the Far East, and the Amur Valley. Occasional individuals were encountered in the southern part of western Siberia and Yakutia (Geptner and Sludskij 1972). In the northern Caucasus the tiger apparently disappeared in the twelfth century. By the 1920s persecution and agricultural development in the Transcaucasus led to a situation in which only occasional tigers came in from Iran. In central Asia tigers survived until the 1950s, when they disappeared completely as a result of direct persecution and the destruction of wild boar. The tiger population steadily declined in the Far East, where as of the early 1980s there were only 180 to 200 individuals enjoying state protection (Zhivotchenko 1981).

Figure 22.9. Skulls of the last Caucasian bison of northern Osetia, killed in the eighteenth and early nineteenth centuries. The Sacred Cave of Digorized. (Photo by N. K. Vereshchagin, 1947)

In the Middle Ages the range of the cheetah (*Acinonyx jubatus* Schreb.) included the plains of the eastern Transcaucasus, where it may have persisted until the eighteenth century (Vereshchagin 1959). In the present century in the USSR, cheetah occurred only in the desert regions of Turkmenia and Uzbekistan. Cheetah became rare as a result of direct persecution and of reduction in the numbers of goitered gazelle. As of 1982, there were probably no cheetah left.

To these examples we can add the well-known extermination of Steller's sea cow (*Hydrodamalis gigas* Zimm.) on the shores of the Commander Islands in 1768; the disappearance of the tarpan (*Equus ferus gmelini* Anton.) and the kulan (*E. hemionus* Pall.) in the mid-nineteenth century on the steppes of the south Russian Plain; and the destruction in the 1920s of the last Caucasian bison (*Bison bonasus caucasicus* Satun.) (fig. 22.9), as well as other data that provide a picture of faunal impoverishment.

The rapid disappearance of wild animals on the planet—of precious genetic resources—has forced biologists to seek ever more decisive measures to conserve them, including the publication of national and global "Red Books" and the development of conservation laws.

When we compare the rate of extinction today with the rate in the Paleolithic, Neolithic, and early metal periods, we are struck by how much it has increased in the age of technology. Besides direct and indirect human destruction of both valuable and harmful mammals, we are now increasingly aware of the strong influence of carnivores and epizootic diseases, and also of various kinds of stressful situations that disrupt population structure. Human exploitation may also disrupt population structure, thereby placing an intolerable burden on exploited animal populations. Hunting practices often inhibit the growth of a wild ungulate or carnivore population by disrupting its structure. Human concentration on large animals and the selection of males for trophies hinders normal reproduction, leading to population exhaustion. Carnivores, epizootic diseases, food shortages, and other factors can have a similar destructive effect on populations.

Conclusion

Analysis of the geography and ecology of extinct and declining Quaternary mammals in the USSR shows clearly that four groups or factors led to the destruction of the mammoth fauna and to the disappearance of a series of other species:

1. Changes in environment, including seasonal and secular climatic oscillations, involving changes in landscapes and biotopes. Here we include tectonic uplift and fluctuation in the level of the world ocean, which not only altered atmospheric circulation, but also weakened natural selection pressures in circumstances of insular and peninsular isolation.

2. The loss by species of resistance and of the ability to adapt quickly to new abiotic and biotic environmental conditions. Here we are talking about ecogenetic and phylogenetic defects, including the concept of excessive specialization and the nonadaptive character of evolution.

3. The disruption of population structure as a result of external factors such as more progressive competitors, carnivores, parasites, epizootic diseases, and stressful situations.

4. The direct destructive influence of people and the indirect action of their economic activity on exploited and unexploited animal species.

Once we postulate these groups of factors, the next step is to evaluate the relative weight of each in time and space, that is, in different stages and geographic zones.

The examples we have presented of extremely rapid decline in some nonexploited mammal species during the Quaternary and at present in northern Eurasia point to the supreme importance of external, abiotic environmental factors in their extinction. The nearly total extinction of large species of the mammoth fauna in the tundra and taiga zones and their partial survival in the forest steppe and steppe confirms the decisive effects of climatic and landscape changes for the life and death of this species group.

The best proof of what has been said is the ubiquitous transformation of the Upper Pleistocene "steppe" fauna into a forest and taiga fauna over huge areas in upper and middle latitudes in Eurasia. In this case, the primitive species were not crowded out by more progressive ones, nor were their population structures disrupted, nor were they destroyed by human activity. Environmental change was so radical and dramatic that morphological evolution simply did not have time to catch up.

The destructive activity of people is often thought to have been decisive in the extinction of Quaternary giants (mammoth, rhinoceros, cave bear, and others). However, while human influence on animal populations steadily increased in prehistory, it became definitive only in the last few millenia and centuries. In addition, the role of man has not been the same in all geographic zones. It has been greatest in the ancient heartlands of civilization in the Mediterranean Basin, western and central Asia, and China, and least in the polar desert.

The remaining causes of extinction have apparently been secondary, merely promoting further reduction in ranges and numbers of species after they had suffered from climatic and landscape changes or from human pressures.

References

Ageeva, E. A., Grichan, YU. V., and Dvodov, N. D. 1978. The Pleistocene hyena in the stone trap (in Russian). *Priroda* 12:114–115.

Alekseeva, L. I. 1977. The theriofauna of the early Anthropogene of eastern Europe (in Russian). *Trudy Geologicheskogo Instituta AN SSSR* 300:214.

Baryshnikov, G. F. 1979. The theriofauna and Upper Pleistocene landscapes of the Gornoe Prikuban'e (in Russian). *Vestnik Leningradskogo Universiteta* 12:53–62.

————. 1981. Order Artiodactyla Owen, 1848 (in Russian). In *Katalog mlekopitayushchikh SSSR (pliotsen—sovremennost')*: 343–408. Moscow, Nauka.

Baryshnikov, G. F., and Dedkova, I. I. 1978. Cave bears of the Greater Caucasus (in Russian). *Trudy Zoologicheskogo Instituta AN SSSR* 75:69–77.

Berglund, B. E., Hakansson, S., and Lagerlund, E. 1976. Radiocarbon dated mammoth (*Mammuthus primigenius* Blumenbach) finds in South Sweden. *Boreas* 5:177–191.

Bibikov, S. N. 1966. An attempt at paleoeconomic modelling in archeology (in Russian). In *Tezisy dokl. Vsesoyuz. cessii, posvyasch. itogam arkheol. i ethnograf. issledovanij.* Kishinev.

————. 1981. *A very ancient musical complex of mammoth bones* (in Russian). Kiev, Naukova dumka.

Bibikova, V. I. 1953. The fauna of the early Tripol'e settlement of Luka-Vrublevitskaya (in Russian). *Mater. i issl. po arkheol. SSSR* 38:411–458.

————. 1973. Bone remains of lion from Eneolithic settlements on the northeastern Black Sea littoral (in Russian). *Vestnik zoologii* 1:57–63.

Bibikova, V. I. and Belan, N. G. 1979. Local variants and groupings of the Upper Paleolithic theriocomplex of southeastern Europe (in Russian). *Byull. Mosk. o-va ispyt. prirody. Otd. biol.* 84 (3):3–14.

Bonfiglio, L. 1978. Resti di Cervide (Megacero) dell' Eutireeniano di Bovetto (RC). *Quaternaria* 20:87–108.

Borgen, U. 1979. Ett fynd av fossil myskoxe i Järntland och nagot om myskoxrnas biologi och historia. *Fauna och flora* (Sver.) 74(1):1–12.

Bouchud, J. 1965. Le *Cervus megaceros* dans la sud et sud-ouest de la France. *Israel J. Zool.* 14 (1/4):24–37.

Brandt, F. F. 1865. On the fossil rhinoceros (in Russian). *Naturalist* 15:1–8.

Budyko, M. I. 1967. The causes of extinction of some animals at the end of the Pleistocene (in Russian). *Izv. AN SSSR, ser. geogr.* 2:28–36.

Cherskij, I. 1891. Description of the collections of post-Tertiary mammals, collected by the Novosibirsk expedition of 1885–1886 (in Russian). In *Prilozhenie k 65 tomy "Zapisok Akad. nauk."* 1.

Dajson, D. L. 1966. *In the world of ice* (in Russian). Leningrad, Gidrometeoizdat.

David, A. I. 1980. *Theriofauna of the Pleistocene of Moldavia* (in Russian). Kishinev, Stiinsta.

Davitashvili, L. SH. 1969. *Causes of the extinction of organisms* (in Russian). Moscow, Nauka.

Devlet, M. A. 1980. *Petroglyphs of Mugur-Sargol* (in Russian). Moscow, Nauka.

Digby, B. 1926. *The mammoth and mammoth-hunting in northeast Siberia.* H. F. and G. Witherby.

Dikov, N. N. 1971. *Rock enigmas of ancient Chukotka* (in Russian). Moscow, Nauka.

Dubrovo, I. A. 1964. Elephants of the genus *Archidiskodon* in the territory of the USSR (in Russian). *Paleontol. Zh.* 3:82–94.

Efremov, I. A. 1950. Taphonomy and the geologic record (in Russian). *I. Trudy Paleontologicheskogo Instituta AN SSSR* 24:1–176.

Erdbrink, D. P. 1953. *A review of fossil and recent bears of the Old World, with remarks on their phylogeny based upon their dentition.* Proefschrift, Deventer.

Ermolova, N. M. 1978. *The theriofauna of the Angara Valley in the late Anthropogene* (in Russian). Novosibirsk, Nauka.

Estevez, J. 1978 (1979). Primer hallazgo del buey almizclado (*Ovibos moschatus* Zimmerman) en el pleistoceno peninsular. *Acta geol. hisp.* 13(2):59–60.

Flerov, K. K. 1979. Systematics and evolution (in Russian). In *Zubr: morfologiya, sistematika, evolutsiya, ekologiya*: 9–27. Moscow, Nauka.

Formozov, A. A. 1969. *Essays on primitive art* (in Russian). Moscow, Nauka.

Formozov, A. N. 1938. On the question of the extinction of some steppe rodents in Upper Quaternary and historic times (in Russian). *Zool. zhurn.* 17(2):260–272.

Gabuniya, L. K. 1969. *Extinction of ancient reptiles and mammals* (in Russian). Tbilisi, Metsniereba.

Garutt, V. E. 1981. Order Proboscidea Illiger, 1811 (in Russian). In *Katalog mlekopitayushchikh SSSR (pliotsen—sovremennost')*: 304–318. Leningrad, Nauka.

Geptner. V. G. and Sludskij, A. A. 1972. *Mammals of the Soviet Union.* Vol. 2. Carnivores (hyenas and cats) (in Russian). Moscow, Vyshaya shkola.

Geptner, V. G., Nasimovich, A. A., and Bannikov, A. G. 1961. *Mammals of the Soviet Union.* Vol. 1. Artiodactyls and perissodactyls (in Russian). Moscow, Vyshaya shkola.

Gernet, E. S. 1981. *Glacial lichens* (in Russian). 2nd ed. Moscow, Nauka.

Goldfuss, G. 1821. Osteologische Beiträge zur Kenntnis verschiedner Säugethiere der

Vorwelt. IV. Über den Schadel des Hohlen-löwen. *Nova Acta Phys. Acad. Casesares Leopold. Carol., Nat. Currosorum, Hall.* 10(2):489–494.

Grave, G. L. 1931. The river beaver in the USSR and its economic significance (in Russian). *Trudy po Lesn. opytn. delu* 14:76–144.

Grichuk, V. P. 1982. The vegetation of Europe in the late Pleistocene (in Russian). In *Paleogeografiya Evropy za poslednie sto tysyach let (atlas-monografiya)*: 92–109. Moscow, Nauka.

Gromov, V. I. 1948. The paleontological and geographical basis of the stratigraphy of the continental deposits of the Quaternary Period in the USSR (mammals, Paleolithic) (in Russian). *Trudy in-ta geol. nauk., geol. ser.* 64(17):1–520.

Gromova, Vera. 1932. New materials on the Quaternary fauna of the Povolzh'e and on the history of the mammals of eastern Europe and northern Asia in general (in Russian). *Trudy Komiss. po izuch. chetv. perioda* 2:69–180.

———. 1965. *A brief review of the Quaternary mammals of Europe (a comparative effort)* (in Russian). Moscow, Nauka.

Grunin, K. YA. 1957. Nasopharyngeal botflies (Oestridae). (in Russian). In *Fauna SSSR. Nasekomye dvukrylye* 19(3):1–147.

———. 1973. The first discovery of the larva of a stomach botfly of mammoth—*Cobbolida (Mammontia,* subgen., n./*russanovi,* sp.n./Diptera—Gasterophilidae) (in Russian). *Entomol. obsr.* 52(1):228–233.

Guenther, E. W. 1960. Funde des Riesenhirsches in Schleswig-Holstein und ihre zeitliche Einordnung. *Steinzeitfragen der Alten und Neuen Welt*: 201–206. Zotz-Festschrift, Bonn.

Hemmer, H. 1967. Fassibhelege zur Verbreitung und Artgeschichte des Löwer. *Panthera leo* (Linne, 1758), mit Gabbild. *Säugethierkundliche Mitteilungen* 15:289–300.

Hilzheimer, M. 1918. Dritter Beiträge zur Kenntnis der Bisonten. *Arch. Naturgesch.,* Abt. A. 84(6).

Howorth, H. 1887. *The mammoth and the flood.* London, Sampson, Low, Marston, Searle, Rivington.

Kahlke, H. D. 1963. Ovibos aus de Kiesen von Süssenborn. Ein Beitrag zur Systematik und Phylogenie der Ovibovinen und zur Stratigraphie des Pleistozäns. *Gedogie* 12(8).

Kirikov, S. V. 1959. *Changes in the animal world in the natural zones of the USSR.* I. Steppe zone and forest steppe (in Russian). Moscow, Izd-vo AN SSSR.

———. 1960. *Changes in the animal world in the natural zones of the USSR.* 2. Forest zone and forest-tundra. Moscow, Izd-vo AN SSSR.

Krause, H. 1977. *The mammoth—in ice and snow?* Stuttgart.

Kuhn, H. 1956. *The rock pictures of Europe.* London.

Kurtén, B. 1958. Life and death of the Pleistocene cave bear. *Acta Zool. Fennica* 95:1–59.

———. 1968. *Pleistocene mammals of Europe.* London, Weidenfeld and Nicolson.

———. 1976. *The cave bear story.* New York, Columbia University Press.

Kurtén, B., and Anderson, E. 1980. *Pleistocene mammals of North America.* New York, Columbia University Press.

Kuz'mina, I. E. 1971. Formation of the theriofauna of the northern Urals in the late Anthropogene (in Russian). *Trudy Zoologicheskogo Instituta AN SSSR* 69:44–122.

Kvasov, D. D. 1977. An increase in moisture at the Pleistocene/Holocene boundary—one cause of mammoth extinction (in Russian). *Trudy Zooligicheskogo Instituta AN SSSR* 73:71–77.

Leidy, Y. 1853. Description of an extinct species of American lion: *Felis atrox. Trans. Amer. Phil. Soc.* 5(10):319–321.

Lindberg, G. U. 1972. *Large-scale fluctuations in ocean levels during the Quaternary Period* (in Russian). Leningrad, Nauka.

Lindner, K. 1937. *Die Jagd der Vorzeit.* Bd. 1. Berlin und Leipzig.

Marikovskij, P. I. 1953. Hunting methods and objects from the motifs of rock depictions of the Chulak Mountains (Kazakh SSR) (in Russian). *Zool. zhurn.* 32(6):1064–1074.

Middendorf, A. 1860. *Journey to the north and east of Siberia.* Part 1 (in Russian). St. Petersburg.

Mitchell, G. F., and Parkes, H. M. 1949. The giant deer in Ireland. *Proc. Royal Irish Acad.* 52(B,7):291–314.

Mowat, F. 1963. *People of the deer country* (translated from English to Russian). Moscow, Izd-vo inostr. lit.

Okladnikov, A. N. 1959. *Shishkin depictions* (in Russian). Irkutsk, Irkutsk knizhn. izd.

Orlova, L. A. 1979. Radiocarbon age of the remains of fossil mammoth in the USSR (in

Russian). *Izv. Sib. otd. AN SSR, ser. ob-shchestv. nauk* 6/2:89–97.

Ovodov, N. D. 1980. Cave sites with mammal remains in Siberia and the Far East (brief review) (in Russian). In: *Kraj Dalnego Vostoka i Sibiri*: 154–163. Vladivostok.

Paaver, K. L. 1965. *Formation of the theriofauna and changeability of the mammals of the Baltic littoral in the Holocene* (in Russian). Tartu, Izd-vo AN Estonskoj SSR.

Pallas, P. S. 1773. De Reliquis animalium exoticorum per Asian borealem repertis complementum. *Nov. Comm. Acad. Sci. Petrop.* 17:576–606.

Pavlova, M. V. 1924. The causes of extinction of animals in past geologic epochs (in Russian). In *Sovremennye problemy estestvoznaniya*: 5–88. Moscow, Pgr. Gosizdat.

Pfizenmayer. E. W. 1926. *Mammutleichen und Urwaldmenschen in Nordost Sibirien.* Leipzig.

Pidoplichko, I. G. 1951a. *The glacial period.* 2 (in Russian). Kiev, Izd-vo AN SSSR.

——. 1951b. *A new method of determining the geologic age of fossil bones from the Quaternary system* (in Russian). Kiev, Izd-vo. AN SSSR.

——. 1969. *Upper Paleolithic dwellings of mammoth bones in the Ukraine* (in Russian). Kiev, Naukova Dumka.

Rashid-Ad-Din 1946. *Collected writings* (translated into Russian from Farsi). 3. Moscow and Leningrad, Izd-vo AN SSSR.

Ryabinin, A. E. 1919. Fossil lions of the Urals and Povolzh'e (in Russian). *Trudy Geol. kom.* 168.

Ryziewicz, Z. 1955. Systematic place of the fossil musk-ox from the Eurasian diluvium. *Prace Wroclawsk. towar. naukow. ser. B.,* 49.

Savvateev, YU. A. 1970. *Zalavruga. Archaeological sites of the lower reaches of the Vyg River.* 1. Petroglyphs (in Russian). Leningrad, Nauka.

Semenov, S. A. 1968. *The development of primitive technology in the stone age* (in Russian). Leningrad, Nauka.

Serebrovskij, P. V. 1935. *A history of the animal world in the USSR* (in Russian). Len. obl. izdat.

Sher, A. V. 1971. *Mammals and stratigraphy of the Pleistocene of the extreme northeast of the USSR and North America* (in Russian). Moscow, Nauka.

Sickenberg, O. 1962. Die Säugetiere aus den elsterszeitlichen Kiesen (Pleistozän) von Bornhausen am Harz. *Geol. Jahrb.* 79.

Skalon, V. N. 1951. River beavers of northern Asia (in Russian). *Mater. k. nozn. fauny i flory SSSR. Nov. ser. Otd. zool.* 25:1–208.

Sludskij, A. A. 1963. Jutes in the Eurasian steppes and deserts (in Russian). *Trudy Instituta Zoologii AN Kazakhskoj SSR* 20:1–88.

Soergel, W. 1922. *Die Jagd der Vorzeit.* Jena.

——. 1936. *Hyaena brevirostris* Aymard und *Hyaena* aff. *crocotta* Erxl. aus den Kiesen von Süssenborn. *Z. Deutsch. Geol. Ges.* 88:52–539.

——. 1942. Die Verbreitung des diluvialen Moschusochsen in Mitteleuropa. *Beitr. Geol. Thüringen* 7:75–95.

Sokolov, V. E. 1964. The past ranges of ungulates in Kazakhstan according to rock depictions (in Russian). *Byull. Mosk. o-va ispyt. prirody 69(1):113–117.*

Tatarinov, K. A. 1977. The vertebrate fauna from the site of Korman IV (in Russian). In *Mnogoslojnaya paleoliticheskaya stoyanka Korman' IV:*112–119. Moscow, Nauka.

Tatishchev, V. N., and Gmelin, I. G. 1730. About bones excavated from the earth (in Russian). *Istorich., geneol. i geograf. primech. v Vedomostyakh* 80–83:319–336 and 88–93:363–372.

Thenius, E. 1980. *Grundzüge der Faunen und Verbreitungsgeschichte des Säugetiere. Eine historische Tiergeographie.* Jena, Gustav Fischer Verlag.

Timchenko, N. G. 1972. *On the history of hunting and stockbreeding in Kievan Russia* (in Russian). Kiev, Naukova Dumka.

Toll', E. V. 1897. Fossil glaciers of the New Siberian Islands and their relationship to the carcasses of mammoth and to the glacial period. *Zap. Russk. geograf. o-va* 32(1):1–139.

Tolmachoff, J. P. 1929. The carcasses of mammoth and rhinoceros found in the frozen ground of Siberia. *Trans. Amer. Philos. Soc. Nov. Ser.* 23(1):1–74.

Tomirdiaro, S. V. 1977. Change in the physical-geographic environment on the plains of northeast Asia on the boundary between the Pleistocene and Holocene as a basis for causing the extinction of the theriofauna of the mammoth complex (in Russian). *Trudy Zool. Instituta AN SSSR* 73:64–72.

——. 1980. *The loess-glacial formation of eastern Siberia in the late Pleistocene and Holocene* (in Russian). Moscow, Nauka.

Tsalkin, V. I. 1962. The history of stockbreeding and hunting in eastern Europe (in Russian). *Mater. i issl. po arkheol. SSSR* 104:1–130.

Tugarinov, A. YA. 1928. The origin of the arctic fauna (in Russian). *Priroda* 7/8:653–680.

———. 1934. Essay on the history of the arctic fauna of Eurasia (in Russian). *Trudy II Konf. Assots. po izuch. chetvertichn. per.* 5:55–65.

Velichko, A. A. 1973. *Natural process in the Pleistocene* (in Russian). Moscow, Nauka.

———. 1982. The basic features of the last climatic macrocycle and the present-day natural environment (in Russian). In *Paleogeografiya Evropy za poslednie sto tysyach let (atlas-monografiya)*: 131–143. Moscow, Nauka.

Vereshchagin, N. K. 1959. *Mammals of the Caucasus: A history of the formation of the fauna* (in Russian). Moscow-Leningrad, Izd-vo AN SSSR.

———. 1971a. The cave lion and its history in the Holarctic and in the limits of the USSR (in Russian). *Trudy Zoologicheskogo Instituta AN SSSR* 69:123–199.

———. 1971b. Prehistoric hunting and the extinction of Pleistocene mammals in the USSR (in Russian). *Trudy Zoologicheskogo Instituta AN SSSR* 69:200–232.

———. 1972. The origin of mammoth "cemeteries" (in Russian). *Prirodnaya obstanovka i fauna proshlogo* (Kiev) 6:131–148.

———. 1979. *Why the mammoths became extinct* (in Russian). Leningrad, Nauka.

———. 1982a. The new Gydansk (Yuribej) mammoth find (in Russian). *Vestnik zoologii* 3:32–38.

———. 1982b. Kizelovskaya Cave—a mammal trap in the Middle Urals (in Russian). *Trudy Zoologicheskogo Instituta AN SSSR* 3:37–44.

Vereshchagin, N. K. and Baryshnikov, G. F. 1980. Remains of mammals in the eastern gallery of Kudaro Cave I (in Russian). In *Kudarskie peshchernye paleoliticheskie stoyanki v Yugo-Osetii*: 51–62. Moscow, Nauka.

Vereshchagin, N. K. and Burchak-Abramovich, N. O. 1948. Rock drawings of southeastern Kabristan (in Russian). *Izv. Gos. geogr. ob-va* 80(5):507–518.

Vereshchagin, N. K. and Kuz'mina, I. E. 1977. Remains of mammals from Paleolithic sites on the Don and Upper Desna (in Russian). *Trudy Zoologicheskogo Instituta AN SSSR* 72:77–110.

Vereshchagin, N. K. and Rusakov, A. N. 1979. *Ungulates of the northwestern USSR* (in Russian). Leningrad, Nauka.

Zhivotchenko, V. P. 1981. The Amur tiger. Success in the restoration of a species (in Russian). *Priroda* (1):91–97.

23

Mammoths in China

LIU TUNG-SHENG AND LI XING-GUO

MAMMOTH FOSSIL LOCALITIES ARE WIDESPREAD in China (fig. 23.1), from Raohe (#33) along the Wusuli River in the east to Tongwei (#146) of Gansgu Province in the west, and from Huma of Da Xingan Mountains (#79) in the north to Lushun (#141) in the south. Fossil localities are especially numerous in the Songliao Plain of northeast China (fig. 23.1). Mammoth is a representative of the *Mammuthus-Coelodonta* (woolly rhinoceros) fauna of the late Pleistocene in northeast China (Pei 1957).

The southern boundary of the distribution of the woolly mammoth (*Mammuthus primigenius*) in the northern hemisphere, Europe, and North America is roughly 40°N latitude (Zhou 1978). The distribution of the mammoth in China is approximately the same, occasionally reaching 35°N latitude.

Mammuthus and Its Distributional Characteristics in China

Mammoth were recorded 2,100 years ago in *The Book of Magic*. They were described in the Emperor Kangxi's *Inquiry on the Physical Law in Leisure*, written by Aixinjueluo Xuanhua of Qing dynasty (1654–1722 A.D.):

> The northern plain near the sea in Russia is the coldest place. There is a kind of beast, which like a mouse, is big as an elephant, crawls in tunnels, and dies as it meets the sun or the moon light. Its teeth are like an elephant's, white, soft and smooth with no crackles. The native people often find it near the river bank. Its bones are used for making bowls, dishes, combs, and double-edged fine-tooth combs. Its meat is chilly and cold in character. Taking it as food, uneasiness and fever can be ridded off and its Russian name is Momentuowa (Chen Zhen 1958).

Thus, as early as 300 years ago, there were narratives on mammoths in China telling of their bones being used for utensils and their meat for food.

According to the study by Zhou Ming-zhen et al. (1974), "The genus *Mammuthus* in China includes *Mammuthus (Parelephas) trogontherii* Pohlig, *Mammuthus (Parelephas) sungari* Chow et Chang, *Mammuthus primigenius* Blumenbach, and one subspecies, *Mammuthus primigenius liupanshanensis* Chow et Chang." From the structure and pattern of the molars it is evident that these species are in a continuous transitional series.

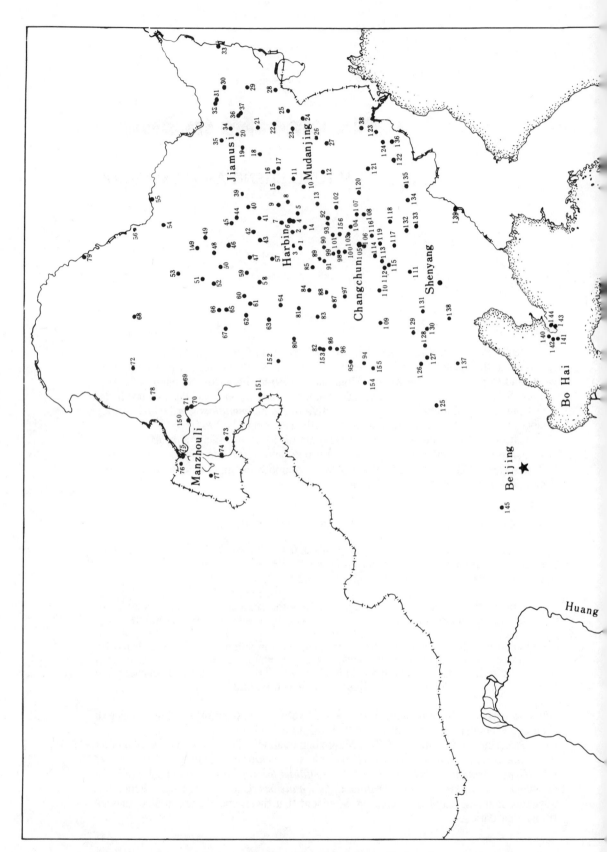

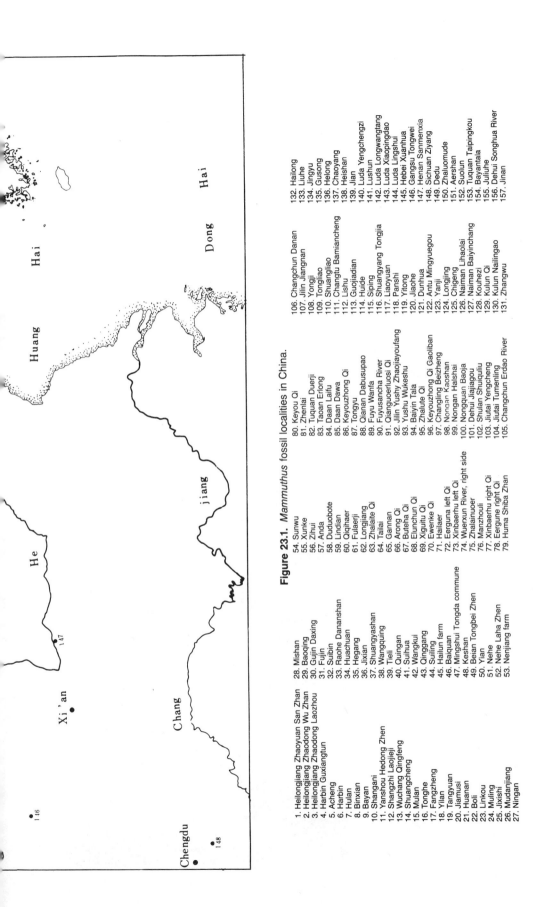

Figure 23.1. *Mammuthus* fossil localities in China.

1. Heilongjiang Zhaoyuan San Zhan
2. Heilongjiang Zhaodong Wu Zhan
3. Heilongjiang Zhaodong Laozhou
4. Harbin Guxiangtun
5. Acheng
6. Harbin
7. Hulan
8. Binxian
9. Bayan
10. Shangani
11. Yanshou Hedong Zhen
12. Shangzhi Laojieji
13. Wuchang Qingfeng
14. Shuangcheng
15. Mulan
16. Tonghe
17. Fangzheng
18. Yilan
19. Tangyuan
20. Jiamusi
21. Huanan
22. Boli
23. Linkou
24. Muling
25. Jixishi
26. Mudanjiang
27. Ningan
28. Mishan
29. Baoqing
30. Gujin Daxing
31. Fujin
32. Suibin
33. Raohe Dananshan
34. Huachuan
35. Hegang
36. Jixian
37. Shuangyashan
38. Wangquing
39. Tieli
40. Quingan
41. Suihua
42. Wangkui
43. Qinggang
44. Suiling
45. Hailun farm
46. Baiquan
47. Mingshui Tongda commune
48. Keshan
49. Beian Tongbei Zhen
50. Yian
51. Nehe
52. Nehe Laha Zhen
53. Nenjiang farm
54. Sunwu
55. Xunke
56. Zihui
57. Anda
58. Duduobote
59. Lindian
60. Qiqihaer
61. Fulaerji
62. Longjiang
63. Zhalaite Qi
64. Tailai
65. Gannan
66. Arong Qi
67. Buteha Qi
68. Elunchun Qi
69. Xiguitu Qi
70. Ewenke Qi
71. Hailaer
72. Eerguna left Qi
73. Xinbaerhu left Qi
74. Wuerxun River, right side
75. Zhalainuoer
76. Manzhouli
77. Xinbaerhu right Qi
78. Eergune right Qi
79. Huma Shiba Zhan
80. Keyou Qi
81. Zhenlai
82. Tuquan Duerji
83. Taoan Erlong
84. Daan Laifu
85. Daan Dawa
86. Keyouzhong Qi
87. Tongyu
88. Qianan Dabusupao
89. Fuyu Wanfa
90. Fuyusancha River
91. Qianguoerluosi Qi
92. Jilin Yushy Zhaojiayoufang
93. Yushu Wukeshu
94. Baiyin Tala
95. Zhalute Qi
96. Keyouzhong Qi Gaoliban
97. Changling Beizheng
98. Noncan Kaoshan
99. Nongan Haitshai
100. Nongguan Baoja
101. Dehui Jiajiagou
102. Shulan Shuiquliu
103. Jiutai Yengcheng
104. Jiutai Tumenling
105. Changchun Erdao River
106. Changchun Danan
107. Jilin Jiangnan
108. Yongji
109. Tongliao
110. Shuangliao
111. Changtu Bamiancheng
112. Lishu
113. Guojiadian
114. Huide
115. Siping
116. Shuangyang Tongjia
117. Liaoyuan
118. Panshi
119. Yitong
120. Jiaohe
121. Dunhua
122. Antu Mingyuegou
123. Yanji
124. Longjing
125. Chigeng
126. Naiman Lihaolai
127. Naiman Baiyinchang
128. Kouhezi
129. Kulun Qi
130. Kulun Nailingao
131. Zhangwu
132. Hailong
133. Liuhe
134. Jingyu
135. Gusong
136. Helong
137. Chaoyang
138. Heishan
139. Jian
140. Luda Yengchengzi
141. Lushun
142. Luda Longwangtang
143. Luda Xiaopingdao
144. Luda Lingshui
145. Hebei Xuanhua
146. Gangsu Tongwei
147. Henan Sanmenxia
148. Sichuan Ziyang
149. Dedu
150. Zhaluomude
151. Aershan
152. Suolun
153. Tuquan Taipingkou
154. Bayantala
155. Juliuhe
156. Dehui Songhua River
157. Jinan

Mammuthus (Parelephas) trogontherii Pohlig and *M. primigenius liupanshanensis* Chow et Chang are relatively older species restricted to western China. Localities of *M. (Parelephas) sungari* Chow et Chang are also rare. These three mammoth species are mainly of paleontological interest, and little other information is available on them.

Distribution and Stratigraphy of the Mammoths

At Heilongjiang, Zhaoyuan, Shan Zhan (fig. 23.1 #1), Zhen et al. (1979) reported that the skeleton of *M. (Parelephas) sungari* Chow et Chang was found buried in a light yellow silt bed, 5 m below the surface (fig 23.2). The skeleton was 5.45 m long and 3.33 m in height with a 2.05-m-long incisor (tusk). It is the most nearly complete fossil mammoth yet discovered in China (fig. 23.3). A ^{14}C age determination of $21,200 \pm 600$ yr B.P. was obtained (Institute of Archaeology 1979).

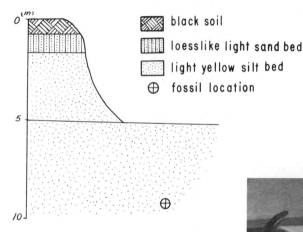

black soil

loesslike light sand bed

light yellow silt bed

⊕ fossil location

Figure 23.2 Stratigraphic section of the mammoth fossil at Zhaoyuan, San Zhan.

Figure 23.3. Skeleton of the fossil mammoth from Zhaoyuan, San Zhan.

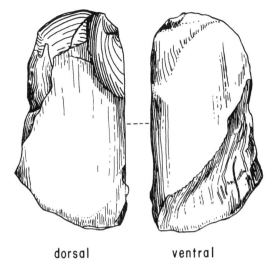

Figure 23.4. Mammoth-tooth spade from Zhoujiayoufang of Yushu, Jiling Province. (After Sun Jian-zhong)

dorsal ventral

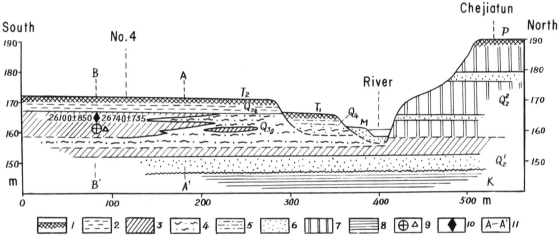

Figure 23.5. Stratigraphic section of the mammoth fossil locality at Zhoujiayoufang of Yushu, Jiling Province (After Sun Jian-zhong). 1. black soil; 2. clayey soil; 3. clay; 4. puddly sand; 5. sandy silt; 6. sandy soil; 7. loess; 8. mudstone; 9. mammoth fossils and cultural relics; 10. ^{14}C ages; 11. surveyed section; K, Cretaceous relics; Q_2^1, lower part of middle Pleistocene; Q_2^2, upper part of middle Pleistocene; Q_3g, Guxiangtun formation of upper Pleistocene; Q_3q, Quanli formation of upper Pleistocene; Q, Holocene; M, valley flat; T_1, first terrace; T_2, second terrace; P, loess platform.

Zhuojiayoufang of Yusha County, Jiling Province in northeast China (fig. 23.1, #92) is a famous locality called "a treasury of fossils." Thirty-five fossil species have been recorded there and a great number of bones of *Mammuthus* and *Coelodonta* were found. In addition, paleoliths and a "spade" (an implement made of mammoth tusk) (fig. 23.4) were discovered (Sun 1977, 1979). This place also may have been a residence of ancient man.

The fossils, identified as *M. primigenius* Blumenbach, were discovered in a black clay bed, 7.7 m below the surface on the second terrace (fig. 23.5). Samples of fossil tree trunks in the same horizon submitted by Li Xing-guo to the radiocarbon laboratory of the Institute of Vertebrate Paleontology and Paleoanthropology yielded ages of 26,100±850 yr B.P. (Li et al. 1979) and 26,740±740 yr B.P. (Institute of Science and Technology 1980).

Figure 23.6. Mammoth coprolites from east opencut coal mine at Zhalainuoer.

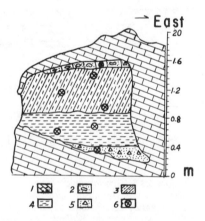

Figure 23.7. Section of mammoth fossil locality at Mingyuigou, Antu, Jiling Province. 1. limestone; 2. brown-gray sandy soil with gravels; 3. grayish-yellow sandy silt; 4. brown-yellow loam; 5. gravels with debris; 6. fossils. (After Jiang-peng)

In May 1980, in the east opencut of the coal mine of Zhalainuoer (#75), Inner Mongolia Autonomous Region, bones of two mammoths were discovered and provisionally identified as *M. primigenius* Blumenbach. One is represented by two remnants of incisors. The other is a nearly complete skeleton. The skeleton, the largest mammoth skeleton found in China, is 8 m in length, 5 m in height, and the incisor (tusk) is 3.05 m in length. In the same fossil locality, mammoth coprolites weighing 3 kg (fig. 23.6) were found and the ^{14}C age of one is $33,400 \pm 1,700$ yr B.P.

Fossil mammoth were discovered in the old river bed deposit 35 to 40 m below the surface. Mammoth tusks and molars identified as *M. primigenius* were discovered in the sandy gravel beds 5 m below the surface at the shipyard of Raohe County (#33), Heilongjiang Province, by Li Xing-guo in June 1980. Implements, possibly Paleolithic, and bone tools were also discovered from this bed.

A limestone cave was discovered at Mingyuegou, Antu County, Jiling Province (Zhou 1978). A number of *Mammuthus* and *Coelodonta* were found in the cave deposits (fig. 23.7). The ^{14}C age of the scapula, which belongs to *M. primigenius*, is $26,560 \pm 550$ yr B.P., while the molar is dated at $35,370 \pm 850$ yr B.P. (Institute of Science & Technology 1980). The discrepancy could be due to contamination of the sample.

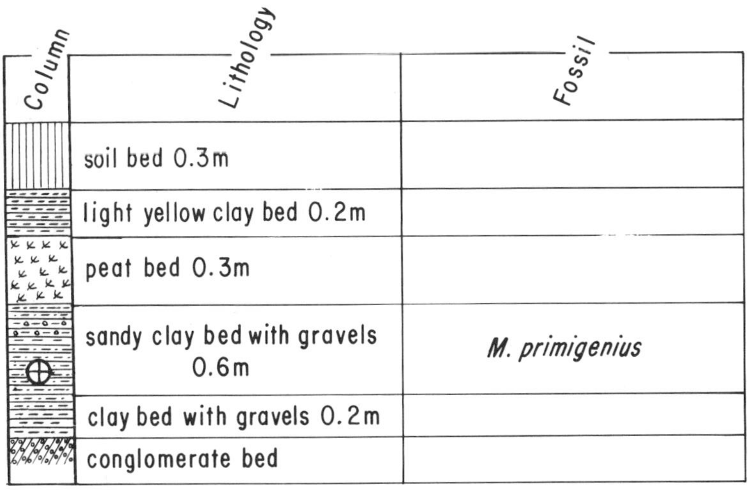

Column	Lithology	Fossil
	soil bed 0.3m	
	light yellow clay bed 0.2m	
	peat bed 0.3m	
	sandy clay bed with gravels 0.6m	*M. primigenius*
	clay bed with gravels 0.2m	
	conglomerate bed	

Figure 23.8. Column of fossil locality of mammoth at Nanshantun of Wangqing, Jiling Province. (After Zhang Zhi-tu)

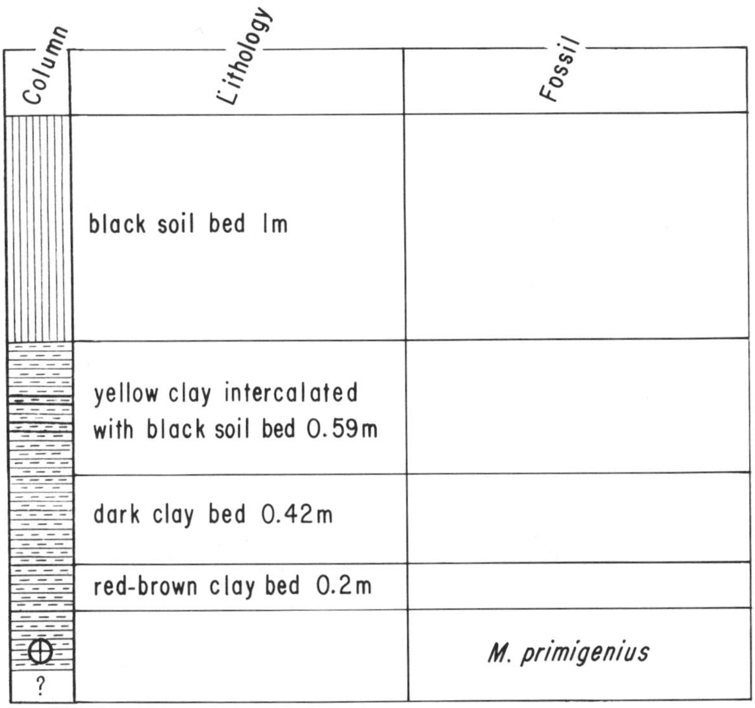

Column	Lithology	Fossil
	black soil bed 1m	
	yellow clay intercalated with black soil bed 0.59m	
	dark clay bed 0.42m	
	red-brown clay bed 0.2m	
?		*M. primigenius*

Figure 23.9. Stratigraphic section of the mammoth fossil locality at Tongbei of Beian, Heilongjiang Province. (After Wei Zheng-yi)

In 1952 a segment of a tusk of *M. primigenius* was found during factory building at the Mudanjiang textile mill, Mudanjiang (Mianto 1967), Heilongjiang Province; it was radiocarbon dated at $21,540 \pm 1,000$ yr B.P. (Institute of Archaeology 1978). A pair of tusks and five molars of *M. primigenius* were found in the clay beds with gravels (Zhang 1964) 1.8 m below the surface at Nanshantun of Wangqing County, Jiling Province (fig. 23.8).

In 1963 a segment of tusk and a complete molar of *M. primigenius* were found in the black-green clay bed 2.5 m below the surface (fig. 23.9) at Tongbei (#49) of Beian, at the upper reaches of Tonggan River, Heilongjiang Province. The geological section was similar to that of the mammoth fossil locality at Hailun farm.

A complete skeleton of fossil mammoth buried in the brownish-yellow silt bed 3 m below the surface was also found at Nehe of Heilongjiang Province, but unfortunately it was not protected in time to avoid loss. Another nearly complete skeleton was found in the yellow-gray loam 1.7 m below the surface, which is preserved in the Heilongjiang Museum.

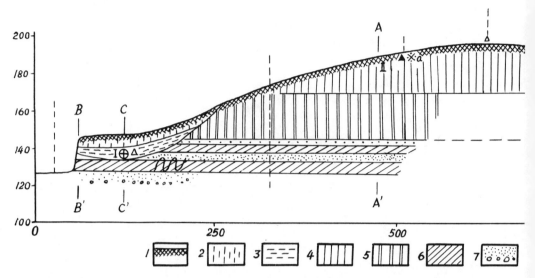

Figure 23.10. General section at Huangshan of Harbin. 1. Tantu black soil; 2. Quanli formation; 3. loess-like soil; 4. Harbin loess and gray-black granular texture soil; 5. loess of Dongfeng formation; 6. clay and gravels of Huangshan formation; 7. angular-granular texture loess; 8. ruins of Mesolithic culture; 9. ruins of Paleolithic culture; 10. fossil locality of mammals; I, *Equus przewalskii, Bos*

Paleolithics were discovered at both Guxiangtun, the type locality of Guxiangtun Formation, and Huangshan of Harbin, Heilongjiang Province (Wei 1963). *M. primigenius, Coelodonta antiquitatis, Equus przewalskii* (Przewalsky's horse), *Bos primigenius* (auroch), and a spore-pollen association were found (Sun 1978). Remains of fossil plants were found, mainly *Picea* cf. *abovata* Ldb. (spruce), *Abies* cf. *nepholepis* Max (fir), *Larix dahurica* Turcz (larch), and *Betula* sp. (birch) (fig. 23.10).

A segment of fossil tree root was collected by Li at Huangshan in 1978 (PV-91); its ^{14}C age is $29,800 \pm 1,340$ yr B.P. (Li 1980).

Discovery of mammoth from localities without a stratigraphic record were also reported. Fossil *Mammuthus* have been found in 157 localities. Apart from *M. primigenius liupanshanensis* Chow et Chang found at Tongwei, Ganshu Province, (a new species) and *M. (Parelephas) trogontherii* found at Sanmenxia, Henan Province, (an early Pleistocene species), most of the mammoth fossils are *M. primigenius* Blumenbach. They were distributed north of 38°N latitude and concentrated in Heilongjiang and Jiling provinces. The geographic pattern of distribution of fossil localities shows dense concentrations in the eastern and northeastern parts of northeast China, which link with the Maritime Province of the Soviet Union where mammoth fossils were found.

Relationship Between Mammoth Distribution and Climatic Factors

The *Mammuthus-Coelodonta* fauna was representative of northeast China's late Pleistocene cold climate. In middle Pleistocene deposits, fossils found in these districts are generally similar to those of the Zhoukoudian (Choukoutien) fauna, for instance *Rhinoceros mercki, Equus sanmeniensis,* and *Bison harbinensis* (Jiang 1977) reflecting the comparatively warm climate.

In the middle Pleistocene, the flora in northeast China was similar to that of north China and mainly consisted of *Pinus, Celtis, Alnus, Ulmus,* and *Carpinus,* indicative of the warm and wet climate. In the late Pleistocene in northeastern China, a cryophilic flora consisting of *Picea, Abies,* and *Betula* replaced the middle Pleistocene thermophilic flora.

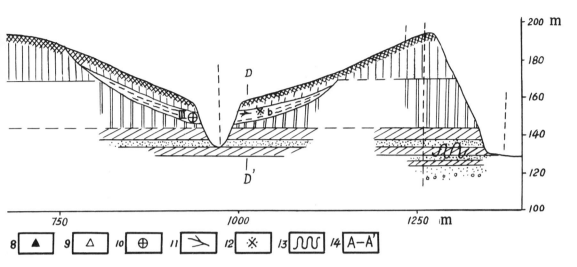

primigenius; II, *Gazella przewalskii;* III, *Mammuthus primigenius, Coelodonta antiquitatis, Bos primigenius;* 11. fossil locality of *Picea* cf., *abovota* Ldb. and other fossil plants; 12. a, 2780 ± 130 yr B.P., b, 29,800 ± 1340 yr B.P.; 13. cryoturbated phenomena; 14. site of surveyed section. (After Sun Jian-zhong)

Age	Layer	m	Column	Lithology	Fossil
	1	0.6		black soil	
	2	1.3		clay	
	3	2.4		clay intercalated with gravels	
Quaternary	4	4.4		clay intercalated with angular detritus	
	5	5.2			*M. primigenius*
	6			gravel bed	

Figure 23.11. Column of mammoth fossil locality at Raohe of Heilongjiang Province.

The mammoth in China usually have been found deeply buried. Generally, a bed of fine sands of 3 to 5 m thick covered the fossils. Some were buried beneath the soil bed and separated from it by loess deposits such as Harbin Huangshan (fig. 23.10) or clay deposits such as Hailun, or Raohe (fig. 23.11). Some were below the clay and peat beds, such as those at Wangqing (fig. 23.8). These deposits indicate a relatively long period of time for burial beneath sediments. Cryoturbation phenomena were observed in some sections which contained cryophilic fossil plants such as *Picea, Abies* and *Betula.*

In Europe and North America, 45,000 to 35,000 yr B.P. was a relatively warm period (Flint 1961). From 35,000 to 18,000 yr B.P. the climate fluctuated, growing colder. At this time mammoth had a wide distribution in northeast China which could have been related to the climatic fluctuation. But no fossil mammoth have been recorded around 18,000 years ago, the coldest part of the last glacial age. No fossil mammoth younger than 20,000 years in age has yet been found in China. From fossils of the Maritime Province of the Soviet Union similar radiocarbon ages have been obtained: Zhalainuoer (#75) (N 49°21', E 117°35') 33,400 years; Huangshan (Wei 1963) (N 45°48', E 126°47') 29,800 years; and Mudanjiang (Mianto 1967) (N 44°38', E 129°35') 21,000 years. Thus it is possible that 18,000 years ago, when the coldest period came, mammoths were already restricted to higher latitudes.

Relationship Between Distribution and Extinction of Mammoth in China and Human Activities

Climatic changes in the late Pleistocene may have forced mammoths to migrate and may account for their distribution in China. At the same time, human hunting activities also affected the migration of mammoth. Mammoth fossils, accumulated as "garbage heaps," were found with stone implements at many places in northeast China. For instance, at the Paleolithic ruin at Zhoujiayoufang, seven individual fossil mammoth were found with spade tools made of mammoth teeth in a 400-square-meter area.

Molars of *M. primigenius* were found with Paleolithic tools in an area of 10 square meters at Raohe, Heilongjiang Province (#33) (N 46°47', E 134°1'). Complete mammoth fossil skeletons were discovered repeatedly at Zhalainuoer (#75) (N 49°21', E 117°35'). In the same strata a great number of Neolithic and Paleolithic tools associated with human and other mammal fossils were discovered, which mainfests the close relation between mammoth extinction and human activities. This supports the view of P. S. Martin (1979) that "the global pattern of extinctions of all large mammals of the continents and islands appears to follow Paleolithic man's footsteps."

The migration and extinction of mammoth had close relationships with the activities of man. As mammoth declined in China, the animals may have persisted farther north, attracting humans in pursuit toward the northeast, allowing the animals to cross the Bering Strait land bridge and continue into North America. Further study on the distribution and extinction of mammoth in northeast China and the relationship with human activities will shed new light in this fascinating and important problem.

Acknowledgments

The authors wish to express their gratitude to Professor Paul S. Martin, University of Arizona, Tucson, Arizona, and Professor Chow Ming Chen, Director of the Institute of Vertebrate Paleontology and Paleoanthropology, for their encourgement and stimulating discussions. To Mrs. Zhang Yu-ping the authors are indebted for her reading of the manuscript. Thanks are also due to Mr. Sun Yi-ying and Mrs. Dai Jia-sheng for their help in translating the manuscript into English and drawing the figures.

References

Chen Zhen. 1958. On the biological history of China. Popular Science Press, pp. 89–90.

Compiling group of Manual of Vertebrate Fos-

sils in China. Institute of Vertebrate Paleontology and Paleoanthropology, Academia Sinica. 1979. Manual of Vertebrate Fossils in

China (revised and enlarged edition). Science Press, pp. 454–455.

Flint, R. F. 1961. Am. Jour. Sci., v. 259, pp. 321–328.

Institute of Archaeology, Chinese Academy of Social Sciences. 1978. A report on age determination of radioactive carbon (5). Archaeology, no. 4, p. 284.

Institute of Archaeology, Chinese Academy of Social Sciences. 1979. A report on age determination of radioactive carbon (6). Archaeology, no. 1, p. 93.

Institute of Science and Technology of Cultural Relic Protection. 1980. A report on carbon 14 age determination (2). Cultural Relics, no. 2, p. 84.

Jiang Peng. 1975. Late Pleistocene cave accumulations at Antu, Jiling. Vertebrata Palasiatica, v. 13, no. 3, pp. 197–198.

———. 1977. The distribution of Late Pleistocene mammal fossils in Jiling. Vertebrata Palasiatica, v. 15, no. 4, pp. 313–316.

Li Shi-zhen. Compendium of Materia Medica, v. 51, The Beast.

Li Xing-guo et al. 1979. C^{14} age determination of some geologic and archaeologic specimens (I). Vertebrata Palasiatica, v. 17, no. 2, pp. 165–171.

———. 1980. A batch of C^{14} age determination of geologic and archaeologic specimens. Vertebrata Palasiatica, v. 18, no. 4, pp. 344–347.

Martin, P. S. 1979. The pattern meaning of Holarctic Mammoth extinction, Paper prepared in advance for participants in Burg Wartenstein Symposium No. 81.

Mianto, M. 1967. On the age of Mammoths in Japan and Siberia. Earth Science, v. 21, pp. 213–217.

Pei Wen-zhong. 1957. The Zoogeographical divisions of Quaternary mammalian faunas in China. Vertebrata Palasiatica, v. 1, no. 1, pp. 9–23.

———. 1957. The geographic distribution of Quaternary mammals in China. Vertebrata Palasiatica, v. 1, no. 1, pp. 23–24.

———. 1957. Ziyang Man. p. 5.

Ren Zhen-ji et al. 1979. The association of spores and pollen of several sections at Zhoukoudian and their correlation, 2nd compilation of papers and abstracts at the commemoration meeting of 50th anniversary of the discovery of the first skull of Peking man.

Shi Yan-shi. 1978. The age of determination of woody specimens in the vicinity of Zhalainuoer. Vertebrata Palasiatica, v. 1, no. 2, pp. 144–145.

Sun Jian-zhong. 1977. On the Guxiangtun formation, Jiling. Geology, no. 2, pp. 27–41.

———. 1978. A preliminary correlation of geology in Dali Glacial ages in China (unpublished).

———. 1979. The ruins of paleolithic culture of Zhoujiayoufang of Yushu, Jiling. Abstract of papers of the third symposium of Quaternary Research Association of China, p. 53.

Wei Zheng-yi. 1963. The investigation of mammoth localities at the upper reaches of Tongken River, Heilongjiang. Vertebrata Palasiatica, v. 7, no. 3, p. 287.

Zhang Zhi-wen. 1964. Quaternary mammal fossils at Wangqing, Jiling. Vertebrata Palasiatica, v. 8, no. 4, pp. 402–407.

Zhen Shuo-nan et al. 1979. The discovery of a complete skeleton of *Mammuthus (Parelephas) sungari* Chow et Chang at Zhaoyuan county, Heilongjiang province, Report of Peking Natural Museum, no. 3, pp. 1–9.

Zhou Ben-xiong. 1978. The geographic distribution of Coelodonta and Mammuthus, paleobioecology and related problem of paleoclimate. Vertebrata Palasiatica, v. 16, no. 1, pp. 47–59.

Zhou Ming-zhen, Zhang Yu-ping. 1974. Elephant fossils in China. Science Press, pp. 57–61.

24

Faunal Turnover and Extinction Rate in the Levant

EITAN TCHERNOV

A BIOGEOGRAPHICAL CROSSROADS, ISRAEL is strategically situated on the Levantine corridor between Euro-Asia and Africa. Geological processes, faunal history, and later cultural events in the Neogene and Pleistocene periods are crucial to understanding the dynamic exchanges of organisms between Eurasia and Africa, the construction of the complicated biomes in the South Palaearctic region, the relatively high species diversity, the faunal turnover, and the rate of extinction in correlation with the abiotic and cultural changes in this region.

Constantly confronted with swarms of Palaearctic species that invaded the Levant from the north, the tropical African elements were slowly eliminated locally throughout the Pleistocene. As yet, very little is known about the animal history of the Levant during the Neogene (Savage and Tchernov 1968), but Pleistocene faunas reflect paleontological events that took place in earlier periods and, to some extent, may clarify the vexing problem of the biotic interconnection between the two realms. Plate tectonic theory (Dewey et al. 1973) suggests that African invasions into Asia resulted from the drawing up of the northeast edge of the Afro-Arabian continent against the margin of the Eurasian continental body by subduction (plate consumption) along the present Anatolian-Iranian tectonic suture line. The first wave of immigrants contained known taxa of the North African early Miocene fauna only (Savage and Hamilton 1973). East African early Miocene representatives such as ambelodont proboscideans and hominoids did not enter Eurasia until the Middle Miocene (Middle Vindobonian). On the other end, Middle Miocene mammals of Eurasia found it difficult to spread into Africa at a time when proboscideans and other African groups found it easy to emigrate to Eurasia. The post-Burdigalian invaders (those that came after the continental passageways were erected), which now live in Africa, are relatively few, and include advanced fissipeds (felids, canids, hyaenids, viverids), equids, advanced rodents, and lagomorphs. None are known yet from beds in eastern Africa older than 12 my. This is also true for the Levant and North Africa. The main waves of immigrants from the north took place during the Pliocene and early and Middle Pleistocene. Few of the newly arrived species could have found open routes to tropical Africa, as the Saharan belt had already created a severe biogeographical barrier.

It is now generally believed that progressive desiccation (fluctuating in correlation with the glacial periods) was the principal climatic trend in the Levant. This shift toward dryness was presumably one factor that caused the extinction of the tropical component of the eastern Mediterranean fauna. The impact of both the northern glacial sequence and the close proximity of a large desert belt upon the terrestrial and aquatic fauna of Israel throughout the Pleistocene is yet to be fully understood.

Pleistocene deposits in Israel have furnished a wealth of fossil material, in particular from the late, but also from the Middle Pleistocene. Most of the Pleistocene bone-bearing deposits came from prehistoric sites in which the deposition of microvertebrates was due to diurnal and nocturnal raptors. Excluding a few cave dwellers and anthropophile organisms, the great majority of animal remains in the prehistoric sites constitute a true thanatocoenosis. Larger vertebrates were mostly collected by man and may largely reflect the hunter's choice at any given time. The ad hoc paleontological information accumulated during this century in Israel, which began mainly with the pioneering work of Bate (1932, 1937, 1942, 1943), enables us to trace the faunal turnover throughout the Pleistocene and try to correlate it with biotic, abiotic, and cultural factors.

The Fossil Record

The Preglacial Pleistocene

The mammalian fauna of Bethlehem constitutes the oldest Pleistocene assemblage known from the Levant and, excluding Ghor-el-Qatar in Jordan (Huckriede 1966), as yet the only one. As shown by Gardner and Bate (1937), Hooijer (1958), Clark (1961), and Horowitz (1974, 1979), this bone-bearing bed should be attributed to the early Pleistocene.

Much like other typical Palaearctic groups (cervids and ursids), *Equus* turned out to be widespread in 'Ubeidiya (Mindel) Formation. None of these groups are represented in the preglacial Pleistocene of the Levant (Table 24.1). Hence, their invasion into the southern part of the eastern Mediterranean region should have taken place after preglacial time (Bar-Yosef and Tchernov 1972), sometime between the Villafranchian and the Mindel, probably during the Günz. It is thus unfortunate that no Günzian fauna is yet known from this region to link the two periods and pinpoint the exact time of the southern expansion of the Palaearctic elements.

The Bethlehem assemblage suggests a flat savannalike landscape crossed with rivers, which agrees with the geological picture presented for this period by Neev (1975) and Horowitz (1979).

Early Middle Pleistocene Fauna

In the Middle East the early Middle Pleistocene generally can be defined first by the wide array of Palaearctic (never Arctic) species, and a minor wave of immigrants from Africa, in addition to the old Ethiopian elements existing in the eastern Mediterranean region since the Miocene. This period can also be typified by the tenacious preservation of older elements, usually from the Villafranchian of Europe. These elements belong to a much older Palaearctic, post-Messinian invasion from the north (Tchernov 1968, 1975).

The source of a large quantity of fossil material is the 'Ubeidiya site, situated at the ingression-regression boundary of the ancient 'Ubeidiya Lake in the Jordan Valley. The 'Ubeidiya Formation is the oldest site in the Levant (Bar-Yosef and Tchernov 1972) and also contains one of the oldest human occupations in the Middle East. The 'Ubeidiya lithic assemblages are basically similar to those of Olduvai Gorge, Bed II, attributed to the developed Oldowan and early Acheulean culture. This suggests a date of early Middle Pleistocene (Stekelis 1966, Leakey 1967, Stekelis et al. 1969, Bar-Yosef and Tchernov 1972).

Table 24.1. Succession of Proboscidean, Ungulate, Primate, and Carnivore Species During the Pleistocene of the Levant

Species	Preglacial Pleistocene	Günz Günz-Mindel	Mindel	Mindel-Riss	Riss	Riss-Würm
Nyctereutes megamastoides	+					
Homotherium sp.	+					
Elephas planifrons	+	?				
Gazellospira torticornis	+					
Hipparion sp.	+	+	?			
Dicerorhinus etruscus	+	+	+			
Sus strozzi	+	+	+			
Giraffa camelopardalis	+	+	+			
Leptobos sp.	+	+	+			
Hippopotamus amphibius	+	+	+	+	+	+
Elephas namadicus		+?	+	+	+	+?
Stegodon mediterraneus		+?	+	+	+	
Macaca sp.			+			
Equus sussenbornensis			+			
Equus stenonis			+			
Equus przewalski			+			
Megantereon megantereon			+			
Ursus etruscus			+			
Oryx sp.			+			
Euctenoceros senezensis			+			
Cervus cf. ramosus			+			
Camelus sp.			+	+		
Metridiochoerus evronensis			+	+		
Megaloceros sp.			+	+		
Bison priscus			+	+	+	+
Crocuta crocuta			+	+	+	+
Vulpes vulpes			+	+	+	+
Dama sp.				+	+	
Dicerorhinus hemitoechus				+	+	+
Hyaena hyaena				+	+	+
Canis aureus				+	+	+
Dicerorhinus merckii					+	
Hemibos sp.					+	
Equus hemionus					+	+
Equus caballus					+	+
Equus hydruntinus					+	+
Sus scrofa					+	+
Phacochoerus garrodae						+
Nyctereutes vinetorum						+
Canis lupaster						+
Bos primigenius						+
Capra aegagros						+
Capra ibex						+
Alcelaphus buselaphus						+
Cervus elaphus						+
Dama mesopotamica						+
Capreolus capreolus						+
Gazella gazella						+
Canis lupus						+
Felis sylvestris						+
Felis pardus						+
Vormella peregusna						+
Meles meles						
Felis leo						
Felis chaus						
Herpestes ichneumon						
Martes foina						
Camelus dromedarius						
Ursus arctos						
TOTAL	10	9	23	13	16	27

| Mousterian | | Würm | | | Holocene | | | | |
Lower	Upper	Upper Paleol.	Aurignacian	Epi-Paleol.	Neolithic	Bronze Age	Iron Age	18th–19th Centuries	Recent
+	+								
+	+								
+	+								
		+	+	+					
+	+	+	+	+	+	+	+	+	+
+	+								
+	+	+	+	+					
		+	+	+	+	+	+	+	+
+	+	+	+	+	+	+	+		
+	+	+	+	+					
+	+	+	+	+	+				
+	+	+	+	+	+	+	+	+	+
+		+	+	+	+	+			
+									
+	+	+	+	+					
+	+	+	+	+	+				
+	+	+	+	+	+				
+	+	+	+	+	+	+	+	+	+
+	+	+	+	+	+	+	+		
+	+	+	+	+	+	+	+		
+	+	+	+	+	+	+	+		
+	+	+	+	+	+	+	+	+	
+	+	+	+	+	+	+	+	+	+
+	+	+	+	+	+	+	+	+	+
+	+	+	+	+	+	+	+	+	+
+	+	+	+	+	+	+	+	+	+
+	+	+	+	+	+	+	+	+	+
+	+	+	+	+	+	+	+		
+	+	+	+	+	+	+	+	+	+
+	+	+	+	+	+	+	+	+	+
+	+	+	+	+	+	+	+	+	+
+	+	+	?						
+	+	+	+	+	+	+	+	+	
32	30	29	29	28	24	21	20	15	13

The lithic assemblage of 'Ubeidiya (Stekelis 1966, Leakey 1967, Stekelis et al. 1969, Bar-Yosef and Tchernov 1972), the roughly dated tectonic movements of the Jordan graben (Picard 1963, 1965; Picard and Baida 1966, 1966a), the correlation with the chronostratigraphy of the Upper Jordan Valley (Horowitz 1973, 1974, 1979; Tchernov 1973), and the radiometric dates for the overlaid and underlaid lava flows (Horowitz et al. 1973, Siedner and Horowitz 1974) suggest a Mindel age of around 1,000,000 years for 'Ubeidiya Formation.

In the absence of Günzian fauna from the eastern Mediterranean area, and with scarce geological data, a large gap exists in our knowledge concerning the faunal succession during this period and the transition to the Mindel. The fauna of 'Ubeidiya clearly shows that great changes took place during the Günz, which can be characterized by two main features: a bloom of Palaearctic elements and a deterioration of tropical ones (Tables 24.1, 24.2, and 24.3). The Mindel glaciation, effective as it was in the Middle East, could have favored better-adapted northern elements, eliminating through ecological exclusion some of the Ethiopian-originating species that were shifted to a lesser adaptive zone. Yet, many of those elements that survived the northern invasion continue to exist in the Middle East, accounting for 15 to 20 percent of the present fauna. Presumably the Günz-Mindel episode saw a fusion of the two faunas with an accelerated rate of extinction of the tropical components. The great diversity of Palaearctic elements in 'Ubeidiya shows that they had entrenched themselves in the Jordan Valley before the Mindel. It is interesting that despite this faunal collision, the ancient Villafranchian elements continued to survive the entire period.

Faunal Succession During the Mindel-Riss and Riss Periods

The faunal succession and the turnover of species after 'Ubeidiya is not completely clear. Mindel-Riss sites—Latamne in Syria (Hooijer 1961; Clark 1967, 1968; Van Liere 1966; Heinzelein 1968) and Evron in northern Israel (Ronen and Amiel 1974, Gilead and Ronen 1977) include only macromammals, while the only known Rissian sites are Benot Ya'akov (Stekelis 1960; Hooijer 1959, 1960, 1961; Horowitz et al. 1973), which contains only macromammalian remains and Giveat Shaul (Tchernov 1968a), which yielded mainly microfaunal remains (Tables 24.1–24.4). A comparative list of the species exposed from these sites is given in Table 24.4

Of the large mammals, excluding the amazing survival of *Stegodon* in Benot Ya'akov, all the Villafranchian elements found in 'Ubeidiya have become extinct. Few ancient forms of rodents survived until the close of the Middle Pleistocene.

The primitive form of *"Meriones" obeidiensis* ('Ubeidiya Formation), which probably represents a transitional stage between *Pseudomeriones* and the modern *Meriones*, is still found in Riss time. Representatives of the genus *Allocricetus* were still extant during the Mousterian.

The Tyrrhenian stage in the Middle East is typified by interpluvial conditions roughly as dry and warm as at present (Horowitz 1979). This probably caused many Palaearctic elements to retreat northward, or through a drastic habitat shrinkage, to be completely eliminated from the scene. The savannalike conditions that prevailed in the area during the Tyrrhenian are well documented, even in the assemblage found as far north as Latamne, northern Syria (Van Liere 1961, 1966; Van Liere and Hooijer 1961), which includes such typical open-land species as *Camelus*, *Equus*, *Dicerorhinus*, and undetermined Antilopini. The fact that human occupation (which scarcely existed during the Tyrrhenian in the eastern Mediterranean region) was concentrated mainly in the vicinity of water bodies explains the association of open-land animals with aquatic or hydrophile species and northward retreat (or extinction) of Palaearctic elements. Yet, this faunal picture can also be due to the paucity of the fossil record.

During the Riss some acceleration of faunal turnover is indicated. Species like *Gazella gazella*, *Equus caballus*, *Sus scrofa*, *Mesocricetus* sp., and *Spalax ehrenbergi*

either first appeared in the area at that time or replaced older forms through autochthonous speciation (like *Jordanomys haasi*). Some of the new Rissian elements were of
Asiatic origin (*Spalax, Mesocricetus*), or they were species that occupied open-land
habitats (*Equus caballus, Dicerorhinus merckii*). The occurrence of *Hemibos,* known
only from Benot Ya'akov Formation, and *Bison priscus,* together with the relative
rareness of cervids and the lack of *Sciurus*, do not indicate the kind of rather cold
climate and pluvial conditions that would be expected in this region during the Riss.

Faunal Rejuvenation During the Riss-Würm Interglacial

Oumm-Qatafa is situated in the Judean Desert south of Jerusalem. It was excavated by Neuville, who described a long sequence of prehistoric cultures from this site
(1951). Following Rust (1950), Howell (1959), Woldstaedt (1962), Tchernov (1968), and
Farrand (1969), the age of Oumm-Qatafa was interpolated at ca. 140,000 years old.

The faunal records indicate that a distinct faunal break took place during the onset
of the Riss and the beginning of the Eem Interglacial. Some ten species of rodents, nine
species of ungulates, and six species of carnivores appeared, causing large-scale recombination of the fauna; notwithstanding the fact that this was a warm, dry period, species
diversity is very high.

Some of the newly arrived species were of Palaeotropic origin (*Felis pardus, Felis
chaus* [Oriental]*, Canis lupaster, Nyctereutes vinetorum* [closely related to the Chinese
form], *Gazella gazella, Alcelaphus* sp., *Hystrix indicus, Rattus haasi* [probably Orien-
tal], and *Phacocheorus garrodae*). Simultaneously, another swarm of immigrants arrived
from the north, like *Talpa, Sciurus, Capreolus, Cervus elaphus, Lepus europaeus, Bos
primigenius, Microtus guentheri, Ursus arctos, Mammuthus primigenius,* and *Felis sylvestris.* Another small group of species invaded the eastern Mediterranean region from
Asia: *Ellobius fuscocapillus,* and *Meriones tristrami* (probably diverged from *M. persicus,* Tchernov 1968). All other species of this time were indigenous, mostly endemic.

The Middle Paleolithic Period

In western European terminology, the Middle Paleolithic of the Middle East can
be broadly defined as a typical Mousterian with Levallois technique. Chronologically,
this period may be placed between 70,000 and 40,000 years ago. The latest phases of
the Mousterian and the transition to the Upper Paleolithic (34,000 to 35,000 years ago;
Farrand 1965, 1969) are scarcely known from caves in Israel (but well established in
Syria and Lebanon), probably due to a nondepositional period, extensive erosion, or
both (Horowitz 1971, Bar-Yosef and Vandermeersch 1972).

The fossil record from the Middle Paleolithic is based on the following sites:
1. Qafzeh (Galilee) (Haas 1972, Bouchud 1974).
2. Tabun (Mt. Carmel) (Garrod and Bate 1937; Bate 1937, 1942, 1943; Farrand 1969).
3. Hayonim (Galilee) (Bar-Yosef and Tchernov 1966).
4. Geula (Mt. Carmel) (Wreschner 1966, Haas 1967, Frenkel 1970, Heller 1970).
5. Kebara (Mt. Carmel) (Schick and Stekelis 1977, Ziffer 1978).
At the dawn of the Würm, as represented in the older Mousterian levels, one
would expect to find a significant change in the faunal list. However, the oldest Mousterian assemblages show neither a "faunal break," large-scale species replacement (only
one species of rodent and a few large mammals were replaced by new species), nor
mass extinction (only one ungulate, one rodent, and one lagomorph became extinct).
Instead, a slow and gradual faunal change occurred (Tables 24.1, 24.2, 24.3). It is also
surprising to find that during Würm 1 species diversity decreased compared to the

Table 24.2 Succession of Rodent Species During the Pleistocene of the Levant

Species	Mindel	Mindel-Riss	Riss	Riss-Würm	Mousterian Lower	Mousterian Upper
Hystrix sp.	+					
Parapodemus jordanicus	+					
Ctenodacylidae	+					
Parallactaga sp.	+					
Peridyromys sp.	+					
Progonomys sp.	+					
Mastomys sp.	+					
Nannocricetus sp.	+					
Cricetus kormosi	+					
Cricetus angustirostris	+					
Jordanomys pusillus	+					
Lagurodon arankae	+					
Arvicola jordanica	+					
Spalax minutus	+					
Myomimus judaicus	+	+	+			
Allocricetus bursae	+	+	+			
Cricetus cricetus	+	+	+			
"Meriones" obeidiensis	+	+	+			
Arvicanthis ectos	+	+	+	+		
Gerbilus sp.	+	+	+	+	+	
Apodemus flavicollis	+		+		+	
Apodemus sylvaticus	+	+	+	+	+	+
Apodemus mystacinus	+	+	+	+	+	+
Mus musculus	+	+	+	+	+	+
Mesocricetus sp.			+			
Cryptomys asiaticus			+			
Jordanomys haasi			+			
Spalax ehrenbergi			+	+	+	+
Allocricetus jesreelicus				+		
Rattus haasi				+		
Mastomys batei				+	+	
Allocricetus magnus				+	+	
Ellobius fuscocapillus				+	+	
Mesocricetus auratus				+	+	+
Sciurus anomalus				+	+	+
Myomimus roachi				+	+	+
Hystrix indica				+	+	+
Meriones tristrami				+	+	+
Microtus guentheri				+	+	+
Cricetulus migratorius						+
Dryomys nitedula						
Rattus rattus						
Eliomys melanurus						
Acomys cahirinus						
Gerbillus dasyurus						
Gerbillus pyramidum						
Gerbillus allenbyi						
Meriones sacramenti						

Riss-Würm. This decrease continued until the end of the Mousterian when a greater faunal change, especially of rodents, took place (Table 24.5).

The Eemian rodent fauna from Oumm-Qatafa includes the maximum number of older elements. One by one, throughout the Upper Pleistocene, they become extinct in the Levant or retreat northward. Mousterian communities which include a higher number of older elements will be dated as older communities. For this reason the rodent

Würm				Holocene			
Upper Paleol.	Aurignacian	Epi-Paleol.	Neolithic	Bronze Age	Iron Age	18th–19th Centuries	Recent
+	+	+					
+	+	+	+	+	+	+	+
+	+	+	+	+	+	+	+
+	+	+	+	+	+	+	+
+	+	+	+	+	+	+	+
+	+	+	+				
+	+	+	+				
+	+	+	+	+			
+	+	+	+	+	+	+	+
+	+	+	+	+	+	+	+
+	+	+	+	+	+	+	+
+	+	+	+	+	+	+	+
?	+	+?	+	+	+	+	+
		+ +	+	+	+	+	+
		+	+	+	+	+	+
		+	+	+	+	+	+
		+	+	+	+	+	+
			+	+	+	+	+
			+	+	+	+	+
			+	+	+	+	+

fauna of Oumm-Qatafa has been used for comparison and correlation of Mousterian (and later) communities.

The Upper Mousterian (with no possibility yet for finer correlations) shows the first appearance of several carnivores: *Martes foina, Herpestes ichneumon, Felis chaus,* and *Felis leo,* and the disappearance of *Talpa chtonia, Mammuthus primigenius,* and *Phacochoerus garrodae.*

Table 24.3. Succession of Insectivore, Lagomorph, Hyracoid,

Species	Mindel	Mindel-Riss	Riss	Riss-Würm	Würm		Upper Paleol.	
					Mousterian			
					Lower	Upper		
cf. *Desmana* sp.	+							
Crocidura	+	+	+					
Suncus sp.	+							
Erinaceus sp.	+							
Hypolagus sp.	+							
Ochotond sp.			+	+				
Procavia capensis			+?	+	+	+	+	
Talpa chthonia				+	+			
Crocidura suaveoleus				+	+	+	+	
Crocidura russula				+	+	+	+	
Crocidura leucodon				+	+	+	+	
Suncus etruscus				+	+	+	+	
Lepus europaeus				+	+	+	+	
Erinaceus europaeus				+	+	+	+	
cf. *Eidolon* sp.								
Rousettus aegyptiacus								
Hemiechinus auratus								

Table 24.4. Comparison of Four Mammalian Assemblages

Mindel-Riss Sites	
Latamne (Orontes, Syria)	**Evron (Northern Israel)**
Stegodon cf. *trigonocephalus*	―――
Elephas trogontherii (= *namadicus*)	*Elephas trogontherii*
Equus sp.	―――
Dicerorhinus cf. *hemitoechus*	―――
―――	―――
―――	*Metridiochoerus evronensis*
Hippopotamus amphibius	*Hippopotamus amphibius*
Megaceros verticornis	―――
―――	*Dama* sp.
	―――
Antilopini (undet.)	―――
Bison priscus	―――
Camelus sp.	―――
Crocuta crocuta	*Crocuta* sp.
Canis cf. *aureus*	―――
―――	―――
―――	―――
―――	―――
―――	―――
―――	―――
―――	―――
―――	―――
―――	―――
―――	―――
―――	―――
―――	―――
―――	―――

and Megachiropteran Species During the Pleistocene of the Levant

	Würm		Holocene				
Aurignacian	Epi-Paleol.	Neolithic	Bronze Age	Iron Age	18th–19th Century	Recent
+	+	+	+	+	+	+
+	+	+	+	+	+	+
+	+	+	+	+	+	+
+	+	+	+	+	+	+
+	+	+	+	+	+	+
+	+	+	+	+	+	+
+	+	+	+	+	+	+
+	+					
	+	+	+	+	+	+
	+	+	+	+	+	+

from Different Mindel-Riss and Riss Bone-bearing Sites in Israel and Syria

Riss Sites	
Benot Ya'akov Formation (Upper Jordan Valley)	Giveat Shaul (Judean Hills, near Jerusalem)
Stegodon mediterraneus	——
Elephas trogontherii	——
Equus caballus	——
Dicerorhinus merckii	——
Sus cf. *scrofa*	——
——	——
Hippopotamus amphibius	——
——	——
Dama cf. *mesopotamica*	——
Cervus cf. *elaphus*	——
——	——
Bison priscus	——
——	——
——	——
——	——
——	——
——	*Cryptomys asiaticus*
——	*Myominus judaicus*
——	*Spalax ehrenbergii*
——	*Apodemus mystacinus*
——	*Apodemus flavicollis*
——	*Apodemus sylvaticus*
——	*Allocricetus bursae*
——	*Mesocricetus* sp.
——	*Cricetus cricetus*
——	*Gerbillus* sp.
——	"*Meriones*" *obeidiensis*
——	*Jordanomys haasi*
——	*Crocidura* sp.

Table 24.5. Representation of Rodent Faunas in Several Mousterian Sites in Israel (Galilee

Rodent Species	Oumm-Qatafa (Haas 1951) (Tchernov 1968) (Eem)	Hayonim Cave (Tchernov)	
		Lower levels	Upper levels
Myomimus roachi	+	+	+
Hystrix indica	+	+	+
Sciurus anomalus	+	+	+
Spalax ehrenbergi	+	+	+
Mesocricetus auratus	+	+	+
Cricetulus migratorius	−	+	+
Allocricetus magnus	+	+	−
Allocricetus jesreelicus	+	−	−
Apodemus mystacinus	+	+	+
Aodemus flavicollis	−	+	−
Apodemus sylvaticus	+	+	+
Mus musculus	+	+	+
Rattus haasi	+	−	−
Mastomys batei	+	+	−
Arvicanthis ectos	+	−	−
Gerbillus sp.	+	+	−
Meriones tristrami	+	+	+
*Psammomys obesus**	+	−	−
Ellobius fuscocapillus	+	+	−
Microtus guentheri	+	+	+
Arvicola sp.	−	−	−

*The presence of this typical deserticulous species proves the vicinity of Oumm-Qatafa to an arid zone.

The Upper Paleolithic

Older phases of the Upper Paleolithic are represented in Kebara Cave (Mt. Carmel) (Schick and Stekelis 1977, Ziffer 1978) and in Qafzeh, which may be at least partially synchronous but probably older (Ronen and Vandermeersch 1972).

Aurignacian fossiliferous beds were exposed in Sefunim Cave (Mt. Carmel) and in Hayonim Cave (layer D), where a wealth of mammalian remains was uncovered.

Upper Paleolithic sites occur in both the Avedat/Aqev and Har Harif areas in Israel's arid Negev zone (Marks 1975), where some faunal remains were identified by Tchernov (1976). The importance of these sites is their location in an arid area, offering the first opportunity for preliminary insight into the climatic and ecological nature of the Israeli desert toward the onset of the Würm. No significant faunal change took place during the "transitional period" (figs. 24.2, 24.3), or throughout the Upper Paleolithic and Epipaleolithic. During the early Upper Paleolithic, but mainly during the Aurignacian, arboreal elements (*Sciurus anomalus* and *Apodemus* spp.) became much more abundant. This was followed by a significant decrease in open-land species (*Microtus quentheri* and *Cricetulus migratorius*). This phenomenon coincides with the palynological picture of the late Würm (Horowitz 1979), which expressed a dominance of oak forests under levels of cool and humid-pluvial vegetation in the Hula Basin and in other places in the Levant (Horowitz and Assaf 1978).

and Mt. Carmel) Compared to the Mindel-Riss Assemblage of Oumm-Qatafa (Judean Hills)

Israeli Sites					
Tabun (Bate 1937, Jelinek et al. 1973)		Qafzeh (Tchernov)		Geula Cave (Heller 1970)	Kebara Cave (Tchernov 1968)
Layer C	Layer D	Layers XII–XV	Layers XVIII–XXII		
+	+	+	+	+	+
+	+	+	+	+	+
+	+	+	+	+	+
+	+	+	+	+	+
–	–	–	–	+	+
+	+	–	–	–	+
+	–	–	–	–	–
–	–	–	–	–	–
+	+	+	+	+	+
?	?	?	?	?	–
+	+	+	+	+	+
+	+	+	+	+	+
–	–	only in XV	–	–	–
–	–	only in XV	+	–	–
–	–	only in XV	+	–	–
–	–	+	+	–	–
+	+	+	+	+	+
–	–	–	–	–	–
–	+	+	–	–	–
+	+	+	+	+	+
–	–	–	–	+	–

The Epipaleolithic

A wealth of Epipaleolithic sites is known from Israel and Sinai, but only a few have yielded reasonable quantities of well-determined faunal remains. Few other sites have been fully studied. Our information on the Kebaran complex and the Natufian period is based mainly on findings from Hayonim Cave layer C (Kebaran) and layer B (Natufian) (Bar-Yosef and Tchernov 1966, Bar-Yosef and Goren 1973) and from Ein-Gev I (Jorand Valley) (Stekelis et al. 1966, Davis 1974).

Natufian deposits from the Negev have been excavated at Rosh-Zin site D16 (Avedat/Aqev area) (Tchernov 1976) and Har Harif (fig. 24.1) (Butler et al. 1977) (Negev, Israel). They have supplied new information on the faunal life of the arid part of the Levant during this period.

During the Epipaleolithic there was again a change in the relative frequencies of rodent species (Tables 24.1, 24.2, 24.3). The relative abundance of arboreal elements (*Sciurus* and *Apodemus*) decreased, and other open-land animal species (*Cricetulus, Spalax*) increased, probably as a consequence of climatic deterioration. This phenomenological trend is clearly expressed by the fact that, for the first time, typical arid to semiarid species (*Acomys cahirinus* and *Gerbillus dasyurus*) that occupy only bare, rocky slopes appeared in the Mediterranean regime of Israel.

Mediterranean elements are still found during the Epipaleolithic in the Negev area (at present a semiarid zone). The presence of *Bos primigenius* during the Epipaleolithic (Harifian) of Har Harif (Tchernov 1976, Butler et al. 1977), indicates that the Mediterranean belt extended far to the south during these stages.

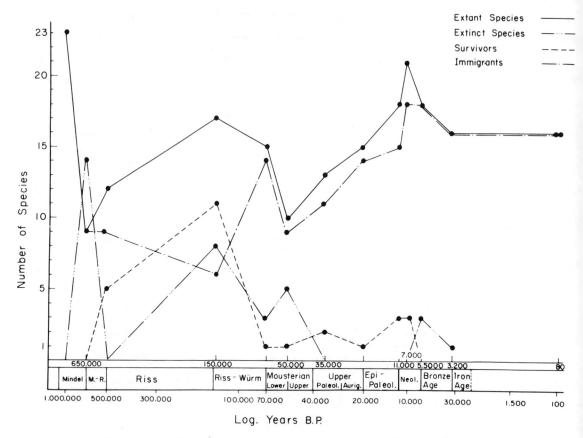

Figure 24.1. The rate of rodent succession during the Middle and Upper Pleistocene of the Levant.

A few other large mammals became extinct or retreated from the Levant, like *Crocuta crocuta*, which completely disappeared from Asia in post-Aurignacian times (Kurtén 1965, Tchernov 1975), and *Dicerorhinus hemitoechus*, which became extinct probably during the Kebaran stage (Tables 24.1, 24.2, 24.3).

The Neolithic and Post-Neolithic Period

The Middle East was heavily settled by Neolithic people, and quite a few sites have been excavated in the region. Microfaunal remains are scarce. The only place where a large microfaunal assemblage has been recovered is from the cave of Rakefet at Mt. Carmel, excavated by Higgs and Noy. Fortunately, recent excavations of Pre-Pottery Neolithic sites in the Jordan Valley (Gilgal) (Noy et al., 1980) and in the southern Sinai (carried out by Bar-Yosef et al.) have enabled us to expand our information on the species diversity of the Neolithic in Israel and the far desert of southern Sinai.

Present information is based mainly on the Neolithic of Jericho (Jordan Valley) (Kenyon 1960; Clutton-Brock 1979), Beidha (Jordan) (Hecker 1975), Nahal Oren (Mt. Carmel), Hagoshrim (Hula Basin), Munhatta (and others) (Ducos 1968), and many post-Neolithic sites (Ducos 1968, Davis 1976).

Faunistically, the Pre-Pottery Neolithic period does not show any change in species diversity. Changes in relative frequencies of animal species due to habitat shrinkage, and dramatic decreases in animal size due to temperature increase (Davis

1977) do occur, however. Yet, the climatic conditions which prevailed in the country, including the extremely arid regions, permitted typical Mediterranean species to survive: *Bos primigenius* in southern Sinai, *Dama dama mesopotamica* (Tchernov 1976) in the Negev, and *Arvicola terrestris* in Gilgal, near the Dead Sea (Noy et al., 1980).

Just after the Pre-Pottery Neolithic, the arid belt of the Levant swiftly extended far north, leaving behind on the mountaintops a trail of Mediterranean and Irano-Turanian relics (Danin 1972, Danin and Orshan 1970). In the late Würm, Middle Eastern environmental conditions were greatly affected by glaciation, which caused a large-scale expansion of the Mediterranean belt to the south, so much so that some species (e.g. *Juniperus*) were shifted as far south as central and southern Arabia and even to the Ethiopian plateau. Few glacial-age elements survived beyond the Pre-Pottery Neolithic. These few survivors were left in the arid belt of the Levant in suitable mesic microhabitats.

A few species of rodents became locally extinct during and immediately after the Neolithic (Tables 24.1, 24.2, 24.3). Some large mammals *(Felis leo, Equus hemionus, Alcelaphus)* became extinct in the Levant in later historical times, and a few others only during the last two centuries.

Faunal Succession and Species Turnover

A community is composed of the following organic components: unchanged survivors, transformed survivors, species added by indigenous speciation, and immigrant elements. At any given period a certain number of species disappears through extinction or temporary elimination from the assemblage. The internal composition of a community, the frequency and abundance of the populations, and the rate of extinction are influenced by many associated factors. Under stable environmental conditions high species diversity is caused by a low rate of extinction, while instability results in quick turnovers of the biota and a lower species diversity from a high extinction rate. It is time to discard the concept that the changing physical environment has an ultimate direct and unique impact on the rate of extinction and the turnover of taxa. If all competitors of a given species are excluded from an area, the resulting ecological amplitude will be far wider than that actually exhibited under natural conditions, allowing it to sustain far more diverse abiotic conditions. The species diversity of a community, the number of specimens in a given population, and their spatial distribution are consequences of the complicated nonequilibrium competitive interplay between the species, and outline the ecological amplitude of each species. Extinction of a species, or its temporary disappearance from an area, results from its inability to compete for the limited resources at a given time (Huston 1979). Less competition gives it more chances, occupation of wider niches, and hence longer survival. Yet, when the steepness of the physical environment increases, intense interspecific competition results, species diversity decreases due to higher rates of extinction, and new elements evolve (either by indigenous speciation or immigration). A reduced rate of competitive displacement allows a longer period of coexistence. In tropical and subtropical realms, if interspecific competition is minimized, survival expectancy is maximized. Floral and faunal turnovers, however, may differ even within the same latitudinal range.

Rate of Extinction and Faunal Succession

To compare the relative rate of extinction (Re) that took place during each period, the equation

$$\mathrm{Re} = \frac{\mathrm{Ne}}{\mathrm{n}} \times \frac{1}{\mathrm{t}} \times 10^5$$

may be adapted, where Ne/n gives the relative number of extinct species, divided by the

Table 24.6 Rate of Extinction of Different Groups of Mammals

Stages (B.P.)	Span of Time Between Two Successive Assemblages	Rodentia	
Mindel to Mindel-Riss	$10^6 - 65 \times 10^4 = 935 \times 10^4$	1.666	
Mindel-Riss to Riss	$65 \times 10^4 - 5 \times 10^5 = 15 \times 10^4$	0	
Riss to Riss-Würm	$5 \times 10^5 - 15 \times 10^4 = 35 \times 10^4$	0.163	
Riss-Würm to early Würm	$15 \times 10^4 - 7 \times 10^4 = 80,000$	1.139	
Early Würm to Upper Mousterian	$7 \times 10^4 - 5 \times 10^4 = 20,000$	1.666	
Upper Mousterian to Upper Paleolithic	$5 \times 10^4 - 35,000 = 15,000$	0	
Upper Paleolithic to Aurignacian	$35,000 - 20,000 = 15,000$	0	
Aurignacian to Epipaleolithic	$20,000 - 11,000 = 9,000$	0	
Epipaleolithic to Neolithic	$11,000 - 7,000 = 4,000$	0.392	
Neolithic to Bronze Age	$7,000 - 5,500 = 2,500$	4.210	
Bronze Age to Iron Age	$5,500 - 3,200 = 2,300$	2.557	
Iron Age to 18th–19th Centuries	$3,200 - 1,900 = 1,300$	0	
18th–19th Centuries to Recent	$1,980 - 1,900 = 80$	0	

NOTE: Rate of extinction is determined by equation described in text.

span of time elapsed between two successive assemblages (Δt), multiplied by 10^5 years (the relative rate of extinction per one hundred thousand years). The index of extinction was calculated separately for different groups of mammals as well as for the whole group of mammals (Table 24.6). The rate of faunal succession and the main periods of faunal turnover are represented in Figures 24.1 and 24.2. The species diversity of rodents (fig. 24.1) attained its maximum during the Mindel and peaked again during the Riss-Würm and the Neolithic due to newly arrived immigrants. The maximum decrease in species deversity is represented in the Upper Mousterian. Species diversity of rodents in the Levant has fluctuated within a relatively large range. Species diversity of the larger mammals (ungulates, primates, proboscideans, and carnivores) attained its maximum during the Riss-Würm (fig. 24.2) and remained high until the end of the Epipaleolithic (Natufian).

The higher rate of extinction during the close of the Riss, the Eem, and the Mousterian periods may have been due to the relatively large numbers of newcomers introduced into the region. This species replacement could have been caused by a competitive exclusion process which decimated the older species. Older species usually show less viability under changing biotic and/or abiotic environments and less adaptability due to genetic senescence. If a climatic shift from a cold Rissian pluvial to a warm and dry Eemian interpluvial regime did not take place, it is difficult to explain how northern and Total Mammals During the Pleistocene of the Levant

Ungulata, Primates, Carnivores, Proboscidea	Insectivora, Hyracoidea, Megachiroptera, Lagomorpha	Total Rate of Extinction
0.169	0.228	0.174
0.154	0	0.087
0.071	0.095	0.026
0.046	0.139	0.099
0.468	0.625	0.727
0	0	0
0	0	0
0.370	0	0.218
3.941	2.857	3.061
4.800	0	3.773
1.976	0	1.811
14.952	0	6.689
156.250	0	60.975

and central Asiatic elements took over and replaced the Mindelian and African forms. As a rule, climatic changes cause habitat shrinkage and heavy competition which may result in total exclusion of a species (Guilday 1967). The shift from a colder, less stable, and less predictable Rissian environment to much warmer, more stable and more predictable interpluvial conditions greatly spurred faunal replacement. But, as expected under milder and more stabilized conditions, this shift also caused a conspicuous increase in species diversity. The extremely high rate of extinction that took place during the Mousterian period may be due to an accelerated rate of immigration rather than environmental changes.

In the Middle East, particularly in the Levant, we mainly consider animals which merely disappeared from the eastern Mediterranean region but still survive elsewhere, either in northern latitudes or in Africa. Rather, we can speak of a nonsynchronous (Epipaleolithic to late historical time) geographical shift of mammals (as well as other groups of animals that underwent a parallel process) that hardly coincides with the swift postglacial climatic deterioration.

Three species of rodents were eliminated from the local fauna during Neolithic time but continue to survive farther north: *Sciurus anomalus* (Sciuridae), *Apodemus flavicollis* (Muridae), and *Mesocricetus auratus* (Cricetidae). During the Iron Age *Myomimus roachi* (Gliridae) became extinct (Haas 1960, Tchernov 1968). On the other

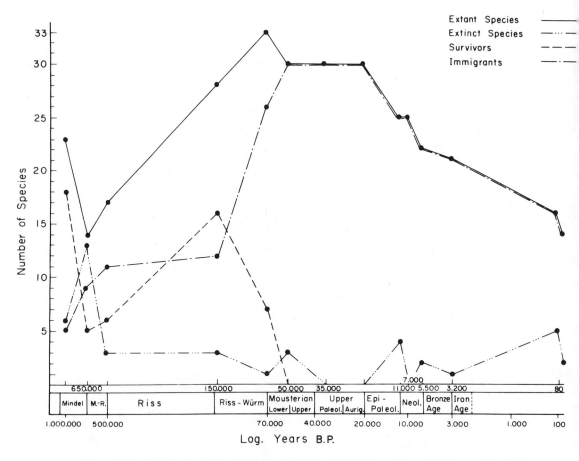

Figure 24.2. The rate of megafaunal succession during the Middle and Upper Pleistocene of the Levant.

hand, a swarm of species well adapted to arid environments emigrated into the deteriorating Mediterranean region of the Levant, either originating in the Arabian desert (*Acomys cahirinus* [Muridae] and *Gerbillus dasyurus* [Gerbillidae], or the Saharan belt *Gerbillus allenbyi*). Autochthonous speciation gave rise to two endemic species (of Saharan ancestors) which also invaded the Mediterranean coastal plain of Israel: *Gerbillus allenbyi* and *Meriones sacramenti* (Gerbillidae). *Rattus rattus* (Muridae) first appeared in the area during the Epipaleolithic together with two glirids: *Dryomys nitedula* and *Eliomys melanurus* (both are rare in the fossil record).

The ungulates represent a different scenario (fig. 24.2). Only two species became extinct towards the end of the Würm: *Rhinoceros hemitoechus* (during the Kebaran time, some five thousand years before the termination of the Würm), and *Equus hyndruntinus* during the Neolithic. The only other large mammal which shifted northward towards the onset of the glacial period is *Capra aegagros*. It was only during much later historical time that other ungulates and large carnivores were finally eliminated from the local fauna (fig. 24.2), in some cases only after the introduction of firearms into the region: *Cervus elaphus, Dama dama mesopotamica,* and *Capreolus capreolus* (all three still coexisted in the Neolithic of Jericho) (Clutton-Brock 1979). *Alcelaphus* sp., *Equus hemionus, Hippopotamus amphibius, Felis leo,* and *Ursus arctos* were doomed to the same fate during historical time. Hence it is evident, that most of the ungulates and larger carnivores withstood the swift climatic deterioration of the postglacial.

An utterly different picture is displayed by the rodents. While the large mammals show a steady decrease in species diversity since the Upper Paleolithic (fig. 24.2), a process that accelerated during historical time, the rodents presented a more complicated dynamic during the same period. A significant increase in species diversity took place among the rodents during the Upper Paleolithic to Neolithic time (fig. 24.1). Species added to the local fauna originated both from the north (*Eliomys melanurus, Dryomys nitedula,* and probably *Rattus rattus*) and from the south (*Acomys cahirinus* and *Gerbillus dasyurus*). It is worth noting that *Hemiechinus auratus* (Insectivora, Erinaceidae) and *Rousettus aegyptiacus* (Megachiroptera) also invaded the eastern Mediterranean region during this period. While some acceleration in species extinction took place during the Neolithic, two other species (*Gerbillus allenbyi* and *Meriones sacramenti*) were added to the local fauna in postglacial time through autochthonous speciation (Zahavi and Wahrman 1957; Tchernov 1968, 1975; Haim and Tchernov 1974).

Taking the mammalian fauna as a whole, there were no drastic biotic changes during the later Würm and early postglacial time, and the fluctuation in species diversity was more gentle than in northern latitudes.

Size Changes in Late Pleistocene and Postglacial

The reaction of a floral or faunal unit to environmental (either spatial or temporal) changes is indirect, driven by many abiotic and organic factors; the consequences are never clear, detectable, and explicable, because no unique correspondence between the outcome of postulated causes and the floral-faunal alteration in the paleontological record can be claimed. A single population, however, will directly reflect specific changes that occurred in its immediate surroundings by changing certain (usually measurable) phenotypic characters. As a rule, the steeper the environmental change, the more abrupt the morphological change. Hence, paleontologists should detect paleoecological changes through the population level, or through certain ecologically associated or closely related species, rather than attempt to deal with a whole assemblage. The environment or selection works on the individual, never on the biota as a whole.

Environmental factors are most often found in gradients rather than as salient changes between adjacent zones. Temporal changes in one locality, or area, will always change more gradually than spatial changes, giving a population more time to adjust itself to new conditions. Morphological variation is controlled by spatial or temporal differences in gene frequencies. A morph-ratio cline is a gradient of phenotypic characters as reflected by a gradient in genotype, controlled either by single major genes or by groups of linked or interacting genes (=supergenes) (Endler 1977). Directed environmental factors will initiate a directional selection, and phenotypes will be shifted away from their relatively stabilized norms. Excluding insular species, three main factors may alter the size of an animal:

1. Ecogeographical changes. The most famous of these is Bergmann's rule which states that "races from cooler climates tend to be larger 'warm-blooded' vertebrates than races of the same species living in warmer climates." The ecophysiological explanation is based on the fact that the volume of the body increases as a cube, and the surface as a square of a linear dimension. The larger the body, the relatively smaller its surface area. In cool climates there is an advantage in the relative reduction of surface area resulting from increased size. Correspondingly, in hot climates, small body size and relatively large surface area is advantageous. The physiological state of this mechanism is not yet fully understood. Yet, even if this is only a small fraction of the adaptive device, it will affect the population, as selective advantages are independent and additive. Since for several cases it has already been concluded that change in body size is a constant factor of the regional mean temperature, it is possible to predict the paleotemperatures of a certain region during a certain period according to the relative body size of a fossil population (Badoux 1964, Kurtén 1965, Tchernov 1968, Davis 1977).

2. Character displacement by competitive exclusion. In order to increase differences between two or more closely related species in a region of sympatry, character

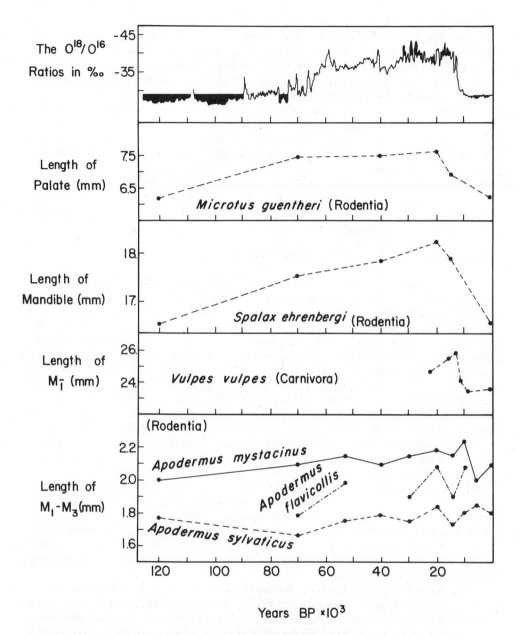

Figure 24.3. Size changes in several species of mammals, represented by mean sizes of populations during the late- and post-Pleistocene period, correlated with data of O^{18} graph from the Camp Century (Greenland) ice sheet (Johnson et al. 1972). Data for *Vulpes vulpes* were taken from Davis 1977.

displacement takes place as a result of selection for reduced competition. Where species compete, changes in linear dimensions, for example, may result (Van Valen 1965). In the time dimension the interaction of two factors affect the body size: interspecific competition of closely related species and climate (Tchernov 1979).

3. Artificial selection through domestication. Changes in organism size are maintained under constant selection pressure for certain favored phenotypes. Different sizes result either from direct selection or as a byproduct of simultaneous selection for many genetic traits.

Ecogeographical Changes. Following the lineage of animals we may apply Bergmann's rule in the time dimension, assuming that size alterations reflect past temperatures in a certain geographical coordinate. Once it is quantitatively scaled for recent populations along a geographical or climatic gradient, it is possible to estimate past climatic shifts (Badoux 1964; Tchernov 1968; Davis 1977, fig. 24.3).

Size change is exemplified most drastically by postglacial radical dwarfing seen in several species of mammals in Israel: *Canis lupus, Ursus arctos, Sus scrofa,* and *Felis sylvestris* (Kurtén 1965); *Spalax ehrenbergi* and *Microtus guentheri* (Tchernov 1968); *Gazella gazella, Vulpes vulpes,* and *Canis lupus* (Davis 1977). Davis showed that the Aurignacian wolf of Israel attained the same dimensions as the present population of southern Sweden. Differences in mean temperature (for the hottest or coldest months) between the two regions is around 15°C. According to Davis (1977), dwarfing commenced within the Natufian period (10,000 to 12,000 B.P.) and was completed approximately during the Neolithic. It is obvious that a conspicuous climatic change took place in the eastern Mediterranean region, an effect not clearly shown by the mammalian fauna turnover.

Character Displacement. Lineages of closely related, competing species will rarely show an overlap in sizes. When all three species of *Apodemus* (*Sylvaemus*) (Muridae) coexist, simultaneous changes in size take place (fig. 24.3) in accordance with Bergmann's rule. All three species, *A. sylvaticus, A. flavicollis,* and *A. mystacinus,* increase in size during colder periods and become smaller in warmer times. When *A. flavicollis* is missing during the warmest periods, the size of both *A. sylvaticus* and *A. mystacinus* is strongly affected independently of climate. Both species occupy the evacuated space by approaching each other's size (enlarging the coefficient of variation or their ecological amplitude) (fig. 24.3). During Würm II (Hayonim D) a size increase took place simultaneously in all three species. During the milder Epipaleolithic period, all three species decreased in size. A short phase of colder and wetter conditions during the Natufian is shown by an abrupt increase of sizes of *A. mystacinus* and *A. sylvaticus.* As the climate grew drier and warmer towards the Neolithic and overall desiccation took place in the Middle East, *A. flavicollis* disappeared. Subsequently a great expansion of *A. mystacinus* included almost the whole space evacuated by *A. flavicollis,* which has not reappeared in the Levant (fig. 24.1). Such northward retreat is also known for several birds (Tchernov 1962, Bar-Yosef and Tchernov 1966): *Pica pica, Pyrrhocorax graculus* (Corvidae), and *Phasianus colchicus* (Phasianidae).

Artificial Selection. When a change in size is intimately connected with human culture, as in the late prehistoric period, it is usually suspected that artificial selection processes are taking place, with direct or indirect reference to climate or other environmental changes. Through the artificial process of domestication, selection is usually simultaneous for various factors, and the results are thus unpredictable; the breed may either increase, decrease, or remain unaltered in size. Davis (1977) showed that a significant decrease in the size of *Canis lupus* took place while the Natufian dog was being domesticated in Israel.

Overview and Discussion

No strict correlation exists between the rate of the mammalian turnover (extinction and immigration) and the Pleistocene environmental changes in the eastern Mediterranean region. Greater correlation is shown between the relative frequencies of species and species diversity of macromammals and the sharp glacial-postglacial transition period. The impact of the climatic changes which took place in this region worked specifically on the population level in a way that each species responded by shifting into another ecological regime, or changed its ecological amplitude (=valence, Endler 1977) within the usually wide range of its environmental tolerance. Body size augmentation during the late Würm and diminution during the postglacial was the usual response of many species to climatic shifts. Changes in spatial distribution were another. Through north-south shifts internal changes in the phenotypical characters of the population possibly can be "avoided." A population may either follow its optimal valence through spatial changes, or stick to its place by adjusting its phenotypic characters to the changing environment. The magnitude of environmental changes in the Levant were steep enough to affect the genetic construction of the populations, but not effective enough to cause an overall faunal turnover.

Guilday (1967) has stated that the main cause for the worldwide extinction of terrestrial megafauna was "post-Pleistocene desiccation," stressing that "the same extinction pattern prevailed throughout the globe." However, it appears that, in the Middle East at least, birds and mammals responded differently in different habitats. Northward and southward retreat from the Mediterranean region was unexpectedly more severe in steppe landscape than in wooded areas. As for the Old World, late Quaternary extinction, Reed (1970) is of the opinion that "climatic change... was important in the extermination of the mammalian megafauna," but in lower latitudes "man...has been the main, and often the only, cause for decreases (and some time eradication) of the Old World Megafauna."

In contrast with Guilday's view (1967) that larger animals are more easily doomed to extermination, micromammals and birds in the Middle East showed no less sensitivity, and responded in still greater magnitude to the postglacial deterioration of climate. With only a few exceptions, it is the Middle East megafauna that showed more stability during the critical period. Their disappearance from the area was caused mainly through the agency of man during historical times; some species disappeared only in the nineteenth and twentieth centuries. These facts coincide with Reed's (1970) view.

In addition, we cannot simply correlate the differential eliminations of species from an entire region with the specific range of environmental tolerance (as suggested by Edwards 1967). A reduced rate of interspecific competition allows a longer period of coexistence and increases species diversity. Less competition gives the species more chances, occupation of wider niches, and hence longer survival. Interassociation complexes between organisms in an ecosystem are significant factors in the stabilization of species or communities. As long as *Tamarix* and *Acacia* trees, for example, are found in varied landscapes and climates in the eastern Mediterranean regions, many Ethiopian species will firmly adhere to them, sometimes in a kind of symbiotic relationship.

Environmental vicissitudes, even if catastrophic, affect organisms differently on various taxonomic levels:

1. On the *population* (intraspecific) level the more adapted sector will be selectively favored. If this sector includes enough individuals to enable the favored deme to breed normally and propagate, a phenotypic change may result that may be measured by the paleontologist. The wider the tolerance (i.e., the wider the range of genetic variability), the greater the chances for at least a narrow section of the population to withstand environmental changes.

Analysis of variability within a population affords insight into its relation to selection value, and distinguishes the effects of environmental factors. Thus, it is now generally accepted that there is a positive correlation between area temperature and the size of endothermal vertebrates, at least. Most of the mammals investigated in the Middle East showed a direct and swift response to the temperature increase in the Pleistocene-Holocene transition period by rather drastic diminution of body sizes. Temperature increase in southwest Asia, as interpolated from the smaller mammals, was roughly 5°C. Using magnitude of changes in animal size as a yardstick for paleotemperatures is still premature, requiring thorough comparative study of the variability of populations under conditions of "natural experiments," especially geographical variability along explicit ecological-environmental gradients.

When dealing with a taxon in which intimate interspecific competition exists, the impact of the temperature changes on the single species is not simple; the competition process may be dominant and avert the direct temperature/body size correlation.

2.   On the *community* level, impact of the postglacial climatic change is less significant, although still not simple. The Levant represents a high species diversity, much higher than in Europe, but lower than in the Ethiopian region as expected because of its geographic position.

The relationship between the diversity and stability of an ecological community should not be taken for granted as positive in all cases, although as a rule highly diversified ecosystems do show more stability than less diversified ones. The correlation is also usually lower when comparing rich—and poor—species communities in space. Higher correlations exist, however, between stability of highly diversified communities and climatic fluctuations in time.

There is only slight "attrition" of the typical Mediterranean communities in postglacial time; species diversity was only slightly affected by climatic change. The deterioration of the Mediterranean ecosystem as a whole in the later Holocene is largely due to the gradual increase in human activity. Where semiarid or more xeric conditions prevailed during the last periglacial, the species surviving were close to their physiological or ecological turning point. Any further increase of temperature and additional desiccation was beyond their adaptive strength, and hence lethal. Indeed, an almost complete collapse of the Palearctic (Mediterranean and Irano-Turanian) communities took place over the entire present Saharo-Arabian belt, leaving behind relicts in enclaves, mostly at higher altitudes. The isolated habitats which include these relict species are found all over the Saharo-Arabian desert belt, few of which are known from Arabia, Sinai, and Israel. These ecological refuges preserve, to a certain degree, the environmental conditions which prevailed in the past and were widespread over vast areas around the south Palearctic arid zone. On the other hand, Saharan and Arabian elements were thrust northward into xeric Mediterranean regions, none of which succeeded in colonizing Mediterranean landscapes north of Israel.

As long as temperature increase and desiccating processes did not bring about a decrease in rainfall below 350 to 400 mm, Mediterranean elements were only slightly affected. Many Mediterranean species survive at present within the range of 150–350 mm. Below the 150 mm isohyete, Arabian and Saharan elements predominate.

References

Badoux, D. M. 1964. Some Remarks on size-trends in mammalian evolution of the Holocene in Sumatra, with some additional notes on the Sampung fauna. *Säugetierkunde Mitteilungen* 1:1–12.

Bar-Yosef, O. and N. Goren. 1973. Natufian remains in Hayonim Cave. *Paléorient* 1(1):49–68.

Bar-Yosef, O. and E. Tchernov. 1966. Archaeological remains and fossil faunas of the Natufian and Microlithic industries at Hayonim Cave (Western Galilee), Israel. *Israel J. Zool.* 15:104–140.

———.1972. *On the palaeoecological history of the site of 'Ubeidiya.* Israel Acd. Sci. Human., Jerusalem. pp. 1–35.

Bar-Yosef, O. and B. Vandermeersch. 1972. The stratigraphical and cultural problems of the passage from Middle to Upper. Palaeolithic in Palestinian caves. In *The Origins of Homo Sapiens*, edited by F. Bordes. UNESCO, Paris. pp. 221–225.

Bate, D. M. A. 1932. On the animal remains obtained from the Mugharet-el-Zuttiyeh in 1925. In *Prehistoric Galilee,* edited by F. Turville-Petre. London: British School of Archaeology. pp. 27–52.

———.1937. *The Stone Age of Mount Carmel* (Vol. 1, Part II, Palaeontology). Oxford, Clarendon. pp. 139–237.

———. 1942. Pleistocene Murinae from Palestine. *Ann. Mag. Nat. Hist.* 9(55):465–486.

———. 1943. Pleistocene Cricetinae from Palestine. *Ann. Mag. Nat. Hist.* 11(10):823–838.

Bouchud, J. 1974. Étude preliminaire de la faune provemant de la grotte du Djebel Qafzeh pres de Nazareth, Israel. *Paléorient* 2(1):87–102.

Butler, B. H., E. Tchernov, H. Hietala and S. Davis. 1977. Faunal exploitation during the late Epipaleolithic in the Har-Harif. In *Prehistory and Paleoenvironments in the Central Nege, Israel*, edited by A. E. Marks. Vol. 2. The Avedat/Aqev area. S.M.U. Press, Dallas. pp. 327–344.

Clark, J. D. 1961. Fractured chert specimens from the Lower Pleistocene Bethlehem levels. *Bull. Brit. Mus.* 5:73–99.

———. 1967. The Middle Acheulian occupation site at Latamne, northern Syria. *Quaternaria* 9:1–68.

———. 1968. The Middle Acheulian occupation site at Latamne, northern Syria, further excavations. *Quaternaria* 10:1–76.

Clutton-Brock, J. 1979. The mammalian remains from the Jericho Tell. *Proc. Prehist. Soc.* 45:135–157.

Danin, A. 1972. Mediterranean elements in rocks of the Negev and Sinai deserts. Notes of the *Royal Botanical Garden* 31(3): 437–440.

Danin, A. and G. Orshan. 1970. Distribution of indigenous trees in the northern central Negev highlands. *La Ya'aran* 20(4):115–119 (in Hebrew).

Davis, S. 1974. Animal remains from the Kebaran site of En Gev I, Jordan Valley, Israel. *Paléorient* 2(2):453–462.

———. 1976. Mammal bones from the early Bronze Age city of Arad, Northern Negev, Israel: Some implications concerning human exploitation. *Journal of Archaeol. Sci.* 3:153–164.

———. 1977. Size variation of the fox, *Vulpes vulpes* in the Palaearctic region today and in Israel during the late Quaternary. *J. Zool. Lond.* 182:343–351.

Dewey, J. F., W. Pitman, B. G. Ryan, and J. Bonnin. 1973. Plate tectonics and the evolution of the Alpine system. *Geol. Soc. Amer. Bull.* 84:3137–3180.

Ducos, P. 1968. *L'Origine de Animaux Domestique en Palestine* Publ. de l'Inst. Prehist. de l'Université de Bordeaux, Mem. 6.

Edwards, W. E. 1967. "The late Pleistocene extinction and diminution in size of many mammalian species." In *Pleistocene Extinctions: The Search for a Cause*, edited by P. S. Martin and H. E. Wright, pp. 141–154. Yale Univ. Press, Hew Haven and London.

Endler, J. A. 1977. Geographic Variation, Speciation and Clines. *Monographs on Population Biology, 10.* Princeton Univ. Press, Princeton.

Farrand, W. R. 1965. Geology, climate and chronology of Yabrud rockshelter I. *Ann. Archeol. Syrie* 15:35–50.

———. 1969. Geological correlation of prehistoric sites in the Levant. *UNESCO Symp. on Environmental Changes and the Origin of Homo sapiens,* Paris. pp. 1–18.

Frenkel, H. 1970. *Hystrix angressi* sp. nov., a large fossil porcupine from the Levalloise-Mousterian of the Geula Cave. *Israel J. Zool.* 19:51–82.

Gardner, E. W. and D. M. A. Bate. 1937. The bone-bearing beds of Bethlehem: Their fauna and industry. *Nature* 140:431–433.

Garrod, D. A. E. and D. M. A. Bate. 1937. *The Stone Age of Mount Carmel, Excavations at the Wadi Mughara*, I. Oxford Univ. Press, Oxford.

Gilead, D. and A. Ronen. 1977. Acheulian industries from Evron on the Western Galilee coastal plain. *Eretz-Israel* 13:56–86.

Guilday, J. E. 1967. Differential extinction during late-Pleistocene and recent time. In *Pleistocene Extinctions: The Search for a Cause*, edited by P. S. Martin and H. E. Wright, pp. 121–140, Yale Univ. Press, New Haven and London.

Haas, G. 1960. Some remarks on *Philistomys roachi* Bate. *Ann. Mag. Nat. Hist.* 13(2):688–690.

———. 1966. *On the vertebrate fauna of the Lower Pleistocene site of 'Ubeidiya.* Israel Acad. Sci. Human. Jerusalem. pp. 1–68.

———. 1967. Bemerkungen ueber die fauna der Geula Hoehle, Carmel. *Quaternaria* 9:97–104.

———. 1972. The microfauna of Djebel Qafzeh Cave. *Paleovertebrata* 5:261–270.

Haim, A. and E. Tchernov. 1974. The distribution of Myomorph rodents in the Sinai Peninsula. *Mammalia.* 38:201–223.

Hecker, H. 1975. The faunal analysis of the primary food animals from Pre-Pottery Neolithic Beidha (Jordan). *Univ. Microfilm Internat.* Ann Arbor, Michigan and London, England. pp. 1–470.

Heinzelein, J. de. 1968. Geological observations near Latamne. In *The Middle Acheulian Occupation Site at Latamne, Northern Syria: Further Excavations*, edited by J. D. Clark, *Quaternaria* 10:3–8.

Heller, J. 1970. The small mammals of Geula Cave. *Israel J. Zool.* 19:1–49.

Hooijer, D. A. 1958. An early Pleistocene mammalian fauna from Bethlehem. *Bull. Brit. Mus. (Nat. Hist.) Geol.* 3(8):267–292.

———. 1959. Fossil mammals from fisr Banat Yaqub, south of Lake Huleh, Israel. *Bull. Res. Counc. Israel* G8:177–179.

———. 1960. A *Stegodon* from Israel. *Bull. Res. Counc. Israel* G9:104–108.

———. 1961. Middle Pleistocene fauna of the Near East. In *Evolution and Hominization*, edited by G. Kurtin. G. Fischer Verlag, Stuttgart. pp. 81–83.

Horowitz, A. 1971. Climatic and vegetational developments in north-eastern Israel during upper Pleistocene-Holocene times. *Pallen et Spores* 13(2):255–278.

———. 1973 Development of the Hula basin, Israel. *Israel J. Earth Sci.* 22:107–139.

———. 1974. *The Late Cenozoic Stratigraphy and Paleogeography of Israel.* Inst. of Archeol. Tel-Aviv University.

———. 1979. *The Quaternary of Israel.* Academic Press, New York.

Horowitz, A. and G. Assaf. 1978. Configuration of interpluvial and pluvial climates in the Levant. *Abst. Int. Colloq. on the History of the Environmental Conditions in South West Asia from the Last Pleniglacial until Today.* Tübingen.

Horowitz, A., G. Siedner, and O. Bar-Yosef. 1973. Radiometric dating of the 'Ubeidiya Formation, Jordan Valley, Israel. *Nature* 242:186–187.

Howell, F. C. 1959. Upper Pleistocene stratigraphy and early man in the Levant. *Proc. Amer. Phil. Soc.* 103(1):1–65.

Huckriede, R. 1966. Das Quartaer des arabischen Jordan-Thales und Beobachtungen ueber "Pebble Culture" und "Prae-Aurignac." *Eiszeitalter Gegenwart* 17:211–212.

Huston, M. 1979. A general hypothesis of species diversity. *The American Naturalist* 113(1):81–101.

Johnson, S. J., W. Dansgaard, and H. B. Clausen. 1972. Oxygen isotope profiles through the Antarctic and Greenland ice sheets. *Nature* 235:429–434.

Kenyon, K. 1960. Excavations of Jericho, 1957–1958. *Palestine Expl. Quart.* 92:1–21.

Kurtén, B. 1965. The carnivora of the Palestine caves. *Acta Zool. Fenn.* 107:1–74.

Leakey, M. D. 1967. Preliminary survey of the cultural material from beds I and II, Olduvac

Gorge, Tanzania. In *Background to Evolution in Africa,* edited by W. W. Bishop and J. D. Clark. Univ. of Chicago Press, Chicago. pp. 417–446.

Liere, W. J. van. 1961. Observation on the Quaternary of Syria. *Proc. State Serv. Archaeol. Invest. Netherlands, The Hague* 10–11:7–69.

———. 1966. The Pleistocene and Stone Age of the Orontes. *Ann. Archeol. Syrie* 16:9–29.

Liere, W. J. van. and D. A. Hooijer. 1961. Paleo-Orontes level with *Archidiskodon meridionalis* (Nesti) at Hama. *Ann. Archeol. Syrie* 11:165–172.

Marks, A. E. 1975. An outline of prehistoric occurrences and chronology in the central Negev, Israel. In *Problems in Prehistory: North Africa and the Levant,* edited by F. Wendorf and A. E. Marks. S.M.U. Press, Dallas. pp. 351–361.

Neev, D. 1975. Tectonic evolution of the Middle East and Levantine basin (easternmost Mediterranean). *Geology* 683–686.

Neuville, R. 1951. *Le Paleolithique et le Mesolithique de Desert de Judée.* Arch. Inst. Paleont. Hum., Mem. 24, Paris.

Noy, T., J. Schulderrein, and E. Tchernov. 1980. Gilgal I. A Pre-Pottery Neolithic site in the south of the Jordan Valley. *Israel Expl. J.*

Picard, L. 1963. The Quaternary in the northern Jordan Valley. *Proc. Israel Acad. Sci. Hum.* 1(4):1–34.

———. 1965. The geological evolution of the Quaternary in the central-northern Jordan graben. *Amer. Geol. Soc. Sp. Papers* 84:337–365.

Picard, L. and U. Baida. 1966. Geological report on the lower Pleistocene deposits of the 'Ubeidiya excavations. *Proc. Israel Acad. Sci. Hum.* 4(1):1–39.

———. 1966a. Stratigraphic position of the 'Ubeidiya Formation. *Proc. Israel. Acad. Sci. Hum.* 4:1–16.

Reed, C. 1970. Extinction of mammalian megafauna in the Old World late Quaternary. *Bioscience* 30:284–288.

Ronen, A. and A. Amiel. 1974. The Evron quarry: a contribution to the Quaternary stratigraphy of the coastal plain of Israel. *Paléorient* 2(1):167–173.

Ronen, A. and B. Vandermeersch. 1972. The upper Paleolithic sequence of the cave of Qafza. (Israel). *Quaternaria* 16:189–202.

Rust, A. 1950. *Die Hohenfunde von Jabrud (Syrien).* Karl Wacholtz Verlag, Neumünster.

Savage, R. J. G. and W. R. Hamilton. 1973. Introduction to the Miocene mammal faunas of Gebel Zelten, Libya. *Bull. Brit. Mus. (Nat. Hist.) Geol.* 22(8):515–527.

Savage, R. J. G. and E. Tchernov. 1968. Miocene mammals of Israel. *Proc. Geol. Soc. Lond.* 1648:98–101.

Schick, T. and M. Stekelis. 1977. Mousterian Assemblages in Kebara Cave, Mt. Carmel. *Eretz-Israel* 13:97–149.

Siedner, G. and A. Horowitz. 1974. Radiometric ages of late Cainozoic basalts from northern Israel: chronostratigraphic implications. *Nature* 250:23–26.

Stekelis, M. 1960. The Paleolithic deposits of Jisr Banat Yaqub. *Bull. Res. Counc. Israel* Gg:61–87.

———. 1966. *Archaeological excavations at 'Ubeidiya 1960–1963.* Israel Acad. Sci. Hum., Jerusalem.

Stekelis, M., O. Bar-Yosef and T. Schick. 1969. *Archaeological excavations at 'Ubeidiya, 1964–1966.* Israel Acad. Sci. Hum., Jerusalem.

Tchernov, E. 1962. Paleolithic avifauna in Palestine. *Bull. Res. Counc. Israel* 11(3):95–131.

———. 1968. *Succession of Rodent Fauna During the Upper Pleistocene of Israel.* Mammalia Depicta, Verlag Paul Parey, Hamburg & Berlin.

———. 1968a. A Pleistocene faunule from a karst fissure filling near Jerusalem, Israel. *Verhandl. Naturf. Ges. Basel.* 79(2):161–185.

———. 1973. *On the Pleistocene Molluscs of the Jordan Valley.* Israel Acad. Sci. Hum., Jerusalem. pp. 1–46.

———. 1975. Rodent faunas and environmental changes in the Pleistocene of Israel. In *Rodents in Desert Environments,* edited by I. Prakash and P. K. Ghosh. Junk, The Hague. pp. 331–362.

———. 1976. Some late Quaternary faunal remains from the Ardat/Aqev area. In *Prehistory and paleoenvironments of the Central Negev, Israel I. The Ardat/Aqev area,* edited by A. E. Marks, S.M.U. Press, Dallas. pp. 68–74.

———. 1979. Polymorphism, size trends and Pleistocene response of the subgenus *Sylvaemus* (Mammalia, Rodentia) in Israel. *Israel J. Zool.* 28:131–159.

Van Valen, L. 1965. Morphological variation and width of ecological niche. *American Naturalist* 99:377–390.

Woldstaedt, P. 1962. Ueber die Gliederung des Quataers und Pleistozaens. *Eiszeitalter Gegenwart* 13:115–124.

Wreschner, E. 1966. The Geula Cave, Mt. Camel. *Quaternaria* 9:69–140.

Zahavi, A. and Y. Wahrman. 1957. The cytotaxonomy, ecology and evolution of the gerbils and jirds of Israel. *Mammalia* 21(4):341–380.

Ziffer, D. 1978. A re-evaluation of the Upper Palaeolithic industries of Kebara Cave and their place in the Aurignacian culture of the Levant. *Paléorient* 4:273–293.

25

Mammalian Extinctions and Stone Age People in Africa

RICHARD G. KLEIN

DISCUSSIONS OF QUATERNARY EXTINCTIONS have commonly ignored Africa because so few large mammals became extinct there compared to the number that disappeared elsewhere. In fact, Africa has been called "the living Pleistocene" because its historic fauna was as abundant and diverse as the now-vanished Pleistocene faunas of other continents, especially North America. Yet, Africa did experience Quaternary extinctions, including end-Pleistocene ones, and Africa cannot be ignored in testing competing explanations for extinctions. It is my purpose here to examine the African case in light of the climatic and human cultural explanations that have been offered for other continents. Since Africa has a longer known record of human occupation than any other continent, I will focus particularly on the possible human role in Quaternary extinctions.

Quaternary Time

As used here, the Holocene is the last 12,000 to 10,000 years (=deep-sea oxygen-isotope stage 1 or the time elapsed since the end of the last glacial). The Pleistocene is the interval between the base of the Olduvai Paleomagnetic Event, roughly 1.8 million years ago, and the beginning of the Holocene. Following Burg Wartenstein Symposium No. 58 (Butzer and Isaac 1975), the Pleistocene may be usefully subdivided into: (1) the Upper Pleistocene, comprising the last interglacial and the last glacial (=deep-sea oxygen-isotope stages 5 through 2), approximately the interval between 130,000 and 12,000 to 10,000 B.P.; (2) the Middle Pleistocene, from the base of the Brunhes Paleomagnetic Epoch to the beginning of the last interglacial, that is, roughly 700,000 to 130,000 B.P.; and (3) the Lower Pleistocene, from the base of the Olduvai Event to the base of the Brunhes Epoch, roughly 1.8 my. to 700,000 B.P.

In Africa, as in most other parts of the world, Upper Pleistocene and Holocene faunal samples are more numerous, often larger, and usually under tighter chronological control than Lower and Middle Pleistocene ones. This is the rationale for discussing Upper Pleistocene and Holocene extinctions separately from earlier ones. One major region—tropical West Africa—has been ignored throughout because its Pleistocene

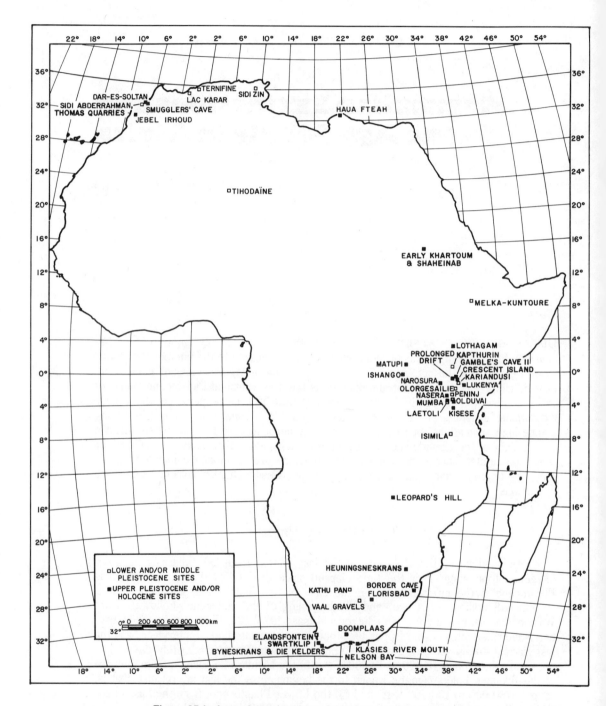

Figure 25.1. Approximate locations of the sites mentioned in the text.

and Holocene faunal history is virtually unknown, reflecting poor preservation conditions, difficulties in finding sites, and a relative lack of interested researchers. Locations of the principal sites mentioned in the text are shown in Figure 25.1.

Upper Pleistocene and Holocene Extinctions

The Upper Pleistocene and Holocene mammal faunas of northern, eastern, and southern Africa include only a handful of extinct species, but there are numerous instances of extant species in sites far outside their historic ranges. Thus local (as opposed to total) extinction was a common event, and in most well-documented instances it has clearly been linked to climatic change.

Northern Africa

The most spectacular examples of local extinctions are from northern Africa, where a stereotypic sub-Saharan savanna fauna spread through much of the Sahara and into the Maghreb on several different occasions during the later Pleistocene and Holocene. Included in this fauna were elephant (*Loxodonta africana*), white rhinoceros (*Ceratotherium simum*), Burchell's zebra (*Equus burchelli*), warthog (*Phacochoerus aethiopicus*), giraffe (*Giraffa camelopardalis*), blue wildebeest (*Connochaetes taurinus*), hartebeest (*Alcelaphus buselaphus*), roan antelope (*Hippotragus equinus*), Bohor reedbuck (*Redunca redunca*), eland (*Taurotragus oryx*), and other species (Vaufrey 1955; Arambourg 1962; Camps 1974; Hays 1975; Wendorf et al. 1977; Wendorf and Schild 1980; Thomas 1977; Geraads 1980, 1981; and Smith 1980).

In the Sahara, bones of the savanna species are consistently associated with ancient lacustrine or spring deposits, indicating a relatively moist climate much different from the historic one. Radiocarbon dates show that the last major Saharan moist period corresponded to the terminal Pleistocene/early Holocene, from 12,000 to 4000 B.P. Earlier moist periods are beyond the range of the radiocarbon method, with the end of the most recent one perhaps just at the method's effective limit of 40,000 to 30,0000 B.P. Between 40,000 and 12,000 B.P., climate was mainly hyperarid, not only in the Sahara but also in the Maghreb, where geomorphic processes dependent upon water (alluviation, colluviation, and pedogenesis) practically ceased. In deeply stratified Maghrebian cave sites, the hyperarid interval shows up as a prolonged period of nonoccupation, when people and other large mammals were either absent or very scarce. The sparse large mammal fauna probably consisted principally of desert-adapted or resistant species such as Barbary sheep (*Ammotragus lervia*), oryx (*Oryx dammah*), addax (*Addax nasomaculatus*), and gazelles (*Gazella* spp.).

Most of the mammal species that occurred in the Sahara and the Maghreb prior to 40,000 to 30,000 B.P. reappeared when moister conditions returned 14,000 to 12,000 years ago, but at least one species did not. This was Kirchberg's "woodland" rhinoceros (*Dicerorhinus kirchbergensis*), a Eurasiatic migrant that may have arrived in the Maghreb only in the late Middle or early Upper Pleistocene (Vaufrey 1955, Jaeger 1975, Thomas 1977). Its failure to reappear there after 14,000 to 12,000 B.P. probably reflects the fact that it had become extinct in Eurasia by this time, leaving no source population for recolonization. Its extinction is basically a Eurasiatic event and will not be discussed further here.

The end of the last major moist interval at 5000 to 4000 B.P. probably left most of the Sahara as arid as it was before 14,000 to 12,000 B.P., but the change in the Maghreb was less drastic and perhaps was reflected mainly in the spread of xerophytic shrubs ("macchia") at the expense of grasses in the regional flora. Thus, while the sub-Saharan fauna disappeared entirely from the Sahara, some prominent elements—elephant, giraffe, and especially hartebeest—survived in places in the Maghreb. Most of the remaining savanna species, of course, survived in substantial numbers immediately below the Sahara, ready to spread north again, should climatic conditions permit. However, at least four species that had been associated with the savanna elements in the Maghreb, the Sahara, or both during previous (Middle and/or Upper Pleistocene) moist phases appear to have become totally extinct at or near the end of the early Holocene one, 5,000 to 4,000 years ago. These are the Atlantic gazelle (*Gazella atlantica*), and more certainly, Thomas' camel (*Camelus thomasi*), the North African "giant" deer (*Megaloceros algericus*), and the large, long-horned African buffalo (*Pelorovis antiquus*).

Eastern Africa

Far fewer Upper Pleistocene and Holocene faunas have been excavated and analyzed in eastern than in northern Africa. The only completely extinct species that has been recorded is the long-horned buffalo (*Pelorovis antiquus*), a nearly complete skeleton of which was found in Upper Pleistocene deposits in the Naivasha-Nakuru Basin of Kenya (Nilsson 1964). Future work will probably show that the buffalo was accompanied by other extinct forms.

Fossiliferous Upper Pleistocene sites are reasonably common in East Africa and have been excavated at Lukenya Hill (site GvJm 22—Gramly and Rightmire 1973, Gramly 1976; site GvJm 46—Miller 1979) in Kenya and at Kisese Shelter II (Inskeep 1962, J. Deacon 1966), Nasera (=Apis) Rock (Mehlman 1977), Mumba Shelter (Mehlman 1979), Olduvai Gorge (in the Naisiusiu Beds—M. D. Leakey et al. 1972), and Loiyangalani River (Bower and Gogan-Porter 1981) in Tanzania. However, comprehensive species lists based on moderate to large bone samples are available for only one of the Lukenya sites (GvJm 22) and for Mumba Shelter (Lehmann as cited in Mehlman 1979). The Mumba list unfortunately requires confirmation, since it includes several seemingly improbable identifications, including the reported but unlikely association of chimpanzee (*Pan troglodytes*) and addax.

East African Holocene faunas are better known and, in combination with geomorphic evidence, suggest a pattern of local extinction or species redistribution broadly comparable to the Holocene pattern in North Africa. Relatively large later Holocene faunas, all postdating 5000 to 3000 B.P. have been described from Shaheinab in the central Sudan (Bate in Arkell 1953) and from Prolonged Drift (Gifford et al. 1980, Gifford 1983), Narosura (Gramly 1974), and Crescent Island (Onyango-Abuje 1977) in south-central Kenya. None of the faunas contain any indigenous species that were not observed near each site historically or at least that would have been out of place in the historic environment.

Medium- and large-sized earlier Holocene faunal samples, dating between 9000 and 5000 B.P., have been described from Early Khartoum in the central Sudan (Bate in Arkell 1949), Lothagam in the Lake Turkana basin of northern Kenya (Fagan in Robbins 1974), Gamble's Cave II in the Lake Nakuru Basin of central Kenya (Hopwood in L. S. B. Leakey 1931), and Ishango on Lake Edward in eastern Zaire (de Heinzelin 1957). In contrast to the later Holocene faunas, each of these earlier ones contains at least one mammal species that was not recorded nearby historically and that suggests a moister climate at the time the site was occupied. The case of Early Khartoum is particularly clearcut, since the wetland species present, water mongoose (*Atilax paludinosus*), cane rat (*Thryonomys* sp.), Nile lechwe (*Kobus megaceros*), and possibly kob (*Kobus kob*), were not found nearer than 450 km historically and are also completely absent in the later Holocene (post-5500 B.P.) fauna of nearby Shaheinab.

Radiocarbon dates related to high lake levels (including ones recorded at Gamble's Cave II and Lothagam) indicate beyond all doubt that the early Holocene was indeed a relatively moist period in East Africa (Butzer et al. 1972), and the local extinctions that have been recorded are readily explained in terms of greater aridity after 5000 B.P. or so. The extent of local extinction appears less dramatic than in North Africa, perhaps partly because many fewer pertinent East African faunas have been studied, but probably mainly because changes in precipitation never led to such extreme aridity in East Africa that large mammals would have nearly disappeared. In particular, in contrast to the situation in the Sahara and the Maghreb, the developing archaeological record of southern Kenya and northern Tanzania suggests that archaeologically visible human populations were present more or less continuously throughout the last 40,000 to 30,000 years, despite geomorphic and palynologic evidence for significant wetter-drier fluctuations.

Southern Africa

The Upper Pleistocene and Holocene faunal record of southern Africa is reminiscent of the North African one, in its evidence for repeated or cyclical local extinction related to climatic change, and in its handful of totally extinct species (Klein 1980).

The best examples of climate-linked local extinctions come from the extreme southwestern part of the subcontinent in what has been termed the Cape Biotic Zone or Cape Ecozone. Historically, the vegetation here was a variable mix of bush, forest, and fynbos (="Cape Macchia"), and the principal ungulates were browsers, including black rhinoceros (*Diceros bicornis*), bushpig (*Potamochoerus porcus*), grysbok (*Raphicerus melanotis*), steenbok (*R. campestris*), bushbuck (*Tragelaphus scriptus*), and duikers (*Sylvicapra grimmia* and *Cephalophus monticola*). The Upper Pleistocene and Holocene witnessed cyclical alternation between this fauna and one including grazers like white rhinoceros, warthog, black wildebeest (*Connochaetes gnou*), and springbok (*Antidorcas australis*) not present historically. The last time that the grassland fauna was prominent was in the very late Pleistocene from before 30,000 until 12,000 B.P., when geomorphic evidence indicates climate was relatively dry (Butzer and Helgren 1972, Butzer et al. 1978).

Interestingly, as in the Maghreb, most deeply stratified cave sites in the Cape record the late Pleistocene dry interval as a period of nonoccupation, suggesting that the aridity depressed human populations, though perhaps not so drastically as in the Maghreb. Some similarity between the records of the Cape and the Maghreb was perhaps to be expected since they are in similar climatic zones today, symmetrically located with respect to the equator.

Six species that are now totally extinct are well documented in Upper Pleistocene sites in southern Africa. The long-horned buffalo (*Pelorovis antiquus*), the "giant hartebeest" (*Megalotragus priscus*), and the "giant Cape horse" (*Equus capensis*) appear to have been present throughout the subcontinent, while Bond's springbok (*Antidorcas bondi*) seems to have occurred everywhere but in the Cape Zone (fig. 25.2). The southern springbok (*Antidorcas australis*) was apparently confined to the Cape Zone, while a large warthoglike pig (*Metridiochoerus* sp. or *Stylochoerus* sp.) is too frequent overall to judge its distribution. Since the giant hartebeest and the Cape horse were so widespread in southern Africa and since they commonly occur in the same sites as the long-horned buffalo, it seems likely they will eventually also be found in Upper Pleistocene sites in East Africa.

Outside the Cape Zone, the extinct species may have disappeared any time from before 25,000 until 5000 B.P., since large faunal samples clearly dating to this interval are lacking. Within the Cape Zone, where such samples are available, the long-horned buffalo, giant hartebeest, southern springbok, and Cape horse apparently made their last appearance between 12,000 and 9,500 years ago. Their disappearance coincides with the initiation of somewhat moister conditions and the replacement of an open grassland fauna by one in which browsing ungulates dominate. This suggests that their extinction in southern Africa was at least partly related to climatic/environmental change.

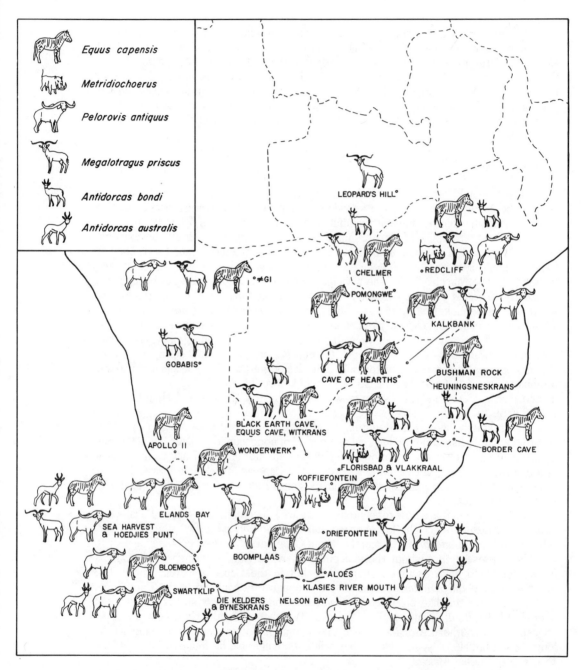

Figure 25.2. Approximate locations of the Upper Pleistocene sites in southern Africa that have provided bones of extinct species.

The Case for Stone Age People as an Ingredient in Upper Pleistocene and Holocene Extinctions

Excluding the special case of Kirchberg's rhinoceros, no mammals appear to have become extinct during the Upper Pleistocene in Africa until the terminal Pleistocene/

early Holocene interval. The profound climatic change that occurred during this period, leading to numerous local extinctions or species redistributions, was almost certainly involved in the total extinctions as well, at least to the extent of diminishing the population size and the territory occupied by each now-extinct species. As a complete explanation for extinctions, however, terminal Pleistocene/early Holocene environmental change is insufficient, since the extinct species clearly survived earlier periods of similar change, occurring from at least the Middle Pleistocene on.

The answer to the dilemma must surely be that the earlier periods of environmental change were not the same as the terminal Pleistocene/early Holocene one. They may have differed in natural features that further paleoenvironmental research will elucidate or perhaps in ones that are too subtle to be detected in such remote periods. Or it may be that the difference was not so much in natural features as it was in cultural ones, that is, that it involved long-term changes in the way Stone Age people interacted with animals. This is the possibility I wish to explore.

Earlier Upper Pleistocene Prehistory

In northern Africa the artifacts found in earlier Pleistocene sites, antedating 40,000 to 30,000 B.P., belong to the Mousterian and Aterian "Cultures," which may be regarded as temporally successive variants of the Middle Paleolithic (Camps 1974, 1975). In both cultures, the principal tools are sidescrapers, points, and denticulates made on flakes. Unlike Mousterian tools, Aterian ones often possess stems or tangs, presumably to facilitate hafting. Neither Mousterian nor Aterian peoples significantly used bone as a raw material for tool manufacture, nor do they seem to have produced art objects or items of personal adornment, such as beads and pendants.

Human remains have been found with Mousterian artifacts at the Haua Fteah in Cyrenaican Libya (Tobias in McBurney 1967) and at Djebel Irhoud in Morocco (Ennouchi 1963, 1968, 1969) and with Aterian artifacts at Dar-es-Soltan II (Debenath 1975, Ferembach 1976a) and the Smugglers' Cave (Temara) (Roche and Texier 1976, Ferembach 1976b), both in Morocco. The remains suggest that both Mousterian and Aterian people belonged to an archaic variety of *Homo sapiens,* different not only from modern *Homo sapiens sapiens,* but also from the Neanderthals who occupied neighboring Europe and southwest Asia in the earlier Upper Pleistocene.

In eastern and southern Africa, the earlier Upper Pleistocene cultures that are contemporaneous with Mousterian and Aterian ones are generally referred to as "Middle Stone Age (MSA)" (Volman 1981). The principal stone tool types were the same as in the Mousterian and Aterian. Also like their North African contemporaries, MSA peoples essentially ignored bone for tool manufacture, and they did not make art objects or items of personal adornment.

There are few well-documented associations between Middle Stone Age artifacts and human remains that are complete enough to diagnose physical type. At the moment it is possible to argue that MSA peoples were archaic *Homo sapiens,* based on the model of the Ngaloba skull from northern Tanzania (Day et al. 1980), or the Florisbad skull from the Orange Free State of South Africa (Rightmire 1978). It is also possible that they were modern *(Homo sapiens sapiens),* especially based on a partial skull and mandibles from Border Cave, Natal, South Africa (de Villiers 1976, Rightmire 1979). It is even possible that an evolution from archaic to modern *Homo sapiens* occurred within the MSA, though many new human fossils from carefully excavated contexts would be necessary to make a serious case.

Later Upper Pleistocene and Holocene Prehistory

In North Africa, the cultures that succeeded Middle Paleolithic ones roughly 40,000 years ago are generally called Upper Paleolithic (Camps 1975, Marks 1975). Like contemporaneous Upper Paleolithic cultures in Eurasia, the North African ones exhibit considerable variability in time and space. They share a common tendency for tools to be made on blades (=flakes that are at least twice as long as wide), while endscrapers, burins, and backed pieces are usually the most important tool types.

Later North African Upper Paleolithic cultures, after 20,000 to 15,000 B.P., are often characterized by large numbers of small, backed blades or bladelets. Such cultures are sometimes referred to as "Epipaleolithic," although continuity with underlying Upper Paleolithic cultures (*sensu stricto*) is almost universally assumed. Terminal Pleistocene/early Holocene Epipaleolithic cultures are often characterized by large numbers of small, geometrically shaped, backed pieces or "geometric microliths." In contrast to Middle Paleolithic peoples, Upper Paleolithic ones (*sensu lato*) regularly manufactured art objects and ornaments, as well as a variety of standardized bone tool types ("awls," "points," "needles," etc.)

In the Maghreb and the Sahara, the earlier Upper Paleolithic is essentially unknown and the earliest well-documented Upper Paleolithic cultures, dating to 15,000 to 14,000 B.P., may postdate the latest Middle Paleolithic ones by 20,000 years or more. The occupation gap probably reflects the extreme aridity of the interval before 30,000 until 15,000 to 14,000 years. Only in Cyrenaican Libya is the record of the Upper Paleolithic from its inception about 40,000 B.P. relatively complete (McBurney 1967), perhaps because Cyrenaica enjoyed a somewhat less arid climate. A complete record may also exist in the Nile Valley, with its perennial water source, though sites clearly dating between the end of the Middle Paleolithic, before 40,000 to 35,000 B.P., and 20,000 to 18,000 B.P. are unknown there. Upper Paleolithic cultures postdating 20,000 to 18,000 B.P. are abundantly documented (Hassan 1980, Wendorf and Schild 1976, Wendorf et al. 1979).

Numerous relatively complete human skeletal remains are known from later Upper Paleolithic (Epipaleolithic) sites in North Africa, and the people were uniformly modern (Camps 1974). Although skeletal evidence is so far lacking, it seems highly probable that earlier Upper Paleolithic people were modern as well.

The most dramatic cultural event following the appearance of the Upper Paleolithic in North Africa was perhaps the appearance of pottery, 9,000 to 8,000 years ago in the Nile Valley and the Sahara, and perhaps 1,000 to 2,000 years later in the Maghreb (Camps 1975, Hays 1975, Smith 1980). Pottery was followed by the appearance of domestic cattle and sheep or goats 8,000 to 7,000 years ago in the Sahara and perhaps 1,000 years later in the Maghreb and along the Nile. The Stone Age cultures that possessed pottery and stock (and in the Nile Valley also domestic cereals) are generally termed "Neolithic." Sometimes the presence of pottery alone, or of ground stone tools and tanged arrowheads, which appeared at about the same time, is thought to justify the use of this term. The Neolithic in the Nile Valley and Mediterranean borderlands was terminated by the development of complex societies with metallurgy, around 5,000 B.P. Over most of the Sahara the Neolithic ended with the onset of hyperarid climate around 4000 to 3000 B.P.

In eastern and southern Africa, the Middle Stone Age was supplanted about 40,000 to 30,000 B.P. by a complex of cultures which are so far poorly known or described. Important sites that have provided early post-MSA industries include Kisese Shelter II (Inskeep 1962, J. Deacon 1966), Nasera (=Apis) Rock (Mehlman 1977), Mumba Shelter (Mehlman 1979), Lukenya site GvJm 46 (Miller 1979), Matupi (van Noten 1977) in East Africa, Leopard's Hill (Miller 1971), Border Cave (Beaumont et al. 1978), Heuningsneskrans (Beaumont and Vogel 1972, Beaumont pers. comm.), and Boomplaas Cave (H. J. Deacon 1979) in southern Africa. The pertinent artifacts from these sites have sometimes been characterized as neither typically Middle Stone Age nor typically Later Stone Age (LSA), but pending much fuller descriptions, I will refer to them collectively here as "early LSA." On the evidence from Border Cave, one important feature the early LSA shared with the later LSA (or LSA *sensu stricto*) was the regular manufacture of standardized bone artifacts and items of personal adornment.

Later LSA assemblages, postdating 24,000 to 20,000 B.P., are relatively well known in both eastern and southern Africa (e.g. M. D. Leakey et al. 1972; van Noten 1971, 1977; Gramly 1976; Miller 1979 for East Africa; and Miller 1971, J. Deacon 1978, and Carter and Vogel 1974 for southern). The later LSA is generally reminiscent of the later Upper Paleolithic of North Africa in that small, backed tools (bladelets, "geometric microliths," or both) are a common element. Later LSA peoples were clearly anatomically modern (Rightmire 1981); the physical appearance of early LSA peoples remains to be established, though it seems highly likely they were modern as well.

Pottery was added to LSA assemblages in East Africa about 8,000 years ago. Domestic cattle and sheep or goats were widespread there by 3000 B.P. and probably by 5000 B.P. or even earlier (Bower et al. 1977, Bower and Nelson 1978). As in North Africa, East African assemblages with domestic stock and pottery are often referred to as "Neolithic." Both pottery and domestic stock reached southern Africa somewhat later, probably only about 2,000 years ago (Philipson 1977, H. J. Deacon et al. 1978, Klein 1983). In southern Africa, stone artifact assemblages with pottery and domestic stock are regarded simply as the latest phase of the local Later Stone Age.

The "Neolithic" in East Africa was terminated by the spread of Iron Age pastoralists and mixed farmers between roughly 2,500 and 1,500 years ago (Phillipson 1977). Iron Age mixed farmers entered southern Africa about 2,000 years ago, but much of the subcontinent was still inhabited by Stone Age hunters-gatherers and pastoralists at the time of European contact, beginning 500 to 400 years ago.

In sum, throughout Africa there is evidence for a dramatic change in artifact traditions and cultural behavior from before 40,000 to 30,000 B.P., perhaps accompanying the appearance of anatomically modern people. The Africans who lived after 40,000 to 30,000 B.P. were the first to manufacture art objects and standardized bone tools, and their cultures exhibit far more variability in time and space than those of their predecessors. To this may be added evidence developed in the Cape Zone of southern Africa that people living before 40,000 to 30,000 B.P. were less effective hunter-gatherers than their successors.

Upper Pleistocene Evolution in Hunting Capability

The Cape evidence for progress in hunting proficiency derives from the analysis of large faunal assemblages from MSA levels in the Klasies River Mouth and Die Kelders Caves and from LSA levels at several sites, especially Nelson Bay Cave and Byneskranskop Cave 1 (Klein 1979, 1980). The analyses indicate that MSA people living along the coast caught fish and flying seabirds much less frequently than their LSA successors. In this context, it is pertinent that artifacts reasonably interpreted as fishing and fowling gear ("gorges" and sinkers) are well known in LSA assemblages but have not been found in MSA ones.

Interesting differences have also been found in the relative abundance of terrestrial species between MSA and LSA sites occupied under similar environmental conditions. In the MSA sites, eland is the most common large ungulate, though it was probably the least common one in the ancient environment. Wild pigs (warthog and bushpig) are barely represented, though one or both was probably reasonably common near the sites. In comparable LSA sites, eland is far less frequent and pigs far more so, and the relative numbers of both eland and pigs probably correspond much more closely to their numbers in the environs of the sites.

In their response to attackers, eland were probably among the least dangerous available large prey, while pigs were among the most dangerous, suggesting that the differences between MSA and LSA faunas may reflect the enhanced ability of LSA hunters to obtain dangerous game. This was perhaps a result of technological innovation—development of projectiles or of snares and traps—that allowed attack from a relatively safe distance. In this regard, it is pertinent that ethnographic analogy suggests that some of the small, backed elements common in at least later LSA sites (but absent in MSA ones) served as arrow armament.

The age profiles of the large bovids in the MSA faunas provide further evidence that MSA peoples were relatively ineffective hunters. In most of the bovid species present, very young and relatively old individuals predominate and prime-age (reproductively active) adults are rare. The pattern is similar to the one that would result from natural attrition in these species and suggests that MSA hunters were largely restricted to taking the most vulnerable (very young and old) individuals, much as lions and other large predators are. The age data for most of the species could even be interpreted to suggest that the MSA people were scavenging carcasses of animals that died naturally or that were killed by other predators. However, active hunting is probably implied by the fact that very young individuals are well represented in the prey samples. In Africa today the carcasses of very young prey are usually completely destroyed by hyenas, lions, and other scavengers/carnivores with whom MSA people had to compete. The people could have obtained so many young carcasses by scavenging only if they were able to locate carcasses before other predators did, which seems unlikely.

In contrast to the other ungulate species, the eland in MSA sites include a large proportion of prime-age adults. The most likely explanation is that MSA people had learned that, unlike most other large African bovids, eland can be driven with relative ease and limited personal risk. An eland herd caught in the right position could be forced over a cliff or into a trap, places in which differences in individual vulnerability due to age would have little meaning.

However, MSA people could not often have driven eland herds to their death or the species would have become extinct, since its reproductive vitality would have been sapped by the continuing loss of a large proportion of the available prime adults. Not only did the eland survive, but there is no evidence that it became less numerous during the long MSA time span. The reason must surely be that eland, living in the same kind of sparsely distributed, wide-ranging herds observed historically, were difficult to locate in a position suitable for drives, and that MSA people were able to obtain only a small proportion of the available herds.

Thus, MSA people were probably not very successful at hunting eland, and this makes it especially interesting that eland is the most abundant large ungulate in the MSA faunas. The clear implication is that MSA people must have been even less successful at hunting the other species that are less common in the sites but were more common in the environment. In short, MSA impact on the large mammal fauna was negligible. By extension, it may be argued that LSA peoples, in whose sites eland and other species are represented more in proportion to their probable live abundance, probably took a higher proportion of game overall. In short, LSA people were almost certainly more proficient hunters.

The combined artifactual and behavioral evidence suggests a qualitative difference between the people who were present during the terminal Pleistocene/early Holocene interval when extinctions took place in Africa and the people who were present during earlier Upper or late Middle Pleistocene times when comparable environmental change took place without extinctions. Perhaps particularly significant is the far greater cultural diversity apparent in the late Upper Pleistocene/early Holocene archaeological record, implying a significantly greater human ability to innovate in the face of environmental change. Therefore I suggest that qualitatively different people were the new ingredient in the terminal Pleistocene/early Holocene situation, directly or indirectly precipitating extinctions after environmental change had reduced species numbers and distributions.

It is unnecessary to hypothesize that each or any extinct species was intensively hunted to its demise, but only that the hunters' adjustment to changing environmental circumstances perturbed a system that was already in tilt, forcing it in a different direction. In the southern African interior, for example, it is possible that the extinct Cape horse, giant hartebeest, and Bond's springbok were interdependent in a grazing succession comparable to that observed among Burchell's zebra, blue wildebeest, and Thomson's gazelle (*Gazella thomsoni*) in modern East Africa (Bell 1971). A technological, social, or ideological innovation that allowed hunters to put more pressure on any one of the three species may have been sufficient to imperil others too, even if they were hunted no more intensively than before.

In northern Africa, where the terminal Pleistocene/early Holocene extinctions appear to have been preceded or accompanied by the introduction of domestic stock, it is possible that the human role was even more subtle, and that it reflects the inability of one or more species to compete with stock for pasture. Although no species disappeared following the introduction of stock to eastern and southern Africa, the archaeological record of the Cape Zone suggests a reduction in numbers in at least two forms—the blue antelope (*Hippotragus leucophaeus*) and the bontebok (*Damaliscus dorcas dorcas*) (Klein 1983). Though the blue antelope is common in sites that antedate the introduction of stock 2,000 years ago, it is infrequent in later sites and was rarely seen by European travelers and colonists. It became extinct about 1800 A.D., the first large African mammal to disappear in historic times. The bontebok survived, but only as a result of deliberate conservation and management.

Clearly, the case for Stone Age people as agents in terminal Pleistocene/early Holocene extinctions is still circumstantial, but it could be substantially improved—or disproved—with more data on the precise timing of extinctions, the relative abundance of extinct species at various times, and the age/sex structure of species samples. With these data, it would be possible to determine the rate of species decline (gradual or precipitous), the true extent of coincidence between extinction and major environmental or cultural change, the degree to which the extinct species may have been ecologically interdependent, and the nature of Stone Age predation on them. It would be particularly critical to determine if the pattern of predation was one that might have impaired reproductive capacity.

Lower and Middle Pleistocene Extinctions

In addition to "extinct" species that survived into the Upper Pleistocene, the Lower and Middle Pleistocene faunas of Africa contained a number of species whose lineages terminated before the beginning of the Upper Pleistocene. The overwhelming majority of these totally extinct species are large mammals, in the same size range as those that became extinct during the terminal Pleistocene/early Holocene. However, there are also some totally extinct small mammals in Lower and Middle Pleistocene samples (Jaeger 1975, 1976; Butler 1978b; Lavocat 1978), as well as extinct birds and reptiles (M. D. Leakey 1979, Tchernov 1976). Since small mammals and birds are particularly uncommon or poorly studied in most Lower and Middle Pleistocene samples, the known number of extinct species is probably only a small fraction of the true number. Because the information is more complete, I will deal here only with large mammals, though I realize that a comprehensive explanation for extinction will eventually have to deal with all species, regardless of size or zoological order.

Table 25.1 lists the large mammal species that do not appear to have survived the Middle Pleistocene in Africa. The list is partially subjective, since the taxonomy of some extinct species is disputed; this is perhaps especially true for the suids (compare White and Harris [1977], and Harris and White [1979], with Cooke and Wilkinson [1978]). More importantly, the list is certainly incomplete, first because Lower and Middle Pleistocene faunas remain far from adequately sampled in Africa, even in major subregions, and second because not all the extinct species that have been found have been described.

Table 25.1. African Lower and Middle Pleistocene Mammal Species With No Apparent Upper Pleistocene or Holocene Descendants

	Northern Africa Pleistocene		Eastern Africa Pleistocene		Southern Africa Pleistocene	
	Lower	Middle	Lower	Middle	Lower	Middle
PRIMATES						
Hominidae						
Australopithecus robustus†			X		X	
Cercopithecidae						
Papioni						
Parapio sp (p)†			X		X	
Dinopithecus ingens†					X	
Gorgopithecus major†					X	
Theropithecus (Simopithecus) oswaldi*	X	X	X	X	X	X
Colobini						
Cercopithecoides williamsi†			?			
CARNIVORA						
Hyaenidae						
Hyaenictis forfex†					X	
Chasmoporthetes nitidula†					X	
Hyaena bellax*					X	
Viverridae						
Pseudocivetta ingens†			X			
Felidae						
Panthera crassidens*			X	X	X	
Dinofelis piveteaui†					X	
Megantereon sp (p)†			X		X	X
Machairodus sp (p)†	X					
HYRACOIDEA						
Procaviidae						
Procavia transvaalensis*					X	
PROBOSCIDEA						
Deinotheriidae						
Deinotherium bosazi†			X			
Elephantidae						
Mammuthus meridionalis†	X					
Loxodonta atlantica*	X	X	X			X
Elephas recki*		X	X	X		X
Elephas iolensis*		X		X		

SOURCES: Butler (1978b), Churcher (1978), Churcher and Richardson (1978), Cooke and Wilkinson (1978), Coppens et al. (1978), Coryndon (1978), Gentry (1978), Geraads (1980, 1981), Hendey (1974b, 1974c, 1978), Meyer (1978), Savage (1978), Simons and Delson (1978), Thomas (1977), and Vrba (1978).

*extinct species in a still extant genus
†extinct species in an extinct genus

Determining the causes of Lower and Middle Pleistocene extinctions is severely hampered by difficulties in dating them precisely. The problem is twofold, involving the inadequacy of sampling (relatively few faunal samples, while most are quite small) and the lack of volcanic extrusives suitable for potassium/argon dating at many important sites. Dating is a problem not just in northern and southern Africa, where there were no active volcanoes, but also in eastern Africa where there were. An additional complica-

Table 25.1. African Lower and Middle Pleistocene Mammal Species With No Apparent Upper Pleistocene or Holocene Descendants

(continued)

	Northern Africa		Eastern Africa		Southern Africa	
	Pleistocene		Pleistocene		Pleistocene	
	Lower	Middle	Lower	Middle	Lower	Middle
ARTIODACTYLA						
Hippopotamidae						
*Hippopotamus gorgops**			x	x	x	
*Hippopotamus aethiopicus**			x			
Suidae						
Notochoerus scotti†			x			
Kolpochoerus limnetes†		x	x			
Kolpochoerus olduvaiensis†			x	x		x
*Phacochoerus modestus**			x	x		
Metridiochoerus nyanzae†			x		x	x
Stylochoerus compactus†	x		x		x	x
Giraffidae						
*Giraffa jumae**	?	?	x	x		
*Giraffa gracilis**			x	x		
*Giraffa pygmaea**			x			
Sivatherium maurusium†	x	x	x	x	x	x
Bovidae						
Tragelaphini						
Tragelaphus (Strepsiceros) sp.*						x
*Tragelaphus nakuae**			x			
Reduncini						
Menelikia lyrocera†			x			
*Kobus ancystrocera**			x			
Thalcerocerus radiciformis†				x		
Hippotragini						
*Hippotragus gigas**	x		x		x	x
Alcelaphini						
Parmularius angusticornis†			x	x		
Parmularius ambiguus†	x					
Parmularius rugosus†			x	x		
Parmularius parvus†				x		x
Beatragus sp.*						x
Ovibovini						
Makapania sp.†					x	
Caprini						
Numidocapra crassicornis†	x					
PERISSODACTYLA						
Chalicotheriidae						
Ancyclotherium hennigi†			x			
Equidae						
Hipparion libycum†	x	?	x	x	x	x

tion is that much of the interval under consideration—particularly the later Middle Pleistocene when many extinctions may have taken place—is too recent for the routine application of the potassium/argon method.

Generally, Lower and Middle Pleistocene extinctions can probably be attributed to one of two basic causes: (1) the failure of species to adapt to changes in climate or physical environment, or (2) their failure to compete successfully in biotic communities that were constantly changing as a result of evolution in local species and the immigration of foreign ones. These two "causes" are not totally separate, since changes in climate or environment probably provided much of the natural selection pressure behind local evolution, as well as new opportunities for migration or species dispersal. The relatively great amount of climatic change that characterized the Pleistocene in Africa is almost certainly the reason that both extinctions and new appearances—"faunal turnover"—appear to have occurred more rapidly than at any other time in the later Cenozoic (Maglio 1978).

Broadly speaking, the same kinds of climatic fluctuations, from drier to wetter and then drier, that have been well documented for the Upper Pleistocene and Holocene probably occurred throughout the Pleistocene. At least within the Middle Pleistocene they led to the same kinds of local extinctions or species redistributions, as shown clearly in the Sahara in northern Africa (Thomas 1977, Wendorf et al. 1977, Wendorf and Schild 1980) and in the Namib Desert in southern Africa (Shackley 1980). However, as in the Upper Pleistocene and Holocene, it seems unlikely that Lower and Middle Pleistocene climatic change by itself was sufficient to cause extinctions, since most, if not all the extinct species probably survived many such changes during their lifespans.

As in the case of the Upper Pleistocene and Holocene, it seems probable that if climatic change was involved, it was in conjunction with evolutionary changes in the biotic environment. Thus, the late Tertiary/early Pleistocene adaptive radiation and obvious evolutionary success of the bovids may be largely responsible for the contemporaneous decline in suid fortunes, while evolutionary advances in the hominids probably account for the end Tertiary/early Pleistocene reduction in terrestrial monkey diversity.

Southern African data can be used to suggest that the successful entry of hominids into a meat-eating niche was perhaps also responsible for a decline in the diversity of carnivores from the late Tertiary into the Pleistocene (Klein 1977). Alternatively, this decline may relate to important developments in the evolution of the carnivores, particularly the appearance and dispersal of large cats of the genus *Panthera,* beginning in the later Pliocene (Hendey 1974a). Finally, some Pleistocene extinctions in northern Africa may have resulted from the failure of indigenous species to compete successfully with Eurasiatic immigrants which periodically appeared there. Immigration seems to have been particularly important during the mid-Pleistocene, when aurochs (*Bos primigenius*), red deer (*Cervus elaphus*), wild boar (*Sus scrofa*), and other Eurasiatic species that were important in North Africa subsequently made their first appearance, probably from southwest Asia (Jaeger 1975).

Hominids were clearly an important element in evolving Lower and Middle Pleistocene communities in Africa, and the possibility that they were directly responsible for Lower or Middle Pleistocene extinctions must certainly be considered. Perhaps the hominid "event" that is easiest to examine from this perspective is the replacement of the Acheulean Earlier Stone Age/Lower Paleolithic cultures by Middle Stone Age/ Middle Paleolithic ones. This occurred sometime in the later Middle Pleistocene, perhaps as many as or more than 200,000 years ago (Wendorf et al. 1975, Klein 1976, Butzer et al. 1978, Szabo and Butzer 1979).

In distinction from Acheulean peoples, MSA/Middle Paleolithic ones rarely made hand axes or cleavers, and when they did, the tools were morphologically distinct from Acheulean ones. The Acheulean tradition began in Africa more than a million years ago (Isaac and Curtis 1974, Isaac 1975, M. D. Leakey 1977), and the rate of change through time appears to have been excruciatingly slow. The early hand-ax makers are clearly assignable to *Homo erectus,* while later ones have been variably assigned to *Homo erectus* or early *Homo sapiens* (Rightmire 1981), a distinction that makes very little difference, insofar as virtually all specialists view *H. erectus* as ancestral to *H. sapiens,* wherever they draw the line between the two. The earliest MSA/Middle Paleolithic peoples were probably all members of archaic *Homo sapiens,* and there is no evidence to suggest a major physical difference from later Acheuleans.

Faunal remains that are stratigraphically associated with Acheulean hand axes and cleavers are known from various African sites, of which the outstanding examples are probably Sidi Abderrahman and Thomas Quarries (Casablanca), Ternifine, Lac Karar, Tihodaïne, and Sidi Zin in northern Africa (Vaufrey 1955; Camps 1974; Freeman 1975; Jaeger 1975; Thomas 1977; Geraads et al. 1980; and Geraads 1980, 1981); Isimila, Peninj, Olduvai Gorge, Olorgesailie, Kariandusi, Kapthurin, and Melka-Kunturé in East Africa (Chavaillon 1976, 1979; Geraads 1979; Westphal et al. 1979; Harris 1977; Howell and Clark 1963; Howell et al. 1972, 1972a; M. D. Leakey 1977; Isaac 1975, 1977); and Kathu Pan, the Vaal "Younger Gravels," and Elandsfontein in southern Africa (Cooke 1963, Helgren 1977, Hendey 1974b, Beaumont and Klein in prep., Klein 1978). The faunal samples from these and other Acheulean sites contain many of the lower and mid-Pleistocene species listed in Table 25.1, while these species are totally unknown in subsequent Middle Stone Age/Middle Paleolithic faunas. It is therefore tempting to conclude, as Martin (1967) did, that advanced Acheulean or early MSA/Middle Paleolithic hunters may have caused a wave of extinctions.

However, the known Acheulean sites with extinct species span at least one million years, and it is by no means clear that many (any) of them date from near the end of the Acheulean. Additionally, there is no evidence to indicate that later Acheuleans were more proficient hunters than earlier ones, or that MSA hunters were even better. It may in fact prove very difficult to evaluate Acheulean hunting capabilities since the stone tools and animal bones found at most (perhaps all) known Acheulean sites in Africa are probably in secondary context. That is, associations may have been partly created (or destroyed) by depositional processes. Even where depositional or postdepositional disturbance is minimal, it is difficult to determine what proportion of the animals present were killed or butchered by people (Klein 1978).

In sum, the lack of evidence for an evolution of hunting proficiency through the Acheulean and into the MSA/Middle Paleolithic, combined with persistent difficulties in dating the relatively small sample of Acheulean sites within the Lower and Middle Pleistocene time range, makes it impossible to argue that early people were directly responsible for any Lower and Middle Pleistocene extinctions, or even that many (any) extinctions coincided with a major cultural event. It remains unknown whether Lower and Middle Pleistocene extinctions occurred episodically (in clumps), as during the terminal Pleistocene/early Holocene transition, or whether the extinct species dropped out gradually over periods of tens or even hundreds of thousands of years. All this, of course, does not mean that Stone Age people were not involved in the extinction process, though I suspect their role was subtle, as part of the wider evolving system to which the extinct species failed to adapt.

Brief Comparisons With Other Continents

On present knowledge, the pace and extent of extinctions that occurred during the Pleistocene and Holocene in Africa seem broadly comparable to the pace and extent of extinctions that occurred during the same interval in Eurasia. On neither continent is there evidence for a major "wave" of extinctions, wiping out a very large proportion of the existing fauna in a relatively brief time. In the context of the potential role played by people, it is notable that after Africa, Eurasia is the continent on which people have been present the longest and that they arrived at a time, more than 700,000 years ago, when they were probably relatively unimportant as predators on other mammals.

In Eurasia as in Africa, the extinctions whose timing is best understood are the ones that occurred in the terminal Pleistocene/early Holocene period, involving especially the woolly mammoth (*Mammuthus primigenius*) and the woolly rhinoceros (*Coelodonta antiquitatis*). Both the numbers and ranges of these species must have been seriously curtailed by terminal Pleistocene/early Holocene reforestation of large areas that had been open steppe throughout most of the preceding last glacial, and it is therefore logical to see environmental change as an important ingredient in their extinction. However, as in Africa, environmental change is an insufficient explanation by itself since both species had survived previous periods of comparable change.

As in Africa, I think the new factor in Eurasia may well have been the kind of people present. Like their African contemporaries, terminal Pleistocene/early Holocene people in Eurasia were anatomically modern, and the archaeological record clearly suggests that their cultures were significantly advanced over those of the nonmodern people who were present during earlier periods of comparable environmental change (Klein 1973). Particularly pertinent in the present context is the fact that these earlier people were unable to colonize the harshest, most continental parts of Eurasia—the northern part of the European USSR and Siberia—during either glacials or interglacials (Klein 1975). It was these areas that probably served as the principal "refuges" for both mammoth and rhinoceros during earlier periods when regions to the south and west became too heavily forested to support viable populations.

The cultural innovations necessary to permit human habitation of northern Eurasia were apparently only developed by anatomically modern people, beginning 40,000 to 30,000 B.P., and thereafter the archaeological record is more or less continuous in the region. It was therefore no longer a faunal refuge in the sense it had been earlier, and people may well have applied the final blow to mammoth and rhinoceros, not necessarily through "overkill" but simply by the important alterations in the north Eurasiatic ecosystem that their presence must have caused. This explanation for terminal Pleistocene/ early Holocene extinctions in Eurasia, combining climatic/environmental change and human actions, is comparable to the one I have offered for like-aged extinctions in Africa.

Unlike Africa and Eurasia, both Australia and the Americas appear to have been swept by "waves" of extinctions during the late Pleistocene, wiping out a very large proportion of the existing fauna in a relatively short period. I see the contrast as support for the argument that the arrival of people in Australia and the Americas was a major factor in causing extinctions, perhaps not so much through "overkill" as through the general disruptive effect on systems that had evolved in the absense of any comparable predator. It is pertinent that the first Australians, and more certainly the first Americans, were anatomically modern, presumably with the same advanced behavioral capabilities that characterized their African and Eurasian contemporaries.

Conclusions

The terminal Pleistocene/early Holocene extinctions of large mammals that have been documented in northern and southern Africa and that may yet be shown in eastern Africa, coincided with a period of relatively dramatic climatic changes. This may have played a role in extinction, at least to the extent of reducing the population numbers and areal distributions of the species that disappeared. Climatic change by itself is an insufficient explanation for extinction, however, because the extinct species survived earlier periods of broadly comparable change.

The difference in the terminal Pleistocene/early Holocene was probably that biotic communities were differently constituted than they had been previously, and, as I have argued, a paramount difference was in the kind of people present. Unlike the people living during earlier periods of climatic change, terminal Pleistocene/early Holocene ones were anatomically modern, and the archaeological record indicates that they were behaviorally advanced over their predecessors. Perhaps especially pertinent is evidence that they were more competent hunters. Even if late Pleistocene/early Holocene people were not directly responsible for extinctions in the sense of "overkill," it is quite possible that their attempts to adjust to changing environmental circumstances perturbed the system sufficiently so that species disappeared that might otherwise have survived.

Evolving people, perhaps again in conjunction with climatic/environmental change, may also have been responsible for Lower and Middle Pleistocene extinctions, but here the case is much more difficult to make. This is partly because there is no direct evidence for an evolution of human hunting abilities during this long time interval, but mainly because there are relatively few Lower and Middle Pleistocene faunal samples and most are not fixed in time very precisely. Thus, it is impossible to say whether Lower and Middle Pleistocene extinctions coincided with major cultural or climatic/environmental events, and more generally whether they occurred episodically or gradually. For the moment, I see no basis for arguing that people, climate, or both were directly involved in extinctions before the Upper Pleistocene or for proposing any other specific cause.

It is perhaps obvious that we are much more likely to obtain reasonably specific explanations for late Pleistocene and Holocene extinctions than for earlier ones, because relevant faunal samples are easier to obtain and to date precisely. Additionally, we will almost certainly always have a fuller appreciation of both cultural and climatic/environmental change during the late Pleistocene and Holocene.

It is my belief, in fact, that while the specific causes of Lower and Middle Pleistocene extinctions may always elude us, the causes of Upper Pleistocene and Holocene ones may be resolved if we devote more attention to obtaining large, well-excavated faunal samples from pertinent sites. Such samples will allow us to deal with such key background questions as whether the extinct species declined gradually or precipitously, whether they were ecologically interdependent with other species that also declined or disappeared, and whether the mortality profiles of the latest representatives of extinct species, as found in archaeological sites, suggest that people may have been hunting them in a way that would impair their reproductive capacity. Clearly then, any specialist interested in advancing our understanding of late Pleistocene/early Holocene extinctions should place the acquisition of fresh, well-excavated, large faunal samples high on his or her list of priorities.

Acknowledgments

I thank J. F. R. Bower, K. W. Butzer, J. D. Clark, K. Cruz-Uribe, D. P. Gifford, Q. B. Hendey, and P. S. Martin for helpful comments on a draft of the manuscript. My research reported here was supported by the National Science Foundation.

References

Arambourg, C. 1962. Les faunes mammalogiques du Pléistocene circum-méditerranéen. *Quaternaria* 6:97–109.

Arkell, A. J. 1949. *Early Khartoum.* London, Oxford University Press.

———. 1953. *Shaheinab.* London, Oxford University Press.

Beaumont, P. B. and Vogel, J. C. 1972. On a new radiocarbon chronology for Africa south of the Equator. *Afr. Stud.* 31:65–89, 155–182.

Beaumont, P. B., de Villiers, H. and Vogel, J. C. 1978. Modern man in sub-Saharan Africa prior to 49,000 years B.P.: a review and evaluation with particular reference to Border Cave. *S. Afr. J. Sci.* 74:409–419.

Bell, R. H. V. 1971. A grazing ecosystem in the Serengeti. *Scient. Amer.* 225 (1):86–93.

Bower, J. R. F. and Gogan-Porter, P. 1981. Prehistoric cultures of Serengeti National Park. *Iowa State Univ. Pap. Anthrop.* 3: 1–56.

Bower, J. R. F. and Nelson, C. M. 1978. Early pottery and pastoral cultures of the Central Rift Valley, Kenya. *Man* 13:554–566.

Bower, J. R. F., Nelson, C. M., Waibel, A. F. and Wandibba, S. 1977. The University of Massachusetts Later Stone Age/Pastoral "Neolithic" comparative study in Central Kenya: an overview. *Azania* 12:119–146.

Butler, P. M. 1978a. Chalicotheriidae. *In* Maglio, V. J. and Cooke, H. B. S. eds. *Evolution of African Mammals*: 368–370. Cambridge (Mass.), Harvard University Press.

———. 1978b. Insectivora and Chiroptera. *In* Maglio, V. J. and Cooke, H. B. S. eds. *Evolution of African Mammals:* 56–58. Cambridge (Mass.), Harvard University Press.

Butzer, K. W. and Helgren, D. M. 1972. Late Cenozoic evolution of the Cape Coast between Knysna and Cape St. Francis. *Quatern. Res.* (N.Y.) 2:143–169.

Butzer, K. W. and Isaac, G. Ll. eds. 1975. *After the Australopithecines*. The Hague, Mouton.

Butzer, K. W., Beaumont, P. B. and Vogel, J. C. 1978. Lithostratigraphy of Border Cave, Kwazulu, South Africa: a Middle Stone Age sequence beginning c. 195,000 B.P. *J. Archaeol. Sci.* 5:317–341.

Butzer, K. W., Isaac, G. Ll., Richardson, J. L. and Washbourn-Kamau, C. 1972. Radiocarbon dating of East African lake levels. *Science* 175:1069–1076.

Butzer, K. W., Stuckenrath, R., Bruzewicz, A. J. and Helgren, D. M. 1978. Late Cenozoic paleoclimates of the Gaap Escarpment, Kalahari Margin, South Africa. *Quatern. Res.* (N.Y. 10:310–339.

Camps, G. 1974. *Les Civilisations préhistoriques de l'Afrique du Nord et du Sahara*. Paris, Doin.

———. 1975. The prehistoric cultures of North Africa: radiocarbon chronology. *In* Wendorf, F. and Marks, A. E. eds. *Problems in Prehistory: North Africa and the Levant*: 181–192. Dallas, Southern Methodist University Press.

Carter, P. L. and Vogel, J. C. 1974. The dating of industrial assemblages from stratified sites in eastern Lesotho. *Man* 9:557–570.

Chavaillon, J. 1976. Mission archéologique Franco-Ethiopienne de Melka-Kunturé. Rapport préliminaire 1972–1975. *L'Ethiopie Avant Histoire* 1:1–11.

———. 1979. Stratigraphie du site archéologique de Melka-Kunturé. *Bull. Soc. Géol. France* 21:225–230.

Churcher, C. S. 1978. Giraffidae. *In* Maglio, V. J. and Cooke, H. B. S. eds. *Evolution of African Mammals*: 509–535. Cambridge (Mass.), Harvard University Press.

Churcher, C. S. and Richardson, M. L. 1978. Equidae. *In* Maglio, V. J. and Cooke, H. B. S. eds. *Evolution of African Mammals:* 379–422. Cambridge (Mass.), Harvard University Press.

Cooke, H. B. S. 1963. Pleistocene mammal faunas of Africa, with particular reference to southern Africa. *In* Howell, F. C. and Bourlière, F. eds. *African Ecology and Human Evolution*: 65–116. Chicago, Aldine Publishing Co.

Cooke, H. B. S. and Wilkinson, A. F. 1978. Suidae and Tayassuidae. *In* Maglio, V. J. and Cooke, H. B. S. eds. *Evolution of African Mammals*: 435–482. Cambridge (Mass.), Harvard University Press.

Coppens, Y., Maglio, V. J., Madden, C. T. and Beden, M. 1978. Proboscidea. *In* Maglio, V. J. and Cooke, H. B. S. eds. *Evolution of African Mammals*: 336–367. Cambridge (Mass.), Harvard University Press.

Coryndon, S. C. 1978. Hippopotamidae. *In* Maglio, V. J. and Cooke, H. B. S. eds. *Evolution of African Mammals*: 483–495. Cambridge (Mass.), Harvard University Press.

Day, M. H., Leakey, M. D. and Magori, C. 1980. A new hominid fossil skull (L. H. 18) from the Ngaloba Beds, Laetoli, northern Tanzania. *Nature* 284:55–56.

Deacon, H. J. 1979. Excavations at Boomplaas Cave—a sequence through the Upper Pleistocene and Holocene in South Africa. *World Archaeol.* 10:241–257.

Deacon, H. J., Deacon, J., Brooker, M. and Wilson, M. L. 1978. The evidence for herding at Boomplaas Cave in the southern Cape, South Africa. *S. Afr. Archaeol. Bull.* 33:39–65.

Deacon, J. 1966. An annotated list of radiocarbon dates for Sub-Saharan Africa. *Ann. Cape. Prov. Mus.* 5:1–84.

———. 1978. Changing patterns in the late Pleistocene/early Holocene prehistory of southern Africa, as seen from the Nelson Bay Cave stone artifact sequence. *Quatern. Res.* (N.Y.) 10:84–111.

Debenath, A. 1975. Découverte de restes humains probablement atériens à Dar es Soltane (Maroc). *C. R. Acad. Sc. Paris* 281: 875–876.

De Heinzelin, J. 1957. Les Fouilles d'Ishango. *Explorations du Parc National Albert*, Fasc. 2. Brussels, Institut des Parcs Nationaux du Congo Belge.

De Villiers, H. 1976. A second adult human mandible from Border Cave, Ingwavuma District, KwaZulu, South Africa. *S. Afr. J. Sci.* 72:212–215.

Ennouchi, E. 1963. Les Néanderthaliens du Jebel Irhoud (Maroc). *C. R. Acad. Sc. Paris* 256:2459–2560.

———. 1968. Le deuxième crâne de l'homme d'Irhoud. *Ann. Paléont.* 54:117–128.

———. 1969. Présence d'un enfant néanderthalien au Jebel Irhoud (Maroc). *Ann. Paléont.* 55:251–265.

Ferembach, F. 1976a. Les restes humains de la grotte de Dar-es-Soltane 2 (Maroc). Campagne 1975. *Bull. Mém. Soc. Anthrop. Paris* 13:183–193.

———. 1976b. Les restes humains atériens de Témara (Campagne 1975). *Bull. Mém. Soc. Anthrop. Paris* 13:175–180.

Freeman, L. G. 1975. Acheulian sites and stratigraphy in Iberia and the Maghreb. In Butzer, K. W. and Isaac, G. Ll. eds. *After the Australopithecines*: 661–744. The Hague, Mouton.

Gentry, A. W. 1978. Bovidae. *In* Maglio, V. J. and Cooke, H. B. S. eds. *Evolution of African Mammals*: 540–572. Cambridge (Mass.), Harvard University Press.

Geraads, D. 1979. La faune des gisements de Melka-Kunturé (Ethiopie): Artiodactyla, Primates. *Abbay* 10:21–49.

———. 1980. La faune des sites à "Homo erectus" des carrières Thomas (Casablanca, Maroc). *Quaternaria* 22:65–94.

———. 1981. Bovidae et Giraffidae (Artiodactyla, Mammalia) du Pléistocene de Ternifine (Algerie). *Bull. Mus. Natn. Hist. Nat. Paris* 3:48–86.

Geraads, D., Beriro, P. and Roche, H. 1980. La faune et l'industrie des sites à Homo erectus des carrières Thomas (Maroc). Précisions sur l'âge de ces Hominidés. *C. R. Acad. Sc. Paris* 291:195–198.

Gifford, D. P. 1983. Behavioral implications of a faunal assemblage from a Pastoral Neolithic site in Kenya. *In* Clark, J. D. and Brandt, S. eds. *Causes and Consequences of Food Production in Africa*. In press. Berkeley, University of California Press.

Gifford, D. P., Isaac, G. Ll. and Nelson, C. M. 1980. Evidence for predation and pastoralism at Prolonged Drift: a pastoral Neolithic site in Kenya. *Azania* 15:57–108.

Gramly, R. M. 1974. Analysis of faunal remains from Narosura. *Azania* 9:219–22.

———. 1976. Upper Pleistocene archaeological occurrences at site GvJm/22, Lukenya Hill, Kenya. *Man* 11:319–344.

Gramly, R. M. and Rightmire, G. P. 1973. A fragmentary cranium and dated Later Stone Age assemblage from Lukenya Hill, Kenya. *Man* 8:571–579.

Harris, J. M. 1977. Mammalian faunas from East African early hominid bearing localities. *In* Wilson, T. H. ed. *A Survey of the Prehistory of Eastern Africa*: 21–48. Nairobi, VIII Pan-African Congress of Prehistory & Quaternary Studies.

Harris, J. M. and White, T. D. 1979. Evolution of the Plio-Pleistocene African Suidae. *Trans. Amer. Phil. Soc.* 69:1–128.

Hassan, F. 1980. Prehistoric settlements along the Main Nile. *In* Williams, M. A. J. and Faure, H. eds. *The Sahara and the Nile*: 421–450. Rotterdam, A. A. Balkema.

Hays, T. R. 1975. Neolithic settlement of the Sahara as it relates to the Nile Valley. *In* Wendorf, F. and Marks, A. E. eds. *Problems in Prehistory: North Africa and the Levant*: 193–204. Dallas, Southern Methodist University Press.

Helgren, D. M. 1977. Geological context of the Vaal River faunas. *S. Afr. J. Sci.* 73:303–307.

Hendey, Q. B. 1974a. Faunal dating of the Late Cenozoic of southern Africa, with special reference to the Carnivora. *Quatern. Res.* (N.Y.) 4:149–161.

———. 1974b. The Late Cenozoic Carnivora of the South-Western Cape Province. *Ann. S. Afr. Mus.* 63:1–69.

———. 1974c. New fossil carnivores from the Swartkrans Australopithecine site (Mammalia: Carnivora). *Ann. Transv. Mus.* 29:27–49.

———. 1978. Late Tertiary Hyaenidae from Langebaanweg, South Africa, and their relevance to the phylogeny of the family. *Ann. S. Afr. Mus.* 76:265–297.

Howell, F. C. and Clark, J. D. 1963. Acheulian hunter-gatherers of Sub-Saharan Africa. In Howell, F. C. and Bourlière, F. eds. *African Ecology and Human Evolution*: 458–533. Chicago, Aldine Publishing Co.

Howell, F. C., Cole, G. H., Kleindienst, M. R., Szabo, B. J. and Oakley, K. P. 1972. Iringa Highlands, Southern Highlands Province, Tanganyika. In Mortelmans, G. and Nenquin, J. eds. *Actes du IVe Congrès Panafricain de Préhistoire et de l'Etude du Quaternaire*: 43–80.

Howell, F. C., Cole, G. H. Kleindienst, M. R., Szabo, B. J. and Oakley, K. P.1972a. Uranium series dating of bone from the Isimila prehistoric site, Tanzania. *Nature,* 237:51–52.

Inskeep, R. R. 1962. The age of the Kondoa rock paintings in the light of recent excavations at Kisese II Rock Shelter. *Ann. Mus. Roy. Afr. Cent.* 40:249–256.

Isaac, G. Ll. 1975. Stratigraphy and cultural patterns in East Africa during the middle ranges of Pleistocene time. *In* Butzer, K. W. and Isaac, G. Ll. eds. *After the Australopithecines*: 495–542. The Hague, Mouton.

————. 1977. *Olorgesailie: Archaeological Studies of a Middle Pleistocene Lake Basin in Kenya.* Chicago, University of Chicago Press.

Isaac, G. Ll. and Curtis, G. H. 1974. The age of early Acheulian industries in East Africa—new evidence from the Peninj Group, Tanzania. *Nature, Lond.* 249: 624–627.

Jaeger, J.-J. 1975. The mammalian faunas and hominid fossils of the Middle Pleistocene of the Maghreb. *In* Butzer, K. W. and Isaac, G. Ll. eds. *After the Australopithecines*: 399–419. The Hague, Mouton.

————. 1976. Les Rongeurs (Mammalia, Rodentia) du Pléistocene Inférieur d'Olduvai Bed I (Tanzanie). 1ère Partie: les Muridés. *In* Savage, R. J. G. and Coryndon, S. C. eds. *Fossil Vertebrates of Africa* 4:57–120. London, Academic Press.

Klein, R. G. 1973. *Ice-Age Hunters of the Ukraine.* Chicago, University of Chicago Press.

————. 1975. The relevance of Old World archaeology to the first entry of man into the New World. *Quatern. Res.* (N. Y.) 5:391–394.

————. 1976. A preliminary report on the "Middle Stone Age" open-air site of Duinefontein 2 (Melkbosstrand, south-western Cape Province, South Africa). *S. Afr. Archaeol. Bull.* 31:12–20.

————. 1977. The ecology of early man in southern Africa. *Science* 197:115–126.

————. 1978. The fauna and overall interpretation of the "Cutting 10" Acheulean site at Elandsfontein (Hopefield), southwestern Cape Province, South Africa. *Quatern. Res.* (N. Y.) 5:391–394.

————. 1979. Stone Age exploitation of animals in southern Africa. *Amer. Scient.* 67:151–160.

————. 1980. Environmental and ecological implications of large mammals from Upper Pleistocene and Holocene sites in southern Africa. *Ann. S. Afr. Mus.* 81:223–283.

————. 1983. The prehistory of Stone Age herders in South Africa. *In* Clark, J. D. and Brandt, S. eds. *Causes and Consequences of Food Production in Africa.* Berkeley, University of California Press.

Lavocat, R. 1978. Rodentia and Lagomorpha. *In* Maglio, V. J. and Cooke, H. B. S. eds. *Evolution of African Mammals*: 69–89. Cambridge (Mass.), Harvard University Press.

Leakey, L. S. B. 1931. *The Stone Age Cultures of Kenya Colony.* Cambridge, Cambridge University Press.

Leakey, M. D. 1977. The archaeology of the early hominids. *In* Wilson, T. H. ed. *A Survey of the Prehistory of Eastern Africa*: 61–79. Nairobi, VIII Pan-African Congress of Prehistory and Quaternary Studies.

————. 1979. *Olduvai Gorge: My Search for Early Man.* London, Collins.

Leakey, M. D., Hay, R. L., Thurber, D. L., Protsch, R. and Berger, R. 1972. Stratigraphy, archaeology, and age of the Ndutu and Naisiusiu Beds, Olduvai Gorge, Tanzania. *World Archaeol.* 3:328–341.

McBurney, C. B. M. 1967. *The Haua Fteah (Cyrenaica) and the Stone Age of the Southeast Mediterranean.* Cambridge, Cambridge University Press.

Maglio, V. J. 1978. Patterns of faunal evolution. *In* Maglio, V. J. and Cooke, H. B. S. eds. *Evolution of African Mammals*: 603–619. Cambridge (Mass.), Harvard University Press.

Marks, A. E. 1975. The current status of Upper Paleolithic studies from the Maghreb to the northern Levant. *In* Wendorf, F. and Marks, A. E. eds. *Problems in Prehistory; North Africa and the Levant*: 439–457. Dallas, Southern Methodist University Press.

Martin, P. S. 1967. Prehistoric Overkill. In Martin, P. S. and Wright, H. E. eds. *Pleistocene Extinctions: the Search for a Cause*: 75–120. New Haven, Yale University Press.

Mehlman, M. J. 1977. Excavations at Nasera Rock, Tanzania. *Azania* 12:111–118.

————. 1979. Mumba-Höhle revisited: the relevance of a forgotten excavation to some current issues in East African prehistory. *World Archaeol.* 11:80–94.

Meyer, G. E. 1978. Huracoidea. *In* Maglio, V. J. and Cooke, H. B. S. eds. *Evolution of*

African Mammals: 284–314. Cambridge (Mass.), Harvard University Press.

Miller, S. F. 1971. The age of the Nachikufan industries in Zambia. *S. Afr. Archaeol. Bull.* 26:143–146.

———. 1979. Lukenya Hill, GvJm 46, Excavation Report. *Nyame Akuma* 14:31–34.

Nilsson, E. 1964. Pluvial lakes and glaciers in East Africa. *Stockholm Contrib. Geol.* 11:2.

Onyango-Abuje, J. C. 1977. Crescent Island: a preliminary report on excavations at an East African Neolithic Site. *Azania* 12:147–159.

Phillipson, D. W. 1977. *The Later Prehistory of Eastern and Southern Africa.* New York, Holmes and Meier.

Rightmire, G. P. 1978. Florisbad and human population succession in Southern Africa. *Amer. J. Phys. Anthrop.* 48:475–486.

———. 1979. Implications of Border Cave skeletal remains for later Pleistocene human evolution. *Curr. Anthrop.* 20:23–35.

———. 1981. Later Pleistocene hominids of Eastern and Southern Africa. *Anthropologie* 19:15–26.

Robbins, L. H. 1974. A Late Stone Age fishing settlement in the Lake Rudolf Basin, Kenya. *Publ. Mus. Mich. State Univ., Anthrop. Ser.* 1:153–216.

Roche, J. and Texier, J.-P. 1976. Découverte de restes humains dans un niveau atérien supérieur de la grotte des Contrabandiers, á Témara (Maroc). *C. R. Acad. Sc. Paris* 285:45–47.

Savage, R. J. G. 1978. Carnivora. *In* Maglio, V. J. and Cooke, H. B. S. eds. *Evolution of African Mammals:* 249–267. Cambridge (Mass.), Harvard University Press.

Shackley, M. 1980. An Acheulean industry with *Elephas recki* fauna from Namib IV, South West Africa (Namibia). *Nature,* 284:340–341.

Simons, E. L. and Delson, E. 1978. Cercopithecidae and Parapithecidae. *In* Maglio, V. J. and Cooke, H. B. S. eds. *Evolution of African Mammals:* 100–119. Cambridge (Mass.), Harvard University Press.

Smith, A. B. 1980. The Neolithic tradition in the Sahara. *In* Williams, M. A. J. and Faure, H. eds. *The Sahara and the Nile*: 451–465. Rotterdam, Balkema.

Szabo, B. J. and Butzer, K. W. 1979. Uranium-series dating of lacustrine limestones from pan deposits with Final Acheulian assemblage at Rooidam, Kimberley District, South Africa. *Quatern. Res.* (N. Y.) 11:257–260.

Tchernov, E. 1976. Crocodilians from the Late Cenozoic of the Rudolf Basin. *In* Coppens, Y., Howell, F. C., Isaac, G. Ll. and Leakey, R. E. F. eds. *Earliest Man and Environments in the Lake Rudolf Basin*: 370–378. Chicago, University of Chicago Press.

Thomas, H. 1977. Géologie et paléontologie du gisement acheuléen de l'erg Tihodaine. *Mémoires du Centre de Recherches Anthropologiques, Préhistoriques et Ethnographiques* 27:1–22.

Van Noten, F. 1971. Excavations at Munyama Cave. *Antiquity* 45:56–58.

———. 1977. Excavations at Matupi Cave. *Antiquity* 51:35–40.

Vaufrey, R. 1955. *Préhistoire de l'Afrique* (Vol. 1: Le Maghreb). Paris, Masson.

Volman, T. P. 1981. *The Middle Stone Age in the Southern Cape.* Ph.D. Dissertation. University of Chicago.

Vrba, E. S. 1978. Problematical alcelaphine fossils from the Kromdraai Faunal Site (Mammalia: Bovidae). *Ann. Transv. Mus.* 31:21–28.

Wendorf, F. and Members of the Combined Prehistoric Expedition. 1977. Late Pleistocene and recent climatic changes in the Egyptian Sahara. *The Geographical Journal* 143:211–234.

Wendorf, F. and Schild, R. 1976. *Prehistory of the Nile Valley.* New York, Academic Press.

———. 1980. *Prehistory of the Eastern Sahara.* New York, Academic Press.

Wendorf, F., Schild, R. and Haas, H. 1979. A new radiocarbon chronology for prehistoric sites in Nubia. *J. Field Archaeol.* 6:219–223.

Wendorf, F., Laury, R. L., Albritton, C. C., Schild, R., Haynes, C. V., Damon, P. E., Shafqullah, M. and Scarborough, R. 1975. Dates for the Middle Stone Age of East Africa. *Science* 187:740–742.

Westphal, M., Chavaillon, J. and Jaeger, J.-J. 1979. Magnétostratigraphie des depôts pléistocenes de Melka-Kunturé (Ethiopie): premiéres résultats. *Bull. Soc. Géol. France* 21:235–239.

White, T. D. and Harris, J. M. 1977. Suid evolution and correlation of African hominid localities. *Science* 198:13–21.

26

Extinctions in Madagascar
The Loss of the Subfossil Fauna

ROBERT E. DEWAR

THE EXTINCTIONS OF LARGE MAMMALS AND BIRDS in Madagascar during the Holocene were severe, comparable in magnitude to the other Quaternary extinctions of North America, Australia, and New Zealand. The fauna of Madagascar today is a remnant of the magnificent bestiary that existed two thousand years ago. Similarities among patterns of faunal extinctions in a number of regions of the world were first highlighted by P. S. Martin (1966). His "overkill" hypothesis remains controversial in two ways: (1) Were these waves of extinction largely the result of human activity? (2) Can each of these waves be explained by the same kind of human activity? Most agree that much of the faunal impoverishment of Madagascar is the result of human activity. Disagreement focuses on the initial causes of the late Holocene crisis and the specific human activity *most* responsible for the extinctions. In this chapter I will outline the evidence concerning, on the one hand, the geographical and chronological patterns of extinction and, on the other, the chronology and nature of the human occupation of the island. I examine competing hypotheses about the relationships between these phenomena and, finally, present a revised model of the human role in the Malagasy extinctions.

First, however, a series of *caveats* are due: (1) No Pleistocene, indeed no Cenozoic, terrestrial fossil deposits are known in Madagascar. Our knowledge of the island's paleozoology is blank from the Mesozoic until the appearance of the subfossil sites described here. As a result, we reconstruct the Pleistocene fauna only from extinct forms from the subfossil deposits and from living specimens. The reconstruction is based upon a much narrower time perspective than is available in other areas of the world. (2) Deposits have not been described in enough detail to allow taphonomic studies and inferences. (3) Geographical distribution of the subfossil sites is nonuniform (see fig. 26.1); the paleofaunas of some regions—in particular the tropical forests of the east and northeast—are not represented. (4) As in all reconstructions, the collections and analysis of the faunal material are probably biased against the smaller forms. In Madagascar, unlike other continents, there is a bias in favor of the primates (Walker 1967a:454). Specifically, a comparison of primate forms recovered from the subfossil sites with surviving genera (see Table 26.1) reveals only two modern genera as yet unrepresented in subfossil collections—*Microcebus* and *Phaner*. Both are small; *Microcebus* is widely distributed. In contrast, fossils of only two of seven living genera of carnivores have been recovered, including a single extinct species. Of the seven en-

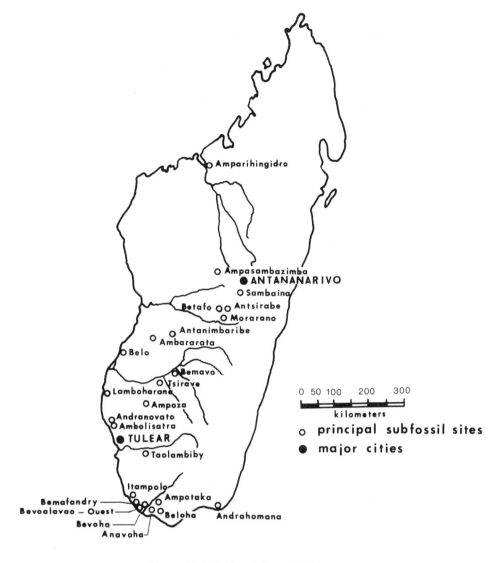

Figure 26.1. Subfossil sites of Madagascar.

demic genera of rodents, only the giant rat of Madagascar (*Hypogeomys antimena*) has been found in a subfossil site. Only one subfossil from the eight or nine endemic genera of insectivores has been found and, again, it is one of the larger forms (Tattersall 1973a; Walker 1967a; Battistini and Vérin 1967, 1972; Battistini 1976; Jungers 1977; Lamberton 1934).

The Subfossils

Lemuroids

The primate fauna of the Malagasy Holocene has been assigned to seventeen genera, of which seven completely disappeared (Tattersall and Schwartz 1974). In addition, two surviving genera are known to have lost a species—*Varecia insignis* and

Table 26.1. Species Reported From Subfossil Localities in Seven Regions of Madagascar

NW = Northwest, PL = Plateau, WC = West Coast, WI = West Interior, SWI = Southwest Interior, SWC = Southwest Coast, SEC = Southeast Coast; * = extinct species; E = reports of extinct species; XP = reports of fossils of extant species in area of modern distribution; XN = reports of fossils of extant species outside areas of modern distribution; −P = no fossils reported, but this area falls within the modern distribution of genus.

Species	NW	PL	WC	WI	SWI	SWC	SEC
MAMMALIA							
Primata							
Cheirogaleinae							
Cheirogaleus sp.	−P	XN			XP		XP
Lemurinae							
Lemur sp.	XP	XP	−P	−P	XP	−P	−P
*Varecia insignis	E	E	E	E	E	E	E
Lepilemur spp.	−P	XP	−P	−P	XP	−P	−P
Hapalemur sp.	−P	XP					
Megaladapinae							
**Megaladapis madagascariensis*			E	E		E	E
**Megaladapis grandidieri*		E					
**Megaladapis edwardsi*			E	E		E	E
Indrinae							
Indri indri		XN					
Propithecus sp.	−P	XN	−P	XP	XP	XP	−P
Avahi laniger	−P	XN					−P
**Mesopropithecus globiceps*			E				
**M. pithecoides*		E					
Archaeolemurinae							
**Archaeolemur majori*	?		E	E	E	E	E
**A. edwardsi*	?	E	?				
**Hadropithecus stenognathus*		E		E		E	E
Paleopropithecinae							
**Paleopropithecus ingens*	E	E	E	E	E	E	
**Archaeoindris fontoyonti*		E					
Daubentonidae							
**Daubentonia robusta*			E	E			

SOURCES: Tattersall (1973a); Walker (1967a); Battistini and Vérin (1967, 1972); Battistini (1976); Jungers (1977); Lamberton (1934).

Daubentonia robusta. Each was the largest form in its genus and was recovered from localities outside the geographical range of the living species.

The extinct lemuroids were all quite large, probably diurnal and probably capable of more locomotor patterns than living lemurs (Walker 1967b). The largest, *Megaladapis edwardsi,* had a skull length of about 30 cm. Estimates of this animal's body weight have varied greatly. One estimate suggests that an adult male weighed between 50 and 100 kg (Jungers 1978). Members of the smallest extinct genus, *Mesopropithecus,* may have overlapped the size range of the largest living species, *Indri indri,* but all other extinct forms were larger than the living lemuroids. Walker used the ratio of orbit diameter to skull length to distinguish diurnal, crepuscular, and nocturnal forms. By this measure, all the extinct species were most likely diurnal (except perhaps *Daubentonia robusta,* for which there is no complete skull). Among the subfossil forms are animals displaying locomotor patterns currently unknown among prosimians, including terrestrial quadrupedalism (*Archaeolemur, Hadropithecus*), arm-swinging (*Paleopropithecus*),

Table 26.1. Species Reported From Subfossil Localities in Seven Regions of Madagascar

NW = Northwest, PL = Plateau, WC = West Coast, WI = West Interior, SWI = Southwest Interior, SWC = Southwest Coast, SEC = Southeast Coast; * = extinct species; E = reports of extinct species; XP = reports of fossils of extant species in area of modern distribution; XN = reports of fossils of extant species outside areas of modern distribution; −P = no fossils reported, but this area falls within the modern distribution of genus.

(continued)

Species	NW	PL	WC	WI	SWI	SWC	SEC
Carnivora							
Viverridae							
Cryptoprocta ferox	−P	XP	−P	−P	−P	−P	XP
*C. spelea		E	E			E	E
Galidictis striata		XP					
Tubulidentata							
*Plesiorycteropus							
madagascariensis		E	E		E	E	
Rodentia							
Hypogeomys sp.		XN	−P			XP	
Insectivora							
Tenrec sp.	−P	XP	−P	−P	XP	XP	XP
Artiodactyla							
Suidae							
Potamochoerus porcus	−P		XP	−P	−P	−?	−?
Sus. sp.	−P	−P	XP	−P	−P	−P	−P
Hippopotamidae							
*Hippopotamus lemerlei		E	E	E	E	E	
Bovidae							
Bos indicus	−P	−P	XP	−P	XP	−P	−P
Capra hircus	−P	−P	−P	−P	−P	−P	XP
AVES							
Aepyorniformes							
Aepyornis spp.		E	E	E	E	E	E
Mullerornis spp.		E	E	E	E	E	
REPTILIA							
Crocodilia							
Crocodilus niloticus	XN	XN	−?	X?	XP		XN
Testudinata							
Geochelone grandidieri } Geochelone abrupta	E	E	E	E	E	E	E

and a ponderous vertical climbing (*Megaladapis*) which has been likened to the locomotor pattern of the koala bear (*Phascolarctos cinereus*) (Walker 1967a, 1967b; Jungers 1977, 1978).

In dental and cranial morphology, *Hadropithecus* was convergent upon the highly terrestrial gelada baboon, *Theropithecus* (Jolly 1970 and Tattersall 1973a). They suggest a broadly similar niche for the two forms. Tattersall has pointed out that *Archaeolemur* may be regarded as broadly convergent upon the *Papio* baboons, and has tentatively suggested that *Archaeolemur* may have preferred a more wooded habitat and frugivorous diet than *Hadropithecus* (1973a:94–97). The archeolemurines apparently exploited ground-level resources more intensively than any living lemuroid (Godfrey 1977, Tattersall 1982). Though *Lemur catta* is sometimes characterized as terrestrial because it moves along the ground, it feeds at all levels of the forest (Sussman 1974).

Once regarded as aquatic (Sera 1950), the largest extinct primate, *Megaladapis*, was apparently a highly arboreal browser, perhaps with a flexible snout (Jungers 1977, Tattersall 1973b, e.g. Mahé 1972).

Walker (1967b) argued that the complete extinction of the terrestrial lemuroids and all other large, diurnal forms, leaving untouched the nocturnal and smaller diurnal, arboreal species, is readily attributable to the coming of human hunters. Nevertheless, the limitations of the fossil record require some caution in accepting Walker's conclusions. Since there is a bias against the discovery of small forms in the subfossil sites, extinctions of as yet unknown small primates may be discovered that would be discordant with the model of high vulnerability to human impact. In addition, the sampling bias in favor of large fossils diminishes the likelihood of finding extinct nocturnal species, since all nocturnal primates are small (Clutton-Brock and Harvey 1977).

With the exception of *Megaladapis,* all the extinct lemur subfossils have been described as ecological vicars of the Anthropoidea. The broad similarities between the anthropoid and lemuroid radiations have often been remarked upon. However, one feature of the pre-extinction fauna is quite remarkable: its diversity, especially of large forms. Table 26.2 lists the primates recovered from Ampasambazimba, a single site in which all deposits are clearly Holocene in date. At the generic level, this is the most diverse community of sympatric primates ever reported. Even more remarkable is the discovery of seven extinct species, all apparently ecological vicars of baboons and hominoids, dwarfing any known similar sympatric assemblage. The original vegetation of the central plateau has usually been envisaged as homogeneous forest, though there may be grounds for doubt. If it was, we may infer that each species exploited a relatively narrow niche and that the population density of each species was low. This is particularly likely given their large body size. Furthermore, studies relating body weight to life history parameters have shown that within lineages of related mammals, increase in body size is strongly correlated with increased lifespan and decreased rates of reproduction (Sacher 1974, Western 1979). The combination of likely demographic and ecological characteristics of the subfossil lemuroids would have rendered them particularly susceptible to extinction, whether by predation or habitat destruction.

Ratites

Next to the lemuroids, the ratites (large, flightless birds) have attracted the most attention among extinct forms. Commonly known as elephant birds, they are usually assigned to two genera, *Aepyornis* and *Mullerornis,* and to between six and more than a dozen species. The largest, *Aepyornis maximus,* resembled a massive robust ostrich, attaining a height of almost three meters; the smallest, *Mulleronis betsilei,* was a little less than half this height. By analogy with surviving ratites, these flightless birds were important terrestrial grazers and browsers. This presumption is strengthened by the absence of endemic medium- or large-sized mammals, save the pygmy hippo. Cracraft

Table 26.2. Primate Remains From Ampasambazimba

Archaeoindris fontoyonti	Lemur sp.
Archaeolemur edwardsi	*Varecia insignis*
Avahi laniger	Lepilemur mustelinus
Cheirogaleus sp.	*Megaladapis grandidieri*
Hadropithecus stenognathus	*Mesopropithecus pithecoides*
Hapalemur griseus	*Paleopropithecus ingens*
Indri indri	Propithecus sp.

SOURCE: Modified from Walker (1967a).
*extinct forms

interprets the robusticity of the Aepyorniformes as evidence of forest habitat (1974:516). However, it has been suggested that, among ratites, robusticity is more a function of absence of predation and reduced need for rapid movement (Amadon 1947).

The debris of ratite eggshells is a common sight along the beaches of the south and southwest coast. Some beaches seem nearly paved with broken eggshells. They have also been reported along northern coasts (Battistini 1965a, 1965b). Most authors have interpreted these beach deposits as remains of colonial nesting sites. While this may be correct for some locations, I suspect that the eggshell debris represents the remains of human meals and broken eggshell vessels transported from elsewhere.

Ratite remains are very common in the subfossil sites of the plateau, south and southwest. Together with the *Geochelone* [*Testudo*] spp., they appear to have been the dominant terrestrial herbivores, with especially high density in the south and southwest (Lamberton 1934).

These ponderous birds would have been easy prey to arriving humans, and it is not hard to imagine them sharing the fate of the moas of New Zealand and the dodo and solitaire of the Mascarenes. Furthermore, their huge eggs, up to eleven liters in capacity, would have invited use both as food and containers.

Nonprimate Mammals

Three other mammals became extinct in Madagascar during the Holocene: a pygmy hippopotamus (*Hippopotamus lemerlei*), a large viverrid (*Cryptoprocta spelea*), and an endemic aardvark (*Plesiorycteropus madagascariensis*). The aardvark is known from several deposits on the central plateau and in the west and southwest (Patterson 1975). *H. lemerlei* had a wide distribution and was probably abundant, judging from the frequency with which its bones appear in subfossil deposits. The pygmy hippo was about two-thirds the length and height of the common African species, *H. amphibius*. Only one other wild artiodactyl is known from Madagascar, the bush pig (*Potomachoerus porcus*). Conspecific with the African population, it is often regarded as a human introduction and has not been recovered from any locations that have not also yielded the bones of domesticated animals.

Cryptoprocta spelea was substantially larger than *C. ferox*, the largest living carnivore on Madagascar. Savage (1978) has suggested that they were conspecific. With a head and body length of about 70 cm, *C. ferox* was a powerful carnivore resembling a short-legged puma; at one time it was classified among the felids (Walker 1975:1262). The close match between the distribution of the subfossils of *C. spelea* and *Megaladapis* led Lamberton to suggest that they were hunter and hunted (1939:155).

Reptiles

The only reptiles known to have died out during the Holocene are the giant land tortoises, *Geochelone* (formerly *Testudo*) *grandidieri* and *G. abrupta* (Auffenberg 1974). Three smaller species of this genus are still found in Madagascar, and one, *G. yniphora*, is on the verge of extinction (Juvik et al. 1981). The carapace of *G. abrupta* attained 80 cm in length and that of *G. grandidieri* could exceed 1.2 m (Blanc 1972). In the past, reference has been made to an extinct crocodile larger than the living species on the island, a taxonomic distinction currently not in vogue (Blanc 1972).

The extinct giant tortoises had a wide distribution (see Table 26.1). At many sites their bones and carapace fragments are among the most common fossils. Presumably, like the giant tortoises of the Galapagos, they were extremely long-lived, slow to mature and reproduce, and certainly very slow moving. Before people arrived, the adults must have had no predators. The Galapagos tortoises seem to be unspecialized herbivores (Van Denburgh 1914), and it is likely *Geochelone* populations were important consumers of ground-level vegetation.

The Subfossil Sites

Types of Sites

The geology of the subfossil sites has been discussed extensively (Walker 1967a:428–460; Tattersall 1973a:10–16; Mahé and Sourdat 1972; Battistini 1976; Battistini and Vérin 1967, 1972; Raison and Vérin 1967). Battistini and Vérin divide them into three categories: a) natural accumulations, b) human habitations with associated subfossil material and c) mixed sites, where a natural accumulation also includes evidence of the contemporaneity of human activity and subfossil species.

Natural accumulations. Natural accumulations are generally small in area, though often rich in subfossils. Early investigators were more concerned with the subfossils than with their context, and we are left with little reliable information about many of the sites or, in some cases, their precise location. A published geological report exists only for Ampasambazimba. Geologists have briefly desribed Itampolo, Andrahomana, Amparihingidro, and Taolambiby; none appears to be very old (Walker 1967a:459–460). Mahé (1965) divided sites into three geological classes: marshes in volcanic regions, marshes in coastal regions, and caves. Walker (1967a:459) questioned the utility of this classification, pointing out that all sites, caves excluded, are located today in areas with active drainage systems, though frequently water flow has diminished. In his view the interior sites were all formed and exposed as a result of local changes, and he found no evidence of island-wide events, such as a major drop in rainfall, which would allow a stratigraphic ordering of the sites. Mahé and Sourdat (1972) have argued that the coastal marsh sites provide evidence of just such a change, but Battistini (1976) suggests that a lowering in sea level, not climatic modification, caused the drying up of coastal marshes in the past three thousand years.

Early investigators attributed subfossils to sites, but almost never recorded stratigraphic, let alone taphonomic, information (Tattersall 1973a:10), so that these death assemblages can only be used to describe generally the local communities. This has created particular difficulties with reconstructing the primeval plant community of the central plateau, which has usually been regarded as having had a continuous forest cover. The subfossils recovered from plateau sites have been seen as one sympatric community. Evidence from two sites, however, seems to point not to homogeneity but to a mosaic of plant formations.

Ampasambazimba, located about 100 km west of Antananarivo, is the marshy remains of a lake formed when a lava flow dammed a stream. Subsequent downcutting breached the dam and drained the lake (Walker 1967a:429–438). The subfossil beds are composed of material washed into the lake from its drainage area. The collections here are rich in species and generically diverse: fourteen genera of primates are associated with the remains of carnivores, an aardvark, insectivores, the crocodile, the pygmy hippo, many ratites, and tortoises. Peat samples from around the subfossil deposits yielded fruits and seeds of *Weinmannia rutenbergeii, W. bojeriana, Colea telfairea, Strychnos* sp., and *Ravensera* sp., suggesting that the local flora resembled that of today's western plateau region (de la Bathie 1927 cited in Tattersall 1973a). The fauna, in contrast, is more characteristic of rain forest or humid seasonal forest (Tattersall 1973a). The second site, also on the central plateau, is Marotampona, west of Antsirabe and about 90 km south of Antananarivo. The floral remains here are unequivocally from a community like the rain forest that today lies at least 70 km east of Antsirabe (de la Bathie 1914 cited in Battistini and Vérin 1967).

Human habitations. Few sites of early human habitation have been excavated, and the only subfossils recovered from them are ratite eggshell debris and, at two southern sites, bones of animals no longer found in the surrounding region: *Crocodilus niloticus* at Beropitikia (Emphoux 1978) and *Hapalemur* sp. at Rezoky (Vérin 1971).

Mixed sites. Mixed sites are difficult to interpert without better understanding their depositional history. Table 26.3 summarizes the evidence for human activity, Table 26.4 the available radiocarbon chronology of these sites.

The identification of mixed sites by the presence of *Aepyornis* eggshell debris (e.g. Battistini 1976) is unreliable, since infertile eggs of the already extinct birds might have been used as containers. The identification of Lamboharana and Ambolisatra as mixed sites is also uncertain since there was apparently no stratigraphic control of the excavations. Three sites remain as important candidates: Ampasambazimba, Taolambiby, and the cave of Andrahomana. Andrahomana, identified as an archaeological site by Walker and Martin (Walker 1967a), is undated and without stratigraphic control. However, the subfossils were found in apparent association with charcoal deposits that may well be human in origin. Walker (1967a) found pottery sherds in the upper strata of Taolambiby, but this is a secondary deposition of uncertain date. The radiocarbon determination is from a lower stratum. Ampasambazimba is also a secondary deposit, with a radiocarbon date that cannot be placed stratigraphically. However, the lateness of the date—around 900 A.D.—and the nature of the artifacts—a *Mullerornis* tibio-tarsus worked into an axhead among them—suggest that this is a probable association of human activity and the now extinct animals.

Dating

Radiocarbon dates of the subfossil and archaeological (human habitation) sites (see Table 26.4) are still limited in number; nonetheless, two interesting points emerge from those available. First, *Megaladapis,* the hippos, and the ratites overlap the radiocarbon-dated epoch of human habitation. Second, the dates present no clear geographical pattern, suggesting that the processes responsible for the extinctions may have had effect largely synchronously over the entire island.

The evidence from the dating indicates that the extinctions of the subfossils were not complete until about 900 B.P., and that they did not occur in any particular order. In the seventeenth century Flacourt recorded reports of an ostrichlike bird in the south

Table 26.3. Some Reported Mixed Sites From Madagascar

Site	Evidence of Human Activity
Ampasambazimba	Worked wooden sticks, *Mullerornis* bone worked into axe-head, earthenware jar, all found mixed in deposits with *Aepyornis* bones and lemuroid remains; a pierced *Archaeolemur* skull
Lamboharana	Many pierced *Daubentonia* teeth associated with *Aepyornis, Megaladapis, Paleopropithecus*; also *Bos* bones and gunflints (or small adzes*)
Taolambiby	Pierced *Geochelone* carapace, potsherds in upper strata; *Bos* bones
Beloha-Anavoha	Pierced *Archaeolemur* skull
Andrahomana	*Archaeolemur* skull fractured by axe-like instrument; burned femur of *A. majori*, ashes, *Capra* bones
Ambolisatra	*Sus* sp. identified with subfossils

*Although the original excavator identified these lithics as identical to gun flints, Battistini and Vérin (1972:322) have suggested they may have been "small adzes of jasper or chalcedony of Indonesian type."

Table 26.4. Radiocarbon Dates Reported for Madagascar

Location	Date (yr B.P.)	Material	Lab. No. Source Note in ( )
		Subfossil Deposits	
Amparihingidro	2850 ± 200	wood	Gif–? (1)
Ampasambazimba	1035 ± 50	*Megaladapis* bone	Pta–739 (2)
Ampoza	1910 ± 120	bone	Gak–2309 (1)
Anavoha	1954 ± 110	wood	Gak–1654 (1)
Bemafandry	1980 ± 90	bone	Gak–1656 (1)
	2060 ± 150	carapace	Gak–1655 (1)
Behavoha	2160 ± 110	carapace	Gak–1658 (1)
Irodo			
(Tafiampatsa)	1150 ± 90	*Aepyornis* eggshell	Gak–? (1)
Itampolo	980 ± 200	hippo bone	Gak–1506 (1)
	2290 ± 90	wood	Gak–1652 (1)
Lamboharana	1220 ± 80	bone	Gak–2310 (1)
	2350 ± 120	bone	Gak–2307 (1)
Taolambiby	2290 ± 90	*Geochelone* bone	Gak–1651 (1)
50 km SW of			
Fort Dauphin	1000 ± 150	*Aepyornis* eggshell	UCLA–1893 (6)
Southern			
Madagascar	1970 ± 90	*Aepyornis* eggshell	? (7)
	2930 ± 85	*Aepyornis* eggshell	? (7)
	5210 ± 140	*Aepyornis* eggshell	? (7)
Anakao, south		Carbonate from	
of Tulear	2375 ± 100	*Aepyornis* eggshell	QC–971 (8)
	2775 ± 95	*Aepyornis* eggshell	QC–970 (8)
	3960 ± 150	*Aepyornis* eggshell	QC–972 (8)
		Archaeological Sites	
Andranosoa	730 ± 90		Gif–4570 (3)
	920 ± 90		Gif–4571 (3)
Beropitikia	750 ± 90	charcoal	Gif–4496 (3)
Irodo	980 ± 100	charcoal	Gak–350b (4)
(Antanimenabe)	1200 ± 40	charcoal	Gak–380 (4)
Irodo			
(Tafiantsirbeka)	1090 ± 90	gastropod shell	Gak–692 (4)
Sarodrano	210 ± 80	charcoal	Gak–927 (5)
	530 ± 80	charcoal	Gak–1057 (5)
	1460 ± 90	charcoal	Gak–928 (5)
Talaky	840 ± 80	charcoal	Gak–276 (4)

SOURCES: 1—Mahé and Sourdat (1972); 2—Tattersall (1973c); 3—Emphoux (1979); 4—Battistini and Vérin (1967); 5—Battistini and Vérin (1971); 6—Berger et al. (1975); 7—Marden (1967); 8—These dates, from the surface of coastal sand dunes, are reported here for the first time.

and an animal the size of a calf with a human face and monkeylike hands and feet, leading to speculation that some of the ratite and primate subfossils may have survived until 300 to 400 yr B.P. Similarly, there have been reports suggesting an even later survival of the hippo (Mahé 1972). However, few as the excavations are, the total absence of any subfossil material beyond ratite eggshell debris in archaeological sites makes it seem unlikely that any but extremely localized populations of the subfossils survived after 900 B.P.

Human Colonization of Madagascar

People colonized Madagascar late in human history. It is the fourth largest island in the world, yet there is no evidence that people arrived before the Christian era. Many scholars have suggested first arrival dates of 500 A.D. or later. By comparison, Australia was settled more than 35,000 years ago and even most prehistoric settlement in the Pacific is older. Three kinds of evidence have been used to estimate the date of human arrival on the island: comparative ethnology, historical linguistics, and the dates of archaeological sites based upon radiocarbon or the appearance of trade goods dated elsewhere along the western margins of the Indian Ocean.

The earliest date for an archaeological site is 1490±90 yr B.P. for Sarodrano, but the sample may have come from a disturbed deposit (Battistini and Vérin 1971). Before the site could be investigated further, it was destroyed by a tropical storm. Similar reservations attach to the dates from Irodo (Antanimenabe), which actually apply to a paleosol underlying a modern sand dune and are only tenuously associated with the nearby archaeological sites (Vérin 1975).

The three next earliest dates, from Irodo (Tafianatsirebeka), Andranosoa, and Talaky, establish an unquestionable pattern of settlements in the ninth or tenth centuries A.D. Each date is well associated with archaeological materials of excellent provenience. The Irodo site can be further dated by a fragment of pottery identified as Sassano-Islamic Ware of the ninth to eleventh centuries (Vérin 1975a, Wright 1979).

Together, these sites show that the tenth and eleventh centuries in Madagascar were already characterized by diverse settlement types and economic patterns. These three sites are geographically dispersed: Irodo is on the northeast coast, Talaky almost a thousand miles distant on the southern coast, and Andranosoa more than 100 km inland in the southeast. Differences in the economic base of each site probably reflect differences in geographical and ecological setting. Talaky appears to have been a coastal fishing village similar to the modern settlements of the Vezo, who are reef fishermen and sailors of outrigger canoes (Battistini et al. 1963). Irodo villagers gathered large amounts of shallow water shellfish, may have farmed the surrounding hills in this hot and humid area, and were already, perhaps, producing chloro-schistite vessels for trade with groups in the Comoros and on the East African coast (Vérin 1975a, Wright 1979). Faunal collections indicate that Andranosoa, located in a region of less than 600 mm annual rainfall today, was a village of cattle pastoralists (Emphoux 1978, 1979).

Wright (1979) has compared the ceramics from Irodo with contemporary material from Mayotte in the Comoros and tentatively suggests that both assemblages are derived from a common ancestral ceramic tradition of uncertain location. On the other hand, the ceramics from Andranosoa resemble those from Talaky, and both are quite distinct from the northern (Irodo) assemblage. None of the sites can be confidently characterized as pioneer settlements, and it is likely that earlier, as yet undiscovered, settlements exist.

Apart from some reports of isolated stone axes (Kellum-Ottino 1972, Bloch and Vérin 1966), evidence of any lithic industry is limited to sinkers and musket flints. This strongly suggests that Madagascar was first colonized by a fully Iron Age people, probably no earlier than the first century A.D., if East African prehistory is a useful guide (Phillipson 1977, van der Merwe 1980).

Vérin, Kottak, and Gorlin (1970) have used linguistic, specifically glottochronological, methods to date the arrival of the first speakers of Malagasy on the island. Their estimate, based upon the divergence of Malagasy dialects, is that initial settlement by speakers of a Proto-Malagasy occurred in the first century A.D. Other scholars have examined the divergence of Malagasy from its most closely related language, Maanyan of Borneo. Dahl's estimate (1951) was that the Malagasy and Maanyan speech communities had split by about 400 A.D. Using glottochronology, Dyen (1953) estimated that that divergence occurred 1,900 years ago. There is no evidence of a pre-Malagasy speech community in Madagascar, but the accuracy of these estimates of colonization is uncertain.

Malagasy culture is a composite of traits traceable to insular Southeast Asia and East Africa, as well as to other areas around the Indian Ocean; tracing traits to homelands, however, is not a reliable method of dating the initial colonization, nor a guide to the original culture of Madagascar. The initial migration of Proto-Malagasy speakers across the Indian Ocean probably accompanied long distance trade. The uninhabited island of Madagascar would not have been an attraction to traders, and the East African coast was likely the terminus of the trade route. It is likely that the initial settlers came from East Africa. The degree of fusion of African and Asian lifeways that took place on the continent before the settlement of Madagascar is controversial (Deschamps 1960, Vérin 1979, Southhall 1975), and archaeological resolution of the issue may be difficult.

Settlement of Madagascar probably occurred during the early part of the first millennium A.D., although we have no archaeological evidence (Vérin 1979, Southhall 1975). Archaeological data suggest that by the ninth century the island was widely settled, and that it had already begun to reflect the ecologic and economic diversity of modern Malagasy society and culture (Rakotoarisoa 1979).

Explanations of the Extinctions

Two distinctive views have emerged concerning the "ultimate cause" of the extinctions of the subfossil species: 1) they were decimated and pushed to the brink of extinction by a severe climatic change in the late Holocene; 2) the arrival of humans was the key to the destabilization and eventual degradation of a Malagasy paradise. There are various versions of the second view, according to the particular aspect of human activity considered most responsible.

Desiccation

I do not believe that strong evidence has been found of a major climatic change 2,000 to 3,000 years ago. In particular, it is difficult to imagine a single climatic event that would have had a uniform effect across the continentally diverse regions of Madagascar (fig. 26.2). The idea of a major desiccation seems to have developed as a result of interest in the sites of the island's arid southwest (Tattersall 1973a). Because these were mainly deposits in the beds of former lakes and marshes, as association was made between extinctions and the drying up of the lakes and marshes. However, the hydrologic modifications that resulted in the emptying of lakes, as at Ampasambazimba and Taolambiby, all apparently have local causes (Walker 1967a).

The most recent attempt to revive the desiccation hypothesis is that of Mahé and Sourdat (1972). They suggest that the radiocarbon dates fall into two clusters: 2300 to 2000 yr B.P. and 1000 to 900 yr B.P. (They did not have access to the dates for Ampasambazimba, Andranosoa, Beropitikia, or Sarodrano.) The second cluster contains dates from the upper strata of some sites and includes all the dates from archeological sites as well as some from mixed sites. The earlier cluster is from lower strata of southwestern coastal sites. The authors argue that the earlier dates indicate a change in hydrology of coastal areas resulting from a drop in sea level and climatic aridification. They regard these changes as the primary cause of the extinctions and suggest that the fauna began to change drastically one thousand years before the dated presence of people.

Responding to Mahé and Sourdat, Battistini (1976) argues that the aridification is almost certainly explained by secular changes in sea level rather than by a change in rainfall. Modern sea level was reached about 4000 B.P., attaining a maximum of about one meter above current height at about 3700 B.P. Since then sea level has been declining slowly, dropping to 40 to 50 cm above modern levels at about 800 B.P. and to just below them in the last two centuries or so (Battistini 1971a:11). Such a drop would

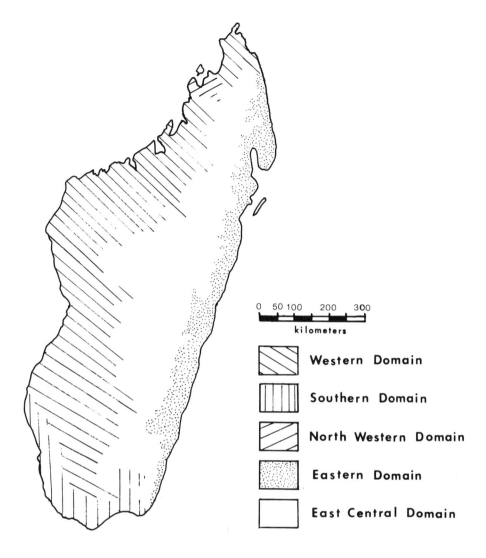

Figure 26.2. Modern vegetation regions in Madagascar. The eastern domain is wet and humid, the southern is dry, and the others are seasonally wet.

account for changes in the water table at the littoral sites of Itampolo, Lamboharana, Amboaboaka, and Ambolisatra, among others. Many fossils were found below the modern water table at Lamboharana, Itampolo, and Amparihingidro (White 1930, Walker 1967a, Tattersall 1973a:10).

To date we have no Holocene pollen diagrams from Madagascar. East African pollen studies indicate no major climatic changes at this time (Livingstone 1975). It should also be noted that the characteristic vegetation of the south and southwest, a xerophytic bush/forest dominated by Euphorbiaceae and Dideriaceae, is perhaps the least degraded of all the regional floras of Madagascar. Along with an extraordinarily high endemicity, more than 90 percent of species, this forest has a high biomass relative to areas of similar rainfall in Africa (Koechlin 1972). It is difficult to accommodate this floral evidence for long-term stability to a very recent, major climatic change.

Human Activity

Among those who view the arrival of *Homo sapiens* as the cause of the disappearance of the subfossil species, various opinions are held about how these animals were exterminated. In general, two means are favored: destruction of habitat *(indirect causation)* and hunting to extinction *(direct causation)*.

Indirect causation. Madagascar is frequently characterized as one of the most degraded areas of the world. Humbert's estimate (1927) that almost 90 percent of Madagascar had suffered from replacement of the indigenous vegetation by a degraded secondary vegetation is widely cited. Soil scientists attribute the extraordinary erosional losses of soil to the loss of plant cover. The hydrology of the island has been radically altered by changes in runoff rates, soil losses, and silting of rivers and streams. This again is attributed to the loss of the primeval vegetation. Our knowledge of much of the original plant cover is largely conjectural, however, and Koechlin (1972), for example, has raised doubts about the commonly held assumption that the now denuded and eroded plateau regions were once completely forested.

Currently, there are two explanations for the deforestation of the island. One assumes that the island flora was particularly vulnerable to fire, based on reports that fire destroyed large regions in the early colonial period (Morat 1973:191–193). According to this view, the areas degraded by humans were far larger than the actual areas of habitation and economic activity. The second explanation is that changes were the direct result of the economic and ecologic activities of agriculturalists (Krantz 1970).

Many observers have commented on the floral degradation caused by the annual burning of grasslands by pastoralists trying to encourage dry-season grass growth. Yet in a modern attempt to document forest loss in the west, by comparing aerial photographs taken in 1949 and 1970 Morat (1973:177–189) found no change in the forest-savanna boundaries over the twenty-year period. While his data provide no evidence that modern savanna plant communities are expanding, they strongly suggest that these communities are maintained by the combined action of grazing and burning. Trying to account for the disappearance of some forests while others, at least today, seem fire-resistant, Morat inferred that the original plant formations of the modern savannas were different from any modern forests, much more fragile and vulnerable to fire. The arrival of fire with humans, he argued, spelled the end to those formations. Morat's explanation is flawed. Natural plant communities vary greatly in the frequency and effects of fire (Vogel 1977), but no natural ecosystem is wholly free of fire.

According to Krantz's model, the extinctions were the result of agricultural habitat transformation together with adventitious hunting. Yet, the island is very large (about 600,000 sq km) and until very recently population density was low with an estimated two and a half million residents in 1900. It seems unlikely that such a sparse population could have transformed the island so completely and so rapidly as Krantz suggests.

Direct causation. Walker's (1976b) reconstruction of the locomotor and activity patterns of the extinct lemuroids led him to believe that the pattern of extinction was best accounted for by the appearance of human hunters. Archaeological findings provide limited support. For example, there is an *Archaeolemur* skull that apparently received a fatal blow from an "axe-like instrument" (1967b:428). At Lamboharana, a large number of fossil *Daubentonia* teeth had been drilled, as if for stringing on a necklace, and Ampasambazimba yielded an ax made from a *Mullerornis* tibio-tarsus (Battistini and Vérin 1967). At Rezoky and Asambalahy, sites of the southwest interior dating to the fourteenth and sixteenth centuries A.D., culinary debris included bones of tenrecs, *Cryptoprocta ferox,* and living lemurs (Battistini 1971b, Vérin 1971).

Some Malagasy groups rely on hunting today. For example, along the southwest coast the Vezo often supplement their fish diet with the endangered species *Geochelone radiata* and tenrecs. To the north of Tulear, the Mikea live exclusively by hunting and foraging, relying on small mammals and reptiles (Fanony n.d., Battistini 1964).

Work on the status of lemur populations suggests that these practices may have rapid effects. For example, *Indri indri,* the largest surviving lemuroid, is now limited to a relatively small section of the east coast rainforest, where its persistence is probably due in part to a local *fady,* or taboo, against killing it (Richard and Sussman 1975). In areas where there has been an influx of migrant workers who do not share in the localized sytems of *fady,* however, these animals are rare or nonexistent. This may be directly attributed to hunting (Randrianasolo personal communication). Local taboo also serves to protect certain populations of land tortoises (Juvik et al. 1981).

Problems with the Models

Models that rely solely on hunting or habitat destruction are insufficient, even though the extinction of certain forms may, indeed, have been due to one or another of these causes. The hippo, perhaps, may have lost much habitat because of hydrologic changes and arboreal forms like *Megaladapis grandidieri, Archaeoindris fontoynonti,* and *Mesopropithecus pithecoides,* known only from now treeless sites on the plateau, may also have been victims of habitat destruction. On the other hand, this explanation does not fit the ratites, giant tortoises, or the archaeolemurines. The fossil distributions of these forms extend over many vegetation zones, including some currently preserved in fairly large patches.

Although no subfossil localities are known in the modern rainforest, the absence of lemuroids larger than *Indri* in this region further suggests the inadequacy of the habitat destruction model. Modern large primates, ourselves excepted, are confined to rain-forests, and it would be surprising if the largest Malagasy primates were not once also to be found (if not exclusively) in the rainforest. Though rapidly decreasing, substantial areas of rainforest remain in Madagascar, so that habitat destruction cannot explain the disappearance of these species. In such cases, human predation ostensibly provides a more plausible explanation. But how could humans living at low densities so rapidly exterminate such a large number of animal populations? This is, of course, a familiar question for analysts of megafaunal extinctions.

P. S. Martin's overkill hypothesis is based upon a pattern found in several areas of the world whereby large animals became extinct almost immediately after the initial arrival of human hunters, i.e. "blitzkrieg" (Martin 1966, 1967, this volume; Mosimann and Martin 1975). The evidence from Madagascar fits this pattern. The earliest dates for human occupation overlap the terminal dates of the subfossil species. Although the length of the overlap period is not known, I would suggest a date of about 500 A.D. for human arrival and about 1,100 A.D. for termination of the extinctions, giving about a 600-year overlap.

Closer examination of the Malagasy evidence in contrast to North America, for example, makes the overkill hypothesis less attractive. Unlike the first Amerindians, who were big game hunters, the first Malagasy were probably traders, agriculturalists, and fishermen. Whereas there is excellent evidence from material culture for nomadic hunting throughout North America, in Madagascar there is none. There are no kill sites in Madagascar. Finally, while in North America the Early Archaic cultures of most areas developed from a shared Paleo-Indian ancestry, in Madagascar the archaeological sites display striking differences in economy and ecology immediately after termination of the extinctions. Differences in ceramics are sufficiently great to doubt that they had been part of a single population in the tenth century A.D.

In summary, the indirect and direct cause models are inadequate because explanations that rely upon habitat destruction cannot account for the absence of some animals today where the habitat is well preserved. In addition, Madagascar was almost certainly not colonized by hunters: whatever brought the first Malagasy to the island, it was not game. The admittedly scanty archaeological evidence indicates that soon after settlement of the island, economic patterns markedly diversified: fishing, farming, and pastoralism all appear at about the same time. This is matched by a geographically dispersed occupation of the island by at least 900 A.D.

Patterns and Processes in the Subfossil Extinctions

The model elaborated here is similar to Krantz's (1970) model, in that it envisages an expanding agricultural population engaged in adventitious hunting. Given the limited archaeological or paleontological data from the rainforest areas of Madagascar, I will focus on the events of the central plateau, western, and southwestern regions. These are, in any case, the most problematic, for here we see the most dramatic ecological changes. The following assumptions underlie my model:

1. The central plateau and western regions of Madagascar were not covered by a uniform forest of unspecified fragile type but by a mosaic of woodland and savanna.

2. The earliest colonization of central and western Madagascar was by people who were primarily cattle pastoralists.

3. These pastoralists were, as are their descendants, adventitious hunters.

4. The subfossil species lived at low densities, had low reproductive rates, and were easy to hunt.

The first assumption is controversial, but I believe reasonable. The fauna at Ampasambazimba is diverse, most particularly the primate fauna. If the surrounding forests were homogeneous, then either the primate species had extremely narrow niches, or else they had an expanded role in the community compared to modern primates. Certainly, the living lemuroids do not display any particular nonprimate roles, with the exception of that odd creature, the aye-aye (*Daubentonia madagascariensis*). A possible exception is *Megaladapis,* which has been described as aquatic (Sera 1950), semiterrestrial (Mahé 1972), and fully arboreal (Jungers 1977). The final interpretation has been well supported through studies of locomotor morphology (Walker 1967a; Jungers 1977, 1978). Thus, Mahé's (1972) suggestion of a diet composed largely of roots and tubers seems to be unfounded. This is the only recent attempt, to my knowledge, to demonstrate an expansion of any of the subfossil forms beyond the normal range or primate diets.

As noted, the large ratites need not require homogeneous forest. The central plateau sites yielded abundant ratite subfossils, with five species at Ampasambazimba. Certainly Madagascar's terrestrial herbivore community diverged considerably from continental patterns. The large- and medium-sized mammalian herbivores of Africa are here replaced by tortoises, ratites, and the terrestrial subfossil lemuroids *Hadropithecus* and *Archaeolemur.*

The archaeolemurines may provide indirected evidence of a heterogeneous environment. While Tattersall (1973a) suggests a mode of life for *Archaeolemur* like that of the forest baboon *Papio,* he and Jolly (1970) place *Hadropithecus* in an environment very similar to the savanna-dwelling *Theropithecus.* The diversity of primate forms in the deposits of Ampasambazimba is best explained as a loose collection of forms drawn not from a single sympatric community, but rather from two or more strongly contrasting habitats and associated ecotones. Undoubtedly forest was here, but I believe it was not continuous and that substantial areas of more open vegetation were also present. It might well have been a savanna-mosaic (Van Couvering 1980). Note that weak support for this interpretation is also provided by the contrasts in the floral collections described by Perrier de la Bathie from Antsirabe and Ampasambazimba. Those from Antsirabe are clearly derived from a forest similar to the modern east coast rainforest. The Ampasambazimba materials are just as clearly not.

The second assumption, that the initial colonization of much of central and western Madagascar was by pastoralists, is supported by the archaeological materials from Andranosoa. Clearly a pastoralist's village, the site is one of the oldest on the island. People in that area today pursue a life reminiscent in many ways of East African pastoralists. All the published reports of sites in the south and west support the third assumption, that they were adventitious hunters.

My scenario is simple: pastoralists settled in Madagascar and began to travel with their herds across the mosaic of the interior, seeking patches of grasslands. If suitable grazing areas were small and scattered, it would have been necessary to travel substantial distances during a season. Along the way these pastoralists killed wild animals they found, and they perhaps already practiced some form of dry-season burning. Cattle, and probably sheep and goats, came to replace the indigenous terrestrial herbivores. It is precisely this replacement of the herbivores, primary consumers of ground-level plants, which I believe is the missing link in the previous scenarios.

A major contrast between the savannas of Madagascar and East Africa is the absence of a single medium- or large-sized herbivore in Madagascar, except for animals domesticated by humans. The Malagasy grasslands are truly eerie in this regard. The ratites, tortoises, and archaeolemurines constituted a terrestrial herbivore community here, to stand as partial vicars of the ungulates of the East African savanna. In Madagascar that community is now completely gone.

Although community structure has long been considered the outcome of processes of interspecific competition, it is now suspected that predation, including plant consumption by herbivores, may play an important role in the evolution of communities (Hutchinson 1978:239). Connell (1975) has discussed the evidence for the importance of predators in determining prey abundance. Among his examples are several which detail the importance of herbivores for plant communities. He suggests that the more benign and predictable an environment is, the more important that predation will be in determining the structure of the community. In Madagascar we contemplate a relatively benign environment that suffered a complete transformation of the terrestrial herbivore community in a very short time. The tortoises, birds, and primates that had coevolved in isolation for millions of years with the indigenous flora were replaced by ruminant herbivores. I believe that this change was devastating, and that the plant community consequently was markedly disturbed.

The subfossil species themselves would have suffered more than just the impact of people gathering eggs or hunting them. If much of the primeval plant cover of the interior was patchy, transformations of the floral communities under attack would have produced a lattice of degradation. Many animals would have suffered a corresponding disruption of range and declining chances of finding mates. And the terrestrial forms would have found themselves in direct competition with the bovids. Other recent arrivals, including *Potamochoerus* and feral *Sus,* would have contributed to this process.

I suggest that this model accounts for the dramatic transformation of the central and western regions more completely than either the hunting or burning hypotheses alone. Changes in the floral community may have altered the response of the indigenous vegetation to fire, either from human or natural causes. This model is not applicable to the best-preserved areas of xerophytic bush or to the east and northeast rainforests. In these regions our data are sparse, but seemingly only one ancient ecological transformation occurred, namely the extinction of large resident animals. No model but direct hunting seems to account for the modern absence of the large subfossil forms in these areas, but without any paleontological or archaeological information this is untestable.

The model presented here invites various tests. It will be supported if palynological evidence discloses a primeval savanna/forest mosaic, followed by a conversion to grasslands without major climatic fluctuation, if evidence of initial settlement of the western and southern interiors was by cattle-raising pastoralists engaged in adventitious hunting, and if the extinctions, establishment of pastoralists, and the floral modification all prove synchronous.

Summary

People coexisted with many species that have been extinct on Madagascar for perhaps a half millennium. We do not know the precise processes responsible for the disappearance of these animals. Climatic change seems an unlikely explanation. Neither habitat destruction nor hunting alone are adequate explanations, and indeed, no single explanation is likely to apply to the entire island. In the modern grasslands of the center and west of the island, I argue that ecological transformation resulted from the substitution of domestic bovids for the native terrestrial herbivores. This was brought about by competition, habitat destruction and fragmentation, and adventitious hunting.

Acknowledgments

The research upon which this article was based was supported in part by grants from the National Geographic Society and the University of Connecticut Research Foundation. I owe debts of gratitude to Jean-Aimé Rakotoarisoa, Jean-Pierre Domeninchini, Ramilisonina, and J. P. Emphoux in Antananarivo; to Fulgence Fanony, Mansare Marikandia, and Noel Gueunier in Tulear; and to R. W. Sussman, A. F. Richard, and Georges Randrianasolo. The final version of this chapter was markedly improved by the critical comments of A. K. Behrensmeyer, A. F. Richard, I. Tattersall, R. MacPhee, P. S. Martin, and, most especially, J. Buettner-Janusch.

References

Amadon, Dean. 1947. An estimated weight of the largest known bird. The Condor 4:159–164.

Auffenberg, Walter. 1974. Checklist of fossil land tortoises, (Testudinidae). Bull. Fla. St. Mus. Biol. Sci. 18(3):121–251.

Battistini, René. 1964. Géographie humaine de la plaine côtiere Mahafaly, 197 pp. Université de Madagascar, Antananarivo.

———. 1965a. Sur la decouverte de l'*Aepyornis* dans le Quaternaire de l'Extrême-Nord de Madagascar. Comptes-rendus somm. Soc. Geol. de France, fasc. 5, p. 174.

———. 1965b. Une datation radio-carbone des oeufs des derniers *Aepyornis* de l'Extrême-Nord de Madagascar. Comptes-rendus somm. Soc. Geol. de France, fasc. 5, p. 309.

———. 1971a. Conditions de gisement des sites littoraux de subfossiles et causes de la disparition de la faune des grands animaux dans le sud-ouest et l'extrême sud de Madagascar. Revue de Musée d'Art et d'Archéologie, Taloha 4:7–18.

———. 1971b. Conditions de gisement des sites de subfossiles et modifications récentes du milieu naturel dans la region d'Ankazoabo. Revue du Musée d'Art et d'Archéologie, Taloha 4:19–27.

———. 1976. Les modifications du milieu naturel depuis 2000 ans et la disparition de la faune subfossile a Madagascar. Assoc. Senegalaise et Quatern. Afr., Bulletin de Liaison, no. 47, pp. 63–76.

Battistini, René and Pierre Vérin. 1967. Ecologic changes in protohistoric Madagascar. pp. 407–424 in Pleistocene Extinctions (P. S. Martin and H. E. Wright, Jr., eds.), Yale University Press, New Haven.

———. 1971. Témoignages archéologiques sur

la cote Vezo de l'embouchure de l'Onilahy à la Baie des Assassins. Revue du Musée d'Art et d'Archéologie, Taloha 4:51–63.

———. 1972. Man and the environment in Madagascar. pp. 311–337 in Biogeography and Ecology in Madagascar (R. Battistini and G. Richard-Vindard, eds.), Dr. W. Junk, B. V., The Hague.

Battistini, René, Pierre Vérin and R. Raison. 1963. La site archéologique de Talaky: cadre géographique et géologique. Annales Malgaches, Faculté des Lettres et Sciences Humaine, Antananarivo, vol. 1, pp. 112–153.

Berger, Rainer, Keith Ducote, Karin Robinson and Hartmut Walter. 1975. Radiocarbon date for the largest extinct bird. Nature 258:709.

Blanc, Charles P. 1972. Les reptiles de Madagascar et des iles voisines. pp. 501–611 in Biogeography and Ecology in Madagascar (R. Battistini and G. Richard-Vindard, eds.), Dr. W. Junk, B. V., The Hague.

Bloch, Maurice and Pierre Vérin. 1966. Discovery of an apparently Neolithic artefact in Madagascar. Man (n.s.) 1:240.

Clutton-Brock, T. H. and Paul Harvey. 1977. Primate ecology and social organization. J. Zool. 183: 1–39.

Connell, Joseph H. 1975. Some mechanisms producing structure in natural communities. pp. 460–490 in Ecology and Evolution of Communities (M. L. Cody and J. M. Diamond, eds.), Harvard University Press.

Cracraft, Joel. 1974. Phylogeny and evolution of the ratite birds. Ibis 116:495–521.

Dahl, Otto. 1951. Malagache et Maanyan. Egede Institutett, Oslo.

Deschamps, Hubert. 1960. Histoire de Madagascar. Berger-Levrault, Paris.

Dyen, Isidore. 1953. [Review of Malagache et Maanyan]. Language 39:578–591.

Emphoux, Jean-Pierre. 1978. Note sur une culture ancienne du XIIeme siecle en pays Antandroy. Communication faite a l'Academie Malgache, December 21, 1978.

———. 1979. Archéologie de l'Androy. Colloque d'Histoire Malgache, Tulear. Mimeo.

Fanony, Fulgence. n.d. A propos des Mikea. Paper delivered at Wenner-Gren Symposium, Burg Wartenstein.

Godfrey, Laurie R. 1977. Structure and function in *Archaeolemur* and *Hadropithecus* (subfossil Malagasy lemurs); the postcranial evidence. Doctoral Dissertation, Harvard University.

Humbert, H. 1927. La destruction d'une flore insulaire par le feu. Mem. Acad. Malg., fasc. 5, pp. 1–78.

Hutchinson, G. Evelyn. 1978. An Introduction to Population Ecology. Yale University Press, New Haven.

Jolly, Clifford. 1970. *Hadropithecus,* a lemuroid small object feeder. Man (n.s.) 5:525–529.

Jungers, William L. 1977. Osteological Form and Function: The Appendicular Skeleton of *Megaladapis,* A Subfossil Prosimian from Madagascar. Doctoral Dissertation. University of Michigan, Ann Arbor.

———. 1978. The functional significance of skeletal allometry in *Megaladapis* in comparison to living prosimians. Am. Jour. Physical Anthro. 49:303–314.

Juvik, J. O., A. J. Andrianarivo and C. P. Blanc. 1981. The ecology and status of *Geochelone yniphora*: a critically endangered tortoise in northwestern Madagascar. Biological Conservation 19:297–316.

Kellum-Ottino. Mari-mari. 1972. Discovery of a Neolithic adze in Madagascar. Asian Perspectives 15:83–186.

Koechlin, Jean. 1972. Flora and vegetation of Madagascar. pp. 145–190 in Biogeography and Ecology in Madagascar (R. Battistini and G. Richard-Vindard, eds.), Dr. W. Junk B.V., The Hague.

Krantz, Grover. 1970. Human activities and megafaunal extinctions. Amer. Scientist 58:164–170.

Lamberton, C. 1934. Contributions à la Connaissance de la Faune Subfossile de Madagascar: Lémuriens et Ratites. Mem. Acad. Malg. fasc. 18.

———. 1939. Contributions à la Connaissance de la Faune Subfossile de Madagascar: Lémuriens et Cryptoproctes. Mem. Acad. Malg., fasc. 27.

Livingstone, D. A. 1975. Late Quaternary climatic change in Africa. Ann. Rev. Ecol. Syst. 5:249–280.

Mahé, Joel. 1965. Les Subfossiles Malgaches, 11 pp. Imprimerie Nationales, Antananarivo.

———. 1972. The Malagasy Subfossils. pp. 339–365 in Biogeography and Ecology in Madagascar (R. Battistini and G. Richard-Vindard, eds.), Dr. W. Junk B.V., The Hague.

Mahé, Joel and Michel Sourdat. 1972. Sur l'extinction des vertébrés subfossiles et l'aridification du climat dans le sud-ouest de Madagascar. Bull. Soc. Geol. France 14:295–309.

Marden, Luis. 1967. Madagascar, island at the

end of the Earth. National Geographic 132:443–487.

Martin, Paul S. 1966. Africa and Pleistocene overkill. Nature 212:339–342.

———. 1967. Pleistocene overkill. pp. 75–120 in Pleistocene Extinctions (P. S. Martin and H. E. Wright, Jr., eds.), Yale University Press, New Haven.

Mosimann, James E. and P. S. Martin. 1975. Simulating overkill by Paleoindians. Amer. Scientist 63:304–313.

Morat, Phillippe. 1973. Les Savanes de sud-ouest de Madagascar. Memoires O.R.S.T.O.M., no. 68, Paris.

Patterson, Bryan. 1975. The Fossil Aardvarks (Mammalia: Tubulidentata). Bull. Mus. Comp. Zool., Harvard Univ., 147(5):185–237.

Perrier de la Bathie, H. 1914. Au sujet des tourbieres de Marotampona. Bull. Acad. Malg. (n.s.) 1:137–138.

———. 1927. Fruits et graines de gisement de subfossiles d'Ampasambazimba. Bull. Acad. Malg. (n.s.) 10:24–25.

Petter, Francis. 1972. The rodents of Madagascar: The seven genera of Malagasy rodents. pp. 661–665 in Biogeography and Ecology of Madagascar (R. Battistini and G. Richard-Vindard, eds.), Dr. W. Junk B.V., The Hague.

Phillipson, D. W. 1977. The later prehistory of eastern and southern Africa. Heinemen, London.

Raison, J. P. and P. Vérin. 1967. Le site de subfossiles de Taolambiby (sud-ouest de Madagascar) dôit-il être attribué a une intervention humaine? Annales de l'Université de Madagascar, Faculté des Lettres et Sciences Humaines, no. 7, pp. 133–142.

Rakotoarisoa, Jean-Aimé. 1979. Principaux aspects des formes d'adaptation de la société traditionelle Malgache. Paper delivered at Wenner-Gren Symposium, Burg Wartenstein.

Richard, Alison F. and Robert W. Sussman. 1975. Future of the Malagasy Lemurs: Conservation or extinction? pp. 335–350 in Lemur Biology (I. Tattersall and R. W. Sussman, eds.), Plenum Press, New York.

Sacher, G. A. 1974. Maturation and longevity in relation to cranial capacity in hominid evolution. pp. 417–441 in Primates Functional Morphology and Evolution (R. Tuttle, ed.), Mouton, The Hague.

Savage, R. J. G. 1978. Carnivora. pp. 249–267 in Evolution of African Mammals (V. J.

Maglio and H. B. S. Cooke, eds.), Harvard University Press, Cambridge, Mass.

Sera, G. L. 1950. Ulteriori osservazioni sui lemuri fossili ed attuali. Paleont. Italica (n.s.) 17:1–113.

Southall, Aidan. 1975. The problems of Malagasy origins. pp. 192–215 in East Africa and the Orient (H. Neville-Chittick and R. I. Rotberg, eds.), Africana Publishing, New York.

Sussman, Robert W. 1974. Ecological distinctions in sympatric species of Lemur. pp. 75–108 in Prosimian Biology (R. D. Martin, G. A. Doyle and A. C. Walker, eds.), Duckworth, London.

Tattersall, Ian. 1973a. Cranial Anatomy of the Archaeolemurines. Anthrop. Pap. Amer. Mus. Nat. Hist. 52(1):1–110.

———. 1973b. Subfossil lemuroids and the "adaptive radiation" of the Malagasy lemurs. Trans. N.Y. Acad. Sci. 35:314–324.

———. 1973c. A note on the age of the subfossil site of Ampasambazimba, Miarinarivo Province, Malagasy Republic. Amer. Mus. Novitiates 2520:1–6.

———. 1982. The Primates of Madagascar. Columbia University Press, New York.

Tattersall, Ian and Jeffrey H. Schwartz. 1974. Craniodental morphology and the systematics of the Malagasy lemurs. (Primates: Prosimii). Anthrop. Pap. Amer. Mus. Nat. Hist. 52(3):139–192.

Van Couvering, J. A. H. 1980. Community evolution in East Africa during the Late Cenozoic. pp. 272–298 in Fossils in the Making: Vertebrate Taphonomy and Paleoecology (A. K. Behrensmeyer and A. P. Hill, eds.), University of Chicago Press, Chicago.

Van Denburgh, John. 1914. The gigantic land tortoises of the Galapagoes Archipelago. Proc. Cal. Acad. Sci. 2(1):203–374.

Van der Merwe, Nikolaas J. 1980. The advent of iron in Africa. pp. 463–506 in The Coming of the Age of Iron (T. A. Wertime and T. D. Muhly, eds.), Yale University Press, New Haven.

Vérin, Pierre. 1971. Les anciens habitats de Rezoky et d'Asambalahy. Revue du Musée d'Art et d'Archéologie, Taloha 4:29–45.

———. 1975a. Les Echelles Anciennes du Commerce sur les Côtes Nord de Madagascar. These du doctorat, Université de Lilles III.

———. 1975b. Austronesian contributions to

the culture of Madagascar: Some archeological problems. pp. 164–191 in East Africa and the Orient (H. Neville-Chittick and R. I. Rotberg, eds.), Africana Publishing, New York.

———. 1979. Le probleme des origines Malagaches. Revue du Musée d'Art et d'Archéologie, Taloha 8:41–55.

Vérin, Pierre, Conrad Kottak and Peter Gorlin. 1970. The glottochronology of Malagasy speech communities. Oceanic Linguistics 8:26–83.

Vogel, Richard J. 1977. Fire: A destructive menace or natural process? pp. 261–289 in Recovery and Restoration of Damaged Ecosystems (J. Cairns, K. Dickson and E. Herricks, eds.), University Press of Virginia, Charlottesville.

Walker, Alan C. 1967a. Locomotor Adaptations in Recent and Subfossil Madagascan Lemurs, Doctoral thesis, University of London.

———. 1967b. Patterns of extinction among the subfossil Madagascar lemuroids. pp. 425–432 in Pleistocene Extinctions (P. S. Martin and H. E. Wright, Jr., eds.), Yale University Press, New Haven.

Walker, Ernest P. 1975. Mammals of the World, 3rd ed. Johns Hopkins Press, Baltimore.

Western, David. 1979. Size, life history and ecology in mammals. Afr. J. Ecol. 17:185–204.

White, E. I. 1930. Fossil hunting in Madagascar. Nat. Hist. 2:209–235.

Wright, Henry. 1979. Early communities on the island of Mayotte and the coasts of Madagascar. Mimeo.

Australia, New Zealand and the Island Pacific: Severe Losses

Australia, New Zealand, and the Island Pacific: Severe Losses

Anyone who has wondered what might emerge from Pleistocene bone beds of Australia can amuse themselves by making friends with Murray and Merrilees's extinct Australian marsupials illustrated in their chapters. Unlike mammoths and mastodonts, the extinct megafauna downunder is rarely reconstructed. Murray's illustrations are thought-provoking. The lumbering diprotodonts must have done something unusual, digging for roots perhaps, with their curious enlarged snouts. The giant extinct kangaroos also look ponderous and less swift than the largest living kangaroos. Murray believes they should have been easy prey for the first human hunters who entered the continent, as Sir Richard Owen concluded a century ago.

However, for reasons he explains in his chapter, David Horton discounts human impact and favors aridity as the main cause of late Pleistocene extinction. As an explanation for marsupial extinctions, the overkill concept has not fared well among many Australians.

For one thing, few archaeological sites have been found to contain any bones of the extinct megafauna, and no kill sites are known. Large bones can be found in vast numbers in Wellington Cave in New South Wales, Victoria Cave at Naracoorte, and Mammoth Cave in Western Australia, which Merrilees treats in his chapter. Large numbers of large bones in shallow burials have been found in spring deposits such as Lancefield, Victoria. On the other hand, when rich archaeological sites are found, such as Lake Mungo, New South Wales, and newly discovered Fraser Cave in southwestern Tasmania, traces of extinct fauna are few or missing. The common bones associated with the hearths, clay ovens, and shell middens left by the earliest inhabitants as early as 15,000 to 30,000 years ago are of modern species. It might be mentioned that neither extinct megafauna nor archaeological sites of pre-Holocene age have been found in the arid interior of the continent northwest of the Darling River. Most fossil deposits of the late Pleistocene megafauna lie within two hundred miles of the coast. Whether Diprotodon *or* Sthenurus *ever reached Alice Springs in the Quaternary remains a subject of speculation.*

At least one extinct genus, Sthenurus, has been reported in late glacial age deposits. However, most megafaunal deposits are apparently much older, and Merrilees finds little evidence of survival in Western Australia after 30,000 years ago. It is not yet certain whether the fifty-three extinct species of large animals Horton recognizes disappeared suddenly and simultaneously, or whether they disappeared gradually in some orderly sequence. This makes Hope's task of utilizing Australian fossil pollen records in testing climatic models more difficult. If one is to search the pollen record for a climatic cause of faunal extinction, is it to be sudden, starting around 26,000 years ago, the age attributed to Lancefield, or gradual, between 40,000 and 15,000 years? Hope makes an important point: the effect of the first significant climatic pulse in the Quaternary sequence should overshadow later ones, if megafaunal extinctions track climatic pulses.

Australia is no different from other continents in the lack of appreciable late Pleistocene plant extinctions. Some plants did, however, experience remarkable range shrinkages, especially rainforest trees, which Kershaw examines. These may have been under way during the time of some faunal extinctions.

If 15,000 years ago was a climatically catastrophic time for large animals in Australia, one wonders if some trace of the event might be detected elsewhere, perhaps in New Zealand 1,600 km to the east. Answering the question will require a much better Pleistocene fossil record in New Zealand. Long beloved by biogeographers for its endemic biota, New Zealand was the largest habitable land mass unoccupied by terrestrial mammals until prehistoric Polynesians arrived, bringing rats and dogs. The extinction of moas and other birds soon followed. In the case of moas there is no doubt of archaeological associations. Anderson and Trotter and McCulloch show how moa extinction followed prehistoric human impact. Anderson estimates the remarkable number of moas found in archaeological sites of the South Island, the richest association of extinct fauna and artifacts to be found anywhere outside Afro-Asia. So much occurred so late that it seems highly unlikely that much moa extinction could have occurred earlier, within say the last 100,000 years. However, the moa fauna of New Zealand during the last glaciation is yet unknown.

Moas were not the only birds lost in New Zealand. Richard Cassels scans the record of extinction of smaller birds there and on other oceanic islands of the western Pacific. Birds from the smaller Pacific Islands once shot and stuffed in the imperial age of "South Sea Expeditions" have disappointed museum curators. As Cassels notes, few endemic genera or species were found. The modern avifauna suggests a retinue of tramp species that survived or reinvaded islands stripped of the native endemics of an earlier time. Whether caused by overhunting, by firestorm and land clearing, by introduction of rats or other alien predators, or by climatic upset, prehistoric island extinctions ran their course. If the losses were severe, it follows that the rich native avifaunas of endemic island birds once sought by museum collectors may yet come to light in the form of fossil bones.

An example of what to expect may be found in the late Pleistocene avifauna of Hawaii. Half of the Hawaiian avifauna became extinct under circumstances that Olson and James believe implicate prehistoric people. The late Pleistocene deposits on other oceanic islands invite an intensified search for other lost endemics. Olson and James go one step further and note how the newly discovered fossil record itself impacts popular models of island biogeography. Paleontologists have sought to explain phyletic extinction in terms of an equilibrium model based in particular on MacArthur and Wilson's famous treatment of origination and extinction of island avifaunas. Since the Holocene island extinctions themselves were not anticipated in most treatments of island biogeography, the application of island biogeography theory to explanations of much older geological extinctions seems especially shaky. One can only ask that biogeographers reexamine their temporal as well as their physical assumptions. Perhaps the past is no less a key to the present as the reverse.

Thus, in this section of Quaternary Extinctions, the recurrent themes of changing climates and colonizing cultures are examined once more. Some of the strongest arguments for extinction at the hand of man are advanced for New Zealand, while in Australia Horton presents one of the strongest cases for extinction entirely independent of man. Those seeking tests of extinction models can be thankful there are so many small oceanic islands to provide prehistoric fossils for their purposes. At the same time for the same reason, the model testers can only mourn the scarcity of continents. Nevertheless, east of Borneo across Wallace's Line there appear to be enough large islands on and off the continental shelf to align any extinction chronology with its possible causes. With further Quaternary paleontology in the South Pacific it should be possible eventually to eliminate one or another theoretical explanations for late Pleistocene extinctions.

27

Extinctions Downunder: A Bestiary of Extinct Australian Late Pleistocene Monotremes and Marsupials

PETER MURRAY

THE EXTINCT FAUNA FROM THE LATE PLEISTOCENE of Australia is not well known. Many forms are represented only by isolated bone fragments, and comparatively little has been accomplished in the way of functional morphology and detailed description of the extinct beasts. Consequently, my interpretations are highly speculative and subject to revision.

A number of attempts to restore Australian marsupials have already been made (Scott 1915, Lord and Scott 1924, Scott and Lord 1920, Gill 1954, Stirton 1959, Bergamini 1964, Martin and Guilday 1967, Tedford 1973, Murray 1978c, Bartholamai 1978, and Archer 1981). Because restorations depend primarily on an anatomical buildup of soft tissues, a restoration is usually a straightforward interpretation based on comparisons with closely related forms with appropriate allometric adjustments made for size differences (fig. 27.1). Some morphologies, especially in the facial region, are difficult to reconstruct because the soft tissues leave little indication of their attachments and because some animals are anatomically unique, ruling out restoration by direct analogy. The various living Australian taxa may be found in Walker (1968), Bergamini (1964), Tindale-Biscoe (1973), Ride (1970), Jones (1969), and Vaughan (1972). Ride (1970) provides the basic taxonomic structure for this discussion.

Monotremes

Tachyglossids, Ornithorhynchids

Large echidnas belonging to the genus *Zaglossus,* once common and widely distributed in Australia, are now confined to the highlands of New Guinea. The first descriptions of fossil *Zaglossus* remains were of isolated postcranial fragments. By the turn of the century a confused synonymy of no less than six species and three genera for only two, or at the most three, species had resulted (Murray 1978b).

Some fossil *Zaglossus* specimens are stouter and have a slightly broader, less downcurved beak than the surviving species, *Zaglossus bruijni* (Murray 1978b). The Pleistocene species is probably specifically distinct, and the use of *Zaglossus ramsayi*

Figure 27.1. Restored skeletons of some Australian Pleistocene mammals: a, *Zygomaturus trilobus;* b, *Thylacoleo carnifex* (after Finch 1981); c, *Zaglossus ramsayi;* d, *Sthenurus occidentalis.*

has been suggested to cover the synonyms *Zaglossus owenii* and *Zaglossus harrissoni* (fig. 27.2b). *Zaglossus robusta* is a large echidna very similar to *ramsayi,* but having some morphological differences in its humerus and a possibly greater antiquity (?Upper Pliocene) (Murray 1978a, 1978b).

Figure 27.2. Reconstructed fossil echidnas: a, *Zaglossus hacketti*, Mammoth Cave, W.A., about 1 m long; b, skull of *Zaglossus ramsayi*, length of skull 165 mm; c, *Zaglossus ramsayi*, Montagu Caves, Tasmania, about 750 mm.

Zaglossus hacketti consists of a tibia, femur, innominate fragment, episternum, coracoid, and a few assigned vertebrae. There are no remains of the skull but the post-cranial elements are unmistakably those of an enormous echidna (fig. 27.2a).

Although a meter in length and half a meter high, the proportions of *Zaglossus hacketti*'s limbs are more like those of the small living echidna, *Tachyglossus* (Murray 1978a). These fossils indicate that the echidnas underwent an adaptive radiation resulting in a range of taxa comparable to the living South American anteaters.

Fossil remains of the platypus are extremely rare (Archer and Bartholomai 1978). The "giant platypus," *Ornithorhynchus maximus* (Dun 1895), turned out on closer inspection to be a *Zaglossus* (Mahoney and Ride 1975, Murray 1978b). The only Pleistocene platypus fossil (*Ornithorhynchus agilis*) is synonomous with the living species, *Ornithorhynchus anatinus* (Archer and Bartholomai 1978).

Marsupials

Dasyurids, Thylacinids

Late Quaternary deposits contain abundant evidence of marsupial carnivores. Isolated skeletal remains of native cats (*D. viverrinus*) appear in Tasmanian cave deposits in the form of fossilized regurgitated owl crop pellets. A large morph of the tiger cat (*Dasyurus maculatus bowlingi*) was present on King Island until European contact (Hope 1973). The *D. maculatus bowlingi* race may have developed as a result of

Figure 27.3. Reconstruction of the Pleistocene devil, *Sarcophilus laniarius,* length of skull, 140 mm; head-body length about 900 mm.

post-Pleistocene isolation of the Bass Strait Islands. The largest living dasyurid, the notorious Tasmanian devil (*Sarcophilus harrisii*), is now confined to the island of Tasmania. Remains of a slightly larger Pleistocene species, *Sarcophilus laniarius,* are plentiful in cave deposits throughout eastern and southern Australia (fig. 27.3). Samples of the Wellington Caves *S. laniarius* can be distinguished from the living devils statistically (Ride 1964). However, this distinction is not always so easy to make in certain other assemblages, i.e. from Tasmania and Western Australia. A dimunitive form of *Sarcophilus* from a 10,000-year-old archaeological deposit in Queensland has also been described (Horton 1977). *Sarcophilus harrisii* survived into Holocene times on the mainland. In Victoria, mid- to late Holocene remains occur in coastal shell middens (Wakefield 1964).

The remains of thylacines or marsupial wolves are not infrequently found in Pleistocene cave deposits. The thylacine was once widespread, ranging from New Guinea to Tasmania, where it has recently become extinct. The Pleistocene fossil species *Thylacinus spelaeus* does not appear to differ from the living or recently extinct Tasmanian wolf (*Thylacinus cynocephalus*) (Ride 1964). Thylacines survived on the mainland until 3,000 or 4,000 years ago (Partridge 1967).

Thylacoleonids

Thylacoleo, the marsupial lion or "giant killer possum," was a leopard-sized carnivorous phalangeroid that weighed between 75 and 100 kg. The permanent premolars (p3/3) were greatly enlarged to function like the shearing carnassials of placental carnivores (fig. 27.4). In many respects *Thylacoleo* resembles the much smaller brushtail possum (*Trichosurus*), which has sectorial permanent premolars that are miniatures of *Thylacoleo*'s. Brushtail possums show an affinity for meat (Archer and Bartholomai 1978) and have been observed to catch and eat insects (Murray 1977). *Thylacoleo carnifex* was a widely distributed species and is relatively common in certain deposits (fig. 27.5). It appears to have survived until the very late Pleistocene (Gill 1954). It is represented by two nearly complete skeletons (Finch 1971 and Rod Wells personal communication) and by numerous crania, dentaries, and isolated teeth.

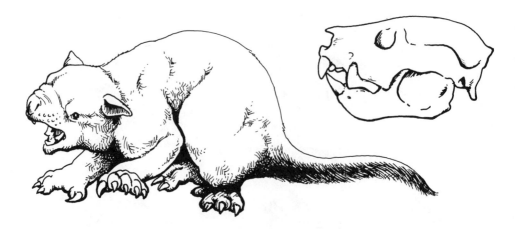

Figure 27.4. Reconstruction of *Thylacoleo carnifex;* length of skull, 240 mm, head-body length about 1,200 mm.

Thylacoleo was a powerful, relatively long-limbed beast that retained a basically phalangeroid hand with a pseudo-opposable pollex bearing an enormous recurved claw (Wells and Nichol 1977) (fig. 27.4). While the pollical claw is proportionally much larger than that of the brushtail possum, as one would expect, the claws on *Thylacoleo*'s digits II–V are relatively small, slender and straight, an arrangement quite different from the scansorial (climbing) adaptation of the brushtail, which has its largest claws on digits II–V. *Thylacoleo*'s foot had the characteristic phalangeroid syndactyly of digits II and III (Wells and Nichol 1977). The elongated limbs and manus and pes morphology of *Thylacoleo* could indicate that it was a tree climber. Of course, its close relationship with arboreal phalangeroids may also be responsible, the characteristics being merely features of heritage. The large pollical claw might conceivably have been employed to help grip its prey, while at the same time could have some way aided the animal in climbing.

Pledge (1977) noted a discontinuous distribution of *Thylacoleo* in South Australian sites that may indicate the absence of suitable trees from the habitat (fig. 27.5). *Thylacoleo* may have resembled the specialized Malagasy viverrid, *Cryptoprocta ferox,* in being an arboreal predator, but its bulk would have limited its activities to the stoutest branches. Horton and Wright (1981) have presented evidence indicating that *Thylacoleo* was a scavenger. Judging from the animal's size and probable efficiency of its kill mechanism (table 27.1), *Thylacoleo* could have killed macropodids up to the size of a grey kangaroo. It is unlikely that it took healthy or adult diprotodontoids for prey. More likely, it was a possum or koala eater.

Thylacoleo remains vary in size and in particulars of the dentition; eventually *T. carnifex* may be broken up into more than one species. A *Thylacoleo* from the west Australian Pleistocene is considered distinct from *T. carnifex* (Merrilees 1967), and a rather small morph with a short premolar has been reported in Tasmania (Murray and Goede 1977).

Vombatids

Late Pleistocene wombats are taxonomically confused and poorly represented as fossils. The emerging picture is that of a minor radiation of huge *Vombatus*-like forms.

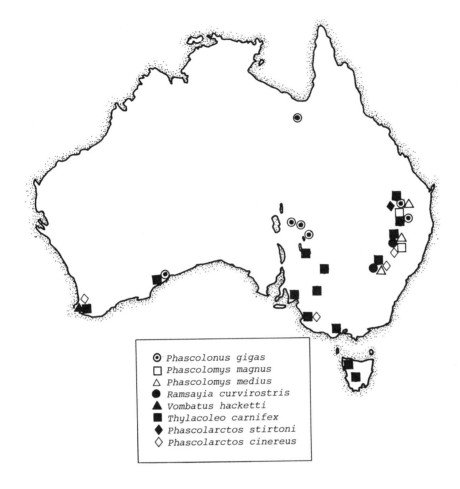

Figure 27.5. Map showing the geographical distribution of *Thylacoleo*, wombats, and koalas in some Australian late Pleistocene localities.

Both *Phascolonus gigas,* a spatulate-incisored species, and *Ramsayia curvirostris,* a narrow-incisored species, stood about a meter high at the shoulder (fig. 27.6) and were at least twice the size of living wombats. Intermediate-sized wombats (*Phascolomys medius* and *P. magnus*) were possibly more closely related to the hairy-nosed wombats (*Lasiorhinus*) (Marcus 1967). Certain smaller species, such as *Vombatus hacketti* of Western Australia, also failed to survive the last glacial. The larger wombats may have relied on burrows for protection which, combined with their great bulk, would have made them easy prey for human hunters.

Phascolarctids

The living koala (*Phascolarctos cinereus*) is the sole survivor of a once diverse family that thrived from mid to late Tertiary times (Archer and Bartholomai 1978). An extinct species *Phascolarctos stirtoni* survived into the late glacial. *P. stirtoni* was substantially larger than *P. cinereus,* perhaps one-third again its size (Bartholomai

**Table 27.1. Summary of Inferred Behavior of Selected Extinct
Australian Pleistocene Marsupials and Monotremes**

Numbers indicate basis for speculation: 1 = based on the most closely related living form; 2 = based on the relative frequency of occurrence in a deposit; 3 = based on inference from dental morphology; 4 = based on inferred dietary requirements; 5 = based on paleoecological data; 6 = based on anatomical inferences.

Species	Habitat	Locomotion & Posture	Diet
Zaglossus ramsayi	forest herbfield (1)	slow quadrupedal gait, rapid digger, can raise up on hind legs to feed (1)	social insects, earth worms (1)
Thylacoleo carnifex	forest edge, open forest (5)	quadrupedal, fast and agile, may have climbed trees (6)	muscle, tissue, and blood (6)
Phascolonus gigas	savanna (5)	quadrupedal, plantigrade, rather slow (6)	leaves, grasses, succulent herbs (3)
Phascolarctos stirtoni	open forest (1)	quadrupedal climber and vertical clinger (1)	folivorous browser (1, 3)
Palorchestes azeal	grassy woodland (5)	plantigrade quadrupedal (1)	selective folivorous browser (3)
Zygomaturus trilobus	grassy woodland (5)	plantigrade, quadrupedal (6), ?swimming/wading	browser, possibly ate roots, tubers, stripped bark, etc. (3)
Diprotodon optatum	light scrub, savanna, bushland (5)	plantigrade, quadrupedal (6)	selective browser, ate leaves, bark, uprooted whole shrubs, etc.
Propleopus oscillans	open forest and forest edge (1, 4)	bipedal, saltator (1)	generalized herbivore: leaves, petioles, flowers, fruit, stems, seeds (1)
Protemnodon anak	forest edge, open forest (1)	bipedal, saltator, fast for short distances (1)	grasses, leaves, stems, petioles (1)
Macropus titan	forest edge, grassland (4), open forest (1)	bipedal, saltator, fast for long distances (1)	grasses (1)
Sthenurus (Sinosthenurus)	forest edge, grassy forest (5)	bipedal, saltator, fast for short distances (6)	leaves, shoots, stems, petioles, fruits, seeds and pods (3)
Procoptodon	forest edge, savanna bushland, open grassland (5)	bipedal, saltator, fast for short distances (6)	leaves, shoots, stems, petioles, grasses, and herbs (3)

1968). Apparently neither species lived in Tasmania even though *P. cinereus* occurred in adjacent southern Victoria (fig. 27.5). *P. cinereus* disappeared from Western Australia despite the continuous presence of seemingly suitable habitat (Merrilees 1968).

Diprotodontoids: Palorchestids, Diprotodontids

No member of the large and varied superfamily Diprotodontoidea survived the late Pleistocene. Two distinct families, the Palorchestidae and the Diprotodontidae, were represented in the last glacial.

Feeding Mechanism	Activity Pattern	Social Behavior
uses long tongue to probe termite's and ant's nests, also extracts worms from soil (1)	crepuscular, out on cool cloudy days (1)	solitary (1, 2)
semiarboreal predator, small prey bitten behind the occiput, large prey strangled by clamping the incisors over trachea (1, 3)	crepuscular and nocturnal (1)	solitary or semigregarious (1)
cropping with incisors (3)	crepuscular and nocturnal (1)	semigregarious (1)
bulk feeding hand-mouth (1)	crepuscular and nocturnal (1)	semigregarious (1)
use of trunk to select leaves and petioles from spiney bushes (6)		solitary (2)
used claws to dig for roots and tubers; stripped bark; uprooted shrubs, etc. (6)		semigregarious (2)
used trunk like upper lip for selective feeding in low scrub and bushland (6)		gregarious (mobs or herds) (2)
used blade like p^3_3 for shearing stems (3)	crepuscular (1)	solitary (1)
used p^3_3 for shearing stems, petioles (1)	diurnal and crepuscular (1)	gregarious (mobs) (1)
incisor cropping mechanism (1, 3)	diurnal, crepuscular, nocturnal (1)	gregarious (mobs) (1)
crushing premolars, trenchant molar lophs for shearing tough vegetation (3)	diurnal, crepuscular (1)	gregarious (mobs) (1, 2)
crushing premolars, low-crowned grinding molars; long arms for reaching branches (3, 6)	diurnal, crepuscular (1)	semigregarious (1, 2)

Palorchestids are found as rare and fragmentary fossils throughout the Tertiary. *Palorchestes azeal* is the only species known from undoubted late Pleistocene deposits, but *P. parvus* may also have been present (figs. 27.7, 27.8). *Palorchestes* was originally thought to have been a gigantic kangaroo (Raven and Gregory 1946) until Woods (1958) demonstrated its probable affinity with the diprotondontids. However, Archer and Bartholomai (1978) consider *Palorchestes* sufficiently distinct to warrant its own family designation.

The enigmatic *Palorchestes* is now beginning to take shape, if recent assignments prove to be correct. A nearly complete cranium of a Miocene form, *P. painei*, has

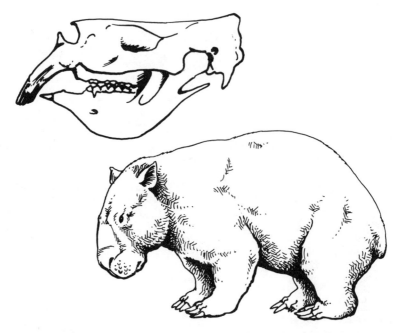

Figure 27.6. Reconstruction of *Phascolonus gigas;* length of skull 472 mm, *P. gigas* was about twice the size of living *Vombatus ursinus,* i.e. about 2,000 mm long and 800 mm high.

markedly retracted nasal bones indicating the presence of a trunk. *Palorchestes azeal* fragments do not include the nasal region, but the long, narrow rostrum is so similar to *P. painei* that it too must have had a tapirlike trunk (Bartholomai 1978). *Palorchestes azeal* also possessed a large kangaroolike tail (Bartholomai 1962 and personal communication). *Palorchestes azeal* was equipped with huge, curved, laterally compressed claws (Archer and Bartholomai 1978) leaving an overall impression of a composite animal with tapir, chalichothere, pantodont, and slothlike features. *P. azeal* is not especially common, but its fossil remains are widely distributed. A Tasmanian specimen from Pulbeena Swamp is dated to 54,200+11,000/−4,500 yr B.P. (Banks et al. 1976).

The family Diprotodontidae contains the two subfamilies Zygomaturinae and the Diprotodontinae. The former is represented in the last glacial by a bizarre-looking, cow-sized animal known as *Zygomaturus trilobus* (figs. 27.1a, 27.9). It was a widely distributed creature that seems to have preferred the coastal and montane regions of Australia (fig. 27.8). Its massive skull is broad, with a curiously narrow, upturned snout. Flaring zygomatic arches dominate the face. Where it originates at the frontal, the nasomaxillary region is narrow and wide with the nasals rising precipitously and abruptly flaring out at the end of the snout where they each terminate in knoblike, roughened growths. These crests may have supported rudimentary horns or cornified pads (Scott and Lord 1920). Alternatively, they may merely mark the attachment of a broad, thick-skinned rhinarium. A high, thin, boney flangelike process projects upwards from the premaxillary, just above the sockets for two diverging, tusklike incisors. The well forward position of the nasals seems to rule out the presence of a trunk. A tapir or elephantlike trunk requires a broad, unrestricted purchase for the facial muscles. In *Zygomaturus* they would have been confined to a relatively small area at the end of the rostrum. The premaxillary process would also have greatly inhibited side-to-side movement of a proboscis. It seems more likely that the premaxillary flange may have supported powerful upper lip musculature that aided *Zygomaturus* in stripping leaves

Figure 27.7. Reconstruction of *Palorchestes azeal,* skull length, approximately 450 mm, head-body length about 1,600 mm.

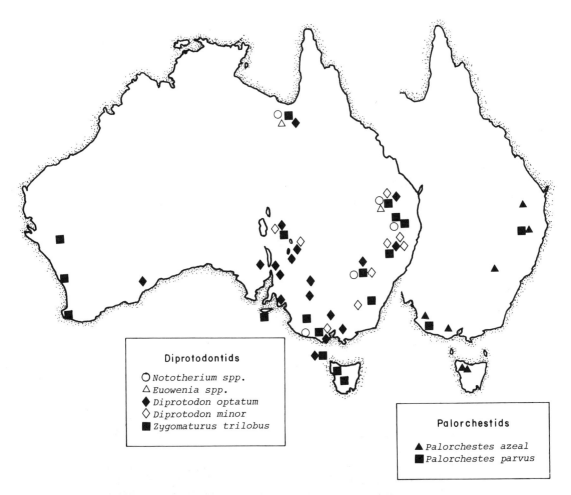

Diprotodontids

○ *Nototherium spp.*
△ *Euowenia spp.*
◆ *Diprotodon optatum*
◇ *Diprotodon minor*
■ *Zygomaturus trilobus*

Palorchestids

▲ *Palorchestes azeal*
■ *Palorchestes parvus*

Figure 27.8. Geographical distribution of palorchestids and diprotodontids from principal late Pleistocene localities.

Figure 27.9. Reconstruction of *Zygomaturus trilobus,* skull length, about 520 mm, head-body length, 2,000 mm.

from shrubs and in grubbing out roots and tubers (Murray 1978c). Merrilees (1968) raised the possibility of *Zygomaturus* being aquatic because of its specialized nasal region.

The two-meter-long *Zygomaturus* was stoutly built, yet its forelimb morphology suggests considerable mobility in contrast to the forelimbs of the more graviportal ungulate placentals of similar mass. The hands were provided with long, slightly curved claws. The hind feet had medially reduced, syndactylous toes, and the hallex may have been vestigial, with the small tarsus being encased in a dense connective tissue pad. The restored posture is rather low and sprawling, suggesting that the animals were not capable of generating much speed.

Diprotodon species were common and widespread, inhabiting drier interior regions as well as coastal margins (Keble 1945). Most finds have been from old lake basins in the interior (Tedford 1973) (fig. 27.8). Two named species of *Diprotodon* occur in late Pleistocene sites: the huge *Diprotodon optatum* (=*D. australis*) and *Diprotodon minor* (fig. 27.10). *D. optatum* stood just under two meters at the shoulder. It had a fairly long neck, a long, narrow head, and a compact body with relatively short legs and proportionally small, strangely formed feet.

Diprotodon optatum has somewhat retracted nasals combined with a large premaxillary process like that of *Zygomaturus,* but it lacks the bony projections from the nasals. It may have had a rudimentary trunk (Thomas Rich personal communication), but as in *Zygomaturus,* the large premaxillary flange would have interfered with its lateral mobility. *Diprotodon optatum* has a long, narrow diastema between its large bilophodont cheek teeth and its extremely long, very procumbent, spatulate upper and chisel-like lower incisors. This seems to have been a selective cropping mechanism enabling it to browse on tough, prickly succulents and shrubs. Archer and Bartholomai (1978) suggest that *Diprotodon optatum* fed on firm, long-stemmed vegetation (Table 27.1).

Figure 27.10. Reconstruction of *Diprotodon optatum,* length of skull about 2,400 mm.

Far less is known about the small species *Diprotodon minor.* It was described from some teeth found at Darling Downs (DeVis 1888), with subsequent finds over a wide area of southeastern Australia (fig. 27.8) contributing a dentary, upper incisors and assorted teeth, and postcranial elements (Gill 1955). *Diprotodon minor* was about one-third smaller than *D. optatum,* thus its postcranial elements could sometimes be confused with those of *Nototherium* and *Zygomaturus.*

The remaining late Pleistocene diprotodontids are also poorly represented, and no attempt has been made to restore them. Among these are various species of *Nototherium* (*N. inerme, N. mitchelli, N. victoriae* and *N. tasmanicum* of which the last two at least, are zygomaturines). *Euowenia* includes two species, one of which is Pliocene (*E. robusta*) and another, *E. grata,* that may have survived the late Pleistocene. These are rare, poorly preserved fossils that may represent some of the still largely unknown fauna of the northern half of the continent.

Macropodids

The kangaroos and their relatives represent the largest radiation of Australian marsupials still in existence, although the diversity of the group was greatly diminished by the beginning of the Holocene (Table 27.2). From the genus *Macropus* alone, at least eight small (wallabies, pademelons) to very large species disappeared (Bartholomai 1975). *Macropus titan* and *Macropus ferragus* were the "giants" of the true kangaroos (Macropodinae) (fig. 27.11). *M. ferragus* may have stood more than 2.5 meters tall and weighed 150 to 250 kg; some individuals may have towered near the 3-meter mark (Table 27.3).

Table 27.2. Extinct Genera and Species of Last Glacial Australian Marsupials and Monotremes

Q = Queensland, NSW = New South Wales, VIC = Victoria,
TAS = Tasmania, SA = South Australia, WA = Western Australia

Genera and Species	Q	NSW	VIC	TAS	SA	WA
Tachyglossidae						
Zaglossus ramsayi	X	X	X	X	X	X
Zaglossus hacketti						X
Dasyuridae						
Sarcophilus laniarius	X	X	X	X	X	?
Thylacoleonidae						
Thylacoleo carnifex	X	X	X	X	X	X
Thylacoleo sp.				?		X
Vombatidae						
Phascolonus gigas	X	X	X		X	
Ramsayia curvirostris		X	X			
Phascolomys magnus	X	X	X			
Phascolomys medius	X	X				
Vombatus hacketti						X
Phascolarctidae						
Phascolarctos stirtoni	X					
Palorchestidae						
Palorchestes azeal	X	X	X	X	X	X
Palorchestes parvus	X		?			
Diprotodontidae						
Diprotodon minor	X	X	X		X	
Diprotodon optatum	X	X	X	X	X	?
Zygomaturus trilobus	X	X	X	X	X	X
Euowenia spp.	X					
Nototherium spp.	X	X	X		X	
Macropodidae						
Propleopus oscillans	X	X	X		X	
Wallabia vishnu	X	X	X			?
Protemnodon anak	X	X	X	X	X	X
Protemnodon brehus	X	X	X		X	X
Protemnodon roechus	X	X	X		X	?
Troposodon minor	X					
Macropus siva	X	X	X			
Macropus rama	X					
Macropus gouldi	X					
Macropus piltonesis	X					
Macropus thor	X					
Macropus birdselli		X			X	
Macropus titan	X	X	X	X	X	?
Macropus ferragus	X	X			X	
Macropus stirtoni	X					
Sthenurus andersoni	X	X			X	
Sthenurus atlas	X	X			X	
Sthenurus oreas	X	X				?
Sthenurus tindalei	X				X	?
Sthenurus pales	X	X				
Sthenurus occidentalis		X	X	X	X	X
Sthenurus brownei			X		X	X
Sthenurus gilli			X	X	X	X
Sthenurus orientalis	X	X	?		?	?
Sthenurus maddocki		?	?		X	
Procoptodon goliah	X	X	X		X	
Procoptodon rapha	X	X	X		X	
Procoptodon pusio	X					
Procoptodon texasensis	X					
Fissuridon pearsoni	X					

Figure 27.11. Reconstruction of the extinct kangaroo species *Macropus (Osphranter) ferragus,* the largest true kangaroo, skull length about 230 mm; standing height about 2,500 mm.

Table 27.3. Estimated Height, Weight, and Length of Selected Extinct Species

SHH = shoulder height, STH = standing height, L = total length, W = weight

Species	SHH (m)	STH (m)	L (m)	W (kg)
Zaglossus ramsayi	.3	—	.6	10
Zaglossus hacketti	.5	—	1.0	30
Sarcophilus laniarius	.4	—	.75	12
Thylacoleo carnifex	.6	—	2.5	75–100
Phascolonus gigas	.8	—	1.7	100–150
Zygomaturus trilobus	1.2	—	2.0	500–1000
Diprotodon optatum	1.8	—	2.4	850–1500
Propleopus oscillans	—	1.0	1.2	30
Protemnodon anak	—	1.5	2.0	40
Macropus ferragus	—	2.5	3.0	150
Sthenurus gilli	—	1.0	1.5	20
Sthenurus occidentalis	—	1.2	2.0	50
Procoptodon goliah	—	3.0	3.5	200–300

Figure 27.12. Diagram showing size reductions of surviving morphs of Pleistocene species. Silhouettes are extinct forms. a, *Macropus titan—Macropus giganteus 33%;* b, *Macropus (Osphranter) cooperi—Macropus (Osphranter) robustus 25%;* c, *Sarcophilus laniarius—Sarcophilus harrisii 25%;* d, *Wallabia indra—Wallabia bicolor 25%;* e, *Phascolarctos stirtoni—Phascolarctos cinereus 33%;* f, *Macropus siva—Macropus agilis,* slight or no reduction.

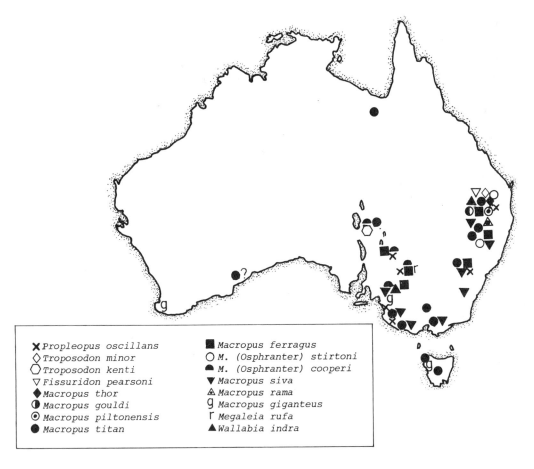

Figure 27.13. Distribution of fossil kangaroos from principal late Pleistocene localities.

Macropus titan was closely related to the living grey kangaroo, *M. giganteus*. In some assemblages it is difficult to determine whether a specimen is a small *M. titan*, possibly a female, or a large *M. giganteus* (Murray and Goede 1977). *Macropus titan* and *Macropus* cf. *titan* are probably nothing more than large Pleistocene morphs of *Marcopus giganteus* and *Macropus fuliginosus*. A similar relationship is found between certain other late Pleistocene kangaroo taxa. *Macropus siva* is a racial form of the living *Macropus agilis* and certain *Macropus* (*Osphranter*) species, i.e. *M. birdselli*, *M. cooperi*, and *M. altus* may represent larger conspecifics of the living euro, *Macropus* (*Osphranter*) *robustus* (fig. 27.12) (Marshall 1973, Main 1978). Because there are only a few fragmentary specimens of the fossil *Osphranter* species, and the characteristics used for their definition are sometimes variable features of which only a few may be present on a fragment, the systematics of this particular group is still fairly tenuous. However, the general pattern is no different than the size reduction phenomenon documented for surviving late Pleistocene faunas throughout Europe (Kurtén 1968).

Several other fossil kangaroo species have been found only in the Darling Downs region in southeastern Queensland (fig. 27.13). *Macropus rama* was similar to the living grey kangaroo in size and morphology (Bartholomai 1975). *Macropus gouldi* is represented only by figures 15–16 of plate 23 in Owen's (1874) description of it, as the type

Figure 27.14. Reconstruction of *Protemnodon brehus*, skull (*Protemnodon* sp.), length 228 mm; standing height about 1,600 mm.

specimen was subsequently lost. *Macropus piltonensis* is a distinctive wallaby with several unique characteristics of the dentition, and *Macropus thor* appears to be very similar to the living *Macropus parryi* (Bartholomai 1975).

The genus *Protemnodon* contains three Australian and one New Guinean species that became extinct during the last glacial (fig. 27.14). *Protemnodon* species are referred to as "giant wallabies," although *Protemnodon anak* was no larger than a female grey kangaroo, and the New Guinean species (*P. otibandus*) is even smaller. *Protemnodon brehus* and *P. roechus* were more substantial animals, the latter having been comparable in mass to *Macropus titan*. At least some of the *Protemnodon* species were compactly built kangaroos with short, stout feet and elongated skulls lacking the characteristic downgrowth of the snout found in modern *Macropus* species (Bartholomai 1973). *Protemnodon* spp. were among the most widely distributed Pleistocene macropodids, especially *P. brehus* which appears to have occurred all over the continent except in the dry center (fig. 27.15).

Protemnodon anak was also a widely distributed species but usually appears in low frequencies in fossil assemblages. The species was probably capable of both browsing and grazing. It retained the long-bladed permanent premolar for shearing stems and petioles. The dentition of *Protemnodon* spp. closely resembles that of the living *Wallabia bicolor*. However, a special taxonomic relationship between *Protemnodon* and the genus *Wallabia* has not been established and the resemblance may only indicate sharing of conservative dental features by the two groups. *Wallabia* is represented in the Pleistocene by *W. vishnu*, probably a large morph of *Wallabia bicolor* (Marshall 1973, Bartholomai 1973).

The Sthenurinae comprise a large group of entirely extinct late Pleistocene browsing kangaroos. The subfamily has been divided into two subgenera, a short-faced *Sthenurus* (*Simosthenurus*) group, and a long-faced *Sthenurus* (*Sthenurus*) group (Tedford 1966) (figs. 27.16, 27.17, 27.18). Both subgenera diversified into a large number of distinct species (Bartholomai 1963; Tedford 1966; Wells and Murray 1979; Merrilees 1965, 1968). Pledge (1980) recommends that the subgenera be given full generic status.

Sthenurines have distinct molar teeth characterized by low crowns and sharp straight lophs with poorly developed midlinks. Some species, particularly the *Simosthenurus* group, have molar "ornamentation" consisting of enamel crenulations in the mid-valley and on the anterior cingulum of each tooth. The premolars are also structurally complex and very distinctive. Rather than being long, bladelike teeth, the perma-

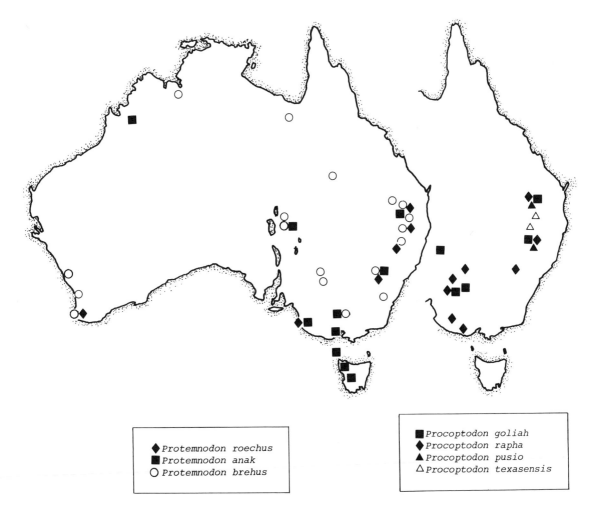

| ◆ Protemnodon roechus |
| ■ Protemnodon anak |
| ○ Protemnodon brehus |

| ■ Procoptodon goliah |
| ◆ Procoptodon rapha |
| ▲ Procoptodon pusio |
| △ Procoptodon texasensis |

Figure 27.15. Distribution of *Protemnodon* and *Procoptodon* species in principal late Pleistocene localities.

nent premolars of *Simosthenurus* are adapted for crushing. Species of *Sthenurus* ranged from small- or medium-sized animals (*S. gilli, S. atlas, S. maddocki*) to large, heavyset forms (*S. orientalis, S. pales*). *Sthenurus gilli* was probably no larger than the living red-necked or Bennett's wallaby and an average size for the genus as a whole would have been comparable to the grey kangaroo.

The long-faced sthenurines represent a trend towards grazing within the subfamily, in parallel with the Macropodinae (Sanson 1976). The short-faced sthenurines were specialized folivorous browsers. The hindlimb morphology of complete skeletons recovered from caves near Tantanoola in South Australia (Pledge 1980) suggest that *Simosthenurus* was a bipedal saltator capable of short bursts of speed. The elongated powerfully built foot of *Sthenurus occidentalis* is composed of a robust fourth metatarsal accompanied by only a rudimentary metatarsal V. The animal's foot would have been effectively composed of a single large toe (Tedford 1966).

Procoptodon goliah was the largest member of the kangaroo family (Table 27.3). At present its relationship within the Macropodidae is in dispute. Tedford (1966) describes many striking similarities of *Procoptodon* to the sthenurines; Sanson (1976)

Figure 27.16. Reconstruction of *Sthenurus (Simosthenurus) occidentalis;* skull length about 185 mm; standing height about 1,200 mm.

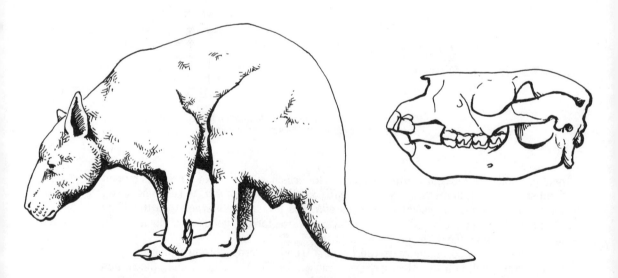

Figure 27.17. *Sthenurus (Sthenurus) atlas,* skull length about 180 mm; standing height about 1,200 mm.

considers these to be parallel developments and places *Procoptodon* directly in the Macropodinae. The larger morphs of *Procoptodon goliah* probably stood between 2.5 and 3 meters tall and would have weighed around 200 to 300 kg (fig. 27.19). The size of individuals from different localities varies considerably. A west to east cline is evident between Lake Menindee, New South Wales (largest) to Darling Downs, Queensland

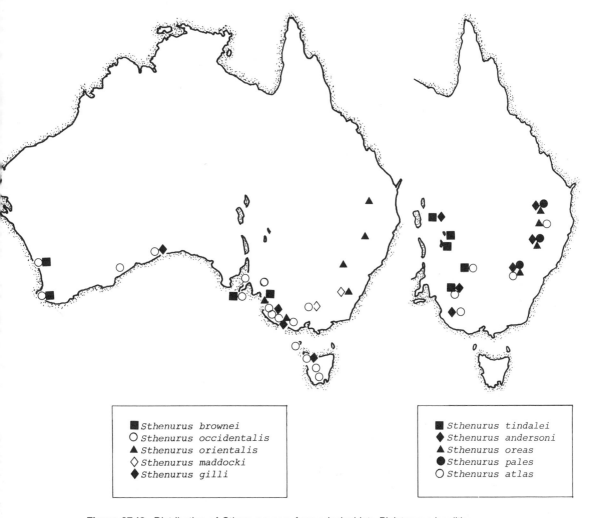

Figure 27.18. Distribution of *Sthenurus* spp. from principal late Pleistocene localities.

(smallest) (Marcus 1967) (fig 27.15). The massive skulls of both *Procoptodon goliah* and *Procoptodon rapha* are as high as they are long. The upper incisors are greatly reduced and peglike, resembling those of koalas. Their short, massive mandibles are remarkably similar to that of *Simosthenurus,* complete with long ascending rami, short diastemata, small blade-shaped incisor crowns, and broad, crushing permanent premolars.

    Procoptodon goliah had long forearms with which it may have pulled down foliage-laden branches (Tedford, 1967). Like *Sthenurus*, the lateral toe was reduced to a splint (Tedford 1966). The heavy jaws, low-crowned molars, and reduced incisors suggest that *Procoptodon* species were folivorous browsers, but Sanson et al. (1980) gives evidence for their having been grazers. *Procoptodon pusio* was smaller than *P. goliah* and *P. rapha.* A recently described species, *P. texasensis,* may also have been present in the late Pleistocene (Archer 1978).

    Propleopus oscillans is the giant muskrat kangaroo. It is represented only by mandibles and a maxillary fragment (fig. 27.20). No postcranial material has been assigned to this form, but doubtless they exist unrecognized in some of the vast Pleistocene vertebrate assemblages held in the larger Australian museums.

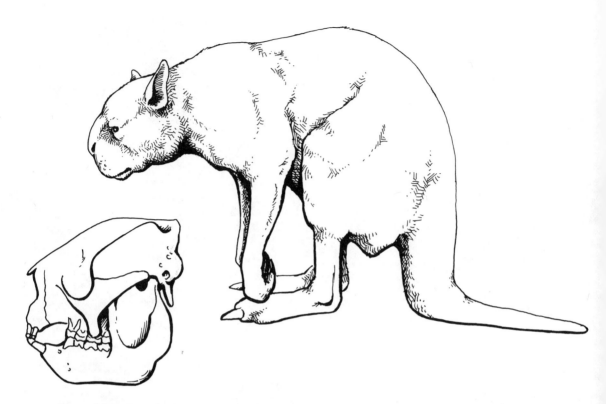

Figure 27.19. Reconstruction of *Procoptodon goliah;* skull length about 225 mm; standing height about 2,600 mm.

Judging from the size of the jaws, *Propleopus* was at least as large as *Macropus rufogriseus,* the red-necked or Bennett's wallaby, and possibly as large as *Macropus giganteus.* However, some forms have proportionally large heads associated with a comparatively small body. Its morphology differs little from its tiny relative, *Hypsiprymnodon moschatus* (Woods 1960), and its appearance has been reconstructed accordingly, having made the appropriate allometric adjustments.

Discussion

From the point of view of taxonomy, there are three kinds of extinction patterns. The first is extinction of complete higher taxonomic units: the subfamily Sthenurinae (including *Procoptodon*), the family Thylacoleonidae, and the superfamily Diprotodontoidea. Second is extinction of genera and species from families and subfamilies that continue to exist, and, third, the disappearance of large morphs of extant species that have retained the status of distinct species, i.e. *Macropus titan* and *M.* cf. *titan*, *Macropus siva* or *Macropus agilis siva*, *Wallabia vishnu* and *Macropus* (*Osphranter*) *cooperi* (possibly synonymous with *M.* [*Osphranter*] *altus* and *M.* [*Osphranter*] *birdselli*) (fig 27.12). Incidentally, in some instances, the converse has occurred, as for example, with *Dasyurus maculatus bowlingi,* possibly the Kangaroo Island *Macropus fuliginosus* and the Tasmanian *Trichosurus vulpecula,* all of which developed into large morphs of their species on islands created by the post-Pleistocene marine transgression.

Adding the surviving genera and species of monotremes and marsupials (only those species known from the fossil record and contemporary with the extinct fauna), it is clear that the faunal diversity during the last glacial of Australia was by no means

Figure 27.20. Reconstruction of *Propleopus oscillans*, the giant musk rat kangaroo, skull length 185 mm; standing height 1,000 mm.

inferior to that of any other continent. While deficient in genera (Australia = 15, U.S. = 32), Australia is comparable to the United States in total number of extinct large species (Australia = 47, U.S. = 51) (fig. 27.21). This is particularly significant when account is taken of the extensive faunal exchange that took place throughout the Tertiary in the American, Eurasian, and African land masses. Because of its isolation, Australia was deprived of significant faunal interchange, receiving only rodents and bats in the Tertiary, man in the late Pleistocene, and the dingo during the early or mid-Holocene. While the rodent invasion appears to have had no significant impact on the marsupial fauna, the arrival of man and the dingo almost certainly did.

It is probably not a coincidence that the larger, more specialized faunal elements seem to have become extinct on the Australian mainland by about 30,000 years ago. In Tasmania, the phenomenon may have been delayed about 10,000 years, probably because man had not arrived there until after the interstadial.

Because Tasmania has been periodically isolated from the mainland by rhythmic marine transgressions, it is possible to say with some assurance when man could or could not have reached it during the last glacial. Evidence for the last interstadial transgression indicates a sea level rise sufficient to sever the land bridge connecting Tasmania and the mainland for the greater part of an interval dating between 30,000 and 50,000 years ago (Chappell 1976). Man either preceded this period, having arrived in Tasmania between 80,000 and 50,000 years ago, or did not arrive until later, less than 30,000 years ago.

On the mainland, evidence of human occupation extends beyond 35,000 years ago (Bowler and Thorne 1976). A consensus is building that at least one and possibly two or more populations of early *Homo sapiens* reached Australia considerably before that time, perhaps 50,000 or more years ago (Birdsell 1977). Certainly 40,000 years ago would not be too early for the first human landfall on the Australian continent.

Full glacial occupation sites in Tasmania, of which there are now three radiometrically dated examples, range from ca. 19,000 to 22,500 yr B.P. (Murray et al. 1980, Bowdler 1975, Rhys Jones personal communication). While too small a sample to be conclusive, the close temporal correspondence of the sites suggests that man entered Tasmania after the interstadial, perhaps around 25,000 years ago.

Figure 27.21. Silhouettes of most of the extinct late Pleistocene Australian vertebrate species drawn to scale (human hunter provides scale).

Row 1, right to left: *Palorchestes azeal, Zygomaturus trilobus, Diprotodon optatum, Diprotodon minor, Euowenia grata (Nototherium* not shown).

Row 2, *Thylacoleo carnifex, Ramsayia curvirostris, Phascolonus gigas, Phascolomys major, Phascolomys medius, Vombatus hacketti, Phascolarctos stirtoni, Propleopus oscillans.*

Row 3, *Procoptodon goliah, Procoptodon rapha, Procoptodon pusio, Sthenurus maddocki, Sthenurus brownei, Sthenurus occidentalis, Sthenurus orientalis (P. texasenis* not shown).

Row 4, *Sthenurus gilli, Sthenurus atlas, Sthenurus tindalei, Sthenurus pales, Sthenurus oreas, Sthenurus andersoni, Troposodon minor, Wallabia indra (not shown, Fissuridon, Troposodon kenti).*

Row 5, *Protemnodon roechus, Protemnodon anak, Protemnodon brehus, Macropus ferragus, Macropus (Osphranter) birdselli, Macropus siva, Macropus titan.*

Row 6, *Macropus rama, Macropus thor, Macropus piltonensis, Macropus gouldi, Macropus stirtoni, Sarcophilus laniarius, Zaglossus hacketti, Zaglossus ramsayi.*

Row 7, *Progura naracoortensis, Progura gallinacea, Genyornis newtoni, Megalania prisca, Wonambi naracoortensis.*

At Beginner's Luck Cave in central Tasmania, there is one cuboid of *Macropus* cf. *titan,* the large Pleistocene morph of *Macropus giganteus* (Murray et al. 1980). Cave Bay Cave contains *Macropus greyi,* a small plains-adapted kangaroo confined to south Australia in the Holocene, where it became extinct early this century (Horton and Murray 1980). A recently excavated cave in southwest Tasmania dates back to about 19,000 years ago and contains only modern species (Rhys Jones personal communication).

The rarity or total lack of extinct marsupial remains in these late glacial ar-chaeological sites is puzzling in light of certain comparatively young dates for fossil remains found in some caves in the northwest and central portions of the island. While the dating of these situations is not without problems, at least some of the determina-tions seem to indicate the persistence of a typical last glacial marsupial/monotreme megafauna well into and beyond the glacial maximum (Murray and Goede 1976, Goede and Murray 1979, Goede et al. 1978).

The Montagu Caves in northwest Tasmania contain abundant remains of *Sthenurus,* but only one or a few individuals of *Thylacoleo, Palorchestes, Zygomaturus, Protemnodon,* and *Zaglossus.* A speleothem incorporated in an underlying stratum gave a U/Th age of 24,000±4,000 yr B.P. indicating that the superposed unit must represent the last glacial maximum or later (Murray and Goede 1976).

Titan Shelter, a site in the Florentine Valley, central Tasmania, contains a similar fauna. Associated charcoal dated to 14,310+2,970/−2,160 yr B.P. (Gak 6875), the large standard error being due to the small sample size. Because charcoal in the lower unit of the excavation may have been reworked, the date must be regarded with circumspec-tion. However, the scarcity of extinct forms in the upper unit suggests that the date may represent a minimum age for the final phase of sedimentation within the cave. The upper unit contains *Sthenurus* and large *Macropus* teeth and appears to be very late last glacial (Goede and Murray 1979).

The evidence, thin though it may be, suggests that critical minimum population sizes of the characteristic Pleistocene fauna in Tasmania were reached between 25,000 and 15,000 years ago. Critical minimum populations on the mainland seem to have occurred between 35,000 and 25,000 years ago (Hope 1978), but this statement requires some qualification because certain extinct forms continue to appear sporadically in late glacial sites to as recently as 19,000 B.P. (McIntyre and Hope 1978). The Lancefield Site in southern Victoria contains *Protemnodon anak, Protemnodon* cf. *brehus, Sthenurus occidentalis, Diprotodon* sp., and *Macropus titan* and is dated to about 26,000 yr B.P. (Gillespie et al. 1978). However, the large grey kangaroo morph, *Macropus titan,* makes up over 90 percent of the total number of individuals, with all other species except *Protemnodon* (7 percent) composing less than 1 percent.

The Lancefield faunal assemblage indicates that certain megafaunal species (e.g. *Sthenurus occidentalis*) that were predominant in earlier sites were in very low numbers by 25,000 years ago. A similar phenomenon is evident from the Willandra Lakes system where the fauna is essentially modern by 30,000 years ago, with a single megafaunal element *(Procoptodon goliah)* hanging on in small numbers until the glacial maximum. The pattern appears to be one of initially rapid population decline before 30,000 years ago, with a slowly dwindling minimal population of a few forms persisting until 25,000 to 20,000 years ago. Because man may not have settled Tasmania until approximately 25,000 years ago, the effect of human presence was not registered until about ten millenia after it occurred on the mainland.

It is possible that the highly nomadic life style of early Australian hunters would leave little in the way of food remains because they would have consumed their game in the open, possibly at or near the kill site rather than returning to limestone caves with all or part of the carcass, where accumulated remains might then be preserved. The generally congenial climate of Australia (even during the Pleistocene) would have made extensive use of caves less likely, except perhaps in the Southeast and Tasmania. Whatever the reason for the lack of faunal remains, the presence of man appears to be associated in some way with the absence of the extinct fauna. Because the majority of these extinct forms were medium- to large-sized, specialized herbivorous species, some having very specialized feeding adaptions, it is possible that they were already under some kind of ecological stress by the time man arrived or that something, either their habits or locomotion, predisposed them to extinction.

Table 27.4. Comparison of Extinct Australian Marsupial

Genus	Dentition	Oral Specializations
Zaglossus	none	long spiny tongue, rostrum beaklike
Thylacoleo	p_3^3-carnassials; long, pointed incisors	large gape, short heavy jaws
Phascolonus	open rooted molars, spatulate incisors	Probably strong mobile lips
Phascolarctos	selenodont molars, peglike upper incisors	small mouth; short, narrow snout; long tongue; short, deep mandible
Palorchestes	selenodont molars, peglike upper incisors	long, narrow rostrum; trunklike proboscis; long, slender tongue
Zygomaturus	bilophodont molars, tusklike upper incisors	muscular, projecting, rhinocerouslike upper lip
Diprotodon	bilophodont molars, long procumbant chisel-like incisors	trunklike upper lip; long narrow tongue, long diastema, buccinator sacculations
Propleopus	bilophodont molars, large bladelike p_3^3	—
Protemnodon	bilophodont molars with weak midlinks; long, bladelike p_3^3	long diastema; long, narrow rostrum
Macropus	bilophodont molars with strong midlinks; reduced p_3^3	long, narrow rostrum; long diastema
Sthenurus	bilophodont molars with low, sharp crowns and weak midlinks; large crushing p_3^3	short diastema; short, narrow rostrum; short, deep mandible; large buccinator sacculations
Procoptodon	bilophodont molars with elaboration of lophs, low crowned; crushing p_3^3; peglike upper incisors	short diastema; narrow, short rostrum; short, deep mandible

The common denominator among the larger extinct species appears to be a browsing habit (Calaby 1976). Grazing species, for example *Macropus*, were affected, but comparatively few Pleistocene grazing genera actually became extinct, undergoing instead only a reduction in body size (Main 1978).

Conclusions

Until the end of the last glacial, the Australian monotreme/marsupial fauna was comparable to other continental faunas in terms of species diversity. It differed in two important aspects, however. The marsupial fauna contains comparatively few genera—the diversity occurring at the species rather than genus level. This, combined with an almost total lack of intercontinental faunal exchange, lends a homogenous character to the fauna not exemplified elsewhere, even in South America in the late Tertiary. These two factors combined may have been a serious liability to a fauna subjected to ecological stress. While many species of large marsupials survived the last glacial, they represent only a few genera and they exhibit a comparatively low degree of morphological diversification. Large browsers were entirely lost.

Genera With Living and Extinct Placentals

Limbs	Trunk and Tail	Comparable Forms
short powerful limbs with large claws	compact body and tail, thick skin muscle layer to erect spines	South American anteaters, aardvark
long, powerful limbs; large, curved pollical claw	long body, probably supple spine	machairodonts felids, *Cryptoprocta ferox*
large claws, short limbs	short tail, thick trunk	caviomorph rodents
large claws and specialized grasping hands and feet	compact body, short tail	sloths, lemuroids
short limbs; huge, curved, laterally compressed claws	long, heavy tail	*Okapia, Tapir* chalicothere, pantodont (*Barylambda*)
mobile forelimbs, short, powerful hnidlimbs	heavy body, short tail	rhinoceros, titanotheres
long limbs, small feet with small claws	heavy short body, short tail, moderately long neck	proboscidians, rhinoceros
—	—	caviomorph, rodents, small browsing antelopines
short to moderately long forelimbs and short, stout feet	compact trunk, moderately long tail	small browsing antelopines and cervids
short forelimbs, large claws	long tail	grazing antelopines, bolids
moderately long forelimbs, large claws, mobile shoulder girdle, heavy hind limbs	heavy trunk and tail	medium-sized ground sloths
long forelimbs; large claws; mobile shoulder girdle; long, powerful hind legs	heavy trunk and tail	large ground sloths, suids (dentition)

The Australian Pleistocene fauna had evolved convergently with several important placental forms, but few marsupial species developed their specializations to the full extent of the placentals (Table 27.4). *Zygomaturus*, for example, appears to have had rudimentary nasal horns, comparable to certain early Tertiary titanotheres or small horned rhinoceroses. *Palorchestes* had a trunklike upper lip, but not as well developed as that of the tapirs. Indeed, the diprotodontoids exhibit trends reminiscent of Tertiary perissodactyls in general, but did not become as specialized or diversified. Of course, the koala, numbat, and the monotremes are exceptions to this generalization. No placentals are as specialized.

The implication of these observations is that the Australian fauna was (and still is) less able to withstand the impact of ecological change in the presence of man, even with the small numbers that were present 50,000 to 30,000 years ago.

References

Archer, M. 1978. Quaternary vertebrate faunas from the Texas caves of southeastern Queensland. *Memoirs of the Queensland Museum* 19(1):61–109.

———. 1981. A review of the origins and radiations of Australian mammals. In A. Keast (ed.). *Ecological Biogeography of Australia.* W. Junk, The Hague.

Archer, M. and Bartholomai, A. 1978. Tertiary mammals of Australia: a synoptic review. *Alcheringa* 2(1):1–19.

Banks, M., Colhoun, E., and Van de Geer, G. 1976. Late Quaternary *Palorchestes azeal* (Mammalia, Diprotodontidae) from northwest Tasmania. *Alcheringa* 1:159–166.

Bartholomai, A. 1962. A new species of *Thylacoleo* and notes on some caudal vertebrae of *Palorchestes azeal*. *Memoirs of the Queensland Museum* 14:33–40.

———.1963. Revision of the extinct macropodid genus *Sthenurus* Owen in Queensland *Memoirs of the Queensland Museum* 14:51–76.

———. 1968. A new fossil koala from Queensland and a reassessment of the taxonomic position of the problematical species *Koalemus ingens* De Vis. *Memoirs of the Queensland Museum* 15(2):65–72.1

———. 1973. The genus *Protemnodon* Owen (Marsupialia, Macropodidae) in the Upper Cainozoic deposits of Queensland. *Memoirs of the Queensland Museum* 16:309–363.

———. 1975. The genus *Macropus* Shaw (Marsupialia, Macropodidae) in the Upper Cainozoic deposits of Queensland. *Memoirs of the Queensland Museum* 17:195–235.

———. 1978. The rostrum in *Palorchestes* Owen (Marsupialia, Diprotodontidae). Results of the Ray E. Lemley Expeditions Part 3. *Memoirs of the Queensland Museum* 18(2):145–150.

Bergamini, D. 1964. *The land and wildlife of Australia.* New York Life Nature Library.

Birdsell, J. 1977. Recalibration of a paradigm for the first peopling of greater Australia. In *Sunda and Sahul,* J. Allen, J. Golson, and B. Jones (eds.). Academic Press, London.

Bowdler, S. 1975. Further radiocarbon dates from Cave Bay Cave, Hunter Island, northwest Tasmania, *Australian Archeology* 3:24–26.

Bowler, J. and Thorne, A. 1976. Human remains from Lake Mungo: discovery and excavation of Lake Mungo III. In *The Origins of the Australians,* R. L. Kirk and A. G. Thorne (eds.). Human Biology Series No. 6, Humanities Press, New Jersey.

Calaby, J. 1976. Some biogeographical factors relevant to the Pleistocene movement of man in Australia. In *The Origins of the Australians,* R. L. Kirk and A. G. Thorne (eds.). Human Biology Series No. 6, Humanities Press, New Jersey.

Chappel, J. 1976. Aspects of Late Quaternary palaeogeography of the Australian East Indonesian region. In *The Origins of the Australians,* R. L. Kirk and A. G. Thorne (eds.). Human Biology Series No. 6, Humanities Press, New Jersey.

DeVis, C. W. 1888. On *Diprotodon minor* Hux. *Proc. Roy. Soc. Qld.* 5:38–44.

Dun, W. 1895. Notes on the occurrence of monotreme remains in the Pliocene of New South Wales. *Records of the Geological Survey of New South Wales* IV:118–126.

Finch, E. 1971. *Thylacoleo* marsupial lion or marsupial sloth? *Australian Natural History* 17:7–11.

Gill, E. D. 1954. Ecology and distribution of the giant marsupial, *Thylacoleo. Victorian Naturalist* 71:7–11.

———. 1955. The range and extinction of *Diprotodon minor* Huxley. *Proc. Roy. Soc. Vict.* 67:225–228.

Goede, A. and Murray, P. 1979. Late Pleistocene bone deposits from a cave in the Florentine Valley, Tasmania. *Pap. Proc. Roy. Soc. Tasm.* 113:39–52.

Goede, A., Murray, P., and Harmon, R. 1978. Pleistocene man and megafauna in Tasmania: dated evidence from cave sites. *The Artefact* 3:139–149.

Hope, J. H. 1973. Mammals of the Bass Strait Islands. *Proc. Roy. Soc. Vict.* 85:163–196.

———. 1978. Pleistocene mammal extinctions: the problem of Mungo and Menindee, New South Wales. *Alcheringa* 2:65–82.

Horton, D. 1977. A 10,000-year-old *Sarcophilus* from Cape York. *Search* 10:374–375.

Horton, D. and Murray, P. 1980. The extinct Toolach wallaby *(Macropus greyi)* from a spring mound in North Western Tasmania, *Records of the Queen Victoria Museum* 71:1–12.

Horton, D. and Wright, R. V. S. 1981. Cuts on Lancefield bones: Carnivorous *Thylacoleo,* not humans, the cause. *Arch. in Oceania* 16(2):73–80.

Jones, F. W. 1969. *The Mammals of South Australia.* Parts I–III. Government Printer, Adelaide.

Keble, R. A. 1945. The stratigraphical range and habitat of the Diprotodontidae in southern Australia. *Proc. Roy. Soc. Vict.* 57 (n.s.):23–48.

Kurtén, B. 1968. *Pleistocene Mammals of Europe.* Aldine, Chicago.

Lord, C. and Scott, H. 1924. *A Synopsis of the Vertebrate Animals of Tasmania.* Hobart, Oldham Beddome and Meredith.

McIntyre, M. L. and Hope, J. H. 1978. *Procop-*

todon fossils from the Willandra Lakes western New South Wales. *The Artefact* 3(3):117–132.

Mahoney, J. and Ride, W. 1975. Index to the genera and species of fossil mammalia described from Australia and New Guinea between 1838 and 1968. *Western Australian Museum Special Publication.* 6:1–250.

Main, A. R. 1978. Ecophysiology: Towards an understanding of late Pleistocene marsupial extinction. In D. Walker and J. C. Guppy (eds.). *Biology and Quaternary environments.* Australian Academy of Science, Canberra.

Marcus, L. 1967. The Bingara fauna. *University of California Publication in Geological Sciences* 114:1–139.

Martin, P. and Guilday, J. 1967. A bestiary for Pleistocene biologists. In *Pleistocene Extinctions,* P. Martin and H. Wright, (eds.). Yale University Press, New Haven.

Marshall, L. G. 1973. Fossil vertebrate faunas from the Lake Victoria Region, S. W. New South Wales, Australia. *Mem. Nat. Mus. Vict.,* 34:151–172.

Merrilees, D. 1965. Two species of the extinct genus *Sthenurus* Owen (Marsupialia, Macropodidae from southeastern Australia including *Sthenurus gilli* sp. nov.). *Journal of the Royal Society of Western Australia* 5(1): 1–24.

———. 1968. Southwestern Australian occurrences of *Sthenurus* (Marsupialia, Macropodidae including *Sthenurus brownei* sp. nov.). *Journal of the Royal Society of Western Australia* 50:65–79.

Murray, P. 1977. Insect predation in the brushtail possum. *Tasmanian Naturalist* 49:5–6.

———. 1978a. Late Cenozoic monotreme anteaters. *The Australian Zoologist* 20(1): 29–55.

———. 1978b. A Pleistocene spiney anteater from Tasmania (Monotremata: Tachyglossidae, *Zaglossus*). *Papers and Proceedings of the Royal Society of Tasmania* 122: 39–68.

———.1978c. Australian megamammals: restorations of some Late Pleistocene fossil marsupials and a monotreme. *The Artefact* 3(2):77–99.

Murray, P. and Goede, A. 1977. Pleistocene vertebrate remains from a cave near Montagu, N. W. Tasmania. *Records of the Queen Victoria Museum* 60:1–30.

Murray, P., Goede, A., and Bada, J. 1980. Pleistocene human occupation at Beginner's Luck Cave, Florentine Valley, Tasmania. *Archaeol. Phys. Anthropol. Oceania* 15: 142–152.

Owen, R. 1874. On the fossil mammals of Australia—Part IX family Macropodidae; Genera *Macropus, Pachysiagon, Leptosiagon, Procoptodon,* and *Palorchestes. Philos. Trans. Roy. Soc. Lond.* 164:783–803.

Partridge, J. 1967. A 3,300-year-old Thylacine (Marsupialia, Thylacinidae) from the Nullabor Plain, Western Australia. *Journal of the Royal Society of Western Australia* 50: 51–59.

Pledge, N. 1977. A new species of *Thylacoleo* (Marsupialia: Thylacoleonidae) with notes on the occurrence and distribution of Thlacoleonidae in South Australia. *Rec. South Austr. Mus.* 17(b):277–283.

———. 1980. Macropodid skeletons including *Simosthenurus* Tedford, from an unusual 'drowned cave' deposit in the southeast of South Australia. *Records of the South Australian Museum* 18(b):131–141.

Raven, H. and Gregory, W. 1946. Adaptive branching of the kangaroo family in relation to habitat. *American Museum Novitates* 1309:1–33.

Ride, W. 1964. A review of Australian fossil marsupials. *Journal of the Royal Society of Western Australia* 47:97–131.

———. 1970. *A Guide to the Native Mammals of Australia.* Oxford University Press, Melbourne.

Sanson, G. 1976. The evolution of mastication in the Macropodinae. Abstract, *The Australian Mammal Society,* 18th General Meeting, Launceston, Tasmania.

Sanson, G., Riley, J., and Williams, M. 1980. A late Quaternary *Procoptodon* fossil from Lake George, New South Wales. *Search* 11(1–2):38–40.

Scott, H. 1915. A monograph of *Nototherium tasmanicum* (Genus—Owen: sp. nov.) *Tasmanian Department of Mines Geological Survey Records* 4:1–46.

Scott, H. and Lord, C. 1920. Studies in Tasmanian mammals living and extinct, Number III *Nototherium mitchelli.* Its evolutionary trend—the skull and such structures as related to the nasal horn. *Papers and Proceedings of the Royal Society of Tasmania* (1920): 76–96.

Stirton, R. 1959. *Time, Life and Man.* John Wiley, New York.

Tedford, R. 1966. A review of the Macropod

genus *Sthenurus. University of California Publications in Geological Science* 57:1–72.

———. 1967. *The Fossil Macropodidae from Lake Menindee, New South Wales.* University of California Press, Berkeley and Los Angeles, 1–165.

———. 1973. The diprotodons of Lake Callabona. *Australian Natural History* 17: 349–54.

Tindale-Biscoe, H. 1973. *Life of Marsupials.* American Elsevier Publishing Company, New York.

Vaughan, T. 1972. *Mammology.* W. B. Saunders, Philadelphia.

Wakefield, N. 1964. Recent mammalian sub-fossils of the Basalt Plain of Victoria. *Proceedings of the Royal Society of Victoria* 77: 419–425.

Walker, E. 1968. *Mammals of the World* (Vol. 1). The Johns Hopkins Press, Baltimore.

Wells, R. and Murray, P. 1979. A new Sthenurine kangaroo (Marsupialia, Macropodidae) from southeastern South Australia. *Transactions of the Royal Society of South Australia* 103(8):213–219.

Wells, R. and Nichol, B. 1977. On the manus and pes of *Thylacoleo carnifex* (Marsupialia). *Transactions of the Royal Society of South Australia* 101(6):139–146.

Woods, J. 1958. The extinct marsupial genus *Palorchestes* Owen. *Memoirs of the Queensland Museum* 13:177–193.

———. 1960. The genera *Propleopus* and *Hypsiprymnodon* and their position within the Macropodidae. *Memoirs of the Queensland Museum* 13:199–212.

28

Comings and Goings of Late Quaternary Mammals in Extreme Southwestern Australia

DUNCAN MERRILEES

IN A LONG, NARROW COASTAL STRIP between Cape Leeuwin and Cape Naturaliste in southwestern Australia, crystalline basement rocks are capped by dune limestone, some of it submerged. The emergent strip of limestone extends about 90 km from north to south, but seldom more than 10 km from the coast inland. In the south, especially, it is riddled with caves, many containing well-preserved bone mainly of mammals. Four such bone deposits are especially relevant, namely those in Mammoth Cave (Archer et al. 1980), Devil's Lair (Baynes et al. 1976, Balme et al. 1978, Dortch 1979), Skull Cave (Porter 1979), and at Turner Brook (Archer and Baynes 1978) (see fig. 28.1). While this region has been investigated intensively, its paleontological possibilities are by no means exhausted.

Descriptions of extant species are given by Ride (1970), whose species concepts I follow for the most part. Illustrations of teeth and other skeletal parts are given by Merrilees and Porter (1979); an artist's impressions of some totally or locally extinct species are shown in figure 28.2. Although these renderings are tentative and not based on rigorous anatomical studies, they do show that the region's mammalian fauna did not include "giant" species. Indeed, no known Australian mammals of any age were much larger than domestic cattle.

In Mammoth Cave two richly fossiliferous deposits were excavated early in this century under conditions that may have led to loss of, or failure to recover, remains of small species, as well as stratigraphic detail. Since the deposits were complex, material from at least two depositional episodes probably has been lumped together, so that any statement about age of the material must be tentative. On the basis of two radiocarbon dates and some tenuous linking of faunal diversity with drying climate due to glaciation, an age either of 70,000 or 150,000 yr B.P. (possibly even earlier) is indicated. Whatever the age, the origin of the remains seem clear—they are the debris of human occupation of the cave.

Numerous radiocarbon dates from several excavations in deep, earthy deposits in Devil's Lair suggest that its mammal remains represent a long stretch of late Quaternary time, from more than 37,000 yr B.P. to about 6500 yr B.P. Most of the radiometric dates are based on redistributed charcoal and may underestimate the age of associated bone (cf. Archer 1974). The one date (SUA-457) on in situ charcoal from a discrete hearth is around 6,000 years older than would be expected from the otherwise self-consistent series, and may indicate the degree to which the other dates underestimate

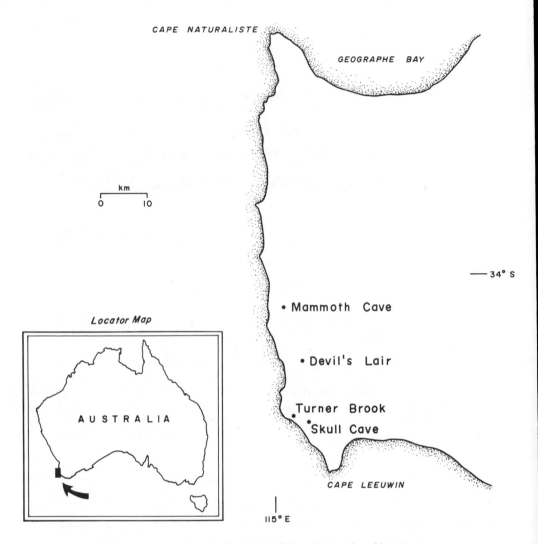

Figure 28.1. Location of principal fossil deposits mentioned in text.

age of bone in the cave fill. If so, it is likely that the Devil's Lair sequence represents the last 40,000 or 50,000 years of Pleistocene time, extending only a little into the early Holocene.

Intermittent deposition in Devil's Lair seems to have taken place throughout this time, but a change of character is marked by what the excavators call Layer 30-lower, with an apparent date of about 33,000 yr B.P. (possibly a true date of about 39,000 yr B.P.). From this layer up, man seems to have been the accumulator of the abundant, highly fragmented bone that is present, probably with some contribution by owls and possibly also by marsupial predators. Below this layer, sedimentation seems to have been more rapid, and there is an incongruous mixture of unbroken bones of small mammals with fragments of bones of large mammals. The unbroken bones are interpreted as owl prey, the large fragments as washed out of some as yet undiscovered primary deposit and secondarily incorporated with the owl prey remains. On this interpretation, the age of the large mammal remains is not known, and it is possible that they are the same order of age as those in Mammoth Cave.

Lagorchestes hirsutus

Sthenurus brownei

Petrogale penicillata

Onychogala (unguifera?)

At least three species of native rats and mice

Sthenurus occidentalis

Vombatus hacketti

Macroderma gigas

Tall Aboriginal man

Potorous tridactylus

Sarcophilus harrisii

Phascolarctos cinereus

At least one, probably two, kangaroo-like species larger than living grey kangaroo

Bettongia penicillata

Perameles (bougainville?)

Protemnodon brehus

Bettongia lesueur

Zygomaturus trilobus

Wallaby of medium size, probably an undescribed species

Thylacoleo carnifex

Thylacinus cynocephalus

Tachyglossus aculeatus

Zaglossus hacketti

Larger Tachyglossus

Wonambi naracoortensis

Figure 28.2. Some species that once lived in the Cape Leeuwin–Cape Naturaliste region, but are no longer found there or, in some cases, anywhere. All to scale of human figure. (Reprinted from the *Journal of the Royal Society of Western Australia*).

Here and in Tables 28.1, 28.2 and 28.3 I will distinguish the upper layers in Devil's Lair (Layer 30-lower and above) from the lower, apparently heterogeneous layers. Broken bones, at least one bone artifact, and stone artifacts in these lower layers suggest that man was the bone accumulator in the undiscovered primary deposit.

**Table 28.1. Late Quaternary Small Mammal Species
Recorded From Southwestern Corner of Australia**

p = species persisting in region into historic time; a = absent from deposit; w = gone from region but persisting on western side of continent; e = surviving only on eastern side of continent; ext = not surviving anywhere into historic time; "small" = not larger than *Rattus rattus*.

Species	Mammoth Cave	Devil's Lair Lower	Devil's Lair Upper	Skull Cave	Turner Brook
Marsupials					
Sminthopsis murina	p	p	p	p	p
Antechinus flavipes	p	p	p	p	p
Phascogale tapoatafa	p	p	p	p	p
Cercartetus concinnus	a	p	p	p	p
Tarsipes spencerae	a	p	p	p	a
Murids					
Pseudomys albocinereus	w	w	w	w	w
Pseudomys occidentalis	a	w	w	a	a
Pseudomys shortridgei	p	p	p	p	p
Pseudomys praeconis	a	a	p	p	p
Notomys mitchellii	a	w	w	a	a
Rattus fuscipes	p	p	p	p	p
Rattus tunneyi	a	a	a	p	p
Bats					
Macroderma gigas	a	a	w	a	a
Small carnivorous bats	p	p	p	p	p

Skull Cave appears to have operated as a pit trap, and two radiocarbon dates, neither representing the lowest layers excavated, suggest that the excavated material may represent most or all of Holocene time. Though artifacts were found, and some of the bone is charred, major human contribution to the deposit is not suspected; any contribution by owls is also thought to be minimal.

By contrast, the bone fillings of two small cavities in a cliff face at Turner Brook are thought to represent owl prey exclusively, and a single radiocarbon date (on mammal hair) of about 400 yr B.P. suggests that the small mammal fauna of immediately prehistoric time is involved. Thus the Turner Brook deposits make a valuable supplement to sketchy early historic records of the mammals of the region.

Together Mammoth Cave, Devil's Lair, Skull Cave, Turner Brook, and early historic records provide some insight into fluctuations in the mammal fauna of the region from some unknown Pleistocene time probably coincident with the second or third last glacial maximum in higher latitudes, through the last glacial maximum of about 20,000 B.P., and through Holocene time into the last century. Modifications of the mammal fauna by the activities of European settlers over the present century have been extreme, but for present purposes these will be ignored. Our end point is the beginning of the historic period in this region, about a century ago.

With one possible exception, all the prehistoric mammal species known from this region are represented in these four deposits. The possible exception is a large macropod, which may or may not be the large kangaroo *Macropus titan*, known from a few limb bone fragments from Strongs Cave near Devil's Lair. Other caves in the region have provided records of mammals, but few are radiometrically dated. Only one case need be considered further; this is Yallingup Cave in the northern part of the region, where remains of *Thylacinus,* the "Tasmanian" tiger, overlie those of the dingo (*Canis*), showing that thylacines persisted in the region until after the undated (but probably

**Table 28.2. Late Quaternary Intermediate-Sized Mammal Species
Recorded From Southwestern Corner of Australia**

Species	Mammoth Cave	Devil's Lair Lower	Devil's Lair Upper	Skull Cave	Turner Brook
Monotremes					
Tachyglossus aculeatus	w*	a	a	a	a
Larger *Tachyglossus*	ext	a	a	a	a
Zaglossus hacketti	ext	a	a	a	a
Marsupials					
Dasyurus geoffroii	p	p	p	p	p
Sarcophilus harrisii	e	e	e	e	e
Thylacinus cynocephalus	e	e	e	a	a
Isoodon obesulus	p	p	p	p	p
Perameles (bougainville?)	w	w	w	w	a
Trichosurus vulpecula	p	a	p	p	p
Pseudocheirus peregrinus	p	p	p	p	p
Phascolarctos (cinereus?)	e	a	e	a	a
Vombatus hacketti	ext	ext	a	a	a
Potorous tridactylus	p	p	p	p	p
Bettongia penicillata	a	a	p	p	p
Bettongia lesueur	a	a	w	a	a
Petrogale (penicillata?)	a	a	w	a	a
Lagorchestes (hirsutus?)	a	w	w	a	a
Onychogalea (unguifera?)	w	a	a	a	a
Macropus eugenii	p	a	p	a	a
Macropus irma	p	p	p	a	a
Setonix brachyurus	p	p	p	p	p
Unidentified wallaby (two species?)	ext	a	a	a	a
Eutherians					
Hydromys chrysogaster	a	a	p	p	p
Canis familiaris	a	a	a	p	a

*see Table 28.1 for meaning of codes

**Table 28.3. Late Quaternary Large Mammal Species
Recorded From Southwestern Corner of Australia**

("large" = comparable to man or larger)

Species	Mammoth Cave	Devil's Lair Lower	Devil's Lair Upper	Skull Cave	Turner Brook
Marsupials					
Thylacoleo (carnifex?)	ext*	a	a	a	a
Macropus fuliginosus	p	p	p	p	a
Sthenurus brownei	ext	ext	ext	a	a
Sthenurus occidentalis	ext	ext	a	a	a
Protemnodon brehus	ext	ext	ext	a	a
(larger *Protemnodon?*)	ext	a	a	a	a
Zygomaturus trilobus	ext	ext	a	a	a

*see Table 28.1 for meaning of codes

Holocene) arrival of dingoes. Throughout the late Quaternary man has been present, but his time of arrival in the region is not known.

Data from the four main mammal-bearing deposits are summarized in Tables 28.1, 28.2 and 28.3. The most easily understandable group of taxa in the tables is those marked "p" throughout. These are the small marsupial carnivores *Sminthopsis,*

Antechinus, and *Phascogale;* the large native mouse *Pseudomys shortridgei;* the bush rat (*Rattus fuscipes*), the carnivorous cat-sized *Dasyurus,* the short-nosed bandicoot (*Isoodon*), the ringtail possum (*Pseudocheirus*), the rat-kangaroo *Potorous,* the small wallabylike quokka (*Setonix*), and small carnivorous bats. The bats are too heterogeneous to warrant further discussion. Of the rest, all but *P. shortridgei* fall into the first of two ecological categories set up by Baynes et al. (1976), namely the "forest" (as opposed to "nonforest") mammals.

Forest mammals are those whose ranges included but were not necessarily confined to forest; some, like *Dasyurus,* were very wide-ranging, and little can be inferred about past habitats from fossils. On the other hand, *Potorous,* and to a lesser extent *Setonix,* seem to have been sufficiently restricted to permit the inference that some forest persisted in the region throughout the period under study, even if only in patches and not in a continuous zone as in historic time.

The only persistent member of the nonforest group, *Pseudomys shortridgei,* may have come from coastal heath that is, and probably always has been, present on the windward side of a high crest in the dune limestone, even in times of lower sea level and a coastline more remote than at present.

It seems likely that at least three forest taxa were persistent, but for reasons of sampling accident do not so appear in Tables 28.1–28.3. These are *Cercartetus* (pigmy possum), *Tarsipes* (honey possum), and *Macropus fuliginosus* (western grey kangaroo), all known in the region from historic records. The first two are very small animals and their remains may well have escaped detection in Mammoth Cave. *Tarsipes* is unusual in that its teeth are rudimentary and its dentaries, as well as maxillae and premaxillae, are extremely fragile. Unlike most mammals, it is more easily recognized from postcranial than cranial remains. Postcranial material from the Turner Brook deposits does not appear to have been studied in detail, and *Tarsipes* may have been overlooked there. Because the Turner Brook material represents owl prey, one would not expect to find even juveniles of so large an animal as *Macropus fuliginosus* represented, even if the species were abundant in the vicinity.

More puzzling is the record of *Trichosurus vulpecula* (brush possum), wide-ranging throughout the continent, abundant in many areas, and able to persist even within large cities. Yet it is absent from the lower layers in Devil's Lair. One would expect it to appear among the secondarily reworked material there, and to find juveniles among the owl prey remains there as at Turner Brook. Balme et al. (1978) established criteria to make the difficult judgment of whether absence of a given taxon from a deposit is due to sampling accident. On these admittedly arbitrary criteria, *Trichosurus* seems to have been absent from the district for thousands of years; the reasons for its absence are obscure.

Two species that more adequate sampling might show to have been persistent are *Pseudomys praeconis* and *Hydromys chrysogaster,* both known from only a few individuals even from the abundant material spanning tens of thousands of years in the Devil's Lair deposit. The specialized aquatic habitat and size of *Hydromys* might have afforded it protection from owls and humans.

The two wallaby species, *Macropus irma* and *M. eugenii,* are known in the region from historic records; *M. eugenii* was perhaps found only in the northern part where the forest is more open. It is not surprising that they are absent from the owl accumulations at Turner Brook, because even juveniles would be rather large for owls to take. Their absence from the pit trap deposit in Skull Cave, however, is surprising since it contains the remains of at least a hundred individual macropods. It is conceivable that both species vanished from the region during the Holocene, only to reinvade after Europeans cleared the forest in historic times.

In many ways the most informative group of species is those marked "w" in Tables 28.1 through 28.3, those species extinct in the region but surviving in other western regions. All of these that occur in Devil's Lair are included by Baynes et al.

(1976) in the nonforest category. This category also would include *Tachyglossus*, the spiny anteater, and *Onychogalea*, the nail-tailed wallaby, known as fossils in this region only from Mammoth Cave.

The occurrence, and in some cases, abundance, of the native mice *(Pseudomys albocinereus* and *P. occidentalis)*, the hopping mouse *(Notomys)*, the ghost bat *(Macroderma)*, the long-nosed bandicoot *(Perameles)*, the burrowing rat-kangaroo *(Bettongia lesueur)*, the rock wallaby *(Petrogale)*, and the hare wallaby, *Lagorchestes*, making up the "w" group in Devil's Lair and also *Pseudomys shortridgei* and *P. praeconis)* has been used to infer that the region was once more sparsely vegetated than in historic time. While virtually no published information on Pleistocene plant fossils from the region exists, the occurrence of *Tachyglossus aculeatus* and perhaps also *Onychogalea* in Mammoth Cave further supports such an inference. In the virtual absence of published evidence from Pleistocene plant fossils, this inference from mammal fossils becomes especially significant for the region. From the record of *Petrogale* I have suggested that a more mosaic vegetational pattern characterized the region for most of the period under study than would be expected from its historic record (Merrilees 1979a). A greater variety of plant formations (implying a drier climate) would account for the greater variety of mammals.

Gradual replacement of a varied vegetational mosaic under a relatively dry climate by dense and more uniform forest (plus a wind-determined strip of coastal heath) under a wet climate may be postulated in the conditions that existed in the transition from late Pleistocene to Holocene. Such a replacement could explain many of the peculiarities in the data in Tables 28.1 through 28.3.

The longest persisting member of the "w" group of Tables 28.1 through 28.3 is *Pseudomys albocinereus*. Among the very few specimens of small species recovered from Mammoth Cave, abundant in Devil's Lair but declining in abundance with time, present in small numbers but not persisting to the top of the Skull Cave deposit, present in very small numbers at Turner Brook, but not known from historic records it presents a picture of steadily declining habitat availability, though the exact nature of this decline is not yet known.

Similarly, *Perameles,* the more abundant of the two bandicoots in Mammoth Cave, persisted throughout Devil's Lair time and through most of Skull Cave time, though in smaller numbers than *Isoodon*. Since owls caught *Isoodon* in Turner Brook time, probably they would have caught the smaller *Perameles* too if it had been available. Its local extinction date therefore seems to be of the order of 1000 B.P.

Absent from the lowest excavated layers in Devil's Lair, *Notomys* became fairly abundantly represented in both the "lower" and "upper" parts of the deposit, but it is absent from the uppermost layers of Devil's Lair, absent from Skull Cave, Turner Brook, and from historic records. Thus, it seems to have arrived in the region perhaps about 35,000 B.P. and become extinct locally before 10,000 B.P.

Likewise, the three macropods, *Bettongia lesueur, Petrogale penicillata,* and *Lagorchestes hirsutus,* seem to have been late arrivals and early departures but to have differed in timing. Although *Lagorchestes* is so scantily represented that little can be safely inferred, *Petrogale* and *Bettongia lesueur* are conspicuous in the Devil's Lair records and occur in other caves in the region. They were not reported in Mammoth Cave, Skull Cave, or Turner Brook, nor are they recorded historically. *Bettongia lesueur* may not have arrived much before the last glacial maximum, preceded some thousands of years by *Petrogale*. Both survived to the end of the Pleistocene but not much beyond, if at all.

Macroderma, a relatively carnivorous bat, and *Pseudomys occidentalis* appear to have briefer records in the region. Both are known only from Devil's Lair (in the region under discussion); *Macroderma* occurs only at high levels in the deposit, and *P. occidentalis* at middle levels. Allowing for some correction of published radiocarbon dates, the incursion of *Macroderma* into the region may have been early Holocene; that of *P. occidentalis* may have occupied the few thousand years before 30,000 yr B.P. The time of arrival of *Tachyglossus aculeatus* and *Onychogalea* in the region are not known, but they seem to have vanished from it early, before, say, 40,000 yr B.P.

Three species seem to have migrated into the region relatively recently but not simultaneously, and to have persisted, namely *Bettongia penicillata, Canis familiaris,* and *Rattus tunneyi,* all known from historic records. Though it is present from the time of the change in character of the Devil's Lair sediments (i.e. throughout the "upper" part), *Bettongia penicillata* is not a conspicuous element in the mammal fauna until a few thousand years before the glacial maximum. The earliest occurrence of *Canis* in the region appears to be a specimen from Yallingup Cave (Merrilees 1979a) from sediments overlying any which contain *Petrogale*. Although radiometric dates are unavailable, the arrival of *Canis* may be taken provisionally as Holocene, perhaps not much before 3,000 B.P. if the Skull Cave evidence is taken at face value. Admittedly the Yallingup Cave specimen has a chalky, fragile appearance, suggesting a rather greater age. *Rattus tunneyi,* on the other hand, seems to be a later Holocene immigrant.

Three taxa are recorded under "e" (all in Table 28.2), those extinct in the southwest but extant on the eastern side of the continent. Curiously, the one most widely distributed in the east in historic time, the koala *(Phascolarctos)*, seems to be the one that vanished earliest from the southwest, thousands of years before the last glacial maximum. *Sarcophilus* and *Thylacinus,* known as the "Tasmanian" devil and "Tasmanian" tiger, because they were confined to Tasmania in historic time, were widespread over the continent for much of the Holocene. *Thylacinus* is commonly believed to have been supplanted by *Canis* (e.g. Archer 1975). While it is possible that *Sarcophilus* also succumbed to the dingo, in neither case is any mechanism clearly evident. *Sarcophilus* may have persisted longer than *Thylacinus* in the southwest, perhaps even up to historic time, though we have no historic records of it.

If Tables 28.2 and 28.3 could be taken at face value, one might say of the totally extinct taxa ("ext") that the extinction processes were spread over tens of thousands of years. It is possible, however, that the upper Devil's Lair extinct entries represent the undiscovered primary deposit, just as in the lower; with the low numbers of individuals, accidents of sampling may well obscure realities of time distribution. This possibility extends also to the differences between the Mammoth Cave and Devil's Lair entries. Conceivably, the Mammoth Cave deposits, although complex, represent only a short time interval and the inferred primary deposit in Devil's Lair dates also from about this time. Specimens from other caves do not help in making this determination because they occur sporadically in low numbers in deposits not yet dated radiometrically.

Several suggestions may be made about this complex picture of immigration and local extinction, even in the face of current ignorance of the ecological requirements and characteristics of Australian mammals. There seems to be one clear case of replacement of one or two marsupial carnivores by a more successful eutherian—*Thylacinus* (and *Sarcophilus?*) by *Canis*—without any indication that climatic or human effects were involved.

Next, it would seem that a long-term climatic swing from drier to wetter marked the passage of late Pleistocene into Holocene time (Wyrwoll 1979). If conceived as fitful and erratic rather than steady, this shift can accommodate the numerous and differently timed arrivals and local extinctions, with one clear overall result—the replacement of a diverse forest and nonforest fauna with a restricted forest fauna. The local extinctions outnumber the new arrivals.

But the crucial question—what happened to the taxa now totally extinct—cannot be answered yet. It does seem that some process or processes bore more heavily on the larger than on the smaller mammals, and that man was present during the operation of these processes. While it is possible that the processes were geologically instantaneous (spread over one or a few thousand years), the evidence taken at face value is for some longer period, say tens of thousands of years. It is unlikely that any major climatic

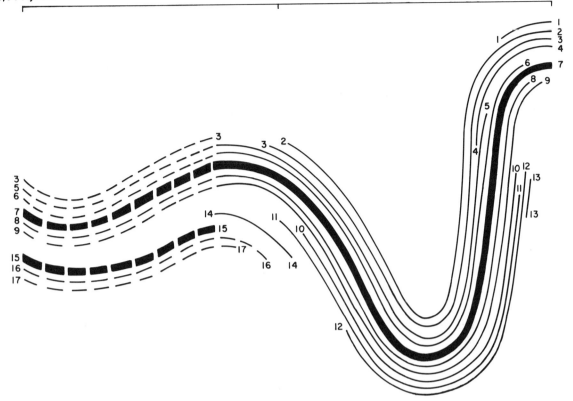

Figure 28.3. Time distribution of mammal species in the Cape Leeuwin–Cape Naturaliste region, southwestern Australia. Schematic only, not to scale. Curves suggest rise and fall of sea level, taken to be correlated with the rise and fall of region's effective rainfall, hence with increasing or decreasing proportion of forest cover. Broken lines indicate uncertain age. 1. *Rattus tunneyi;* 2. *Bettongia penicillata;* 3. *Trichosurus vulpecula;* 4. *Canis familiaris;* 5. *Thylacinus cynocephalus;* 6. *Sarcophilus harrisii;* 7. Man and probably fourteen other species, certainly those entered under "p" in tables 28.1–28.3. May also include *Macropus irma, M. eugenii,* and some small carnivorous bats; 8. *Perameles (bougainville?);* 9. *Pseudomys albocinereus;* 10. *Petrogale (penicillata?);* 11. *Notomys (mitchellii?);* 12. *Bettongia lesueur;* 13. *Macroderma gigas;* 14. *Pseudomys occidentalis;* 15. Six species entered only under "Mammoth Cave" in Tables 28.2 and 28.3; perhaps includes macropod from Strongs Cave; 16. *Phascolarctos (cinereus?), Protemnodon brehus,* and *Sthenurus brownei;* 17. *Vombatus hacketti, Sthenurus occidentalis,* and *Zygomaturus trilobus.*

swing is responsible for the extinctive processes, for they seem to antedate the last major swing, and nothing indicates that they represent the next to last. Therefore, it is tempting to suggest that some human activity underlies these extinctions, whether habitat modification by fire (Hallam 1975), selective hunting or indiscriminately wasteful methods of hunting, or some other activity as yet undiscerned. Further studies, such as those by Balme (1979 and in press) or Archer et al. (1980), of specifically human effects on bone and other paleontological material might be expected to resolve some uncertainties, and radiometric or other absolute dating of known deposits would resolve others. Based on present knowledge of the Cape Leeuwin–Cape Naturaliste region, the most promising expectation seems to be location and systematic excavation and study of the postulated primary deposit near the old entrance to Devil's Lair.

Pending further investigations, present knowledge of the late Quaternary record of the Cape Leeuwin–Cape Naturaliste mammal fauna is summarized in Figure 28.3. In this, it is assumed that neither the postulated old primary deposit in Devil's Lair nor any of the Mammoth Cave sequences is much older than the owl prey in Devil's Lair "lower." It is assumed also that the published sequence of radiocarbon dates from Devil's Lair generally underestimates the true age of associated bones by several thousand years, and that the occurrence of *Phascolarctos, Protemnodon,* and *Sthenurus brownei* in Devil's Lair "upper" reflects later survival of these taxa than *Vombatus, Sthenurus occidentalis,* and *Zygomaturus.*

As read from left (onset of next to last major glaciation) to right (early historic period in this region), Figure 28.3 expresses first the impoverishment in stages of a once varied mammal fauna, the incursion before the last glacial maximum, and the local extinction after it of a nonforest suite, two new arrivals during the Holocene, and a large measure of faunal stability in the persistent suite. But reasons for the extinction, total in many cases, local in a few, before the last glacial maximum, of a substantial suite, mainly of the large mammals, remains obscure.

References

Archer, M. 1974. Apparent association of bone and charcoal of different origin and age in cave deposits. Memoirs of the Queensland Museum. v. 17, pp. 37–48.

———. 1975. New information about the Quaternary distribution of the Thylacine (Marsupialia, Thylacinidae) in Australia. Journal of the Royal Society of Western Australia. v. 57, pp. 43–50.

Archer, M. and Baynes, A. 1973. Prehistoric mammal faunas from two small caves in the extreme south-west of Western Australia. Journal of the Royal Society of Western Australia. v. 55, pp. 80–89.

Archer, M., Crawford, I.M. and Merrilees ,D. 1980. Incisions, breakages and charring, some probably man-made, in fossil bones from Mammoth Cave, Western Australia. Alcheringa v. 4, pp. 115–131.

Balme, J. 1979. Artificial bias in a sample of kangaroo incisors from Devil's Lair, Western Australia. Records of the Western Australian Museum. v. 7, pp. 229–244.

———. in press. An analysis of charred bone from Devil's Lair, Western Australia. Records of the Western Australian Museum.

Balme, J., Merrilees, D. and Porter, J. K. 1978. Late Quaternary mammal remains, spanning about 30,000 years, from excavations in Devil's Lair, Western Australia. Journal of the Royal Society of Western Australia. v. 61, pp. 33–65.

Baynes, A., Merrilees, D. and Porter, J. K. 1976. Mammal remains from the upper levels of a late Pleistocene deposit in Devil's Lair, Western Australia. Journal of the Royal Society of Western Australia. v. 58, pp. 97–126.

Dortch, C. 1979. Devil's Lair, an example of prolonged cave use in south-western Australia. World Archaeology. v. 10, pp. 258–279.

Hallam, S. J. 1975. Fire and hearth. A study of Aboriginal usage and European usurpation in south-western Australia. Australian Institute of Aboriginal Studies, Canberra.

Merrilees, D. 1979a. Prehistoric rock wallabies (Marsupialia, Macropodidae. *Petrogale*) in the far south west of Western Australia. Journal of the Royal Society of Western Australia. v. 61, pp. 73–96.

———. 1979b. The prehistoric environment in Western Australia. Journal of the Royal Society of Western Australia. v. 62, pp. 109–128.

Merrilees, D. and Porter, J. K. 1979. Guide to the identification of teeth and some bones of native land mammals occurring in the extreme south west of Western Australia. Western Australian Museum, Perth.

Porter, J. K. 1979. Vertebrate remains from a stratified Holocene deposit in Skull Cave, Western Australia, and a review of their significance. Journal of the Royal Society of Western Australia. v. 61, pp. 109–117.

Ride, W. D. L. 1970. A guide to the native mammals of Australia. Oxford University Press, Melbourne.

Wyroll, K.-H. 1979. Late Quaternary climate of Western Australia: evidence and mechanisms. Journal of the Royal Society of Western Australia. v. 62, pp. 129–142.

29

Red Kangaroos: Last of the Australian Megafauna

D. R. HORTON

THE FRAMEWORK WITHIN WHICH HYPOTHESES about the extinction of the Australian megafauna must be set is a relatively simple one. Hypotheses that the extinctions were humanly caused were based on the premise that there was a coincidence in the time of the extinctions and the first arrival of people on the Australian continent. An early date for the extinction was important for two reasons. First, it suggested that the extinctions were *not* correlated with the major climatic change of the late Pleistocene (Horton 1979, 1980). Second, a close correlation between extinction and time of human arrival provided a *mechanism* for extinctions, in the sudden presence of man in a country where the fauna and flora had not previously been exposed to man's activities. The megafauna was unable to adapt to either human predation or modification of the environment by fire and so became extinct.

Human arrival in Australia is presently believed to have occurred at least 40,000 years ago. The finding of megafauna in relatively young sites, notably at Lancefield (Gillespie et al. 1978, Ladd 1976, Horton 1976, Horton and Samuel 1978), a site dated at 26,000 yr B.P., had two implications. The timing of the extinctions now correlated with the late Pleistocene climatic changes and the extinctions could not have resulted from the impact of man on a virgin landscape and a naive fauna. "Overkill" *by itself* was therefore no longer tenable, and if fire was the mechanism then either man did not use fire to modify the environment for a long time after arrival or environmental change in response to Aboriginal firing was much slower than had been supposed.

The recognition that the extinctions coincided in time with a period of major climatic change has influenced thinking on the subject in two divergent ways. It has been seen as rescuing the human causation hypothesis because it is suggested that although neither overkill nor fire-induced habitat change is sufficient to cause extinction during normal times, they are effective during periods of climatic stress. Based on this hypothesis, the megafauna had survived earlier periods of climatic stress in the absence of man, but man's presence at the end of the Pleistocene tipped the balance. The corollary is that the megafauna would have also survived through this period if man had not been present. The other point of view is that the proposed mechanisms for human causation are implausible even in times of climatic stress. The economics of hunting are that, other things being equal, it is the most numerous species of a given taxonomic category that is caught most frequently. A species will form the bulk of the prey only until it becomes less numerous than an alternative species so predation cannot cause extinction irrespective of the climatic conditions. Burning is also not a viable mechanism because even if Aborigines changed the Australian fire regime (and I think that they caused little if any change) (see Horton 1982), such changes can only have had the effect of *improving* conditions for the megafauna. If human causation is unlikely, we must look seriously at the hypothesis that climatic change caused extinctions.

When examining causation it is not sufficient simply to find a correlation in the timing of two events. Testable mechanisms must be proposed which are the simplest that fit the facts presently known. I hope the model presented here meets these criteria.

The idea that we should aim to produce the simplest possible model seems arbitrary because the extinction of species must be a complex event and the late Pleistocene in Australia is a complex period. If the extinctions are due to the combined effects of hunting, burning and climatic change, then the actual mechanism would be too complex to be deciphered with the kind of data available from the fossil and archaeological records; it is, then, nonsense to talk about *the* cause of the extinctions.

The question of causation, however, can be framed very simply—had Aborigines not colonized Australia, would the megafauna have become extinct? If the answer is no, then man is the cause; if yes, then climatic change is the cause. The question cannot, of course, be tested directly—we cannot conduct a controlled experiment. Thus, two approaches are open to us. One suggests that the controlled experiment has been conducted for us—the timing of the extinctions in different parts of the world is correlated with the timing of man's arrival in different parts of the world. The Americas, for example, provide a control for Australia: had man not arrived here until 12,000 B.P., the megafauna would have survived until 12,000 B.P. This approach is unacceptable because other variables are involved in addition to the time of human colonization, some of which will be discussed.

Because the continents are so different, I suggest that the causes of extinction on each will also vary and must be considered individually. On each continent we must look for the kind of data that can test various mechanisms. If there is a mechanism by which climatic change can cause extinctions in Australia without man's influence, and we can find evidence that the mechanism has operated, then climatic change is *the* cause of the extinctions in Australia. A separate but related proof would be to show that the climatic change of the late Pleistocene was different in either kind or intensity, or both, to anything that had occurred previously in the Pleistocene. The difference need not necessarily be great, but if *no* difference exists except the presence of man, then even if the proposed climatic change mechanism can be shown to *affect* the megafauna, man is still the *cause* of the extinctions. Even without evidence for the climatic change being different, however, it would be difficult to sustain the human causation argument without supporting evidence such as kill sites or humanly induced vegetation changes coincident with extinctions.

Against this background I present this model for climatic change as the causal agent in the extinction of the Australian megafauna.

Defining the Australian Megafauna

Before considering the distribution of "the megafauna," we must clarify which species are included in the group. The answer carries implications about how we view the extinction process, and we must consider it carefully. Martin (this volume) uses an arbitrary weight of 44 kg as the lower limit in size for the megafauna, but in Australia this would result in including the modern kangaroos (grey, antilopine, wallaroo and red) in the megafauna, and probably excluding such genera as *Thylacoleo, Propleopus,* and the smaller *Sthenurus* from the megafauna.

Table 29.1. The Extinct Megafauna

Family	Genus	Species	Family	Genus	Species
Mammals			Mammals (continued)		
Macropodidae	Propleopus	chillagoensis	Dasyuridae	Sarcophilus	laniarius
		oscillans			
	Macropus	altus	Thylacoleonidae	Thylacoleo	carnifex
		cooperi			
		ferragus	Diprotodontidae	Diprotodon	optatum
		gouldi			minor
		mundjabus		Euowenia	grata
		piltonensis		Euryzygoma	dunense
		rama		Nototherium	inerme
		stirtoni		Zygomaturus	trilobus
		thor		Palorchestes	azael
		titan			
	Wallabia	indra	Vombatidae	Lasiorhinus	medius
	Protemnodon	anak		Phascolonus	gigas
		brehus		Ramsayia	magna
		roechus			
	Troposodon	kenti	Tachyglossidae	Zaglossus	hacketti
		minor			ramsayi
	Fissuridon	pearsoni			
	Sthenurus	andersoni	Birds		
		atlas			
		brownei	Dromornithidae	Genyornis	newtoni
		gilli			
		maddocki	Megapodiidae	Progura	gallinacea
		occidentalis			naracoortensis
		oreas			
		orientalis	Reptiles		
		pales			
		tindalei	Meiolaniidae	Meiolania	oweni
	Procoptodon	goliah			
		pusio	Varanidae	Megalania	prisca
		rapha			
		texasensis	Boidae	Wonambi	naracoortensis

To my knowledge, no real attempt has been made to define the Australian megafauna, although agreement does exist about its composition. The unstated definition would be that it consists of animals that became extinct before the Holocene and are large, either in an absolute sense or relative to other members of some taxonomic rank, or are part of a taxonomic category all of whose members became extinct and some of whose members are large. Table 29.1 lists the megafaunal species covered by this definition. The second part of the definition is necessary to cope with the difficult cases of *Sthenurus* (some members of which were smaller than living kangaroos), *Macropus* (most species of which survived, only the larger ones becoming extinct), and *Propleopus* (which although very large relative to the other Potoroine or rat kangaroos was only about the same size as the living kangaroos). Marshall (1973) suggests treating *Macropus* species as distinct from the rest of the megafauna because most of the larger species appear to have dwarfed living relatives, i.e. the species evolved smaller-sized descendants that did not become extinct, although this distinction is somewhat clouded by disagreements over the relationships between the living and "extinct" species of *Macropus*.

The red kangaroo appears in the title of this chapter because it symbolizes the argument that climatic change caused the extinction of the Australian megafauna. First, it is an excellent symbol for the fauna of the arid areas of Australia, a reminder that this continent is the driest in the world. Second, the presence of its fossil remains in what are now well-watered farming areas in Victoria graphically illustrates the dramatic changes in climate that have occurred during the Pleistocene and Holocene in Australia. Finally, and most importantly, considering the red kangaroo to be a megafaunal species highlights the problem of defining "the megafauna," and the implications of defining that group in a particular way.

Defining "the megafauna" as "species which became extinct near the end of the Pleistocene," the implicitly accepted practice, and then using this grouping to investigate the causes of extinction result in confusion of cause and effect. The megafauna tends to be seen retrospectively as having always been a group, and this can be used as an argument against the climatic extinction model, because surely some members of this group should have been able to find refuge areas and survive. The model presented here, in a sense, reverses the question by asking whether all animal species would have been expected to find refuge during the climatic fluctuations of the late Pleistocene. The answer obviously is no, and the extinct species are those that could not. Similarly, if we think that what we are discussing is the extinction of an *entire group*, this implies that some extraordinary process must be involved because natural processes are not seen as being capable of removing *every* member of a group. The extinction of the dinosaurs has aroused similar feelings. If we were to define the megafauna by an arbitrary body weight, perhaps species weighing more than 20 to 25 kg, clearly a number of megafaunal species did survive. Their survival tells us that particular kinds of "refuges" were available and may also result in a more ready acceptance of the proposition that natural processes are involved in the extinctions.

The most interesting thing about Table 29.1, then, is not the species which are included in it, but those which are not. Some other possible inclusions are the modern kangaroos (*M. giganteus, M. fuliginosus, M. antilopinus, M. robustus, M. rufus*), the wombats, the emu, the cassowary, and the amethyst python. I believe the reasons for the survival of these species are greatly relevant to understanding the cause of extinction of the other megafauna. *Macropus rufus*, the red kangaroo, is of particular importance in achieving this understanding.

Distribution Patterns

The reconstruction of paleoclimates by Jones and Bowler (1980) dramatically shows the change that occurred between 30,000 B.P. and 18,000 B.P. At 30,000 B.P. the continent was much wetter than at present, with all the climatic zones shifted towards the center. Jones and Bowler (1980:9) describe this as the "mega-lake phase," during which the presently dry or near-dry lakes were ten times larger and full of water. Around 18,000 years ago, however, most of the lakes had entirely disappeared and dune building had extended south and north well beyond present limits of mobility (1980:11).

Their map (Jones and Bowler 1980:fig. 5) shows that the arid zone expanded across southern Victoria and also northern Tasmania. Pollen analysis (Dodson 1974, 1975) and geomorphology (Bowler 1975) indicate that the arid period in western Victoria lasted until about 10,000 B.P. Fossil work in the area gives us some additional information about the paleoenvironment. From fossil red kangaroos (*Macropus rufus*) at Keilor (Marshall 1974), Lake Corangamite (Wilkinson 1972) and Spring Creek (Flannery and Gott in prep.), and more recently at Lake Bolac (Coutts et al. in prep.) in a site dated at 12,000 yr B.P., I infer that the rainfall in southern Victoria was less than 250 mm at this time, since this isohyet marks the present southern limit of red kangaroo distribution.

There is evidence for a considerable expansion of the arid core of Australia during the period approximately 25,000 B.P. to 10,000 B.P. What was happening to the megafauna while the range of the red kangaroo was expanding? The answer to this question depends on where the megafauna was living, that is, its distribution patterns and adaptations. If it was widespread and adapted to different conditions, then expansion of conditions suitable for red kangaroos would have little effect. If adapted to more mesic conditions, then expansion of the arid area could have great effects. Jones (1975:29) claims that "there is a positive suggestion that, for example, diprotodonts could live in diverse environments, fossils being found in cyclical lakes in semi-arid Central Australia, wooded hills of the Great Dividing Range in the east, and in the tropics of New Guinea."

In spite of the evident importance of documenting the distribution of extinct species in Australia, we have few examples. Gill (1954), Tedford (1966), Rich (1979) and Wells (1978) have made contributions, but these are limited in scope. In this paper I present range maps of all the more common extinct megafaunal species.

A number of problems exist in analyzing the distribution of extinct species that have inhibited analysis in the past:

1. Identification of species may not be accurate and full lists of site contents are frequently unavailable. The maps presented here, then, simply summarize the present state of knowledge as recorded in the literature.

2. Different groups of animals vary in their propensity to become fossilized and, conversely, sites differ in their chances of containing fossils of particular kinds of animals. Some parts of the range of extinct species may have contained no sites in which fossilization could occur. Because of these problems, mapping distributions may seem futile. The situation resembles a modern ecological survey where a range of different devices—pit traps, baited traps, snares—are set out in an area and the presence of different species recorded. A different set of species will be obtained from each kind of device, and some may not be obtained from any. However, if pit traps in one area consistently catch a particular species while those in another area remain empty, it is a reasonable assumption that the species does not occur in the second area. We could not, of course, conclude this if the first area contained pit traps and the second only baited traps. With fossils the main difference is that we have no control over the position or the kind of traps, and so our conclusions about absence must be more general. Wells (1978) adopted this principle in analyzing South Australian sites. Arguing that there were similar kinds of sites spread across the state, he went on to show that while *Diprotodon* was found in all areas, *Procoptodon* was not found in the most northern sites. The difference could only be due to different distribution patterns of the living species. The same approach is used here.

3. Relatively few fossil sites in Australia have been studied stratigraphically; even fewer have been dated. Undoubtedly the maps presented here include sites ranging from early to late Pleistocene, although since the number of sites decreases through time because of destruction of the remains, the great majority are likely to be late Pleistocene in age. In any case, the maps are suitable for the purposes of the model presented here because I am concerned only with the *maximum* extent of the distribution of the total megafauna. The existence of a mixture of sites of different ages would matter only if we were trying to determine the *minimum* extent of distribution at any one time. However, the only minimum distribution that concerns us is that at the end of the Pleistocene, and we already know that this is zero for the extinct megafauna.

4. Since we cannot measure population densities for fossil species, we can map distributions only by using a line to map limits of occurrence. This line is roughly equivalent to that which could be drawn for a living species, but it will not correspond exactly because we have not chosen the sampling points. The only way to decide the core areas (and hence adaptations) of fossil species is to compare the distributions of a number of species. If the distributions of two species greatly overlap, their core areas are also likely to have overlapped and their adaptations to have been similar. Conversely, small areas of overlap imply different adaptations.

Figure 29.1. Pleistocene fossil sites. Included are all localities referenced in the literature. Relatively few sites have been described in detail, and many are simply chance surface finds. The sites are likely to range from early to late Pleistocene, but there seems to be no way to quantify this with presently available information. An approximate division could be made between cave and swamp sites, which are likely to be late Pleistocene, and river or creek sites, which generally involve fossils eroding out of silts which could have been laid down at any time during the Pleistocene.

```
OPEN SITE    ▼
CAVE         ●
UNKNOWN      ◆
```

1. Abercrombie Caves
2. Albert River, Queensland
3. Attunga (9 miles west of)
4. Barwon River (Brewarrina)
5. Bearbung, Gilgandra
6. Beginner's Luck Cave,
   Florentine Valley + Titans Shelter
7. Billybillong
8. Bingara
9. Breeza
10. Brothers Island, Coffin Bay
11. Wanneroo
12. Calca Station (Calca Hill)
13. Cement Mills, Gore
14. Chinchilla (Charley Creek,
    Condamine River)
15. Cloggs Cave + 'Buchan Caves'
16. Colac
17. Coreena
    (between Barcaldine and Aramac)
18. Cunningham Creek, County Harden
19. Curramulka—'Town' Cave
20. Devils Lair + Strongs Cave
21. Dingo Rock, Karonie
22. Douglas Cave
23. Duck Ponds
24. Eastern Darling Downs—Warra Warra
    Station, St. Ruths Station,
    Condamine River
25. Egg Lagoon, King Island
26. Floraville
27. Footscray + North Melbourne
28. Ginna Gullah
29. Reddestone Creek, near Glen Innes
30. Greenough River
31. Huntsgrave, Keepit
32. Keilor
33. Kirban, Mendooran (Castlereagh River)
34. Knapps Creek (near Beaudesert)
35. Koala Cave
36. Lake Callabonna
37. Lake Colongulac
38. Lake Darlot (40 miles west of)
39. Lake George
40. Lake Menindee
41. Lake Omeo

42. Lake Tandou
43. Lake Victoria
44. Lancefield
45. Landsdowne Station, near Tambo +
    Tambo Station
46. Macintyre Gully
47. Madura Cave
48. Mammoth Cave
49. Marmor
50. Goodravale Caves
51. Maryvale Creek, Clarke River
52. Montagu Cave + Scotch Town Cave
53. Monto
54. Mowbray Swamp + Pulbeena Swamp
55. McEacherns Cave
56. Nepean Peninsula
57. Nogoa River, Rawbelle
58. Normanville (= Salt Creek)
59. Oakover River
60. Parsens Hill Plain, Warrah
61. Peak Downs
62. Planet Downs near Gregory Downs
63. Quanbun
64. Near Roma
65. Rubyvale, Anakie
66. Russenden Rear Cave + The Joint
67. Scone
68. Surprise Bay, King Island
69. Strathdownie + Puralka?
70. Forbes
71. Tambar Springs
72. Talbot, Back Creek

73. Tantanoola Caves + Submerged Cave
    (= Green Waterhole?) + Glencoe +
    Mt. Burr Cave
74. Tocumwal
75. Victoria Cave, Naracoorte +
    Haystall Cave + Hensckes Quarry +
    Jones Quarry + Alexandra Cave +
    Specimen Cave + Fox Cave
76. Warburton River, near Cowarie Station
77. Warrnambool
78. Watch Hill (Lake Corangamite)
79. Weetalabah Creek, near Coolah
80. Wellington Cave
81. Willandra Lakes
82. Wonberna
83. Yalpara (= Orroroo) + Hillpara Creek
    30 km NE Orroroo
84. South East Lagoon, King Island
85. Seton Rock Shelter + Emu Caves +
    Kelly Hill Caves + Mt. Taylor Cave +
    Fossil Cave
86. Rocky River
87. Labyrinth Cave
88. Marshall Ponds Creek
89. Wombeyan Caves
90. Pejark Marsh
91. Mount Hamilton Lava Cave

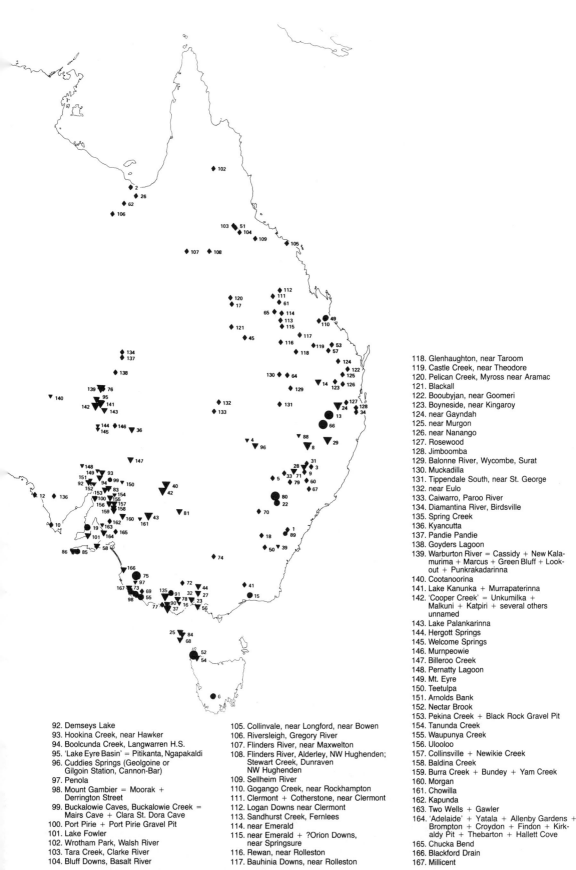

118. Glenhaughton, near Taroom
119. Castle Creek, near Theodore
120. Pelican Creek, Myross near Aramac
121. Blackall
122. Booubyjan, near Goomeri
123. Boyneside, near Kingaroy
124. near Gayndah
125. near Murgon
126. near Nanango
127. Rosewood
128. Jimboomba
129. Balonne River, Wycombe, Surat
130. Muckadilla
131. Tippendale South, near St. George
132. near Eulo
133. Caiwarro, Paroo River
134. Diamantina River, Birdsville
135. Spring Creek
136. Kyancutta
137. Pandie Pandie
138. Goyders Lagoon
139. Warburton River = Cassidy + New Kala-
     murima + Marcus + Green Bluff + Look-
     out + Punkrakadarinna
140. Cootanoorina
141. Lake Kanunka + Murrapaterinna
142. 'Cooper Creek' = Unkumilka +
     Malkuni + Katpiri + several others
     unnamed
143. Lake Palankarinna
144. Hergott Springs
145. Welcome Springs
146. Murnpeowie
147. Billeroo Creek
148. Pernatty Lagoon
149. Mt. Eyre
150. Teetulpa
151. Arnolds Bank
152. Nectar Brook
153. Pekina Creek + Black Rock Gravel Pit
154. Tanunda Creek
155. Waupunya Creek
156. Ulooloo
157. Collinsville + Newikie Creek
158. Baldina Creek
159. Burra Creek + Bundey + Yam Creek
160. Morgan
161. Chowilla
162. Kapunda
163. Two Wells + Gawler
164. 'Adelaide' + Yatala + Allenby Gardens +
     Brompton + Croydon + Findon + Kirk-
     aldy Pit + Thebarton + Hallett Cove
165. Chucka Bend
166. Blackford Drain
167. Millicent

92. Demseys Lake
93. Hookina Creek, near Hawker
94. Boolcunda Creek, Langwarren H.S.
95. 'Lake Eyre Basin' = Pitikanta, Ngapakaldi
96. Cuddies Springs (Geolgoine or
    Gilgoin Station, Cannon-Bar)
97. Penola
98. Mount Gambier = Moorak +
    Derrington Street
99. Buckalowie Caves, Buckalowie Creek =
    Mairs Cave + Clara St. Dora Cave
100. Port Pirie + Port Pirie Gravel Pit
101. Lake Fowler
102. Wrotham Park, Walsh River
103. Tara Creek, Clarke River
104. Bluff Downs, Basalt River

105. Collinvale, near Longford, near Bowen
106. Riversleigh, Gregory River
107. Flinders River, near Maxwelton
108. Flinders River, Alderley, NW Hughenden;
     Stewart Creek, Dunraven
     NW Hughenden
109. Sellheim River
110. Gogango Creek, near Rockhampton
111. Clermont + Cotherstone, near Clermont
112. Logan Downs near Clermont
113. Sandhurst Creek, Fernlees
114. near Emerald
115. near Emerald + ?Orion Downs,
     near Springsure
116. Rewan, near Rolleston
117. Bauhinia Downs, near Rolleston

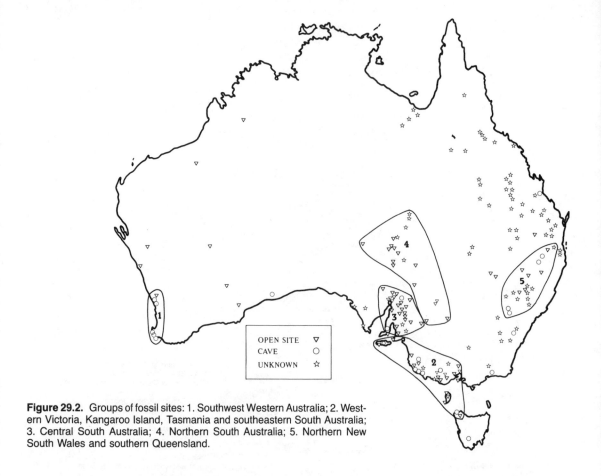

Figure 29.2. Groups of fossil sites: 1. Southwest Western Australia; 2. Western Victoria, Kangaroo Island, Tasmania and southeastern South Australia; 3. Central South Australia; 4. Northern South Australia; 5. Northern New South Wales and southern Queensland.

For all these reasons the individual maps presented here are obviously not, nor are they intended to be, definitive representations of the distribution of particular species at a particular time. They are simply summary statements of published knowledge now available. I hope that they are used in the spirit in which they are offered, that is, to provide a focus on the great gaps which exist in our knowledge about the past distribution of species. If they provide a starting point from which a series of increasingly more accurate maps can be drawn, they will have served their purpose.

It is obvious that modern species of animals are distributed in regular patterns in the landscape, those patterns being determined by environmental factors. It is equally clear that extinct species would also have had regular patterns of distribution in the past. If enough fossil sites are investigated, such patterns should be apparent even within the limitations of the fossil evidence.

Literature records disclose Pleistocene megafauna in 167 fossil sites (fig. 29.1). These can be divided (fig. 29.2) into five main areas: 1. southwestern Western Australia; 2. western Victoria, Kangaroo Island, Tasmania, southeastern South Australia; 3. central South Australia; 4. northern South Australia; and 5. northern New South Wales, southeastern Queensland. Areas 2, 3, and 5 contain a similar range of sites covering all of the major kinds of Australian fossil deposits. Area 1 consists almost entirely of cave sites, while area 4 includes no cave sites.

An interpretation of the living distributions that these fossil distributions represent is given in figure 29.3. Figures 29.4 through 29.8 show distributions of all species, divided into far southern, southern, northern, and wide-ranging species.

Far Southern Species

Figure 29.4(a–d) shows that the distribution of these species closely matches the former distribution of the historically extinct toolach wallaby *M. greyi* (Horton and

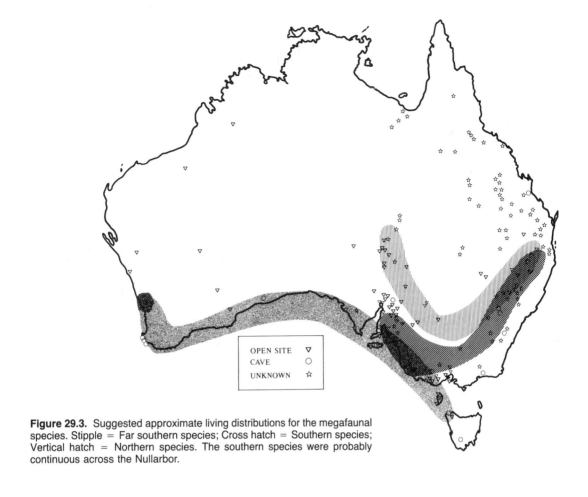

Figure 29.3. Suggested approximate living distributions for the megafaunal species. Stipple = Far southern species; Cross hatch = Southern species; Vertical hatch = Northern species. The southern species were probably continuous across the Nullarbor.

Murray 1980) and the related *M. irma,* western brush wallaby (fig. 29.6a), suggesting that all may have lived in the grassland habitat that occurred in northwestern Tasmania, on the Bass Strait land bridge, and in southern South Australia (Hope 1978). The grassland seems to have been lost from northwestern Tasmania around 11,000 B.P. (Colhoun 1978). The modification of the foot structure in *Sthenurus* suggests adaptation for quick movement in open habitat; their skull structure suggests browsing adaptation (Wells 1978). Perhaps, like the modern swamp wallaby, they ate the coarser sedges and rushes rather than grass. It is interesting that *M. greyi* also exhibited a much modified foot structure, suggesting the need to move fast and erratically, perhaps in response to predators in an open environment (Finlayson 1927).

Since *S. brownei* and *S. occidentalis* reached southwestern Western Australia, *S. gilli* extended at least to the Nullarbor, and *M. greyi* is represented by *M. irma* in Western Australia, this grassland habitat probably extended right across the Nullarbor and the fringing dry land exposed by the fall in sea level.

Southern Species

The distribution of these species (fig. 29.5a–g) broadly matches the present-day combined distribution of *M. giganteus/M. fuliginosus* (eastern and western grey kangaroos) (fig. 29.6d) and the distribution of *M. rufogriseus* (red-necked wallaby) (fig. 29.6b) and *Wallabia bicolor* (swamp wallaby) (fig. 29.6c), although the red-necked and swamp wallabies do not extend into South Australia except in the very southernmost portion. All three species are found in coastal areas of New South Wales, and all inhabit woodland and dry sclerophyll forest (*W. bicolor* being also found in rainforest). The three species have relatives, or at least share morphological similarities, with a number of extinct forms. *Macropus titan* and *M. giganteus/fuliginosus* are extremely closely related, while *M. agilis siva* and the *Protemnodon* species are similar to the large wallabies.

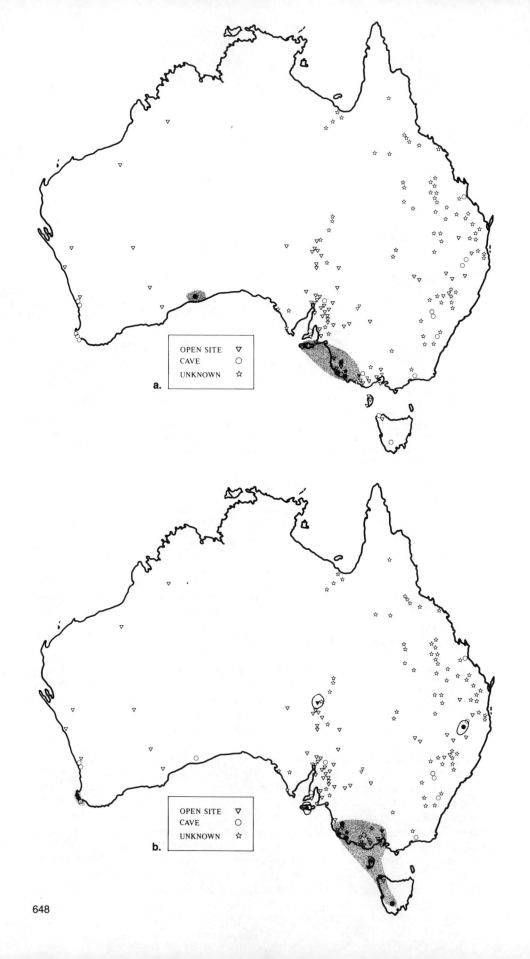

OPEN SITE ▽
CAVE ○
UNKNOWN ☆

a.

OPEN SITE ▽
CAVE ○
UNKNOWN ☆

b.

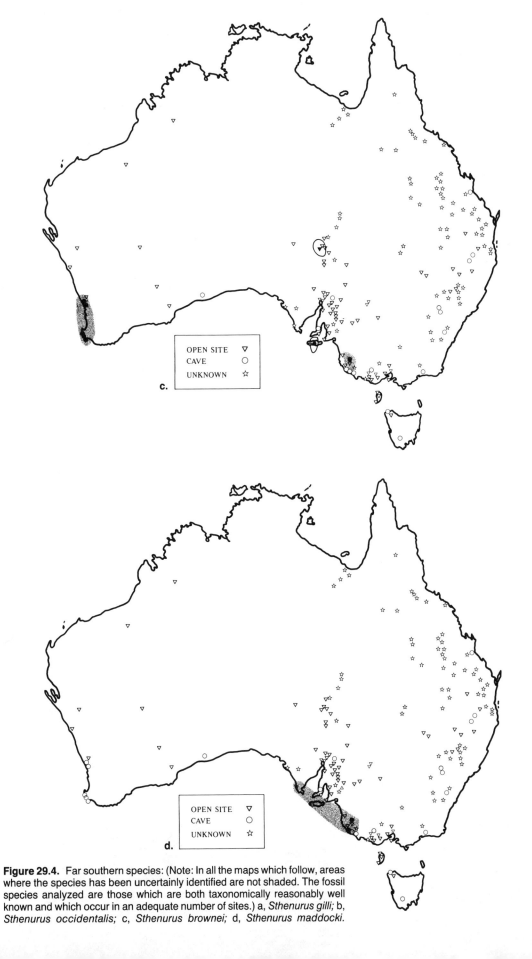

Figure 29.4. Far southern species: (Note: In all the maps which follow, areas where the species has been uncertainly identified are not shaded. The fossil species analyzed are those which are both taxonomically reasonably well known and which occur in an adequate number of sites.) a, *Sthenurus gilli;* b, *Sthenurus occidentalis;* c, *Sthenurus brownei;* d, *Sthenurus maddocki.*

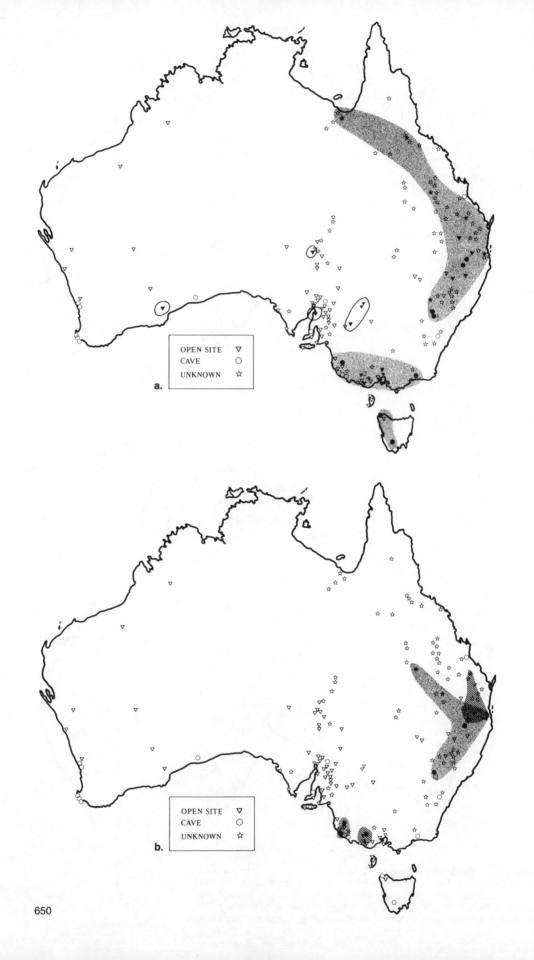

Figure 29.5. Southern species: a, *Macropus titan;* b, *Macropus agilis siva* (hatched) and *Protemnodon roechus* (stippled); c, *Protemnodon anak;* d, *Procoptodon rapha*.

c.

OPEN SITE	▽
CAVE	○
UNKNOWN	☆

d.

OPEN SITE	▽
CAVE	○
UNKNOWN	☆

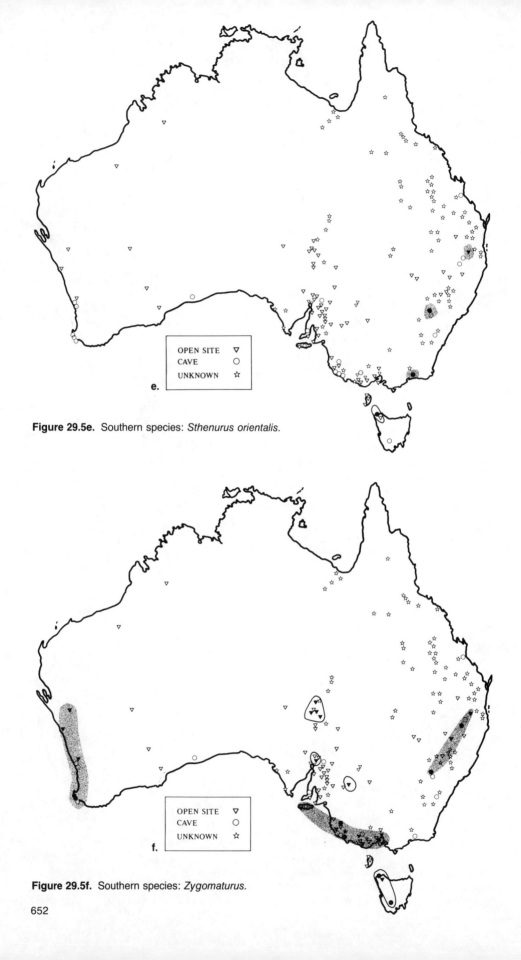

Figure 29.5e. Southern species: *Sthenurus orientalis.*

Figure 29.5f. Southern species: *Zygomaturus.*

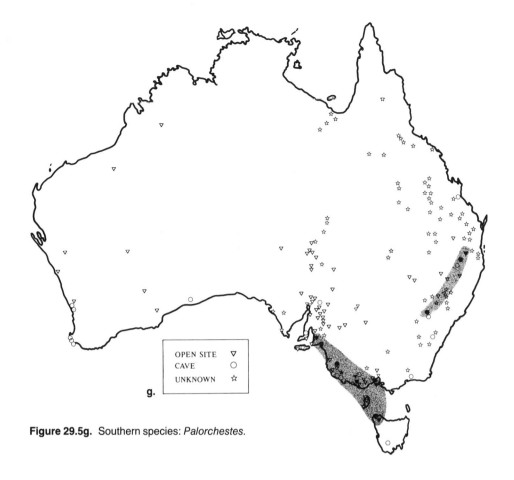

g.

Figure 29.5g. Southern species: *Palorchestes.*

Although *M. titan* could be spread over its fossil distribution with conditions like to-day's, South Australia would have to be wetter in the Flinders Range area to provide conditions suitable for species like *M. rufogriseus* and *W. bicolor*. Similarly, wetter conditions on the Nullarbor area are indicated by the fact that *Zygomaturus* and *P. anak* reached Western Australia. Wyroll (1979) points out that although paleoclimatic evidence from the Nullarbor Plain is inconclusive a number of studies do imply wetter conditions in the past. Main, Lee, and Littlejohn (1957) pointed out that permanent water supplies must have been available across the Plain for some frog species to have reached the southwest.

Northern Species

No living species have distributions similar to these (fig. 29.7a–g). The overlap in distribution between these and the previous group, particularly in northern New South Wales, implies adaptation to a similar kind of habitat, but their more inland distribution suggests somewhat warmer and drier conditions. The distribution of both these groups implies a considerably wetter environment, or at least one with considerably more water, in central and northern South Australia.

An increase in the number of frontal systems reaching South Australia could cause increased rainfall in the Flinders Range area, but such a mechanism does not account for the filling of lakes and the presence of northern species around Lake Eyre. Increased precipitation, lower temperatures, or both, in the catchments of rivers supplying the area would provide water in this system, just as it did for Lakes Mungo and Menindee. The *Sthenurus* and *Procoptodon* species could live in the woodland area around these lakes and rivers, just as grey kangaroos extend across western New South Wales by using the Murray River corridor.

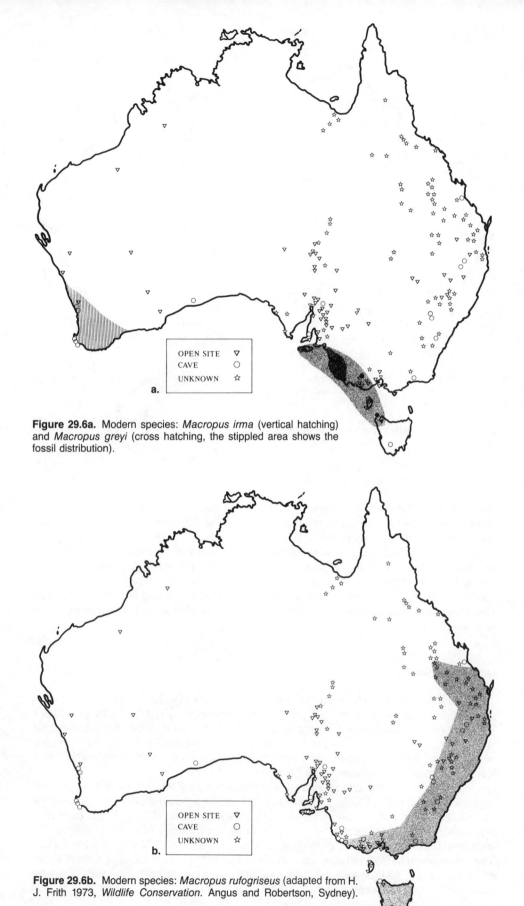

Figure 29.6a. Modern species: *Macropus irma* (vertical hatching) and *Macropus greyi* (cross hatching, the stippled area shows the fossil distribution).

Figure 29.6b. Modern species: *Macropus rufogriseus* (adapted from H. J. Frith 1973, *Wildlife Conservation*. Angus and Robertson, Sydney).

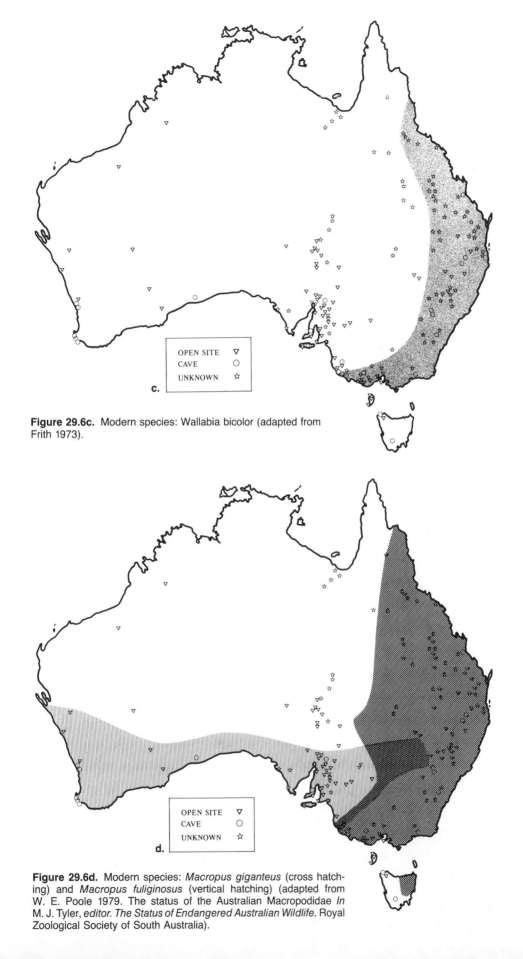

Figure 29.6c. Modern species: Wallabia bicolor (adapted from Frith 1973).

Figure 29.6d. Modern species: *Macropus giganteus* (cross hatching) and *Macropus fuliginosus* (vertical hatching) (adapted from W. E. Poole 1979. The status of the Australian Macropodidae *In* M. J. Tyler, *editor. The Status of Endangered Australian Wildlife.* Royal Zoological Society of South Australia).

OPEN SITE ▽
CAVE ○
UNKNOWN ☆

a.

OPEN SITE ▽
CAVE ○
UNKNOWN ☆

b.

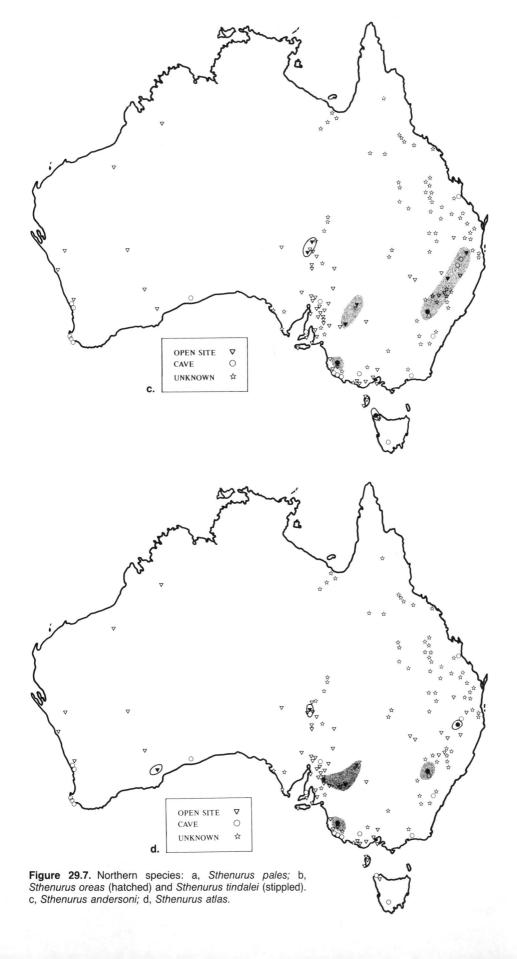

OPEN SITE ▽
CAVE ○
UNKNOWN ☆

c.

OPEN SITE ▽
CAVE ○
UNKNOWN ☆

d.

Figure 29.7. Northern species: a, *Sthenurus pales;* b, *Sthenurus oreas* (hatched) and *Sthenurus tindalei* (stippled). c, *Sthenurus andersoni;* d, *Sthenurus atlas.*

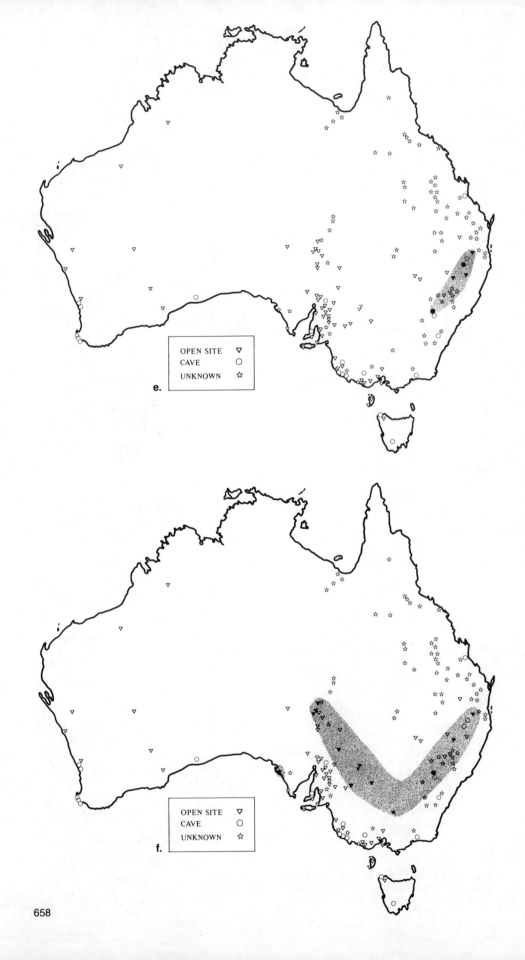

OPEN SITE ▽
CAVE ○
UNKNOWN ☆

e.

OPEN SITE ▽
CAVE ○
UNKNOWN ☆

f.

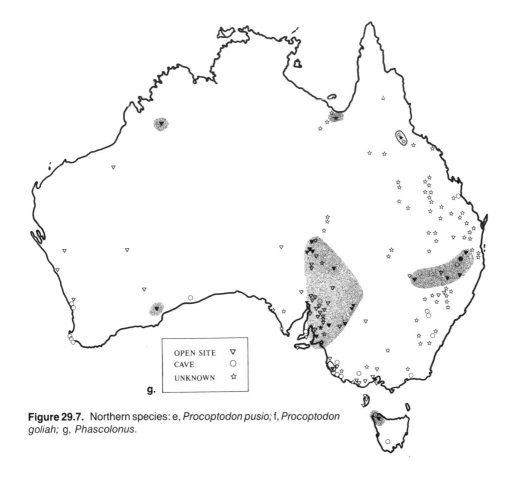

Figure 29.7. Northern species: e, *Procoptodon pusio;* f, *Procoptodon goliah;* g, *Phascolonus.*

Distribution of the southern and northern megafauna groups suggests adaptation to a mesic environment, one intermediate between the aridity of central Australia and the wetness of the mountains and coastal areas. The likely habitat in such areas is woodland. To check this proposition, we can analyze the modern species, whose habitat preferences are known, which are commonly found in association with the megafaunal species.

Both modern fauna and megafauna occur and have been described at twenty-three sites. *Thylacinus cynocephalus*, the thylacine, occurs in all these sites. Guiler (1961:209) says "thylacines were caught in all types of country ranging from the coast to the mountains, but with the greatest number being caught in the drier parts of Tasmania . . . animals were taken in savannah woodland or open forest." Other common woodland species found at between thirteen and sixteen sites are *Sarcophilus harrisii*, Tasmanian devil; *Dasyurus viverrinus,* Eastern quoll; *Trichosurus vulpecula,* common brushtail possum; and *Vombatus ursinus,* common wombat.

These associated modern species confirm the hypothesis based on distribution patterns—the two main groups of megafaunal species were adapted to woodland or open forest conditions. This is not an unexpected finding for such large herbivores. Merrilees (1968:3) notes that "the African savannas, which are anything but 'lush' in the conventional meaning, carry a far greater biomass of large mammals than any other African environment."

Wide-Ranging Species

These species do not fit into any other category. The two that clearly belong here are *Diprotodon* (fig. 29.8a), and *Genyornis* (fig. 29.8b), but *Propleopus* (fig. 29.8c), *Thylacoleo* (fig. 29.8d), and *Protemnodon brehus* (fig. 29.8e) probably also belong here.

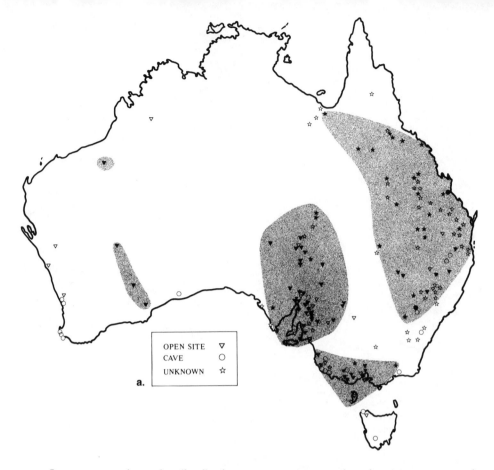

OPEN SITE ▽
CAVE ○
UNKNOWN ☆

a.

In summary, then, the distribution patterns suggest that there were two main groups of megafaunal species, one adapted to the southern grasslands and the other to woodlands, with a further subdivision of the woodland group into warmer and drier woodlands and cooler and wetter.

The only previous attempt to provide a detailed analysis of megafaunal distribution and relate it to adaptation was by Tedford (1966). He divided *Sthenurus* into two subgenera, *S. (Sthenurus)* and *S. (Simosthenurus)*. Whereas members of both subgenera could be found in inland sites, only *Simosthenurus* species were found in sites on the coastal fringes of the south. Suggesting that the inland sites would have been characterized by grassland and savanna habitats, while the coastal ones would have savanna and woodland habitats, he equated the differences in distribution with morphology. The *Sthenurus* species, with a "grazing" morphology, would be more suited to habitats with grassland; *Simosthenurus* "browsing" species would be more adapted to woodland. Overlap between the two groups would occur in savanna habitats. Three new *Simosthenurus* species have been described since Tedford's analysis: *S. brownei, S. gilli, S. maddocki*. All occupy the coastal fringe areas.

As Tedford has noted, a clear division exists between species of *Sthenurus* which occur in the southern coastal fringe (my far southern species group), and those which occur in inland areas (my northern species group), only one species falling into the intermediate southern group (*S. orientalis*). However, the distribution of living macropods does not suggest a difference between grazing and browsing species (see Calaby 1966). In addition, if the division were correct, we would expect *Procoptodon,* evidently a browser, to be found in the same area as the morphologically similar *Simosthenurus* species. In fact, two *Procoptodon* species are inland (northern), one has the intermediate (southern) pattern, while none have the coastal fringe distribution. Probably all these species are adapted to open country, which can be found either inland or coastally, and both browsers and grazers can find their food requirements in such areas. An alternative explanation for the various megafaunal distribution patterns is obviously needed.

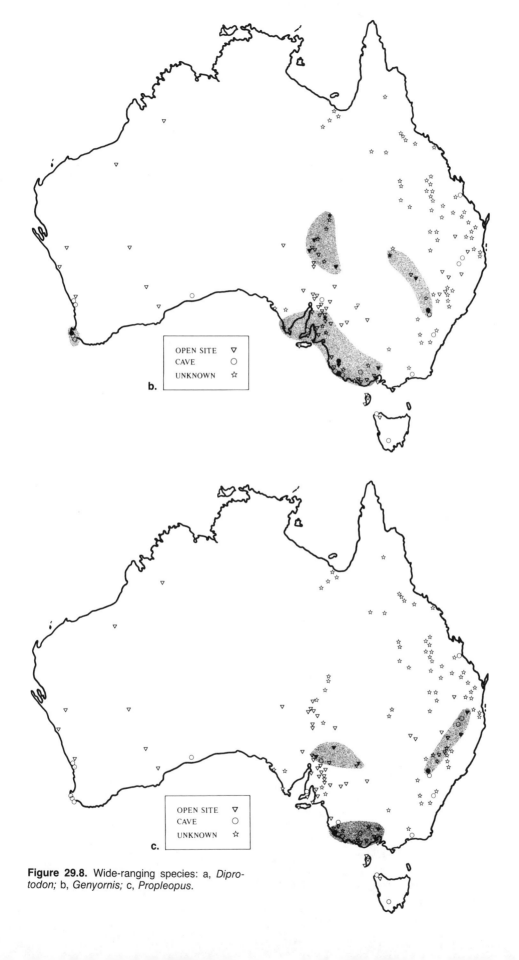

Figure 29.8. Wide-ranging species: a, *Diprotodon;* b, *Genyornis;* c, *Propleopus.*

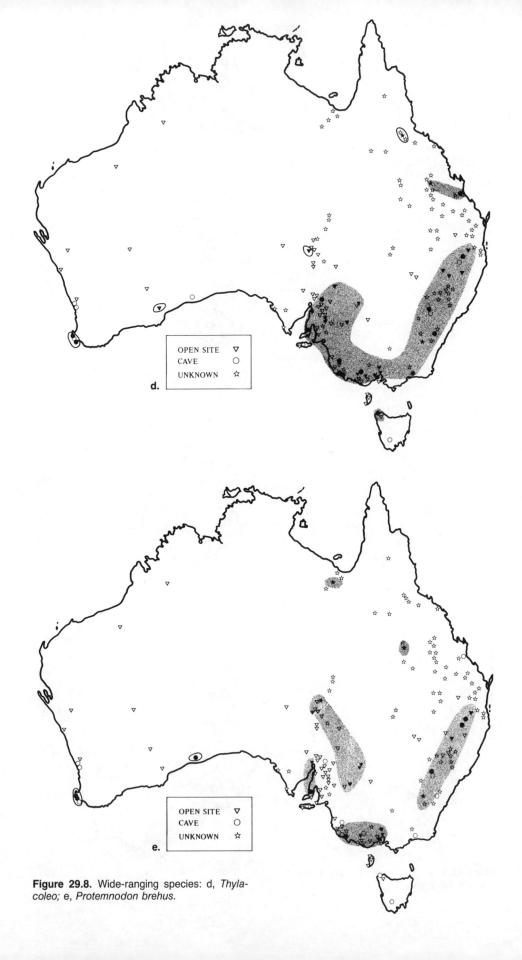

Figure 29.8. Wide-ranging species: d, *Thyla-coleo;* e, *Protemnodon brehus.*

OPEN SITE ▽
CAVE ○
UNKNOWN ☆

d.

OPEN SITE ▽
CAVE ○
UNKNOWN ☆

e.

Distribution patterns of *living* Australian macropods can be related to rainfall patterns. The four areas identified are (1) those with more than 750 mm annually, inhabited, for example, by *Thylogale thetis* (red-necked pademelon), *T. stigmatica* (red-legged pademelon), *Macropus parma* (parma wallaby); (2) those with more than 500 mm annually, inhabited by *Wallabia bicolor, M. rufogriseus, M. irma* and *M. greyi*; (3) those with more than 250 mm annually, inhabited by *M. giganteus* and *M. fuliginosus*; (4) those with less than 250 mm annually, inhabited by *Macropus rufus*. The lack of overlap in distribution between the third and fourth areas is almost complete, but grey and red kangaroos can be found together in a small area of southern Queensland, a fact at least partially related to environmental changes caused by the grazing industry. The distribution of these groups forms a roughly concentric pattern, group (1) on the coastal fringe and group (4) in central Australia.

The concentric rainfall pattern (fig. 29.9) is responsible for the concentric habitat arrangement in Australia, for example, woodland (fig. 29.10), which in turn determines animal distribution patterns. The fossil distributions suggest the same pattern as for the modern macropods, but with the wetter zones expanded and the arid zone contracted. If this is the case, then it could be suggested that the equivalence between the fossil and modern distributions is as in Table 29.2.

Distribution and Extinction

Although in the previous section particular modern species were listed as if they occurred only within a single range of rainfall, the first three rainfall categories are not exclusive. *Thylogale, Wallabia, Macropus rufogriseus*, and *M. giganteus*, for example,

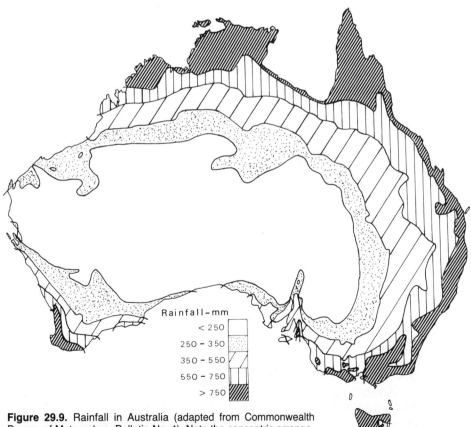

Rainfall – mm

< 250
250 – 350
350 – 550
550 – 750
> 750

Figure 29.9. Rainfall in Australia (adapted from Commonwealth Bureau of Meteorology Bulletin No. 1). Note the concentric arrangement of zones surrounding the arid central area.

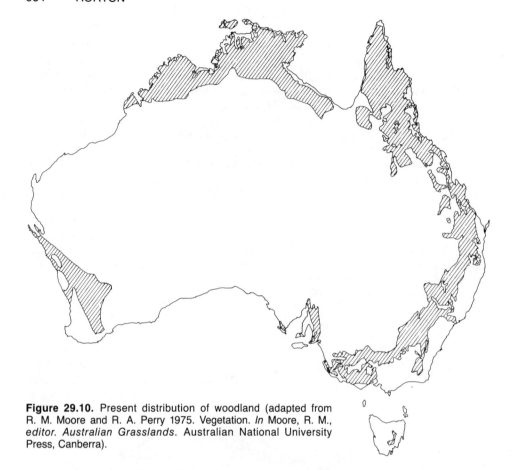

Figure 29.10. Present distribution of woodland (adapted from R. M. Moore and R. A. Perry 1975. Vegetation. *In* Moore, R. M., editor. *Australian Grasslands.* Australian National University Press, Canberra).

can all be found together in northeastern New South Wales (Calaby 1966). Similarly, it is clear from the fossil record that overlaps would have occurred in distribution between megafaunal species from different groups. Among the modern macropod fauna, however, one species, *Macropus rufus*, does occupy an exclusive category. If red kangaroos occur in an area, then modern species from the other three categories will not be found. If the equivalence between modern and megafaunal species presented here is correct, then we can argue by analogy that areas in which red kangaroos were found in the past would not have been suitable for the megafauna. The fossil record, then, allows construction of the following hypothesis to describe the distribution of red kangaroos and megafauna during the late Pleistocene (figs. 29.11, 29.12):

1. From 40,000 to 26,000 B.P. the range of red kangaroos was considerably reduced, being inland from the area covered by megafaunal sites. Admittedly red kangaroos could have been found quite close to the most inland of these sites, in the same way their present distribution abuts the present distribution of grey kangaroos.

2. From 26,000 to 10,000 B.P. red kangaroos considerably expanded their range, almost, if not completely, covering all areas in which megafaunal sites have been found. The megafaunal species became extinct.

3. From 10,000 B.P. to modern times the range of red kangaroos contracted once again. Species favoring more humid ranges such as grey kangaroos and swamp wallabies expanded back into some of the area which they originally occupied jointly with megafaunal species. This hypothesis would require that at sites yielding both red kangaroos

Table 29.2. Rainfall and Large Living and Extinct Marsupial Distributions

Modern			Fossil		
Group	Annual Rainfall*	Species	Group	Annual Rainfall	Species
1	>750mm	Thylogale thetis, T. stigmatica, M. parma	—	—	—
2	>500mm	(1) W. bicolor, M. rufogriseus M. irma	2	>500mm (woodland)	Zygomaturus, Palorchestes, M. titan, M. agilis siva, P. roechus, P. anak, P. rapha, S. orientalis
		(2) M. greyi	1	>500mm (grassland)	S. occidentalis S. brownei, S. gilli, S. maddocki
3	>250mm	M. giganteus, M. fuliginosus	3	>250mm	S. pales, S. oreas, S. tindalei, S. andersoni, S. atlas, P. pusio, P. goliah, Phascolonus
			4	>250mm	Diprotodon, Thylacoleo, Propleopus, Genyornis, Protemnodon brehus
4	<250mm	M. rufa	—	<250mm	(M. rufa)

*For the modern species the rainfall figures are derived from the isohyets marking the limits of their distribution. In the cases of the extinct species, the values are also inferred from their distribution patterns at a time when Australia was much wetter than at present. However, "rainfall" for these species should be understood as effective water availability. A lower temperature can reduce evaporation and therefore increase available water even if rainfall remains constant. Similarly, the extinct species may have lived in areas of lower rainfall than those suggested above, if lakes fed by rivers from high rainfall areas were present.

and megafauna (Spring Creek, Keilor, Menindee, Tandou), the red kangaroos should be more recent in age than the megafauna; this is the case at least at Keilor. However, the extinctions and the range extension could be so closely connected in time as to be stratigraphically inseparable.

This arid expansion model requires some additional elaboration. If concentric rainfall zones exist, they will endure whether the arid zone is large or small, although it could be argued that the zones suitable for megafauna may have become so attenuated that they could not support sufficiently large populations to ensure the survival of species. Jones (1975) implies that the megafaunal species could stay ahead of the expanding arid front, remaining in refuge areas around the fringe ready to reexpand when the climate improved.

The model I propose for the selective death of megafauna is based on its presumed need for free water. Smaller species may obtain sufficient water from their food (Horton 1978, 1980). Normally, the large species lived where water supplies were spaced at a distance no greater than that which the animals could cover in a time no greater than the longest interval they could survive without drinking. During drought, less reliable water supplies dried up, and those remaining were spaced farther apart. Therefore, animals at one water source could not travel far enough to reach another water source. Food supplies, reduced in quantity and quality by the drought, were eaten out within a radius equal to half the distance that the animals could travel in a time no greater than the longest interval they could survive without drinking. Ecologically they became tethered (White and O'Connell 1979). When the food supplies were exhausted, animals died even though a water source was still available, as Shipman (1975) has described in the case of recent droughts in Africa.

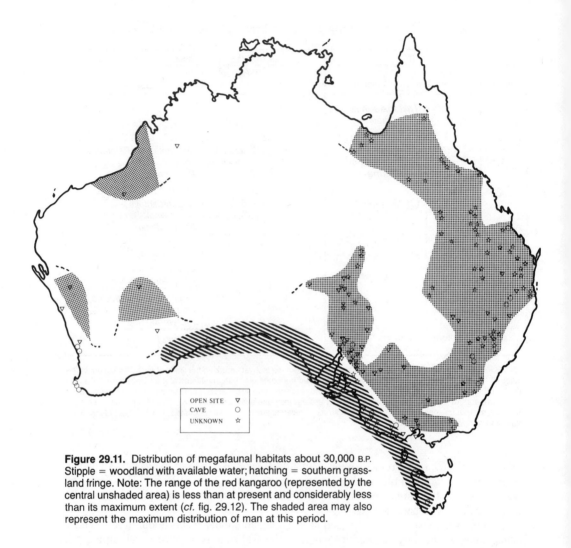

Figure 29.11. Distribution of megafaunal habitats about 30,000 B.P. Stipple = woodland with available water; hatching = southern grassland fringe. Note: The range of the red kangaroo (represented by the central unshaded area) is less than at present and considerably less than its maximum extent (*cf.* fig. 29.12). The shaded area may also represent the maximum distribution of man at this period.

This process would have prevented megafaunal species retreating in advance of the arid zone, and would have killed species even in refuge areas of otherwise suitable habitat. The megafaunal species could not respond to the change and move ahead of it because it was *the change itself* that killed them, not the final result. Surrounding the expanding arid zone would be a zone in which megafauna were dying, rather in the way that in a hotel fire, people die from smoke asphyxiation in advance of the flames.

The megafaunal species had never occupied the more heavily wooded areas of the Great Divide, perhaps because such habitats yielded little forage, or movement was difficult because of the animals' size. The late Pleistocene arid expansion at its height produced an effect as shown in Figure 29.13.

When conditions improved, species from these areas could reinhabit the woodland area, which was itself expanding. In any case, some species may well have been able to survive there throughout this period, not needing permanent water.

Since four species of *Sthenurus* (*S. gilli, S. occidentalis, S. brownei,* and *S. maddocki*) have a set of shared, unique characteristics—small size, a pattern of distribution different from the rest of the megafauna, and probably a much later survival than the rest of the megafauna—we must consider a two-stage process for megafaunal extinc-

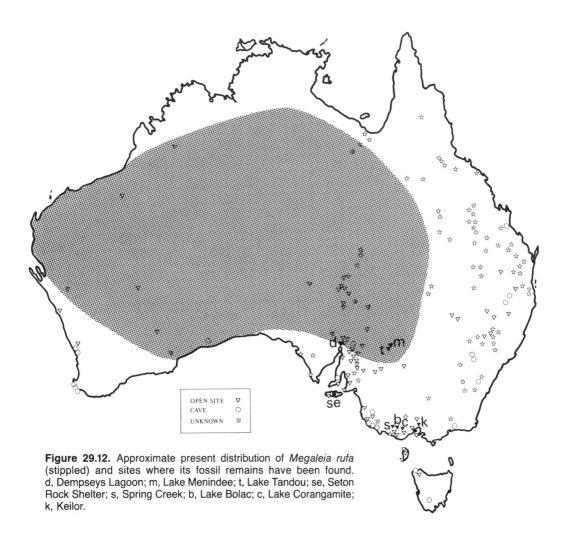

Figure 29.12. Approximate present distribution of *Megaleia rufa* (stippled) and sites where its fossil remains have been found. d, Dempseys Lagoon; m, Lake Menindee; t, Lake Tandou; se, Seton Rock Shelter; s, Spring Creek; b, Lake Bolac; c, Lake Corangamite; k, Keilor.

tion. In the first, beginning about 26,000 B.P., the woodland megafauna was placed under enormous ecological pressure by a series of droughts. Each drought may have been short, but the cumulative effect, over 7,000 to 8,000 years, was enough to cause the extinction of all these large species. The four *Sthenurus* species survived this early period, perhaps because the area they occupied along the southern fringes of the continent remained somewhat better watered than more inland areas. The rain-bearing frontal systems, although forced much farther south than at present, may have occasionally brought rain to this part of Australia. These species, being somewhat smaller than the rest of the megafauna, may also have had less stringent water requirements. Presumably these species survived the earlier fluctuations, but then became extinct between 17,000 to 15,000 B.P., when aridity reached a peak.

Alternatively, perhaps because of their small size, these *Sthenurus* species were not affected by changes in water regime but by habitat changes. Hope (this volume) points out that the area occupied by these species and *Macropus greyi* was "a shrub-rich grassland with abundant grass and daisies, and only scattered eucalypts," a habitat that has no equivalent today. It may be that, unlike *Macropus greyi*, these *Sthenurus* species could not adapt to the climatically induced change in this habitat around 17,000 to 15,000 B.P. or even as late as 11,000 B.P.

a

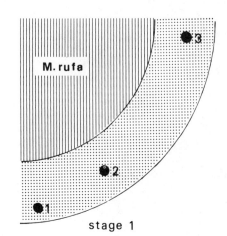

M.rufa

stage 1

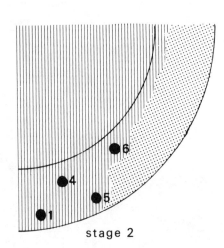

stage 2

b

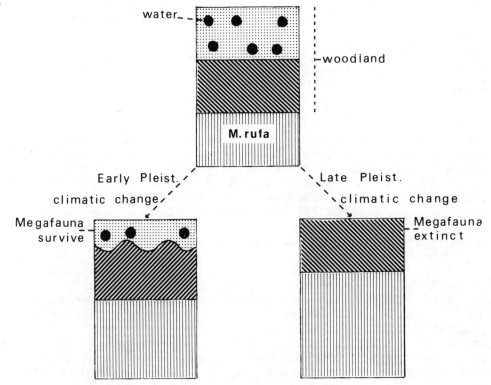

water

woodland

M.rufa

Early Pleist.
climatic change

Megafauna
survive

Late Pleist.
climatic change

Megafauna
extinct

There have been three major criticisms of the kind of climatic extinction model presented here, the first dealing specifically with this proposal, the others with climate models in general.

Hope (this volume) suggests that "The evidence for very severe droughts...is unfortunately invisible in the present geomorphic and vegetational records. Severe droughts would have had their greatest effect after 20,000 B.P., when the dunefields were at their most extensive." This comment reveals some misunderstanding both of the model presented here and of the geomorphic record. It is not the arid expansion, as such, that causes extinction, at least of the woodland species, but fluctuations in water availability in the woodland areas fringing the arid zone. The geomorphic record contains clear evidence for such fluctuations (droughts).

At Lake Mungo, Bowler (1975:76) notes: "At 26,000 B.P. the first appearance of clay pellets in aeolian sediments on the Walls of China records the beginning of a major water level oscillation. Levels in L. Mungo and Outer Arumpo fell, overflow from the system ceased, and the environment underwent a major change. This low, saline oscillatory phase persisted for several thousand years." This period is followed by a short-lived, high water level at around 22,000 B.P., then another period of several thousand years of saline oscillations, and finally the lakes completely dried up between 17,000 and 15,000 B.P. As Bowler (1975:79) also notes: "To explain the events recorded from southeastern Australia we must turn to the highland catchments of the streams that fed the lakes and which exerted a significant control of the palaeohydrologic events recorded on the plains." In fact, the record shows changes not only in water in the highland catchment areas of Mungo (and other lakes in southeastern Australia), but also in the megafaunal areas, since these rivers all flow across the western slopes and plains of New South Wales and other areas in which the megafauna were found. The fluctuations at Mungo provide evidence of the changes in water availability postulated in the model presented here.

It is difficult to know what kind of vegetational change would provide evidence for these droughts. Hope's own discussion (this volume) notes that the presence of drier adapted vegetation types after 25,000 B.P. supports the geomorphological evidence. It is hard to see how the occurrence of frequent droughts during the period to, say, 20,000 B.P. could have had any *additional* effects. The model proposed here can cause extinction of animals by recurrent drought while having little effect on the plant communities whose reproductive processes ensure species' survival in a way that those of animals do not.

Earlier episodes of aridity during the Pleistocene did occur. Bowler (1976) notes that both the lake-full and lake-dry phases of the upper sequence at Mungo overlie an older unit (Golgol), which probably represents a previous dry-lake episode of clay lunette building. Bowler suggests that this probably relates to the glacial maximum around 120,000 years ago. Below this layer is another unit that Bowler suggests may represent an earlier episode of aridity perhaps 200,000 years ago. Bowler also notes that dunes in northwestern Victoria may represent a 300,000-year-old event.

Figure 29.13. Diagrammatic representation of the extinction model: (a) In *stage 1* the range of *M. rufa* has expanded and is fringed by an area in which water resources have been reduced, leading to the death of megafauna at sites such as Spring Creek (1), Lancefield (2), and Reddestone Creek (3). All three sites show the Lancefield characteristics of large numbers of mature animals and few young animals. In *stage 2* the range of *M. rufa* has further expanded, entering part of the area in which the megafauna has now become extinct. Its fossilized remains are found at Spring Creek (1), Lake Bolac (4), Lake Corangamite (5) and Keilor (6). Note that the extinctions do not depend on the arid zone covering all parts of the megafaunal range; (b) This diagram illustrates the threshold factor in climatic change in Australia. As long as some refuges with water remain, few or none of the megafauna will become extinct. Once the threshold is crossed, large numbers of extinctions will result.

A number of authors have interpreted the occurrence of these earlier events as evidence against the climatic models of extinction, suggesting that since the megafauna had survived the earlier events they should also have survived the most recent one, since there is no evidence that it was more severe. This argument overlooks the fact that the Willandra Lakes are geographically so located as to be sensitive indicators of climatic change, being near the arid core of Australia, and more importantly, a long way from their water sources in the eastern highlands. Climatic changes will be felt in this area first, whether they are extensive enough to affect more southern and eastern areas or not.

The hypothesis presented here suggests that sites, for example in southern Victoria and the western slopes and plains of New South Wales, representing the fringe area (stippled in fig. 29.13) will show features relating to aridity representing only the *last* glacial period. That is, they will not have the equivalent of the Golgol unit or the earlier unnamed units. Only in the last glacial period did the arid core expand far enough, and the water sources in the highlands fluctuate greatly enough, to eliminate the megafauna in its refuges.

The question of refuge areas has been influenced by concepts of drought in Australia. Some have assumed that drought may affect only a few areas in Australia at any one time; in effect, refuge areas would shift. It has also been assumed that some areas are never drought affected and would invariably have had food and water for megafauna.

Neither concept can be shown to be correct even in the last 150 years of Australian climatic records. At least six major droughts have affected all or almost all of Australia in this period (Foley 1957): 1864–68, 1880–88, 1895–1903, 1911–16, 1918–20, and 1939–45.

In all these major droughts stock losses were massive and would have been greater without bores and hand-fed fodder. The consistent pattern in every period has been a great reduction in water sources—dams and waterholes dry up all over Australia. Some may assume that rivers would still be available to provide water but this is not the case. Foley lists numerous occasions when even major rivers like the Murray, Lachlan, Namoi, and Hunter have ceased to flow during major droughts.

It should be noted that all these droughts have occurred during a much wetter climatic regime than that of 25,000 to 15,000 years ago. Water supplies at that time would have been much less extensive and short-term droughts would have made water very scarce indeed. In that period it was not cattle and sheep that were dying by the millions but megafaunal species.

Various geomorphological studies confirm that in the late Pleistocene no potential refuge areas existed even in southwestern Australia (Wyroll 1980), northern Australia (Webster and Streten 1972), and Tasmania (Colhoun 1978). The climatic reconstruction by Jones and Bowler (1980) shows the expansion and contraction of the arid center and the concurrent contraction and expansion of the better-watered areas arranged concentrically around it. A concentric arrangement ensures that extinctions will occur because no part of the continent can remain unaffected.

The findings that the extinct megafaunal species lived mainly in open forest or woodland and that they required free water to drink regularly should not generate controversy. The change from wet to dry conditions in the late Pleistocene, with the arid center expanding outwards, is well established from geomorphological and palynological evidence. The new proposal I offer is that not only are these factors related, but that together they provide a mechanism for extinctions that will selectively affect only the megafauna and which will cause their extinctions to be concentrated at a particular time.

The essential problem with a climatic extinction model has always been to explain why climatic change, which proceeds gradually and consecutively in adjacent areas, should cause extinction rather than changes in distribution patterns. Jones has elegantly recognized this point: "Even if there were climatic changes of some magnitude, this would only have meant a shift of ecological zones laterally a few hundred kilometers, with survival of large refuge zones. . . ." (1975:29). The model presented here answers this objection by showing that extinctions can occur *in advance* of the shift in ecological zones because of reduction in water availability.

The controversial proposition is that there should be sites in the fringe area (fig. 29.13) that contain evidence that drought caused the deaths of large numbers of individuals of megafaunal species. Conversely, if the model proposed here is incorrect, and if man caused the extinctions, we would expect to find evidence for this in the sites where megafaunal fossils occur.

Fossil Sites

The search for proof of association between man and megafauna has been the result of two different hypotheses (Horton 1979, 1980). At first, association was sought to demonstrate the antiquity of man's occupation of Australia. Later, following the work of Martin and Wright (1967), Jones (1968), and Merrilees (1968), demonstration of man's exploitation of the megafauna was sought. In neither case was the nature of the sites felt to be an integral part of the problem—if megafauna was present it was possible, in theory, to find examples of association and perhaps exploitation.

While Wells (1978) produced a classification of Australian fossil sites based on geographical considerations, an alternative classification is possible based on the characteristics of the animals fossilized: mainly young animals, mainly young animals but with a sizable proportion of old animals, and mainly mature animals.

The first type reflects a random sampling of an unstressed living population. This would apply to caves acting as pit traps, like Victoria Cave (see Wells 1978) and to situations where a large part of a population has been killed simultaneously, as at Edith Creek (Horton and Murray 1980).

In the second type the animals fossilized are those dying naturally, yielding a random sample of the mortality curve. This could apply in the case of a cave that has functioned as a carnivore den, since the inexperienced young and the slow, weak, old animals are those normally caught by predators. Victoria Cave also exemplifies this type (Smith 1972). It could also apply to cases in which animals died because of weaning problems, old age, or disease and were fossilized by chance. In both cases the number of old animals is likely to be low, but the lack of mature animals in the second type is distinctive.

The third type is characterized by the presence of large numbers of mature animals and the virtual absence of young and old individuals. This would be the case mainly in swamp sites, but possibly also in lacustrine and fluviatile sites. Examples are known from Lancefield (Gillespie et al. 1978), Reddestone Creek (Horton and Connah 1981), and Spring Creek (Flannery and Gott in prep.).

It occurs at times of drought as a result of a two-stage process. One response to drought is the cessation of breeding (e.g. Newsome 1977). If the drought continues for only a short time, breeding can resume when it ends and the population can recover. A very long drought, however, of fifteen to twenty years, would result in the extinction of local populations. If such an event occurred in the past, the fossil remains would consist entirely of old animals.

At sites such as Lancefield, however, the remains consist of mainly mature animals, which have clearly not died at the end of their normal lifespan. Some additional mechanism must be involved—this is the loss of food by overgrazing. Alternatively, and with the same result, the water supply could dry up before this stage was reached. In this case evidence is likely to be found of animals being mired as they tried to reach the remaining water.

All three kinds of sites may have large numbers of individuals of megafaunal species present, and attention has therefore been focused on them in the hope of finding associated artifacts. It is apparent, however, that this has been the wrong strategy. In none of the three types is man the cause of death of these animals—any association can only be fortuitous.

The possible exception to this is sites where mature animals are the most numerous age group. There are two alternative strategies man could adopt in drought conditions that I believe account for this age distribution. He could move away from the drought conditions to the coastal fringe areas of Australia and therefore *away* from areas where the megafauna was dying. Alternatively, he may have retreated *to* the remaining sources of water that were attracting the megafauna. This would not have been a very good strategy for two reasons: first, there can have been no certainty, and indeed there must have been evidence to the contrary, that these water sources would remain where others had failed; second, the destruction of vegetation in a wide area around the water, and pollution of the water by carcasses, would have made these sites unsuitable for man's occupation. It seems unlikely that any substantial evidence of man will be found at these sites, but it is possible that occasional groups of people exploited the "captive" mammal populations at such sites.

In general, then, we should be looking not for "megafaunal" sites, to see if man is present, but for *archaeological* sites, greater than 26,000 years old, to see if megafauna is present. It seems that three possible patterns of exploitation exist, hence three different kinds of sites at which such evidence may be found.

Megafaunal species may have simply formed part of the overall exploitation pattern at a site. In this case megafaunal species occasionally would be present among the normal rockshelter food refuse of wallabies, possums, birds, reptiles, and others. At all sites some species were commonly eaten while others were not. It would be possible in theory, then, to find a site at which the most common species was a member of the megafauna, but this is by no means a straightforward expectation.

Jones (1980) has noted an uneven relationship between the probability of obtaining a resource and the return from that resource for the Gidjingali of Arnhem Land. While this is clearly not an invariable rule it is a useful guide to analysis.

The implication of Jones's model is that there also tends to be a relationship between animal size and frequency of hunting success: the larger the species the less likely it is to be caught. There is evidence for this from archaeological sites, where kangaroo remains are rarely found but those of small wallabies such as bettongs, pademelons, and rock wallabies are abundant. The reason for this difference lies in the biology of these species. Kangaroos tend to be nomadic in small groups in open country, and therefore unpredictable, difficult to approach, and difficult to obtain in large numbers even when successfully approached. Small wallabies tend to be sedentary in particular kinds of habitat, thus easier to find. Their tendency to live in dense cover means they can be driven out by fire, or trapped, or ambushed on the runways they use. Thus, it seems likely that the megafauna would have been even more rarely caught than the modern kangaroos, and only occasional fragments will be found among other food refuse in archaeological sites.

The second possibility is that the megafauna was not treated the same as smaller prey. People may have followed the game to where it was killed and set up camp until the meat was consumed, much as Kalihari bushmen do when they wound elephants and follow them until they die. Alternatively, hunters could follow large game until it was exhausted, and they could get close enough to kill it. Freshness of meat then becomes a consideration. If killing a *Diprotodon* was a long and difficult process, a single man would be able to carry so little meat that the effort would not be worthwhile and the distances traveled may be so great that the meat would have "gone off" by the time he returned to camp. A small group of people following the hunter(s), and thus the game, would therefore seem to be a good strategy. Such activity would be almost invisible archaeologically, but it is possible that remains of a single individual of a megafaunal species could be found in relation to a hearth.

If megafauna was as difficult to catch and kill as I have suggested, then another alternative strategy might involve many men dealing with many animals in a single event such as driving them over the edge of a cliff. In the northern hemisphere such a strategy would mean hunting herd animals since these occur in groups and behave predictably. We have no reason to think that any of the megafauna were herd animals since none of the modern macropods are. The modern species do, however, congregate around food and water sources, and similar behavior in the megafauna, particularly congregation at water sources, could have allowed large numbers of them to be trapped.

The remains of such an activity should be archaeologically highly visible since many people would have been required to ambush, kill, and butcher animals, and large numbers of people to consume the meat would be needed to make the operation worthwhile. While results would superficially resemble those sites where drought-caused deaths occurred, sufficient burnt bones and artifacts should be present to distinguish them (Horton 1976). In addition, the age structure should be distinctive. A human kill of this kind would almost inevitably involve young animals.

Examples of this third kind of exploitation have not yet been found in Australia, but examples of the other two kinds are beginning to appear. These will always be debatable because of the low intensity of use, but the presence of megafauna among other faunal debris is reported from Seton Rock Shelter (Hope et al. 1977) and Beginner's Luck Cave (Geode et al. 1978), while Menindee has long been claimed as an example of exploitation (Tedford 1955). Undoubtedly, clearer examples will be found as more sites with fauna are discovered.

In spite of considerable searching, and the discovery of numerous sites with megafauna, no sites have been found that have the characteristics of a mass-kill by humans. Although it is dangerous to argue from negative evidence, there is now a good deal of it. It is hard to disagree with Bowdler's (1977:232) ironic remark that "In America, the kill site came first and the overkill hypothesis was generated subsequently; in Australia, we have an overkill hypothesis, but are still waiting for the kill site."

In summary, evidence that the climatic extinction model presented here has operated in late Pleistocene times is beginning to be found. Evidence is also now being found that the Aboriginal economic system (Horton 1981) in operation in recent times in Australia has had a long history. During late Pleistocene times this economic system encompassed the megafaunal species but the low level of exploitation of large species inherent in this system could not have caused megafaunal extinctions. There is no evidence that a "mass-killing" system, the only one which could conceivably have affected the megafauna (although this is debatable), was ever in operation in Australia.

Some authors, notably Jones (1968) and Merrilees (1968), reject overkill and turn to the effects of fire. Neither say specifically how fire might have affected megafauna, but both generally assume that man's arrival in Australia resulted in increased fire leading to habitat change and therefore extinction.

I have suggested elsewhere that after their arrival in Australia, Aborigines used fire in the context of an already existing natural fire regime so that it caused no environmental change (Horton 1982). Since such change has been strongly argued as a mechanism for extinction, however, let us briefly consider hypothetically what effect increased fire frequency *could* have had on the megafauna.

A considerable amount of data now exists on the effects of fire on modern Australian mammals (e.g. Fox and Mackay 1981, Catling and Newsome 1981, Fox 1978, Christensen and Kimber 1975, Bolton and Latz 1978, Sampson 1971) and the trends are clear. Many species can find food, often in increased amounts, in areas burnt one or two years previously. Shelter is the main factor. Generally, large animals can survive well without shelter but smaller animals cannot because they need structural features such as large grass tussocks, dense shrubs, leaf litter, logs, and hollow trees, not found in newly burnt areas and which take a very long time to develop. Had fire frequency been increased, the amount of shelter would have been reduced and growth of new grass and shrubs encouraged. Had Aborigines changed the Australian fire regime, small fauna, not megafauna, would have become extinct.

Discussion

In this study I have divided the megafaunal species into a number of groups. First are the arid species, such as the red kangaroo, which survived to the present because they had adapted to arid conditions early in the Pleistocene and lived in Central Australia. The range of these species would have been reduced during wetter climatic periods, although their survival is evidence that arid areas were never completely lost in Australia. Second are the "dwarfed" species, such as grey kangaroos, that survived (or evolved) either because there were smaller individuals within populations of the related megafaunal species, or because there were whole populations of smaller individuals in particular areas. Third are the woodland species, which became extinct soon after the change to arid conditions, and, fourth, the southern fringe species, which survived somewhat longer but became extinct before the Holocene.

I have suggested two biological attributes to account for the extinction of woodland megafauna: Adaptation to woodland habitats ensured that these species were forced into reduced peripheral areas by the expansion of the arid interior during dry periods in the Pleistocene. Their water requirements ensured that even the woodland habitat that remained as a refuge for smaller species could not support them. Small species could travel between water sources; the megafauna, with a greater need for water, could not.

Water requirements, then, explain the size selection which is a notable feature of late Pleistocene extinctions in Australia. The concentric habitat arrangement ensures that no refuges for large species can occur if the arid core expands beyond a certain point.

The concentric arrangement has also produced the curious concentration in time of the extinctions. The mechanism ensured that extinctions would not be gradual because there was a built-in threshold. The arid pulses could keep occurring with little extinction as long as their magnitude was small enough to leave some woodland areas with adequate water supplies. A few species may have become extinct under the pressure but most were able to survive. The arid expansion of 26,000 to 15,000 B.P. went beyond this threshold and, therefore, resulted in massive extinctions. It was by chance that this threshold was crossed after man had arrived in Australia, and it is this coincidence that caused the debate. Had it been crossed half a million years earlier the event would have been seen as just another of the many extinction episodes in Australian (and world) history (fig. 29.14).

The model proposed here has a number of implications for Australian prehistory. Man, like the megafauna, is dependent upon free water. If the model here is correct, then the map showing the maximum extent of megafaunal distribution (fig. 29.10) also shows the maximum possible extent of human distribution before 26,000 B.P. (Horton 1981).

These areas would have had an economy based on the principal resources, the proportions determined by a balance between economic return and effort required (Jones 1980), as well as more intangible factors such as palatability. It seems likely that the megafauna fit into this economy at the top end of the scale, that is, offering high yield at high cost. Their remains will, therefore, be rare in archaeological sites. In areas close to lakes and rivers, the most frequent food remains are likely to be the high certainty, low effort, low return elements such as shellfish and fish; away from these resources,

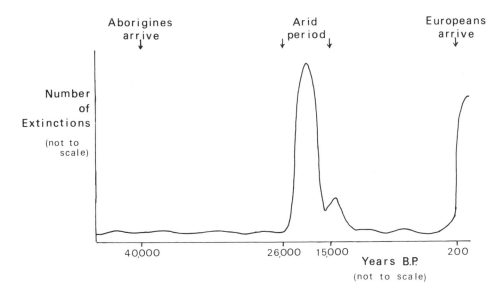

Figure 29.14. Timing of the extinctions in relation to climatic change and human occupation of Australia. The extinctions in the last two hundred years have been mainly of small species in contrast to the extinction of large species in the earlier period. The small peak at around 15,000 B.P. represents the extinction of the four "far southern" species of *Sthenurus*.

small mammals are likely to be the common food. Under this regime, hunting would have had no impact on the megafauna, just as it had none on the smaller species.

As knowledge of the magnitude of late Pleistocene climatic change has increased, the tendency has been to grant it a role in extinction but to see man as the decisive element in the equation. The arguments I have put forward against both the suggested mechanisms for human causation however remain *unaffected* by changes in climate. A more subtle variant of this hypothesis (White and O'Connell 1979) takes the extinction model presented here and adds human predation as the decisive factor. To sustain this view, I think, requires a belief that man was somehow immune from the effects of climatic change. I have discussed the possible human responses during the period 26,000 to 15,000. I think it likely that man, relying on good water resources, retreated to the fringe of the continent during the period of arid expansion. Some people may have been able to survive in what had been the major megafaunal areas because they were not restricted in their range of food the way megafaunal species were. They may even have killed some of the animals clustering around the remaining water resources, but to suggest that this had any impact on the extinction process, as do White and O'Connell (1979), is illogical. The argument is that since the "ecological tethering" resulted in large numbers of animals being assembled, then man could have had a much greater impact on such assemblages than he would have had in normal circumstances. This overlooks the fact that the mechanism I have proposed that produces such "tethering" also kills the animals concerned—the final assemblage is the immediate prelude to death. Man can have no impact without such assemblages, but once they occur any animals killed by man were about to die in any case. The megafaunal species were destined to become extinct whether man was inhabiting the continent or not.

Hope (this volume) suggests that the hypothesis presented here to explain extinctions in mainland Australia cannot be used to explain megafaunal extinction in New Guinea and Tasmania, and certainly these two geographic extremes have always seemed to pose problems for any climatic model.

There are only two sites in New Guinea, Nombe and Pureni Swamp, which seem to be late Pleistocene megafaunal sites. This is clearly insufficient evidence with which to decide whether the proposed extinction model for Australia also has application in New Guinea. However, the fact that both localities are at relatively low altitudes in the highlands (1500 to 1660 meters) does not indicate a montane adaptation in the New Guinea forms, and it is of interest in this context that *Thylacinus* is found with *Protemnodon* at Nombe. I suggest that these may well be fringe localities in what was basically a lowland megafaunal distribution.

Climatic changes were extensive during the late Pleistocene in New Guinea. Hope and Hope (1976) note that the movement of the ice caps produced vegetation changes in the highlands. Such effects would also have been felt at lower altitudes, not least because such changes in ice cover would have considerable impact on the water regimes in the lowlands. Nix and Kalma (1972) suggested that the presence of dry land in what is now the Arafura Sea would also have had a great impact on the climate of southern New Guinea. It is too early to say whether such changes caused megafaunal extinction in New Guinea or whether an alternative explanation is more appropriate for that country.

In Tasmania most of the megafaunal finds have been made in the northwest. I suggest that the four megafaunal species that occurred in the state are likely to have also been present in eastern Tasmania, but not in the western highlands, although work presently is insufficient to demonstrate this. There is now good evidence for extensive climatic change in these areas. Colhoun (1978:6) noted the "widespread occurrence of lunettes and sandsheets in eastern Tasmania" and that lunettes were found "from near Smithton in the northwest to Flinders Island in the northeast, and to Bruny Island in the southeast." The few dates available "correspond closely with the time of maximum aridity in southeastern Australia." It seems, then, that with some reservations due to poverty of evidence, the model for mainland Australia can be applied to New Guinea and Tasmania.

I have not attempted to apply this model to other parts of the world, but I think that, again with some reservations, it probably can be applied. The reservations are that Australia had a very restricted area affected by glacial and periglacial activity, and this is more important on some continents; the relationship between climatic change and the shape of the continent seems to be a crucial factor in Australia; the concentric climatic arrangement has been important in extinction here; the relationship between fire, vegetation, and animal species will vary between continents; the nature of the animal species themselves may well be important. If herds are present, hunting may well have a greater impact than in Australia where gregarious species are lacking.

It is clearly important to examine each case individually. The Australian evidence suggests that the extent of late Pleistocene climatic change has been underrated and that the possible effects of climatic change on large mammals should be determined. Climate has clearly been the dominant agent in the most recent set of Australian extinctions, just as it was in earlier periods.

The apparent differences in the timing and effects of extinction in Africa, North America, and Australia have been used as an argument against the climatic mechanism and, because humans arrived in the different areas at different times, in favor of the human mechanism.

However, given the widely different geography of these three areas, some lack of climatic synchrony is to be expected. Completely synchronous extinctions would be evidence in favor of some unique, perhaps extraterrestrial event, rather than of the climatic model. The extinctions are, however, all clearly linked to the last glacial change, those in Australia occurring relatively early in the period while those in North America were relatively late. The timing in Africa is uncertain (Klein this volume).

The hypothesis presented here for Australia suggests a two-stage process in which the earlier extinctions in the period (say 26,000 to 20,000 B.P.) were caused by water-related problems, while the later ones (of the small *Sthenurus* species) may have been caused in the period 18,000 to 15,000 by additional water stress, or at the same time or a little later by habitat change. Because of either the nature of the climatic change or the animal species concerned, it may be that in North America all extinctions were due to habitat change that occurred late in the Pleistocene and that were roughly synchronous with the *Sthenurus* extinctions. Webb (this volume) essentially suggests this, correlating the extinctions in North America with the furthest glacial advance, and noting that some North American species were able to survive in Central America. For Africa, Klein (this volume) correlates some extinctions with habitat change between 12,000 to 9,500 years ago.

The difference in the extent of the extinctions is also consistent with the climatic change model. In both North America and Australia the rate is high. In Australia this is because, as I have suggested, the central arid expansion in a concentric habitat arrangement leaves no possible refuge areas. In North America, Webb (this volume) has suggested that the glacial advances from the north essentially produce a bottleneck effect in Central America because "extinction events appear to have swept many North American large mammals southward" and "the rapidly decreasing area of Central America has an overriding effect."

In Africa the extinction rate is much lower, and this is clearly because Africa, astride the equator, always has a range of refuges available. The wet forests of western Africa have remained in place, although fluctuating in size, since the Tertiary. Klein (this volume) makes the point that, while local extinctions are apparent in the fossil record, complete extinction is rare. An area of suitable habitat is always available somewhere in the continent. The extinction rate, however, is higher, and complete extinctions do occur in the late Pleistocene. In contrast to those of North America and Australia, these extinctions all seem to be species that have surviving relatives in Africa. It may be that this late Pleistocene episode is simply a special case of earlier ones. It may also be that a threshold mechanism is operating in southern Africa similar to the one suggested here for Australia. That is, normally some humid areas survive as refuges in southern Africa, but the last glacial change was just a little more severe and removed these refuges, causing extinction of some species that occurred only in southern Africa, but leaving their relatives surviving in other parts of the continent.

Thirty thousand years ago a biologist studying Australian mammals would have considered red kangaroos to be very much the poor relations of the large mammal fauna: sheltering to conserve water and therefore forced to subsist on poor quality food, or endlessly moving to follow the rains to obtain water and good quality food, and existing in low numbers in the harshest part of Australia.

But this was the species destined to survive, and for about 10,000 years to considerably expand its range. Australia was not, in the long term, a continent that was to maintain well-watered woodlands supporting many species of large mammals, but this member of the megafauna was preadapted to the long drought that began around 26,000 B.P. The other species could not survive long enough to see the woodland and the water resources return to at least part of their original area.

Cows and sheep took their place on the western slopes and plains when Europeans colonized the continent. And though conditions had ameliorated in the previous 15,000 years, these introduced species could be supported on a *permanent* basis only by providing artificial waterholes and by adding new grass species to the existing ones.

Summary

The normal Aboriginal economic system, as it operates in modern times, in which small mammals are staple food items and large animals are rarely caught, could not have caused the extinction of the megafauna. Archaeological evidence suggests that this economic system operated in the Pleistocene, and sites are beginning to be found in which a few specimens of megafaunal species are found among more numerous smaller species. While it is difficult to imagine an Aboriginal economic system which included the mass killing of large animals, it is equally difficult to see how even such an extreme form of predation could have caused extinction. No archaeological evidence for this kind of activity has been found in Australia.

The woodland habitat was one with a potential for frequent, low intensity fires. In most areas this potential would have been achieved by lightning strikes igniting the fuel. In some areas with low frequency of lightning the woodland may have remained in climax form longer. Neither situation would have been disadvantageous to the megafauna in terms of shelter, but the higher frequency fire areas may have been slightly advantageous in terms of food. The arrival of man in Australia can at most have only increased the rate of ignition so that all the woodland areas could achieve their potential fire regimes. If man altered fire regimes at all, the effect can only have been favorable to the megafauna.

Not surprisingly, the time of man's arrival in Australia, whether at 40,000 B.P. or earlier, bears no relationship to the timing of the extinctions (fig. 29.14). The extinctions, however, do coincide with a period of major climatic change in Australia. The model I have put forward to explain how this climatic change caused the extinctions relies on the following propositions: Most Australian megafaunal species were woodland inhabitants, but one small group of species occupied the fringing grasslands of south-eastern Australia, and two other species occupied arid central Australia. Only those two species survived to the present.

The megafaunal species that became extinct relied on the availability of water much more heavily than the species that have survived. The combination of these two biological attributes was the cause of the extinctions, and it also resulted in their being concentrated in time. The concentric arrangement of habitats in Australia creates a threshold effect (fig. 29.13b). The arid area of central Australia expanded and contracted throughout the Pleistocene without crossing the threshold because woodland areas with adequate water supplies always remained, and few extinctions were caused. The threshold was crossed between 26,000 and 15,000 yr B.P. when the arid area expanded further than usual and water resources in the woodland areas were severely reduced. In some cases populations were trapped around drying-up water supplies, in others food supplies were eaten out even though the water remained. Smaller species with less water dependence could travel far enough to find food.

The evidence for this extinction model lies in the occurrence of sites of the Lancefield type with a preponderence of mature animals and few young animals, and of sites where red kangaroos are found outside their present range. The sites of concern for this model will range in age between about 26,000 and 20,000 B.P. Some earlier sites of this type may serve to record earlier episodes of arid expansion.

That the extinctions coincide with a period of major climatic change is not in itself evidence that the two events are causally related. It has been argued that the megafauna could survive both human activity and climatic change as individual effects, but only the two combined were capable of causing extinction. I have found no need for the hypothesis that human activity is involved in the extinctions. The model proposed here would operate to cause extinctions whether man was present or not. It also explains all the data available at present without proposing human actions, which are invisible in the archaeological and fossil records. Finally, the model is testable, and that is perhaps the best attribute that any model can have.

Acknowledgments

I thank Geoff Hope, John Mulvaney, Peter White, and Richard Wright for reading and commenting on an earlier draft of this manuscript.

References

Bolton, B. L. and Latz, P. K. 1978. The western hare-wallaby, *Lagorchestes hirsutus* (Gould) (Macropodidae), in the Tanami Desert. *Australian Wildlife Research* 5:285–293.

Bowdler, S. 1977. The coastal colonization of Australia. *In* Allen, J., Golson, J. and Jones, R., editors. *Sunda and Sahul:* 205–246. Academic Press, London.

Bowler, J. 1975. Deglacial events in southern Australia: their age, nature, and palaeoclimatic significance. *In* Suggate, R. P. and Cresswell, M. M., editors. *Quaternary Studies:* 75–82. Royal Society of New Zealand, Wellington.

———. 1976. Aridity in Australia: age, origins and expression in aeolian landforms and sediments. *Earth-Science Reviews* 12:279–310.

Calaby, J. H. 1966. Mammals of the Upper Richmond and Clarence rivers, New South Wales. *CSIRO, Division of Wildlife Research Technical Paper No. 10.*

Catling, P. and Newsome, A. 1981. Responses of the Australian vertebrate fauna to fire: an evolutionary approach. *In* Gill, A. M., Grove, R. H. and Noble, I. R., editors. *Fire and the Australian Biota:* 273–310. Australian Academy of Science, Canberra.

Christensen, P. and Kimber, P. 1975. Effects of prescribed burning on the flora and fauna of south-west Australian forests. *Proceedings of Ecological Society of Australia* 9:85–107.

Colhoun, E. R. 1978. Recent Quaternary and geomorphological studies in Tasmania. *Australian Quaternary Newsletter* 12:2–15.

Coutts, P., Horton, D. R., Kershaw, A. P. and Peterson, J. in prep. An archaeological site at Lake Bolac, western Victoria.

Dodson, J. R. 1974. Vegetation history and water fluctuations at Lake Leake, south-eastern South Australia I: 10,000 B.P. to present. *Australian Journal of Botany* 22:719–741.

———. 1975. Vegetation history and water fluctuations at Lake Leake, south-eastern South Australia II: 50,000 B.P. to 10,000 B.P. *Australian Journal of Botany* 23:815–831.

Finlayson, H. H. 1927. Observations on the South Australian members of the subgenus *Wallabia. Transactions of the Royal Society of South Australia* 15:363–377.

Flannery, T. F. and Gott, B. in prep. Late Pleistocene megafaunal extinction and the Spring Creek locality of southwestern Victoria.

Foley, J. C. 1957. Droughts in Australia. *Bulletin No. 43,* Bureau of Meteorology, Melbourne.

Fox, A. M. 1978. The '72 fire of Nadgee Nature Reserve. *Parks and Wildlife* 2:5–24.

Fox, B. J. and MacKay, G. M. 1981. Small mammal responses to pyric successional change in eucalypt forest. *Australian Journal of Ecology* 6:29–41.

Gill, E. D. 1954. Ecology and distribution of the extinct giant marsupial, *Thylacoleo. Victorian Naturalist* 71:18–35.

Gillespie, R., Horton, D. R., Ladd, P., Macumber, P. G., Thorne, R. and Wright, R. V. S. 1978. Lancefield Swamp and the extinction of the Australian megafauna. *Science* 200:1044–1048.

Goede, A., Murray, P. and Harmon, R. 1978. Pleistocene man and megafauna in Tasmania: dated evidence from cave sites. *The Artefact* 3:139–149.

Guiler, E. R. 1961. The former distribution and decline of the thylacine. *Australian Journal of Science* 25:207–210.

Hope, G. S. 1978. The late Pleistocene and Holocene vegetational history of Hunter Island, north-western Tasmania. *Australian Journal of Botany* 26:493–514.

Hope, J. H. and Hope, G. S. 1976. Palaeoenvironments for man in New Guinea. *In* Kirk, R. L. and Thorne, A. G., editors. *The Origin of the Australians:* 29–54. Australian Institute of Aboriginal Studies, Canberra.

Hope, J. H., Lampert, R. J., Edmondson, E., Smith, M. and Van Tets, G. F. 1977. Late Pleistocene faunal remains from Seton Rock Shelter, Kangaroo Island, South Australia. *Journal of Biogeography* 4:363–385.

Horton, D. R. 1976. Lancefield: the problem of proof in bone analysis. *The Artefact* 1:129–143.

———. 1978. Extinction of the Australian megafauna. *Australian Institute of Aboriginal Studies Newsletter* 9 (n.s.):72–75.

———. 1979. The great megafaunal extinction debate: 1879–1979. *The Artefact* 4:11–25.

———. 1980. A review of the extinction question: man, climate and megafauna. *Archaeology and Physical Anthropology in Oceania* 15:86–97.

———. 1981. Water and woodland: the peopling of Australia. *Australian Institute of Aboriginal Studies Newsletter* 16:21–27.

———. 1982. The burning question: Aborigines, fire and Australian ecosystems. *Mankind* 13:237–251.

Horton, D. R. and Connah, G. E. 1981. Man and megafauna at Reddestone Creek, near

Glen Innes, northern New South Wales. *Australian Archaeology* 13:35–52.

Horton, D. R. and Murray, P. 1980. The extinct Toolach wallaby *(Macropus greyi)* from a spring mound in north western Tasmania. *Records of the Queen Victoria Museum* 71:1–12.

Horton, D. R. and Samuel, J. 1978. Palaeopathology of a fossil macropod population. *Australian Journal of Zoology* 26:279–292.

Jones, R. 1968. The geographical background to the arrival of man in Australia and Tasmania. *Archaeology and Physical Anthropology in Oceania* 3:186–215.

———. 1975. The Neolithic, Palaeolithic and the hunting gardeners; man and land in the Antipodes. *In* Suggate, R. P. and Creswell, M. M., editors. *Quaternary Studies:* 21–34. Royal Society of New Zealand, Wellington.

———. 1980. Hunters in the Australian coastal savanna. *In* Harris, D., editor. *Human Ecology in Savanna Environments.* Academic Press, London.

Jones, R. and Bowler, J. M. 1980. Struggle for the savanna: northern Australia in ecological and prehistoric perspective. *In* Jones, R., editor. *Northern Australia: Options and Implications:* 3–31. Research School of Pacific Studies, Australian National University, Canberra.

Ladd, P. G. 1976. Past and present vegetation of the Lancefield area, Victoria. *The Artefact* 1:113–127.

Main, A. R., Lee, A. K. and Littlejohn, M. J. 1957. Evolution in three genera of frogs. *Evolution* 12:224–233.

Marshall, L. G. 1973. Fossil vertebrate faunas from the Lake Victoria region, southwestern New South Wales, Australia. *Memoirs of the National Museum of Victoria* 34:151–281.

———. 1974. Late Pleistocene mammals from the Keilor cranium site, southern Victoria. *Memoirs of the National Museum of Victoria* 35:63–86.

Martin, P. S. and Wright, H. E. Eds. 1967. *Pleistocene Extinctions: The Search for a Cause.* Yale University Press, New Haven.

Merrilees, D. 1968. Man the destroyer: late Quaternary changes in the Australian marsupial fauna. *Journal of the Royal Society of Western Australia* 51:1–24.

Newsome, A. E. 1977. Imbalance in the sex ratio and age structure of the red kangaroo, *Macropus rufus*, in central Australia. *In* Stonehouse, R. and Gilmore, D., editors.

The Biology of Marsupials: 221–223. Macmillan, London.

Nix, H. A. and Kalma, J. D. 1972. Climate as a dominant control in the biogeography of Northern Australia and New Guinea. *In* Walker, D., editor. *Bridge and Barrier: the natural and cultural history of Torres Strait:* 61–92. Australian National University Press, Canberra.

Rich, P. V. 1979. The Dromornithidae, an extinct family of large ground birds endemic to Australia. *Bureau of Mineral Resources, Geology and Geophysics Bulletin* 84:1–196.

Sampson, J. C. 1971. The Biology of *Bettongia penicillata* Gray, 1937. Unpublished Ph.D. Thesis, University of Western Australia.

Shipman, P. 1975. Implications of drought for vertebrate fossil assemblages. *Nature* 257:667–668.

Smith, M. J. 1972. Small fossil vertebrates from Victoria Cave, Naracoorte, South Australia 2. Peramelidae, Thylacinidae and Dasyuridae (Marsupialia). *Transactions Royal Society of South Australia* 96:125–137.

Tedford, R. H. 1955. Report on the extinct mammalian remains at Lake Menindee, New South Wales. *Records of the South Australian Museum* 11:299–305.

———. 1966. A review of the macropodid genus *Sthenurus. University of California Publications in Geological Science* 57:1–72.

Webster, P. J. and Streten, N. A. 1972. Aspects of late Quaternary climate in tropical Australia. *In* Walker, D., editor. *Bridge and Barrier:* 39–60. Australian National University, Canberra.

Wells, R. T. 1978. Fossil mammals in the reconstruction of Quaternary environments with examples from the Australian fauna. *In* Walker, D. and Guppy, J. C. editors. *Biology and Quaternary Environments:* 103–124. Australian Academy of Science, Canberra.

White, J. P. and O'Connell, J. F. 1979. Australian prehistory: new aspects of antiquity. *Science* 203:21–28.

Wilkinson, H. E. 1972. *Macropus rufus*, Victorian fossil find. *Victorian Naturalist* 191:95–99.

Wyroll, K-H. 1979. Late Quaternary climates of Western Australia: evidence and mechanisms. *Journal of the Royal Society of Western Australia* 62:129–142.

———. 1980. Quaternary studies group of Western Australia. *Australian Quaternary Newsletter* 14:9–11.

30

Australian Environmental Change
Timing, Directions, Magnitudes, and Rates

GEOFFREY HOPE

An impression of stability and continuity through time is conveyed by much of the Australasian region. This impression arises from many causes. The strangeness and endemicity of the biota seem to argue for a long history of undisturbed specialization and adaptation in isolation. The ubiquitous genera *Eucalyptus, Acacia,* and *Casuarina* lend a superficial uniformity to a large proportion of the Australian continent, and this seems too extensive for change. Finally the subdued topography with its widespread basement of Paleozoic, or older, rocks, suggests a continent on which environmental change has been minor. The Australian continent, however, does have an extremely active orogenic zone on its northern edge, forming the island of New Guinea separated by the shallow intermittent sea of Torres Strait; also, worldwide changes of climate in the ice age did not bypass Australia.

Our understanding of the historical development of the present Australian environment is the result of several advances over the past twenty years. Since the first major review in 1959 (Keast et al. 1959), the larger scale biogeography has been changed by the application of plate tectonic theories. These provide a basis for interpreting modern distributions when combined with the increasingly well-documented fossil record. A reliable microfloral zonation for southeastern Australia is now available for the Tertiary (e.g. Stover and Partridge 1973, Truswell and Harris 1982).

After a long period of neglect, more geological work is now being carried out on the continental Cenozoic history in Australia. Based on geomagnetics, volcanic rocks, Tertiary weathering, and the interpretation of southern ocean cores, a patchy record of landscape evolution is emerging to complement the paleontologic record. Late Quaternary studies have been more vigorous, and there is now a framework of vegetational and geomorphic histories that provide fair coverage back to the last interglacial. Despite a lack of support for taxonomy and field ecology because they are "unfashionable," knowledge of modern Australian–New Guinean biota and communities has increased rapidly, to the benefit of paleoecologists.

The general consensus now is that, at least during the late Pleistocene, Australia and New Guinea experienced great change of a scale comparable to that in Europe and North America. This is based on evidence of environmental change, mostly from sites on the eastern and northern margins of the continent. The consensus does not extend to details, and various interpretations of the paleoenvironments are possible. Ideas on the evolution and dynamics of communities are only just beginning to be developed, so it is still not possible to describe the environments in which faunal extinctions took place. There are several reviews which present the general data on environmental change in Australia (e.g. Bowler et al. 1976, Jones and Bowler 1980, Kershaw 1981, Singh 1982, Walker 1981, Walker and Hope 1982, Walker and Singh 1981).

This chapter summarizes the general trends of climatic and vegetational change, and then analyzes them for influences which might have affected vertebrate extinction rates. It is necessary to briefly examine the evolutionary history of the biota in order to consider whether particular environmental changes should result in extinction.

The modern biota of the Australian kingdom can be allocated to five rather distinctive biogeographic provinces, as defined for plants by Burbidge (1960) (fig. 30.1). The provinces correlate with the climatic zones of Australia–New Guinea, and thus the history of environmental change also reflects the history of these provinces. If extinction is correlated with environmental change, it should also be linked to the relative fortunes of biogeographic provinces; in the longer term the decline of a province reflects an opportunity for expansion of others.

Have the rates or directions of change altered markedly over the period of extinction? Is the reaction of plant communities to change (their stability or resilience) likely to have contributed to extinction in their dependent fauna? Are there marked differences in the histories of the biogeographic provinces that might have contributed to the timing or location of extinctions?

The general trends of biotic change can be summarized in four time periods:

1. Late Tertiary   —Appearance of many modern niches and genera

2. Early-Mid Pleistocene   —Evolution of modern species/environmental relationships

3. Late Pleistocene   —Reaction of biota to environmental change

4. Pleistocene-Holocene transition   —Appearance of modern environmental pattern and population structures (communities)

Late Tertiary

The first appearances of modern plant genera and families now widespread in Australia and New Guinea fall readily into two major groups, those that were already present in the early Tertiary and those that appear within the last 25 my. The degree to which the latter migrated from southeast Asia is controversial (Webb and Tracey 1981), but certainly a large proportion developed autochthonously in response to slow but massive environmental changes. These reflected orogenic uplift in eastern Australia, the rafting of the continent into lower latitudes, the opening of the southern ocean, and the worldwide increase in thermal gradient between the equator and the poles, leading to the major climatic fluctuations of the Pleistocene. Some of these changes, such as movement into lower latitude seas and the creation of moist mountain habitats in New Guinea, offset the general climatic cooling and drying to maintain habitats that resembled those of higher latitude–warmer ocean of the early Tertiary. For the bulk of Australia the movement into the subtropics and cooling of the oceans brought a seasonality and unreliability in rainfall and the reduction of precipitation/evaporation balances. The late Miocene is marked in parts of southern Australia by deep weathering to a degree now confined to parts of northern Australia and New Guinea. The late Tertiary vegetation histories (e.g. Kemp 1978, Martin 1978) are not well known, but fairly widespread closed forests with taxonomic and possibly structural affinities to the modern montane rainforests of New Guinea, New Zealand, and western Tasmania did exist.

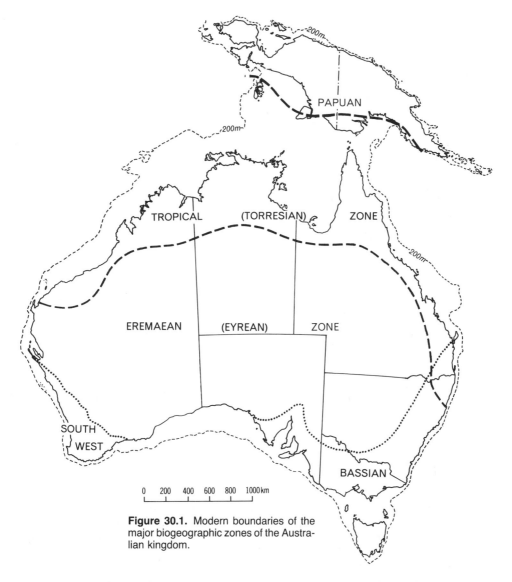

Figure 30.1. Modern boundaries of the major biogeographic zones of the Australian kingdom.

But more open communities also had been present from the early Tertiary, and perhaps xerophytic taxa had already developed. There are very few Pleistocene records; rainforest may have become more restricted while grasses, composites, Myrtaceae, *Casuarina*, and Proteaceae became widespread. In the north the lowland forests were presumably still extensive but gradually became isolated from New Guinea. In terms of modern floristic and faunal provinces, the Papuasian was widespread, the southwest Australian, Bassian, and Torresian were already forming, but the Eremaean (Eyrean) arid zone had hardly appeared. The transition to the Pleistocene is not apparently marked by a great increase in extinction and replacement rates, and the boundary is not yet definable on fossil grounds. Although the presence of certain pollen types, such as *Nothofagus brassii*, in southeast Australia is taken as indicating pre-Pleistocene age, the actual time of extinction of these floristic elements is unknown.

Early and Mid-Pleistocene

Although there are no records of floristic change, it is during the first half of the Pleistocene that massive biological radiation must have taken place or at least continued from the Pliocene. This resulted in the broad categories of vegetation and fauna that occur today. The retreat of rainforests to restricted high rainfall areas, and their replacement by open eucalypt forest in eastern Australia, was matched by the spread of arid communities and the appearance of grasslands, shrub steppe, and probable sclerophyll-dominated woodlands. The Eremaean province came into being, but there are no direct records of this event. Volcanic activity occurred in Victoria and Queensland, and the cycles of minor glaciation may have commenced.

Although no date has been accepted for creation of the dryland and desert sandridge country, some circumstantial evidence indicates that it was actively spreading in the middle Pleistocene. The Blanchetown clays, laid down by Lake Bungunnia, are thought to be more than 350,000 years old (based on paleomagnetic reversal dating at Lake Tyrrell, in central Victoria). These clays are overrun by the red sandridges, and indicate relatively saline or brackish conditions. This record is matched by the presumed mid- or early-Pleistocene drying of Lake Dieri in central Australia, the sediments of which also underlie the red dunes, which in many places rest on gypseous clays blown from a drying lake floor (Loeffler and Sullivan 1979). Problems arise in dating these occurrences and the associated faunas preserved in the alkaline sediments. In New Guinea widespread orogenic uplift led to extensive vulcanism and the appearance of the first upper montane habitats.

Although a comparison between the environments of the late Pleistocene and those of the Pliocene tells us what must have happened, we have little detail of the spread of arid communities or of the decline of closed forests. Central Pacific and Southern Ocean cores (Frakes 1978) seem to indicate that the five climatic oscillations of the Pleistocene prior to about 400,000 B.P. were of lower magnitude and frequency than the last three. On this slender basis, it may be that climatic variation was less extreme in the early Pleistocene than in the last third.

Late Pleistocene

Here the data are much more extensive, although only three pollen records extend back to the last interglacial. Two opposing themes emerge from a consideration of all the data. On the one hand, it is difficult to find records of communities of modern vegetation over the time of the last glacial. In contrast many relict distributions seem to predate the available records, and so may represent communities of considerable antiquity. Although plant extinctions take place (see Kershaw this volume) at the generic level, there are few appearances or replacements in the flora. From this it follows that over much of Australia and New Guinea, taxa formed different communities in late Pleistocene environments. For example, upland sites in eastern Australia and lowland sites in Tasmania and coastal Victoria were occupied by a shrub-rich grassland with abundant grass and daisies, and only scattered eucalypts during the coldest period about 18,000 B.P. (Hope 1978, Dodson 1979). Moisture-demanding taxa, restricted to dense forest today, occurred in open grasslands. Channel morphology and source-bordering dunes suggest that woodlands and riparian trees were absent from the inland plains 18,000 years ago, although they were quite extensive at 30,000 B.P.

From Lake George in New South Wales (Singh et al. 1981) it is possible to compare previous glacial periods with the last. The modern eucalypt-dominated woodland is not found at earlier times, instead sharing dominance with *Casuarina* and some moist-forest elements. At Lake George only sparse vegetation grew near the site at 18,000 B.P., while many wet-forest taxa and subalpine herbs were present before 25,000. Much of the Australian flora has poor representation in pollen diagrams, e.g. *Acacia*, so it is possible that other shrub and tree genera besides *Casuarina* had an important role in the Pleistocene. While these floristically and structurally distinctive communities undoubtedly reflect different climatic controls, Singh et al. (1981) point out that reduced fire frequencies may have contributed to these patterns. Both floras and faunas indicate associations of taxa which today are separated into open (semiarid) or closed (moist forest and heath) communities (e.g. Hope et al. 1978).

In a continent as little affected by glaciation as Australia, it seems surprising that few sites record late Pleistocene vegetation communities that could have given rise to the present pattern. Exceptions include the arid Nullarbor Plain, which appears to have had a chenopod steppe 25,000 years ago similar to that occurring inland of the site today (Martin 1973). The basaltic plains of western Victoria still preserve an open grassland with scattered trees, which has persisted despite great changes in the hydrology (which can be related to effective precipitation) of lakes in the area (Dodson 1979). Dense eucalypt forest was present on the eastern coast ranges at 14,000 B.P. (Ladd 1979). The existence of many isolated patches of warm and cool temperate rainforest along the eastern coast suggests that this relatively humid area has acted as a refuge for many taxa, possibly since the last interglacial or longer.

In humid Tasmania, Colhoun (1980) interprets a deposit as probably interglacial, based on the presence of a *Nothofagus* forest which was subsequently forced from the area until about 10,000 B.P. Kershaw (1978) has shown that an interglacial rainforest in northeast Queensland also resembled that of the present day.

In New Guinea cold-indicating taxa reached lower altitudes before 35,000 B.P. than during the last major depression of the treeline between 27,000 and 14,000 B.P., when a large area of subalpine grasslands existed (Walker and Hope 1982). Similarly, a relative extension of cool wet forest follows a less pronounced cold or dry period in Tasmania, Lake George, and northeast Queensland about 70,000 to 50,000 B.P. But after about 35,000 B.P the wet forest declined and became absent from most records, being replaced by associations of xerophytes and, in some cases, alpine herbs.

The decline of rainforest and spread of shrub steppe took place between 35,000 and 25,000 B.P., and was a prelude to widespread active slopes and some glaciation, particularly in Tasmania and New Guinea. The extent of treelessness, hence the degree of associated cold and aridity, seems to have been greater than during earlier periods of open conditions within the last 100,000 years, at least in Australia. Figure 30.2 indicates probable vegetation boundaries in southeastern Australia during this period of extreme conditions. The vegetation boundaries show a reasonable agreement with models of precipitation/evaporation provided by Jones and Bowler (1980) for Australia for 18,000 B.P. This model indicates that there was a change from a considerable water surplus relative to today at 35,000 B.P. to the deficit indicated after 25,000 B.P. Kershaw (1978) found no evidence for a marked surplus in northeast Queensland, but expanded former lake shores in central and northern Australia suggest a period of increased surface water, stable dunefield, and presumably denser vegetation (Jones and Bowler 1980, Hughes and Lampert 1980).

The latter part of the glacial period also reflects a change in fire regime across the continent. In the New Guinea highlands burning commences at 30,000 B.P. near the time of first occupation. Although fires had been common in Australia, a change in regime is apparent in the last glacial period. Singh et al. (1981) attribute some changes in vegetation composition to the relative absence of fire-sensitive taxa today and suggests the arrival of humans as the cause. This raises a question for assessing climatic change from vegetation histories which will also be responding to changing fire regimes. In general, the shift from dense, moist vegetation to open grasslands/shrublands will favor more frequent and extensive firing.

Pleistocene-Holocene Transition

Warming of the Australian continent took place about 16,000 B.P. or earlier. Combined with a cold and lowered sea surface, this gave rise to substantial water deficits and widespread aridity. Most sites across Australia record the driest time in their record for the period 16,000 to 12,000 B.P. Slope instability in the Derwent Valley in Tasmania at about this time (Wasson 1977) is only one of many areas in which sand mobilization, lunette formation or total drying of lakes, sand sheet formation, fluvial changes, and vegetation histories indicate very open conditions. The interest in the Tasmanian deposits arises because buried sand sheets above slope deposits indicate similar events earlier in the last glacial period. The longitudinal dune system of inland Australia was active, but it is not known if it became more extensive than in earlier dry phases.

After 10,000 B.P. the very high rate of change continued; in some areas the direction reversed. Along the coast and through Torres Strait the rising and warming sea brought increases in rainfall which offset the increased land temperatures. Many lakes within 150 km of the modern coastline reached their highest levels in the period 8600 to 6000 B.P., and closed forest was more extensive. In general the modern vegetation distributions began to be established about 10,500 B.P. and were complete with some exceptions by 8500 B.P. In detail, most Holocene communities appear to be still adjusting in species composition and range, and relict distributions persist in many areas. This is not surprising, as considerable migrations had taken place, but it also reflects the novelty of the present climate pattern.

The semiarid and arid winter rainfall areas of the Eremaen (Eyrean) region did not recover from the late Pleistocene aridity. Evaporation increases presumably outstripped any increase in rainfall. There is a significant difference between high evaporation/ medium precipitation and medium evaporation/low precipitation regimes, the latter favoring shrub steppe and the former grasslands and savannah. Higher evaporation increases the influence of unreliability of rainfall, and the occurrence of long periods of severe drought favors ephemeral rather than perennial shrubs and trees. The flooding of Torres Strait brought increasing and relatively reliable rain, partly due to cyclones, to wide areas of northern Australia (Nix and Kalma 1972) and allowed the spread of woodlands and savannah to the south.

Also around 11,000 B.P. the closed forest in New Guinea invaded the extensive mountain grasslands and spread out onto much of the southern lowlands, replacing savannah (Hope and Hope 1976). In both Australia and New Guinea the Holocene/ Pleistocene boundary is marked by the onset of burning regimes similar to those endured at present, and the plant migrations which took place seem, in some cases, to have been hindered or prevented due to fire. The lack of many long records and the large changes in vegetation types prevent a useful comparison of late Pleistocene and early Holocene fire regimes.

Discussion

Biotic extinction with replacement can occur in isolated continental biota as a result of environmental change. If the change occurs in a single direction in evolutionary time, then the replacing taxa may be directly related to the extinct populations. If the change is multidirectional and rapid, the replacing taxa may be selected from other taxonomic groups to suit different ecological niches. Only if the change is so sudden and complete that virtually nothing of an original niche is left can extinction without replacement be expected. The Australian Quaternary contains examples of extinction of fauna and flora both with replacement and without, and the latter seems to be concentrated in the period from ca. 100,000 to 10,000 years ago. This poses a general question: Does the timing, direction, magnitude, and rate of change evidenced by vegetation and geomorphic responses in Australia and New Guinea help explain the pattern of faunal extinction there? The magnitude cannot be separated from the timing or rate; the first substantial cold/arid cycle may, for example, be expected to trigger more extinction than subsequent, more extreme, climatic oscillations. Similarly, a rapid environmental

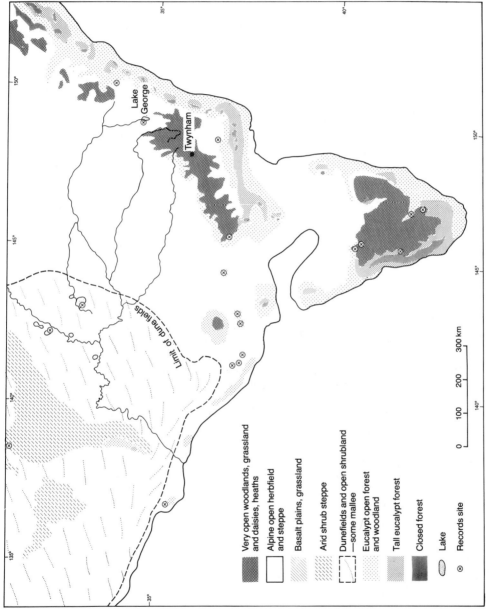

Figure 30.2. A reconstruction of the vegetation boundaries in Southeast Australia at 22,000–16,000 B.P. This represents the most extreme variation from cool moist conditions prior to 30,000 B.P. and the warm conditions after 10,000 B.P.

Very open woodlands, grassland and daisies, heaths

Alpine open herbfield and steppe

Basalt plains, grassland

Arid shrub steppe

Dunefields and open shrubland —some mallee

Eucalypt open forest and woodland

Tall eucalypt forest

Closed forest

Lake

⊗ Records site

Lake George

Twynham

Limit of dune fields

0   100   200   300 km

change may have a greater impact on damaged populations than would slower shifts of greater magnitude. Finally, the role of very extreme isolated events, such as record drought or cold, can be important and extremely difficult to discern in the fossil record.

The appearance and spread of the Eyrean and Torresian provinces, the fragmentation of the closed forest biota, and limited migration of new taxa allowed many faunal groups to radiate and replace ancestral marsupial species in the Pliocene–early Pleistocene. The glacial-interglacial cycles of the Pleistocene imposed a faster rate of change on biota, in that they required the ability to migrate or tolerate a wide climatic range.

These cycles may have become more intense with time, but general records such as those derived from sea levels in New Guinea (Chappell et al. 1974) suggest that the rate of change is gradual in terms of the time scales of populations. The advent of new fire regimes was presumably far more sudden, and brought change on time scales shorter than the life cycles of many longer-lived plant species. Nonreplaced extinction of herbivores, particularly the Pleistocene browsers, also changed the competitive balance between plant taxa, but in a fairly gradual way.

The general question can be rephrased if only the late Pleistocene, nonreplaced extinctions are considered.

Are there environmental changes in the period 50,000 to 10,000 B.P. that are remarkable in comparison with previous events? The answer, based on sea levels and the Lake George record (Singh et al. 1981), is that events similar to the very open vegetation of 25,000 to 10,000 years ago have occurred before, though this period is as extreme as any for which we have evidence and more extreme than cool periods earlier in the last glacial. With the exception of the closed forests and possibly some shrub steppe, no present community seems to have had a long history in its present range, and the degree of vegetation migration that took place 35,000 to 25,000 B.P. has been surprising. The subsequent changes have resulted in modern plant communities different in many respects from any seen before.

The rate of environmental changes with the onset of the cool-arid maximum cannot yet be compared with earlier analogues. In terms of resource changes the Eremaean province expanded, although its central core became substantially less vegetated. The Torresian province migrated north onto Torres Strait and was possibly restricted. The Bassian province remained extensive, although the ranges of individual communities changed greatly. The Papuan and southwest Australian provinces were relatively unchanged.

In fact it seems that only two extensive Pleistocene niches have disappeared. First, the steppe-grassland of the Bassian region, dominated by daisies, grass, and probably snowgum, with some heaths, has been replaced by eucalypt woodlands or forest. Remnants occur as subalpine or montane woodland. Similarly in New Guinea, the montane grasslands have been replaced by newly evolved closed forest communities (Hope 1980). These changes are the result of Holocene warming, not the onset of cold-arid conditions.

The timing of the onset of cold-arid conditions can be postulated as occurring at around 30,000 B.P., and many of the vegetation and hydrological changes continue until 23,000 B.P. The adjustment of vegetation boundaries was helped by fire in some instances, which tends to remove marginal or relict stands incapable of replacing themselves. Surface water would have become considerably scarcer over this time, and rainfall reliability in the last 30,000 years never seems to have exceeded that of the time of pronounced water surplus before then.

If niche changes are the sole explanation for extinction, the above results suggest that we can expect more extinctions to occur in the period 30,000 to 20,000 B.P. We can further expect relatively higher extinction rates in the Bassian Province than elsewhere, although very few extinctions can be attributed to niche disappearance in any province. Extinction should also be concentrated on the wet forest and other reliable moisture-demanding vegetation communities, and be least for pyrophytic grasslands and woodlands. This does not appear to be the case, either chronologically or geographically. There is also a rather low level of plant extinction (see Kershaw this volume) compared with faunal extinction, and situations of remarkably high floral diversity and endemicity, for example the southwest province (Carlquist 1974), contrast with a rather cosmopolitan disharmonious vertebrate fauna in which extinction has continued up to the present day (Merrilees this volume).

If niche change is additional to some other disturbance, one could predict a drawn-out process of extinction, with large taxa of open environments dying out (or surviving) independently of the climatic variation. Browsing and forest edge taxa would have become more vulnerable as the cold-arid changes continued. Predators would follow a similar pattern, except they may be still more sensitive to drought intensification and would have more restricted ranges after 25,000 B.P. Increased fire frequencies and restriction of permanent water supplies would favor taxa capable of rapid or long-range movement and would restrict taxa of dense habitats.

This suggests that climatic variation was not a major element in nonreplaced extinction. Individual plants cannot migrate as such, they can only propagate, and this would seem to place them at greater risk from environmental change than large mobile animals. This lack of migratory ability by plants in the face of periodic climatic change results in resilience that may be missing from fugitive taxa. This resilience, developed over the Pleistocene, has obviously benefited the Australian and montane New Guinean vegetation faced with abruptly imposed human disturbance. In some cases vegetation structure and taxa seem to have remained as or reverted to glacial-period types, more open and more capable of regenerating after disturbance than those which could potentially develop in the present climate. Perhaps some animal groups have not developed this resilience because the ability to migrate had been their major adaptation to changing environments. This strategy proved insufficient when the environmental changes of the late Pleistocene were compounded by disturbance at a rate never experienced before.

Acknowledgments

I thank Joe Jennings, Judy Owen, and Gurdip Singh for commenting on this paper and Tony Dare-Edwards, Jeannette Hope, Phil Hughes, Peter Kershaw, Mick Macphail, John Magee, and Marjorie Sullivan for helpful discussions of landscape evolution in Australia.

References

Bowler, J. M., G. S. Hope, J. N. Jennings, G. Singh and D. Walker. 1976. Late Quaternary climates of Australia and New Guinea. *Quaternary Research* 6:359–394.

Burbidge, N. 1960. The phytogeography of the Australian region. *Aust. J. Bot.* 8:75–212.

Carlquist, S. 1974. *Island biology.* Columbia Univ. Press, New York.

Chappell, J. M., A. L. Bloom, W. S. Broecker, R. R. Mathews and K. J. Mesolella. 1974. Quaternary sea level fluctuations on a tectonic coast; new Th^{230}/U^{234} dates from the Huon Peninsula, New Guinea. *Quaternary Research* 4:185–205.

Colhoun, E. A. 1980. Quaternary fluviatile deposits from the Pieman Dam site, Western Tasmania. *Proc. R. Soc. Lond.* B 207:355–384.

Dodson, J. R. 1975. Vegetation history and water fluctuations of Lake Leake, southeastern South Australia. II. 50,000 B.P. to 10,000 B.P. *Aust. J. Bot.* 23:815–831.

———. 1979. Late Pleistocene vegetation and environments near Lake Bullenmerri, Western Victoria. *Aust. J. Ecol.* 4:419–427.

Frakes, L. A. 1978. Cenozoic climates: Antarctica and the Southern Ocean. In *Climatic Change and Variability*, pp. 53–68, A. B. Pittock, L. A. Frakes, D. Jenssen, J. A. Peterson and J. W. Zillman eds. Cambridge Univ. Press, Cambridge.

Hope, G. S. 1978. The late Pleistocene and Holocene vegetational history of Hunter Island, north-western Tasmania. *Aust. J. Bot.* 26:493–514.

———. 1980. Historical influences on the New Guinea flora. In *Alpine Flora of New Guinea*, pp. 223–248. P. van Royen, ed. Cramer Verlag, Stuttgart.

Hope, J. H. and G. S. Hope. 1976. Palaeoenvironments for man in New Guinea. In *The Origin of the Australians*, pp. 29–54, R. L. Kirk and A. G. Thorne eds. Aust. Inst. Aboriginal Studies, Canberra.

Hope, J. H., R. J. Lampert, L. Edmonson, M. J. Smith and G. F. van Tets. 1978. Late Pleistocene remains from Seton rock shelter, Kangaroo Island, South Australia. *J. Biogeog.* 4:363–385.

Hughes, P. J. and R. J. Lampert. 1980. Pleistocene occupation of the arid zone in southeast Australia: research prospects for the Cooper Creek–Strzelecki Desert region. *Aust. Archaeol.* 10:52–67.

Jones, R. M. and J. M. Bowler. 1980. Struggle for the savanna: northern Australia in ecological and prehistoric perspective. In *Northern Australia: Options and Implications*, pp. 3–31, R. Jones, ed. R. S. Pac. S., Australian National University, Canberra.

Keast, A., R. L. Crocker and C. S. Christian. 1959. *Biogeography and Ecology in Australia*, Monographiae Biologicae, 8. Junk, The Hague.

Kemp, E. M. 1978. Tertiary climatic evolution and vegetation history in the southeast Indian Ocean region. *Palaeogeography, Palaeoclimatology and Palaeoecology* 24:169–208.

Kershaw, A. P. 1978. Record of last interglacial-glacial cycle from northeastern Queensland. *Nature* London 272:159–161.

———. 1981. Quaternary Vegetation and environments. In *Ecological Biogeography of Australia*. pp. 81–101. A. L. Keast, ed. Monographiae Biologicae. Junk, The Hague.

Ladd, P. G. 1979. A Holocene vegetation record from the eastern side of Wilson's Promontory, Victoria. *New Phytologist* 82:265–276.

Loeffler, E. and M. E. Sullivan. 1979. Lake Dieri resurrected: an interpretation using satellite imagery. *Z. Geomorph.* 23:233–242.

Martin, H. A. 1973. Palynology and historical ecology of some cave excavations in the Australian Nullarbor. *Aust. J. Bot.* 21:283–316.

———. 1978. Evolution of the Australian flora and vegetation through the Tertiary: evidence from pollen. *Alcheringa* 2:181–202.

Nix, H. A. and J. D. Kalma. 1972. Climate as a dominant control in the biogeography of northern Australia and New Guinea. In *Bridge and Barrier: The Natural and Cultural History of Torres Strait*, pp. 61–92, D. Walker ed. Department of Biogeography and Geomorphology Publication BG/3, Australian National University, Canberra.

Singh, G., A. P. Kershaw, and R. L. Clark. 1981. Quaternary vegetation and fire history in Australia. In *Fire and the Australian Biota*. pp. 23–54. A. M. Gill, R. H. Groves, and I. M. Noble eds. Academy of Science, Canberra.

Singh, G. 1982. Environmental upheaval: vegetation of Australasia during the Quaternary. In *History of Australian Vegetation*. pp. 90–108, J. M. B. Smith, ed. McGraw-Hill, Roseville.

Stover, L. E. and A. D. Partidge. 1973. Tertiary and Late Cretaceous spores and pollen from the Gippsland Basin, southeastern Australia. *Proc. Roy. Soc. Victoria* 85:237–286.

Truswell, E. M. and W. K. Harris. 1982. The Cainozoic palaeobotanical record in arid Australia. In *Evolution of the Flora and Fauna of Arid Australia*. pp. 67–76, W. R. Barker and P. J. M. Greenslade, eds. Peacock, Adelaide.

Walker, D. 1981. Speculations on the origins and evolution of Sunda-Sahul rainforests. In *Biological Diversification in the Tropics*. pp. 554–575, G. Prance ed. Columbia Univ. Press, New York.

Walker, D. and G. S. Hope. 1982. Late Quaternary vegetation history. In *Biogeography and Ecology in New Guinea*, pp. 263–287, J. L. Gressitt ed. Monographiae Biologicae. Junk, The Hague.

Walker, D. and G. Singh. 1981. Vegetation history. In *Australian Vegetation*. pp. 26–43, R. H. Groves ed., Cambridge Univ. Press, Cambridge.

Wasson, R. 1977. Catchment processes and the evolution of alluvial fans in the lower Derwent Valley, Tasmania. *Zeitschr. f. Geomorphologie* 21:147–168.

Webb, L. J. and J. G. Tracey. 1981. Australian rainforest: patterns and change. In *Ecological Biogeography of Australia*, pp. 605–694, A. L. Keast ed. Monographiae Biologicae. Junk, The Hague.

Late Cenozoic Plant Extinctions in Australia

A. PETER KERSHAW

THE LIMITED SPACE DEVOTED TO PLANTS in this volume reflects the degree of interest in and knowledge about Pleistocene plant extinctions. Even within the area of environmental conservation, where literature abounds on rare and endangered animal species, only recently have attempts been made to assess the status of plants, despite their fundamental position in the biosphere.

General Record of Plant Extinction

One sound reason for the lack of evidence of plant extinctions is that there haven't been many, at least at higher taxonomic levels. In contrast to animals, where large-scale turnovers are well documented and have formed the basis of geological stratigraphy, the plant record is much less spectacular, with newly evolved groups superimposed on existing ones rather than totally replacing them.

No extinctions are known in the families of Pteridophyta (Banks et al. 1967a, b), and only one taxon of possibly similar taxonomic status has become extinct in the Gymnospermae (Alvin et al. 1967) within the Cenozoic. This is the ginkgoalian, *Torellia*, which provisionally has been given a generic name in the absence of a formal family classification for the group. No extinctions at the family level in the Angiospermae have been recorded (Chesters et al. 1967), though botanic affinities of early angiosperms are totally unproven (Muller 1970). Part of the explanation for the relatively low extinction rate of higher plants undoubtedly is related to the slow evolution rate which results from the long generation time of many members. With the group of shortest-lived higher plants, the annual and biennial herbs, evolution has been relatively rapid. Environmental conditions since their development, however, have tended to lead to continual radiation and areal expansion. Plants, with their sedentary habit and general lack of dependence on other macroorganisms, are capable also of surviving as small populations in localized refugia during unfavorable conditions.

Quality of Evidence

Apart from the apparent evolutionary complacency of plants, the nature of fossil evidence severely limits determination of plant extinctions. First, most information is derived from assemblages of pollen and spores, which provide an incomplete picture of existing vegetation. The pollen of many taxa is not preserved, and the degree of morphological variation tends to prevent identification to a refined taxonomic level. Within the late Tertiary it is often impossible or unwise to determine botanic affinities below the family level, while even in the late Quaternary the family or genus is normally the lowest identifiable taxonomic unit.

A second problem is the way in which information is presented. A great many palynologists working in the Cenozoic are trained in stratigraphic geology, and their main aim often is to separate pollen and spore types into form taxa, with little concern for botanic affinities or vegetation reconstructions. Therefore it becomes extremely difficult for others to incorporate this material into regional pictures of vegetation change.

Third, palynological information is limited to areas where suitable preservation sites exist, usually in less dynamic or highly stressed environments, where evolutionary changes, including extinctions, are less likely to occur.

By far the most useful evidence for extinctions has been derived from the study of macro-plant remains (see Leopold 1967). These can often be identified to genus or species level, although the restricted number of sites and local nature of the evidence generally prohibit the determination of actual times and places of extinction. Unfortunately, little emphasis has been placed on this kind of study in Australia.

Modern Vegetation of Australia

In the absence of complete fossil records for all plant taxa, realistic assessment of the degree and causes of plant extinction can be made only by reference to gross variations in the nature and distribution of major vegetation types. A useful starting point is an appreciation of the present-day vegetation pattern, which is shown in simplified form in Figure 31.1. Canopy dominants are used to define vegetation types, except in the case of closed forests (rainforests), which are characterized normally by a mixture of canopy species. Structural complexity of the vegetation generally decreases along gradients of effective rainfall from the coast inland, with variations being produced largely by soil factors and fire regimes. Elevation on this continent of subdued relief is of little importance except along the east coast, where the wetter ranges provide suitable conditions for the maintenance of a variety of closed forest types.

Apart from the patches of closed forest, the remainder of the continent is characterized by woody sclerophyll or grassland communities. Eucalypts, forming a forest or woodland cover, have an extensive distribution in more humid areas and dominate some shrub communities in semiarid regions. Major understory types are heath or open shrubland on soils of lower nutrient status with tussock grassland on better soils. Tussock grasslands without a eucalypt canopy occur on some heavy alluvial soils of northern and eastern Australia. The arid heart of the continent is dominated by acacias, which form an open woodland and shrubland cover above a sparse understory usually composed of hummock grasses commonly referred to as spinifex. Some calcareous or saline soils in southern arid regions are covered with chenopodiaceous shrublands.

Superimposed on this general pattern are patches of other vegetation types whose distributions cannot easily be explained by reference to present-day environmental factors. These include small pockets of closed forest in low rainfall areas, relatively high rainfall *Acacia* communities such as Brigalow in eastern Australia, and scattered occurrences of *Casuarina*- and *Callitris*-dominated forests.

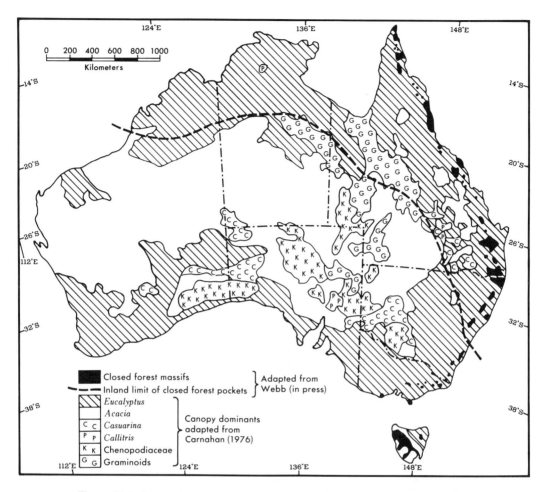

Figure 31.1. Present-day distribution of major vegetation types in Australia.

Late Cenozoic Vegetation Changes

Late Tertiary

The broad picture of late Tertiary environments in Australia is presented by Hope (this volume). The contribution of pollen information to this picture is critical despite the paucity of sites. Figure 31.2 shows the distribution of major late Tertiary pollen sequences, restricted to the eastern half of the continent. All sites indicate that rainforest was dominant at the beginning of the Miocene, about 20 million years ago. The first significant indication of more open vegetation is found in the early Miocene at two inland sites: the Lake Eyre Basin (Harris, quoted in Callen and Tedford 1976) and southern Northern Territory (Kemp 1978). Here the presence of Poaceae pollen suggests that grasslands may have existed in interfluve areas between gallery forests.

In mainland southeastern Australia a number of taxa, including *Ilex, Nothofagus* subsection *Bipartitae* (with *brassii*-type pollen), *Austrobuxus-Dissilaria,* and Cupanieae, now confined to lower-latitude rainforests, disappear within the mid-late Miocene, but the majority of rainforest taxa survive here until at least the end of the Tertiary (Stover and Partridge 1973, Martin 1978). A similar situation would appear to

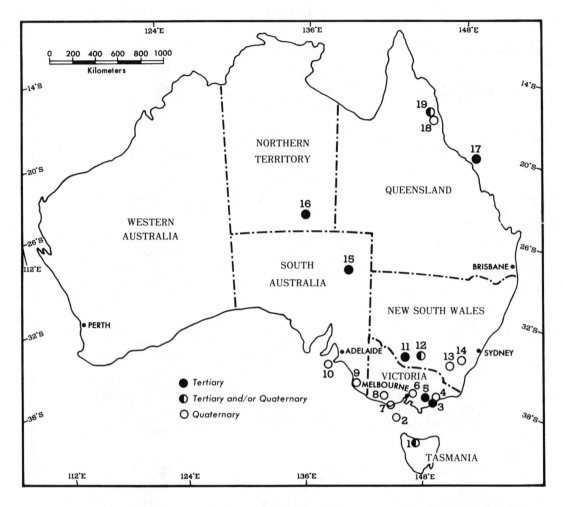

Figure 31.2. Approximate localities for Late Cenozoic fossil studies. 1. Forth Valley (Paterson et al. 1967) 2. King Island (Jennings 1959) 3. Gippsland Basin (Stover and Partridge 1973) 4. Sperm Whale Head (Hooley et al. 1980) 5. Latrobe Valley (Luly et al. 1980, Blackburn 1980) 6. Dandenongs (Churchill and Dodson 1980) 7. Lake Hordern (Head 1979) 8. Western Plains (Churchill and Dodson 1980) 9. Lake Leake (Dodson 1975) 10. Kangaroo Island (Clark in Singh et al. 1981a) 11. Murray Basin (Martin 1977) 12. Lachland Valley (Martin 1973) 13. Kosciusko (Caine and Jennings 1968) 14. Lake George (Singh et al. 1981a, Singh et al. 1981b) 15. Etadunna and Namba formations (Harris in Callen and Tedford 1976) 16. Hale River (Kemp 1978) 17. Central Queensland (Heckel 1972) 18. Lynch's Crater (Kershaw 1978, Singh et al. 1981) 19. Butcher's Creek (Kershaw 1980, Kershaw and Sluiter 1982).

hold for central Queensland (Heckel 1972) with the last appearances of pollen of *Nothofagus brassii*-type, *Austrobuxus-Dissilaria,* and Cupaniae in the mid-Miocene. More information, however, is needed to complete the picture. If the *N. brassii*-type is an indicator of the decline of ombrophilous rainforest in the eastern part of Australia (fig. 31.3), the trend that began in inland areas extended along the present rainfall gradient, leaving New Guinea as the only part of the landmass where the taxon is still living. Despite the reduction in range and regional extinction of a number of rainforest taxa in the late Tertiary, there is no proof that any taxon present became totally extinct or disappeared completely from Australia during this time.

Expansion of more open vegetation, particularly in the Pliocene, is indicated by increases in the abundance of the cosmopolitan taxa Poaceae, Asteraceae, and Chenopodiaceae (Truswell and Harris 1982). Information is scant on the status of the sclerophyll forest, woodland, and heath that now dominate much of the continent between the humid rainforests and semiarid grasslands and shrublands. (No fossil evidence has yet been found for the evolution of truly arid communities.) The mere presence of *Acacia* pollen from the late Tertiary is significant, as dispersal from parent plants is extremely poor and representation always low. Important also are significant values for Cupressaceae (probably *Callitris*) as early as the mid-late Eocene in the center of Australia (Kemp 1976), and from the late Miocene to Pliocene in southeastern Australia (Martin 1978). It is possible that both these taxa, which dominate restricted communities within extensive eucalypt forests and woodlands under subhumid conditions, formed a greater proportion of the "sclerophyll" element in the past than they do today.

Hypotheses on the status of heathlands (Specht 1979, Westman 1978) suggest an ancient origin, but they are based on an almost complete lack of fossil evidence. Recent macrofossil and palynological studies (Blackburn 1980, Luly et al. 1980) do provide some support for the presence of heath-type communities in late Tertiary brown coal deposits of the Latrobe Valley in Victoria, but structural or floristic evidence is lacking to indicate the existence of similar communities at earlier times.

Determination of the time of origin for eucalypt-dominated communities provides a particular problem because the pollen of *Eucalyptus* cannot be easily separated from that of some species in other myrtaceous genera including the *Eugenia* complex, which is almost completely restricted to rainforest. Macrofossils considered to be derived from *Eucalyptus* have been recorded as early as Oligocene (Gill 1965) or even late Eocene (Ambrose et al. 1979), but pollen falling within the range of morphological variation of the genus and outside that of other Myrtaceae occurs only in small amounts throughout the Tertiary. The most convincing evidence of the presence of *Eucalyptus* within the Tertiary comes from an arid-zone site in South Australia (Lange 1978). A whole range of silicified fruits suggests high diversification of the genus and its existence with other sclerophyll taxa such as *Casuarina*, cf. *Melaleuca*, and cf. *Leptospermum*. Although probably late Tertiary, the site is not firmly dated. Eucalypts possibly were originally rainforest plants fulfilling the role of secondary species in the same way as *E. deglupta,* one of the two truly extra-Australian eucalypt species, does today (Carr 1972). A similar origin can be postulated for another major sclerophyll genus, *Casuarina*. Pollen from *Casuarina* is conspicuous in almost all pollen spectra from the Paleocene to present. The genus is divided into two divisions: the Cryptostomae, largely confined to sclerophyll vegetation in Australia, and the Gymnostomae, generally associated with rainforest outside Australia but containing one species *(C. nodiflorum)* with restricted distribution in northeast Australia. The vegetation relationships of *Casuarina* have always been in doubt due to the difficulty of separating the two divisions on pollen evidence (Kershaw 1970). Examination of macrofossil material has revealed that all recorded early Tertiary occurrences of the genus fall within the present-day morphological range of Gymnostomae and that there is no evidence of Cryptostomae before the Miocene (Christophel 1980). Detailed examination of community relationships of the pollen flora of the Latrobe Valley coal deposits strongly suggests the presence of two *Casuarina* components, one associated with dry-land temperate and subtropical rainforest vegetation, the other forming part of the local swamp vegetation (Luly et al. 1980). The second may represent the emergence of the Cryptostomae as sclerophyll taxa.

Quaternary

There is no information from Australia that can be positively related to the earlier part of the Quaternary. Pollen spectra from New South Wales have been assigned tentatively to this period (Martin 1973), indicating that the vegetation may have been little different from today. Other evidence that may relate to the Quaternary or very late Tertiary is restricted to the high-rainfall areas of Tasmania and northeastern Queensland. Pollen samples from the Lemonthyme deposits of northwestern Tasmania were originally assigned a Pleistocene age (Paterson et al. 1967) but may be older (Eric Colhoun personal communication). A similar situation exists with a pollen sequence from an oil shale and coal deposit beneath basalt at Butcher's Creek on the Atherton Table-land (Kershaw 1980, Kershaw and Sluiter 1982). Dates on basalt flows within the Volcanic Province (Stephenson et al. 1980) provide a minimum age of 0.8 million years, but the sequence may have originated some time before this. A maximum age is indicated by the presence of *Polygonum*, which has not been recorded in Australia before the Pliocene (Martin in press). Regardless of the actual age of these deposits, they are undoubtedly younger than any Tertiary sequence previously discussed. Both contain *Nothofagus brassii*-type pollen in subtantial numbers (see Figure 31.3) and a variety of other rainforest genera, including *Dacrydium* and *Phyllocladus*. Apart from the presence of these taxa together with *Dacrycarpus*, the pollen samples from the Atherton Tableland closely match surface samples from montane rainforest, which oc-curs at somewhat higher altitudes than the site today. The only other major difference is the almost complete absence of pollen which could be referred to as *Eucalyptus*, suggesting either that open eucalypt vegetation had not yet evolved or that it had a much more restricted distribution than at present.

More continuous records of vegetation do not exist until the late Quaternary and are almost entirely confined to coastal and subcoastal areas. Lake George, situated within open eucalypt woodland on the Southern Tablelands of New South Wales, pro-vides a record from before 300,000 yr B.P. to present (Singh et al. 1981a, Singh et al. 1981b). Age control is provided by radiocarbon dates in the uppermost section, and the position of the Brunhes/Matuyama magnetic reversal in the sediment core well below the base of the pollen-analyzed section. Between these dates, a time scale has been estimated largely from comparisons of fluctuations with those from the deep-sea core record of Shackleton and Opdyke (1973). Pollen and stratigraphic evidence indicate the presence of a number of periods characterized by high pollen values of forest or wood-land plants existing under high effective rainfall, which alternates with phases of open herbaceous vegetation occurring under dry and cold conditions. Rainforest elements were important during wetter times until within the last 100,000 years, when the major taxa, *Nothofagus* cf. *cunninghamii* (of the *N. menzesii* group) and *Phyllocladus*, disap-pear. *Casuarina* shows a trend similar to that for rainforest taxa in the lower part of the diagram, and it survives with reduced representation in the upper part. Conversely, *Eucalyptus* and myrtaceous shrubs are poorly represented until about 100,000 yr B.P., when they increase during periods of higher rainfall. They have obviously replaced *Casuarina* and more mesic taxa.

Support for the contraction in range of rainforest taxa and *Casuarina* within the late Quaternary comes from other sites in southeastern Australia. In situ remains of *Nothofagus* cf. *cunninghamii* wood from the Snowy Mountains southwest of Lake George demonstrate the survival of rainforest here until at least 35,000 yr B.P. (Caine and Jennings 1968). No later record of *Phyllocladus* than that at Lake George is found in New South Wales, but there is some evidence for its recent decline within Victoria and on King Island in Bass Strait. Churchill and Dodson (1980) report consistent trace values for *Phyllocladus* pollen in lake sediments of a volcanic crater on the western plains of Victoria until about 1100 yr B.P. They rule out long-distance transport from Tasmania, where the plant still survives, as its pollen has not been found anywhere in more recent sediments on the mainland. They suggest that the most likely sources were cool-temperate rainforest patches in the nearby Otway Ranges. The evidence, though, of such a late survival of the plant in Victoria is still equivocal. A recent pollen study from a site that includes most of the Otway Ranges in its catchment shows no sign of *Phyllocladus* within the last 4,000 years (Head 1979). The decline of *Phyllocladus* on King Island is dramatic. A pollen sample dated at about 38,000 yr B.P. (Jennings 1959) yielded over 90 percent *Phyllocladus*, yet in the 1940s the plant was restricted to a few

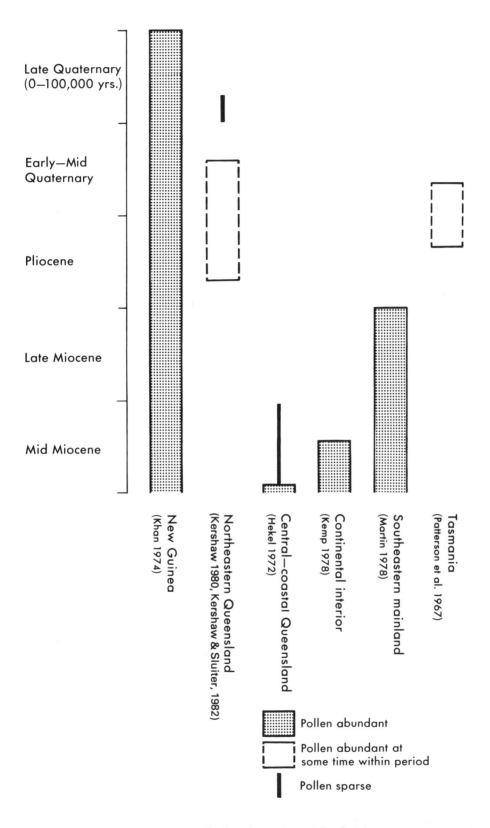

Figure 31.3. Late Cenozoic records of *Nothofagus* section *brassii* in Australia.

moist gully sites. Subsequent clearing has apparently brought about its extinction on the island. Two other cool-temperate forest taxa, *Drimys* and *Nothofagus*, were recorded in the same fossil deposit, with no evidence of either having survived until the present.

Within the casuarinas, there is information on the decline of two pollen-morphological types. Dodson (1975) records from a pollen diagram in southeastern South Australia the disappearance of a very small type of grain about 10,000 years ago. Between 13μm and 18μm, it falls below the size range of all modern taxa (Kershaw 1970) and of measured Tertiary grains (Cookson and Pike 1954). Within the last 10,000 years, a number of diagrams show major and sustained reductions in representation of large-size classes of *Casuarina,* referable to the major tree or large-shrub species *C. stricta* and *C. littoralis*, with a concomitant increase in *Eucalyptus* (Hooley et al. 1980). In some instances, further declines in very recent times have led to local disappearance.

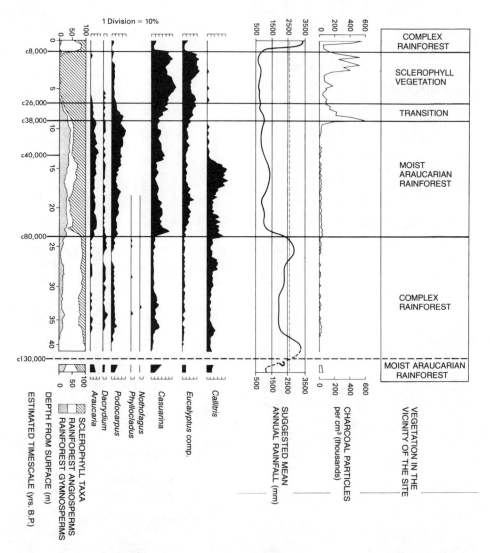

Figure 31.4. Selected attributes of the pollen diagram from Lynch's Crater. Pollen values are expressed as percentages of total pollen of dry land plants excluding *Callitris*.

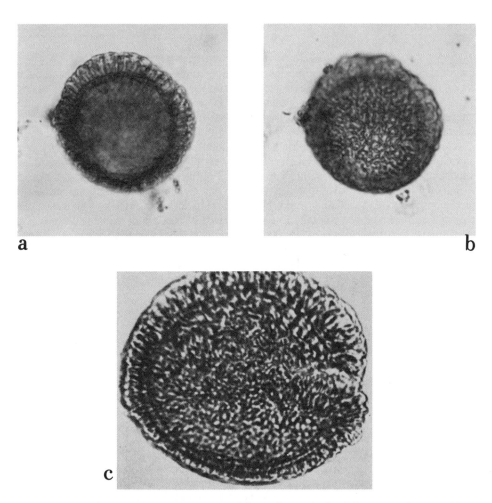

Figure 31.5. Comparisons of pollen types of *Dacrydium*. a, b: fossil *Dacrydium* from Lynch's Crater core; c: *D. guillauminii* from New Caledonia (photo from Erdtman 1957). Magnification approximately x1000.

In northeastern Queensland, Lynch's Crater on the Atherton Tableland provides an almost continuous pollen record from an estimated 140,000 yr B.P. to present (Kershaw 1978, 1980; Singh et al. 1981a) (fig. 31.4). As at Lake George, there appears to be a good correlation with deep-sea core records, with times of high sea level and sea-surface temperatures corresponding with high effective precipitation on land. An explanation for this correlation is provided by Kershaw and Nix (1975). Of the regionally extinct taxa, which were present in the Butcher's Creek samples, only *Dacrycarpus* is not represented here. Occasional grains of *Nothofagus brassii*-type and of *Phyllocladus* are recorded prior to about 80,000 years ago. These suggest, but do not prove, the presence of parent plants in north Queensland. It is a remote possibility that they were derived from New Guinea, but high pollen influx from the regional vegetation at this time makes the detection of any long-distance component unlikely. By contrast, *Dacrydium* was a common component of the complex rainforest phase prior to about 80,000 years ago and maintained consistent representation during the succeeding moist *Araucaria*-dominated rainforest phase. It is last recorded about 28,000 years ago, when all rainforest within the area was replaced by sclerophyll vegetation characterized by *Eucalyptus* and Casuarina, and it may now be totally extinct. The pollen is unlike that of any extant species in Tasmania and New Guinea, but it resembles the "primitive" New

Caledonian species *D. guillauminii* described by Erdtman (1957) (see fig. 31.5). Similarity includes the possession of a vesiculate frill in place of two bladders, which is normal for pollen of the genus, but the two grains differ in size and fine morphological detail. It is interesting to note that *D. guillauminii* has a very restricted distribution in New Caledonia today (De Laubenfels 1972).

Sustained reductions in the ranges of the other conifers, *Araucaria, Podocarpus,* and *Callitris,* appear to have taken place also in the late Quaternary. *Araucaria,* an important component of the rainforest that was replaced by sclerophyll vegetation, is found only in small, isolated patches today in the northeast of the continent, and it probably was unable to expand from refuges occupied during the sclerophyll phase when the rainforest increased within the last 10,000 years. *Podocarpus,* which suffered a fate similar to that of *Araucaria* some 28,000 years ago, is now a significant member of some rainforest communities, but a reduction in the range of pollen-morphological types suggests that extinction at the species level is a possibility. *Callitris* declined slightly earlier than other taxa, but in high-rainfall areas it now has a restricted distribution similar to that of *Araucaria.*

Causes of Vegetation Changes

Without doubt the late Cenozoic has been characterized by deteriorating climatic conditions (Hope this volume). These have had a marked influence on the contraction of complex rainforest to eastern coastal areas and the expansion of sclerophyll vegetation. It has often been assumed that rainforest was replaced directly by the major types of sclerophyll vegetation that occur today. The lack of fossil evidence within the Tertiary for eucalypt-dominated communities in particular tends to negate this assumption. There is evidence, however, for the rise of the division Cryptostomae within the genus *Casuarina* and for high percentages of *Callitris* pollen in some places, and these taxa may have dominated some early sclerophyll vegetation types. The presence of *Acacia* pollen, although in low numbers, could also suggest the development of *Acacia* communities such as Brigalow, which tend to exclude sclerophyllous grasses and to contain understory species with strong rainforest affinities. Drier rainforest types related to the present-day relict pockets found within a large area of northern and northeastern Australia (fig. 31.1) are likely to have evolved and become extensive within the late Tertiary. A lack of fossil evidence here could be attributed to a number of factors, including a floristic similarity to wetter rainforest types, poor pollen production as indicated by Kershaw (1973) for deciduous vine thickets in northeastern Queensland, and the absence of suitably located pollen-bearing sedimentary deposits. In fact, the absence of pollen-bearing sediments in drier environments generally precludes the full elucidation of the origin, nature, and distribution of sclerophyll and drier rainforest vegetation.

Not only did the range of rainforest contract because of decreasing rainfall but its components were also reorganized in response to a steepening of the climatic gradient from the equator to the pole, a result of intensification of the atmospheric circulation pattern (Kemp 1978). The range of a number of taxa present in the late Tertiary of mainland southeastern Australia contracted northwards, while a few others became restricted to Tasmania. The most marked retreat was in *Nothofagus brassii*-type, that dominated pollen spectra from New Zealand and Australia during much of the Tertiary and is now restricted to the low-latitude islands of New Guinea and New Caledonia. As this taxon generally prefers higher altitude there, it could be that increasing climatic variability rather than low temperatures were the prime cause of its extinction within Australia.

An additional factor traditionally put forward to help explain the demise of the existing rainforest element and the strong floristic latitudinal gradient is the invasion of a Laurasian or Indo-Malaysian component from rainforests to the north of Australia after contact was made between the two landmasses within the last 20 million years (see

Burbidge 1960, Westman 1978). Fossil evidence indicates, however, that any northern influence cannot have been great enough to alter the gross floristic composition of rainforest communities. Apart from the now relict status of *Nothofagus* and many of the conifers, the pollen dominants of Quaternary spectra from northeast Queensland and late Tertiary spectra from Victoria are very similar, and the majority of minor taxa are shared (Kershaw and Sluiter 1982).

With increased climatic variability during the Quaternary, communities dominated by *Eucalyptus* probably increased in number and extent, but the trend appears to have accelerated within the last 100,000 years. In the absence of evidence indicating more extreme climatic conditions during this period than during previous glacial-interglacial cycles, other factors must be examined.

At Lynch's Crater, the major expansion of sclerophyll vegetation at the expense of rainforest took place between about 38,000 and 26,000 years ago. Effective rainfall at this time was most likely reduced, but drier conditions alone would have tended to cause a change to a less complex closed forest community such as deciduous vine thicket rather than to sclerophyll vegetation. The replacement of rainforest by shade-intolerant eucalypts demands that the forest first be removed. Fire is the most likely cause, since eucalypts are fire-adapted and, in fact, actively promote the spread of fire through their volatile oils and heavy litter production. Support for this hypothesis is provided by a dramatic and sustained increase in the charcoal curve and by the composition of the earlier *Araucaria*-dominated phase, which is considered to represent the latter part of the penultimate "glacial" period. Here sclerophyll vegetation did not replace drier rain-forest as it did in the period 40,000 to 10,000 B.P.

A similar explanation can be put forward to explain the sharp decline in fire-sensitive *Callitris* vegetation some 2,000 years earlier than the beginning of the transition from *Araucaria* forest to sclerophyll vegetation. Charcoal in the core does not increase at the same time, but this may be because *Callitris* existed on soils of lower nutrient status surrounding the Tableland, and charcoal was not carried to the site. It has been suggested that only local evidence of fire is recorded by high levels of charcoal in swamp and lake sediments (Robin Clark personal communication). It would be expected that communities with open canopies, such as *Callitris* forests that cannot maintain a humid microclimate, would succumb to fire pressure before the more complex rainforests. As areas of eucalypt vegetation grew, fires could become more intense and increase the rate of destruction of surrounding vegetation. Once established, eucalypts could be successfully maintained by frequent fires, and under favorable climatic conditions could have prevented the reinvasion of drier rainforest and fire-sensitive sclerophyll communities. Further studies need to be undertaken to assess the extent of vegetation change within the area during this time.

A change in the fire regime in northeastern Queensland could be considered to be the end of a process that developed millions of years earlier in more inland areas, were it not for the developing knowledge of charcoal concentrations at other sites. At Lake George, a sharp and sustained increase in charcoal accompanies the replacement of *Casuarina* and closed forest communities by *Eucalyptus* and myrtaceous shrubs within the last 100,000 years (Singh et al. 1981a). The later replacements of *Casuarina* by *Eucalyptus* at Sperm Whale Head in Victoria (Hooley et al. 1980) and on Kangaroo Island in South Australia (Clark in Singh et al. 1981a) were accompanied by peaks in charcoal curves. Levels of charcoal particles as high as those recorded in the late Quaternary have not been found in older deposits, except for local concentrations associated with fire holes in Tertiary brown-coal deposits in Victoria (Kemp 1981).

Major vegetation changes, including regional and total plant extinctions within the late Quaternary, can be attributed largely to the influence of fire, and it is no coincidence that changes in fire regimes come within the time when aboriginal man, whose use of fire is widely known, may have been in Australia. The earliest well-documented record of the presence of man is 32,000 yr B.P. (Barbetti and Allen 1972), and undated evidence exists that suggests an earlier presence (Bowler 1976). The beginning of apparently irreversible vegetation changes at Lynch's Crater dates from about 40,000 yr B.P., and there is no problem advocating man's involvement there. The suggested date of 100,000 yr B.P., or even slightly older, for changes at Lake George provides more of a problem. If man was present by this time, a large discrepancy in dates remains between Lake George and Lynch's Crater for the first likely human impact on vegetation. This can best be resolved by considering the replacement of fire-sensitive communities by open eucalypt vegetation as a gradual process, with drier areas dominated by more open vegetation succumbing before rainforests, as explained previously with *Callitris* and *Araucaria* forests. In addition, climatic conditions in southeastern Australia may have been more conducive to the spread of fire than those in the northern part of the continent. This hypothesis is supported by the different fire regimes that presently exist in these areas (Gill 1975), and by differences in present-day vegetation. In the southeast, fires are generally infrequent but of high intensity, and they have resulted in the massive invasion of rainforests by eucalypts and associated sclerophylls to produce extensive areas of wet sclerophyll and mixed forests. By contrast, fires in north Queensland tend to be more frequent but less effective, a greater variety of rainforest and other fire-sensitive communities is preserved even under low rainfall, and rainforest/eucalypt woodland ecotones are generally narrow.

Decreasing effective rainfall through the late Tertiary caused the regional extinction of complex evergreen rainforest in many inland areas. Fossil evidence is lacking to allow any realistic reconstruction of drier vegetation types that replaced it, but many fire-sensitive sclerophyll taxa such as *Acacia, Casuarina,* and *Callitris* were probably much more extensive than at present. All three taxa can exist in rainforest or on rainforest margins today, and these may originally have been secondary plants that contained sufficient genetic plasticity to take advantage of the deteriorating climatic conditions. Grasses and composites, with very effective dispersal ability, intruded to form an understory in many open-canopied forests. Heath communities, which may have evolved at various times from rainforest ancestors in response to low nutrient soils (Specht 1979, Westman 1978), would have had the opportunity to expand also.

There is no fossil evidence for the arrival of an Indo-Malaysian rainforest element within the last 15 million years, and the latitudinal separation of rainforest communities can be explained by a steepening of the temperature gradient from equator to pole within the late Cenozoic.

Increasing climatic variability in the Quaternary probably led to the emergence of the now-dominant eucalypt communities, although the major spread of *Eucalyptus* may have resulted from increased fire frequency since the time of arrival of aboriginal man. High fire frequencies have been maintained through at least the last 40,000 years and have led to the gradual replacement of many rainforest and fire-sensitive sclerophyll communities by eucalypt forests and woodlands. Major changes generally relate to times of decreasing rainfall, when fire-sensitive communities would have been under stress and succumbed more easily to fire.

Late Pleistocene Vegetation Changes and Megafaunal Extinctions

Major changes in the vegetation landscape of Australia occurred at the same time or just before the main phase of megafaunal extinction (sometime from 50,000 to 5000 B.P. [Hope 1978]). The regional, continental, or total extinction of such taxa as *Nothofagus* cf. *brassii, Nothofagus* cf. *cunninghamii, Phyllocladus* spp., and *Dacrydium* spp., all large, long-lived trees, could be regarded as "megafloral" extinction that is equivalent in status to megafaunal extinction, except for the very different histories of the two groups. Whereas the megafauna evolved in response to Quaternary environments, the megaflora are representatives of an ancient element that has declined throughout the Quaternary period. In the megaflora the existing trend was accelerated by late Quaternary events; in the megafauna, the evolutionary trend was severely and abruptly reversed.

Some causal relationships must be examined. It is unlikely that dependence of any animal on a particular plant type that became extinct would have been sufficiently great to bring about its own extinction, or vice versa, but the impact of major vegetation changes on animal populations could have been substantial. Possible vegetation changes resulting from an increase in fire frequency or significant changes in fire regimes have been put forward at various times as contributing to the decline of Pleistocene megafauna, but this hypothesis has been rejected recently on the grounds that fire regimes implemented by aboriginal man would lead to greater vegetation patchiness, with a consequent increase in numbers and diversity of fauna (White and O'Connell 1979). This may well be the case within particular ecosystems, but would increased numbers and diversity necessarily have included those taxa that became extinct? More information is needed on the ecology, especially the feeding habits, of extinct species to assess the direct impact of changes in the nature and distribution of vegetation types.

An increase in burning combined with the expansion of fire-promoting vegetation at the expense of more stable and structurally more complex communities may also have had an indirect effect on the decline of the megafauna, through changing characteristics of the water table. It has been suggested that the extent of surface waters could have been a critical factor in the survival of fauna during the late Pleistocene (White and O'Connell 1979), which was dry throughout much of Australia (Bowler et al. 1976). The changes in the vegetation and fire regimes probably would have led to increased variation in water tables and possibly also an increase in water salinity, as suggested by the rise of salt-tolerant Chenopodiaceae in the pollen diagram from Lake George (Singh et al. 1981a).

Regardless of the actual mechanisms involved in the decline of megafauna, it is felt that sustained vegetation changes and related environmental changes experienced during the late Pleistocene must have been a contributing factor. If the megafauna can be shown to have been under severe stress during previous similar dry periods, as suggested by Witter (1978), then habitat changes alone may have been sufficient to tip the balance in the direction of extinction.

Comparisons Between Late Cenozoic Vegetation Changes in Australia and Other Parts of the World

Climatic deterioration has characterized much of the earth's surface within the late Cenozoic. Progressively drier climatic regimes during this period have led to the development of increasingly more xeric vegetation in a number of areas, for example the southwestern United States (Axelrod 1979) and southwestern Africa (Tankard and Rogers 1978). In line with Australia, acceleration of the trend took place in the Quaternary. By contrast, temperature has exerted major control over vegetation changes in Europe, with the gradual replacement of warm-temperate/sub-tropical taxa by a more temperate flora (van der Hammen et al. 1971). Again, increased climatic fluctuations during the Quaternary accelerated evolutionary changes, but by this time most taxa belonging to the more southern element had disappeared. The degree of vegetation change was greater in Europe than in similar latitudes of North America and eastern Asia, a phenomenon partly attributed to the presence of an east-west chain of mountains in southern Europe, preventing the free retreat of thermophilous plant taxa during cooler phases. Differences in rates and magnitudes of floristic changes are also noted between oceanic and continental climates of temperate parts of the Northern Hemisphere (Leopold 1967).

Studies have tended to concentrate on the reconstruction of vegetation patterns and on the evolution of emerging vegetation types in response to changing environmental conditions, rather than on plant extinctions. Even in those studies that do examine extinction, little attention has focused on taxa without established present-day botanic affinities. Consequently there is a strong bias towards the documentation of taxa that have become regionally extinct and against those that have suffered total extinction. An examination of the Australian data shows that a great number of extinctions at the species level is likely, though this cannot be proven, at least from pollen evidence. A number of angiosperm pollen types die out within the late Quaternary record from Lynch's Crater, but without a complete reference collection for rainforest plants on the continent, along with a great deal of time spent on pollen morphology and investigation of present-day distributions and ecology, their present status cannot be assessed. Evidence is, therefore, restricted to selected taxa like the gymnosperms, which have distinctive and well-dispersed pollen and are respresented by few species.

Despite this severe limitation, some evidence shows that regional differences in the timing and magnitude of extinction have occurred. Regional extinctions are best documented from northwestern Europe, where some eighty-three taxa disappeared between the late Miocene and present (van der Hammen et al. 1971). Of these, only ten are considered to be totally extinct; it is appropriate that all extinctions occurred before the Quaternary, and all were identified from macrofossil rather than pollen evidence. Few regional extinctions took place after the earlier part of the Quaternary. Similar though less dramatic trends are suggested by Leopold (1967) for other parts of temperate latitudes in the Northern Hemisphere where data exist.

Data on extinctions from Southern Hemisphere continents are sparse. The emphasis in South America has been on first appearances of Northern Hemisphere taxa in the flora of the gradually rising Andes (van der Hammen 1974), and only recently has any attention been paid to possible extinctions there (T. van der Hammen personal communication). No literature is known on plant extinctions in Africa, at least in the Pleistocene (E. M. van Zinderen Bakker personal communication).

In contrast to these Southern Hemisphere continents, the islands of New Zealand provide a more comprehensive record of late Cenozoic vegetation changes. From pollen evidence, about twenty-five extinctions are recorded, with the majority in the Pliocene and early Pleistocene (Mildenhall 1980). That the *brassii* type became extinct in the mid-Pleistocene (Fleming 1975) reinforces the argument that a loss of climatic equability was probably the reason for its disappearance in all but low latitudes within the Australasian region. No late Pleistocene exinctions are known from New Zealand (Neville Moar personal communication) and, like Tasmania, the country retains a number of coniferous taxa lost within the Quaternary of mainland Australia.

In general terms the nature of vegetation changes appears to differ between temperate northern latitudes and New Zealand, where the major climatic influence has been a decrease in temperature, and other parts of the world, where a reduction in precipitation has been a dominant feature. In the northern latitudes, coolings have caused a southern retreat of warmer elements, and many failed to return during intervening warmer periods. Glacial advances would have decreased the chances of survival in local, favorable areas. By contrast, with increasing aridity in other areas, survival of moisture-demanding taxa is likely to have been facilitated by the presence of locally moist environments. During wetter periods of the Quaternary, these taxa could expand temporarily from retreats, rather than having to migrate long distances. Consequently, extinction rates would be lower. It is possible that extinction in Australia has been greater than in South America and Africa, which have retained larger areas of high precipitation.

The presence of the fire-promoting genus *Eucalyptus* could explain apparently higher extinction rates in Australia than elsewhere during the late Pleistocene.

Conclusions

The late Cenozoic plant record for Australia, as for most other parts of the world, is extremely patchy. Information that does exist indicates a general change in the continent's vegetation from complex rainforest to structurally more simple communities with a concomitant reduction in species diversity. Traditional ideas of a direct replacement of rainforest by eucalypt vegetation, similar to that which is dominant today, are challenged in light of recent fossil evidence. Instead it is envisaged that complex rainforest was first replaced by a variety of open sclerophyll and drier rainforest communities in response to increasing aridity and climatic variability. Only with increased fire activity in the late Pleistocene, probably resulting from the activities of aboriginal man, did eucalypt communities come to occupy their present dominant position. The active role of fire-promoting sclerophylls in the creation of vegetation change is put forward as an explanation for higher rates of regional plant extinctions in the late Pleistocene compared to earlier times and in Australia at this time compared to other parts of the world. Vegetation and resulting environmental changes are likely to have contributed to the demise of megafauna in the late Pleistocene of Australia, though an apparent lack of any relationship between times of major plant and animal extinction in other areas suggests that vegetation may not have been a critical factor.

Acknowledgments

I thank Herb Wright, Liz Truswell, and Garry Werren for critically reading the paper, Gary Swinton for drawing the figures, and Helen MacDonald for typing the manuscript.

References

Alvin, K. L., Barnard, P. D. W., Harris, T. M., Hughes, N. F., Wagner, R. H. and Wesley, A. 1967. Gymnospermophyta. In W. B. Harland et al. (eds.), *The Fossil Record,* Geol. Soc. London, pp. 247–268.

Ambrose, G. J., Callen, R. A., Flint, R. B. and Lange, R. T. 1979. *Eucalyptus* fruits in stratigraphic context in Australia. *Nature* 280:387–389.

Axelrod, D. I. 1979. Age and origin of Sonoran Desert vegetation. *California Acad. Sci., Occasional Papers* 132, 74 pp.

Banks, H. P., Chaloner, W. G. and Lacey, W. S. 1967a. Pteridophyta—1. In W. B. Harland et al. (eds.), *The Fossil Record,* Geol. Soc. London, pp. 219–231.

Banks, H. P., Collett, M. G., Gnauck, F. R. and Hughes, N. F. 1967b. Pteridophyta—2. In W. B. Harland et al. (eds.), *The Fossil Record,* Geol. Soc. London, pp. 233–245.

Barbetti, M. and Allen, H. 1972. Prehistoric man at Lake Mungo, Australia by 32,000 years B.P. *Nature* 240, 46–48.

Blackburn, D. T. 1980. Floristic, environmental and lithotypic correlations in the Yallourn Formation, Victoria. *Bur. Min. Res. Geol. Geophys. Record* 1980/67, p. 9.

Bowler, J. M. 1976. Recent developments in reconstructing late Quaternary environments in Australia. In R. L. Kirk and A. G. Thorne (eds.), *The Origin of the Australians,* Aust. Inst. Aboriginal Studies, Canberra, pp. 55–77.

Bowler, J. M., Hope, G. S., Jennings, J. N., Singh, G. and Walker, D. 1976. Late Quaternary climates of Australia and New Guinea. *Quat. Res.* 6:359–394.

Burbidge, N. T. 1960. The phytogeography of the Australian region. *Aust. J. Bot.* 8:75–212.

Caine, N. and Jennings, J. N. 1968. Some blockstreams of the Toolong Range, Kosciusko State Park, New South Wales, *J. Proc. Roy. Soc. N.S.W.* 101:93–103.

Callen, R. A. and Tedford, R. H. 1976. Late Cainozoic environments of part of northeastern South Australia. *J. Geol. Soc. Aust.* 21:151–169.

Carnahan, J. A. 1976. Natural vegetation *Atlas of Australian Resources,* Second Series, Canberra.

Carr, S. G. M. 1972. Problems in the geography of the tropical eucalypts. In D. Walker (ed.), *Bridge and barrier: the natural and cultural history of Torres Strait,* Publication B/G3, Dept. of Biogeography and Geomorphology, Australian National University, pp. 153–182.

Chesters, K. I. M., Gnauck, F. R. and Hughes, N. F. 1967. Angiospermae. In W. B. Harland et al. (eds.), *The Fossil Record,* Geol. Soc. London, pp. 269–288.

Christophel, D. C. 1980. Occurrence of *Casuarina* megafossils in the Tertiary of south-eastern Australia, *Aust. J. Bot.* 28:249–259.

Churchill, D. M. and Dodson, J. R. 1980. The occurrence of *Phyllocladus asplenifolius* (Labill.) Hook. F. in Victoria prior to 1100 B.P. *Muelleria* 4:277–284.

Cookson, I. C. and Pike, K. M. 1954. Some dicotyledonous pollen from Cainozoic deposits in the Australian region. *Aust. J. Bot.* 2:197–219.

De Laubenfels, D. J. 1972. *Flore de la Nouvelle Caledonie et Dependances. 4 Gymnosperms,* Museum National D'histoire Naturelle.

Dodson, J. R. 1975. Vegetation history and water fluctuations at Lake Leake, south-eastern South Australia II 50,000 to 10,000 B.P. *Aust. J. Bot.* 23, 815–831.

Erdtman, G. 1957. *Pollen and spore morphology/plant taxonomy. Gymnospermae, Pteridophyta, Bryophyta (illustrations),* Almqvist and Wiksell, Stockholm.

Fleming, C. A. 1975. The geological history of New Zealand and its biota. In G. Kuschel (ed.), *Biogeography and Ecology in New Zealand.* Junk, The Hague, pp. 1–86.

Gill, A. M. 1975. Fire and the Australian flora: a review. *Aust. For.* 38:4–25.

Gill, E. D. 1965. The palaeogeography of Australia in relation to the migration of marsupials and men. *Trans. N.Y. Acad. Sci.* 28:5–14.

Head, L. M. 1979. Change in the aire: a palaeoecological study in the Otway Region of south-western Victoria. Unpublished B.A. (Hons) thesis, Monash University.

Hekel, H. 1972. Pollen and spore assemblages from Queensland Tertiary sediments. *Publs. Geol. Surv. Qld.* 355 (Palaeont. Pap. 30), 1–33.

Hooley, A. D., Southern, Wendy and Kershaw, A. P. 1980. Holocene vegetation and environments of Sperm Whale Head, Victoria, Australia. *J. Biogeog.* 7:349–362.

Hope, J. H. 1978. Pleistocene mammal extinctions: the problem of Mungo and Menindee, New South Wales. *Alcheringa* 2:65–82.

Jennings, J. N. 1959. The coastal geomorphology of King Island, Bass Strait, in relation to changes in the relative level of land and sea. *Records, Queen Vict. Mus., Launceston,* New Series No. 11:1–39.

Kemp, E. M. 1976. Early Tertiary pollen from Napperby, Central Australia. *J. Aust. Geol. Geophys.* 1:109–114.

——. 1978. Tertiary climatic evolution and vegetation history in the southeastern Indian Ocean region. *Palaeogeog. Palaeoclimatol. Palaeoecol.* 24:169–208.

——. 1981. Pre-Quaternary fire in Australia. In A. M. Gill, R. A. Groves and I. R. Noble (eds.), *Fire and Australian Biota,* Australian Academy of Science, Canberra, pp. 3–21.

Kershaw, A. P. 1970. Pollen morphological variation within the Casuarinaceae, *Pollen et Spores* 12:145–161.

——. 1973. Late Quaternary vegetation of the Atherton Tableland, north-east Queensland, Australia. Unpubl. Ph.D. thesis, Australian National University.

——. 1978. Record of last interglacial-glacial cycle from north-eastern Queensland. *Nature* 272:159–161.

——. 1980. Evidence for vegetation and climatic change in the Quaternary. In R. A. Henderson and P. J. Stephenson (eds.), *The Geology and Geophysics of north-eastern Australia,* Geol. Soc. Aust. (Qld Div.), Brisbane, pp. 398–402.

Kershaw, A. P. and Nix, H. A. 1975. The regional climatic significance of Late-Quaternary vegetation changes in north-eastern Queensland, Australia. *Aust. Conf. on Climate and Climatic Change, Melbourne.*

Kershaw, A. P. and Sluiter, I. R. 1982. Late Cenozoic pollen spectra from the Atherton Tableland, northeastern Australia. *Aust. J. Bot.* 30:279–295.

Khan, A. M. 1974. Palynology of Neogene sediments from Papua (New Guinea) stratigraphic boundries. *Pollen et Spores* 16:265–284.

Lange, R. T. 1978. Carpological evidence for fossil *Eucalyptus* and other Leptospermae from a Tertiary deposit in the South Australian arid zone. *Aust. J. Bot.* 26:221–233.

Leopold, E. B. 1967. Late-Cenozoic patterns of plant extinction. In P. S. Martin and H. F. Wright, Jr. (eds.), *Pleistocene Extinctions, The Search for a Cause,* Yale University Press, New Haven, pp. 203–246.

Luly, J., Sluiter, I. R. and Kershaw, A. P. 1980. Pollen studies of Tertiary brown coals: preliminary analyses of lithotypes within the Latrobe Valley, Victoria. *Monash Publications in Geography No. 23,* Geography Department, Monash University, 78 pp.

Martin, H. A. 1973. Upper Tertiary palynology in southern New South Wales. *Geol. Soc. Aust. Special Publ.* 4:35–54.

――――. 1977. The Tertiary stratigraphic palynology of the Murray Basin in New South Wales. 1. The Hay–Balranald–Wakool Districts. *J. Proc. Roy. Soc. N.S.W.* 110:41–47.

――――. 1978. Evolution of the Australian flora and vegetation through the Tertiary: evidence from pollen. *Alcheringa* 2:181–202.

――――. in press. Changing Cainozoic barriers and the Australian palaeobotanical record. *Annals, Missouri Botanical Garden.*

Mildenhall, D. C. 1980. New Zealand late Cretaceous and Cenozoic plant biogeography: a contribution. *Palaeogeog. Palaeoclimatol. Palaeoecol.* 31:197–233.

Muller, J. 1970. Palynological evidence on early differentiation of angiosperms. *Biol. Rev.* 45:417–450.

――――. 1981. Fossil pollen records of extant angiosperms. *Bot. Rev.* 47:1–142.

Paterson, S. J., Duigan, S. L. and Joplin, G. A. 1967. Notes on the Pleistocene deposits of Lemonthyme Creek in the Forth Valley. *Papers Proc. R. Soc. Tasmania* 101:221–225.

Shackleton, N. J. and Opdyke, N. D. 1973. Oxygen isotope and palaeomagnetic stratigraphy on equatorial Pacific Core V28-238: oxygen isotope temperatures and ice volumes on a 10^5 year and 10^6 year scale. *Quat. Res.* 3:39–55.

Singh, G., Kershaw, A. P. and Clark, Robin. 1981a. Quaternary vegetation and fire history in Australia. In A. M. Gill, R. A. Groves and I. R. Noble (eds.), *Fire and Australian Biota,* Australian Academy of Science, Canberra, pp. 23–54.

Singh, G., Opdyke, N. D. and Bowler, J. M. 1981b. Late Cenozoic stratigraphy, paleomagnetic chronology and vegetational history from Lake George, Australia. *J. Geol. Soc. Aust.* 28: 435–452.

Specht, R. L. 1970. Vegetation. In G. W. Leeper (ed.), *The Australian Environment,* 4th edn., C.S.I.R.O. and Melbourne Univ. Press, Melbourne.

――――. 1979. Heathlands and related shrublands of the world. In R. L. Specht (ed.), *Heathlands and Related Shrublands A.* Lange and Springer, Berlin, pp. 1–18.

Stephenson, P. J., Griffin, T. J. and Sutherland, F. L. 1980. Cainozoic volcanism in northeastern Australia. In R. A. Henderson and P. J. Stephenson (eds.), *The Geology and Geophysics of north-eastern Australia.* Geol. Soc. Aust. (Qld Div.), Brisbane, pp. 349–374.

Stover, L. E. and Partridge, A. D. 1973. Tertiary and Late Cretaceous spores and pollen from the Gippsland Basin, southeastern Australia. *Proc. Roy. Soc. Vict.* 85:237–286.

Tankard, A. J. and Rogers, J. 1978. Cenozoic palaeoenvironments on the west coast of South Africa, *J. Biogeog.* 5:319–338.

Truswell, E. M. and Harris, W. K. 1982. The Cainozoic Palaeobotanical record in arid Australia: fossil evidence for the origins of an arid-adapted flora. In W. R. Barker and P. J. M. Greenslade (eds.), *Evolution of the Flora and Fauna of Arid Australia,* Peacock Publications, South Australia, pp. 67–76.

van der Hammen, T. 1974. The Pleistocene changes in vegetation and climate in tropical South America, *J. Biogeog.* 1:3–26.

van der Hammen, T., Wijmstra, T. A. and Zagwin, W. H. 1971. The floral record of the Late Cenozoic of Europe. In K. K. Turekian (ed.), *The Late Cenozoic Glacial Ages,* Yale University Press, pp. 391–424.

Webb, L. J., Tracey, J. G. and Williams, W. T. in press. An ecological framework of Australian rainforests II. Floristic classification. *Aust. J. Bot.*

Westman, W. E. 1978. Evidence for the distinct evolutionary histories of canopy and understory in the *Eucalyptus* forest-heath alliance of Australia. *J. Biogeog.* 5:365–376.

White, J. P. and O'Connell, J. F. 1979. Australian prehistory: new aspects of antiquity. *Science* 203:21–28.

Witter, D. C. 1978. Late Pleistocene extinctions: a global perspective. *The Artefact* 3:51–65.

32

Moas, Men, and Middens

MICHAEL M. TROTTER AND BEVERLEY McCULLOCH

FOR A MILLENNIUM PEOPLE HAVE BEEN interested in moas. To the Polynesians, the word simply means common fowl, and in fact for the first five to six hundred years, these giant New Zealand ratites of the order Dinornithiformes aroused mainly culinary interest. During the last century and a half, this interest has been replaced by scientific curiosity regarding an order of birds that, by the time the first Europeans arrived in New Zealand, had become extinct. The questions of exactly when they died out and for what reasons have engendered much discussion, study, and controversy for more than 140 years, since Polack discovered subfossil bones in the 1830s (Polack 1838). Were they "living fossils"—animals unfitted for survival in the post-Pleistocene world under any circumstances—or were they the victims of ancient ecocide?

Undoubtedly man's relatively recent colonization of New Zealand has had a marked effect on the flora and fauna. Thirty-four species of birds that were living here when Polynesians first arrived about a thousand years ago were extinct by the time of European contact some eight hundred years later (Kinsky et al. 1970; Williams 1962 gives forty species), thus providing a Recent parallel to the worldwide extinctions that occurred during the late Pleistocene. Populations of marine mammals, shellfish, crustaceans, and a terrestrial reptile were seriously depleted. Vast areas of New Zealand forest had been burned as well. Virtually the entire eastern side of the South Island, and even parts of the high-rainfall western side, were fired, as were eastern, central and northern regions of the North Island.

Many of the bird species that were exterminated were highly vulnerable to any environmental changes that involved predation, since they had evolved in a predator-free environment for millions of years. The situation that prevailed at the time of man's arrival in New Zealand can best be explained in the context of the country's geological and biogeographical background.

The islands of New Zealand, almost 270,000 square kilometers in area, lie in the South Pacific Ocean between latitudes 30°20' and 47°20'. Although often regarded as part of Polynesia, New Zealand is geologically quite different from the central Pacific islands, as it is the emergent portion of an extensive area of mainly submerged continental crust—the largest submerged area of such crust in the world. This piece of continental material was originally part of the great southern continent of Gondwana, along with Antarctica and the other southern hemisphere continents. Thus, New Zealand terrestrial flora and fauna contain elements common to other southern continents derived from the original Gondwana biota before that land mass broke into separate continents. The more recently formed and much smaller central Pacific islands lack many of the groups represented in New Zealand.

Until man's arrival in New Zealand, the fauna was characterized by many birds and a few reptiles; the only mammals were marine forms and two small bats. Of this fauna, the reptile tuatara (*Sphenodon punctatus*) and the ratite birds clearly had a Gondwana origin. Cracraft (1973, 1974) has assessed the relationships and distribution of the ratites relative to the break-up of Gondwana. Since the tuatara narrowly survived human predation, it does not concern us further here.

In the isolation of New Zealand's temperate forest environment and in the complete absence of large predators, birds prospered. Many species, which had arrived with wings and power of flight, developed what Roger Duff in 1951 described as that fatal New Zealand tendency to adopt a pedestrian habit. The already flightless, autochthonous ratites were able to colonize niches usually filled in other lands by ground-dwelling reptiles and mammals.

The New Zealand ratites comprise the orders Apterygiformes and Dinornithiformes (Kinsky et al. 1970). They have probably been separated from their relatives on other continental land masses since at least the end of the Cretaceous (see e.g. Stevens 1980). Although Cracraft (1974) does not recognize separate ordinal status for these groups, the division remains widely accepted. The Apterygiformes remained small, nocturnal, and unobtrusive, surviving into the present to become New Zealand's national bird, the kiwi, of which there are three species. The Dinornithiformes, or moas, grew large, often cumbersome, and obtrusive, the tallest having a reaching height estimated to exceed three meters.

Moas developed into numerous species, aided by the later division of New Zealand into separate islands (principally the North and South Islands), and possibly also by localized isolation during periods of Tertiary marine transgression and Pleistocene glaciation. The exact number of species is yet to be agreed upon. The classification most widely used at present is that of Oliver (1949) who listed twenty-nine species, although some were based on isolated, even fragmentary, bones. In 1976 Cracraft reclassified the moas into only thirteen species, regarding some of those previously described as not valid and others to be sexually dimorphic pairs. We accept neither of these classifications as being wholly correct, although we believe Cracraft's study to be more realistic; considerably more taxonomic study, using reliable samples and modern techniques, is required before a really workable classification can be developed.

The topic of moa speciation was particularly important a few decades ago, when it was widely believed that only one or two species had survived into the human era. Now, however, remains of most (if not all) species have been recognized in primary association with other archaeological evidence of prehistoric human activity, and the moas can largely be considered together as a single group when discussing their extinction. Those species that have not been shown indisputably to be contemporaries of early man are, in any case, doubtful species.

It is possible that some species of moa were rare at the time of the first human settlement in New Zealand, either because they had never been abundant or because their numbers had been depleted before man arrived. Little research has yet been undertaken on the distribution and size of New Zealand ratite populations in either prehuman or human times, and it is not possible to say that any given species might not have survived to the present had man not arrived. For dates of prehuman moa bone deposits see McCulloch and Trotter 1979.

Biology and Ecology

One of the greatest problems arising from any attempt to interpret data relating to moa extinction is the paucity of accurate, up-to-date information on the natural history of moas. Quite apart from the question of systematics, information is still sparse or largely speculative on diet, distribution, reproduction, population size, and basic behavior.

Regarding diet, it has been widely believed that, with the possible exception of some of the smaller species, moas were grass eaters. The main proponent of this idea was Roger Duff, who likened the moas to grazing animals of present-day farmers, and suggested that the largest variety probably required as much grass per day as a bullock (Duff 1951). This persistent myth endures, for example in recent books on geology (Stevens 1980:252) and conservation (Halliday 1978: 150), although earlier published lists of identified plants from gizzard contents do not contain any grass or tussock species (e.g. Oliver 1949). The grass-eating theory was put forward before it was realized that prior to European settlement the forests that had once covered most of the country had been burned off.

Analyses of the preserved contents of a number of moa gizzards from swamps at Pyramid Valley (fig. 32.1) and from inland Otago have confirmed earlier identifications. The diet of the South Island *Dinornis* consisted mainly of twigs, and to a lesser extent leaves, fruits and seeds of woody plants, and occasional herbs that grew in forests and forest margins (Burrows et al. 1981). It has been suggested by Greenwood and Atkinson (1977) that the coevolutionary relationship between browsing moas and shrubs may account for the high proportion of small-leaved divaricating plants in the New Zealand flora.

Included in the plant fragments from the moas' last meals were quantities of gizzard stones that functioned in grinding ingested food. At Pyramid Valley some stones were more than 50 millimeters long, and the total weight of stones in a gizzard exceeded 5 kilograms in some cases. As noted by several investigators (e.g. Haast 1872), gizzard stones almost invariably could have been collected in the immediate vicinity where moa bones were found. This suggests that moas did not normally roam widely and hence may have been more vulnerable to the effects of a depleted population.

As more becomes known of moa dietary requirements and feeding habits, we may be able to better determine their distribution. They obviously survived the Pleistocene ice advances in New Zealand; these events must have affected the extent of forest, thus restricting the available habitat, unless moas were able to adapt to changing conditions. The probable distribution of the prehuman moa population can be estimated only from the evidence of fortuitously preserved remains, mainly in swamps, sand dunes, caves, and loess deposits. While the fossil record may not reflect their natural distribution, moas appear to have ranged from the coast to more than 1,000 meters above sea level—roughly the same as the prehuman forest distribution—with isolated, but largely unconfirmed, reports of moa bones being found up to 2,000 meters, well above the maximum altitude for present-day tree growth. High elevation records might suggest that the moa was not restricted to a forest habitat, but we believe they should be treated with caution. Assuming the reports are accurate, it is not impossible, for instance, that the remains were of birds driven upslope by forest fires.

Figure 32.1. Gizzard contents (indicated by arrow) lie beneath the 3,600-year-old skeleton of a moa *(Dinornis)* in a natural swamp deposit at Pyramid Valley.

Moa populations probably fluctuated, but evidence suggests that they occurred in considerable numbers in the millenia prior to man's arrival. Regardless of the classification used, there can be little doubt that there were a number of moa species in at least five genera; such a number is unlikely to have been maintained unless there were adequate breeding populations. The numbers preserved in favorable situations indicate a very large population. At Pyramid Valley, for instance, Duff (1951) estimated that there were approximately 800 skeletons of four genera per acre of swamp (nearly 2,000 per hectare). All radiocarbon dates for this site fall around 3600 B.P. (McCulloch and Trotter 1979), indicating a very short period when conditions were suitable to trap the birds and preserve their remains.

Reproduction rates are not safely extrapolated from the living continental ratites such as ostriches and emus. These groups have successfully survived in an environment containing predators, presumably by such strategies as laying a large clutch of eggs. The moas' closest relatives, the kiwis, incubate single eggs.

The only specific study to have been made of moa nests is that of an amateur observer, W. H. Hartree, in the Hawkes Bay district of the North Island. He found that nests were mostly scooped-out hollows in deposits of volcanic ash in rockshelters, and as a rule contained remains of only one egg per nest. Some shelters had been used by more than one species of moa. Results of Hartree's study have not been published in detail, but are contained in correspondence at the National Museum, Wellington, and in the summary reports of Golson (1957) and Hartree (1960). Our own investigations in the South Island, where olive, blue, and white eggshell of varying thickness was found in an inland Canterbury nesting shelter (dated as 633±51 yr B.P.), also suggest that only one egg was incubated at a time. Evidence of moa nesting is found frequently in rockshelters that were subsequently used by humans. Presumably, moas did not nest only in rockshelters; large quantities of eggshell also occur in open-country loess deposits in Marlborough. It seems likely, then, that the moas had a fairly low rate of reproduction, and that if depleted their population recovery could have been slow at best (cf. Hamel 1979).

Considering the anatomical differences between moas and living ratites, it is probably unwise to assume that their behavior was closely similar. Nor can archaeology yet provide much information on aspects of moa behavior, or even on what methods the Polynesians used to hunt and kill them. Imaginative suggestions such as trapping them by fire, driving them into swamps or to the ends of sandspits, rounding them up with dogs, trapping them in pits—even the possibilities of "farming" them and of breaking their legs to prevent their escape—have been made, all without acceptable supportive evidence. The only possible weapon to be found in archaeological sites, and which could have been used in their capture, is a harpoon (Trotter 1979:217–18). Other aids, made of materials that have not survived, were doubtless used.

Their diminutive heads and ponderous legs have led to the supposition that moas were stupid and clumsy. From a morphometric analysis of moa leg bone, Cracraft (1976b) suggests that the genus *Dinornis* may have been adapted for running, while the other genera had an essentially graviportal locomotion. Whatever the reason for its legs being proportionately more slender than those of other genera, the archaeological record (especially in the South Island) shows that *Dinornis* had no survival advantages when faced with human predators (see e.g. Simmons 1968, Scarlett 1974).

We infer that because the moas had evolved in a predator-free environment, they were highly vulnerable to man's predation. There is one hint that they might have made some attempt at self protection. On the limestone wall of a prehistoric rockshelter in South Canterbury is what is possibly the only surviving depiction of moas as seen by man. Here, a prehistoric Polynesian drew in charcoal and hematite three moas, one of which appears to be kicking forward with its clawed feet in a manner similar to that used by emus and ostriches (fig. 32.2). Like all South Island prehistoric rock art, the moa drawing is somewhat stylized (Trotter and McCulloch 1971), and others may prefer to interpret it differently.

Scientific interest in moas has enjoyed a resurgence in the past few years; not only are studies being conducted in the contentious field of systematics, but also, for the first time, serious attempts are being undertaken to determine through taphonomy the probable biology and ecology of these birds.

The Archaeological Record

When man, represented by a few Polynesian canoeists, arrived in New Zealand about a thousand years ago, a very large number of moas lived here, especially on the South Island. They must have represented an easily gathered, high-protein food source unequaled on any Pacific island from which the settlers may have come. Details of the increase in the human population and the corresponding decrease in moas have been the subject of much disagreement. Relevant archaeological evidence has been frequently misinterpreted, often as a result of attempts to make it fit into a preconceived framework.

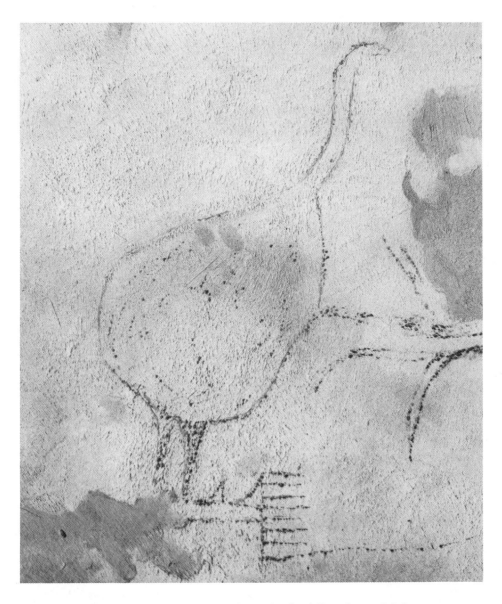

Figure 32.2. Prehistoric drawing of a moa in a South Canterbury rockshelter. (Copy by Theo Schoon)

Archaeological sites in which evidence of moa hunting occurs—mostly in the form of middens—are found all over New Zealand (fig. 32.3). Some sites are large, covering many hectares; others are small, and may indicate only where an isolated "kill" was cooked and eaten. The data do not necessarily accurately reflect the distribution of prehistoric moa populations, since they are affected by the intensity of field research in some areas, the variable degree of preservation of remains, and other cultural and natural factors. Many moa hunter sites have been destroyed during the past century, both by European development and by those digging for prehistoric artifacts.

Research on moas, with or without associated cultural remains, has been conducted mostly on the eastern side of the South Island. This is where the most important and largest sites have been found, and where some of the principal research workers have been based. While most of the available data are derived from this region, they in no way conflict with those obtained elsewhere in the country.

Figure 32.3. Map of New Zealand showing locations of sites referred to in text.

Investigators were first attracted to early South Island archaeological sites by the large quantities of moa bones often present on the surface. As early as 1851 Charles Torlesse noted that at Redcliffs the abundant remains of moas to be seen in the sand dunes were "apparently the refuse of former native feasts"; 1875 Julius Haast noted numerous large heaps of broken bones together with eggshells and artifacts at this same site (Torlesse 1851, Haast 1875a). Haast also described a moa hunter site at Rakaia, exposed by early European agriculture, as comprising forty acres (about sixteen hectares) more or less covered with earth ovens and kitchen middens, the greater portion of which consisted of bones of several species of moas, "so that in some instances they must have offered some difficulty to the plough" (Haast 1870:54, 1872:82). The same gentleman wrote of evidence of the wholesale slaughter of moas at Shag River where the occupants apparently had "such abundance of game that they selected for their food only the most valued portions of the birds killed" (Haast 1875b:94–95). This was supported by Augustus Hamilton (1904:24), who reported finding between fifty and sixty moa necks, mostly with skulls attached, apparently thrown into midden heaps as useless. Later in the nineteenth century several wagonloads of moa bones from this site were sent to Dunedin bone mills (Skinner 1924:3). At Puketoi numerous ovens each contained great quantities of moa bones, when they were examined in 1865, together with moa eggshell which indicated that "a vast number of eggs must have been consumed as food" (Murison 1872:122). When the famous Wairau Bar site was first plowed, the ground is said to have been "white with great thick bones"; they were first thought to be bullock bones, but subsequently were found to be moa (Duff 1951:25). Similarly, plowing at the Waitaki Mouth site left the ground "covered with thousands of moa-bones" reduced by weathering to "a few scattered fragments" in about twelve years (Teviotdale 1939:168). David Teviotdale, who dug the site in 1936 and 1937, records personally excavating the bones of large numbers of moas (fig. 32.4). Other early excavations, recorded in summaries by Buick (1931), Teviotdale (1932), and Duff (1977), also produced ample evidence of large-scale moa hunting.

Figure 32.4. Moa bone midden at the mouth of the Waitaki River, 1936.

Figure 32.5. Part of a moa butchery at Tai Rua, a moa-hunter site occupied some 500 years ago.

Subsequent investigators have not been able to repeat the discoveries of such large quantities of moa remains at hunter sites, thus some modern writers have not fully appreciated the extent of moa-hunting activity (see Green 1975, Halliday 1978). Archaeological deposits in New Zealand typically occur at shallow depths and are often located on arable land where agricultural operations have damaged perishable materials and exposed them to weathering. Others, located on sandspits well away from threat of the plow, have been much disturbed by artifact collectors, with similar results. Other sites, because of the same features that made them attractive to moa hunters, have now become European settlements with consequent destruction of archaeological deposits. Yet sufficient evidence has survived, below plow level or in areas not frequented by artifact collectors or house builders, to confirm the veracity of the early reports of large quantities of moa bones on sites.

At Wakanui, for example, a single earth oven hollow was packed tightly with moa bones when investigated in 1972 (Trotter 1977); aerial photography indicated that a large number of apparently similar ovens occurred on the site. In a moa-butchering area excavated at Tai Rua in the early 1960s (fig. 32.5) complete sections of neck vertebrae in position of articulation showed that moa necks, some with heads attached, had been discarded whole, a practice previously reported at Shag River (Trotter 1979). At Wairau Bar, the remains of sixteen whole moa eggs, with a hole in one end so that they could be used as water containers, and fragments of many more were found, as well as quantities of plow-disturbed bones over a large area (Duff 1977). At Kaupokonui, to give a North Island example, all the signs were present of "large scale slaughter over a short period of time, accompanied by considerable waste of flesh and bone by the early settlers" (Cassels n.d.:46).

Reinvestigations of the remnant portions of earlier-discovered sites confirm the original reports. At Rakaia, fragmentary weathered moa bone was found scattered over more than eleven hectares, and an undisturbed oven depression conformed to Haast's 1872 description (Trotter 1972). At Redcliffs, almost a hundred years of artifact hunting

Figure 32.6. A moa egg perforated at one end for use as a water container. Wairau Bar.

had left little ground undug, but by good chance an undisturbed area discovered in 1969 contained great quantities of moa (and seal) bones, along with a large earth oven (Trotter 1975). The story is much the same at the large moa hunter sites at Waitaki, Shag River, and Pleasant River. Where archaeologists have located undisturbed ground, there is evidence that moas were plentiful when the sites were occupied. A possible exception is a site at Awamoa, where investigation has shown that a natural prehuman deposit of moa bones occurred, as well as the culturally derived deposit, a fact not recognized by Walter Mantell when he investigated it in 1852. Nevertheless, evidence at Awamoa is indisputable, even though it may not have been on as large a scale as previously thought (see Trotter 1980b).

Moas were useful to prehistoric New Zealanders for more purposes than food. Although little skin has survived, the incorporation of a small fragment in a bird-skin cloak preserved in a rockshelter in the dry Central Otago region (Hamilton 1893) reminds us of the probable usefulness of this material, considering New Zealand's cold climate and the general absence of other terrestrial animals (apart from a dog that the Polynesians brought with them) from which skins could be obtained. Moa leg bones were made into useful and ornamental objects—fishhooks, harpoon heads, beads for necklaces, pendants, and miscellaneous tools. And a practice that must have had considerable effect on moa populations was the taking of their eggs both for food and for use as containers for liquids (fig. 32.6).

When all the evidence is examined, a clear picture emerges. When most moa hunter sites were occupied—particularly the large ones—moas were plentiful; they were hunted for food, their nests were robbed, and the utilization of their carcasses was probably wasteful. Man was a voracious predator of the moas. It seems likely that the dogs and rats he introduced may also have had a detrimental effect on these ground-nesting birds.

Dating and Its Interpretation

The development of radiocarbon dating in the 1950s was hailed by archaeologists as a major advance in absolute dating methods, one that could be applied to the majority of archaeological sites. Its accuracy and reliability have been improved, and it is still the foremost dating method available. In most countries, with many millenia of prehistory, comparatively few problems have been encountered in its use. In New Zealand, however, where we are dealing with only a few centuries of prehistory, relatively small errors and inaccuracies are more significant.

The most common material to be used for radiocarbon dating is charcoal, yet the death of the wood cells rarely coincides exactly with the lighting of a fire for cooking or for warmth. Where available, dead branches and sticks would have been used, and even in living trees the inner portions of the trunks and branches not actively growing have a greater radiocarbon age than the outer portion. Also, the porous nature of charcoal is ideal for the absorption of contaminating substances from the soil; for example, dated archaeological charcoal from Redcliffs and Tai Rua has been found to contain 47.5 percent and 46.4 percent, respectively, of soil-derived organic carbon (Goh and Molloy 1979). Such degrees of contamination must lead to a grossly inaccurate date being obtained where the absorbed carbon is of a different age from the event being dated. None of the many New Zealand archaeological charcoal samples have been tested or treated for such contamination before being dated.

The collagen portion of moa bone, on the other hand, appears to be relatively immune to contamination from the ground or the atmosphere. Since the moa obtained its food from terrestrial sources, there are no problems with variation in ^{14}C uptake. Hence, moa bone should be an excellent material for radiocarbon dating, and in fact results are highly consistent. Coastal marine shell has also been found to give similar results, although the dates do not appear to be quite as consistent with other data as those for moa bone collagen. (See McCulloch and Trotter 1975, Rafter et al. 1972.)

Where an archaeological site is radiocarbon dated by both moa bone collagen (or marine shell) and charcoal, the charcoal tends *on average* to give results some 300 years earlier than the faunal material; the difference, however, is not consistent and varies from nearly the same age to many hundreds of years earlier. In the short time span of New Zealand prehistory, some 800 years, discrepancies on the order of 300 years can vastly affect any synthesis or reconstruction of the sequence of events. In the matter of moa extinction, accurate dating of man's arrival in New Zealand, his population increases, and movements throughout the country are fairly crucial. A difference of a few hundred years could lead to an entirely different interpretation of a series of data.

In discussing the date of moa extinction and the part played by man in that extinction, we would prefer to consider only samples of moa bone collagen. Unfortunately, for one reason or another, collagen dates are not yet available for all moa hunter sites—some have not been radiocarbon dated—and we have found it necessary to use marine shell dates in some instances, avoiding dates obtained from samples in which the shell material has significantly recrystallized, and also of *Amphibola*, a pulmonate snail of questionable reliability as a dating material.

At the time of this writing, twenty-seven radiocarbon dates from moa bone collagen and twenty-two from coastal marine shell were available for moa hunter sites throughout the country, though very few were from the North Island (see Table 32.1). For ease of reference and comparison, all those for the eastern part of the South Island—the majority of those available—are shown graphically in Figure 32.7, where the sites are arranged in geographical order from north to south, as if their positions were projected normally on to the long axis of the South Island (their exact locations are shown in Figure 32.3). These dates are calculated with reference to local standards

Table 32.1. Radiocarbon Dates for Moa Hunter Sites

Site	Dates (yr B.P.)	
(North to South)	Moa Bone	Marine Shell
Tairua		570 ± 60
Hotwater Beach		453 ± 40
		524 ± 40
Kaupokonui	610 ± 50	
	660 ± 60	
Foxton		550 ± 70
Titirangi	830 ± 90	
Wairau Bar	590 ± 60	680 ± 50
Heaphy		570 ± 60
Clarence		750 ± 50
Avoca Point	1010 ± 50	860 ± 40
	740 ± 90	880 ± 30
Buller		465 ± 25
Hurunui	730 ± 80	
Redcliffs	581 ± 40	617 ± 34
	615 ± 40	
	735 ± 56	
Takamatua		666 ± 52
Rakaia	585 ± 64	
	518 ± 80	
Wakanui	421 ± 55	
	596 ± 59	
	672 ± 56	
Woolshed Flat	493 ± 70	
Awamoa	660 ± 54	678 ± 58
Ototara		528 ± 32
Tai Rua	543 ± 32	485 ± 32
	503 ± 32	
	503 ± 32	
Hampden	538 ± 70	
	554 ± 53	
Waimataitai	717 ± 171	626 ± 30
Shag River	679 ± 85	657 ± 34
Pleasant River	446 ± 57	
Purakanui		562 ± 30
		571 ± 34
Hawkesburn	440 ± 55	
Cannibal Bay		450 ± 60
Pounawea		500 ± 60
Papatowai	640 ± 60	300 ± 40

and to the "old" half life of ^{14}C; tree-ring corrections have not been made because of some lingering doubts about the exact relationship between radiocarbon years and calendar years.

For present purposes a moa-hunter site is defined as one that contains definite evidence of hunting, preparation, or use of moas for food, and which can from artifact typology and materials be considered to be culturally "early," i.e. corresponding to what has been termed the Moa-hunter Period of Maori culture or the Archaic Phase of New Zealand Polynesian culture (Duff 1977, Golson 1959). (Note that "Moa-hunter" is a cultural term used in studies of New Zealand prehistory; it is applied to the early period of Polynesian settlement, but does not necessarily imply actual moa-hunting activities.)

Despite the relative reliability of moa bone collagen and marine shell dates, some sites show a spread of dates that is not in accord with the archaeological evidence. In some cases explanations for the discrepancy can be suggested. The younger Wairau Bar date, for example (Table 32.1), was obtained from material kept in open storage for some years that may have become contaminated. The two Papatowai dates, on the other hand (Table 32.1), were obtained by different people at different times from different parts of the site (Hamel 1978), and while there is apparent confusion over stratigraphic correlation, this does not wholly account for the considerable difference in dates.

For the most part, all the listed sites were places of occupation for well-established groups that attended to various food-gathering and preparation activities (besides moa hunting) and to the manufacture of a variety of tools and ornaments.

It can be seen from Figure 32.7 that the dates obtained for South Island moa hunter sites are not widely scattered in time. The greatest ranges, in fact, occur on single sites believed to have been occupied for much less time than the dates indicate; in other words, the greatest variations are apparently due to peculiarities associated with the dating method rather than to a long period of occupation on the individual sites. At only one, Papatowai, has it strongly been suggested that there was a long human occupation (Lockerbie 1959:80–82). The original evidence for this, however, was mainly from charcoal dates and moa bone carbonate (not collagen) dates. While charcoal tends to produce dates that are too old, bone carbonate has been found to give the opposite effect. A number of carbonate dates indicate that moas are still living! A new series of radiocarbon dates for this site has been published (Hamel 1978); these also cover a long period, though they do not correlate with the previous series. However, all are from charcoal samples except one of shell (the one included in Figure 32.7) which the collector considers to be anomalous, presumably because it is younger than charcoal from the same level. It should be noted that many of the radiocarbon dates obtained in the 1950s and 1960s (e.g. some of those used by Fleming 1962 and Duff 1963) are not acceptable in light of present knowledge.

There appears to be a trend of earlier sites occurring towards the north and later sites towards the south of the South Island—the regression line drawn through the scatter plot diagram (fig. 32.7) emphasizes this point. In general terms this could be interpreted as indicating a moving peak in a moa-hunting activity which occurred in the north of the South Island some 800 years before present, gradually becoming more recent to about 500 years before present in the south. Further it might suggest a gradual population movement down the South Island, hunting out local communities of moas on the way. As game became scarce in one region, the hunters shifted farther south where their prey was more plentiful, over the course of 300 years. This echoes, on a smaller scale, Martin's (1973, 1974) hypothesis of prehistoric overkill. That the man/moa overlap existed much longer than the application of the North American model would suggest may be attributed to considerable cultural and environmental differences: the Polynesians grew crops, had an abundance of coastal food always on hand, and because of heavy forest and rugged terrain probably found mass slaughter techniques increasingly difficult.

The suggestion of a population peak moving southward is not based on strong evidence; only a few early dates from the far south would upset it. We must stress that the dates do not represent the first occurrence of man in each area. That there was prior exploration, probably by small groups, is evident on most of the main sites by the well-developed use of stone resources transported from different parts of the country, some quite remote.

Less hospitable climate and terrain make it unlikely that population movements would have been identical on the western side of South Island. Only two moa-hunter sites have been recorded there, with shell dates of 570±60 yr B.P. for Heaphy and 465±25 yr B.P. for Buller. While such minimal data do not provide proof, they could suggest utilization of this coast after the more-favored northern half of the eastern side of the island had been hunted out.

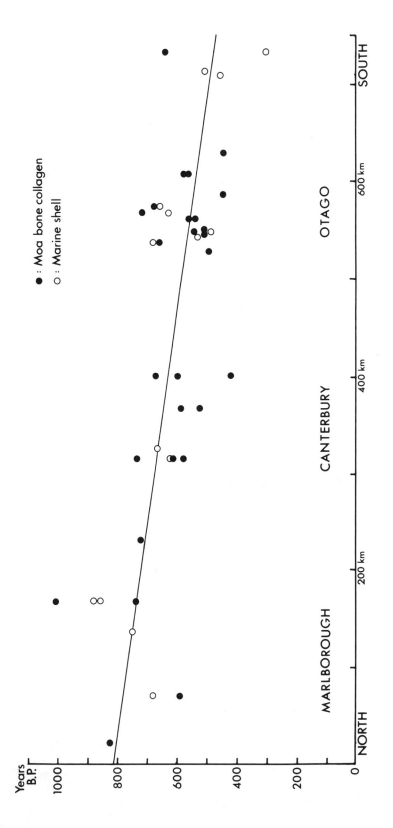

Figure 32.7. Moa bone and marine shell radiocarbon dates for moa-hunter sites on the east coast of the South Island.

Incredibly, moa bone collagen dates have been obtained for only one North Island archaeological site, that of Kaupokonui which has dates of 610±50 and 660±60 yr B.P. (Cassels pers. comm. 1980). A few others have shell dates (Moore and Tiller 1975), but these do not form any obvious pattern, and some are at variance with stratigraphic data, e.g. at Foxton (McFadgen 1978:40; questionable dates are not included in Table 32.1). We believe that most North Island moa-hunting activity occurred at about the same time as it did on the South Island. Although archaeological research is uneven throughout the country, there do appear to be fewer large moa-hunter sites in the North Island. One explanation is that moas were fewer there during the early Polynesian era; it seems unlikely that people would have chosen to live in the colder South Island if there had been a good supply of moas in the North Island.

During the millenia before man's arrival in New Zealand, a number of volcanic eruptions occurred in the central North Island, the largest being Taupo some 1,800 years ago; an estimated 26 cubic kilometers of pumice were thrown over a wide area, burning forests and completely changing the appearance of the central part of the island. It has been suggested that one would have to have been at least 150 kilometers away from Taupo to be safe (Clark 1974:864). Such an event could well have decimated the moas and other terrestrial fauna, leaving only isolated communities that may not have increased to any great extent by the time man arrived.

It has been suggested by Helen and Foss Leach (1979:236) that forest destruction by high winds, which were associated with periods of glacial advance, may have been responsible for the local extinction of moas in southern Wairarapa in the North Island before human settlement. It is now known, however, that moas lived in the area just before man's arrival in New Zealand, a natural moa bone deposit having been dated at 1470±50 yr B.P. (McCulloch and Trotter 1977); Bruce McFadgen has reported a natural deposit of the bones of mature and immature moas (suggesting a breeding population) dating well into the human era (McFadgen pers. comm. 1980). Occupation times of the Leach's sites are based wholly on charcoal radiocarbon dates and may be open to question.

Sites containing evidence of moa hunting are also known from the general area (McFadgen pers. comm. 1980). It seems likely that moa material is absent from the Leach's Wairapa sites because the sites were occupied after man killed out the moas.

Taking into account all available evidence in both South and North islands—from "dirt archaeology" and from radiocarbon dates—there is a strong suggestion that prolonged periods of intensive moa hunting did not occur in any district.

As mentioned in our opening paragraphs, man burned vast areas of forest, apparently at about the same time as moa extinction; this burning was a major contributing factor in that extinction. Many have observed that forests had been destroyed by the time Europeans had arrived. But not until numerous radiocarbon dates were obtained was it widely appreciated that most forest destruction had probably occurred during the Polynesian era. The evidence is most compelling for the eastern part of the South Island, where dates for subfossil logs and buried charcoals are mostly between 500 and 1,000 years before present (Molloy et al. 1963, Molloy 1969). Analysis of pollen grains from swamp deposits in both islands indicates vegetation changes compatible with forests having been burned off during the last 1,000 years (e.g. Moar 1970, McGlone 1978).

Summarizing evidence of nonarchaeological charcoal dates and pollen analysis (some of it unpublished), palynologist M. S. McGlone tells us that around 1000 B.P. the first signs of permanent forest clearance appears in both islands outside Central Otago (where there had been some earlier clearance by natural fires). At about 700 B.P. the rate of forest clearance accelerated, and within about 200 years nearly all the areas that were to be cleared by prehistoric man had been burned. Subsequent fires merely maintained this treeless situation or only marginally increased the area in bracken, scrub, and grassland (McGlone pers. comm. 1980). The problems associated with dating charcoal—and to a lesser extent wood—do not allow us to match the dates of wide-spread burning against those of major moa-hunting activity with much accuracy, but certainly both occurred at approximately the same time. With the recognition of moas as forest dwellers and consumers of forest plants, it follows that besides man's direct predation, his use of fire must have resulted in the destruction of much of their habitat.

The Role of Man—Conservationist or Killer?

No moa, no moa,
In old Ao-tea-roa.
Can't get 'em.
They've et 'em;
They've gone and there aint no moa!
                              A popular New Zealand song

In light of the evidence, the conclusion as to man's role in the extinction of the moas seems inevitable. We know that moas survived major environmental changes throughout the Cenozoic and that they existed, apparently in large numbers, im-mediately before the first human settlement of New Zealand about a thousand years ago. We know, too, that they were then hunted and eaten in quantity, their eggs were collected, and their natural habitat was depleted by forest fires. We also know that they became extinct within the first few centuries of the human era.

Yet there remains great and not unemotive reluctance to blame the moas' demise on man. Although the Noble Savage concept has taken a battering in recent decades, it lives on strongly enough to color our reasoning on such matters. Despite a very short recorded history and only about 800 years of prehistory (much of it still unknown), New Zealand has become a great country for traditions. While many of these have been laid to rest, one that persists, probably for the same reasons that led to its conception, is that the prehistoric Maoris, the Polynesian inhabitants of New Zealand, were great conser-vationists. Unlike the later European settlers, we are often told, these people did not destroy their natural environment, cut down the forests, kill off the animals, and otherwise deplete the country's natural resources. These ideas were fostered by the writings of such people as ethnologist Elsdon Best (e.g. 1924) and economist Raymond Firth (1929:238). Naturalist Robert Cushman Murphy observed that "like most primi-tive folk the Maori were effective conservationists" (Murphy 1951), and even recent researchers such as archaeologist Peter Coutts (1969) and historian Ann Parsonson (1980) imply deliberate conservation of resources by Polynesians, though not explicitly saying so. Not surprisingly, general writings embodying this sentiment are manifold. Two examples will suffice here: Andrew Clark (1949), in his book on the invasion of the South Island by people, plants, and animals, states that "Essentially the Maori are believed to have lived in harmony with the region and to have altered its pristine character little if at all." Don Sinclair (1976), whose publication was written with the assistance of the Department of Maori Affairs, tells us that "the ancient Maori was a natural conservationist." However, geographer Kenneth Cumberland had few illusions when in 1962 he put forward the somewhat radical argument of cultural interference rather than naturally occurring changes being responsible for forest destruction and moa extermination.

The general confusion surrounding this issue is perhaps typified by Tim Halliday (1978) who on page 38 of his *Vanishing Birds* states that the moas were "hunted for food [by the Polynesians] and gradually exterminated." Later in the book, however, (page 124) he refers (somewhat obscurely) to the Maoris' "conservationist policies," and still later (page 126) states that "the Maoris are commonly blamed for the demise of the moas. While there is no doubt that they did hunt and eat them it is most improbable that hunting alone could have exterminated these once numerous birds." He then goes on to suggest that the activities of Maori-introduced alien animals were the principal cause of extinction, and concludes (on page 150) that "we shall never know precisely why the moas declined."

The general acceptance that moa extinction occurred within the human era has not wholly banished the old bogies of climatic change and genetic weakness, now offered as if by extenuation and excuse (see Green 1975). Although minor climatic changes have occurred during the last thousand years, they cannot be compared with those that must have taken place during the ice advances of the Pleistocene; and the inability of moas to radically alter their reproduction rate and their defensive behavior within a few generations, as well as to rapidly adapt to a greatly modified habitat, cannot seriously be ascribed to genetic weakness.

A number of stories are told of the possible survival of some of the smaller species in the more remote, still forested, areas of New Zealand well into the nineteenth century. Possibly some may have been true; the odd, small, isolated community of moas could well have survived the main wave of extinction by several centuries, although it is most unlikely that any are now alive.

Such relict survival, if it did occur, does not detract from the thesis that extinction was due to the activities of man. Nor does it mitigate against the facts of widespread extermination five hundred and more years ago.

An obvious parallel can be drawn with the extinction of much of the large land fauna of Madagascar. Like New Zealand, Madagascar is an island remnant of Gondwana, from which it inherited ratites along with distinctive primate mammals, the lemurs. Anthropological research indicates that man was responsible for the extinction of the entire ratite population as well as other large fauna on the island, either by direct predation or by destruction of their forest habitat by fire (Battistini and Verin 1972, Mahe 1972, Dewer this volume). Richard and Sussman (1975) note that the fourteen species of lemurs that became extinct after man's arrival were all large, slow-moving, diurnal, ground-living forms; smaller nocturnal species were the most successful survivors. In this, too, we can see a parallel with the survival of the small nocturnal Apterygiformes (kiwis) in New Zealand.

Although moas were the most interesting and abundant New Zealand birds to fall victim to Polynesian predation, a number of other species suffered a similar fate at the same time. These included a flightless goose (*Cnemiornis calcitrans*), a giant rail (*Aptornis otidiformis*), a swan (*Cygnus sumnerensis*), and several other flighted species, among them the giant eagle (*Harpagornis moorei*), and indirect victim of man's invasion. Eagle bones are rare in archaeological sites, and it seems unlikely that this species was important as food. It was, however, largely a carrion eater with diminished powers of flight. Its remains are found in swamps and sinkholes along with those of the dead moas on which it fed. When its main food supply was exterminated, it too perished.

During the European period of occupation in New Zealand, at least five more bird species have become extinct (Williams 1962). A number more are so endangered that their survival depends on the success of special protective measures and breeding programs. The basic causes of this recent depletion of the indigenous fauna have been land clearing for timber and agriculture and the introduction of predatory mammals, but it is also due in part to the deliberate killing of birds for museum displays and for scientific study. The extreme collectors' attitude of a bygone era is typified by Walter Buller (1888), whose appreciation of the rarity of a species appears to have made him all the more determined to shoot every specimen he could find.

Undoubtedly the changes to New Zealand's environment wrought by European activities of the last two centuries greatly exceed those caused by a few thousand Stone-Age Polynesians in the preceding centuries. Nevertheless, the impact of that initial colonization is more obvious in its effect on a predator-free island environment. By the time Europeans arrived, the most vulnerable species—including moas—had been exterminated.

We have no illusions about the role of man in their demise, nor that the arguments about that role will continue for years to come. We can find no evidence of any convincing "natural" causes for moa extinction, and more importantly, we see no compelling need to continue to look for them.

Contemplating the monumental loss of native forms, not only in New Zealand but elsewhere on oceanic islands, we are inclined to agree with W. S. Gilbert, that "man is Nature's sole mistake."

Acknowledgments

We wish to record our thanks to the Institute of Nuclear Sciences, Lower Hutt, for radiocarbon dates and pertaining to them; to our colleagues who supplied us with unpublished information, particularly Mat McGlone, Bruce McFadgen, Richard Cassels, and Atholl Anderson; to Paul Martin, Colin Burrows, and Matt McGlone for helpful criticism of our draft manuscript; and to the landowners of natural and archaeological sites who allowed us to carry out excavations in the course of our investigations of moas, men, and middens.

References

Battistini, R., and Verin, P. 1972. Man and the Environment in Madagascar. In *Biogeography and Ecology in Madagascar* (R. Battistini and G. Richard-Vinard, editors). Dr. W. Junk, The Hague, pp. 311–37.

Best, Elsdon. 1924. *The Maori as He Was.* Dominion Museum, Wellington. (Reprinted 1934, 1952.)

Buick, L. T. 1931. *The Mystery of the Moa.* Thomas Avery & Sons, New Plymouth.

Buller, Walter L. 1888. *A History of the Birds of New Zealand* (2nd edition). Published by the author, London. Vol. 1 (Earlier edition 1873).

Burrows, C. J., McCulloch, B., and Trotter, M. M. 1981. The Diet of Moas, Based on Gizzard Contents Samples from Pyramid Valley, North Canterbury, and Scaifes Lagoon, Lake Wanaka, Otago. *Records of the Canterbury Museum* 9(6).

Cassels, Richard. n.d. Kaupokonui N128/3B; A Moa Butchery and Cooking Site (Mimeographed) Anthropology Department, University of Auckland.

Clark, A. H. 1949. *The Invasion of New Zealand by People, Plants and Animals.* Rutgers University Press, New Brunswick.

Clark, R. H. 1974. Volcanoes. *New Zealand's Nature Heritage* 3(31):862–70.

Coutts, P. J. F. 1969. Merger or Takeover. *Journal of the Polynesian Society* 78(4): 495–516.

Cracraft, Joel. 1973. Continental Drift, Paleoclimatology, and the Evolution and Biogeography of Birds. *Journal of Zoology* 169(4): 455–545.

———. 1974. Phylogeny and Evolution of the Ratite Birds. *Ibis* 116(4):494–521.

———. 1976a. The Species of Moas (Aves: Dinornithidae). In *Collected Papers in Avian Paleontology Honoring the 90th Birthday of Alexander Wetmore* (Storrs L. Olson, editor). Smithsonian Contributions to Paleobiology 27:189–205.

———. 1976b. The Hindlimb Elements of Moas (Aves, Dinornithidae): A Multivariate Assessment of Size and Shape. *Journal of Morphology* 150:495–526.

Cumberland, Kenneth B. 1962a. 'Climatic Change' or Cultural Interference? In *Land and Livelihood* (M. McCaskill, editor). New Zealand Geographic Society, Christchurch, pp. 88–142.

———. 1962b. Moas and Men; New Zealand About A.D. 1250. *The Geographical Review* (New York) 52(2):151–73.

Duff, Roger. 1951. *Moas and Moa-Hunters.* Government Printer, Wellington. (Also issued as *Post Primary School Bulletin* 5(7):129–63.) (Reprinted 1957).

———. 1963. *The Problem of Moa Extinction* The Cawthron Institute, Nelson.

———. 1977. *The Moa-Hunter Period of Maori Culture* (3rd edition). Canterbury Museum Bulletin 1. Government Printer, Wellington. (Earlier editions in 1950, 1956.)

Firth, Raymond. 1929. *Primitive Economics of the New Zealand Maori.* George Routledge, London.

Fleming, C. A. 1962. The Extinction of Moas and Other Animals During the Holocene Period. *Notornis* 10:113–17.

Goh, K. M. and Molloy, B. P. J. 1979. Contaminants in Charcoals Used for Radiocarbon Dating. *New Zealand Journal of Science* 22(1):39–47.

Golson, J. 1957. New Zealand Archaeology 1957. *Journal of the Polynesian Society* 66(3):271–90.

———. 1959. Culture Change in Prehistoric New Zealand. In *Anthropology in the South Seas* (J. D. Freeman and W. R. Geddes, editors). Thomas Avery & Sons, New Plymouth, pp. 29–74.

Green, R. C. 1975. Adaptation and Change in Maori Culture. In *Biogeography and Ecology in New Zealand.* Monographiae Biologicae 27. Dr. W. Junk, The Hague, pp. 591–641. (Reprinted by Stockton House, Albany, pp. 2–48.)

Greenwood, R. M. and Atkinson, I. A. E. 1977. Evolution of Divaricating Plants in New Zealand in Relation to Moa Browsing. *Proceedings of the New Zealand Ecological Society* 24:21–33.

Haast, J. 1870. Letter to Professor Owen in *Proceedings of the Zoological Society of London* 1870:53–56.

———. 1872. Moas and Moa-hunters. *Transactions of the New Zealand Institute* 4:66–107.

———. 1875a. Researches and Excavations carried on in and near the Moa-bone Point Cave, Sumner Road, in the Year 1872. *Transactions of the New Zealand Institute* 7:54–85.

———. 1875b. Notes on the Moa-hunter Encampment at Shag Point, Otago. *Transactions of the New Zealand Institute* 7:91–98.

Halliday, Tim. 1978. *Vanishing Birds.* Sidgwick & Jackson, London.

Hamel, G. 1978. Radiocarbon Dates from the Moa-hunter Site of Papatowai, Otago, New Zealand. *New Zealand Archaeological Association Newsletter* 21(1):53–54.

———. 1979. The Breeding Ecology of Moas. In *Birds of a Feather* (A. Anderson, editor). British Archaeological Reports International Series 62:61–66.

Hamilton, A. 1893. Notes on Some Old Flax Mats Found in Otago. *Transactions of the New Zealand Institute* 25:486–88.

———. 1904. Notes on the Southern Maori. *In Dunedin and its Neighbourhood* (A. Bathgate, editor). Otago Daily Times and Witness Co. Ltd., Dunedin.

Hartree, W. H. 1960. A Brief Note on the Stratigraphy of Bird and Human Material in Hawkes Bay. *New Zealand Archaeological Association Newsletter* 3(4):28.

Kinsky, F. C. 1970. (Convener). *Annotated Checklist of the Birds of New Zealand.* A. H. & A. W. Reed, Wellington.

Leach, H. M. and Leach, B. F. 1979. Environmental Change in Palliser Bay. In *Prehistoric Men in Palliser Bay* (B. F. Leach and H. M. Leach, editors). National Museum of New Zealand Bulletin 21:229–40.

Lockerbie, L. 1959. From Moa-hunter to Classic Maori in Southern New Zealand. In *Anthropology in the South Seas* (J. D. Freeman and W. R. Geddes, editors). Thomas Avery & Sons, New Plymouth, pp. 75–110.

McCulloch, B. and Trotter, M. M. 1975. The First Twenty Years: Radiocarbon Dates for South Island Moa-hunter Sites. *New Zealand Archaeological Association Newsletter* 18(1):2–17.

———. 1979. Some Radiocarbon Dates for Moa Remains from Natural Deposits. *New Zealand Journal of Geology and Geophysics* 22(2):277–79.

McFadgen, B. G. 1978. *Environment and Archaeology in New Zealand.* Unpublished Ph.D. thesis, Victoria University of Wellington.

McGlone, M. S. 1978. Forest Destruction by Early Polynesians, Lake Poukawa, Hawkes Bay, New Zealand. *Journal of the Royal Society of New Zealand* 8(3):275–81.

Mahe, J. 1972. The Malagasy Subfossils. In *Biogeography and Ecology in Madagascar* (R. Battistini and G. Richard-Vinard, editors). Dr. W. Junk, The Hague, pp. 339–65.

Mantell, W. B. D. 1853. News from the South. *New Zealand Spectator and Cook's Strait Guardian*, 27 August 1853.

Martin, Paul S. 1973. The Discovery of America. *Science* 179:969–74.

———. 1974. Paleolithic Players on the American Stage: Man's Impact on the Late Pleistocene Megafauna. *Arctic and Alpine Regions* (J. D. Ives and R. G. Barry, editors). Methuen & Co. London, pp. 669–700.

Moar, N. T. 1970. A New Pollen Diagram from Pyramid Valley Swamp. *Records of the Canterbury Museum* 8(5):455–61.

Molloy, B. P. J. 1969. Recent History of the Vegetation. In *The Natural History of Canterbury* (G. A. Knox, editor). A. H. & A. W. Reed, Wellington, pp. 340–60.

Molloy, B. P. J., Burrows, C. J., Cox, J. E., Johnston, J. A. and Wardle, P. 1963. Distribution of Subfossil Forest Remains, Eastern South Island, New Zealand. *New Zealand Journal of Botany* 1(1):68–77.

Moore, P. R. and Tiller, E. M. 1975. Radiocarbon Dates for New Zealand Archaeological Sites. *New Zealand Archaeological Association Newsletter* 18(3):98–107.

Murison, W. D. 1872. Notes on Moa Remains. *Transactions of the New Zealand Institute* 4:120–24.

Murphy, Robert Cushman. 1951. The Impact of Man Upon Nature in New Zealand. *Proceedings of the American Philosophical Society* 95(6):569–82. [Reprinted in abridged form in the *New Zealand Geographer* 8(1):1–14].

Oliver, W. R. B. 1949. *The Moas of New Zealand and Australia.* Dominion Museum Bulletin 15, Wellington.

Parsonson, Ann R. 1980. The Expansion of a Competitive Society. *The New Zealand Journal of History* 14(1):45–60.

Polack, J. S. 1838. *New Zealand.* Richard Bentley, London.

Rafter, T. A., Jansen, H. S., Lockerbie, L. and Trotter, M. M. 1972. New Zealand Radiocarbon References Standards. *Proceedings of the 8th International Radiocarbon Dating Conference.* Royal Society of New Zealand, Wellington.

Richard, A. F. and Sussman, R. W. 1975. Future of the Malagasy Lemurs; Conservation or Extinction. In *Lemur Biology* (I. Tattersall and R. Sussman, editors). Plenum Press, New York.

Scarlett, R. J. 1974. Moa and Man in New Zealand. *Notornis 21* (1):1–12.

———. 1979. Avifauna and Man. In *Birds of a Feather* (A. Anderson, editor). British Archaeological Reports International Series 62:75–90.

Simmons, D. R. 1968. Man, Moa and the Forest. *Transactions of the Royal Society of New Zealand* 2(7):115–27.

Sinclair, Don. 1976. *The Maori in Colour.* Bascands Ltd., Christchurch.

Skinner, H. D. 1924. Results of the Excavations at the Shag River Sandhills. *Journal of the Polynesian Society* 33(1):11–24.

Stevens, Graeme R. 1980. *New Zealand Adrift. The Theory of Continental Drift in a New Zealand Setting.* A. H. & A. W. Reed, Wellington.

Teviotdale, D. 1932. The Material Culture of the Moa-hunters in Murihiku. *Journal of the Polynesian Society* 41(2):81–120.

———. 1939. Excavation of a Moa-hunters' Camp Near the Mouth of the Waitaki River. *Journal of the Polynesian Society* 48(4):167–85.

Torlesse, C. O. 1851. Report on the Canterbury Block. *Lyttelton Times,* 5 July 1851, p. 7.

Trotter, Michael M. 1972. A Moa-hunter Site Near the Mouth of the Rakaia River, South Island. *Records of the Canterbury Museum* 9(2):129–50.

———. 1975. Archaeological Investigations at Redcliffs, Canterbury New Zealand. *Records of the Canterbury Museum* 9(3):189–220.

———. 1977. Moa-hunter Research since 1956. In *The Moa-hunter Period of Maori Culture* (by R. S. Duff). Canterbury Museum Bulletin 1, Government Printer, Wellington, pp. 348–75.

———. 1979. Tai Rua: A Moa-hunter Site in North Otago. In *Birds of a Feather* (A. Anderson, editor). British Archaeological Reports International Series 62:203–30.

———. 1980a. Archaeological Investigations at Avoca Point, Kaikoura. *Records of the Canterbury Museum* 9(4):277–88.

———. 1980b. Radiocarbon Dates for Awamoa, North Otago. *New Zealand Archaeological Association Newsletter* 23(3):184–85.

Trotter, M. M. and McCulloch, B. 1971. *Prehistoric Rock Art of New Zealand.* A. H. & A. W. Reed, Wellington. (Second edition published by Longman Paul, Auckland, 1981.)

Williams, G. R. 1962. Extinction and the Land and Freshwater-Inhabiting Birds of New Zealand. *Notornis* 10:15–32.

33

The Extinction of Moa in Southern New Zealand

ATHOLL ANDERSON

THE LARGE RATITES KNOWN COLLECTIVELY AS MOA were found throughout the main islands of New Zealand but, judging from the distribution of natural bone deposits and moa-hunting sites, moa were particularly common in southern New Zealand, the region south of Banks Peninsula (fig. 33.1). Cracraft (1976) has assigned moa to the family Dinornithidae. The eight species that he regards as certainly having been present on South Island (i.e. excluding *Megalapteryx benhami* and *Dinornis struthoides*) were living in the southern region at the time the first people, the Maori, arrived about 1,000 years ago. By the time of European settlement in the nineteenth century, however, all moa had apparently become extinct. I will discuss the course and causes of their disappearance, first by looking at their ecology.

Aspects of Moa Ecology in Southern New Zealand

For more than a century it was widely assumed that, with the possible exception of the smallest genera, moa inhabited mainly open grasslands (Duff 1956, Oliver 1955). This view arose substantially from the broad correlation of natural bone beds and moa-hunting sites with the historically tussock-covered lands of eastern South Island. Recent investigations have shown this hypothesis to be no longer tenable. First, it was demonstrated, especially by Lockerbie (1959), that archaeological sites rich in moa bone were also to be found in districts such as the Catlins (fig. 33.1), where closed forest remained throughout the prehistoric era. Second, distributional studies of fossil logs and other forest phenomena such as soil charcoals, podsols, and "tree dimples," together with the radiocarbon dating of the charcoal, demonstrated that extensive forest grew in eastern South Island until well into the last millennium (Molloy et al. 1963, Molloy 1969, Mark 1974). Third, it has been argued that the divaricating habit of numerous indigenous shrubs evolved as a response to intensive moa browsing (Greenwood and Atkinson 1977), although alternative explanations exist for this (McGlone and Webb 1981). Fourth, analyses of preserved moa crops (Burrows 1980, Burrows et al. 1981), and some early studies (e.g. Hamilton 1891), indicate that larger moa (*Dinornithinae*) were mainly consuming twigs, fruits, and seeds of woody plants from shrubland and forest-fringe microenvironments. These data are not yet conclusive, but they

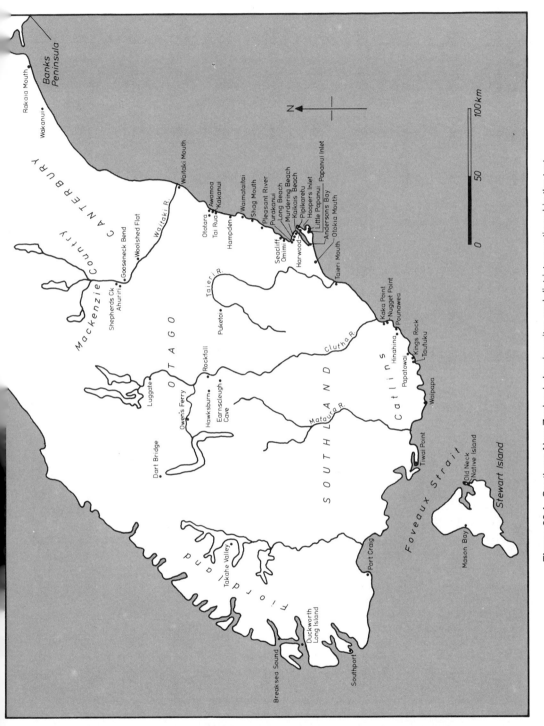

Figure 33.1. Southern New Zealand showing sites and districts mentioned in the text.

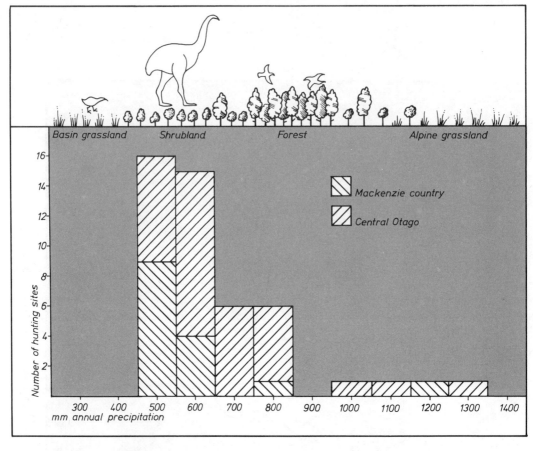

Figure 33.2. Correlation of interior moa hunting sites with annual precipitation and prehistoric vegetation patterns.

do point confidently to a preferred moa habitat of shrublands and forest, particularly the margins, rather than open grasslands.

An interesting case study of this hypothesis can be formulated regarding moa-hunting site locations in the interior districts of central Otago and the Mackenzie country (fig. 33.1). Most of the land there was already under tussock before the Maori arrived around 900 B.P. (MgGlone pers. comm.). Forest was confined to bands or patches on the midslopes of the ranges, with the lower fringe generally defined by the 600-mm annual isohyet (Mark pers. comm.). Below the forests, shrublands opened out downslope into tussock. If it is assumed that moa-hunting sites were usually located at or close to the places in which moa were most commonly found, then they ought to be clustered about the predicted forest-fringe/shrubland ecotone. In fact, this seems to have been largely the case (fig. 33.2) (for further details see Anderson 1983, n.d.a.).

This is not to say that moa were necessarily sedentary, although that argument has been advanced (Haast 1871, Ambrose 1968, Scarlett 1974). Rather, if moa, like other ratites (Davies 1976), required high quality forage, it is likely that they were continually moving around patchy resource territories to locate each flush of suitably nutritious food. In upland areas some transhumant movements of the kind that occur there today among large herbivores may be envisaged, but it is unlikely that wide-ranging migrations occurred as they do among emu (*Dromaius novaehollandiae*), for example.

Related to the need for high quality food are rapid growth rates of ratite young (Davies 1976). In environments with the comparatively marked seasonality of southern New Zealand, this was probably reflected in a breeding peak immediately prior to the main growth flush of woody plants in spring. How productive moa breeding was, however, is almost impossible to say. The only extant New Zealand ratites, the kiwis (*Apteryx* sp.), usually lay only one egg at a time, and Falla (1962) reported evidence of a similarly small clutch size for *Anomalopteryx* sp. On these admittedly slim grounds (see also Trotter and McCulloch, this volume), moa reproductive rates may have been considerably lower than those of large ratites living today.

How many moa were living in southern New Zealand at the time the Maori arrived is also a matter of conjecture. It seems most unlikely, given the large size of these birds and their probably precise dietary requirements, that they ever approached the numbers suggested by Duff (1956:280). Could they, however, have been "numerous," "plentiful," or in "large numbers" as others have suggested (e.g. Lockerbie 1959, Simmons 1968)? Reference to the population densities of the large Australian ratites of today suggests that these too may be overestimates. The emu, probably as common today as it was before the arrival of Europeans (Davies 1976), and the cassowaries (*Casuarius* sp.) are distributed at densities of 5 to 15 km² per individual (Crome 1976, Davies 1976, Caughley et al. 1980). These figures, applied to the area of southern New Zealand below the alpine zone, would produce an estimated "standing crop" of only 3,000 to 10,000 birds. This range is almost certainly too low because, on the one hand, there were no significantly competing herbivores and, on the other, a gross population of this size could hardly support more than a few species. Nevertheless, there does not seem to be any reason to imagine that the total moa population south of Banks Pensinsula at about 1000 B.P. was any greater than some tens of thousands of individuals.

The Course of Extinction

One hundred seventeen moa-hunting sites have been recorded in southern New Zealand, undoubtedly only a minor proportion of the actual number that still exist. In the interior, for example, another forty-nine sites are uninvestigated, from which all the distinctive characteristics of moa hunting, except moa bone, have been recorded. More than 100 painted rockshelters are also thought to be largely attributable to moa-hunting groups (Trotter and McCulloch 1981). Along the coast the largest archaeological sites, by an order of magnitude, are moa butchery sites (e.g. Waitaki Mouth at 60 ha in area) and both there and in the interior moa are the most frequently represented resource in the early prehistoric era. Despite the manifest importance of moa hunting, comparatively few systematic site excavations have been conducted, and for only fifty-six sites in the region have moa remains been identified to the species level (Table 33.1); in many cases these amount to little more than "grab" samples.

Assuming these data reasonably reflect actual availability of moa, then the most common species was *Euryapteryx geranoides* followed by *Emeus crassus*. These were medium-sized moa with estimated liveweights of 61 kg and 51 kg, respectively (Smith n.d.). *E. geranoides* was found in about 60 percent of the sites, while *E. crassus* was confined mainly to the coast. In the interior *Megalapteryx didinus* (estimated liveweight 25 kg, Smith n.d.) is as common as *E. geranoides* (Table 33.1). The three largest species of Dinornithinae ranged from 125 kg to 230 kg liveweight (Smith n.d.). They are not well represented overall and are missing in coastal sites south and west of the Catlins (Table 33.1). Insofar as the chronology of moa hunting can be regarded as precise, it would seem that *M. didinus* survived longest, although only in remote western districts; at the other extreme, *Dinornis giganteus* may have been regionally extinct as early as 600 B.P.

The radiocarbon chronology of southern New Zealand prehistory is difficult to interpret for several reasons. Many dates are unpublished, and exact details of their provenance are unknown. The same may be said of too many published dates. Worse, there is frequently a discrepancy of up to 300 years between radiocarbon estimates from charcoal samples and those from other materials such as bone collagen and marine shell.

Table 33.1. Archaeological Sites With Identified Moa Species

Site	SPECIES*								
	Anomalopteryx didiformis	*Megalapteryx didinus*	*Pachyornis elephantopus*	*Euryapteryx geranoides*	*Emeus crassus*	*Dinornis torosus*	*Dinornis novaezealandiae*	*Dinornis giganteus*	Radiocarbon Dates (yr B.P.) (Old T½, uncorr., charcoal)
COASTAL									
Rakaia Mouth	x		x	x	x				
Wakanui				x	x				
Waitaki Mouth	x		x	x	x	x	x		980 ± 70, 642 ± 34
Awamoa			x		x	x		x	984 ± 37
Ototara				x					
Tai Rua			x	x					831 ± 33
Hampden				x					
Waimataitai			x	x	x				708 ± 40
Shag Mouth	x		x	x	x	x	x		823 ± 55, 802 ± 55
Pleasant River			x	x	x				
Seacliff	x	x		x	x	x		x	
Omimi	x				x				
Purakanui				x				x	1030 ± 60
Long Beach				x					868 ± 80†
Murdering Beach					x				
Kaikai's Beach				x					900 ± 60
Pipikaretu					x				
Harwood	x		x	x	x				
Papanui Inlet	x		x					x	
Little Papanui			x	x	x		x	x	
Hooper's Inlet				x	x				
Anderson's Bay				x				x	
Otokia Mouth	x			x	x				
Taieri Mouth								x	
Kaka Point				x	x				
Nugget Point			x						
Pounawea	x	x	x	x	x			x	815 ± 80, 584 ± 77†
Hinahina					x				740 ± 75
Papatowai	x	x	x	x	x	x	x	x	910 ± 80, 560 ± 60†
King's Rock				x			x		
Tautuku Peninsula						x			
Tautuku Point	x		x	x	x		x		
North Tautuku						x			
Waipapa					x				
Tiwai Point (X)				x	x				660 ± 40, 640 ± 40†
Old Neck				x					ca. 700
Native Island				x					
Mason Bay					x				

Table 33.1. Archaeological Sites With Identified Moa Species

(continued)

Site	Anomalopteryx didiformis	Megalapteryx didinus	Pachyornis elephantopus	Euryapteryx geranoides	Emeus crassus	Dinornis torosus	Dinornis novaezealandiae	Dinornis giganteus	Radiocarbon Dates (yr B.P.) (Old T½, uncorr., charcoal)
Port Craig			x						
Duckworth				x					
Long Island				x					590 ± 150, 455 ± 78
Southport		x							
Breaksea Sound		x							
INTERIOR									
Shepherd's Creek Flat		x		x					
Gooseneck Bend			x						850 ± 150
Ahuriri		x							625 ± 65
Woolshed Flat		x	x	x					860 ± 32, 810 ± 50
Luggate			x					x	
Puketoi		x					x	x	
Rockfall II				x					949 ± 59, 674 ± 60†
Earnscleugh Cave				x	x	x	x	x	
Hawksburn	x	x	x	x	x	x	x		742 ± 34, 590 ± 50†
Owen's Ferry	x			x				x	763 ± 35, 607 ± 29†
Dart Bridge	x	x		x					723 ± 57, 337 ± 56†
Takahe Valley		x							860 ± 60, 230 ± 60
Number of occurrences	14	12	19	33	27	9	9	13	

SOURCES: Moa species data from Hamel (1977) with additions and radiocarbon dates from Anderson (1982b) and unpublished data.

*Follows Cracraft (1976).

†Indicates dates on samples screened for species and lifespan composition (see text).

NOTE: Where two dates are shown, those are the extremes; in many cases estimates exist between them. Only latest dates are shown.

The difference is inconsistent, and the cause is yet unclear. The earliest charcoal dates extend to about 1000 B.P., those on collagen and marine shell generally to about 800 B.P., while the latest dates for moa-hunting sites come into closer agreement at about 500 B.P. (with a few exceptions). In several cases almost exactly the same result has been obtained upon charcoal, collagen, and marine shell samples from the same context (for further discussion see McCulloch and Trotter 1975; Anderson 1982; and Trotter and McCulloch this volume). For present purposes I prefer the estimates obtained from charcoal, especially the recent dates in which the samples have been treated for soil contamination and screened for species and lifespan composition, since these are most comparable to the radiocarbon dates used in the reconstruction of vegetational changes.

The radiocarbon dates shown in Table 33.1 and from other sites in which moa remains have not been identified to species (Anderson 1982, n.d.b.), indicate that moa hunting extended from about 1000 B.P. to 500 B.P. in southern New Zealand, but that it was mainly concentrated from 900 to 600 B.P. Within that time moa hunting seems to have significantly declined, although data are still scarce. At Papatowai, for instance, Hamel (1977, 1978) recovered six individuals of moa (five species) from the lowest layer dated to ca. 850 yr B.P., three of two species in the middle layer (ca. 700 yr B.P., and one individual in the upper layer (ca. 600 yr B.P.). Found at Pounawea (Hamel 1980) were six individuals of five species in the lower layer (ca. 800 yr B.P.), and three individuals of three species in the upper layer (ca. 600 yr B.P.). Intensive moa hunting seems to have been fairly brief in the Catlins, as in two interior districts. In the lower Mackenzie country the charcoal dates for moa hunting span a period of less than two centuries (about 850 to 700 B.P.), and Ambrose (1970) regards this activity as having been brief and destructive. In east-central Otago most charcoal dates also fall within a 150-year span from about 700 yr B.P. to 550 yr B.P. (Anderson n.d.a.). These cases suggest that at a district level intensive moa hunting ran its course comparatively quickly and that, at a regional level, it was over within four centuries of the Maori's arrival.

How long sporadic moa hunting continued after that depends on the credence ascribed to a variety of indications. Charcoal radiocarbon dates are associated with primary moa bone or eggshell refuse (i.e. nonindustrial material) up to about 500 yr B.P. along the coast and to between 400 yr B.P. and 300 yr B.P. at some interior sites (Italian Creek, Ritchie n.d., Dart Bridge, unpublished dates). In addition there is a date of 230 ± 60 yr B.P. obtained from tussock underlying *Megalapteryx* sp. remains at Takahe Valley (Duff 1952) which need not be discounted, although Trotter and McCulloch (1973) regard it as dubious. Younger still are Coutts's (1972:71, 190) records of moa bone and feathers in apparently post-European contexts in several Fiordland rockshelters. Finally, an extensive list of reported sightings of live moa or fresh remains date from the arrival of Europeans to the end of the nineteenth century (summary in Beattie 1958). Many of these accounts are quite fantastic but some are plausible, including Meurant's alleged observations of moa being consumed at Molyneux (Clutha) Mouth in 1823 (Taylor 1855:238) and the surveyor Thomson's report (1959:333) of fresh moa bone strewn among newly abandoned Maori storehouses at a settlement in western Southland in 1857.

The course of moa extinction in southern New Zealand may be summarized thus: hunting began with the earliest arrival of the Maori, around 1000 B.P., and became an intensive subsistence activity over the whole region from 900 to 600 B.P., although it was apparently much shorter than that in each district. Opportunistic hunting continued along the east coast until about 500 B.P. and in the western interior until 300 to 200 B.P. Some moa could have survived in the most remote western districts until the advent of European settlement.

Causes of Extinction

Moa extinction in southern New Zealand has been put down to a wide variety of causes. Of them we can eliminate gross climatic changes such as episodes of severe cold (Booth 1874) or severe drought (Pyke 1890), neither of which occurred during the last millennium. Deterioration in moa habitats resulting from climatic changes was probably not a significant factor either. Since 1000 B.P. there have been fairly regular but minor temperature fluctuations of the magnitude of $\pm 0.7°C$. The lowest temperatures, and east of the Southern Alps probably the lowest precipitation levels, were reached between 300 to 100 B.P. (Burrows and Greenland 1979, Burrows 1982, but these seem to have had little effect upon forest regeneration patterns; even if they had, moa were probably already close to extinction by that time.

The late persistence of all the moa species and their subsequent rapid disappearance, on a geological time scale, also renders unlikely hypotheses that attribute extinction to evolutionary age. Even if, as McDowall (1969) has argued, positive correlation exists between extinction and endemism among New Zealand birds, it seems improbable that the large ratites and other birds present throughout the Pleistocene should have coincidentally reached evolutionary senescence only within the last millennium.

Left with cultural factors, the problem is to try to sort them into an order of relative importance. Some simply cannot be assessed because of lack of evidence. This is the case with the possible introduction of fowl pests or diseases by the Maori (Williams 1973) and introduced rat (*Rattus exulans*) predation (Fleming 1969). It is substantially true of introduced dog (*Canis familiaris*) predation as well, although there is at least historical evidence of feral dogs preying upon ground-dwelling birds (Anderson 1981). The two hypotheses considered most likely are that moa became extinct because of human interference with their preferred habitats or that they did so as a direct result of human predation.

The first requires demonstration that forest and shrubland were cleared extensively during the last millennium. Evidence is now abundant that this did, in fact, occur and that the primary agency was cultural firing (Molloy et al. 1963, Molloy 1969, Wells 1972, Mark 1974, Molloy 1977, Burrows and Greenland 1979, McGlone 1982 pers. comm., Anderson 1983). Since woody vegetation seems to have provided the main moa habitats, and since the firing was particularly destructive of the podocarp forests in eastern South Island (fig. 33.3), which provided richer niches for moa and other birds than the western beech (*Nothofagus* sp.) forests, it was conceivably a phenomenon of major significance. As a satisfactory explanation of moa extinction, however, it is open to question on the grounds of when the forest burning took place. McGlone's (1982 pers. comm.) reconstruction of the sequence of firing indicates that the main phase of burning began about 700 B.P. Thus, burning postdated the most intensive moa hunting, and moa populations were probably already significantly depleted. Forest firing may have been a deliberate response to lowered moa availability in southern New Zealand. Whatever the intent, any attempt to increase the incidence of shrubland and forest margins by burning was doomed to failure for reasons peculiar to the nature of this vegetation and its limited regenerative potential. However, large tracts of forest remained, especially in the southernmost districts. There is room, therefore, for an alternative explanation.

Although widely acknowledged, the direct human predation hypothesis is not one archaeologists have considered in much depth. Some have assumed that moa were so plentiful that, as Green (1975:613) has argued, "nothing like sufficient numbers of moa seem to have been killed and eaten by man to have directly brought about their extinction." In my view this approximate calculation does not hold for southern New Zealand. It is very difficult to estimate how many moa were slaughtered by the southern Maori, but since the question is such a crucial one it is worth attempting. In addition to the impressive historical evidence of abundant moa remains in southern New Zealand butchery sites (Trotter and McCulloch, this volume), it is possible, by the use of a few data from recent excavations, to suggest very approximately the number of moa killed in this region. The density of moa individuals per square meter can be calculated as 0.06 at Tiwai Point (area X, Sutton and Marshall 1980), 0.07 at Pounawea (Hamel 1980), 1.4 at Hawksburn (Anderson n.d.a.), and 2.2 at Papatowai (Hamel 1977). Using 0.1 to 0.5 moa individuals per square meter as a conservative estimate and extending these figures to the approximately 10×10^5 square meters which the 117 known moa-hunting sites cover, then 100,000 to 500,000 moa are represented. This may be regarded as a low estimate because the density of moa remains was probably significantly higher in many of the large specialized moa-hunting sites along the east coast not represented in these figures; undoubtedly there are many more moa-hunting sites besides the 117 presently

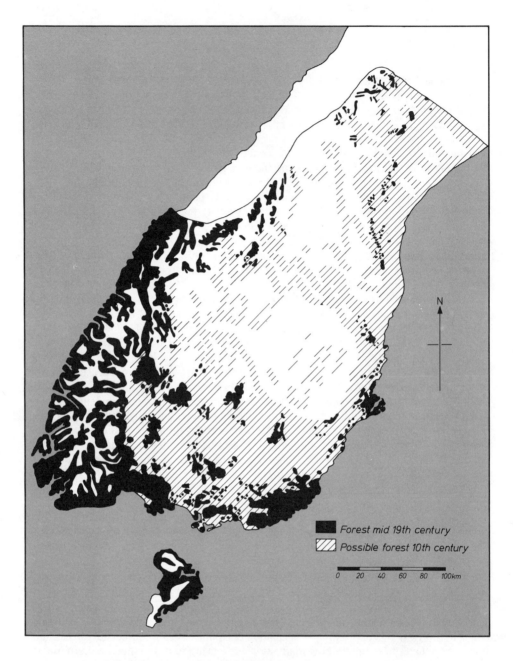

N

| | Forest mid 19th century |
| | Possible forest 10th century |

0   20   40   60   80   100km

Figure 33.3. The approximate extent of deforestation in southern
New Zealand during the last millenium.

recorded. The coastline along which the largest sites are situated (Waitaki Mouth to
Banks Peninsula) has been receding at rates of 0.5 to 1.0 m per year in the historical era
(Gibb 1977), and this erosion has probably removed some major butchery settlements.
However, all I wish to suggest is that such few data as can be mustered would support
the view that whereas the size of the moa population available to the early Maori may
have been measured in tens of thousands, the number of moa that were ultimately
slaughtered was an order of magnitude greater. In other words, the arithmetic of
exploitation might well remove the necessity for envisaging any other significant agency
of extinction.

Few archaeological sites in southern New Zealand have, however, been analyzed in a way that would provide direct evidence of overexploitation, although various early investigators certainly formed this impression from the remains they observed (Trotter and McCulloch this volume). Accounts of great quantities of eggshell at Puketoi (Murison 1871), Awamoa (McDonnell 1888), and some other sites (e.g. Hocken 1871) are consistent with that view. Analysis of the moa remains from Hawksburn (Anderson n.d.a.) provides a similar case. There, two-thirds of the 430 moa in the site were represented by trimmed leg joints, not whole carcasses, a pattern indicating that between one-third and one-half of the available meat weight slaughtered was discarded in the field.

Despite the scarce evidence, the predator-prey relationship of man and moa was certainly one likely to lead to overexploitation. Moa had few, if any, significant predators prior to the arrival of people, had probably evolved various characteristics of the "K strategy" of adaptation, including a low reproductive rate (Halliday 1980:34), and were, in the absence of large terrestrial mammals, the most conspicuous targets for the Maori in an otherwise impoverished resource array. Moreover, the hunters possessed preadapted fowling technology of spears, nets, and snares that apparently was sufficient (there is only one possible trapping pit recorded, Skinner 1934) and had no knowledge of the sustained-yield techniques of big-game management.

A likely consequence of such intensive exploitation would be rapid human population growth. This does seem to be reflected in southern New Zealand by the fact that the largest sites which include the widest range of faunal remains and artifacts (including domestic structures) date to the early settlement period, ca. 950 to 650 B.P. After that, ca. 650 to 450 B.P., settlement functions were apparently fragmented among smaller and less numerous sites, which may indicate a declining population in the face of depleted resources (Anderson 1982).

Conclusions

Recent evidence bearing upon the question of moa extinction in southern New Zealand has refined earlier views. First, it is now assumed that moa generally preferred woody plant to grassland habitats and were especially attracted to species-rich forest fringes and mixed shrubland. Second, the chronological evidence of extinction, although still subject to problems of interpretation, suggests that the depletion of moa populations occurred more or less contemporaneously along the coast and in the interior as a rapid and devastating phase that resulted in declining yields as early as within three centuries of the arrival of the Maori. Third, paleoclimatic data have disposed of climatic change during the last millennium as a direct agent of extinction and have reduced it, as an indirect agent through vegetational changes, to a minor role at most. Fourth, by ascribing the destruction of the forest to Maori burning, the environmental hypothesis of extinction has now been firmly cast in a cultural mold.

Taken with the longstanding evidence for remarkably intensive moa hunting in southern New Zealand, these conclusions point strongly to a primary role for human interference as the explanation of moa extinction. If moa populations were smaller and the number of moa eventually slaughtered greater than has been assumed, and if the intensive phase of habitat destruction began some centuries later than the intensive phase of moa hunting, then overexploitation would appear to have been the most effective agent of severe depletion, though not necessarily of ultimate extinction.

The scenario I propose for southern New Zealand is this: moa hunting entered an intensive and overexploitive phase about a century after the arrival of the Maori, and it was accompanied by rapid human population growth. Within several centuries forest burning had become sufficiently widespread to accelerate the decline of the moa, which by 500 to 400 B.P. had become so scarce as to no longer be systematically hunted.

Continued reduction of preferred moa habitats, sporadic hunting, and possibly depredations of feral dogs administered the *coup de grâce* immediately before European settlement.

Acknowledgments

My thanks to Alan Mark, Matt McGlone, and Ian Smith for valuable discussions on various matters discussed in this paper, to Martin Fisher for the illustrations, and to Ann Trappitt for the typing.

References

Ambrose, W. 1968. The unimportance of the inland plains in South Island prehistory. *Mankind* 6:585–593.

———. 1970. Archaeology and rock drawings from the Waitaki gorge, central South Island. *Records of the Canterbury Museum* 8:383–437.

Anderson, A. J. 1981. Pre-European hunting dogs in the South Island, New Zealand. *New Zealand Journal of Archaeology* 3:15–20.

———. 1982. A review of economic patterns during the Archaic phase in southern New Zealand. *New Zealand Journal of Archaeology* 4:45–75.

———. 1983. Habitat preferences of moa in central Otago, A.D. 1000–1500, according to palaeobotanical and archaeological evidence. *Journal of the Royal Society of New Zealand* 12:321–336.

———. n.d.a. The prehistoric hunting of moa (Aves: Dinornithidae) in the high country of southern New Zealand. Paper presented at the Fourth International Conference of the International Council for Archaeozoology, London, April 1982.

———. n.d.b. Faunal depletion and subsistence change in the early prehistory of southern New Zealand. Paper presented at the 52nd Congress of the Australian and New Zealand Association for the Advancement of Science, Macquarie University, May 1982.

Beattie, H. 1958. The Moa: when did it become extinct? *Otago Daily Times and Witness*, Dunedin.

Booth, B. D. 1874. Description of the moa swamp at Hamilton. *Transactions of the New Zealand Institute* 7:123–138.

Burrows, C. J. 1980. Diet of the New Zealand Dinornithiformes. *Naturwissenschaften* 67:S.151.

———. 1982. On New Zealand climate within the last 1000 years. *New Zealand Journal of Archaeology* 4:157–167.

Burrows, C. J. and Greenland, D. E. 1979. An analysis of the evidence for climatic change in New Zealand in the last thousand years: evidence from diverse natural phenomena and from instrumental records. *Journal of the Royal Society of New Zealand* 9:321–373.

Burrows, C. J., McCulloch, B., and Trotter, M. M. 1981. The diet of moas, based on gizzard contents samples from Pyramid Valley, north Canterbury, and Scaifes Lagoon, Lake Wanaka, Otago. *Records of the Canterbury Museum* 9:309–336.

Caughley, C., Griggs, G. C., Caughley, J., and Hill, G. J. E. 1980. Does dingo predation control the densities of kangaroos and emu? *Australian Wildlife Research* 7:1–12.

Coutts, P. J. F. 1972. The emergence of the Foveaux Strait Maori from prehistory: a study of culture contact. Ph.D. dissertation, University of Otago.

Cracraft, J. 1976. The species of moas (Aves: Dinornithidae). *Smithsonian Contributions to Paleobiology* 27:189–205.

Crome, F. H. J. 1976. Some observations on the biology of the cassowary in northern Queensland. *The Emu* 76:8–14.

Davies, S. J. J. F. 1976. The natural history of the emu in comparison with that of other ratites. *In* Frith, H. J. and Calaby, J. H. (Eds.), *Proceedings of the 16th International Ornithological Conference*. Australian Academy of Science, Canberra, pp. 109–120.

Duff, R. S. 1952. Recent Maori occupation of Notornis Valley, Te Anau. *Journal of the Polynesian Society* 61:90–119.

———. 1956. *The Moa-hunter period of Maori culture*. Government Printer, Wellington.

Falla, R. A. 1962. The moa, zoological and archaeological. *Newsletter of the New Zealand Archaeological Association* 5:189–191.

Fleming, C. A. 1969. Rats and moa extinction. *Notornis* 16:210–211.

Gibb, J. G. 1977. Historical shoreline changes in New Zealand. *In* Neall, V. E. (Ed.), *Soil Groups of New Zealand, Part 2: Yellow-brown sands.* Government Printer, Wellington, pp. 182–196.

Green, R. C. 1975. Adaptation and change in Maori culture. *In* Kuschel, G. (Ed.), *Biogeography and Ecology in New Zealand.* Junk, The Hague, pp. 591–641.

Greenwood, R. M. and Atkinson, I. A. E. 1977. Evolution of divaricating plants in New Zealand in relation to moa browsing. *Proceedings of the New Zealand Ecological Society* 24:21–33.

Haast, J. von. 1871. Moas and moa hunters. *Transactions of the New Zealand Institute* 4:66–107.

Halliday, T. 1980. *Vanishing Birds: Their Natural History and Conservation.* Penguin, London.

Hamel, G. E. 1977. Prehistoric man and his environment in the Catlins, New Zealand. Ph.D. dissertation, University of Otago.

———. 1978. Radiocarbon dates from the moa-hunter site of Papatowai, Otago, New Zealand. *Newsletter of the New Zealand Archaeological Association* 21:53–54.

———. 1980. Pounawea: the last excavation. Unpublished report to the New Zealand Historic Places Trust.

Hamilton, A. 1891. Notes on moa gizzard stones. *Transactions of the New Zealand Institute* 24:172–175.

Hocken, T. M. 1871. Fourth meeting of the Otago Institute, 1871. *Transactions of the New Zealand Institute* 4:413–415.

Lockerbie, L. 1959. From moa-hunter to classic Maori in southern New Zealand. *In* Freeman, J. D. and Geddes, W. R. (Eds.), *Anthropology in the South Seas.* Avery, New Plymouth, pp. 75–110.

McCulloch, B. and Trotter, M. M. 1975. The first twenty years: radiocarbon dates for South Island moa-hunter sites. *Newsletter of the New Zealand Archaeological Association* 18:2–17.

McDonnell, Lt. Col. 1888. The ancient moa-hunters at Waingongoro. *Transactions of the New Zealand Institute* 21:438–441.

McDowall, R. M. 1969. Extinction and endemism in New Zealand land birds. *Tuatara* 17:1–12.

McGlone, M. S. 1982. Polynesians and the late Holocene deforestation of New Zealand. *In* Ross, A. (Ed.), *Environment and People in Australasia: Abstracts of the 52nd Congress of the Australian and New Zealand Association for the Advancement of Science.* Macquarie University, p. 43.

McGlone, M. S. pers. comm. Botany Division, Department of Scientific and Industrial Research, Christchurch, New Zealand.

McGlone, M. S. and Webb, C. J. 1981. Selective forces influencing the evolution of divaricating plants. *New Zealand Journal of Ecology* 4:20–28.

Mark, A. F. 1974. Old Man Range. *New Zealand's Nature Heritage* 2:524–531.

———. pers. comm. Botany Department, University of Otago.

Molloy, B. P. J. 1969. Recent history of the vegetation. *In* Knox, G. A. (Ed.), *The Natural History of Canterbury.* Reed, Wellington, pp. 340–360.

———. 1977. The fire history. *In* Burrows, C. J. (Ed.), *Cass: History and Science in the Cass District, Canterbury, New Zealand.* University of Canterbury, Botany Department, pp. 157–170.

Molloy, B. P. J., Burrows, C. J., Cox, J. E., Johnston, J. A. and Wardle, P. 1963. Distribution of subfossil forest remains, eastern South Island, New Zealand. *New Zealand Journal of Botany* 1:68–77.

Murison, W. D. 1871. Notes on moa remains. *Transactions of the New Zealand Institute* 4:120–124.

Oliver, W. R. B. 1955. *New Zealand Birds.* 2nd Edition. Reed, Wellington.

Pyke, V. 1890. The moa (Dinornis) and the probable causes of its extinction. *Otago Witness*, September 18th.

Ritchie, N. A. n.d. The excavation of a Maori campsite, Italian Creek, central Otago. Unpublished report to the New Zealand Historic Places Trust.

Scarlett, R. J. 1974. Moa and man in New Zealand. *Notornis* 21:1–12.

Simmons, D. R. 1968. Man, moa and the forest. *Transactions of the Royal Society of New Zealand (General)* 2:115–127.

Skinner, H. D. 1934. A pit in peat, central Otago. *Journal of the Polynesian Society* 43:293.

Smith, I. W. G. n.d. Estimated meat weights of moa. Unpublished manuscript, Anthropology Department, University of Otago.

Sutton, D. G. and Marshall, Y. M. 1980. Coastal hunting in the subantarctic zone.

New Zealand Journal of Archaeology 2:25–49.

Taylor, Rev. R. 1855. *Te Ika A Maui, or New Zealand and its Inhabitants.* Wertheim and Macintosh, London.

Thomson, J. T. 1959. Extracts from a journal kept during a reconnaissance survey of the southern districts of the province of Otago. *In* Taylor, N. M. (Ed.), *Early Travellers in New Zealand.* Clarendon Press, Oxford, pp. 325–348.

Trotter, M. M. and McCulloch, B. 1973. Radiocarbon dates for South Island rock shelters. *Newsletter of the New Zealand Archaeological Association* 16:176–178.

———. 1981. *Prehistoric Rock Art of New Zealand.* 2nd Edition. Longman Paul, Auckland.

Wells, J. A. 1972. Ecology of *Podocarpus hallii* in central Otago, New Zealand. *New Zealand Journal of Botany* 10:399–426.

Williams, G. R. 1973. Birds. *In* Williams, G. R. (Ed.), *The Natural History of New Zealand.* Reed, Wellington, pp. 304–333.

34

The Role of Prehistoric Man in the Faunal Extinctions of New Zealand and Other Pacific Islands

RICHARD CASSELS

EVER SINCE THE DISCOVERY OF THE MOAS of New Zealand, over 100 years ago, debate has been lively about the cause or causes of their extinction. Academic opinion has remained as divided as it ever was. The question remains, "Who killed cock robin?" Was it prehistoric man, or were the moas dying out anyway? Was it climatic change or a change in forest composition? Or was it something to do with the dogs and rats that were introduced to the Pacific islands by the Polynesians?

While focusing on the New Zealand extinctions, I also consider some other Pacific islands. The geographical scope is limited to islands east of New Guinea and west of Easter Island (fig. 34.1). The extinctions discussed are of those species that were never recorded alive by Europeans, and are known to be no older than late Pleistocene in age. The extinctions that occurred following the arrival of Europeans (see Greenway 1958) are not described, although they may best be viewed as a continuation of the process that started in prehistory. Dramatic as they are, post-European extinctions do not compare to the scale of the pre-European ones.

In the case of New Zealand, there is no doubt that most, if not all, known Pleistocene-Holocene extinctions occurred during the period of prehistoric Polynesian occupation, that is, within the last 1,000 years. Elsewhere in the Pacific these extinctions are less precisely dated. Three chronological alternatives exist: first, that they took place soon after Europeans arrived and before extensive ornithological records were made; second, that they occurred during the period of prehistoric settlement; and third, that they predate the arrival of man, possibly going back as far as the late Pleistocene. Increasingly, archaeological evidence favors the second alternative: that they occurred during the period of prehistoric settlement. Theoretical explanations are, however, hampered by the chronological uncertainty that still remains in the Pacific islands other than New Zealand.

At present the pre-European extinctions of New Zealand provide one of the best cases for arguing that prehistoric man was capable of causing the extermination of fauna on a catastrophic scale. It is no surprise that this should have occurred in the remote oceanic islands of the Pacific where, in the absence of land mammals (Gressitt 1963), birds evolved into endemic forms extraordinarily vulnerable to man.

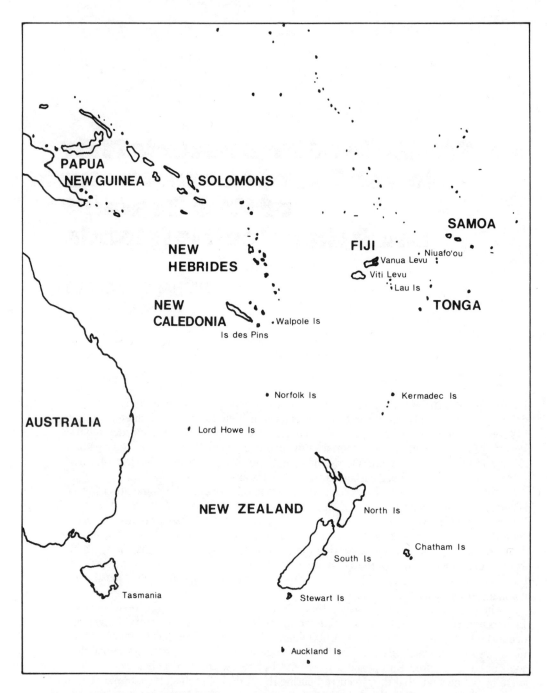

Figure 34.1. Islands of the western Pacific, including those discussed in the text. The Auckland Islands and Lord Howe Island are among the few that were not inhabited prehistorically.

Perhaps the one surprise in the New Zealand evidence is that some very vulnerable species, notably the moas, seem to have coexisted with man for roughly five centuries. Even more remarkable is the survival to the present day of such large, flightless curiosities as the kiwi. Such survivals offer insights into the prehistoric extinction process. Following analysis of the extinct prehistoric avifaunas of New Zealand and the Pacific archipelagos, I will compare characteristics of the living and extinct birds.

Table 34.1. Characteristics of Extinct New Zealand Bird Species

KEY: Characteristics: 1 = flightlessness; 2 = flightless and survived by flighted relative (N.Z., same genus); 3 = ground-nesting; 4 = colonial nesting; 5 = primarily nocturnal; 6 = location of closest surviving relative; 7 = larger than closest surviving relative (N.Z.); 8 = largest form in its family (N.Z.); 9 = level of endemism; + = definite; (+) = probable; ? = possible; − = unlikely/negative; u = unique; * = no archaeological association with man.

Bird Species	Characteristics								
	1	2	3	4	5	6	7	8	9
Species Extinctions									
Moas (13 + species)	+	−	+			N.Z.	+	+	order
Aptornis otidiformis	+	−	+			N.C.		u	family
Gallinula hodgeni	+	−	+			Aus		−	species
Capellirallus karamu	+	−	+			—		−	genus
Gallirallus minor (uncertain sp.)	+	−	+			N.Z.	−	−	species/genus
Fulica chathamensis	+	−	+			Aus		−	species
Cygnus sumnerensis	−	−	(+)	(+)		Aus	+	+	species
Cnemiornis (2 species)	+	−	+			Aus	+	−	genus
Euryanas finschi	(+)	−	(+)			Aus		−	genus
Palaeocorax moriorium	−	−	?	?		Aus	+	u	species
Harpagornis moorei	−	−				Aus	+	+	genus
"Circus" (Accipiter?) (2 species)	−	−				Aus?	?	−	species
*Megaegotheles novaezealandiae	?	−	(+)		+	Aus		u	genus
*Malacorhynchus scarletti	−	−	(+)			Aus		−	species
Max. Poss. Total 28	22	0	25	2	1		20	15	
(Percent)	78.6	0	89.3	7.1	3.6		71.4	53.6	
Subspecies Extinctions									
Pelecanus conspicillatus	−	−	(+)	?	−	Aus	+	u	subspecies
Biziura lobata	−	−	(+)	−	−	Aus		−	subspecies
Max. Poss. Total 30	22	0	27	3	1		21	15	
(Percent)	73.3	0	90	10	3.3		70	50	
Regional Extinctions									
Notornis mantelli	+	−	+	−	−	N.Z.			genus/species
*Coenocorypha aucklandica	−	−	+	−	−	N.Z.			genus
Mergus australis	−	−	(+)	−	−	N.Z.			species
Nestor meridionalis subsp.	−	−	−	−	−	N.Z.			subspecies

SOURCES: Owen 1879; Rothschild 1907; Falla 1953; Oliver 1955; Fleming 1962b; 1974, 1979; McDowall 1969; Kinsky 1970; Falla et al. 1970; Cracraft 1973, 1974, 1976; Olson 1977a, b; Millener ms; Slater 1970; Frith 1967; Hamel 1977, 1979; Simmons 1968.

Bird Extinctions on the New Zealand Mainland

The extinct avian species of the New Zealand Quaternary that were never recorded alive by Europeans consist of all the moas and approximately twenty other birds: fourteen to sixteen endemic species and five forms very similar to living Australian species. These are listed below and with their characteristics in Table 34.1. Nomenclature follows Kinsky (1970) unless otherwise specified.

The most famous of the extinct New Zealand birds are the moas, Dinornithiformes (see fig. 34.2), classified into twenty species by Archey (1941); twenty-seven by Oliver (1949); and thirteen by Cracraft (1976, see also Fleming 1977). In addition there were at least fifteen other extinct endemic species, namely *Aptornis otidiformis* or Giant "Rail"; *Gallinula hodgeni*, New Zealand Gallinule (Olson 1975); *Capellirallus karamu*, Cave Rail; *Gallirallus minor*, Little Weka, uncertain species (Olson 1975); *Fulica chathamensis*, New Zealand Coot (Olson 1975, Millener 1981);

Figure 34.2. The moa, *Dinornis,* as it is usually depicted, standing erect like an ostrich. Emu feathers were used in the reconstruction.

Cygnus sumnerensis, New Zealand Swan; *Cnemiornis calcitrans*, South Island Extinct Goose; *Cnemiornis gracilis*, North Island Extinct Goose; *Euryanas finschi*, Finsch's Duck; *Palaeocorax moriorium*, Extinct New Zealand Crow or Raven; *Harpagornis moorei*, Extinct Eagle; *Circus teauteensis* and *Circus eylesi*, extinct North Island and extinct South Island Harriers (probably goshawks, *Accipiter* spp. [R. J. Scarlett pers. comm.]); *Megaegotheles novaezealandiae*, New Zealand Owlet Nightjar; and *Malacorhynchus scarletti*, New Zealand Pink-eared Duck (Olson 1977b). Among possible additions to this list are "*Gallirallus insignis*" (Olson 1977a), "*Gallirallus harteei*" (Scarlett 1970a, 1976; but see Olson 1975), and a large shag (Forbes 1891).

The extinct birds of Australian affinity include *Biziura lobata*, Musk Duck, (Olson 1977b, Dawson 1958); *Pelecanus conspicillatus novaezealandiae*, New Zealand Pelican (Scarlett 1957); *Fulica atra*, Australian Coot (Scarlett 1979:84, Teal 1975); *Dupetor flavicollis*, Australian Black Bittern (Horn 1980); and *Oxyura australis*, Australian Blue-billed Duck (Millener ms.). Scarlett's (1967) identification of a Barn Owl has not been reconfirmed (Millener pers. comm.). It is difficult to tell whether these Australian species are simply short-lived populations that became extinct in ecological time or even vagrants picked up by chance in the paleontological record, or whether they are established local species that subsequently became extinct (e.g. Williams 1962:17–18, 1964:142). *Pelecanus* and *Biziura* at least seem to have formed local subspecies.

Several other birds disappeared from part of their previous ranges by the time of European recording. *Notornis mantelli*, the Takahe (Williams 1960a), became extinct in the North Island and survived only in a remote part of the South Island. *Apteryx oweni*, the Little Spotted Kiwi (Reid and Williams 1975); *Strigops habroptilus*, the Kakapo (Williams 1956); and perhaps also *Podiceps cristatus*, the Southern Crested Grebe (Bull and Whitaker 1975:264) were seriously reduced in the North Island, finally disappearing there in historic times. They survive in the South Island. *Coenocorypha aucklandica*, the Subantarctic Snipe, became extinct in both the North and South islands, but survived on some offshore islands (Bull and Whitaker 1975:265). *Mergus australis*, the Auckland Island Merganser, used to occur in the North, South, Stewart, and Chatham islands (Kear and Scarlett 1970, Davidson 1978b:334; Marshall et al. n.d.), but was found alive by Europeans only on the Auckland Islands, the last record being in 1905.

Unquestionably some seabirds also once had wider distributions. Bones of petrels (Procellariiformes) are commonly found in inland caves and other deposits of both main islands (Falla 1957:19, Scarlett 1979:90, Williams n.d., Millener ms.). Today only a few of these birds breed inland, and the main breeding grounds are now on offshore islands. Penguin breeding patterns may have altered. Hamel (1977:57) suggested that, in southern South Island, "with the arrival of human predators, there seems to have been a change from penguin species which nest in exposed colonies to those which nest cryptically in pairs and small scattered groups," but direct evidence of colonial-nesting penguins breeding on the New Zealand mainland is still lacking. Similarly it is unproven but possible that albatrosses used to breed on the mainland until the arrival of man (see Leach and Hamel 1978:245).

In the South Island a number of species such as kiwis, kakapo, and takahe disappeared from the east, to survive only in the forested west. For further discussion of regional avian extinctions within each island see Williams 1956, 1976; Reid and Williams 1975; Hamel 1977; Fleming 1979:99; and Millener ms.

Some reptile and amphibian distributions also seem to have changed since the *Rallus philippensis*, and *Ninox novaeseelandiae* which appear occasionally in archaeological sites (such as Ototara, Mangakaware, Washpool Midden, Kaupokonui, Paremata, Hotwater Beach, Ohawe; see Table 34.2) lack older fossil records (Millener ms.). Possibly they first colonized New Zealand during the prehistoric period, or became more numerous then as a result of the ecological changes associated with man.

Not only birds suffered range reductions or extinctions during prehistoric settlement. Other species were affected also, including sea mammals, reptiles (tuataras and lizards), amphibians (frogs), insects, and to a lesser extent, fish, mollusca, and crustacea. The identification of pups of the New Zealand Fur Seal *(Arctocephalus forsteri)* at two northern North Island "Archaic" sites (Mt. Camel [Shawcross 1972:608] and Tairua [Smith 1978]) shows that this species formerly bred considerably north of its present breeding areas (see Gibb and Flux 1973). Evidently this reduction of breeding range occurred during the prehistoric period. Human predation is the most likely explanation, since known climatic changes during the prehistoric period would, in fact, have favored more northern distribution (see Smith 1978). The southern elephant seal's *(Miroungia leonina)* occurrence in a number of northern archaeological sites (e.g. Tairua, Kaupokonui, Washpool Midden) suggests that it too may have been more common in northern waters in prehistory than at present, although these locations are not outside its present nonbreeding range. Although breeding evidence is lacking, the occurrence of the southern sea lion *(Phocarctus hookeri)* at a number of North Island sites (e.g. Kaupopkonui, Washpool Midden) also indicates a past distribution that extended farther

Table 34.2. References for New Zealand Archaeological Sites Cited in Text

Site	References
Avoca Point, S49/46	Trotter 1980
Black Rocks, Wairarapa, N168/77	Anderson 1979
Hot Water Beach, Coromandel, N44/69	Leahy 1974
Kaupokonui, Taranaki, N128/3	Buist 1963, Cassels n.d., Foley 1980
King's Rock, Otago, S184/6	Hamel 1977: table 4:12
Mangakaware, Waikato, N63/35	Bellwood 1978b: 32–33
Mount Camel, Northland, N6/4	Pike 1973, Shawcross 1972
Ohawe, Taranaki, N128/77 (Waingongoro)	Buist 1962, Buist and Yaldwyn 1960
Ototara, Otago, S136/2	Trotter 1965
Paremata, Wellington, N160/50	Davidson 1978a
Pleasant River, Otago, S155/2	Teal 1975
Port Jackson, Coromandel, N35–6/88	Davidson 1979a
Rakaia, Canterbury, S93/20	Trotter 1972
Redcliffs, Canterbury, S84/76 & 118	Trotter 1975
Rotokura, Nelson, S14/1	Butts 1977
Sunde, Motutapu, N38/24	Scott 1970
Tai Rua, Otago, S136/1	Trotter 1965
Tairua, Coromandel, N44/2	Rowland 1975
Tiwai Point, Southland, S181–2/16	Hamel 1969, Sutton and Marshall 1980
Waimataitai, Otago, S146/2	Trotter 1965
Wakapatu, Southland, S176/4	Higham 1968
Washpool, Wairarapa, N168/22	Leach 1979, Smith 1979
Whakamoenga, Taupo, N94/7	Leahy 1976

north than the present one (see Smith 1979:217). In particular, the frequency of this species at Kaupokonui in Taranaki (Foley 1980) indicates a major difference from the modern situation.

Various archaeologists have studied the effects of human predation on marine shellfish, fish, and crayfish (e.g. Shawcross 1972, Swaddling 1977, Rowland 1976, Leach and Anderson 1979, Anderson 1979). While no cases of extinction have been discovered, individuals in the largest size classes disappear soon after prehistoric fishing begins.

Some reptile and amphibian distributions also seem to have changed since the arrival of man, presumably as a result of the introduction of rats (Whitaker 1978, Bull and Whitaker 1975). The tuatara, *Sphenodon punctatus,* an archaic lizardlike reptile up to two feet long, has been found in early archaeological sites on the mainland; it now survives only on certain offshore islands (fig. 34.3). *Sphenodon* is reduced or extinct on islands where the Polynesian rat is present, and common where rats are absent. Other lizards are today more common on New Zealand offshore islands that lack the Polynesian rat, and some species occur only on such islands. A large subfossil gecko occurred on the mainland; a similar-sized species occurs today only on offshore islands (fig. 34.3). An extinct large frog also occurred on the mainland; the largest surviving frog today occurs only on two rat-free islands. The same pattern is observable among insects, where several large, flightless, and ground-dwelling species now have disjunct or insular distributions (Ramsay 1978); rats are probably responsible for this.

Concerning moa and other avian extinctions, recent explanations in the published literature have ranged from the view that they died out naturally, largely unaffected by the arrival of man (Duff 1956:280, Williams 1962), to Fleming (1962a) who laid the blame fairly at man's door. There are theories of a natural decline aided by a final coup de grace from prehistoric man (Scarlett 1974:12). Cautious reviews hesitate to exclude any of a wide range of possible factors: natural environmental change, particularly in forest composition, burning of the forest by Polynesians, hunting with or

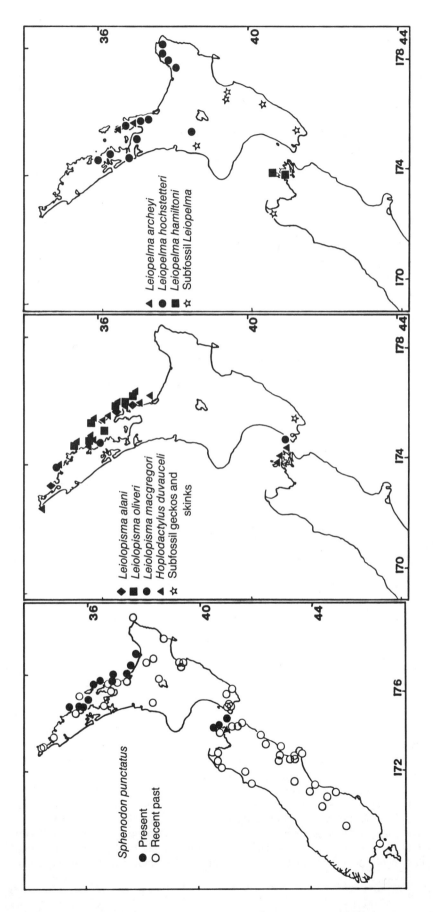

Figure 34.3. The present and past distribution of small terrestrial vertebrates. From left, the tuatara (*Sphenodon*), skinks (*Leiolopisma*) and geckos (*Hoplodactylus*), and frogs (*Leiopelma*). Range contraction is attributed to impact of introduced *Rattus* by prehistoric people.

without dogs, and predation by rats (e.g. Davidson 1979b:230–31). In addition to these, Green (1975:21) suggested an inherited tendency to extinction. More recently Fleming (1969) has expanded on the possible importance of rats, and Anderson (1981) has stressed the role of prehistoric dogs. The abundance of explanations leads one to wonder that anything managed to survive. Except for Fleming's dismissal of "natural" factors, few possibilities have been firmly ruled out.

If we now turn to the evidence presently available, certain facts are well established. It is now clear that virtually all known pre-European Quaternary avian extinctions of New Zealand occurred after the arrival of prehistoric man. An association with man has been documented for twenty-one out of twenty-four possible moa species (Fleming 1974:68). Apart from the moas, most of the other extinct New Zealand birds previously listed are known in archaeological associations (Foley 1980, Kinsky 1970, Scarlett 1979, Davidson 1978a, Scarlett 1970b, Davidson 1979a, Teal 1975, Horn 1980, Scarlett 1969). There are three exceptions: *Cnemiornis gracilis,* North Island Goose, *Malacorhynchus,* Pink-eared Duck, and *Megaegotheles,* the Owlet Nightjar. These can be dismissed as imperfections of the fossil record, although *Megaegotheles* has been cited as an extinction not related to man (Scarlett 1968, Rich and Scarlett 1977).

The virtual absence of a pre-Holocene avian fossil record (Millener and Cassels in press) does not alter the evidence that many large, vulnerable birds were present when man arrived. The lack of Pleistocene avifaunas means that it is difficult to put the prehistoric extinctions into perspective. It appears that the prehistoric extinctions were unprecedented in scope and scale, but it would be gratifying to have something to compare them with.

Claims that species like the moas actually were dying out at the time of human settlement (e.g. Scarlett 1974) are not based on quantifiable evidence. By world standards the number of archaeological sites containing extinct avifauna, particularly moas (e.g. Trotter 1977, A. Anderson this volume), is substantial, with no serious reason to doubt the viability of these species at the time of first human colonization. The first human settlement of New Zealand, the last large habitable land mass in the world to be discovered and colonized, took place about 1000 A.D. The settlers were east Polynesians with a Neolithic technology, who hunted birds, burned the forest, and introduced the domesticated dog (but no other domesticated animals), the small Polynesian rat, *Rattus exulans,* and root-crop horticulture based on the sweet potato (Green 1975). In all these activities, one can see possible causes of bird extinction.

Turning to the question of hunting, little doubt exists that direct human predation, particularly of moas, was widespread and intensive. Many hunting sites are known that occasionally attained enormous size (Anderson 1982; Trotter 1972:131, 1977). Moa bones, especially of some middle-sized species (Scarlett 1974), are frequent in archaeological sites (see figs. 34.4, 34.5). Moa-hunting sites are most common in regions like the far North, Coromandel, the Taranaki-Wellington coast, and the east coast of the South Island. In other areas such as Northland, Auckland, Bay of Plenty, Waikato, Hawke's Bay, and the South Island west coast, the sites are rare or small. There are many possible explanations for this, one being that moa hunting was a regional phenomenon.

Radiocarbon dates make it clear that moas were still being hunted in large numbers several centuries after the first settlement of the country (Trotter 1977; McCulloch and Trotter 1975, this volume; Anderson this volume). It is certain that the moas were not wiped out in the first wave of human colonization, as Martin (1973) has postulated for the late Pleistocene extinct mammals in America. From earliest times the prehistoric Maoris based their subsistence on horticulture, fishing, and gathering (Davidson 1979b); they were not dependent on hunting moas for their livelihood.

The date of the final extinction of the last moa is unknown. Several Europeans claim to have sighted moas in the nineteenth century (Falla 1964), and it is possible that one species did survive until then. However, archaeological evidence suggests that

Figure 34.4. Moa skull and articulated neck vertebrae associated with a tibia from the Kaupokanui moa-hunting site (N128/3), North Island, New Zealand.

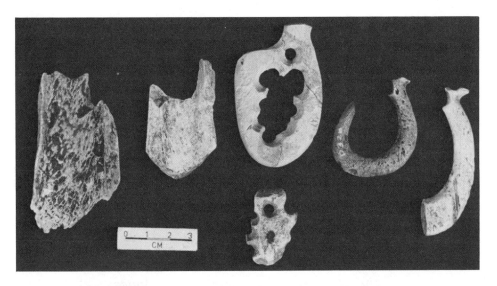

Figure 34.5. Moa bone used to manufacture fishhooks by drilling and filing.

moas had become rare, if not extinct, several centuries before this (Davidson 1979a; Hamel 1978, 1979). Their absence from late-period archaeological sites cannot prove that the moas were extinct; nevertheless, it is at least strongly suggestive.

Apart from moas, the most common land birds in archaeological sites (see Table 34.2) are species that survive to the present day. These include spotted shags, *Stictocarbo punctatus* (e.g. at Redcliffs, Rotokura, Ototara, Tai Rua, Waimataitai, Rakaia, Avoca Point); kakas, *Nestor meridionalis* (at Paremata, Port Jackson, Sunde); parakeets, *Cyanoramphus* spp. (at Black Rocks, Washpool Midden, Tiwai Point, Wakapatu); tuis, *Prosthemadera novaeseelandiae* (Wahakamoenga, Washpool Midden); and wekas, *Gallirallus australis* (Kaupokonui).

The date of the extinction of the smaller avian species is less well established than that of moas. The sites of Paremata, King's Rock, Rotokura, Hot Water Beach, and Kaupokonui (see Table 34.2) have extinct species in a context that may postdate the archaeological disappearance of moas. Some of the regional extinctions, such as the Kakapo and Little Spotted Kiwi in the North Island (Williams 1956:32, Turbott 1967:251, Falla et al. 1970:113), were not completed until after the arrival of Europeans. This suggests that the post-European extinctions should perhaps be seen as a continuation of the process that started in prehistory.

The period of moa extinction (1500 to 1800 A.D.) coincides with minor climatic oscillations and some associated changes in forest composition (Simmons 1969:22–25; Wardle 1963, 1973; Burrows and Greenland 1979). While these changes have been held accountable for the extinctions, the New Zealand fauna had survived climatic and environmental changes of incomparably greater magnitude during the Pleistocene (e.g. Fleming 1979:79). The minor fluctuations of the Holocene cannot seriously be invoked as a major cause of extinction.

Whether intentionally or by accident, the prehistoric Polynesians caused the deforestation of large areas of New Zealand by fire, which must have affected the native birds. Burning was extensive in both islands (see McLintock 1959:maps 14 and 15), although it has been more fully studied in the south than the north (Molloy et al. 1963, Cumberland 1962, McGlone 1978). Despite this burning, however, substantial areas of forest remained intact at the time of European settlement, notably on the west coast of the South Island and interior of the North Island (McLintock 1959:maps 14 and 15). The existence of these refugia from fire and human settlement must be taken into account in any theory of extinction. It is generally assumed (perhaps erroneously, see Cassels 1972:198) that these remaining forests are similar to those burned by Polynesians, and thus would have provided suitable habitats for species like the moa (Hamel 1977, Simmons 1968). If this is the case, then some agency other than fire has to be invoked to account for the extinctions. One imagines processes like the relentless pressure of human hunting parties seeking the prestige of capturing increasingly rare prey, or some form of internal population collapse resulting from the pressures applied to the peripheries of moa populations.

In conclusion, the known Quaternary avian extinctions of New Zealand are dated, with only a few possible exceptions, to the human period, that is, after 1000 A.D. On present evidence, by far the greatest number of extinctions took place before the arrival of Europeans. Undoubtedly, the arrival of prehistoric man, in one way or another, caused these extinctions. This view is widely held among the New Zealand public, and it is academic writers who have spent time and energy attempting to sustain alternative explanations that do not involve man.

Bird Extinctions on Pacific Archipelagos

Chatham Islands

Large assemblages of faunal remains have recently been recovered from archaeological sites in the Chatham Islands (Sutton and Marshall 1977, Sutton 1979, Marshall et al. n.d.). New records have been established, and for the first time the

Table 34.3. Extinct Chatham Island Birds: Possible Chracteristics

KEY: Columns 1–5 same as Table 34.1. Column 6A = same species survives on New Zealand mainland; 6B = same species occurred in N.Z. but extinct there in prehistory; 7 = larger than closest surviving relative (Chats); 8 = largest in its family (on Chathams); 9 = level of endemism (Chathams); * = no archaeological association with man.

Bird Species	Characteristics									
	1	2	3	4	5	6A	6B	7	8	9
Species Extinctions										
*Icthyophaga australis	−	−	?	−	−	−	−	+	+	species?
Pachyanas chathamica	+	−	+	−	−	−	−	+	−	genus
Diaphorapteryx hawkinsi	+	−	+	−	−	−	−	+	+	genus
Coenocorypha chathamica	−	−	+	−	−	−	−	+	+	species
Max. Poss. Total 4	2	0	4	0	0	0	0	4	3	
(Percent)	50		100					100	75	
Local Extinctions										
Fulica chathamensis	+	(+)	+	−	−	−	+	+	−	
Cygnus sumnerensis	−	−	+	?	−		+	+	+	
Palaeocorax moriorium	−	−	?	?	−	−	+	+	+	
Gallirallus australis	+	−	+	−	−	+	−	+	−	
Gallirallus minor	+	−	+	−	−	+	−	+	−	
Circus sp.	−	−	(+)	−	−	−	+	+	−	
Mergus sp.	−	−	(+)	−	−	−	+			
Nestor meridionalis	−	−	−	−	−	+	−	+	+	
Strigops habroptilus	+	−	+	−	+	+	−	+	+	
Falco novaeseelandiae	−	−	+	−	−	+	−	−	−	
Sceloglaux albifacies	−	−	+	−	(+)	+	−	+	+	
Anas gibberifrons	−	−	+	−	−	+	−	−	−	
Aythya novaeseelandiae	−	−	+	?	−	+	−			
Chlidonias hybrida	−	−	+	+	−	+	−	−	−	
Sterna nereis	−	−	+	−	−	+	−	−	−	
Procellariiformes (7 spp.?)	−	−	(+)	+	−	?	?			−
*Sula sp. (booby)	−	−	?	?	−	−	−			
Max. Poss. Total 27	6	1	26	6	2	9+	6+	13	8	
(Species & Local)										
(Percent)	22.2	3.7	96.3	22.2	7.4	33.3+	22.2+	48.1	29.6	
Regional Extinctions (within Chathams)										
Cabalus modestus	−	−	+	−	−	−	−		−	species?
Coenocorypha aucklandica	−	−	+	−	−	−	−		−	
Thinornis novaeseelandiae	−	−	−	−	−	+	−		−	
Petroica traversi	−	−	−	−	−	−	−		−	species

SOURCES: Oliver 1955, Kinsky 1970, Bourne 1967, Marshall et al. n.d., Harrison and Walker 1973, Millener ms., Falla et al. 1970, Olson 1977a.

assemblages are well dated. The Chathams were settled within the last 1,000 years, probably by people from New Zealand (Sutton 1980). For their size they lost an astounding number of species, probably at least nineteen and maybe as many as twenty-seven (Table 34.3). This occurred sometime in the Holocene period, before the advent of adequate European ornithological recording in the late nineteenth century (Forbes 1893). Fifteen of the nineteen species have been recovered from certain archaeological contexts, thus excluding the possibility of prehuman extinction.

Following Kinsky (1970), unless otherwise specified, the extinctions can be divided into four groups:

Endemic forms:
   1. Sea Eagle, *Haliaeetus* sp. (Olson pers. comm.) or *Icthyophaga australis* (Harrison and Walker 1973); no human association

2. *Pachyanas chathamica*, Chatham Island Goose Duck (human association recently reported by Marshall et al. n.d.)
3. *Diaphorapteryx hawkinsi*, Giant Chatham Island Rail
4. *Coenocorypha chathamica*, Extinct Chatham Island Snipe

Species that also became extinct on the New Zealand Mainland:
5. *Fulica chathamensis*, New Zealand Coot
6. *Cygnus sumnerensis*, New Zealand Swan
7. *Palaeocorax moriorium*, Extinct New Zealand Crow or Raven
8. *Gallirallus minor*, Little Weka (uncertain species, see Olson 1977a)
9. Extinct Hawk *(Circus eylesi?)* (Dawson 1960)
10. Merganser Duck (Marshall et al. n.d.)

Land birds never recorded alive on the Chathams since European contact, although living at that time on the New Zealand mainland:
11. *Gallirallus australis*, Weka, presumably introduced by prehistoric man from New Zealand; apparently extinct by the nineteenth century; reintroduced subsequently, now abundant (Oliver 1955:370)
12. *Nestor meridionalis*, Kaka, perhaps two species (Dawson 1952, 1959) or possibly a local subspecies (Scarlett pers. comm.). Dawson (1959) disputes Oliver's (1955:542) claim for a Kea, *Nestor notabilis,* in the Chathams.
13. *Strigops habroptilus*, Kakapo, very rare (Dawson 1959), presumably prehistoric human introduction
14. *Falco novaeseelandiae*, New Zealand Falcon, vagrant from New Zealand? (Dawson 1957)
15. *Sceloglaux albifacies*, Laughing Owl
16. *Anas gibberifrons*, Grey Teal (Marshall et al. n.d.)
17. *Aythya novaeseelandiae*, New Zealand Scaup (Marshall et al. n.d.)

Two seabirds not recorded by Europeans on the Chathams (Marshall et al. n.d.):
18. *Chlidonias hybrida albostriatus*, Black-fronted Tern
19. *Sterna nereis*, Fairy Tern

Four species found archaeologically on the main Chatham Island, but historically recorded since European contact only on smaller islands of the Chatham group:
20. *Cabalus modestus*, Chatham Island Rail (Oliver 1955:356)
21. *Coenocorypha aucklandica*, Subantarctic Snipe (Marshall et al. n.d.)
22. *Thinornis novaeseelandiae*, New Zealand Shore Plover (Marshall et al. n.d.)
23. *Petroica (Miro) traversi,* Black Robin (Marshall et al. n.d.)

In addition, breeding colonies of pelagic birds were reduced. Millener (ms.) cites Van Tets (pers. comm.) as recording a booby (*Sula* sp.) from the Chathams, a species not present today. Marshall et al. (n.d.) mention evidence for prehistoric breeding populations of a Crested Penguin, *Eudyptes pachyrhynchus,* and Mottled Petrel, *Petrodroma inexpectata,* neither of which breed in the Chathams at present. They also suggest that further extinct forms will be identified. Bourne (1967) claimed that up to twenty-one petrels may once have bred in the Chatham Islands, compared to the fourteen species recorded there in historic times. Most of his specimens probably came from prehistoric archaeological contexts; unfortunately this will never be known for certain. A total of twenty-one breeding species would make the Chathams, at the time of human settlement, "the most important breeding station for petrels in the world" (Bourne 1967:7).

Table 34.4 compares the avifauna extinctions of New Zealand and the Chathams. It can be seen that the Chathams have a higher proportion of smaller-sized birds, most of them volant. This suggests that the extinction process had gone further in the

**Table 34.4. Comparison of Characteristics
of Extinct Birds of New Zealand and Chathams**

(Figures are percentages)

		Characteristics						
		1*	2	3	4	5	7	8
Species	New Zealand	78.6	0	89.3	7.1	3.6	71.4	53.6
Extinctions	Chathams	50	0	100	0	0	100	75
	Difference	28.6	0	10.7	7.1	3.6	28.6	21.4
Species +
Subspecies
Extinctions (N.Z.)	New Zealand	73.3	0	90	10	3.3	70	50
+ Island
Extinctions
(Chathams)	Chathams	22.2	3.7	96.3	22.2	7.4	48.1	29.6
	Difference	51.1	3.7	6.3	12.2	4.1	21.9	20.4

*See Table 34.1 for key

Chathams, extending to species that survived the prehistoric period in New Zealand. These greater losses could also be a result of their smaller size and greater isolation, making the Chathams a more vulnerable ecosystem.

So far there is no definite association with man for four species, *Haliaeetus*, *Strigops*, *Sceloglaux*, and the extinct hawk. The near-flightless *Strigops* was almost certainly a human introduction from New Zealand. The other three may have become extinct before the arrival of the Morioris, or in the period between first settlement and the sixteenth century, to which the excavated sites belong.

Three of Sutton's sites—CHA, CHB, and CHC—may be as late as the eighteenth century A.D. If so, the persistence of *Diaphorapteryx*, a large flightless rail, is particularly remarkable, suggesting that this species survived at least several centuries of prehistoric human occupation. Four other Chatham species present but rare in the sixteenth century are absent from later sites: *Pachyanas*, *Cygnus*, *Palaeocorax*, and *Mergus*. Possibly this reflects late prehistoric extinction.

The abundance of the magenta petrel (*Pterodroma magentae*) in late prehistoric sites notably contrasts with its present near-extinction. This species is one of several whose decline seems likely to have occurred as a result of European contact.

The date of settlement of the Chathams becomes crucial to understanding these avian extinctions. The only evidence for the date of colonization is the stylistic affinities of Chatham Island material culture, which is definitely with the New Zealand Archaic Phase and not with the Classic Phase (Sutton 1980). This narrows the possible date of settlement to between 1000 and 1400 A.D. Thus, the major sixteenth-century faunal assemblage, from Waihora, represents conditions sometime between 100 and 500 years after initial human colonization.

Unlike most other Polynesians, the Moriori, the prehistoric inhabitants of the Chathams, did not practice agriculture (Sutton and Marshall 1980). Nevertheless, they did burn the forests, contributing to the disappearance of certain birds. The dog was apparently not introduced to the islands, and therefore cannot be a factor in the extinctions. This indirectly strengthens the case that it may not have been a factor on the New Zealand mainland. Rats were introduced prehistorically (Sutton 1979:131) and may have affected some of the smaller birds.

The two most likely explanations for the bird extinctions of the Chathams are, therefore, burning of the forest, direct human predation, or both. Regarding human predation, it is noteworthy that sea mammals did not become extinct, despite being a major prehistoric food source (Sutton 1979); neither did the albatrosses and mollymawks, whose colonies were regularly exploited (Sutton 1977, but see also Bourne 1967:2). This apparent paradox can be approached in several ways. It may mean that the seals were seen as economically important and therefore intentionally conserved, while the terrestrial birds were not. Or it may simply reflect the much greater vulnerability of terrestrial birds to man, in contrast to the seabirds and sea mammals whose stocks could be replenished from subantarctic colonies beyond the range of human influence.

The Hawaiian Islands

Discoveries over the last few years (Stearns 1973; Pacific Science Assn. 1972. 1974, 1976; Olson and Wetmore 1976; Olson and James 1982, this volume) have suddenly and dramatically demonstrated that the Hawaiian Islands, like New Zealand, used to have a much larger avifauna than they did when Europeans arrived. A significant proportion of these newly discovered extinct species were contemporaries of prehistoric man. Olson and James (1982, this volume) argue that direct human predation and habitat destruction—particularly of ecologically diverse arid lowland forest—by the prehistoric Polynesians were the main causes of the extinctions. They also consider that dogs, pigs, and rats may have played a part.

The bird extinctions are still not well dated; some may have occurred before the arrival of the Polynesians around 500 A.D. (Cordy 1974, Bellwood 1978a:324–25). However, as research progresses, more and more species are demonstrated to be contemporary with man (Olsen and James this volume). Whether the Hawaiian bird extinctions were completed before or after the arrival of Europeans is not yet known; the difficulties involved in proving pre-European extinctions are the same as in New Zealand. At present, however, pre-European extinction seems the more plausible hypothesis.

While only preliminary reports are available, the following extinctions have been identified. First there is a group of totally extinct species, among them flightless geese, including *Thambetochen chauliodus* from Molokai (Olson and Wetmore 1976) and *Geochen rhuax* from Hawaii (Wetmore 1943); flightless ibises, including *Apteribis glenos* (Olson and Wetmore 1976); flightless rails; a new genus of long-legged owl; a species of *Haliaeetus* sea eagle; a large meliphagid; a number of species of honeycreepers; at least one extinct species of crow; and a new species of *Pterodroma* petrel. Second is a group of species that may have living or recent representatives on other Hawaiian Islands, but not on the island where they have been found subfossil. These include a *Branta* goose, a crow, a hawk *(Buteo)*, some honeycreepers, and several petrels.

In the extinction of birds of prey, a crow, flightless rails, and flightless geese, one sees remarkable parallels between Hawaii, New Zealand, and the Chathams. The discovery of the extinct Hawaiian passeriformes is yet unparalleled in New Zealand. Perhaps man-induced ecological changes were more severe in the Hawaiian Islands; or it may be that new, small species remain to be discovered in New Zealand. The extinction of petrels may be paralleled in the Chathams but has not been documented on the New Zealand mainland.

Marquesas

Kirch (1973) documents the disappearance through time of seabirds from archaeological sites in the Marquesas. There is, however, no indication of extinction. Possible explanations include changes in human gathering strategies or bone-disposal techniques, elimination of easily accessible breeding grounds, or alteration of the bird's behavior.

Fiji

In the Lau Islands, near Fiji, Best (pers. comm.) has recovered the remains of extinct birds from a lower layer in a rockshelter that probably dates to the period of first human settlement, about 1000 B.C. The birds are provisionally identified as a parrot, a pigeon larger than modern species, and a megapode *(Megapodius?)*. An excavation on Naigani Island just off the east coast of Vitu Levu revealed another site of the Lapita tradition with remains of a megapode, a turkey-size mound-builder probably of a new species (Best pers. comm.). The dimensions of the bones strongly suggest it was flightless. Two individuals were identified, both probably juvenile; one bone had been made into an awl.

Megapodes have not been recorded during the post-European period anywhere between the New Hebrides to the west, and Niuafoo in the Tongan group to the east (Mayr 1945:57, Olson 1980). The new Fijian finds show that megapodes were once more continuously distributed. In 1926 Christian claimed that the native names for megapodes were practically identical from Malaysia to Samoa and Tonga. R. Clark (pers. comm.) has confirmed that similar names are used in Tonga, New Hebrides (Mele-Fila and pidgin), Luanguia, and North Celebes. The possibility that the megapode used to live in the Kermadecs (Cheeseman 1890:219) also supports the argument for a previously wider distribution. In Australia one of the few examples of avian extinction known to have occurred during the period of prehistoric settlement involves a megapode (Van Tets 1974).

New Caledonia

New Caledonia was settled by at least 1000 B.C. and possibly several thousand years earlier (White 1979). Occurring in one of the driest parts of the tropical Pacific, the New Caledonian vegetation must have been very vulnerable to burning. With 3,000 to 6,000 years of settlement leading up to extensive horticulture (Bellwood 1978a:147), prehistoric environmental modification must have been extensive.

A large and flightless extinct bird identified as a ratite *(Sylviornis neocaledoniae)* has recently been found on the Isle des Pins off the southern end of the main island (Dubois 1976, Poplin 1980). The age of this find is not established but is almost certainly late Pleistocene or Holocene. While an association with man was not demonstrated, Poplin (1980) commented that it was without doubt man who exterminated the New Caledonian ratite. Prehistoric human impact may also be held accountable for the extinction of a horned tortoise, *Meiolania mackayi,* known from guano deposits on Walpole Island near New Caledonia (Mittermeier 1972).

At present New Caledonia has only one really old bird form, namely the kagu, *Rhynochetos jubatus,* probably related to *Aptornis* of New Zealand (Olson 1977a) and perhaps ultimately to the sun-bitterns (Eurypygidae) of South America. There is an "amazingly small" number of endemic bird genera (Mayr 1940:209), in contrast to the rich endemic flora of New Caledonia (Mayr 1965). One plausible explanation for the lack of endemic birds is that most became extinct as a result of the impact of prehistoric man on the landscape and fauna, with the kagu being an exception.

Lord Howe, Norfolk, and the Kermadecs

The avifaunas of Lord Howe Island, Norfolk Island, and the Kermadec group offer an instructive natural experiment. The three island groups are comparable in latitude, area, and geology. Lord Howe is the smallest (Douglas 1969:455–57). Prehistoric Polynesian settlement occurred on Norfolk and the Kermadecs but is not known on Lord Howe (Specht 1978, Sykes 1975, Pope 1960). Hence, it is noteworthy that at the time of European contact only Lord Howe possessed flightless species—a woodhen,

Tricholimnas sylvestris, and a large rail, *Porphyrio albus,* like the New Zealand takahe (Recher 1974, Turner et al. 1968, Cheeseman 1890). A small woodhen is reported from subfossil remains on Norfolk (Van Tets et al. 1981).

A marine turtle, probably the marine green turtle, was present when Europeans landed on Lord Howe; this breeding population was soon exterminated (Recher and Clark 1974). Although evidence for the effect of prehistoric man on Pacific turtles is scanty, these reptiles must have been very vulnerable; there are hints of a decline in frequency in the Pacific archaeological record (Green 1979:37).

Potentially equally vulnerable, the horned tortoise, *Meiolania platyceps,* was not present on Lord Howe at the time of historic contact. It is known only from late Pleistocene deposits (Sutherland and Ritchie 1974); its extinction must be attributed to natural causes. Bones of seabirds recovered from Lord Howe indicate the former presence of several species not recorded during the period of European ornithological study. Natural or early post-European extinction seems likely in the case of the White-faced Storm Petrel, *Pelagodroma marina* (Van Tets and Fullagar 1974), and another petrel, *Cookilaria* sp. (Bourne 1974).

Other Pacific Islands

It is highly likely that in prehistoric time flightless rails occupied many Pacific islands, particularly the smaller ones. The flightless rails surviving until recently (see Ripley 1977) occur either on islands that were uninhabited or only temporarily settled in prehistory, such as Lord Howe, Henderson, Laysan, and Wake, or on the larger islands providing better cover such as New Caledonia, Guam, Samoa, Fiji, Solomons, Hawaii, and New Zealand (see Ripley 1977, Dupont 1976, Mayr 1945). Bulmer (1976) argued on linguistic grounds that the Barn Owl, *Tyto alba* (or some other owl), may have been more widely distributed in east Polynesia in early prehistoric times than now.

Following McDowall's (1969) argument that endemic forms, particularly forms at a high level of endemism, are most extinction prone, one could infer that islands like New Caledonia with very few endemics today may have suffered extinctions in the past. The whole Tongan group of islands with a long history of prehistoric settlement had, at European contact, only one endemic species and no endemic genera (Mayr 1940:200). Easter Island, high and volcanic, appears to be a more attractive place for birds than the small, waterless raised coral atoll of Henderson. Yet the latter has four species, two endemic (Williams 1960b), while Easter Island has no land birds. The key may lie in prehistoric settlement, lengthy on Easter Island and minor on Henderson (Sinoto n.d.). The Galapagos group, with at best limited prehistoric human occupation (Heyerdahl and Skjolsvold 1956), had twenty-eight endemic species (Harris 1974).

Mayr listed a number of birds that were surprisingly absent in Samoa (1940:208). He also remarked on the absence in Micronesia of parrots and honeyeaters of such widespread genera as *Lalage* and *Pachycephala,* and on the small number of pigeons. In view of the controversies concerning the composition of island faunas (e.g. Lack 1976), I feel it would be premature to attribute all such absences to the hand of man. On the other hand, the subfossil record of islands is yielding increasing evidence of extinct endemic land birds, and the opportunity for testing various extinction models by searching out fossil deposits appears much more promising than has been realized.

Patterns of Extinction and Survival

So far discussion has been confined to the traditional issues of the extinction debate—the dating evidence, the question of an association with man, and the possible causes of extinction. However, another, and complementary, approach can be followed along the lines of Walker's (1967) study of the Madagascar lemuroids. This is to recon-

Table 34.5. Patterns of Extinction and Survival

Extinct Species	Surviving Species
1. Large, often largest	———
2. Many flightless	A few flightless
3. Flightless species were diurnal	Flightless species mostly nocturnal
4. Flightless species mainly forest-ecotone	Flightless species mostly primary forest
5. Flying species are large/largest	———
6. Mainly ground-nesting	Ground- and tree-nesting
7. Dispersed and colonial-nesting	Few colonial nesting
8. Small clutches, large eggs	———
9. Highly endemic	Lesser degree of endemism

struct the characteristics of the species that became extinct prehistorically and to compare them with the species that survived (see Table 34.5). The contrast then sheds light on the type of selection that was taking place, and hence on the different possible causes of extinction that have been proposed.

The characteristics of the extinct species (see Tables 34.1 and 34.3) are drawn from the large body of literature cited in the table captions. Deductions are made primarily from the morphology of the species, the ecological and other characteristics of their nearest surviving relatives, and other archaeological, paleontological, and geographical data. The characteristics of the surviving fauna have been extracted from the following standard Pacific ornithological sources: Berger 1972, Cayley 1931, Delacour 1966, Dupont 1976, Falla et al. 1970, Frith 1967, Mayr 1945, Munro 1960, Oliver 1955, Sharland 1958, Slater 1970.

The first striking characteristic of the extinct species is that many of them come from five groups, notably the ratites, rails, waterfowl, birds of prey, and crows. Except within the crow family, Corvidae, any extant species in the same group are smaller than their extinct relatives. Evidently it is not a case of entire groups of birds being unsuited to an ecosystem with man in it, but rather that the larger representatives were more vulnerable than smaller ones.

Large size was a general characteristic of the extinct species; for example, the nineteen largest land birds of New Zealand (thirteen moas, six others) disappeared. However, large size is not inevitably linked with extinction. While the largest waterfowl disappeared, such as the swan of New Zealand and the Chathams, and the flightless geese of New Zealand, Hawaii, and the Chathams, the next largest species, the Paradise, Grey, and Blue Ducks of New Zealand survived. In turn the Finsch's Duck, which was smaller still, became extinct. While the largest rail in the Chathams (*Diaphorapteryx*) became extinct, the largest in New Zealand (*Notornis*) survived—if only marginally. All the extinct rails of New Zealand were smaller than the surviving weka (*Gallirallus*) and pukeko (*Porphyrio*), and some of the extinct Hawaiian rails were tiny. These constitute exceptions to the general rule that large size was disadvantageous.

Two other apparent exceptions to the generalization are provided by the cases of the extinct "small weka" (*Gallirallus minor*), survived by the larger modern *G. australis,* and the small extinct South Island Kaka (*Nestor meridionalis* subsp.) survived by the larger modern form. Both small extinct forms are dubious species, or even subspecies. In the case of the weka, we may simply be seeing a reversal in selection with the elimination of smaller individuals in the post-European populations by European-introduced predators such as cats, weasels, stoats, and the large European rats.

The reduction of size range by elimination of smaller individuals is in distinct contrast to the main tendency evident among prehistoric avifauna. The disappearance of the largest birds can be explained by human or dog predation or natural change, which might select against the largest, and hence rarest species (see Guilday 1967:123).

This range of options is somewhat reduced when one takes account of the frequency of flightlessness among the extinct species, notably among the ratites, rails, and geese (see Table 34.1). Clearly flightlessness made species more vulnerable. This is further supported by the number of cases where extinct or nearly extinct flightless species are survived by volant relatives such as most contemporary rails and waterfowl, and the parrots of New Zealand. Furthermore, flightless species frequently survive only on islands that suffered little or no prehistoric impact, for example the Auckland Island flightless duck (*Anas aucklandica*), the Galapagos flightless cormorant (*Nannopterum harrisi*), and the flightless rails of Lord Howe (*Porphyrio alba* and *Tricholimnas sylvestris*) and on Henderson, Laysan, and Wake (Ripley 1977). It is hard to imagine a reason why selection against flightlessness would result from a change in climate.

Some flightless species did survive prehistoric island occupation, and these exceptions to the general rule require closer examination. Examples are the kiwis of New Zealand, the kagu of New Caledonia, the woodhens of New Zealand (*Gallirallus*) and New Caledonia (*Tricholimnas*), and the flightless rails of Hawaii and Samoa. Unlike most of the extinct species, kiwis and kagus are mainly dwellers of primary forest. Here they are nocturnal insectivores, in contrast to the diurnal herbivores that make up most of the extinction lists (notably the moas, geese, and swans). Survival of kiwis and kagus could be interpreted several ways. It could be taken as a sign that predation by humans, dogs, or both, falls heavily in the diurnal, forest-edge species, sparing nocturnal birds of the deeper forests. Alternatively, it could be taken as support for the argument that burning of the forest edge was a primary cause of extinction, permitting only deep-forest dwellers to survive.

At least two extinctions are unlikely to be explained by destruction of habitat. These are the extinctions of the weka, *Gallirallus australis,* in the Chatham Islands, and of the swan in New Zealand. In both cases the same species has been successfully reintroduced in historic times (Dawson 1959, Bull and Whitaker 1975:258). Their prehistoric extinction may have resulted from overhunting.

Although some flightless birds survived prehistory, a number of flying birds did not: eagles, goshawks, crows, pelicans, ducks, and others. Extinction of these species seems most unlikely to be related to the role of dogs and rats, and fairly unlikely to result from forest burning. As the largest and hence rarest members of the ecosystem, or in the case of the birds of prey, the top of the ecological pyramid, these are the species that would be most vulnerable to any environmental change. They, their nests, eggs, feathers and bones, would also have been sought by human predators, and their absolute numbers would have been sufficiently small for overhunting to have occurred more easily.

Many of the extinct species nested on the ground. Obviously all the flightless species did, and the snipe, pelican, and swan may have, as do their surviving relatives. In the absence of ground-living predators, other New Zealand birds may also have originally nested on the ground. This would have made them especially vulnerable to predation or forest burning.

However, ground-nesting was by no means an automatic route to extinction once man had arrived; the survival of the New Zealand Quail (*Coturnix*) into the historic period is notable, as is the contemporary survival of kiwis and wekas in New Zealand, megapodes in Melanesia, etc. Many other ground nesters survive even today, such as plovers, dotterels, stilts and harriers, the grassland owls of Oceania, gulls, terns, and some ducks. Not all of these can defend themselves against rats. Clearly it is some factor other than ground-nesting alone that made some species more vulnerable than others. Pierce's (1979) interesting work on the relative breeding successes of Pied and Black Stilts (*Himantopus* spp.) in New Zealand demonstrates the importance of minor differences in choice of nesting locations, notably the distinction between streambank and wetland nesting, to successful escape from predators. Such a distinction may explain the success and failure of various ground nesters.

If predation by humans was important, colonial-nesting species may have been at a disadvantage compared to dispersed nesters. By analogy with surviving Australian forms, the extinct New Zealand swan, crow, and pelican may have been colonial nesters. It is notable that very few New Zealand birds today nest in colonies. One that does, the White Heron (*Egretta alba*), survives only in one colony in remote Westland (Falla, Sibson, and Turbott 1970); previously it had a much wider breeding range (Millener ms.).

A major exception is seen in the shags (Phalacrocoracidae), colonial nesters in trees and cliffs, that have survived despite heavy prehistoric and historic human exploitation. Similarly colonial-nesting seabirds also survived within the reach of prehistoric man—notably the petrels and gannets of New Zealand, and albatrosses, mollymawks, and petrels of the Chathams. This indicates either conservationist "culling" of colonies of some species, or that human predation was not severe enough to lead to extinction of all populations.

A characteristic of extinct birds compared to their surviving relatives may be that they laid smaller clutches of larger eggs. This has been argued for the moas as opposed to other surviving large ratites (Davies 1976:114), and it is true of the nearly extinct takahe in contrast to the flourishing pukeko (Oliver 1955:373, 377). Lack (1970:9) has demonstrated that small clutches of larger eggs are a general trend in island species of ducks, in contrast to mainland species. This is one of various examples of the "uncompetitiveness" of island species when they encounter introduced continental predators. Many of the extinct birds have very similar relatives, usually smaller and more mobile (see Fleming 1974), that survive on neighboring continents. The robust extinct island ratites of New Zealand and Madagascar are survived by their more gracile relatives in continental Africa, Australia, and South America (e.g. Cracraft 1974). The New Zealand eagle *Harpagornis* is survived by the comparable Wedge-tailed Eagle of Australia (Harrison and Walker 1973); the New Zealand "goshawk" by the goshawks of Australia and the Pacific; the New Zealand Swan, by the very similar Australian Black Swan (Bull and Whitaker 1975:258); the New Zealand Pelican by the Australian Pelican (Scarlett 1957); the New Zealand Owlet Nightjar by an Australian Nightjar (Scarlett 1968); Finsch's Duck by a comparable Australian species. *Chenonetta jubata* (Falla 1953); and the Chatham Sea Eagle by other sea eagles (Harrison and Walker 1973).

Most of the extinct species are highly endemic (see McDowall 1969, Fleming 1974:68). However, endemism did not inevitably lead to extinction. In New Zealand one order, the moas; one family, *Aptornis* (see Olson 1977a); eight to nine genera; three species; and at least one subspecies represent extinction of endemics. On the other hand one order, the kiwis; three families (Callaeatidae, Turnagridae, Xenicidae); sixteen genera; thirteen species; and seventeen subspecies represent survivals of endemic groups.

Despite these qualifications, it is clear that the prehistoric extinctions are characterized by the loss of highly endemic species. Such a pattern is most unlikely to result from a climatic change. The highly endemic species are inevitably the oldest inhabitants of New Zealand, and were unavoidably exposed to all Pleistocene climatic changes including those of the last glaciation. The rapid loss in the last 1,000 years of many highly endemic forms, particularly the moas, is strong evidence that this period saw the impact of totally new selective forces, quite different from those operating over the previous millions of years.

Five hypotheses were evaluated in light of the various patterns of extinction and survival (see Table 34.6):

1. That human predation was an important factor in eliminating some species but not others is not contradicted by any of the evidence. It is supported by the selection against large and vulnerable species in both wetland and forest, the loss of the largest flying species, and the survival of nocturnal species. The main difficulties with this hypothesis are raised by the survival of shags, seabirds, and flightless rails.

Table 34.6. Suggested Implications of Pattern of Extinction and Survival

Pattern*	Hypotheses				
	Hunting	Burning	Dogs	Rats	Natural Change
1	√		?	X	√
2	√	√	√		X
3	√				X
4	?	√	X		
5	√	X	X	X	√
6	√	√	√	√	X
7	√				X
8	√		?	X	
9	√	√	√	√	X

√ = compatible with hypothesis; X = not compatible with hypothesis
*Same as Table 34.5

2. The domestic dog, as a hunting aid, could well have played a part in eliminating the larger, more vulnerable New Zealand species. The survival of kiwis, which are notoriously vulnerable to dogs, suggests that if domesticated dogs of the prehistoric Polynesians in New Zealand became feral, they were not numerous and did not penetrate far into the bush. The survival of many ground-nesting birds also suggests the ineffectiveness of canine predation. Neither domestic nor feral dogs are likely to have had any major part in the extinction of the flying species. Dogs were absent in the Chathams and New Caledonia (Urban 1961, Cassels in press).

3. The Polynesian rat could have played a part in the extinction of a few of the smaller species, but it probably did not affect anything larger than chicken-sized birds (Imber 1978, Atkinson 1978). As with dogs, it seems very unlikely that they played any part in the extinction of large flying species like eagles and crows.* However, the fate of certain insects, reptiles, and amphibians seems to have been determined by rats.

4. The possibility that burning by Polynesians was a major factor in the extinction of moas and various land birds has not been ruled out. It is unlikely to account for the extinction of pelicans and ducks. The fire hypothesis is supported by the survival of deep-forest species like kiwis, by the survival and perhaps even spread of grassland species like owls and quail, and by the loss of flightless and ground-nesting species.

5. The likelihood that extinctions of endemic species were caused by a natural process such as climatic deterioration encounters the objection that these birds experienced and survived previous climatic changes of greater magnitude. The apparent shift in selection against characteristics such as flightlessness and colonial nesting is not obviously related to climatic change.

Table 34.6 summarizes the argument so far. The "extinction by hunting and burning" hypotheses are compatible with most of the present evidence. The "natural change" hypothesis is contradicted by some of the patterns that have been demonstrated. The dog and rat hypotheses also are contradicted by some of the evidence.

Summary

The pre-European faunal extinctions of the Pacific consist mainly of endemic terrestrial birds, along with some seabirds. Terrestrial mammals are virtually unknown

---

*The well-known case of rat attacks on nesting albatrosses (see Kepler 1967) did not lead to extinction, and occurred only in an extreme environmental situation.

on the islands concerned. Some sea mammals, reptiles, amphibians, and insects were affected. The greater number of extinctions are known from New Zealand, the Chatham Islands, and the Hawaiian Islands. Recent finds in New Caledonia and Fiji strongly suggest that similar discoveries will be made elsewhere in the Pacific as the search for fossils continues.

In New Zealand, the Chathams, and Fijian Islands the majority of the extinctions coincided with the settlement of the islands by prehistoric human populations, and there is new evidence that this was also the case in the Hawaiian Islands. Prehuman extinction has rarely been demonstrated for any of this late Quaternary fauna; more examples may be expected. Nevertheless most large and vulnerable species apparently survived until the arrival of man. It is conceivable that some extinctions thought of as prehistoric actually occurred in the first years of European contact, before extensive ornithological observations were made. However, it seems certain that pre-European extinctions were considerably more drastic than post-European ones.

Certain regularities are apparent in the type and characteristics of the extinct avifauna. Ratites, rails, large waterfowl, large birds of prey, and crows are commonly represented. The insular syndrome includes large size, lowered fecundity, flightlessness, and a high degree of endemism. These characteristics were a serious handicap when man arrived but did not invariably lead to extinction. What was critical for one species was not necessarily so for another, even a relative (see Halliday 1978). Despite this, enough regularities may be seen by contrasting the characteristics of extinct and surviving species to make some useful generalizations that shed light on the causes of the extinctions.

The hypothesis that hunting by humans was a major factor in the prehistoric extinctions fits well with the existing evidence. It would account for the loss of the moas, despite the persistence of large forest refugia as in the South Island of New Zealand. It would account for the selection against large and vulnerable species, both flightless and volant. It would account for the rarity today of colonial-nesting species, or their restriction to inaccessible locations. As a totally unprecedented type of selective force, it would account for the loss of the older, highly endemic species.

Certainly there are difficulties with an overkill model. These are raised in particular by the survival of kiwis, wekas, kagu, and other large flightless birds, the persistence of some accessible colonial-nesting shags and seabirds, and the perseverance of the sea mammal colonies of the South Island. Most of these were important prehistoric food sources, and some conservationist-type culling may be inferred.

In general it seems that, before human settlement, the islands of the Pacific commonly possessed large, vulnerable, and remarkable endemic birds and seabird colonies. These are becoming better known as the subfossil record is explored. While one cannot predict the accidents of colonization that determined whether an ibis, a goose, or a rail became flightless, large and vulnerable species in these groups once existed on various Pacific islands. Such birds had a low chance of surviving extensive human settlement. They may have formed a significant proportion of an island's fauna, and this should be taken into account in any modern study of island biogeography.

By comparison with Australia or North America, the fossil evidence for prehistoric contact with man, and the massive killing of extinct species is overwhelming. The time period was short, less than 1,000 years, but not instantaneous, involving at least several hundred years in New Zealand. Presumably, none of the human populations of the Pacific depended on the extinct species for their livelihood. Burning and hunting had a catastrophic effect. Competition with introduced animals was probably not a major factor. The relationship between prehistoric human settlement of the Pacific islands and evidence of unprecedented faunal extinction opens a new field—human ecology during colonization.

Acknowledgments

For their assistance I would like to thank Simon Best, Ralph Bulmer, Ross Clark, Roger Green, Paul Martin, Phil Millener, Ron Scarlett, Douglas Sutton, and Alan Ziegler.

References

Anderson, A. J. 1979. Prehistoric exploitation of marine resources at Black Rocks Point, Palliser Bay. pp. 49–65 in B. F. and H. M. Leach (eds.) Prehistoric Man in Palliser Bay. *Bulletin of the National Museum of New Zealand* 21, 272 pp.
———. 1981. Pre-European hunting dogs in the South Island, New Zealand. *New Zealand Journal of Archaeology* v. 3, pp. 15–20.
———. 1982. North and Central Otago. pp. 112–128 in Nigel Prickett (ed.) *The First Thousand Years. Regional Perspectives in New Zealand Archaeology.* N.Z. Archaeological Association Monograph 13. Palmerston North, Dunmore Press, 204 pp.
Archey, G. 1941. The moa: a study of the Dinornithiformes. *Bulletin of the Auckland Museum* v. 1, 119 pp.
Atkinson, I. A. E. 1978. Evidence for effects of rodents on the vertebrate wildlife of New Zealand islands. pp. 7–30 in P. R. Dingwall et al. (eds.) *The Ecology and Control of Rodents in New Zealand Nature Reserves.* Dept. of Lands and Survey, Wellington, Information Series no. 4, 237 pp.
Bellwood, Peter. 1978a. *Man's Conquest of the Pacific. The Prehistory of Southeast Asia and Oceania.* Auckland, Collins, 462 pp.
———. 1978b. Archaeological research at Lake Mangakaware, Waikato, 1968–1970. *New Zealand Archaeological Association Monograph* No. 9, 79 pp.
Berger, Andrew J. 1972. *Hawaiian Birdlife* Honolulu, University Press of Hawaii, 270 pp.
Bourne, W. R. P. 1967. Subfossil petrel bones from the Chatham Islands. *Ibis* 109:1:1–7.
———. 1974. Notes on some subfossil petrels from the New Zealand region, including a cranium of the subgenus *Cookilaria* from Lord Howe Island. *Emu* 74:4:257–58.
Buist, A. G. 1962. Archaeological evidence of the Archaic phase of occupation in South Taranaki. *New Zealand Archaeological Association Newsletter* 5:4:233–37.

———. 1963. Kaupokonui midden, South Taranaki: N128/3. Preliminary report. *New Zealand Archaeological Association Newsletter* 6:4:175–83.
Buist, A. G. and J. C. Yaldwyn. 1960. An articulated moa leg from an oven excavated at Waingongoro, South Taranaki. *Journal of the Polynesian Society* 69:2:76–87.
Bull, P. C. and A. H. Whitaker. 1975. The amphibians, reptiles, birds and mammals. pp. 231–76 in G. Kuschel (ed.) *Biogeography and Ecology in New Zealand.* The Hague, Dr. W. Junk, 689 pp.
Bulmer, Ralph. 1976. Review of Bruner, P. L. Birds of French Polynesia. *Journal of the Polynesian Society* 85:1:119–22.
Burrows, C. J. and D. E. Greenland. 1979. An analysis of the evidence for climatic change in New Zealand in the last thousand years: evidence from diverse natural phenomena and from instrumental records. *Journal of the Royal Society of New Zealand* v. 9, pp. 321–73.
Butts, David J. 1977. *Seasonality at Rotokura, Tasman Bay.* Unpublished B. A. (Hons.) thesis, University of Otago, 102 pp.
Cassels, Richard. 1972. Human ecology in the prehistoric Waikato. *Journal of the Polynesian Society* v. 81, no. 2, pp. 196–247.
———. n.d. Kaupokonui N128/3B: a moa butchery and cooking site. A preliminary report on 1974 rescue excavations. Mimeographed. Anthropology Dept., University of Auckland.
———. in press. Prehistoric man and animals in Australia and Oceania. To appear in World Animal Science Series, v. 1, *Domestication, Conservation and Use of Animal Resources.* L. J. Peel (ed.), Amsterdam, Elsevier.
Cayley, Neville W. 1931. *What Bird Is That? A Guide to the Birds of Australia.* Sydney, Angus and Robertson, 315 pp.
Cheeseman, T. F. 1890. On the birds of the Kermadec Islands. *Transactions and Proceedings of the New Zealand Institute* 23:216–26.

Christian, F. W. 1926. The megapode bird and the story it tells. *Journal of the Polynesian Society* 35:260.

Cordy, Ross H. 1974. Cultural adaptation and evolution in Hawaii: a suggested new sequence. *Journal of the Polynesian Society* 83:2:180–91.

Cracraft, Joel. 1973. Continental drift, palaeoclimatology and the evolution and biogeography of birds. *Journal of Zoology* (London) 169:455–545.

———. 1974. Phylogeny and evolution of the ratite birds. *Ibis* 116:494–521.

———. 1976. The species of moas (Aves: Dinornithidae). In Storrs L. Olson (ed.) Collected Papers in Avian Palaeontology. *Smithsonian Contributions to Paleobiology* 27, 211 pp.

Cumberland, K. B. 1962. Climatic change or cultural interference? pp. 88–142 in M. McCaskill (ed.) *Land and Livelihood.* Christchurch, New Zealand Geographical Society, 280 pp.

Davidson, Janet. 1978a. Archaeological salvage excavations at Paremata, Wellington, New Zealand. *National Museum of New Zealand Records* 1:13:203–36.

———. 1978b. The prehistory of Motutapu Island, New Zealand: five centuries of Polynesian occupation in a changing landscape. *Journal of the Polynesian Society* 87:4:327–37.

———. 1979a. Archaic middens of the Coromandel region: a review. pp. 183–202 in A. J. Anderson (ed.) Birds of a Feather: A Tribute to R. J. Scarlett. *British Archaeological Reports, International Series* v. 62, 295 pp.

———. 1979b. New Zealand. pp. 222–48 in Jesse D. Jennings (ed.) *The Prehistory of Polynesia.* Cambridge and London, Harvard University Press, 399 pp.

Davies, S. J. F. 1976. The natural history of the emu in comparison with that of other ratites. pp. 109–20. *Proceedings of the 16th International Ornithological Congress,* Canberra, 765 pp.

Dawson, Elliot W. 1952. Subfossil *Nestor* (Psittacidae) from New Zealand. *Emu* 52:259–72.

———. 1957. Falcon in the Chatham Islands. *Notornis* 7:113.

———. 1958. Rediscoveries of the New Zealand subfossil birds named by H. O. Forbes. *Ibis* 100:2:232–37.

———. 1959. The supposed occurrence of kakapo, kaka and kea in the Chatham Islands. *Notornis* 8:4:106–15.

———. 1960. New evidence of the former occurrence of the kakapo *(Strigops habroptilus)* in the Chatham Islands. *Notornis* 9:3:65–7.

Delacour, Jean. 1966. *Guide des oiseaux de la Nouvelle-Calédonie.* Neuchatel, Delachaux and Niestlé, 175 pp.

Douglas, Gina. 1969. Draft checklist of Pacific oceanic islands. *Micronesica* 5:2:327–463.

Dubois, M. J. 1976. Trouvailles à l'Ile des Pins, nouvelle-Calédonie. *Journal de la Société des Océanistes* 32:51–52:233–41.

Duff, Roger. 1956. *The Moa-hunter Period of Maori Culture.* Wellington, Government Printer, 400 pp.

Dupont, John E. 1976. South Pacific Birds. *Delaware Museum of Natural History Monograph Series* No. 3, 218 pp.

Falla, R. A. 1953. The Australian element in the avifauna of New Zealand. *Emu* 53:36–46.

———. 1957. Discussion. *Proceedings of the New Zealand Ecological Society* 4:19.

———. 1964. Moa. pp. 477–79 in A. L. Thomson (ed.) *A New Dictionary of Birds.* London, Nelson, 928 pp.

Falla, R. A., R. B. Sibson and E. G. Turbott. 1970. *A Field Guide to the Birds of New Zealand.* 2nd edn. London, Collins, 256 pp.

Fleming, C. A. 1962a. The extinction of moas and other animals during the Holocene period. *Notornis* 10:113–7.

———. 1962b. History of New Zealand land bird fauna. *Notornis* 9:270–74.

———. 1969. Rats and moa extinction. *Notornis* 16:210–11.

———. 1974. The coming of the birds. *New Zealand's Nature Heritage* v. 1, part 3:61–8.

———. 1977. Wishbones for Wetmore, review of Smithsonian Contributions to Paleobiology 27. *Notornis* 24:144–47.

———. 1979. *The Geological History of New Zealand and Its Life.* Auckland University Press, Auckland, 141 pp.

Foley, Diane. 1980. *Analysis of Faunal Remains from the Kaupokonui site (N128/3B).* Unpublished master's thesis, Anthropology, University of Auckland, 239 pp.

Forbes, H. O. 1891. Preliminary notice of additions to the extinct fauna of New Zealand. Abstract. *Transactions and Proceedings of the New Zealand Institute* 24:185–89.

———. 1893. A list of birds inhabiting the Chatham Islands. *Ibis* 5:521–46.

Frith, H. J. 1967. *Waterfowl in Australia.* Angus and Robertson, Sydney, 328 pp.

Gibb, J. A. and J. E. C. Flux. 1973. Mammals. pp. 334–71 in G. R. Williams (ed.) *The*

Natural History of New Zealand: an Ecological Survey. Wellington, A. H. and A. W. Reed, 434 pp.

Green, Roger C. 1975. Adaptation and change in Maori culture. pp. 591–641 in G. Kuschel (ed.) *Biogeography and Ecology in New Zealand,* The Hague, Dr. W. Junk.

——. 1979. Lapita. pp. 27–60 in J. D. Jennings (ed.) *The Prehistory of Polynesia,* Cambridge, Harvard University Press, 399 pp.

Greenway, James. 1958. *Extinct and Vanishing Birds of the World.* Special Publication 13. New York, American Committee for International Wild Life Protection, 518 pp.

Gressitt, J. L. (ed.) 1963. *Pacific Basin Biogeography. A Symposium.* 10th Pacific Science Congress. Honolulu, Bishop Museum Press, 563 pp.

Guilday, John E. 1967. Differential extinction during Late-Pleistocene and Recent times. pp. 121–40 in P. S. Martin and H. E. Wright, Jr. (eds.) *Pleistocene Extinctions: The Search for a Cause.* New Haven and London, Yale University Press, 453 pp.

Halliday, Tim. 1978. *Vanishing Birds. Their Natural History and Conservation.* London, Sedgwick and Jackson, 296 pp.

Hamel, G. E. 1969. Ecological method and theory: Tiwai Peninsula. *New Zealand Archaeological Association Newsletter* 12:3: 147–63.

——. 1977. *Prehistoric Man and His Environment in the Catlins, New Zealand.* Unpublished Doctoral thesis, Anthropology, University of Otago, 347 pp.

——. 1978. Radiocarbon dates from the Moa-hunter site of Papatowai, Otago, New Zealand. *New Zealand Archaeological Association Newsletter* 21:1:53–4.

——. 1979. Subsistence at Pounawea: an interim report. *New Zealand Archaeological Association Newsletter* 22:3:117–21.

Harris, Michael. 1974. *A Field Guide to the Birds of Galapagos.* London, Collins, 160 pp.

Harrison, C. J. O. and C. A. Walker. 1973. An undescribed fish-eagle from the Chatham Islands. *Ibis* 115(2):274–77.

Heyerdahl, Thor and Arne Skjolsvold. 1956. Archaeological evidence of pre-Spanish visits to the Galapagos Islands. Memoirs of the Society for American Archaeology No. 12. *American Antiquity* 22(2) part 3.

Higham, Charles. 1968. Prehistoric research in western Southland. *New Zealand Archaeological Association Newsletter* 11(4):155–64.

Horn, P. L. 1980. Probable occurrence of the Black Bittern *Dupetor flavicollis* (Linnaeus) in New Zealand. *Notornis* 27:401–03.

Imber, M. J. 1978. The effects of rats on breeding success of petrels. pp. 67–72 in P. R. Dingwall, I. A. E. Atkinson, and C. Hay (eds.) *The Ecology and Control of Rodents in New Zealand Nature Reserves.* Dept. of Lands and Survey, Wellington, Information Series No. 4, 237 pp.

Kear, J. and R. J. Scarlett. 1970. The Auckland Island Merganser. *Wildfowl* 21:78–86.

Kepler, Cameron B. 1967. Polynesian rat predation on nesting Laysan Albatrosses and other Pacific seabirds. *Auk* 84(3):426–30.

Kinsky, F. C. (Convenor). 1970. *Annotated Checklist of the Birds of New Zealand.* Wellington, A. H. & A. W. Reed for the Ornithological Society of New Zealand, 96 pp.

Kirch, P. V. 1973. Prehistoric subsistence patterns in the northern Marquesas Islands, French Polynesia. *Archaeology and Physical Anthropology in Oceania* 8(1):24–40.

Lack, David. 1970. The endemic ducks of remote islands. *Wildfowl* 21:5–10.

——. 1976. *Island Biology, Illustrated by the Land Birds of Jamaica.* Oxford, Blackwell, 445 pp.

Leach, B. F. 1979. Excavations in the Washpool Valley, Palliser Bay. pp. 67–136 in B. F. and H. M. Leach (eds.) Prehistoric Man in Palliser Bay. *Bulletin of the National Museum of New Zealand* 21, 272 pp.

Leach, B. F. and A. J. Anderson. 1979. Prehistoric exploitation of crayfish in New Zealand. pp. 141–64 in Atholl Anderson (ed.) Birds of a Feather. *British Archaeological Reports, International Series* 62, 295 pp.

Leach, H. M. and G. E. Hamel. 1978. The place of Taiaroa Head and other Classic Maori sites in the prehistory of East Otago. *Journal of the Royal Society of New Zealand* 8(3):239–51.

Leahy, Anne. 1974. Excavations at Hot Water Beach (N44/69), Coromandel Peninsula. *Records of the Auckland Institute and Museum* 11:23–76.

——. 1976. Whakamoenga Cave, Taupo, N94/7. *Records of the Auckland Institute and Museum* 13:29–75.

McCulloch, Beverley and Michael Trotter. 1975. The first twenty years: radiocarbon dates for South Island moahunter sites 1955–1974. *New Zealand Archaeological Association Newsletter* 18:1:2–17.

McDowall, R. M. 1969. Extinction and endemism in New Zealand land birds. *Tuatara* 17(1):1–12.

McGlone, M. S. 1978. Forest destruction by early Polynesians, Lake Poukawa, Hawkes Bay, New Zealand. *Journal of the Royal Society of New Zealand* 8:3:275–81.

McLintock, A. H. (ed.). 1959. *A Descriptive Atlas of New Zealand*. Wellington, Government Printer, 109 pp.

Marshall, Y. M., R. J. Scarlett and D. G. Sutton. ms. Bird species present on the southwest coast of Chatham Island in the sixteenth century A.D. Paper submitted for publication.

Martin, Paul S. 1973. The discovery of America. *Science* 179:969–74.

Mayr, Ernst. 1940. Origins and history of the bird fauna of Polynesia. *Proceedings of the 6th Pacific Science Congress,* v. IV:197–216.

———. 1945. *Birds of the Southwest Pacific.* New York, Macmillan, 316 pp.

———. 1965. Avifauna turnover on islands. *Science* 150:1587–88.

Millener, P. R. 1981. The subfossil distribution of extinct New Zealand Coots. *Notornis* 28:1–9.

———. ms. The Origin and Evolution of New Zealand's Avian Fauna. Unpublished manuscript.

Millener, P. R. and R. J. S. Cassels. in press. Moa. *A New Dictionary of Birds.* Revised edition, B. Campbell (ed.), British Ornithologists Union.

Mittermeier, Russell A. 1972. Zoogeography of fossil and living turtles *Australian Natural History* 17:8:265–69.

Molloy, B. P. J., C. J. Burrows, J. E. Cox, J. A. Johnston, P. Wardle. 1963. Distribution of subfossil forest remains, eastern South Island, New Zealand. *New Zealand Journal of Botany* 1:68–77.

Munro, George C. 1960. *Birds of Hawaii.* Rutland, Vermont, Charles E. Tuttle, 192 pp.

Oliver, W. R. B. 1949. The moas of New Zealand and Australia. *Dominion Museum Bulletin* 15. Wellington, 205 pp.

———. 1955. *New Zealand Birds.* Wellington, A. H. and A. W. Reed, 661 pp.

Olson, Storrs L. 1975. A review of the extinct rails of the New Zealand region. *Records of the National Museum of New Zealand* 113:63–79.

———. 1977a. A synopsis of the fossil Rallidae. pp. 339–74 in S. Dillon Ripley, *Rails of the World,* Boston, Godine, 406 pp.

———. 1977b. Notes on subfossil Anatidae from New Zealand, including a new species of pink-eared duck *Malacorhynchus. Emu* 77:3:132–35.

———. 1980. The significance of the distribution of the Megapodiidae. *Emu* 80:21–24.

Olson, S. L. and H. F. James. 1982. Fossil birds from the Hawaiian Islands: evidence for wholesale extinction by man before western contact. *Science* 217:633–35.

Olson, Storrs L. and Alexander Wetmore. 1976. Preliminary diagnosis of two extraordinary new genera from Pleistocene deposits in the Hawaiian Islands. *Proceedings of the Biological Society of Washington* 89(18):247–58.

Owen, Richard. 1879. *Memoirs on the Extinct Wingless Birds of New Zealand.* London, J. Van Voorst, 465 pp. + vol. 2 plates.

Pacific Science Association. 1972. *Information Bulletin* 24(2):9. Extinct Goose, Hawaiian Islands.

———. 1974. *Information Bulletin* 26(4–6):10. Remains of flightless Ibis found in Hawaiian Islands.

———. 1976. *Information Bulletin* 28(6):7. Further fossil information on the evolution of Hawaiian birds.

Pierce, R. 1979. Differences in susceptibility to predation between Pied and Black Stilts *(Himantopus* spp.*).* p. 334, v. 1, Abstracts, 49th Congress, Australian and New Zealand Association for the Advancement of Science. Auckland, The University, 583 pp.

Pike, Brigid. 1973. *A Prehistoric Ornithology.* Unpublished Master's Research Essay. Anthropology, University of Aukland, 86 pp.

Pope, Elizabeth C. 1960. The natural history of Lord Howe Island. *Australian Museum Magazine* 13:7:207–10.

Poplin, François. 1980. *Sylviornis neocaledoniae* n.g., n. sp. (Aves). Ratite éteint de la Nouvelle-Calédonie. *Compte Rendus Academie des Sciences,* Paris, 290:Serie D; 691–94.

Ramsay, G. W. 1978. A review of the effects of rodents on the New Zealand invertebrate fauna. pp. 89–98 in P. R. Dingwall, I. A. E. Atkinson and C. Hay (eds.) *The Ecology and Control of Rodents in New Zealand Nature Reserves.* Dept. of Lands and Survey, Wellington, Information Series No. 4, 237 pp.

Recher, H. F. 1974. Colonization and extinction: the birds of Lord Howe. *Australian Natural History* 18:2:64–9.

Recher, H. F. and S. S. Clark. 1974. A biological survey of Lord Howe Island with recommendations for the conservation of the island's wildlife. *Biological Conservation* 6:4:263–73.

Reid, B. and G. R. Williams. 1975. The Kiwi. pp. 301–30 in G. Kuschel (ed.) *Biogeography and Ecology in New Zealand*. The Hague, Dr. W. Junk.

Rich, P. V. and R. J. Scarlett. 1977. Another look at Megaegotheles, a large owlet nightjar from New Zealand. *Emu* 77:1–8.

Ripley, S. Dillon. 1977. *Rails of the World*. Boston, Godine, 406 pp.

Rothschild, Walter. 1907. *Extinct Birds*. London, Hutchinson, 244 pp.

Rowland, Michael J. 1975. *Tairua and Offshore Islands in Early New Zealand Prehistory*. Unpublished master's thesis, Anthropology, University of Auckland, 274 pp.

Rowland, Michael J. 1976. *Cellana denticulata* in middens on the Coromandel coast, N.Z.—possibilities for a temporal horizon. *Journal of the Royal Society of New Zealand* 6:1:1–15.

Scarlett, R. J. 1957. Sub-fossil bones of the Australian Pelican from the South Island. *Notornis* VII:114.

———. 1967. A sub-fossil record of a Barn Owl in New Zealand. *Notornis* 14:218–20.

———. 1968. An Owlet-Nightjar from New Zealand. *Notornis* 15(4):254–66.

———. 1969. The occurrence of the Musk Duck, *Biziura lobata* (Shaw), in New Zealand. *Notornis* 16(1):57–9.

———. 1970a. A small woodhen from New Zealand. *Notornis* 17:68–74.

———. 1970b. The Genus *Capellirallus*. *Notornis* 17:303–19.

———. 1974. Moa and man in New Zealand. *Notornis* 21(1):1–12.

———. 1976. Extinct rails. *Notornis* 23 (1):78.

———. 1979. Avifauna and man. pp. 75–90 in Atholl Anderson (ed.) Birds of a Feather. *British Archaeological Reports, International Series* 62, 295 pp.

Scott, Stuart D. 1970. Excavations at the Sunde Site, N38/24, Motutapu Island, New Zealand. *Records of the Auckland Institute and Museum* 7:13–30.

Sharland, Michael. 1958. *Tasmanian Birds*. Sydney, Angus and Robertson, 175 pp.

Shawcross, Wilfred. 1972. Energy and ecology: thermodynamic models in archaeology. pp. 577–622 in D. L. Clarke (ed.) *Models in Archaeology*. London, Methuen, 1055 pp.

Simmons, D. R. 1968. Man, moa and forest. *Transactions of the Royal Society of New Zealand, General* 2:7:115–27.

———. 1969. Economic change in New Zealand prehistory. *Journal of the Polynesian Society* 78:3–34.

Sinoto, Y. H. n.d. Polynesian migrations based on the archaeological assessments. Paper read at IX Congress, Union International des Sciences Prehistoriques et Protohistoriques, Nice, 1976.

Slater, Peter. 1970. *A Field Guide to Australian Birds: Non-Passerines*. Adelaide, Rigby, 428 pp.

Smith, Ian W. G. 1978. Seasonal sea mammal exploitation and butchering patterns in an Archaic site (Tairua N44/2) on the Coromandel Peninsula. *Records of the Auckland Institute and Museum* 15:17–26.

———. 1979. Prehistoric sea mammal hunting in Palliser Bay. pp. 215–24 in Leach, B. F. and H. M. (eds.) Prehistoric Man in Palliser Bay. *National Museum of New Zealand Bulletin* 21, 272 pp.

Specht, Jim. 1978. The early mystery of Norfolk Island. *Australian Natural History* 19:7:218–23.

Stearns, Harold T. 1973. Geologic setting of the fossil goose bones found on Molokai Island, Hawaii. *Occasional Papers, Bernice P. Bishop Museum*, Honolulu, v. 24, no. 10:155–63.

Sutherland, Lin and Alex Ritchie. 1974. Defunct volcanoes and extinct horned turtles. *Australian Natural History* 18:2:44–9.

Sutton, Douglas G. 1977. The archaeology of the Little Sister, Chatham Islands. *Working Papers in Chatham Islands Archaeology* 10. Dunedin, Anthropology Dept. University of Otago.

———. 1979. *Polynesian Coastal Hunters in the Subantarctic Zone*. Unpublished Doctoral thesis, Anthropology, University of Otago. 411 pp.

———. 1980. A culture history of the Chatham Islands. *Journal of the Polynesian Society* v. 89, pp. 67–93.

Sutton, D. G. and Y. M. Marshall. 1977. Archaeological bird bone assemblages from Chatham Island: an interpretation. *Working Papers in Chatham Islands Archaeology* 12, Dunedin, Anthropology Department, University of Otago.

———. 1980. Coastal hunting in the subantarctic zone. *New Zealand Journal of Archaeology* 2:25–50.

Swaddling, Pamela. 1977. The implications of shellfish exploitation for New Zealand prehistory. *Mankind* 11:11–18.

Sykes, W. R. 1975. The Kermadecs. *New Zealand's Nature Heritage* 4:59:1641–47.

Teal, F. Jane. 1975. *Pleasant River Excavations 1959–62*. Unpublished Research Essay, Post-graduate Diploma, Anthropology, University of Otago, Dunedin, 53 pp.

Trotter, Michael M. 1965. Avian remains from North Otago archaeological sites. *Notornis* 12(3):176–78.

———. 1972. A Moa-hunter site near the mouth of the Rakaia River, South Island. *Records of the Canterbury Museum* 9(2): 129–50.

———. 1975. Archaeological investigations at Redcliffs, Canterbury, New Zealand. *Records of the Canterbury Museum* 9(3):189–220.

———. 1977. Moa-hunter research since 1956. pp. 348–375 in Duff, Roger, *The Moa-hunter Period of Maori Culture*, 3rd ed. Wellington, Government Printer, 433 pp.

———. 1980. Archaeological investigations at Avoca Point, Kaikoura. *Records of the Canterbury Museum* 9:4:277–88.

Turbott, E. G. (ed.) 1967. *Buller's Birds of New Zealand*, Christchurch, Whitcombe and Tombs, 261 pp.

Turner, John S., C. N. Smithers and R. D. Hoogland. 1968. *The Conservation of Norfolk Island*. Special Publication 1, Australian Conservation Inc., 41 pp.

Urban, Manfred. 1961. *Die Haustiere der Polynesier*. Band 2, Volkerkundliche Beitrage zur Ozeanistick. Gottingen, Hantzschel, 367 pp.

Van Tets, G. F. 1974. A revision of the fossil Megapodiidae (Aves), including a description of a new species of *Progura* De Vis. *Transactions of the Royal Society of South Australia* 98(4):213–24.

Van Tets, G. F. and P. J. Fullagar. 1974. Of sketches, skins and skeletons. *Australian Natural History* 18:2:72–3.

Van Tets, G. F., P. V. Rich, P. J. Fullagar, and P. M. Davidson. 1981. Fossil, subfossil and early historic birds of Lord Howe and Norfolk Islands. pp. 29–31 in Lord Howe Island, *Occasional Reports of the Australian Museum* No. 1, Sydney.

Walker, Alan. 1967. Patterns of extinction among sub-fossil Madagascan lemuroids. pp. 425–32 in P. S. Martin and H. E.

Wright, Jr. (eds.) *Pleistocene Extinctions: The Search for a Cause.* New Haven and London, Yale University Press, 453 pp.

Wardle, P. 1963. The regeneration gap of New Zealand gymnosperms. *New Zealand Journal of Botany* 1:301–15.

———. 1973. Variations of the glaciers of Westland National Park and the Hooker Range, New Zealand. *New Zealand Journal of Botany* 11:349–88.

Wetmore, A. 1943. An extinct goose from the island of Hawaii. *Condor* 45:146–48.

Whitaker, A. H. 1978. The effects of rodents on reptiles and amphibians. pp. 75–88 in P. R. Dingwall, I. A. E. Atkinson and C. Hay (eds.) *The Ecology and Control of Rodents in New Zealand Nature Reserves.* Dept. of Lands and Survey, Wellington, Information Series No. 4, 237 pp.

White, J. Peter. 1979. Melanesia. pp. 352–77 in J. D. Jennings (ed.) *The Prehistory of Polynesia.* Cambridge, Harvard University Press, 399 pp.

Williams, G. R. 1956. The kakapo *(Strigops habroptilus* Gray*).* A review and reappraisal of a near-extinct species. *Notornis* 12(2): 29–56.

———. 1960a. The takahe *(Notornis mantelli* Owen 1848): A general survey. *Transactions of the Royal Society of New Zealand* 88(2):235–58.

———. 1960b. The birds of Pitcairn Island, central South Pacific Ocean. *Ibis* 102:58–70.

———. 1962. Extinction and the land and freshwater-inhabiting birds of New Zealand. *Notornis* 10:15–32.

———. 1964. Extinction and the Anatidae of New Zealand. *Wildfowl Trust Annual Report* 15:140–46.

———. 1976. The New Zealand wattlebirds (Callaeatidae). pp. 161–70 *Proceedings of the 16th Ornithological Congress,* Canberra.

Williams, Paul W. (ed.) n.d. *Metro Cave. A Survey of Scientific and Scenic Resources.* Unpublished report for Dept. of Lands and Survey.

35

The Role of Polynesians in the Extinction of the Avifauna of the Hawaiian Islands

STORRS L. OLSON AND HELEN F. JAMES

BECAUSE A GREAT PROPORTION of the native land birds of the Hawaiian archipelago became extinct or very rare in the period since European discovery of the islands, it has often been assumed that the "serious degradation of the Hawaiian environment... began in earnest within a few years after the arrival of Captain Cook and his successors," whereas the "new balance [that] was established between the Hawaiians and their environment during the centuries following the first colonization of the islands will never be known" (Berger 1972:7). Discoveries of major deposits of fossil and subfossil bird bones on four of the main Hawaiian Islands, as well as lesser finds on another island, drastically change these assessments. Contrary to the stated belief that "no serious inroads were made on the native birds by the Hawaiians" (Amadon 1950:210), we find that the destruction of the greater part of the avifauna took place well before Cook's arrival.

The Fossil Sites

The sites that have produced bones of extinct birds (fig. 35.1), along with bones of extant species, encompass a surprising variety of geologic settings on five of the eight main Hawaiian Islands. Detailed information on these sites and the composition of the fossil faunas is provided elsewhere (Olson and James 1982b); here we limit ourselves to a brief recapitulation, adding new information from our latest excavations on Maui and Oahu.

On the island of Molokai a large collection of fossil material has been recovered from numerous individual sites in aeolian sand dunes at Ilio Point, on the northwestern tip of the island, and near Moomomi Beach. Aeolian dunes at the southern end of the island of Kauai are the source of another extensive fossil avifauna. Four samples of land snail shells and crab claws from the dunes on Kauai and from Ilio Point, Molokai, yielded radiocarbon ages ranging from 5,145 ± 60 to 6,740 ± 80 yr B.P. (Olson and James 1982b), indicating that the extinct species in these deposits persisted well into the Holocene. At least one specimen of fossil bird from the Molokai dunes appears to be Pleistocene in age, however, as land snails associated with the holotype of the flightless goose *Thambetochen chauliodous* yielded a radiocarbon age of 25,150 ± 1000 yr B.P. (Stearns 1973).

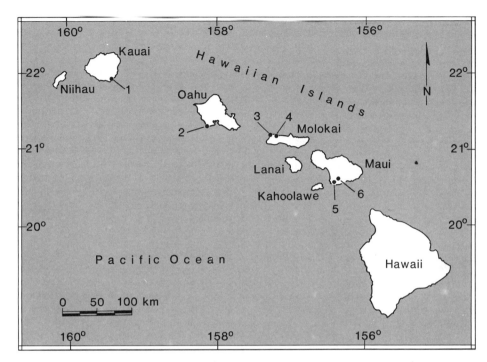

Figure 35.1. Outline map of the main Hawaiian Islands to show the location of principal fossil sites: 1, Makawehi dunes; 2, Barber's Point sinkholes; 3, Ilio Point dunes; 4, Moomomi Beach dunes; 5, Pau Naio lava tube; 6, Auwahi lava tube.

On Oahu we obtained a very rich avifauna from various sinkholes and caverns in an elevated limestone reef at Barber's Point, on the southwestern corner of the island. A unique site here is a flooded cavern, exposed during limestone quarrying operations, in which associated skeletons of extinct birds were found under 5 or more meters of fresh water. These skeletons must have been deposited at a time when most of the cavern floor was dry, presumably when lowered sea levels depressed the water table, indicating that fossils from this site are probably at least Wisconsinan in age. Apart from this site, however, the bones from Barber's Point were recovered from sediment-filled sinkholes and appear to be Holocene in age.

Recently, a new fossil bird site on Oahu was located by G. Paulay near Mokapu Point. Well-mineralized bird bones are exposed here in a sea cliff in sediment that was apparently deposited when a crater lake filled the center of the Ulupau tuff cone. The bird-bearing sediments underlie a beach deposit that has been identified as part of the Waimanalo Formation, dated radiometrically at 120,000 yr B.P. (Ku et al. 1974). Fossils from this deposit will add a new dimension to our knowledge of evolution in Hawaiian birds, but because they greatly antedate the arrival of man in the archipelago, they are not relevant to the present discussion and are not included in our calculations.

Until recently, few bird remains had been found on Maui (Olson and James 1982a,b), but two richly fossiliferous lava tubes were discovered in 1982 by R. M. Severns on the southeast slope of Haleakala on east Maui. Both are at considerably greater elevations than any of the productive sites on other islands. The first of these, the Auwahi Tube, is situated at 1145 m in a once-forested area that is now cattle pasture. Access to this tube was through two roof collapses about 58 m apart, both of which probably acted as pitfall traps. By far the majority of bird remains recovered here were those of flightless geese, ibises, and rails. Bones were concentrated near the entrances but were also scattered throughout the remainder of the tube. Many were exposed at the surface, while others were buried in the shallow sediment that had accumulated near the entrances. Preservation was usually excellent. The bones are unmineralized and do not appear to be of great age.

The second site, the Puu Naio Tube, is at roughly 300 m elevation. The entrance to this tube is at the bottom of a sinkhole about 10 m in depth. Both the uphill and downhill portions of this lava tube had in the past been used for Polynesian burials. Bird remains were found only in the somewhat drier upslope portion of the tube. Bones of geese and ibises were found among the coarse rockfall at the rear of this cave; for the most part these were more friable and poorly preserved than those obtained in the Auwahi Tube, due presumably to greater moisture and degree of exposure. Extensive deposits of sediment occur back from the entrance of the upslope cave for a distance of at least 25 m. A small test pit here yielded remains of several species of extinct birds, and more extensive excavations should yield bones of a greater variety of small passerines than previously recovered from Maui.

A few bird bones have also been found in lava tubes on Hawaii, and it is hoped that further exploration will expand our knowledge of the former birdlife of that island. Archaeological sites are another source of remains of extinct birds, and on Oahu, Molokai, and Hawaii they have provided unequivocal evidence of contemporaneity of Hawaiians and extinct species of birds.

The Extinct Birds

The prehistorically extinct species of birds in our faunal samples (Table 35.1) include one small petrel, ten or eleven species of geese, most of them bizarre flightless forms, three species of flightless ibises, eight species of flightless rails, at least three species of a new genus of long-legged owl, an extinct eagle (*Haliaeetus*), an extinct hawk (*Accipiter*), two large species of extinct crows (*Corvus*), one species of large meliphagid (*Chaetoptila*), and as many as fifteen species of the so-called "Hawaiian honeycreepers," or drepanidines, which are actually finches belonging to the subfamily Carduelinae of the Fringillidae. Among the last mentioned are numerous taxa of finch-billed forms, as well as extinct genera not known historically.

We recognize thirty-six endemic species of land birds in the historically known avifauna of the main Hawaiian Islands and five from the Leeward Islands. Sixty-eight endemic species of land birds occur in fossil deposits, forty-four of which became extinct before they could be recorded in life by ornithologists. Fourteen of the historically known species have not yet been collected as fossils, giving a total of eighty-two endemic species of land birds known from the main Hawaiian Islands. Of this total, 53 percent became extinct prehistorically. These figures are constantly being updated, and the addition of taxa recovered during our excavations on Maui accounts for the difference between these figures and those in Olson and James (1982a,b).

Although the historic period in the Hawaiian Islands began with Captain Cook in 1778, to avoid the tedious repetition of such phrases as "before thorough ornithological collecting took place," we define the historic avifauna as comprising all species that have been described as living birds in the scientific literature. Many of these descriptions were published a century or more after 1778, and it is quite possible that some of the species we count as prehistoric extinctions may have survived into the early part of the historic period. It is extremely unlikely, however, that an appreciable portion of the extinct fossil species survived this long.

Many of the extinct species are found in fossil deposits on more than one island. For instance, one species of fossil crow occurs both on Molokai and on Oahu; thus, two separate island populations of this crow became extinct. By analyzing the avifauna at the

Table 35.1. Species or Populations of Fossil Land Birds that Became Extinct Prehistorically in the Hawaiian Islands

KAUAI
  Anatidae
    Branta sandvicensis (Hawaiian Goose)
    Medium Kauai goose
    Large Kauai goose
  Rallidae
    Medium Kauai rail
  Strigidae
    Long-legged Kauai owl
  Fringillidae
    Drepanidini
      Psittirostra (Telespyza) (a finch), medium sp.
      Psitirostra (Chloridops) (a grosbeak finch), Kauai sp.
      Psittirostra (Subgenus incertae sedis) Cone-billed Finch
      Psittirostra (Subgenus incertae sedis) Additional Kauai Finch
      Hoopoe-like sickle-bill
      Ciridops sp. (related to Ula 'ai hawane), Kauai

OAHU
  Anatidae
    †*Branta* sp. (a goose)
    Supernumerary Oahu goose
    Oahu *Thambetochen* sp. (a flightless goose)
  Accipitridae
    Haliaeetus sp. (an eagle)
    Accipiter sp. (a hawk)
  Rallidae
    Small Oahu rail
    Medium-large Oahu rail
  Strigidae
    Long-legged Oahu owl
  Corvidae
    Corvus sp., slender billed (a crow)
    Corvus sp., deep billed (a crow)
  Meliphagidae
    Chaetoptila sp. (similar to Kioea)
  Fringillidae
    Drepanidini
      Psittirostra (Telespyza) cf. *cantans* ("Laysan" finch)
      Psittirostra (Telespyza) medium sp. (a finch)
      Psitirostra (Loxioides) bailleui (Palila)
      Psittirostra (Rhodacanthis) flaviceps (Lesser Koa Finch)
      Psittirostra (Chloridops) Lesser Oahu sp. (a grosbeak finch)
      Psittirostra (Chloridops) Giant Oahu sp. (a grosbeak finch)
      Psittirostra (Subgenus incertae sedis) Ridge-billed Finch
      Hoopoe-like sickle-bill
      Icterid-like gaper, Oahu
      Sickle-billed gaper
      Ciridops sp., Oahu (related to Ula 'ai hawane)

MOLOKAI
  Plataleidae
    Apteribis glenos (a flightless ibis)
  Anatidae
    Branta sandvicensis (Hawaiian Goose)
    Thambetochen chauliodous (a flightless goose)
  Accipitridae
    Haliaeetus sp. (an eagle)
    Buteo solitarius (Hawaiian Hawk)

**Table 35.1. Species or Populations of Fossil Land Birds
that Became Extinct Prehistorically in the Hawaiian Islands**

(continued)

Rallidae
    Very small Molokai rail
Strigidae
    Long-legged Molokai owl
Corvidae
    Corvus sp., slender billed (a crow)
Fringillidae
    Drepanidini
      Psittirostra (*Telespyza*) cf. *cantans* ("Laysan" Finch)
      Psittirostra (*Telespyza*) cf. *ultima* ("Nihoa" Finch)
      Psittirostra (*Telespyza*) small sp. (another finch)
      Psittirostra (Subgenus incertae sedis) Ridge-billed Finch
      Pseudonestor xanthophrys (Maui Parrotbill)
      Heterorhynchus lucidus (Nukupu'u)
      Hemignathus obscurus ('Akialoa)
      Paroreomyza maculata ('Alauwahio)
      Icterid-like gaper, Molokai
      Ciridops sp., Molokai (related to Ula 'ai hawane)

MAUI
    Plataleidae
      Maui *Apteribis* sp. A (a flightless ibis)
      †Maui *Apteribis* sp. B (a flightless ibis)
    Anatidae
      Maui *Thambetochen* sp. A (a flightless goose)
      Maui *Thambetochen* sp. B (a flightless goose)
      †Maui *Branta* sp. A (a goose)
      †Maui *Branta* sp. B (a goose)
    Rallidae
      Smallest Maui rail
      Small Maui rail
      Larger Maui rail
    Strigidae
      †Long-legged owl
    Muscicapidae
      Phaeornis sp. (a thrush)
    Meliphagidae
      Moho sp. ('O'o)
    Fringillidae
      Drepanidini
      †*Psittirostra* sp. (a finch)

HAWAII
    Anatidae
      Geochen rhuax (a goose)
      Large Hawaii goose
    Rallidae
      Large Hawaii goose

NOTE: Nomenclature follows Olson and James (1982b)
*Taxa that survived into the historic period, but not on island where listed.
†Taxon may not be specifically distinct from one occurring on another island.

level of individual island populations rather than species, we find that a minimum of eighty-nine populations of endemic species of land birds occur as fossils, sixty-seven (75 percent) of which became extinct prehistorically.

We can also document through fossils that many extant taxa with historically restricted distributions were, once, more widespread in the archipelago. Some of the better examples are among the species known historically only from the Leeward Islands or the island of Hawaii. A number of species absent from the historically known avifauna of Maui are known historically from islands on either side of Maui [e.g. the thrush (*Phaeornis*), flycatcher (*Chasiempis*), meliphagid (*Moho*), and drepanidine (*Hemignathus obscurus*)]. Our recent excavations have now begun to fill in some of these gaps, although even when prehistoric human disturbance is considered it is still difficult to explain the absence from the large island of Maui of many taxa that managed to survive on smaller islands.

So far we have limited our discussion of the fossil evidence to endemic land birds. Seabirds, particularly petrels (Procellariidae), which come to land to breed and are essentially helpless in their nesting burrows, also suffered extinctions. Among the bones from Oahu are remains of a new extinct species of small *Pterodroma*. Another species, the Bonin Petrel, *Pterodroma hypoleuca*, is found in the Hawaiian chain today only in the Leeward Islands, yet it occurs in the deposits from the main islands and is particularly well represented in a site at Moomomi on Molokai that consists in part of cultural midden. The Dark-rumped Petrel, *Pterodroma phaeopygia*, once occurred on all the main islands, where the Polynesians used it extensively for food. It is now found as dwindling populations only in the higher parts of Maui and Hawaii, its decline being attributed to predation by mammals introduced after European colonization. Although unknown historically from Oahu, the Dark-rumped Petrel is by far the most abundant species represented in each of the many sinkholes at Barber's Point. Munro (1960) has suggested that Polynesians exterminated the Oahu population.

The number of Holocene extinctions in the Hawaiian avifauna was undoubtedly considerably greater than the figures presented here suggest. Certainly, many extinct species and populations remain to be discovered. Only from three of the eight main islands (Kauai, Oahu, and Molokai) are fossil collections comprehensive enough to include most of the species that were present on the island before human disturbances took place. But even on these islands, the samples do not include the entire naturally occurring avifauna (Olson and James 1982b). Niihau and Kahoolawe have not been searched for fossils yet, and efforts to find fossils on Lanai have so far produced only some eggshell fragments, probably of a goose. Fossil collections from the island of Hawaii are meager at present. Productive fossil sites on Maui eluded us for years, but new discoveries are beginning to increase the number of species known from fossils on that island.

Contemporaneity of Prehistoric Humans and Extinct Fossil Birds

What can account for the massive extinction of species of birds as diverse as flightless geese, eagles, petrels, and small insectivorous and granivorous passerines? While Pleistocene changes in climate and habitat can be correlated with extinctions of vertebrates in the West Indies (Pregill and Olson 1981), they cannot be invoked to explain the massive extinction of Hawaiian birds, since in most cases the extinct species persisted well into the Holocene. Further, we can show that many of the prehistorically extinct birds survived long enough to be contemporaneous with the Polynesians. The best evidence for this comes from finding bones of extinct birds in archaeological sites.

The Hawaiian Goose, *Branta sandvicensis*, is known historically only from the island of Hawaii. Remains of *Branta* (the systematics of the fossil populations have not yet been resolved) have been found in cultural midden deposits on Molokai and Oahu. A small sinkhole on Barber's Point, Oahu, contained bones of hundreds of individuals of the Dark-rumped Petrel, and, in addition to *Branta*, the following taxa that are either extinct altogether or extinct on Oahu: Bonin Petrel, *Pterodroma hypoleuca*; Harcourt's

Storm Petrel, *Oceanodroma castro*; a small flightless rail; a flightless goose, *Thambetochen* sp.; an extinct crow, *Corvus* sp.; and a large passerine, probably the extinct meliphagid, *Chaetoptila* sp. These were in association with a grindstone and bones of fish, chicken (*Gallus gallus*), Pacific rat (*Rattus exulans*), and a larger mammal (probably dog), as well as shells of marine mollusks commonly taken as food by Hawaiians. Of the myriad petrel bones from this site, all elements of the skeleton are represented, and none is broken or burned, indicating that cooking must have been accomplished by steaming or boiling. Thus, bird bones from Hawaiian middens will not necessarily exhibit either cut marks or charring.

In contrast to the Barber's Point cooking site, seabird bones from archaeological sites on southeastern Oahu almost always have the ends broken off. Bones of an extinct goose (*Thambetochen* sp.), an extinct rail, and locally extirpated seabirds were also recovered from these sites (Olson and James 1982b).

On the island of Hawaii, remains of a prehistorically extinct rail were found in cultural deposits high on Mauna Kea. A partial cranium of a large extinct goose was recovered from an occupation site in a lava tube at a level slightly below one dated at 606 ± 90 radiocarbon yr B.P. (Olson and James 1982b), but the association of this specimen with cultural deposits we now consider doubtful.

Two of the fossil sites in the Moomomi dunes on Molokai are composed at least partially of cultural midden, including bones of fish and chickens (*Gallus*), limpet shells ("opihi"), other edible mollusks, and crab claws. However, because of the unstable nature of dune deposits, we cannot be certain that the midden material was not deposited in an area that already contained bones of extinct birds. At one of these sites there is an unusually large concentration of bones of the Bonin Petrel, *Pterodroma hypoleuca*, a species now extinct in the main Hawaiian Islands. Along with the petrel bones are those of the flightless goose *Thambetochen chauliodous* and the flightless ibis *Apteribis glenos* (Olson and Wetmore 1976), an extinct long-legged owl, an extinct crow (*Corvus* sp.), a nene-like goose (*Branta* sp.), and a hawk (*Buteo solitarius*), the last two being extinct on Molokai, although still present on the island of Hawaii.

One of the most significant sites on Oahu is a very large, deep sinkhole with signs of considerable human modification (fig. 35.2). Beautifully preserved bones of extinct birds were obtained here from a deposit of fine, dry limestone dust that had accumulated under a sheltered overhang. Additional bones were recovered from the sediments in the central part of the sinkhole. Excavations here indicated that the sinkhole may have been used by Polynesians for agricultural purposes. A hearth uncovered in the central area contained charred bones of *Branta* sp., a larger extinct goose, Dark-rumped Petrel, and *Rattus exulans*. Charcoal from the hearth gave a radiocarbon age of 770± 70 yr B.P. (Olson and James 1982b), indicating that at least some species of extinct birds may have survived well after the original human colonization of the archipelago about 1,500 years ago.

In the dusty sediments of the sheltered portion of the sinkhole, bones of *Rattus exulans* were found in place in the same levels as bones of extinct birds. These were disassociated, isolated elements, all of which lay in a horizontal plane and thus were evidently not deposited as a result of burrowing by the rats. Thus, bones of extinct birds and *Rattus exulans*, which was imported by Polynesians, were being deposited simultaneously. These deposits must, therefore, postdate the arrival of man in Hawaii.

The remains of "marker" taxa such as *Rattus exulans*, that indicate Polynesian introduction, offer great potential for showing contemporaneity of extinct birds and Polynesians. Land snails also may be important indicators (Kirch and Christensen 1981, Christensen and Kirch 1982). At least one species, *Lamellaxis gracilis*, is believed to have been introduced to the Hawaiian Islands before European contact. It now appears likely that most of the species of lizards that occur at present in the Hawaiian archipelago were introduced prehistorically by Polynesians (G. K. Pregill in prep.), so that lizard bones, too, should provide good indicators.

Figure 35.2. The largest of the fossiliferous sinkholes in the raised limestone reef at Barber's Point, Oahu. The large stone cairn at the left and the rock wall extending from it are prehistoric Polynesian modifications of the site. Bones of extinct birds were recovered in association with Polynesian-introduced marker taxa in the fine dusts beneath the overhang at the top of the picture. Additional bones of extinct birds, cultural remains, a radiocarbon-dated hearth, and evidence of agricultural use of the sinkhole were uncovered in the unsheltered sediments. A test pit, later greatly expanded, may be seen in these sediments at the right of the picture.

In another of the larger soil-filled sinks excavated at Barber's Point, bones of extinct birds were found in the upper 5 cm of sediment and to depths of nearly 40 cm. Disassociated bones of *Rattus exulans* occurred throughout the deposit, although they were more abundant in the upper portions. In cores taken from this site, shells of *Lamellaxis* were distributed throughout the column. In another sinkhole with deeper sediments, *Lamellaxis* was found two-thirds of the way down into the layer that produced the most bones of extinct birds. It is evident that much of the deposition of bones of extinct birds at Barber's Point took place during the prehistoric Polynesian period. Kirch and Christensen (1981) also document the decrease in abundance and eventual extinction of a number of species of endemic land snails up through the stratigraphic column.

Causes of Extinction

Having shown that the majority of prehistorically extinct species of birds were not victims of Pleistocene climatic changes or other natural causes, and having implicated man in the demise of these birds, we now turn to possible mechanisms of extinction. Direct predation for food could have been responsible for the reduction in numbers or total obliteration of some species. The large flightless geese, the flightless ibises, and ground-nesting or burrowing species of seabirds had no defenses against man or the dogs, pigs, and rats the Polynesians brought with them. To account for the loss of population after population of small forest birds, which could hardly have been hunted to extinction by any means available to the Polynesians, some other mechanism must have been at work. The most likely explanation is that relatively dry lowland forests in the Hawaiian islands were largely destroyed by clearing, mostly by fire, for agriculture.

In general, the pattern of extinctions within the avifauna corroborates the idea that human disturbances were involved. Possibly as many as twenty-two of the extinct fossil species of birds found so far were flightless or nearly so, and it is obvious that conditions in the archipelago were ideally suited to the evolution of flightlessness. However, only one of these flightless species is known certainly to have survived into the historic period. It is hardly likely that climatic changes or any other natural force would have been so intolerant of an adaptation that had been selected for repeatedly throughout the evolutionary history of the Hawaiian avifauna.

Except for some of the lava tubes, all of our fossil sites are from lowland areas; most are within a few hundred meters of the present shoreline. Although the Maui sites are at higher elevations, they are on the drier, leeward slope of Haleakala. During the historic period, native forest birds have been found almost exclusively in wet montane forests, and in fact some authors have considered them to be "montane species," unable to exist in lowland habitats (e.g. Juvik and Austring 1979). Fossil evidence conclusively shows not only that many prehistorically extinct species once flourished in the lowlands, but also that virtually all extant species occurred there as well.

With such exceptions as recent lava flows, active sand dunes, and the alpine regions of the highest mountains, there is no reason to believe that the vegetation in any part of the Hawaiian Islands before man's arrival consisted of anything other than forest of one kind or another. This was definitely not the case at the time of European contact, however. Early explorers described much of the drier lowlands as being barren or grassy and destitute of trees. On the leeward slopes of Kauai, Cook (1784) reported that on his visit "no wood can be cut at any distance convenient to bring it from," and fourteen years later Vancouver (1798) confirmed that this area was periodically burned when he "observed the hills to the eastward of the river to be on fire to a considerable height, in particular directions, down towards water's edge." From shipside, King (1784) could discern no tree growth on southwest Molokai or on the island of Kahoolawe. Unpublished journals of other members of Cook's expedition confirm these observations and add that Lanai and Niihau also appeared to be barren of trees (Wilson 1977).

On Vancouver's expedition, Menzies (1790–1792) found continuous plantations reaching six or seven miles inland from Kealakekua Bay on the west coast of Hawaii, stretching in a broad band along the coast as far as he could see in either direction. Of the land farther north, he reported that "from the northwest point of the island [of Hawaii] the country stretches back for a considerable distance with a very gradual ascent and is destitute of trees or bushes of any kind, but it bears every appearance of industrious cultivation by the number of small fields into which it is laid out. . . ."

Of leeward Oahu, Chamisso (1830:316) gave the following account: "The culture of the vallies which lay behind Hanaruru [Honolulu] is really astonishing. Artificial irrigations enable the natives to form, even upon the hills, large aquatic plantations of *Tarra* [taro], which are at the same time employed as fish-ponds, while all kinds of useful plants grow on the banks which form their borders." Archaeological studies of prehistoric land use on Molokai and Oahu provide further evidence that the natural vegetation of many lowland regions was removed, probably for agricultural purposes, and that an accelerated rate of erosion and the extermination of native land snail faunas ensued (Yen et al. 1972, Kirch and Kelly 1975).

The nature of the original vegetation of lowland regions is a matter of conjecture, as the forests here were virtually destroyed before botanists arrived to collect in them. Fosberg (1972) believes that "large areas of dry coastal slopes and higher rain shadows, probably most of the relatively dry areas below 1500 m, were originally covered by an open scrub forest." In sampling a few remnants of lowland forest, Rock (1913:15) was surprised to find a much greater diversity of species of trees than in wet montane forests, and he remarked that "not less than 60 percent of all the species of indigenous trees growing in these islands can be found in and are peculiar to the dry regions or lava fields of the lower forest zone...." This suggests that there was a greater diversity of feeding niches in these lowland forests, a view supported by the number of new species of drepanidines that we found in the lowland fossil deposits.

In the West Indies, arid scrub forest in Puerto Rico was found to have twice the species diversity and three times the number of individuals of birds as montane rain forest (Kepler and Kepler 1970). For this reason, and with support from the fossil record, it has been postulated that loss of arid habitats in the West Indies since the last glaciation probably caused the extinction of many species (Pregill and Olson 1981). By removing such habitats from the Hawaiian Islands, the Polynesians wrought a greater change in the total biota of the archipelago than has been accomplished by all post-European inroads in the wet montane forests.

Implications for Island Ecosystems Elsewhere

There is no reason to regard the prehistoric fate of the Hawaiian Islands as exceptional. It is known that practically the entire flora of Easter Island was eliminated by Polynesians, and in fact the eventual absence of wood strongly influenced the cultural development of the islanders (McCoy 1976). The Maoris in New Zealand were responsible for extensive deforestation by burning shortly after they colonized those islands (McGlone 1978). Zimmerman (1938) lamented that at least three quarters of the forest had been eliminated from Rapa by fire and grazing animals. When queried whether there were any islands in the Pacific that are "sufficiently virgin to enable a study of the primitive conditions" Zimmerman (1963:63) responded that: "From my experience, I would say there are very few, almost no such islands as a matter of fact, and that in the eastern or central south Pacific there are none."

Entomologists, however, can do little more than speculate on what the consequences of this deforestation might have been for insects, as few insects are preserved as fossils under the usual conditions of deposition met with in these islands. With the fossil record of birds from Hawaii we get a better picture of the magnitude of the destruction. With the possible exception of some Galapagos islands (D. W. Steadman pers. comm.), probably no islands in the Pacific have a relatively intact fauna with a full complement of naturally occurring species. Preliminary studies of a few archaeological remains from the Lakeba Islands in the Fiji group, for example, have shown that megapodes (Megapodiidae) and a very large pigeon were eliminated from the avifauna since the arrival of man (D. W. Steadman pers. comm.).

Exercises based on MacArthur and Wilson's (1967) "island biogeography," using statistics derived from present island ecosystems without considering the human history of the islands involved, are now seen as unlikely to be biologically meaningful. As an example, Juvik and Austring (1979) calculated species-area curves for native land birds of the Hawaiian Islands on the basis of historically known taxa. They found a high correlation between number of species and island area. It seemed reasonable to conclude that the Hawaiian avifauna was in natural equilibrium when the islands were first visited by Europeans. The fossil record, however, has shown that the historically known

Table 35.2. Area, Elevation, and Number of Endemic Species of Land Birds Known Historically and as Fossils from the Five Largest Hawaiian Islands

		Molokai	Kauai	Oahu	Maui	Hawaii
Endemic Species	Fossil	21	21	32	12 + *	3*
of Land Birds	Historic	9	13	11	10	23
Island Area (km²)		676	1,422	1,536	1,880	10,464
Maximum Elevation (m)		1,515	1,598	1,227	3,056	4,206

*Fossil samples from these islands are still too small to reflect true prehistoric species diversity.

avifauna of the archipelago constitutes only a fraction of the natural species diversity of the islands (Table 35.2). For example, thirty-two endemic species of land birds occur in the fossil deposits on Oahu, yet only eleven species are known historically. Even with the inclusion of the available data from fossils, there would be little chance of deriving a realistic correlation based on numbers of species per island because of the deficiencies of the fossil record.

Biotas of oceanic islands outside the Pacific region have also been depleted by human-caused extinctions. Fossils indicate that the small islands of the South Atlantic underwent a period of extinction immediately after their discovery by Europeans early in the sixteenth century. A flightless rail was exterminated on Ascension Island (Olson 1973), a rail and a rodent disappeared from Fernando de Noronha (Olson 1981), and two rails, a cuckoo, a hoopoe, and at least five species of seabirds, including one endemic, vanished from St. Helena since the coming of man (Olson 1975). In the North Atlantic, fossil evidence of man-caused extinctions is as yet known only from Bermuda (Olson et al. in prep.), whereas the nature of the prehuman faunas of the Azores, Canaries, Madeira, and Cape Verde islands, which have longer histories of occupation by man, are yet unknown. The absence of any endemic species of birds in the Azores is almost certainly an artifact of human interference. In the Indian Ocean, the ill fate of the biota of the Mascarene Islands since their discovery in the fifteenth century is renowned (e.g. Greenway 1958), although the fauna of these islands is still incompletely documented in the fossil record.

We have only begun to appreciate the effects of the Polynesian invasion on the biotas of Pacific islands. Heretofore, the best evidence, though often not recognized as such, came from New Zealand. We have now shown in the Hawaiian Islands that the devastation was more widespread and comprehensive than previously imagined (Olson and James 1982a, b). If the Hawaiian archipelago, which was colonized by humans considerably later than most of Polynesia, can be taken as an indication of the extent to which other Pacific Islands suffered from prehistoric human-caused alterations in environment, then the period of the original peopling of the diverse islands of Oceania, with their highly endemic biotas, may have been marked by one of the greatest waves of rapid extinction of species of animals and plants in the history of the earth. As a consequence, at least for vertebrate zoologists, scientific exploration of the Pacific Islands must be conducted anew. Until these islands are investigated paleontologically, we will not know what their natural diversity may have been.

Acknowledgments

We would like to express our indebtedness to all the individuals who helped with the collection and curation of fossils, especially to Alan Ziegler, Joan Aidem, C. J. and Carol P. Ralph, and the staff of the B. P. Bishop Museum. Specimens are housed at the B. P. Bishop Museum in Honolulu and the National Museum of Natural History, Smithsonian Institution.

Figure 35.3. Extinct flightless ibis (left) and flightless rail from the Hawaiian Islands. (Courtesy of B. P. Bishop Museum, Honolulu. Painting by H. Douglass Pratt)

References

Amadon, D. 1950. The Hawaiian honeycreepers (Aves, Drepanidae) *Bull. Amer. Mus. Nat. Hist.*, v. 95, pp. 151–262.

Berger, A. 1972. *Hawaiian birdlife*. Honolulu, The University Press of Hawaii, 270 pp.

Chamisso, Admiral de. 1830. Notices respecting the botany of certain countries visited by the Russian voyage of discovery under the command of Capt. Kotzebue. (Translated from the German edition of the voyage), pp. 305–323 *in* W. J. Hooker, *Botanical Miscellany*, v. 1, London, John Murray.

Christensen, C. C. and Kirch, P. V. 1982. (Abstract) Land snails and environmental change at Barbers Point, Oahu, Hawaii. *Bull. Amer. Malacol. Union* [for 1981], p. 31.

Cook, J. 1784. *A voyage to the Pacific Ocean.* v. 2, London, G. Nicol and T. Cadell, 549 pp.

Fosberg, F. R. 1972. *Guide to Excursion III. 10th Pacific Science Congress.* rev. ed., Honolulu, University of Hawaii, 249 pp.

Greenway, J. C., Jr. 1958. Extinct and vanishing birds of the world. New York, *American Comm. Int. Wild Life Protection*, Spec. Publ. v. 13, 518 pp.

Juvik, J. O. and Austring, A. P. 1979. The Hawaiian avifauna: biogeographic theory in evolutionary time. *J. Biogeogr.*, v. 6, pp. 205–224.

Kepler, C. B. and Kepler, A. K. 1970. Preliminary comparison of bird species diversity and density in Luquillo and Guanica Forests. pp. E-183–E-191, *in* H. T. Odum, *editor, A tropical rain forest; a study of irradiation and ecology at El Verde, Puerto Rico*, Oak Ridge, Tenn., U. S. Atomic Energy Comm. Div. Tech. Inform.

King, J. 1784. *A voyage to the Pacific Ocean*. v. 3, London, G. Nicol and T. Cadell, 558 pp.

Kirch, P. V. and Christensen, C. C. 1981. Nonmarine molluscs and paleoecology at Barbers Point, O'ahu. pp. 242–286 *in* H. Hammatt and W. H. Folk. *Archaeological and Paleontological investigation at Kalaeloa (Barbers Point), Honouliuli, 'Ewa, O'ahu, Federal Study Area 1a and 1b, and State of Hawaii Optional Area 1.* Unpublished report prepared for U.S. Army Corps of Engineers, U.S. Army Engineer District, Honolulu, by Archaeological Research Center Hawaii, Inc., MS No. ARCH 14–115, 398 pp.

Kirch, P. V. and Kelly, M., *editors.* 1975. Prehistory and ecology in a windward Hawaiian valley: Haelawa Valley, Molokai. *Pacific Anthropol. Rec.*, v. 24, 203 pp.

Ku, T., Kimmel, M. A., Easton, W. H., and O'Neil, T. J. 1974. Eustatic sea level 120,000 years ago on Oahu, Hawaii. *Science*, v. 183, pp. 959–962.

MacArthur, R. H. and Wilson, E. O. 1967. *The theory of island biogeography.* New Jersey, Princeton University Press, 203 pp.

McCoy, P. C. 1976. Easter Island settlement patterns in the late prehistoric and protohistoric periods. *Easter Island Comm., Intern. Fund for Monuments, Inc. Bull.,* v. 5, 164 pp.

McGlone, M. S. 1978. Forest destruction by early Polynesians, Lake Poukawa, Hawkes Bay, New Zealand. *J. Royal Soc. N. Z.,* v. 8, pp. 275–281.

Menzies, A. 1790–1792. *Journal of Vancouver's voyage.* Transcript of the original in the British Museum. (Xerographic copy of 874 pages, bound in 2 volumes, in Smithsonian Institution Libraries.)

Munro, G. C. 1960. *Birds of Hawaii.* rev. ed., Rutland, Vt., Charles E. Tuttle, 192 pp.

Olson, S. L. 1973. Evolution of the rails of the South Atlantic islands (Aves: Rallidae). *Smithson. Contr. Zool.*, v. 152, 53 pp.

———. 1975. Paleornithology of St. Helena Island, South Atlantic Ocean. *Smithson. Contr. Paleobiol.*, v. 23, 49 pp.

———. 1981. Natural history of vertebrates on the Brazilian islands of the mid South Atlantic. Nat. Geogr. Soc. Res. Repts., v. 13, pp. 481–492.

Olson, S. L. and James, H. F. 1982a. Fossil birds from the Hawaiian Islands: evidence for wholesale extinction by man before Western contact. *Science*, v. 217, pp. 633–635.

———. 1982b. Prodromus of the fossil avifauna of the Hawaiian Islands. *Smithson. Contr. Zool.*, v. 365, 59 pp.

Olson, S. L. and Wetmore, A. 1976. Preliminary diagnoses of two extraordinary new genera of birds from Pleistocene deposits in the Hawaiian Islands. *Proc. Biol. Soc. Wash.*, v. 89, pp. 247–258.

Pregill, G. K. and Olson, S. L. 1981. Zoogeography of West Indian vertebrates in relation to Pleistocene climatic cycles. *Ann. Rev. Ecol. Syst.*, v. 12, pp. 75–98.

Rock, J. F. 1913. The indigenous trees of the Hawaiian Islands. Honolulu, published under patronage, 518 pp.

Stearns, H. T. 1973. Geologic setting of the fossil goose bones found on Molokai Island, Hawaii. *Occ. Pap. B. P. Bishop Mus.*, v. 24, pp. 155–163.

Vancouver, G. 1798. *Voyage of discovery to the North Pacific Ocean and round the world.* v. 1, Amsterdam, N. Israel, 432 pp. (Facsimile reprint of 1967 published as Bibliotheca Australiana No. 30, New York, Da Capo Press.)

Wilson, E. 1977. Observations of Hawaiian avifauna during Cook's expeditions. *'Elepaio, J. Hawaii Audubon Soc.*, v. 38, pp. 13–18.

Yen, D. E., Kirch, P. V., Rosendahl, P., and Riley, T. 1972. Prehistoric agriculture in the upper valley of Makaha, Oahu. pp. 59–94, *in* E. J. Ladd and D. E. Yen, *editors,* Makaha Valley Historical Project, Interim Report No. 3. *Pacific Anthropol. Rec.*, v. 18.

Zimmerman, E. C. 1938. Cryptorhynchinae of Rapa: *Bernice P. Bishop Mus. Bull.*, v. 151, 75 pp.

———. 1963. Nature of the land biota. pp. 57–64, *in* F. R. Fosberg, *editor, Man's place in the island ecosystem. 10th Pacific Science Congress, 1961,* Honolulu, Bishop Mus. Press, 264 pp.

An Overview

An Overview

The task of responding to Quaternary Extinctions *was given to three principals from three different fields—vertebrate paleontology, zooarchaeology, and ecology. Larry Marshall leads off with a comparative glossary of the concepts used by various chapter authors including some new ones he proposes on his own, like "coextinction" for tightly linked symbionts. He helps readers determine what various authors are saying as well as where their philosophical position may lie. He expands the theoretical range of the book, noting that different explanations for late Pleistocene extinctions seem to apply on different land masses. Marshall identifies his own views as "middle of the road," falling between the more extreme paradigms of man-the-big-game hunter and natural-climatic catastrophe. A multidimensional paradigm has been popular since the time of Lyell, is likely to be the least offensive, and (alas) seems to be the least falsifiable.*

Falsifications are dealt with in Don Grayson's brilliant historical and structural analysis, in places so contrary to some of our own thinking that we anticipate heightened interest among readers in what Grayson is up to. He notes that climatic modelers have been effective in improving their falsifiability as illustrated in various chapters, especially in the third section on geologic-climatic models. This is welcome news. Grayson also finds that the overkill champions have painted themselves into a corner through excessive application of ad hoc explanations to accommodate imperfections in their models. That remains to be seen. For example, it is by no means absolutely clear that hunters were in America prior to 12,000 years ago, or that most North American bird extinction does not represent coextinction, or that accepting all radiocarbon dates "as is" is preferable to a judicious selectivity. Selectivity for dates of high quality avoids such idiotic results as concluding that mastodonts survived in Michigan to 6,000 years ago; selectivity may be especially important now in the crucial case of determining the chronology of extinction in Australia. Patience and forbearance are needed. The radiocarbon time machine has been "on line" for a few decades only. Without radiocarbon dates there would be no struggle between climatic, cultural, and other paradigms and far fewer opportunities for tests or falsifications of models.

Finally Jared Diamond scrutinizes historic extinctions, a detailed attempt to fathom their meaning in the context of late Pleistocene losses. Historic disappearances of modern birds and mammals can be laid to a variety of effects, some climatic and some cultural. Diamond turns to the "extinctions that did not occur" with damaging results for some models. Losses of island animals in historic time were considerable; in prehistoric time they were more so, but not everywhere and not for all groups. Islands of various sizes, some on and others beyond the once emergent shelf of the continents, have yielded some of the best models for biogeographers and evolutionary biologists. Darwin, Wallace and many more biologists since their time (Diamond included) have done their best work there. Will the newly emerging late Pleistocene-Holocene fossil record of oceanic islands provide the turning point in the great debate about what took place throughout the world at the end of the Pleistocene? Diamond's chapter and our book end with question marks embracing the disembodied bones and yet unknown extinction chronologies of those extracontinental lands.

36

Who Killed Cock Robin?
An Investigation of the Extinction Controversy

LARRY G. MARSHALL

THE EXTINCTION CONTROVERSY is analogous to a trial in which court has been held for sixteen years. The first session, *Pleistocene Extinctions: The Search for a Cause,* was adjourned so that the principals could accumulate more facts. Two prime suspects, *overkill* and *climate,* remain in custody. Both plead guilty, yet some witnesses swear they saw only one at the scene of the crime. Was the death of Cock Robin (alias *the megafauna*) truly the work of a single fiend, or was it a conspiracy? And who, if anyone, fired that second shot reported by other witnesses? For a second session, *Quaternary Extinctions: A Prehistoric Revolution,* additional evidence has been amassed by nearly four dozen specialists and presented to the court. The verdict? None; it's a hung jury! (Table 36.1).

The collected chapters in this book are reminiscent of chapters in an Agatha Christie novel. The clues are apparently there, but as in any good whodunit some are pertinent and others simply lead the unwary reader astray. Unfortunately, and this is the basis for the dilemma, Agatha never wrote *the* final chapter. That the real villain(s) remains a mystery surprises none. Moreover, we may never be certain whom Agatha intended as the elusive culprit(s). There is still "no solution and no consensus" (Martin chap. 17).

Stimulated by Don Grayson's line of thought (chap. 37), I focus on the failure to define *the rules of the game* that are formidable obstacles blocking the search for conviction. As will become evident, the trial does not involve a single crime. A number of charges have been made; some warrant separate convictions, while others will be tried jointly. To this end I address ten issues—terminology; scientific method; chronology, calibration, correlation, causation; historical perspective; extinction; turnover; megafauna; dwarfing; overkill; and climate.

Terminology

Many terms are static in meaning, or nearly so, and are used by one or a few authors (or coauthors)—for example, *individualistic or biotic reorganization, vegetational mosaic, normalcy of the present, fallacy of the primeval, secular climatic deteriorations, disharmonious associations, coevolutionary disequilibrium, ripple effect, threshold density, ecological stress, and plaids vs. stripes* (italics here and elsewhere are mine).

Other terms, such as *megafauna* and *overkill*, are used by most authors but often with slightly or significantly different meanings [e.g. *megafauna* of Martin (chap. 17) is not the same as *megafauna* of Graham and Lundelius (chap. 11)]. Consistent definitions of basic terms are essential, especially when they serve for defining data sets. It is only when the units of measure are equivalent that comparisons and contrasts are meaningful.

Scientific Method

The terms *debate, hypothesis, model, scenario,* and *theory* are used to describe extinction processes. As defined by *Webster's New Collegiate Dictionary* (1977), a *debate* is "a regulated discussion of a proposition between two matched sides"; a *hypothesis* is "a tentative assumption made in order to draw out and test its logical or empirical consequences"; a *model* is "a description or analogy used to help visualize something that cannot be directly observed"; a *scenario* is "an account or synopsis of a projected course of actions or events"; a *theory* is "a plausible or scientifically acceptable general principle or body of principles offered to explain phenomena"; and *phenomena* are "facts or events of scientific interest susceptible to scientific description and explanation." I realize that there are variations in defining these terms and that even among philosophers of science there is no consensus. For this reason I follow the definitions in *Webster's,* which is an easily accessible and neutral source.

Late Pleistocene/Holocene extinction phenomena have themselves become subject to scrutiny and debate in recent years. As in any debate, persistence is an important tactic determining who triumphs (i.e. who continues to argue after his opponents have tired and gone off to attend to other issues). Assertion and repetition are other virtues, and the position of "If I've said it thrice, it's true" can take its toll on one's opponents.

Conflicting views exist as to whether hypotheses (referred to as *models* by some authors) are or are not *testable* or *falsifiable*. Predictions are made employing *deductive* reasoning (i.e. from the general to the particular). For example, Martin (chap. 17) states, "One striking advantage of the *overkill model* over others is its testability," and "the [blitzkrieg] model can be rejected in Australia... if the large extinct mammals are found to have maintained sizable populations long after initial human invasion of the continent." In contrast, Grayson (chap. 37) states: "the overkill hypothesis became so resilient that it could withstand virtually any factual onslaught" and "... has proven so resilient... that... it cannot be falsified." Horton (chap. 29) proposes a *climatic model* that alone explains megafaunal extinctions in Australia. "The model is testable, and that is perhaps the best attribute that any model can have." Dewar (chap. 26) offers a *scenario* for vertebrate extinctions on Madagascar which he regards as synonymous with a *model* and details how it can be tested. By the above definitions this makes his *model a hypothesis.* Had it been tested and shown to be correct, then it would be a *theory.*

Hypotheses are *tested* and *falsified* by the method called the *null hypothesis*. "The null hypothesis is like the assumption of innocence in a court of law, which assumption is maintained until the weight of evidence is overwhelmingly against it" (Simpson et al. 1960, p. 175). Yet it appears that what we are dealing with are not hypotheses, but what Hempel (1966, p. 31) calls *pseudo-hypotheses:* hypotheses in appearance only:

> There may be no conceivable way of adjudicating these conflicting views.... Neither of them yields any testable implications; no empirical discrimination between them is possible. Not that the issue is "too deep" for scientific decision: the two verbally conflicting interpretations make no assertions at all. Hence, the question whether they are true or false makes no sense, and that is why scientific inquiry cannot possibly decide between them.

The term *paradigm* is probably most appropriate for positions expressed by participants in the extinction controversy. As defined by Kuhn (1970, p. 10), paradigms share two essential characteristics: "their achievement was sufficiently unprecedented to attract an enduring group of adherents away from competing modes of scientific activity. Simultaneously, it was sufficiently open-ended to leave all sorts of problems for the redefined groups of practitioners to resolve." A paradigm is thus resilient—new data permit reevaluation of previous views and its redesign for further scrutiny, yet it has not reached the stage of being testable. Nevertheless, "acquisition of a paradigm and of the more esoteric type of research it permits is a sign of maturity in the development of any given scientific field" (Kuhn 1970, p. 11). Martin (chap. 17) promotes the virtues of resiliency, Grayson (chap. 37) does not. Ironically none of the "hypotheses" discussed here have been tested, and possibly none of them can be.

It must be emphasized that the human and climatic paradigms are *not exclusive*; if anything, they are complementary. The exclusion of one paradigm clearly does not warrant acceptance of another, although such an approach is often implied or inferred by avid proponents. This is where much controversy arises. Although evidence may suggest preference of one paradigm, that fact does not eliminate *coacceptance* of another. As aptly put by Martin (chap. 17), "absence of evidence is not evidence of absence." The null hypothesis for alternative or complementary paradigms cannot be rejected.

Chronology, Calibration, Correlation, Causation

A *chronology* of events and factors believed relevant in extinction is the foundation of our understanding of the process (Mead and Meltzer chap. 19). Late Pleistocene/Holocene extinctions are *calibrated* by the ^{14}C isotope dating of organic materials, and the age is given in years before present (yr B.P.). A precise chronology permits determination of whether or not events are *synchronous* (Table 36.2). If they are, then a *correlation* (a mutual interrelationship) between two or more may exist. This last process permits inference about the *causation* of one of the observed events—in this case the extinction.

Marcus and Berger (chap. 8) survey ^{14}C dates available for the late Pleistocene/Holocene fauna of Rancho La Brea. They conclude that "extinction of the... megafauna [apparently] occurred over a relatively short time span." Likewise, Mead and Meltzer (chap. 19) review ^{14}C dates associated with twenty-three genera of extinct large mammals in North America. Reliable ^{14}C dates exist only for *Camelops, Equus, Mammut, Mammuthus, Nothrotheriops,* and *Panthera,* indicating that all were extinct by 10,000, possibly 10,800 yr B.P.: "at least among these genera, the extinction process may contain an element of synchroneity." These authors are concerned with calibrating the extinction process, not with identifying a cause.

Mead and Meltzer (chap. 19) caution that although a peak in the curve of reliable ^{14}C dates occurs at about 11,000 yr B.P., this fact "cannot and does not reflect *how* the extinctions took place; it is simply the best available age of the latest survivors of the extinction process." Nevertheless, it is conventional practice to interpret such data as indicating *when*, and if a synchronous event is identified (e.g. the appearance of man) then it may be interpreted as the *how*.

The step in this *chain of inference* of equating synchroneity of events (when) with causation (how) is the source of much debate (Table 36.2). *Inductive reasoning*, from particular cases to general principles, is employed. Data from a study of a species at a particular site(s) (King and Saunders chap. 15) or of a species over an entire continent (Agenbroad chap. 3) are used to document and calibrate an extinction event. These data and the inferences which they generate provide the bases for formulating models and hypotheses.

Table 36.1. Tabulation of Favored Paradigms for Extinctions as Concluded by Chapter Authors (Categories are Defined in Text).
Parentheses ( ) indicate secondary or inferred paradigms.

Cause / Chapter & Authors	Man — Blitzkrieg Direct	Man — Blitzkrieg Associated	Man — Innovation	Man — Attrition Indirect	Man — Attrition Competition	Climate Direct	Climate Linkage	Climate Coextinction	Not Relevant X / No Conclusions 0
1. Grayson									X
2. E. Anderson									X
3. Agenbroad	X								X
4. Agenbroad									X
5. Gruhn & Bryan									0
6. Gilbert & Martin						X			
7. Phillips									0
8. Marcus & Berger									0
9. Webb						X			
10. Gingerich						X			
11. Graham & Lundelius						X		X	
12. Guilday		X				X			
13. Guthrie						X		X	0
14. Kiltie							(X)		0
15. King & Saunders						X			
16. Haynes		X							
17. P. Martin	X								
18. McDonald	X								
19. Mead & Meltzer									X
20. Whittington & Dyke	X								
21. Steadman & Martin	(X)							X	
22. Vereshchagin & Baryshnikov						X			
23. Liu & Li	X								
24. Tchernov	X	X		X	X	X			
25. Klein		X	X	X					
26. Dewar				X					
27. A...									

28. Merrilees		X		X	
29. Horton			X		
30. Hope					0
31. Kershaw			(X)		0
32. Trotter & McCulloch		X			
33. A. Anderson		X			
34. Cassels		X			
35. Olson & James		X			
37. Grayson			X		

Table 36.2. Calibration in Years Before Present (yr B.P.) of Human, Climatic, and Faunal Events on Major Land Masses as Documented by Authors in This Book

Event / Land Mass	Man		Climatic Change	Dwarfing	Extinction
	First Appearance	Big-game Hunting Technology			
Africa	>2 million	12,000–8,000	12,000–8,000	12,000	12,000–8,000
Australia	>40,000	(not known)	26,000–15,000 16,000–12,000	26,000–15,000	26,000–20,000 (some 18,000–15,000)
Europe	>1 million	60,000–10,000	14,000–10,000	15,000–12,000 (in Levant)	12,000–10,000
North America	12,000–10,000*	12,000–10,000 11,500–10,500	12,000–10,000 12,000–11,000 11,500–10,500	12,000–10,000 12,000–11,000 11,000–10,000	12,000–10,000 11,500–10,500
South America	12,000–8,000*	12,000–8,000	10,000	(examples known but time not well documented)	12,000–8,000
Madagascar	1500–900	(not known)	(not known)	(no examples)	by 500
New Zealand	1000	(minor)	(not reviewed in this book)	(no examples)	by 500 (concentrated 900–600)

*Much earlier arrivals are claimed by some archaeologists.

It must be remembered that kill sites document only the death of individual animals. A convincing kill is one in which a projectile point is found embedded in a fossil. Such cases are rare, and for North America only fourteen are recorded for *Mammuthus* (Agenbroad chap. 3). Some kill sites document the human-related death of a population (i.e. a herd of bison wedged in an arroyo and showing evidence of butchering). Such data are used to infer causes for the local or total extinction of a species. Other species which become extinct at about the same time are inferred to have been influenced by the same or similar causes, even if there is no direct evidence for such a causal relationship.

I do not mean to undermine the quality of the fossil record. On the contrary, I agree with Martin's (chap. 17) observation that if blitzkrieg did happen we should be surprised not at the small number of kill sites known, but at the fact that we have as many as we do since discovery of these sites was chiefly by serendipity. There are numerous known historical extinctions or near-extinctions related to man-the-hunter, and I predict that for many of them (i.e. Steller's sea cow) one would be hard pressed to produce an irrefutable kill site.

Most workers are sensitive to the tenuous nature of this inference process and are exceedingly cautious in proceeding from correlation to causation. Haynes (chap. 16), for example, states that "the stratigraphic coincidence of the first visible evidence of hunters and the last skeletons of the... megafauna is intriguing."

In a different vein, Phillips (chap. 7) demonstrates that extinction of the Shasta ground sloth in Rampart Cave occurred about 11,000 yr B.P., "at a time... which should have been nearly ideal for its continued existence." There is no evidence of dietary, climatic, or environmental stress that could account for the extinction of this animal. Favorable or similar environmental conditions existed for about 3,000 years after its disappearance. However, Phillips offers no alternative explanation of *why* it became extinct and does not turn to another cause (e.g. man) in his process of eliminating climate.

The establishment of synchroneity between events is the first step in the inference process. Yet establishment of a correlation does not dictate a causation relationship. "When examining causation it is not sufficient simply to find a correlation in the timing of two events. Testable mechanisms must be proposed which are the simplest that fit the facts presently known" (Horton chap. 29). To put it another way, "the mechanisms that functionally link the specified... changes with the extinctions must be provided" (Grayson chap. 37). As an example, "a close correlation between extinction and the time of human arrival provided a mechanism for extinctions, in the sudden presence of man in a country where the fauna and flora had not previously been exposed to man's activities" (Horton chap. 29).

Thus, a particular cause is deemed a viable explanation only if it is linked with a testable mechanism. Such linkage permits the identification of synchronous events which are spurious (irrelevant) and nonspurious (relevant). A causation relationship is considered nonexistent if two events are shown to be diachronous.

If an extinction event can be confidently correlated with only one factor (e.g. climatic change), then the probability of a causation relationship is increased. In such cases it is necessary to consider whether it is *the* or *the only* cause. For example, "the correlation is so good between the timing of the extinctions and the early Clovis culture... that the burden of proof has probably shifted from the overkill theorists to those who would argue for nonhuman causes" (Guthrie chap. 13).

However, multiple synchronous events can usually be identified, and the causes have been specified as *primary, ultimate, interrelated, complex,* or *mosaic.* The trick is to identify, if possible, the independent effects of each specified event on the fauna.

A good example of multiple correlations is provided by King and Saunders (chap. 15). Between 12,000 to 10,000 yr B.P. in northeastern North America there occurred four phenomena: 1) the retreat of the ice sheet north of the Great Lakes, 2) the replacement of spruce woodland and tundra by pine and deciduous species, 3) the first undisputed evidence of man in New World, and 4) the extinction of *Mammut americanum*. In the Southwest between 11,500 to 10,500 yr B.P. Haynes (chap. 16) documents 1) megafaunal extinction, 2) Clovis culture, and 3) the greatest vegetational-climatic change since the Sangamon interglacial. Diverse interpretations for the relevance of these synchronous events exist. One review points out that both climatic change and the arrival of man "obviously had an effect upon the ecosystems they encountered. What effect one would have had in the absence of the other is untestable, therefore their relative importance must remain unknown. Did one set up the punch that the other delivered, or was it vice versa?" (Guilday chap. 12). On the other hand, "it is possible to argue that both megafaunal extinctions and the expansion of humans [into the Arctic and hence the Americas] are features of the same climatic event" and were related only secondarily to each other (Guthrie chap. 13). It has also been argued that "the megafauna could survive both human activity and climatic change as individual effects, but only the two combined were capable of causing extinction" (Horton chap. 29).

Historical Perspective

Martin and Neuner (1978) remark that it "seems fruitless to study earlier extinctions when the one at the end of the Pleistocene still eludes us." Yet, viewed in perspective of other extinction events during the Phanerozoic, that for the late Pleistocene/Holocene is insignificant (Raup and Sepkoski 1982). A feature of earlier extinction events (e.g. Permo-Triassic, Cretaceous-Tertiary) is that both marine and continental biotas show major losses. For the late Pleistocene/Holocene there was conspicuous loss of large land mammals, but little else (Martin chap. 17).

Gingerich (chap. 10) asks, "in the context of the entire Cenozoic era, are Pleistocene extinctions unusual?" He concludes that "56 percent of large mammalian species disappeared in Rancholabrean-3 and were not replaced by new large mammals." Furthermore, "the rate of late Pleistocene extinctions was the highest of any subdivision of the Cenozoic, ... [yet] when viewed in terms of genera per total genera, the rate of late Pleistocene extinctions was about average for earlier parts of the Cenozoic."

Webb (chap. 9) analyzes the last ten million years of mammalian history in North America. He concludes that the greatest extinction episode, involving sixty-two genera (thirty-five large ones) was in the late Hemphillian (five million years ago), while forty genera (thirty-nine large ones) went extinct during the Rancholabrean. Webb cautions that the Rancholabrean extinctions are "synchronized within a few thousand years, [yet for the late Hemphillian it is not known whether] extinctions all occurred near the end of that interval within a few thousand years, in which case it was at least as great an extinction episode as the Rancholabrean," or are protracted through the late Hemphillian, in which case it is a considerably smaller episode than the Rancholabrean. While inconclusive, his suggestion is provocative.

Extreme caution must be observed when comparing and contrasting the above data sets. Raup and Sepkoski (1982) use stratigraphic range data of families, Webb (chap. 9) uses genera, and Gingerich (chap. 10) uses species. In studies of this sort it is assumed, but has never been adequately demonstrated, that different taxonomic levels track one another, and that aspects of taxonomic evolution shown by one rank (e.g. genus) mirror that of another (e.g. species). In one study (Marshall et al. 1982) the pattern shown by analysis of families and genera were, in part, different. Stratigraphic range data for Pleistocene mammal genera and species in North America are available Kurtén and Anderson 1980), and a comparative study focusing on tracking would be most informative.

Many chapter authors argue that the old axiom—*the present is a key to the past*—no longer stands. Guthrie (chap. 13) speaks of the *standards tied to normalcy of present* as being erroneous when looking at the Pleistocene. The present can no longer be regarded as the norm. The appropriate question is not, "Where were today's biotic provinces during the late Pleistocene?", but rather, "Why were the late Pleistocene biotic provinces so different from those of today?" (chap. 13). He also speaks of the *fallacy of the primeval*—the assumption that biomes as we know them today are primordial units. Not true; biomes of today are products of the Holocene. The present disharmonious associations are a result of a filtering and separation of communities and species which had a long history of association. Graham and Lundelius (chap. 11) believe that "modern community patterns began to evolve from these disharmonious associations between 12,000 and 10,000 years ago throughout North America." Hope (chap. 30) found a similar situation in Australia and New Guinea, where plant "taxa formed different communities in late Pleistocene environments." Most authors (Graham and Lundelius, Guilday, Guthrie, Gingerich) agree that since the late Pleistocene there has been a decrease in habitat diversity. That there is also a decrease in faunal diversity is expected.

Extinction

Various types of extinctions are recognized and are defined by either the mechanism causing the extinction or the magnitude of the extinction itself.

1. *Taxonomic* or *phyletic* extinctions occur when one taxonomic unit (i.e. species) evolves into another of equal or higher rank. For example, species A in the middle Pleistocene gives rise to species B in the late Pleistocene. There is only one phyletic lineage, but with evolutionary change it appears that we are dealing with multiple-origination (first appearance) and extinction (last appearance) events. *Phyletic* extinctions are inherent in any data set, and they increase the number of apparent extinctions for any given time interval.

2. *Accidental* extinctions (*sensu* MacArthur and Wilson 1967) are "the departure or extirpation of undifferentiated island populations of plants and animals" (Martin chap. 17).

3. *Differential* extinction, in the context of the late Pleistocene/Holocene, refers to disproportionately greater extinction in larger- than in smaller-sized animals (Guilday 1967). "It does seem that some process or processes bore more heavily on the larger than on the smaller mammals" (Merrilees chap. 28). Guilday (chap. 12) discusses why animals with large bodies go extinct while smaller-sized forms survive. In contrast, Graham and Lundelius (chap. 11) argue that late Pleistocene/Holocene extinctions in North America include "all size categories of mammals, and three of the four size groups [they recognize] (small, medium, and large) contributed about equally to the extinction."

4. *Local* or *regional* extinctions (i.e. range reductions) result from the temporary or permanent disappearance of a species population from a given area. Martin (chap. 17), following Stenseth (1979), defines *local extinctions* as "repeatable phenomena that operate in ecological time" (also see Klein chap. 25, Guilday chap. 12).

5. *Biological, blanket, total,* or *global* extinctions occur when a species death rate exceeds birth rate for a prolonged period of time. This, the permanent disappearance of a taxonomic lineage, is what Martin (chap. 17) calls *extinction forever*. Guilday (chap. 12) discusses *biological* extinctions, and Klein (chap. 25) *total* extinctions. Martin (chap. 17), following Stenseth (1979), defines *global* extinctions as those which are "irreversible and operate in geological time."

6. *Extinction episodes* or *events* occur when, within a relatively short period of time, extinction rates exceed origination rates; the result is a decrease in taxonomic diversity.

7. *Coextinction* occurs when the extinction of one taxon is the direct result of extinction of another (i.e. loss of one taxon is dependent upon, not independent of, loss of another; Steadman and Martin chap. 21). Two types are recognized: *animal-animal* and *animal-plant*. *Animal-animal* is divisible into two categories: *predator-prey* coex-

tinction which results when the extinction of a predator (e.g. *Smilodon*) follows the extinction of its prey (i.e. mastodont) (E. Anderson chap. 2); and coextinction of animals and their *commensals, scavengers,* and *parasites*. As examples, most bird extinctions in North America are attributed to disappearance of mammal megafauna and the loss of niches for bird scavengers and commensals (Steadman and Martin chap. 21), and the subcutaneous botfly *Pallasiomyia antilopum* virtually disappeared with its saiga host (Vereshchagin and Bryshnikov chap. 22).

Animal-plant coextinction results from "disequilibrium created by the disruption of coevolutionary interactions between plants and animals" (Graham and Lundelius chap. 11). The disappearance of a plant food species or disruption (i.e. simplification) of plant communities may result in the disappearance of herbivores which feed on a particular plant species or are dependent on a unique vegetational community (Guthrie chap. 13).

Turnover

Taxonomic turnover is a function of absolute and relative differences in rates of origination and extinction of taxa as documented by stratigraphic range data. The Equilibrium Hypothesis of MacArthur and Wilson (1967) predicts that, in a saturated fauna, a stable equilibrium exists in which origination and extinction rates are stochastically constant. That is, equilibrium persists until disrupted by appearance of new faunal groups or environmental change. The former disruption results from biological interactions of taxa, the latter from changing ecologies. Random or chance phenomena can also cause disequilibrium. Aspects of turnover are easy to document in the fossil record. Yet attempts to attribute the turnover episodes in general, or origination and/or extinction events in particular, biological or environmental causes are largely speculative.

Gingerich (chap. 10), using the range data in Kurtén and Anderson (1980), analyzes turnover patterns in North American Pleistocene mammals. "Given that the extinction rate remains in close equilibrium with origination rate, the high rate of late Pleistocene extinctions can be viewed as a natural equilibrium ending a one- to two-million-year interval of high saturation that followed an unusually high rate of early Pleistocene originations. In this context, *what requires explanation is not late Pleistocene extinctions but the very high rate of early Pleistocene originations.*" He concludes that late Pleistocene extinction is "best regarded as a natural consequence of high faunal turnover caused by major oscillations in climate and environmental heterogeneity" (i.e. a climatically or environmentally induced turnover episode as described by Webb chap. 9). At the same time Gingerich (chap. 10) entertains the possibility that the most recent wave of immigrants from Asia is in some way causally related to late Pleistocene extinctions in North America (i.e. an immigration- or invasion-induced turnover episode of Webb chap. 9).

There are two fundamental types of turnover, *with replacement* and *without replacement*. Both are relevant to late Pleistocene/Holocene extinctions and each can be divided into mutually exclusive ranks.

Turnover With Replacement

With replacement turnover occurs when originations follow extinctions, and diversity remains relatively constant. Two types are distinguishable. The first, *active or competitive replacement,* may ensue between native and invading *vicar:*

[groups which have a similar role in nature or occupy the same trophic level within an adaptive zone (Van Valen 1971, Van Valen and Sloan 1966)]. The vicars need not be closely related taxonomically and may be in different families, orders, or classes. If the native form is more efficient (i.e. competitively superior) in exploiting the available resources, the invader will either be expelled or be prevented from entering.... If the invader is victorious, the native becomes extinct (Marshall 1981, p. 146).

A causal relationship for inferring competitive replacement is permitted if the appearance of an exotic (*allochthonous*) taxon is synchronous (or nearly so) with the disappearance of a native (*autochthonous*) vicar.

As a general example of active replacement, Klein (chap. 25) attributes early and middle Pleistocene extinctions in Africa to failure of species "to compete successfully in biotic communities that were constantly changing as a result of evolution in local species and the immigration of foreign ones."

Active replacement may be accomplished by either *invasions* or *introductions*. E. Anderson (chap. 2) provides numerous examples of invasions: extinction of *Arctodus* due to competition with invading brown and grizzly bears, extinction of *Canis dirus* due to competition with *C. lupus*, and so forth. Guthrie (chap. 13) also notes that "it is possible... the sheer increase in numbers of cervid browsers hastened the mastodont's extinction."

Examples of introductions may be found in Madagascar, where the native terrestrial herbivores (tortoises, birds, primates) were replaced by domestic bovids (cattle, sheep, goats) and suids introduced by man. "This was brought about by competition, habitat destruction and fragmentation, and adventitious hunting" (Dewar chap. 26). "In northern Africa... terminal Pleistocene/early Holocene extinctions appear to have been preceded or accompanied by the introduction of domestic stock,...[reflecting] the inability of one or more [native] species to compete with stock for pasture" (Klein chap. 25).

Passive replacement, the second type of turnover with replacement, as stated by Marshall (1981, p. 146),

> may occur when chance phenomena or environmental factors cause extinction of a lineage and its role is assumed by a native group or timely invader. The successor fills a vacated niche or adaptive zone, but it need not necessarily be competitively superior to the group it replaces. In such instances, "it is not physical or behavioral limitations which guide a group's evolutionary potential or success, but merely the opportunity to exploit an available adaptive zone, which, because of the nature of the fauna, was open" (Hecht 1975, p. 248).

Klein (chap. 25) provides examples of passive replacement by *autochthonous (native) groups*: the "evolutionary" success of the bovids [in Africa] may be largely responsible for the contemporaneous decline in suid fortunes, while evolutionary advances in the hominids probably account for... reduction in terrestrial monkey diversity.... The successful entry of hominids into a meat-eating niche was perhaps also responsible for a decline in the diversity of carnivores." Webb (chap. 9) documents

> an apparent correlation between late Cenozoic decreases in large mammalian herbivore genera and increases in small mammalian herbivore genera.... These data suggest that the replacement of ungulate diversity by rodent diversity occurred not because of direct competition, but as a consequence of most major extinctions more adversely affecting large ungulates. In the late Cenozoic the net result was that small grazing herbivores filled the vacancies left by large grazing herbivores (Webb 1969).

As for passive replacement by *allochthonous (timely) invaders*, Webb (chap. 9) shows that for the late Pleistocene, "the record suggests that immigrations generally tracked extinctions. In effect, immigrant taxa filled the vacuums produced by prior extinctions." As an example, Graham and Lundelius (chap. 11) conclude that "with the extinction of numerous herd herbivores, other preadapted species such as *Rangifer*, *Ovibos*, and *Antilocapra* were able to invade the newly formed communities."

Turnover Without Replacement

Without replacement turnover occurs when extinctions are not followed by originations. The net result is a decrease in diversity as a result of local and/or total extinctions (Gilbert and Martin chap. 6). As an example of *climate-related* turnover of this type, E. Anderson (chap. 2) states that the disappearance of *Tapirus* from North America is "probably due to climatic change," and *"Glyptotherium* was so specialized and lived in such a restricted environment that local populations were easily wiped out by climatic change."

A second category of turnover without replacement is *man-related*. Extinction of *Myotragus* and *Hydrodamalis* was caused by human predation (E. Anderson chap. 2). The disappearance of lizards, frogs, and insects on islands is related to activities of rats and mice introduced by man (Cassels chap. 34, E. Anderson chap. 2).

A third category is a *combination of man and climate*. E. Anderson (chap. 2) attributes the extinction of *Mammut* and *Mammuthus* to both climatic change and hunting.

The Replacement-Nonreplacement Debate

Martin (chap. 17), argues that late Pleistocene/Holocene extinctions exemplify turnover without replacement, and that the niches once filled by these now extinct megafauna are vacant and still existing today. If, however, some extinctions resulted from competition with invading taxa, then there was at least partial replacement. Furthermore, Graham and Lundelius (chap. 11) and Guthrie (chap. 13) feel that many of the niches are no longer in existence; if so, there is no possibility for replacement to occur.

Megafauna

Megafauna is not a taxonomic group nor is there a standard definition. Horton (chap. 29), for example, defines megafauna as consisting of "animals that became extinct before the Holocene and are large, either in an absolute sense or relative to other members of some taxonomic rank, or are part of a taxonomic category all of whose members became extinct and some of whose members are large." Attempts have been made to define megafauna in terms of *average adult body weight*, although *large* (megafauna) and *small* (non-megafauna) have been interpreted differently. Some definitions are fraught with circular reasoning (i.e. to be large and to be extinct is to be megafauna; to be small and to be extinct is something else), and the problem of arbitrary selection of size (i.e. what is large and what is small) is open to wide interpretation. Four chapter authors define megafauna in terms of kilograms (kg), and each uses a different criterion and weight category.

Webb (chap. 9) define megafauna as species >5 kg, based on Bourliere (1975) who "explains 'the bimodal distribution of body weights among terrestrial mammals' and recognizes as large mammals those with average adult body weight over 5 kg."

Horton (chap. 29) defines "megafauna by an arbitrary body weight, perhaps species weighing more than 20 to 25 kg."

Martin (chap. 17) defines megafauna as follows: "Since most genera of continental herbivores which became extinct in the late Pleistocene approximated or exceeded 44 kg (100 lbs.) in adult body weight, animals of this size or larger will be assigned to the 'megafauna'; those under 44 kg will be considered 'small.'"

Graham and Lundelius (chap. 11) and Gingerich (chap. 10) use the stratigraphic range data of Kurtén and Anderson (1980) and follow the size categories recognized by these authors. *Small* includes the "small" (1 g to 907 g) and "medium" (908 g to 181 kg) categories of Kurtén and Anderson, and *large* (megafauna) includes the "large" (182 kg to 1.9 tons) and "very large" (>2 tons) categories of Kurtén and Anderson.

Thus megafauna as defined in this book begins at a size range between 5 and 182 kg. It is possible that other authors, who simply use the term without defining it in terms of kilograms, may have had an even different weight category in mind. In 1984 the concept of megafauna is considerably different from what it was seventeen years ago. It has become general practice to divide megafauna of earlier workers into various categories.

Marshall (1973, 1974) divides Australian megafauna into two groups: 1) those species which became extinct; and 2) those species which underwent a diminution in body size, with living forms smaller than their late Pleistocene ancestors (e.g. species of *Macropus, Megaleia, Osphranter, Petrogale, Sarcophilus, Dasyurus*). The latter group exemplify *phyletic dwarfism* (Marshall and Corruccini 1978), a Quaternary phenomenon that apparently represents "a *general evolutionary trend*" (Hooijer 1950, p. 360).

Horton (chap. 29) divides Australian megafauna into four groups: 1) arid species (i.e. red kangaroo) which survive to the present, 2) dwarfed species which survive to the present (i.e. grey kangaroo, euro, Tasmanian devil), 3) "woodland species" which die out in a first wave of extinctions 26,000 to 20,000 yr B.P. (i.e. most megafauna), and 4) "southern fringe species" (i.e. four species of *Sthenurus*), which survive the first wave but die out in a second wave of extinctions 18,000 to 15,000 yr B.P. Thus Horton regards megafaunal extinction in Australia as a two-stage process. [Actually the red kangaroo (*Megaleia rufa*) was also dwarfed (see Marshall 1974, p. 79), reducing Horton's categories to three.]

Extinction and dwarfing are documented for megafaunal species on all continents. In North America alone dwarfing is reported in species of *Alces, Ovis, Ovibos, Rangifer, Bison, Felis, Hydrochoerus, Panthera, Tapirus,* and *Ursus* (Guthrie chap. 13, Guilday chap. 12, McDonald chap. 18, Kurtén and Anderson 1980). In fact, all large species in North America which are living today underwent some degree of size decrease during the Holocene (Gilbert and Martin chap. 6).

Extinction and dwarfing were concurrent processes (Guthrie chap. 13) and probably resulted from a common causative factor(s) (Gilbert and Martin chap. 6, Guthrie chap. 13, McDonald chap. 18). This correlation is highlighted by detailed studies of extinct species that indicate they underwent dwarfing just prior to their extinction [i.e. dwarfed populations of *Mammuthus, Equus,* and *Coelodonta* occur around 12,000–11,000 yr B.P. in Europe (Guthrie chap. 13); dwarfed *Mammuthus* (with small tusks) in Clovis (and other) sites around 12,000–11,000 yr B.P. in North America (Guthrie chap. 13); dwarfed *Mammut americanum* between 12,000–10,000 yr B.P. in North America (King and Saunders chap. 15)]. There is good evidence that man was killing these dwarfed animals just prior to their extinction (Guthrie chap. 13).

Two unresolved issues exist with regard to *megafauna*. First, is it a useful term and is there good reason to continue to use it? I have my doubts. Much debate and effort has gone into defining the term and in trying to pigeonhole taxa into or out of it. At such times the term is a hindrance to understanding the extinction process itself. Why not simply list which taxa go extinct, which dwarf, and which do neither? These three groups will have different size components for each landmass, for the "extinct group" for North America is not the same as "extinct group" for Australia. This method should help clear the air and enable a less restricted approach to understanding extinctions both on an individual landmass and a worldwide basis. Second, if the term is retained, then it is evident that two categories need to be recognized. The term *dwarfed megafauna* may be used for those large-body-sized species which dwarf but survive, and *extinct megafauna* for those large-body-sized species which go extinct, some after they dwarf. I use these terms conceptually and not operationally and am at a loss to provide a definition incorporating body weights for each that will be useful for a particular land-mass, much less the world.

Dwarfing

The knowledge that extinction and dwarfing are concurrent processes may provide new insight into extinction phenomena. If we can identify mechanisms causing a decrease in body size, we can, so to speak, approach extinction "through the back door"—assuming that both extinction and dwarfing are indeed linked to a common causal factor(s). An explanation for the dwarfing may then be extrapolated to explain the extinctions.

Guthrie (chap. 13), Guilday (chap. 12), and McDonald (chap. 18) develop elaborate paradigms to explain extinction and dwarfing, yet Guthrie and Guilday attribute these causes to climate, and McDonald attributes them to man.

Guthrie (chap. 13), approaching extinctions from the view of *paleo range management*, concludes that a changing seasonal regime resulted in a decrease in habitat complexity (his *plaids vs. stripes*) and an increase in homogeneous plant communities, a decrease in the quality and quantity of available plant resources, climatic deterioration, dwarfing, and extinction. The key feature affecting the dwarfing and extinction was a decrease in the length of the mammalian growing season. The "major brunt of the extinctions was borne by those ungulates with conservative life histories that do not do well in short growing seasons" (chap. 13). Upon deglaciation, habitats available to the large herbivores became less favorable (Haynes chap. 16).

McDonald (chap. 18), however, believes that upon deglaciation, 18,000 to 8,000 yr B.P., more favorable habitats became available to the large herbivores as a result of increased environmental heterogeneity. Approaching the issue from the view of *paleo life history strategies*, he concludes that milder climates and more diverse habitats resulted in an increase in primary production; dwarfing and extinction he relates to human activity.

Many causes have been proposed to explain dwarfing, and some are discussed by Marshall and Corruccini (1978, pp. 113–116). A representative but not exhaustive list of these is given below.

1. *Man.* McDonald (chap. 18) and Edwards (1967) attribute dwarfing to human hunting. The premise is that an increase in predation by man results in a decrease in average size of his prey.

> With a downward shift in the age distribution, adults that mature faster and reproduce faster are genetically selected, and these tend to be smaller. Human technology... greatly reduces the counterattacking defense advantages of larger size and emphasizes concealment and speed of flight. At this point of increased pressure of human predation, the genetically selected optimum body size of many forms declines sharply. Also, humans may intentionally bypass smaller prey individuals for more efficient collecting or for game conservation; thus the genetically determined, mature dwarfs may escape predation by being considered less productive of food or by being mistaken for immature animals (Edwards 1967, p. 149).*

McDonald (chap. 18) views human hunting as an ecological process, causing a restructuring of the North American large mammal selection regime. The long gestation periods of large-bodied mammals were selected against. Human hunting either lowered

---

*Reprinted by permission from *Pleistocene Extinctions: The Search for a Cause,* edited by P. S. Martin and H. E. Wright, Jr. (New Haven: Yale University Press), copyright 1967.

population age structures to levels where viability was lost and extinction occurred, or favored phyletic evolution to a more secure position within the new selection regime. One of the five "measures of fitness" in this new regime was a "relatively small body size."

Davis (1981) and Tchernov (chap. 24) also document instances of Holocene dwarfing in Europe which they attribute to artificial selection accompanying domestication.

2. *Resource limitation and climate*. Guthrie (chap. 13) believes dwarfing to be the result of a decrease in *animal quality* or *fitness* imposed by a reduction in growth season and concurrent vegetational change. King and Saunders (chap. 15) correlate dwarfing in *Mammut americanum* to a decrease in *habitat quality*: coniferous (pine-dominated) vegetation began to break up into scattered, island-like areas about 10,000 yr B.P., and "these islands would have provided short-term habitats for mastodont concentrations." Thus dwarfing may reflect response to insular distributions of habitat. Kurtén (1968, p. 252) likewise suggests that the primary cause for dwarfing on continents "is the same that resulted in the evolution of dwarf forms on islands; the necessity to keep up an adequate population density in spite of severe limitation of habitat and/or food supply." By the process of dwarfing, a species can shift from K-selected to more r-selected life-history strategies (McDonald chap. 18). A feature of r-selected strategies is an increase in rate of reproduction (Dewar chap. 26). The smaller body size permits maintenance of higher population levels and hence greater genetic variability, features which favor survival despite *ecologic accidents* (Kurtén 1965, p. 62). Dwarfing is thus viewed by Kurtén to be a factor of *population density*. "In other words, these adaptive arguments apply to groups, not to individuals" (Marshall and Corruccini 1978, p. 115). They are thus group-selectionist.

King and Saunders (chap. 15) found a disproportionately high percentage of prime or mature individuals and few young and old ones in a population of *Mammut americanum*. This bias in age structure they interpret as *imposed self-regulation*. "As the coniferous habitat was collapsing,... mastodonts were undergoing self-regulation, evidenced by low recruitment and high modal age class analogous to the African evidence..., during a time of environmental stress." The ultimate extinction of *M. americanum* is attributed to its inability to successfully self-regulate or otherwise adapt in response to the rapidly changing climates and vegetation of the terminal late-Wisconsin. This and other studies of age structure of populations (e.g. see Horton chap. 29) indicate low recruitment rates in forms going extinct.

Contrasting with the study by King and Saunders is that of Agenbroad (chap. 4), who describes a 26,000-yr-B.P. death assemblage of *Mammuthus columbi*. This population was a sample at a time when there is no evidence of an unfavorable climate. The population demonstrates a balanced age structure and is composed of animals of normal size (i.e. they are not dwarfed).

3. *Temperature or ecogeography*. "There seems to be general agreement that this size reduction was the result of climatic change (Bergmann's rule)" (Gilbert and Martin chap. 6). "Most of the mammals investigated in the Middle East show a direct and swift response to the temperature increase in the Pleistocene/Holocene transition period by rather drastic diminution of body sizes" (Tchernov chap. 24).

4. *Character displacement and competition* is discussed by Tchernov (chap. 24).

Overkill

Overkill embraces all aspects of human intervention as causal factors in extinction processes. As defined by Martin (chap. 17), *overkill* is "human destruction of native fauna either by *gradual attrition* over many thousands of years *or* suddenly in as little as a few hundred years or less. Sudden extinction following initial colonization of a landmass inhabited by animals especially vulnerable to the new human predator represents, in effect, a prehistoric faunal *'blitzkrieg.'*" The overkill paradigm is appealing "because of its resemblance to the dire effects that Recent human cultures are inflicting on many surviving species of large mammals" (Webb chap. 9).

In Europe, as Kurtén (1968, p. 273) points out, "it seems fairly certain that modern man has played a dominant role in the wiping out of many species, although perhaps by *indirect influences* as much as by *actual hunting*" (see also Martin chap. 17). In Australia "it is tempting to suggest that some human activity underlies these extinctions, whether habitat modification by fire..., selective hunting or indiscriminately wasteful methods of hunting, or some other activity as yet undiscerned" (Merrilees chap. 28). On the Hawaiian Islands the extinction of some 54 percent of endemic species of land birds is attributed to both man's use of fire in slash-burn agriculture and direct hunting (Olson and James chap. 35). The same case has been made for all other continents and islands (Martin chap. 17).

Dewar (chap. 26) addresses the question "were these waves of extinctions largely the result of human activity?... [and if so], can each of these waves be explained by the same kind of human activity?" He proposes two favored means of human participation: *direct causation* (i.e. hunting) and *indirect causation* (i.e. destruction of habitat). Other authors (this volume) also define overkill to include multiple aspects of human-related activity (e.g. *overkill, Pleistocene overkill, slow overkill, rapid overkill*). Below I propose a hierarchical classification for overkill and define each rank using examples from this book. They are not all mutually exclusive, and boundaries are somewhat arbitrary.

Blitzkrieg (Man-the-Hunter, Exterminator)

Martin (1973) defines *blitzkrieg* as "*a special case of overkill.*" The three basic ingredients are 1) rapid deployment of human populations into an area not previously inhabited by man, 2) man's possession of a big-game-hunting technology, and 3) the virtual simultaneous and synchronous extinction of megafauna resulting from direct hunting by humans (Martin chap. 17). Blitzkrieg results in man's overexploitation of "a virgin landscape and a naive fauna" (Horton chap. 29). Extinctions will occur in such a short period of time that there will be an absence or dearth of kill sites recording this process (Martin chap. 17). "The swifter the job was done, the narrower the archaeological window, therefore the fewer the sites that might be expected to record the event" (Guilday chap. 12).

The blitzkrieg paradigm is attractive "because of its widespread efficiency (across diverse habitats and various climatic regimes)" (Webb chap. 9). The *direct causation* of Dewar (chap. 26) and *rapid overkill* of Grayson (chap. 37) represents blitzkrieg. I recognize two types of blitzkrieg: *direct* and *associated*.

Direct blitzkrieg. Man-the-big-game-hunter is sole cause of extinction of unstressed populations of megafauna (Martin chap. 17, Whittington and Dyke chap. 20). Man's presence alone is not enough to invoke direct blitzkrieg; he must also have a hunting technology (e.g. Clovis culture) which enables him to effectively kill megafauna.

Associated blitzkrieg. Man-the-big-game-hunter and concurrent natural climatic change contribute hand-in-hand to megafaunal extinctions. Megafaunal populations are stressed and vulnerable as a result of climatic change. *Associated blitzkrieg* thus relies on "...a particular set of cultural and environmental conditions" (Grayson chap. 37). In Alaska, for example, there is good evidence that man dealt the coup de grace (*sensu* Guilday 1967) by killing stressed megafauna (i.e. the animals he was killing were dwarfed before they became extinct; Guthrie chap. 13).

As Guilday states (chap. 12) "It is possible, indeed probable, that human predation was the event that forced the apparently geologically concordant extinction of so many large mammals... [but] only if those same species had previously been reduced numerically and geographically [by climatic change] to relict status." Haynes (chap. 16) puts it this way: "Animals that concentrated about the remaining watering places were more vulnerable to predation. Adding Clovis hunters to the scene may have been the determining factor in bringing about extinctions 11,000 years ago."

Innovation

The three basic ingredients of innovation are 1) man's long-established presence in an area (i.e., Africa, Asia, Europe), 2) natural climatic change stressing megafauna, and 3) man's innovation of a big-game-hunting technology. This technology results "... from the development of a new, hunting-oriented subsistence base by inhabitants of an area who previously only occasionally took large mammals" (Grayson chap. 37).

Klein (chap. 25) provides an excellent example of innovation in Africa. "I suggest that qualitatively different people were the new ingredient in the terminal Pleistocene/ early Holocene situation, directly or indirectly precipitating extinctions after environmental change had reduced species numbers and distribution.... It is unneccessary to hypothesize that each or any extinct species was intensively hunted to its demise, but only that the hunters' adjustment to changing environmental circumstances perturbed a system that was already in tilt, forcing it in a different direction."

Extinction resulting from innovation will predictably be less abrupt and severe than that resulting from blitzkrieg, because man and megafauna had been coevolving only in the former situation. Megafauna in an innovation situation became more wary of man and their chances for survival greater. This situation is recognized by Kurtén (1971, p. 221): "most of the Eurasian invaders to North America, the moose, wapiti, caribou, musk ox, grizzly bears, and so on—were able to maintain themselves, perhaps because of their long previous conditioning to man." Also, Klein (chap. 25) notes that in neither Africa nor Eurasia "is there evidence for a major 'wave' of extinctions, wiping out a very large proportion of the existing fauna in a relatively brief time." Humans have been on these continents for a long time and arrived at a time "when they were probably relatively unimportant as predators on other mammals."

Attrition (Man-the-Farmer, Occasional Hunter)

The four basic ingredients of attrition are 1) man's previous presence in an area, 2) man's manipulation of his environment and alteration of ecologies through farming practices, 3) occasional hunting, and 4) the slow disappearance of megafauna (i.e., extinctions not rapid as in blitzkrieg). Attritional extinctions will result in many recoverable kill sites. In its broadest sense, attrition is equivalent to *indirect causation* (i.e., human participation in habitat destruction) of Dewar (chap. 26) and Grayson's *slow overkill* (chap 37).

Indirect attrition. This form of attrition involves man-the-farmer and habitat manipulator (activities which stress populations of native animals) and occasional hunting. Included are *fire stick farming* (Horton chap. 29) and the *extinction-by-hunting-and-burning* hypothesis (Cassels chap. 34). For example, some authors propose that in Australia man's arrival "resulted in increased fire leading to habitat change and therefore extinction" (Horton chap. 29, a thesis he rejects).

The burning and clearing of woodlands by pastoralists resulted in destruction of areas larger than those needed for habitation and economic activity (Dewar chap. 26). This activity may account for extinction of small-body-sized animals, such as forest birds on the Hawaiian Islands (Olson and James chap. 35).

A possible example of indirect attrition is available in Dewar's (chap. 26) description of the disappearance of large native mammals in Madagascar:

pastoralists settled in Madagascar and began to travel with their herds across the mosaic of the interior, seeking patches of grasslands. If suitable grazing areas were small and scattered, it would have been necessary to travel substantial distances during a season. Along the way these pastoralists killed wild animals they found, and they perhaps already practiced some form of dry-season burning. Cattle, and probably sheep and goats, came to replace the indigenous terrestrial herbivores. It is precisely this replacement of the herbivores, primary consumers of ground-level plants, which I believe is the missing link in the previous scenarios.

Cassels (chap. 34) offers another example of indirect attrition for South Island, New Zealand, where there is evidence of mass killings of moas. "The time period [for the moa killings] was short, less than 1000 years, but not instantaneous, involving at least several hundred years.... Presumably, none of the human populations of the Pacific depended on the extinct species for their livelihood. Burning and hunting had a catastrophic effect."

A similar view is taken by Trotter and McCulloch (chap. 32). Large numbers of ratites lived in a predator-free environment in New Zealand until the coming of man about 1,000 yr B.P. Within a few centuries of man's arrival, ratites and a number of other birds became extinct due to hunting, egg-gathering, and the destruction of their forest habitat by fire.

It must be emphasized that *contra* Martin (chap. 17), moa extinctions in New Zealand are not a case of blitzkrieg (as defined above). "It is certain that the moas were not wiped out in the first wave of human colonization.... From earliest time the prehistoric Maoris based their subsistence on horticulture, fishing, and gathering...; they were not dependent on hunting moas for their livelihood" (Cassels chap. 34).

Competitive attrition. Competition may have occurred between man and other animals such as terrestrial monkeys, carnivores, and suids in Africa (Martin chap. 17, Klein chap. 25). Such attrition would have resulted from man's competitive exclusion of other animals due to overlap in diet preference, feeding strategies, or habitat utilization.

Another type of competitive attrition is due to man's present inability to tolerate native taxa. "Most recent extinctions... are due to habitat reductions rather than human overhunting.... The bears, wolves, bison, bobcats, pumas, wapiti, and others could be easily restocked in a matter of years—but they aren't because the 'habitat' in which we can tolerate them living, amongst sheep ranchers, row crops, and suburbia, has all but been eliminated" (Guthrie chap. 13).

Climate

The two basic ingredients are 1) climatic change and 2) concurrent megafaunal extinctions. A proponent of climate-caused extinctions is Horton (chap. 29), who believes that even if the Aborigines had never reached Australia, "that continent's fauna and flora still would have been almost identical to [that] which the first European explorers found around two hundred years ago." Other workers (e.g. Gilbert and Martin chap. 6) view climate as an "attractive alternative to overkill."

Guthrie (chap. 13) attempts to provide a *central biotic explanation* for his *ecological model*, which he summarizes as follows:

1. Climatic changes in seasonal regimes decreased diversity, increased zonation of plant communities, and caused a shift in net antiherbivory defense strategies. 2. This change in the plant community resulted in a shorter, less diverse growing

season for ungulates, decreasing net annual quality and quantity of resources available to many large mammal species. 3. These restrictions in available resources decreased local faunal diversity, body size, distributional ranges and frequently resulted in extinctions.

The components of climatic change may be divided into two categories, *direct aspects* and *linking mechanisms*.

1. *Direct aspects* are those primary consequences of climatic change affecting fauna. For example, *intensifying seasonality* can result in drought conditions. Megafauna will be stressed by lack of water, and this stress can contribute to or directly cause extinction of local populations or, if widespread, can cause extinction of species (see Haynes, chap. 16, for discussion of such events in the southwestern United States).

An excellent example is offered by Horton (chap. 29) for megafaunal extinctions in Australia.

> The model I propose for the selective death of megafauna is based on its presumed need for free water. Small species may obtain sufficient water from their food. . . . Normally, the larger species lived where water supplies were spaced at distance no greater than that which the animals could cover in a time no greater than the longest interval they could survive without drinking. During drought, less reliable water supplies dried up, and those remaining were spaced farther apart. Therefore, animals at one water source could not travel far enough to reach another water source. Food supplies, reduced in quantity and quality by the drought, were eaten out within a radius equal to half the distance that the animals could travel in a time no greater than the longest interval they could survive without drinking. Ecologically they became tethered. . . . When the food supplies were exhausted, animals died even though a water source was still available as Shipman (1975) has described in the case of recent droughts in Africa.

2. *Linking mechanisms* are those secondary consequences of climatic change affecting fauna. For example, climate-linked changes in vegetation quality and type constrain herbivore growth potential (Guthrie chap. 13). Guilday (chap. 12) also links climatic change with range readjustments, dwarfing, and extinctions.

Gilbert and Martin (chap. 6) list four assumptions in the climate paradigm: 1) that vegetational change should reflect climatic change, 2) that large and small animals will be affected in any given area, 3) that climatic changes at the end of the Pleistocene will be qualitatively and quantitatively different from those of earlier interglacials, and 4) that living animals should show evidence of this climatic change through changes in distribution and/or morphology (i.e. dwarfing).

McDonald (chap. 18) lists five possible changes attributed to climate-related causes: 1) decreasing equability and increasing continentality of climate, 2) desiccating post-glacial ecosystem, 3) inability of some taxa to adjust reproductive habits to changing climates, 4) ecological disequilibrium initiated by rapid glacial retreat, and 5) competitive exclusion.

Graham and Lundelius (chap. 11) and Guthrie (chap. 13) believe that unlike earlier interglacials, the present one is more severe, with clear evidence of intensifying seasonality.

Theoretically extinction can also result if the young of large animals with long gestation periods are born out of season (Slaughter 1967, Axelrod 1967). This mechanism is no longer given much credence, and it "clearly cannot account for all extinctions of herbivorous mammals in extratropical regions at the end of the Pleistocene" (Kiltie, chap. 14). McDonald (chap. 18) includes birth-to-first parturition intervals of three years or less in herbivores or five years or less in carnivores, gestation periods less than twelve months, and annual biotic potentials equal to or greater than one per fertile female among his six points of "measure of fitness in the new selection regime." This selection regime is imposed by man-the-hunter and is not linked directly or otherwise to climate.

Conclusions

Chapters in this book disclose the diverse views regarding late Pleistocene/ Holocene extinctions. Included are a mixture of old and new explanations for the disappearance of many large terrestrial vertebrates in the late Cenozoic. As Cassels (chap. 34) notes, "The abundance of explanations leads one to wonder that anything managed to survive."

Pleistocene Extinctions: The Search for a Cause (Martin and Wright 1967) was successful in defining controversies and in stimulating research. We have come a long way since 1967 in documenting and in establishing chronologies for these extinctions; yet, the interpretation of these data remains the primary issue in identifying the cause(s) of the extinctions.

Late Pleistocene/Holocene extinctions are the result of one or multiple causes, depending upon which author is followed and which landmass (and time period) is considered (Table 36.1). In some cases a reasonable argument is made that extinction on landmass *A* resulted from cause 1, or extinction on landmass *B* from cause 2. On landmass *C*, however, causes 1 and 2 may both appear reasonable or even 1 and 2 and 3. The number of landmass-cause combinations are large, and for ten landmasses and three basic causes alone there are at least seventy possibilities. To rank these in a simple fashion is inconceivable, and the categories I used above are somewhat arbitrarily simplified, and the boundaries between them are ill-defined. But this is the state of art of late Pleistocene/Holocene extinctions, and I applaud the contributors of this book for attempting to deal with such a miasmatic problem.

There are two extreme paradigms for these extinctions—one favors man-the-big-game-hunter, the other natural climatic catastrophes. Many authors have taken the middle-of-the-road view and favor a multidimensional paradigm. I too prefer to be in this category.

Such paradigms are expressed by four of the authors as follows:

1) "An informal survey of colleagues... revealed that the majority did not believe that a single factor... was responsible for the disappearance of many genera; instead a mosaic of factors... combined to cause extinction" (E. Anderson chap. 2).

2) "Large mammal taxa did not simply vanish from a scene of ecological composure, but instead disappeared during a time of great ecological ferment when biotas were adjusting, dissolving, and reforming under new climatic parameters to emerge into the Holocene in greatly altered aspect.... All of this ecological fermentation served as a backdrop behind the spread and increasing technological sophistication of man-the-hunter.... Hunting pressure becomes just one of many interrelated 'causes' acting in consort to drive the species to extinction none of which could be singled out as the ultimate, or even the primary cause" (Guilday chap. 12).

3) "This explanation for terminal Pleistocene/early Holocene extinctions in Eurasia, combining climatic/environmental change and human actions, is comparable to the one I have offered for like-aged extinctions in Africa" (Klein chap. 25).

4) "Although neither overkill nor fire-induced habitat change are sufficient to cause extinction during normal times, they are effective during periods of climatic stress" (Horton chap. 29).

Multidimensional paradigms can be visualized using continuum diagrams (fig. 36.1). In the two-dimensional continuum (fig. 36.1a) the antipodes are man-the-big-game-hunter and natural climatic catastrophes (e.g. volcanic eruption causes total extinction of an island biota). In the three-dimensional continuum (fig. 36.1b) the antipodes

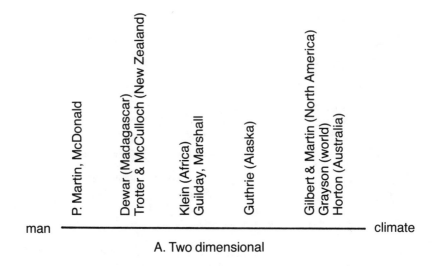

A. Two dimensional

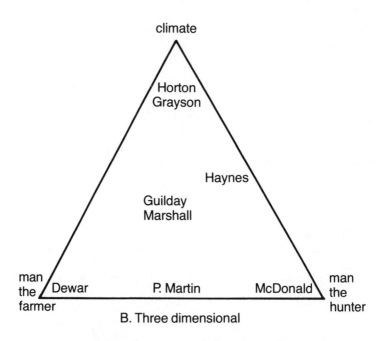

B. Three dimensional

Figure 36.1. Continuum extinction diagrams showing positions favored by chapter authors.

are man-the-big-game-hunter, man-the-farmer-habitat-manipulator, and natural climatic catastrophes. All of the paradigms for late Pleistocene/Holocene extinctions proposed in this book can be plotted with some degree of confidence on these diagrams. I indicate the position of paradigms favored by some chapter authors to illustrate this approach. (The positions and views indicated for the chapter authors are based on my interpretation of their works with the assistance of Paul Martin. The actual intended positions of the authors may be somewhat different.)

Rather than close with conclusions, I offer the following cautions:

1. Extinction on each landmass should be viewed as a discrete event—each *may be unique*.
2. Elimination of one cause for extinctions on one landmass does not warrant its exclusion as a possible cause on another landmass.
3. An extinction paradigm reasonable for one landmass need not be reasonable for another.
4. Extension of a paradigm from the specific (e.g. landmass *A*) to the general (i.e. the world) is risky and may result in its loss of credibility even for the area for which it was conceived.
5. A *which-came-first, the-chicken-or-egg* dilemma is often evident in the late Pleistocene/Holocene extinction debate. Could man have done all he is credited for had it not been for the direct or indirect help of climate, and where does the influence of one stop and the other start? A unique set of climatic events was responsible for letting man into the New World in the first place. Do I detect a conspiracy?

Acknowledgments

I wish to thank Julio Betancourt, Owen Davis, Cynthia Lindquist, Everett Lindsay, Paul S. Martin, Jim Mead, and Tinco van Hylckama for their help on various aspects of this chapter and Deborah Gaines for typing it.

References

Axelrod, D. I. 1967. Quaternary extinctions of large mammals. *Univ. Calif. Publ. Geol. Sci.*, V. 74, pp. 1–42.

Bourliere, F. 1975. Mammals, small and large: the ecological implications of size. pp. 1–8 *in* E. B. Golley et al., eds., *Small Mammals: Their Productivity and Population Dynamics*. Cambridge Univ. Press, London.

Davis, S. J. M. 1981. The effects of temperature change and domestication on the body size of late Pleistocene to Holocene mammals in Israel. *Paleobiology*, V. 7, pp. 101–114.

Edwards, W. E. 1967. The late-Pleistocene extinction and diminution in size of many mammalian species. pp. 141–154 *in* P. S. Martin and H. E. Wright, Jr., eds., *Pleistocene Extinctions: The Search for a Cause*. Yale Univ. Press, New Haven, 453 pp.

Guilday, J. E. 1967. Differential extinction during late-Pleistocene and Recent times. pp. 121–140 *in* P. S. Martin and H. E. Wright, Jr., eds., *Pleistocene Extinctions: The Search for a Cause*. Yale Univ. Press, New Haven, 453 pp.

Hecht, M. K. 1975. The morphology and relationships of the largest known terrestrial lizard, *Megalania prisca* Owen, from the Pleistocene of Australia. *Proc. Roy. Soc. Vict,*, V. 87, pp. 239–249.

Hempel, C. G. 1966. *Philosophy of Natural Science*. Prentice-Hall, Inc., Englewood Cliffs, N.J., 116 pp.

Hooijer, D. A. 1950. The study of subspecific advances in the Quaternary. *Evolution*, V. 4, pp. 360–361.

Kuhn, T. S. 1970. *The Structure of Scientific Revolutions,* Univ. of Chicago Press, Chicago, 210 pp.

Kurtén, B. 1965. The carnivora of the Palestine caves. *Acta Zool. Fenn.*, V. 107, pp. 1–74.

———. 1968. *Pleistocene Mammals of Europe*. Aldine Publ. Co., Chicago. 317 pp.

———. 1971. *The Age of Mammals*. Columbia Univ. Press, New York.

Kurtén, B. and E. Anderson. 1980. *Pleistocene Mammals of North America*. Columbia Univ. Press, New York. 442 pp.

MacArthur, R. H. and E. O. Wilson. 1967. *The Theory of Island Biogeography*. Princeton Univ. Press, Princeton, New Jersey, 203 pp.

Marshall, L. G. 1973. Fossil vertebrate faunas from the Lake Victoria Region, S.W. New

South Wales, Australia. *Mem. Natl. Mus. Victoria*, V. 34, pp. 151–172.

———. 1974. Late Pleistocene mammals from the "Keilor Cranium Site," Southern Victoria, Australia. *Mem. Natl. Mus. Victoria*, V. 35, pp. 63–86.

———. 1981. The Great American interchange—an invasion induced crisis for South American mammals. pp. 133–229, *in* M. H. Nitecki, ed., *Biotic Crises in Ecological and Evolutionary Time*, Academic Press, New York.

Marshall, L. G. and R. S. Corruccini. 1978. Variability, evolutionary rates, and allometry in dwarfing lineages. *Paleobiology*, V. 4, pp. 101–119.

Marshall, L. G., S. D. Webb, J. J. Sepkoski, Jr. and D. M. Raup. 1982. Mammalian evolution and the Great American interchange. *Science*, V. 215, pp. 1351–1357.

Martin, L. D. and A. M. Neuner. 1978. The end of the Pleistocene in North America. *Trans. Nebraska Acad. Sci.*, V. 6, pp. 117–126.

Martin, P. S. 1967. Prehistoric overkill. pp. 75–120 *in* P. S. Martin and H. E. Wright, Jr., eds., *Pleistocene Extinctions: The Search for a Cause*. Yale Univ. Press, New Haven, 453 pp.

———. 1973. The discovery of America. *Science*, V. 179, pp. 969–974.

Martin, P. S. and H. E. Wright, Jr. (eds.).

1967. *Pleistocene Extinctions: The Search for a Cause*. Yale Univ. Press, New Haven, 453 pp.

Raup, D. M. and J. J. Sepkoski, Jr. 1982. Mass extinctions in the marine fossil record. *Science*, V. 215, pp. 1500–1503.

Shipman, P. 1975. Implications of drought for vertebrate fossil assemblages. *Nature*, V. 257, pp. 667–668.

Simpson, G. G., A. Roe and R. C. Lewontin. 1960. *Quantitative Zoology*, Harcourt, Brace and World, Inc., New York. 440 pp.

Slaughter, B. H. 1967. Animal ranges as a clue to late-Pleistocene extinction. pp. 155–167 *in* P. S. Martin and H. E. Wright, Jr., eds., *Pleistocene Extinctions: The Search for a Cause*, Yale Univ. Press, New Haven, 453 pp.

Stenseth, N. C. 1979. Where have all the species gone? On the nature of extinction and the red queen hypothesis. *Oikos*, V. 33, pp. 196–227.

Van Valen, L. 1971. Adaptive zones and the orders of mammals. *Evolution*, V. 25, pp. 420–428.

Van Valen, L. and R. E. Sloan. 1966. The extinction of the multituberculates. *Syst. Zool.*, V. 15, pp. 261–278.

Webb, S. D. 1969. Extinction-origination equilibria in Late Cenozoic land mammals of North America. *Evolution*, V. 23, pp. 688–702.

37

Explaining Pleistocene Extinctions
Thoughts on the Structure of a Debate

DONALD K. GRAYSON

OVER A CENTURY HAS PASSED SINCE Charles Lyell concluded that "the growing power of man" may have played a role in causing the extinction of many species of Pleistocene mammals. Dedicated scientists can accumulate a lot of facts in a century, and the dedicated scientists who have labored since Lyell's time have done just that. We have accumulated facts on the nature of ancient floras and faunas, on past climates, on human prehistory, and on the chronology of it all. These are precisely the kinds of facts that scientists have assumed all along are needed to provide an adequate explanation of late Pleistocene extinctions.

Nonetheless, from an historical perspective one of the most interesting lessons to be learned from this volume is that we are apparently no closer to that adequate explanation, or at least to agreement as to what that adequate explanation is, than we were when Owen's theology forced him to exclude the horse from his list of animals driven to extinction by prehistoric Americans (Grayson chap. 1). We still debate the differential merits of overkill and climatic change much as those merits were debated during the 1860s and 1870s. The accumulation of facts, it would seem, has been of surprisingly little help in resolving one of the major problems that confronts the student of mammalian history. There is a question that can hardly be avoided, one that is nearly as interesting as the causes of the extinctions themselves: why has the huge increase in our knowledge of the past failed to move the issue detectably closer to resolution? Why has so much time made so little difference?

Is the Empirical Record Adequate?

The existence of an unsolved scientific problem often suggests that we simply do not have enough factual information about the phenomena involved to find the solution to that problem. We are, for instance, ignorant of the origin of the archaeological phenomenon known as Clovis, and it may well be that this is due to our lack of knowledge of pre-Clovis archaeology. In seeking the origin of Clovis, we are stopped cold by factual gaps.

Some have suggested that factual gaps also prevent us from understanding the causes of Pleistocene extinctions. "The overkill issue," Webster (1981:556) argues, "is one of those scientific controversies fueled by the paucity or ambiguity of relevant data;

the amount written on the subject...is in inverse proportion to the hard evidence."
Similarly, J. H. Hope (1978:81) concluded her excellent discussion of Australian mammalian extinctions by suggesting that "before speculation on the cause of the extinctions can be profitable, a rigorous examination of the dating and taphonomy of all
relevant sites is needed, in order to set up a basic chronology of the late Pleistocene extinctions." It is, of course, impossible to quantify whether we know a lot or
a little about the late Pleistocene, and until we have satisfied ourselves that we know
the causes of late Pleistocene extinctions, it will always be possible that we are missing
some crucial piece of information, some crucial fact or set of facts whose possession
would make all the difference.

Hope (1978) is on strong ground in observing the weakness of Australian extinction chronologies. In contrast to the North American situation, with its more than 150
dated Pleistocene sites with extinct taxa (Mead and Meltzer chap. 19), there are only
some two dozen dated sites in Australia. In addition, the dates available for many of
these sites may be—indeed, have been—reasonably questioned. Thus, Balme (1978),
Balme et al. (1978), Dortch (1979), and Merrilees (chap. 28) all note that the remains of
extinct Pleistocene mammals from Devil's Lair, Western Australia, in deposits dating to
between roughly 25,000 and 32,000 yr B.P., may have been redeposited from levels
dating to at least 38,000 yr B.P. Goede and Murray (1977) feel that *Sthenurus* remains
from Beginner's Luck Cave, Site M, Tasmania, dated to 14,400 yr B.P., may imply late
Pleistocene survival of this animal, but Murray et al. (1980) suggest that an aspartic acid
racemization date of ca. 80,000 yr B.P. for the same deposits provide a more likely
estimate of the age of this material. Milham and Thompson (1976) argue that the
remains of extinct mammals in Holocene deposits in Madura Cave, Western Australia,
have been redeposited. Gillespie et al. (1978), Horton and Wright (1981), and White and
O'Connell (1979, 1982) accept an age of 26,000 yr B.P. for the Lancefield, New South
Wales, bone bed, but Martin (chap. 17) observes that this assessment depends on two
dates, while material from beneath the Lancefield bones proved modern in age. Many
other radiometrically dated Pleistocene Australian sites have similar problems (e.g.
Lake Colungulac [Errey and Flannery 1978], Tandou Lunette 1 [Hope 1978], Titans
Shelter [Goede et al. 1978, Goede and Murray 1979, Murray et al. 1980], and the Nulla
Nulla Sand fauna, Lake Victoria [Gill 1973, Marshall 1973, Hope 1978]).

As a result, it is possible to construct a chronology of Australian extinctions that
fits either climatic or overkill hypotheses. Martin (chap. 17) argues that the extinctions
occurred prior to 30,000 B.P. and that the cause of the extinctions lies in human
predation, while Horton (1981, chap. 29) places the extinctions after 26,000 B.P. and
attributes them to climatic change. Both positions depend on different interpretations of
the same dates, and both can be made to appear reasonable. Indeed, the Australian
extinction chronology is sufficiently weak that we do not know if all the mammals
involved became extinct at about the same time, or in piecemeal fashion across the
millennia.

A case, then, can be made that we are factually deprived in at least some settings,
and that the source of the continuing debate lies in this deprivation. If only we knew
precisely when the Australian mammals became extinct, the solution would be at hand.
It could even be argued that this situation is good, since it allows hypotheses to be
clearly stated prior to the availability, or even recognition, of the data needed to test
them. "After about 26000 years ago," Horton (1981:26) hypothesized, "Australia began to dry up, reaching its most arid phase at around 18000 years ago...during this
period the megafauna became extinct...conditions began to improve after about 15000
years ago, and by 12000 years ago the present climatic regime was established." This
hypothesis requires the extinctions to have occurred roughly between 18,000 and
25,000 years ago. "The blitzkrieg...model can be rejected in Australia," Martin (chap.
17) suggests, "if the large extinct mammals are found to have maintained sizeable

populations long after initial human invasion of the continent." The hypotheses, it would appear, have set the stage; the basic chronology, once established, will dictate the choice.

Although a case for factual deprivation can be made, such a case can always be made prior to the emergence of a satisfying explanation for any scientific question. The last two decades of debate over the causes of Pleistocene extinctions, however, suggest that the reason for the persistence of this debate lies elsewhere. In particular, that reason seems to lie in how we have gone about searching for an answer.

The Structure of the Debate

Elsewhere (Grayson 1980), I have suggested that attempts to explain Pleistocene extinctions have been characterized by what Wilson (1975:28) has called the advocacy method of developing science:

> Author X proposes a hypothesis to account for a certain phenomenon, selecting and arranging his evidence in the most persuasive manner possible. Author Y then rebuts X in part or in whole, raising a second hypothesis and arguing his case with equal conviction. Verbal skill now becomes a significant factor. Perhaps at this stage author Z appears as *amicus curiae*, siding with one or the other or concluding that both have pieces of the truth that can be put together to form a third hypothesis—and so forth seriatim through many journals and over years of time.

Although Wilson was describing the sociobiological literature, it would appear that he could just as well have been describing attempts to explain Pleistocene extinctions. There are, for instance, few studies in which competing hypotheses are arrayed alongside one another, differentially supportive test implications drawn from those hypotheses, and the factual record examined in detail in order to shed serious doubt on those that do not meet the empirical test (for an exception see Van Valen 1969). Instead, as the factual record has grown more detailed, it has generally been used to build more detailed statements by supporters of one or another account of Pleistocene extinctions, not more penetrating tests of alternative hypotheses. The debate has been enlivened by the exceptional force and skill of the partisans, but that very skill and force has helped to entrench the members of each camp and has made it less likely that a cherished hypothesis would be abandoned and more likely that a masterful piece of advocacy on one side would evoke an equally, or more, masterful statement on the other. In many ways, the process at times has seemed more akin to an election campaign than to the deductive hypothesis-testing procedures of science.

Arguments made by Paul Martin, the foremost overkill partisan, and his adversaries illustrate the process well. At one time, for example, Martin accepted dates of 13,000 to 16,000 yr B.P. for the initial human occupation of Australia; rejecting Holocene dates for the extinct Pleistocene Australian mammals, he argued that those extinctions had occurred by about 14,000 yr B.P., and concluded that the extinctions swiftly followed the human arrival (Martin 1967). An ensuing explosion of well-documented information demonstrated that people arrived in Australia much earlier than 30,000 yr B.P., and climatic partisans skillfully argued that the apparent overlap of many thousands of years between the first human presence and the last megafaunal extinction falsified overkill (e.g. Horton 1980, White and O'Connell 1979). Overkill partisans responded with equal skill, rejecting once-acceptable terminal Pleistocene dates for the extinct mammals, and calling instead for extinction prior to 30,000 B.P. (Martin chap. 17, Martin and Murray n.d.); as the dates for the earliest Australians receded deeper into the

Pleistocene, so did the dates Martin was willing to accept for the timing of the extinctions. Continued argument is unlikely to resolve this issue in the near future, since the method of debate has brought us to a position where adequate resolution requires facts we do not have.

Similarly, Martin (1967) at one time argued that late Pleistocene North American extinctions were primarily confined to large herbivorous mammals and their ecological dependents. In response, a member of the climatic school asserted that the magnitude of terminal Pleistocene bird extinctions in North America did not support the overkill position (Grayson 1977). An expert avian paleontologist entered on the overkill side to assist in the argument that the birds that became extinct were ecologically dependent on the mammals that were lost (Steadman and Martin chap. 21). Are the adaptations of extinct birds known—or knowable—with sufficient precision to be used to clear these waters? "The extinct genera of icterids regarded by paleontologists as 'cowbirds,' had an ecology unknown and unknowable to us," Paul Martin once wrote to me, a point with which I then agreed and with which I still agree. Continued argument is unlikely to resolve this issue in the near future; again, the method of debate has brought us to a position where adequate resolution requires facts we do not have. Indeed, in this case, we may never have them.

Likewise, Martin's initial detailed development of the overkill hypothesis maintained that late Pleistocene North American extinctions were caused by human predation, while passing by the fact that only mammoths were known prey (Martin 1967). Insightful supporters of other hypotheses responded by carefully underscoring the missing kill sites (e.g. Guilday 1967, Krantz 1970). Overkill partisans replied that the process of overkill occurred so rapidly that the kill sites are not to be found (Martin 1973, chap. 17; Mosimann and Martin 1975b). "Unanswerable and untestable" responds one of the best in the climatic school (Guilday chap. 12). Once again, this issue is not likely to be solved in the near future, since the process of debate has worked us into a position where adequate resolution requires unavailable information.

Examples could be multiplied, but the structure would remain the same. The strategy on all sides has routinely involved weaving a flexible wall of argument from fact and conjecture, the usual materials of science, and rallying around the weak point when an opponent succeeds in piercing, or even denting, the wall. To date, the outcome of successful attacks has been to reweave, often by maneuvering to a position where resolution requires facts that are unknown and perhaps unknowable, and thus gaining at least a draw. One role of hypotheses is to define what the relevant facts are, but partisans of both overkill and climatic change have often employed this feature to create factual gaps, either as by-products of their arguments, or on purpose, as sanctuaries for positions difficult to maintain in other ways. In this sense, those who maintain that we do not known enough to explain Pleistocene extinctions are in part correct, but their correctness comes more from the way the search has been conducted than from deficiencies in our knowledge of crucial aspects of the Pleistocene.

It might appear from my examples that the creative reweaving has been done primarily by overkill partisans. In fact, this has been the case. This situation is ironic; it occurred only because overkill advocates, and in particular Paul Martin, had a hypothesis that seemed to provide readily testable implications. Even though overkill has proven difficult to test, this is a result of the structure of the debate that has occurred during the past fifteen years or so. When first proposed in detailed form in 1967, overkill appeared readily testable. It was the first detailed explanation of Pleistocene extinctions ever to have been proposed in testable form (Grayson 1980), and was so powerfully framed that its appearance marked the end of routine publication of multicausal explanations of those extinctions. Henceforth, most scientists were to argue that either climate or people, but not both, provided the driving force behind the extinctions.

Furthermore, until very recently even the best of the climatic accounts appeared extremely difficult to test, depending largely on the sheer chronological correlation between environmental change and the extinctions, or dealing largely with proximate causes that required major assumptions concerning intricate relationships between organism and environment, or both (e.g. Guilday 1967, Slaughter 1967). Most climatic accounts are carefully constructed narratives that depend on gestalt-like assessments of goodness-of-fit between the argument line and the perceived facts of the Pleistocene, rather than leading immediately to sets of inferences with clear empirical import that could be used to falsify the account, or to differentially support it as compared to overkill. As a result, advocates of climatic explanations have often attempted to show their position correct by showing overkill incorrect. Were overkill falsified, it has been assumed, a climatic account would ascend by default, though it has not been at all clear how one might choose among the various climatic hypotheses now available for different parts of the world and for different periods of time. In short, overkill partisans have been forced to reweave more often because they are the ones with the visible structure to attack. The results of this situation have been unfortunate, as I shall discuss.

My earlier suggestion that the advocacy approach to explaining Pleistocene extinctions be abandoned in favor of judging hypotheses through examination of mutually exclusive test implications (Grayson 1980) has been criticized by Merrilee Salmon (1982). Salmon does not defend advocacy, but suggests that it is not "always possible to set up mutually exclusive test implications... when hypotheses one wishes to test are enmeshed in a network of other hypotheses" (Salmon 1982:43). Salmon's suggestion that mutually exclusive test implications might not exist in this setting may well be correct, even though those attempting to explain the extinctions often behave as if such implications do exist. The late Pleistocene extinctions of North American birds, I optimistically concluded in 1977, were too great to support the overkill hypothesis (Grayson 1977). Martin (chap. 17) notes that Australian blitzkrieg must be rejected if the extinct mammals survived human entry for a substantial period of time. Other authors in this volume provide similar statements (see, for instance, Graham and Lundelius chap. 11). We certainly talk as if mutually exclusive test implications exist, at least at the level of climatic versus overkill accounts, and we certainly talk as if both overkill and climatic hypotheses have been conducted in such a way as to be falsifiable. Do such mutually exclusive test implications exist?

Can Overkill be Falsified?

The breadth and resilience of both overkill and climatic accounts of Pleistocene extinctions may be evident to anyone who has followed the debate over the past two decades, or over the past century. I believe that one of the reasons that overkill has proven so resilient is that, in its current form, it cannot be falsified, regardless of the causes of the extinctions. To show why this is the case, I will begin with Martin's blitzkrieg model, and will then examine the more general overkill hypothesis.

The blitzkrieg model was first offered by Martin in 1973, applied to the New World situation. The model was not meant to argue that overkill had occurred, but was instead meant to provide a sequence of events demonstrating how overkill *could* have happened *if* it happened. "Simulations can prove nothing about prehistory. If extinctions were caused by overkill, simulations can suggest ways in which it might have happened" (Mosimann and Martin 1975a:381).

Now expanded to include Australia (Martin chap. 17, Martin and Murray n.d.), the blitzkrieg model, in barest outline, introduces people to a large territory unoccupied by human big-game hunters but stocked with many easily hunted large game mammals. The human predators take immediate advantage of this unwary prey, soon reach and then sustain high growth rates, and rapidly spread outwards from their original point of entry, leaving behind extinct populations, and ultimately extinct genera (see Whittington and Dyke, chap. 20, for a more detailed review).

Martin (1973) and Mosimann and Martin (1975b) provided a computer simulation of North American blitzkrieg overkill, using various values for such critical parameters as initial prey biomass, initial human population density, prey destruction rate, and maximum human growth rate. They demonstrated that, if their values were acceptable, overkill could have occurred in the blitzkrieg mode.

The logical status of this particular blitzkrieg model in terms of the overkill hypothesis is important. Because the overkill hypothesis does not logically entail the blitzkriegs quantified by Martin (1973) and Mosimann and Martin (1975b), falsification of these simulations would not affect the status of the more general hypothesis. If no version of the simulated blitzkriegs occurred, then some other kind of overkill could have taken place. Blitzkrieg is simply a "special case of overkill" (Martin chap. 17), and the general overkill hypothesis can accommodate many special cases. Of course, if any version of the simulated blitzkriegs took place, overkill took place, but the simulations themselves do not tell us that. They were meant only to demonstrate that overkill could have occurred, that the phenomenon is plausible.

Thus, while fascinating, the blitzkrieg simulations in no way affect the underlying, more major issues. No matter how much the details of the simulations are explored (see Webster 1981 for a thoughtful exploration), the outcomes tell us nothing about whether or not overkill occurred.

More important, however, is the demonstration by Whittington and Dyke (chap. 20) that the general model on which the specific blitzkrieg simulations were based is sufficiently robust that selection of different values for various crucial parameters allows North American extinctions to be accounted for by human predation "in as few as 250, or as many as 22,000 years (or more)" (Whittington and Dyke chap. 20).

I have noted that one of the strengths of the overkill hypothesis is that it has appeared testable. One of the reasons it has appeared testable is that it seems to lead to strong inferences about the chronology of extinction. Martin (1967) suggested that overkill would be falsified by the discovery of many clear-cut cases of extinct North American mammals having survived into the Holocene. He has also suggested that the notion of sudden overkill would be falsified if archaeological sites with extinct mammals are found scattered across thousands of years and if human predation on extinct New World Pleistocene mammals is found to have occurred much before 11,000 B.P. (Martin chap. 17), as Gruhn and Bryan (chap. 5) assert. While it is true that these chronological implications of sudden and complete overkill—true blitzkrieg—still stand, the conclusion reached by Whittington and Dyke removes the critical chronological test of the general overkill hypothesis. Extinctions well prior to, and well after, the end of the Pleistocene in the New World can clearly be accommodated by that hypothesis. So, presumably, can post-30,000 B.P. extinctions in Australia, assuming that similar calculations yield similar results for that continent. Since we lack theory that will provide uncontroversial values for the crucial parameters in the overkill simulations, the argument may well become whether or not empirical observations of the archaeological and modern records support the figures used by Whittington and Dyke or some other calculations (though certainly there can be no surprise in the demonstration that overkill could have occurred either slowly or rapidly). Since such parameters as population density cannot be accurately quantified archaeologically, and since it is unclear which—if any—modern settings provide proper analogues for the late Pleistocene settings, continued argument would be unlikely to resolve the issue.

In short, given the general chronology of late Pleistocene extinctions—most New World extinctions at the very end of the Pleistocene, most Australian extinctions between 40,000 and 10,000 years ago—it would appear that overkill can be made sufficiently resilient to withstand most chronological assaults. While true blitzkrieg—some three dozen North American genera driven to extinction within a thousand years, for example—has strong chronological implications, there is a smooth transition between this phenomenon and slower overkill that suggests there is resilience here as well.

Although chronology is usually presented as having the potential of providing the most clear-cut, decisive source of falsification of overkill accounts, other types of data are also seen as potential falsifiers. Of these, archaeological sites containing evidence for human predation on the extinct mammals loom large on both sides of the issue.

Indeed, as I have discussed in chapter 1, it was acceptance of the associations of human remains with those of extinct Pleistocene mammals that led to the rapid adoption of overkill into explanations of Pleistocene extinctions during the 1860s, though the associations themselves were not provided by kill sites. When Martin provided his powerful development of the overkill hypothesis in 1967 (see also Martin 1958), the presence or absence of North American kill sites played little role, except to support the notion that there were big-game hunters here at the time of the extinctions. He took a very different approach for New Zealand, however, observing that "twenty-two of the extinct moa species have now been found in association with prehistoric man...In addition to the moas, a number of other birds became extinct in the same general period; half of these have been found in cultural association" (Martin 1967:102).

No time was lost by supporters of other hypotheses in pointing out the dearth of North American kill sites. "Examination of the archaeological evidence for the reduction of animal populations by the *hunting* of early man," Hester (1967:186) observed, "reveals that evidence to be inadequate and ambiguous." Krantz (1970:165) concurred: "in North America at least, early hunting man did not significantly prey upon those animals he supposedly exterminated.... Only two genera were hunted in significant numbers, mammoth and bison, one extinct and the other not." Even those favorably disposed to overkill saw the lack of kill sites as a problem (Jelinek 1967). The passage of time has not altered the archaeological situation very much; of the three dozen or so genera that became extinct at the end of the Pleistocene, only three—mammoth, horse, and camel—have been found convincingly associated with human remains (see E. Anderson chap. 2).

The lack of kill sites thus appeared to falsify overkill; given the difficulty that authors of climatic hypotheses had in deriving predictions whose confirmation would support their own accounts, above and beyond the chronological coincidences that suggested the accounts in the first place, it is not surprising that so much emphasis was placed on the slim archaeological record for human predation on the extinct taxa. These arguments became sufficiently compelling that Martin (1973) and Mosimann and Martin (1975b) addressed them directly in their development of the blitzkrieg model. Overkill occurred so rapidly, they argued, that the wonder is not that there are so few kill sites, but that there are any at all. "The dozen well-documented cases of human artifacts associated with extinct mammoths may be a rich, not a poor, record of extinction by overkill" Mosimann and Martin (1975b:313) concluded. "Scarcity of evidence," Martin and Murray (n.d. p. 10) note, "is a requirement of the blitzkrieg model."

Can the archaeological record for kill sites provide evidence against overkill? While the special case of true blitzkrieg seems to require that most kill sites be confined to a narrow period of time, the more general overkill argument does not. Overkill, Whittington and Dyke (chap. 20) suggest, could have taken many thousands of years in North America, and there is no reason to deny the possibility for Australia. Thus, while "scarcity of evidence" is presented as a test of the blitzkrieg model, it cannot be presented as a test of slower overkill. Although Martin (1973, chap. 17) argues that low archaeological visibility follows from blitzkrieg overkill, it is clear that rapid extinction associated with human activities can leave impressive evidence behind. Although the precise mechanism whereby people caused the extinction of moas in New Zealand is still debated, that people were involved is not at issue (see A. Anderson chap. 33, Trotter and McCulloch chap. 32, and Cassels chap. 34). On South Island, there are 117 known moa-hunting sites confined to a period of about 500 years, and strong indications that many more exist. Here, rapid extinction associated with human predation has left many kill sites. Indeed, the very visibility of moa kill sites was employed by Martin (1967) in support of overkill.

Thus, the blitzkrieg models allow the argument that rapid overkill will leave few kill sites; New Zealand shows that it can leave many. Either many kill sites or no kill sites can be associated with this special case of overkill, so no decisive test of blitzkrieg can be had on these grounds. Slow overkill would presumably generate a number of kill sites involving a sample, or perhaps even all, of the taxa that became extinct, scattered across the millennia. Given that the North American extinctions seem to have been roughly synchronous, and may have been so in Australia as well, the discovery of kill sites scattered across time and across taxa would be consistent with both overkill and climatic accounts. Indeed, while such a pattern would falsify blitzkrieg caused by the entry of people into a previously unoccupied area, it would not discount rapid overkill that resulted from the development of a new, hunting-oriented subsistence base by inhabitants of an area who previously only occasionally took large mammals, since Martin's blitzkrieg model predicts that only scattered kill sites would now remain to be found from the blitzkrieg itself.

Most associational arguments against North American overkill proceed by asserting that there are too few extinct Pleistocene taxa represented in archaeological settings here, although there is no theoretical structure that informs us when "not many" in the empirical sense has become "too few" in the overkill setting. This argument assumes that we have many terminal Pleistocene (12,000 to 10,000 B.P.) records for the missing taxa, and thus that the scanty archaeological record is meaningful (for a similar assertion in the Australian setting, see Bowdler 1977; see also the responses to Bowdler by White and O'Connell 1979, 1982). However, in an examination of a large sample of sites with extinct taxa radiocarbon dated to less than 12,000 yr B.P., I found that while mammoth occurred more often than would be expected by chance (p $<$.01) in the subset of those sites with secure archaeological associations, no other taxa were significantly over- or under-represented (Grayson 1983). I have reconducted that analysis, armed with the larger sample of demonstrably terminal Pleistocene sites provided by combining my initial sample with that amassed by Meltzer and Mead (1983, see also Mead and Meltzer chap. 19). This larger sample (151 occurrences dated to less than 12,000 yr B.P., and excluding *Smilodon* because of its presumed dependence on the herbivores that became extinct) shows that while mammoth occur in secure archaeological settings more often than would be expected by chance (p $<$.001), mastodont occur less frequently (p $<$.05); other taxa occur neither more nor·less than chance will allow (see Grayson [1983] for details of the analysis). The taxa that are unknown from North American archaeological contexts may be unknown simply because they are so poorly represented in terminal Pleistocene contexts as a whole.

Does the mastodont, which is significantly under-represented, pose a problem for blitzkrieg, or for overkill in general? The blitzkrieg model predicts that few mastodont will ever be found in archaeological contexts, so this discovery is, in fact, in accord with that model. But how about the fact that so many mammoth kills are known? There are about the same number of dated terminal Pleistocene mammoth sites in my sample (37) as there are mastodont sites (33), but while my sample includes eight mammoth kill sites, it includes none for mastodont. Stepping outside my sample and adding the one possible dated mastodont kill of which I am aware (Manis, Washington: Gustafson et al. 1979; Kimmswick, Missouri is undated: Graham et al. 1981) would still leave mastodont significantly under-represented. Even this discordancy can be accounted for by overkill, by positing a faster kill rate, and thus lesser archaeological visibility, for mastodont than for mammoth. Martin (chap. 17), for instance, suggests that the greater number of moa kill sites on South Island, New Zealand, than on North Island may be accounted for by slower attrition on the former island (A. Anderson, chap. 33, attributes the difference to archaeologists' search time). Thus, the disparity between numbers of mammoth and mastodont kill sites can readily be accommodated by the blitzkrieg model and by overkill.

Chronology and archaeology provide the bulk of what have appeared to be robust test implications of the overkill hypothesis, implications that if met would support overkill at the expense of climatic accounts, and that if not met would do the opposite. I suggest that overkill can be made so resilient that it cannot be falsified through these implications. In addition, other statements generated by supporters of overkill appear to be tests, but are not, and would not be even if the hypothesis had not become so resilient. Martin's observation that overkill would be falsified by the "extinction of late Pleistocene faunas under circumstances that preclude an anthropogenic effect, i.e. before prehistoric humans arrived" (Martin chap. 17) is one such instance. A wave of extinction that swept across the Americas during, say, the early Wisconsin and that was marked by the same attributes that marked North American terminal Pleistocene extinctions—for example, the loss of many large mammals in a very short time—would certainly reduce the impact of the correlation between a new human presence and the extinctions that provides the overkill hypothesis with so much of its driving force, although still not precluding overkill toward the very end of the Pleistocene. But it is a fact that in both the New World and in Australia, the only large land masses to which the tests could be applied, the extinctions followed the arrival of people. Were this not the case, the debate presumably would have taken a very different direction. The lack of earlier late Pleistocene extinctions does not differentially support either overkill or climatic accounts, nor, given what we know of the late Pleistocene, can it be used to test either account in the future. There are many nonexistent features of the Pleistocene that could have falsified overkill had they occurred, but listing them does not increase the likelihood that overkill is correct, nor does it increase the falsifiability of that hypothesis, although it may increase the appearance of falsifiability.

It is not my purpose to survey all aspects of late Pleistocene extinctions that appear to be *consistent* with overkill (see Martin, chap. 17, for such a survey) and to argue whether these aspects show such consistency. It is no surprise that supporters of overkill find that hypothesis consistent with those things it was initially advanced to explain. It is also no surprise that supporters of other explanations feel that many of the consistencies, including the size of the animals involved, the speed of extinction, and the fact that massive extinctions appear to have been roughly synchronous, can be more convincingly accounted for by those other explanations. Instead, my purpose is to argue that the structure of the debate over the causes of Pleistocene extinctions has made the overkill hypothesis exceedingly difficult to test. Like a highly successful organism, it has evolved defenses for most, if not all, challenges; even if true blitzkrieg can be falsified, the parental hypothesis probably cannot. Martin's statement that overkill would be weakened by "the extinction of late Pleistocene faunas under circumstances that preclude an anthropogenic effect, i.e. before prehistoric human arrival" is telling. An anthropogenic effect can be precluded only for those cases in which extinction occurred prior to a human presence. Only extinction in the absence of people would seem to falsify overkill, yet that did not occur.

Here, then, is a major reason why the debate has persisted the way it has. One side would seem to be armed with an account that cannot be shown to be wrong. Where does the other side stand?

Can Climatic Explanations be Falsified?

While overkill has appeared testable, in the past even well-developed climatic models have often seemed to lack clear implications that would allow them to be falsified. At their heart, all climatic accounts rely on the fact or the presumption that the North American and Australian extinctions occurred at a time of climatic change, just as overkill accounts rely on the correlation between the extinctions and the first appearance of human predators. Most participants on the climatic side of the debate have realized—often because supporters of overkill made them realize—that the simple correlation between the time of the extinctions and the time of climatic change in these areas does not establish the cause of the extinctions. Accordingly, proponents of climatic change have attempted to establish that causal relationship.

Two general approaches have been taken in attempting to turn the general climatic correlation into climatic causation. First, there are those who have attempted to falsify overkill without providing climatic models of their own (e.g. Grayson 1977, 1983; Webster 1981). The underlying assumption of these arguments is simple: if people did not cause the extinctions, climate must have, and research should be directed along these more promising lines. If, however, overkill is as difficult to falsify as I have suggested, such approaches may lead to adjustments of the overkill hypothesis, but they will not be successful in establishing a climatic account by default. Indeed, the last twenty years have shown us that as the attempts to falsify overkill in this fashion become more ingenious, the resulting adjustments in the overkill hypothesis become more ingenious, and the chances of establishing a climatic account by default decrease.

Second, there are those who provide their own climatic hypotheses usually with, but sometimes without, specific attempts to falsify overkill. Within this large category of arguments, some depend largely on detailed assessments of environmental change at the time of extinction, attempting to document the particular aspect or aspects of climatic change that caused the extinctions (e.g. Guilday chap. 12, Lundelius 1983). Others combine such detailed analyses with specifications of equally detailed linking mechanisms that attempt to explain how a given set of environmental changes caused the extinctions (e.g. Horton 1978, Guthrie chap. 13, Slaughter 1967, see also Kiltie chap. 14). A few studies have even hypothesized linking mechanisms that could have caused the extinctions under a particular set of environmental conditions without examining the paleoenvironmental record itself (e.g. Main 1978).

An adequate explanation of Pleistocene extinctions in climatic terms must contain three components. It must, of course, be demonstrated that the extinctions coincided with an episode of climatic change. Without such a correlation, there is no reason to call on climate as a cause. Second, if the specific climatic cause(s) invoked (for instance, loss of equability) is meant to possess explanatory value at least as general as that provided by overkill, and thus to account for the extinctions whenever and wherever they occurred, it must be shown that the magnitude of the invoked climatic cause is correlated with the magnitude of extinction. Third, some mechanism must be provided that links the invoked climatic cause with the extinctions to establish that, given the kind of climatic change that occurred, extinction was an expected result. The order in which these components are provided is unimportant. If, for instance, we had the third component in hand, we could predict the nature and timing of past climatic changes at times of Pleistocene extinctions from knowledge of the extinctions alone. If our predictions were confirmed, we would certainly consider our climatic hypothesis strongly confirmed.

Each of these components differs in the power it provides a climatic account. The simple correlation between an episode of climatic change and an episode of extinction carries little weight, since all the contested sets of extinctions occurred at a time of change in both the archaeological and climatic records.

Correlation counts more heavily in the second component, for here some specific attribute of climatic change, often inductively derived from examination of one set of extinctions, is predicted to correlate positvely with the magnitude of other sets of extinctions, whether or not people were present. Because of this strongly predictive aspect, this component may present many opportunities for falsification, as I shall discuss.

What is lacking in the second component, however, is any explanation as to why extinction occurred when the hypothesized climatic cause occurred. All that has been established is the empirical demonstration that when a specified climatic change happened, as from an equable to a continental climatic regime, extinction followed, the magnitude of the extinction event depending on the magnitude of the climatic event. Such a demonstration does not explain why the extinctions followed, only that they did follow. The "why" can be answered only by specifying the mechanism that functionally links the climatic changes to the extinctions. Once this component has been supplied, the extinctions can be said to have been fully explained climatically.

Where do climatic accounts stand now? Since 1967, when Martin and Wright published their critically important survey of Pleistocene extinctions, much progress has been made toward specifying the particular climatic factor or factors that may have caused the extinctions, and in specifying them in such a way that their relevance to questions of the causes of the extinctions can be tested.

This was not the case in 1967. At that time, most proponents of climatic change relied heavily on establishing the simple correlation of an episode of climatic change with an episode of extinction, usually coupled with attempts to falsify overkill, in order to establish their thesis (e.g. Guilday 1967, Hester 1967). As a result, it was quite difficult for them to answer a simple and appropriate question: "why, after surviving a succession of four or more glacial advances, do most elements of the large mammalian faunas of the Northern Hemisphere disappear during the last glaciation?" (Jelinek 1967:194). Why extinction at that time, when the animals had been through similar changes before? This was a difficult question for members of the climatic school to answer because, although they fully believed climatic change was the cause of the extinctions, they did not have a clear idea as to exactly what the crucial, extinction-causing climatic changes were. Accordingly, they were unable to examine the ends of previous glacials to see if they had been characterized by such changes, as their accounts required. Left without a satisfactory answer, some took refuge in factual gaps to resolve the problem: our poorer records for earlier extinctions might simply be a function of the decreasing completeness of the fossil record as one travels farther back in time (e.g. Guilday 1967).

It is telling that this problem was not a difficult one for those armed with hypotheses that specified in detail the environmental attributes felt to have caused the extinctions, since those hypotheses also specified what earlier interglacials should not have been like were their accounts correct. Axelrod (1967), for example, argued that lower climatic equability had caused terminal Pleistocene/early Holocene extinctions in North America, and called not on our lack of knowledge to account for the lack of equivalent extinctions earlier in the Pleistocene, but instead on evidence that earlier interglacials had been more equable than the early Holocene, just as the hypothesis required. Axelrod had specified the environmental parameters thought to have caused the extinctions; as a result, he was able to turn what had been a puzzlement for others into potentially important support for his thesis.

In many ways, Axelrod's work was a portent of things to come. Many newer climatic hypotheses specify in detail the kinds of environmental changes felt to have caused the extinctions, and in so doing have become readily testable. The hypothesis provided by Graham and Lundelius (chap. 11) provides an example. These authors hypothesize that Pleistocene extinctions were caused by the loss of climatic equability. They argue, for instance, that late Pleistocene climates were "equable and seasonal extremes in moisture and effective precipitation were reduced" (Graham and Lundelius chap. 11); the loss of equability caused the North American extinctions. While Graham and Lundelius discuss a number of predictions derived from their hypothesis, I note simply that their account requires that late Pleistocene extinctions occurred at the same time that disharmonious faunal associations were disappearing. Many disharmonious pairs must have existed in North America up to the time of extinction, the number decreasing rapidly at the time of extinction. This seems to have occurred, but forms the basis of the hypothesis in the first place (see also Guilday chap. 12, Guthrie chap. 13). More importantly, then, there must have been fewer disharmonious pairs at the end of earlier North American glacials, when extinctions were fewer. Disharmonious pairs must have been high in Australia up to the time of extinction, low or absent after that time. Where terminal Pleistocene extinctions were fewer, as in Africa, the number of disharmonious pairs must have been low at the time of extinction, still lower afterwards. Thus, Graham and Lundelius do not have to hide from the reasonable challenge posed by Klein in his tightly argued survey of African extinctions. Klein argues that climatic change alone cannot account for the teminal Pleistocene/early Holocene large mammal extinctions in Africa "because the extinct species survived earlier periods of broadly comparable change" (Klein chap. 25). Graham and Lundelius have a way of testing whether or not the mammals had been through earlier periods of comparable change, because their hypothesis provides them with a definition of "comparable change," and allows them to place the terminal Pleistocene African extinctions in global context. Indeed, the predictions made by their hypothesis are sufficiently robust that they can proceed in the absence of a precise absolute chronology. They can even be applied to Australia now, where the extinction chronology is poorly developed, since their account predicts that sizeable Pleistocene faunas that contain extinct mammals will also contain disharmonious pairs, and that sizeable faunas postdating the time of extinction will not contain them in any number, no matter what the exact time of extinction was. Such tests have the power to lend strong support, or to entirely refute, the Graham and Lundelius climatic hypothesis. Although the outcome of these tests is for the future, Lundelius has begun to apply the tests both to Australia and to earlier North American interglacials (Lundelius, 1983, and personal communication).

The gains that have been made in this area since 1967 can be seen in the work of single scholars. In 1967, for instance, Guilday (1967) took refuge in the incompleteness of the interglacial fossil record to defend his climatic account against the argument that the late Pleistocene mammals had been through it all before. Now, Guilday (chap. 12) uses the interglacial fossil record as support for his account, since his more precise—and testable—hypothesis defines what is and what is not important about earlier fauna in explaining the North American terminal Pleistocene extinctions. Although Guilday cites a current reference to make his point, the same point could have been made in 1967; indeed, it could have been made as early as 1960 (see Guthrie chap. 13). In 1967 Guilday took refuge in a factual gap; now, armed with a more powerful and predictive hypothesis, he no longer needs the gap to defend his position, and it is gone. The changes in Guilday's approach during the past fifteen years parallel developments in climatic explanations as a whole during that time. We have moved from a set of generally ambiguous hypotheses that relied heavily on the simple temporal correlation between episodes of climatic change and episodes of extinction, to a set of hypotheses that specifies the precise climatic attributes thought to have caused the extinctions. These new hypotheses have clear empirical import and, as a result, can be supported or refuted by what we know or will learn about the Pleistocene.

If all the predictions derived from the equability portion of the Graham and Lundelius hypothesis were to be met, we would have strong reasons to think that whenever a specific kind of climatic event occurred during the Pleistocene, extinctions followed, but we would not know why they followed. To explain the extinctions, the mechanisms that functionally link the specified climatic changes with the extinctions must be provided.

"The difficulty most paleoecologists have had with extinction theories based on climatically induced events," Guthrie (chap. 13) observes, "is their ambiguity and diffuseness in the proximate coup." This criticism has not been leveled at overkill because scientists have generally been willing to accept human predation as providing a precise mechanism whereby extinction occurs, a mechanism that links the first appearance of people with the loss of the mammals in a causal way. Mechanisms of this apparent precision are much harder to come by in climatic accounts. Many suspect that episodes of rapidly changing climate could cause extinction, but the links between such an episode and the extinctions—be they gestation time, diet, temperature tolerance, or something else—are complex indeed, and the kinds of links required are not the kind that leave obvious traces in the ground. Unfortunately, ambiguity and diffuseness in the proximate coup have come with the territory. Biologists lack theory that can predict the responses of organisms to climatic change in general, and because they lack such theory, it does not seem surprising that many climatic accounts of Pleistocene extinctions have focused on the search for provocative correlations.

The role of linking mechanisms in climatic hypotheses is analogous to the role of blitzkrieg in the overkill hypothesis. Just as blitzkrieg is meant to demonstrate that given a particular set of cultural and environmental conditions, extinction would ensue, such mechanisms as gestation time and dietary tolerance are meant to show that given a particular set of environmental conditions alone, extinctions would ensue. True blitzkrieg is falsifiable because it makes heavy chronological demands, but falsifying blitzkrieg does not falsify overkill, unless it could be shown that overkill could have occurred only in this fashion. Similarly, falsification of any particular linking mechanism or set of mechanisms would not falsify the parental climatic account, unless it could be shown that climatically induced extinctions could only have been caused by the falsified mechanisms. But are there ways of falsifying linking mechanisms in the climatic setting?

It would appear that there are, though the process is not likely to be a simple one. Among the hypotheses presented in this volume that offer linking mechanisms (e.g. Graham and Lundelius, Guilday, Guthrie, Horton, and Kiltie), those offered in Guthrie's fascinating contribution are perhaps the most complex.

Guthrie maintains that changed seasonal regimes at the end of the North American Pleistocene led to increasingly homogeneous plant communities, which decreased the quality and quantity of plant resources available and decreased the length of the mammalian growing season, causing extinction. Here, the climatic cause of extinction is hypothesized to reside in the development of a more intensely seasonal climatic regime. This hypothesized cause has empirical import and can be falsified if it is false. The development of vegetational zonation and simplification links the climatic cause to the hypothesized proximate causes of extinction, and is also falsifiable. The crucial linking mechanisms that provide Guthrie's account with its explanatory power, however, assert that the mammalian growing season shortened and that the structure of critically important food resources changed. How can we know that this was so, or that it was not?

Guthrie's linking mechanisms do, in fact, direct us toward the examination of a number of attributes of late Pleistocene faunas. They predict, for instance, altered bone growth at the time of onset of intensely seasonal climates, and direct us to look for such phenomena as Harris's lines in the bones of the taxa that became extinct, to compare such phenomena through time (they must increase as the time of extinction approached) and across taxa (they must be common in those animals that became extinct, less so in those that did not). They direct us to examine such demographic variables as recruitment rates through time among the animals that became extinct (they must have decreased as the time of extinction approached; see King and Saunders chap. 15), and between taxa that did and did not become extinct at the end of the Pleistocene (recruitment rates must have been greater among the latter than among the former). They direct us to look at specific kinds of dietary information available from those rare, stratified accumulations of the dung of extinct animals, and to compare that information with data derived from accumulations of dung of the survivors. They direct us to look at bone chemistry, since differential utilization of plants of very different chemical composition is basic to Guthrie's thesis. None of these tests is simple, but all are tests and all can lend support to, or refute, Guthrie's thesis.

Where, then, do current climatic hypotheses stand now? They appear to stand on firm ground as regards their ability to be profitably tested. As exemplified by a number of contributions to this volume, many of these hypotheses carefully specify climatic causes and often carefully specify linking mechanisms as well, and do so in such a way that both components have empirical import that will allow them to be supported or refuted. This situation is markedly different from that which existed only a decade ago.

Conclusions

When first proposed in detail (Martin 1958 and, especially, 1967), the overkill hypothesis not only seemed to account for many previously perplexing facts presented by late Pleistocene extinctions, particularly in the New World, but also seemed to make many predictive statements about the nature of the archaeological, paleontological, and paleoclimatic records. At the same time, the climatic hypotheses that were available to account for the same facts were vague and diffuse, gaining support almost entirely from the chronological coincidence of extinctions and climatic change. While attractive to a number of scientists dealing with Pleistocene extinctions, these climatic accounts made few predictions about the nature of the past, and were thus extremely difficult to test.

Because overkill partisans had a testable hypothesis while climatic partisans did not, the process that took place may have been almost inevitable. With little to seek in the late Pleistocene record that would corroborate their own hypotheses, members of the climatic school took aim instead at overkill. Virtually every statement with empirical import made by the overkill position was subjected to detailed scrutiny. Are there enough kill sites? No. Did other organisms become extinct at the same time? Yes. Were there people in North America and Australia well prior to the time of extinction? Yes. Did overkill adequately account for the large mammals, including bison, musk oxen, and large kangaroos, that survived? No. Proponents of overkill did not shy from the debate. Lacking the opportunity to compare their detailed accounts with any similar account provided by their opponents, champions of overkill posed general and unanswerable questions to the champions of climatic causes. In addition, and in the process of responding to the critics of overkill, the overkill hypothesis became so resilient that it could withstand virtually any factual onslaught. There are so few kill sites in North America because it all happened so quickly. Where there are many kill sites (as in New Zealand), it is because overkill took longer. Birds did become extinct in number, but they were dependent on the mammals that became extinct. The large mammals that survived did so because only they had behavioral patterns that provided them protection. Overkill partisans were, at first, like blind boxers fighting in the dark; they could tighten their defense so nothing could get through, but they could only return roundhouse blows covering so much territory that they had to hit something. Why didn't comparable extinctions occur at the end of earlier glaciations? In the end, the process seems to have made overkill unfalsifiable, but the fault cannot simply be attributed to overkill advocates. The fault lies in the temporal asymmetry of the development of overkill and climatic accounts, and in the responses by scientists to this asymmetry.

As the process of debate moved overkill from a readily testable to an apparently unfalsifiable statement about the way the world has worked, climatic accounts moved in the opposite direction. While overkill partisans became frustrated at being attacked in the dark, climatic partisans became frustrated at their own inability to answer the general and valid questions being asked of them. Why did sloths become extinct even though they fed on plants still available in their ancient homeland? Why did horses become extinct in North America at the end of the Pleistocene, yet thrive when reintroduced by Europeans thousands of years later? Given that climate was the driving force, what exactly was it at the end of the last glacial that caused the extinctions? There were leads to be followed here, leads provided by the insightful subtleties in the work of Axelrod (1967), Guilday (1967), Lundelius (1967), and Slaughter (1967). Members of the climatic school built their own, detailed explanations of Pleistocene extinctions, explanations with clear empirical import that can be readily falsified if they are false. The best of the proposed explanations are to be seen in this volume. But by the time they built them, overkill's wall was virtually, if not entirely, impregnable.

There are ways in which the situation I have described is similar to Lakatos's view of the march of science (Lakatos 1978; I am indebted to Merrilee Salmon for pointing this similarity out to me). Central to Lakatos's analysis is his distinction between "progressive" and "degenerating" research programs. Each step of a progressive program increases its empirical content, predicting new facts as it moves along, and seeing those predictions corroborated as time passes. In contrast, degenerating programs are marked by the accretion of ad hoc hypotheses designed to protect the heart of the program from important inconsistencies, while failing to predict new and unexpected phenomena. Progress in science, Lakatos suggests, resides in the diversification of competing research programs, and in the continued growth of progressive ones. In this light, it may well be that the overkill program is best characterized as a degenerating one, the climatic program as progressive. However, if such distinctions are apt, they are certainly best made in deep retrospect. In addition, there is certainly no way to detect whether climatic accounts will have to be shored by auxiliary hypotheses as ad hoc as those that now shore overkill, generated in response to the criticisms overkill partisans are sure to make of this apparently revitalized research program.

It is clear that we now have two problems with which to contend. The first is the problem of explaining Pleistocene extinctions, a problem made more difficult because we have no theory that relates organisms to their environment in such a way as to enable us to explain any set of ancient vertebrate extinctions, be they of Cretaceous dinosaurs or of Pleistocene mammals. The second is the problem of having to select between two sets of accounts, either (or neither) of which may be correct, but only one of which is not currently burdened by auxiliary hypotheses that protect it from falsification. There are, of course, philosophers of science and scientists who would automatically opt for that set of accounts that can be shown wrong. If that is the situation we are in, and I believe it is, then the situation is an ironic one. The overkill hypothesis, as framed by Paul Martin, was the first detailed explanation of Pleistocene extinctions ever to have been forwarded. It was presented in such a powerful way that its appearance marked the end of most attempts to attribute the underlying causes of the extinctions to diverse phenomena acting in concert. It, and he, stimulated much of what followed, but what followed was an uneven dialectical process that saw overkill become surrounded by protective ad hoc hypotheses with little predictive power, and that made it, at least as it stands now, unfalsifiable, while leading to the development of climatic accounts ready for the test. Salmon (1982) certainly appears to have been correct. There are no mutually exclusive test implications to be had here. The problem, however, does not stem simply from the complexity of the hypotheses involved. Instead, the problem stems from the fact that overkill now excludes very little, a result of the process of debate that has taken place over the past two decades.

Acknowledgments

My sincere thanks to the friends and colleagues who provided critical comments on the substance of this paper: R. C. Dunnell, R. D. Guthrie, G. T. Jones, R. G. Klein, E. L. Lundelius, Jr., P. S. Martin, M. Salmon, M. B. Schiffer, R. S. Thompson, and J. F.

O'Connell. I must especially thank R. C. Dunnell, E. L. Lundelius, Jr., and M. Salmon for their critical efforts, and J. P. White and J. F. O'Connell for making page proofs of their new book on Australian prehistory available to me. I thank D. J. Meltzer for providing me with the date list described elsewhere in this volume. Above all, I thank Paul S. Martin for soliciting this paper long ago, when neither of us knew what it would say, and for sticking with it after we both found out.

References

Axelrod, D. I. 1967. Quaternary extinctions of large mammals. *University of California Publications in Geological Sciences* 74:1–42.

Balme, J. 1978. An apparent association of artifacts and extinct fauna at Devil's Lair, Western Australia. *The Artefact* 3:111–116.

Balme, J., D. Merrilees, and J. K. Porter. 1978. Late Quaternary remains, spanning about 30,000 years, from excavations in Devil's Lair, Western Australia. *Journal of the Royal Society of Western Australia* 61(2):33–65.

Bowdler, S. 1977. The coastal colonisation of Australia. In *Sunda and Sahul: prehistoric studies in Southeast Asia, Melanesia and Australia*, edited by J. Allen, J. Golson, and R. Jones. pp. 205–246. Academic Press, London.

Dortch, C. 1979. Devil's Lair, an example of prolonged cave use in southwestern Australia. *World Archaeology* 10:258–279.

Errey, K. and T. Flannery, 1978. The neglected megafaunal sites of the Colungulac region, Western Victoria. *The Artefact* 3:101–106.

Gill, E. D. 1973. Geology and geomorphology of the Murray River region between Mildura and Renmark, Australia. *Memoirs of the National Museum of Victoria, Melbourne* 34:1–98.

Gillespie, R., D. R. Horton, P. Ladd, P. G. Macumber, T. H. Rick, R. Thorne, and R. V. S. Wright. 1978. Lancefield Swamp and the extinction of the Australian megafauna. *Science* 200:1044–1048.

Goede, A. and P. Murray. 1977. Pleistocene man in south central Tasmania: evidence from a cave site in the Florentine Valley. *Mankind* 11:2–10.

———. 1979. Late Pleistocene bone deposits from a cave in the Florentine Valley, Tasmania. *Papers and Proceedings of the Royal Society of Tasmania* 113:39–52.

Goede, A., P. Murray, and R. Harmon. 1978. Pleistocene man and megafauna in Tasmania: dated evidence from cave sites. *The Artefact* 3:139–149.

Graham, R., C. V. Haynes, D. L. Johnson, and M. Kay. 1981. Kimmswick: a Clovis-mastodon association in eastern Missouri. *Science* 213:1115–1117.

Grayson, D. K. 1977. Pleistocene avifaunas and the overkill hypothesis. *Science* 195:691–693.

———. 1980. Vicissitudes and overkill: the development of explanations of Pleistocene extinctions. *Advances in Archaeological Method and Theory* 3:357–403.

———. 1983. Some tests of the overkill hypothesis. In *Man in late glacial North America*, edited by D. J. Meltzer and J. I. Mead. *University of Maine Center for the Study of Early Man, Publication Series* No. 1. (in press).

Guilday, J. 1967. Differential extinction during late-Pleistocene and Recent times. In *Pleistocene extinctions: the search for a cause*, edited by P. S. Martin and H. E. Wright, Jr., pp. 121–140. Yale University Press, New Haven.

Gustafson, C. E., D. Gilbow, and R. D. Daugherty. 1979. The Manis Mastodon Site: early man on the Olympic Peninsula. *Canadian Journal of Archaeology* 3:157–164.

Hibbard, C. 1960. An interpretation of Pliocene and Pleistocene climates in North America. *Annual Report of the Michigan Academy of Science, Arts, and Letters* 62:5–30.

Hester, J. J. 1967. The agency of man in animal extinctions. In *Pleistocene extinctions: the search for a cause*, edited by P. S. Martin and H. E. Wright, Jr., pp. 170–192. Yale University Press, New Haven.

Hope, J. H. 1978. Pleistocene mammal extinctions: the problem of Mungo and Menindee, New South Wales. *Alcheringa* 2:65–82.

Horton, D. R. 1978. Extinction of the Australian megafauna. *Australian Institute of Aboriginal Studies Newsletter* n.s. 9:72–75.

———. 1980. A review of the extinction question: man, climate and megafauna. *Archaeology and Physical Anthropology in Oceania* 15:86–97.

————. 1981. Water and woodland: the peopling of Australia. *Australian Institute of Aboriginal Studies Newsletter* n.s. 16: 21–27.

Horton, D. R. and R. V. S. Wright. 1981. Cuts on Lancefield bones: carnivorous Thylacoleo, not humans, the cause. *Archaeology in Oceania* 16:73–80.

Jelinek, A. J. 1967. Man's role in the extinction of Pleistocene faunas. In *Pleistocene extinctions: the search for a cause*, edited by P. S. Martin and H. E. Wright, Jr., pp. 193–200. Yale University Press, New Haven.

Krantz, G. 1970. Human activities and megafaunal extinctions. *American Scientist* 58:164–170.

Lakatos, I. 1978. Falsification and the methodology of scientific research programs. In *The methodology of scientific research programmes*, edited by J. Worrall and G. Currie, pp. 8–101. Cambridge University Press, Cambridge.

Lundelius, E. L., Jr. 1967. Late-Pleistocene and Holocene faunal history of central Texas. In *Pleistocene extinctions: the search for a cause*, edited by P. S. Martin and H. E. Wright, Jr., pp. 287–319. Yale University Press, New Haven.

————. 1983. Climatic implications of late Pleistocene and Holocene faunal associations in Australia. *Alcheringa* 7:125–149.

Main, A. R. 1978. Ecophysiology: towards an understanding of late Pleistocene marsupial extinctions. In *Biology and Quaternary environments*, edited by D. Walker and J. C. Guppy, pp. 169–183. Australian Academy of Science, Canberra.

Marshall, L. G. 1973. Fossil vertebrate faunas from the Lake Victoria region, S. W. New South Wales, Australia. *Memoirs of the National Museum of Victoria, Melbourne* 34:151–172.

Martin, P. S. 1958. Pleistocene ecology and biogeography of North America. In *Zoogeography*, edited by C. L. Hubbs. *American Association for the Advancement of Science Publication* 51:375–420.

————. 1967. Prehistoric overkill. In *Pleistocene extinctions: the search for a cause*, edited by P. S. Martin and H. E. Wright, Jr., pp. 75–120. Yale University Press, New Haven.

————. 1973. The discovery of America. *Science* 179:969–974.

Martin, P. S. and P. F. Murray. n.d. Australia's giant marsupials: victims of overkill? Unpublished manuscript, Department of Geosciences, University of Arizona, Tucson.

Meltzer, D. J. and J. I. Mead. 1983. Radiocarbon dating and late Pleistocene extinctions. In *Man in late glacial North America*, edited by D. J. Meltzer and J. I. Mead. *University of Maine Center for the Study of Early Man, Publication Series* No. 1. (in press).

Milham, P. and P. Thompson. 1976. Relative antiquity of human occupation and extinct fauna at Madura Cave, southeastern Western Australia. *Mankind* 10:175–180.

Mosimann, J. E. and P. S. Martin. 1975a. [Letter]. *American Scientist* 63:381.

————. 1975b. Simulating overkill by Paleoindians. *American Scientist* 63:303–313.

Murray, P. F., A. Goede and J. L. Bada. 1980. Pleistocene human occupation at Beginner's Luck Cave, Florentine Valley, Tasmania. *Archaeology and Physical Anthropology in Oceania* 15:142–152.

Salmon, M. 1982. Models of explanation: two views. In *Theory and explanation in archaeology: the Southampton Conference*, edited by C. Renfrew, M. J. Rowlands, and B. A. Seagraves, pp. 35–44. Academic Press, New York.

Slaughter, B. H. 1967. Animal ranges as a clue to late-Pleistocene extinction. In *Pleistocene extinctions: the search for a cause*, edited by P. S. Martin and H. E. Wright, Jr., pp. 155–167. Yale University Press, New Haven.

Van Valen, L. 1969. Evolution of communities and late Pleistocene extinctions. *Proceedings of the North American Paleontological Convention*, Part E: 469–485.

Webster, D. 1981. Late Pleistocene extinctions and human predation: a critical overview. In *Omnivorous primates: gathering and hunting in human evolution*, edited by R. S. O. Harding and G. Teleki, pp. 556–594. Columbia University Press, New York.

Wilson, E. O. 1975. *Sociobiology: the new synthesis*. Belknap Press, Harvard.

White, J. P. and J. F. O'Connell. 1979. Australian prehistory: new aspects of antiquity. *Science* 203:21–28.

————. 1982. *A prehistory of Australia, New Guinea, and Sahul*. Academic Press, Sydney.

38

Historic Extinctions: A Rosetta Stone for Understanding Prehistoric Extinctions

JARED M. DIAMOND

THOUSANDS OF SPECIES AND LOCAL POPULATIONS have become extinct in recent centuries, under the eyes of biologists and other literate observers. For many of these modern extinctions the timing and causes are known with relative certainty. The demise of the Great Auk (*Pinguinus impennis*), for example, can be traced in detail, through such episodes as the visit of Jacques Cartier to the Funk Island breeding colony on 21 May 1534 to feed his crew on auk meat; the Icelander Latra Clemens filling his boat with auks at Gunnbjorn Rocks off eastern Greenland in 1590; the destruction of the Funk Island colony by feather collectors beginning in 1785; the killing of the last individual on Orkney in 1812, on St. Kilda in 1821, on Faeroe in 1828, and in Ireland in 1834; and the killing of the last two individuals of the species, and the smashing of their single egg, on June 3, 1844 at Eldey Rock off Iceland, by two fishermen Jón Brandsson and Siguror Islefsson. Obviously, this degree of detail and confidence about the course of extinction is unattainable for late Pleistocene events.

It seems appropriate to conclude a book on prehistoric extinctions with a chapter on historic ones. This will provide a clear picture of witnessed events possibly similar to the unwitnessed events that preceding chapters have attempted to decipher. I will demonstrate that studies of natural extinction on any time scale show uncommon species (including most large species) to be the ones most at risk. I will illustrate the various mechanisms by which extinctions have resulted from environmental effects independent of man, or from effects related to man. My own observations of how modern hunter-gatherers in New Guinea affect populations of large mammals, together with the history of recent man-related extinctions, indicate how diverse man's impact on prehistoric faunas is likely to have been. Hence the critical problem in understanding prehistoric extinctions is not just to understand how they could have occurred: there are plausible theories in abundance. Instead, we need to understand why extinctions did befall some beasts in some places at some times, but not other beasts in the same places nor similar beasts at other places or times. Such paradoxes will be used in this chapter to formulate challenges to climate-based theories and to man-linked theories of prehistoric extinctions. Finally, I shall suggest some directions for future research.

Patterns of Natural Extinction

How does a population's risk of extinction vary with the area of habitat and with the species' abundance? To answer this question five time scales or types of communities will be examined—communities at approximate equilibrium of species number, over a decade, the same communities over many decades, communities undergoing a large excess of extinctions over many decades, communities with many extinctions since the end of the Pleistocene, and communities on a time scale long enough for the evolution of new, higher taxa.

Communities Near Equilibrium
Viewed Over a Decade

If one follows species abundances for a decade or two on a small island or within a small study area on a continent, one is likely to encounter repeated local disappearances and recolonizations. Table 38.1 illustrates this phenomenon, using censuses of breeding land birds from 1954 to 1969 on Bardsey Island, an island of 1.8 km² lying 3 km off the west coast of Britain. The number of breeding species in a single year fluctuated between twenty-four and thirty, but thirty-seven species bred at least once during this time span. Of these thirty-seven, four were initially absent and became regular breeders (Kestrel, *Falco tinnunculus;* Starling, *Sturnus vulgaris;* Yellowhammer, *Emberiza citrinella*; Reed Bunting, *Emberiza schoeniclus*); perhaps two were initially regular breeders and then disappeared (Sparrowhawk, *Accipiter nisus*; Corncrake, *Crex crex*); about fifteen flickered in and out, breeding in some years but not others (e.g. Wood Pigeon, *Columba palumbus;* Cuckoo, *Cuculus canorus;* Song Thrush, *Turdus philomelos*; Robin, *Erithacus rubecula;* Chaffinch, *Fringilla coelebs*); and sixteen species bred in all sixteen census years (e.g. Moorhen, *Gallinula chloropus*; Oystercatcher, *Haematopus ostralegus*; Little Owl, *Athene noctua*; Swallow, *Hirundo rustica*).

Communities Near Equilibrium,
Viewed Over Many Decades

From this flickering of small populations on a small island over sixteen years, it is seemingly a long jump to the total extinction of mammoths in North and Central America (chaps. 3 and 4), the Soviet Union (chap. 22), and China (chap. 23) or of mastodonts in South America (chap. 5) and North America (chap. 15). The "extinctions" on Bardsey were not the end of genetically distinct populations, nor will they even necessarily prove to be the final end for these species on Bardsey itself. Table 38.2 starts to bridge the gap in time and space between temporary extinctions on Bardsey and permanent worldwide ones by summarizing changes in species abundance for Britain, a much larger island (230,737 km²), over a longer time span, 1900 to 1975. During this period there were 157 land bird species that bred at least once in Britain. Of these 157 species, 125 bred every year, and 32 bred in some but not all years, yielding 134 species breeding in an average year. Of the 32 species that did not breed in every year, two initially bred regularly and then disappeared; two initially bred regularly, disappeared for many years, then resumed breeding regularly; 11 bred irregularly (from 1 to 20 times each) but each time disappeared promptly or eventually; and 17 were initially absent but eventually bred regularly, after up to four false starts ending in temporary abandonment.

Figures 38.1 and 38.2 show the two patterns underlying the short-term flickerings of species documented in Tables 38.1 and 38.2. On a given set of islands the probability that a particular species will disappear within a given time is highest for the species with

Table 38.1. Annual Censuses of Breeding Land Birds on Bardsey Island, 1954–69

+ = species bred but total number of pairs undetermined; S = number of breeding species in a year.

	ESTIMATED NUMBER OF BREEDING PAIRS															
	1954	1955	1956	1957	1958	1959	1960	1961	1962	1963	1964	1965	1966	1967	1968	196
Sparrowhawk (*Accipiter nisus*)	1	1	1													
Kestrel (*Falco tinnunculus*)								1	1	1	1	1	1	1	1	1
Corncrake (*Crex crex*)	2	2														
Moorhen (*Gallinula chloropus*)	1	1	1	1	1	2	2	4	5	4	5	4	4	7	6	8
Oystercatcher (*Haematopus ostralegus*)	37	35	35	35	27	35	40	44	39	44	44	34	40	52	55	55
Lapwing (*Vanellus vanellus*)		10	8	10	8	7	20	11	10	5	4	3	6	7	6	10
Ringed Plover (*Charadrius hiaticula*)															1	1
Curlew (*Numenius arquata*)										1	1	1	1	1	1	
Wood Pigeon (*Columba palumbus*)													1		1	2
Cuckoo (*Cuculus canorus*)		1						2	1		1	2				1
Little Owl (*Athene noctua*)	3	3	3	3	4	4	5	6	7	6	5	5	5	6	6	5
Skylark (*Alauda arvensis*)	5		2	3	3	4	4	4	4	6	7	5	6	7	7	7
Swallow (*Hirundo rustica*)	10	6	6	6	6	6	6	5	4	7	5	2	4	5	7	7
Raven (*Corvus corax*)	2	3	2	3	1	2	1	3	1	1	1	2	2	3	3	2
Carrion Crow (*Corvus corone*)	3	4	4	6	7	7	6	6	4	6	5	4	4	5	4	4
Jackdaw (*Corvus monedula*)	35	30	30	25	25	30	30	20	20	25	30	50	45	50	50	47
Chough (*Pyrrhocorax pyrrhocorax*)	2	2	2	2	2	1	2	2	2	3	4	4	4	4	4	3
Wren (*Troglodytes troglodytes*)	20	19	20	20	21	30	30	20	30	4	7	10	14	20	20	20

the lowest population density (fig. 38.1). If one calculates for a given island the percentage of its breeding land bird species that disappear each year, this percentage decreases with island area, from 5 to 20 percent per year for northern European islands less than 2 km² in area, to 0.5 to 2 percent per year for Britain and Iceland, the largest European islands below the Arctic Circle (fig. 38.2). Both patterns—the inverse dependence of population extinction on population density and on area—combine to mean that probability of disappearance is observed to decrease with population size (= population density × area). This finding is intuitively reasonable and follows from probabilistic models of birth and death rates (MacArthur and Wilson 1967, Richter-Dyn and Goel 1972, Leigh 1975).

Table 38.1. Annual Censuses of Breeding Land Birds on Bardsey Island, 1954–69

+ = species bred but total number of pairs undetermined; S = number of breeding species in a year.

(continued)

	ESTIMATED NUMBER OF BREEDING PAIRS															
	1954	1955	1956	1957	1958	1959	1960	1961	1962	1963	1964	1965	1966	1967	1968	1969
Song Thrush *(Turdus phiolomelos)*					2	2	4	2				1				
Blackbird *(Turdus merula)*	3	2	2	3	2	5	7	10	14	12	12	16	16	45	47	40
Wheatear *(Oenanthe oenanthe)*	6	6	2	4	5	9	3	12	8	12	5	6	10	17	20	20
Stonechat *(Saxicola torquata)*	10	8	3	4	6	8	15	12	6	5		2	3	7	8	6
Robin *(Erithacus rubecula)*	10	7	4	1	1		3	3	3		1	1	2	2	1	
Sedge Warbler *(Acrocephalus schoenobaenus)*	2			1								1	1	1	1	1
Whitethroat *(Sylvia communis)*	5	5	5	5	7	10	15	12	14	5	5	6	6	9	10	
Willow Warbler *(Phylloscopus trochilus)*														1		
Chiffchaff *(Phylloscopus collybita)*																1
Dunnock *(Prunella modularis)*	15	13	14	15	15	15	15	15	12	5	12	13	15	20	22	35
Meadow Pipit *(Anthus pratensis)*	100	80	55	60	80	100	100	100	95	25	42	45	47	50	48	45
Rock Pipit *(Anthus spinoletta)*	50	50	50	30	20	25	25	30	30	13	20	26	20	40	40	45
Pied Wagtail *(Motacilla alba)*	3	2	3	2	2	3	2	3	3	3					1	1
Starling *(Sturnus vulgaris)*									3	3	4	12	20	21	36	50
Linnet *(Acanthis cannabina)*	35	27	20	25	27	30	30	27	30	14	20	23	15	15	15	15
Chaffinch *(Fringilla coelebos)*					1							1				
Yellowhammer *(Emberiza citrinella)*				1	2	4	4	2	3	4	4	3	4	4	3	3
Reed Bunting *(Emberiza schoeniclus)*					2	2	2	2	4	3	4	1	1		2	2
House Sparrow *(Passer domesticus)*	12	+	+	10	10	15	30	30	28	30	30	25	25	20	20	12
S	24	24	23	25	25	25	25	27	27	26	26	30	28	27	30	29

SOURCE: Annual reports of Bardsey Bird and Field Observatory, as evaluated by Dr. Timothy Reed.

Communities Undergoing Net Losses of Species, Viewed Over Many Decades

For the islands discussed so far, the short-term record does not show a net loss of species, but instead the local departures of some species are roughly balanced by establishment of other species. However, in the late Pleistocene there was net loss of

Table 38.2. Turnover of Breeding Land Birds in Britain, 1900–75

(Bred in some but not all years between 1900 and 1975).

For species that bred irregularly, the number of years of breeding is given (e.g. "3x"). E = regularly bred in the years after 1900 and then stopped breeding (became extinct); I = did not breed in the years after 1900 and then began breeding regularly (immigrated); EIE, IEI, etc. = extinction or immigration temporarily reversed (e.g. IEIEIEIEI = began breeding and then stopped breeding four successive times before finally beginning to breed regularly).

Became extinct:
White-tailed Eagle	*Haliaeetus albicilla*	E
Kentish Plover	*Charadrius alexandrinus*	EIE

Became extinct, then re-immigrated:
Osprey	*Pandion haliaetus*	
Golden Oriole	*Oriolus oriolus*	

Bred irregularly:
Little Bittern	*Ixobrychus minutus*	3x
Black-winged Stilt	*Himantopus himantopus*	1x
Green Sandpiper	*Tringa ochropus*	2x
Bee-eater	*Merops apiaster*	2x
Hoopoe	*Upupa epops*	20x
Shore Lark	*Eremophila alpestris*	1x
Moustached Warbler	*Acrocephalus melanopogon*	1x
Icterine Warbler	*Hippolais icterina*	1x
Bluethroat	*Luscinia svecica*	1x
Brambling	*Fringilla montifringilla*	1x
Serin	*Serinus serinus*	2x

Immigrated:
Bittern	*Botaurus stellaris*	I
Honey Buzzard	*Pernis apivorus*	IEI
Hen Harrier	*Circus cyaneus*	I
Goshawk	*Accipiter gentilis*	IEI
Little Ringed Plover	*Charadrius dubius*	IEI
Temminck's Stint	*Calidris temminckii*	IEIEIEIEI
Ruff	*Philomachus pugnax*	IEIEIEIEI
Black-tailed Godwit	*Limosa limosa*	I
Whimbrel	*Numenius phaeopus*	IEI
Wood Sandpiper	*Tringa glareola*	I
Collared Dove	*Streptopelia decaocto*	I
Cetti's Warbler	*Cettia cetti*	I
Savi's Warbler	*Locustella luscinioides*	I
Firecrest	*Regulus ignicapillus*	I
Black Redstart	*Phoenicurus ochruros*	I
Redwing	*Turdus iliacus*	I
Fieldfare	*Turdus pilaris*	I

species; numerous authors have discussed the hypothesis that this loss was related to the shrinkage or fragmentation of certain habitats on continents. Do the patterns of Figures 38.1 and 38.2, observed when species number is near equilibrium, also apply during times of reduction of species numbers?

The classic case study of short-term loss of species associated with habitat fragmentation is for Barro Colorado Island, a Panamanian hilltop that became a 16-km² island when construction of the Panama Canal dammed surrounding valleys, creating Gatun Lake around 1914. Of the approximately 200 land bird species known to have bred on Barro Colorado, about 47 had disappeared by 1981: 26 due to decline or disappearance of their habitat through successional changes, 21 due to insularization (Willis 1974, Karr 1982).

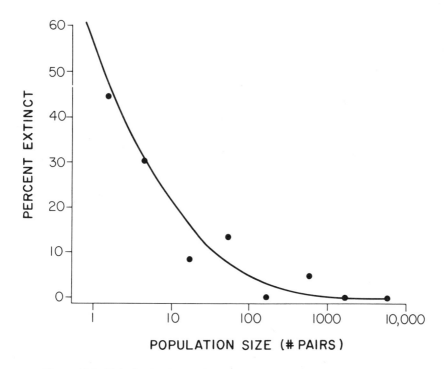

Figure 38.1. Risk of extinction as a function of abundance, for breeding land birds of the California Channel Islands, on a time scale of up to seventy-nine years. Breeding populations of each species on each island were grouped into abundance classes (1–3 pairs, 4–10 pairs, 11–30 pairs, etc.). For each class (abscissa) the ordinate gives the percentage of the populations in that class that have become extinct during the twentieth century. Note that risk of extinction is highest for the rarest species. From Jones and Diamond (1976).

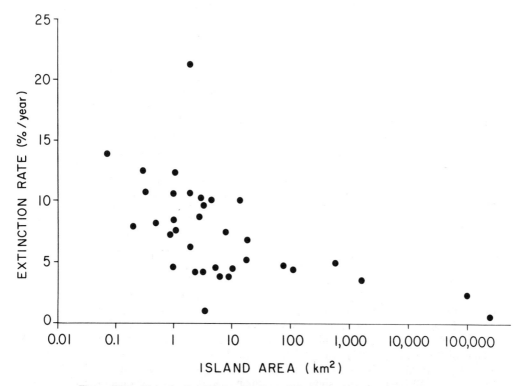

Figure 38.2. Risk of extinction as a function of island area for breeding land birds of northern European islands, on a time scale of one year. Note that risk of extinction decreases with island area.

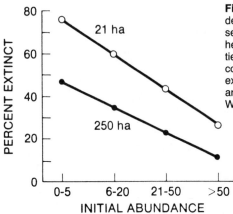

Figure 38.3. Risk of extinction as a function of population density and habitat area, for Brazilian forest birds over several decades. For two forest patches, one of area 21 hectares, the other of 250 hectares, initial population densities were estimated and expressed as bird individuals encountered per 100 hours of observation. Note that risk of extinction decreases with population density for each patch and is higher for the smaller patch. From Terborgh and Winter (1980), based on data of Willis (1980).

A further study of short-term faunal impoverishment due to habitat fragmentation is for three different-sized patches of Brazilian subtropical forest isolated by clearing of forests for agriculture (Willis 1980). Within a few decades the 1,400-hectare forest patch lost 14 percent of its breeding bird species, the 250-hectare patch lost 41 percent, and the 21-hectare patch lost 62 percent. On each patch the probability of extinction was greater for the least abundant species (fig. 38.3).

Thus, in short-term studies the pattern of impoverishment is the same whether the losses are balanced (figs. 38.1 and 38.2) or not balanced (fig. 38.3) by immigrations. In either case, population extinction rates decrease with a species' population density and with available area.

Communities Undergoing Net Losses of Species, Viewed Since the End of the Pleistocene

The patterns of extinction due to habitat fragmentation can also be studied on a long time scale directly relevant to the theme of this book, namely, the 10,000 years since the end of the Pleistocene. Rising late-Pleistocene sea levels from melting glaciers flooded low-lying land areas throughout the world, severing land bridges and carving modern islands off the edges of the Pleistocene continents. This process was qualitatively identical to the creation of Barro Colorado Island by the rising water of Gatun Lake, but occurred much more slowly and on a grander scale. Examples of such land-bridge islands created at the end of the Pleistocene include Britain carved from Europe, Trinidad from South America, Fernando Po from Africa, Tasmania from Australia, and Ceylon, Borneo, Hainan, and Japan from Asia.

Continental species, unable to cross the water gaps, were left stranded on the islands and became subject to a process of differential extinction. These extinctions have been documented directly from the fossil record for mammals on islands of the Bass Straits platform joining Tasmania to Australia (Hope 1973), and of the Sunda Shelf (Hooijer, in Terborgh 1974). In other cases the extinctions are inferred by comparing the modern faunas of a land-bridge island and of a comparable area on the adjacent mainland, as in studies of birds on New Guinea land-bridge islands (Diamond 1972, 1976), birds on neotropical land-bridge islands (Terborgh 1974), birds on land-bridge islands of the Solomon archipelago (Diamond 1983), mammals on islands of the Sunda Shelf (Soulé et al. 1979), and lizards on islands off Baja California (Case 1975; Wilcox 1978, 1980). All these studies showed that small islands lost species faster, and ended up losing more species, than did large islands (fig. 38.4). Bird and mammal populations

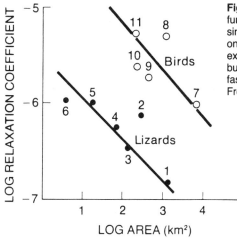

Figure 38.4. Risk of extinction for birds and lizards as a function of island area over 10,000 years. Extinction rates since the end of the Pleistocene were calculated for species on land-bridge islands of the New World. Note that risk of extinction decreases with area for both birds and lizards, but that bird populations go extinct more than ten times faster than lizard populations on an island of the same area. From Wilcox (1980).

disappeared more rapidly than did lizard populations (fig. 38.4), because lizards are generally smaller and have lower metabolic requirements and higher population densities.

Of particular interest in the context of this volume are the great differences among mammal species of the same continent in their proneness to extinction following areal fragmentation. For instance, in Hope's (1973) study of Bass Straits, large carnivores were more susceptible to extinction than small carnivores, carnivores generally were more susceptible than similar-sized herbivores, and habitat specialists were more susceptible than habitat generalists. These three patterns are each special cases of the general pattern that proneness to extinction decreases with population density, for late-Pleistocene fragmentation as for habitat fragmentation in the twentieth century (fig. 38.3). For example, the largest Bass Strait carnivores (the thylacine, *Thylacinus cynocephalus*, and the Tasmanian devil, *Sarcophilus harrisii*) survived only on the largest land-bridge island (Tasmania, 67,900 km²) and disappeared on all smaller islands. The smallest carnivores (*Sminthopsis leucopus* and *Antechinus minimus*), however, survived on islands as small as 6 and 9 km², respectively. Among mammals weighing over 1 kg, three of the four marsupial carnivores (*Thylacinus cynocephalus*, *Sarcophilus harrisii*, and *Dasyurus viverrinus*) survived only on Tasmania, while seven of the nine marsupial herbivores in this weight range survived on five to seventeen islands, including islands as small as 2 km². Two small rodents of specialized habitats, *Mastacomys fuscus* and *Pseudomys higginsi*, also survived only on Tasmania. *Homo sapiens*, because of large body size and low population density, also survived only on Tasmania and disappeared from the smaller islands (Jones 1977). The explanation of these patterns is that herbivores are generally more abundant than carnivores, small mammals are more abundant than big mammals, and habitat generalists are more abundant than habitat specialists.

Besides carving real islands off the continents, late Pleistocene changes in temperature and rainfall also created virtual islands within the continents by fragmenting habitats. For example, the mountains rising from the Great Basin of western North America are inhabited by numerous small mammal species confined to pinyon-juniper, meadow, and riparian habitats above 7,500 feet. Today these species occur as disjunct mountaintop populations on "islands in the sky," separated by the unsuitable habitats of the basin floor. In the Pleistocene, however, the distributions of these species were continuous across the Great Basin. Rising late Pleistocene temperatures drove these habitats and their mammals up the mountain slopes, fragmenting their ranges and subjecting them to differential disappearance of populations (Brown 1971, 1978). Just as in Hope's study of Bass Strait mammals, small mountain "islands" lost more species than large islands, while herbivores, habitat generalists, and small species survived on more and smaller islands than did carnivores, habitat specialists, and large species (Table 38.3). Similar late Pleistocene shifts in altitudinal zones occurred on mountains throughout the world, as discussed for New Guinea by Hope (chap. 30) and as conveniently documented for Arizona by plant remains in fossil packrat middens (chap. 7).

Communities Viewed Over Geological Time

To detect patterns in extinction rates on a time scale much longer than the 10,000 years since the end of the Pleistocene, let us examine the indirect evidence from faunal endemism (Mayr 1965, Diamond 1980). Some oceanic islands, such as Hawaii and New Zealand, are famous for their biotas consisting largely of endemic species. Other islands, such as Palau and the New Hebrides, lack such fame and share most of their species with neighboring archipelagoes. It can be shown that the percentage of an archipelago's species (within some natural group of species, such as birds) that are endemic to the archipelago varies inversely with the average extinction rate for the group's species in the archipelago. (The reason is that if populations go quickly extinct, few survive long enough to evolve into an endemic species or genus. Let q be the probability per unit time that a species present in an archipelago will go extinct. Then the fraction of an archipelago's species that survive for time t [e.g. for the time necessary to evolve to a distinct species] is simply e^{-qt}. On the other hand, if the colonist pool outside the archipelago consists of P species with probability p per unit time of colonizing the archipelago, then the number of species on the archipelago at any instant is $Pp/(p+q)]$. Thus, species number depends on both immigration and extinction rates, but degree of endemism directly reflects only extinction rates. However, extinction rates are influenced by immigration through competition. In addition, continually arriving colonists will inhibit differentiation of a founder population until it has diverged to the point of reproductive isolation.)

Figure 38.5 shows how this percentage of endemics increases with archipelago area and isolation for Pacific island avifaunas. The Pacific islands with a high percentage of endemics are the huge, somewhat isolated islands of New Zealand (69 percent), Australia (49 percent), and New Guinea (47 percent), and the medium-sized, remote Hawaiian archipelago (83 percent). There is only a modest percentage of endemics on the large but close Solomons (15 percent) and Bismarcks (5 percent), and on the remote but small Societies (17 percent) and Marquesas (9 percent). The reason why endemism increases with area is that extinction rates decrease with area over geological time, just as over short times (figs. 38.2, 38.3) and over postglacial times (fig. 38.4). Endemism increases with isolation because more remote archipelagoes have fewer species, hence less competition, hence lower extinction rates.

The percentages in Figure 38.5 take into account known subfossil extinct species of New Zealand (chap. 34). The figure was prepared before the discoveries of subfossil Hawaiian species by Olson and James (chap. 35); these discoveries would raise the value for Hawaii from 83 percent to nearly 100 percent. As Olson and James point out, subfossil bird species probably await discovery in other Pacific archipelagoes and are important for biogeographic analyses. How does this gap in our knowledge affect the conclusions to be drawn from Figure 38.5? The quantitative values will certainly be subject to upwards revision. However, it is difficult to see how the qualitative pattern of endemism being greatest on large and remote archipelagoes could arise as an artifact of our present ignorance of subfossils. If anything, subfossil discoveries may steepen the increase in endemism with isolation, because isolated, originally mammal-free archipelagoes like Hawaii and New Zealand are the ones most likely to have suffered extinctions of endemic birds when man arrived with his rats, cats, and pigs.

Table 38.3. Modern Distribution of Small Flightless Mammal Species of Pinyon-Juniper Woodland on Nineteen Mountain Ranges in the Great Basin of North America

Habitat fragmentation since the Pleistocene has subjected mammal populations to differential extinction. H = herbivore; C = carnivore; G = habitat generalist; S = confined to specialized habitat such as streams, meadows, or talus; "—" sign = modern absence of species from mountain range.

Species	Number of Ranges per Species	Weight (grams)	Diet	Habitat Preference	Toiyabe	Ruby	White-Inyo	Snake	Toquima	Schell Creek	Deep Creek	White Pine	Desatoya	Spring	Stansbury	Oquirrh	Grant	Diamond	Spruce	Roberts Creek	Sheep	Pilot	Panamint
Uinta chipmunk (Eutamias umbrinus)	17	60	H	G	+	+	+	+	+	+	+	+	+	+	+	+	+	+	+	+	+	−	−
Bushy-tailed woodrat (Neotoma cinerea)	17	300	H	G	+	+	+	+	+	+	+	+	+	+	+	+	−	+	+	−	+	−	+
Cliff chipmunk, Panamint chipmunk (Eutamias dorsalis), (E. panamintinus)	16	55	H	G	+	+	+	+	+	+	+	+	+	−	+	+	+	−	+	−	+	+	+
Golden-mantled ground squirrel (Spermophilus lateralis)	14	170	H	G	+	+	+	+	+	+	+	+	+	+	+	−	+	−	+	−	−	+	+
Long-tailed vole (Microtus longicaudus)	13	45	H	G	+	+	+	+	+	+	+	+	+	+	+	+	+	−	−	−	−	+	−
Nuttall's cottontail (Sylvilagus nuttallii)	12	800	H	G	+	+	−	+	+	+	+	+	+	+	+	+	+	−	−	−	−	−	+
Yellow-bellied marmot (Marmota flaviventris)	10	3,000	H	G	+	+	+	+	+	+	+	+	−	+	−	−	+	−	−	−	−	+	+
Vagrant shrew, Inyo shrew (Sorex vagrans), (S. tenellus)	8	7	C	G	+	+	+	+	+	−	+	−	+	−	−	−	+	−	−	−	−	−	−
Northern water shrew (Sorex palustris)	6	14	C	S	+	+	+	+	+	−	+	−	−	+	−	−	−	−	−	+	−	−	−
Pika (Ochotona princeps)	5	120	H	S	+	+	+	+	+	−	−	+	−	−	−	−	−	−	−	−	−	−	−
Western jumping mouse (Zapus princeps)	4	25	H	S	+	−	−	+	−	−	−	−	−	−	+	−	−	−	−	+	−	−	−
Ermine (Mustela erminea)	4	50	C	G	+	−	+	+	−	−	−	−	−	−	−	−	−	−	−	+	−	−	−
Belding's ground squirrel (Spermophilus beldingi)	3	300	H	S	+	+	−	−	−	−	−	−	−	−	+	−	−	−	−	−	−	−	−
White-tailed jack rabbit (Lepus townsendii)	1	3,000	H	S	−	+	−	−	−	−	−	−	−	−	−	−	−	−	−	−	−	−	−
Number of species per mountain					13	12	11	10	10	8	8	7	7	6	6	6	5	4	4	4	3	3	3
Mountain area above 7,500 feet (sq. miles)					684	364	738	417	1,178	1,020	223	262	83	125	56	82	150	159	49	52	54	12	47

SOURCE: Brown (1971, 1978)

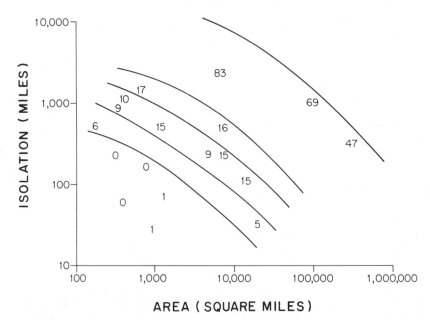

PERCENTAGE OF AVIFAUNA
ENDEMIC AT THE SPECIES LEVEL OR HIGHER

AREA (SQUARE MILES)

Figure 38.5. Numbers on the graph give, for some Pacific island or archipelago, the percentage of its breeding land bird species that are endemic at the level of full species or higher. Location of number gives the island's area (square miles, abscissa) and distance from nearest major colonization source (miles, ordinate). For instance, 16 percent of the species of Fiji, with an area of 7,055 square miles and distance of 520 miles from the New Hebrides, are endemic full species or belong to endemic genera. Curves are drawn by hand to group islands with similar endemism. Note that endemism increases with island area and isolation. Since percentage of endemics is an inverse measure of the risk of extinctions, this means that extinction rates over evolutionary time are lowest on the largest, most remote islands. From Diamond (1980).

Conclusion: The Main Pattern of Natural Extinction

An inverse relation between extinction rates and population size (proportional to area and to population density) appears at every time scale examined, from a decade to extended evolutionary times. (In addition, several studies have shown that, for species with similar population size, the one showing wider fluctuations in abundance is more susceptible to extinction [cf. Karr 1982].) This main pattern of natural extinction includes a preferential susceptibility of megafauna, since large animals have on the average lower population densities than small mammals. Could the megafaunal extinction waves of the late Pleistocene just be one more example of this "normal" pattern of extinction, with extinction rates accelerated by late Pleistocene climatic changes? I shall return to this question.

Extinctions Related to Climate

During the last millenium numerous cases have been recorded of local expansions and contractions of populations that are convincingly related to changes in climate. These shifts have even involved populations of man: notably, the colonization of Iceland,

Greenland, and Newfoundland by Vikings during the warm period A.D. 900 to 1300; the subsequent disappearance of the Greenland Vikings and southward expansion of the Greenland Eskimos during the colder centuries A.D. 1300 to 1500; and the occasional arrivals of Greenland Eskimos (as well as hooded seals, *Cystophora cristata*, and other Arctic mammals and birds) in Britain during the Little Ice Age from A.D. 1550 to 1800.

The literature on these range shifts is most detailed for Europe. Papers discussing the relative effects of climate and of man on range shifts of European birds include Kalela (1949), Williamson (1975), von Haartmann (1978), and Järvinen and Ulfstrand (1980). Corresponding papers for European butterflies include Heath (1974) and Burton (1975).

The climate-linked population changes documented in recent centuries have consisted almost entirely of flickerings of local populations in and out of existence, and longer-term advances or retreats, rather than complete extinctions of species. The reason is that species are always being tested by fluctuating climates, and the fluctuations in recent centuries have been modest compared to those that extant species must have survived in their earlier histories. Thus, the paucity of modern complete extinctions linked to climate is not an argument against hypotheses attributing many complete extinctions to the drastic climate shifts of the Pleistocene. For example, Webb (chap. 9) describes five extinction waves of North American mammals since the Miocene, four of which preceded man's arrival in North America and were thus surely due to climate or other natural causes. At least three of these four climate-linked extinction waves coincided with glacial terminations. Of the mammalian genera eliminated from North America by the most recent of Webb's five extinction waves, the Rancholabrean, nearly half (e.g. *Tapirus*, *Tremarctos*, and various species of *Felis* and *Panthera*) do not constitute total extinctions but instead range retreats to tropical America. Similarly, Guilday (chap. 12) describes numerous range shifts of small mammals as well as extinctions of large mammals in late Pleistocene Appalachia; Liu and Li (chap. 23) suggest that the disappearance of mammoths from China may partly have involved a temperature-driven migration; and Tchernov (chap. 24) documents range shifts of mammals through the Levant in response to Pleistocene rainfall and temperature cycles. The modern climate-linked population shifts that I review are less drastic, but they do illustrate the types of processes that may have culminated in climate-linked extinctions in the Pleistocene.

Effects of Temperature

Consider first the effects of changing temperatures. An exceptionally clear study is for Iceland, whose breeding birds are particularly sensitive to temperature changes, since Iceland lies at the northern range limit of numerous southern bird species and is also at the southern limit of some northern species (Gudmundsson 1951).

The century 1870 to 1970 was a period of rising land and sea temperatures in Iceland, the rise being greatest (up to 2½°C) in the winter months and least in June. Southern species that first bred in Iceland during the twentieth century and now breed regularly are the Shoveler (*Anas clypeata*), Pochard (*Aythya ferina*), Tufted Duck (*Aythya fuligula*), Black-headed Gull (*Larus ridibundus*), Herring Gull (*Larus argentatus*), Lesser Black-backed Gull (*Larus fuscus*), Common Gull (*Larus canus*), Short-eared Owl (*Asia flammeus*), Starling (*Sturnus vulgaris*), and Blackbird (*Turdus merula*). Other southern species began to breed but failed to become established: Coot (*Fulica atra*), Lapwing (*Vanellus vanellus*), Collared Dove (*Streptopelia decaocto*), Wood Pigeon (*Columba palumbus*), Fieldfare (*Turdus pilaris*), Swallow (*Hirundo rustica*), and House Martin (*Delichon urbica*). Black-tailed Godwit (*Limosa limosa*) and Oyster-catcher (*Haematopus ostralegus*) formerly bred largely or solely in southern Iceland but now also breed in northern Iceland. The Merlin (*Falco columbarius*), Widgeon (*Anas penelope*), Snipe (*Gallinago gallinago*), Redshank (*Tringa totanus*), and Redwing (*Turdus iliacus*) have always bred in Iceland in the summer but formerly migrated south for the winter; increasing numbers now winter in southern Iceland. Curlews (*Numenius arquata*), Herons (*Ardea cinerea*), Lapwings (*Vanellus vanellus*), Fieldfares (*Turdus pilaris*), Common Gulls (*Larus canus*), and Blackbirds (*Turdus merula*) that bred on the European mainland began to winter regularly in Iceland by 1950, and the latter four of these species have now bred, the last two regularly. Wrens (*Troglodytes troglodytes*), which are resident in Iceland, were formerly rare because of winter deaths but are now commoner. On the negative side, two arctic species that reach their southern breeding limit in Iceland, the Long-tailed Duck (*Clangula hyemalis*) and Little Auk (*Alle alle*), decreased greatly in numbers; the Little Auk is now almost extinct in Iceland.

The importance of the particular season at which a short-term temperature trend becomes most marked is further illustrated by changes in breeding birds of the European mainland from 1830 to 1940 (Kalela 1949). In that period the winter and early spring temperatures rose in central Europe and especially in northern Europe, but summers grew cooler in central Europe. As a result, southern breeding bird species that are permanent residents and generally limited by winter conditions tended to expand northward; species that winter in southern Europe and northern Africa and then migrate northward in early spring and breed in April and May also expanded their breeding range northward, being especially sensitive to temperature swings during migration and breeding; but species that migrate northward later in the spring and breed in May through July retreated southward. Thus, different species in the same area underwent opposite population trends due to opposite temperature trends in the months of greatest importance—i.e. altered patterns of seasonality. As summarized by Grayson (chap. 1), there is a century-old school of thought that relates seasonality and climatic equability to Pleistocene extinctions. In the present volume this approach is exemplified in the chapters by Graham and Lundelius (chap. 11), Guthrie (chap. 13), and Kiltie (chap. 14), who develop different models to explain how increased seasonality at the end of the Pleistocene could have precipitated the late Pleistocene extinction wave. Gilbert and Martin (chap. 6) discuss increased seasonality as a likely cause of late Pleistocene extinctions at Natural Trap Cave, Wyoming.

In cases of a northern species and a southern species that compete and tend to replace each other geographically, temperature trends may shift the border between the two (Merikallio 1951, Williamson 1975, Järvinen and Väisänen 1979). Cases include the northward advance of the Crested Tit (*Parus cristatus*), Chaffinch (*Fringilla coelebs*), Jay (*Garrulus glandarius*), and Curlew (*Numenius arquata*) in Scandinavia at the expense of their boreal counterparts the Siberian Tit (*P. cinctus*), Brambling (*F. montifringilla*), Siberian Jay (*Perisoreus infaustus*), and Whimbrel (*N. phaeopus*); the displacement of the Long-tailed Duck (*Clangula hyemalis*) by the Tufted Duck (*Aythya fuligula*) in Iceland; and the northward advance of the Red-neck Phalarope (*Phalaropus lobatus*) and retreat of the Grey Phalarope (*P. fulicarius*) in Greenland. Such temperature-dependent range shifts had led to hybridization between northern and southern species in the case of the Redpoll (*Acanthis flammea*) and Arctic Redpoll (*A. hornemanni*) in Greenland, Iceland, and Scandinavia and the Herring Gull (*Larus argentatus*) and Glaucous Gull (*L. hyperboreus*) in Iceland (see summary by Williamson 1975). E. Anderson (chap. 2) points out that some Pleistocene extinctions may have involved competitive replacement, such as the replacements of the short-faced bear (*Arctodus simus*), the moose (*Cervalces scotti*), the dire wolf (*Canis dirus*), and the giant beaver (*Castoroides*) by the brown bear (*Ursus arctos*), moose (*Alces alces*), timber wolf (*Canis lupus*), and beaver (*Castor*). If so, these extinctions may have been the culmination of range shifts similar to those just discussed for northern and southern bird species in Europe.

While the examples of the preceding three paragraphs involve short-term trends in temperature and hence in populations, unusual temperatures during a single year can also produce temporary population consequences. Well-studied examples are the local extinctions of British birds during the exceptionally cold winters of 1916–17, 1939–40, 1946–47, and 1962–63 (Dobinson and Richards 1964, Magee 1965, Batten 1980). The winter of 1939–40 exterminated Stonechat (*Saxicola torquata*) throughout northwest Suffolk and the Orkneys, while the winter of 1946–47 exterminated it throughout southern Wales and much of eastern and central Scotland. The winter of 1962–63, which was the coldest in England since 1740, clobbered British populations of Stonechat, Wren (*Troglodytes troglodytes*), Song Thrush (*Turdus philomelos*), Kingfisher (*Alcedo atthis*), Barn Owl (*Tyto alba*), Green Woodpecker (*Picus viridis*), Grey Wagtail (*Motacilla cinerea*), Goldcrest (*Regulus regulus*), and Long-tailed Tit (*Aegithalos caudatus*) (note in Table 38.1 the drop in breeding pairs of Wren on Bardsey from thirty in 1962 to four in 1963). It took several years for these populations to recoup their losses.

Effects of Rainfall

Having considered these cases of temperature-linked population shifts, I now examine rainfall-linked shifts. A good example of a short-term trend is the decline of the Whitethroat (*Sylvia communis*), and to a lesser extent the Sedge Warbler (*Acrocephalus schoenobaenus*), Redstart (*Phoenicurus phoenicurus*), Spotted Flycatcher (*Musicapa striata*), Garden Warbler (*Sylvia borin*), and Yellow Wagtail (*Motacilla flava*), in Britain as a result of the Sahel drought (Winstanley et al. 1974). British populations of these species winter in West Africa south of the Sahara. Rainfall in the Sahel in 1968 was 25 percent to 70 percent (locally) below normal, shifting the isohyets up to 400 km southwards. Sahel rains recovered slightly in 1969 but then resumed their decline. From the summer of 1968 to the summer of 1969 the breeding Whitethroat population declined by 77 percent throughout Britain and by 60 percent in western and central Europe. With the 1969 Sahel rains the British Whitethroat population rose slightly, but the decline resumed with continuation of the drought, and the population had not recovered by the early 1980s.

A more transient effect of rainfall is provided by the unusually wet summer of 1927 on the north German island of Helgoland. Exceptionally for Helgoland in that period, standing fresh water was available to birds during the breeding season, and the number of breeding species jumped from three to at least eight with the addition of Skylark (*Alauda arvensis*), White Wagtail (*Motacilla alba*), Whitethroat (*Sylvia communis*), Great Tit (*Parus major*), and Tree Sparrow (*Passer montanus*) and possibly House Martin (*Delichon urbica*) and Chaffinch (*Fringilla coelebs*). All but two of these species failed to breed in the following year.

For species of arid habitats with marked year-to-year variation in rainfall, wide fluctuations in abundance and frequent local extinctions are the norm. Many bird species of the Australian desert, such as Grey Teal (*Anas gibberifrons*), are nomads that breed when it rains and then move elsewhere. An example for the arid southwestern United States is a pine-oak woodland community that Cody (1981) studied in the Chiricahua Mountains at an elevation of 1,650 m. Annual rainfall varies three-fold between years, and with it insect abundance varies eight-fold and bird abundance three-fold. In a very dry year five bird species of lower elevations or drier habitats joined the community, one species increased in abundance, six species declined greatly in abundance, and seven species disappeared completely, compared to average years. In very wet years two species of higher elevations or riparian habitats joined the community, and eleven species increased in abundance compared to average years.

Other clear modern examples of local extinctions due to fluctuations in rainfall are described by Vereshchagin and Baryshnikov (chap. 22). In dry years the mole, *Talpa caucasica*, becomes extinct over large areas of the western Caucasus, while the vole, *Microtus socialis*, is locally eliminated in the eastern Transcaucasus after dry winters.

For the Pleistocene and early Holocene, Klein (chap. 25) describes how savanna mammals now confined to the sub-Sahara occupied the Sahara during the moist period ending 30,000 to 40,000 B.P., then retreated as the Sahara dried out, reoccupied the Sahara again in the moist period beginning 12,000 to 14,000 B.P., and retreated again with the return of dry conditions to the Sahara at 4000 to 5000 B.P. Most of the population changes that Klein discusses are expansions and contractions rather than extinctions, but the camel *Camelus thomasi* and at least three other large savanna mammals became totally extinct around 4000 to 5000 B.P. Horton (chap. 29) argues that the arid phase at 15,000 to 26,000 B.P. in Australia similarly eliminated Australia's woodland megafauna, while Merrilees (chap. 28) links some extinctions in southwest Australia to rainfall changes.

Extinctions Related to Man

In considering modern effects of climate, we had to content outselves with examining range contractions and local extinctions, because almost no modern cases exist of total extinction due clearly to climate. In contrast, there is a rich modern data base of total extinctions due to man, who is believed in one way or another responsible for most or nearly all recorded extinctions of vertebrates in modern times. Standard reference works on modern extinct and endangered species include: for mammals, Allen (1942), Harper (1945), Goodwin and Goodwin (1973), Corbet and Hill (1980), and Thornback and Jenkins (1982); for birds, Greenway (1967), Halliday (1978), and King (1977, 1980, 1981, 1983); for vertebrates generally, Ziswiler (1967), Fisher et al. (1969), McClung (1976), and Day (1981); for Australian vertebrates, Ovington (1978), Tyler (1979), and Frith (1979); and for general discussion of mechanisms of man-related extinctions, Frankel and Soulé (1981, their chaps. 2 and 5), Ehrlich and Ehrlich (1981, their chaps. 6 and 7), and Soulé (1983).

Of the approximately 4,200 modern species of mammals, an estimated 63 have become extinct since A.D. 1600, plus 52 subspecies. As summarized in Table 38.4, 81 extinctions are of continental populations, 1 of a pelagic whale, and 33 of remote island populations. Among small mammals, Europe, Africa, and Asia suffered only 1 or 2 extinctions each, while North America suffered 6, Australia 22, and islands 33.

It would be valuable to compare these figures with the number of extant forms. For example, there are far more island birds than mammals, both extinct and extant, and it is unclear whether the fractional extinction rate for island birds has exceeded that for island mammals. On all continents except South America the fractional extinction rate for mammals has considerably exceeded that for birds. It may be significant that the continents show the same patterns in their modern susceptibility to small mammal extinction (Table 38.4) and in their late Pleistocene sucsceptibility to large mammal extinction (Table 17.1 of chap. 17). The lack of modern large mammal extinctions on Australia and remote islands is for the trivial reason that all those formerly present had already gone extinct before A.D. 1600, except for 4 large kangaroo species of Australia.

Of the approximately 8,500 modern species of birds, about 88 have become extinct since A.D. 1600, plus 83 subspecies. The overwhelming brunt of these extinctions (155 out of 171) has befallen island populations (Table 38.4). About three times this number are close to extinction. These figures are underestimates and do not include dozens of other taxa have not been observed in recent decades and whose status is unknown.

No comparable estimates are available for modern extinctions among reptiles, amphibia, and fish.

Modern man has proven versatile as an exterminator, with at least six major methods long at his disposal (and a seventh, chemical pollution, recently added): overkill; habitat destruction by logging, fire, introduced browsing and grazing animals, and

Table 38.4. Species and Subspecies of Mammals and Birds Extinct Since 1600

( ) = number of large mammal taxa (>44 kg)/number of small taxa

	Mammals	Birds
Continents		
Africa	11 (10/1)	0
Asia	11 (9/2)	6
Australia	22 (0/22)	0
Europe	7 (6/1)	0
North America	22 (17/5)	8
South America	0	2
Total	73 (42/31)	16
Pelagic	1 (1/0)	0
Islands		
Continental		
Africa	0	2
Asia	4 (4/0)	0
Australia	0	2
North America	4 (3/1)	3
Oceanic		
Pacific Ocean		
Galapagos	4 (0/4)	0
Baja California islands	0	8
Hawaii	0	24
New Zealand	0	16
Chatham	absent	5
Lord Howe	0	8
Norfolk	absent	6
Cebu (Philippines)	0	11
Bonin, Ryukyu	0	10
Other	0	21
Indian Ocean		
Madagascar	1 (0/1)	2
Christmas	3 (0/3)	0
Mascarenes	0	14
Seychelles	0	2
Atlantic Ocean		
West Indies	22 (0/22)	15
Other	1 (0/1)	5
Mediterranean	2 (0/2)	1
Total, all islands	41 (7/34)	155
Total, all locations	115 (50/65)	171

NOTE: This tabulation omits numerous island birds believed to have become extinct after 1600 but known only from recently reported subfossil bones (e.g. Olson 1977).

SOURCES: Mammal data are from Goodwin and Goodwin (1973), except that two extinct named races of deer from remote Pacific islands are assumed to represent populations introduced by man and are omitted. Bird data are from King (1981, 1983).

draining; introduction of predators; introduction of competitors; introduction of diseases; and extinctions secondary to other extinctions. All six modes were probably effective in prehistoric extinctions as well. Let us consider these mechanisms in turn.

Overkill

This is the sole mechanism by which marine mammals have been exterminated (Steller's sea cow, *Hydrodamalis gigas*; Atlantic gray whale, *Eschrichtius gibbosus*; Japanese sea lion, *Zalophus califorianus japanicus*; Caribbean monk seal, *Monachus tropicalis*) or decimated (numerous other pinnipeds and cetaceans; dugong, *Dugong dugon*; sea otter, *Enhydra lutris*) by man, and by which man has decimated other marine species (turtles, crocodiles, and numerous fish, shark, and ray populations: cf. Brander 1981). Overkill has also been the main cause or a major cause in virtually all the forty-six modern extinctions of large terrestrial mammals (e.g. the thylacine, *Thylacinus cynocephalus*; Quagga, *Equus quagga*; Burchell's zebra, *Equus burchelli burchelli*); in about 15 percent of bird extinctions (e.g. Great Auk, *Pinguinus impennis*; Passenger Pigeon, *Ectopistes migratorius*; Dodo, *Raphus cucullatus*; Arabian Ostrich, *Struthio camelus syriacus*); in exterminations of populations of giant tortoise (*Geochelone*) on the Seychelles, Galapagos, and Mascarenes; and in the extinction of trees logged for their wood (various *Santalum* species) or sap (the wine palm, *Pseudophoenix ekmanii*). For a few decimated bird species the overkill has been of eggs, not of adults (e.g. the Jackass Penguin, *Spheniscus demersus*).

Most of the species exterminated by overkill were large. However, the list includes numerous bird species weighing one pound or less, such as the Wake Island Rail, *Rallus wakensis* (exterminated by starving soldiers during World War II), Passenger Pigeon, and Carolina Parakeet, *Conuropsis carolinensis*.

Interestingly, while numerous taxa of fish and mammals have been decimated at sea, there have been no complete extinctions at sea (with the possible exception of the Atlantic gray whale), due to man's difficulties in searching down every last individual. Three of the four exterminations of marine mammals, those of Steller's sea cow, Japanese sea lion, and Caribbean monk seal, took place on land or in shallow water at their breeding grounds.

We tend to associate overkill with hunting for meat, and this was certainly true for the Passenger Pigeon, Steller's sea cow, giant tortoises, and numerous other victims. However, there have been at least five other modern motives for hunting, and collectively these outweigh hunting for meat. One is the economic value of body parts, such as fur (pinnipeds and sea otters), skins (crocodile), feathers for bedding (destruction of the Great Auk colony on Funk Island and of albatross colonies on many Pacific islands), and oil (whales). A second motive is the cultural value of body parts, such as rhinoceros horn, elephant ivory, butterfly wings, and the display plumes of egrets, ostrich, and birds of paradise. A third motive is to protect gardens (extermination of the Carolina Parakeet) and men's domestic animals (cf. extermination of the thylacine; the Guadalupe Caracara, *Polyborus lutosus;* the Falkland Island fox *Dusicyon australis,* and large predators such as lion and wolf populations). A fourth motive is capture as pets (parrots, primates, and tropical fish), and a fifth is hunting simply for fun, without economic motive.

Among prehistoric extinction discussed in this book, the most incontrovertible case of overkill involves New Zealand's moas (chaps. 32 and 33). At least some extinctions on Hawaii (chap. 35), other Pacific islands (chap. 34), North America (chaps. 16 and 18), Madagascar (chap. 26), and Africa (chap. 25), and possibly many more, provide other examples. Overkill is reviewed on a worldwide basis in Chapter 17 and modeled theoretically in Chapter 20.

Habitat Destruction by Man

This is now the most important cause of extinction, especially because of accelerating destruction of the earth's most species-rich habitat, tropical rainforest. During the past four centuries habitat destruction is believed to have played the (or a) major role in about half of continental bird extinctions and about one-fifth of island bird extinctions (King 1980). The causes of habitat destruction have been varied.

Deforestation for timber, agriculture, or stock grazing has been the most significant form of habitat destruction. Examples include the nearly complete clearing of Mauritius and Rodriguez, resulting in the extinction of at least twelve bird species; the complete clearing of Cebu Island in the Philippines, resulting in extinction of all ten of Cebu's endemic birds; the clearing of the forests of southeast Brazil and consequent near-extinction of dozens of mammals and birds; the clearing of much of eastern Australia and consequent extinctions of mammals; and the clearing of St. Helena and consequent extinction of eighty of St. Helena's more than 100 endemic plant species, most of its endemic land snails, and three of its four known native land birds. Bachman's Warbler (*Vermivora bachmanii*), which bred in the southeastern United States, is probably extinct due to conversion of its wintering habitat in Cuba to sugar cane fields. The Ivory-billed Woodpecker (*Campephilus principalis principalis*) of the United States required old dead trees in alluvial forest, and logging brought its disappearance.

A second type of habitat destruction has resulted from introduced grazing and browsing animals, notably goats and rabbits. The continent on which this factor has contributed most to extinctions is Australia (Ovington 1978, Tyler 1979, Frith 1979). Famous island examples are the extinctions of many plant species and half the endemic bird species on Guadalupe Island off Baja California, due to destruction of most of the vegetation by introduced goats, and extinction of the endemic rail and two of the three songbirds of Laysan Island following removal of vegetation by introduced rabbits.

Drainage and transformation of wetlands contributed to or caused the extinction of the ducks, *Rhodonessa caryophyllacea* in the Ganges delta and *Anas oustaleti* in the Marianas; the deer, *Cervus schombergki*, in the swamps and wet grasslands of Thailand; the grackle, *Cassidix palustris*, in marshes near Mexico City; and the frog, *Discoglossus nigriventer*, of the Near East; the extinction or decimation of populations of Seaside Sparrows in North America; the decimation of the Hawaiian Duck (*Anas wyvilliana*), Stilt (*Himantopus mexicanus knudseni*), Coot (*Fulica americana alai*), and Gallinule (*Gallinula chloropus sandvicensis*); the next extinction of the two most famous plants of ancient Egypt, the lotus, *Nymphaea lotus*, and the papyrus, *Cyperus papyrus hadidii*; and the extinction of numerous fish that lost their habitats to dam-building or drainage of springs.

Fire eliminated the Song Sparrow race, *Melospiza melodia graminea*, of Santa Barbara Island off California, and restricted the Noisy Scrubbird (*Atrichornis clamosus*) of southwest Australia to one unburnt area.

Among prehistoric extinctions, habitat destruction by man is discussed elsewhere in this book as contributing to those on Hawaii (chap. 35), New Zealand and other Pacific islands (chap. 34), and Madagascar (chap. 26).

Introduced Predators

The effect of introduced predators on birds is well documented and is thought to be responsible for about half of island bird extinctions, but for no continental bird extinctions. Introduced predators have also had important but less well analyzed effects on continental freshwater fish, Australian mammals, and island amphibia, reptiles, and invertebrates. Victims of extermination or decimation by predation include prey species much larger than the predator. Most such examples involve predation on small juveniles (e.g. on young Galapagos giant tortoises by introduced pigs, dogs, cats, and rats, and on chicks of the Atitlan Grebe, *Podilymbus gigas*, by introduced bass). However, Kepler (1967) documented killing of nesting adult Laysan Albatross (*Diomedea immutabilis*) by *Rattus exulans*.

There are striking differences among biotas in their susceptibility to introduced predators. The effect of introduced European mammals has been serious in Australia, nonexistent in South America. Introduced rats have exterminated birds on twenty-six islands, including catastrophic extinction waves that eliminated a large fraction of native bird species within a few years on Hawaii, Midway, Lord Howe, and Big South Cape. Yet other islands that share a Pacific location and related avifaunas with the catastrophically decimated islands received introduced rats without suffering extinctions—e.g. Fiji,

Tonga, Samoa, Marquesas, Rennell, and the Solomons, plus Aldabra and Christmas in the Indian Ocean, the Galapagos in the eastern Pacific, and of course the continents of North and South America and Australia. Why these differences in susceptibility?

An important contribution to understanding this difficult problem has been made by Atkinson (1983). He points out that rat-related extinctions of birds are virtually confined to islands lacking native rats and land crabs. The Solomons, Christmas, Galapagos, and North and South America and Australia have (or had) native rats, while the other "immune" islands all have native land crabs, nocturnal scavengers that climb trees, enter holes, and are the invertebrate ecological equivalent of rats. Thus, birds on islands with native rats and land crabs evolved to be able to coexist with such predators. Only predator-naive birds succumbed to introduced rats.

Even on islands that suffered extinctions, some birds continued to thrive. For example, when *Rattus rattus* reached Lord Howe in 1918, five native songbirds were quickly exterminated, but three remain common to this day. Atkinson has interpreted such differences in terms of nest position in relation to the particular rat species introduced. *Rattus norvegicus* mainly affects birds nesting at a height of 0 to 3 m; *Rattus exulans* gets birds nesting in burrows, on the ground, or in the canopy; and *Rattus rattus*, the species responsible for the greateat number of rat-related extinctions, gets birds nesting at any height.

Atkinson's analysis of the effect of rats on birds naive to rats and ratlike predators furnishes a model for the effect of introduced cats and foxes on small mammals of Australia but of no other continent. Australia is unique among the continents in its paucity of cursorial carnivorous mammals cat-sized or larger: only five modern species, and no other extinct Pleistocene ones (chap. 27), fewer even than Madagascar (seven viverrid species extant and at least one extant)! Why Australia evolved so few swift mammalian carnivores is a mystery, but the resulting naiveté of mammalian prey undoubtedly explains the unique impact of introduced cats and foxes in Australia. Atkinson's analysis also furnishes a model for Martin's discussion (chap. 17) of the varying effects of man the hunter on late-Pleistocene biotas with different histories of exposure to man.

Among predators introduced by man, the ones responsible for the largest number of extinctions are the rats, *Rattus rattus* and *R. norvegicus*, carried around the world by Europeans in the last five centuries, and *R. exulans*, carried across the Pacific by Polynesians and Micronesians in preceding millenia. Notorious exterminations by rats include the previously mentioned destruction of five bird species on Lord Howe Island within a few years after rats landed from a ship in 1918; destruction of four endemic birds, the bat *Mysticina tuberculata robusta*, and numerous invertebrates after rats reached New Zealand's Big South Cape Island in 1964 (Atkinson and Bell 1973); and extermination of the finch, *Telespyza cantans*, and the rail, *Porzana palmeri*, on Midway Island within a few years of rats' arrival. In addition, the timing of their arrival and bird extinctions on the various main Hawaiian islands between 1873 and 1932 suggests that rats played a major role in the historic extinction wave of birds on Hawaii (Atkinson 1977), as well as of Hawaiian achatinellid land snails. In New Zealand, rats eliminated mainland populations of the tuatara (*Sphenodon punctatus*) and numerous endemic lizards, frogs, and large insects, now virtually confined to rat-free offshore islands (cf. chap. 34). On Mauritius, rats exterminated most ground snakes and lizards.

Cats are second only to rats in destructiveness among introduced predators. The classic case is the Stephen Island Wren, *Xenicus lyalli*, the only flightless songbird in the world, exterminated single-handedly by the lighthouse keeper's cat in 1894. (Its earlier disappearance from the New Zealand mainland was due to predation by *Rattus exulans* introduced by Polynesians.) Other cat victims include the Guadalupe Island Petrel (*Oceanodroma macrodactyla*), the Crested Pigeon (*Microgoura meeki*) of Choiseul Island, and the Saddleback (*Plilesturnus caruculatus*) population of Little Barrier Island. A controlled experiment on effects of cats is presently underway on Little

Barrier, where cats were recently eradicated by the New Zealand Wildlife Service: populations of the three species of honey-eaters (Meliphagidae) on Little Barrier have already begun to rise (Veitch 1983). On Herekopare Island off New Zealand introduced cats reduced the seabird population from 400,000 to a few thousand and eliminated six resident species.

Other introduced mammalian predators with numerous avian victims include pigs (sharing responsibility with man for exterminating the Dodo), mongoose (also responsible for exterminating several colubrid snakes in the West Indies), monkeys, foxes, and weasels. Introduced predatory birds with avian victims include the Australian Harrier, *Circus approximans*, which was brought to Tahiti to control introduced rats and proceeded to eliminate the parrot, *Vini peruviana*, there; and the Weka (*Gallirallus australis*), a flightless rail of New Zealand, which has exterminated or decimated breeding petrels on several islands where it was introduced.

As already mentioned, cats and foxes are thought to have played an important role in exterminating or decimating small native Australian mammals, although it is difficult to disentangle the predators' effects from those of habitat destruction and introduced competitors. Similarly, introduced fish are thought to have exterminated native fish in numerous lakes and rivers, although again the effects of predation and competition by introduced species have usually not been disentangled. A clear case is the extermination of the trout, *Salvelinus namaycush*, and the ciscos, *Coregonus nigripennis* and *C. johannae*, from the Great Lakes by introduced sea lamprey, *Petromyzon marinus*. Ironically, *Salvelinus namaycush* was itself introduced to Lake Titicaca of the Andes and exterminated the largest endemic fish species of genus *Orestias*, *O. cuvieri*. Other possible exterminations of fish by introduced predatory fish involve effects of the introduced serranid, *Lates niloticus*, on endemic fish of East African lakes (Fisher et al. 1969), and of introduced trout and bass on freshwater fish of the Cape Province of South Africa and on the now-extinct New Zealand grayling (*Prototroctes oxyrhynchus*).

Introduced predators have not yet been implicated in prehistoric extinctions. However, I suspect that *Rattus exulans* introduced by Polynesians was responsible for the extinctions of some of the subfossil birds of Hawaii discussed by Olson and James (chap. 35) and of other Pacific islands discussed by Cassels (chap. 34). The introduction of the dog and its consequent feral establishment as the dingo in Australia several thousand years ago are also likely to have contributed to extinctions.

Introduced Competitors

Available evidence suggests that introduced competitors can drastically reduce the abundance and distribution of native species, but rarely lead to complete extinction, at least in terrestrial continental habitats (Frankel and Soulé 1981, Soulé 1983). The reason pointed out by Soulé is that the relative competitive ability of two species generally varies among habitats. It is rare that one species is superior to another everywhere. Thus, an introduction may at most cause a native species to shrink into favored refuge habitats, where the native might then disappear slowly and from other causes. For example, the introduced North American squirrel, *Sciurus carolinensis*, replaced the native squirrel, *S. vulgaris*, in most of southern Britain; the introduced European House Sparrow (*Passer domesticus*) and Starling (*Sturnus vulgaris*) drove the native Purple Finch (*Carpodacus purpureus*) and Eastern Bluebird (*Sialis sialis*) out of eastern North American suburbs and farms; and the introduced ant, *Iridomyrmex humilis*, drove out North American ants. However, the invasions have not yet led to extinctions in these examples.

A further complication is that, even when extinction does occur, it would be more difficult to document competition than predation as its cause. Predation can be documented dramatically just by observing one animal sinking its teeth into the throat of another. Competition must be documented more laboriously, through studies of complementary range shifts or shared resource items.

A well-documented example is the role of the introduced red deer (*Cervus elaphus*) in the recent range contraction of New Zealand's flightless gallinule (*Notornis mantelli*, the Takahe). Deer and Takahe utilize the same species of tussock grasses with approximately the same order of relative preference, and both deer and Takahe select individual plants on the basis of nutrient and mineral content. Since the introduction of deer to the Takahe's range between 1901 and 1910, Takahe have disappeared from most of their former range and are declining in parts of the remainder (Mills and Mark 1977).

Perhaps the most marked effects of introduced competitors on native species have been in freshwater habitats (Fisher et al. 1969, Anon. 1981). The introduced mosquito-fish, *Gambusia affinis*, exterminated or decimated numerous native fish populations in Australia and North America. Introduced whitefish (*Coregonus*) eliminated the trout, *Salvelinus alpinus*, in Scandinavia. There are many other examples involving introduced trout, bass, and tilapia in North America, Europe, Africa, and New Zealand, but the relative impact of the introduced species as predators and as competitors on native species has not been unraveled.

Competition with introduced placental herbivores and carnivores is assumed to have contributed to declines and extinctions of their native marsupial and rodent equivalents in Australia. Again, effects of competition, predation, and habitat destruction are hard to untangle (Ovington 1978, Tyler 1979, Frith 1979). In the Alice Springs area numerous marsupials that have become locally extinct since 1930 within the range of cattle survive beyond the range of cattle but in the presence of foxes. The disappearance of the burrowing kangaroo, *Bettongia lesueur*, formerly one of the most abundant marsupials, from the Australian mainland is attributed to competition with rabbits for food and burrows. Competition from sheep and rabbits is held responsible for the decline of the wombat, *Lasiorhinus kreffti*.

Declines of native New Zealand and Hawaiian forest songbirds are often attributed partly to competition from the abundant introduced songbird species in forest. However, in New Zealand forests that remain free of introduced mammalian browsers and predators, native songbirds are able to exclude competing introduced bird species. Thus, browsing and predation were the cause of the decline in native songbirds, and abundance of introduced songbirds is a result rather than a cause (Diamond and Veitch 1981).

Among prehistoric extinctions, the disappearance of the thylacine and possibly the Tasmanian devil (*Sarcophilus harrisii*) from the Australian mainland, and their survival on Tasmania, are often attributed to competition from the introduced dingo on the mainland. Competition with invading North American mammals three million years ago, following emergence of the Panamanian land bridge, has been proposed and denied as playing a major role in the slow wave of subsequent mass extinctions that befell South American mammals (Simpson 1953, Webb 1976, Marshall 1981, Marshall et al. 1982, chap. 9 of this book). E. Anderson (chap. 2) attributes some late Pleistocene extinctions of large mammal species partly to competition with related forms.

Introduced Diseases

This has been the chief means by which European man decimated numerous human populations of North and South America and Pacific islands, and exterminated the last Tasmanians on Flinders Island, during the past five centuries. In these cases the exterminating disease was carried by colonizing humans, and the victims were resident humans. When the victims, however, were resident plants or other animals, the vectors were almost always not man himself but instead species that man introduced. Clear examples of decimations due to introduced disease were those of the American chestnut (*Castanea dentata*) due to chestnut blight from Europe, the American elm (*Ulmus americana*) due to Dutch elm disease introduced from Europe, and Swayne's hartebeest, *Alcelaphus buselaphus swaynei*, due to rinderpest introduced from Asia with cattle. Diseases carried by introduced poultry and parrots contributed respectively to the declines of the Heath Hen (*Tympanuchus cupido cupido*) and Norfolk Island Parakeet (*Cyanorhamphus novaezelandiae cookii*). Other suggested but less well documented are the decline of several dasyurid marsupials, including the extinction of the formerly abundant *Dasyurus viverrinus* on the eastern Australian mainland around 1900, due to an unidentified disease (Tyler 1979); the possible role of introduced avian malaria in the nineteenth-century decline of native Hawaiian birds (Warner 1968); and the role of disease spread from introduced *Rattus rattus* in the extinctions of the two endemic rats of Christmas Island.

Secondary Extinctions ("Trophic Cascades")

Since species abundances depend on each other in numerous ways, disappearance of one species is likely to produce cascading effects on abundances of species that use it as prey, pollinator, or fruit disperser. Four modern examples illustrate the range of such phenomena. Puffins (*Fratercula arctica*) feed their young on small fat-rich fish, especially sand eel and sprat. When North Sea fishermen, having fished out herring in the 1960s, switched to sand eel and sprat in the 1970s, all puffin chicks in the breeding colony of a million puffins on Rost Island starved to death (Mills 1981). A two-step trophic cascade has been eliminating ground-nesting birds on Barro Colorado Island in this century: insularization led to the loss of the largest predators (jaguar, puma, Harpy Eagle), leading to a population explosion of smaller predators such as monkeys, peccaries, coatimundis, and possums that served as their prey and that in turn rob bird nests. All five species of the endemic Hawaiian plant genus *Hibiscadelphus* are extinct or nearly extinct due to the disappearance of their pollinators, Hawaiian honeycreepers, whose long curved bills match the plants' narrow tubular curved flowers. The failure of the tree *Calvaria major* on Mauritius to establish seedlings in recent centuries, despite producing numerous seeds, has been attributed to its adaptation for seed dispersal by the now-extinct Dodo (Temple 1977).

Prehistoric examples of secondary extinctions have been suggested. E. Anderson (chap. 2) attributes late Pleistocene extinctions of the scimitar cat (*Homotherium*) and sabertooth (*Smilodon*) to extinctions of their prey. Steadman and Martin (chap. 21) show that among the few late Pleistocene extinctions of birds in North America, large predators and carrion feeders predominated, suggesting extinction secondary to disappearance of mammalian prey. Further examples of prehistoric trophic cascades may include the extinction of New Zealand's giant eagle, *Harpagornis moorei*, during man's extermination of the moas (chaps. 32 and 34); extinctions of an eagle and a hawk (*Accipiter*) in Hawaii during the Polynesian-linked extermination of half of Hawaii's avifauna (chap. 35); and extinctions of several giant owls and hawks in the West Indies, linked to extinctions of numerous mammals (Table 17.13 of chap. 17, and Arredondo 1976).

Multiple Factors

We have been considering mechanisms of man-related extinctions as if the mechanisms operated separately. In some cases one mechanism is so overwhelmingly important that this approach is proper. For instance, one has only to read accounts of how Passenger Pigeons (*Ectopistes migratorius*) were shot en masse and their nesting colonies destroyed to appreciate that overkill alone was sufficient to exterminate the Passenger Pigeon. In many other cases, however, several factors contributed significantly. Consider two typical examples.

The Heath Hen (*Tympanuchus cupido cupido*) was common in open country of eastern North America at the time of European discovery. In the following centuries it was shot by the thousands for food, preyed on by introduced cats, and afflicted with diseases of introduced poultry, all while its grassland habitat was being converted to farmland. By 1830 it had been exterminated on the mainland and was confined to the island of Martha's Vineyard, where its numbers rose to 20,000 by 1916. In that year its numbers were decimated by a fire in the summer, followed by a harsh winter and invasion of Goshawks. Cats, inbreeding, and a disease introduced with turkeys reduced its number to 13 in 1928, 2 in 1929, and in 1930 one, which died in 1932. Thus, four man-related factors—overkill, habitat destruction, and introduced predators and diseases—brought the Heath Hen to the edge of extinction, and two man-related factors—introduced predators and diseases—combined with climate to deliver the final blow (Greenway 1967, Halliday 1978).

The Norfolk Island Parakeet (*Cyanorhamphus novaezelandiae cookii*) is endemic to Norfolk Island, an isolated 3.5-km² island lying between New Zealand and Australia. Visited but not permanently settled by Polynesians, the island was rapidly deforested following European settlement until only 15 percent of the original forest remained. As the parakeet is confined to native forest, its numbers dropped drastically; predation by introduced cats and rats also contributed to the drop. The aggressive introduced Australian parrot, *Platycercus elegans*, is now abundant on Norfolk, eating the same foods and occupying the same remnant forest as the endemic parakeet. In 1976 disease broke out in the introduced species and spread in 1977 to the parakeet. Some thirty individuals remained in the early 1980s. Thus, four man-related factors caused the decline: habitat destruction and introduced competitors, predators, and disease (see chapter by Forshaw in Pasquier 1980).

Impact of Modern Hunters in New Guinea

In the course of ecological field work in many areas of New Guinea since 1964, I have spent about two years living with Papuan hunters. Their impact may offer an instructive model for understanding the impact of prehistoric neolithic hunters.

The people I observed include several lowland groups of nomadic hunter-gatherers in the Meervlakte, foothills of the van Rees Mountains, and foothills of the Bewani Mountains, and highland agriculturalists on the southern watershed of the Central Dividing Range. Their main weapon for hunting is a bow and arrow made entirely of wood, plus some use of nooses and wooden spears. These weapons would be invisible in the archaeological record and are probably less effective than the stone-tipped weapons widespread in the late Pleistocene. In the tropical rainforest habitat in which the Papuan hunters operate, it is much harder to detect prey than in open habitats or in temperate-zone forest. The largest prey are tree kangaroos (*Dendrolagus*), wallabies (*Thylogale, Macropus, Dorcopsis, Dorcopsulus*), feral pigs, the flightless cassowary (*Casuarius*), and large flying birds. None of these prey species occurs in herds or flocks. Thus, weaponry, habitat structure, and prey habits all make decimation of prey more difficult for these modern New Guinea hunters than it would have been for Clovis hunters and moa-hunters.

While virtually all of New Guinea is within the hunting territory of some human group, the hunting impact would seem nevertheless to be minimal in much of New Guinea. Human population density often averages less than one person per square mile, and parts of a territory are visited only every year or two for a short period by a small band of hunters. Not until I entered the Gauttier (Foja) Mountains by helicopter in 1979 and 1981 was I able to appreciate the impact that even this low hunting pressure exerts. The Gauttier Mountains are an isolated range rising steeply from the swamps of the Meervlakte and the north New Guinea coastal plain. Today no humans live in these mountains, and except in the foothills they are never visited by people from the adjacent swamps. Missionary pilots who have flown over these mountains on regular flight paths for the past ten years have never seen signs of humans. Except for my visits in 1979 and 1981, the only evidence that a human had previously been in the Gauttier Mountains is the existence of three skins of the bowerbird, *Amblyornis flavifrons*, shipped to Europe in 1895 from an unspecified New Guinea location by a feather merchant dealing with New Guinea hunters. This bowerbird is now known to be confined to higher elevations in the Gauttier Mountains and must have been collected there (Diamond 1982). The reasons for the absence of humans from the Gauttiers are multiple. The mountains are too steep, isolated, and small in area to support mid-montane agriculture or specialized montane hunter-gatherers; they rise from lowland swamps with few human inhabitants; and they lack sago, the most important food of the lowland hunter-gatherers. These same factors presumably operated throughout the 30,000+ years that man has occupied New Guinea. Thus, the Gauttier Mountains may be one of the few forested areas in the modern world whose animals are still naive to man.

Elsewhere in New Guinea the largest native mammals, tree kangaroos, are nocturnal, uncommon, and extremely shy. I have never seen one in the wild outside the Gauttiers. In these mountains the tree kangaroo *Dendrolagus matschiei* is common and diurnal and permitted me to approach it openly within 10 meters. Wallabies elsewhere in New Guinea are also very shy. In twenty-four months I had glimpsed about six individuals as they fled after being surprised. In the Gauttier Mountains I found the wallaby *Dorcopsulus vanheurni* abundant, saw it daily, and was again able to approach within 10 meters. Displays of *Amblyornis* bowerbirds elsewhere in New Guinea have been witnessed only by concealed observers. In contrast, a male *A. flavifrons* in the Gauttiers displayed for twenty minutes to a female, while I stood in full view at the bower.

The contrast between my experience in the uninhabited Gauttier Mountains and everywhere else in New Guinea suggests that even infrequent visits by hunters eventually transform the behavior of surviving prey species. Until I had worked in the Gauttiers, I was mystified to understand how the few Maoris in the vastness of New Zealand's South Island could have killed *all* the moas, and how anyone could take seriously the Mosimann-Martin hypothesis of Clovis hunters eliminating most large mammals from North and South America in a millenium or so. I no longer find this at all surprising when I recall the large kangaroo *Dendrolagus matschiei* remaining on a tree trunk at a height of 2 meters, watching my field assistant and me as we talked nearby in full sight. The low densities of these mammals elsewhere in New Guinea, even in areas visited annually only by nomadic hunters, illustrate how susceptible large, K-selected mammals with low reproductive rates are to hunting pressure.

Two other observations about hunters and hunted in New Guinea may also be relevant to understanding Quaternary extinctions: the use of dogs, and the motives for hunting and destruction.

Hunting large terrestrial mammals and flightless cassowaries in New Guinean rainforest depends heavily on the use of dogs. The hunter's dog (or dogs) finds the prey and trees it or brings it to bay, until the hunter catches up with the sound of the barking and spears the prey or dispatches it with bow and arrow. Without a dog, it is impossible to detect silent prey at any distance in rainforest, or to overtake the prey once detected. All seven known specimens of the Goodenough Island Wallaby (*Dorcopsis atrata*) were collected and sold to the Fourth Archbold Expedition by a single New Guinean hunter with his dog (Van Deusen 1957). Van Deusen and the other expedition biologists never saw this wallaby alive during the forty-one days they spent on Goodenough Island in its habitat. During my field work in New Guinea's Karimui Basin in the early 1960s, the basin's population of wild pig was being decimated through the skill of one particular small dog at bringing pigs to bay. I wonder whether a contributing factor to the extinction wave at the end of the Pleistocene may have been the domestication of the dog and consequent jump in efficiency of man the hunter. (However, clear evidence that Clovis and Folsom hunters of North America had dogs is lacking [Olsen 1977], and the first human settlers of Australia did not have dogs).

Meat is a major motive for exploitation of animals in New Guinea. Vertebrate prey cover the whole size spectrum from pigs and cassowaries down to rats and wrens. Nearly as important a motive is acquisition of bird plumes for decoration, the main targets being birds of paradise, lorikeets, the New Guinea Harpy Eagle (*Harpyopsis novaeguinea*), and Pesquet's Parrot (*Psittrichas fulgidus*). The last two are large-bodied, long-lived species that are among the first animals to disappear locally with the spread of shotguns. Another motive, underrated by biologists but widespread among New Guineans as among Amerindians with whom I have lived, is acquisition of live young birds and mammals as pets. Finally, some destruction is purely for fun, notably the firing of alpine grassland to enjoy the sight of spectacular blazes.

Lessons of Modern Man-Related Extinctions

It is sometimes assumed that an extinction wave due to man can be equated with rapid extermination of big animals by hunting for meat. The preceding discussion of modern man's impact reveals that the actual situation is far more complex:

Hunting is merely one of many ways by which man exterminates. Thus, the choice of explanations is not overkill versus climate, but man-related explanations versus man-independent explanations generally.

Meat is merely one of several motives for hunting today, even among the protein-starved hunter-gatherers of New Guinea. The most endangered species in New Guinea today are hunted for decoration, not for meat. Excavated late Pleistocene necklaces of *Sarcophilus* teeth from Australia, Cromagnon cave paintings, and flower remains around Neanderthal burials suggest that man's drive to sweat, risk, kill, and exterminate for decoration is of long antiquity.

Man's impact is not only on the megafauna. On New Zealand and Hawaii, both the modern European wave and the prehistoric Polynesian extinction wave claimed as victims small frogs, lizards, insects, and birds in addition to large flightless birds.

Some witnessed cases of overkill did happen very rapidly. From discovery to extinction lasted less than 1 year for the Stephen Island Wren (*Xenicus lyalli*), 2 years for Solander's Petrel (*Pterodroma solandri*) on Norfolk Island, 27 years for Steller's sea cow in Bering Straits, 174 years for the Dodo on Mauritius, and 275 years for the Labrador Duck (*Camptorhynchus labradorius*) in North America.

Population collapses tend to be especially rapid for gregarious species that breed colonially and whose reproductive behavior is stimulated socially. Prime examples are the unanticipatedly rapid extinction of Passenger Pigeons (*Ectopistes migratorius*) and of Tasmanian humans after cessation of the shooting that had decimated their numbers.

In contrast, other witnessed cases of overkill happened extremely slowly. The Great Auk was hunted at least 20,000 years ago and did not succumb until 1844. Other species, races, or populations that survived thousands or hundreds of thousands of years of hunting to succumb in modern times include the wolf (*Canis lupus*), bear (*Ursus arctos*), and beaver (*Castor fiber*) populations of Britain; the aurochs (*Bos primigenius*), Caucasian wisent (*Bison bonasus caucasicus*), and tarpon (*Equus gmelini*) of Europe; the races of ostrich (*Struthio camelus syriacus*), lion (*Panthera leo persicus*), tiger (*Panthera tigris virgata*), and leopard (*Panthera pardus jarvisi*) of the Near and Middle East; the Bali tiger (*Panthera tigris balica*); and the wolves (*Canis lupus hodophilax, C.l. hattai*) and sea lion (*Zalophus californianus japonicus*) of Japan.

Some species survived one or even two human colonization waves, only to succumb quickly later to human colonists. Examples include the numerous Hawaiian species that survived the Polynesians, New Zealand species that survived the Maoris, and South African species (*Equus quagga, Equus b. burchelli, Alcephalus buselaphus caama, Hippotragus leucophaeus*) that survived the Hottentots and Bushmen, only to succumb within a century or two to European invaders. The thylacine population of the Australian mainland survived the first human invasion of Australia before 30,000 B.P. to disappear with the second invasion that brought the dingo (ca. 3000 to 4000 B.P., chap.

27), while several dozen other Australian species survived these first two invasions but not the third invasion by Europeans beginning in 1788. Other populations coexisted abundantly with man for millenia, only to disappear quickly when the same human population acquired a new hunting technology (e.g. the rapidly spreading local extinctions in Papua-New Guinea today following introduction of the shotgun). For this reason the possible presence of humans in the New World before Clovis huntings seems to me not to argue against a Clovis-linked interpretation of late Pleistocene New World extinctions.

While the extermination of the moas left behind thousands of butchered carcasses, some species succumbed to modern overkill so quickly that museums today have found only a few skeletons a century or two later (e.g. the solitaires of Reunion and Rodriguez), and other species continue to thrive under heavy hunting that leaves abundant carcasses. For instance, introduced possum (*Trichosurus vulpecula*) and deer (*Cervis elaphus*) remain abundant in New Zealand, despite the best efforts of government-hired hunters using helicopters and poison. Possum carcasses in particular are a regular feature of the New Zealand landscape. There are similar prehistoric examples: eland (*Taurotragus oryx*) survived despite being the most abundant human prey recorded archaeologically for Middle Stone Age Africa (chap. 25); albatross and marine mammals survived heavy Polynesian exploitation on the Chatham Islands (chap. 34); and cormorants, parrots, and honeyeaters survived Polynesian exploitation on New Zealand (chap. 34).

Coexisting and closely related species differ greatly in their susceptibility to modern extinction waves. The coyote (*Canis latrans*) prospered in North America while the timber wolf (*C. lupus*) vanished. Why is the Amakihi (*Hemignathus virens*) today the second commonest Hawaiian honeycreeper, while its congeners, the Greater Amakihi (*H. sagittirostris*), Akialoa (*H. obscurus*), and Kauai Akialoa (*H. procerus*) are extinct and the Nukupuu (*H. lucudus*) nearly so? Why did the kangaroo *Macropus greyi* become extinct and *M. parma* nearly so, while *M. giganteus*, *M. fuliginosus*, and *M. rufus* prospered and *M. robustus* became an abundant pest? Why have no more than one or two of Australia's 571 breeding bird species, but about 22 of its 237 mammals become extinct? Similar problems perplex students of prehistoric extinctions, as discussed in many chapters of this book.

In short, witnessed modern extinctions that are certainly linked to man are very diverse. Overkill occurs out of many motives, exterminates very quickly or very slowly, and is but one of many means by which man exterminates. Some species succumbed to the first human wave, other similar species succumbed to the second, and still others prosper to this day. Some species disappeared and left few butchered carcasses, others disappeared and left many carcasses, and still others survived abundantly and continue to leave many carcasses. Grayson (chap. 37) objects that the overkill hypothesis of prehistoric extinctions has become so varied and resilient as to be unfalsifiable. Expressed alternatively, this hypothesis has become increasingly faithful to the facts of modern man-related extinctions, which are so varied as to make any rigid hypothesis of past man-related extinctions unrealistic.

The Extinctions That Did Not Occur: Challenges to Theories of Prehistoric Extinctions

[Inspector Gregory]: "Is there any point to which you would wish to draw my attention?"
[Sherlock Holmes]: "To the curious incident of the dog in the night-time."
[Inspector Gregory]: "The dog did nothing in the night-time."
"That was the curious incident," remarked Sherlock Holmes.
(From the story "Silver Blaze," in *Memoirs of Sherlock Holmes,* by A. C. Doyle)

There were too many plausible theories to explain what did happen when Colonel Ross's valuable horse Silver Blaze disappeared from his stable shortly before the race for the Wessex Cup. The decisive clue that Sherlock Holmes appreciated, and that Inspector Gregory missed, was what did not happen. (The stable dog did not bark, because the horse was stolen by the horse's trainer himself, not by a stranger).

The problem in explaining late Pleistocene extinctions is similar. There are numerous variants on climate-based or man-based hypotheses that *might* explain extinctions. Considering the extinctions alone, it has proven difficult to decide what cause was actually responsible. The decisive clues, it seems to me, will be all the dogs that did nothing in the nighttime: the species and biotas that survived while other species and biotas were disappearing. Let us examine how these nonextinctions challenge climate-based and man-based hypotheses.

Challenges to Climate-Based Hypotheses

If late Pleistocene extinctions of animals were due ultimately to altered climate, the animal extinctions should have been mediated at least partly, perhaps largely, by plant extinctions. Where is the evidence for the plant extinctions that were supposedly responsible?

In fact, we have megafaunal extinction waves without floral extinction waves. It is not that we would be unable to detect fossil extinctions of plant species if they had occurred. About twenty-five extinctions of plant species are known for New Zealand, mostly Pliocene and early Pleistocene, none late Pleistocene (chap. 31). A similar pattern applies to western Europe (Leopold 1967). How could climate have eliminated mammoths and ground sloths, while the plant species represented in their gut contents and dung remain abundant and widespread?

Climate should have had the least effect on those animals best buffered against climate, the homeotherms, and should have had the most devasting effect on the animals least buffered against climate, the poikilotherms. In fact, it was one of the two classes of homeotherms, the mammals, that suffered worst. Why was there no devastation of beetles (Coope 1979), fish, amphibia, and reptiles?

Might the studies of Brown (1971, 1978) and Hope (1973) explain through a climate-based mechanism why large mammals suffered far heavier extinctions than small mammals on the continents? As discussed, (cf. Table 38.3), these authors studied the end-of-the-Pleistocene local extinctions of mammals associated with climate-driven shrinking habitat area on Great Basin mountains and shrinking land area in Bass Straits. Local extinctions should, therefore, provide excellent models for continent-wide extinctions of megafauna, if the latter were similarly due to climate-driven shrinkage of habitat. At first, the pattern seems encouragingly similar: large mammals were indeed preferentially extinction prone in the Great Basin and in Bass Straits. But closer examination shows that this pattern is part of a broader pattern of preferential susceptibility of species occurring at low population density for any reason—the "main pattern of natural extinction." Thus, carnivores suffered proportionately more extinctions than herbivores, and habitat specialists proportionately more than habitat generalists, both in the Great Basin and in Bass Straits. Do these two patterns seen in undoubtedly climate-driven extinctions also describe the continent-wide extinctions of North America, South America, and Australia?

The glacial retreat at the end of the Pleistocene greatly increased the productivity and habitable land area of North America (by about 75 percent, chap. 18), as well as in Europe and the palearctic zone of Asia. Similar increases occurred at the end of other glacial periods and corresponding decreases occurred at the end of interglacials. One therefore expects maximal extinction rates at the ends of interglacials, minimal extinction rates at the ends of glacials. In fact, only in Australia do megafaunal extinctions appear to come at a time of deteriorating rather than ameliorating climate (chap. 29). Why does this straightforward reasoning fail except perhaps in Australia? Where are the extinction waves that should have come at the ends of interglacials in Europe and North America, as glaciers advanced and habitats shrank? It does not solve this dilemma to conclude that this proves the climate-based hypothesis to be wrong and the late Pleistocene extinctions to be man related, for Webb (chap. 9) describes extinction waves in North America coincident with earlier glacial terminations before man's arrival.

One could refine the argument of the preceding paragraph by noting that, while some habitats expanded at the end of the Pleistocene, others contracted (e.g. boreal habitats on mountaintops). Perhaps the extinctions were climate-based but confined to species of those habitats that shrank. Consider that the North American extinctions whose dating is reviewed in Chapter 19 included mammals of desert (Shasta ground sloth, chap. 7), spruce forest and pine forest (mastodont, chap. 15), montane conifer parkland (chap. 6), grassland and open steppe (mammoth, chaps. 3 and 4), and pinyon-juniper woodland (Table 38.3). If extinctions struck in this diverse range of habitats, what were the habitats that expanded and of which the faunas were spared?

Several chapters in this book analyze possible mechanisms that could produce late Pleistocene extinction waves without human intervention. For instance, Gingerich (chap. 10) sees the high rate of late Pleistocene extinctions as related to the high rate of early Pleistocene speciation; Graham and Lundelius (chap. 11) discuss the disruption of coevolutionary equilibria between plants and animals; Guthrie (chap. 13) considers the trend towards homogeneous plant communities presenting mammalian herbivores with higher loads of toxic compounds; and Kiltie (chap. 14) argues that increased seasonality at the end of the Pleistocene stressed large animals with gestation periods exceeding a year. Each of these proposed mechanisms rests on general biological considerations and invokes no specific properties of North and South America. Why, then, was late Pleistocene megafaunal extinction so much more severe in North and South America than in Europe and Asia? Why was there not an even more severe extinction wave eliminating moas 10,000 years before the Maori reached New Zealand, where late Pleistocene habitat changes were at least as drastic as those in North America (Fleming 1975)? How could the arid period of 25,000 to 10,000 B.P. in Australia have dispatched Australia's woodland megafauna so completely (chap. 29), while the arid period of 40,000 to 12,000 B.P. in Africa left the African megafauna largely intact?

These questions pose challenges to the future development of climate-based extinction models.

Challenges to Man-Based Hypotheses

When Martin and Wright's (1967) *Pleistocene Extinctions* was published sixteen years ago, there was not yet an instance of an extinction wave provenly due to prehistoric man. We now have much more information about late Pleistocene extinction on the continents, plus four undoubted examples of waves due to prehistoric man: Hawaii, New Zealand, Chatham, and Madagascar. Three of these four man-linked waves are heterogeneous in their victims and mechanisms.

Table 38.5 summarizes the diverse patterns of extinction waves (see "Observed Extinctions" columns). In late Pleistocene North America the victims consisted of most of that continent's large mammals, few small mammals, and some birds that were large (by avian standards). This also seems true of late Pleistocene Australia and South America, except that their avifaunas are too little known to assess extinction patterns. The Hawaiian extinction pattern following Polynesian settlement is very different: the victims include approximately as many species of small flying songbirds as large flight-less birds (chap. 35). Known New Zealand victims of the Polynesians included all of New Zealand's giant flightless birds, local populations of marine mammals, large and small flightless and flying birds, and frogs, lizards, and flightless insects (chaps. 32–34).

Table 38.5. An Attempt to Account for

Location	Time of Human Arrival[b]	Man's Arsenal[c]						
		Dogs	Weaponry	Rats	Stock	Agriculture	Swift predators	Pigs
Australia	ca. 40,000 B.P.							
Australia	1788 A.D.	+	+	+	+	+	+	+
North America	ca. 11,000 B.P.	?		?				
Eurasia, Africa	ca. 11,000 B.P.	?						
Madagascar	ca. 500 A.D.	+		+?	+	+		+
West Indies	ca. 5000 B.P.	?		?		+		
West Indies	1492 A.D.	+	+	+	+	+	+	+
New Zealand	ca. 1000 A.D.	+		+		+		
New Zealand	1790 A.D.	+	+	+	+	+	+	+
Chatham	ca. 1000 A.D.	+		+				
Hawaii	ca. 500 A.D.	+		+		+		+
Hawaii	1778 A.D.	+	+	+	+	+	+	+
Mediterranean islands	ca. 6000 B.P.	+?		?	+	+	?	?
New Caledonia	ca. 4000 B.P.	+		+		+		
Bismarcks, Solomons	by 11,000 B.P.	?		+		+		+
Remote Islands, RC+[a]	1500–1800 A.D.	+	+	+	+	+	+	+
Remote Islands, RC−[a]	1500–1800 A.D.	+	+	+	+	+	+	+

NOTE: Preliminary test of a theory that relates differences in intensity of extinction waves to differences in two sets of controlling variables: man's arsenal of extermination devices, and the fauna's susceptibility.

[a]RC+ = remote islands with native rats or land crabs (e.g. Fiji, Marquesas, Christmas, Aldabra, Galapagos); RC− = remote islands without native rats or land crabs (e.g. Lord Howe, Norfolk, Tristan).

[b]Approximate time of arrival of first humans, or else of subsequent European settlement (Australia, West Indies, New Zealand, Hawaii). The dates are given simply to identify the arriving human wave by its currently estimated arrival date (e.g. West Indies, 5000 B.P., = means first Amerindian settlers of West Indies, whenever they actually arrived). The date given for Eurasia and Africa is the end of the Pleistocene, not an arrival time for humans.

[c]Extermination devices introduced by man. Stock = domestic grazing and browsing animals; swift predators = cat, mustelids, fox, mongoose. A " + " in the weaponry column means firearms; better would be a graded scale of weapon effectiveness.

[d]Properties of the fauna, island, or continent predisposing to man-related extinction. Fire risk = extensive arid areas highly susceptible to burning. Australia is assigned ½+ for "native swift predators absent" because it had so few.

[e]The score is taken as the sum of the following factors: 3 = no previous humans, ½ = domestic dogs, ½ = firearms, ½ (Australia) or 1 (elsewhere) = native swift predators absent.

Known Chatham victims of the Polynesians include several dozen bird species, large and small, flying and flightless (chap. 34). The Madagascar victims were mammals, flightless birds, and tortoises, all large or giant. Some caveats about apparent extinction patterns for small birds and small mammals will be discussed at the end of this chapter.

Thus, of the four prehistoric extinction waves definitely attributable to man, only one (Madagascar) resembles the late Pleistocene continental waves in drawing its victims overwhelmingly from the megafauna. The other three waves resemble what literate Europeans have been doing on numerous oceanic islands, in the New World, and in Australia for the past several centuries: exterminating species of any size.

Man's Varying Impact on Faunas

Faunal Susceptibility[d]				Predicted Resulting Impact		Predicted Extinctions[g]			Observed Extinctions[h]		
Fire risk	Native rats and land crabs absent	Native swift predators absent	No previous humans	Hunting overkill[e]	Habitat destruction[f]	Megafauna	Small mammals	Small birds	Megafauna	Small mammals	Small birds
+	½+	+		3½	1	4½	1	1	+++	0	?
+	½+			1½	3	4½	3½	4	+	++	0
+		+		3	1	4	1	1	+++	0	0?
				0	0	0	0	0	+	0	0
+			+	3½	3	6½	3	3	+++	?	?
		+	+	4	1	5	1	1	+++ }	++	?
		+		2	2	na	3	4	na }		+
	+	+	+	4½	1	5½	na	4	+++	na	+
	+	+		2	2	na	na	8	na	na	++
	+	+	+	4½		4½	na	3	+++	na	?
	+	+	+	4½	1	5½	na	5	+++	na	++
	+	+		2	2	na	na	8	na	na	++
		?	+	3½	2	5½	2	2	+++	+	?
+	?	+	+	4½	2	6½	na	5	?	na	?
		+	+	4	1	5	1	1	?	?	?
		+	+	5	2	na	3	4	na	++	+
	+	+	+	5	2	na	na	8	na	na	++

[f]The score is taken as the sum of the following factors: 1 = stock, 1 = agriculture, 1 = fire risk.

[g]Predicted intensity of extinctions relative to that for the same group of animals elsewhere or at another time. Megafaunal score: sum of columns 14 and 15 (na = not applicable, because there was no megafauna to exterminate). Small mammal score = column 15 (habitat destruction), plus ½ (Australia) or 1 if swift predators were introduced to a land lacking native ones; na = no native flightless land mammals. Small bird score = column 15 (habitat destruction) plus 1 (Australia) or 2 if swift predators were introduced to a land lacking native ones, plus 3 if rats were introduced to a land lacking native rats and land crabs, plus 1 if pigs were introduced to a land lacking native rats and land crabs and swift predators.

[h]Observed intensity of extinctions, based on fraction of the fauna exterminated. Observed extinctions of small mammals in the West Indies were intense (+ +), but it is not known how many occurred on the arrival of Amerindians (~ 5000 B.P.) as opposed to those that occurred with the arrival of Europeans (~ 1492 A.D.). The main discrepancies with predictions are that small birds and the surviving megafauna in Australia have survived the arrival of Europeans better than predicted (perhaps because column 15 exaggerates degree of habitat destruction in Australia compared to other sites). The scoring systems for observed and predicted extinctions are obviously crude, and readers are encouraged to devise better scoring systems.

If the late Pleistocene continental extinction waves are to be attributed to man, the main challenge for future research is to fit them into the framework of a general theory of man's impact on faunas. Such a theory would attempt to explain why witnessed impacts are so diverse. The theory would need to incorporate three sets of variables (Table 38.5):

1. *Man-linked independent variables*, specifying man's arsenal of mechanisms for extermination:

- introduced grazing and browsing animals (leading to habitat destruction);
- crop farming (leading to habitat destruction);
- introduced cursorial predators (especially cat, fox, mongoose, mustelids);

- introduced pigs (acting as predators on naive ground animals, and also leading to habitat destruction);
- dogs (possibly increasing hunting efficiency, especially for hunters without firearms);
- weaponry (firearms, or quality of stone weapons visible archaeologically).

In addition, commensal rats have accompanied all human colonists since the late Pleistocene, possibly with the exceptions of the first colonists of Australia and the New World. Available information suggests that introduced diseases contributing to extinctions have mostly been transmitted from introduced mammals and birds to similar native species. Hence the disease variable may be approximately proportional to the sum of those above-listed variables involving introduced species.

   This list makes it clear that improvements in weaponry, from the stone tools of Australia's first settlers to Clovis points to metal-tipped spears to firearms, only partly explain why the deadliness of man the exterminator has increased with time. The exterminating arsenal of the first Australians consisted solely of their hunting weapons plus fire. The arsenal of early Amerindians was improved by Clovis and Folsom points, and perhaps by commensal canids. The Polynesians added rats (*Rattus exulans*) and dogs, farming, and pigs. Bronze-age European colonists of Mediterranean islands and the first settlers of Madagascar added grazing and browsing stock, and the former probably added *Rattus rattus*. Modern Europeans added cursorial predators and *Rattus norvegicus*.

   2. *Island- or continent-linked independent variables* that facilitate man-linked extinctions:

- lack of prior exposure to man (hence fauna naive to human hunters);
- lack of native flightless mammals or land crabs (hence small birds, reptiles, amphibia, and terrestrial invertebrates susceptible to commensal rats);
- lack of native cursorial mammalian predators (hence native birds and small mammals susceptible to introduced cursorial predators; may also make fauna more naive to human hunters);
- long arid seasons (hence habitats especially susceptible to destruction by fire).

   3. *Dependent variables:*

- widespread destruction of habitat (leading in turn to possible extinction of any species);
- widespread extinctions of megafauna;
- widespread extinctions of small mammals (as in the European extinction wave in Australia and the West Indies, but not in the late Pleistocene waves in Australia and the New World);
- widespread extinctions of small volant birds (as in the Polynesian and European waves in Hawaii, but not in the European wave in Australia and North America).

   Table 38.5 summarizes values of these variables for late Pleistocene, prehistoric Holocene, and witnessed modern extinction waves. The table is not intended to prove that all these waves were due to man. It is simply an attempt to see how far one can go with current knowledge in constructing a general theory of human impact, and how closely the predictions of such a theory match the facts. Independent variables assumed to contribute to habitat destruction are stock, farming, pigs, and aridity (= susceptibility to fire). Variables contributing to extinction of megafauna are habitat destruction, dogs, quality of weaponry, lack of prior exposure to man, and lack of native cursorial mammals. Variables contributing to extinctions of small mammals are habitat destruction, and introduced cursorial predators in combination with lack of native cursorial predators. Variables contributing to extinctions of small birds are habitat destruction, introduced cursorial predators in combination with lack of native cursorial predators, and introduced rats in combination with lack of native flightless mammals or land crabs.

Let us summarize extinction waves in terms of Table 38.5. The reason why the late Pleistocene extinctions in Australia and North America were largely confined to the megafauna is that hunting was then the major man-related mechanism of extermination, with habitat destruction minor. Two factors leading one to expect worse damage to the megafauna of Australia than of North America are that Australia's aridity meant greater habitat destruction by fire (chap. 29 gives another view), and that Australia lacked large native cursorial predators other than the thylacine. Offsetting these two advantages of early Australians, Paleo-Indians had one or two advantages as exterminators: superior weapons, and (questionably) canid commensals. In practice, the fraction of the megafauna exterminated was slightly higher in Australia than in the New World (Table 17.1 of Chap. 17). The only independent variable in Table 38.5 that differs between late Pleistocene Eurasia/Africa and North America is prior exposure to man on the former continents, and this factor would have to be largely responsible for the slightness of megafaunal extinction in Eurasia/Africa compared to North America.

Prehistoric Madagascar, uniquely among oceanic islands, resembles late Pleistocene North America in having native mammals including numerous cursorial predators (at least eight viverrids). This may be why the Madagascar extinction wave was the only island wave to afflict mainly the megafauna and hence to offer a model for the late Pleistocene extinction wave in North America. The extinction patterns apparently fail (see below) to reflect greater habitat destruction in prehistoric Madagascar than in late Pleistocene North America, due to greater fire risk, introduced stock, and farming. The prehistoric Hawaiian, New Zealand, and Chatham extinction patterns share with North America the overkill of a naive megafauna, but differ conspicuously in the additional extinctions of small vertebrates and invertebrates for two reasons: susceptibility to rats, due to lack of native mammals and land crabs, and habitat destruction by agriculture (Hawaii and New Zealand but not Chathams).

Table 38.5 also lists differences among the independent variables relevant to understanding differences in the modern extinction waves of Hawaii, New Zealand, Australia, and various oceanic islands.

While understanding these differences in man's impact offers the major challenge to man-based hypotheses of prehistoric extinctions, an additional challenge comes from differences in abundance of butchered skeletons. As already noted, there are cases of man-linked extinctions that left few skeletons, and cases of hunting that left abundant skeletons without causing extinction. The *paucity* of butchered skeletons in the New World stimulated Mosimann and Martin (1975) to develop their blitzkrieg hypothesis. As Chapter 33 describes, the New Zealand moas offer a likely case of blitzkrieg, with progressively more recent radiocarbon dates from north to south end of South Island, exactly as Mosimann and Martin postulate for New World overkill. Yet this blitzkrieg reveals itself in a superabundance of skeletons, exactly opposite to the Mosimann-Martin predictions! As A. Anderson discusses in Chapter 33, the estimated number of moa skeletons at the 117 known moa-hunter archaeological sites is an order of magnitude greater than the standing crop of moas at any moment. Similarly, the Chathams have yielded abundant remains of slaughtered seabirds (Bourne 1967). In Madagascar a similar blitzkrieg carried off another group of moa-like ratites, the elephant birds, along with numerous large mammal species (chap. 26). Yet skeletal remains of these extinct forms in archaeological contexts are extremely few for Madagascar. The eggshells of the elephant birds are superabundant, but moa eggshells are not. Where are the elephant-bird hunter sites in Madagascar, and the goose-hunter sites in Hawaii (chap. 35), that correspond to New Zealand's moa-hunter sites?

These enormous differences in archaeological visibility of blitzkriegs escape present understanding. Could part of the explanation be site differences in large scavenging mammals and birds (absent from New Zealand) that would tend to scatter bones from kill sites?

The Future for Understanding Prehistoric Extinctions

In this final section I offer some comments on the strength of proof that we can expect for reconstructions of prehistoric extinctions. I then consider three directions for future research: extinct minifaunas, species-level analyses, and island studies as the potential basis of a broadened perspective.

The Strength of Proof Attainable

Some modern extinctions, such as that of the Great Auk, were so simply caused, visible, and well documented that we have no doubt about why they happened. Other modern extinctions are much harder to understand. In North America we do not know why the Labrador Duck (*Camptorhynchus labradorius*) became extinct, how to rate several plausible causes for extinction of the Carolina Parakeet (*Conuropsis carolinensis*), and even whether the Ivory-billed Woodpecker (*Campephilus principalis*) and Bachman's Warbler (*Vermivora bachmanii*) still exist. North American biologists have been studying Kirtland's Warbler (*Dendroica kirtlandi*) intensively for several decades and still do not understand the reason for its decline. In Australia the relative roles of habitat destruction and introduced predators, competitors, and diseases in the extinctions or declines of several dozen mammal species are unknown.

How can we expect the fossil record to yield unambiguous interpretations of extinctions that happened 10,000 to 30,000 years ago, when we are often unable to understand declines of species being studied in the field today? For the answers to many questions about prehistoric extinctions, we will be lucky if we can find enough evidence to support plausible guesses.

Have We Overlooked Minifaunal Extinctions?

Most of this book is about prehistoric extinction of megafaunas. To what extent is this emphasis an artifact of the problem that large bones are more likely to be preserved than small bones, hence the minifaunal extinction tends to be underestimated?

For mammals of North America, South America, Europe, and Australia most extant small-bodied genera are known as Pleistocene fossils (chap. 17). This argues against much mammalian minifaunal extinction having been overlooked due to sample bias—unless this conclusion is invalidated by the species-level problem (see next section).

For Madagascar mammals, however, the question of size-related sample bias cannot be dismissed (chap. 26). Of fifteen extant rodent species in seven endemic genera, only the largest is known as subfossil. Of more than twenty extant insectivore species in eight or nine endemic genera, only one large species is known as subfossil. Of ten extant primate genera, all except the smallest and third smallest are known as subfossils. Thus, the sample of the fossil minifauna is much too poor to judge whether it suffered extinctions comparable to those of the megafauna.

For birds the problem is more acute, as there are so few scientists studying fossil birds (Olson 1981). In Hawaii Olson and James (1982; see also chap. 35) recovered 81 percent of modern avian genera as fossils, implying fairly good generic sampling of the fossil record. They found that 5 percent of passerine genera and 47 percent of nonpasserine genera known as prehistoric fossils became extinct before modern times. Since nonpasserines are generally larger than passerines, this implies differential extinctions of large-bodied genera (but see the species-level analysis in the next section). A similar pattern emerges for North America, but on shakier evidence. Of the nineteen known extinct Pleistocene genera of North American birds (chap. 21), fifteen are large-bodied, four medium-small (jay-sized), none small. However, fewer than half of the extant genera of passerines are known as fossils, so that sampling of the fossil avifauna is evidently quite incomplete. From New Providence in the Bahamas 66 percent of known avian Pleistocene genera have become extinct: 38 percent of the nonpasserines, 25 percent of the passerines (Olson and Hilgartner 1982). No avian Pleistocene genus known from Europe has become extinct. In New Zealand much more is known about large than small fossil birds. For Australia we know little, and for Madagascar practically nothing, about small fossil birds.

Thus, there are few parts of the world where the fossil avifauna is sufficiently well known to compare the fates of the megafauna and the minifauna. For cold-blooded vertebrates our ignorance is worse. Extinct fossil reptiles large and small are known from the Bahamas (Pregill 1982), and small ones are known from New Zealand (chap. 34). Of the species identified as fossils at Rancho La Brea, many of the mammals, some of the birds, but almost none of the insects and fish, are now extinct (chap. 8). There are almost no studies of cold-blooded vertebrates from any part of the world that document good sampling of the fossil record by recovering a high proportion of modern genera as fossils, and then analyze fossil extinction patterns.

In brief, a major task for future research will be to show whether large mammals really were the primary victims of prehistoric extinctions, or whether that apparent pattern is in some cases an artifact of selective preservation and study of large mammalian fossils.

The Species-Level Problem

Many studies of prehistoric extinction use the genus as the unit of analysis. This procedure has the advantage that it avoids problems of deciding whether an "extinct" fossil species in an extant genus really represents an extinction or just the ancestor of a modern species. However, the genus-level analysis has two crippling disadvantages: it tends to underestimate extinction, and it may yield biased conclusions about patterns of extinction.

Both problems are illustrated by the fossil Hawaiian avifauna (Olson and James 1982 and chap. 35). Of the thirty-five defined genera of fossil Hawaiian nonmarine birds, only nine (i.e. 26 percent of the total) are extinct. This high rate of generic survival conceals mass extinction at the species level: forty-two of the sixty-nine known fossil species, or 61 percent.

As mentioned previously, the extinction rate at the generic level is much higher for fossil Hawaiian nonpasserines (47 percent) than for passerines (5 percent), implying that large-bodied birds were at an extreme disadvantage. For extinctions at the species level this difference is much less extreme: 76 percent for the nonpasserines, 50 percent for the passerines. The reason is that relatively more passerine than nonpasserine genera suffered multiple extinctions of species. More generally, it is dangerous to compare the frequency of generic extinctions for the megafauna and minifauna, or for birds and mammals, if the species/genus ratio among extinct relative to extant species differs among the groups compared.

Thus, a further challenge for the future is to shift the focus from the genus level to the species level in studying prehistoric extinctions.

Broadening the Geographical Perspective

In Martin and Wright (1967) discussion of prehistoric extinctions was based mainly on information from North America, the Palearctic, and Madagascar. There was slight discussion of some other areas, and no discussion of the oriental region and New Guinea.

In the present volume North America still receives the most discussion, while the oriental region, New Guinea, and South America are still largely neglected. The most striking advances in knowledge since 1967 have been for Australia, New Zealand, Chatham, and Hawaii. The expanded geographical perspective has been a revelation. Australian paleontology has revealed an extinction wave even more complete than that of North America, but the Australian wave apparently antedates the end of the Pleistocene. This fact must be incorporated into any theory that interprets the North American extinctions in terms of end-of-the-Pleistocene climate changes: either they were not the cause of the North American extinctions, or else there were important differences between the end of the Pleistocene in North America and Australia. New Zealand, Chatham, and Hawaii have yielded clear examples of prehistoric extinction waves linked to man, and New Zealand offers a clear case of overkill.

The insights gained from Australia, New Zealand, Chatham, and Hawaii illustrate the advantages of a broadened geographic perspective. It is difficult to draw unambiguous interpretations from studies of extinction on a single continent. The present state of this field resembles the position of archaeologists who open three tombs, disagree about the meaning of the contents, open three more tombs, find their understanding transformed, and realize that thirty more tombs remain unopened. Continental extinctions may be clarified by extinctions on many islands that await paleontological exploration. I give four sets of examples (also see discussion in chap. 17):

1. Dozens of islands had not been reached by man by the end of the Pleistocene, and they remained uninhabited until within the last 3,000 years. Here are dozens of chances to see the effects of late Pleistocene climatic changes in the absence of human influence. Among the many islands of particular interest are Fiji, New Caledonia, New Zealand, Hawaii, the Galapagos, Norfolk, and Lord Howe in the Pacific; Madagascar, the Seychelles, Mascarenes, and Aldabra in the Indian Ocean; and St. Helena, the Canaries, Azores, and Madeiras in the Atlantic. This set includes two small island-continents, New Zealand and Madagascar, that evolved megafaunas and that should provide examples to be compared to the major continents. If climate and habitat changes associated with deglaciation caused the late Pleistocene megafaunal extinctions of North America and the Palearctic, extinctions should have been more drastic in New Zealand. At the end of the Pleistocene, New Zealand lost most of its grassland and tundra habitat, while its forests expanded in area by an order of magnitude. A variety of large flightless birds survived to be butchered by the Maori; how many more had been eliminated by the climatic upheaval 10,000 years earlier? Similarly, if changes in rainfall caused the extinction wave around 30,000 B.P. in Australia, there should also have been widespread late-Pleistocene extinctions in Madagascar (with a known megafauna), and possibly in New Caledonia and Fiji. It would be striking support for the climatic interpretation of late Pleistocene continental extinctions if evidence were discovered for similar waves on New Zealand and Madagascar. Conversely, I find it hard to see how that interpretation could be rescued for the major continents if there were not similar waves in late Pleistocene New Zealand and Madagascar.

2. New Caledonia and Fiji have low-rainfall areas susceptible to burning. When man arrived a few thousand years ago, did these two islands suffer heavy faunal losses due to habitat destruction, as did Hawaii? Paleontologically these two islands are nearly unexplored. Remains of three extinct birds have been found on Fiji, and bones of a large bird dubiously claimed to be a ratite have been reported for New Caledonia (see references in chap. 34). Both islands have endemic genera of lizards, neither has native

flightless mammals, and Fiji has two questionably native frogs. The modern avifaunas are undistinctive: only four endemic genera in Fiji and five in New Caledonia, one flightless species in Fiji and two in New Caledonia. (The Kagu, *Rhynochetos jubatus*, of New Caledonia, formerly placed in a monotypic family, is now considered related to the recent fossil *Aptornis* of New Zealand [Olson 1977].) On the other hand, no one would have guessed from the single flightless modern bird species of Hawaii that Hawaii had at least ten other flightless bird species when the Polynesians arrived. My guess is that extinct birds, including large flightless ones, await discovery both in Fiji and New Caledonia, but not in Hawaii's variety. I would not be surprised, however, if the surviving endemic lizard genera of New Caledonia are the remnants of a rich herpetofauna (including megafauna) reduced by the rats, dogs, fires, and forest clearing of the first Melanesians.

3. The West Indies and the islands of the Mediterranean are of interest because both lacked man at the end of the Pleistocene, both were colonized by man by about 5000 B.P., and both supported fossil megafaunas (Table 17.12 and 17.13 of chap. 17; see Olson 1978 and Pregill and Olson 1981 for summary of West Indian fossils). The dates of disappearance are still unknown for most of the extinct species. Probably there were some extinctions at the end of the Pleistocene, another extinction wave when neolithic agriculturists arrived, and (in the case of the West Indies) a third wave when Europeans arrived in A.D. 1492. At least for the West Indies, changes in rainfall and emergent land area at the end of the Pleistocene were on a scale likely to cause extinctions (Pregill and Olson 1981). The West Indies, Mediterranean islands, and Madagascar may be the only islands that had numerous large flightless birds and mammals, and that nevertheless were not reached by man until so long after the end of the Pleistocene as to make the climate-linked and man-linked extinction waves unequivocally separable by fossil dates.

4. New Guinea, the Bismarck archipelago, and the Solomon archipelago lie near each other at the same latitude, are climatically similar, and potentially offer interesting comparisons for studies of extinction. All have high mountains: New Guinea to 16,500 feet, the Bismarcks to 8,000 feet, the Solomons to 8,500 feet. New Guinea was joined to Australia for much of the Pleistocene, but the Bismarcks and Solomons were not joined to each other or to New Guinea or Australia. New Guinea and the Solomons had much greater land area during Pleistocene periods of low sea level. All three were subject to major expansions and compressions of montane vegetation zones during the Pleistocene, as documented for New Guinea in Chapter 30.

All three support montane faunas and floras. All three have native flightless mammals: about 130 species on New Guinea, including marsupial carnivores up to the size of cats extant and the thylacine extinct; about 11 species in the Bismarks, including rats and marsupial omnivores and herbivores; and about 5 species in the Solomons, including moderate-sized rats. All have native flightless bird species: 4 in New Guinea, 1 or 2 in the Bismarcks, 2 in the Solomons. All have snakes and large carnivorous lizards. The Bismarcks and Solomons are paleontologically unexplored, while some mammals are known as fossils from New Guinea. Man reached New Guinea at least by 30,000 B.P., the Bismarcks and Solomons apparently only after the end of the Pleistocene.

What extinctions did the climate changes of the Pleistocene produce in these three island groups? Such changes may be confounded by the presence of man in late Pleistocene New Guinea, but the Bismarcks and Solomons may offer controls for climatic effects on similar tropical mountainous islands in the absence of man. Did New Guinea share the rich extinct mammalian megafauna of Australia? Were there more flightless birds before man arrived, especially in the Solomons, which have the fewest mammals? Are the few flightless mammals of the Bismarcks and Solomons the survivors of a larger mammal fauna that man reduced? Did prehistoric man and his commensal pigs and dogs have differing impacts on the faunas of these island groups, related to the differing variety of mammalian carnivores with which the faunas had evolved? Will an understanding of island extinctions offer a key to understanding prehistoric continental extinctions?

Acknowledgments

It is a pleasure to record my debt to Paul Martin and Michael Soulé for valuable discussion, and to Richard Cassels, Robert Dewar, and Warren King for providing important unpublished manuscripts.

References

Allen, G. M. 1942. *Extinct and Vanishing Mammals of the Western Hemisphere.* American Committee for International Wildlife Protection, New York.

Anon. 1981. Conservation of the genetic resources of fish: problems and recommendations. *F.A.O. Fisheries Technical Paper* no. 217.

Arredondo, O. 1976. The great predatory birds of the Pleistocene of Cuba. *Smithsonian Contributions to Paleobiology*, no. 27, 169–188.

Atkinson, I. A. E. 1977. A reassessment of factors, particularly *Rattus rattus* L., that influenced the decline of endemic forest birds in the Hawaiian Islands. *Pacific Sci.* 31:109–133.

———. 1983. Effect of rodents on islands. *Proc. 18th World Conf. ICBP*, in press.

Atkinson, I. A. E. and B. D. Bell. 1973. Offshore and outlying islands. In: *The Natural History of New Zealand* (G. R. Williams, ed.), pp. 372–392. Reed, Wellington.

Batten, L. A. 1980. Some recent bird population changes in Britain. *Ibis* 122:420.

Berger, A. J. 1981. *Hawaiian Birdlife.* 2nd ed. University Press of Hawaii, Honolulu.

Bourne, W. R. P. 1967. Subfossil petrel bones from the Chatham Islands. *Ibis* 109:1–7.

Brander, K. 1981. Disappearance of common skate *Raia batis* from Irish Sea. *Nature* 290:48–49.

Brown, J. H. 1971. Mammals on mountaintops: nonequilibrium insular biogeography. *Amer. Natur.* 105:467–478.

———. 1978. The theory of insular biogeography and the distribution of boreal birds and mammals. *Great Basin Nat. Mem.* 2:209–227.

Burton, J. 1975. The effects of recent climatic changes on British insects. *Bird Study* 22:203–204.

Case, T. J. 1975. Species numbers, density compensation, and colonizing ability of lizards on islands in the Gulf of California. *Ecology* 56:3–18.

Cody, M. L. 1981. Habitat selection in birds: the roles of vegetation structure, competitors and productivity. *BioScience* 31:107–113.

Coope, G. R. 1979. Late Cenozoic fossil Coleoptera: evolution, biogeography, and ecology. *Ann. Rev. Ecol. Syst.* 10:247–267.

Corbet, G. B. and J. E. Hill. 1980. *A World List of Mammalian Species.* Cornell University Press, Ithaca.

Day, D. 1981. *The Doomsday Book of Animals.* Ebury Press, London.

Diamond, J. M. 1972. Biogeographic kinetics: estimation of relaxation times for avifaunas of southwest Pacific islands. *Proc. Nat. Acad. Sci. USA* 69:3199–3203.

———. 1976. Relaxation and differential extinction on land-bridge islands: applications to natural preserves. *Proc. 16th Intern. Ornith. Congr.* 616–628.

———. 1980. Species turnover in island bird communities. *Proc. 17th Intern. Ornith. Congr.* 777–782.

———. 1982. Rediscovery of the Yellow-fronted Gardener Bowerbird. *Science* 216:431–434.

———. 1983. Report of a 1974 ornithological expedition to the Solomon Islands: survival of bird populations stranded on land-bridge islands. *Nat. Geog. Soc. Research Reports*, in press.

Diamond, J. M. and C. R. Veitch. 1981. Extinctions and introductions in the New Zealand avifauna: cause and effect? *Science* 211:499–501.

Dobinson, H. M. and A. J. Richards. 1964. The effects of the severe winter of 1962–63 on birds in Britain. *British Birds* 57:373–434.

Ehrlich, P. R. and A. Ehrlich. 1981. *Extinction.* Random House, New York.

Fisher, J., N. Simon and J. Vincent. 1969. *Wildlife in Danger.* Viking Press, New York.

Fleming, C. A. 1975. The geological history of New Zealand and its biota. In: *Biogeography and Ecology in New Zealand* (G. Kuschel, ed.), pp. 1–86. Junk, The Hague.

Frankel, O. H. and M. E. Soulé. 1981. *Conservation and Evolution.* Cambridge University Press, Cambridge.

Frith, H. J. 1979. *Wildlife Conservation*. Angus and Robertson, London.

Goodwin, A. J. and J. M. Goodwin. 1973. List of mammals which have become extinct or are possibly extinct since 1600. *I.U.C.N. Occasional Paper* no. 8.

Greenway, J. C., Jr. 1967. *Extinct and Vanishing Birds of the World*. Dover Publications, New York.

Gudmundsson, F. 1951. The effects of the recent climatic changes on the bird life of Iceland. *Proc. 10th Intern. Ornith. Congr.* 502–514.

Halliday, T. 1978. *Vanishing Birds*. Holt, Rinehart, and Winston, New York.

Harper, F. 1945. *Extinct and Vanishing Mammals of the Old World*. American Committee for International Wildlife Protection, New York.

Heath, J. 1974. A century of changes in the Lepidoptera. In: *The Changing Flora and Fauna of Britain* (D. L. Hawksworth, ed.), pp. 275–292. Systematics Association, London.

Hope, J. H. 1973. Mammals of the Bass Straits Islands. *Proc. Roy. Soc. Victoria* 85:163–196.

Järvinen, O. and F. Ulfstrand. 1980. Species turnover of a continental bird fauna: northern Europe, 1850–1970. *Oecologia* 46:186–195.

Järvinen, O. and R. A. Väisänen. 1979. Climatic changes, habitat changes, and competition: dynamics of geographic overlap in two pairs of congeneric bird species in Finland. *Oikos* 33:261–271.

Jones, R. 1977. Man as an element of a continental fauna: the case of the sundering of the Bassian Bridge. In: *Sunda and Sahul* (J. Allen, J. Golson, and R. Jones, eds.), pp. 317–386. Academic Press, London.

Jones, H. L. and J. M. Diamond. 1976. Short-time-base studies of turnover in breeding bird populations on the California Channel Islands. *Condor* 78:526–549.

Kalela, O. 1949. Changes in geographic ranges in the avifauna of northern and central Europe in relation to recent changes in climate. *Bird Banding* 20:77–103.

Karr, J. R. 1982. Avian extinction on Barro Colorado Island, Panama: a reassessment. *Amer. Nat.* 119:220–239.

Kepler, C. B. 1967. Polynesian rat predation on nesting Laysan Albatrosses and other Pacific seabirds. *Auk* 84:426–430.

King, W. B. 1977. Endangered birds of the world and current efforts toward managing them. In: *Endangered Birds* (S. A. Temple, ed.), pp. 9–18. University of Wisconsin Press, Madison.

———. 1980. Ecological basis of extinction in birds. *Proc. 17th Intern. Ornith. Congr.* 905–911.

———. 1981. *Endangered Birds of the World*. The ICBP Bird Red Data Book. Smithsonian Institution Press, Washington.

———. 1983. Island birds: will the future repeat the past? *Proc. 18th World Conf. ICBP*, in press.

Leigh, E. G., Jr. 1975. Population fluctuations, community stability, and environmental variability. In: *Ecology and Evolution of Communities* (M. L. Cody and J. M. Diamond, eds.), pp. 51–73, Harvard University Press, Cambridge, Mass.

Leopold, E. B. 1967. Late-Cenozoic patterns of plant extinction. In: *Pleistocene Extinctions* (P. S. Martin and H. E. Wright, Jr., eds.), pp. 203–246. Yale University Press, New Haven.

MacArthur, R. H. and E. O. Wilson. 1967. *The Theory of Island Biogeography*. Princeton University Press, Princeton.

McClung, R. M. 1976. *Lost Wild World*. William Morrow, New York.

Magee, J. D. 1965. The breeding distribution of the Stonechat in Britain and the cause of its decline. *Bird Study* 12:83–89.

Marshall, L. G. 1981. The great American interchange: an invasion-induced crisis for South American mammals. In: *Biotic Crises in Ecological and Evolutionary Time* (M. H. Nitecki, ed.), pp. 133–229. Academic Press, New York.

Marshall, L. G., S. D. Webb, J. J. Sepkoski, Jr., and D. M. Raup. 1982. Mammalian evolution and the great American interchange. *Science* 215:1351–1357.

Martin, P. S. and H. E. Wright, Jr. 1967. *Pleistocene Extinctions*. Yale University Press, New Haven.

Mayr, E. 1965. Avifauna: turnover on islands. *Science* 150:1587–1588.

Merikallio, E. 1951. Der Einfluss der letzten Wärmeperiode (1930–49) auf die Vogelfauna Nordfinnlands. *Proc. 10th Intern. Ornith. Congr.* 484–493.

Mills, S. 1981. Graveyard of the puffin. *New Scientist* 91:10–13.

Mills, J. A. and A. F. Mark. 1977. Food preferences of Takahe in Fiordland National Park, New Zealand, and the effect of competition from introduced red deer. *J. Anim. Ecol.* 46:939–958.

Mosimann, J. E. and P. S. Martin. 1975.

Simulating overkill by Paleoindians. *Amer. Scientist* 63:304–313.

Olsen, S. J. and J. W. Olsen. 1977. The Chinese wolf: ancestor of New World dogs. *Science* 197:533–535.

Olson, S. L. 1977. A synopsis of the fossil Rallidae. In: *Rails of the World* (S. D. Ripley, ed.), pp. 339–373. Godine, Boston.

———. 1978. A paleontological perspective of West Indian birds and mammals. *Acad. Nat. Sci. Phil. Special Publ.* 13 (Zoogeography in the Caribbean), pp. 99–117.

———. 1981. The museum tradition in ornithology: a response to Rickleffs. *Auk* 98:193–195.

Olson, S. L. and W. B. Hilgartner. 1982. Fossil and subfossil birds from the Bahamas. *Smithsonian Contributions to Paleobiology* no. 48, pp. 22–56.

Olson, S. L. and H. F. James. 1982. Prodromus of the fossil avifauna of the Hawaiian Islands. *Smithsonian Contributions to Zoology* no. 365.

Ovington, D. 1978. *Australian Endangered Species*. Cassell, Australia.

Pasquier, R. F. 1980. *Conservation of New World Parrots*. Smithsonian Institution Press, Washington.

Pregill, G. K. 1982. Fossil amphibians and reptiles from New Providence Island, Bahamas. *Smithsonian Contributions to Paleobiology* no. 48, pp. 8–21.

Pregill, G. K. and S. L. Olson. 1981. Zoogeography of West Indian vertebrates in relation to Pleistocene climatic cycles. *Ann. Rev. Ecol. Syst.* 12:75–98.

Richter-Dyn, N. and N. S. Goel. 1972. On the extinction of a colonizing species. *Theor. Pop. Biol.* 3:406–433.

Simpson, G. G. 1953. *The Major Features of Evolution*. Columbia University Press, New York.

Soulé, M. E. 1983. What do we really know about extinction? In: *Genetics and Conservation* (T. Schonewald-Cox, ed.), in press. Addison-Wesley, Massachusetts.

Soulé, M. E., B. A. Wilcox and C. Holtby. 1979. Benign neglect: a model of faunal collapse in the game reserves of East Africa. *Biol. Conservation* 15:259–272.

Temple, S. A. 1977. Plant-animal mutualism: coevolution with dodo leads to near-extinction of plants. *Science* 197:885–886.

Terborgh, J. 1974. Preservation of natural diversity: the problem of extinction-prone species. *BioScience* 24:715–722.

Terborgh, J. and B. Winter. 1980. Some causes of extinction. In: *Conservation Biology* (M. E. Soulé and B. A. Wilcox, eds.), pp. 119–134. Sinauer, Sunderland, Massachusetts.

Thornback, J. and M. Jenkins. 1982. *The IUCN Mammal Red Data Book*. Part 1. IUCN, Gland, Switzerland.

Tyler, M. J. 1979. *The Status of Endangered Australasian Wildlife*. Royal Zoological Society of South Australia, Adelaide.

Van Deusen, H. M. 1957. Results of the Archbold Expeditions. No. 76. A new species of wallaby (genus *Dorcopsis*) from Goodenough Island, Papua. *Amer. Mus. Novitates* no. 1826.

Veitch, C. R. 1983. Cat eradication on Little Barrier Island. *Proc. 18th World Conf. ICBP*, in press.

von Haartman, L. 1978. Changes in the bird fauna in Finland and their causes. *Fennia* 150:25–32.

Warner, R. E. 1968. The role of introduced diseases in the extinction of the endemic Hawaiian avifauna. *Condor* 70:101–120.

Webb, S. D. 1976. Mammalian faunal dynamics of the great American interchange. *Paleobiology* 2:220–235.

Wilcox, B. A. 1978. Supersaturated island faunas: a species-age relationship for lizards on post-Pleistocene land-bridge islands. *Science* 199:996–998.

———. 1980. Insular ecology and conservation. In: *Conservation Biology* (M. E. Soulé and B. A. Wilcox, eds.), pp. 95–118. Sinauer, Sunderland, Massachusetts.

Williamson, K. 1975. Birds and climatic change. *Bird Study* 22:143–164.

Willis, E. O. 1974. Populations and local extinctions of birds on Barro Colorado Island, Panama. *Ecol. Mongr.* 44:152–169.

———. 1980. Species reduction in remanescant wood lots in southern Brazil. *Proc. 17th Intern. Ornith. Congr.* 783–786.

Winstanley, D., R. Spencer and K. Williamson. 1974. Where have all the Whitethroats gone? *Bird Study* 21:1–14.

Ziswiler, V. 1967. *Extinct and Vanishing Animals*. Springer, New York.

About the Contributors

Larry D. Agenbroad was trained as an engineer, geologist, and archaeologist. He has excavated at several archaeological sites with extinct megafauna in the western United States.

Atholl Anderson has researched prehistoric hunting and fishing patterns in various parts of Europe and the Pacific. An archaeologist at the University of Otago, New Zealand, he has worked primarily on the big-game hunting activities of the early Maori people.

Elaine Anderson, widely known Pleistocene mammalogist, is coauthor of *Pleistocene Mammals of North America*. Her special interests have been mustelids, faunal studies, biogeography, and of course, extinction.

Rainer Berger has been engaged for more than twenty years in radiocarbon studies of the archaeology and environment of the late Quaternary.

Gennadij F. Baryshnikov has been a scientific collaborator in the Zoological Institute of the Academy of Sciences of the USSR, where he has studied the Quaternary history of alpine mammals in the Crimea, the Caucasus, and central Asia. He has participated in field work in Yakutia and the Caucasus.

Alan L. Bryan has undertaken archaeological field work from Beringia (Pribilof Islands) to Patagonia in an attempt to resolve questions concerning the early peopling of the Americas.

Richard Cassels was trained in England as a paleolithic archaeologist. In New Zealand his interest in hunter-gatherer economics led him naturally to study moa extinction, a dramatic example of a man-fauna relationship that had "failed."

Robert E. Dewar has conducted archaeological research in Madagascar since 1977. His principal research interest has been the impact of human economies on natural communities.

Jared M. Diamond has focused on the ecology and evolution of New Guinean birds, membrane biophysics, and applications of island biogeographic studies to conservation ecology.

Bennett Dyke has studied primate behavior at the Southwestern Foundation for Research in San Antonio, Texas. He has worked extensively with anthropological computer simulations, particularly in connection with human population studies.

B. Miles Gilbert has extensive experience with both late Pleistocene and Holocene mammalian faunas and directed the excavations at Natural Trap Cave, Wyoming, from 1974 to 1980. His published books and articles have covered a wide spectrum of osteological topics from anthropology to zoology.

Philip G. Gingerich has been interested primarily in the rapid evolutionary diversification of early Cenozoic mammals and the dynamics of faunal turnover.

Russell W. Graham has studied the evolution of Quaternary mammalian communities and the interaction between early man and environment.

Donald K. Grayson has conducted extensive research on the biogeographic history of mammals in the arid western United States. His strong interest in the history of archaeology and paleontology is illustrated in his book, *The Establishment of Human Antiquity*.

Ruth Gruhn has carried out archaeological research at early sites in western Canada, the western United States, Central America, Venezuela, and Brazil.

John E. Guilday, until his death in November 1982, was a specialist in the biota of the Appalachian Mountain area and an expert on rodents. He was particularly interested in the effects of Pleistocene climatic change on individual mammalian species and on mammal communities.

R. Dale Guthrie, a vertebrate paleontologist, has worked for twenty years on Pleistocene mammals from interior Alaska. The reasons for large mammal extinction in the north have always been an important issue in his research.

C. Vance Haynes has devoted twenty-five years to investigating the origin and geochronology of the Clovis culture and the paleoecology of the Pleistocene-Holocene transition. He stresses the coincidence between Pleistocene extinction and the appearance of the Clovis culture in the stratigraphic framework.

Geoffrey Hope has used pollen analysis to study late Quaternary vegetational and climatic change in Australia and New Guinea.

David R. Horton has applied his interests in paleoecology and prehistory to the investigation of Quaternary sites throughout eastern Australia.

Helen F. James has devoted six years to collecting and studying the fossil avifaunas of the Hawaiian Islands. One outcome of this research has been evidence for a link between early human activities and extinction.

A. Peter Kershaw has researched the vegetational history of the Australian region. He has been particularly interested in elucidating the nature of past rainforest types and in explaining the development of present rainforest/sclerophyll patterns.

Richard A. Kiltie has been strongly interested in the evolutionary determinants of reproductive seasonality in mammals.

James E. King has studied vegetational response to Pleistocene glaciation in central and eastern North America for many years. His special interest has been vegetational changes south of the ice limits and their relationship to the extinct megafauna.

Li Xing-guo has been director of the Laboratory of Radiocarbon Dating at the Institute of Vertebrate Paleontology and Paleoanthropology at the Academia Sinica in Peking.

Liu Tung-sheng has served as president of the Chinese Quaternary Association, while engaged in his lifelong career studying loess and aeolian geology.

Ernest L. Lundelius, Jr. has worked on Pleistocene and Holocene mammals from the United States and Australia.

Beverley McCulloch has studied the paleoecology of man, moa, and other species in New Zealand at the time people first arrived there, a thousand years ago.

Jerry N. McDonald a biogeographer, has specialized in Quaternary megafaunal zoogeography. His particular interests have been in identifying realistic biological bases for late Quaternary megafaunal extinctions and the adjustments of the megafaunal community to extinctions.

Leslie F. Marcus has been associated with Rancho La Brea research for more than forty years. He has also delved into biometrics and computer applications in natural science.

Larry G. Marshall studied Quaternary and Recent mammals in Australia for two years. For the past seven years, he has focused on the Cenozoic land mammals of South America. His research interests have included the reasons for faunal turnover, including extinction events and processes.

Larry D. Martin has cooperated with B. Miles Gilbert in the study of the Natural Trap Cave in Wyoming and has been engaged in a study of late Cenozoic extinctions in North America.

Jim I. Mead has undertaken research on the late Pleistocene paleoenvironments of the arid western United States.

David J. Meltzer has been involved in research on late Pleistocene human adaptations in eastern North America, including the relationship of human populations to the extinct megafauna.

Duncan Merrilees, until his retirement, worked at the Western Australian Museum, concentrating on prehistoric interactions of man, other mammals, and climate in the extreme southwest of the Australian continent.

Peter Murray has used his special anatomical training to flesh out the unknown Australian megafauna. He has been curator of anthropology at the Museums and Art Galleries of the Northern Territory, Darwin, Australia.

Storrs L. Olson has been studying the extinct avifauna of islands in the South Atlantic, Caribbean, and Pacific since 1970, in connection with other studies of the systematics and paleontology of birds. He has also been concerned with the biogeographical implications of man-caused extinctions of vertebrates on islands.

Arthur M. Phillips, III has studied the modern vegetation and flora and the late Pleistocene paleoecology of the Grand Canyon for many years. He has been especially interested in the former plant and animal inhabitants of the canyon.

Jeffrey J. Saunders has excavated numerous mastodont sites, pursuing their remains in spring site depositional environments and bone beds. With others, he has studied mastodont adaptive strategies.

David W. Steadman has studied Quaternary extinctions of vertebrates, especially birds and rodents, in North America, the West Indies, and the Galapagos.

Eitan Tchernov has pursued interests in biostratigraphy, paleoecology, and faunal evolution in the Middle East. He was involved for many years in the Jordan Valley project and has participated in numerous prehistoric excavations in Israel.

Michael M. Trotter has studied the archaeology of New Zealand for nearly twenty years. He has begun writing a human prehistory of the South Island of New Zealand, a study in which the economic significance of moas and moa extinction plays an important role.

Nikolaj K. Vereshchagin has studied the ecology, zoogeography, and paleontology of mammals for more than fifty years at the Zoological Institute of the Academy of Sciences of the USSR. He has published frequently on the acclimatization of fur-bearing animals and on the Quaternary history of the fauna of the Russian Plain, the Crimea, the Caucasus, and Siberia.

S. David Webb undertook a study of *Camelops* from the La Brea Tar Pits a quarter century ago. Since then, he has worked with many groups of extinct, large mammals, especially in Florida, but has not yet satisfied his curiosity.

Stephen L. Whittington has pursued his doctoral research at Pennsylvania State University in New World archaeology and computer simulation of social systems.

Index